Proceedings of the 8th International
Congress of Immunology Budapest 1992

Progress in Immunology
Vol. VIII

Editors: J. Gergely

M. Benczúr, Anna Erdei, A. Falus, Gy. Füst,
G. Medgyesi, Gy. Petrányi, Éva Rajnavölgyi

Springer-Verlag
Berlin Heidelberg New York
London Paris Tokyo Hong Kong Barcelona Budapest

János Gergely
Department of Immunology
Eötvös Loránd University
Jávorka S. u. 14.
2131 Göd, Hungary

Names and addresses of the other editors can be found on page XXVII

ISBN 978-3-642-51481-4   ISBN 978-3-642-51479-1 (eBook)
DOI 10.1007/978-3-642-51479-1

© Springer Hungarica 1993
Softcover reprint of the hardcover 1st edition 1993

Printing: Gyomai Kner Nyomda

CONTENTS

1. B AND T CELL ONTOGENY, REPERTOIRE, SELECTION AND TOLERANCE

2. GENETICS, STRUCTURE, FUNCTION AND EVOLUTION OF IMMUNOGLOBULINS AND T CELL RECEPTORS

3. MHC, MOLECULAR AND CELLULAR ASPECTS OF ANTIGEN PROCESSING AND PRESENTATION

4. SIGNAL TRANSDUCTION PATHWAYS

7. CYTOKINES AND CYTOKINE RECEPTORS IN HEALTH AND DISEASE

9. Fc RECEPTORS AND Ig BINDING FACTORS

1O. COMPLEMENT SYSTEM; STRUCTURE AND FUNCTION

13. THERAPEUTIC APPLICATION OF REGULATORY NETWORKS

14. IMMUNOLOGY OF INFECTIOUS DISEASES AND VACCINATION

15. TRANSPLANTATION IMMUNOLOGY

16. REPRODUCTIVE IMMUNOLOGY

17. TUMORIMMUNOLOGY

18. LATE ARRIVALS

Opening Speeches

On behalf of the Organizers I have the pleasure to welcome you at the 8th International Congress of Immunology.

Six years ago on the IUIS Council Meeting after thorough consideration Budapest became the first candidate for this important conference. Its fortunate location, its easy accessibility from East and West, the well known Hungarian hospitality all were in favour of this decision. In addition, the tradition of immunology in Hungary, the good cooperation between Hungarian immunologists and colleagues abroad, moreover, the successful congresses (among others the 4th European Immunology Meeting) held in Hungary were also of great importance. Everybody agreed that Hungary is the place where a relatively high number of participants from the Eastern-block countries can be expected and that the time had come for a break-through in this respect. At that time nobody could predict that history will support our ideas so successfully.

Three years ago in Berlin, I was in the lucky situation to invite you to this Congress. Keeping this intention of the IUIS in mind, we have chosen a bridge, as the symbol of the 8th International Congress of Immunology. A bridge which helps the interaction of cells facilitating information transfer during various levels of immune responses. A bridge which connects people from East and West, South and North enabling easy understanding for all. A bridge which connects Buda and Pest the wonderful capital of our friendly country which is now ready to host you and offer you a pleasant environment for a stimulating and memorable conference.

Now, when opening this meeting, I am convinced that concerning the symbol, our choice was good: the participants at this conference represent 75 countries (and only God knows, how many nationalities), and the number of colleagues from the former eastern-block countries could be mentioned in the next issue of the Guinness book of records. More than 600, mainly young colleagues from these, until now isolated countries, have now the opportunity to shake hands with fellow-scientists of the Western world. And we can be proud of the high representation from the developing countries as well.

The "bridge-function" of Hungary seems to be working. However, I have to emphasize also, the significance of international cooperation in the preparation of this meeting. Not only arranging the scientific program results from the effort of hundreds of excellent immunologists from all over the world, but without the enthusiasm of the IUIS officers and generous financial support of several national societies we could not have this relatively high number of new friends among us.

Immunology is greatly acknowledged in Hungary. It is therefore not accidental, that the government and our scientific organizations are represented on the highest level at this opening ceremony. It is a great pleasure for me to announce that His Excellency Mr. Árpád Göncz, President of our Republic, Mr. Miklós Marschall the representative of the Chief Mayor of Budapest, Professor Béla Halász, the Vice President of the Hungarian Academy of Sciences, are present in our Presidium. His Excellency Prime Minister József Antall, is greeting the Congress by letter. Professor Jacob Natvig, President of the International Union of Immunological Societies represents the Board of IUIS. The Hungarian Organizers are very glad that the doyen of Hungarian immunologists, Professor Zoltán Ováry has accepted our invitation and is also in our midst. Finally,

Professor Győző Petrányi, President of the Hungarian Society for Immunology represents the HSI.

And now, on behalf of the Organizers I wish you a very successful meeting.

János Gergely

Immunology for all

Address of the IUIS President at the 8th International Congress of Immunology, Budapest, Hungary, August 1992

Jacob B. Natvig

Institute of Immunology and Rheumatology Rikshospitalet, The National Hospital, University of Oslo
Oslo, Norway

INTRODUCTION

Let me first at this splendid gathering in the old Town Hall of Budapest thank the Republic of Hungary, represented by the Minister of Welfare and Social Affairs, László Surján, the city of Budapest represented by the Deputy Mayor, Miklós Marschall and the Hungarian Society for Immunology represented by the President, Győző Petrányi and the Congress-president, János Gergely for making it possible for almost 5000 immunologists from all over the world to have this reunion for the 8th International Congress of Immunology in your beautiful country and city.

You are fortunate in Hungary to have an active scientific community based in universities and other academic institutions which has contributed actively to this meeting. But the main burden has after all been on the Hungarian immunologists, more than 200, who have spent most of the last years preparing for this meeting. In thanking you, let me also point out that our Hungarian immunologists represent an active, scientific community on a high, international level.

My second point deals with the question, what is immunology and what do immunologists really do? Our field is the science of how biological systems discriminate between self (own body constituents) and nonself (foreign cells and particles). The common denominators to do this are the very specialized cells such as T and B lymphocytes as well as the antibodies and other effector and signal molecules which they produce (See Hodgkin and Kehry 1993, Miller 1993). They promote tissue reactions of animals and man which can either be beneficial, as seen in the fields of immunity to infections or harmful, as seen in allergic reactions to allergens such as pollen. In addition, autoimmune diseases which can be regarded as a war against our own body components, result from lack of tolerance to self. Such autoimmune diseases can be life threatening, destroying organs and tissues.

Examples are chronic kidney diseases, rheumatoid arthritis, gastrointestinal autoimmune diseases, diabetes and many other affections. Immunology is of course also essential to the understanding and management of immunodeficiencies, transplantation and hopefully malignancies.

Now, thirdly, what is our aim as immunologists? As was just outlined by our President elect, Henry Metzger in our new journal The Immunologist (Metzger 1993), the most fundamental goal of the International Union of Immunological Societies (IUIS) has always been, and should continue to be, to actively promote the advancement of immunology in all its aspects. This is the work and goal for approximately 30.000 immunologists in the world, and involves 2 very important actions.

1. To increase our knowledge of immunology.
2. To distribute this knowledge and the application of it to all.

IN WHÁT AREA CAN WE INCREASE THE KNOWLEDGE OF IMMUNOLOGY AND WHERE IS IMMUNOLOGY MOVING IN THE NEXT CENTURY?

Let us focus on three areas.

1. Our past president Gustav J. V. Nossal just pointed out as one major trend the modulation of immunological processes through surface molecules on immunological cells and signal substances acting between cells (Nossal 1993). According to Nossal "Immunology has been particularly successful in its reductionist mode. Our understanding of the organs, cells, molecular products and key genes of the immune system has improved immensely, and we have in vitro techniques for manipulating a large number of important variables. Although this "anatomical" approach will be even further developed, it seems to me that tomorrow will be more "physiological," that is, much of our research work in the future will deal with the regulation of the immune system more directly. Immunoregulation is effected by cytokines, transcription factors and CD molecules at the surfaces of T, B and other lymphoid cells. "We shall, therefore, have to meticulously avoid the temptation of having everyone working on their pet interleukine, CD molecule, or gene knock-out mice, and instead have to struggle hard to integrate the many disparate finding from the various models into a framework that makes sense at the level of the entire animal" (Nossal 1993).

2. Major advanced will be expected in new vaccination strategies and a World Health Organization (WHO) program for expanded immunization has so far become a great success. A lot of work is made to develop new vaccines including principles for depot effects with release of antigen at various times after one single injection.

Paul Henri Lambert who is the head of the Immunology Unit of WHO is working very actively in this area. The hope for an HIV vaccine is also gradually increasing (Wigzell 1993).

3. Biotechnology - and gene technology will take a new lead also in immunology. In a recent overview termed "A revolution in medicine like no other", Reed R. Pyeritz describes the revolution of molecular and genetic medicine based on biotechnology and gene technology (Pyeritz 1992). This revolution which began in 1953 by the identifiction of the structure and function of DNA will bring about

new understanding of pathogenesis, management of molecular disease, diagnostics and screening methods for these diseases, disorders of development, inborn errors of metabolism, inflammation and immunity and potentially even disorder of cognition, behaviour and emotions (see Pyeritz 1992).

HOW CAN WE DISTRIBUTE THE KNOWLEDGE OF IMMUNOLOGY AND THE APPLICATION OF IT TO ALL?

Unfortunately, the distribution of immunology to all is hampered by the fact that the 30.000 scientists who are the immunologists of the world are not equally distributed. Recent information shows that 94% of all scientists in the world live and work in the developed, or industrialized countries in Europe, North America, or Japan and other limited parts of Asia. In contrast, only 6% of the worlds scientists live and work in the developing areas representing 80% of the world population (ASCEND 1991). Now this would actually indicate that out of our 30.000 IUIS members about 28.000 belong to the developed countries mostly in the Western world and Japan. Only 1.800-2.000 would be in the enormous areas represented by the developing countries. These data fit quite well with our current membership list (Arnon 1990), even after the creation of new immunological societies and regional federations in Africa and Asia-Oceania (see Esa and Natvig 1992). Let me give you one typical example of this inequality. During this congress a young immunologist from an African country came to me to receive a travel bursary. I asked him how many immunologists from his country were participating. " I am the only one" he replied. "How many people live in your country?" I asked. "25 million" was the answer. In comparison to this would indicate that only about 12 out of almost 300 million Americans would participate in our congress. In contrast, approximately 1.300 are actually here. The comparison would be similar for most western countries.

We must therefore make the progress in immunology accessible to all as soon as possible. This includes its preventive measures of vaccination, its potential in diagnosis and treatment and its application to a broad group of basic and applied sciences. In addition, it means generating a better knowledge of immunology and making it available all over the world through medical and veterinary schools, in science faculties (see Reveillard 1993) and other academic curricula, for nurses, technical staff and others to reach the ambitious aim "Immunology for all".

We really need many parallell actions. This includes promoting new immunological societies with a series of new engagements in areas with modest or no immunology. We must encourage the involvement of new devoted people both in member societies, regional federations (see Natvig and Arnon 1993) and the IUIS itself. This deals with congresses, meetings, symposia and a variety of committee activities within education, specialization, standardization and others (see Natvig and Arnon 1993).
We also need to improve the distribution of information about immunology and IUIS activities which we have recently among other things done by the new journal, The Immunologist, published by the international publishers, Hogrefe

and Huber located in Germany, Switzerland, Canada and the USA. Finally, we should tighten our interactions with international bodies such as the International Council of Scientific Unions (ICSU), The World Health Organization (WHO) as well as with international affiliated members of IUIS (see Natvig and Arnon 1993).

ARE WE WILLING TO RISK FAILURE ON THE WAY TO OUR GOALS?

How are we going to manage all these endeavours and reach our goals? We have to proceed by trial and errors which is the framework of science. The well known Nobel Laureate T. H. Morgan according to his Norwegian student and friend Otto Lous Mohr once divided his experiments into "fool experiments, damned fool experiments, and those that are still worse" (Mohr 1952). When we try hard to reach our goals we should not be afraid that some of these activities might actually fail and not cause the wanted effects. In my view we need the willingness to risk failure recognizing that, the man or woman who cannot do a mistake cannot do anything. I have with great encouragement read an editorial by Harold T. Shapiro where he points out "We must also be aware of raising the flag of failure to quickly. The world too often calls it failure if we do not immediately reach our goals, ... failure lies, rather, in giving up on our goals. When 10.000 experiments with a storage battery failed to produce results, Thomas Edison said, "... I have not failed. I've just found 10.000 ways that won't work". Few battles result in immediate and full victory. For example, instead of being frustrated and immobilized by the continued hardship experienced by around the world, we must keep alive the idea of eventual deliverance from suffering and oppression by demonstrating our willingness to move forward along those paths actually available to us.

FROM KNOWLEDGE TO WISDOM

The question is how we apply this knowledge in a wise way. This has been focused by Nicholas Maxwell in his book "It is what we do and what we are that ultimately matters, our knowledge is but an aspect of our life and being" and "This basic Socratic idea has been betrayed, and as a result, to put it at its most extreme, we now stand on the brink of self-destruction. In the circumstances, there can scarcely be any more urgent task for all those associated in any way with the academic enterprise - scientists, technologists, scholars, teachers, administrators, students, parents, providers of funds - than to help put into practise the new kind of inquiry, rationally devoted to the growth of wisdom" (Maxwell 1984).

If we are willing to work along these lines, to transform the best of our knowledge in the field of immunology to wisdom, I have the hope for the next decade, the next century and the new millennium that immunology will win the war against infections, help to conquer cancer, provide tools for fertility control and to influence the aging processes. I have the hope that immunology will eventually solve the problems of autoimmune diseases, immunodeficiencies and transplantation (Natvig 1990).

I have the hope that immunology together with many other scientices will bring safer, happier and more meaningful life to all mankid.

REFERENCES

Arnon R (1990) International Union of Immunological Societies (IUIS), Weizmann Institute, Rehovot, Israel

ASCEND (1991) Agenda for Science and development for the 21 century, organized November 1991 in Vienna by the International Council of Scientific Union (ICSU), 51 Boulevard de Montmorency, 75016 Paris, France

Esa A, Natvig JB (Eds) (1992) Proceedings of the First African Immunology Meeting, Harare, Zimbabwe 1992 Scand J Immunol Supplement 11: pp 218

Hodgkin PD, Kehry MR (1993) The Immunologist 1:5-8

Maxwell N (1984) From knowledge to wisdom. Revolution in the aims & methods of science, Basil Blackwell, Oxford & New York, pp 298

Miller JFAP (1993) The Immunologist 1:9-15

Mohr OL (1952) Taler (Speeches) Gyldendal Norsk Forlag, Oslo, pp. 145

Natvig JB, Arnon R (1993) The Immunologist 1:29-31

Metzger H (1993) The Immunologist 1:3-4

Natvig JB (1990) Immunology Today 11:72-73

Nossal GJV (1993) The Immunologist 1:3-4

Pyeritz RE (1992) FASEB J 6:2761-2766

Reveillard J-P (1993) The Immunologist 1:22-28

Shapiro HT (1990) Science 250:609

Wigzell H (1993) The Immunologist 1:20-21

OPENING ADDRESS
Béla Halász, Vice-President of the Hungarian Academy of Sciences

On behalf of the Hungarian Academy of Sciences, Domonkos Kosáry it is a great priviledge and a special pleasure for me to warmly welcome all participants of the Eight International Congress of Immunology. To its deepest regret, President Kosáry could not be present, because he had to go abroad to participate in a meeting.

The Hungarian Academy of Sciences is very pleased that this important event takes place in Hungary, in Budapest. We consider it as a recognition of the significant scientific contribution of Hungarian researchers to immunology.

Immunological science is of fundamental importance, both in basic science as well as in everyday clinical practice. Immunology has made a spectacular progress in the last years and its field has expanded to a great extent. As an example of the expansion I would like to refer to my own research field which is neuroendocrinology. Twenty or thirty years ago there were practically no links between neuroendocrinology and immunology. But now-a-days these desciplines are closely connected with each other, and a joint research field called neuro-immuno-endocrinology developed. You developed have solved several basic problems in the past, but new crucial questions have risen. I am sure the immunologists, who have several Nobel Laureates which is an indication of their outstanding contribution to science, will give the right answers to the questions they are presently faced with.

Wish you all a very successful meeting, lively and constructive discussions and a very pleasant stay in our city, which we like very much.

LADIES AND GENTLEMEN

I was asked to welcome the participants of the 8th International Congress of immunology. I thank the organizers for this exceptional honor. I think this choice was made because I am Hungarian and perhaps the most senior of the actively working immunologists. Moreover, I am quite international as I was born Hungarian in Kolozsvár, Transsylvania and I worked in France, Italy, Brasil, Japan and the United States.

When I started to work it was the golden age of Immunochemistry dominated by one of the most prominent scientists: Michael Heidelberger. Now, many different fields and techniques are needed in Modern Immunological Research, especially the powerful technics of Molecular Biology.

Immunology has become a multidisciplinary field and we all look forward to what the future will bring.

I wellcome the participants and I hope that we will all profit of the exchange of ideas at the Congress of Budapest.

I great you with the old Hungarian welcome:

ISTEN HOZTA!

(Good brought you)

Zoltán Ováry

EDITORS

Benczur, M.
National Institute of Haematology and Blood Transfusion
Daróczy út 24,
1113 Budapest, Hungary

Erdei, Anna
Department of Immunology
Eötvös Loránd University
Jávorka S. u. 14,
2131 Göd, Hungary

Falus, A.
National Institute of Rheumatology
Frankel Leo út. 17-19.
1027 Budapest, Hungary

Füst, Gy.
National Institute of Haematology and Blood Transfusion
Daróczy út 24,
1113 Budapest, Hungary

Gergely, J.
Department of Immunology
Eötvös Loránd University
Jávorka S. u. 14,

Medgyesi, G. A.
National Institute of Haematology and Blood Transfusion
Daróczy út 24,
1113 Budapest, Hungary

Petrányi, Gy. G.
National Institute of Haematology and Blood Transfusion
Daróczy út 24,
1113 Budapest, Hungary

Rajnavölgyi , Éva
Department of Immunology
Eötvös Loránd University
Jávorka S. u. 14,
2131 Göd, Hungary

CONTRIBUTORS

The number in brackets following the author's name indicates the first page of the paper he/she authored/coauthored.
The bold indicates the author who presented the paper at the congress.

Adamczewski, M. [435]
Molecular Allergy and Immunology Section
NIAD, NIH
12441 Parklawn Drive
Rockville, MD 20852
USA

Aichele, P. [65]
Institute of Experimental Immunology
University of Zürich
Sternwartstrasse 2
8091 Zürich
Switzerland

Al-Ramadi, B. [181]
Section of Immunobiology
Yale University Sch. of Medicine
New Haven, CT 06510
USA

Al-Sabbagh, A. [627]
Center for Neurologic Diseases
Bringham and Women s Hospital
Harvard Medical School
75 Francis Street
Boston, MA 02115
USA

Alber, G. [27]
Max-Planck Institut für Immunbiologie
Stübeweg 51
7800 Freiburg
Germany

Alber, G. [473]
Section on Chemical Immunology
Arthritis and Rheumatism Branch
NIAMSD, NIH
Bethesda, MD 20892
USA

Almerigogna, F. [239]
Division of Clinical Immunology and Allergy
University of Florence
Instituto di Clinica Medica 3
50134 Firenze
Italy

Amadori, A. [699]
Institute for Oncology
Interuniversity Center for Res. on Cancer
University of Padova
Via Gattamelata 64
I-35128 Padova
Italy

Amemiya, C.T. [107]
Lawrence Livermore Natl. Laboratory
Div. of Biomedical Science
Livermore, CA 94551
USA

Amigorena, S. [457]
Laboratoire d'Immunologie
Cellulaire et Clinique
INSERM U255, Institut Curie
26 rue d'Ulm
75231 Paris, Cedex 05
France

Anand, R. [91]
Dept. of Biotechnology, ICI Pharmaceuticals
Alderley Park, Macclesfield, Cheshire
SK10 4TG
UK

Anderson, W.F. [553]
NCI, NHLBI
National Institutes of Health
Bethesda, MD
USA

Arch, R. [289]
Kernforschungszentrum Karlsruhe
Institut für Genetik und Toxicologie
PO Box 3640
D-7500 Karlsruhe 1
Germany

Armant, M. [465]
Notre-Dame Hospital Res. Center
1560 Sherbrooke Street East
Montreal, Quebec, H2L 4M1
Canada

Arnon, R. [767]
Weizmann Institute of Science
Dept. of Chemical Immunology
Rehovot, 76100
Israel

Arunan, K. [849]
National Institute of Immunology
New Delhi 110067
India

Bach, J.F. [595]
INSERM U25
Hopital Necker
161 rue de Sevres
75015 Paris
France

Baggiolini, M. [385]
Theodor-Kocher Institut
University of Berne
Switzerland

Baker, C.L. [893]
The Johns Hopkins University
Dept. of Medicine
Oncology and Mol. Biology
Baltimore, MD 21205
USA

Banwatt, D. [841]
McMaster University
Hamilton, Ontario
Canada

Barrett, T. [261]
Ben May Institute and Committee
on Immunology MC 1089
University of Chicago
Chicago, IL 60637
USA

Bartlott,T. [269]
Max-Planck Institut für Biologie
Abteilung Immungenetik
7400 Tübingen
Germany

Batchelor, R. [793]
Dept. of Immunology
Royal Postgraduate Medical School
Hammersmith Hospital
Du Cane Road
London, W12 ONN
UK

Battegay, M. [659]
Institute for Experimental Immunology
Dept. of Pathology
University of Zürich
Sternwartstrasse 2
CH-8091 Zürich
Switzerland

Baumann, H. [377]
Roswell Park Cancer Institute
Dept. of Molecular and Cellular Biology
Buffalo, NY 14263
USA

Ben-Sasson, S.Z. [347]
Lautenberg Center for General and
Tumor Immunology
Hebrew University
Hadassah Medical Center
Jerusalem 91010
Israel

Bentley, A.M. [403]
Dept. of Allergy and Clinical Immunology
National Heart and Lung Institute
Dovehouse Street
London, SW3 6LY
UK

Berberich, I. [305]
Institute of Virology and Immunobiology
University of Würzburg
Versbacherstrasse 7
D-8700 Würzburg
Germany

Berg, D. [355]
Dept. of Immunology
DNAX Research Institute
901 California Ave.
Palo ALto, CA 94304
USA

Beverley, P.C.L. [231]
Imperial Cancer Research Fund
Human Tumor Immunology Group
UCMSM
91 Riding House Street
London W1P 8BT
UK

Biagiotti, R. [239]
Division of Clinical Immunology and Allergy
University of Florence
Instituto di Clinica Medica 3
50134 Firenze
Italy

Bierer, B.E. [275]
Division of Pediatric Oncology
Dana Farber Cancer Institute
Boston, MA
USA

Bischoff, S.C. [411]
Institute of Clinical Immunology
Inselspital
Ch-3010 Bern
Switzerland

Blaese, R.M. [553]
NCI, NHLBI
National Institutes of Health
Bethesda, MD
USA

Bloom, B.R. [761]
Dept. of Microbiology and Immunology
Albert Einstein College of Medicine
Bronx, NY 10461
USA

Bluestone, J.A. [261]
Ben May Institute and Committee
on Immunology MC 1089
University of Chicago
Chicago, IL 60637
USA

Boehmer, H. von [81],[129]
Basel Institut for Immunology
487 Grenzacherstrasse
CH 4005 Basel
Switzerland

Bonnerot, C. [457]
Laboratoire d'Immunologie
Cellulaire et Clinique
INSERM U255, Institut Curie
26 rue d'Ulm
75231 Paris, Cedex 05
France

Boog, C.J.P. [587]
Dept. of Infectious Diseases and Immunology
University of Utrecht
Yalelaan 1, P.O. Box 80165
3508 TD, Utrecht
The Netherlands

Boon, T. [871]
Ludwig Institute for Cancer Research
74 Ave. Hippocrate
B-1200 Brussels
Belgium

Borgulya, P. [129]
Basel Institut for Immunology
487 Grenzacherstrasse
CH 4005 Basel
Switzerland

Bowen, M.A. [255]
Dept. of Microbiology and Immunology
University of Miami Sch. of Medicine
P.O.Box 016960
Miami, FL 33101
USA

Bowne, D.B. [213]
Howard Hughes Medical Institute
Dept. of Pathology
Washington University Medical School
660 South Euclid
St. Louis, MI 63110
USA

Boyer, V.[509]
INSERM U271
69424 Lyon Cedex 03
France

Brandle, D. [65]
Institute of Experimental Immunology
University of Zürich
Sternwartstrasse 2
8091 Zürich
Switzerland

Brasseur, F. [871]
Ludwig Institute for Cancer Research
74 Ave. Hippocrate
B-1200 Brussels
Belgium

Brinkworth, R. [451]
Centre for Drug Design
University of Queensland
Australia

Brod, S.A. [627]
Center for Neurologic Diseases
Bringham and Women s Hospital
Harvard Medical School
75 Francis Street
Boston, MA 02115
USA

Bruggen, P. van der [871]
Ludwig Institute for Cancer Research
74 Ave. Hippocrate
B-1200 Brussels
Belgium

Bründler, M. [659]
Institute for Experimental Immunology
Dept. of Pathology
University of Zürich
Sternwartstrasse 2
CH-8091 Zürich
Switzerland

Brunner, M.C. [613]
Deutsches Rheumaforschungszentrum Berlin
Robert Koch Institut
Nordufer 20
D-1000 Berlin
Germany

Brunner, T. [411]
Institute of Clinical Immunology
Inselspital
Ch-3010 Bern
Switzerland

Bruno, L. [129]
Basel Institut for Immunology
487 Grenzacherstrasse
CH 4005 Basel
Switzerland

Buell, R.D. [107]
Dept. of Pediatrics
University of South Florida
801 Sixth Street S.
St. Petersburg, FL 33701
USA

Burakoff, S.J. [275]
Division of Pediatric Oncology
Dana Farber Cancer Institute
Boston, MA
USA

Bürki, K. [65],[255]
Preclinical Research
Sandoz Pharma Ltd.
4002 Basel
Switzerland

Cabanas, C. [283]
Macrophage Laboratory
Imperial Cancer Research Fund
44 Lincoln s Inn Fields
London, WC2A 3PX
UK

Callahan, J. [57]
Dept. of Medicine
Natl. Jewish Center for Immunology and
Respiratory Medicine
1400 Jackson Street
Denver, CO 80206
USA

Cambier, J.C. [199]
Division of Basic Sciences
Dept. of Pediatrics
Natl. Jewish Center for Immunology and
Respiratory Medicine
1400 Jackson Street
Denver, CO 80206
USA

Campbell, K.S. [199]
Division of Basic Sciences
Dept. of Pediatrics
Natl. Jewish Center for Immunology and
Respiratory Medicine
1400 Jackson Street
Denver, CO 80206
USA

Campbell, R.D. [525]
MRC Immunochemistry Unit
South Parks Road
Oxford OX1 3QU
UK

Campos, S.P. [377]
Roswell Park Cancer Institute
Dept. of Molecular and Cellular Biology
Buffalo, NY 14263
USA

Capron, A. [675],[739]
Centre d Immunologie et de
Biologie Parasitaire
Unité Mixte INSERM U167- CNRS 624
Institute Pasteur
1 rue du Calmette
59019 Lille Cedex
France

Capron, M. [675]
Centre d Immunologie et de
Biologie Parasitaire
Unité Mixte INSERM U167- CNRS 624
Institute Pasteur
1 rue du Calmette
59019 Lille Cedex
France

Caput, D. [613]
Laboratoire de Biologie Moleculaire
Sanofi Elf Bio-Recherches
Labege
France

Carli, M. De [239]
Division of Clinical Immunology and Allergy
University of Florence
Instituto di Clinica Medica 3
50134 Firenze
Italy

Carpenter, Ch. [627]
Center for Neurologic Diseases
Bringham and Women s Hospital
Harvard Medical School
75 Francis Street
Boston, MA 02115
USA

Carrera, A.C. [893]
The Johns Hopkins University
Dept. of Medicine
Oncology and Mol. Biology
Baltimore, MD 21205
USA

Carter, R.H. [495]
Johns Hopkins University Sch. of Medicine
Division of Molecular and Clinical Rheumatology
Baltimore, MD 21205
USA

Casamayor-Palleja, M. [21]
Dept. of Immunology
University of Birmingham Med. School
Birmingham, B15 2TT
UK

Cavazzana-Calvo, M. [557]
INSERM U132
Hopital des Enfants Malades
149, rue de Sevres
75730 Paris Cedex 15
France

Cerundolo, V. [175]
Institute of Molecular Medicine
John Radcliffe Hospital
Oxford
UK

Chan, A.C. [205]
Dept. of Medicine
University of California
San Francisco U426
3rd and Parnassus Ave.
San Francisco, CA
USA

Chandrasekhar, S. [849]
National Institute of Immunology
New Delhi 110067
India

Chaouat, G. [825],[841],[861]
Service de Gynecologie et
d Obstetrique
Hopital Antoine Beclere
Clamart 92140
France

Chatterjee, M. [619]
Roswell Park Cancer Center
Buffalo, NY 14263
USA

Chatterjee, N. [849]
National Institute of Immunology
New Delhi 110067
India

Chieco-Bianchi, L. [699]
Institute for Oncology
Interuniversity Center for Res. on Cancer
University of Padova
Via Gattamelata 64
I-35128 Padova
Italy

Chomez, P. [871]
Ludwig Institute for Cancer Research
74 Ave. Hippocrate
B-1200 Brussels
Belgium

Chua, K.Y. [427]
The Western Australian Research
Institute for Child Health
Princess Margaret Hospital
GPO Box D184
Perth, WA 6001
Australia

Ciernik, I.F. [659]
Institute for Experimental Immunology
Dept. of Pathology
University of Zürich
Sternwartstrasse 2
CH-8091 Zürich
Switzerland

Claas, F. [785]
University Hospital
Leyden
The Netherlands

Clark, D.A. [841]
McMaster University
Hamilton, Ontario
Canada

Clerici, M. [707]
Experimental Immunology Branch
NIC, NIH
Bethesda, MD
USA

Coffman, R.L. [707]
Dept. of Immunology
DNAX Research Institute of
Molecular and Cellular Biology Inc.
Palo Alto, CA
USA

Cohen, I.R. [579]
Dept. of cell Biology
The Weizmann Institute of Science
Rehovot 76100
Israel

Coley, J. [651]
Dept. of Immunology
Unilever Research
Colworth Laboratory
Sharnbrook, Bedford MK4 1LQ
UK

Collins, T. [275]
Division of Pediatric Oncology
Dana Farber Cancer Institute
Boston, MA
USA

Colten, H.R. [483]
Dept. of Pediatrics
Washington University Medical School
660 South Euclid
St. Louis, MI 63110
USA

Conceicao-Silva, F. [775]
WHO Immunology Res. and Training Centre
Institute of Biochemistry
University of Lausanne
Switzerland

Conley, M.E. [545]
Dept. of Immunology
St. Jude Children's Hospital
332 North Lauderdale
Memphis, TN 38105
USA

Convit, J. [761]
Instituto de Biomedicina
Caracas
Venezuela

Cooper, M.D. [535]
Howard Hughes Med. Istitute
Divisions of Developmental and
Clin. Immunology and Rheumatology
University of Alabama at Birmingham
Birmingham, AL 35294
USA

Corrigan, C.J. [403]
Dept. of Allergy and Clinical Immunology
National Heart and Lung Institute
Dovehouse Street
London, SW3 6LY
UK

Cory, S. [51]
The Walter and Eliza Hall
Institute of Med. Research
Royal Melbourne Hospital
Victoria 3050
Australia

Coulie, P.[871]
Ludwig Institute for Cancer Research
74 Ave. Hippocrate
B-1200 Brussels
Belgium

Coutinho, A. [603]
Unite d'Immunobiologie
CNRS URA 359
Institut Pasteur
25 rue du Docteur Roux
75724 Paris Cedex 15
France

Cronin, K. [361]
Dept. of Haematologic Oncology
Memorial Sloan Kettering Cancer Center
New York, NY 10021
USA

Csiszár, T. [861]
Institute Microbiology
University Medical School
H-7643 Pécs
Hungary

Culver, K.W. [553]
NCI, NHLBI
National Institutes of Health
Bethesda, MD
USA

D Andrea, E. [699]
Institute for Oncology
Interuniversity Center for Res. on Cancer
University of Padova
Via Gattamelata 64
I-35128 Padova
Italy

Daeron, M. [457]
Laboratoire d'Immunologie
Cellulaire et Clinique
INSERM U255, Institut Curie
26 rue d'Ulm
75231 Paris, Cedex 05
France

Dahinden, C.A. [411]
Institute of Clinical Immunology
Inselspital
Ch-3010 Bern
Switzerland

David, F. [825]
Service de Gynecologie et
d Obstetrique
Hopital Antoine Beclere
Clamart 92140
France

David, J.R. [753]
Dept. of Tropical Public Health
Harvard Medical School
Bringham and Women s Hospital
Boston, MA 02115
USA

Day, A.J. [525]
Dept. of Biochemistry
South Parks Road
Oxford OX1 3QU
UK

Deacock, S. [793]
Dept. of Immunology
Royal Postgraduate Medical School
Hammersmith Hospital
Du Cane Road
London, W12 ONN
UK

Deans, R.J. [877]
Kenneth Norris Jr. Cancer Center
University of Southern California
School of Medicine
Los Angeles, CA
USA

Deist, F. Le [557]
INSERM U132
Hopital des Enfants Malades
149, rue de Sevres
75730 Paris Cedex 15
France

Del Giudice, G. [683]
Microbiology and Immunology
World Health Organization
CH-1211 Geneva 27
Switzerland

Del Mistro, A. [699]
Institute for Oncology
Interuniversity Center for Res. on Cancer
University of Padova
Via Gattamelata 64
I-35128 Padova
Italy

Delespesse, G. [465]
Notre-Dame Hospital Res. Center
1560 Sherbrooke Street East
Montreal, Quebec, H2L 4M1
Canada

Delibrias, C. [509]
INSERM U28
Hopital Broussais
75014 Paris
France

Deng, G. [255]
Dept. of Microbiology and Immunology
University of Miami Sch. of Medicine
P.O.Box 016960
Miami, FL 33101
USA

Desgranges, C. [509]
INSERM U28
Hopital Broussais
75014 Paris
France

Dierich, M. [691]
Institute für Hygiene
Leopold Franzens Universitat
Fritz Pregl Strasse 3
A-6020 Innsbruck
Austria

Dietrich, G. [643]
INSERM U28
Hopital Broussais
Paris
France

Dinarello, Ch.A. [339]
Tufts University Sch. of Medicine
New England Medical Center
750 Washington Street
Boston, MA 02111
USA

Disanto, J. [557]
INSERM U132
Hopital des Enfants Malades
149, rue de Sevres
75730 Paris Cedex 15
France

Djian, V. [825]
Service de Gynecologie et
d Obstetrique
Hopital Antoine Beclere
Clamart 92140
France

Durham, S.R. [403]
Dept. of Allergy and Clinical Immunology
National Heart and Lung Institute
Dovehouse Street
London, SW3 6LY
UK

Duvaux-Miret, O.[675]
Centre d Immunologie et de
Biologie Parasitaire
Unité Mixte INSERM U167- CNRS 624
Institute Pasteur
1 rue du Calmette
59019 Lille Cedex
France

Eden, W. van [587]
Dept. of Infectious Diseases and Immunology
University of Utrecht
Yalelaan 1, P.O. Box 80165
3508 TD, Utrecht
The Netherlands

Ehrfeld, A. [269]
Max-Planck Institut für Biologie
Abteilung Immungenetik
7400 Tübingen
Germany

Eichmann, K. [269]
Max-Planck Institut für Biologie
Abteilung Immungenetik
7400 Tübingen
Germany

Eisenbarth, G. [627]
Center for Neurologic Diseases
Bringham and Women s Hospital
Harvard Medical School
75 Francis Street
Boston, MA 02115
USA

Elbe, A. [305]
Dept. of Dermatology I.
University of Vienna VIRCC
Brunnerstrasse 59
A-1235 Vienna
Austria

Elliott, T. [175]
Institute of Molecular Medicine
John Radcliffe Hospital
Oxford
UK

Erb, K. [305]
Institute of Virology and Immunobiology
University of Würzburg
Versbacherstrasse 7
D-8700 Würzburg
Germany

Eynde, B. van den [871]
Ludwig Institute for Cancer Research
74 Ave. Hippocrate
B-1200 Brussels
Belgium

Fearon, D.T. [495]
Johns Hopkins University Sch. of Medicine
Division of Molecular and Clinical Rheumatology
Baltimore, MD 21205
USA

Feinstein, A. [651]
Dept. of Immunology
AFRC Institute of Animal Physiology
Babraham, Cambridge CB2 4AT
UK

Finkelman, F.D. [419]
Dept. of Medicine
Uniformed Services
University of the Health Sciences
4301 Jones Bridge Rd.
Bethesda, MD
USA

Fischer, A. [557]
INSERM U132
Hopital des Enfants Malades
149, rue de Sevres
75730 Paris Cedex 15
France

Fischer, E. [509]
INSERM U28
Hopital Broussais
75014 Paris
France

Fischer-Lindahl, K. [145]
Dept. of Microbiology and Biochemistry
University of Texas
Health Science Center at Dallas
5323 Harry Hines Blvd.
Dallas, TX 75235-9050
USA

Fitch, F.W. [247]
Dept. of Pathology
The Ben May Institute
University of Chicago MC6027
5841 South Maryland Ave.
Chicago, IL 60637
USA

Flanders, K.C. [841]
Laboratory of Chemoprevention
NIH
Bethesda, MD
USA

Flaswinkel, H. [27]
Max-Planck Institut für Immunbiologie
Stübeweg 51
7800 Freiburg
Germany

Flores, E. [213]
Howard Hughes Medical Institute
Dept. of Pathology
Washington University Medical School
660 South Euclid
St. Louis, MI 63110
USA

Foon, K.A. [619]
Scripps Clinic and Res. Foundation
La Jolla, CA 92037
USA

Fournier, S. [465]
Notre-Dame Hospital Res. Center
1560 Sherbrooke Street East
Montreal, Quebec, H2L 4M1
Canada

Frank, M.M. [501]
Duke University Medical Center
Box 3352
Durham, NC 27710
USA

Freed, J.H. [57]
Dept. of Medicine
Natl. Jewish Center for Immunology and
Respiratory Medicine
1400 Jackson Street
Denver, CO 80206
USA

Fridman, W.H. [457]
Laboratoire d'Immunologie
Cellulaire et Clinique
INSERM U255, Institut Curie
26 rue d'Ulm
75231 Paris, Cedex 05
France

Fruth, U. [775]
WHO Immunology Res. and Training Centre
Institute of Biochemistry
University of Lausanne
Switzerland

Fujimoto, K. [369]
Dept. of Molecular Pathology
Cancer Research Institute
Kanazawa University
13-1 Takaramachi
Kanazawa 920
Japan

Fukita, Y. [91]
Center for Molecular Biology and Genetics
Faculty of Medicine, Kyoto University
Sakyo-ku, Kyoto 606
Japan

Gajewski, T.F. [247]
Dept. of Pathology
The Ben May Institute
University of Chicago MC6027
5841 South Maryland Ave.
Chicago, IL 60637
USA

Gani, M. [651]
Dept. of Immunology
Unilever Research
Colworth Laboratory
Sharnbrook, Bedford MK4 1LQ
UK

Gansbacher, B. [361]
Dept. of Haematologic Oncology
Memorial Sloan Kettering Cancer Center
New York, NY 10021
USA

Garchon, H.J. [595]
INSERM U25
Hopital Necker
161 rue de Sevres
75015 Paris
France

Giudizi, M.G. [239]
Division of Clinical Immunology and Allergy
University of Florence
Instituto di Clinica Medica 3
50134 Firenze
Italy

Goldman, C. [333]
Metabolism Branch
NCI, NIH
Bethesda, MD 20892
USA

Goldman, J. [793]
Dept. of Haematology
Royal Postgraduate Medical School
Hammersmith Hospital
Du Cane Road
London, W12 ONN
UK

Good, R.A. [107]
Dept. of Pediatrics
University of South Florida
801 Sixth Street S.
St. Petersburg, FL 33701
USA

Gordon, J. [21]
Dept. of Immunology
University of Birmingham Med. School
Birmingham, B15 2TT
UK

Groettrup, M. [129]
Basel Institut for Immunology
487 Grenzacherstrasse
CH 4005 Basel
Switzerland

Guarini, R. [361]
Dept. of Haematologic Oncology
Memorial Sloan Kettering Cancer Center
New York, NY 10021
USA

Guilbert, L.J. [857]
Dept. of Immunology
University of Alberta
8-65 Medical Sciences Bldg.
Edmonton, Alberta T6G 2H7
Canada

Gulizia, R.J. [725]
The Scripps Research Institute
La Jolla, CA 92037
USA

Gupta, H. [849]
National Institute of Immunology
New Delhi 110067
India

Gyódi, É. [807]
National Institute of Haematology
Blood Transfusion and Immunology
Daróczi 24
H-1113 Budapest
Hungary

Haas, W. [561]
Hoffmann-La Roche Inc.
340 Kingsland Street
Bldg. 102, Rm. 116d
Nutley, NJ 07110-1199
USA

Haasner, D. [11]
Basel Institut for Immunology
487 Grenzacherstrasse
CH 4005 Basel
Switzerland

Haeffner-Cavaillon, N. [509]
INSERM U28
Hopital Broussais
75014 Paris
France

Hafler, D.A. [627]
Center for Neurologic Diseases
Bringham and Women s Hospital
Harvard Medical School
75 Francis Street
Boston, MA 02115
USA

Hagen, M. [443]
Dept. of Pathology
University of Iowa
College of Medicine
Iowa City, Iowa 52242
USA

Hahn, W.C. [275]
Division of Pediatric Oncology
Dana Farber Cancer Institute
Boston, MA
USA

Haino, M. [91]
Center for Molecular Biology and Genetics
Faculty of Medicine, Kyoto University
Sakyo-ku, Kyoto 606
Japan

Haire, R.N.[107]
Dept. of Pediatrics
University of South Florida
801 Sixth Street S.
St. Petersburg, FL 33701
USA

Hamid, Q. [403]
Dept. of Allergy and Clinical Immunology
National Heart and Lung Institute
Dovehouse Street
London, SW3 6LY
UK

Hardie, D. [21]
Dept. of Immunology
University of Birmingham Med. School
Birmingham, B15 2TT
UK

Hardy, R.R. [43]
Institute for Cancer Research
Fox Chase Cancer Center
Philadelphia, PA 19111
USA

Harel, W. [877]
Kenneth Norris Jr. Cancer Center
University of Southern California
School of Medicine
Los Angeles, CA
USA

Harris, A.W. [51]
The Walter and Eliza Hall
Institute of Med. Research
Royal Melbourne Hospital
Victoria 3050
Australia

Harvey,J. [283]
Macrophage Laboratory
Imperial Cancer Research Fund
44 Lincoln s Inn Fields
London, WC2A 3PX
UK

Hashim, G. [635]
Dept. of Microbiol. and Surgery
St. Luke s Roosevelt Hospital Center
Columbia University
New York, NY
USA

Hatekeyama, M. [321]
Institute for Molecular and Cellular Biology
Osaka University
Yamadaoka 1-3
Suitashi, Osaka 565
Japan

Haury M. [603]
Unite d'Immunobiologie
CNRS URA 359
Institut Pasteur
25 rue du Docteur Roux
75724 Paris Cedex 15
France

Hayakawa, K. [43]
Institute for Cancer Research
Fox Chase Cancer Center
Philadelphia, PA 19111
USA

Heap, R.B. [651]
Dept. of Immunology
AFRC Institute of Animal Physiology
Babraham, Cambridge CB2 4AT
UK

Heath, A.W. [327]
DNAX Research Institute
901 California Ave.
Palo Alto, CA 94304
USA

Heckels, J.E. [667]
Molecular Microbiology Group
University of Southampton Medical School
Southampton General Hospital
Southampton, SO9 4XY
UK

Hedrick, S. [261]
University of California
San Diego, CA
USA

Heider, K.-H. [289]
Kernforschungszentrum Karlsruhe
Institut für Genetik und Toxicologie
PO Box 3640
D-7500 Karlsruhe 1
Germany

Hein, W.R. [121]
Basel Institut for Immunology
487 Grenzacherstrasse
CH 4005 Basel
Switzerland

Helbert, M.R. [613]
Deutsches Rheumaforschungszentrum Berlin
Robert Koch Institut
Nordufer 20
D-1000 Berlin
Germany

Hengartner, H. [65],[255]
Institute of Experimental Immunology
University of Zürich
Sternwartstrasse 2
8091 Zürich
Switzerland

Herrlich, P. [289]
Kernforschungszentrum Karlsruhe
Institut für Genetik und Toxicologie
PO Box 3640
D-7500 Karlsruhe 1
Germany

Higgins, P. [627]
Center for Neurologic Diseases
Bringham and Women s Hospital
Harvard Medical School
75 Francis Street
Boston, MA 02115
USA

Hinds-Frey, K.R. [107]
Dept. of Pediatrics
University of South Florida
801 Sixth Street S.
St. Petersburg, FL 33701
USA

Hivroz, C. [557]
INSERM U132
Hopital des Enfants Malades
149, rue de Sevres
75730 Paris Cedex 15
France

Hofmann, M. [289]
Kernforschungszentrum Karlsruhe
Institut für Genetik und Toxicologie
D-7500 Karlsruhe 1
Germany

Hogarth, P.M. [451]
Schutt Laboratory for Immunology
Austin Research Institute
Heidelberg 3084
Australia

Hogg, N. [283]
Macrophage Laboratory
Imperial Cancer Research Fund
44 Lincoln s Inn Fields
London, WC2A 3PX
UK

Holland, G. [355]
Dept. of Immunology
DNAX Research Institute
901 California Ave.
Palo ALto, CA 94304
USA

Honjo, T. [91]
Center for Molecular Biology and Genetics
Faculty of Medicine, Kyoto University
Sakyo-ku, Kyoto 606
Japan

Horak, I. [305]
Institute of Virology and Immunobiology
University of Würzburg
Versbacherstrasse 7
D-8700 Würzburg
Germany

Horuzsko, A. [807]
National Institute of Haematology
Blood Transfusion and Immunology
Daróczi 24
H-1113 Budapest
Hungary

Houssiau, F. [313]
Ludwig Institute for Cancer Research
Catholic University of Louvain
74 avenue Hippocrate
B-1200 Brussels
Belgium

Howard, M.C. [327]
DNAX Research Institute
901 California Ave.
Palo Alto, CA 94304
USA

Hu-Li, J. [347]
Laboratory of Immunology
NIAID, NIH
Bethesda, MD 20892
USA

Hudak, S. [355]
Dept. of Immunology
DNAX Research Institute
901 California Ave.
Palo ALto, CA 94304
USA

Hugo, P. [57]
Dept. of Medicine
Natl. Jewish Center for Immunology and
Respiratory Medicine
1400 Jackson Street
Denver, CO 80206
USA

Hulett, M.D. [451]
Schutt Laboratory for Immunology
Austin Research Institute
Heidelberg 3084
Australia

Hulst, M.A. [107]
Dept. of Pediatrics
University of South Florida
801 Sixth Street S.
St. Petersburg, FL 33701
USA

Humphreys, A.S.[651]
Dept. of Immunology
AFRC Institute of Animal Physiology
Babraham, Cambridge CB2 4AT
UK

Hünig, T. [305]
Institute of Virology and Immunobiology
University of Würzburg
Versbacherstrasse 7
D-8700 Würzburg
Germany

Hunt, B. [355]
Dept. of Immunology
DNAX Research Institute
901 California Ave.
Palo ALto, CA 94304
USA

Hurez, V. [643]
INSERM U28
Hopital Broussais
Paris
France

Ierio, F. [451]
Schutt Laboratory for Immunology
Austin Research Institute
Heidelberg 3084
Australia

Ignatowitz, L. [57]
Dept. of Medicine
Natl. Jewish Center for Immunology and
Respiratory Medicine
1400 Jackson Street
Denver, CO 80206
USA

Igras, V. [275]
Division of Pediatric Oncology
Dana Farber Cancer Institute
Boston, MA
USA

Imai, T. [91]
Gene Bank, Tsukuba Life Science Center
Inst. of Physical and Chemical Research
Koyadai, Tsukuba, Ibaraki 305
Japan

Imhof, B.A. [121]
Basel Institut for Immunology
487 Grenzacherstrasse
CH 4005 Basel
Switzerland

Indraccolo, S. [699]
Institute for Oncology
Interuniversity Center for Res. on Cancer
University of Padova
Via Gattamelata 64
I-35128 Padova
Italy

Irving, B. [205]
Dept. of Physiology
University of California
San Francisco U426
3rd and Parnassus Ave.
San Francisco, CA
USA

Ishida, H. [327]
DNAX Research Institute
901 California Ave.
Palo Alto, CA 94304
USA

Ishihara, H. [465]
Notre-Dame Hospital Res. Center
1560 Sherbrooke Street East
Montreal, Quebec, H2L 4M1
Canada

Iwashima, M. [205]
Howard Hughes Medical Institute
University of California
San Francisco U426
3rd and Parnassus Ave.
San Francisco, CA
USA

Janeway Jr., C.A. [181]
Section of Immunobiology
Yale University Sch. of Medicine
New Haven, CT 06510
USA

Johnson, G.D. [21]
Dept. of Immunology
University of Birmingham Med. School
Birmingham, B15 2TT
UK

Jones, V.E. [377]
Roswell Park Cancer Institute
Dept. of Molecular and Cellular Biology
Buffalo, NY 14263
USA

Jouvin, M.-H. [435]
Molecular Allergy and Immunology Section
NIAD, NIH
12441 Parklawn Drive
Rockville, MD 20852
USA

Kagi, D. [255]
Dept. of Experimental Pathology
University of Zurich
Sternwartstrasse 2
Zurich
Switzerland

Kan-Mitchell, J. [877]
Kenneth Norris Jr. Cancer Center
University of Southern California
School of Medicine
Los Angeles, CA
USA

Kapovic, M. [861]
Dept. of Physicology an Immunology
University of Rijeka
Croatia

Kappler, J.W. [57]
Dept. of Medicine
Natl. Jewish Center for Immunology and
Respiratory Medicine
1400 Jackson Street
Denver, CO 80206
USA

Karasuyama, H. [11]
Basel Institut for Immunology
487 Grenzacherstrasse
CH 4005 Basel
Switzerland

Karvelas, M. [3]
The Walter and Elisa Hall Institute
The Royal Melbourne Hospital
Victoria 3050
Australia

Kassai, M. [807]
Blood Center
Country Hospital
Hódmezővásárhely
Hungary

Kaushic, C. [849]
National Institute of Immunology
New Delhi 110067
India

Kaveri, S.V. [643]
INSERM U28
Hopital Broussais
Paris
France

Kay, A.B. [403]
Dept. of Allergy and Clinical Immunology
National Heart and Lung Institute
Dovehouse Street
London, SW3 6LY
UK

Kazatchkine, M.D. [509]
INSERM U28
Hopital Broussais
75014 Paris
France

Kelvin, D.J. [297]
Laboratory of Molecular Immunoregulation
Natl. Cancer Institute - Frederick
Cancer and Res. Development Center
Frederick, MD 21702-1201
USA

Kent, U.M. [473]
Section on Chemical Immunology
Arthritis and Rheumatism Branch
NIAMSD, NIH
Bethesda, MD 20892
USA

Khoury, S.J. [627]
Center for Neurologic Diseases
Bringham and Women s Hospital
Harvard Medical School
75 Francis Street
Boston, MA 02115
USA

Kim, K.-M. [27]
Max-Planck Institut für Immunbiologie
Stübeweg 51
7800 Freiburg
Germany

Kinet, J.-P. [435]
Molecular Allergy and Immunology Section
NIAD, NIH
12441 Parklawn Drive
Rockville, MD 20852
USA

Kinsky, R. [825],[861]
Service de Gynecologie et
d Obstetrique
Hopital Antoine Beclere
Clamart 92140
France

Kirberg, J. [129]
Basel Institut for Immunology
487 Grenzacherstrasse
CH 4005 Basel
Switzerland

Kishimoto, T. [887]
Dept. of Medicine III.
Osaka University Med. School
1-1-50 Fukushima, Fukushima-ku
Osaka 553
Japan

Kisielow, P. [81], [129]
Basel Institut for Immunology
487 Grenzacherstrasse
CH 4005 Basel
Switzerland

Klein, D. [137]
Dept. of Microbiology and Immunology
University of Miami Sch. of Medicine
Miami, FL 33101
USA

Klein, J. [137], [153]
Max-Planck Institut für Biologie
Abteilung Immungenetik
7400 Tübingen
Germany

Knight, K.L. [99]
Dept. of Immunology and Microbiology
Stritch School of Medicine
Loyola University
Chicago, Maywood, IL 60135
USA

Kobayashi, N. [321]
Institute for Molecular and Cellular Biology
Osaka University
Yamadaoka 1-3
Suitashi, Osaka 565
Japan

Kohler, H. [619]
IDEC Pharmaceuticals Corporation
La Jolla, CA 92037
USA

Kono, T. [321]
Institute for Molecular and Cellular Biology
Osaka University
Yamadaoka 1-3
Suitashi, Osaka 565
Japan

Koulmanda, M. [815]
Transplantation Unit
The Walter and Eliza Hall Institute of
Medical Research
Parkville, 3050 Victoria
Australia

Kramer, S. [305]
Institute of Virology and Immunobiology
University of Würzburg
Versbacherstrasse 7
D-8700 Würzburg
Germany

Krieger, M. [411]
Institute of Clinical Immunology
Inselspital
Ch-3010 Bern
Switzerland

Kubagawa, H. [535]
Howard Hughes Med. Istitute
Divisions of Developmental and
Clin. Immunology and Rheumatology
University of Alabama at Birmingham
Birmingham, AL 35294
USA

Kuhn, R. [355],[561]
Universitat Köln
Institute für Genetik
Weyertal 121
D-5000 Köln
Germany

Kullberg, M. [731]
Dept. of Immunology
Stockholm University
S-106 91 Stockholm
Sweden

Kumar, S. [731]
Malaria Section
NIAID, NIH
Bethesda, MD 20892
USA

Kyburz, D. [659]
Institute for Experimental Immunology
Dept. of Pathology
University of Zürich
Sternwartstrasse 2
CH-8091 Zürich
Switzerland

Lalor, P.A. [3]
The Walter and Elisa Hall Institute
The Royal Melbourne Hospital
Victoria 3050
Australia

Lambert, P.-H. [683]
Microbiology and Immunology
World Health Organization
CH-1211 Geneva 27
Switzerland

Lancki, D.W. [247]
Dept. of Pathology
The Ben May Institute
University of Chicago MC6027
5841 South Maryland Ave.
Chicago, IL 60637
USA

Landis, R.C. [283]
Macrophage Laboratory
Imperial Cancer Research Fund
44 Lincoln s Inn Fields
London, WC2A 3PX
UK

Lanzavecchia, A. [189]
Basel Institut for Immunology
487 Grenzacherstrasse
CH 4005 Basel
Switzerland

Larcher, C. [691]
Institute für Hygiene
Leopold Franzens Universitat
Fritz Pregl Strasse 3
A-6020 Innsbruck
Austria

Lassoued, K. [535]
Howard Hughes Med. Istitute
Divisions of Developmental and
Clin. Immunology and Rheumatology
University of Alabama at Birmingham
Birmingham, AL 35294
USA

Lea, R.G. [841]
McMaster University
Hamilton, Ontario
Canada

Lechler, R. [793]
Dept. of Immunology
Royal Postgraduate Medical School
Hammersmith Hospital
Du Cane Road
London, W12 ONN
UK

Lethe, B. [871]
Ludwig Institute for Cancer Research
74 Ave. Hippocrate
B-1200 Brussels
Belgium

Levelt, C. [269]
Max-Planck Institut für Biologie
Abteilung Immungenetik
7400 Tübingen
Germany

Levy, J.A. [725]
Cancer Research Institute
School of Medicine
University of California-San Francisco
San Francisco, CA 94143
USA

Lichtenheld, M.G. [255]
Dept. of Microbiology and Immunology
University of Miami Sch. of Medicine
P.O.Box 016960
Miami, FL 33101
USA

Lider, O. [627]
Center for Neurologic Diseases
Bringham and Women s Hospital
Harvard Medical School
75 Francis Street
Boston, MA 02115
USA

Lin, H. [857]
Dept. of Immunology
University of Alberta
8-65 Medical Sciences Bldg.
Edmonton, Alberta T6G 2H7
Canada

Lisowska-Grospierre, B. [557]
INSERM U132
Hopital des Enfants Malades
149, rue de Sevres
75730 Paris Cedex 15
France

Litman, G.W. [107]
Dept. of Pediatrics
University of South Florida
801 Sixth Street S.
St. Petersburg, FL 33701
USA

Litman, R.T. [107]
Dept. of Pediatrics
University of South Florida
801 Sixth Street S.
St. Petersburg, FL 33701
USA

Liu, Y.-J. [21]
Dept. of Immunology
University of Birmingham Med. School
Birmingham, B15 2TT
UK

Lloyd, A.W. [297]
Laboratory of Molecular Immunoregulation
Natl. Cancer Institute - Frederick
Cancer and Res. Development Center
Frederick, MD 21702-1201
USA

Lombardi, G. [793]
Dept. of Immunology
Royal Postgraduate Medical School
Hammersmith Hospital
Du Cane Road
London, W12 ONN
UK

Lorquin, C. [871]
Ludwig Institute for Cancer Research
74 Ave. Hippocrate
B-1200 Brussels
Belgium

Louis, J.A. [683],[**775**]
Microbiology and Immunology
World Health Organization
CH-1211 Geneva 27
Switzerland

Lynch, R.G. [443]
Dept. of Pathology
University of Iowa
College of Medicine
Iowa City, Iowa 52242
USA

MacIsaac, P.D. [725]
Immunology Division
Medical Biology Institute
La Jolla, CA 92037
USA

Maclennan, I.C.M. [21]
Dept. of Immunology
University of Birmingham Med. School
Birmingham, B15 2TT
UK

Maggi, E. [239]
Division of Clinical Immunology and Allergy
University of Florence
Instituto di Clinica Medica 3
50134 Firenze
Italy

Mandel, T.E. [815]
Transplantation Unit
The Walter and Eliza Hall Institute of
Medical Research
Parkville, 3050 Victoria
Australia

Manetti, R. [239]
Division of Clinical Immunology and Allergy
University of Florence
Instituto di Clinica Medica 3
50134 Firenze
Italy

Mao, S.-Y. [473]
Section on Chemical Immunology
Arthritis and Rheumatism Branch
NIAMSD, NIH
Bethesda, MD 20892
USA

Marcos, M.A.R. [603]
Unite d'Immunobiologie
CNRS URA 359
Institut Pasteur
25 rue du Docteur Roux
75724 Paris Cedex 15
France

Margittai, M. [107]
Dept. of Pediatrics
University of South Florida
801 Sixth Street S.
St. Petersburg, FL 33701
USA

Markiewicz, S. [557]
INSERM U132
Hopital des Enfants Malades
149, rue de Sevres
75730 Paris Cedex 15
France

Marrack, P. [57]
Dept. of Medicine
Natl. Jewish Center for Immunology and
Respiratory Medicine
1400 Jackson Street
Denver, CO 80206
USA

Marschang, P. [691]
Institute für Hygiene
Leopold Franzens Universitat
Fritz Pregl Strasse 3
A-6020 Innsbruck
Austria

Matis, L. [261]
Medicine Branch
NCI, NIH
Bethesda, MD
USA

Matsuda, F. [91]
Center for Molecular Biology and Genetics
Faculty of Medicine, Kyoto University
Sakyo-ku, Kyoto 606
Japan

Matsui, M. [627]
Center for Neurologic Diseases
Bringham and Women s Hospital
Harvard Medical School
75 Francis Street
Boston, MA 02115
USA

Matsumura, R. [91]
Center for Molecular Biology and Genetics
Faculty of Medicine, Kyoto University
Sakyo-ku, Kyoto 606
Japan

Matthews, R.J. [213]
Howard Hughes Medical Institute
Dept. of Pathology
Washington University Medical School
660 South Euclid
St. Louis, MI 63110
USA

McCormack, J. [57]
Dept. of Medicine
Natl. Jewish Center for Immunology and
Respiratory Medicine
1400 Jackson Street
Denver, CO 80206
USA

McDowall, A. [283]
Macrophage Laboratory
Imperial Cancer Research Fund
44 Lincoln s Inn Fields
London, WC2A 3PX
UK

McFarland, E.C. [213]
Howard Hughes Medical Institute
Dept. of Pathology
Washington University Medical School
660 South Euclid
St. Louis, MI 63110
USA

McHeyzer-Williams, M.G. [3]
The Walter and Elisa Hall Institute
The Royal Melbourne Hospital
Victoria 3050
Australia

McIntyre, J.A. [833]
Center for Reproduction and
Transplantation Immunology
Methodist Hospital of Indiana
Indianapolis, IN 46202
USA

McKenzie, I.F.C. [451]
Schutt Laboratory for Immunology
Austin Research Institute
Heidelberg 3084
Australia

McKisic, M.D. [247]
Dept. of Pathology
The Ben May Institute
University of Chicago MC6027
5841 South Maryland Ave.
Chicago, IL 60637
USA

McLean, M. [3]
The Walter and Elisa Hall Institute
The Royal Melbourne Hospital
Victoria 3050
Australia

Melchers, F. [11]
Basel Institut for Immunology
487 Grenzacherstrasse
CH 4005 Basel
Switzerland

Menu, E. [825]
Service de Gynecologie et
d Obstetrique
Hopital Antoine Beclere
Clamart 92140
France

Metzger, H. [473]
Section on Chemical Immunology
Arthritis and Rheumatism Branch
NIAMSD, NIH
Bethesda, MD 20892
USA

Migita, K. [73]
Dept. of Immunology and Med. Genetics
University of Toronto
Mount Sinai Hospital
600 University Ave.
Toronto, Ontario, M5G 1X5
Canada

Mihalik, R. [807]
National Institute of Haematology
Blood Transfusion and Immunology
Daróczi 24
H-1113 Budapest
Hungary

Miller, A. [627]
Center for Neurologic Diseases
Bringham and Women s Hospital
Harvard Medical School
75 Francis Street
Boston, MA 02115
USA

Miller, A.D. [553]
Fred Hutchinson Cancer Center
Seattle, WA
USA

Milon, G. [775]
Institut Pasteur
Paris
France

Minami, Y. [321]
Institute for Molecular and Cellular Biology
Osaka University
Yamadaoka 1-3
Suitashi, Osaka 565
Japan

Mion, M. [699]
Institute for Oncology
Interuniversity Center for Res. on Cancer
University of Padova
Via Gattamelata 64
I-35128 Padova
Italy

Mitchell, M.S. [877]
Kenneth Norris Jr. Cancer Center
University of Southern California
School of Medicine
Los Angeles, CA
USA

Mitchison, N.A. [613]
Deutsches Rheumaforschungszentrum Berlin
Robert Koch Institut
Nordufer 20
D-1000 Berlin
Germany

Mitomo, K. [369]
Dept. of Molecular Pathology
Cancer Research Institute
Kanazawa University
13-1 Takaramachi
Kanazawa 920
Japan

Modlin, R.L. [761]
Division of Dermatology
Dept. of Microbiol. and Immunology
UCLA School of Medicine
Los Angeles, CA 90033
USA

Moore, K.W. [327]
DNAX Research Institute
901 California Ave.
Palo Alto, CA 94304
USA

Moqbel, R. [403]
Dept. of Allergy and Clinical Immunology
National Heart and Lung Institute
Dovehouse Street
London, SW3 6LY
UK

Morella, K.K. [377]
Roswell Park Cancer Institute
Dept. of Molecular and Cellular Biology
Buffalo, NY 14263
USA

Moser, B. [385]
Theodor-Kocher Institut
University of Berne
Switzerland

Mosier, D.E. [725]
The Scripps Research Institute
La Jolla, CA 92037
USA

Moskophidis, D. [659]
Institute for Experimental Immunology
Dept. of Pathology
University of Zürich
Sternwartstrasse 2
CH-8091 Zürich
Switzerland

Mosmann, T.R. [857]
Dept. of Immunology
University of Alberta
8-65 Medical Sciences Bldg.
Edmonton, Alberta T6G 2H7
Canada

Mozes, E. [573]
Dept. of Chemical Immunology
Weizmann Institute of Science
Rehovot, 76100
Israel

Müller, Ch. [65]
Dept. of Pathology
University of Bern
3010 Bern
Switzerland

Müller, I. [775]
WHO Immunology Res. and Training Centre
Institute of Biochemistry
University of Lausanne
Switzerland

Müller, S. [619]
San Diego Regional Cancer Center
San Diego, CA 92121
USA

Müller, W. [355]
Universitat Köln
Weyertal 121
D-5000 Köln
Germany

Murphy, D.B. [181]
Section of Immunobiology
Yale University Sch. of Medicine
New Haven, CT 06510
USA

Nagaoka, H. [91]
Center for Molecular Biology and Genetics
Faculty of Medicine, Kyoto University
Sakyo-ku, Kyoto 606
Japan

Nakamura, T. [535]
Howard Hughes Med. Istitute
Divisions of Developmental and
Clin. Immunology and Rheumatology
University of Alabama at Birmingham
Birmingham, AL 35294
USA

Nakayama, K. [369]
Dept. of Molecular Pathology
Cancer Research Institute
Kanazawa University
13-1 Takaramachi
Kanazawa 920
Japan

Natsuume-Sakai, S. [517]
Dept. of Immunobiology
Cancer Research Institute
Kanazawa University
Kanazawa
Japan

Neefjes, J.J. [167]
The Netherlands Cancer Institute
Plesmanlaan 121
1066 CX Amsterdam
The Netherlans

Nelson, D. [333]
Metabolism Branch
NCI, NIH
Bethesda, MD 20892
USA

Nemazee, D. [35]
Dept. of Pediatrics
Natl. Jewish Center for Immunology and
Respiratory Medicine
1400 Jackson Street
Denver, CO 80206
USA

Neta, R. [297]
Dept. of Experimental Haematology
AFRRI
Bethesda, MD 20814

Nishimoto, N. [535]
Howard Hughes Med. Istitute
Divisions of Developmental and
Clin. Immunology and Rheumatology
University of Alabama at Birmingham
Birmingham, AL 35294
USA

Noben, N. [443]
Dept. of Pathology
University of Iowa
College of Medicine
Iowa City, Iowa 52242
USA

Nobrega, A. [603]
Unite d'Immunobiologie
CNRS URA 359
Institut Pasteur
25 rue du Docteur Roux
75724 Paris Cedex 15
France

Nonaka, M. [517]
Dept. of Immunology
Kanazawa Medical University
Uchinada, Ishikawa
Japan

Nossal, G.J.V. [3]
The Walter and Elisa Hall Institute
The Royal Melbourne Hospital
Victoria 3050
Australia

Nunez, C. [535]
Howard Hughes Med. Istitute
Divisions of Developmental and
Clin. Immunology and Rheumatology
University of Alabama at Birmingham
Birmingham, AL 35294
USA

Nunez, R. [443]
Dept. of Pathology
University of Iowa
College of Medicine
Iowa City, Iowa 52242
USA

Nussenblatt, R.B. [627]
Center for Neurologic Diseases
Bringham and Women s Hospital
Harvard Medical School
75 Francis Street
Boston, MA 02115
USA

O'hUigin, C. [137]
Max-Planck Institut für Biologie
Abteilung Immungenetik
7400 Tübingen
Germany

Ochi, A. [73]
Dept. of Immunology and Med. Genetics
University of Toronto
Mount Sinai Hospital
600 University Ave.
Toronto, Ontario, M5G 1X5
Canada

Odermatt, B. [659]
Institute for Experimental Immunology
Dept. of Pathology
University of Zürich
Sternwartstrasse 2
CH-8091 Zürich
Switzerland

Oehen, S. [65]
Institute of Experimental Immunology
University of Zürich
Sternwartstrasse 2
8091 Zürich
Switzerland

Offner, H. [635]
Neuroinmmunology Research 151D
V.A. Medical Center
Dept. of Microbiol. and Immunology
Portland, OR 97201
USA

Ohashi, P. [65]
Institute of Experimental Immunology
University of Zürich
Sternwartstrasse 2
8091 Zürich
Switzerland

Ohnheiser, R. [115]
Institut für Physiologische Chemie
Universitat München
Schillerstrasse 44
D-8000 München
Germany

Ohta, Y. [107]
Dept. of Pediatrics
University of South Florida
801 Sixth Street S.
St. Petersburg, FL 33701
USA

Olsen, K.J. [255]
Dept. of Microbiology and Immunology
University of Miami Sch. of Medicine
P.O.Box 016960
Miami, FL 33101
USA

Ono, H. [137]
Max-Planck Institut für Biologie
Abteilung Immungenetik
7400 Tübingen
Germany

Oppenheim, J.J. [297]
Laboratory of Molecular Immunoregulation
Natl. Cancer Institute - Frederick
Cancer and Res. Development Center
Frederick, MD 21702-1201
USA

Padányi, Á. [807]
National Institute of Haematology
Blood Transfusion and Immunology
Daróczi 24
H-1113 Budapest
Hungary

Pajovic, S. [377]
Roswell Park Cancer Institute
Dept. of Molecular and Cellular Biology
Buffalo, NY 14263
USA

Pal, R. [849]
National Institute of Immunology
New Delhi 110067
India

Pals, S.T. [289]
Academic Medical Center
Dept. of Pathology
Meibergdreef 9
NL-1105 AZ Amsterdam
The Netherlands

Paolini, R. [435]
Molecular Allergy and Immunology Section
NIAD, NIH
12441 Parklawn Drive
Rockville, MD 20852
USA

Pardoll, D.M. [893]
The Johns Hopkins University
Dept. of Medicine
Oncology and Mol. Biology
Baltimore, MD 21205
USA

Park, J.K. [275]
Division of Pediatric Oncology
Dana Farber Cancer Institute
Boston, MA
USA

Parolini, O. [545]
Dept. of Immunology
St. Jude Children's Hospital
332 North Lauderdale
Memphis, TN 38105
USA

Parronchi, P. [239]
Division of Clinical Immunology and Allergy
University of Florence
Instituto di Clinica Medica 3
50134 Firenze
Italy

Paul, W.E. [347]
Laboratory of Immunology
NIAID, NIH
Bethesda, MD 20892
USA

Pecht, I. [221]
Dept. of Chemical Immunology
The Weizmann Institute of Science
Rehovot 76100
Israel

Perlaza, B. [775]
WHO Immunology Res. and Training Centre
Institute of Biochemistry
University of Lausanne
Switzerland

Perlmann, H. [731]
Dept. of Immunology
Stockholm University
S-106 91 Stockholm
Sweden

Perlmann, P. [731]
Dept. of Immunology
Stockholm University
S-106 91 Stockholm
Sweden

Perner, F. [807]
1st Clinical of Surgery
Semmelweis Medical School
Budapest
Hungary

Petrányi, G.Gy. [807]
National Institute of Haematology
Blood Transfusion and Immunology
Daróczi 24
H-1113 Budapest
Hungary

Piccinni, M.-P. [239]
Division of Clinical Immunology and Allergy
University of Florence
Instituto di Clinica Medica 3
50134 Firenze
Italy

Pingel, J.T. [213]
Howard Hughes Medical Institute
Dept. of Pathology
Washington University Medical School
660 South Euclid
St. Louis, MI 63110
USA

Pircher, H.P. [65], [659]
Institute of Experimental Immunology
University of Zürich
Sternwartstrasse 2
8091 Zürich
Switzerland

Plaen, E. de [871]
Ludwig Institute for Cancer Research
74 Ave. Hippocrate
B-1200 Brussels
Belgium

Pócsik, É. [807]
National Institute of Haematology
Blood Transfusion and Immunology
Daróczi 24
H-1113 Budapest
Hungary

Podack, E.R. [255]
Dept. of Microbiology and Immunology
University of Miami Sch. of Medicine
P.O.Box 016960
Miami, FL 33101
USA

Ponta, H. [289]
Kernforschungszentrum Karlsruhe
Institut für Genetik und Toxicologie
PO Box 3640
D-7500 Karlsruhe 1
Germany

Potocnik, A. [269]
Max-Planck Institut für Biologie
Abteilung Immungenetik
7400 Tübingen
Germany

Powell, M.S. [451]
Schutt Laboratory for Immunology
Austin Research Institute
Heidelberg 3084
Australia

Preston-Hurlburt, P.[181]
Section of Immunobiology
Yale University Sch. of Medicine
New Haven, CT 06510
USA

Prete, G.F. Del [239]
Division of Clinical Immunology and Allergy
University of Florence
Instituto di Clinica Medica 3
50134 Firenze
Italy

Pribluda, V.S. [473]
Section on Chemical Immunology
Arthritis and Rheumatism Branch
NIAMSD, NIH
Bethesda, MD 20892
USA

Pulendran, B. [3]
The Walter and Elisa Hall Institute
The Royal Melbourne Hospital
Victoria 3050
Australia

Quilliam, A.J. [451]
Schutt Laboratory for Immunology
Austin Research Institute
Heidelberg 3084
Australia

Raman, Ch. [99]
Dept. of Immunology and Microbiology
Stritch School of Medicine
Loyola University
Chicago, Maywood, IL 60135
USA

Rast, J.P. [107]
Dept. of Pediatrics
University of South Florida
801 Sixth Street S.
St. Petersburg, FL 33701
USA

Reich, E.-P. [181]
Section of Immunobiology
Yale University Sch. of Medicine
New Haven, CT 06510
USA

Reimann, A. [269]
Max-Planck Institut für Biologie
Abteilung Immungenetik
7400 Tübingen
Germany

Renauld, J.-C. [313]
Ludwig Institute for Cancer Research
Catholic University of Louvain
74 avenue Hippocrate
B-1200 Brussels
Belgium

Rennick, D. [355]
Dept. of Immunology
DNAX Research Institute
901 California Ave.
Palo ALto, CA 94304
USA

Reth, M. [27]
Max-Planck Institut für Immunbiologie
Stübeweg 51
7800 Freiburg
Germany

Réti, M. [807]
National Institute of Haematology
Blood Transfusion and Immunology
Daróczi 24
H-1113 Budapest
Hungary

Reynaud, C.-A. [121]
Institut Necker
156 rue de Vaugiard
75730 Paris Cedex 15
France

Rieux-Laucat, F. [557]
INSERM U132
Hopital des Enfants Malades
149, rue de Sevres
75730 Paris Cedex 15
France

Riley, J.H. [91]
Dept. of Biotechnology, ICI Pharmaceuticals
Alderley Park, Macclesfield, Cheshire
SK10 4TG
UK

Roberts, T.M. [893]
The Johns Hopkins University
Dept. of Medicine
Oncology and Mol. Biology
Baltimore, MD 21205
USA

Robinson, D.S. [403]
Dept. of Allergy and Clinical Immunology
National Heart and Lung Institute
Dovehouse Street
London, SW3 6LY
UK

Rolink, A. [11]
Basel Institut for Immunology
487 Grenzacherstrasse
CH 4005 Basel
Switzerland

Romagnani, S. [239]
Division of Clinical Immunology and Allergy
University of Florence
Instituto di Clinica Medica 3
50134 Firenze
Italy

Romero, P. [775]
Ludwig Institute for Cancer Research
1066 Epalinges
Switzerland

Ronda, N. [643]
INSERM U28
Hopital Broussais
Paris
France

Rood, J.J. van [785]
University Hospital
Leyden
The Netherlands

Rosat, J.P. [775]
WHO Immunology Res. and Training Centre
Institute of Biochemistry
University of Lausanne
Switzerland

Rosenstein, Y. [275]
Division of Pediatric Oncology
Dana Farber Cancer Institute
Boston, MA
USA

Rosenthal, F.M. [361]
Dept. of Haematologic Oncology
Memorial Sloan Kettering Cancer Center
New York, NY 10021
USA

Rothbard, J. [181]
Section of Immunobiology
Yale University Sch. of Medicine
New Haven, CT 06510
USA

Roussev, R.G. [833]
Center for Reproduction and
Transplantation Immunology
Methodist Hospital of Indiana
Indianapolis, IN 46202
USA

Rowen, D. [643]
INSERM U28
Hopital Broussais
Paris
France

Roy, G. [213]
Howard Hughes Medical Institute
Dept. of Pathology
Washington University Medical School
660 South Euclid
St. Louis, MI 63110
USA

Rubensky, A.Y. [181]
Section of Immunobiology
Yale University Sch. of Medicine
New Haven, CT 06510
USA

Rudy, W. [289]
Kernforschungszentrum Karlsruhe
Institut für Genetik und Toxicologie
PO Box 3640
D-7500 Karlsruhe 1
Germany

Rülicke, T. [65]
Institute of Experimental Immunology
University of Zürich
Sternwartstrasse 2
8091 Zürich
Switzerland

Sacco, R. [443]
Dept. of Pathology
University of Iowa
College of Medicine
Iowa City, Iowa 52242
USA

Sachs, D. [801]
Transplantation Biology Research Center
Massachusetts General Hospital
Harvard Medical School
Charlestown, MA 02129
USA

Sad, S. [849]
National Institute of Immunology
New Delhi 110067
India

Sadlack, B. [305]
Institute of Virology and Immunobiology
University of Würzburg
Versbacherstrasse 7
D-8700 Würzburg
Germany

Sahai, P. [849]
National Institute of Immunology
New Delhi 110067
India

Saint Basile, G. de [557]
INSERM U132
Hopital des Enfants Malades
149, rue de Sevres
75730 Paris Cedex 15
France

Saizawa, K.M. [269]
Max-Planck Institut für Biologie
Abteilung Immungenetik
7400 Tübingen
Germany

Salgame, P. [761]
Dept. of Microbiology and Immunology
Albert Einstein College of Medicine
Bronx, NY 10461
USA

Salunke, D. [849]
National Institute of Immunology
New Delhi 110067
India

Sampognaro, S. [239]
Division of Clinical Immunology and Allergy
University of Florence
Instituto di Clinica Medica 3
50134 Firenze
Italy

Sandor, M. [443]
Dept. of Pathology
University of Iowa
College of Medicine
Iowa City, Iowa 52242
USA

Sarfati, M. [465]
Notre-Dame Hospital Res. Center
1560 Sherbrooke Street East
Montreal, Quebec, H2L 4M1
Canada

Satta, Y. [153]
Max-Planck Institut für Biologie
Abteilung Immungenetik
7400 Tübingen
Germany

Sautes, C. [457]
Laboratoire d'Immunologie
Cellulaire et Clinique
INSERM U255, Institut Curie
26 rue d'Ulm
75231 Paris, Cedex 05
France

Sayegh, M. [627]
Center for Neurologic Diseases
Bringham and Women s Hospital
Harvard Medical School
75 Francis Street
Boston, MA 02115
USA

Schalcher, Ch. [659]
Institute for Experimental Immunology
Dept. of Pathology
University of Zürich
Sternwartstrasse 2
CH-8091 Zürich
Switzerland

Scharton, T. [747]
Dept. of Pathobiology
School of Veterinary Medicine
University of Pennsylvania
3800 Spruce Street
Philadelphia, Pennsylvania 19104
USA

Schimpl, A. [305]
Institute of Virology and Immunobiology
University of Würzburg
Versbacherstrasse 7
D-8700 Würzburg
Germany

Schmidt, B. [807]
National Institute of Haematology
Blood Transfusion and Immunology
Daróczi 24
H-1113 Budapest
Hungary

Schorle, H. [305]
Institute of Virology and Immunobiology
University of Würzburg
Versbacherstrasse 7
D-8700 Würzburg
Germany

Schwarer, T. [793]
Dept. of Haematology
Royal Postgraduate Medical School
Hammersmith Hospital
Du Cane Road
London, W12 ONN
UK

Scott, P. [747]
Dept. of Pathobiology
School of Veterinary Medicine
University of Pennsylvania
3800 Spruce Street
Philadelphia, Pennsylvania 19104
USA

Seder, R.[347]
Laboratory of Immunology
NIAID, NIH
Bethesda, MD 20892
USA

Sela, M. [573]
Dept. of Chemical Immunology
Weizmann Institute of Science
Rehovot, 76100
Israel

Sercarz, E.E. [159]
Dept. of Microbiol. and Mol. Genetics
University of California
5304 Life Sciences Bldg.
405 Hilgard Ave.
Los Angeles, CA 90024-1489
USA

Shamblott, M.J. [107]
University of Maryland, C.O. M. B.
600 East Lombard Street
Baltimore, MD 21202
USA

Sharabi, Y. [801]
Dept. Of Life Sciences
Bar-Ilan University
Ramat-Gan 52100
Israel

Shaw, A. [213]
Howard Hughes Medical Institute
Dept. of Pathology
Washington University Medical School
660 South Euclid
St. Louis, MI 63110
USA

Shearer, G.M. [707]
Experimental Immunology Branch
NIC, NIH
Bethesda, MD
USA

Shenoi, H. [213]
Howard Hughes Medical Institute
Dept. of Pathology
Washington University Medical School
660 South Euclid
St. Louis, MI 63110
USA

Shibuya, H. [321]
Institute for Molecular and Cellular Biology
Osaka University
Yamadaoka 1-3
Suitashi, Osaka 565
Japan

Shimizu, H. [369]
Dept. of Molecular Pathology
Cancer Research Institute
Kanazawa University
13-1 Takaramachi
Kanazawa 920
Japan

Shin, E.K. [91]
Center for Molecular Biology and Genetics
Faculty of Medicine, Kyoto University
Sakyo-ku, Kyoto 606
Japan

Shoemaker, C. [753]
Dept. of Tropical Public Health
Harvard Medical School
Bringham and Women s Hospital
Boston, MA 02115
USA

Sidhu, S. [793]
Dept. of Immunology
Royal Postgraduate Medical School
Hammersmith Hospital
Du Cane Road
London, W12 ONN
UK

Sieper, J. [613]
Deutsches Rheumaforschungszentrum Berlin
Robert Koch Institut
Nordufer 20
D-1000 Berlin
Germany

Simon, K. [613]
Deutsches Rheumaforschungszentrum Berlin
Robert Koch Institut
Nordufer 20
D-1000 Berlin
Germany

Sims, M.J. [651]
The Wellcome Res. Laboratories
Langley Court
South Eden Rd.
Beckenham, Kent BR3 3BS
UK

Singh, M. [849]
National Institute of Immunology
New Delhi 110067
India

Singh, O. [849]
National Institute of Immunology
New Delhi 110067
India

Sleckman, B.P. [275]
Division of Pediatric Oncology
Dana Farber Cancer Institute
Boston, MA
USA

Sobel, R. [627]
Center for Neurologic Diseases
Bringham and Women s Hospital
Harvard Medical School
75 Francis Street
Boston, MA 02115
USA

Soeda, E. [91]
Gene Bank, Tsukuba Life Science Center
Inst. of Physical and Chemical Research
Koyadai, Tsukuba, Ibaraki 305
Japan

Soudais, C. [557]
INSERM U132
Hopital des Enfants Malades
149, rue de Sevres
75730 Paris Cedex 15
France

Souza, V.R. de [643]
INSERM U28
Hopital Broussais
Paris
France

Sperling, A. [261]
Ben May Institute and Committee
on Immunology MC 1089
University of Chicago
Chicago, IL 60637
USA

Stadler, B.M. [395]
Institute of Clinical Immunology
University of Bern
Sahlihaus, Inselspital
3010 Bern
Switzerland

Stanley,P. [283]
Macrophage Laboratory
Imperial Cancer Research Fund
44 Lincoln s Inn Fields
London, WC2A 3PX
UK

Stewart, M. [283]
Macrophage Laboratory
Imperial Cancer Research Fund
44 Lincoln s Inn Fields
London, WC2A 3PX
UK

Stingl, G. [305]
Dept. of Dermatology I.
University of Vienna VIRCC
Brunnerstrasse 59
A-1235 Vienna
Austria

Strasser, A. [51]
The Walter and Eliza Hall
Institute of Med. Research
Royal Melbourne Hospital
Victoria 3050
Australia

Straus, D. [205]
Howard Hughes Medical Institute
University of California
San Francisco U426
3rd and Parnassus Ave.
San Francisco, CA
USA

Sundblad, A. [603]
Unite d'Immunobiologie
CNRS URA 359
Institut Pasteur
25 rue du Docteur Roux
75724 Paris Cedex 15
France

Suzuki, S. [269]
Max-Planck Institut für Biologie
Abteilung Immungenetik
7400 Tübingen
Germany

Swat, W. [81],[129]
Basel Institut for Immunology
487 Grenzacherstrasse
CH 4005 Basel
Switzerland

Sykes, M. [801]
Transplantation Biology Research Center
Massachusetts General Hospital
Harvard Medical School
Charlestown, MA 02129
USA

Szekeres-Barthó, J. [861]
Institute Microbiology
University Medical School
H-7643 Pécs
Hungary

Szelényi, J. [807]
National Institute of Haematology
Blood Transfusion and Immunology
Daróczi 24
H-1113 Budapest
Hungary

Szigetvári, I. [807]
Dept. of Gynecology
Postgraduate Medical Schoool
Budapest
Hungary

Taga, T. [887]
Division of Immunology
Institute for Mol. and Cell. Biology
1-3 Yamada-oka, Suita
Osaka 565
Japan

Taka-ishi, Sh. [91]
Center for Molecular Biology and Genetics
Faculty of Medicine, Kyoto University
Sakyo-ku, Kyoto 606
Japan

Takahashi, M. [517]
Dept. of Immunobiology
Cancer Research Institute
Kanazawa University
Kanazawa
Japan

Takahata, N. [153]
National Institute of Genetics
Mishima 411
Japan

Talwar, G.P. [849]
National Institute of Immunology
New Delhi 110067
India

Tamir, I. [221]
Dept. of Chemical Immunology
The Weizmann Institute of Science
Rehovot 76100
Israel

Tanaka, T. [347]
Laboratory of Immunology
NIAID, NIH
Bethesda, MD 20892
USA

Taniguchi, T. [321]
Institute for Molecular and Cellular Biology
Osaka University
Yamadaoka 1-3
Suitashi, Osaka 565
Japan

Tatsumi, Y. [261]
Ben May Institute and Committee
on Immunology MC 1089
University of Chicago
Chicago, IL 60637
USA

Taub, D.D. [297]
Laboratory of Molecular Immunoregulation
Natl. Cancer Institute - Frederick
Cancer and Res. Development Center
Frederick, MD 21702-1201
USA

Taussig, M.J. [651]
Dept. of Immunology
AFRC Institute of Animal Physiology
Babraham, Cambridge CB2 4AT
UK

Teeraratkul, P. [443]
Dept. of Pathology
University of Iowa
College of Medicine
Iowa City, Iowa 52242
USA

Teillaud, J.L. [457]
Laboratoire d'Immunologie
Cellulaire et Clinique
INSERM U255, Institut Curie
26 rue d'Ulm
75231 Paris, Cedex 05
France

Thieblemont, N. [509]
INSERM U28
Hopital Broussais
75014 Paris
France

Thomas, M.L. [213]
Howard Hughes Medical Institute
Dept. of Pathology
Washington University Medical School
660 South Euclid
St. Louis, MI 63110
USA

Thomas, W.R. [427]
The Western Australian Research
Institute for Child Health
Princess Margaret Hospital
GPO Box D184
Perth, WA 6001
Australia

Titus, R.G. [753]
Dept. of Tropical Public Health
Harvard Medical School
Bringham and Women s Hospital
Boston, MA 02115
USA

Tölg, C. [289]
Kernforschungszentrum Karlsruhe
Institut für Genetik und Toxicologie
PO Box 3640
D-7500 Karlsruhe 1
Germany

Top, L. [333]
Metabolism Branch
NCI, NIH
Bethesda, MD 20892
USA

Torbett, B.E. [725]
The Scripps Research Institute
La Jolla, CA 92037
USA

Townsend. A. [275]
Institute of Molecular Medicine
John Radcliffe Hospital
Oxford
UK

Traversari, C. [871]
Ludwig Institute for Cancer Research
74 Ave. Hippocrate
B-1200 Brussels
Belgium

Troye-Blomberg, M. [731]
Dept. of Immunology
Stockholm University
S-106 91 Stockholm
Sweden

Tsicopoulos, A. [403]
Dept. of Allergy and Clinical Immunology
National Heart and Lung Institute
Dovehouse Street
London, SW3 6LY
UK

Upadhyay, S.N. [849]
National Institute of Immunology
New Delhi 110067
India

Uyttenhove, C.[313]
Ludwig Institute for Cancer Research
Catholic University of Louvain
74 avenue Hippocrate
B-1200 Brussels
Belgium

Van Pel, A. [871]
Ludwig Institute for Cancer Research
74 Ave. Hippocrate
B-1200 Brussels
Belgium

Van Snick, J. [313]
Ludwig Institute for Cancer Research
Catholic University of Louvain
74 avenue Hippocrate
B-1200 Brussels
Belgium

Vandenbark, A.A. [635]
Neuroinmmunology Research 151D
V.A. Medical Center
Dept. of Microbiol. and Immunology
Portland, OR 97201
USA

Varga, P. [861]
Institute Microbiology
University Medical School
H-7643 Pécs
Hungary

Vassilev, T. [643]
Centre for Infectious and Parasitic Dis.
Sofia
Bulgaria

Veronese, M.L. [699]
Institute for Oncology
Interuniversity Center for Res. on Cancer
University of Padova
Via Gattamelata 64
I-35128 Padova
Italy

Veronesi, A. [699]
Institute for Oncology
Interuniversity Center for Res. on Cancer
University of Padova
Via Gattamelata 64
I-35128 Padova
Italy

Villartay, J.P. de [557]
INSERM U132
Hopital des Enfants Malades
149, rue de Sevres
75730 Paris Cedex 15
France

Vink, A. [313]
Ludwig Institute for Cancer Research
Catholic University of Louvain
74 avenue Hippocrate
B-1200 Brussels
Belgium

Volanakis, J.E. [535]
Howard Hughes Med. Istitute
Divisions of Developmental and
Clin. Immunology and Rheumatology
University of Alabama at Birmingham
Birmingham, AL 35294
USA

Waldmann, T.A. [333]
Metabolism Branch
NCI, NIH
Bethesda, MD 20892
USA

Wang, J.-M. [297]
Laboratory of Molecular Immunoregulation
Natl. Cancer Institute - Frederick
Cancer and Res. Development Center
Frederick, MD 21702-1201
USA

Wang, M.-W. [651]
Ligand Pharmaceuticals
9393 Towne Centre Dr. Suite 100
San Diego, CA 92121
USA

Warnier, G. [313]
Ludwig Institute for Cancer Research
Catholic University of Louvain
74 avenue Hippocrate
B-1200 Brussels
Belgium

Watts, C. [189]
Dept. of Biochemistry
Medical Sciences Institute
University of Dundee
Dundee, DD1 4HN
UK

Wauben, M.H.M. [587]
Dept. of Infectious Diseases and Immunology
University of Utrecht
Yalelaan 1, P.O. Box 80165
3508 TD, Utrecht
The Netherlands

Weck, A.L. de [411]
Institute of Clinical Immunology
Inselspital
Ch-3010 Bern
Switzerland

Wegmann, T.G. [857]
Dept. of Immunology
University of Alberta
8-65 Medical Sciences Bldg.
Edmonton, Alberta T6G 2H7
Canada

Weichhold, G.M. [115]
Institut für Rechtmedizin
Universitat München
Frauenlobstrasse 7a
D-8000 München
Germany

Weill, J.-C. [121]
Institut Necker
156 rue de Vaugiard
75730 Paris Cedex 15
France

Weiner, H.L. [627]
Center for Neurologic Diseases
Bringham and Women s Hospital
Harvard Medical School
75 Francis Street
Boston, MA 02115
USA

Weiser, P. [27]
Max-Planck Institut für Immunbiologie
Stübeweg 51
7800 Freiburg
Germany

Weiser, W.Y. [753]
Dept. of Rheumatology
Harvard Medical School
Bringham and Women s Hospital
Boston, MA 02115
USA

Weiss, A. [205]
Howard Hughes Medical Institute
University of California
San Francisco U426
3rd and Parnassus Ave.
San Francisco, CA
USA

White, J. [333]
Metabolism Branch
NCI, NIH
Bethesda, MD 20892
USA

Witort, E. [451]
Schutt Laboratory for Immunology
Austin Research Institute
Heidelberg 3084
Australia

Wu, P. [613]
Deutsches Rheumaforschungszentrum Berlin
Robert Koch Institut
Nordufer 20
D-1000 Berlin
Germany

Wu, Z. [255]
Dept. of Microbiology and Immunology
University of Miami Sch. of Medicine
P.O.Box 016960
Miami, FL 33101
USA

Würch, A. [269]
Max-Planck Institut für Biologie
Abteilung Immungenetik
7400 Tübingen
Germany

Yamaguchi, N. [517]
Dept. of Immunobiology
Cancer Research Institute
Kanazawa University
Kanazawa
Japan

Yamamoto, K. [369]
Dept. of Molecular Pathology
Cancer Research Institute
Kanazawa University
13-1 Takaramachi
Kanazawa 920
Japan

Yamamura, M. [761]
Division of Dermatology
Dept. of Microbiol. and Immunology
UCLA School of Medicine
Los Angeles, CA 90033
USA

Ying, S. [403]
Dept. of Allergy and Clinical Immunology
National Heart and Lung Institute
Dovehouse Street
London, SW3 6LY
UK

Yoneyama, M. [321]
Institute for Molecular and Cellular Biology
Osaka University
Yamadaoka 1-3
Suitashi, Osaka 565
Japan

Zachau, H.G. [115]
Institut für Physiologische Chemie
Universitat München
Schillerstrasse 44
D-8000 München
Germany

Zakarija, M. [255]
Dept. of Microbiology and Immunology
University of Miami Sch. of Medicine
P.O.Box 016960
Miami, FL 33101
USA

Zamarchi, R. [699]
Institute for Oncology
Interuniversity Center for Res. on Cancer
University of Padova
Via Gattamelata 64
I-35128 Padova
Italy

Zawadzki, V. [289]
Kernforschungszentrum Karlsruhe
Institut für Genetik und Toxicologie
PO Box 3640
D-7500 Karlsruhe 1
Germany

Zee, R. van der [587]
Dept. of Infectious Diseases and Immunology
University of Utrecht
Yalelaan 1, P.O. Box 80165
3508 TD, Utrecht
The Netherlands

Zgaga-Griesz, A. [269]
Max-Planck Institut für Biologie
Abteilüng Immungenetik
7400 Tübingen
Germany

Zhang, Z.J. [627]
Center for Neurologic Diseases
Bringham and Women s Hospital
Harvard Medical School
75 Francis Street
Boston, MA 02115
USA

Zheng, J.-H. [517]
Dept. of Immunobiology
Cancer Research Institute
Kanazawa University
Kanazawa
Japan

Zilch, A.C. [107]
Dept. of Pediatrics
University of South Florida
801 Sixth Street S.
St. Petersburg, FL 33701
USA

Zimmer, S. [289]
Lucille P. Markey Cancer Center
Dept. of Microbiology and Immunology
800 Rose Street
Lexington, Kentucky 40536-0093
USA

Zinkernagel, R.M. [65], [659]
Institute for Experimental Immunology
Dept. of Pathology
University of Zürich
Sternwartstrasse 2
CH-8091 Zürich
Switzerland

Zisman, E. [573]
Dept. of Chemical Immunology
Weizmann Institute of Science
Rehovot, 76100
Israel

Zöller, M. [289]
Deutsches Kerbsforsungszentrum Heidelberg
Institut für Radiologie und Pathophysiologie
Im Neuenheimer Feld 280
D-6900 Heidelberg 1
Germany

1. B and T Cell Ontogeny, Repertoire, Selection and Tolerance

Immunological Tolerance Revisited in the Molecular Era

G. J. V.Nossal, M. G. McHeyzer-Williams, B. Pulendran, M. McLean,
P. A. Lalor and M.Karvelas

The Walter and Eliza Hall Institute of Medical Research, Post Office, The Royal Melbourne Hospital, Victoria 3050, Australia

ACKNOWLEDGEMENTS

This work was supported by the National Health and Medical Research
Council, Canberra, Australia; by Grant AI-03958 from the National
Institute of Allergy and Infectious Diseases, United States Public
Health Service; and by a grant from the Human Frontiers Science
Program, Principal Investigator Professor K Rajewsky.

We would also like to thank Ms Amanda Light for excellent technical
assistance.

INTRODUCTION

It is indeed a great honour to be asked to deliver the first of these
six Ehrlich Lectures at this historic 8th International Congress of
Immunology. It is a happy coincidence that my 35 years work in
immunology has concerned itself, to a great extent, with the subject
of "horror autotoxicus" that so fascinated Paul Ehrlich (1900).
Immunological tolerance at the cellular level clearly reflected
functional or actual repertoire purging for both T and B cells as
studied by single cell cloning of specific repertoire elements
(Nossal, 1983). Even before the present exciting period, clonal
abortion or negative selection within the primary lymphoid organs and
clonal anergy had been fully established as concepts. What has made
the last five years so special has been the extensive use of
transgenic models and anti-clonotypic antibodies to lend a great deal
of precision and detail to the cellular mechanisms involved.

IGNORANCE AND TOLERANCE IN THE T CELL REPERTOIRE

As regards the T cell repertoire, negative selection within the thymus
clearly represents a powerful and sensitive mechanism for the deletion
of cells reactive to such self peptides as may be presented within the
thymic microenvironment. However, transgenic experiments may not
always be as simple as they seem. Some of the complexities have
recently been revealed by an elegant collaboration between Miller's
group in Melbourne and Arnold's group in Heidelberg (Heath et al.,

1992). This involved judicious combination of three transgenic strains. The first were mice rendered transgenic for the class I major histocompatibility complex (MHC) antigen K^b placed under the control of the rat insulin promoter (RIP-K^b mice). The second involved the creation of T cell receptor transgenic mice expressing an anti-K^b reactivity that could be marked by the monoclonal antibody Desiree (Des-TCR transgenic mice). When these two mouse strains were crossed, double transgenics resulted in which virtually all the mature T cells were CD8$^+$ and expressed the Des clonotypic marker. While the K^b antigen, as judged by conventional criteria such as immunohistology or Northern analysis, appeared to be confined to the β cells of the pancreatic islets of Langerhans, PCR analysis did detect a small amount of K^b in the thymus. Careful flow cytometric investigation of CD8 single positive cells was performed on the thymus and lymph nodes of Des-TCR single transgenic and Des-TCR x RIP-K^b double transgenic mice. This clearly showed that the double transgenics had centrally deleted those T cells with the highest TCR expression (high avidity T cells) whereas lower TCR expressors had been permitted to enter the peripheral lymphoid tissues. Such double transgenics were not tolerant of K^b skin grafts, in fact, they rejected them promptly within about 10 days, while accepting littermate skin grafts. So, the lower avidity anti-K^b T cells that had reached the periphery in double transgenic mice were capable of destroying K^b skin grafts. Nevertheless, they appeared entirely to ignore the β cells in the pancreatic islets of Langerhans. However, when double transgenic mice were mated to a third transgenic strain carrying the IL-2 gene under the control of the RIP promoter, the triple transgenics (RIP-K^b-TCR-Des-RIP-IL-2) developed diabetes rapidly, half the mice being diabetic at 2 weeks of age and all being diabetic by 18 days. The most straightforward explanation of these findings is that the strong localised help provided by IL-2 secretion in the islet provided the help which the CD8$^+$ T cells in the double RIP K^b-TCR-Des mice had not been receiving. If this is the correct explanation, it suggests that even low avidity T cells directed to a differentiation antigen present in a peripheral tissue remain a threat, capable of provoking autoimmunity, if help is delivered through a cross reacting antigen of some kind. These studies support a position long promoted by Zinkernagel (Ohashi et al., 1991).

MAINTENANCE OF TOLERANCE IN THE SECONDARY B CELL REPERTOIRE

For the primary B cell repertoire, single cell studies on hapten specific B lymphocytes had shown that while some degree of clonal abortion or central deletion could be induced by high doses of multivalent antigen, much lower doses could act as toleragens by causing clonal anergy (Nossal, 1983). More recently, a considerable number of transgenic studies have validated both central deletion within the bone marrow and clonal anergy induction (Reviewed in Nossal, 1992). Cell membrane anchored neo-self antigens preferentially induce the former, whereas soluble antigens (perhaps aggregating in vivo) will cause anergy at an appropriate dose. For many self antigens present at only low concentration, B cells escape tolerance and thus display a state of ignorance towards the self antigen in question.

In contrast to T lymphocytes, B lymphocytes display the capacity for extensive and rapid V gene mutation following antigenic stimulation. This somatic hypermutation process occurs predominantly in germinal centres and is followed by rapid death unless the mutation has been an affinity-raising one. In other words, selection of "improved" clonotypes within the germinal centre appears to be the mechanism of affinity maturation in the antibody response. One might, therefore, ask what happens if immunoglobulin V gene hypermutation is such as to confer anti-self reactivity by chance? If the mutation happened to confer reactivity to some self antigen present at the surface of the antigen-retaining follicular dendritic cells, the risk of further proliferation of the anti-self B cell clone would appear to be quite real. Two mechanisms, not mutually exclusive, have been proposed to negate this possibility of "horror autotoxicus". The first is a "second window" of tolerance susceptibility in a special lineage of B cells destined to become memory B cells. If such pre-memory B cells were to pass through a tolerance-sensitive phase shortly after their activation by antigen, the potential auto-immune mutation would be "nipped in the bud" (Linton and Klinman, 1991). The second hypothesis suggests that mutated, potentially anti-self B cells would not progress because of T cell tolerance to the self antigen concerned and therefore lack of T cell help. In our recent studies, we have sought to mimic the situation in question by the introduction of soluble deaggregated antigen before or after challenge immunisation of an animal with the same antigen adsorbed onto alum and given with a B pertussis adjuvant.

Two model situations have been explored, namely the anti-human serum albumin (HSA) response (Nossal and Karvelas, 1992) and, more recently, the response to the hapten NP attached to HSA as a carrier. In the latter case, soluble, freshly deaggregated NP_2-HSA has acted as the standard toleragen. In both of these models, which support the same conclusions and will thus be discussed together, the development of memory cells in immunised animals has been studied by in vitro analysis of the splenic repertoire following immunisation. An empirical trick has been used to identify memory B cells. It is the fact that the spleen of a normal unimmunised animal has very few cells indeed with a sufficiently high affinity for either HSA or NP presented as an oligovalent conjugate on an ELISA plate to permit detectable binding of its product immunoglobulin when attention is focused solely on IgG rather than IgM antibody. It is now possible to clone splenic B lymphocytes with high efficiency in clonal cultures, and to achieve a switch to IgG1 and IgE antibody production. The system that we use relies on E coli lipopolysaccharide as a polyclonal mitogen, 3T3 fibroblasts to support continued proliferation, and a mixture of three interleukins, IL-4, IL-5 and IL-2, each at optimal concentration, to drive proliferation and isotype switching. Limit dilution cloning shows that from the very low background prior to immunisation, high affinity cells can be detected in increasing numbers following immunisation, peaking approximately 2 weeks after challenge. Typically, the numbers of relevant B cells may rise from about 100 per total spleen to about 50,000 per total spleen over this period.

When the antigen concerned is introduced in soluble form several days before the challenge immunisation, this appearance of high affinity clonable B cells is virtually ablated. Dose response parameters have been worked out, and the threshold for an effect appears to be between 1 and 10 μg of soluble deaggregated antigen per animal. Somewhat to our surprise, the soluble antigen can also achieve this effect when introduced a considerable time after primary immunisation. In studies where mice were killed at 14 days after challenge, a reduction of more than 20-fold could be achieved 6 days after challenge and in one experiment out of two a significant reduction was achieved 8 days after challenge. Both of these times are well after the initiation of the germinal centre process.

The cellular targets for tolerance induction could be investigated in two ways, namely through adoptive transfer analysis or through the use of toleragens representing only the carrier or only the hapten. Adoptive transfer studies have shown the almost complete ablation of any adoptive response when T cells from tolerant animals and B cells from tolerant animals are co-transferred. An almost equally profound reduction in the adoptive response was shown when T cells from tolerant animals were co-transferred with B cells from control animals. When T cells were harvested from normal unimmunised control mice and mixed with B cells from tolerant animals, a reduction in adoptive response was also seen, but it was less pronounced.

Hapten-carrier studies on the basis of this tolerance similarly suggest that both the T cell and the B cell are the target for tolerance induction. For example, when mice were immunised with NP_{18}-HSA, given a soluble deaggregated antigen 6 days later, and were then finally killed 8 days later again, that is, 14 days after challenge immunisation, soluble NP_2-HSA reduced the response to approximately 2% of normal. When the carrier alone, HSA, was introduced 6 days post challenge, the reduction was down to 10% of control values and when a conjugate of NP_3-horse serum albumin, thus an irrelevant carrier, was administered 6 days post challenge, the response was reduced to an almost equal extent as with HSA. Interestingly, when NP_8-cytocrome c was injected it lowered memory cell generation only slightly and not to a statistically significant extent. This may have represented a conjugate that was too heavily substituted to remain in the serum and extra cellular fluids for a sufficient length of time and we are re-investigating this point with a more lowly substituted NP-cytochrome c conjugate.

As far as these studies have gone to date, they do not support the view that B cells destined to become memory cells are exquisitely more sensitive to tolerance induction than B cells in their unimmunised state. Rather, they demonstrate that B cells can be rendered tolerant with a sufficient concentration of deaggregated antigen but also that T cells are absolutely necessary for the correct development of the germinal centre reaction, even after it has begun. If those T cells are silenced, for example through the induction of clonal anergy by the soluble antigen, then memory cell generation is frustrated. These findings clearly implicate an important role for the $CD4^+$ T cells resident in germinal centres. The capacity to "switch off" the germinal centre mutation and selection process by deaggregated antigen offers an intriguing further level of control of immune responses.

It is now possible in the C57 Bl mouse to sort out this putative memory cell population by 6 parameter flow cytometry (McHeyzer-Williams et al., 1991). Our method relies on the identification of $IgG1^+$ λ light chain positive IgM^- and NP^+ B cell populations which, in the particular case of the anti-NP response in this strain of mice, will usually use the VH 186.2 gene. Such sorted cells can be analysed for mutations by subjecting single cells to PCR analysis using sets of nested primers from the 5'- end of the VH 186.2 gene and from the heavy chain constant region. The method has shown very few mutations in immunised cells at day 7 of the response but an average of 3.4 mutations per V gene at days 12 to 14 of the immune response (Lalor et al., 1992; McHeyzer-Williams et al. [manuscript in preparation]). A tryptophan to leucine interchange at position 33 in the first hypervariable region is known to raise the affinity of anti-NP antibody by a factor of about 10. This mutation was never encountered at day 7 or earlier but was present in 55% of cells harvested on days 12 or 14. This is a dramatic indication of the rapid mutation rate and the effectiveness of the selection process.

When similar FACS analysis to that which yielded these highly selected cells from immunised mice is performed on mice that had received 1 mg of soluble, deaggregated NP_2-HSA before challenge, a marked deficit in NP^+ λ^+ $IgG1^+$ cells is found. This deficit as judged by flow cytometry is at least 20-fold, in other words, it parallels exactly the deficit noted through clonal analysis. We are in the process of studying the V gene sequence of the very few cells that meet the criteria for high affinity cells in these tolerant animals, but even before the results of this study are available, we can conclude that the deaggregated toleragen has frustrated in large measure the appearance of high affinity memory B cells. Most probably this is due to effects on the germinal centre V gene hypermutation and selection process, the inactivation of helper T cells being crucial and a direct tolerance (perhaps anergy) effect on germinal centre B cells also contributing.

In view of these results suggesting a profound and continuing role for helper T cells within germinal centres, it is important to briefly address current discussions concerning helper T cell subsets. Kelso and her colleagues (Kelso et al., 1991) have suggested two opposing models for the determination of the lymphokine secretion profile of T cell clones. On the one hand, antigenic stimulation may differentially select cells predetermined for their lymphokine secretion phenotype as in the strictest version of the Th1-Th2 postulation. On the other hand, lymphokine profiles might be determined through differential instruction of a given T cell which is, in fact, capable of secreting all possible T cell lymphokines but "chooses" to manufacture only certain ones when signalled in a particular way. The Kelso group have now completed an analysis of three different T cell stimulatory systems. These are the graft-versus-host reaction, in vitro recall of an anti-KLH response, and polyclonal stimulation of unimmunised splenic T cells by a combination of signals involving anti-CD3, anti-CD4 or 8 and anti-LFA1 with concomitant support of cultures by IL-2. While the first system is interferon γ-biased and the second IL-4-biased, detailed examination of positive clones does not support the Th1-Th2 hypothesis for these early clones. One model which Kelso is currently exploring is that interferon γ secretion is the "default" pathway when the T cell receptor is ligated appropriately, but that the production of IL-4 (with or without interferon γ) depends upon receipt of a specific instructional signal.

We do not know what the cytokines are that are important in the germinal centre reaction. As CD40 is clearly important, and as the CD40 ligand has recently been cloned, we should soon know more about its possible role. Soluble CD23 has been implicated, and it is by no means excluded that follicular dendritic cells may secrete yet other stimulatory molecules. The detailed role of cytokines in the germinal centre reaction is clearly a fertile field for further study.

SUMMARY AND CONCLUSIONS

At the cellular level, the big picture in tolerance is emerging. Thymic negative selection is clearly a dominant mechanism for both CD4 and CD8 positive T cells. Peripherally, concerning antigens not expressed in the thymus, T cell anergy is clearly a mechanism which exists but whether it is really relevant to tolerance for antigens expressed only in very specialised locations is now in some doubt. Certainly, in some cases, ignorance of the T cell concerning such antigens is documented, making autoimmunity through cross-reactive help a real possibility. B cell tolerance must now be regarded as proven, central deletion being the mechanism for the strongest cross-linking signals and anergy for weaker signals, but ignorance clearly prevails with respect to many self antigens, permitting some autoantibody formation even in healthy individuals. Whether these low affinity, largely IgM autoantibodies are the substrate for the formation of pathogenic autoantibodies is not clear, but evidence has been presented that many pathogenic auto-antibodies display numerous mutations, and thus presumably result from antigenic stimulation. Lack of T cell help is clearly a factor in preventing B cell hypermutation towards anti-self but, at the same time, B cells themselves are capable of being turned off, i.e. rendered anergic, when they encounter a sufficient concentration of soluble antigen. Through these two mechanisms, soluble de-aggregated antigens can "switch off" the germinal centre V gene hypermutation and selection mechanism, and presumably self antigens in the extracellular fluid can do likewise.

REFERENCES

Ehrlich P (1900) On immunity with special reference to cell life. Proc Roy Soc Lond B 66:424-448

Heath WR, Allison J, Hoffmann MW, Schönrich G, Hämmerling G, Arnold B, Miller JFAP (1992) Autoimmune diabetes as a consequence of locally produced interleukin-2. Nature (in press)

Kelso A, Troutt AB, Maraskovsky E, Gough NM, Morris L, Pech MH, Thomson JA (1991) Heterogeneity in lymphokine profiles of CD4[+] and CD8[+] T cells and clones activated in vivo and in vitro. Immunol Rev 123:85-114

Lalor PA, Nossal GJV, Sanderson RD and McHeyzer-Williams MG (1992) Functional and molecular characterisation of single, NP-specific, IgG$_1$[+] B cells from antibody-secreting and memory B cell pathways in the C57BL/6 immune response to (4-hydroxy-3-nitrophenyl) acetyl (NP). Eur J Immunol (In press)

Linton P-J, Rudie A, Klinman NR (1991) Tolerance susceptibility of newly generating memory B cells. J Immunol 146:4099-4104

McHeyzer-Williams MG, Nossal GJV, Lalor PA (1991) Molecular characterization of single memory B cells. Nature 350:502-505

Nossal GJV (1983) Cellular mechanisms of immunological tolerance. Ann Rev Immunol 1:33-62

Nossal GJV (1992) Cellular and molecular mechanisms of B lymphocyte
tolerance. Adv Immunol 52 (In press)

Nossal GJV, Karvelas M (1990) Soluble antigen abrogates the appearance
of anti-protein IgG$_1$-forming cell precursors during primary
immunisation. Proc Natl Acad Sci USA 87:1615-1619

Ohashi PS, Oehen S, Buerki K et al. (1991) Ablation of "tolerance" and
induction of diabetes by virus infection in viral antigen transgenic
mice. Cell 65:305-317

Birth, Life and Death of a B Cell

F. Melchers, H. Karasuyama, D. Haasner, and A. Rolink

Basel Institute for Immunology[*], Grenzacherstrasse 487, CH-4005 Basel, Switzerland

INTRODUCTION

The generation of B lymphocytes from progenitors and precursors, first during embryonic development and later continuously throughout life, is characterized by the expression of lineage-related markers, by successive rearrangements of the Ig loci, by development of cells with different proliferative and differentiating potentials, and by differential potentials of B lineage-cells to populate different B cell compartments in severe combined immunodeficient (SCID) hosts. Recent reviews have dealt with the molecular processes which accompany the Ig gene rearrangements (Alt et al., 1992), with selective expression of markers and their potential functions, and proliferative and differentiating capacities of progenitors and precursors of the B-lymphocytic lineages (Kincade et al., 1989; Dorshkind, 1990; Rolink and Melchers, 1991, 1993) of $Ly1^+$ and $Ly1^-$ B cells (Hardy, 1992).

Once sIg^+ B cells have been generated they are subjected to selective processes. Negative selection against self antigens either deletes or anergizes self reactive B cells (Nemazee and Bürki, 1989; Goodnow et al., 1989; see also Nemazee, this volume). The peripheral B cell repertoire appears also positively selected by antigen (Förster and Rajewsky, 1990; Gu et al., 1991a; Schittek and Rajewsky, 1992). This might occur in germinal centers in helper T cell-dependent, antigen-specific responses, in which Ig class switching and hyper-mutations of V-regions of rearranged Ig genes occur (reviewed by Liu et al., 1992; see also MacLennan et al., this volume). Positive selection appears to prevent cells of the B-lineage to die by apoptosis, and the induction of expression of the oncogene bcl-2 appears to be one of the molecular modes by which this is achieved (Vaux et al., 1988; McDonnell et al., 1989, 1990; Strasser et al., 1990, 1991a,b; Sentman et al., 1991; Nuñez et al., 1992; Hockenberry et al., 1991; Hardie et al., 1991; see also Strasser et al., this volume). It has been estimated that a mouse continues to newly generate 5×10^7 B-lineage cells every day (Osmond, 1991) but the fate of most newly generated cells is to die in a few days unless positive selection induces a larger life expectancy in them.

DIFFERENT EXPRESSION OF LINEAGE-RELATED MARKERS DURING B CELL DEVELOPMENT

A predictably oversimplified summary of our current view of early steps in B cell development before and during successive rearrangements of first the IgH-, and then the Ig-L loci is shown in Figure 1. It remains to be seen whether all stages of this development occur in this way prenatally in embryonic blood, yolk sac, embryonic placenta, liver and omentum, and postnatally in spleen, blood and bone marrow (Solvasan et al., 1991; reviewed in Rolink and Melchers, 1991). At least the enzyme terminal desoxynucleotidyl-transferase (TdT), which is responsible for N-region insertions at the D_H to J_H and V_H to $D_H J_H$ joints is not expressed prenatally in liver, so that the majority of all joints of the rearranged H chain loci are devoid of N-regions in fetal liver-derived pre B and B cells (Holmberg et al., 1989; Gu et al., 1990, 1991b; Feeney, 1990; Meek, 1990). Fetal liver-derived pre B cells can also be distinguished from bone marrow derived pre B cells by the differential expression of a novel regulatory myosin light chain in the bone marrow (Oltz et al., 1992).

[*] The Basel Institute for Immunology was founded and is supported by F. Hoffmann-La Roche Ltd., Basel, Switzerland

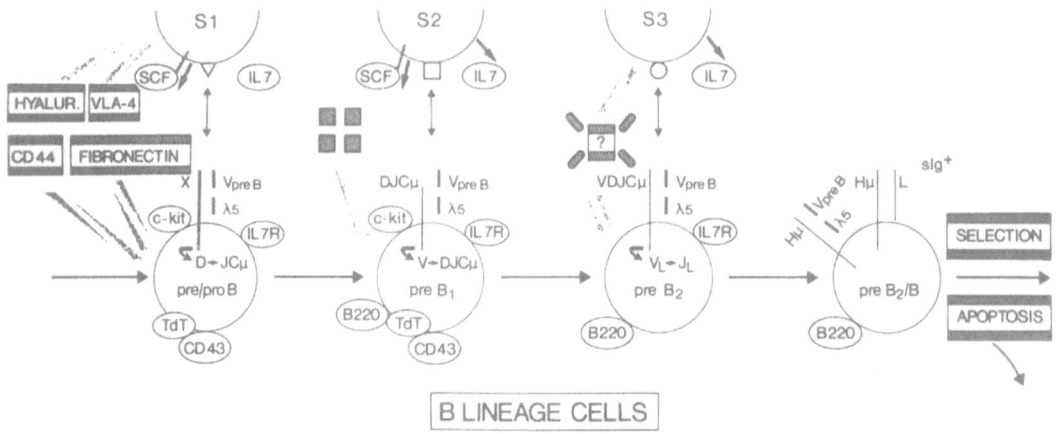

Fig. 1. B cell development in the mouse. S1, S2, and S3 are ligands on stromal cells proposed to react with the Ig-like complexes containing the surrogate L chain on pre B cells at different stages of their development. For details see text.

The vast majority of V_L to J_L joints are devoid of N-regions, since the enzyme TdT is already shut off at these late stages of pre B cell development. It is not known what turns on or off the expression of TdT, and whether productive rearrangements in the H chain loci and a subsequent expression of μH chains on the surface of pre B cells contributes to the downregulation of expression of TdT. Nevertheless, earlier stages (B220⁻, TdT⁺ and B220⁺, TdT⁺ cells) can be distinguished (Osmond, 1991; early and intermediate pro B cells) from later stages (B220⁺ TdT⁻, late pro B, large and small pre B cells) in bone marrow. The late stages are expected to have their H chain gene rearrangements completed, which makes some of these cells detectable by the expression of cytoplasmic μH chains. The later B220⁺, TdT⁻ stages of pre B cells are also expected to be in the process of (maybe continuous and multiple) rearrangements of their L chain loci (Feddersen and Van Ness, 1990; Harada and Yamagishi, 1991; see also Nemazee, this volume).

B cell differentiation has also been ordered by the differential expression of the high molecular weight form of the common leukocyte antigen B220 (CD45), of leukosialin (CD43) and of the tyrosine kinase c-kit (Figure 1, Hardy et al., 1991; Rolink and Melchers, 1993). Not shown in Figure 1 are the heat-stable antigen and BP-1 which are also differentially expressed during B cell development (Hardy et al., 1991). As expected, RAG-1 and RAG-2 are expressed throughout pre B cell development as is the protooncogene N-myc, MHC class I antigens and the B lineage-related alkaline phosphatase PB-76 (Blackwell et al., 1986; Lieber et al., 1987; Schatz and Baltimore, 1988; Strasser, 1988; Zimmerman and Alt, 1990; Ma et al., 1992).

Many B220⁻ and B220⁺, CD43⁺, c-kit⁺, TdT⁺ cells are expected to be $D_H J_H$-, but not $V_H D_H J_H$-rearranged pre B-I cells, while many of the B220⁺, CD43⁻, c-kit⁻, BP-1⁺, TdT⁻ cells might be already $V_H D_H J_H$-rearranged (Rolink et al., 1991a; Hardy et al., 1991; Hardy and Hayakawa, 1991), but that, again, is likely to be an oversimplification of a much more complex set of pre B cell-subpopulations differing in all these parameters (Figure 1).

DIFFERENTIAL CAPACITIES OF PRO AND PRE B CELL SUBPOPULATIONS TO PROLIFERATE "IN VITRO" AND TO POPULATE B CELL COMPARTMENTS OF SCID MICE "IN VIVO"

Proliferation and differentiation of pro and pre B cells require the contact with a microenvironment of stromal cells present in the primary organs where they develop (Kincade et al., 1989; Dorshkind, 1990; Rolink and Melchers, 1991). A variety of stromal cells, often preadipocytic fibroblast lines from bone marrow and even thymus, support long-term proliferation of progenitors and precursors of the B-lineage. IL-7 has been identified as a major cytokine which costimulates this proliferation (Namen et al., 1988), but it is likely that other cytokines such as IGF-I (Landreth et al., 1992), IL-6, IL-11, GM-CSF (Hirayama et al., 1992) and "Steel" encoded SCF (McNiece et al., 1991) have costimulatory activities, maybe with different types of stromal cells, and maybe on different stages and subpopulations of pre B cell development.

Some of the molecular contacts that regulate pro and pre B-I cell proliferation in contact with stromal cells, and that may control the performance of stromal cells to provide contacts and cytokines, have been identified. Monoclonal antibodies specific for c-kit inhibit the "in vitro" proliferation of some (Rolink et al., 1991b), but not all (Collins and Dorshkind, 1987; Kodama et al., 1992) long-term proliferating cells. This indicates that c-kit regulates pro and pre B cell proliferation, in line with the observation that the majority of all long-term proliferating pro and pre B cells express c-kit on their surface. Other contacts include VLA-4 and fibronectin (Miyake et al., 1991) and CD44 and hyaluronate (Miyake et al., 1990 a,b).

We have established lines and clones of pro and pre B cells from fetal liver, blood, spleen and bone marrow, which proliferate in serum-substituted media in the presence of exogenously added IL-7 and in contact with stromal cells for long periods of time (Rolink et al., 1991a). They can be cloned and recloned with efficiencies near 100% and retain their stage of differentiation. They are capable of differentiation to sIg^+ B cells "in vivo" and "in vitro", and they can populate pre B and B cell compartments of SCID mice for long periods of time. Therefore, they have properties of B-lineage committed stem cells, which undergo equal divisions into two cells at the same stage of differentiation when they proliferate on stromal cells in the presence of IL-7, and they can undergo unequal, or differentiating divisions into one or two differentiated, eventually mature B cells when removal from their environment of stromal cells and IL-7 induces their differentiation. Nishikawa and his colleagues (Nishikawa et al., 1988; Hayashi et al., 1990) have defined three stages of pro and pre B cell development by colony assays of early cells, which need only stromal cells, a subsequent stage which needs stromal cells and IL-7, and a late stage which only needs IL-7 to proliferate (Figure 1).

Frequencies and numbers of clonable cells change in the different organs during life. A wave of clonable cells appears before birth and disappears after birth in liver (Melchers, 1979). Up to two weeks after birth, high frequencies of clonable cells are present in spleen, and also detectable in blood, but become undetectable at 6-8 weeks in these sites. In bone marrow, up to 2% of all cells are clonable early, but decrease 10-20 fold within 6 months of age.

Clonable pro and pre B cells are enriched in the $B220^-$ $c\text{-}kit^{low}$ as well as the $B220^+$ $c\text{-}kit^+$ (and $B220^+$ $CD43^+$) cell populations of bone marrow. They are depleted from $B220^+$ $c\text{-}kit^-$ and $B220^+$ $CD43^-$ populations, and are absent in $B220^-$ $c\text{-}kit^-$ and $B220^-$ $c\text{-}kit^{high}$ populations of bone marrow. $B220^+$ $c\text{-}kit^+$ and $B220^+$ $CD43^+$ cells are likely to be the same, since long-term proliferation of $B220^+$ $c\text{-}kit^+$ cells in culture also express $CD43^+$ (Figure 1). The absolute numbers of clonable $B220^-$ $c\text{-}kit^{low}$ and of $B220^+$ $c\text{-}kit^+$ cells drop 5-20 fold within 6 months of life.

Clonable pro and pre B cells are present in fetal liver and bone marrow of a wide variety of different inbred strains of mice, transgenic mice and mice in which genes with functions in the B-lineage pathway of differentiation have been inactivated by targeted integration of a defective gene. The frequencies and absolute numbers of such clonable cells are equal, if not higher, in severe combined immunodeficient (SCID) (Bosma et al., 1988), and RAG-2^{ko} mice (Shinkai et al., 1992), in B cell deficient μH chain-trans-membrane^{ko} mice (Kitamura et al., 1991), and in λ_5^{ko} mice (Table I) (Kitamura et al., 1992).

13

Clonable pro and pre B cells are decreased in frequencies in mice expressing transgenic μH chains. Since they have B220+ c-kit- cells in their pre B cell compartments, it is likely that the lower numbers of clonable cells are a result of accelerated differentiation (Reichman-Fried et al., 1990; Era et al., 1991) into a more mature compartment which is no longer capable of extended proliferation on stromal cells in the presence of IL-7. Finally, mice expressing transgenic κL chains (Carmack et al., 1991), λL chains (Bogen and Weiss, 1991; Vasicek et al., 1992), are either delayed or severely depressed in their B cell development. The numbers of clonable pro and pre B cells, as well as the B220+ c-kit- cells, are also reduced. It remains to be seen how the transgenic L chains, prematurely expressed in pro and pre B cells block B cell development at very early stages and whether competition with the surrogate L chain (see below) plays a role in this inhibition.

Differentiation along the B-lineage pathway, therefore, appears accompanied by a change in proliferative capacities of cells which express different markers, some with functions controlling the performance of pro and pre B cells (Figure 1; Rolink and Melchers, 1993). Progenitor (pro) B cells, originally with all Ig genes in germline configuration, proliferate on stromal cells in the presence of IL-7, and probably also in response to other cytokines. "In vitro", as well as "in vivo" during early B cell development in fetal liver and bone marrow, pro B cells begin D_H to J_H rearrangements which they continue as pre B-I cells by successive, secondary D_H to J_H rearrangements, maybe until they have reached the most 5' located D_H segment or the most 3' located J_H segment on both H chain loci since clones grown from single pro/pre B cells have more than two forms of $D_H J_H$-rearranged H chain loci, and thus, appear not to be clonal in the H chain gene configurations.

It remains to be investigated in greater detail whether the earliest B220- c-kit+ progenitors are only B lineage-committed. In fact, bipotent precursors for B lineage cells and macrophages have been found early in fetal liver development which derive their dual potential from a cell with all Ig-loci still in germline configuration (Cumano et al., 1992; Cumano and Paige, 1992). In addition, B220- progenitors have been found which give rise to T and B lineage cells (Takai et al., 1992).

H chain loci in germline configuration can still be found after two or three weeks in tissue cultures of early fetal liver- or bone marrow-derived pro and pre B cells, i.e. at a time when a single cell could have expanded to 2-5 x 10^6 cells. This might indicate that these early pro B cells keep their germline H chain gene conformation by asymmetric divisions in which one germline pro B cell gives rise to one germline and one $D_H J_H$-rearranged cell.

Precursor B cells which do not proliferate on stromal cells in the presence of IL-7 for long periods of time are the vast majority in bone marrow. They are B220+ CD43- c-kit-. They might be reactive to IL-7 for a limited period of time (Hayashi et al., 1990; Rolink et al., 1991a). These pre B-II cells are likely to still express RAG-1 and RAG-2, as well as the surrogate L chain, but they are probably all TdT- and have their $V_H D_H J_H$-rearrangements at least on one of the two H chain loci completed (Osmond, 1991; Hardy et al., 1991). Deposition of a productive $V_H D_H J_H$-rearranged μH chain together with the surrogate L chain, may in fact, signal the cell to turn off CD43 and c-kit expression, but that needs to be investigated in greater detail. Removal of IL-7 from "in vitro" cultures of pre B-I cells certainly allows unproductive rearrangements of either H and/or L chain gene loci, and this differentiation also leads to downregulation of CD43 and c-kit, to a loss of stromal cells/IL-7-reactivity and to apoptosis of sIg-negative, as well as sIg positive differentiated cells (Rolink et al., 1991).

It is possible that pre B-II cells represent just a transitory state on the way to mature sIg+ or sIg- B cells with no interactive capacity with stromal cells, and no special B cell generating functions. Since these pre B-II cells are, however, so frequent, as they represent over 90% of all B220+ pre B cells (Osmond, 1991), it is conceivable that they are occupying a bone marrow pre B cell compartment controlled by stimuli different from those controlling pre B-I cells. Pre B-II cells could divide once every day (see below) in asymmetrical divisions to produce one new pre B-II cell and one cell on its way to L chain rearrangements and to a mature B cell (Figure 1). "In vivo" this might occur in situations when all sites in bone marrow are occupied and when further divisions crowd the marrow too much, so that only one of the two cells produced in the division remains attached to stroma while the other leaves the local area of IL-7-production and, thereby, is induced to differentiate by initiating L chain gene rearrangements. In this scenario, the contacts of the pre B cells with its ligands on stroma and with the cytokine IL-7 keeps them at their given stage of differentiation, while loss of ligands induces differentiation. Pre B-I cells are kept at their stage by stromal cells and IL-7, pre B-II cells maybe by IL-7 alone.

It has repeatedly been suggested that early differentiation along the B lymphocyte-lineage pathway, marked by successive rearrangements of Ig gene segments, is guided by the expression of productive rearrangements of the H-chain locus, either as $D_H J_H C_\mu$-proteins or as $V_H D_H J_H$-rearranged μH chains. Deposition of the μH chains on the surface of the pre B cells selects cells along a pathway that leads to rearrangments of the next Ig gene segment along the differentiation pathway. Successful cells are thereby selected over unsuccessful cells along their way toward a B cell. Deposition on the surface is mediated by the disulphide-bonded association of the different forms of the μH chains with surrogate L chain (Melchers et al., 1989; Misener et al., 1992). Signaling for either survival or rearrangements is likely to occur via the Ig-associated molecules Ig-α and Ig-β encoded by the mb-1 and B29 genes (Sakaguchi et al., 1988; Hombach et al., 1988; Hermanson et al., 1988). Signaling along the B lineage-differentiation pathway has been reviewed in detail elsewhere (Reth et al., 1991; Alés-Martinez et al., 1991).

It is evident from mice in which the λ_5 gene has been inactivated by targeted integration of a defective form of the gene (Kitamura et al., 1992) that normal expression and function of the λ_5 gene is critical for normal B cell development. While the heterozygous littermates of the $\lambda_5$ko mice appear to develop their B cells normally, the homozygous $\lambda_5$ko mice have an altered precursor pool and a delayed appearance of both Ly1+ as well as Ly1- B cells (Table I). While the Ly1+ B cells compartment appears normalized after one month of life, the Ly1- compartments fill up much more slowly, so that even after 6 months of life, only around half of the normal numbers of Ly1- B cells are present in the peripheral lymphoid organs. Even with the lower number of B cells, however, $\lambda_5$ko mice mount normal immune responses to T-independent as well as T-dependent antigens.

Table 1. Alterations in the pre B and B cell compartments of $\lambda_5^{k.o}/_{k.o}$ mice

	$\lambda_5^{k.o}/_{k.o}$	normal
B cell development Ly1+ sIg+ B	delayed	normal
Ly1- sIg+ B	delayed	normal
B cell responses to mitogen	normal	normal
to T independent antigens	normal	normal
to T dependent antigens	normal	normal
Bone marrow		
B220+ CD43- μ^- pre B-II	40 fold lower	normal
B220+ c-kit- μ^- pre B-II		
B220+ CD43+ μ^- pre B-I	normal	normal
B220+ c-kit+ μ^- pre B-I		
Frequencies of pro/pre B-I cells clonable on stromal cells in IL7	higher at early times 1 in 15-30 at 4 weeks 1 in 600 at 23 weeks	normal 1 in 100 at 4 weeks 1 in 500 at 28 weeks
Surrogate L chain VpreB/λ_5 on the surface of pre B-I cell clones	absent	present
Reading frame distribution in D_H-J_H joints of fetal liver pre B-I clones	I:II:III = 20:11:7 (38) no N rfII represented	I:II:III = 24:3:10 (37) no N rfII suppressed
VJ-rearranged k chain loci in pre B-I clones	low (~1%)	low (~1%)
Development of sIg+, mitogen-reactive B cells from pre B-I-clones in vitro	normal	normal

When pro and pre B cells rearrange D_H-segments to J_H-segments these rearrangements can occur in three reading frames (rf). RfII, but not rfI or III, allows the expression of a $D_HJ_HC_\mu$-protein, since the reading frame of J_H is in frame with the promotor and the start codon found upstream of most D_H elements within the H-chain gene locus. This is particularly so for D_HJ_H-joints made in fetal liver, where no N-region diversity is inserted (Holmberg et al., 1989; Feeney, 1990; Meek, 1990; Gu et al., 1990). An analysis of the representation of rfI, II and III within the repertoire of pre B cells of fetal liver has been shown that rfII is suppressed in Ly1$^-$, but not Ly1$^+$ B cells (Gu et al., 1990, 1991b). This suppression has been interpreted to result from a suppression of all those pre B cells expressing $D_HJ_HC_\mu$-protein on their surface, and consequently, a lack of expansion on stromal cells in the presence of IL-7.

If surface deposition was mandatory for a signaling to stop further expansion of rfII-D_HJ_H-rearranged pre B cells then pre B cells of λ_5^{ko} mice should be unable to do so, because $D_HJ_HC_\mu$-protein cannot be inserted into the surface membrane in the absence of surrogate L chain. An analysis of the repertoire of D_HJ_H joints in pre B cells of λ_5^{ko} mice, in fact, shows that rfII is present is now represented normally, i.e. appears not suppressed (Table I). Similarly, $V_HD_HJ_H$-rearranged pre B cells may be suppressed for further proliferation and expansion in the presence of IL-7, as experiments with μH-chain transgenic mice might suggest (Era et al., 1991), as soon as the surrogate L chain deposits the μH chain in the surface membrane. Again, λ_5^{ko} pre B cells should not be able to do this and, therefore, accumulate productively $V_HD_HJ_H$-rearranged pre B cells in the stromal cell/IL-7-reactive, c-kit$^+$ compartment. This is under investigation.

A delayed appearance of mature Ly1$^-$ B cells in the periphery of λ_5^{ko} mice could be expected to be the result of an abnormally low rate of L chain gene rearrangements in $V_HD_HJ_H$-rearranged pre B-II cells which are incapable of inserting μH chains into their surface membrane. This assumes that rearrangements of L chain gene segments are induced in pre B cells by the deposition of mH chains on their surface. However, our "in vitro" experiments with pre B cell lines and clones support the notion that V_L to J_L-rearrangements can occur without previous productive $V_HD_HJ_H$-rearrangement and insertion of mH chain into the surface membrane. The mere removal of IL-7 from the culture induces V_H to D_HJ_H and V_L to J_L rearrangements, and even pre B cells do so that never are able to produce a productive V_H to D_HJ_H-rearrangement (clone 18, in Rolink et al., 1991a). These findings make it unlikely that a slower rate of L chain-rearrangements in λ_5^{ko} pre B cells could explain the delayed appearance of sIg$^+$ B cells in the periphery of λ_5^{ko} mice.

We expect that the surrogate L chain plays yet another role in B cell development. In fact, the surrogate L chain is found on the surface of progenitors and pre B cells before μH chains are ever expressed (either as $D_HJ_HC_\mu$ proteins or as $V_HD_HJ_H$-rearranged μH chains) (Misener et al., 1992; Karasuyama et al., sumitted). Pre B-I cells from fetal liver with D_HJ_H-rearranged H chain loci in rfI or III and stromal cells/IL-7-reactive progenitors from RAG-2^{ko} mice (Shinkai et al., 1992; Melchers et al., in preparation) are such cells on which the surrogate L chain has been detected with the aid of monoclonal antibodies specific for V_{preB} and λ_5. Immunoprecipitation with these monoclonal antibodies detects a complex of protein molecules, p130/p64/p46, associated with the surrogate L chain, of which p64 is disulphide-bonded to the protein. An example of such an analysis is shown in Figure 2. A delayed B cell development in the λ_5^{ko} mice might be the result of the inability of the progenitor and pre B cells of these mice to deposit this complex into the surface membrane. This might slow down either the proliferation or further development of these early cells along the B lymphocyte-lineage pathway of differentiation.

On the other hand, pre B-I-type stromal cell/IL7-reactive cells from λ_5^{ko} mice can be cloned in normal, if not elevated frequencies, from fetal liver or bone marrow of λ_5^{ko} mice (Table I). Even if they had their Ig-H chain loci in germline configuration at the time of cloning "ex vivo", they are D_HJ_H-, and often also $V_HD_HJ_H$-rearrranged, when they have grown up from one precursor to 10^6-10^9 cells, i.e. within 30 divisions in three to four weeks "in vitro". These observations do not point to a delayed capacity to rearrange D_H to J_H gene segments within the IgH locus.

The most dramatic defect of λ_5^{ko} mice is their 40 fold reduced compartment of B220$^+$ CD43$^-$ c-kit$^-$ pre B-II cells, leading to a delayed appearance of Ly1$^-$ sIg$^+$ B cells that is only 40% of normal, even at 6 months of age (Kitamura et al., 1992; Table I). These Ly1$^-$ sIg$^+$ B cells could well be generated from the approximately 5×10^5 D_HJ_H-rearranged pre B-I cells present in the λ_5^{ko} mice, if they generated that number of $V_HD_HJ_H$-rearranged, V_LJ_L-rearranged cells by asymmetric divisions each day, if 10% of all of these fully rearranged cells had a productive H and a productive L chain gene and, therefore, became sIg+, and if these B cells became long lived in the process. Half of the normal level of Ly1⁻ sIg+ B cells (i.e. 5×10^7 cells) would thus be reached in 100 days.

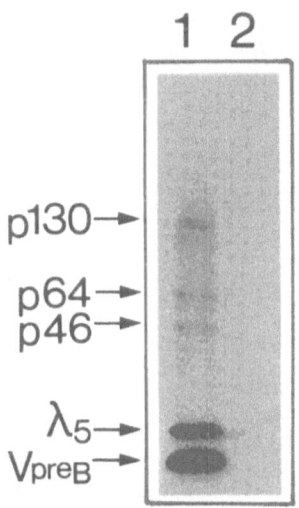

Fig.2. A pre B cell line 40E1, which has IgL chain loci in germline configuration and D_HJ_H-rearrangements on both alleles of the IgH chain locus, was surface labeled with ^{125}I and lysed with NP40 lysis butter. Detergent soluble lysates were reacted with either a λ_5-specific mAb (lane 1) or a µH chain-specific mAb (lane 2). Immunoprecipitates were electrophoresed under reducing conditions on 4-15% gradient SDS-PAGE. Note that while neither µH or $D_HJ_HC\mu$-protein was detectable, proteins of 130kD, 64kD and 46kD (p130, p64 and p46) were found coprecipitated wit V_{preB}/λ_5 surrogate L chains by the λ_5-specific mAb. The association of the molecules (p130, p64 and p46) with surrogate L chain was also detectable on the cell surface of other precursor B cell lines not yet producing mH chain or $D_HJ_HC\mu$-protein, including pro B cell lines in which both IgH and IgL loci are in germline configuration. The analysis by nonreducing/reducing two-dimensional SDS-PAGE revealed that p64 was disulphide-linked to λ_5, whereas V_{preB}, p130 and p46 proteins were noncovalently associated with them (not shown).

The lack of pre B-II cells in the λ_5^{ko} mice underlines the importance of the B220+ CD43⁻ c-kit- pre B-II compartment for the development of Ly1⁻ sIg+ B cell compartment in mice. It suggests that, in normal mice the surrogate L chain plays the role of selecting from all $V_HD_HJ_H$-rearranged cells those which have a productive rearrangement and, therefore, express µH chain protein. If the µH chain cannot be deposited in the surface membrane, as it is the case in pre B cells from RAG-2^{ko}, µH chain-transmembrane^{ko}, and λ_5^{ko} mice, this pre B-II-compartment is greatly reduced or totally missing (Shinkai et al., 1991; Kitamura et al., 1991, 1992; Table I). It is tempting to speculate that this positive selection is brought about by the binding of the µH chain/surrogate L chain complex to (a) ligand(s) in the environment.

A transitory cell expressing µH chains together with surrogate L chains and with normal L chains, even as hybrid molecules (Cherayil and Pillai, 1991; Karasuyama et al., submitted) is expected to be the next, and maybe the final site where the surrogate L chain is expressed. It may also be the first cellular stage which is selectable by antigens which fit the µH chain/normal L chain receptors. Antigen binding might turn down surrogate L chain expression, select the cells for exit from the primary lymphoid organ into the periphery and induce a longer lifespan.

In summary, the most probable function of the V_{preB}/λ_5 surrogate L chain is comparable to that of normal L chains, namely the capacity to transport disulphide-linked µH chains and heavy chain-like proteins through the cell and onto the surface of pre B cells. Binding to the environment might effect the proper lodging of the pre B cells and result in the proper filling of the pre B compartments. When the pre B-II compartment of bone marrow is filled with 5×10^7 cells, it suffices to produce the 5×10^7 cells from which 3×10^6 sIg+ are selected into the periphery that the mouse makes every day (Osmond, 1991; Rolink and Melchers, 1993).

REFERENCES

Alt FW, Oltz EM, Young F, Gorman J, Taccioli G, Chen J (1992) Immunol Today 13:306-313
Alés-Martínez JE, Cuende E, Martínez-A C, Parkhouse RME, Pezzi L, Scott DW (1991) Immunol Today 12:201-205
Blackwell TK, Moore MW, Yancopoulos GD, Suh H, Lutzker S, Selsing E, Alt FW (1986) Nature 324:585-589
Bogen B, Weiss S (1991) Eur J Immunol 21:2391-2395
Bosma M, Schuler W, Bosma G (1988) Curr Topics Microbiol Immunol 137:197-202
Carmack CE, Camper SA, Mackle JJ, Gerhard WU, Weigert MG (1991) J Immunol 147:2024-2033
Cherayil BJ, Pillai S (1991) J Exp Med 173:111-116
Collins LS, Dorshkind K (1987) J Immunol 138:1082
Cumano A, Paige CJ, Iscove NN, Brady G (1992) Nature 356:612-615
Cumano A, Paige C (1992) EMBO J 11:593-601
Dorshkind K (1990) Annu Rev Immunol 8:111-137
Era T, Ogama DL, Nishikawa SI, Okamoto M, Honjo T, Akaji K, Miyasaki JI, Yamamura K (1991) EMBO J 10:337-342
Feeney AJ (1990) J Exp Med 172:1377-1390
Förster I, Rajewsky K (1990) Proc Natl Acad Sci USA 87:4781-4784
Goodnow CC, Crosbie J, Jorgensen H, Brink RA, Basten A (1989) Nature 342:385-391
Gu H, Förster I, Rajewsky K (1990) EMBO J 9:2133-2140
Gu H, Tarlinton D, Müller W, Rajewsky K, Förster I (1991a) J Exp Med 173:1357-1371
Gu H, Kitamura D, Rajewsky K (1991b) Cell 65:47-54
Hardie DL, Gordon J, MacLennan ICM (1991) Eur J Immunol 21:1905-1919
Hardy RR, Hayakawa K (1991) Proc Natl Acad Sci USA 88:11550-11554
Hardy RR, Carmack CE, Shinton SA, Kemp JD, Hayakawa K (1991) J Exp Med 173:1213-1225
Hardy RR (1992) Curr Op In Immunol 4:181-185
Hayashi SI, Kunisada T, Ogawa M, Sudo T, Kodama H, Suda T, Nishikawa S, Nishikawa SI (1990) J Exp Med 171:1683-1695
Hermanson GG, Eisenberg D, Kincade PW, Wall R (1988) Proc Natl Acad Sci USA 85:6890-6894
Hirayama F, Shih J-P, Awgulewitsch A, Warr GW, Clark SC, Ogawa M (1992) Proc Natl Acad Sci USA 89:5907-5911
Hockenberry DM, Zutter M, Hickey W, Nahm M, Korsmeyer S (1991) Proc Natl Acad Sci USA 88:6961-6965
Holmberg D, Anderson A, Carlson L, Forsgren S (1989) Immunol Rev 110:889-903
Hombach J, Leclercq L, Radbruch A, Rajewsky K, Reth M (1988) EMBO J 7:3451-3456
Kincade PW, Jyonouchi H, Landreth KS, Lee G (1982) Immunol Rev 64:81
Kitamura D, Roes J, Kühn R, Rajewsky K (1991) Nature 350:423-426
Kitamura D, Kudo A, Schaal S, Müller W, Melchers F, Rajewsky K (1992) Cell 69:823-831
Kodama H, Nose M, Yamaguchi Y, Tsunoda J-i, Suda T, Nishkawa S, Nishikawa S-i (1992) J Exp Med 176:351
Landreth KS, Narayanan R, Dorshkind K (1992) Blood, in press
Lieber MR, Hesse JE, Mizuuchi K, Gellert M (1987) Genes Dev 1:751-761
Liu Y-J, Johnson GD, Gordon J, MacLennan ICM (1992) Immunol Today 13:17-21
Ma A, Fisher P, Dildrop R, Oltz E, Rathbun G, Achacoso P, Stall A, Alt FW (1992) EMBO J 11:2727-2734
McDonnell TJ, Deane N, Platt FM, Nunez G, Jaeger U, McKearn JP, Korsmeyer SJ (1989) Cell 57:79-88
McDonnell TJ, Nunez G, Platt JM, Hockenberry D, London L, McKearn JP, Korsmeyer SJ (1990) Mol Cell Biol 10:
McNiece IK, Langley KE, Zsebo KM (1991) J Immunol 156:3785-3790
Meek K (1990) Science 250:820-823
Melchers F (1979) INSERM Symp 10:281-289
Melchers F, Strasser A, Bauer SR, Kudo A, Thalmann P, Rolink A (1989) Cold Spring Harbor Symp Quant Biol LIV:183-189
Miyake K, Medina KL, Obo S, Hamaoka T, Kincade OW (1990a) J Exp Med 171:477-488
Miyake K, Underhill CB, Lesley J, Kincade PW (1990b) J Exp Med 172:69-75
Miyake K, Weissman IL, Greenberger JS, Kincade PW (1991) J Exp Med 173:599-608
Namen AE, Lupton S, Hjerrild K, Wagnall J, Mochuzuki DY, Schmierer A, Mosley B, March C, Urdal D, Gillis S, Cosman D, Goodwin RG (1988) Nature 333:571-573

Nemazee DA, Bürki K (1989) Nature 337:562-566
Nishikawa SI, Ogawa M, Nishikawa S, Kumisada T, Kodama H (1988) Eur J Immunol 18:1767
Nuñez G, Hockenbery D, McDonnell TJ, Sorensen CM, Korsmeyer SJ (1991) Nature 353:71-73
Oltz EM, Yancopoulos GD, Morrow MA, Rolink A, Lee G, Wong F, Kaplan K, Ollies S, Melchers F, Alt FA (1992) EMBO J 11:2759-2767
Osmond DG (1991) Curr Op In Immunol 3:179-185
Reichmann-Fried M, Hardy RR, Bosma MJ (1990) Proc Natl Acad Sci USA 87:2730-2734
Reth M, Hombach J, Wienands J, Campbell KS, Chien N, Justement LB, Cambier JC (1991) Immunol Today 12:196-201
Rolink A, Melchers F (1991) Cell 66:1081-1094
Rolink A, Kudo A, Karasuyama H, Kikuchi Y, Melchers F (1991a) EMBO J 10:327-336
Rolink A, Streb M, Nishikawa SI, Melchers F (1991b) Eur J Immunol 21:2609
Rolink A, Melchers F (1993) Adv Immunol 53, in press
Sakaguchi N, Kashiwamura SI, Kimoto M, Thalmann P, Melchers F (1988) EMBO J 7:3457-3464
Schatz DG, Baltimore D (1988) Cell 53:107-115
Schittek B, Rajewsky K (1992) J Exp Med 176:427-438
Sentman CL, Shutter JR, Hockenbery D, Kanagawa O, Korsmeyer SJ (1991) Cell 87:879-885
Shinkai Y, Rathbun G, Lam KP, Oltz EM, Stewart V, Mendelson M, Charron J, Datta M, Young F, Stall AM, Alt FM (1992) Cell 68:855-867
Solvason N, Lehuen A, Kearney JF (1991) Internatl Immunol 3:543-550
Strasser A (1988) Eur J Immunol 18:1803
Strasser A, Harris AW, Vaux DL, Webb E, Bath ML, Adams JM, Cory S (1990) Curr Top Mircrobiol Immunol 166:175-181
Strasser A, Harris AW, Cory S (1991a) Cell 67:888-899
Strasser A, Whittingham S, Vaux DL, Bath ML, Adams JM, Cory S, Harris AW (1991b) Proc Natl Acad Sci USA 85:5881-5885
Takai Y, Sakata T, Iwagami S, Tai XG, Kita Y, Hamaoka T, Sakaguchi N, Yamagishi H, Tsuruta T, Teraoka H, Fujiwara H (1992) J Immunol 148:1329-1337
Vasicek TJ, Levinson DA, Schmidt EV, Campos-Torres J, Leder P (1992) J Exp Med 175:1169-1180
Vaux DL, Cory S, Adams JM (1988) Nature 335:440-442
Zimmerman K, Alt FW (1990) Crit Rev Oncogenesis 2:75-95

Germinal Centres in the Affinity Maturation of T Cell-Dependent Antibody Responses

Ian C. M. Maclennan, Yong-Jun Liu, Deborah Hardie, Gerald D. Johnson, Montserrat Casamayor-Palleja and John Gordon

The Department of Immunology, University of Birmingham Medical School, Birmingham B15 2TT, England

INTRODUCTION

Germinal centre formation is a hallmark of T cell-dependent antibody responses. These structures are intimately associated with the expansion of B-cell clones and the introduction of mutations into their rearranged immunoglobulin (Ig) V-region genes. The B cells produced in germinal centres are selected on the basis of their affinity for and speed of combination with antigen held on follicular dendritic cells (FDC). Those cells which are positively selected survive to receive signals which induce them to leave the germinal centre as plasmablasts or memory B cells. Non-selected cells die by apoptosis.

THE DEVELOPMENT OF GERMINAL CENTRES

Germinal centres develop in B-cell follicles found in all secondary lymphoid tissues - the lymph-nodes, the spleen, and mucosal lymphoid tissues. In the absence of antigen-driven proliferation follicles consist of a network of FDC with the spaces in the network filled with small recirculating B cells (Nieuwenhuis and Ford 1976). These are termed primary follicles, as opposed to secondary follicles which contain activated B cells. In germ-free rodents there are no germinal centres in the follicles (Thorbecke 1959) and these structures are very unusual in congenitally athymic rodents (Jacobsen et al 1974) indicating the antigen and T cell dependence of these structures. Although germinal centres are always found in Peyer's patches of healthy individuals it is not uncommon to find only primary follicles in the spleen, for this organ is relatively protected from exogenous antigen. Consequently the spleen is particularly suitable for studying the development of germinal centres following immunisation with T cell-dependent antigens. The germinal centres which develop during the first three weeks of these responses can be clearly distinguished from the small level of B cell proliferation which continues throughout the remaining months of the response (Liu et al 1991a). An essential feature of follicles is the capacity of FDC to take up antigen in the form of immune complex and hold this in a non-degraded form for many months (Szakal et al 1989). It seems likely that this stored antigen is important both for cell selection in germinal centres and the maintenance of plasma cell and memory cell production during the months of established T cell-dependent antibody responses. Late secondary responses to T cell-dependent antigens can only be transferred to syngeneic naive animals if the recipient is given both cells and antigen (Gray and Skarvall 1988, Askonas and Williamson 1972).

Studies of the way in which germinal centres develop have been reviewed by Nieuwenhuis and Ospstelten (1984), Szakal et al (1989) MacLennan et al (1990) and Liu et al (1992). We have used double immunoenzyme techniques (van Rooijen et al 1986) to identify antigen-specific B cells in sections of tissues taken during simultaneous antibody responses in rats to two different T cell-dependent antigens (Liu et al 1991a). The two antigens used were protein conjugates with the haptens: 2,4-dinitrophenyl (DNP) and phenyl oxazalone (Ox). If rats are immunised with the carrier protein to provide T cell help then subsequent responses to combined i.v. challenge with DNP-protein and Ox-protein is associated with the rapid formation of large germinal centres in all follicles in the spleen. From the second day after challenge hapten-specific blasts can be seen proliferating within the follicular dendritic cell network. It is striking that the blasts are exclusively specific for one of the haptens in 6 to 31% of follicles, i.e. some follicles contain only DNP-specific cells, others only Ox-specific cells while the remainder are of mixed specificity. This reflects the oligoclonality of these follicular responses first recognised by Kroese et al (1987). Statistically if 12.5 per cent of the follicles were monospecific one would expect an average of 3 B cells had colonized each follicle. This only applies after a single dose of antigen. If immunisations are repeated or there is a continued supply of new antigen further B cells are induced to migrate to follicles. The oligoclonality of the response is maintained throughout the 3 weeks of the germinal centre reaction (Liu et al 1991a). The initial stage of the follicular reaction is associated with the exponential growth of the B blasts so that 72 hours after entering the follicle the three or so blasts colonizing the follicle have become 1-1.5x10^4 blasts. These fill the FDC network and the small recirculating B cells which were initially in the follicle centre are displaced to form the follicular mantle. Cell cycle times of around 7 hours are required to achieve this rapid clonal expansion. This rate of proliferation has been confirmed using stathmokinetic analysis (Zhang et al 1988) and studies of the uptake of the thymidine analog 5-bromo-2'-deoxyuridine (BrdUrd) (Liu et al 1991a).

This phase of exponential growth of B blasts ends when the blasts fill the FDC network. During the following few hours the classical zonal pattern of the germinal centre develops. The blasts move to one edge of the FDC network and reduce their amount of sIg. These cells, now termed centroblasts, continue in rapid cell cycle but do not increase in numbers, for as well as reproducing themselves they are also continually producing daughter cells which come out of cell cycle. The non-dividing progeny are centrocytes, which increase their level of sIg and re-enter the dense part of the FDC network (Fliedner et al 1964, Liu et al 1991a). There is a high death rate among centrocytes but some survive to become memory B cells (Klaus et al 1980, Coico et al 1983) or plasma cells (Liu et al 1991b, Tew et al 1992).

The zone of proliferating centroblasts at one edge of the FDC network is conventionally termed the dark zone. For orientation purposes the position of the dark zone serves to identify the base of the follicle. The central part of the follicle which contains the dense FDC network and centrocytes is termed the light zone. Immunohistological analysis of germinal centres has shown that compartmentalisation of the germinal centre is more complex (Johnson et al 1986, 1989, Liu et al 1992).

The dark zone

The proliferating cells in the dark zone in human tonsil can be identified by a monoclonal antibody Ki67 which binds to a nuclear antigen associated with cells in active cell cycle. The expression of this correlates well with the area with the highest frequency of mitoses seen in germinal centres. In germinal centres in experimental animals most cells in this zone are labelled in vivo within seven hours by an infusion of tritiated thymidine (Fliedner et al 1964) or BrdUrd (Liu et al 1991a). By contrast only a small proportion of cells in the light zone are labelled during this period. The Ki67[+] cells at the base of germinal centres in tonsil show the highest level of CD77 expression among germinal centre cells. In tonsils CD77 is expressed by germinal centre cells and vascular endothelium but not by extra-follicular lymphocytes and follicular mantle B cells. The dark zone defined conventionally on sections stained with methyl green pyronin is: that part of the germinal centres with closely-packed large and intermediate sized cells with pyroninophilic cytoplasm and nuclei with open chromatin. By this definition the Ki67[+] cells occupy the lower part of the dark zone. The upper part of the histologically-defined dark zone contains pyroninophilic cells, which are not in cycle. These are the centrocytes in what we have termed, perhaps confusingly, the basal light zone.

The basal and apical light zones

The dense part of the FDC network can be divided into two zones. The FDC in both zones hold antigen in the form of immune complex, and express CD21, CD54, and the pan FDC antigens recognised by the monoclonal antibodies BU10 and R4/23 (Johnson et al 1989). The FDC nearer the follicular mantle i.e. furthest from the dark zone express CD23 strongly while those nearer the dark zone show little CD23 expression. This is the main feature which allows the discrimination of the apical light zone from the basal light zone. The B lineage cells in both the apical and basal light zone are mainly centrocytes in that they are germinal center B lineage cells which are not in cell cycle. Tritiated thymidine or BrdUrd labelling experiments indicate, however, that these cells have been in cell cycle within the previous 7-12 hours (Fliedner et al 1964, Liu et al 1991a). It has long been recognised that there is a high death rate among cells in germinal centres (reviewed Nieuwenhuis and Opstelten 1984). This is seen by the presence of intensely condensed chromatin and nuclear fragmentation which characterises cells undergoing apoptosis (Wyllie et al 1984). These condensed nuclear fragments, tingible bodies, which are so characteristic of germinal centres are derived from cells which have recently been in cycle. The tingible bodies have been found to become labelled within a few hours of starting an infusion of tritiated thymidine (Fliedner 1967). Although tingible bodies can be found throughout tonsillar germinal centres, in a study using detailed quantitative microscopy the majority of apoptotic nuclei were found in the basal light zone.

There is a narrow fourth zone within the germinal centre which surrounds both the dense part of the FDC network and the dark zone and abuts on its outer surface to the small lymphocytes of the follicular mantle. This is termed the outer zone and is characterised by a relatively sparse FDC network which is CD23[-]. Some of the lymphoid cells in this outer zone contain cytoplasmic immunoglobulin, others are morphologically similar to the centrocytes of the apical light zone and occasional cells are Ki67[+] blasts. Most share the common feature of the cells of this zone by expressing high levels of CDw75 antigen. These cells share the germinal centre B cell characteristics of expressing CD77, CD38 and not expressing CD39 or extra-nuclear bcl-2 protein. The significance of cells in this outer zone remains obscure.

SIGNALS INVOLVED IN CELL SELECTION AND DIFFERENTIATION

Virgin B cells are recruited into T cell-dependent antibody responses only in the first few days after exposure to antigen; the response is subsequently maintained by memory B-cell clones (Gray, et al 1986). Somatic mutation occurring in Ig V-region genes during responses to Ox-protein in mice, does not start until after the onset of antibody production (Griffiths et al 1984 and Bereck et al 1985). These two sets of data taken together point to the Ig V-region-directed mutagenic process being induced in cells that have already been activated by antigen. In addition the peak time of accumulation of mutations in Ig V-region genes occurs when germinal centres are present. On the basis of these data it was proposed that the mutational process might be taking place in centroblasts and that centrocytes are selected on the basis of their ability to receive appropriate antigen-dependent signals when they pass into the FDC network (MacLennan and Gray 1986). Recently direct evidence has been provided for the hypermutation mechanism being switched on in germinal centres (Jacob et al 1991). The high death rate of cells in the basal light zone points to this being the possible site of selection. Evidence pointing to a mechanism by which selection of cells is associated with successful interaction with antigen has been provided by isolating germinal centre cells from human tonsil. These cells can be obtained by negative selection of sIgD[-] and sCD39[-] tonsil B cells (Liu et al 1989). Provided these cells are isolated in the cold they show good viability. When they are cultured at 37°C, however, they rapidly start to kill themselves by apoptosis (Liu et al 1989). It has been found that cross-linking the sIg of these cells with anti-immunoglobulin delays the onset of apoptosis by some 24-48 hours. Subsequently a number of co-signals have been identified which stabilise rescue from entry to apoptosis and induce the rescued germinal centre cells to differentiate.

The protection of isolated germinal centre cells from entering apoptosis appears to be linked to the induction of expression of the protein encoded by the proto-oncogene bcl-2. Analysis of genetic defects associated with germinal centre cell tumours identified a frequent translocation between the bcl-2 gene on chromosome 18 and the IgH genes on chromosome 14 (Tsujimoto et al 1985). Bcl-2 encodes a 26kD protein that is located partially in mitochondria (Hockenbury et al 1990, Liu et al 1991c and partially in other extra-nuclear sites (Liu et al 1991c). This protein is not expressed in appreciable amounts in germinal centre cells, but it is found in most if not all small lymphocytes, and extra-follicular B-cell blasts. The temporary rescue of germinal centre cells from apoptosis by cross-linkage of sIg is associated with the induction of bcl-2 expression within four hours (Liu et al 1991c). The link between bcl-2 expression and cell survival was first identified when the bcl-2 gene with an appropriate expression system was transfected into IL-3-dependent haemopoietic cell lines. This was found to prevent cell death when IL-3 was withdrawn (Vaux et al. 1988). Several other systems including transgenic mice have served to demonstrate a link between bcl-2 expression and cell survival (reviewed Korsmeyer 1992). As neoplastic germinal centre cells with t14;18 express the bcl-2 gene inappropriately it may be that this leads to the survival of cells which normally would have died in germinal centres.

Three groups of signals have been identified which prolong germinal centre cell-survival and induce differentiation: (i) exposure to the CD40 antibody G28-5 induces cells to leave cell cycle, express bcl-2 and acquire the phenotype of small lymphocytes (Liu et al 1989 and 1991c). Cells treated in this way increase their level of sIg expression and start to express sCD23. This is consistent with the production of memory B cells, which are known to arise in germinal centres (Klaus et al 1980, Coico et al 1983). If IL-4 is added with CD40 antibody the cells remain in cycle and the sIg and CD23 expression of these cells is greater than that of cells treated with CD40 alone. This phenotype resembles extra-follicular B blasts and is quite distinct from centroblasts. (ii) A combination of a recombinant 25 kDa fragment of the CD23 molecule plus IL-1α also prevents germinal centre cells from entering apoptosis (Liu et al 1991b). Together these induce germinal centre cells to acquire features of plasmablasts, including the expression of cytoplasmic IgG, the development of endoplasmic reticulum and a prominent Golgi apparatus. (iii) A proportion of germinal centre cells respond to IL-2 (Holder et al 1992). A proportion of cells with low sIg expression proliferate over periods of several days in response to IL-2. These cells continue to resemble centroblasts in that they do not express bcl-2. The second effect of IL-2 is on a small fraction of germinal centre cells which express sIgM. These cells start to produce cytoplasmic IgM and also fail to express bcl-2. These cells may reflect the IgM-producing plasma cells found in tonsil germinal centres which also do not express bcl-2. It remains to be seen which if any of these signals represent differentiation pathways which act in vivo.

REFERENCES

Askonas BA and Williamson AR (1972) Eur J Immunol 2:487-493
Berek C Griffiths GM and Milstein C (1985) Nature 316 412-418
Coico RF Bhogal BS and Thorbecke GJ (1983) J Immunol 131 2254-2257
Fliedner TM Kress M Cronkite EP and Robertson JS (1964) Ann N Y
    Acad Sci 113:578-594

Fliedner TM (1967) in Germinal centres in immune responses Cottier
   H et al eds pp 218-224 Springer Berlin
Gray D MacLennan ICM and Lane PJL (1986) Eur J Immunol 16, 641-448
Gray D and Skarvall H (1988) Nature 336:70-73
Griffiths GM Berek C Kaartinen M and Milstein C (1984) Nature 312:
   271-275
Hockenbery D Nunez G Milliman C et al (1990) Nature 348: 334-336
Holder M Liu Y-J de France T et al (1991) Internat Immunol 12:1243-
   1251
Jacob J Kelsoe G and Rajewsky K (1991) Nature 354:389-392
Jacobson EB Caporale IH and Thorbecke GJ (1974) Cell Immunol
   13:416- 430
Johnson GD MacLennan ICM Khan M et al (1989) in Leucocyte typing IV
   Knapp W et al eds pp 183-184 Oxford University Press
Johnson GD MacLennan ICM Ling NR et al (1986) in Leucocyte typing
   II Reinherz L et al eds pp 289-297 Springer-Verlag New York
Klaus GGB Humphrey JH Kunkle A and Dongworth DW
   (1980) Immunol. Rev. 53, 3-28
Korsmeyer SJ (1992) Ann Rev Immunol 10:785-808
Kroese FGM Wubenna AS Seijen HG and Nieuwenhuis P (1987) Eur J
   Immunol 17:1069-1072
Liu Y-J Cairns JA Abbot SD et al (1991b) Eur J Immunol 21:1107-
   1114
Liu Y-J Johnson GD Gordon J and MacLennan ICM (1992) Immunol Today
   13:17-21
Liu Y-J Joshua DE Williams GT et al (1989) Nature 342:929-931
Liu Y-J Mason DY Johnson GD et al (1991c) Eur J Immunol 21:1905-
   1910
Liu Y-J Zhang J Chan E et al (1991a) Eur. J. Immunol. 12:2951-
   2962
MacLennan ICM and Gray D (1986) Immunol Rev 91:61-85
Maclennan ICM Liu Y-J Oldfield S et al (1990) Curr Top Microbiol
   Immunol 159:37-63
Nieuwenhuis P and Ford WL (1976) Cell Immunol 23:254-267
Nieuwenhuis P and Opstelten D (1984) Amm J Anat 170:421-435
Szakal AK Kosco MH and Tew JG (1989) Ann Rev Immunol 7:91-109
Tew JG DiLosa RM Burton GF et al (1992) Immunol Rev 126:99-112
Thorbecke GJ (1959) Ann N Y Acad Sci 78:237-246
Tsujimoto Y Cossman J Jaffe E and Croce C (1985) Science 228:1440-
   1443
Van Rooijen N Classen E and Eikelenboom P (1986) Immunol Today 7:
   193-196
Vaux DL Cory S and Adams JM (1988) Nature 335:440-442
Vonderheid,R.H. and Hunt S.V. (1990) Eur. J. Immunol. 20, 79-86
Wyllie AH Morris RG Smith AL and Dunlop D (1984) J Path 142:67-77
Zhang J MacLennan ICM Liu Y-J and Lane PJL (1988) Immunol Letters
   18:297-299

Structure and Function of the B Cell Antigen Receptors

G. Alber; H. Flaswinkel; K.-M. Kim; P. Weiser and M. Reth

Max-Planck-Institut für Immunbiologie, Stübeweg 51, 7800 Freiburg, Germany

INTRODUCTION

The B cell antigen receptor is a complex consisting of the membrane-bound immunoglobulin (mIg) molecule and two B cell-specific transmembrane proteins (Ig-α and Ig-β) which form a covalently linked α/ß heterodimer (for review see Reth et al., 1991; Reth, 1992). All five different mIg classes are associated with the same α/ß heterodimer (Venkitaraman et al., 1991) although, depending on the Ig class, the Ig-α molecule can differ in the glycosylation of its extracellular Ig domain (Campbell et al., 1991). The Ig-α and the Ig-β proteins both contain in their cytoplasmic tail a sequence motif (Reth 1989), which is important for their signaling function (for review see Cambier, 1992; Chan et al., 1992).
In this report we describe the analysis of the surface expression of the Ig-α/Ig-β heterodimer using specific antibodies and the functional analysis of the Ig-α and Ig-β tail sequence.

Surface expression of the Ig-α/Ig-β heterodimer without mIg molecule

The Ig-α and Ig-β proteins are both members of the Ig-superfamily (Williams and Barclay, 1988) and are encoded by the B cell-specific genes mb-1 and B29, respectively (Sakaguchi et al., 1988; Hermanson et al., 1988). They consist of one extracellular Ig domain, a transmembrane part and a cytoplasmic tail of 61 and 48 aminoacids.
We generated antibodies against the Ig domain and cytoplasmic tail of Ig-α and Ig-β by immunizing mice and rabbits with fusion proteins. The tail sequences appended to glutathion-S-transferase (GST) were expressed in bacteria We obtained monoclonal antibodies (mAb) against the GST-Ig-α and GST-Ig-β proteins.
The mIgM molecule requires assembly with the Ig-α/Ig-β heterodimer for transport to the cell surface (Hombach et al., 1990).
The Ig-α/Ig-β heterodimer itself, however, can be expressed without mIg molecule on the B cell surface. As has previously been shown for Ig-α by the group of Sakaguchi (Nomura et al., 1990) we found that pre-B cell lines which do not produce any Ig molecule carry Ig-α and Ig-β on the cell surface. Using our specific rabbit antisera we detected Ig-α and Ig-β expression on the surface of two Ig-negative Abelson pre-B cells (Figure 1). The α/β heterodimer may exist on these cells either alone or in association with another pre-B cell-specific proteins replacing the mIgM molecule.

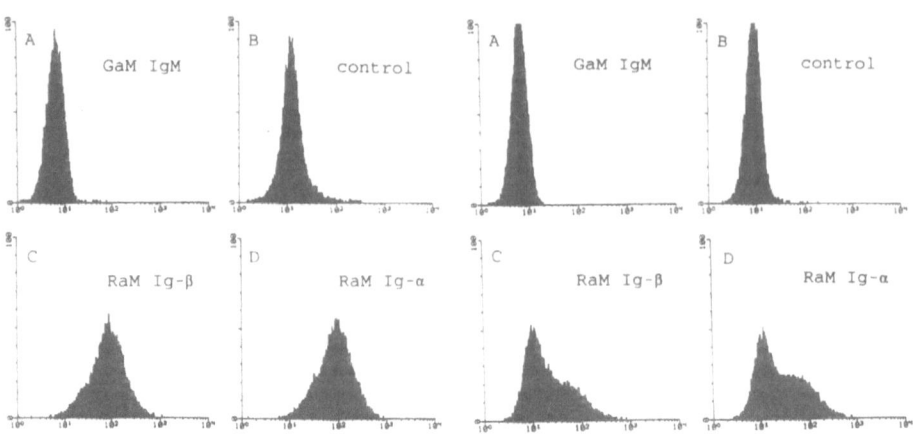

Fig. 1. FACScan analysis of the pre-B cell lines P17-27 and 33.11 (Iglesias et al., 1991). The cells were incubated with either a rabbit anti-IgM(A), a normal rabbit (B), a rabbit anti-Ig-β (C) or a rabbit anti-Ig-α antiserum. Bound antibodies were detected with FITC-labelled goat anti-rabbit antibodies.

To test for the former possibility we transfected the J558L myeloma line producing only Ig-β with an mb-1 expression vector. In a FACScan analysis the Ig-α and Ig-β protein were detected on the cell surface of the J558Lmb-1 transfectant but not on the J558L parental line (data not shown). This experiment suggests, that the α/β heterodimer can be expressed alone on the B cell surface. It furthermore demonstrates that Ig-β cannot be transported to the cell surface alone but only as an Ig-α/Ig-β heterodimer.

Functional dissection of the B cell antigen receptor

The isolated Ig-α/Ig-β heterodimer expressed on the surface of null pre-B cell can already have a signaling function in these cells. Indeed, Nomura et al. demonstrated a Ca^{++} mobilisation after treatment of these cells with anti-Ig-α antibodies. This finding is in line with experiments of Irving and Weiss, 1992 and Romeo and Seed, 1991 showing that the cytoplasmic tail of isolated components of the T cell receptor or FcεRI receptor can already have some signaling function. All these receptor chains share with the Ig-α and Ig-β proteins a cytoplasmic sequence motif. E/D x_7 E/D x_2 Y x_2 L x_7 Y x_2 L/I containing two tyrosines (Reth 1989).
We have analysed the signaling function of isolated cytoplasmic tails of the IgG2a antigen receptor. The IgG2a antigen receptor contains the same Ig-α/Ig-β heterodimer as the mIgM class. The heavy chain of the mIgG2a molecule contains, in contrast to the μm chain, an evolutionarily conserved cytoplasmic tail of 28 amino

28

acids. We have expressed hybrid molecules between the murine CD8α (Lyt2a) surface protein and the cytoplasmic tail of Ig-α, Ig-β or the membrane form of γ2a, as minimal receptors on the surface of the K46 B lymphoma line (Fig. 2).

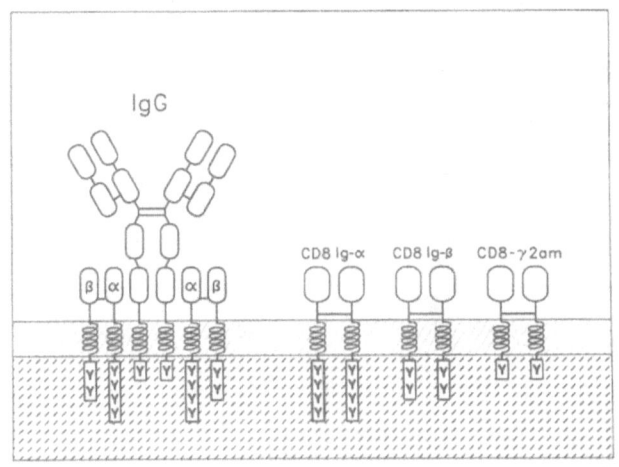

Fig. 2. Presumed structure of the IgG2a antigen receptor and CD8 fusion protein carrying the isolated cytoplasmic tails of the antigen receptor.

The K46 line also carries the complete IgG2a antigen receptor on its cell surface. Cross-linking of the antigen receptor results in a rapid release of Ca ions from internal cellular stores (Fig. 3d). Cross-linking of the CD8/Ig-α and CD8/Ig-β homodimer with anti-CD8 antibodies results in Ca mobilisation albeit in only 5% of the transfectants and with slower kinetics (Fig. 3a and 3b). No rise in intracellular Ca was seen after crosslinking of the CD8/γ2a homodimer (Fig. 3c) which thus serves as negative control for these experiments.

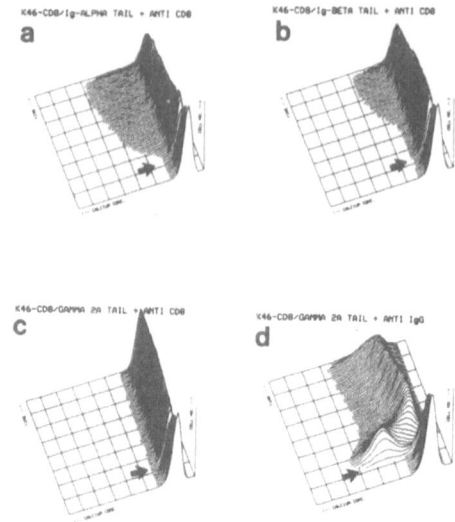

Fig. 3. Rise of the intracellular Ca-concentration in K46 transfectant carrying CD8/Ig-α(a); CD8/Ig-β(b) or CD8/γ2a (c and d) homodimers after crosslinking with either anti-CD8 (a-b) or anti-IgG(d) antibodies. Measurements were done with Indo-AM1-loaded cells in a FACStar-plus.

Activation of tyrosine phoshorylation after antigen receptor cross-linking

The Ig-α and Ig-β proteins have been detected as Ig-associated phosphoprotein in normal B cells (Campbell and Cambier). Indeed, cross-linking of the antigen receptor results in a rapid activation of tyrosine kinases resulting in increased phosporylation of intracellular substrate proteins (Campbell and Sefton, 1990; Gold et al., 1990) including the Ig-α and to a weaker extent the Ig-β protein (Gold et al., 1991). Four src-related tyrosine kinases (lyn,blk,fyn,and lck) can be activated via the antigen receptor and become partially associated with the antigen receptor (Yamanashi et al., 1991; Burkhardt et al., 1991; Campbell and Sefton, 1992). These src-related tyrosine kinases are bound via myristic acid to the inner leaflet of the plasma membrane. Another type of tyrosine kinase, which becomes activated via and connected to the antigen receptor is the cytoplasmic tyrosine kinase syk (Taniguchi et al., 1991) also known as PTK72 (Hutcroft et al., 1992).
We have analysed substrate proteins of tyrosine phosphorylation in B lymphoma and myeloma lines before and after antigen receptor cross-linking. An increase in tyrosine phosphorylation was seen not only in B lymphoma lines but, to our surprise, also in the myeloma line J558Lμm expressing IgM antigen receptor on its surface. Thus, cells specialized to secrete Ig molecules can still have functional components of the B cell signaling machinery. The tyrosine kinases lyn and syk are active in this myeloma line. Dominant substrate proteins were among others the 34-36 Kd Ig-α

protein and to a lesser extent the 39 Kd Ig-β protein (Fig. 4).
The activation of the tyrosine phosphorylation after antigen
receptor cross-linking was dependent on the presence of the
Ig-α/Ig-β heterodimer in the antigen receptor complex. We have
obtained a J558L variant (J558Lδm2.6) which expresses the IgD
molecule not as a transmembrane (mIg) protein but as a GPI-linked
molecule on the cell surface (Wienands and Reth, 1992). The
GPI-linked form of surface IgD is not connected to the Ig-α/Ig-β
heterodimer and does not activate tyrosine phosphorylation after
cross-linking (Fig. 4, lane 5-7). After transfection of the
J558Lδm2.6 line with an mb-1 expression vector, most IgD molecules
are expressed on the cell surface as transmembrane proteins in
association with the Ig-α/Ig-β heterodimer. In these cells, the
ability of the cross-linked IgD antigen receptor to activate
tyrosine phosphorylation is restored (Fig. 4, lane 8-10).

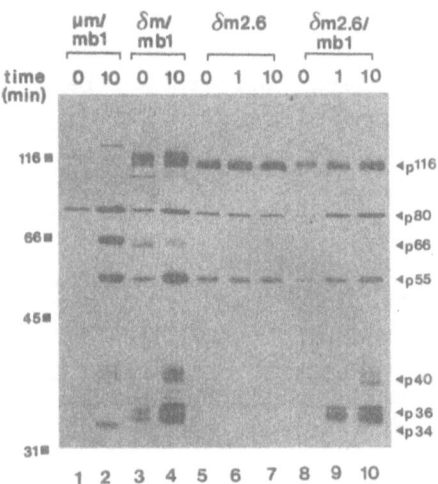

Fig 4. Induced tyrosine phosphorylation of substrate proteins in
J558L transfectants expressing membrane-bound IgM (1,2), IgD (3,4
and 8-10) and GPI-linked IgD(5-7) on their cell surface. The
different antigen receptors on these cells were crosslinked for the
indicated time with the antigen NIP-BSA. Phosphorylated substrate
proteins are detected on a Western blot with the
anti-phosphotyrosine antibody 6410.
In conclusion, our data suggest that while mIg molecules require
association with the α/β heterodimer, the heterodimer itself can
also be transported alone on the B cell surface. Furthermore, we
have shown that the conserved cytoplasmic tail of Ig-α and Ig-β
have an important signaling function. Cross-linking of hybrid
molecules carrying the cytoplasmic tails results in the activation
of tyrosine kinases and Ca^{++} mobilisation.

ACKNOWLEDGEMENT

We thank Dr. C. Gee for providing us with the 6410 antibody.

REFERENCES

Burkhardt A, Brunswick M, Bolen J, Mond J (1991) Anti-immunoglobulin stimulation of B lymphocytes activates src-related protein-tyrosine kinases. Proc Natl Acad Sci USA 88, 7410-7414.

Campbell MA, Sefton BM (1990) Protein tyrosine phosphorylation is induced in murine B lymphocytes in response to stimulation with anti-immunoglobulin. EMBO J 9:2125-2132.

Campbell KS, Cambier JC (1990) B-lymphocyte antigen receptors (mIg) are non-covalently associated with a disulfide-linked, inducibly phosphorylated glycoprotein complex. EMBO J 9:441-448.

Campbell K, Hager E, Friedrich R, Cambier J (1991) α-chains of IgM and IgD antigen receptor complexes are differentially N-glycosylated mb-1-related molecules. J Immunol 147:1575-1580.

Campbell MA, Sefton BM (1992) Association between B lymphocyte membrane immunoglobulin s and multiple members of the src-family of protein tyrosine kinases. Mol Cell Biol, 12:2315-2321.

Chan AC, Irving BA, Weiss A (1992) New insights into T-cell antigen receptor structure and signal transduction. Curr Opin Immunol 4:246-251.

Gold MR, Law DA, DeFranco AL (1990) Stimulation of protein tyrosine phosphorylation by the B-lymphocyte antigen receptor. Nature 345:810-812.

Gold MR, Matsuuchi L, Kelly RB, DeFranco AL (1991) Tyrosine phosphorylation of components of the B-cell antigen receptors following receptor crosslinking. Proc Natl Acad Sci USA 88:3436-3440.

Hermanson, GG, Eisenberg D, Kincade PW, Wall R (1988) A member of the immunoglobulin gene superfamily exclusively expressed on B-lineage cells. Proc Natl Acad Sci USA 85:6890-6894.

Hombach J, Tsubata T, Leclercq L, Stappert H, Reth M (1990) Molecular components of the B cell antigen receptor complex of the IgM class. Nature 343:760.

Hutchcroft JE, Harrison ML, Geahlen RL (1992) Association of the 72-kDa protein-tyrosine kinase PTK72 with the B cell antigen receptor. J Biol Chem 267:1-7.

Iglesias A, Kopf M, Williams GS, Bühler B, Köhler G (1991) Molecular requirements for the μ-induced light chain gene rearrangement in pre-B cells. EMBO J 10:2147-2156.

Irving BA, Weiss A (1991) The cytoplasmic domain of the T cell receptor ζ chain is sufficient to couple to receptor-associated signal transduction pathways. Cell 64:891-902.

Nomura J, Matsuo T, Kubota E, Kimoto M, Sakaguchi N (1990) Signal transmission through the B cell-specific MB-1 molecule at the pre-B cell stage. Int Immunol 3:198-199.

Reth M (1989) Antigen receptor tail clue. Nature 338:383-384.

Reth M, Hombach J, Wienands J, Campbell KS, Chien N, Justement LB, Cambier JC (1991) The B-cell antigen receptor complex. Immunol Today 12:196-201.

Romeo C, Seed B (1991) Cellular immunity to HIV activated by CD4 fused to T cell or Fc receptor polypeptides. Cell 64:1037-1046.

Sakaguchi N, Kashiwamura SI, Kinoto M, Thalmann P, Melchers F (1988) B lymphocyte lineage restricted expression of mb-1, a gene with CD3-like structural properties. EMBO J 7:3457-3464.

Taniguchi T, Kobayashi T, Kondo J, Takahashi K, Nakamura H, Suzuki J, Nagai K, Yamada T, Nakamura S, Yamamura H (1991) Molecular cloning of a porcine gene syk that encodes a 72-kDa protein-tyrosine kinase showing high susceptibility to proteolysis. J Biol Chem 266:15790-15796.

Venkitaraman AR, Williams GT, Dariavach P, Neuberger MS (1991) The B-cell antigen receptor of the five immunoglobulin classes. Nature 352:777-781.

Wienands J, Reth M (1992) Glycosyl-phosphatidylinositol linkage as a mechanism for cell-surface expression of immunoglobulin D. Nature 356:246-248.

Williams AF, Barclay AN (1988) The immunoglobulin superfamily - domains for cell surface recognition. Ann Rev Immunol 6:381-405.

Receptor Editing in Immature Self-Reactive B-cells

D. Nemazee

Division of Basic Sciences, Dept. of Pediatrics, National Jewish Center for Immunology and Respiratory Medicine, 1400 Jackson St., Denver, CO 80206, USA[1]

CLONAL SELECTION VS RECEPTOR SELECTION

The clonal selection theory (Talmage 1957; Burnet 1957) is the basis for much of the way we think about the immune system. Lymphocytes are seen as an inherently diverse population, containing cells that are born with differing specificities for antigens, upon which Darwinian antigenic selection subsequently eliminates unfit, autoreactive clones, and allows the survival and maturation of the fittest clones. Here we review our data showing that in immature, self-reactive B-cells, encounter with autoantigen can induce secondary immunoglobulin light chain gene rearrangements (Tiegs et al., submitted). These rearrangements effectively alter the B-cell's antigen specificity and, in extinguishing the autoreactive specificity of the cell, allow it to develop further and populate the peripheral lymphoid organs. This type of selection is is radically different from the classical clonal selection concept because it permits the environment to direct relevant genetic change.

Our conclusions are based on analyses of a useful genetic system to study B-cell tolerance, the 3-83$\mu\delta$ line mouse (Russell et al. 1991), which is transgenic for IgM and IgD anti-MHC class I immunoglobulin genes. The transgene-encoded 3-83 antibody recognizes H-2K molecules of all allelic forms tested except for d and f. In 3-83$\mu\delta$ mice that are H-2^d, and therefore lack antigen to which the transgenic B-cells react, anti-H-2K B-cells develop and populate the peripheral lymphoid organs. In these H-2^d 3-83$\mu\delta$ mice (non-deleting mice) B-cells are abundant and are essentially monoclonal with respect to specificity, with >95% of the B-cells bearing the 3-83 clonotype, as detected with a specific monoclonal antibody. By contrast, in F1 crosses between 3-83$\mu\delta$ mice and H-2^k or H-2^b mice (centrally-deleting mice), which bear antigen that reacts with the transgenic B-cells, autoreactive B-cells are eliminated from the peripheral lymphoid organs.

---

1   Department of Microbiology and Immunology, University of Colorado Health Sciences Center, Denver, CO 80206. This work was supported by a grant from the National Institutes of Health.

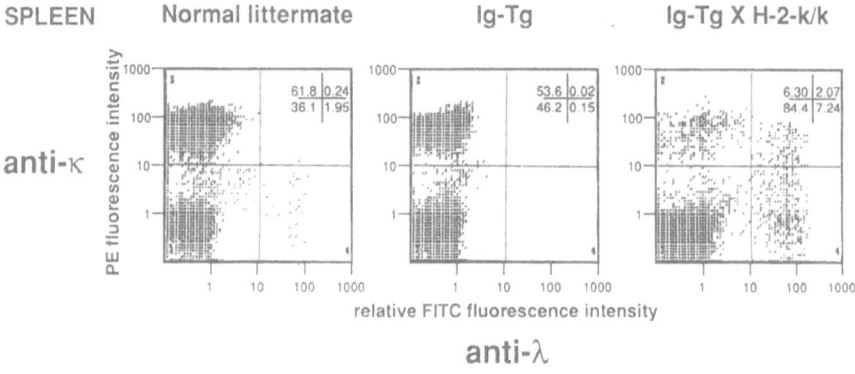

Figure 1. Increase in λ-light chain-bearing B-cells in the spleen of a centrally-deleting transgenic mouse (right panel) relative to non-deleting transgenic mouse (center panel). Spleen cells were stained with phycoerythrin-labeled monoclonal rat anti-mouse kappa (Becton Dickinson) and FITC-labeled Goat anti-mouse lambda (Fisher) and analyzed with a Profile (Coulter) flow cytometer. All mice were six weeks old. The percentages of cells falling into each of the four quadrants is indicated in the upper right hand corner of each histogram.

The extremely good allelic exclusion found in the non-deleting 3-83μδ mice and the unexpectedly large numbers of variant B-cells that escaped deletion in the centrally-deleting mice is striking. One of the remarkable properties of the B-cells of centrally-deleting mice is their frequent expression of λ-light chains (Figure 1). B-cells bearing λ-light chains must arise by the rearrangement and expression of endogenous λ-light chain genes because the transgenic light chain gene is κ.

In light of the extremely low frequency of λ-bearing B-cells in the non-deleting mice, we reasoned that the large number of such cells appearing in the spleens of the centrally-deleting mice could be explained by one of three mechanisms: (1) The selective outgrowth of the λ-bearing B-cells, which is made possible by the "space" provided by elimination of the predominant anti-H-2K-specific population; (2) The induction of increased bone marrow B-cell output, which indirectly increases the number of variant B-cells, including those bearing λ. This induction would be a feedback phenomenon resulting from the decrease in the numbers of peripheral B-cells caused by clonal elimination of the anti-H-2K^{k,b} cells; or (3) the direct induction of light chain gene rearrangement by antigen crosslinking of sIg on B-cells, which would presumably increase both the proportion of B-cells bearing λ-light chains and the numbers of variant B-cells lacking autoreactive receptors.

In order to distinguish among these possibilities, we took advantage of our previously described peripheral B-cell deletion model (Russell et al. 1991). In 3-83μδ mice crossed to MT-K^b transgenic mice (Morahan et al. 1989), which express H-2K^b exclusively in the periphery (i.e. not in the bone marrow), autoreactive B-cells do not encounter antigen in their bone marrow birthplace, but are deleted later, after they have entered the circulation (Russell et al. 1991). When we compared (3-83μδ X H-2$^{k \text{ or } b}$)F1 mice (centrally-deleting mice) with the (3-83μδ X MT-K^b)F1 mice (peripherally-deleting mice) we found a number of important and informative differences. First, the peripherally-deleting mice usually had fewer B-cells than age-matched centrally-deleting mice. Second, the peripherally-deleting mice lacked the large (~10-fold) increase in λ-bearing B-cells seen in the centrally-deleting mice. Finally, we observed an increase in immunoglobulin light chain rearrangement activity in the bone marrows of centrally-deleting mice, but not in peripherally-deleting mice (Table 1). Rearrangement activity was assessed by measuring the levels of RAG-1 and RAG-2 recombinase gene mRNA (Schatz et al. 1989; Oettinger et al. 1990), and by quantifying Ig gene rearrangement excision products in bone marrow DNA (Tiegs et al., submitted). These results suggested that in centrally-deleting bone marrow the *rate* of assembly of κ and λ light chain genes, but not heavy chain genes, was greatly elevated.

Table 1. Increased Ig light chain gene rearrangement in centrally-deleting bone marrow.

Mouse type	RAG expression[a]		Light chain gene rearrangement[b]		
	RAG-1	RAG-2	V_H-to-D	D-to-J	V_L-to-J[c]
Ig-Tg (non-deleting)	-/+	-/+	-	-/+	-
Ig-Tg X MT-K^b (peripherally-deleting)	-/+	-/+	-	-/+	-
Ig-Tg X H-2$^{b \text{ or } k}$ (centrally-deleting)	+	+	-	+	+
Non-Tg littermates	+	+	+	+	+

Mice were two months old.

a RAG mRNA detected with a quantitative PCR assay.

b Ig gene rearrangements quantitated by the detection of rearrangement excision   products in bone marrow DNA.

c Both κ and λ were analyzed with similar results.

These results are incompatible with the simple model of selective outgrowth of λ-bearing cells (mechanism [1] above) because this model assumes no change in the rate of generation of λ^+ B-cells, just a difference in their subsequent growth. These data are also incompatible with a simple feedback-induction model (mechanism [2] above) because the peripherally-deleting mice are equally, or more, B-cell deficient than the centrally-deleting mice, yet they fail to demonstrate increased rearrangement activity in the bone marrow. We therefore favor the alternative possibility to explain the data-- namely that in immature, self-reactive B-cells, encounter with autoantigen can induce secondary immunoglobulin light chain gene rearrangements.

One of the attractive features of our hypothesis is that it helps to explain the arrangement of the mouse (and human) immunoglobulin genes. After VDJ assembly on the heavy chain locus, further rearrangements using conventional signal sequences are not possible on the same allele, because of the deletion of unused D elements and the inability of V_H elements to join to downstream J_H's (Cory and Adams 1980; Brodeur et al. 1988). By contrast, the κ-locus, which lacks D elements and in which V's join directly to J's, appears to encourage nested rearrangements that can delete and replace functionally rearranged V-J coding regions (e.g. Van Ness et al. 1982, Feddersen and Van Ness 1985). Products of such secondary rearrangements that delete functional $V_κ$-$J_κ$ coding segments have recently been demonstrated to exist in normal splenic B-cells (Harada and Yamagishi 1991), but the cause of their production has been unclear. We propose that nested $V_κ$-$J_κ$ or $V_κ$-RS rearrangements are induced in the κ-locus of immature IgM+ B-cells encountering autoantigens that crosslink their Ig receptors. (RS is the mouse kappa deleting element, Durdik et al. 1984; Moore et al. 1985). These rearrangements remove and replace active $V_κ$-$J_κ$ elements with different light chain specificities-- κ in the case of nested $V_κ$-$J_κ$ rearrangements (Feddersen and Van Ness 1985), and λ in the case of nested $V_κ$-RS rearrangements, which are believed to play a role in the activation of the λ loci (Moore et al. 1985). Thus, secondary rearrangements in the κ-locus, but not in the heavy-chain gene locus, are both possible and effective in changing a B-cell's specificity. The ability to take advantage of receptor editing at what must be a late, sIg+ stage in B-cell development could be a driving force in the evolution of both the heavy and light chain genes. In the heavy chain locus the relative transcriptional orientations of the V and J elements and the use of D elements should prevent effective receptor editing, while in the κ-locus the inverted orientation of about half the V genes, relative to the J's (Shapiro and Weigert 1987), and the absence of D elements, permits repeated attempts at receptor editing.

MULTIPLE MECHANISMS OF B-CELL TOLERANCE

To further explain the differences between centrally-deleting and peripherally-deleting mice we assume that the peripheral deletion of autoreactive B-cells in (3-83μδ X MT-K^b)F1 mice proceeds by a mechanism of rapid programmed cell death. Programmed cell death has been shown to occur in autoreactive B-1 (CD5) B-cells challenged with autoantigen (Murakami et al. 1992) and in germinal center B-cells that fail to bind to antigen with sufficient affinity (Liu et al. 1989). We think it likely that only B-cells at an early developmental stage are capable of receptor editing, while at later stages of their development B-cells can be deleted or anergized (Goodnow et al. 1988). Even in B-cells capable of receptor editing, we would guess that cells that fail to replace their autoreactive Ig receptors with appropriate receptors must eventually turn over by a death program. However, we have failed to detect apoptotic cells among sorted, bone marrow sIg⁺ B-cells from centrally-deleting mice, only a small subset of which eventually populate the peripheral lymphoid organs.

RECEPTOR SELECTION IN LYMPHOCYTES

Our results provide a third example in the immune system of *receptor selection*, which is distinguished from *clonal selection* by the feature that the fate of the cell is uncoupled from the fate of its antigen-specific receptor. In the case described here, antigen in the bone marrow can act to eliminate an autoreactive B-cell's anti-self specificity without eliminating the cell itself. The other two examples of receptor selection are the well-documented induction of somatic hypermutation by antigenic stimulation in B-cells (Weigert et al. 1970; Möller 1987), and the recent observation of down-regulation of ongoing TCR-α rearrangement in positively-selected immature TCR⁺ thymocytes (Turka et al. 1991; Borgulya et al. 1992). In our particular case this receptor selection plays an important role in self-tolerance and therefore should be considered as a potential target of defects leading to autoimmunity.

REFERENCES

Borgulya P, Kishi H, Uematsu Y, von Boehmer H (1992) Exclusion and inclusion of α and β T cell receptor alleles. Cell 69: 529-537

Brodeur PH, Osman GE, Mackle JJ, Lalor TM (1988) The organization of the mouse *Igh-V* locus. Dispersion, interspersion, and the evolution of the V_H gene family clusters. J Exp Med 168: 2261-2278

Burnet FM (1957) A modification of Jerne's theory of antibody production using the concept of clonal selection. Aust J Sci 20 : 67-69

Cory S, Adams JM (1980) Deletions are associated with somatic rearrangement of immunoglobulin heavy chain genes. Cell 19: 37-51

Durdik J, Moore MW, Selsing E (1984) Novel kappa light-chain gene rearrangements in mouse lambda light chain-producing B lymphocytes. Nature 307: 749-752

Feddersen, R.M. and Van Ness, B.G. (1985). Double recombination of a single immunoglobulin κ-chain allele: Implications for the mechanism of rearrangement. Proc. Natl. Acad.Sci.USA *82*, 4793-4797.

Goodnow CC, Crosbie J, Adelstein S, Lavoie TB, Smith-Gill SJ, Brink RA, Pritchard-Briscoe H, Wotherspoon JS, Loblay RH, Raphael K, Trent RT, Basten A (1988) Altered immunoglobulin expression and functional silencing of self-reactive B lymphocytes in transgenic mice. Nature 334: 676-682

Harada K, Yamagishi H (1991) Lack of feedback inhibition of Vκ gene rearrangement by productively rearranged alleles. J Exp Med 173: 409-415

Liu YJ, Joshua DE, Williams GT, Smith CA, Gordon J, MacLennan ICM (1989) Mechanism of antigen-driven selection in germinal centres. Nature (London) 342:929-931

Moore MW, Durdik J, Persiani DM, Selsing E (1985) Deletions of κ chain constant region genes in mouse λ chain-producing B cells involve intrachromosomal DNA recombinations similar to V-J joining. Proc Natl Acad Sci USA 82: 6211-6215

Morahan G, Brennan FE, Bhathal PS, Allison J, Cox KO, Miller JFAP (1989) Expression in transgenic mice of Class I histocompatibility antigens controlled by the metallothionene promoter. Proc Natl Acad Sci USA 86: 3782-3786

Möller G, ed (1987) Role of somatic mutation in the generation of lymphocyte diversity. Immunol Rev vol. 96

Murakami M, Tsubata T, Okamoto M, Shimizu A, Kumagai S, Imura H, Honjo T (1992) Antigen-induced apoptotic death of Ly-1 B cells responsible for autoimmune disease in transgenic mice. Nature 357:77-80

Nemazee D, and Bürki K (1989) Clonal deletion of B lymphocytes in a transgenic mouse bearing anti-MHC Class I antibody genes. Nature 337: 562-566

Nemazee D, Bürki K (1989) Clonal deletion of autoreactive B lymphocytes in bone marrow chimeras Proc Natl Acad Sci USA 86: 8039-8043

Oettinger MA, Schatz DG, Gorka C, Baltimore D (1990) RAG-1 and RAG-2, adjacent genes that synergistically activate V(D)J recombination. Science 248: 1517-1522

Russell DM, Dembic' Z, Morahan G, Miller JFAP, Bürki K, Nemazee D (1991) Peripheral deletion of self-reactive B-cells. Nature 354: 308-311.

Schatz DG, Oettinger MA, Baltimore D (1989) The V(D)J recombination activating gene, RAG-1. Cell 59: 1035-1048

Shapiro, M.A. and Weigert, M. (1987). How immunoglobulin Vκ genes rearrange. J. Immunol. *139*, 3834-3839.

Talmadge DW (1957) Allergy and immunology. Ann Rev Med 8: 239-256

Turka LA, Schatz DG, Oettinger MA, Chun JJ, Gorka C, Lee K, McCormack WT, Thompson, CB (1991) Thymocyte expression of RAG-1 and RAG-2: termination by T cell receptor crosslinking. Science 253: 778-781

Van Ness BG, Coleclough C, Perry RP, Weigert M (1982) DNA between variable and joining gene segments of immunoglobulin κ light chain is frequently retained in cells that rearrange the κ locus. Proc Natl Acad Sci USA 79: 262-266

Weigert M, Cesari IM, Yonkovich SJ, Cohn M (1970) Variability in the lambda light chain sequences of mouse antibody. Nature 228:1045-1047

A Developmental Switch in B Lymphopoiesis: A Fetal Origin for CD5+ B Cells

R. R. Hardy and K. Hayakawa

Institute for Cancer Research, Fox Chase Cancer Center, Philadelphia, PA, 19111, USA

INTRODUCTION

Since their identification a decade ago (Hayakawa et al. 1983) as an infrequent subset of B cells in both mouse and man, CD5+ B cells have posed a number of questions. One of the most important is their relationship to the major population of IgD++/CD5- B cells ("conventional" B cells). Although early data showing differences in the capacity of fetal liver and adult bone marrow to repopulate this subset in irradiated allotype-congenic recipients suggested a relatively distinct origin (Hayakawa et al. 1985), the ability to induce CD5 on conventional B cells by treatment with phorbol ester (Miller and Gralow 1984) has led others to suggest that these cells arise by a novel activation pathway from bone marrow B cells. To decide between these two explanations for the origin of CD5+ B cells we have first carefully discriminated intermediate stages of early B cell differentiation, both in bone marrow of adults and in fetal liver. This analysis allowed us to define a fraction of cells in both locations that is B-lineage committed at a comparable differentiation stage (possessing D-J, but not V-D-J rearrangements) and capable of generating large numbers of B cells after transfer into immunodeficient SCID mice. Our results show that the progeny of this Pro B cell fraction isolated from fetal liver gives rise to a phenotypically and functionally distinct B cell population when compared with B cells arising from bone marrow Pro B cells. We identify this fetal-derived population as CD5+ B cells and the adult-derived population as conventional B cells.

Distinctive properties of CD5+ B cells.

CD5+ B cells possess a number of properties that distinguish them from the bulk of CD5-IgD++ B cells, often referred to as "conventional" B cells (reviewed in (Hayakawa and Hardy 1988); see Table 1). Besides their distinctive phenotype, their enrichment in the peritoneal cavity is useful in analyzing this relatively rare B cell population. Further, transfer studies have demonstrated their capacity to expand and reconstitute irradiated recipients, in contrast with conventional B cells (Hayakawa et al. 1986). This ability to self-renew may also be responsible for the observed expansions of CD5+ B cells in aged normal animals and in adult mice of certain autoimmune strains (Stall et al. 1988). Indeed many B cell neoplasms in mice bear CD5 and the CH series of B lymphomas are uniformly CD5+. The CH-series lines also exhibit another novel feature ascribed to CD5+ B cells, that of restricted immunoglobulin repertoire, with a bias toward certain autoreactive specificities (Haughton et al. 1986).

Table 1. Distinct Features of CD5+ and "Conventional" B Cells

	CD5+ B cells	CD5-IgD++ B cells
Phenotype	IgM^{high}IgD^{low}	IgM^{low}IgD^{high}
Localization	Peritoneal Cavity, Spleen	Lymph Nodes, Spleen
Lifespan	Self-renewing; no IgM- precursors in adult	Replaced by IgM- precursors in bone marrow
Specificities	enriched for self-reactivities	largely to foreign antigens
Growth Properties	propensity to expansion	dies easily in vitro
Ig Genes	low N-region addition	N-regions common
Origins	in fetal liver	in adult bone marrow

Several years ago we showed that reactivity with a determinant on mouse red blood cells exposed by treatment with the proteolytic enzyme bromelain (anti-BrMRBC) was strikingly enriched in the CD5+ B cell subset (Hayakawa et al. 1984). More recently, we and others found that the V genes encoding this reactivity belonged to a small V_H family, V_H11 (Reininger et al. 1987; Hardy et al. 1989). Utilizing a PCR assay, we showed that most of the BrMRBC-specific CD5+ B cells possessed V_H11 gene rearrangements (Carmack et al. 1990). Since the frequency of cells with this specificity is 10% in normal mouse strains, it was easy to show an increased frequency of V_H11 rearrangement in the CD5+ B cell subset of both spleen and peritoneal cavity. We have suggested that this enrichment is due to selection of a fetally-generated B cell population, based on germline-encoded self-reactivity, resulting in accumulation of certain specificities in the adult long-lived CD5+ B cell subset.

Stem cells from adult bone marrow are ineffective in reconstituting CD5+ B cells.

Recently we have repeated our earlier transfer experiments, but utilizing precursor fractions enriched for hematopoietic stem cells using the procedures developed by Muller-Seiburg and Weissman (Muller-Sieburg et al. 1986; Spangrude et al. 1988). Thus we transferred Thy-1+Lin− fractions sorted from bone marrow of adult mice and from fetal liver of day 16 gestation into immunodeficient SCID recipients that were irradiated the previous day (350R). Recipient animals were sacrificed 4 or 8 weeks after transfer and analyzed for lymphoid engraftment by flow cytometry. As is clearly shown in Fig. 1, there was a striking difference in the relative repopulation of CD5+ B cells in the peritoneal cavity, similar to the earlier findings utilizing unfractionated precursor sources. These new experiments suggest that the difference in repopulating capacity occurs in a very early lymphocyte precursor, possibly a lymphoid cell progenitor or even the hematopoietic stem cell itself.

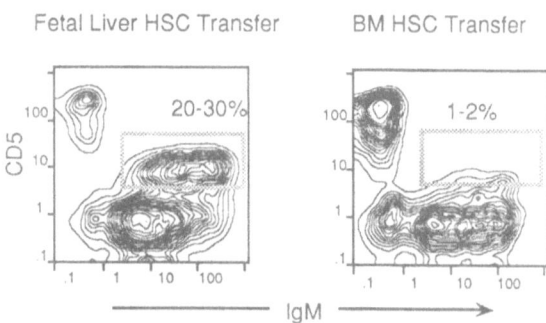

Fig. 1. Analysis of B cell subsets reconstituted in peritoneal cavity of SCID mice 2 months after transfer of 10^4 Thy-1+Lin− cells sorted from either fetal liver or adult bone marrow. "Lin−" means that cells in this fraction lack differentiation lineagemarkers such as B220, Mac-1, CD4, CD8, etc.

Resolution of intermediate stages of early B lineage cells

In order to more clearly define the stage at which a cell becomes restricted to generating CD5+ B cells versus conventional B cells, we have initially focused on carefully resolving intermediate stages of early B lineage cells in bone marrow. We have taken advantage of the S7 monoclonal antibody, initially shown to be expressed at late stages of B cell differentiation (Gulley et al. 1988). This antibody is now known to recognize the murine homolog of CD43, also known as sialophorin, a highly O-glycosylated molecule present on diverse cell types (Baecher et al. 1988). However it is found on only a subset of B lineage cells in bone marrow that are recognized by expression of the high molecular weight form of the common leukocyte antigen, B220. We showed that these B220+CD43+ cells could proliferate rapidly in stromal cell culture and in response to interleukin-7, a characteristic of an early B lineage cell. We also found that we could subdivide the B220+CD43+ fraction (3% of bone marrow) into three subsets based on correlated expression of the heat stable antigen (HSA) and BP-1, a molecule expressed preferentially on early B lineage cells (Hardy et al. 1991).

We wished to correlate our phenotypic intermediates with classical stages of B cell differentiation so we sought to determine the immunoglobulin rearrangement status of these fractions. We chose to design a polymerase chain reaction (PCR) assay, due to the difficulty of obtaining significant numbers of these cells. We selected a pair of oligonucleotide primers that amplify a segment of DNA that is deleted upon rearrangement of any D to any J, another pair that amplify a segment deleted upon V to D rearrangement

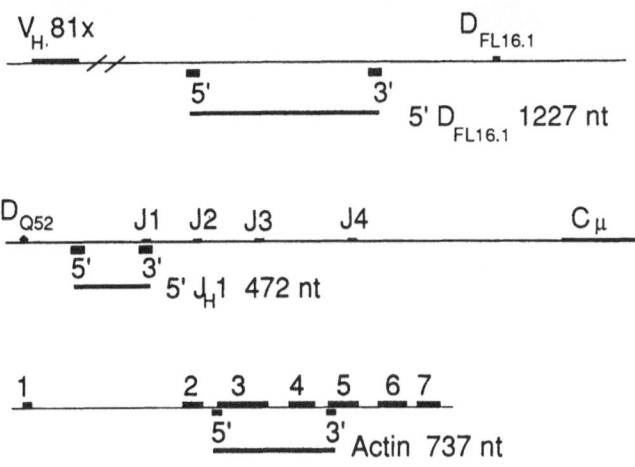

Fig. 2. Diagram of PCR assay for determining degree of rearrangement in sorted B lineage subsets. Three fragments are amplified simultaneously, with the ratio of 5′$D_{FL16.1}$ to actin and 5′ J_H1 to actin determining the degree of V to D and D to J rearrangement, respectively.

and a third pair that amplify a segment from a gene unaffected by Ig rearrangement, α-actin. The basic scheme is diagrammed in Fig. 2.

Our results are summarized in Table 2. Thus, we found that we could resolve a set of cells in which D to J rearrangement was ongoing, but before the onset of V to D-J rearrangement. This "Pro-B" cell fraction was recognized as $B220^+CD43^+HSA^+$. We were also able to resolve a similar fraction of cells in fetal liver, and PCR analysis showed ongoing D-J, but not V-DJ rearrangement in this population also (see Fig. 3). Interestingly, the latter stages of B cell differentiation, recognized as $B220^+CD43^-$, were absent from fetal liver at day 16 of gestation.

Table 2. Heavy chain rearrangement in B lineage cells determined by PCR

| Cell fraction | Fractional Retention of Germline | | | |
| | fragment 5' of D | | fragment 5' of J | |
	mean	(SE)	mean	(SE)
BM Pro B	0.98	(0.03)	0.39	(0.02)
Pre-B	0.50	(0.01)	0.05	(0.01)
B cell	0.50	(0.04)	0.05	(0.01)
FL Pro B	0.95	(0.04)	0.55	(0.04)

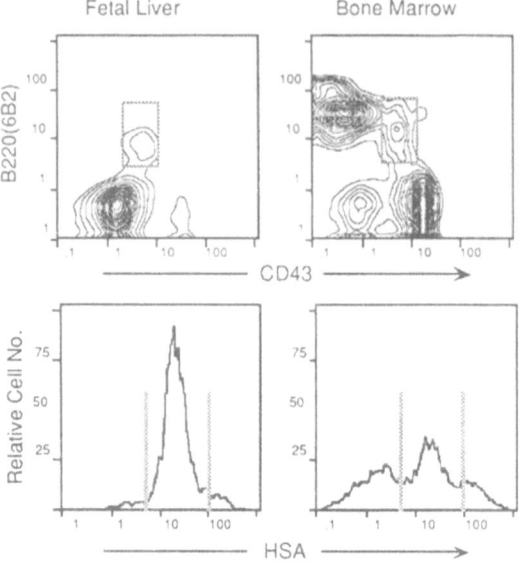

Fig. 3. Analysis of fetal liver and adult bone marrow for correlated expression of B220, CD43 and HSA. The gates used in sorting cells are marked on the plots.

Repopulation of SCID mice with Pro B cells of different developmental origins.

We next sought to assess the differentiation potential of Pro B cells when provided with an essentially complete microenvironment, an adult SCID mouse (Bosma 1989). Thus we sorted B220+CD43+HSA+ cells from fetal liver and adult bone marrow and transferred these cells into comparable adult SCID recipients that were irradiated the previous day (350R). Such transfers repopulated B cells in both spleen and peritoneal cavity when analyzed three weeks or more after injection. Analysis of bone marrow and thymus showed no permanent lymphocyte precursor pool and there were no T cells in the periphery, in contrast with hematopoietic stem cell (HSC) transfers (Fig. 4). Therefore we conclude that the Pro B cell fraction is largely B lineage restricted and differentiates, rather than extensively self-renews, when placed in a SCID microenvironment.

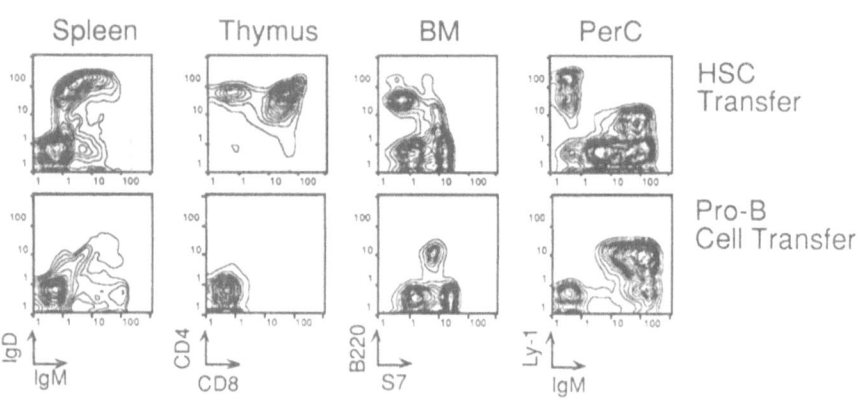

Fig. 4. Analysis of lymphoid subsets reconstituted in SCID mice two months following transfer of either HSC (Thy-1+LIN−) or Pro B (B220+CD43+HSA+) sorted from fetal liver.

46

We found that the phenotype of B cells generated from fetal liver differed from that arising from adult bone marrow: most B cells generated from fetal liver Pro B had low to negligible levels of IgD and may bore CD5; in contrast, B cells derived from adult bone marrow were CD5⁻ and most expressed high levels of IgD (Hardy and Hayakawa 1991). Thus, essentially all Pro B cells found in fetal liver at day 16 of gestation follow a distinct differentiation program compared to the B cells in adult bone marrow, even when placed into an identical adult microenvironment. Fig. 5 presents flow cytometric analyses of the two different B cell phenotypes generated in spleen of such SCID Pro B cell recipients.

Finally, we also wished to determine whether any functional activity associated with the normal CD5⁺ B cell subset could be found in the engrafted population. As mentioned above, a significant fraction of the normal CD5⁺ B cell subset is specific for a determinant present on bromelain-treated mouse red blood cells. These cells can be revealed by their capacity to bind phosphatidyl-choline(PtC)-containing lipid vesicles that have incorporated a fluorescent dye as analyzed by flow cytometry (Mercolino et al. 1988). Accordingly, we stained peritoneal washout cells from SCID mice reconstituted two months earlier with Pro B cells from bone marrow and fetal liver with PtC-vesicles together with reagents specific for IgM and CD5. Fig. 6 shows the contour maps of IgM⁺ gated cells displaying CD5 versus PtC-binding. As was shown above, B cells in the bone marrow transferred mouse largely lack CD5 expression and few bind PtC. In contrast, many of the cells in the fetal liver transferred mouse bear CD5 and about 5% bind PtC. Thus one of the functional features of the CD5⁺ subset in intact animals is also found in Pro B cell reconstituted SCID mice.

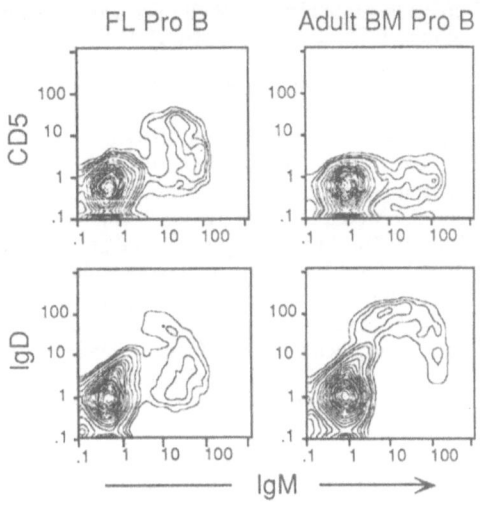

Fig. 5. B cell phenotypes repopulated in spleen of SCID mice three weeks after transfer of Pro B cell fractions (10^5 cells) sorted from either fetal liver or adult bone marrow.

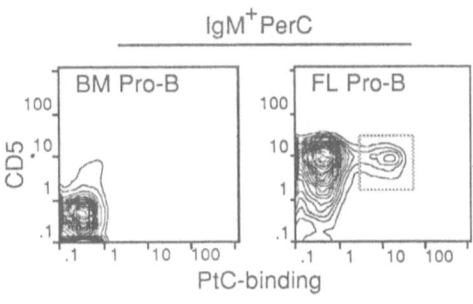

Fig. 6. Generation of a phosphatidyl-choline-specific population of CD5⁺ B cells in SCID mice receiving fetal liver Pro B cells, but not in mice receiving bone marrow Pro B cells. Analysis of B cells in peritoneal cavity of SCID recipients two months after transfer.

SUMMARY AND CONCLUSIONS

Since the initial description of CD5 expression on a subset of B cells in normal mice, there has been an ongoing debate as to the relation of this population to the bulk of CD5⁻ B cells. Although early transfer experiments suggested that their origins might be largely independent of conventional B cell differentiation, the potential of CD5⁻ B cells to express this molecule after certain types of treatment in vitro has led others to view CD5 as an "activation antigen" (Miller and Gralow 1984; Cong et al. 1991). In opposing this "activation model", our results from transfers of sorted populations of defined precursors into identical adult microenvironments provides some of the clearest evidence that the origins of most CD5+ B cells are distinct from most CD5⁻IgD++ (conventional) B cells.

In an attempt to reconcile these data with an activation model, some have proposed that CD5 expression is induced only by certain antigens (possibly those known as "type II") and that such specificities are uniquely restricted to the fetal-derived B cell population, possibly due to low N region addition. While it is clear that N region addition is low in the progeny of fetal Pro B cells (Feeney 1990) it remains unclear how their lack would result in such a biased reactivity (or alternately how the presence of N addition would preclude it). Furthermore, not all CD5+ B cells show an absolute lack on N-addition (Gu et al. 1990). Another suggestion is that preferential V_H/V_L expression restricted to fetal B cells might give the type of reactivity that preferentially induces CD5 expression. However such preferential usage has been reported in early B lineage cells of both fetal liver and adult bone marrow (Perlmutter et al. 1985; Yancopoulos et al. 1988). Even if it turns out that fetal and adult B cells do express distinct repertoires, then the argument that CD5 is simply an activation antigen on B cells becomes in essence semantic, since CD5 expression would still serve to mark the progeny of B cells largely restricted to fetal development.

We favor a model wherein numerous differences in cellular physiology distinguish fetal and adult B cells. In addition to expression of TdT, the recent report of a difference in expression of a novel pre-B specific myosin light chain between fetal liver and adult bone marrow marks another molecular distinction between B lineage cells at these two developmental timings (Oltz et al. 1992). We hypothesize that crosslinking of surface Ig transduces a different signal in fetal B cells compared to adult B cells such that germline-encoded self-reactivity is positively selected into an adult long-lived pool. Further, we predict that this population should lack molecules important for interaction with T cells normally leading to somatic hypermutation and affinity maturation since this could lead to high affinity, likely pathogenic, autoreactivity. It should be possible to test these ideas in the near future.

In summary, even though CD5 expression can be induced on bone marrow derived B cells by a type of "activation", it appears likely that CD5 expression *in vivo* normally serves to mark the progeny of a distinctive wave of B cell development separate from adult bone marrow derived B cells. Thus, as suggested long ago, CD5+ B cells constitute a separate developmental pathway or lineage with novel molecular and functional properties. The associations with autoreactivity and B cell neoplasia make this an important population to understand more fully. Finally, the functional significance of a fetally-generated subset of B cells selected for germline-encoded self-reactivity remains an intriguing puzzle for further investigation.

REFERENCES

Baecher CM, Infante AJ, Semcheski KL, Frelinger JG (1988). Identification and characterization of a mouse cell surface antigen with alternative molecular forms. Immunogenetics 28: 295-302.
Bosma MJ (1989). The scid mutation: occurrence and effect. Curr Top Microbiol Immunol 152: 3-9.
Carmack CE, Shinton SA, Hayakawa K, Hardy RR (1990). Rearrangement and selection of V_H11 in the Ly-1 B cell lineage. J Exp Med 172: 371-374.
Cong YZ, Rabin E, Wortis HH (1991). Treatment of murine CD5⁻ B cells with anti-Ig, but not LPS, induces surface CD5: two B-cell activation pathways. Int Immunol 3: 467-76.
Feeney AJ (1990). Lack of N regions in fetal and neonatal mouse immunoglobulin V-D-J junctional sequences. J Exp Med 172: 1377-1390.
Gu H, Forster I, Rajewsky K (1990). Sequence homologies, N sequence insertion and J_H gene utilization in V_HDJ_H joining: implications for the joining mechanism and the ontogenetic timing of Ly1 B cell and B-CLL progenitor generation. Embo J 9: 2133-2140.
Gulley ML, Ogata LC, Thorson JA, Dailey MO, Kemp JD (1988). Identification of a murine pan-T cell antigen which is also expressed during the terminal phases of B cell differentiation. J Immunol 140: 3751-3757.
Hardy RR, Carmack CE, Shinton SA, Kemp JD, Hayakawa K (1991). Resolution and characterization of pro-B and pre-pro-B cell stages in normal mouse bone marrow. J Exp Med 173: 1213-1225.

Hardy RR, Carmack CE, Shinton SA, Riblet RJ, Hayakawa K (1989). A single V$_H$ gene is utilized predominantly in anti-BrMRBC hybridomas derived from purified Ly-1 B cells. Definition of the V$_H$11 family. J Immunol 142: 3643-3651.

Hardy RR, Hayakawa K (1991). A developmental switch in B lymphopoiesis. Proc. Natl. Acad. Sci. USA 88: 11550-11554.

Haughton G, Arnold LW, Bishop GA, Mercolino TJ (1986). The CH series of murine B cell lymphomas: neoplastic analogues of Ly-1+ normal B cells. Immunol Rev 93: 35-51.

Hayakawa K, Hardy RR (1988). Normal, autoimmune, and malignant CD5+ B cells: the Ly-1 B lineage? . Annu Rev Immunol 6: 197-218.

Hayakawa K, Hardy RR, Herzenberg LA, Herzenberg LA (1985). Progenitors for Ly-1 B cells are distinct from progenitors for other B cells. J Exp Med 161: 1554-1568.

Hayakawa K, Hardy RR, Honda M, Herzenberg LA, Steinberg AD, Herzenberg LA (1984). Ly-1 B cells: functionally distinct lymphocytes that secrete IgM autoantibodies. Proc Natl Acad Sci U S A 81: 2494-2498.

Hayakawa K, Hardy RR, Parks DR, Herzenberg LA (1983). The "Ly-1 B" cell subpopulation in normal immunodefective, and autoimmune mice. J Exp Med 157: 202-218.

Hayakawa K, Hardy RR, Stall AM, Herzenberg LA, Herzenberg LA (1986). Immunoglobulin-bearing B cells reconstitute and maintain the murine Ly-1 B cell lineage. Eur J Immunol 16: 1313-1316.

Mercolino TJ, Arnold LW, Hawkins LA, Haughton G (1988). Normal mouse peritoneum contains a large population of Ly-1+ (CD5) B cells that recognize phosphatidyl choline. Relationship to cells that secrete hemolytic antibody specific for autologous erythrocytes. J Exp Med 168: 687-698.

Miller RA, Gralow J (1984). The induction of Leu-1 antigen expression in human malignant and normal B cells by phorbol myristic acetate (PMA). J Immunol 133: 3408-3414.

Muller-Sieburg CE, Whitlock CA, Weissman IL (1986). Isolation of two early B lymphocyte progenitors from mouse marrow: a committed pre-pre-B cell and a clonogenic Thy-1-lo hematopoietic stem cell. Cell 44: 653-662.

Oltz EM, Yancopoulos GD, Morrow MA, Rolink A, Lee G, Wong F, Kaplan K, Gillis S, Melchers F, Alt FW (1992). A novel regulatory myosin light chain gene distinguishes pre-B cell subsets and is IL-7 inducible. Embo J 11: 2759-2767.

Perlmutter RM, Kearney JF, Chang SP, Hood LE (1985). Developmentally controlled expression of immunoglobulin V$_H$ genes. Science 227: 1597-1601.

Reininger L, Ollier P, Poncet P, Kaushik A, Jaton JC (1987). Novel V genes encode virtually identical variable regions of six murine monoclonal anti-bromelain-treated red blood cell autoantibodies. J Immunol 138: 316-323.

Spangrude GJ, Heimfeld S, Weissman IL (1988). Purification and characterization of mouse hematopoietic stem cells. Science 241: 58-62.

Stall AM, Farinas MC, Tarlinton DM, Lalor PA, Herzenberg LA, Strober S, Herzenberg LA (1988). Ly-1 B-cell clones similar to human chronic lymphocytic leukemias routinely develop in older normal mice and young autoimmune (New Zealand Black-related) animals. Proc Natl Acad Sci USA 85: 7312-7316.

Yancopoulos GD, Malynn BA, Alt FW (1988). Developmentally regulated and strain-specific expression of murine V$_H$ gene families. J Exp Med 168: 417-435.

The Role of *bcl* - 2 in Lymphoid Differentiation and Transformation

A. Strasser, A. W. Harris and S. Cory

The Walter and Eliza Hall Institute of Medical Research, Post Office, Royal Melbourne Hospital, Victoria 3050, Australia

INTRODUCTION

Stringent selection mechanisms ensure the establishment of a repertoire of mature B and T cells that is potentially useful and not harmful (for reviews see von Boehmer and Kisielow, 1990; Blackman et al., 1990; Rolink and Melchers, 1991). Lymphocytes appear to be continuously checked by their environment for functional antigen receptor expression. Appropriate engagement of the antigen receptor, be it surface imunoglobulin (sIg) or a T cell receptor (TCR) heterodimer, determines cell survival and induces proliferation and differentiation. Inappropriate engagement evokes cell suicide by apoptosis (Murphy et al., 1990; Swat et al., 1991).

The bcl-2 protein has recently been recognized as one determinant of lymphoid cell survival. The proto-oncogene *bcl*-2 was initially identified by virtue of its involvement in the t(14;18) chromosomal translocation, the hallmark of follicular center B cell lymphoma. The translocation recombines the *bcl*-2 gene into the immunoglobulin (Ig) heavy chain locus (Tsujimoto et al., 1984; Bakhshi et al., 1985; Cleary et al., 1986), perturbing its regulation but not its coding region. The *bcl*-2 gene encodes a 26 kD non-glycosylated cytoplasmic protein (Tsujimoto et al., 1987; Chen-Levy et al., 1989) which associates via its hydrophobic C-terminus with the plasma membrane and perinuclear endoplasmic reticulum (Chen-Levy et al., 1989; Chen-Levy and Cleary, 1990) and with the inner mitochondrial membrane (Hockenbery et al., 1990). The first indication of the function of *bcl*-2 was the unexpected finding that enforced *bcl*-2 expression in growth factor-dependent cell lines permitted their survival in the absence of factor (Vaux et al., 1988). This observation raised the possibility that a normal role of *bcl*-2 is to govern the life and death of lymphocytes during the generation and function of the immune system. Transgenic mice have been developed to test this hypothesis and to evaluate the oncogenic potential of *bcl*-2 for lymphoid cells (McDonnell et al., 1989,1990; Strasser et al., 1990a,1991a,1991b; Sentman et al., 1991).

CONSEQUENCES OF DEREGULATING *BCL*-2 EXPRESSION IN THE B CELL LINEAGE

Constitutive *bcl*-2 expression significantly perturbed homeostasis of the B lymphoid lineage (McDonnell et al., 1989,1990a; Strasser et al., 1990a, 1991a). The *bcl*-2 transgenic mice accumulated a 3- to 5-fold excess of small, non-cycling, mature conventional (IgM+, IgD+, CD23+, class II MHC+) B cells in all organs where B cells are normally found (peripheral blood, spleen, bone marrow, lymph nodes and peritoneal cavity). After 6-8 weeks of age, there was essentially no further increase, consistent with the idea that the rate of B cell production from progenitor cells declines with age (Rolink and Melchers, 1991). The number of terminally differentiated Ig-secreting cells was markedly increased (30- to 200-fold), producing a substantial increase in serum Ig levels. Immunization provoked an amplified and drastically prolonged humoral immune response, presumably due to an increase in the lifespan of plasma cells (Strasser et al., 1991a; Nunez et al., 1991). B lymphoid cells of all differentiation stages (pre-B, virgin B, mature B, activated B and plasma cell) exhibited prolonged survival in simple tissue culture medium (McDonnell et al., 1989, 1990; Strasser et al., 1990a, 1991a). Significantly, all of the survivors were small and non-cycling, but they could readily be induced to re-enter the cell cycle by the addition of antigens or mitogens.

Our transgenic mice expressing *bcl*-2 in the B lineage often developed a fatal autoimmune disease resembling human systemic lupus erythematosus (Strasser et al., 1991a). The genetic background of the mice has recently been shown to significantly influence the incidence of this disease (Strasser et al., manuscript submitted). We speculate that the long-lived B cells and plasma cells presumably included those with anti-self reactivity, causing autoreactive antibodies to accumulate to pathologic concentrations. Perturbation of deletion mechanisms that normally operate in germinal centres may have increased the frequency of self-reactive clones.

The inference from these experiments with transgenic mice is that *bcl*-2 may play an important role in determining the lifespan of activated and terminally differentiated B lineage cells during a humoral immune response *in vivo*. Intriguingly, bcl-2 protein is readily apparent in long-lived mature resting B cells but is not detectable in most rapidly dividing centroblasts and centrocytes, which are prone to apoptotic cell death (Pezzella et al., 1991; Hockenbery et al., 1991; Liu et al., 1991). It would therefore appear that differentiation from the centrocyte to the memory B cell stage requires the induction of *bcl*-2. Consistent with this hypothesis, treatment of centrocytes *in vitro* with anti-Ig antibody induced *bcl*-2 expression, delayed cell death and induced progression to a memory B cell phenotype (Liu et al., 1991). This observation suggests that when germinal center B cells bind to the antigen/antibody complexes retained by follicular dendritic cells, *bcl*-2 expression is induced and differentiation proceeds to the memory B cell stage. Only those B cells which express high affinity antigen receptors generated by somatic hypermutation would be able to bind and hence be positively selected.

In view of the extended lifespan of transgenic pre-B cells *in vitro*, it seemed likely that *bcl*-2 expression also plays a role in determining cell survival during differentiation from the progenitor to the mature B cell stage. However, only a modest increase in B220$^+$sIg$^-$ B lineage cells was evident in lymphoid organs of *bcl*-2 transgenic mice (McDonnell et al., 1989,1990; Strasser et al., 1991a and our unpublished observations). While this might imply that *bcl*-2 cannot confer survival on B cells which have not undergone productive Ig gene rearrangements, an alternative interpretation was that the survival advantage for surface Ig$^-$ B lymphoid cells was concealed by the selective advantage *in vivo* of surface Ig$^+$ B cells. We therefore analysed specific-pathogen free newborn mice, in which antigen driven expansion is minimal, and found a significant excess of pre-B cells in the *bcl*-2 transgenic mice (Strasser at al., in preparation).

To further address this issue, we have combined our *bcl*-2 transgene with the *scid* mutation by cross-breeding. As expected, *bcl*-2 transgene expression was unable to correct the *scid* defect in generating productive antigen receptor gene rearrangements (Schuler et al., 1986). There was, however, a notable increase in the number of B220$^+$ surface Ig$^-$ B lineage cells in the bone marrow (3- to 5-fold) and spleen (20- to 50-fold) (Strasser et al., manuscript in preparation). Surprisingly, the *bcl*-2/ *scid* B lineage cells exhibited properties normally only associated with mature long lived sIgM$^+$ IgD$^+$ B cells. Most were small and in G$_0$, and they expressed CD23 and class II MHC, which are hallmarks of mature resting B cells but are not found on B220$^+$ *scid* cells. The *bcl*-2/*scid* B lineage cells survived as well in culture as those from normal *bcl*-2 transgenic mice, whereas *scid* B220$^+$ cells died even more rapidly than those from normal mice, probably because of their defect in DNA repair (Fulop and Phillips, 1990). Taken together, these experiments suggest an important role for *bcl*-2 in mediating survival of immature B lymphoid cells.

It has been postulated that B lymphoid proliferation and differentiation is controlled by the engagement of a succession of receptor complexes (Rolink and Melchers, 1991). Pro-B cells, which lack any Ig gene rearrangement, express λ$_5$ and VpreB surrogate light chains which are presumed to be associated on the cell surface with an as yet unidentified 'surrogate heavy chain protein X'. Pre-B cells, having undergone IgH rearrangement and achieved μ expression, express a μ/λ5/VpreB complex. Finally, after productive rearrangement of either κ or λ light chain genes, B cells express IgM and then IgM plus IgD. We speculate that the engagement of each of these receptors is important not only for proliferation and differentiation, but also to ensure cell survival by inducing *bcl*-2 synthesis.

CONSEQUENCES OF DEREGULATING *BCL*-2 EXPRESSION IN THE T CELL LINEAGE

During T cell development, *bcl*-2 is expressed in mature T cells - medullary thymocytes and peripheral T cells - but not in immature cortical thymocytes (Pezzella et al., 1991; Hockenbery et al., 1991), implying an important role for *bcl*-2 during clonal selection. Transgenic mice expressing a *bcl*-2 transgene in the T lymphoid compartment were developed to address this issue. T cells from these mice are remarkably long-lived when cultured in simple medium and the usually hypersensitive CD4+CD8+ cortical thymocytes are relatively resistant to a wide range of cytotoxic agents (Strasser et al., 1991b, Sentman et al., 1991; Siegel et al., 1992). Significant resistance was also observed *in vivo* to treatment with γ-radiation, hydrocortisone and anti-CD3 antibody (Strasser et al., 1991b; Sentman et al., 1991; Siegel et al., 1992). The enhanced and sustained immune response of peripheral T cells to superantigen (Strasser et al., 1991b) suggested, in addition, that activated T cells had an extended lifespan. Surprisingly, however, T cell homeostasis was not perturbed (Strasser et al., 1991b). Both the number and relative proportions of the major thymic and peripheral T cell subsets remained normal throughout the first year of life, and thymic involution was unaffected in aging animals.

Several mouse strains that overexpress a *bcl*-2 transgene in the T lineage have been tested for defects in clonal deletion of autoreactive T cells. In peripheral lymphoid organs, self-superantigen reactive T cells were either not detectable or very rare (Siegel et al., 1992) . In the thymus, low levels of autoreactive cells were detected by two groups (Strasser et al., 1991b; Siegel et al., 1992) but not another (Sentman et al., 1991). Since the transgene regulatory sequences varied between strains, this apparent discrepancy might be the result of differences in the timing and level of *bcl*-2 transgene expression. Alternatively, they might be a reflection of differences between the self-superantigens that were studied. In an attempt to resolve this issue and to extend the analysis to investigate the effect of *bcl*-2 on deletion of thymocytes recognizing a non-Mls self antigen, we have generated anti-HY TCR/*bcl*-2 double transgenic mice. Anti-HY TCR transgenic mice express a TCR α/β heterodimer that recognizes the male antigen HY in the context of class I MHC H-2D^b (reviewed by von Boehmer, 1990). Our experiments with doubly transgenic hybrids have confirmed that deregulated *bcl*-2 expression interferes with the deletion of autoreactive thymocytes but does not totally block this process (Strasser et al., submitted for publication).

Our current view of the role of *bcl*-2 in T cell development is the following.

(i). Immature thymocytes that have undergone productive TCRβ gene rearrangement express a TCRβ-β homodimer on the cell surface (Kishi et al., 1991). Signalling through this receptor has been postulated to induce proliferation and further differentiation (e.g. initiation of TCRα rearrangement). While it might also induce *bcl*-2, by itself this is not sufficient to ensure survival at this stage of T cell differentiation. Unlike their counterparts in the B lymphoid lineage (see above), *bcl*-2/*scid* thymocytes do not accumulate *in vivo* and rapidly undergo apoptotic death *in vitro*, despite expressing high levels of *bcl*-2 protein (Strasser et al., manuscript in preparation). Thus, cells with non-functional TCRβ rearrangements are destined to die and *bcl*-2 cannot protect them from this fate. The reason for the unexpected difference in the response of very early T and B lineage cells to *bcl*-2 expression is unclear.

(ii). CD4+8+CD3^{low} thymocytes that have undergone functional TCRβ and α gene rearrangements are checked for binding to self MHC molecules. Those that fail to bind presumably die by apoptosis because they fail to receive a positive signal. Their death may be due primarily to a lack of *bcl*-2 expression, since bcl-2 protein is singularly lacking in cortical thymocytes (Pezzella et al., 1991; Hockenbery et al., 1991) and it is clear from the transgenic studies that *bcl*-2 expression can greatly enhance their survival in the face of a variety of insults.

(iii). CD4+8+CD3^{low} thymocytes that can bind to self MHC molecules are selected for development and/or further scrutiny. The mechanism of positive selection is not known but we speculate that up-regulation of *bcl*-2 expression is a critical component. We are currently generating female anti HY TCR/*bcl*-2 double transgenic mice on H-2D^b /*scid* and H-2D^d /*scid* backgrounds to test the role of *bcl*-2 in positive selection. In the absence of H-2D^b, thymocytes expressing the transgenic TCR cannot be positively selected and, because of the *scid* mutation, thymocytes expressing endogenous TCR cannot be generated. Thus, female H-2D^d transgenic TCR mice are deficient in CD4$^+$CD8$^+$ and CD4$^-$CD8$^+$ T cells (von Boehmer, 1990). If *bcl*-2 transgene expression can replace the positive selection signal, female mice should display ample numbers of these cells irrespective of the H-2D background.

(iv). TCRα/β^{low} CD4$^+$8$^+$ thymocytes that bind to self MHC with high affinity are deleted and die by a form of apoptosis (Murphy et al., 1990 Swat et al., 1991) that can be inhibited but not ablated by constitutive *bcl*-2 expression (see above).

In summary, we conclude that thymic differentiation invokes more than one type of death-inducing mechanism and that *bcl*-2 expression can protect developing T cells against some but not all of these. It seems likely that *bcl*-2 expression is critical for positive selection. However, down-regulation of *bcl*-2 is insufficient to account for negative selection of autoreactive clones. This process must occur via a pathway relatively insensitive to *bcl*-2. The prime candidate for the killing mechanism is interaction of the FAS/APO-1 receptor with its putative ligand the *gld* gene product. However, since negative selection is largely intact in *lpr* mice (Sidman et al., 1992), which lack a functional FAS/APO-1 receptor (Watanabe-Fukunaga et al., 1992), down-regulation of *bcl*-2 may also play a critical role. We are currently investigating the effect of introducing the *bcl*-2 transgene into *lpr* mice.

THE ROLE OF *BCL*-2 IN LYMPHOMAGENESIS

Studies of large cohorts of several independent strains of *bcl*-2 transgenic mice (Strasser et al., submitted) have revealed that *bcl*-2 can facilitate the spontaneous transformation of B lymphoid cells, in agreement with results obtained with a different *bcl*-2 transgene by Korsmeyer and colleagues (McDonnell et al., 1991), but displays very little if any lymphomagenic potential for T cells. The principal types of B lymphoid tumors were plasmacytomas and novel lymphomas exhibiting a phenotype consistent with an origin early in B cell differentiation (Strasser et al., submitted). All types of tumours were of low incidence and developed only after long latent periods. This indicates that constitutive *bcl*-2 expression was not sufficient for transformation and that somatic mutation must have played an important role. Rearrangement of the *myc* gene was common in the plasmacytomas, as reported previously for *bcl*-2 tumours designated large cell lymphoma (McDonnell et al., 1991), implying a synergistic role for *myc* and *bcl*-2 in their etiology. Progression of some cases of follicular lymphoma to an aggressive variant have been correlated with acquisition of a *myc*/IgH t(8;14) translocation (de Jong et al., 1988; Lee et al., 1989). Collaboration of *myc* and *bcl*-2 in lymphomagenesis has been confirmed by cross-breeding experiments. Doubly transgenic *myc*/*bcl*-2 mice succumbed at an early age to tumors with a cell surface phenotype and gene expression pattern indicating that they are neoplastic counterparts of normal primitive stem or progenitor cells (Strasser et al., 1990b).

In conclusion, *bcl*-2 is a novel oncogene which promotes cell survival rather than proliferation. Its primary role in follicular lymphoma appears to be to enable the cells which have acquired a t(14;18) translocation to resist apoptosis in the germinal centre. The ability of the *bcl*-2 expressing cells to survive a range of adverse circumstances, e.g. lymphokine deprivation, ensures the continued survival of the affected clone. Should cells in this clone subsequently acquire more frankly oncogenic mutation(s) which confer a.proliferative advantage, the enhanced survival capacity becomes a potent force ensuring the ultimate emergence of a fully malignant clone. The recent demonstration that constitutive *myc* expression induces apoptosis under limiting growth conditions (Askew et al., 1991; Evan et al., 1992) underlines the significance of the survival signal for transformation. Cytokine concentrations *in vivo* are usually limiting, so cells acquiring a *myc* mutation would normally drive themselves to suicide unless they also acquired either a survival function or a mutation conferring growth factor independence. Significantly, germinal center B cells are uniquely exposed to mutagenic processes. Somatic hypermutation introduces multiple point mutations into the variable region genes of rearranged Ig genes and DNA deletion ensures Ig class switch recombination. It is possible that the enzymes responsible for these processes sometimes choose aberrant substrates and thereby introduce oncogenic mutations.

ACKNOWLEDGEMENTS

We thank Dr. Harald von Boehmer for kindly providing anti HY TCR transgenic mice; Dr. Jerry Adams for stimulating discussions; Maureen Stanley, Fiona Horsburgh and Jenny Beaumont for expert technical assistance; Kim Patane and Jo Parnis for animal husbandry; and Iras Collins for preparing the manuscript. A.S. was supported by fellowships from the Swiss National Science Foundation and the Leukemia Society of America. This work was supported by the National Health and Medical Research Council of Australia and the U.S. National Cancer Institute (CA43540).

References

Askew D, Ashman R, Simmons B, Cleveland J (1991) Oncogene 6:1915-1922

Bakhshi A, Jensen JP, Goldman P, Wright JJ, McBride OW, Epstein AL, Korsmeyer SJ (1985) Cell 41:899-906

Blackman M, Kappler J, Marrack, P Science 248:1335-1341 (1990)

Chen-Levy Z, Nourse J, Cleary ML (1989) Mol Cell Biol 9:701-710

Chen-Levy Z, Cleary ML (1990) J Biol Chem 265:4929-4933

Cleary ML, Smith SD, Sklar J (1986) Cell 47:19-28

de Jong D, Voetdijk MH, Beverstock GC, van Ommen GJB, Willemze R, Kluin PM (1988) New Eng J. Med. 318:1373-1378

Evan GI, Wyllie AH, Gilbert CS, Littlewood TD, Land H, Brooks M, Waters CM, Penn LZ, Hancock DC (1992) Cell 63:119-125

Fulop GM, Phillips RA (1990) Nature 347:479-482

Hockenbery D, Nunez G, Milliman C, Schreiber RD, Korsmeyer SJ (1990) Nature 348:334-336

Hockenbery DM, Zutter M, Hickey W, Nahm M, Korsmeyer SJ (1991) Proc Natl Acad Sci USA 88:6961-6965

Kishi H, Borgulya P, Scott B, Karjalainen K, Traunecker A, Kaufman J, von Boehmer H (1991) EMBO J 10:93-100

Lee JT, Innes DJ, Williams ME (1989) J Clin Invest 84:1454-1459

Liu YJ, Mason DY, Johnson GD, Abbot S, Gregory CD, Hardie DL, Gordon J, MacLennan ICM (1991) Eur J. Immunol 21:1905-1910

McDonnell TJ, Deane N, Platt FM, Nunez G, Jaeger U, McKearn JP, Korsmeyer SJ (1989) Cell 57:79-88

McDonnell TJ, Nunez G, Platt FM, Hockenberry D, London L, McKearn JP, Korsmeyer SJ (1990) Mol Cell Biol 10:1901-1907

McDonnell TJ, Korsyemer SJ (1991) Nature 349:254-256

Murphy KM, Heimberger AB, Loh DY (1990) Science 250:1720-1723

Nunez G, Hockenbery D, McDonnell TJ, Sorensen CM, Korsmeyer SJ (1991) Nature 353:71-73

Pezzella F, Tse AG, Cordell JL, Pulford KA, Gatter KC, Mason DY (1990) Am J Pathol 137:224-232

Rolink A, Melchers F (1991) Cell 66:1081-1094

Schuler W, Weiler IJ, Schuler A, Phillips RA, Rosenberg N, Mak T, Kearney JF, Perry RP, Bosma MJ (1986) Cell 46;963-972

Sentman CL, Shutter JR, Hockenbery D, Kanagawa O, Korsmeyer SJ (1991) Cell 67:879-888

Sidman CL, Marshall JD, von Boehmer H (1992) Eur J Immunol 22:499-504

Strasser A, Harris AW, Vaux DL, Webb E, Bath ML, Adams JM, Cory S (1990a) Curr Top Microbiol Immunol 166:175-181

Strasser A, Harris AW, Bath ML, Cory S (1990b) Nature 348:331-333

Strasser A, Whittingham S, Vaux DL, Bath ML, Adams JM, Cory S, Harris AW (1991a) Proc Natl Acad Sci U S A 88:8661-8665

Strasser A, Harris AW, Cory S (1991b) Cell 67:889-899

Swat W, Ignatowicz L, von Boehmer H, Kisielow P (1991) Nature 351:150-153
Tsujimoto Y, Finger LR, Yunis J, Nowell PC, Croce CM (1984) Science 226:1097-1099
Tsujimoto Y, Ikegaki N, Croce CM (1987) Oncogene 2:3-7
Vaux DL, Cory S, Adams JM (1988) Nature 335:440-442
von Boehmer H (1990) Ann Rev Immunol 8:531-556
von Boehmer H, Kisielow, P. (1990) Science 248:1369-1373
Watanabe-Fukunaga R, Brannan CI, Copeland NG, Jenkins NA, Nagata S (1992) Nature 356:314-317

Control of the T Cell Repertoire

P. Marrack[*], J. H. Freed, L. Ignatowitz[*], J. McComarck[*], J. Callahan, P. Hugo[*] and J. W. Kappler[*]

Howard Hughes Medical Institute[*], Departement of Medicine, National Jewish Center for Immunology and Respiratory Medicine, 1400, Jackson Street, Denver, CO 80206, USA

INTRODUCTION

T cell receptors for antigen bound to Major histocompatibility complex proteins (MHC) are made up of a number of different variable elements, $V\alpha$, $J\alpha$, $V\beta$ and so on. Although any given mouse or man can probably make at least 10^{10} different versions of these receptors it is certain that not all possible combinations are expressed in any given animal. In part the selection is limited by the collection of germ line genes for the variable elements available in that animal. For example, some mice have a large deletion at their $V\beta$ locus and consequently can make 50% or less of all receptors possible in animals which have an intact β locus. Since somatic mutation does not occur for $\alpha\beta$ T cell receptor genes, in theory deletions of this type could severely limit the ability of the T cells in such animals to recognize particular antigens. This does not usually seem to be the case, however. The T cell response to most antigen/MHC complexes is so degenerate that in fact it is quite difficult to demonstrate "holes" in the T cell repertoire due to problems of this type, although it can be done.

The various $\alpha\beta$ combinations in any given animal are also limited by 2 important somatic processes, tolerance to self and positive selection. The T cells of a healthy animal cannot react with the tissues of that animal itself. Some years ago we and others showed that T cell tolerance was in large part due to the fact that thymocytes go through a stage of development during which contact with antigen is lethal (Kappler et al 1987). This seems to be caused by the increased intracellular Ca^{++} levels which ensue from receptor engagement. High intracellular Ca^{++} is a lethal event for thymocytes. Therefore any thymocyte which encounters antigen at this stage in its life history will be eliminated. This phenomenon ensures self tolerance for any self components which can reach the appropriate regions in the thymus.

Induction of T cell tolerance is not limited to the thymus, however. Mature peripheral T cells may also die or become inactivated if they encounter antigen under certain circumstances (Rammensee et al, 1989; Kawabe and Ochi, 1990, Webb et al 1990). The precise conditions which favor tolerance over response will be discussed in further detail below.

Although the phenomenon of positive selection was discovered 15 years ago, many aspects of the process are still not understood (Zinkernagel et al, 1978, Bevan and Fink, 1978). It is known that positive selection involves interaction between the receptors on developing thymocytes and MHC proteins on thymus cortical epithelial cells. The accessory molecules CD4 or CD8 are also critical. Several laboratories have reported that peptides bound in the grooves of the selecting MHC also influence which T cells can be positively selected by a given MHC allele. The principles which underlie positive selection are not understood, however. Since the process depends upon reaction between thymocyte $\alpha\beta$ receptors and MHC, have can it be reconciled with self tolerance? Several theories have been advanced to deal with the paradox. Currently the most popular of these states that positive selection involves low affinity reaction between T cell receptor and MHC, an affinity which is too low to drive negative selection or response in mature T cells. Another hypothesis, which we have suggested, posits that positively selecting thymus cortical epithelial cells bear a collection of peptides bound to MHC which is different than that found elsewhere. Hence positive selection can occur on an MHC/peptide ligand not found in other parts of the animal.

This latter theory depends upon the assumption that developing thymocytes go through a stage of their existence at which engagement of their receptors leads not to death (see above) or activation, but rather to maturation. Data from $\alpha\beta$ receptor transgenic mice does not support this notion, however, we have found that in normal animals a population of immature, receptor-bearing thymocytes cannot be killed be receptor engagement. We have suggested that these are the cells in normal animals which are subject to positive selection.

PEPTIDES ASSOCIATED WITH THYMUS AND SPLEEN CLASS II

For several reasons there is some interest in a comparison between the peptides bound to MHC proteins in the thymus and those found associated with the same proteins elsewhere. First tolerance to self peptides can only be established in the thymus if the peptides are actually present in that organ. Secondly, if positive selection does depend upon the peptides expressed plus MHC on thymus cortical epithelial cells, then an examination of these peptides is important. Recently we have begun studies to find out what peptides are bound to Class II MHC proteins in the thymus and spleen.

These 2 organs were taken from C3H/HeJ mice. After of the cells in NP40 IA^k and IE^k were isolated by passage over 17/227 or 14-4-4 columns respectively. Class II proteins were eluted from the columns at high pH and were concentrated on prewashed Centricon 30 columns. Peptides were eluted from Class II with 2.5M acetic acid, lyophilized and passed over C8 HPLC in an acetonitrile gradient. Peptide elution was monitored by OD at 214. Peaks were sequenced by standard techniques. A list of the peptides we have found associated with spleen Class II, and the proteins from which they were probably derived, is shown in Table 1.

Table 1

Peptides bound to Class II in C3H/HeJ Spleens

MHC	Peptide	Origin
IA^k	VVKKGTDFQLNQLE	Transferrin 97-
IA^k	KGDFQLNQLEGKKG	Transferrin 100-
IA^k	YVRFDSFVGEYRAVT	Aβ^k 37-51
IA^k	ENLRFDSDVGEFRAV	Eβ^k 33-
IA^k	EDENLYEGLNLDDCSMYE	MB1 177-
IA^k	YILYNKGIMGEDSYPY	Cathepsin H 77-
IA^k	SYLDAWVCEQLAT	FcE Receptor II 298-
IA^k	SQRHFVHQFQPFCYF	Aβ^k 3-
IA^k	QFQPFCYFTNT	Aβ^k 10-
IE^k	HPPHIEIQMLKNG	β2 Microglobulin 42-
IE^k	VNKEIQNAVQGVK	C' Cytolysis Inh. 41-
IE^k	DNRMVNHFIAEFKRK	Cognate HSP 70 234-
IE^k	TPTLVEAARNLGRVG	Serum Albumin 347-

These are all derived from the sources expected to give rise
to Class II-associated peptides, ie serum, plasma membrane
and lysosomal and endosomal proteins.

They have several other noteworthy features. For example,
the peptides derived from transferrin and the N terminal
proximal end of IA^k are closely related to each other. It
is likely that each pair uses the same sequence to bind to
Class II, but overhangs the Class II protein to different
extents at the N terminal end. Another pair of peptides
differ at their C terminal ends since the $\beta 2$ microglobulin
sequence show above and derived from IE^k elutes on HPLC at 2
different positions (not shown). It is likely that these 2
peptides differ in length, and, of course, share their N
terminal sequences. These data suggest that Class II
binding peptides, unlike those associated with Class I, can
overhang the ends on the Class II groove to variable extents
at either end. Such a result means that it probably will
not be profitable to search bulk peptides isolated fom Class
II for patterns of amino acids found at particular position
within the peptide.

These peptides are not all expressed to the same degree on
thymus Class II. Although in many cases the same peptides
were isolated from the 2 locations, indicating that peptide
processing and loading pathways are to some extent shared
between spleen and thymus, some differences were seen.
Table II shows a comparison of some of the results from the
2 organs.

Table 2

Thymus and Spleen Class II do not have
exactly the same spectrum of peptides bound

MHC	Peptide	Origin	% MHC bound in Spleen	Thymus
IA	VVKKGTDFQLNQLEGKKG	Transferrin	2.5	<0.1
IA	EDENLYEGLNLDDCSMYE	MB1	9.2	<0.1
IA	SYLDAWVCEQLAT	FcE Rec. II	1.0	<0.1
IA	YVRFDSFVGEYRAVT	$A\beta^k$	9.3	1.1
IA	xPNALQFAELPVNKG	Unknown	0.5	0.2
IE	HPPHIEIQMLKNG	$\beta 2M$	3.5	2.8
IE	TPTLVEAARNLGRVG	Serum alb.	1.1	2.6

Some of the peptides were not found in the thymus at all. Two of these were derived from proteins expected to be at reasonably high concentration in spleen and absent from thymus, the surface immunoglobulin associated protein, MB1 and the FcE receptor II protein. This result is not therefore surprising and presumably simply reflects the geographical distribution of the donor proteins. Some peptides were found at drastically different concentrations in the 2 locations, for example, that derived from $A\beta^k$, a result reminiscent of that previously described for another Class II associated peptide (Murphy et al, 1989).

Overall, although there are differences between thymus and spleen it is at present difficult to assess the significance of these differences. Presumably tolerance to self peptides which are high concentration peripherally and absent from the thymus cannot be mediated by mechanisms with pertain to the thymus. It will be interesting to find out whether or not tolerance can be broken more easily to self peptides which are absent from the thymus.

PERIPHERAL TOLERANCE

We and others have shown that tolerance can be induced in mature peripheral T cells. For example, chronic infection with a superantigen-encoding exogenous mouse mammary tumor virus (MMTV) causes the disappearance of cells bearing the target $V\beta$ for the superantigen not only in the thymus but also in the periphery. This model for peripheral tolerance is not easy to manipulate experimentally, however, since dose and expression of the MMTV superantigen cannot be controlled.

In an attempt to create a better system we decided to inject mice continuously with another powerful superantigen, staphylococcal enterotoxin A (SEA) which reacts with mouse T cells bearing $V\beta3$. Adult mice were therefore injected every other day with low doses of SEA. We were surprised to find that very low doses of the superantigen, in the range of 0.01 ug/injection, caused the almost complete deletion of $V\beta3$-bearing cells from all peripheral organs of the mouse within about 9 days. This deletion occurred in the absence of any apparent prior proliferation of the target T cells suggesting that death of the mature T cells was not dependent, as others have previously suggested, on massive activation of the cells.

Several mechanisms can be suggested which would account for this profound deletion. One of these suggests that presentation of the superantigen by "nonprofessional" antigen presenting cells, ie B cells, is a major contributory factor. Alternatively presentation might occur in an inappropriate environment, the wrong part of the lymph node, or without associated lymphokines required for T cell stimulation. To test these and other ideas we examined the ability of various reagents to interfere with tolerance induction. Complete Freund's adjuvant turned out to be an interesting material. This did appear to interfere with tolerance induction, perhaps by changing the nature of the antigen presenting cell, though further tests will be required to prove this conjecture.

CELL LINES AND POSITIVE SELECTION

One of the problems in studying positive selection has been the absence of good _in vitro_ systems in which the phenomenon can be examined. So far, in fact, even cell lines which can catalyze the process are not known. We have tested several continuous thymus epithelial cell lines for their ability to mediate positive selection. In our experiments B10.BR, $H2^k$ mice were irradiated and reconstituted with (B10.BR x B10)F1, $H2^{kxb}$ bone marrow. Three weeks later the thymuses of some of these animals were injected with an epithelial cell line derived from B10, $H2^b$ mice. At intervals after this, the mice were challenged with (TG)AL, a synthetic copolymer known to be presented by IA^b and not IA^k. T cells from control mice, irradiated and reconstituted as described above, but given intrathymic injections of balanced salt solution alone, did not respond well to (TG)AL. T cells from chimeras made by injection F1 bone marrow into irradiated F1 recipients, responded well, presumably because T cells in these animals were selected on thymus epithelium bearing IA^b. Cells from B10.BR mice reconstituted with F1 bone marrow and injected intrathymically with an epithelial cell line bearing IA^b responded intermediately.

These results suggest that the epithelial cell line does have some activity in positive selection. We hope that this cell line will allow various experimental manipulations, including the introduction and deletion of various genes encoding proteins thought to be important in positive selection, and thus improve our capacity to investigate this important and still mysterious phenomenon.

CONCLUSION

Although much has been learned about the processes of positive and negative selection, much remains to be learned. Tolerance in particular appears to occur in several different ways. Self tolerance is probably so important that a number of backup mechanisms have developed to prevent potentially autoreactive T cells from "sneaking through". In fact, T cells seem to have been set up so that their default response is death or inactivation rather than stimulation. Only under special circumstances, when infectious agents or their products are present, do T cells respond, to conventional or super- antigens.

ACKNOWLEDGEMENTS

The authors would like to thank Joel Boymel, Ella Kushnir, Helena Morales and Ava Serges very much for their excellent assistance. This work was supported by granst from the USPHS.

REFERENCES

Bevan M and Fink P (1978) The influence of thymus H-2 antigens on the specificity of maturing killer and helper cells. Immunol. Rev. 42:3-19.

Kappler J, Roehm N and Marrack P (1987) T cell tolerance by clonal elimination in the thymus. Cell 49:273-280.

Kawabe Y and Ochi A (1990) Selective anegry of $V\beta8+$ T cells in Staphylococcal enterotoxin-B primed mice. J. Exp. Med. 172:1065-1070.

Murphy D, Lo D, Rath S, Brinster R, Flavell R, Slanetz A, and Janeway CA Jr (1989) A novel MHC class II epitope expressed in thymic medulla but not cortex. Nature 338:765-768.

Rammensee HG, Kroschewski R and Frangoulis B (1989) Clonal anergy induced in mature $V\beta6+$ T lymphocytes on immunizing Mls-1^b mice with Mls-1^a expressing cells. Nature 339:541.

Webb, S, Morris C and Sprent J (1990) Extrathymic tolerance of mature T cells: Clonal elimination as a consequence of immunity. Cell 63:1249.

Zinkernagel, R., Callahan, G., Althage, A., Cooper, S., Klein, P., and Klein, J. (1978) On the thymus in the differentiation of H-2 self-recoginition by T cells: Evidence for dual recognition?. J. Exp. Med. 147:882-896.

Tolerance, Autoimmunity and Immunopathology

H. Hengartner, P. Ohashi, P. Aichele, S. Oehen, D. Braendle,
Ch. Müller[+], T. Rülicke, K. Bürki[*], H.P. Pircher, R. M. Zinkernagel

Institute of Experimental Immunology, University of Zurich, Sternwartstrasse 2, 8091 Zurich, Switzerland
[+]*Departement Pathology, University of Berne, 3010 Berne*
[*]*Preclinical Research, Sandoz Pharma Ltd, 4002 Basel*

INTRODUCTION

Many viral infections are efficiently defeated by a vigorousy cyto-toxic T cell mediated early immune response. Before T cells in the periphery can be activated they have to undergo thymic positive and negative selection. During positive selection T cells acquire the capacity to H-2 restrictedly recognize the foreign antigens whereas negative selection prevents the development of self reactive T cells. However not all self antigens are presented to the developing T cells in the thymus to establish the so called central tolerance by physical elimination of self reactive T cells (Kappler *et al.* 1988; MacDonald *et al.* 1988). Therefore potentially autoreactive T cells may reach the periphery. These cells are presumable involved in many autoimmune disease processes (Sinha *et al.* 1990; Zamvil and Steinman, 1990). Several studies have shown that potentially self reactive T cells may be rendered tolerant to self antigens not ex-pressed in the thymus by extrathymic mechanisms of tolerance in-duction (Webb and Sprent, 1990; Hanahan *et al.* 1990; Ramsdell and Fowlkes, 1990; Schönrich *et al.* 1991; Schoenrich *et al.* 1992). How-ever other models are consistent with the idea that T cell tolerance to certain antigens does not occur because the self antigen is nei-ther tolerogenic nor immunogenic and therefore may be ignored by the immune system (Adams *et al.* 1987; Böhme *et al.* 1989; Murphy *et al.* 1989; Götz *et al.* 1990; Schild *et al.* 1990; Ohashi *et al.* 1991; Oldstone, 1990; Lo *et al.* 1992). If not all peripheral T cells have been tolerized against all self antigens one has to question why au-toimmune disease induction does not occur more frequently and which are the processes important in triggering such potentially self re-active T cells.

We have generated transgenic mice expressing lymphocytic chori-omeningitis virus glycoprotein (LCMV-GP) under the control of rat insulin promoter (RIP) in ß islet cells of the pancreas (Ohashi *et al.* 1991). These animals do not spontaneously become diabetic. How-ever approximately 11 days after infection with LCMV-WE the RIP-GP transgenic mice develop diabetes. The pancreatic islets are infil-trated predominantly by CD8[+] T cells and to a lesser extent by CD4[+] T cells. Because the self antigen that is the target of immunologi-cal attack is known, this model is amenable to detailed analysis. In addition we also developed single and double chain transgenic mice with the TCR α ($V_{\alpha2}$) and β ($V_{\beta8.1}$) chain genes derived from a LCMV-GP (peptide 33-42) specific H-2D[b] restricted cytotoxic T cell clone P14

(Pircher et al. 1989). By crossing the TCR and RIP-GP transgenic animals the precursor frequency of potentially self reactive transgenic cytotoxic T cells against the transgenic RIP-GP neo-self antigen expressed exclusively on pancreatic islet cells was drastically increased. Also under these circumstances no spontaneous induction of hyperglycemia occurred unless the animals were infected with LCMV.

The TCR α or β or $\alpha\beta$ double transgenic and RIP-GP transgenic animals allowed us to study the following problems: The controlled expression of the TCR during thymic development of the T cells (Brändle et al. 1992) and the induction of tolerance and immunopathology leading to diabetes. The induction of autoreactivity in this model can be studied by the use of different LCMV isolates and GP recombinant vaccinia viruses in transgenic mice of various haplotypes. In addition the transgenic TCR uses $V_{\beta 8.1}$ which is known to react both with Mls[a] and SEB. What is the role of superantigens, the number of autoreactive precursor T cells or which other cell populations are involved in the disease?

The induction of immunopathological processes in this animal model for diabetes is based on a shared or crossreactive self T cell epitope with the infectious agent LCMV. We therefore studied possibilities to actively vaccinate against diabetes or to induce tolerance in newborns or in adults by establishing bone marrow chimerism with stem cells expressing the transgenic neo-self antigen as efficient tolerogen.

POSITIVE SELECTION OF T CELLS IN THE THYMUS DOWN-REGULATES RAG-1 EXPRESSION

The recombination activating genes RAG-1 and 2 are crucially involved in the TCR gene rearrangement processes (Fenton et al. 1988; Uematsu et al. 1988; Pircher et al. 1990b; Oettinger et al. 1990; Schatz et al. 1989). We have examined the expression of the recombination activating gene RAG-1 by in situ hybridization of thymi from mice bearing either TCR α, TCR β or both, TCR $\alpha\beta$ chain transgenes. RAG-1 transcription was found in the thymic cortex of single TCR α and single TCR β chain transgenic mice comparable to normal mice. However, RAG-1 transcription was strikingly reduced in the thymic cortex from transgenic mice carrying both TCR α and β chain genes and expressing H-2[b] MHC class I molecules necessary for positive selection of the transgenic TCR. In contrast, thymi of transgenic mice also carrying both TCR α and β chain genes but expressing H-2[d] MHC molecules which did not positively select the transgenic TCR displayed high levels of RAG-1 transcription. The low thymic RAG-1 expression coincided with high transgenic TCR α chain surface expression and with inhibition of endogenous TCR α chain rearrangement. Our findings suggest that binding of the TCR to self MHC molecules during positive selection signals downregulation of RAG-1 transcription in cortical thymocytes and thereby prevents further TCR α chain rearrangement.

FACTORS INFLUENCING THE ONSET OF AUTOREACTIVE IMMUNOPATHOLOGICAL PROCESSES.

As already mentioned before our RIP-GP transgenic H-2^b mice expressing the viral glycoprotein on β-islet cells exhibit GP specific T cell immune responsiveness and are not tolerized by clonal deletion or anergy induction (Ohashi *et al.* 1991). Diabetes is rapidly induced by infection with LCMV-WE, but also with LCMV-strains Armstrong, Pasteur and clone 13 within around 11 days. However infection with LCMV-GP recombinant vaccinia virus or immunization with isolated LCMV-glycoprotein, or with the peptide 32-42 of LCMV-GP or with a polyclonal allogeneic activation of T cells by the injection of allogeneic spleen cells does not induce T cell mediated diabetes. Since our transgenic LCMV-GP specific TCR utilizes the V$_{\beta8.1}$ gene segment RIP-GP/TCR double transgenic animals (H-2$^{b/k}$, Mls^b) were challenged with CBA/J (H-2^k, Mls^a) and SEB. But neither one of the two superantigens did induce efficient cytotoxic T effector cells leading to hyperglycemia within 60 days. Anergy induction by this treatment with Mls^a was excluded by a subsequent infection with LCMV-WE. These observations demonstrate that only an effective activation of a large cytotoxic T cell population leads to immunopathological damage of the β-cells.

Extensive studies have shown, that predisposition to certain autoimmune diseases are MHC associated (Sinha *et al.* 1990; Todd *et al.* 1988). We therefore analysed the induction of hyperglycemia in our RIP-GP transgenic animals in dependence of the MHC-haplotype. Mice of H-2^k and H-2^q haplotype show a poor glycoprotein specific cytotoxic T cell immune response when compared to an H-2^b restricted response (Schulz *et al.* 1989; Hany *et al.* 1989). After infection with LCMV hyperglycemia was detected around day 24 in H-2^k, whereas H-2^q RIP-GP transgenic animals were already diabetic on day 9. Therefore although the H-2^k and H-2^q haplotypes respond poorly to LCMV-GP, some GP-specific T cell reactivity was induced and diabetes still occurred, albeit with altered kinetics.

In H-2^b RIP-GP transgenic animals, two different lines Bln and Brx were analyzed to study the CD4 and CD8 dependence of diabetes induction by monoclonal antibody treatment. Anti-CD8 treatment could block the onset of diabetes in both lines, while the anti-CD4 treatment only blocked in RIP-GP(Brx) but not in RIP-GP(Bln) transgenic animals.In the RIP-GP(Bln) transgenic mice, CD4$^+$ T cells played a role in the induction of diabetes only in the H-2^k but not H-2^q haplotype. This suggests that CD4$^+$ T cells may be necessary for sufficient CD8$^+$ T cell activation to induce diabetes in low responders.

The onset of diabetes might also be influenced by the precursor frequency of self-reactive cytotoxic T cells. The two GP specific TCR transgenic mouse lines 327 and 318 express the GP specific transgenic TCR on approximately 90% and 50% of CD8$^+$ T cells respectively. The RIP-GP transgenic mouse as such or crossed with the TCR transgenic lines 318 or 327 were used to demonstrate the influence of the frequency of self reactive T cells on the induction of hyperglycemia after GP specific immunization. For this purpose we made use of the special property of recombinant vaccinia-GP virus, which is able to prime for LCMV-GP specific cytotoxicity in vivo but fails to induce

diabetes in the RIP-GP single transgenic lines Brx and Bln as stated
before. However in the two double transgenic mouse line with the intermediate (line RIP-GP/TCR 318, 50%) and high (line RIP-GP/TCR 327,
90%) self reactive T cell precursor frequency, infection with vaccinia-GP resulted within 8-18 days in hyperglycemia with an intermediate blood glucose level (15-20mM) or within 4 days in high (30-
35mM) blood glucose levels respectively. These results indicated
that the frequency of responding GP specific T cells is directly determining the severity and the outcome of immunopathological disease.

Since RIP-GP transgenic animals did not get diabetes upon infection
with recombinant vaccinia-GP virus, histological analysis were performed. Six days after infection with vaccinia-GP, both CD4[+] and
CD8[+] T cell infiltrates were seen in the pancreatic islets of RIP-GP
transgenic mice, but not in normal mice. However, unlike histological sections of RIP-GP infected with LCMV-WE, where all islets were
infiltrated, lymphocytes were only detected in a few islets upon infection with vaccinia-GP.

Because infiltrating cells are present after infection of RIP-GP
transgenic mice with vaccinia-GP, it is possible that some pathological damage is done to islet cells that is probably insufficient to
result in overt hyperglycemia. Several other treatments were combined with vacc-GP infection to provoke the onset of diabetes. RIP-
GP(Bln) single transgenic animals were immunized with vaccinia-GP
and treated with IFN-γ. Histological analysis of the islet cells
clearly indicated that an increase in MHC class-I antigen expression
had occurred. Hyperglycemia was detected only in 1 of 5 animals that
have undergone this treatment. Interestingly, the blood glucose levels in the mouse that did develop hyperglycemia returned to normal
levels within a period of 10 days, suggesting that the islets had
regenerated.

VACCINATION TO PREVENT DIABETES

Our viral antigen expressing transgenic mice allowed to consider the
development of vaccination protocols against the autoimmune disease
induced by infectious LCM virus which shares crossreactive T cell
epitopes expressed on the pancreatic islet cells (Ohashi et al.
1991). LCMV infections are most efficiently defeated by a CD8[+] T
cell mediated immune response against epitopes on the viral glycoprotein (GP) and the nucleoprotein (NP) (Lehmann-Grube, 1971;
Buchmeier et al. 1980; Schulz et al. 1989; Traub, 1936). It is
therefore not surprising that RIP-GP transgenic animals could not be
protected against diabetes by anti-LCMV specific polyclonal antiserum treatment prior to LCMV infection (Wright and Buchmeier,
1991).

The specificity of cytotoxic T cell activity against the GP or NP is
H-2 haplotype dependent: Against GP mainly in H-2[b], and against NP
in H-2[d] or H-2[q] (Hany et al. 1989; Schulz et al. 1989). Recombinant
vaccinia-GP does not induce diabetes in RIP-GP transgenic animals.
The two available recombinant vaccinia viruses vaccinia-GP and vac-

cinia-NP were therefore used to prime RIP-GP transgenic animals for a CD8$^+$ cytotoxic T cell activity at different time points prior to the diabetes inducing LCMV infection.

RIP-GP-trangenic mice of H-2^b haplotype primed either with vaccinia-GP or vaccinia-NP were not protected at any time point. However RIP-GP transgenic mice of the haplotype H-2$^{b/q}$ vaccinated with recombinant vaccinia-NP were protected up to 32 days against hyperglycemia upon infection with LCMV in contrast to H-2$^{b/d}$ which were not protected anymore after day 15. Therefore, vaccination with LCMV epitopes differing from the neo-self antigen shared by the host and the virus prevented immunopathology and disease in an MHC-dependent fashion.

These experiments show that vaccination to prevent immunopathology can be achieved, however by an extremely delicate protocol. Vaccination with non-GP LCMV vaccine was able to eliminate LCMV efficiently enough to delay or even prevent immunopathology. This vaccine strategy depends on the immunogenicity of the non-GP determinant in a given H-2 haplotype. Ideally a vaccine providing all possibly recognized non-GP antigens should prove to be most effective; obviously this possibility is limited to H-2 haplotypes where non-GP epitopes are protective. Such a vaccine could be composed of a cocktail of recombinant vaccines or of virus expressing mutated GP-epitopes (Pircher et al. 1990a). The obvious limitations of such "self-epitope excluding" vaccines are given by the MHC-polymorphism and one would require a different vaccine for each MHC haplotype. Although seemingly rather complicated, this perspective may nevertheless not be too discouraging. The known MHC-dependence of juvenile diabetes may well signal that potential vaccines may have to be sought only for a few HLA types. A more reliable method to prevent diabetes in our animal model may therefore be to induce central tolerance to neo-self antigen by thymic negative selection. Complete abrogation of the specific autoimmune reaction by induction of tolerance either neonatally by infection or in adults by therapy with transfected bone marrow stem cells may then be more general approach to prevent diabetes.

CONCLUSIONS

Maintenance of tolerance towards self antigens and the induction of selfreactive T cells to self antigens to induce immunopathological processes leading to autoimmune disease was studied in several transgenic animals. In such models even some peripheral antigens expressed under the control of identical promoters have extremely variable effects on immunological manifestations (Adams et al. 1987; Oldstone et al. 1991; Ohashi et al. 1991; Zinkernagel et al. 1991; Lo et al. 1992).

The presented results of our animal model, where LCMV-GP is expressed on the pancreatic β-islet cells showed, that this antigen is ignored by peripheral GP-specific T cells. Upon LCMV infection peripheral GP-specific T cells, activated via professional antigen presenting cells in the secondary lymphoid organs, induces immunopathology in the transgenic islets resulting in hyperglycemia. As shown this effect is dependent on the properties of the infectious agent such as the ability to replicate in the host and the ability of certain antigens being presented by antigen presenting cells, an MHC haplotype dependent phenomenon, to the responding T cells. The affinity of the TCR of these T cells with a crossreactivity to tissue specifically expressed neo-self antigen itself plays in addition a crucial role.

Experiments with our RIP-GP and GP-specific TCR double transgenic mice further showed, that immunopathology mediated diabetes cannot be that easily induced by either activation via crossreactive determinants such as Mls^a, SEB, allogeneic stimulator cells together with interleukins or even by direct GP specific stimulation with the GP-peptide aa32-42 or the glycoprotein itself. Some immunological damage could be observed in RIP-GP transgenic animals by vaccinia-GP infection induced GP specific T cells, but these cells apparently are not perpetually stimulated by GP expressed on the islet cells and cumulative damage resulting in diabetes does not occur. This finding is relevant in at least 2 important situations: With respect to induction of CTL mediated autoimmune disease, these results suggest that a strong primary CTL response is required, and that the existence of "memory" CTL in the presence of the GP antigen in the islets is not able to cause sufficient immunopathological damage to result in diabetes. These findings also have implications in potential immunotherapy against tumors. One approach would be to develop vaccines against tumor specific antigens. In the case where tumor growth may be controlled by CTLs, our model would predict that even if an individual were vaccinated against a tumor specific antigen, the effector CTL function may not be sufficient to control tumor growth.

Why does the expression of peripheral neo-self antigens in the various transgenic mouse models result in such variety of immunological manifestations? There is no unique type of a so called peripheral tolerance induction pathway. Many obvious factors are determining whether potentially self reactive T cells towards tissue specifically expressed antigens react by ignorance, downregulation of the expression of the TCR or of accessory molecules, by anergy or by depletion. Peripheral antigen density and its tissue specific expression pattern, as well as the affinity of the MHC/neo-self antigen complex interaction with the TCR determined by the H-2 molecule density and the TCR itself are of crucial importance. Accessibility of lymphocytes towards the antigen expressing tissue (size and the environment of the organ) is determining whether a large proportion of the lymphoid cells will get in contact with the self antigen. The expression of neo-self antigens expressed on either professional or nonprofessional antigen presenting cells may also determine the type of tolerance induction.

The chance to develop immunopathologically mediated autoimmune disease is often strongly MHC class-II haplotype associated. However in our animal model the induction of diabetes occurred via MHC class-I antigen restricted cytotoxic T cells. Nevertheless the outcome of/ the disease was more or less CD4$^+$ T cell dependent. It is also known that many patients prior to the outbreak of certain autoimmune diseases, e.g. juvenile diabetes, suffered from a virus infection. A possible scenario to explain the involvement of the CD8$^+$ and CD4$^+$ T cells would therefore be, that virus activated CD8$^+$ T cells with either crossreactivity for β-cell related antigens or a specificity to virus infected pancreas are mediating tissue cell destruction followed by a release of β-cell characteristic proteins. Tissue macrophages attracted by lymphokines would process and present such β-cell antigens to CD4$^+$ T cells. These T cells were not confronted with such antigens during their thymic education, since these β-cell characteristic antigens under normal circumstances are not present in the thymus. Such β-cell antigen specific CD4$^+$ T cells would maintain a self perpetuating inflammatory process. Future studies especially on the antigen specificity of the T cells in animal models should help to clarify these speculations.

Adams, T.E., Alpert, S. and Hanahan, D. (1987) *Nature*, **325**, 223-228.

Böhme, J., Haskins, K., Stecha, P., van Ewijk, W., LeMeur, M., Gerlinger, P., Benoist, C. and Mathis, D. (1989) *Science*, **244**, 1179-1183.

Brändle, D., Müller, C., Rülicke, T., Hengartner, H. and Pircher, H. (1992) *Proc.Natl.Acad.Sci.U.S.A.*, in press,

Buchmeier, M.J., Welsh, R.M., Dutko, F.J. and Oldstone, M.B.A. (1980) *Adv.Immunol.*, **30**, 275-312.

Fenton, R.G., Marrack, P., Kappler, J.W., Kanagawa, O. and Seidman, J.G. (1988) *Science*, **241**, 1089-1092.

Götz, J., Eibel, H. and Köhler, G. (1990) *Eur.J.Immunol.*, **20**, 1677-1683.

Hanahan, D., Jolicoeur, C., Alpert, S. and Skowronski, J. (1990) *Cold Spring Harbor Symp.Quant.Biol.*, **54**, in press.

Hany, M., Oehen, S., Schulz, M., Hengartner, H., Mackett, M., Bishop, D.H.L. and Zinkernagel, R.M. (1989) *Eur.J.Immunol.*, **19**, 417-424.

Kappler, J.W., Staerz, U.D., White, J. and Marrack, P. (1988) *Nature*, **332**, 35-40.

Lehmann-Grube, F. (1971) *Virol.Monogr.*, **10**, 1-173.

Lo, D., Freedman, J., Hesse, S., Palmiter, R.D., Brinster, R.L. and Sherman, L.A. (1992) *Eur.J.Immunol.*, **22**, 1013-1022.

MacDonald, H.R., Schneider, R., Lees, R.K., Howe, R.C., Acha-Orbea, H., Festenstein, H., Zinkernagel, R.M. and Hengartner, H. (1988) *Nature*, **332**, 40-45.

Murphy, K.M., Weaver, C.T., Elish, M., Allen, P.M. and Loh, D.Y. (1989) *Proc.Natl.Acad.Sci.USA*, **86**, 10034-10038.

Oettinger, M.A., Schatz, D.G., Gorka, C. and Baltimore, D. (1990) *Science*, **248**, 1517-1523.

Ohashi, P.S., Oehen, S., Bürki, K., Pircher, H.P., Ohashi, C.T.,

Odermatt, B., Malissen, B., Zinkernagel, R. and Hengartner, H. (1991) *Cell*, **65**, 305-317.

Oldstone, M.B.A. (1990) *J.Exp.Med.*, 171, 2077-2089.

Oldstone, M.B.A., Nerenberg, M., Southern, P., Price, J. and Lewicki, H. (1991) *Cell*, **65**, 319-331.

Pircher, H.P., Bürki, K., Lang, R., Hengartner, H. and Zinkernagel, R. (1989) *Nature*, **342**, 559-561.

Pircher, H.P., Moskophidis, D., Rohrer, U., Bürki, K., Hengartner, H. and Zinkernagel, R.M. (1990a) *Nature*, 346, 629-633.

Pircher, H.P., Ohashi, P.S., Miescher, G., Lang, R., Zikopoulos, A., Bürki, K., Mak, T.W., Zinkernagel, R.M., MacDonald, H.R. and Hengartner, H. (1990b) *Eur.J.Immunol.*, 20, 417-424.

Ramsdell, F. and Fowlkes, B.J. (1990) *Science*, **248**, 1342-1348.

Schatz, D.G., Oettinger, M.A. and Baltimore, D. (1989) *Cell*, **59**, 1035-1048.

Schild, H.J., Rötzschke, O., Kalbacher, H. and Rammensee, H.-G. (1990) *Science*, **247**, 1587-1589.

Schoenrich, G., Momburg, F., Malissen, M., Schmitt-Verhulst, A.M., Malissen, B., Hammerling, G. and Arnold, B. (1992) *Int.Immunol.*, **4**, 581-590.

Schönrich, G., Kalinke, U., Momburg, F., Malissen, M., Schmitt-Verhulst, A.-M., Malissen, B., Hämmerling, G.J. and Arnold, B. (1991) *Cell*, **65**, 293-304.

Schulz, M., Aichele, P., Vollenweider, M., Bobe, F.W., Cardinaux, F., Hengartner, H. and Zinkernagel, R.M. (1989) *Eur.J.Immunol.*, 19, 1657-1667.

Sinha, A.A., Lopez, M.Th. and McDevitt, H.O. (1990) *Science*, **248**, 1380-1388.

Todd, J.A., Acha-Orbea, H., Bell, J.I., Chao, N., Fronek, Z., Jacob, C.O., Timmermann, L., Steinmann, L and McDevitt, H.O. (1988) *Science*, **240**, 1003-1009.

Traub, E. (1936) *J.Exp.Med.*, **63**, 847-861.

Uematsu, Y., Ryser, S., Dembic, Z., Borgulya, P., Krimpenfort, P., Berns, A., von Boehmer, H. and Steinmetz, M. (1988) *Cell*, **52**, 831-841.

Webb, S.R. and Sprent, J. (1990) *Science*, **248**, 1643-1646.

Wright, K.E. and Buchmeier, M.J. (1991) *J.Virol.*, **65**, 3001-3006.

Zamvil, S.S. and Steinman, L. (1990) *Annu.Rev.Immunol.*, **8**, 579-621.

Zinkernagel, R.M., Pircher, H.P., Ohashi, P.S., Oehen, S., Odermatt, B., Mak, T.W., Arnheiter, H., Bürki, K. and Hengartner, H. (1991) *Immunol.Rev.*, **122**, 133-171.

T Cell Receptor Signalling Defects in Anergic T Cell Induced *in vivo* by Bacterial Superantigen

Atsuo Ochi, and Kiyoshi Migita

Division of Molecular Immunology and Neurobiology Samuel Lunenfeld Research Institute, and The Department Immunology and Medical Genetics University of Toronto, Mount Sinai Hospital, 600 University Ave. Toronto, Ontario, Canada M5G 1X5

INTRODUCTION

Although mice inherit a large number of gene segments of T cell receptor (TCR) α- and β- chain variable domains (Vα, Vβ), generally not every V gene takes part in the repertoire of T cells in the periphery. This development of a limited TCR repertoire results from thymic negative selection where apoptic cell death plays a major role in the loss of self-reactive immature T cells (Kappler 1987). This type of thymus dependent "self-tolerance" is under the striking influence of self-genome integrated mouse mammary tumour virus encoded superantigens (previously denoted as minor lymphocyte stimulating antigens) that react with T cells based on the particular Vβ phenotypes (Kappler 1988; MacDonald 1988). In contrast, peripheral T cell tolerance against foreign antigens is caused, at least in part, by functional unresponsiveness (anergy). Mice tolerized to allogenic viral superantigens or bacterial superantigens give rise to anergized T cells expressing specific TCR Vβ's (Rammensee 1989; Kawabe and Ochi 1990). Anergic T cells are also implicated in transgenic mice expressing nonself antigens in the periphery (Burkly 1989; Murphy 1989: Wieties 1990), or self reactive TCRs (Blackman 1991), and mice tolerized orally to nonself protein antigens (Whitacre 1991). Anergic T cells fail to proliferate or produce the T cell growth factor, interleukin-2 (IL-2) in response to antigenic stimulation. Though the fate of these functionally inactive cells is not yet clear, circumstantial evidence suggests that most but perhaps not all of the anergic T cells recover from anergy (Migita & Ochi submitted). A fundamental role of anergic T cells in T cell memory has also been suggested (Bandeira 1991). Assuming that anergy represents an antigen induced transient state of mature T cells in the periphery, and that anergic T cells represent a novel, functional T cell subset, it appears to be important to elucidate the mechanism of T cell anergy. Understanding of the role of the anergic state in T cell function will help to elucidate the mechanisms that underlie T cell-repertoire development in antigen challenged animals. In this paper we will present results on TCR signal transduction and associated protein tyrosine kinase (PTK) activation of *in vivo* bacterial superantigen, Staphylococcal Enterotoxin B (SEB) induced anergic T cells.

RESULTS

The initiation of T cell tolerance in SEB-injected adult mice has been previously demonstrated with data from this laboratory and others (Kawabe and Ochi 1990; Rellahan 1990; MacDonald; 1991). This tolerance was established because SEB-reactive $V\beta8^+$ T cells had been anergized. Because of the activation induced death of SEB-stimulated T cells which takes place after SEB-injection, the number of $V\beta8^+$ cells decreased to 25 % of total T cells in SEB-treated mice within 5 days whereas 30 % of T cells are $V\beta8^+$ in normal mice (Kawabe and Ochi 1991). The reduction of $V\beta8^+$ T cells occurred mainly within the $CD4^+$ subset. When 100 μg of SEB was injected to mice, previously anergized with SEB, both spleen CD4 and CD8 T cells proliferated poorly and production of IL-2 and γIF were suppressed. Established T cell anergy remained for almost two months without changing the number of $V\beta8^+$ T cells in spleen. We have used such anergic $V\beta8^+$ T cells as an invaluable source of cells for the study of unresponsiveness established in normal cells.

Proliferation and IL-2 production by anergic $V\beta8^+$ T cells after stimulation by phorbol 12-myristate 13-acetate (PMA) and Ca^{2+} ionophore

In order to obtain insight into the defects in anergic T cells, purified anergic $V\beta8^+$ T cells were co-stimulated with a phorbol ester and ionomycin. We reasoned that these agents might reverse the anergic state if the signalling defects was upstream to protein kinase C (PKC) or intracellular Ca^{2+}. While stimulation with SEB resulted in unresponsiveness of proliferation and IL-2 production by SEB anergized $V\beta8^+$ T cells, both PMA and ionomycin induced both proliferation and IL-2 production in anergic T cells. The level of the responses were undistinguishable from non-anergic purified $V\beta8^+$ T cells. Therefore this result indicated that SEB-anergic $V\beta8^+$ T cells are normal with respect to mechanisms downstream of PKC and intracellular Ca^{2+} metabolism. This focused our attention on possible defects in the signal transduction cascade upstream of PKC and intracellular Ca^2+ source.

Induction failure of protein phosphorylation

We first examined the phosphorylation events in SEB-tolerized spleen T cells after stimulation with SEB. As shown in Fig. 1, SEB in the presence of the fixed class II major histo-compatibility complex (MHC) antigen-positive B lymphoma, A20-2J, did not induce a significant increase in the amount of tyrosine phosphorylated proteins. In contrast when spleen T cells of PBS-treated control mice were stimulated similarly, an increase in the amount of tyrosine phosphorylation was readily detected. Thus TCR stimulation does not induce detectable increases in tyrosine phosphorylation in anergic $V\beta8^+$ T cells.

Fig 1. Failure to increase intra-cellular protein tyrosine phosphorylation in SEB-anergized spleen T cell with SEB. Balb/c ByJ mice (H-2^d) were injected with PBS or 100 μg SEB (Sigma) by the intraperitoneal route. A week later spleen T cells were enriched (>90% CD3$^+$) by depletion of surface immunoglobulin (Ig)-positive cells on goat anti-mouse Igs (100 μg/ml)-coated petri dishes and with anti-mouse Igs coated magnetic beads. They were sorted by a magnetically activated cell sorter (MACS). These purified T cells were incubated for 2 min at 37°C with formaldehyde fixed murine B lymphoma, A20-2J (H-2^d, class II MHC$^+$) in either the presence or absence of 10 μg/ml SEB. Incubations were terminated by adding lysis buffer (50 mM Hepes pH 7.5 / 150 mM NaCl / 2mM EDTA / 10 μg ml^{-1} aprotinin / 10 μg ml^{-1} leupeptin / 1 mM phenylmethylsulfonyl fluoride (PMSF) / 500 μM sodium orthovanadate). After 20 min of incubation on ice, nuclei were pelleted and supernatants were subjected

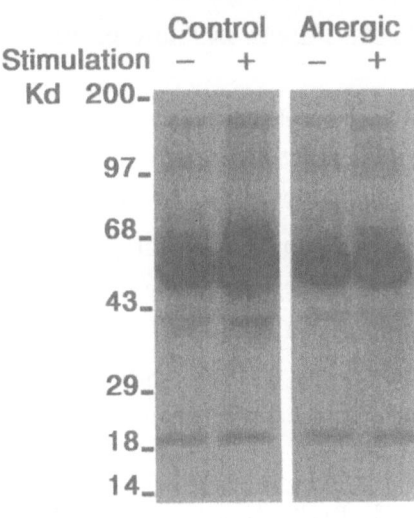

to SDS-PAGE on 8 to 16 % gradient polyacrylamide gels (Novex). The fractionated proteins were transferred to Immobilon (Millipore) and the membrane was incubated for 16 hours in 5 % BSA / 10 mM Tris pH 8.2 / 140 mM NaCl / 0.01 % NaN$_3$ to block nonspecific protein binding. The blots were incubated with phosphotyrosine (ptyr)-specific rabbit antibodies (Koch 1989) for 2 hours. The membranes were then incubated with 1 μCi of ^{125}I-protein A (Amersham) for 30 min. The filters were then washed and autoradiographed.

Constitutively phosphorylated ζ

In order to examine tyrosine phosphorylation events in TCR-stimulated anergic T cells more specifically, we immunoprecipitated a TCR component, CD3-ζ (Fig. 2). The phosphorylation of CD3-ζ occurs subsequent to antigen stimulation of normal T cells (Samelson 1986). In control cells, phosphorylation of CD3-ζ was increased when the TCR was stimulated by SEB whereas the same treatment did not alter phosphorylation of CD3-ζ in anergic Vβ8$^+$ T cells. Interestingly, unlike the case in normal T cells, CD3-ζ was phosphorylated at a high level in anergic Vβ8$^+$ T cells prior to stimulation.

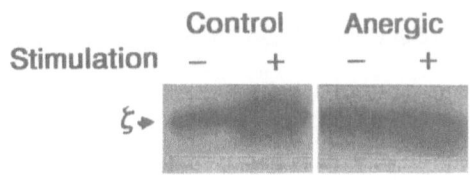

Fig. 2. Tyrosine phosphorylation of CD3-ζ in anergic spleen T cells. Purified spleen T cells from normal and SEB-anergic mice were stimulated with SEB for 2 min as described in Fig. 1. Cells were lysed in 0.5 % Triton-X lysis buffer and a 5x10^6 cell equivalent extract (50 μl) was immunoprecipitated with Vβ8-specific antibody (Staerz 1985) fixed on protein A conjugated beads (Parmacia) on ice for 1 hour. Beads were washed and the immuno-precipitated proteins were solubilized in loading buffer for SDS-PAGE analysis. The blots were probed for phosphotyrosine (ptyr) as described above.

Normal phosphatase activity of anergic Vβ8⁺ T cell membrane

The above observations prompted us to investigate the activity of tyrosine phosphatase, CD45, in anergic Vβ8⁺ T cells since this enzyme is known to be essential for responses to TCR stimulation (Koretzky 1991). We found that the anergic Vβ8⁺ T cells expressed levels of membrane CD45 equivalent to control cells (Migita & Ochi submitted). Furthermore, when the membrane fraction of purified anergic Vβ8⁺ cells was assayed for tyrosine phosphatase activity it was indistinguishable from that of control cells (Fig. 3) and was equally sensitive to the action of specific tyrosine phosphatase inhibitor, Na-orthovanadate (Mustelin 1989). Therefore, abnormal functional phosphatase activity cannot account for our results on aberrant CD3-ζ tyrosine phosphorylation.

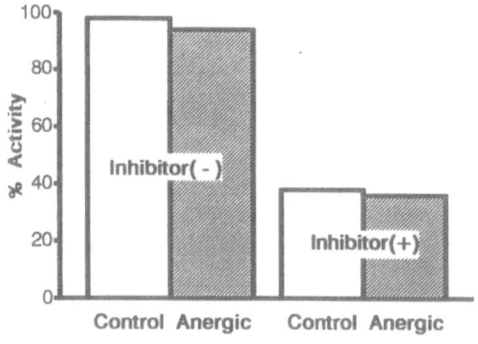

Fig. 3. Tyrosine phosphatase activity of anergic Vβ8⁺ T cells. Normal or SEB-anergized spleen T cells were incubated with biotinylated anti-Vβ8 antibody and the cells were then washed and mixed with Streptavidin-magnetic beads (Dynal) for 30 min at room temperature. The magnetically coated cells were isolated by a magnetic field generated by MACS. Membrane fractions of Vβ8⁺ T cells were prepared according to the method of Mustelin (1989). Briefly, purified Vβ8⁺ T cells from PBS or SEB (100 μg) treated spleen cells were sonicated in a hypotonic lysis buffer [25 mM Tris-HCl pH 7.5 / 25 mM sucrose / 0.1 mM EDTA / 5 mM dithiotreitol / 1 mM PMSF / leupeptin (10 μg/ml) / aprotinin (10 μg/ml)], in the presence and absence of 1 mM Na₃VO₄. After nuclei were removed by low-speed centrifugation, the membranes were sedimented at 100,000 g for 60 min at 4°C. The resulting pellet was suspended in lysis buffer by sonication. Protein concentration was determined using the Protein assay kit (Bio-Rad). The phosphatase activity was measured according to previously described methods (Zanke 1992). 20 μg of membrane protein was incubated in a reaction mixture of 5 mM P-nitrophenyl phosphate (Sigma), 80 mM 2-[N-morpholino]ethane sulfonic acid (pH 5.5), 10 mM EDTA, and 10 mM DTT, at 37°C for 15 minutes. The reactions were stopped by the addition of 0.2 N NaOH. Absorbance was measured at 410 nm and expressed for each value as a percentage of the maximum observed. The experiment was performed in triplicate.

Normal amount of *fyn* and CD3-ζ but abortive induction of TCR associated PTK in anergic T cells

fyn is a member of the *src* family of tyrosine kinases and represents the only characterized PTK known to associate with TCR (Samelson 1990). Whereas the functional role of *fyn* is not fully understood, this PTK and CD3-ζ appear to be crucial components necessary for signal transduction through the TCR (Cooke 1991; Irving and Weiss 1991). It was therefore possible that abnormalities in CD3-ζ or *fyn* expression or action were responsible for the failure to transduce signals through the

TCR in anergic $V\beta8^+$ T cells. In order to examine this possibility, we postulated three potential alternatives: 1) that there was a decrease in the absolute amounts of CD3-ζ and/or *fyn*; 2) that CD3-ζ and/or *fyn* did not associate with the TCR; or 3) that *fyn* was not activated by the TCR in anergic T cells. We initially approached these alternatives by performing western blot analysis of purified anergic $V\beta8^+$ T cell extracts using specific antibodies against CD3-ζ and *fyn*. We found that the amounts of CD3-ζ which co-precipitated with $V\beta8$-TCR were almost the same in control and anergic T cells. Levels of *fyn* which co-precipitate with $V\beta8$-TCR were comparable between control and anergic T cells. Thus, T cell anergy cannot be due to inadequate amounts of CD3-ζ or *fyn*, or insufficient association of these molecules to TCR.

Finally, in order to examine the third possibility for aberrant activation of TCR-associated *fyn* is aberrant in anergic T cells, we examined the function of TCR-associated *fyn*. This enzyme auto-phosphorylates as well as phosphorylates exogenous substrates and these activities are measurable in *in vitro* kinase assays (Samelson 1990). We measured the kinase activity of TCR-associated *fyn* in samples immunoprecipitated with anti-$V\beta8$ antibody from mild detergent (digitonin) cell lysates. Activation of control cells with SEB dramatically increased TCR-associated *fyn*-kinase activity in anti-$V\beta8$ immunoprecipitates compared to resting levels (Fig. 4). In contrast, TCR-associated *fyn*-kinase activity increased only slightly following TCR mediated activation of purified anergic $V\beta8^+$ T cells. The TCR-activated kinase was confirmed to be *fyn* by reprecipitation of the *in vitro* kinase with anti-*fyn* antibody. As demonstrated in Fig. 4, only TCR-associated *fyn* and coprecipitating proteins were demonstrably phosphorylated following activation of the TCR. When the protein blots were transferred to immobilon and incubated with KOH to examine tyrosine-specific phosphorylation (Cooper 1983), we found that activation of normal T cells with SEB resulted in a marked increase in TCR-associated *fyn*-kinase activity (data not shown). In contrast, *fyn* activity was not detectably increased by SEB activation of anergic T cells. Thus TCR associated *fyn* remains inactive in response to stimulation in anergic T cells.

Fig. 4. *In vitro* kinase assay of $V\beta8$-TCR-associated protein tyrosine kinase. Spleen T cells were solubilized at 10^7 /ml cells in 50 mM Hepes / 150 mM NaCl / 2 mM EDTA / 1 % digitonin (w/v), 10 μg ml^{-1} aprotinin / 10 μg ml^{-1} leupeptin / 1 mM PMSF / 0.5 mM Sodium orthovanadate. The lysates were centrifuged at 12,000g for 10 min and supernatants were pre-cleared for 30 min with 50 μl of 10 % Staphylococcus Cowan⁻ Strain I (Calbiochem). The supernatants were then incubated with anti-$V\beta8$ antibody for 1 hour at 4°C. The immune complex was then collected by incubation with Protein A-Sepharose 4B (Pharmacia) 4°C for 30 min. After immuno-precipitation, the immune complex was washed three times in lysis buffer without EDTA and 3 times with kinase buffer (50 mM Hepes pH 7.5 / 100 mM NaCl / 5 mM MnCl$_2$ / 5 mM MgCl$_2$ / 1 μM ATP) and suspended in a 50 μl of kinase buffer containing 10 μ Ci of γ-^{32}P-ATP (3×10^3 Ci / mM, Amersham). After 15 min at 30°C, the reaction was terminated by the addition of an equal volume of 2 x SDS sample buffer (20 % glycerol, 10 % 2-mercaptoethanol, 4.6 % SDS, 125 mM Tris pH 6.8, 0.004 % Bromophenol blue). The samples were boiled and analyzed by electrophoresis through a 8 % SDS-polyacrylamide gel. This experiment was performed four times with similar results.

fyn : After the kinase reaction was performed with immunoprecipitated TCR, the immune complex was incubated in 1 % SDS at 95°C for 10 min and diluted 10 fold in buffer with 1 % Nonidet P-40. The samples were re-immunoprecipitated with anti-*fyn*.

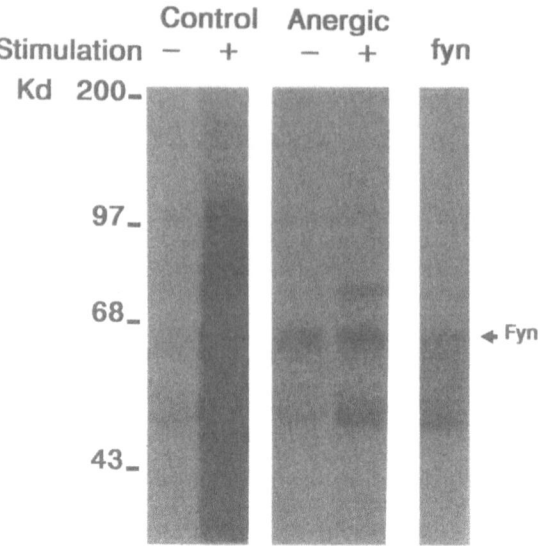

DISCUSSION

In studies described in this report we have shown that TCR stimulation does not activate TCR-associated *fyn* in anergic $V\beta 8^+$ T cells. Thus, unresponsiveness to TCR stimulation of anergic T cells *in vivo* may be due to a defect(s) in activation of TCR-associated PTK(s). These cells however produce IL-2 and proliferate in response to co-stimulation with PMA and ionomycin (Migita and Ochi manuscript in preparation). Thus, they seem to carry a functional IL-2 gene which can be regulated by downstream events initiated by increases in cytosolic calcium and activation of PKC. Previously, a similar phenomenon was observed when cells were treated with a PTK-inhibitor, Herbimycin A, in which TCR stimulation failed to induce IL-2 production. Co-stimulation with PMA and ionomycin induced IL-2 production in the presence of Herbimycin A (June 1990). Blackman reported that TCR $V\beta 8.1$ transgenic mice generated in an Mls-1^a carrying host develop anergic $CD4^+$, $V\beta 8.1^+$ cells in the periphery (Blackman 1990). These cells are unable to mobilize intracellular calcium when stimulated with a TCR-specific antibody. However, similar to our SEB-induced anergic $V\beta 8^+$ T cells, they proliferated in response to PMA and ionomycin. Because Ca^{2+} mobilization is triggered by inositol 1,4,5-triphosphate created by hydrolysis of inositol phospholipids by phospholipase C-γ (PLC-γ), and since a TCR-associated PTK activates PLC-γ by phosphorylation of tyrosine residues (Mustelin 1990), anergic T cells in transgenic mice may have defect(s) in the signalling circuit: TCR-associated PTK -PLC-γ -intracellular Ca^{2+} release and/or PKC activation.

Anergic T cells have constitutively phosphorylated CD3-ζ. Interestingly, $CD4^-$, 8^- T cells of MRL*lpr/lpr* autoimmune-prone mice also respond poorly to TCR-activation and also posses similarly constitutively phosphorylated CD3-ζ (Samelson 1986; Katagiri 1987). Similarly, $CD45^-$ cell lines also fail to signal through the TCR (Koretzky 1991). However, CD45 levels and tyrosine phosphatase activities were normal in SEB-induced anergic T cells. Taking the observations together persistent phosphorylation of TCR components may result in anergy in T cells. The observation that phosphorylation of the terminal tyrosine residue of *src* family tyrosine kinases negatively regulates their enzyme activities is of interest in terms of this latter hypothesis (Davidson 1992). Constitutive tyrosine phosphorylation of this residue of *fyn* may impair *fyn* dependent TCR signal transduction and result in T cell anergy.

Although *fyn* and CD3-ζ are essential for signalling, and known to associate with the TCR, it is unclear which component of the TCR is responsible for *fyn* association and whether this interaction requires additional molecules. Furthermore, it has been suggested that additional tyrosine kinase are involved in the TCR-signal transduction system (Chan 1991). The molecular mechanism leading to an uncoupling of PTK activation from stimulation of the TCR is under further investigation.

REFERENCES

Bandeira A, Mengel J, Burlen-Defranoux O, Coutinho A (1991) Int Immunol 3:923-931

Baniyash M, Garcia-Morales P, Luong E, Samelson LE, Klausner RD (1988) J Biol Chem 263: 18225-18230

Blackman MR, Gerhard-Burgert H, Woodland DL, Palmer E, Kappler JW, Marrack P (1990) Nature 345:540-542

Blackman MA, Finkel TH, Kappler J, Cambier J, Marrack P (1991) Proc Natl Acad Sci USA 88: 6682-6686

Burkly LC, Lo D, Kanagawa O, Brinster RL, Flavell RA (1989) Nature 342:564-566

Chan AC, Irving BA, Fraser JD, Weiss A (1991) Proc Natl Acad Sci USA 88:9166-9170

Cooke MP, Abraham KM, Forbush KA, Perlmutter RM (1991) Cell 65:281-291

Cooper JA, Sefton BA, Hunter T (1983) Methods Enzymol 99:387-402

Davidson D, Chow LML, Fournel M, Veillette A (1992) J Exp Med 175:1483-1492

Irving BA, Weiss A (1991) Cell 64:891-901

June CH, Fletcher MC, Ledbetter JA, Schieven GL, Siegel JN, Phillips AF, Samelson LE (1990) Proc Natl Acad Sci USA 87:7722-7726

Kappler JW, Roehm N, Marrack P (1987) Cell 49:273-280

Kappler JW, Staerz UD, White J, Marrack P (1988) Nature 332:35-40

Katagiri K, Katagiri T, Eisenberg RA, Ting J, Cohen PL (1987) J Immunol 138:149-156

Kawabe Y, Ochi A (1990) J Exp Med 172:1065-1070

Kawabe Y, Ochi A (1991) Nature 349:245-248

Koch CA, Moran M, Sadowsky I, Pawson T (1989) Mol Cell Biol 9:4131-4140

Koretzky GA, Picus J, Schultz T, Weiss A (1991) Proc Natl Acad Sci USA 88:2037-2041

MacDonald HR, Schneider R, Lees RK, Howe RC, Acha-Orbea H, Festenstein H, Zinkernagel RM, Hengartner H (1988) Nature 332:40-45

MacDonald HR, Baschieri S, Lees RK (1991) Eur J Immunol 21:1963-1966

Murphy KM, Weaver CT, Elish M, Allen PM, Loh DY (1989) Proc Natl Acad Sci USA 86:10034-10038

Mustelin T, Coggeshall MK, Altman A (1989) Proc Natl Acad Sci USA 86:6302-6306

Mustelin T, Coggeshall KM, Isakov N, Altman A (1990) Science (Wash. DC) 247:1584-1587

Nakayama T, Singer A, Hsi ED, Samelson LE (1989) Nature 341:651-654

Pullen AM, Marrack P, Kappler JW (1988) Nature 335:796-801

Rammensee H, Kroschewski R, Frangoulis B (1989) Nature 339:541-544

Rellahan BL, Jones LA, Kruisbeek AM, Fry AM, Matis LA (1990) J Exp Med 172:1091-1100

Samelson LE, Davidson WF, Morse III HC, Klausner RD (1986) Nature 324:674-676

Samelson LE, Patel MD, Weissman AM, Harford JB, Klausner RD (1986) Cell 46:1083-1090

Samelson ES, Phillips AF, Luong ET, Klausner RD (1990) Proc Natl Acad Sci USA 87:4358-4362

Staerz UD, Rammensee H-G, Benedetto DJ, Bevan JM (1985) J Immunol 134:3994-4000

Veillette A, Bookman MA, Horak EM, Bolen JB (1988) Cell 55:301-308

Whitacre CC, Gienapp IE, Orosz CG, Bitar DM (1991) J Immunol 147:2155-2163

Wieties K, Hammer RE, Jones-Youngblood S, Forman J (1990) Proc Natl Acad Sci USA 87:6604-6608

Zanke B, Suzuki H, Kishihara K, Mizzen L, Minden M, Pawson A, Mak T-W (1992) Eur J Immunol 22:235-239

ACKNOWLEDGEMENTS

We wish to thank Drs. G. B. Mills, A. J. Pawson, L. Samelson and A. Veillette for invaluable discussions and provision of antibodies specific for ptyr, CD3ζ and *fyn*. We also thank Drs. M. J. Bevan for the monoclonal antibody, F23.1. We thank Drs. D. E. Spaner and L. Siminovitch for the critical readings of the manuscript. We also thank D. S. Hill for his help in preparing the manuscript. This work was supported by grants from the Medical Research Council of Canada, the National Cancer Institute of Canada and the Arthritis Society of Canada.

Intrathymic Selection and Maturation of $\alpha\beta$ T Cells

P. Kisielow, W. Swat and H. von Boehmer

Basel Institute for Immunology*, Grenzacherstrasse 487, CH-4058 Basel, Switzerland

INTRODUCTION

Intrathymic development of CD4+ and CD8+ ('single positive', SP) T cells from CD4-8- ('double negative', DN) precursors can be divided into two phases controled by the products of genes encoding $\alpha\beta$ T cell receptor (TCR). The TCR β chain - whose surface expression in the homodimeric form preceeds in ontogeny the expression of $\alpha\beta$ heterodimer (Groettrup and von Boehmer- submitted) - controls the transition of DN thymocytes into CD4+8+ ('double positive', DP) thymocytes (von Boehmer 1990, P. Mombaerts 1992, personal communication). The $\alpha\beta$ heterodimer in turn controls the selection and development of DP thymocytes into SP T cells (von Boehmer et al. 1989, Philpott et al. 1992, P. Mombaerts 1992, personal communication). In contrast to the TCR β genes, the expression of functional TCR α gene and protein does not prevent further rearrangements whithin TCR α locus (von Boehmer 1990, Borgulya et al. 1992, Mallissen et al. 1992). It is therefore possible that more than one $\alpha\beta$ TCR with different specificities can be expressed on the same clone of developing DP thymocyte thus increasing the pool of receptors avaliable for selection without increasing the pool of selectable cells.

Thus, the long lasting controversy concerning the role of CD4+8+ thymocytes in T cell development has been resolved and it is now generally accepted that it is at this stage where positive and negative selection processes shaping the repertoire of $\alpha\beta$TCR specificities expressed on mature T cells take place. The best evidence for this conclusion has been provided by experiments using TCR transgenic mice in which the fate of T cells expressing TCR of known specificity for antigen and MHC restricting molecule could be followed throughout all stages of development in positively (MHC-minus-antigen) and negatively (MHC-plus-antigen) selecting environments (von Boehmer 1990, Kisielow and von Boehmer 1990). However, the population of $\alpha\beta$TCR+ thymocytes expressing both CD4 and CD8 molecules is very heterogenous (Table 1) and the responsiveness of the various subpopulations of DP thymocytes to TCR mediated signals responsible for positive and negative selection is not clear. In order to investigate this problem we begun to study the in vitro differentiation potential and responsiveness to antigen and/or to anti-TCR antibody of isolated subsets of DP and SP thymocytes at different stages of development.

* The Basel Institute for Immunology was founded and is supported by F. Hoffmann-La Roche Ltd., Basel, Switzerland

Table 1. Phenotypic heterogenity of CD4+8+TCR+ thymocytes

| | | Marker: | | | |
SIZE	TCR	CD4	CD8	HSA	CD69
	low >95%	high	high	high	-
		high	high	high	-
					+
				low	?
small >95%	high <5%	high	low	high	-
					+
		low	high	low	-
					+
large <5%	low	high	high	high	?
	high	high	high	high	?

HETEROGENEITY OF DPTCR+ THYMOCYTES AND THE SEQUENCE OF PHENOTYPIC CHANGES ACCOMPANYING THEIR MATURATION

Most of DP thymocytes are small and express low levels of αβ TCR but a minor population expresses high levels of TCR and is heterogenous with regard to size and expression levels of CD4, CD8, HSA and CD69 molecules (Hugo et al. 1990, Bendelac et al. 1992, Swat et al. 1992). It is generally believed that in the normal course of maturation DP thymocytes proceed from large HSA^{hi}CD4^{high}TCR^{lo} to small HSA^{hi}CD4^{high}TCR^{lo} and then without cell division (Huesmann et al. 1991, Shortman et al. 1991) to small HSA^{hi} CD4^{high}TCR^{hi} and then to either mature HSA^{lo}CD4^{hi}8-TCR^{hi} cells through an intermediate stage of small HSA^{hi}CD4^{hi}8^{lo}TCR^{hi} thymcytes or to mature HSA^{lo}CD4-8^{hi} TCR^{hi} cells through an intermediate stage of small HSA^{hi}CD4^{lo}8^{hi}TCR^{hi} thymocytes (Shortman et al. 1991). The critical changes are thought to include the upregulation of TCR level as the earliest indication of positive selection event (Borgulya et al. 1991, Ohashi et al. 1990) and downregulation of HSA as indication of the transition of immature to mature (competent) stage of development. However, the identification of large HSA^{hi}CD4^{high}TCR^{hi} thymocytes in both normal and TCR transgenic mice expressing as well as lacking selecting MHC molecules (Swat et al. 1992) and unraveling the heterogeneity of HSA expression within CD4/CD8 subpopulations which can be further subdivided on the basis of expression of CD69 ('early activation marker') molecule (Bendelac, 1992, Swat et al. 1992, submitted), necessitates a reevaluation of this picture.

IDENTIFICATION OF PRE AND POSTSELECTION STAGES OF DPTCR+ THYMOCYTES

In order to distinguish stages in development of DP thymocytes which precede positive selection from those which result from it we analysed in detail the phenotype of these cells from mice in which thymocytes expressing transgenic HY-specific H-2D^b (class I MHC) restricted TCR developed in selecting (H-2^b) or nonselecting (H-2^d) environment (Teh et al. 1988). The results have showed that small TCR^{hi} thymocytes with the transgenic TCR could be detected only in H-2^b but not in H-2^d female mice indicating that they represent the postselection stages whereas large HSA^{hi}TCR^{hi} and small HSA^{hi}CD69$^-$TCR^{lo} thymocytes represent nonselected pool of cells (Swat et al. 1992 and Swat et al. 1992-submitted). Thus the small but not large CD4^{highi}TCR^{hi} thymocytes were identified as the earliest product of positive selection. Recently, it has been shown that at this stage the rearrangement whithin TCR α locus ceases due to the termination of activity of recombination activating (RAG) genes (Borgulya et al. 1992). The fact that in contrast to small CD4^{highi}TCR^{lo} thymocytes small CD4^{highi}TCR^{hi} thymocytes contain a small proportion of HSA^{lo} and CD69$^+$ cells (Bendelac et al. 1992, Swat et al. 1992 and Swat et al. 1992-submitted) suggests that positive selection activates and induces downmodulation of HSA expression at least in some selected cells. We found that when placed into culture the isolated small but not large CD4^{highi}TCR^{hi} or small CD4^{highi}TCR^{lo} cells from normal and TCR transgenic mice differentiated towards SP thymocytes. However, in contrast to small DPTCR^{hi} cells from normal mice, which generated both CD4 and CD8 SP cells by gradual downmodulation of either CD4 or CD8 molecules, the small DP thymocytes expressing high level of class I MHC restricted TCR downmodulated only CD4 molecule and generated only CD8 SP thymocytes. No downmodulation of CD8 was observed in the latter cultures. (Swat et al. 1992). These results confirm the conclusion that small but not large DPTCR^{hi} thymocytes are the immediate product of positive selection and in addition are consistent with an instructive model of positive selection (von Boehmer 1986, Robey et al. 1990) whereby TCR mediated, MHC class specific signal inducing selective downmodulation of CD4/CD8 coreceptors is received at CD4^{highi} stage of development.

IDENTIFICATION OF SMALL HSA^{hi}CD4^{highi}TCR^{lo}CD69$^-$ THYMOCYTES AS THE TARGET OF BOTH POSITIVE AND NEGATIVE SELECTION

The immediate precursors of positively selected DPTCR^{hi} thymocytes have not been so far directly identified and it is still the matter of controversy whether they include small DPTCR^{lo} thymocytes (Guidos et al. 1989). To obtain direct evidence that cells with this phenotype contain precursors of SP cells and can be the target of positive selection, highly purified population of nonselected small HSA^{hi}CD4^{highi}TCR^{lo}CD69$^-$ thymocytes from H-2^d (Thy1.2) transgenic mice were injected intrathymically into normal H-2^b (Thy1.1) female mouse and three to four days later the phenotype of recovered donor (Thy1.2$^+$) thymocytes was analysed. The results are shown in Fig.1. The great majority of recovered cells were CD8 SP, some remained DP but all expressed high levels of transgenic TCR. From this result we conclude that small HSA^{hi}CD4^{highi}TCR^{lo}CD69$^-$ are the target of positive selection.

Phenothype of thymocytes from H-2d (Thy 1.2) TCR (T3.70) transgenic mice injected i.t. into normal H-2b (Thy 1.1) recipient

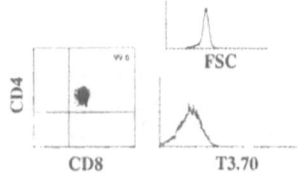

Phenotypes of recovered donor thymocytes

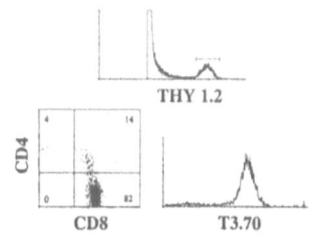

Fig.1. Differentiation of small DPTCR^{lo} into SPTCR^{hi} thymocytes after intrathymic (i.t.) transfer in vivo.

The results shown in Fig. 2. demonstrate that when the same cells are cultured in the precence of male but not female antigen presenting cells (APCs) from H-2^b mice they become specifically deleted. Thus small HSA^{hi}CD4^{high}TCR^{lo}CD69$^-$ thymocytes are the target of both positive and negative selection.

Another conclusion which can be drawn from the fact that small DPTCR^{lo} thymocytes from H-2^d transgenic mice can be deleted by APCs from H-2^b mice is that susceptibilty to deletion by antigen does not depend on prior positive selection by restricting MHC molecules on thymic epithelium. The above results indicate that during the lifetime of DP thymocytes, stages susceptible to negative and positive selection do overlap.

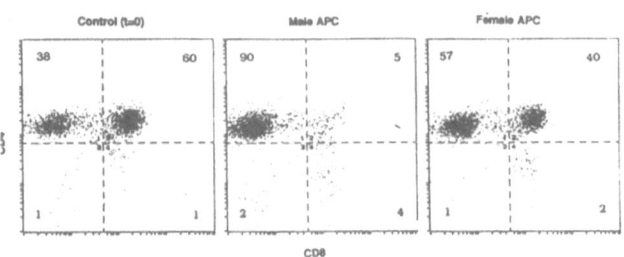

Fig. 2. Male (HY) antigen specific deletion of CD4$^+$8$^+$ thymocytes from H-2^d αβ TCR transgenic mice expressing TCR specific for HY antigen in the context of H-2D^b molecule. Sorted CD4$^+$8$^-$ and CD4$^+$8$^+$ thymocytes were cocultured for 24 hrs with APCs from H-2^b males or females. Expression of CD4 and CD8 was analysed on cell populations from which cell debris and apoptotic cells were excluded as described (Swat et al. 1991).

POSITIVE SELECTION RESULTS IN COUPLING OF TCR TO CD69 EXPRESSION PATHWAY

To get insight into the mechanism of positive selection of small DP thymocytes by MHC molecules we have analysed the inducibility of protein kinase C (PKC) dependent expression of CD69 molecule in TCR transgenic thymocytes developing in the presence (H-2^b) or absence (H-2^d) of selecting class I (D^b) MHC molecules.

To determine whether there is a difference in TCR mediated signalling in pre- and postselection stages in T cell development we studied the induction of CD69 expression on small DPTCR^{lo} and TCR^{hi} thymocytes by triggering TCR with idiotype specific monoclonal antibody (T3.70) in vitro.

As shown in Fig.3. stimulation with anti TCR antibody induced expression of CD69 on most CD8 SP thymocytes but only on the small fraction of DPTCR^{lo} thymocytes from the H-2^b (selecting) but not from the H-2^d (nonselecting) mice. Small DP thymocytes from H-2^d mice could not be induced to express CD69 by TCR crosslinking even after spontaneous in vitro upregulation of TCR level which resulted in enhanced Ca^{++} flux. (not shown). Unlike the anti-TCR antibody, PMA - a direct activator of PKC - induced the expression of CD69 on all thymocytes (Fig.3). These results suggest that TCR signalling in nonselected small DP cells is not coupled to PKC-dependent pathway leading to CD69 expression (PKC$_{CD69}$ pathway) and that coupling results from positive selection of DPTCR^{lo} thymocytes.

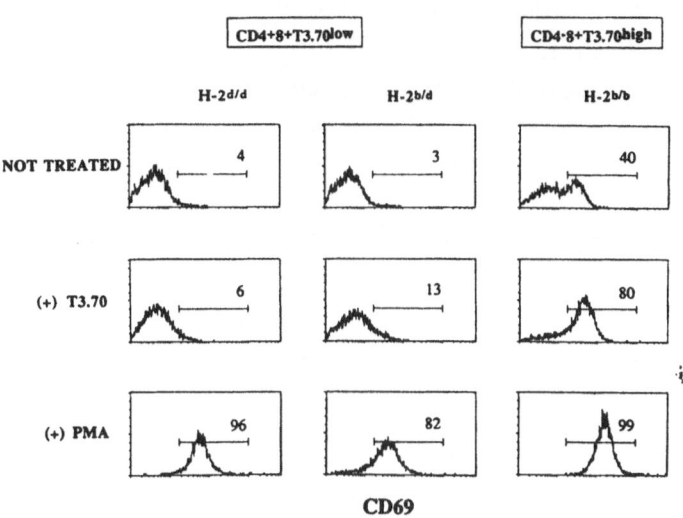

Fig. 3. TCR versus PMA mediated induction of CD69 expression on sorted populations of CD4+8+TCR^{lo} and CD4-8+TCR^{hi} thymocytes. CD4+8+T.370^{lo} thymocytes from females of H-2^d (nonselecting) and H-2$^{b/d}$ (selecting) TCR transgenic mice and CD4-8+T3.70^{hi} thymocytes from females of H-2^b TCR transgenic mice (purity >98%) were cultured without or with T3.70 mab or with PMA and then stained with anti-CD69 antibody. Numbers represent percentages of CD69$^+$ cells.

Because ligation of TCR on nonselected DPTCR^{lo} thymocytes by an antibody does not induce expression of CD69 but nevertheless results in cell death (see below) we suggest that deletion of immature thymocytes does not require activation of the PKC$_{CD69}$ pathway. The above results show that signals generated by crosslinking of TCR on small DP thymocytes are different before and after positive selection and provide a hint that TCR mediated signals involved in negative and positive selection of thymocytes may be different.

MATURATION OF POSITIVELY SELECTED DP THYMOCYTES

To determine at which stage of maturation positively selected thymocytes become resistant to deletion and begin to respond by proliferation to TCR mediated signals we compared the responsiveness of sorted subpopulations of TCR^{lo} and TCR^{hi} thymocytes and CD8 T cells to stimulation by anti-TCR antibody in vitro. The results (Swat et al. 1992- submitted) are summarized in Fig.4. which shows that the large proportions of DPTCR^{lo} and DPTCR^{hi} thymocytes were specifically deleted both in the absence and in the presence of IL-2 indicating that coupling of TCR to CD69 expression pathway *per se* does not influence the ability of the TCR to deliver a signal resulting in deletion. In contrast to DP thymocytes, CD8 SP thymocytes as well as CD8 T cells contained an increased proportion of cells which, although deletable in the absence of IL-2, respond with proliferation in its presence. However, even in the presence of IL-2 a significant proportion of CD8 SP thymocytes and smaller proportion of T cells could be deleted by anti-TCR antibody. These results indicate that positive selection of DP thymocytes does not immediately result in resistance to deletion and that the switch from deletion to proliferation in response to TCR triggering occurs at the late SP stage of maturation.

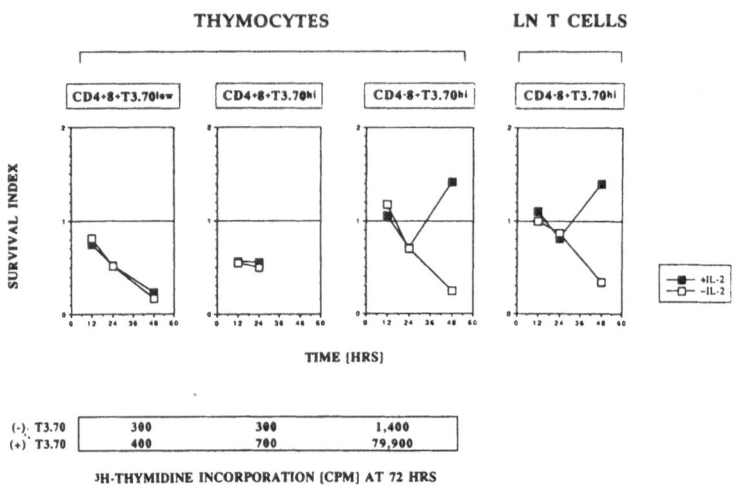

Fig. 4. Susceptibility to TCR-mediated deletion and to TCR-mediated induction of proliferation of developing T cells before and after positive selection. The indicated sorted subsets of thymocytes and lymph node T cells were from H-2^b TCR transgenic female. Survival index <1 indicate deletion, >1 indicate proliferation.

We know nothing about the nature of the signals promoting the maturation of selected DP thymocytes and whether TCR mediated signalling is involved in this process. Therefore because the expression of CD69 molecule on a T cell reflects the sustained signalling via TCR (Testi et al. 1989) it was interesting to determine the maturational status of CD69$^+$ and CD69$^-$ cells among HSA^{hi}CD8 SP and HSA^{lo}CD8 SP thymocytes. The maturational sequence suggested by the analysis of the frequency of cells responding by proliferation (which may be taken as an indication of the functional maturity of the given population) (Swat et al. 1992-submitted) is compatible with the view that HSA^{hi}CD69$^+$ and HSA^{lo}CD69$^+$ cells represent sequential stages of MHC-induced-PKC$_{CD69}$-dependent maturation ending with mature HSA^{lo}CD69$^-$ cells. Thus our results suggest that sustained interaction of TCR with selecting ligands is needed to promote maturation of positively selected DP thymocytes which involves gradual loss of the sensitivity to deletion and acquisition of the ability to respond by proliferation to TCR triggering.

Fig. 5 presents our working hypothesis on the involvement of αβTCR during selection and maturation of thymocytes.

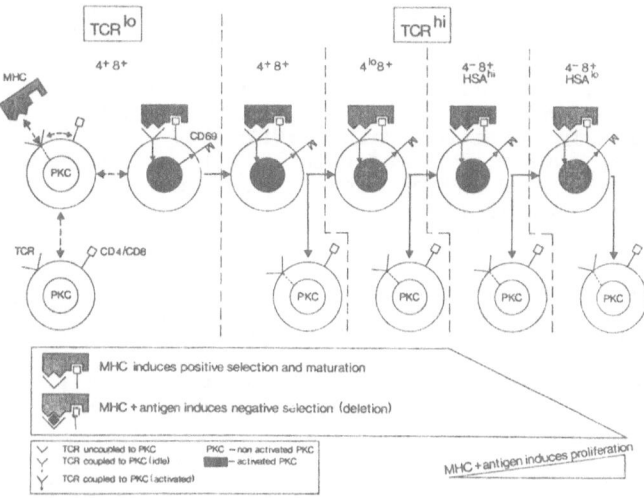

Fig. 5. Hypothetical model of intrathymic selection and maturation of αβ T cells.

In unselected small DPTCR^lo thymocytes, the TCR is not coupled to PKC. The coupling is established by interaction of TCR with MHC molecules probably by crosslinking of the TCR with CD4 or CD8 coreceptor by class I or class II MHC molecules depending on the specificty of TCR. The coupling could be required to shut off the activity of RAG-1 and RAG-2 genes in DP thymocytes after binding of their TCR to MHC ligands (Borgulya et al. 1992). This idea is supported by recent data showing that activation of PKC by PMA leads to inhibition of RAG-1 gene expression in immature thymocytes (Turka et al. 1991). In this way coupling of TCR to PKC would provide the mechanism preventing the generation of new receptors due to the ongoing rearrangements within TCR α locus once the selectable receptor becomes expressed on developing thymocytes.

The CD69+ cells among positively selected thymocytes could represent cells undergoing the process of MHC-induced-PKC-dependent maturation ending with mature HSA^loCD69- cells. According to this scenario most of the HSA^hiCD69- cells would represent 'dropouts' from the maturation process which may either die or become mature HSA^loCD69- cells if rescued by interaction with MHC ligand.

REFERENCES:

Adams, J. M., Harris, A.W., Pinkert, S., Cory, S., Palmiter, R.D., and Brinster, R.L. (1985). The c-myc oncogene fused with the immunoglobulin enhancers induces lymphoid malignancy in transgenic mice. Nature 318:533-538

Becker, R.S., and Knight, K.L. (1990). Somatic diversification of immunoglobulin heavy chain VDJ genes: evidence for somatic gene conversion in rabbits. Cell 63:987-997

Becker, R.S., Suter, M., and Knight, K.L., (1990) Restricted utilization of VH and DH genes in leukemic rabbit B cells. Eur. J. Immunol. 20:397-402

Currier, S.J., Gallarda, J.L. and Knight, K.L. (1988) Partial molecular genetic map of the rabbit VH chromosomal region J. Immunol. 140:1651-1659

DiPietro, L.A., and Knight, K.L. (1990). Restricted utilization of germ-line V_H genes and diversity of D regions in rabbit splenic Ig mRNA. J. Immunol. 144:1969-1973

Frence, D. L., Laskov, R. and Scharff, M.D. (1989) The role of somatic hypermutation in the generation of antibody diversity. Science 244:1152-1157

Gallarda, J. L., Gleason, K.S. and Knight, K.L. (1985) Organization of rabbit immunoglobulin genes I. Structure and multiplicity of germ-line V_H genes. J. Immunol. 135:4222-4228

Kelus, A.S., and Weiss, S. (1986). Mutation affecting the expression of immunoglobulin variable regions in the rabbit. Proc. Natl. Acad. Sci. USA 83:4883-4886

Kindt, T.J. and Capra, J.D., (1984). The Antibody Enigma. Plenum Press, New York.

Knight, K.L., and Becker, R.S. (1990). Molecular basis of the allelic inheritance of rabbit immunoglobulin V_H allotypes: implications for the generation of antibody diversity. Cell 60:963-970

Knight, K.L., Spieker-Polet, H., Kazdin, D. S., and Oi, V.T. (1988). Transgenic rabbits with lymphocytic leukemia induced by the c-myc oncogene fused with the immunoglobulin heavy chain enhancer. Proc. Natl. Acad. Sci., USA 85:3130-3134

Mage, R.G., Bernstein, K.E., McCartney-Francis, N., Alexander, C.B., Young-Cooper, G.O., Padlan, E.A., and Cohen, G.H. (1984). The structural and genetic basis for the expression of normal and latent V_Ha allotypes of the rabbit. Mol. Immunol. 21:1067-1081

McCormack, W.T., and Thompson, C.B. (1990). Somatic diversification of the chicken immunoglobulin light-chain gene. Adv. Immunol. 48:41-67.

Raman, C. and Knight, K. L. CD5[+] B cells predominate in peripheral tissues of rabbit, J. Immunol., in press.

Reynaud, C.-A., Dahan, A., Anquez, V. and Weill, J.-C, (1989). A Somatic hyperconversion diversifies the single V_H gene of the chicken with a high incidence in the D region. Cell 59:171-183

2. Genetics, Structure, Function and Evolution of Immunoglobulins and T Cell Receptors

Organization of the Human Immunoglobulin Heavy-Chain Locus

Fumihiko Matsuda, Euy Kyun Shin, Hitoshi Nagaoka, Ryusuke Matsumura, Makoto Haino, Yosho Fukita, Shigeo Takaishi, Takashi Imai[1], John H. Riley[2], Rakesh Anand[2], Eiichi Soeda[1], and Tasuku Honjo

Center for Molecular Biology and Genetics, and Department of Medical Chemistry, Faculty of Medicine, Kyoto University, Sakyoku, Kyoto 606, Japan.

INTRODUCTION

Immunoglobulin (Ig) heavy-chain (H) genes cluster at three loci in the human genome; the distal region of chromosome 14(1), chromosome 15(2) and chromosome 16(3). Among these only the chromosome 14 locus has been shown to generate a functional Ig gene by recombination of the variable (V_H), diversity (D) and joining (J_H) segments. V_H segments on chromosome 16 and D segments on chromosome 15 are putative orphons as these loci do not seem to contain any J_H and constant (C_H) genes, although interchromosomal recombination might occur to associate the orphon V_H and D segments with the J_H segments on chromosome 14.

Our long standing efforts to construct the complete physical map of the human V_H locus on chromosome 14 resulted in mapping of the 3' 200-kb region(4) by using yeast artificial chromosome (YAC) system. The construction of the physical map was extended to the further upstream region covering the 0.8-Mb DNA with 64 V_H segments. We determined the complete nucleotide sequences of the V_H segments, transcriptional polarities of V_H segments, and location of most frequently used V_H segments. We estimated the total number of the V_H segments in the human genome to be about 120 by comparison of the characterized YACs and total DNA by Southern blot hybridization.

RESULTS

Organization of the 3' 0.8-Mb V_H region

We have screened two YAC libraries of human lymphoblastoid cell lines CGM1 and GM1416 using V_{H-I}- and V_{H-III}- specific primers. The longest contig of 8 YACs overlapped with the previously characterized 3' most YACs, Y103 and Y20(4) and together covered

[1] Gene Bank, Tsukuba Life Science Center, the Institute of Physical and Chemical Research (RIKEN), Koyadai, Tsukuba, Ibaraki 305, Japan and [2] Department of Biotechnology, ICI Pharmaceuticals, Alderley Park, Macclesfield, Cheshire, SK10 4TG, United Kingdom

91

830 kb from the Cδ gene. Four YAC clones Y24, Y6, Y21, and 17H covering the 620-kb region upstream of Y103 and Y20 were extensively characterized by subcloning into cosmids and phages.

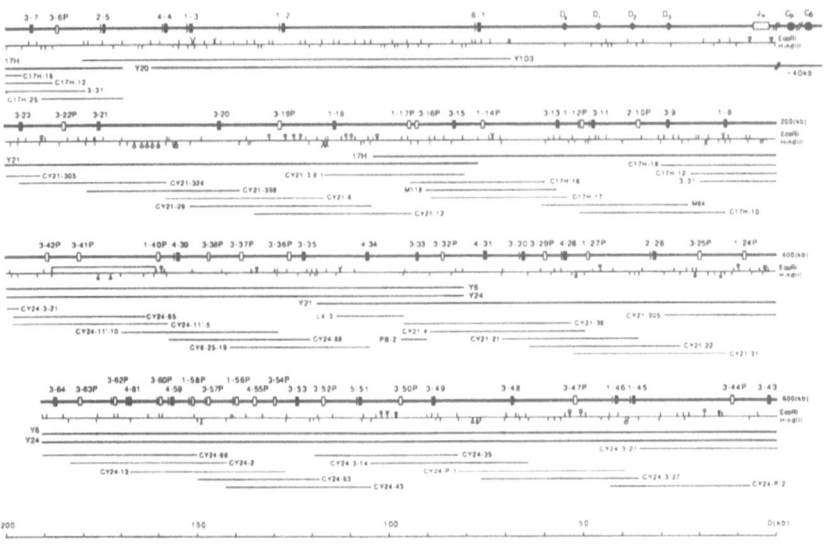

Fig. 1. Organization and restriction map of the 0.8-Mb region of the human IgH locus. The 0.8-Mb DNA is shown by 4 thick horizontal lines with the 3' end at the top right corner. The 3' end region is partially deleted. Restriction sites of EcoR I and Hind III are shown by vertical lines on the thin lines below. Vertical lines with open circles represent sites whose order are not determined and dotted box indicates the region of which EcoR I sites are not determined. Mlu I (∇) and Not I (▲) sites are also shown. DNA insert contained in each clone is shown by horizontal line further below. Locations of functional and pseudo V_H segments are indicated by filled and open rectagnles, respectively. 21 V_H segments whose transcriptional orientations are determined are shown by the addition of 5' leader exons. Cosmid subclones from YAC DNAs are indicated by C followed by the name of parental YAC clone. Cosmid clone 3-31 was isolated from a human placenta DNA(5). Cosmid M84 and M118 were obtained from FLEB14-14(3). Phage subclone L4-3 and plasmid subclone PB-2 were from Y6 DNA.

The cosmid and phage libraries were screened with V_H segments as well as total human DNA as probes. Isolated subclones were digested with restriction endonucleases *EcoR* I and *Hind* III, and aligned by overlap of the restriction maps (Fig. 1). The V_H segments belonging to six different families were highly interspersed among each other in agreement with our earlier observation(5). We named all the V_H segments in this region by the family number, followed by the order from the 3' end as we proposed in the previous report(4). Slight modifications are; only arabic numbers are used and pseudogenes are shown by P at the end. The average distance between neighboring V_H segments was about 12 kb and the longest region without V_H segments was between V6-1 and V1-2 segments (about 50 kb).

Only 33 of the 64 V_H segments were apparently functional (Table 1). In V_{H-I} and $_{-III}$ families, more than half (8 out of 14 for V_{H-I} and 21 out of 37 for V_{H-III}) of V_H segments are pseudogenes. On the other hand, all but one of the 8 V_{H-IV} segments were structurally non-defective. Two other V_{H-IV} segments (V11 and V58) whose locations are not known also turned out to be potentially functional genes(6). This family has several unique features as follows: 1)high homology of nucleotide sequences (>90%) between the family members, 2) remarkably few polymorphisms among individuals, and 3) high proportion of functional V_H segments. It is probably, at least in part, due to recent expansion of the V_{H-IV} family as evident from internal duplication (V4-28/V4-31 and V4-59/V4-61). Extensive comparison among 64 V_H sequences will be described elsewhere.

TABLE 1 Summary of V_H segments in the 0.8 Mb region[a]

V_H	Kb from J_H	Known V_H (homology %)
6-1	75	15P1
		M71
1-2	125	20P3
1-3	150	
4-4	160	
2-5	175	P46 (99.3), M44
3-6P	185	
3-7	190	HUMHHG19G
1-8	215	TH9* (97.6)
3-9	230	3D6 (97.3), P4 (98.7)
2-10P	235	
3-11	245	22-2B, N54P3, P2
1-12P	250	22-1

3-13	260	38P1 (97.6)
1-14P	275	
3-15	280	20P1 M26 (99.7), SB5/D6 (97.3),4B4*(99.7)
3-16P	290	
1-17P	295	
1-18	315	VH1GRR* (99.6)
3-19P	330	
3-20	345	
3-21	375	HUMWHG16G (99.7)
3-22P	385	
3-23	395	30P1, M43 (99.7), 18/2*, IGMV (99.7), VH26 (99.3), Ab18* (99.0)
1-24P	410	
3-25P	420	
2-26	430	
1-27P	450	
4-28	455	V12G-1, 1.9II, 4.13
3-29P	460	
3-30	470	56P1 (98.3), M72 (99.0), FL2-2(99.3), 1.9III, HV3005 (98.7), Kim4.6* (99.7), RFSJ2* (98.0), RFTS2* (97.6)
4-31	475	P30 (99.0)
3-32P	485	
3-33	490	
4-34	505	HT112 (99.7), LS2*, AB63* (99.7), HUMTOUVH* (99.7), Ab44* (97.7)
3-35	520	
3-36P	525	
3-37P	540	
3-38P	545	
4-39	555	V2-1, 4.18, HUMVH4R, μF7 (98.3), Ab61* (98.0)
1-40P	560	
3-41P	580	
3-42P	590	
3-43	600	
3-44P	610	
1-45	635	7-2
1-46	640	7-1, VHG3 (99.7)
3-47P	650	
3-48	670	N80P1m (99.7), HUMWHG26G
3-49	690	
3-50P	695	
5-51	710	M61, VH251, 5AU,
3-52P	715	H16BR (99.7), VH105
3-53	725	63P1
3-54P	730	
4-55P	735	
1-56P	740	V71-7
3-57P	745	V71-6
1-58P	750	V71-5

4-59	755	V71-4, 58P2 (99.7), 4.11 (99.7), OMM, Pag-1* (98.6)
3-60P	760	V71-3
4-61	770	V71-2
3-62P	775	V71-1
3-63P	780	
3-64	790	

a, identical known V_H segments are listed. Homologous but not identical V_H segments are shown by % homology. DNA sequences of coding regions except for leader sequences were aligned to maximize homology. Asterisks indicate V_H segments identified as autoantibody.

TRANSCRIPTIONAL POLARITY

We already showed that the transcriptional polarities of $D-J_H$ proximal five V_H segments are identical to those of the D and J_H segments(4). Comparison of the restriction maps indicate that V_H segments at 775(V3-62P), 710(V5-51), 640(V1-46), 555(V4-39), 470(V3-30), 430(V2-26), 250(V1-12P) and 175(V2-5) kb from the J_H segments had the same transcriptional orientations as those of the D and J_H segments. Therefore, inversion of a large chromosomal region containing many V_H segments as reported in the human V_k locus(7) is unlikely in this 0.8-Mb region although we can not exclude the possibility that a few V_H segments may have opposite polarities.

Homology of isolated V_H segments with known V_H segments

Computer-assisted search of known V_H segments homologous to 64 V_H segments was carried out using GENBANK, EMBL and DDBJ databases. We listed all the known V_H sequences including cDNA which displayed more than 97% nucleotide sequence homology with the 64 germline V_H segments analyzed in this study (Table 1). Among 33 functional V_H segments, 24 V_H segments were identical or almost identical to known V_H sequences. V_H segments often used preferentially in early stages of ontogeny were V6-1, V1-2, V3-13, V3-15, V3-23, V3-30, V5-51, V3-53 and V4-59 segments(8,9). Germline counterparts for several other V_H cDNA found in fetal liver could not be identified in the 0.8-Mb region, probably because they are located further upstream region or absent from this particular haplotype. The results indicate that V_H segments preferentially used in the early stage of ontogeny do not necessarily cluster in the J_H-proximal region.

We could identify several germline V_H segments which were often used for auto-antibodies. The V3-30 sequence was homologous to cDNA for rheumatoid factors, RF-TS2, RF-SJ1 and RF-SJ2(10). In

addition, this germline V_H segment was 99.7% identical to cDNA
for Kim 4.6 auto-antibody(8). Similarly, the V3-15 segment,
which is the germline gene of 20P1 cDNA expressed in fetal
liver(8), is 99.7% identical to cDNA for 4B4, an anti-Sm
antibody(11). The V3-23 segment is identical to 18/2 (an anti-
DNA auto-antibody)(12) and 30P1 cDNA found in fetal liver(8).

It is worth noting that more than half (about 100) of known V_H
genes in databases failed to identify their germline counterparts
in our sequences. These V_H sequences may be encoded by V_H
segments in more J_H-distal region as more not-yet sequenced V_H
segments should be present in the upstream portion within the
chromosome 14 locus. Allelic polymorphism could be another
reason as we showed in the case for the additional V_{H-I} segment
(V1-4.1b) in one haplotype(4). Accumulation of somatic mutations
also makes it difficult to find out germline counterparts of V_H
segments.

TABLE 2 Estimation of the total number of the V_H segments

Families	No. of V_H in the genone	No. of known Orphon V_H	No. of V_H on Chr. 14	V_H segments in isolated DNA
I	29	4	25	25
II	8	1	7	4
III	68	2	66	55
IV	12	0	12	11
V	2	0	2	2
VI	1	0	1	1
Total	120	7	113	98

Estimation of the total number of the human V_H segments

To assess how many of the total V_H segments in the human genome
have been cloned, quantitative hybridization analysis was
performed using the four YAC clones, namely Y6, Y21, 17H and Y20.
The numbers of V_H bands in an equimolar mixture of the four YACs,
and human germline (FLEB14-14) DNA were compared by Southern
hybridization using probes specific to V_H family. The estimated
total number of V_H segments on chromosome 14 is 113 including one
V_{H-VI} segment while we have already isolated at least 98 V_H
segments in YACs and cosmids (Table 2). All the results taken
together, we estimate that the total number of the human V_H
segments on chromosome 14 may be no more than 120.

References

(1) Croce C. M. et al.   Proc. Natl. Acad. Sci. U.S.A. 76, 3416-3419 (1979)

(2) Cherif, D. & Berger, R.   Genes Chromosomes & Cancer 2, 103-108 (1990)

(3) Matsuda, F. et al.   EMBO J. 9, 2501-2506 (1990)

(4) Shin, E. K. et al.   EMBO J. 10, 3641-3645 (1991)

(5) Kodaira, M. et al.   J. Mol. Biol. 190, 529-541 (1986)

(6) Lee, K. H., Matsuda, F., Kinashi, T., Kodaira, M. & Honjo, T.   J. Mol. Biol. 195, 761-768 (1987)

(7) Lorenz, W., Straubinger, B. & Zachau, H. G.   Nucleic Acids Res. 15, 9667-9676 (1987)

(8) Schroeder, H. W. Jr., Hillson, J. L. & Perlmutter, R. M.   Science 238, 791-793 (1987)

(9) Schroeder, H. W. Jr. & Wang, J. Y.   Proc. Natl. Acad. Sci. U.S.A. 87, 6146-6150 (1990)

(10) Pascual, V. et al.   J. Clin. Invest. 86, 1320-1328 (1990)

(11) Sanz, I., Dang, H., Takei, M., Talal, N. & Capra, J. D.   J. Immunol. 142, 883-887 (1989)

(12) Dersimonian, H., Schwartz, R. S., Barrett, K. J. & Stollar, B. D.   J. Immunol. 139, 2496-2501 (1987)

Restricted V_H Gene Usage and Generation of Antibody Diversity in Rabbit

Katherine L. Knight and Chander Raman

Departement of Microbiology and Immunology Stritch School of Medicine, Loyola University, Chicago Maywood, IL 60153, USA

INTRODUCTION

Through the years, the rabbit has offered researchers a diverse and rich source of antibodies. At the same time, it left us with an enigma regarding the presence of genetic markers in the variable region of the heavy chain. In the process of resolving this enigma, we've expanded our understanding of the unique ways in which immune systems create antibody diversity. Specifically, in the rabbit we have found that there is preferential use of $\underline{V_H1}$ in VDJ gene rearrangements and that antibody diversity occurs by gene conversion in the V_H region and by somatic mutation in the D region.

PROBLEM OF V_H ALLOTYPES

The V_H region allotypes, a1, a2, and a3, are controlled by allelic genes, and they are present on 80 to 90% of serum Ig molecules (reviewed by Kindt and Capra, 1984; Mage et al., 1984). The 10 to 20% of Ig molecules that do not have these V_Ha allotypes are called V_Ha-negative. Concerning the V_H genes, we know that the rabbit heavy chain chromosomal region contains multiple V_H genes, perhaps 200 or so (Gallarda et al, 1985). And, based on nucleotide sequence analysis, we estimate that approximately 50% of them are functional (Currier et al., 1988). We assumed that most of the functional V_H genes would be used in VDJ gene rearrangements. And because most serum Ig molecules have V_Ha allotypic specificities, we expected that most of the V_H genes would encode V_Ha allotypes. So, the question arose as to how these V_H allotypes could be inherited as alleles. One would expect that during the generation of rabbits a cross-over might have occurred among the V_H genes at meiosis. If so, then the $\underline{a}^1$ and $\underline{a}^2$ V_H genes would be linked to each other and inherited as linked genes rather than as alleles. So the question we asked was: What is the genetic basis for the allelic inheritance of the allotypes?

B-CELL LEUKEMIAS IN TRANSGENIC RABBITS

To begin to understand the genetic basis for these allotypes, we needed VDJ gene rearrangements. Normally one might use myelomas or hybridomas as a source of VDJ genes. Yet, because rabbit myelomas have not been identified and because at the time we began these experiments, we did not have stable rabbit-mouse hybridomas, we decided to develop transgenic rabbits using c-myc conjugated to the heavy chain enhancer. Adams et al., (1985) showed that transgenic mice carrying c-myc driven by the heavy chain enhancer developed B-lymphomas. When we used a similar construct to make transgenic rabbits, we found that they developed B cell leukemia within 3 weeks of birth (Knight et al., 1988). These leukemias were polyclonal and allowed us to clone multiple VDJ gene rearrangements.

After cloning and sequencing several of the VDJ genes from the leukemic rabbits, we were quite surprised to find that the V regions of these VDJ gene rearrangements were essentially identical to each other if they were derived from rabbits of the same allotype (Becker et al., 1990). This suggested to us that the leukemic B cells from rabbits of the same allotype had all used the same V_H gene. To determine which germline V_H gene was used by the leukemic B cells, we began to map the V_H chromosomal region, starting at the D-proximal end. We cloned the 3'-most V_H genes from the three different haplotypes, a^1, a^2 and a^3 (Knight and Becker, 1990). The V_H genes are numbered starting at the 3'-end with V_H1. The nucleotide sequences of several of the 3'-most V_H genes were determined, and when we compared these sequences with the sequences of the VDJ genes cloned from the leukemic rabbits, we found that the sequences of the leukemia-derived VDJ genes were identical to the sequence of the 3'-most gene, V_H1 from the respective allele. The fact that the sequence of the V_H1 gene was identical to the sequence of the genes utilized in the B-cell leukemia suggested to us that the B-cell leukemias had used V_H1 in their VDJ gene rearrangements.

V_H1 GENE USAGE IN B-CELL LEUKEMIAS

To determine if germline V_H1 was used in the VDJ gene rearrangements of the B-cell leukemias, we screened a cosmid library from one of the leukemic rabbits, identified a VDJ gene rearrangement, and compared it with germline V_H1 and with the region immediately upstream of V_H1. Indeed, we found that the region upstream of the VDJ gene from the leukemia was identical to that upstream of germline V_H1, showing that V_H1 had been used in the VDJ gene rearrangement. This observation led us to the idea that there could be restricted use of V_H1 in VDJ gene rearrangements. Based on nucleotide sequence analysis, we know that: The V_H1 gene from the a^1 allele encodes a prototypic a1 allotype V_H region, the V_H1 gene from the a^2 allele encodes a prototypic a2 allotype molecule, and likewise, the 3'-most gene V_H1 gene from the a^3 allele encodes a prototypic a3 molecule. Since the V_H1 genes are truly allelic in the different haplotypes, and since they encode the prototypic V_Ha allotype, we hypothesized that the allelic inheritance of V_Ha allotypes could be explained by preferential usage of V_H1 in VDJ gene rearrangements (Knight and Becker, 1990).

LOSS OF V_Ha ALLOTYPE EXPRESSION CORRELATES WITH LOSS OF V_H1 IN MUTANT ALICIA RABBITS

If the allelic inheritance of the V_H allotypes was the result of preferential usage of V_H1 then a rabbit with a defective V_H1 might have altered expression of the V_Ha allotypes. Kelus and Weiss (1986) identified a mutant rabbit, Alicia that was genotypically a^2/a^2, but phenotypically it had primarily V_Ha-negative Ig molecules. To determine if the Alicia rabbit had a defect in V_H1, we constructed a cosmid library from Alicia DNA and cloned the region around V_H1. We compared the restriction map of this region with the restriction map surrounding V_H1 in a normal a2 rabbit (Fig. 1). We found that in the Alicia DNA there was a 10 Kb deletion that included V_H1 as well as V_H2; V_H2 being a non-functional gene. Therefore, the loss of the V_H1 gene correlated with the loss of a2 allotype molecules in Alicia rabbits, further supporting the idea that the allelic inheritance of V_Ha allotypes may be due to preferential usage of V_H1 in VDJ gene rearrangements.

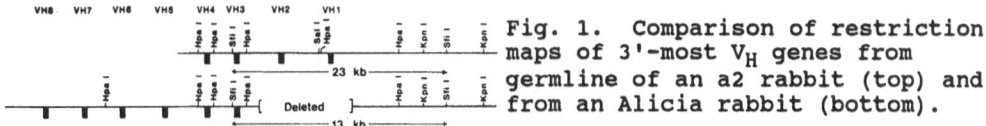

Fig. 1. Comparison of restriction maps of 3'-most V_H genes from germline of an a2 rabbit (top) and from an Alicia rabbit (bottom).

V_H1 USAGE IN STABLE RABBIT/MOUSE HYBRIDOMAS

In a recent study to examine V_H gene usage in normal rabbits, we developed several stable rabbit-mouse hybridomas to further examine the use of V_H genes by normal rabbit B cells. We fused mouse Sp2/0 myeloma cells with rabbit spleen cells and selected hybrids that secreted rabbit V_Ha allotype Ig. We cloned the VDJ gene rearrangements from eight of these hybridomas and then examined the restriction map 5' of the VDJ gene to determine which V_H gene was used in the VDJ gene rearrangements. We found that in each of the 8 hybridomas, V_H1 was utilized in the VDJ gene rearrangements (unpublished data). Taken together, the data from the hybridomas and from the Alicia rabbit convinced us that V_H1 is preferentially utilized.

IS V_H1 PREFERENTIALLY REARRANGED?

Does the restricted use of V_H1 reflect preferential rearrangement and expression of V_H1, or selective expansion of B cells that express surface V_H1 Ig? To distinguish between these two possibilities, we chose to PCR amplify VDJ gene rearrangements from RNA and from DNA of newborn to 10 day old rabbits. The RNA was taken from bone marrow, while the DNA came from bone marrow and spleen. We chose young rabbits for this experiment because pre-B cells are generally abundant in bone marrow of newborn animals. Also, the VDJ genes would not yet have undergone somatic diversification. This would allow us to unequivocally identify the V_H gene used in the rearrangement. The genes were sequenced and we analyzed a total of 25 sequences from cDNA and 40 sequences from genomic DNA (Table 1).

Table 1.   Rearranged V_H Genes in Newborn to 10 day old rabbits.

Utilized V_H gene	RNA	DNA
$\underline{V_H}1$	19	27
$\underline{V_H}x$	3	1
$\underline{V_H}y$	1	4
$\underline{V_H}z$	2	0
Other	0	8 (6 new V_H genes)

From the 25 sequences cloned from the RNA, we found a total of four different V_H gene sequences (Friedman and Knight, unpublished data). Nineteen of the 25 sequences, approximately 75%, were identical to $\underline{V_H}1$. The remaining six were identical to $\underline{V_H}x$, $\underline{V_H}y$ and $\underline{V_H}z$. These three genes encode the 10-20 % of serum Ig molecules that do not have the V_Ha allotype, that is the V_Ha-negative molecules (Short et al., 1991). To determine if there were genes other than these four that were rearranged, we examined the DNA. In this case we again found that 27, or most, of the VDJ gene rearrangements were $\underline{V_H}1$. However, we also found six other genes which we have never seen before, but which encoded a functional V_H region. If the restricted use of $\underline{V_H}1$ is due to the selective expansion of $\underline{V_H}1$-expressing B-cells, then V_H genes other than V_H1 would have to be rearranged and expressed. Although we found several new V_H genes that were rearranged, we did not find any of them expressed in mRNA.

To rule out the possibility that these novel genes were expressed at low levels, we attempted to directly PCR amplify them from the cDNA using PCR primers specific for these genes. Indeed, no PCR product corresponding to an expressed gene was found. Thus, we conclude that preferential use of $\underline{V_H}1$ is due to its preferential rearrangement.

GENERATION OF ANTIBODY DIVERSITY

Somatic Gene Conversion of V_H regions

Given the limited use of V_H genes and consequently, the limited extent to which combinatorial joining can contribute to antibody diversification, how is antibody diversity generated? To examine the generation of antibody diversity, we needed to compare the nucleotide sequence of the diversified V_H gene with its germline counterpart (Becker and Knight, 1990). The nucleotide sequence of FR1 of a diversified $\underline{V_H}1$-utilizing VDJ gene is compared in Figure 2 with the sequence of germline $\underline{V_H}1$. The diversified gene differs from the germline gene by multiple nucleotides. Most striking is the insertion of a codon at position two in the VDJ gene. Such a codon insertion is not characteristic of the hypermutation process for generating antibody diversity in mouse (reviewed in French et al., 1989). Instead, it is similar to what one might expect for a gene conversion event as Reynaud et al., (1989) showed for chicken Ig genes. If rabbit VDJ genes diversify by gene conversion, then we should be able to identify donor V_H genes with sequences identical to the diversified regions.   We searched our Gene-bank of rabbit germline V_H gene sequences for ones with sequences similar to the diversified region of the VDJ gene. We identified a gene, $\underline{V_H}6$, that has a sequence identical to that of the diversified region of the VDJ gene (Fig.2)(Becker and Knight, 1990). We suggest that this region of the

VDJ gene was diversified by gene conversion using V_H6 as the donor gene.

```
        FR 1
        1                                       10
V_H1    CAG \\\ TCG TTG GAG GAG TCC GGG GGA GAC CTG GTC AAG CCT GGG GCA
VDJ     --- GAG CA- C-- -T- --- --- ---     --- -G- --- --- C-- --- -A- -G-
V_H6    --- GAG CA- C-- -T- --- --- ---     --- -G- --- --- C-- --- -A- -G-
```

Fig. 2. Somatic gene conversion. Regions of diversification of a V_H1-utilizing VDJ gene and potential donor gene, V_H6. Numbers refer to codon numbers. Sequence of the utilized V_H1 gene is shown (top).

Another example of gene conversion is from a recent study analyzing the rabbit hybridomas which I previously mentioned. In one of those hybridomas there were multiple nucleotide differences between the VDJ gene and the utilized germline V_H gene, V_H1 (Fig. 3).

```
        Leader Intron
V_H1    cacagacagtgtgagtgacag/tacctgaccatgtcgtctgtgttttcag  GT GTC
VDJ     -c----g----------------c-t------------t---------g---  -- ---
V_H8    -c----g----------------c-t------------t---------g---  -- ---
```

```
        FR1
        1                                       10
V_H1    CAG TGT CAG /// TCG TTG GAG GAG TCC GGG GGA GAC CTG GTC AAG
VDJ     --- --- --- GAG CA- C-- -T- --- --- --- --- -G- --- --- C--
V_H8    --- --- --- GAG CA- C-- -T- --- --- --- --- -G- --- --- C--
```

```
                                20
V_H1    CCT GGG GCA TCC CTG ACA CTC ACC TGC ACA GCC TCT
VDJ     --- -A- -G- --- --- --- --- --- --- --- --T ---
V_H8    --- -A- -G- --- --- --- --- --- --- --- --T ---
```

Fig. 3. Somatic gene conversion. Regions of diversification of a V_H1-utilizing VDJ gene and potential donor gene, V_H8. Sequence of the utilized V_H1 gene is shown (top). Numbers refer to codon numbers.

A similar search of our rabbit Gene-bank for these sequences led us to identify that the entire sequence from the leader-intron, through FR1 was identical to the sequence of the germline gene, V_H8. We conclude that the VDJ gene was diversified by somatic gene conversion using V_H8 or a gene very similar to it, as the donor gene. Thus we conclude that rabbit VDJ genes diversify extensively by somatic gene conversion; this observation represents the first clear example of somatic gene conversion within mammals.

Somatic Mutation of D regions

While much of our work has concentrated on the generation of antibody diversity in the V region, we have also carefully analyzed the D region. By examining the VDJ gene rearrangements from the spleen of adult rabbits, we found the D regions were extremely different from one VDJ gene to another (DiPietro et al., 1990). To study the diversification of D regions we need to examine VDJ gene rearrangements early in ontogeny, prior to extensive diversification. We analyzed the D regions from VDJ genes cloned from 3, 6 and 9 week old rabbits (Short et al., 1991) (Fig. 4). At three weeks of age the D regions are relatively undiversified. However, at 6 weeks of age, we begin to see some diversification of the D regions and by 9 weeks, the D regions are quite diversified, to the point where the germline origin of the D regions can only be identified with difficulty. In these diversified D regions, we find no evidence for codon insertions. And, we do not believe these D regions are diversified by somatic gene conversion. Instead, diversification of D regions appears to be an accumulation of somatic point mutations. Therefore, we suggest that the D regions diversify by a somatic mutation process, perhaps by a hypermutation process similar to that which diversifies mouse V_H and V_L genes.

CLONE	D-REGION	GERMLINE D
3 WKS		
2	------------------------------------	D2b
3	----------------------------ACC---	D1a
4	-----------------------	D3
5	------------C-	D3
6	--------------------------------	D2b
8	--------------------------------	D2b
9	-----------------------------------	D2b
10	--------------------------------------	D2
6 WKS		
21	--------------A----------C----C-----A--	D2a
22	----------------C------	D3
23	-A--------G---C-------	D3
24	----------	D3
25	--TAG-------G--	D3
26	------G----------------	D3
28	-G-----TA--------A---------------	D2a
29	------------------AA----C-----	D2b
30	--------------------	D1a
9 WKS		
41	--G-A------G------	D2b
42	--------G----T-----	D2a
43	------AT-A--AC----CC------GG-	D2b
44	-----C-C-------AT----GG--GC-T---	D2a
45	-G------C----G----------GG-T-------	D2a
46	---C--GA-G--G---C-------G--	D3
47	-------G-----GA-G-----	D3
48	----GGC-G--------ATAGTGCG-	D2a
49	-C----AT----GG-TT----	D2b

Fig. 4. D region diversity of VDJ genes from 3-week, 6-week and 9-week-old rabbits. The nucleotide sequences of D regions from VDJ genes from PCR amplified splenic-derived cDNA are compared to germline D genes. The germline D gene used in the rearrangement is indicated at the right; dashes indicate identity between the cDNA clone and the corresponding germline D gene.

To summarize our findings, we believe that...One, the rabbit preferentially uses one V_H gene, $\underline{V_H}1$. Two, preferential usage results from preferential rearrangement of $\underline{V_H}1$, and three, antibody diversity is generated in the V region by gene conversion and in the D region by somatic mutation. The V region may also be diversified by somatic mutation, but once the V region is diversified by somatic gene conversion, we cannot know if it is further diversified by point mutations.

PERIPHERAL CD5[+] B-CELLS IN RABBIT

Recently, we developed a monoclonal antibody to rabbit CD5 molecules and tested the B-cell population for CD5[+] B-cells (Raman and Knight, in press). In a two-color immunofluorescence experiment with adult rabbit spleen cells stained with anti-CD5 and anti-L chain antibodies, we made the surprising finding that essentially all Ig[+] cells are also CD5[+] (Fig. 5).

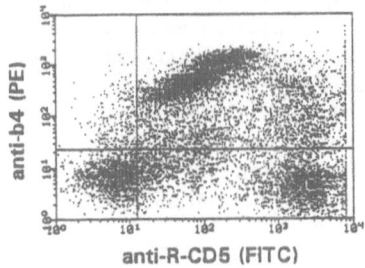

anti-R-CD5 (FITC)

Fig. 5. Two-color FACS analysis of adult rabbit spleen cells stained by indirect immunofluorescence with anti-b4 κ-light chain allotype antibodies and monoclonal anti-rabbit CD5 antibody.

These data indicate that essentially all rabbit B cells are CD5[+]. If these CD5[+] B cells belong to the same lineage as CD5 B cells in mice, they probably develop early in ontogeny and are maintained in the periphery by a self-renewing proces. We suggest that rabbit B cells, like chicken B cells (McCormack and Thompson, 1990), develop early in ontogeny and are maintained throughout life by a self-renewing proces. Further, we suggest that rabbits develop a primary antibody repertoire by somatic gene conversion and/or mutation, in an antigen independent manner. Additional diversification probably occurs following stimulation with antigen.

CONCLUSIONS: COMPARISON OF SPECIES

To appreciate the unique immune system of the rabbit, we can compare it to other species. The rabbit, even though it has many V_H genes that appear by nucleotide sequence analysis to be functional, rearranges predominantly one of these, V_H1. Chicken also rearranges only one V_H gene — but it has only one functional V_H gene — all of the others are pseudogenes (Reynaud et al., 1989). In contrast, mouse and human have multiple functional V_H genes and they both rearrange many of them, giving rise to a large number of VDJ rearrangement combinations.

In terms of the generation of antibody diversity, mice and humans- species that utilize multiple V_H genes in VDJ gene rearrangements- diversify their Ig genes predominantly by somatic hypermutation; chickens, which have only one functional V_H gene, diversify V_H and V_L by gene conversion, and rabbits, in spite of having multiple apparently functional V_H genes, use primarily one of them, but use both gene conversion and somatic mutation to diversify their VDJ genes. Clearly, nature offers many solutions for generating antibody diversity. And, we are left to wonder how many other mechanisms may be used in species not yet studied.

REFERENCES

Bendelac A, Matzinger P, Seder R, Paul WE, Schwartz RH (1992) Activation events during thymic selection. J Exp Med. 175:731

Borgulya PH, Kishi H, Müller U, Kirberg J von Boehmer H (1991) Development of the CD4 and CD8 lineage of T cells: instruction versus selection. EMBO J 10:913

Borgulya P, Kishi H, Uematsu Y, von Boehmer H (1992) Exclusion and inclusion of alpha and beta T cell receptor alleles. Cell 69:529

Guidos CJ, Weissman IL, Adkins B (1989) Intrathymic maturation of murine T lymphocytes from CD8+ precursors. PNAS USA 86:7542

Huesmann M, Scott B, Kisielow P, von Boehmer H (1991) Kinetics and efficacy of positive selection in the thymus of normal and T cell receptor transgenic mice. Cell 66:533

Hugo P, Boyd RL, Waanders GA, Petrie HT,Scollay R (1990) Timing of deletion of autoreactive Vβ 6+ cells and down-modulation of either CD4 or CD8 on phenotypically distinct CD4+8+ subsets of thymocytes expressing intermediate or high levels of T-cell receptor. Int Immunol 3:265

Kisielow P, von Boehmer H (1990) Negative and positive selection of immature thymocytes: timing and role of the ligand for αβ T cell receptor. Sem Immunol 2: 5

Malissen M, Trucy J, Jouvin-Marche E, Cezanave PA, Scollay R, Malissen B (1992) Regulation of TCR a and b gene allelic exclusion duriong T-cell development. ImmToday 13:315

Philpott KL, Viney JL, Kay G, Rastan S, Gardiner EM, Chae S, Hayday AC, Owen MJ (1992) Lymphoid development in mice congenitally lacking T cell receptor αβ-expressing cells. Science 256:1448

Ohashi PS, Pircher H, Burki K, Zinkernagel RM, Hengartner H (1990) Distinct sequence of negative or positive selection implied by thymocyte T cell receptor densities. Nature 346:861

Robey EA, Fowlkes BJ,Pardoll DM (1990) Molecular mechanisms for lineage commitment in T cell development. Sem Immunol 2:25

Shortman K, Vremec D, Egerton M (1991) The Kinetics of T Cell Antigen Receptor Expression by Subgroups of CD4+8+ Thymocytes: Delineation of CD4+8+3++ Thymocytes as Post-selection Intermediates Leading to Mature T Cells. J Exp Med 173:323

Swat W, Dessing M, Baron A, Kisielow P, von Boehmer H (1992) Phenotypic changes accompanying positive selection of CD4+8+ thymocytes. Eur J Immunol, in press

Teh HS, Kisielow P, Scott B, Kishi H, Uematsu Y, Blüthmann H, von Boehmer H (1988) Thymic major histocompatibility complex antigens and the specificity of the αβ T cell receptor determine the CD4/CD8 phenotype of T cells. Nature 335:229

Testi R, Phillips JH, Lanier LL, (1988) Constitutive expression of a phosphorylated activation antigen (Leu 23) by CD3bright human thymocytes. J Immunol 141:2557

Turka LA, Schatz DG, Oettinger MA, Chun JJ, Gorka C, Lee K, McCormack WT, Thompson CB (1991) Thymocyte expression of RAG-1 and RAG-2: termination by T cell receptor cross-linking. Science 253:778

von Boehmer H (1986) The selection of the α,β heterodimeric T cell receptor for antigen. Imm Today 7:333

von Boehmer H (1990) Developmental biology of T-cells in T cell receptor transgenic mice. Ann Rev Immunol 8:531

von Boehmer H, Kishi H, Borgulya P, Scott B, van Ewijk W, Teh HS, Kisielow P (1989) Control of T cell development by the TCR αβ for antigen. Cold Spring Harbor Symp Quant Biol 54:111

Evolutionary Origins of Immunoglobulin Gene Diversity

G. W. Litman, J. P. Rast, M. A. Hulst, R. T. Litman,
M. J. Shamblott[*], R. N. Haire, K. R. Hinds-Frey, R. D. Buell,
M. Margittai, Y. Ohta, A. C. Zilch, R. A. Good, and C. T. Amemiya[**]

*Departement of Pediatrics, University of South Florida, 801 Sixth Street S., St.
Petersburg, FL USA 33701
University of Maryland, C.O.M.B., 600 East Lombard Street, Baltimore, MD USA
21202[*]
Lawrence Livermore National Laboratory, Division of Biomedical Science,
Livermore, CA USA 94551[**]*

INTRODUCTION

During the past several years, our laboratory has identified immunoglobulin genes in species that represent major phylogenetic groups of vertebrates (Litman et al. 1989). Taken together, the descriptions of immunoglobulin genes in chondrichthyes (Litman et al. 1985a; Hinds and Litman 1986; Kokubu et al 1988a), bony fishes (Amemiya and Litman 1990, 1991), amphibians (Schwager et al. 1988b, 1989; Haire et al. 1990, 1991), avians (Reynaud et al. 1987, 1989) and mammals (Blackwell and Alt 1988), as well as ongoing studies in our laboratory involving other phylogenetically important species are revealing, the overall patterns in the evolution of immunoglobulin gene structure and diversity. In all of these species, segmental organization and DNA sequence-mediated selective rearrangement in somatic tissues, the most distinctive features of the immunoglobulin gene system, are conserved. The nucleotide and predicted peptide sequences of individual V_H, V_L, D_H, J_H, J_L and some C_H exons, in most instances, are highly conserved as are the recombination signal sequences (RSSs) flanking the V_H, V_L, D_H, J_H and J_L segmental elements (Litman et al. 1985a; Hinds and Litman 1986; Kokubu et al. 1988a). In addition, the exon-intron organization of the C_H exons in *Heterodontus* (horned shark), a phylogenetically primitive vertebrate, as well as other chondrichthyes (see below), is equivalent to that described in mammals (Kokubu et al. 1988b). To a certain degree, some of these findings are not unexpected since in many cases the genes in lower vertebrates were detected by cross-hybridization with a mammalian immunoglobulin heavy chain variable region gene-specific probe (Litman et al. 1983, 1985a). Typically, any gene detected using this procedure would have to be at least 60% related at the nucleotide sequence level. Although it is informative to compare the sequences of these genes to one another in an overall sense, it must be emphasized that the V_H genes are members of extensively diversified multigene families and few guidelines can be applied to determine whether similarities and/or differences in gene sequence reflect orthologous evolutionary relationships (Litman et al. 1985b). For this reason, the emphasis in this discussion is on the organization of these gene segments, the relationships of gene organization to function and the phylogenetic development of mechanisms which generate genetic diversity.

Unique Patterns of Gene Organization are Associated with Major Pathways of Evolution

As indicated above, the individual segmental elements in different vertebrate species are remarkably related at the DNA and predicted peptide sequence levels, even in cases where the segmental elements, e.g., D_H and J_H, have not been selected by cross-hybridization directly but rather have been detected by sequencing of an adjacent DNA segment(s), i.e., the gene segment(s) is in close chromosomal proximity to a gene that was localized by cross-hybridization or has been identified in a cDNA (detected with a probe complementing different segmental elements) (Amemiya and Litman 1990; Hinds and Litman 1986; Harding et al. 1990a). The genomic organization of these genes in different vertebrates, however, varies markedly. In *Heterodontus*, immunoglobulin heavy chain gene segmental elements (regions) are in ~20,000 bp "clusters" consisting of a single V_H, two D_H, one J_H and one C_H region, encoded by six exons (Hinds and Litman 1986; Kokubu et al. 1988a, b). We have resolved the complete nucleotide sequences of such a cluster and diagrammatic representation of the germline and somatically rearranged forms are shown in Fig. 1. Hundreds of these clusters are present in the genome of an individual, contrasting markedly with the organization of mammalian immunoglobulin genes, which are organized at a single locus, although some human V_H segments have been shown to reside outside the primary locus (Matsuda et al. 1990). Preliminary studies using *in situ* hybridization have shown immunoglobulin gene clusters in both *Heterodontus* and *Raja* (skate) to be present at different chromosomal locations (C. Amemiya, J. Rast and G. Litman unpublished observations). The particular features of the detection system suggest strongly that multiple clusters also exist in close proximity. In *Heterodontus*, antibody gene diversity presumably is generated by the same rearrangement mechanism(s) that is employed by higher vertebrate species, but recombination does not occur between segmental elements in different clusters. First detected in *Heterodontus*, this form of gene organization also has been found in representatives of a different elasmobranch order (Harding et al. 1990a) as well as in a chimaera, *Hydrolagus* (ratfish), a species that represents an independent evolutionary line of the chondrichthyes (J. Rast, C. Amemiya and G. Litman unpublished observations).

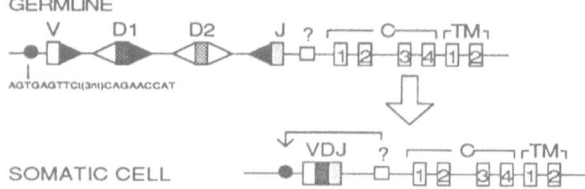

Fig. 1. Germline organization of a cartilaginous fish ("cluster")-type immunoglobulin heavy chain gene locus and the specific rearrangement product which would occur in a somatic cell (B lymphocyte). Variable (V), diversity (D) and joining (J) segments only rearrange within clusters. The constant region (C) consists of six exons, including two transmembrane (TM) exons. ▶◀ reflect 22/23 nucleotide spacers in recombination signal sequences; ▷◁ represent 12 nucleotide spacers in recombination signal sequences. A decamer-(spacer)-nonamer sequence (●), closely resembling that associated with mammalian T cell antigen receptor Vβ is designated in uppercase letters, where identical to the mammalian consensus vs. lowercase letter (t) where these differ. The spacer distance is three nucleotides ($3nt$). By analogy, an enhancer(s) (☐) may (?) be present in the heavy chain J-C intron.

A second, markedly different heavy chain constant region isotype has been described in *Raja* (Kobayashi et al. 1984), and we have characterized it at the gene level (Harding et al. 1990b). Recent studies of the segmental organization of this gene, which we have designated IgX, have shown it to exhibit the same pattern of segmental organization associated with the other IgM (μ)-type elasmobranch heavy chain genes (M. Hulst and G. Litman unpublished observations) and likewise, to be chromosomally dispersed (C. Amemiya, J. Rast and G. Litman unpublished observations). The V_H segment of IgX-type genes differs markedly at the nucleotide and predicted peptide levels from the V_H segments associated with IgM clusters (Harding et al. 1990b). In early development, both IgM and IgX (encoded at separate loci) are expressed on single lymphocytes, in contrast to allelic exclusion that is associated with immunoglobulin gene expression in higher vertebrates. These findings suggest that in one major group of vertebrates a unique form of immunoglobulin gene organization and regulation is found, which has not been observed in any other vertebrate species. The relationship between this type of gene organization and the generation of antibody diversity will be discussed later in this report.

A second, unusual feature of immunoglobulin genes in cartilaginous fishes is the germline-joining of segmental elements. Approximately one-half of the clusters in *Heterodontus* (Kokubu et al. 1988a; Harding et al. 1990a; Fig. 2) are joined to some extent. In all cases thus far examined, fully joined (VDJ) segments maintain a correct reading frame (Kokubu et al. 1988b, unpublished observations); if joining were random, only one sequence in three would be expected to correctly encode the J_H segment, which has a well-conserved peptide structure. For $V_H D_H$-J_H-joined genes in *Heterodontus* (Kokubu et al. 1988a) and $V_H D_1$-$D_2 J_H$-joined genes in *Raja erinacea* (little skate) (Harding et al. 1990a; see below), the RSSs, found at the 3′ of D_H in *Heterodontus* and at the 3′ of D_1 and 5′ of D_2 in *Raja*, are similar to those found in the non-joined clusters that have been shown to be expressed, suggesting that these can potentially rearrange; none of the germline-joined genes contain stop codons, frameshift mutations, etc. These alternative forms of gene organization are shown in Fig. 2. The V_H coding segments of joined genes are not related appreciably more to one another than to the non-joined clusters (Kokubu et al. 1988a; K. Hinds-Frey and G. Litman unpublished observations). However, the D regions of joined genes, by comparison to genomic D elements found in non-joined clusters, differ markedly, suggesting that these may be encoding unique "predetermined" specificities that could be important components of defense in these species. This form of organization may represent a "regression" from the segmental rearrangement system and may be of decided survival advantage in species where the kinetics of antigen driven cell selection/proliferation are not as efficient or otherwise differ from that observed in higher vertebrates. The chimaera, *Hydrolagus*, also exhibit germline-joining and in one case, unequivocal evidence has been obtained (using PCR) that a cDNA transcript of a joined gene is expressed (J. Rast, C. Amemiya and G. Litman unpublished observations). Similarly, it was noted recently that all of the members of a major family of light chain genes, recovered from a *Raja* genomic DNA library, are "joined" (M. Shamblott, M. Hulst and G. Litman unpublished observations), in marked contrast to the major expressed family of light chain genes in *Heterodontus*, which are organized in a cluster configuration (V_L-J_L-C_L) but show no evidence of "germline-joining", despite an exhaustive effort to detect such a configuration (Shamblott and Litman 1989). Although studies are not yet completed, we have been unable to detect an unmodified transcript of the *Raja* germline-joined light chain genes, suggesting that there are other members of this family or the joined elements are targets of somatic change. Based on comparisons of light chain peptide sequences and the sequences of cDNAs that are related to the germline-joined gene, it is likely that *Raja* possesses additional light chain gene families (M. Hulst, M. Shamblott and G. Litman unpublished observations) as does *Heterodontus* (J. Rast, M. Shamblott and G. Litman unpublished observations).

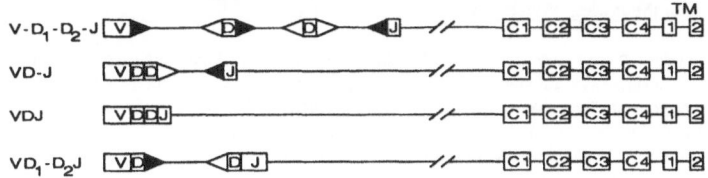

Fig. 2. Alternative patterns of cluster-type gene organization in cartilaginous fish. From top: cluster-type (*V-D₁-D₂-J*) *Heterodontus*, germline-joined (*VD-J*) *Heterodontus*, germline-joined (*VDJ*) *Heterodontus* and germline-joined (*VD-DJ*) *Raja*. *V*=variable, *D*=diversity, *J*=joining, *C*=constant region exon 1-4, *TM* (1-2)=constant region transmembrane exons. ▶◀ recombination signal sequences with 22/23 nucleotide spacers; ▷◁ recombination signal sequences with 12 nucleotide spacers. Intersegmental distances not to scale

A third, unique feature of the organization of heavy chain genes in a representative cartilaginous fish involves the absence of the regulatory octamer (ATGCAAAT or ATTTGCAT) of the B cell-specific immunoglobulin promoter, an invariant component of immunoglobulin gene expression in higher vertebrates (see below). The absence of this sequence motif at its characteristic position 5′ of the "TATA"-equivalent sequence in cartilaginous fishes suggests that the tissue specific regulation of immunoglobulin gene expression in these species may be unique. The decamer/nonamer sequence (Fig. 1) that is associated with regulation of the β T cell antigen receptor (TCR) and is homologous to a CRE (Lee and Davis 1988) has been detected upstream of the initiation codon in *Heterodontus* immunoglobulin heavy chain genes (Kokubu et al. 1988a; E. Davidson personal communication). It should be noted, however, that *Heterodontus* light chain genes possess the octamer sequence in a 5′ position that corresponds to the location of this regulatory sequence in higher vertebrates (Shamblott and Litman 1989), whereas in *Raja* at least some light chain genes lack an octamer. Similarly, higher vertebrate enhancer-like sequences have been identified in the J$_L$-C$_L$ intron of *Heterodontus* (M. Shamblott and G. Litman unpublished observations). A recently completed *Heterodontus* ~9 kb J$_H$-C$_H$ intron sequence exhibits evidence for several immunoglobulin enhancer-like sequences but none of these match the established "mammalian" consensus to the same degree as that observed with other mammalian immunoglobulin enhancer sequences. In summary, there are major differences in the organization and presumably, in the regulation of immunoglobulin heavy (and light) chain genes in the most phylogenetically primitive jawed vertebrates. Strong interspecies sequence identity in V$_H$, the exon organization of C$_H$ and the differential processing of 3′ secretory and transmembrane sequences are characteristics of higher vertebrate immunoglobulin and specifically μ-type heavy chains. However: 1) the close linkage of segmental elements, 2) presence of two D$_H$ segments and 3) absence of an immunoglobulin octamer with the presence of a TCR nonamer/decamer (in *Heterodontus*) suggest that the rearranging genes in the cartilaginous fish also are related closely (in terms of overall organization) to TCRs. It appears that the heavy chain locus reflects properties of both TCR and immunoglobulin gene loci (Fig. 3). In the absence of evidence either for or against the presence of TCRs in cartilaginous fishes, one interpretation of this chimeric gene structure/organization is that the cluster-type immunoglobulin heavy chain gene(s) in *Heterodontus* reflects the structure of the ancestral genes from which the immunoglobulin and TCR systems evolved. The light chain locus, on the other hand, is more mammalian-like, despite the dominant, close linkage form of gene organization. In subsequent phylogenetic development, a recombination event may have imposed the octamer-type transcriptional regulation on the heavy chain gene system.

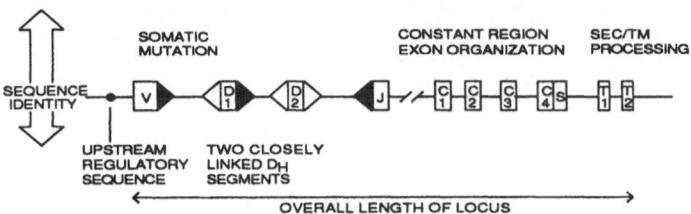

Fig. 3. Relationships of "immunoglobulin" gene clusters in cartilaginous fish to mammalian immunoglobulin vs. T cell antigen receptors. Notations *above* the horizontal axis of the figure imply immunoglobulin-like characteristics and *below* the figure imply T cell antigen receptor-like characteristics. Overall, the V-D-J segmental elements share a greater degree of *sequence identity* with immunoglobulin than with T cell antigen receptors, reflected in the length of the respective vertical arrows. The exons of a μ-type constant region (shown after —//—) are indicated as constant region (*C*) 1-4; C4 contains the secretory (*S*) sequence and two transmembrane exons (*T1-2*). ▶◀ recombination signal sequences with 22/23 nucleotide spacers; ▷◁ recombination signal sequences with 12 nucleotide spacers. Intersegmental distances not to scale

Immunoglobulin Gene Homologs in the Jawless Vertebrates

Two important findings that may relate to the nature of immunoglobulin gene structure (if homologs exist) in the more phylogenetically primitive vertebrates (jawless vertebrates) emerge from the studies of these genes in jawed vertebrates: 1) at least four distinct sequence regions are hyperconserved: variable region FR2 and FR3, J_H and the constant region secretory/transmembrane regions and 2) close segmental linkage appears to be the phylogenetically "earlier" form of gene organization. Thus, "immunoglobulin" gene structure and organization in jawless vertebrates (lampreys and hagfishes) may be homologous to the patterns noted in the cartilaginous fishes. Alternatively, the gene structure and organization may be more closely akin to that observed in the bony fishes and more recent vertebrate radiations or the "antibodies" will not be significantly (or will be only distantly) related to immunoglobulin, as defined by the "shared" criteria indicated above. It would seem that relatedness to the cartilaginous fish pattern is the most likely; however, exhaustive efforts to detect immunoglobulin gene homologs in both lampreys and hagfishes, including the screening of cDNA libraries constructed from tissue sources in which histologically defined lymphoid-like elements are detected, have been uniformly unsuccessful using a variety of probes that complement the regions of intense interspecies (jawed vertebrates) sequence relatedness indicated above. In addition, varying stringency-degenerate primer polymerase chain reaction (PCR) analyses directed at sequence regions exhibiting elevated levels of nucleotide sequence identity have proven unsuccessful, although these same methods yield predicted products among a broad representation of jawed vertebrates, including those which did not factor in primer design. Although we (Varner et al. 1991) and others (Kobayashi et al. 1985; Hanley et al. 1990) have identified immunoglobulin-like heterodimers that exhibit limited amino acid sequence identity with other vertebrate immunoglobulins, large numbers of peptides isolated from these heterodimers lack sequence identity with either heavy or light chain genes (Varner et al. 1991). Recently, it has been shown that this molecule represents a complement polypeptide (Ishiguro et al. 1992).

These findings come as no surprise given the peculiar nature of the "antibody" response in lamprey. Antibody is recovered in an ~300-500 kD, highly labile serum fraction. Immunization with antigenically complex human group "O" erythrocytes results in high titer antibody directed exclusively to the H surface antigen; no cross-reactivity with type-A or -B cells is evident. Similarly, high titer antibody to a *Brucella abortus* vaccine, does not crossreact with "O"-erythrocytes and exhibits coincidental chromatographic (gel filtration and ion exchange) behavior with the anti-"O" antibody, i.e., there is an indication of an antibody "repertoire". The physicochemical properties of the serum protein fraction possessing antibody activity are dissimilar to the properties associated with conventional antibody, most notably, a lack of intersubunit disulfide bonding and stability of antibody found in the most phylogenetically distant, modern forms of jawed vertebrates. While homologous forms of immunoglobulin may be present in basal protochordate species, the extant jawless vertebrates may lack "immunoglobulin" that is homologous to that seen in representatives of the more recent vertebrate radiation. A completely different system of inducible, specific immunity that may not involve members of the immunoglobulin gene superfamily, as presently recognized, may exist in these species.

Restriction in the V_H Repertoire in the Early Vertebrates

Estimates of the size of the V_H repertoire in lower vertebrate species are based on comparisons of both germline and rearranged, expressed (cDNA) immunoglobulin genes. Our initial analyses of V_H gene sequences in *Heterodontus*, including genes that exhibit germline-joining, indicate that with a single exception, all V_H genes are members of a single family (Type 1), i.e., the genes possess $\geq 70\%$ absolute nucleotide identity (typically V_H genes in *Heterodontus* are ~90% related). A single exception (Type 2) is only 61% identical at the nucleotide sequence level to the most closely related members of the Type 1 *Heterodontus* V_H family (Kokubu et al. 1988a). An iterative-type screening procedure as described previously (Haire et al. 1990) has been used to attempt to detect additional V_H families in *Heterodontus*; to date, this effort has been unsuccessful. Thus, the primary antibody repertoire in *Heterodontus* appears to derive from a sizeable number of genes in the V_H-D_H-D_H-J_H configuration that are closely related members of a single gene family and a single gene cluster that appears to be the monotypic representative of a second V_H gene family.

Despite the limitation in V_H gene diversity, N-region, junctional and complementarity determining region (CDR) diversity are extensive (Kokubu et al. 1988a; K. Hinds-Frey and G. Litman unpublished observations). Using an oligonucleotide-based approach that relates "parental" clusters to specific mRNA transcripts, we recently have been able to provide unequivocal evidence for somatic variation in both CDR and FR segments of rearranged, presumably functionally translated *Heterodontus* genes. Similar observations have been made with the rearrangement products of the second (monotypic) type of V_H gene (not illustrated). While it can be argued that all of the differences that we observe arise from mutated germline variants, which we are not able to detect or are not integrated in the libraries, the number of such gene clusters that would be required is inconsistent with our current understanding of the number of the heavy chain genes in *Heterodontus* (Kokubu et al. 1987). Furthermore, the recent, successful application of direct PCR amplification of genomic V_H segments has failed to lead to the identification of any additional germline V_H sequences that would correspond to the presumably somatically-"mutated" cDNAs. Thus, *Heterodontus* exhibits two significant restrictions in the immune repertoire: 1) sequence diversity in V_H, D_H and J_H segments is considerably lower than in higher vertebrates and 2) combinatorial rearrangement between segmental elements found in different clusters has not been detected; these points are discussed further (below). In view of the marked restrictions in the D_H and J_H repertoires, this latter effect may be insignificant, i.e. no selective advantage would be realized in recombination between clusters. The near absence of variation in D_H is particularly intriguing, as these segments, which are located within 350 nt of V_H, maintain sequence stability in different chromosomal locations, while the V_H genes exhibit appreciable sequence variation in CDR2. Based on analyses of rearrangement products, it is evident that the role of D_H segments in this species is in mediating/facilitating joining rather than in contributing unique coding specificities.

The Origins of the Single Extended Locus Form of Immunoglobulin Gene Organization

Having established the presence of a rearranging immunoglobulin gene system in the cartilaginous fish, it would be expected that the system is present in all subsequent vertebrate radiations, including the bony fishes, dipnoi, amphibians, reptiles, avians and mammals. This essentially is the case; however, there are major variations in the organization of the rearranging gene system, which are remarkable for a system that is associated with only a single primary function, i.e. antibody specificity. In the bony fishes, we and others have shown that V_H genes are tandemly linked upstream of J_H (and presumably D_H segments) and the constant region exons (mammalian-type organization) (Amemiya and Litman 1990). Preliminary evidence from other bony fishes, including a representative chondrostean and holostean, indicates variation in the numbers of V_H and C_H segments (Amemiya and Litman 1991), suggesting that characterization of the heavy chain genes in these species may be worthwhile. Diversification of V_H regions into specific families has been established in teleost fish (Amemiya and Litman 1990; Ghaffari and Lobb 1991), in marked contrast to *Heterodontus* which possesses one large V_H gene family and a monotypic second V_H family. In *Latimeria*, the living coelacanth, two distinct forms of gene organization are encountered. Variable regions either are tandemly linked or are associated with a single D segment in the same close linkage arrangement as seen in *Heterodontus* but lacking the closely linked $D_H 2$ and J_H segments as well as the constant region exons. The tandemly-linked variable regions are pseudogenes which may have lost recognizable, flanking D_H segments. This has led us to propose that *Latimeria* may occupy a phylogenetic point intermediate between the cartilaginous and bony fishes, a conclusion that is consistent with many anatomical, locomotive, physiological and endocrinological considerations (C. Amemiya and G. Litman unpublished observations). Extensive diversification of V_H families is evident in *Xenopus*, with eleven distinct, highly interspersed families now identified (Haire et al. 1990, 1991). *Xenopus* also possesses three constant region isotypes (Schwager et al. 1988b; Haire et al. 1989, 1990). Extensive V_H gene families also are found in reptiles with evidence for V_H genes in four different chromosomal linkage groups (C. Amemiya and G. Litman unpublished observations).

A unique variation on the general theme of segmental rearrangement is found in avians. For both the heavy and light chain genes in *Gallus*, a single, functional gene is the target for extensive gene conversion by flanking pseudogenes (Reynaud et al. 1987, 1989). Similar findings have been made with another avian (McCormack et al. 1989). The evolutionary mechanism(s) that gave rise to this unique variation may relate to extensive chromosomal dispersal of the immunoglobulin gene loci as noted above for an ancestral reptilian species and loss of chromosomes during speciation. Alternatively, the modern chicken immunoglobulin loci may reflect a highly derived and perhaps degenerate form of the mammalian type locus. To a certain degree, the rabbit which preferentially uses its most 3' V_H element as a gene conversion acceptor and upstream V_H segments as donors (Becker and Knight 1990), may represent a functional intermediate between the avian and mammalian heavy chain arrangements. With the exception of the avians, two primary patterns of gene organization are found throughout the vertebrates: the cluster and extended, single tandem locus (mammalian-) types. Furthermore, it is interesting to note that in each of the three major types of gene organization [cluster, extended locus and single gene (avian-type)], the organization and mechanisms for generating diversity are identical for the heavy and light chain gene families, within the representative taxa. Recent studies in our laboratory suggest that cartilaginous fish possess joined as well as unjoined light chain gene families. The joined and nonjoined light chain gene families are not closely related at the nucleotide sequence level, suggesting another organizational parallel. Collectively, these observations strongly suggest that these independent loci are coevolving, which we have attributed to an obligatory need for utilization of a common recombination mechanism(s) (Shamblott and Litman 1989).

Relative Diversity of Lower Vertebrate Antibody

One of the foremost questions that has arisen from studies of immunoglobulin genes in lower vertebrate systems involves whether restrictions in antibody heterogeneity are associated with species which emerged early in phylogenetic development. Claims regarding restricted diversity in lower vertebrates have been based largely on biological observations and some immunochemical findings (Du Pasquier 1982). It now is possible to examine this broad question in terms of the large body of data which has accumulated from studies of immunoglobulin gene structure and function in phylogenetically distant species. Clearly, there are indications for limitations in V_H diversity. A single, minimally diversified (~90% interrelatedness between members) V_H family and a second monotypic gene family are found in

·

Heterodontus; efforts to identify additional gene families have been unsuccessful. Even more striking is the minimal diversification of the coding sequences of D_H and J_H segments. Although it is not certain whether non-lymphoid germline-joining of gene clusters, found in all of the cartilaginous fish, are expressed, their putative transcription products would exhibit restrictions since potential somatic variation at V/D, D/D, D/J junctions would be restricted to a single junction in a VD-J-joined gene. Unequivocal evidence has been found for an absence of combinatorial joining in *Heterodontus* and the presence of only a single μ-type constant region in this species represents a further restriction. However, consideration of immunoglobulin genes in a representative of a different order of cartilaginous fish limit such generalizations. In *Raja* (skate) a unique, divergent V_H family and second heavy chain isotype have been described. Despite the V_H family and constant region isotype restrictions in *Heterodontus*, N region diversity is extensive and there is compelling evidence for somatic mutation. Teleost fish (Amemiya and Litman 1990; Ghaffari and Lobb 1991) as well as *Xenopus laevis* have extensively diversified V_H families (Haire et al. 1990) and the levels of putative D_H and J_H diversification in *Xenopus* are significant (Schwager et al. 1988a; Hsu et al. 1989; Haire et al. 1990). Somatic variation in the single V_H and V_L genes found in an avian, *Gallus*, is extensive and recent work reported elsewhere in this volume shows that a gene conversion mechanism functions to diversify immunoglobulin genes in mammals. Thus, many of the earlier conclusions reached about antibody restriction need to be modified in terms of our current understanding of the molecular genetics in these select but representative systems. There are elements of restriction in some components of the antibody gene system, particularly in *Heterodontus*; however, the overall concept is in serious need of redefinition. The recent findings of somatic mutation (or gene conversion) in even the earliest jawed vertebrates is particularly significant in this regard, since this process acts on the committed (rearranged) gene and generally was considered to represent the final step in the progressive evolution of gene diversifying mechanisms. We now propose that the mutation (hypermutation) or gene conversion mechanisms arose early in vertebrate evolution, prior to the emergence of the cellular mechanisms that effect selection and expansion of mutated forms. Interestingly, germinal centers where mutation/selection occurs are not histologically recognizable until later stages of vertebrate phylogeny (above the levels of the cartilaginous and bony fishes). In summary, if one applies objective molecular genetic criteria, there is ample evidence for complex antibody gene heterogeneity in lower vertebrates. A summary of these conclusions is shown in Fig. 4.

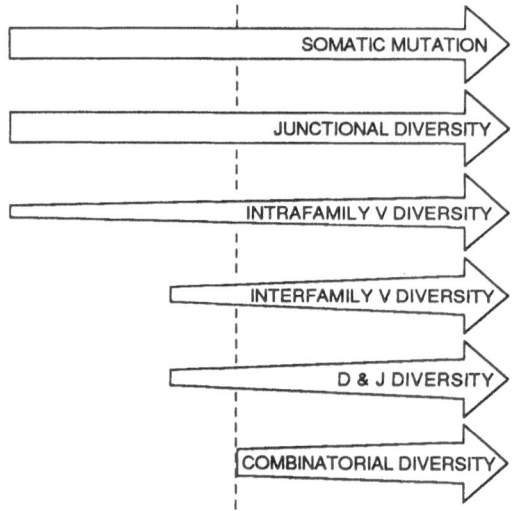

Fig. 4. Hypothetical progressive phylogenetic expansion of six features of immunoglobulin gene diversity. The vertical line separates the cartilaginous fishes (left side) and the bony fishes and the more advanced vertebrates (right side). Arrows are in the direction of increasing phylogenetic development. In this scheme, the extent of somatic and junctional diversity are the same in all species, intrafamily V diversity expands during vertebrate development; interfamily V diversity (multigenic) originates in the cartilaginous fishes as does D and J diversity but expands in subsequent phylogenetic development. Combinatorial diversity does not arise until after the emergence of the bony fishes (and acquisition of an extended multiple segment locus)

CONCLUSIONS

Based on these observations and additional data from our laboratory, hypotheses about the principal events that gave rise to the immunoglobulin gene systems of contemporary vertebrates can be made. The first event in the evolution of this rearranging gene system presumably involved the segmentation of a single exon encoding a protein that possessed the basic conformation of an immunoglobulin domain. This may have been mediated by a transposon containing an RSS-like sequence or by intragenic recombination. We have described such a transposon-like structure within an immunoglobulin intervening sequence (Litman et al. 1985b). In functional terms, rearrangement of the segmented gene could introduce sequence variation at the joining junction analogous to junctional and N-type diversity seen in higher vertebrates. This segmented exon may have evolved into a V_H-J_H-C_H- or V_H-D_H-J_H-C_H-type structure through duplication and/or recombination; however, this discussion should not be taken to suggest that a heavy chain gene necessarily preceded a light chain gene in the evolution of this rearranging gene system.

Indeed, there are reasons to believe that a light chain-like structure may have preceded the heavy chain in the evolution of the rearranging gene system; notably, light chains can form a homodimer that simulates the folding patterns of immunoglobulin heterodimers. Without readdressing whether TCRs preceded immunoglobulins in phylogeny (discussed above), the most conservative interpretation is that a primordial, segmented V-D-J-type structure duplicated, giving rise to V_1-J_1-C_1 and V_2-J_2-C_2. One of these, V_2, underwent an internal rearrangement, recombination or secondary inversion by a transposon, yielding V_2-D_2-J_2-C_2. Notably, both the immunoglobulin and TCR heterodimers are encoded by pairs of D- containing and non D-containing segmental rearranging genes, e.g., heavy (D_H) and light chain, α and β (D_β) TCR and γ and δ (D_δ) TCR. The modern-type transcriptional regulatory sequences detected in *Heterodontus* light chain genes and absence of such in the TCR-like heavy chain genes argues that the light chain may have evolved from a TCR-like heavy chain cluster and through recombination imposed the "modern" forms of gene regulation on the heavy chains in future stages of evolutionary development.

In their most fundamental role, the D_H segments, which in at least one phylogenetically early vertebrate (*Heterodontus*) lack appreciable sequence variation, serve as mediators of genetic change through the recombination process rather than contributing sequence variability to the antibody repertoire. Further development of this system presumably involved unit duplication of the V-D-J clusters. Sequence diversification in the CDRs of the V_H segments would expand the range of combining site specificities. An alternative pathway of gene evolution resulted in duplication (and diversification) of the individual V_H, D_H and J_H segments in a single, extended locus. Significant chromosomal distances between the individual elements would serve to promote nonrestrictive rearrangement of segmental elements, facilitating combinatorial diversity. Further diversification of V_H FRs gave rise to the separate V_H gene families found in contemporary vertebrates and duplication and diversification of C_H exons expanded constant region diversity. Throughout this process, somatic mutation (and hypermutation) targeted to the immunoglobulin locus most likely became an increasingly important factor in the generation of antibody diversity. Alternatively, somatic mutation may have been an early component in the system and this process co-evolved with cellular selection mechanisms that could improve the quality (affinity and specificity) of antibody reactivity to a defined antigen. Experiments to distinguish these competing hypotheses are presently underway.

It may never be possible to project the time frame in which these various changes took place, other than to state that the first point in vertebrate evolution where heterodimeric (higher vertebrate-type) immunoglobulin genes can be recognized is in the phylogenetically primitive cartilaginous fishes. The recent findings that the heterodimeric structure found in an Agnathan, the hagfish, has only limited amino acid sequence identity with immunoglobulin (and many other peptides from the hagfish heavy and light chains have no identity with immunoglobulin) is not unexpected since extensive efforts using molecular genetic approaches to detect immunoglobulin genes in this species (and lamprey, another Agnathan) have been unsuccessful as have been efforts to identify plasma cells, the terminally differentiated B lymphocyte form. The search for the origins of the B cell repertoire may need to be restricted to the earliest cartilaginous fishes, as the jawless vertebrates either may lack recognizable immunoglobulin genes or rely on other mechanisms and molecules, which may or may not be related to contemporary members of the immunoglobulin gene superfamily, to mount effective challenges to invading foreign pathogens. The Agnathans may possess highly derived mechanisms for antigen reactivity but the more phylogenetically primitive protochordates may retain characteristics that are related more closely to immunoglobulin found in contemporary species. It is apparent that phylogenetic investigations into the evolution and phylogenetic origins of this complex system have and will continue to provide essential information as to the mechanisms of evolution of this and other multigenic systems which are comprised of multiple, interacting cell types.

ACKNOWLEDGEMENT

The editorial assistance of Barbara Pryor is appreciated. This work was supported by grants from the National Institutes of Health, AI-23338 and GM-38656.

REFERENCES

Amemiya CT, Litman GW (1990) Proc Natl Acad Sci USA 87:811-815
Amemiya CT, Litman GW (1991) Amer Zool 31:558-569
Becker RS, Knight KL (1990) Cell 63:987-997
Blackwell TK, Alt FW (1988) Molecular immunology. IRL Press Ltd, Oxford
Du Pasquier L (1982) Nature 296:311-313
Ghaffari SH, Lobb CJ (1991) J Immunol 146:1037-1046
Haire RN, Shamblott MJ, Litman GW (1989) Nuc Acids Res 17:1776
Haire RN, Amemiya, CT, Suzuki D, Litman GW (1990) J Exp Med 171:1721-1737
Haire RN, Ohta Y, Litman RT, Amemiya CT, Litman GW (1991) Nuc Acids Res 19:3061-3066

Hanley PJ, Seppelt IM, Gooley AA, Hook JW, Raison RL (1990) J Immunol 145:3823-2828

Harding FA, Cohen N, Litman GW (1990a) Nuc Acids Res 18:1015-1020

Harding FA, Amemiya CT, Litman RT, Cohen N, Litman GW (1990b) Nuc Acids Res 18:6369-6376

Hinds KR, Litman GW (1986) Nature 320:546-549

Hsu E, Schwager J, Alt FW (1989) Proc. Natl. Acad. Sci. USA 86:8010-8014

Ishiguro H, Kobayaski K, Suzuki M, Titani K, Tomonaga S, Kurosawa Y (1992) EMBO J 11:829-837

Kobayashi K Tomonaga S, Kajii T (1984) Mol Immunol 21:397-404

Kobayashi K, Tomonaga S, Hagiwara K (1985) Mol Immunol 22:1091-1097

Kokubu F, Hinds K, Litman R, Shamblott MJ, Litman GW (1987) Proc Natl Acad Sci USA 84:5868-5872

Kokubu F, Litman R, Shamblott MJ, Hinds K, Litman GW (1988a) EMBO J 7:3413- 3422

Kokubu F, Hinds K, Litman R, Shamblott MJ, Litman GW (1988b) EMBO J 7:1979-1988

Lee NE, Davis MM (1988) J Immunol 140:1665-1695

Litman GW, Berger L, Murphy K, Litman RT, Hinds KR, Jahn CL, Erickson BW (1983) Nature 303:349-352

Litman GW, Berger L, Murphy K, Litman R, Hinds KR, Erickson BW (1985a) Proc Natl Acad Sci USA 82:2082-2086

Litman GW, Murphy K, Berger L, Litman R, Hinds K, Erickson BW (1985b) Proc Natl Acad Sci USA 82:844-848

Litman GW, Shamblott MJ, Haire R, Amemiya C, Nishikata H, Hinds K, Harding F, Litman R, Varner J (1989) Progress in Immunology VII. Springer-Verlag, Berlin

Matsuda F, Shin EK, Hirabayashi Y, Nagaoka H, Yoshida MC, Zong SQ, Honjo T (1990) EMBO J 9:2501-2506

McCormack WT, Tjoelker LW, Carlson LM, Petryniak B, Barth CF, Hunphries EH, Thompson CB (1989) Cell 56:785-791

Reynaud C-A, Anquez V, Grimal H, Weill J-C (1987) Cell 48:379-388

Reynaud C-A, Dahan A, Anquez V, Weill J-C (1989) Cell 59:171-183

Schwager J, Grossberger D, Du Pasquier L (1988a) EMBO J 7:2409-2415

Schwager J, Mikoryak CA, Steiner LA (1988b) Proc Natl Acad Sci USA 85:2245-2249

Schwager J, Burckert M, Courtet M, Du Pasquier L (1989) EMBO J 8:2989-3001

Shamblott MJ, Litman GW (1989) EMBO J 8:3733-3739

Varner J, Neame P, Litman GW (1991) Proc Natl Acad Sci USA 88:1746-1750

The Human Immunoglobulin κ Locus: Facts and Artifacts in Establishing a PFGE Map

G. M. Weichold[1], R. Ohnheiser, H. G. Zachau

Institut für Physiologische Chemie, Universität München Schillerstrasse 44, D-8000 München 2, Germany

INTRODUCTION

The structure and various functional aspects of the human κ locus have been established in the past decade (reviews Zachau 1989,1990). Some structural features have been compiled in a recent study on polymorphisms and haplotypes in the κ locus (Pargent 1991). First attempts to establish a long-range map of the κ locus by pulsed field gel electrophoresis (PFGE) were published by Lorenz et al. (1987). PFGE was then used to analyze the products of $V_κ$-$J_κ$ rearrangement mechanisms, which involve deletions or inversions of Mb sized fragments (Weichhold et al. 1990). Recently, a detailed study of the long-range structure of the κ locus was completed (Weichhold 1992). The main results of this study were submitted as a publication (Weichhold et al. 1993a). In addition to the 2 Mb of the locus and the 1.5 Mb to each side of it, another 2 Mb, including the CD8α locus, were mapped and linked to the κ locus (Weichhold et al. 1993b). In the present paper three groups of experiments will be described, which were carried out in the course of our PFGE studies on the κ locus but which may be of some general interest.

RESULTS AND DISCUSSION

The Effect of 5-Azacytidine on DNA Methylation

The PFGE mapping of human loci with the help of rare cutter restriction nucleases is influenced by the fact that CpG sequences are methylated to different extents in the DNAs of various individuals or cell lines. The recognition sites of several rare cutters contain one or two CpGs whose methylation prevents cleavage at those sites. It was of interest to explore the possibility of producing DNA samples with reduced m^5C content.

It was shown that the growth of the cells in the presence of 5-azacytidine leads to undermethylated DNA (Dobkin et al. 1987; Selig et al. 1988 and earlier literature). Therefore the lymphoid cell line GM607, which was used in several of our studies (Weichhold et al. 1990), was grown in the presence of several concentrations of 5-azacytidine and the DNA studied in PFGE experiments. An example is shown in Fig. 1. It is seen that the two NotI sites whose cleavage leads to fragments of 190 and 340 kb become accessible only after 5-azacytidine treatment of the cells. The accessibility of a SalI site giving rise to a 370 kb fragment was strongly increased by the pretreatment. The results show that the 5-azacytidine in principle also works for our type of experiments. For two reasons the method

---

[1] Present address: Institut für Rechtmedizin der Universität München, Frauenlobstrasse 7a, D-8000 München 2, Germany

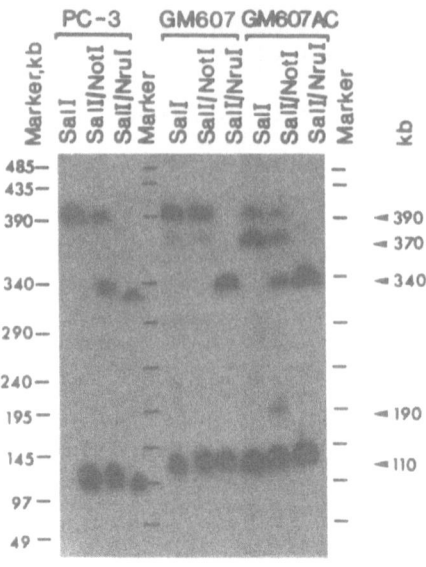

m217-1

Fig. 1. Alteration of cleavage patterns by cell growth in the
presence of 5-azacytidine. GM607 cells were grown in the absence
and, for two weeks, i.e. until the number of cells was triplicated,
in the presence of 2 μM 5-azacytidine (Selig et al. 1988) and DNA
digests of GM607 and GM607-AC (= 5-azacytidine treated) were
compared to those of PC-3. The conditions of restriction nuclease
digestions, PFGE separation (150 V with 30 s pulses for 40 h) and
hybridization (with the probe m217-1) were as described in Weichhold
et al. (1993a). The cleavage patterns of GM607 and GM607-AC can be
interpreted on the basis of the restriction map (Fig. 2 in Weichhold
et al. 1993a): all fragments start from the SalI site at position
440 and reach to the SalI site at 550 (110 kb) to the NotI sites at
positions 630 and 780, which become cleavable in GM607 only after
5-azacytidine treatment (190 kb, 340 kb), to a previously not observed SalI site at position 810 (370 kb) and to the NotI site at
position 830 (390 kb).

was not widely employed in our work: the cells grow very slowly
depending on the 5-acazytidine concentration, and it will probably
not be easy to obtain sufficient amounts of DNA with reproducible
cleavage properties. The main reason was, however, that the aim of
establishing a narrowly spaced restriction map could eventually be
reached without the 5-azacytidine treatment, i.e. by employing
several different DNAs and a large panel of restriction nucleases.
But there may be situations where cell growth in the presence of
5-azacytidine is the method of choice for restriction mapping of
m5CpG containing DNA, e.g. if one wants to render a certain rare
cutter site cleavable, which is known to exist in cosmid clones but
is not seen in genomic DNA.

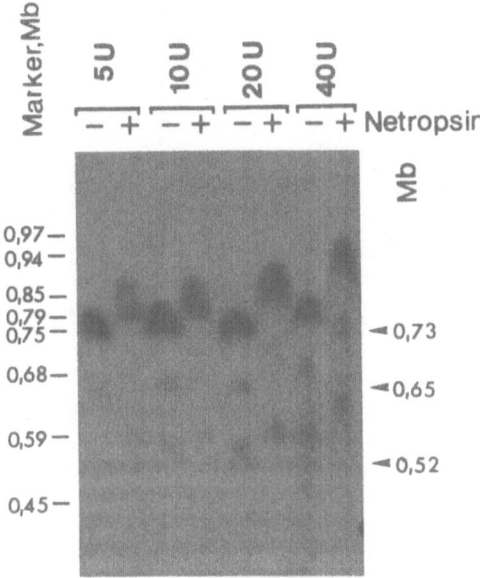

SgrAI

m21-2

Fig. 2. The effect of netropsin on the cleavage of DNA with SgrAI
and on the separation of the cleavage products. Cleavage of PC-3 DNA
was with 5 - 40 units of SgrAI in the absence (-) and the presence
(+) of 0.8 mM netropsin in the buffer systems of Tautz et al. (1990)
and Laue et al. (1990). Separation was with 120 V and 150 s pulses
for 40 h and 100 s pulses for another 24 h. Hybridization was with
the L-region probe m21-2. The 730 kb fragment is derived from the p
copy (positions 590 to -140 of Weichhold et al. 1993a), the 650 kb
and 520 kb fragments are from the d copy (positions 1380 to 2040 and
1900, respectively).

DNA Cleavage with SgrAI in the Presence and Absence of Netropsin

The rare cutter restriction nuclease SgrAI,which cleaves at CRCCGGYG
sites (Laue et al. 1990), proved to be a very useful enzyme for our
purposes since it generates different large fragments from the p and
d copies of the κ locus (Fig. 2 in Weichhold et al. 1993a). The
strong 0.73 Mb band from the p copy and the weak 0.65 and 0.52 Mb
bands from the d copy can be seen in Fig. 2; the major cleavage
product of the d copy is a 1.4 Mb fragment and is therefore not resolved under the conditions of Fig. 2. Laue et al. (1990) reported
that in λ DNA digestions with ten and more units of SgrAI a side
activity termed SgrAI[*] is seen, which leads to additional defined
smaller restriction fragments. The same type of fragments is also
obtained with low nuclease concentrations in the presence of
glycerol or of Mn^{++} in stead of Mg^{++}. The appearance of the minor
fragments could be prevented when the incubations were carried out
in the presence of 0.2 mM netropsin. We wanted to find out whether

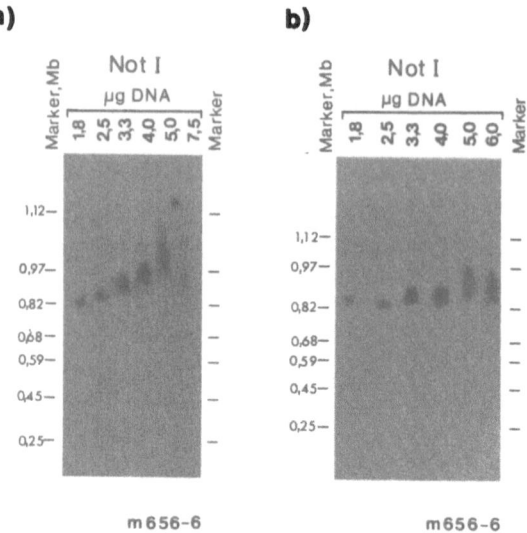

Fig. 3. Effects of the amounts of DNA per track and of the separation conditions on the mobility of fragments. Increasing amounts
of NotI digests of PC-3 DNA per track were applied in both experiments. Hybridization was with the d-copy specific probe m656-6.
a) 1.1 % agarose gel, 150 V with 90 s pulses for 40 h.
b) 1.3 % agarose gel, 180 V with 110 s pulses for 30 h.
0.25xTEB buffer was used instead of the normally applied 0.5xTEB
buffer.

the small amounts of the two d-copy derived fragments were possibly
due to the SgrAI* activity in the enzyme. The experiment of Fig. 2
shows that this is probably not the case: the intensities of the two
bands did not increase when the SgraAI concentration was raised beyond 10 units. The bands rather seem to be derived from SgrAI sites
that are only partially cleavable. Netropsin under the conditions
applied here seems to lead to smaller amounts of the d- and p-copy
derived fragments at 5 and 10 units SgrAI. But this may be due to a
general inhibition of the enzyme which is overcome at higher SgrAI
concentrations. In addition, it was observed that in PFGE netropsin
at a concentration of 0.8 mM slows down the speed of migration and,
in some experiments, leads to the artificial appearance of double
bands. It is concluded that the d-copy derived fragments arose from
SgrAI and not from SgrAI* cleavage. The SgrAI* activity probably
plays no role under the conditions of digestion of agarose embedded
DNA.

Influence of the DNA Concentration and the Separation Conditions on the Mobility of DNA Fragments in PFGE

The mobility of DNA fragments in PFGE relative to marker DNAs and
consequently the apparent fragment sizes are influenced by a number
of parameters (Cantor et al. 1988; Birren et al. 1988; Anand and
Southern 1990). A systematic study of the DNA concentration on the
mobility of fragments was published recently (Doggett et al. 1992).

Since the variation of apparent fragment sizes depending on the details of the experimental conditions was a serious problem in our work, some comparative experiments were carried out. Two of them are shown in Fig. 3 a and b. A fragment size of 0.85 Mb was assigned to the d-copy derived NotI fragment, as it is observed at low DNA concentrations under optimal separation conditions (Fig. 3a). This value may be 10 % too high, since it was found that up to the size of our largest cloned contig (0.6 Mb) fragment sizes determined by PFGE with 3 - 4 µg DNA/track were about 10 % larger than the ones derived from the addition of the sizes of cloned fragments determined by conventional gel electrophoresis (Weichhold et al. 1993a). The apparent fragment sizes keep increasing with increasing amounts of DNA (Fig. 3a; Doggett et al. 1992). Under different conditions. which are perhaps less well suited for size calibrations in the 0.8 - 1.0 Mb range, the apparent size variation with the amounts of DNA applied is smaller, but at higher DNA concentrations a double band appears which is clearly an artifact (Fig. 3b).

Our résumé was and is that fragment sizes should be determined, whenever possible under the conditions of optimal separation in the respective size range. In order to avoid artifacts the separations should be carried out under a number of different conditions.

ACKNOWLEDGEMENTS

We thank Drs. W. Ankenbauer and G.G. Schmitz for samples of SgrAI prior to publication and marketing. Our work was supported by Bundesministerium für Forschung and Technologie, Center Grant 0316200A, and Fonds der Chemischen Industrie.

REFERENCES

Anand R, Southern EM (1990) Pulsed field gel electrophoresis. In: Rickwood D, Hames BD (eds) Gel electrophoresis of nucleic Acids. IRL Press, Oxford / Washington D.C., p 101
Birren BW, Lai E, Clark SM, Hood L, Simon MI (1988) Nucleic Acids Res 16:7563-7582
Cantor CR, Smith CL, Mathew MK (1988) Ann Rev Biophys Biophys Chem 17:287-304
Dobkin C, Ferrando C, Brown WT (1987) Nucleic Acids Res. 15:3183
Doggett Na, Smith, CL, Cantor CR (1992) Nucleic Acids Res 20:859-864
Laue F, Schmitz GG, Kessler C (1990) Nucleic Acids Res 18:3421
Lorenz W, Straubinger B, Zachau HG (1987) Nucleic Acids Res 15:9667-9676
Pargent W, Schäble KF, Zachau HG (1991) Eur J Immunol 21:1829-1835
Selig S, Ariel M, Galtein R, Marcus M, Cedar H (1988) EMBO J 7:419-426
Tautz N, Kaluza K, Frey B, Jarsch M, Schmitz GG, Kessler C (1990) Nucleic Acids Res 18:3087
Weichhold GM (1992) Doctoral Thesis, Fakultät für Chemie und Pharmazie der Universität München
Weichhold GM, Ohnheiser R, Zachau HG (1993a) Submitted
Weichhold GM, Huber C, Parnes JR, Zachau HG (1993b) Submitted
Zachau HG (1989) Immunglobulin light chain genes of the κ type in man and mouse. In: Honjo T, Alt FW, Rabbitts, TH (eds) The Immunoglobulin Genes. Academic Press, London / San Diego, p 91
Zachau HG (1990) Biol. Chem. Hoppe-Seyler 371:1-6

Diversity Is Generated with Diversity

C.- A. Reynaud, W.R. Hein*, B.A. Imhof* and J.-C. Weill

*Institut Necker - 156 rue de Vaugirard - 75730 Paris Cedex (France)
Basel Institute for Immunology - 487 Grenzacherstrasse - CH 4005 Basel (Switzerland)

INTRODUCTION

Whereas the thymic differentiation step appears universal for all species studied so far, different developmental pathways of B cell differentiation have evolved in divergent species (Weill and Reynaud 1992). Accordingly, rules that were thought to be of general value for B cell development have to be replaced by new ones as these new models are being described.

The first B cell system studied was the mouse system. In mice the B cell repertoire is generated by a mechanism which consists of assorting randomly a small number of immunoglobulin (Ig) coding elements to give rise to a very large number of different Ig molecules. This rearrangement process occurs at first in fetal liver and later on in bone marrow, generating throughout life a diversified population of B cells. The mouse B cell repertoire is generated in bone marrow by a population of B cell progenitors which most probably have their Ig genes in a germline configuration and which undergo a continuous rearrangement process while maintaining their potential by self renewal. When and where these progenitors are made is not precisely known at the moment and this may require the definition of specific markers to trace back these cells during early embryonic development (reviewed in Phillips 1989 ; Rolink and Melchers 1991). When looking at the avian model of B cell development in the bursa of Fabricius, the situation appears totally different. There is, as in the mouse bone marrow, accumulation of B cell committed progenitors at a precise stage of development in a precise lymphoid micro-environment (at day 8 to 14 in the embryonic bursa), but the phenotype of these progenitors is different. They have an IgM molecule at the surface and have therefore already rearranged their Ig genes (Pink et al 1985 ; Weill et al. 1986). Once having colonized bursal follicles, chicken B cell committed progenitors divide extensively within these follicles and diversify their Ig receptor by an ongoing post-rearrangement process: gene conversion (Reynaud et al. 1987 ; Thompson and Neiman 1987).

SHEEP ILEAL PEYER'S PATCHES BEHAVE AS A PRIMARY ORGAN OF B CELL DIVERSIFICATION.

When we decided to study the B cell system of ruminants, it had been shown that sheep ileal Peyer's patches (IPP), as opposed to jejunal Peyer's patches, possess many histological and physiological similarities with the avian bursa (Reynolds 1987). They are organized in segregated follicles which are colonized by progenitors around 110 days of

Table 1. Somatic mutation in Ig light chain sequences isolated from sheep ileal Peyer's patches at fetal and adult stages

	FR	CDR	Total
Rate of modifications			
Fetal stage	1.4 per 10^3 nucl.	6.6 per 10^3 nucl.	0.9 per V sequence
Adult stage	7.4	110	11.5
R/S ratio			
Theoretical	2.9	3.6	
Fetal stage	1.5	4.2	
Adult stage	1.1	8.1	

Number of modifications are calculated from 49 rearranged light chain sequences at 144 days of fetal life ("fetal stage") and 15 at 4 months after birth ("adult stage") compared to the corresponding germline gene (taken from Reynaud et al., 1991a). Total numbers are corrected for the errors of the Taq polymerase. FR: framework region ; CDR: complementarity-determining region ; the total V sequence emcompasses 473 nucleotides from the beginning of the leader segment to 12 nucleotides of the J segment.

fetal life (birth is at 150 days) which thereafter divide extensively and generate a large population of surface IgM B cells. This active lymphopoiesis (3×10^9 B cells produced per hour) goes on until 6 to 8 months after birth at which time IPP start to involute. During this massive proliferation cells migrate to the periphery and generate the B cell compartment of the animal. If IPP are surgically ablated during fetal life, the animal is severely B cell deficient (< 10% of circulating B cells when compared to the normal situation, our unpublished data). Based on these features, Reynolds and colleagues have proposed that IPP of sheep have the properties of a primary B cell organ and could therefore be considered as a mammalian bursa equivalent (Reynolds and Morris 1983). When looking at this B cell model at the molecular level, we confirmed that sheep IPP behave as a bursa-like primary organ of B cell diversification (Reynaud et al. 1991a). The progenitors which accumulate in the IPP follicles during fetal life (2-3 per follicles, 10^5 follicles) have already rearranged their Ig genes and as they proliferate in IPP, diversify their Ig receptors by an ongoing post-rearrangement process. However while chicken B cells diversify their Ig receptors by gene conversion, sheep B cells use another molecular mechanism to generate diversity: somatic hypermutation. The process starts *in utero* and there is on the average one mutation per V sequence at birth ; it then goes on gradually, with up to 12 modifications per sequence at 4 months. Somatic mutations are highly clustered and are strongly selected for replacement substitutions (R) in CDRs, whereas silent substitutions (S) are favored in framework regions (Table 1). These biased R/S ratios are the landmark of a strong selective pressure acting on the binding site of the mutant Ig molecules. Immunologists have studied somatic hypermutation for a long time. It occurs during secondary immune responses to T dependent antigens within germinal centers. In sheep IPP follicles however, hypermutation is triggered during fetal development in the absence of external antigens and lasts for several months in an IgM B cell population. Moreover less than 1% T cells are present within IPP follicles and only a few per cent recirculating T cells are present in the interfollicular areas.

fig. 1: <u>Construction of isolated ileal segments.</u> Transsection of a segment of the terminal ileum is performed at 120 days of fetal life. The ileum is reanastomosed and the isolated segment ("ileal loop") is maintained *in situ* after ligation at both ends with its nervous, lymphatic and vascular connections. Such segments are then recovered at various stages (144 days of fetal life ; 4 and 8 weeks after birth), together with adjacent normal tissue as a control, for histological staining and analysis of Ig diversification (Reynaud et al., in preparation)

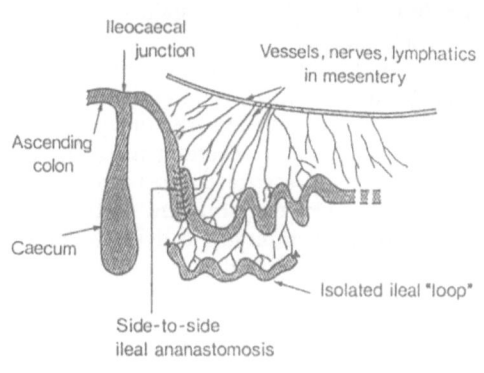

In order to find out what is the role of food and bacterial antigens present in the gut, we followed the development of IPP in segments which have been surgically separated from the ileal portion of the gut during fetal life. These loops remain attached through vascular and lymphatic connections but are never in contact with gut associated products (fig.1). IPP follicles develop in these isolated segments rather normally until several weeks after birth, at which time their growth is impaired when compared to the normal situation (Reynolds 1976). At the molecular level, the diversification of rearranged light chain genes within these isolated loop follicles at different stages of post-natal development appears very similar to the normal situation (fig.2). Preliminary results obtained with germ-free sheep fed with a sterilized natural diet seem to show the same result, implying that there is no reduction in the number of mutations and probably in the level of selection applied to the mutant B cells in these two experimental situations (unpublished data). Along with these experiments we also performed thymectomy *in utero* at day 70 of fetal development. The thymectomized animals showed at 4 weeks after birth a 90% reduction in the number of T cells and had practically no γδ T cells

fig. 2: <u>CDR 1 diversification of light chain Ig sequences in an isolated segment of the sheep ileum.</u> Rearranged light chain sequences (corresponding to one of the two major light chain rearrangements, Reynaud et al., 1991a) were amplified by PCR from an isolated ileal segment in parallel with a fragment of IPP from the same animal at 8 weeks after birth. The CDR 1 portion of these somatic sequences ("8 week loop" vs. "8 week control") is compared to the corresponding germline gene (Reynaud et al., in preparation)

```
                                    CDR  1

          TCT GGA AGC TAC ATC GGT AGT AGT GGT GTA GGC
   8      --- --- --- --- --- --- --- --- --- --- ---
   W      --- --- --- --- --- -A- --C --- --- --- ---
          --- --- --- --- --- --- --- --- -C- --- ---
   L      --- --- --- --- --- --- --- --- --- --- AA-
   o      --- --- --- --- --- AA- -A- -A- -A- -C- ---
   o      --- --- --- --- --- --- -C- G-- --- --- ---
   P      --- --- --- G-- --- --- --- --- -A- --- ---
          --- --- --- --- --- --- --- --- -C- --- ---

   8      --- --- --- --- --- A-- C-- --- A-- --T ---
   W      --- --- --G --- --- --- --- --- --- --- ---
          --- --- --- --- --- --- --- --- -C- --- ---
   C      --- --- --- AGT --- --- G-- --- --- --- ---
   o      --- --- --- --- --- -A- --- -A- --- --- -A-
   N      --- --- --- --- --- --- --- --- --- --- ---
   T      --- -A- --- --- --- --- --- --- -C- --- ---
   R      --- --- --- --- --- G-- --- --- --A --A --- ---
   o      --- --- --- --- --- --- --- --- -C- --- ---
   L
```

123

(Hein et al. 1990). Here again IPP Ig genes show the same pattern of somatic mutations and selection.

These preliminary results confirm our earlier proposition that B cells dividing in IPP follicles are engaged in an antigen-independent program of development which consists in diversifying their Ig receptor by hypermutation, the mutants being selected as the cells are produced before seeding the periphery (90% of B cells produced in IPP die *in situ*). The selective pressure is most probably due to internal epitopes present in the precise lymphoid micro-environment of the IPP. This selection process of B cells as they are made in the primary B cell organ of ruminants may be reminiscent of the selection applied on B cells as they are produced in the bone marrow of mice before they migrate as "naive" B cells to the spleen (Freitas et al. 1990 ; Gu et al. 1991).

DIVERSITY IS GENERATED WITH DIVERSITY.

The primary antibody repertoire can be generated by ongoing rearrangement, ongoing gene conversion or ongoing hypermutation. It is striking that post- rearrangement processes take place in lymphoid follicles of gut-associated lymphoid tissues (GALT) in which clones of B cells expand for weeks and months and therefore can accumulate with time somatic modifications on their Ig receptors. These events take place in bursal follicles in birds and in IPP follicles in ruminants. The rabbit has been shown to generate part of its immune repertoire by gene conversion (Becker and Knight 1990). Whether B cell diversification takes place in IPP and appendix of this animal (as originally proposed 30 years ago (Archer et al. 1963 ; Cooper et al. 1966)) remains to be defined. In contrast, rearrangement of Ig genes in mouse B cell progenitors requires only a few cell divisions, and takes place as an ongoing process in the mouse bone marrow (Osmond 1990). The young B cells as they are produced will not diversify further their Ig receptor and therefore migrate as such to the periphery.

One mechanism does not exclude the other and it is probable that in the rabbit ongoing rearrangement taking place in the bone marrow, where pre-B cells have been described (Hayward et al 1978), contributes to generate the antibody repertoire. In humans one cannot exclude that GALT may produce in the perinatal period a diversified population of B cells. In all cases B cells as they are produced in primary B cell organs seem to undergo a selective process before they migrate to the periphery. In sheep IPP the pattern of mutation indicates strongly that there is selection during B cell production and our recent data suggest that it may be caused by internal ligands. Recent results obtained in transgenic mice coding for an autoantibody also imply that newly-formed B cells recognizing self antigens may either be anergized or deleted, or induced to rearrange further their Ig genes in order to escape self recognition (Nemazee et al. ; Weigert et al. ; this volume).

EMERGENCE OF COMMITTED PROGENITORS IN THE EMBRYONIC BURSA.

As already discussed in the first part, committed B cell progenitors have a unique phenotype in the chicken, having their Ig genes in a rearranged configuration and presenting a surface IgM molecule. This unique property allows us to ask where and when these progenitors emerge during embryonic development by following at the single cell level the rearrangement pattern of embryonic cells (Reynaud and al. 1992). This can be done at all stages of development, taking advantage of the easy access to the embryo, the simplicity of chicken Ig loci with single VH, VL and J elements and the use of the PCR technique allowing the detection of one rearrangement event in 10^5 cells. The results are summarized in table 2.

The first DJ rearranged cells are detected at day 5-6 of embryonic development in the chicken yolk sac. At that stage 10 to 100 cells may have undergone that event. These cells can be seen 1 or 2 days later in the blood, in the spleen (day 6, 7, 8) and later in the bursa, bone marrow and thymus (day 9, 10). The time of appearance of DJ_H rearrangements in an organ corresponds to its morphogenesis and to the time at which it becomes colonized by lymphoid progenitors, the spleen being the earliest formed. One or two days after DJ_H rearrangement, the B cell progenitors start to rearrange simultaneously their single V_H and V_L genes in the yolk sac, blood, spleen (day 8, 9) and in the bursa, bone marrow and thymus (day 10, 11). Quantitatively, the picture is as follows (table 3): approximately 1000 DJ_H progenitors are present at day 8 in the yolk sac, then DJ_H progenitors dominate in the blood at day 10 (100 000-200 000 cells). This population remains very high in blood at day 13 as well as in the spleen and bone marrow (100 000-500 000 cells), while it starts to accumulate in the bursa. At day 17 these populations have declined in blood, spleen and bone marrow but expand strikingly in the bursa. We can therefore, based on these and previous results (Moore and Owen 1967; Dieterlen-Lièvre and Martin 1981 ; Lassila et al. 1978, 1982 ; Ratcliffe et al. 1986 ; Houssaint et al. 1991), recapitulate the development of the B cell system in the early chicken embryo (fig.3). Hematopoietic precursors can first be detected in the intra-embryonic para-aortic region at day 3-4. These progenitors then migrate to the yolk sac at day 5-6 at which stage the specific B cell progenitors start to perform D to J_H rearrangement, thus segregating from the other hematopoietic lineages (chicken T cells as opposed to mouse T cells do not show any DJ_H rearrangement). This DJ_H rearrangement could be triggered either by the yolk sac environment or by the removal of a differentiation block as cells migrate out from intra-embryonic sites. Later on B cell progenitors circulate back to the embryo and seed the various lymphoid organs. At that stage, whether in the blood or in a lymphoid organ, they start to rearrange their heavy and light chain V genes, this event being most probably part of an intrinsic cellular

Table 2: <u>Onset of Ig gene rearrangement in various chicken embryonic tissues</u>

Ig gene rearrangement	DJ_H	$V_H DJ_H$	$V\lambda J\lambda$
<u>Tissue</u>	days of development		
Yolk sac	5/6	9	N.D.
Periaortic region	9	(-)	(-)
Blood	8	9	9
Spleen	6/7	8/9	8/9
Bursa	9/10	10/11	10/11
Bone marrow	10	11	11
Thymus	9	11	11

The first day of detection of the three Ig gene rearrangements is indicated with a one-day variation reflecting individual variations of development. N.D.: not done : (-): negative over the period studied (up to day 9 for the para-aortic region) (Reynaud et al.,1992).

Table 3: Quantification of DJ$_H$ committed progenitors in various embryonic compartments

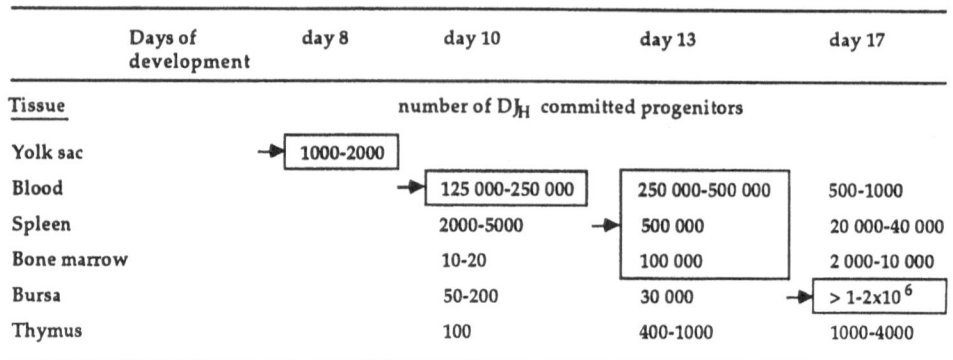

Days of development	day 8	day 10	day 13	day 17
Tissue		number of DJ$_H$ committed progenitors		
Yolk sac	→ 1000-2000			
Blood		→ 125 000-250 000	250 000-500 000	500-1000
Spleen		2000-5000	→ 500 000	20 000-40 000
Bone marrow		10-20	100 000	2 000-10 000
Bursa		50-200	30 000	→ > 1-2x10^6
Thymus		100	400-1000	1000-4000

The total number of DJ$_H$ committed progenitors in various compartments at various development stages was estimated from quantitative PCR analysis (performed on serial dilutions of 10^5, 10^4, 10^3... cells) reported to the size of each organ at the stage (Reynaud et al., 1992). Boxed figures pointed by an arrow represent the dominant DJ$_H$-progenitor compartment(s) at each developmental stage.

program rather than being induced in any particular site. Once in spleen or in bone marrow the B cell progenitors, whether they have rearranged their Ig genes productively or not, cease to divide and slowly disappear. To support this view we have shown that the percentage of in-frame sequences (50% in-frame being the outcome of V$_H$DJ$_H$ rearrangement) does not increase with time in the spleen while they go from 50 to 90 % in the bursa in 5 days (Reynaud et al. 1991b ; as described previously for bursal light chain sequences, McCormack et al. 1989). Moreover light chain circular excision products remain approximately at a constant level in the embryonic spleen B cell progenitors while they gradually disappear in the bursa during the same stage, both these results implying that there is no cellular division and selection for functional sequences in the spleen. The maximal concentration of B cell progenitors in the spleen occurs at a stage when most of the bursal colonization is accomplished, suggesting that most B cell progenitors ending outside the bursa are somehow lost for the system. It remains difficult to understand why so many cells (1-2x10^6) are engaged in the B lymphoid lineage while approximately only 3x10^4 productively rearranged committed B cell progenitors have to be formed in bursal follicles between day 10 to day 14 of embryonic development.

Overall, our results suggest that entry into the B cell progenitor population takes place at a very early stage of embryonic development. Once these progenitors are formed, they need inductive signals from the particular bursa environment in order to further differentiate into a diversified B cell population (cf .also McCormack et al 1989). In the mouse system, committed B cell progenitors may emerge during early embryonic development as in the chicken (Palacios and Imhof, submitted), and once they are formed they will only generate the B cell repertoire of the animal in the specific stromal environment of the bone marrow. Thus if one takes into account the differences in the molecular mechanism used to generate B cell diversity, the precise

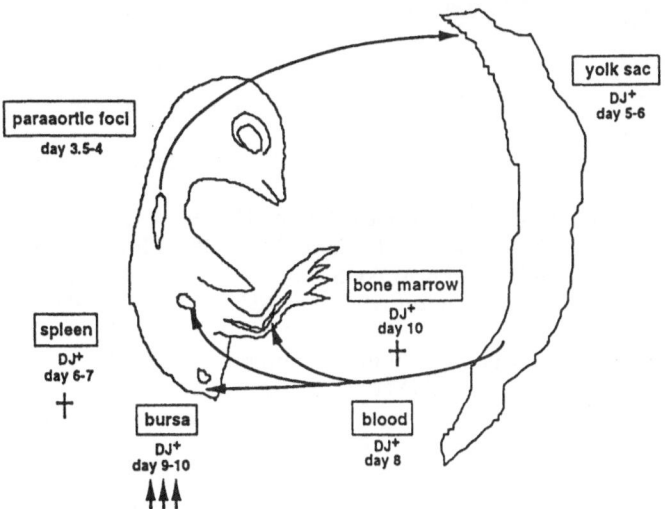

fig . 3: <u>A proposed scheme for the emergence of B cell progenitors in the chicken embryo</u> (Reynaud et al., 1992). Hematopoietic progenitors are first detected within the embryo in the para-aortic region (Lassila et al. (1978); Dieterlen-Lièvre and Martin (1981)); they further migrate through the yolk sac, where DJ_H committed progenitors segregate and seed *via* the general circulation the various lymphoid organs. These progenitor populations regress thereafter in spleen and bone marrow (marked with a † symbol) and only expand in the bursa (↑↑↑). The first day of detection of DJ_H rearrangement is indicated for each compartment.

lymphoid compartment where primary B cell repertoires are made may simply be the tissue site where committed progenitors -whatever their phenotype- can differentiate and maintain their self renewing capacity.

CONCLUSION

The study of B cell differentiation in various species, despite the great differences it has revealed, nevertheless allows the formulation of some general rules. B cell committed progenitors emerge most probably during very early embryonic development. They then migrate to a specific lymphoid microenvironment where they generate the antibody primary repertoire. These specific diversity generator lymphoid compartments (bone marrow, GALT) allow the progenitors to divide actively and to self-renew and therefore to produce large quantities of B cells. The molecular mechanisms used to generate the repertoire vary among different species as well as the phenotype of the B cell committed precursors. As naive diversified B cells are produced in these primary B cell organs, they undergo a selection before migrating to the periphery. This selection is in part due to internal ligands which most probably allow the system to eliminate potentially harmful anti-self B cells.

REFERENCES

Archer OK, Sutherland DER, Good RA (1963) Nature 200: 337-339

Becker RS, Knight KL (1990) Cell 63: 987-997.

Cooper MD, McKneally MF, Gabrielsen AE, Sutherland DER, Good RA (1966) Lancet 1: 1388-1391.

Dieterlen-Lièvre F, Martin C (1981) Dev Biol 88: 180-191

Freitas AA, Andrade L, Lembezat MP, Coutinho A (1990) Int Immunol 2: 15-23

Gu H, Tarlinton D, Müller W, Rajewsky K, Förster I (1991) J Exp Med 173: 1357-1371

Hayward AR, Simons MA, Lawton AR, Mage RG and Cooper MD (1978) J Exp Med 148: 1367-1377

Hein WR, Dudler L, Morris B (1990) Eur J Immunol 20: 1805-1813

Houssaint E, Mansikka A, Vainio O (1991) J Exp Med 174: 397-406

Lassila O, Eskola J, Toivanen P, Martin C, Dieterlen-Lièvre F (1978) Nature 272: 353-354

Lassila O, Martin C, Dieterlen-Lièvre F, Gilmour DG, Eskola J, Toivanen P (1982) Scand J Immunol 16: 265-268

McCormack WT, Tjoelker LW, Barth CF, Carlson LM, Petryniak B, Humphries EH and Thompson CB (1989) Genes Dev 3: 838-847

Moore MAS and Owen JJT (1967) Nature 215: 1081-1082

Osmond DG (1990) Semin Immunol 2: 173-180

Pink JRL, Ratcliffe MJH, Vainio O (1985) Eur J Immunol 15: 617-620

Phillips RA, in Melchers F. (ed), Progress in Immunol., 7th International Congress of Immunology (1989), Springer-Verlag, Berlin: 305-315

Ratcliffe MJH, Lassila O, Pink JRL, Vainio O (1986) Eur J Immunol 16: 129-133

Reynaud CA, Anquez V, Grimal H, Weill JC (1987) Cell 48: 379-388

Reynaud CA, MacKay C, Mueller RG, Weill JC (1991a) Cell 64: 995-1005

Reynaud CA, Anquez V, Weill JC (1991b) Eur J Immunol 21: 2661-2670

Reynaud CA, Imhof BA, Anquez V, Weill JC (1992) EMBO J in press

Reynolds JD (1976) PhD thesis, Australian National University, Canberra, Australia

Reynolds JD (1987) Curr Top Microbiol Immunol 135: 43-56

Reynolds JD, Morris B (1983) Eur J Immunol 13: 627-635

Rolink A, Melchers F (1991) Cell 66: 1081-1094

Thompson CB, Neiman P (1987) Cell 48: 369-378

Weill JC, Reynaud CA, Lassila O, Pink JRL (1986) Proc Natl Acad Sci USA 83: 3336-3340

Weill JC, Reynaud CA (1992) Curr Op Immunol 4: 177-180

Exclusion and Inclusion of α and β T Cell Receptor Alleles in Different T Cell Lineages

H. von Boehmer, L. Bruno, M. Groettrup, J. Kirberg, W. Swat, and P. Kisielow

Basel Institute for Immunology*, Grenzacherstrasse 487, CH-4058 Basel, Switzerland

INTRODUCTION

The introduction of $\alpha\beta$ T cell receptor genes into rearrangement deficient and normal mice has shown that T cell development is very tightly controlled by these genes (von Boehmer 1990, Scott et al. 1989, Teh et al. 1988, Kisielow et al. 1988). It is probably useful to distinguish an early developmental phase controlled by the β TCR chain only from a later phase controlled by the complete $\alpha\beta$ TCR (von Boehmer 1990). The early phase of development has been less extensively discussed simply because the expression of the β TCR chain in the absence of the α TCR chain on developing T cells has escaped detection for many years. The later phase of development has been extensively reviewed and consists of positive and negative selection of lymphocytes according to the specificity of their receptor such that nonselected cells die, positively selected cells survive and negatively selected cells undergo rapid apoptosis (Huesmann et al. 1991, Swat et al. 1991). The delineation of the main developmental pathway of $\alpha\beta$ T cells in $\alpha\beta$ TCR transgenic mice has also resulted in the recognition of an unusual $\alpha\beta$ lineage which is subject to different selection mechanisms (von Boehmer et al. 1991). In the following we will focus on the expression and allelic exclusion of α and β TCR genes in early development as well as in different $\alpha\beta$ T cell lineages.

THE EXPRESSION OF AN $\alpha\beta$ TCR HOMODIMER ON PRE T CELLS CORRELATES WITH CELLULAR EXPANSION, CD4 AND CD8 CORECEPTOR EXPRESSION AND ALLELIC EXCLUSION

The introduction of a β TCR transgene only into rearrangement deficient SCID mice has regularly resulted in an increased cell number of thymocytes, CD4 and CD8 coreceptor expression and increased transcription of the α TCR locus (von Boehmer 1990, Kishi et al. 1991). This was correlated with the cell surface expression of the β TCR chain as a dimer as well as a monomer on developing thymocytes in the apparent absence of CD3 proteins (Kishi et al. 1991). In follow up studies by Groettrup et al. (1992), who analyzed the expression of various TCR genes in immature T cell lines obtained from rearrangement deficient mice, the initial picture was qualified: the surface expression of the β TCR chain in these cells was restricted to the homodimeric form and it could be shown that the β TCR homodimer was associated with all CD3 components even though

* The Basel Institute for Immunology was founded and is supported by F. Hoffmann-La Roche Ltd., Basel, Switzerland

this association was of different quality than that of the αβ TCR. Nevertheless, crosslinking of the β TCR by monoclonal antibodies resulted in a strong Ca++ mobilization (Groettrup et al. 1992). These new results then raised the questions whether the surface expression of the β TCR chain represented a transgenic artifact or whether the homodimeric form or the monomeric form represented the physiologically relevant pre-T cell receptor. This was analyzed by cell surface staining and immunoprecipitation in fetal thymocytes from normal mice (Groettrup and von Boehmer, submitted). The results showed that fetal thymocytes from normal mice express the β TCR only as a homodimer on the surface and that thymocytes from transgenic mice which exhibit a much brighter surface staining than normal thymocytes have much of the β TCR chain as a GPI linked monomer which can be removed from the surface by phospholipase C. The latter probably reflects a mechanism by which cells can get rid of abundant proteins and it is this form which initially obscured the physiologically relevant homodimeric form which is associated with CD3 components.

It remains to be seen whether the pre T cell receptor is responsible for inducing allelic exclusion of the β TCR locus, cellular expansion, CD4 and CD8 coreceptor expression as well as increased transcription of the α TCR locus. In this context it is important to note that, in contrast to mature α TCR chain deficient T cell lines, immature T cells have means to prevent degradation of the β TCR chain in the endoplasmatic reticulum such that the chain can reach the cell surface. It is likely that this is mediated by specific chaperones present in immature T cells only. Their identification and elimination in gene targeting experiments will eventually address the biological role of the β TCR pre-T cell receptor.

ALLELIC EXCLUSION OF THE α TCR LOCUS IN DIFFERENT αβ LINEAGES

When α and β TCR transgenes from an HY-specific, D^b MHC restricted CD8+ T cell clone were introduced into normal D^b positive female mice the transgenic receptor was expressed early in ontogeny on CD4-8- cell and CD4+8+ thymoblasts. This resulted in the suppression of rearrangement at the endogenous β TCR loci such that no or very few productive β TCR rearrangements were seen in T cell clones expressing the β TCR transgene (Uematsu et al. 1988, Kisielow et al. 1988). In contrast, inhibition of rearrangement at the endogenous α TCR locus was much less pronounced such that on small CD4+8+ T cells, endogenous α TCR chains could readily be detected and became more prominent on mature T cells (Borgulya et al. 1992). An analysis of mature T cells at the RNA level showed that all CD4+8- and most CD4-8+ cells expressed endogenous α TCR loci and that even among CD4-8+ T cells which expressed the transgenic α TCR chain on the cell surface, more than fifty percent transcribed one of the endogenous TCR loci (Borgulya et al. 1992, Bruno et al. unpublished). Consistent with rearrangement of endogenous TCR loci was the finding that CD4+8+ thymocytes contained high levels of RAG1 and RAG2 RNA in spite of their expression of the transgenic αβ TCR. RAG1 and RAG2 expression was only shut off in CD4+8+ cells with high levels of TCR (Borgulya et al. 1992), a population which results from positive selection (Swat et al. 1992). These experiments then indicate the αβ TCR on developing T cells is initially not fixed but can be changed by the elimination of an old and the formation of a new α TCR gene until the cell dies or undergoes positive selection (Malissen et al. 1988, von Boehmer 1990).

This scenario raises the question whether deletion of immature T cells affects, in addition to pre-selection stages of CD4+8+ cells (von Boehmer 1990), also the post selection stages represented by CD4+8+ TCRhi cells whose TCR can no longer be changed because of inactive RAG genes. Appropriate experiments were conducted in the in vitro system developed by Swat et al. (1991) which allows the study of deletion of thymocyte subsets in suspension culture. As reported in this volume by Kisielow et al. (1992) it was clear that positive selection and negative selection did not affect distinct developmental stages as deletion clearly affected stages prior to and after positive selection, thus eliminating self reactive developing cells after their αβ TCR is fixed.

Further studies in the αβ TCR transgenic mice revealed a lineage of αβ T cells which is subject to different selection mechanisms in that it does not require positive selection of MHC molecules on thymic epithelium to exit from the thymus and become functionally mature. In this lineage, cells with TCRs specific for self antigens are not deleted (von Boehmer et al. 1991). This could be shown by transferring T cell-depleted bone marrow cells from male H-2b αβ transgenic nu/nu mice, which lack mature T cells with the transgenic receptor in bone marrow and peripheral lymph organs, into lethally x-irradiated B10.D2 mice which lack in the thymus Db MHC molecules required for the selection of cells with the transgenic TCR of the conventional αβ TCR lineage. Several weeks after transfer these animals contained CD4-8- as well as CD4-8low T cells with the transgenic αβ TCR, a phenotype which could also be easily detected in the periphery of male H-2b αβ TCR transgenic mice. Thus, these cells do not result from positive selection by H-2b MHC molecules on thymic epithelium with subsequent down regulation of CD8 coreceptors but represent a different lineage. In fact, cells of this lineage do recognize self antigen and persist or slowly expand as activated cells over long time periods (von Boehmer et al. 1991). Since these cells are not undergoing positive selection one might expect that the status of endogenous α TCR genes may be different in those cells compared to those from CD4-8+ T cells resulting from positive selection by Db MHC molecules on thymic epithelium. This was indeed the case: while none of thirteen CD4-8- or thirteen CD4-8low T cell clones from male αβ TCR transgenic H-2b positive mice showed transcription of the endogenous α TCR loci, seven out of thirteen CD4-8+ T cells from female H-2b αβ TCR transgenic mice contained transcripts from the endogenous α TCR loci.

DISCUSSION

The experiments reported here suggest that, similar to B lymphocyte development, a receptor on pre-T cells control development of αβ T cells. This receptor does not contain any disulfide linked chains in addition to the β TCR chain and thus differs from the pre-B cell receptor which contains a disulfide linked surrogate light chain in addition to other noncovalently linked components (Rolink and Melchers 1991). It appears that the β TCR chain itself can substitute for the α TCR chain and the formation of this homodimer is sufficient to be transported to the cell surface together with the various proteins of the CD3 complex. This does not happen in mature T cell lines as the loss of the α TCR chain or transfection of a β TCR chain into an α and β TCR deficient mature T cell line (Yagüe et al. 1985, Groettrup et al. 1992) does not result in cell surface expression of the β TCR chain only even when all CD3 components are present in that cell. It is thus logical to assume that the β TCR on the surface of pre T cells serves a purpose and is responsible for progress in T cell development observed after the introduction of a β TCR gene into rearrangement deficient mice including allelic exclusion of the β TCR locus.

Allelic exclusion of the α TCR locus works differently, at least in the conventional αβ lineage in that there may be no suppression of α-rearrangement at all by the αβ TCR unless the receptor binds to MHC ligands in the thymus. There exists, however, a different lineage of αβ T cells subject to different selection mechanisms and it appears that in this lineage the expression of the αβ TCR prevents rearrangement of both TCR loci. This latter lineage does not represent a transgenic artefact as CD4-8- αβ T cells with the same properties have been described in the thymus (Egerton and Scollay, 1990) as well as in the periphery of normal mice (Huang and Crispe, 1992). It remains to be seen what role cells of this lineage have in the immune system and especially whether activated and expanded clones have an enhancing or suppressing effect on autoimmunity.

REFERENCES

Borgulya P, Kishi H, Uematsu Y, von Boehmer H (1992) Exclusion and inclusion of α and β T cell receptor alleles. Cell 69:529-537

Egerton M, Scollay R (1990) Intrathymic selection of murine TCR αβ+ CD4-8- thymocytes. Int Immunology 2:157-163

Groettrup M, Baron A, Griffiths G, Palacios R, von Boehmer, H (1992) T cell receptor β chain homodimers on the surface of immature but not mature αγδ deficient T cell lines. EMBO J 11:2735-2740

Huang L, Crispe IN (1992) Distinctive selection mechanisms govern the T cell receptor repertoire of peripheral CD4-8+ αβ T cells. J Exp Med, in press

Huang L, Crispe N (1992) Distinctive selection mechanisms govern the T cell receptor repertoire of peripheral CD4-8- αβ T cells. J Exp Med, in press

Huesmann M, Scott B, Kisielow P, von Boehmer H (1991) Kinetics and efficacy of positive selection in the thymus of normal and T cell receptor transgenic mice. Cell 66:533-540

Kishi H, Borgulya P, Scott B, Karjalainen K, Traunecker A, Kaufman J, von Boehmer H (1991) Surface expression of the β T cell receptor (TCR) chain in the absence of other TCR or CD3 proteins on immature T cells. EMBO J 10:93-98

Kisielow P, Bluthmann H, Staerz UD, Steinmetz M, von Boehmer H (1988) Tolerance in T cell receptor transgenic mice involves deletion of nonmature CD4+8+ thymocytes. Nature 333:742-746

Kisielow P, Swat W, von Boehmer H (1992) Intrathymic selection and maturation of αβ T cells. Progress in Immunology VIII (this volume)

Malissen M, Trucy J, Letourneur F, Rebai N, Dunn DE, Fitch FW, Hood LE, Malissen B (1988) A T cell clone express two T cell receptor α genes but uses one αβ heterodimer for allorecognition and self MHC restricted antigen recognition. Cell 55:49-59

Rolink A, Melchers F (1991) Molecular and cellular origins of B lymphocyte diversity. Cell 66:1081-1094

Scott B, Bluthmann M, Teh H, von Boehmer H (1989) The generation of mature T cells requires interaction of the αβ T cell receptor with major histocompatibility complex. Nature 338:555-558

Swat W, Dessing M, Baron A, Kisielow P, von Boehmer H (1992) Phenotypic changes accompanying positive selection of CD4+8+ thymocytes. Eur J Immunol, in press

Swat W, Ignatowicz L, von Boehmer H, Kisielow P (1991) Clonal deletion of immature CD4+8+ thymocytes in suspension culture by extrathymic antigen presenting cells. Nature 351:150-153

Teh H, Kisielow P, Scott B, Kishi H, Uematsu Y, Bluthmann H, von Boehmer H (1988) Thymic major histocompatibility complex antigens and the αβ T cell receptor determine the CD4/CD8 phenotype of T cells. Nature 335:229-233

Uematsu Y, Ryser S, Dembic Z, Borgulya P, Krimpentort P, Berns A, von Boehmer H, Steinmetz M (1988) In transgenic mice the introduced functional T cell receptor β gene prevents expression of endogenous β genes. Cell 52:831-841

von Boehmer H (1990) Developmental biology of T cells in T cell receptor transgenic mcie. Ann Rev Immunol 8:531-545

von Boehmer H, Kirberg J, Rocha B (1991) An unusual lineage of αβ T cells that contains autoreactive cells. J Exp Med 174:1001-1008

Yagüe J, White J, Coleclough C, Kappler J, Palmer E, Marrack P (1985) The T cell receptor: the α and β chains define idiotype, and antigen and MHC specificity. Cell 42:81-87

3. Mhc, Molecular and Cellular Aspects of Antigen Processing and Presentation

The Accordion Model of *Mhc* Evolution

Jan Klein[*+], Hideki Ono[*], Dagmar Klein[+], and Colm O'hUigin

[*]Max-Planck-Institut für Biologie, Abteilung Immunogenetik, 7400 Tübingen, Germany
[+]Departement of Microbiology and Immunology, University of Miami School of Medicine, Miami, FL 33101, USA

FROM FISH TO PHILOSOPHER

Up until a few years ago, the supposition that the major histocompatibility complex (*Mhc*) existed in vertebrates other than the mouse, human, and chicken, rested largely on hunches -- educated hunches, but hunches just the same. Not so any longer. Now that *Mhc* genes have been cloned from representatives of all vertebrate classes except jawless fish (Agnatha), more than 400 million years (my) of *Mhc* evolution, from fish to philosopher (Smith 1953), have been documented. One can now safely hazard a guess that *all* vertebrates possess an *Mhc* and seriously entertain notions about the existence of *Mhc* in primitive chordates -- the cephalochordates and tunicates. What's more, the search for invertebrate roots of the *Mhc* genes is no longer a foolhardy proposition.

The purpose of this communication is to sketch out a rough outline of the *Mhc* evolution as it emerges from the accumulating data of our own laboratory as well as others. The discussion will focus on four phylogenetic trees (Figures 1 through 4) depicting the relationships among representative sequences of the major vertebrate groups. The trees are based on genetic distances between sequences, calculated from nonsynonymous substitutions (amino acid replacements) differentiating a given pair of sequences. They were constructed using the neighbor-joining method of Saitou and Nei (1987). The main observation issuing from the consideration of the trees is that like *Janus*, the god of beginnings and ends, the *Mhc* presents us with two faces -- the backward-looking face of stability and the foresighting face of instability.

Mhc STABILITY

For lack of space, we are able to present here only a summary of the evolutionary studies, without detailed documentation. The summary is based on the reports listed in the figure legends. The principal conclusions afforded by the evolutionary data are these:
-- All vertebrates studied thus far have two classes, I and II, of genuine *Mhc* genes and no evidence for the existence of additional classes has been found in any vertebrate species.
-- The two subclasses in each class of *Mhc* genes -- that is those coding for the α and ß polypeptide chains -- were established as long as 425 my ago, before the emergence of cartilaginous fishes.
-- The basic structure of the Mhc molecule (α,ß heterodimer) has been conserved in all vertebrates, without any principal change in its design during the entire period of *Mhc* evolution known to us.
-- The organization of the class I and class II genes (the number and disposition of exons and introns) has not changed essentially for more than 400 my. The documented variations seem minor and are probably not significant functionally.
-- The distribution of nucleotide (amino acid) variability along coding parts of the *Mhc* genes (polypeptides) lends itself to the interpretation that the characteristic structure of the peptide-binding region (PBR), with its two ridges of α-helices rising above the

floor of ß-pleated sheets, was established early in vertebrate evolution, not later than in the ancestors of the jawed fish (Ono et al. 1992, D. Klein et al. 1992), but probably even earlier than that, and has been retained by all vertebrate classes. The implication of this observation is that the prime function of the Mhc molecules (i.e., the binding of self and nonself peptides) is the same in jawed fishes, amphibians, reptiles, birds, and mammals.

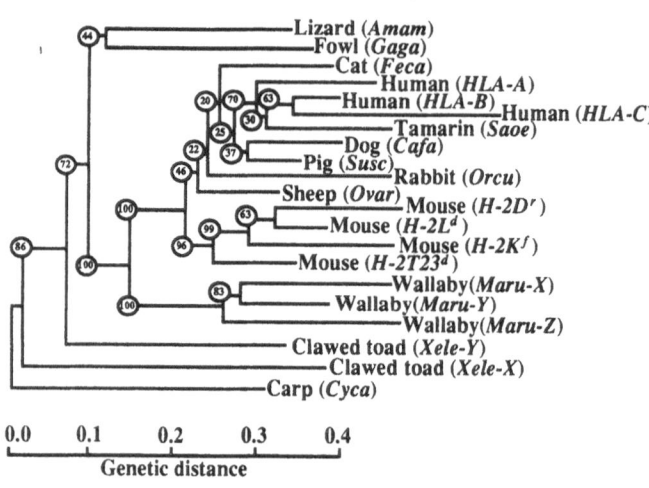

Fig. 1. Phylogenetic relationships of class I α chains. References for sequences used: *Amam*, Grossberger & Parham, Immunogenetics *36*:166, 1992; *Gaga*, Kroemer et al., Immunogenetics *31*: 405, 1990; *Feca*, Yuhki et al. J. Immunol. *142*:3676, 1989; *HLA-A*, Mayer et al., EMBO J. 7:2765, 1988; *HLA-B*, Little & Parham, Tissue Antigens *38*: 186, 1991; *HLA-C*, Boccoli et al., Immunogenetics *29*:80, 1989; *Saoe*, Watkins et al., J. Immunol. *144*: 1136, 1990; *Cafa*, Sarmiento & Storb, Immunogenetics *31*: 400, 1990; *Susc*, Ehrlich et al., Mol. Cell. Biol. 8:695, 1988; *Orcu*, Marche et al., Immunogenetics 21:71, 1985; *Ovar*, Grossberger et al. Immunogenetics 32:77, 1990; *H-2D^r*, Cai & Pease, Immunogenetics 32:456, 1990; *H-2L^d*, Moore et al. Science 215:679, 1982; *H-2K^f*, Horton et al. J. Immunol. 145:1782, 1990; *H-2T23^d*, Lalanne et al. Cell 41:469, 1985; *Maru*, Mayer et al., unpubl., *Xele*, Flajnik et al., Proc. Natl. Acad. Sci. USA *88*:537, 1991; and unpubl., *Cyca*, Hashimoto et al. Proc. Natl. Acad. Sci. USA *87*:6867, 1990.

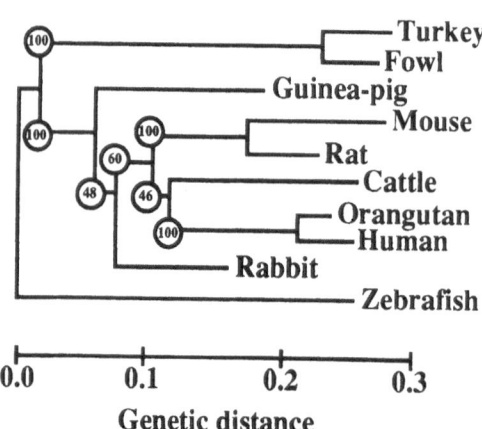

Fig. 2. Phylogenetic relationships of class I ß chains (ß₂-microglobulins). References for sequences used: *Turkey*, Welinder et al., Mol. Immunol. 28:177, 1991; *fowl*, Kaufman et al., J. Immunol. *148*:1532, 1992; *guinea pig*, Wolfe & Cebra, Mol. Immunol. *17*: 1493, 1980; *mouse*, Daniel et al., EMBO J. 2:1061, 1983; *rat*, Mauxion & Kress, Nucleic Acids Res. *15*:7638, 1987; *cattle*, Groves & Greenberg, J. Biol. Chem. 25:2619, 1982; *orangutan*, Lawlor et al., Immunol. Res. *113*:147, 1990; *human*, Güssow et al., J. Immunol. *139*:3132, 1987; *rabbit*, Gates et al., Biochem *18*:2267, 1979; *zebrafish*, Ono et al., Immunogenetics, in press 1992.

All these findings lead us to the conclusion that the evolution of the *Mhc* has been remarkably conservative: it has strived to preserve the basic *Bauplan* of the *Mhc* genes and molecules once it has been established and tested by natural selection, which must have taken place in the early vertebrates, nonvertebrate chordates, or even in the invertebrate ancestors of the chordates. From the tempo of *Mhc* gene evolution we calculate that the *Mhc* classes and subclasses began to diverge more than 700 my ago, hence well before the current estimated date of the emergence of vertebrates (Y. Satta and J. Klein, unpublished data).

This conservative mode of *Mhc* evolution in terms of class, subclass, molecule, and gene organization contrasts sharply with the evolutionary mode of evolution experienced by the immunoglobulin (Ig) genes (reviewed by Amemiya and Litman 1991). Not only do different vertebrate taxa possess different Ig classes, they also differ fundamentally in the organization of their genes, in the manner in which they generate antibody diversity, and in the specificity of their antibody molecules. Many other genes not involved in immune response have also undergone substantial structural and functional adaptations during the period in which the *Mhc* genes have basically not changed at all. Take the globin genes as an example (reviewed by Dickerson and Geis 1983). Although their three-exons, two introns organization was established early in the vertebrate evolution, a series of duplications produced many new genes that have diversified the globin molecules in a major way, both structurally and functionally. The original monomeric molecules (which can still be found in jawless fishes) were replaced by tetramers, some of the monomeric molecules adapted themselves to oxygen storage (instead of exchange) in muscle tissue and became myoglobins, and different hemoglobin molecules adapted themselves to exchanging oxygen at different stages of development -- embryonic, fetal, and adult. The many other examples of this sort all serve to underpin the unusual stability of vertebrate *Mhc* genes and molecules.

INSTABILITY

The *Mhc* Janus has, however, also a second face characterized by a gargantuan *in*stability. This feature is manifested at two levels -- that of the gene sequence and that of the genomic organization. At the sequence level, the divergence of *Mhc* genes from the primitive and advanced vertebrates is so great that only a combination of perspicacity and perspiration on the part of the involved investigators has led to cloning of these genes from representatives of the different vertebrate classes, with the jawless fish still defying all efforts in this regard. While there are only a few positions at which all vertebrates share an amino acid, at some positions more than half of the 20 amino acids known to occur in proteins have been found. The overall amino acid sequence similarity between homologous Mhc proteins can be as low as 27%. There is thus a tremendous propensity for diversification during the evolution of the *Mhc* genes. It is, however, at the level of genomic organization where *Mhc* instability manifests itself to the greatest extent. The sequencing and mapping data, however fragmentary, allow us to glean an unprecedented number of reorganizations in the *Mhc* region during its evolution. The reorganizations are implied by phylogenetic trees (Figures 1, 3, and 4) and have been documented in those instances in which *Mhc* regions have been mapped by contig analysis. A good example of implied instability is the phylogenetic tree of the *Mhc* genes encoding the class II ß chains in Figure 4. In Figures 1 and 3, the trees have been much simplified because of space limitations and the implied instability is therefore less obvious.

Turning to Figure 4 and starting with the familiar organization of our own species, we find the *HLA-D* region to consist of five ß-chain encoding gene families, *DOB, DPB, DQB, DRB*, and the recently discovered *DMB*. Ignoring *DMB* for the moment, we meet with the same families in most, if not all, the orders of eutherian (placental) mammals. (There may be additional *DB* families, possibly not represented in humans, in some of the orders, but this is beside the point.) That, however, is as far back as we can safely trace the families -- to the common ancestor from which all the living orders of eutherian mammals radiated some 80 my ago. Already in marsupials, the second of three major subdivisions of mammals exemplified by the wallaby, we lose track of the eutherian families entirely (Schneider et al. 1991). Instead of them, we find new families of expressed genes that do not seem to have orthologous relationships to the eutherian *DB* families: The marsupial genes are more related to one another than any of them is to any of the eutherian *DB* families (Schneider et al. 1991). We surmise, therefore, that the marsupial *DB* genes originated from a common ancestor which was not identical with the common ancestor of the eutherian *DB* genes.

When we move from mammals to birds, we encounter the same phenomenon. The avian class II ß-encoding genes form families that stand in no direct relationship either to the marsupial or eutherian families. The fowl class II ß-encoding genes, too, seem to be derived from a common ancestor, but one distinct from that of either the marsupial or the eutherian mammals.

In the jawed fishes, represented by the carp (Hashimoto et al. 1990), zebrafish (*Brachydanio rerio*; see Ono et al. 1992), and the cichlid *Aulonocara hansbaenschi* (D. Klein et al. 1992), the class II ß-encoding genes once again seem to have arisen from yet another ancestor.

The most straightforward interpretation of these observations is that the class II region has been undergoing repeated cycles of expansion through gene duplications from a single ancestral element, followed by contraction back to a single functional gene that becomes the ancestral element of a new expansion (Fig. 5). A similar argument can be put forward for the α-encoding class II genes (Fig. 2) and the α-encoding class I genes (Fig. 1). For these three subclasses, the evolution of the *Mhc* region resembles the bellow movements of an accordion played by a virtuoso musician. Each contraction does not necessarily mean deletion of all duplicated genes save one; it merely indicates that only one of the duplicated genes becomes the ancestor in the new duplication cycle. The other genes may linger on for a while, perhaps a long while even by evolutionary standards, without being used functionally. The *DM* might be such a lingering group of genes, since the phylogenetic trees indicate that the *DM* elements diverged from all the other class II genes before the different members of the latter diverged from one another (Figures 3 and 4). The reason why the *DM* genes may have remained protected from the accordion effect could be their close linkage to the *TAP* genes (Transporter associated with antigen processing), known to evolve conservatively (Kelly et al. 1991; Cho et al. 1991).

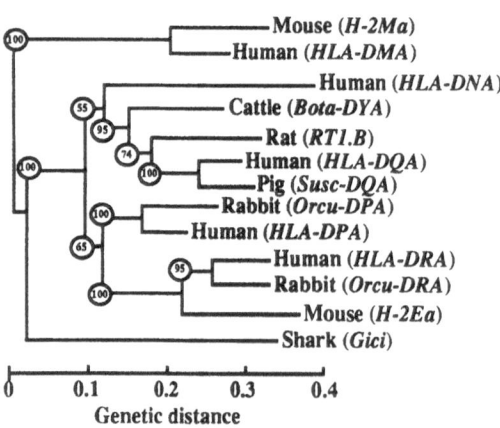

Fig. 3. Phylogenetic relationships of class II α chains. References for sequences used: *H-2Ma*, Cho et al., Nature *353*:573, 1991; *HLA-DMA*, Kelly et al., Nature *353*:571, 1991; *HLA-DNA*, Jonsson & Rask, Immungenetics *29*:411, 1989; *Bota-DYA*, der Poel et al., Immunogenetics *31*:29, 1990; *RT1.B*, Barran & McMaster, Immunogenetics *26*:56, 1987; *HLA-DQA*, Jonsson et al., Immunogenetics *30*:232, 1989; *Susc-DQA*, Hirsch et al., Immunogenetics *31*:52, 1990; *Orcu-DPA*, Sittisombut et al., J. Immunol. *140*:3237, 1988; *HLA-DPA*, Lawrance et al., Nucleic Acids Res. *13*:7515, 1985; *HLA-DRA*, Schamboeck et al., Nucleic Acids Res. *11*:8663, 1983; *Orcu-DRA*, Laverriere et al., Immunogenetics *30*:137, 1989; *H-2Ea*, Hyldig-Nielsen et al., Nucleic Acids Res. *11*:5055, 1983.

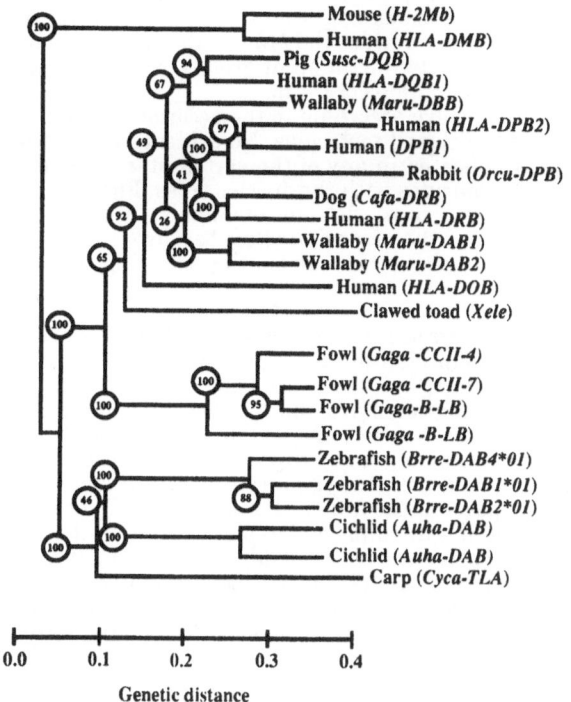

Fig. 4. Phylogenetic relationship of class II ß chains. References for sequences used: *H-2Mb*, Cho et al., Nature *353*:573, 1991; *HLA-DMB*, Kelly et al., Nature *353*:571, 1991; *Susc-DQB*, Gustafsson et al., J. Immunol. *145*:1946, 1990; *HLA-DQB1*, So et al., J. Immunol. *139*:3506, 1987; *Maru*, Schneider et al., Mol. Biol. Evol. *8*:753, 1991; *HLA-DPB2*, Gustafsson et al., J. Biol. Chem. *262*:8778, 1987; *HLA-DPB1*, Kelly & Trowsdale, Nucleic Acids Res. *13*:1607, 1985; *Orcu-DPB*, Sittisombut et al., J. Immunol. *140*:3237, 1988; *Cafa-DRB*, Sarmiento & Storb, Immunogenetics *31*:396, 1990; *HLA-DRB*, Young et al., Proc. Natl. Acad. Sci. USA *84*:4929, 1987; *HLA-DOB*, Johnsson & Rask, Immunogenetics *29*:411, 1989; *Xele*, Kasahara et al., Proc. Natl. Acad. Sci. USA, in press 1992; *Gaga-CCII*, Xu et al., J. Immunol. *142*:2122, 1989; *Gaga-B-LB*, Zoorob et al., Immunogenetics *31*: 179, 1990; *Brre*, Ono et al., Proc. Natl. Acad. Sci. USA, in press 1992; *Aauha*, D. Klein et al., Proc. Natl. Acad. Sci. USA, submitted 1992; *Cyca-TLA*, Hashimoto et al., Proc. Natl. Acad. Sci. USA *87*:6867, 1990.

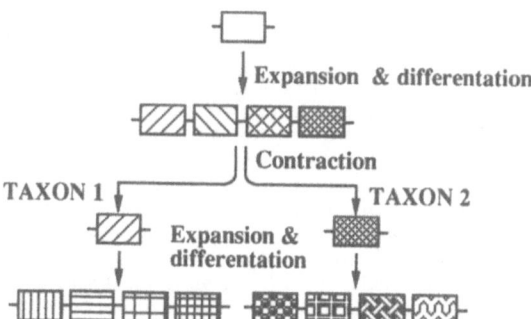

Fig. 5. Diagrammatic representation of expansion-contraction cycles of the sort presumably occurring in the *Mhc*. Each rectangle represents a single *Mhc* locus; functional differentiation is indicated by different shadings. Contraction to a single locus could be functional rather than physical.

It may not be a coincidence that expansion cycles seem to be associated with periods of major evolutionary radiation, where the foundations for higher taxa are laid down (the radiations of eutherian mammals, marsupials, birds, amphibians, and of major fish taxa). In this respect, the documented expansions might only be the proverbial tip of the iceberg; more expansions might be revealed when the *Mhc* studies are extended to include additional vertebrate taxa, particularly the different groups of fishes, many of which have been separated for hundreds of millions of years.

Only one *Mhc* subclass does not seem to evolve according to the accordion model -- the ß$_2$ microglobulin (ß2m), the ß-chain of the class I molecules. No evidence has yet been obtained to show that the *B2m* gene duplicated in any of the vertebrates studied, and the phylogenetic tree of the *B2m* genes coincides with the species trees (Fig. 2), so that the *B2m* genes of the different species, from fish to mammals, appear to be orthologous. Perhaps the *B2m* gene escaped the accordion effect because it transposed early in vertebrate evolution to a different chromosome from that carrying the rest of the *Mhc* genes.

The apparent association of the expansion-contraction cycles with the emergence of major taxonomic groups suggests that adaptive radiations provide conditions for the fixation of the rearrangements in the *Mhc* region. There are two such, not necessarily exclusive, conditions. First, if adaptive radiation starts, as is generally believed, from a relatively small founding population, the fixation of rearrangements could be the result of a purely stochastic process -- random genetic drift. The supposition here is that rearrangements occur all the time and are always present at low frequencies in the population, which is true for some parts of the *Mhc* at least (see below). In a large population, they do not have a chance of spreading and replacing the predominant chromosomal forms, but in a small population, the odds of their fixation are reasonably good. Second, adaptive radiation occurs when a population colonizes an entirely new environment and so it is driven by intense selection for adaptation to the diverse niches. An important part of this adaptation is, undoubtedly, coming to terms with new breeds of parasites. Since the surmised main function of the *Mhc* is protection against parasites, the expansion-contraction process might have an adaptive value. It might provide means of remolding the *Mhc* quickly in response to the need for protecting the emerging taxa from the dangers looming over them in the new nonself world.

Evidence that the expansion-contraction process is not *limited* to periods of adaptive radiation is available. It is provided, for example, by the variation in length of the primate *DRB* haplotypes (Klein et al. 1991) and by the homogenization of the *C4* and *CYP21* genes residing in the middle of the *Mhc* region (Kawaguchi et al. 1991). In the latter case, the multiple copies of the *C4* (or *CYP21*) genes in each of the primate species examined resemble each other in their sequence more than they do any of the *C4* genes of even a closely-related species. The most likely explanation for this observation is that unequal crossing-over frequently reduces the number of *C4* (and *CYP21*) copies to one and then expands it again in each species separately.

RESOLUTION

The contradiction between the two aspects of the *Mhc* -- the stability and instability aspects -- can be reconciled by assuming that the complex has evolved to accommodate two opposing needs. One is the need to nurture the T lymphocytes, and this requires a certain stability of the Mhc molecules, otherwise the two interacting components would not be able to recognize each other. (If we are right on this point, we would expect a similar evolutionary stability on the part of the T-cell receptor molecules.) The other is the need to deal with the extremely variable and ever-changing realm of the parasites, and this requires flexibility and hence instability. These are then the two faces of the *Mhc* Janus, one face turned to the self world (that of T cells) and the other to the nonself world (the parasites). Or, using the second of our two metaphors, the framework of the accordion must remain fixed so that the bellows can move freely and so allow the musician to play any melody he or she desires.

ACKNOWLEDGMENTS. We thank Ms. Lynne Yakes for editorial assistance and Ms. Anica Milosev for execution of the computer graphics.

REFERENCES

Amemiya CT, Litman GW (1991) Amer Zool 31:558-569
Cho S, Attaya M, Monaco JJ (1991) Nature 353:573-576
Dickersen RE, Geis I (1983) Hemoglobin: structure, function, evolution, and pathology. Benjamin/Cummings, Menlo Park, CA
Flajnik MF, Canel C, Kramer J, Kasahara, M (1991) Proc Natl Acad Sci USA 88:537-541
Grossberger D, Parham P (1992) Immunogenetics 36:166-174
Hashimoto K, Nakanishi T, Kurosawa Y (1990) Proc Natl Acad Sci USA 87:6863-6867
Hashimoto K, Nakanishi T, Kurosawa, Y (1992) Proc Natl Acad Sci USA 89:2209-2212
Kasahara M, Vazquez M, Sato K, McKinney EC, Flajnik MF (1992) Proc Natl Acad Sci USA, in press
Kawaguchi H, O'hUigin C, Klein J (1991) Evolution of primate *C4* and *CYP21* genes. In: Klein J, Klein D (eds) Molecular evolution of the major histocompatibility complex. Springer-Verlag, Heidelberg, p 357
Kelly AP, Monaco JJ, Cho S, Trowsdale J (1991) Nature 353:571-573
Klein D, Ono H, Vincek V, O'hUigin C, Klein J (1992) Proc. Natl. Acad. Sci. USA, submitted
Klein J, O'hUigin C, Kasahara M, Vincek V, Klein D, Figueroa F (1991) Frozen haplotypes in *Mhc* evolution. In: Klein J, Klein D (eds) Molecular evolution of the major histocompatibility complex. Springer-Verlag, Heidelberg, p 261
Ono H, Klein D, Vincek V, Figueroa F, O'hUigin C, Tichy H, Klein J (1992a) Proc Natl Acad Sci USA, in press
Ono H, Figueroa F, O'hUigin C, Klein J (1992b) Immunogenetics, in press
Saitou N, Nei M (1987) Mol Biol Evol 4:406-425
Schneider S, Vincek V, Tichy H, Figueroa F, Klein J (1991) Mol Biol Evol 8:753-766
Smith HW (1953) From fish to philosopher. The story of our internal environment. Little, Brown, Boston

Are Medial Class I Histocompatibility Antigens Coming of Age?

K. Fischer Lindahl

Howard Hughes Medical Institute, Departements of Microbiology and Biochemistry, University of Texas Health Science Center at Dallas, 5323 Harry Hines Blvd., Dallas, TX 75235-9050, U.S.A.

INTRODUCTION

The major histocompatibility complex (MHC) encodes one to three classical class I, or class Ia, molecules. These are expressed in all tissues at relatively high levels, present endogenous minor histocompatibility antigens and foreign viral antigens on the cell surface, and are usually highly polymorphic. In addition to the class Ia genes, vertebrates have other class I genes, the total ranging from seven in the miniature swine (Singer *et al.* 1987) and some ten in humans (Koller *et al.* 1989) to more than 45 in mice (Stroynowski 1990; Fischer Lindahl *et al.* 1991) and over 60 in rats (Jameson *et al.* 1992), the record so far. The function of these other genes has long been a matter of debate, reflected in their many names: *nonclassical*, reflecting their functional difference from the classical ones, *nonconventional*, reflecting their often peculiar expression patterns, *medial*, reflecting their lower antigenic strength, the neutral *class Ib*, and *class I-like*, which I dislike.

The class Ib genes and their products have been most intensively studied in the mouse, and authoritative and comprehensive reviews have appeared in recent years (Chen *et al.* 1987; Stroynowski 1990; Flaherty *et al.* 1990). My aim in this brief essay will be to discuss some selected aspects and recent advances in our understanding of class Ib genes and molecules and, in the process, to address some of the arguments for the notion that the class Ib genes are "rotting hulks" (Howard 1987) and evolutionary debris (Parham *et al.* 1989) in the "junkyard of the MHC" (Klein & Figueroa 1986).

A BRIEF HISTORY

The thymus-leukemia (TL) antigen was the first class Ib molecule to be discovered, as Old, Boyse, and Stockert were searching for leukemia-specific cell surface molecules (1963). Over the next ten years, biochemical analysis showed that TL was associated with ß$_2$-microglobulin (ß2m) (Vitetta *et al.* 1975) and related to the classical H-2 antigens. Considerable ingenuity went into the genetic and serological analysis of the expression of TL on thymocytes of certain strains and on leukemias, even from strains that did not express the antigen in the thymus (recently reviewed by Chorney *et al.* (1991)). The next class Ib molecules defined were Qa-1 (Stanton & Boyse 1976) and Qa-2 (Flaherty 1976), which were discovered on peripheral lymphoid cells and showed genetic linkage to *Tla*. A few years after the serological definition of these antigens, several laboratories independently observed cytotoxic T lymphocytes (CTL) specific for Qa-1 and Qa-2 that were not restricted by H-2 (reviewed by Fischer Lindahl & Langhorne (1981)).

So far, the study of class Ib antigens had been an esoteric interest of a few immunogeneticists and biochemists, but in 1982, molecular biologists entered the field with the demonstration of a surprising plethora of class I genes in the mouse MHC (Steinmetz *et al.* 1982). These genes now became a respectable object for study of evolution, regulation, tissue-specific expression, and alternative splicing, and of class I proteins with unusual or missing membrane associations; but, apart from a slowly growing number of protein products and their recognition by alloreactive CTL, function was unknown.

In 1983 came the first demonstrations, in rats (Livingstone 1983) and mice (Fischer Lindahl et al. 1983), that genes in the class Ib region of the MHC could affect the presentation of a minor histocompatibility antigen to CTL that were not restricted by a class Ia molecule; in the mouse, the antigen was mitochondrially inherited and ß2m was also involved. Since 1988 has followed demonstrations that T cell clones with γδ receptors recognize class Ib molecules (Bluestone et al. 1988; Bonneville et al. 1989), and Qa-1^b was shown to present a synthetic polypeptide, GT (Vidović et al. 1989). This immediately fostered the notion that presenting antigen to γδ T cells is what class Ib molecules are for (Raulet 1989; Strominger 1989).

Because they are weak, class Ib antigens can be studied functionally only in species with inbred strains and recombinant MHC haplotypes; cloned reagents, such as T cells, monoclonal antibodies, and transfected cell lines, will be required to learn about human class Ib molecules. From the bias in the literature, one might gain the impression that only mice have class Ib molecules. It is therefore worth pointing out that CTL directed at MHC products, but not restricted by class Ia, were discovered first in rats (Marshak et al. 1977), that the high number of class Ib genes in rats was noted early (Palmer et al. 1983), that the serological definition and CTL analysis of their products proceeded apace in rats (Stock & Günther 1982), and that class Ib restriction was discovered simultaneously in rats and mice (reviewed by Wang et al. (1991a)). However, fewer groups work with rats, and, although discoveries were made simultaneously, the wealth of reagents in mice facilitated subsequent analysis and exploration. The field owes a great debt to its pioneer, Dr. Edward A. Boyse, who began the development of recombinant inbred strains for the dissection of class Ib antigens, and his heir, Dr. Lorraine Flaherty, who continued the study and for many years has provided everybody else with essential mice and antisera.

ANTIGEN PRESENTATION

Cell surface presentation of processed peptides to CD8$^+$ T cells is now understood to be *the* biological function of class Ia molecules. The earliest demonstrations of MHC restriction employed strains that shared many class Ib antigens, and these did not allow T cell recognition of the viral or minor histocompatibility antigens that were being studied (Zinkernagel & Doherty 1975; Gordon et al. 1975; Bevan, 1975). Specific tests for the ability of Qa-1 to present minor antigens soon confirmed this conclusion (Kastner et al. 1979; Fischer Lindahl & Langhorne, 1981). However, examples of antigen presentation by class Ib molecules have emerged and accumulated over the years; as I have recently reviewed these in detail (Wang et al. 1991a; Fischer Lindahl 1992), I shall discuss only a few examples here.

The Maternally Transmitted Antigen and *H-2M3*

The maternally transmitted antigen (Mta) of mice was discovered as a transplantation antigen, detected by H-2-unrestricted CTL, that showed strict maternal inheritance (Fischer Lindahl et al. 1980), and it was later realized that an *H-2*-linked gene, presumably a class Ib gene, was also required for its recognition (Fischer Lindahl et al. 1983). We now know that the maternally inherited factor, MTF, is the amino-terminal end of the mitochondrially encoded protein, ND1, a subunit of respiratory complex I located in the mitochondrial inner membrane. A polymorphism in the sixth codon causes conservative substitutions (Ile, Ala, Val, and Thr) that make this peptide act as a minor histocompatibility antigen (Loveland et al. 1990).

MTF is presented on cells by a ß2m-associated, class Ib molecule, H-2M3, which is encoded at the distal end of the mouse MHC (Wang et al. 1991b). *H-2M3* is expressed throughout life, from before day 8 of gestation, in all nucleated cells examined and perhaps at highest levels in the thymus (Wang & Fischer Lindahl 1992). The *H-M3* mRNA is rare, and it has not appeared in random screens of class I cDNA clones from various sources, nor is the protein common, and its surface expression may also be limited by the supply of the mitochondrial ligand (Loveland et al. 1990). The H-2M3 protein has only a short cytoplasmic anchor, but its extracellular domains have the same length and share many consensus residues with class Ia molecules, including disulphide bridges and a single site for N-glycosylation at Asn86. A substitution of Gln for Leu95, expected to point into the peptide-binding groove, prevents recognition of the H-2M3^{cas3} form by CTL (Wang & Fischer Lindahl 1992).

Mitochondria, like the prokaryotes they originated from, initiate protein synthesis with *N*-formyl-methionine, and MTF therefore carries a formyl-group that distinguishes it from any peptide synthesized on cytoplasmic ribosomes (Loveland *et al.* 1990). Shawar and his colleagues (1990) showed that this formyl-group is essential for binding of ND1 peptides to H-2M3, and that H-2M3 can also bind other peptides, provided they have a N-terminal formyl-group. This finding provided a structural foundation for the notion that H-2M3 might present antigens of intracellular parasites (Fischer Lindahl *et al.* 1989) and thus play a selective role in host defense against prokaryotes (Shawar *et al.* 1991b).

Definitive evidence for such a role has come from studies of the immune response to *Listeria monocytogenes*. It has been known for some time that mice make CD8$^+$ CTL upon *Listeria* infection, and that some of these are not restricted by H-2 class Ia molecules (Kaufmann *et al.* 1988). Pamer *et al.* (1992) have now shown that one such clone recognizes a peptide produced by the bacteria, rather than the infected cells, and that this peptide, which has a blocked amino-terminus, is presented by H-2M3. An *H-2M3^{cas3}* fibroblast cell line cannot present the peptide, unless it is transfected with a cosmid that carries the *H-2M3^d* gene (as its only class I gene), and presentation of this peptide can be prevented by competition with an ND1 peptide and some other formylated peptides, but not by peptides with free aminotermini. Kurlander *et al.* (1992) have reached the same conclusion based on similar competition experiments, using partially purified *Listeria* membranes.

RMA-S, Qa-1, and Heat Shock Proteins

The lack of polymorphism of class Ib molecules has made it harder to detect their products. T cells restricted by such molecules may appear MHC-unrestricted, but cell lines and mice deficient in ß2m are now available to test involvement of a class I molecule in such responses. Furthermore, mutant cell lines, such as the mouse RMA-S cells (Townsend *et al.* 1989), have led to the realization that class Ia molecules are unstable at 37°C and impeded in their transport to the cell surface, unless they can associate with a peptide in addition to the ß2m light chain (Townsend *et al.* 1990; Ljunggren *et al.* 1990). These cell lines are deficient in MHC-encoded transporter proteins required for normal peptide processing in the class I presentation pathway (Monaco 1992); surface display of class I molecules on RMA-S can be restored by transfection of a normal *Tap-2* gene (Powis *et al.* 1991; Attaya *et al.* 1992). The RMA-S defect affects not only class Ia molecules, but also Mta and Qa-1 (Hermel *et al.* 1991). As expected, Mta can be restored by incubation of RMA-S cells with a synthetic ND1 peptide that mimics MTF.

RMA-S cells have been useful for dissecting the specificity of alloreactive, anti-Qa-1^b CTL responses into at least four types of clones (Aldrich *et al.* 1992). About half the clones generated in a secondary in vitro response recognize RMA-S cells at least as well as the parent RMA line; they most likely recognize a signal peptide that enters the endoplasmic reticulum without the aid of the TAP transporters (Wei & Cresswell 1992; Henderson *et al.* 1992). The rest of the clones do not recognize RMA-S cells, unless these are transfected with a functional *Tap-2* gene. The peptide-specific clones can be further subdivided. Some of them do not recognize cells homozygous for *H-2D^k*; a gene, *Qdm*, in this region has long been known to affect Qa-1 expression (Fischer Lindahl 1983; Aldrich *et al.* 1988), and a likely explanation is that all H-2D molecules, except D^k, provide a peptide that binds to Qa-1. And finally, some, but not all, *Tap-2*-dependent, *Qdm*-independent clones recognize RMA-S treated with oligomycin or ionomycin; the increase in intracellular calcium caused by these drugs may activate proteases or cause selective leakiness of the ER membranes, and it does not affect Mta or H-2 display to nearly the same degree (Hermel *et al.* 1992).

Surface expression of Qa-1 on L cells transfected with the $T23^d$ gene, which encodes Qa-1^b, is increased when these L-g37 cells are incubated with a tryptic digest of mycobacterial 65 kDa heat shock protein, HSP60 (Imani & Soloski 1991). A synthetic peptide corresponding to amino acids 181-195 of HSP60 binds to and is immunoprecipitated with Qa-1 (M. Soloski, pers. comm.); and heat-shocked L-g37 cells display more Qa-1^b (Imani & Soloski 1991). Heat shock can also stimulate γδ T cell recognition of TL (Eghtesady & Kronenberg 1992) (see below). The 180-196 peptide of mycobacterial HSP60 and the homologous mouse peptide are also favoured antigens for T cells with Vγ1 and, usually, Vδ6 receptor chains (O'Brien et al. 1992); though appearing unrestricted, these γδ T cells may also require presentation by a ß2m-associated, class Ib molecule (Harshan et al. 1992).

In the Groove

If indeed class Ib molecules are adapted to the binding of conserved, bacterial peptides, one would expect them to have correspondingly conserved grooves with special features to accommodate these peptides. Four conserved tyrosines in class Ia molecules are implicated in the coordination of the first amino acid in the bound peptides (Madden et al. 1991; Matsumura et al. 1992); a Phe in place of Tyr171 in H-2M3 may accommodate the formyl-group (Wang et al. 1991b). The C-terminus of a bound peptide is coordinated by conserved residues Tyr84, Thr143, Lys146, and Trp147. In H-2M3, Trp147 is replaced by Leu (M3^d) or Phe (in *Mus spretus*) (Wang et al. 1991a), and in Qa-1 both 143 and 147 are replaced by Ser (Connolly et al. 1992). The loss of consensus residues for binding of peptide termini may make the groove more selective for peptides with more correctly fitting side chains in the middle.

An ortholog of *H-2M3* is abundantly expressed in rats; the protein encoded by this gene differs in several residues pointing up from the α1 and α2 helices, but not in any residue on the bottom of the groove, and it shares the signature residues Gln34, Leu95 and Phe171 (Wang et al. 1991a). The TL antigen can be encoded by nonallelic genes in different haplotypes, e.g. $T18^d$ or $T3^b$, but when the predicted structures are compared, the groove is conserved and the differences cluster on the outside loops (Chen et al. 1987; Eghtesady et al. 1992), accounting for their many serological epitopes. Qa-1^b and Qa-1^a are also encoded by pseudoallelic genes, but their α1 and α2 domains differ in only two residues (9 and 24) inside the groove and twelve outside (Connolly et al. 1992).

The $T22^b$ gene encodes a class Ib molecule with a deletion at the beginning of the α2 helix; yet, it is recognized by the KN6 γδ T cell hybridoma and thus clearly functional (Ito et al. 1990); $T10^d$, presumably a pseudoallele (see below), has the same deletion. Until we know the peptides bound by class Ib molecules and understand their affinity, we must be careful not to judge Ib molecules nonfunctional solely because they deviate from a class Ia consensus.

The peptide elution that has been so useful for obtaining sequences of peptides in class Ia molecules is hampered by the shortage of class Ib material. However, work on Qa-2 is now in progress in two laboratories, which have identified a small set of nonamer peptides with His in the seventh position (I. Stroynowski, pers. comm.) and Leu, Ile, or Phe in the ninth (Rötzschke et al. 1992).

T Cell Receptors

Some class Ib molecules are indeed recognized by γδ T cell receptors, and the limited tissue distribution of some class Ib molecule may match the idiosyncratic distribution of T cells with receptors of particular Vγ and Vδ classes. TL antigen in the small intestinal epithelium stimulates Vγ5 T cells (Eghtesady & Kronenberg 1992); a class Ib molecule expressed in keratinocytes might be the ligand for Vγ3 T cells (Allison & Havran 1991). Qa-1 is seen by both γδ (Vidović et al. 1989) and αß receptors (Germana & Shinohara 1991; C.J. Aldrich, pers. comm.). Qa-2 (L. Lowen & J. Forman, pers. comm.) and M3 (Fischer Lindahl et al. 1989; Shawar et al. 1991a; Pamer et al. 1992) specific clones and lines have αß receptors, but this does of course not exclude that these antigens might also be seen by γδ T cells.

T cell interaction with Qa-1 and M3 depends on CD8 (Aldrich *et al.* 1992; Shawar *et al.* 1991a). The α3 domain of TL can also bind CD8 (Teitell *et al.* 1991), but that of Qa-2 cannot (Aldrich *et al.* 1991). A role for class Ib molecules in positive selection in the thymus is yet to be demonstrated, but selection of the KN6 γδ receptor (specific for T22^b) does require expression of ß2m, and hence class I molecules, in the thymus (Pereira *et al.* 1992).

NONCONVENTIONAL EXPRESSION

Twenty seven years after its discovery, the thymus-leukemia antigen was found to be expressed in the epithelium of the small intestine (Hershberg *et al.* 1990; Wu *et al.* 1991). It had been argued that because certain strains of mice did not express TL in the thymus, the function of TL, if any, could not be essential; however, *all* strains of mice express TL in the gut. Furthermore, intraepithelial T lymphocytes isolated from the small intestine respond to epithelial cells by release of serine esterases, an early sign of activation of cytolytic T cells. This response requires expression of ß2m on the epithelial cells, it can be inhibited by antibodies against TL, γδ T cell receptors in general, or Vγ5 specifically, and it can be stimulated by heat-shocking the epithelial cells (Eghtesady *et al.* 1992; Eghtesady & Kronenberg 1992). This first evidence of a T cell response to TL, coupled with expression of the antigen in all strains of mice, suggests that we may finally be on the trail of TL's function. It follows that any search for a TL homolog in the thymus of other species was looking in the wrong place.

Other class Ib molecules have similarly nonconventional expression patterns. Q10 is secreted in the liver (Kress *et al.* 1983; Devlin *et al.* 1985); Q4 (Qb-1) is also secreted (Robinson 1987); Qa-2 may be linked to the membrane by phosphoinositol, which can be cleaved (Stroynowski *et al.* 1987); I shall not venture a guess as to the function of such soluble class I molecules in serum, found also in other species (Güssow & Ploegh 1987). Q5^k is expressed in postimplantation embryos, in the thymus, and in the uterus of pregnant mice (Schwemmle *et al.* 1991). HLA-G is expressed only in placenta and in the eye (Ishitani & Geraghty 1992). Given these idiosyncracies, we must be cautious about the failure to find expression of other full-length, seemingly intact class Ib genes from the *M* region, such as *M1* (Singer *et al.* 1988), *M5* (Fischer Lindahl *et al.* 1991), and *Mb2* and *Mb3* (R. Arepalli, D.S. Singer, & S. Rudikoff, pers. comm.). Similarly, low levels of expression may be compatible with presentation of a few species of peptides that bind with high affinity to an adapted groove.

PSEUDOGENES - A PSEUDOPROBLEM?

The class Ib region is unquestionably littered with fragments of class I genes whose only remaining function could be as a source for segmental exchange of genetic information (Geliebter, Nathenson 1987). Other class Ib genes have awkward stop codons or faulty splice sites, but different forms of membrane association may overcome such defects (Stroynowski *et al.* 1987). Alternative splicing is observed for class Ib transcripts (Lalanne *et al.* 1985; Ulker *et al.* 1990; R. Ehrlich & D.S. Singer, pers. comm.) and it can lead to protein products (Ishitani & Geraghty 1992), for which a function remains to be defined, however.

The high number of tandemly repeated, homologous genes in the central, class Ib region of the MHC is both cause and result of unequal recombination, which continuously shapes the MHC, deleting and duplicating class I genes (for examples, see Stephan *et al.* (1991a)). Haplotypes lacking certain genes may suggest these are dispensable, but the plasticity of the class Ib gene family creates a redundancy that can compensate for loss of gene function. Thus the allele of the *T23^d* gene that encodes Qa-1^b is a pseudogene (Nakayama *et al.* 1990), and Qa-1^a is encoded by a pseudoallele (Connolly *et al.* 1992), whose location has yet to be identified. The *d* allele of *T22^b* is a pseudogene, but its function may be assumed by the closely related *T10^d* gene, whose *b* allele is a pseudogene (Ito *et al.* 1990; Pereira *et al.* 1992). In the *d* haplotype, TL is encoded by *T18*, missing in the *b* haplotype, which uses *T3*; the *a* haplotype, which has the highest level of TL may use more than one gene (Chen *et al.* 1987). It would be of great interest to know the organization of *T* region genes in further haplotypes and to test (by PCR analysis) whether the known pseudogenes are pseudogenes in all haplotypes.

Of the class Ia genes, *H-2K*, *RT1.A*, and *HLA-B* are relatively isolated from other class I genes at one end of the MHC, unlike *H-2D* in its frequently changing environment (Stephan *et al.* 1986) and *HLA-A* with its entourage of class Ib genes. At the other end of the mouse MHC, the expressed class Ib genes, *H-2M2* and *M3*, which are both conserved in various species of wild mice and in rats (Brorson *et al.* 1989; Wang *et al.* 1991a; Wang *et al.* 1991b), also reside in splendid isolation; we have isolated each on a yeast artificial chromosome of about 250 kb with no other class I genes (E.P. Jones & K. Fischer Lindahl, unpublished).

Polymorphism is a hallmark of class Ia genes, which are distinguished both by the high number of alleles and by the divergence of these alleles (Klein 1986). There are good arguments (Klein 1991) and direct evidence (Hill *et al.* 1991) that this polymorphism, which is unique among genetic systems, is maintained through selection by parasites. In contrast, the class Ib genes are oligomorphic; this does not imply that they have no function, but it does argue for a function different from class Ia, perhaps the presentation of conserved motifs as discussed above. I believe that some class Ib genes, such as *H-2M3*, have more alleles (or pseudoalleles) than currently identified, but these alleles are not as divergent as the class Ia alleles.

CONCLUSION

None of the arguments presented heɪe will convince Jan Klein that class Ib molecules are important, or anybody else that they are essential (Hedrich 1992), but they do show that class Ib molecules can present antigen to both $\alpha\beta$ and $\gamma\delta$ T cells. This was discovered in fifteen years or less, whereas it took close to forty years to reach the same conclusion for class Ia molecules (Zinkernagel & Doherty 1975). At the very least, the class Ib molecules should teach us some humility; if indeed TL was studied for twenty seven years in the wrong place, who knows whether we are barking up other wrong trees?

REFERENCES

Aldrich CJ, Rodgers JR, Rich RR (1988) Immunogenetics 28:334-344

Aldrich CJ, Hammer RE, Jones-Youngblood S, Koszinowski U, Hood L, Stroynowski I, Forman J (1991) Nature 352:718-721

Aldrich CJ, Waltrip R, Hermel E, Attaya M, Fischer Lindahl K, Monaco JJ, Forman J (1992) J Immunol, in press

Allison JP, Havran WL (1991) Annu Rev Immunol 9:679-705

Attaya M, Jameson S, Martinez CK, Hermel E, Aldrich C, Forman J, Fischer Lindahl K, Bevan MJ, Monaco JJ (1992) Nature 355:647-649

Bevan MJ (1975) Nature 256:419-421

Bluestone JA, Cron RQ, Cotterman M, Houlden BA, Matis LA (1988) J Exp Med 168:1899-1916

Bonneville M, Ito K, Krecko EG, Itohara S, Kappes D, Ishida I, Kanagawa O, Janeway CA,Jr., Murphy DB, Tonegawa S (1989) Proc Natl Acad Sci USA 86:5928-5932

Brorson K, Richards CS, Hunt SW,III, Cheroutre H, Fischer Lindahl· K, Hood L (1989) Immunogenetics 30:273-283

Chen Y-T, Obata Y, Stockert E, Takahashi T, Old LJ (1987) Immunol Res 6:30-45

Chorney MJ, Mashimo H, Chorney KA, Vasavada H (1991) *TL* genes and antigens. In: Srivastava R (ed) Immunogenetics of the Major Histocompatibility Complex. VCH, New York, p 177

Connolly DJ, Dyson PJ, Hederer RA, Thorpe CJ, Travers PJ, McVey JH, Robinson PJ (1992) Submitted

Devlin JJ, Lew AM, Flavell RA, Coligan JE (1985) EMBO J 4:369-374

Eghtesady P, Kronenberg M (1992) Submitted

Eghtesady P, Panwala C, Teitell M, Kronenberg M (1992) Submitted
Fischer Lindahl K (1983) Transplant Proc 15:2042-2044
Fischer Lindahl K (1992) Semin Immunol, in press
Fischer Lindahl K, Langhorne J (1981) Scand J Immunol 14:643-654
Fischer Lindahl K, Bocchieri M, Riblet R (1980) J Exp Med 152:1583-1596
Fischer Lindahl K, Hausmann B, Chapman VM (1983) Nature 306:383-385
Fischer Lindahl K, Hermel E, Loveland BE, Richards S, Wang C-R, Yonekawa H (1989) Cold Spring Harbor Symp Quant Biol 54:563-569
Fischer Lindahl K, Hermel E, Loveland BE, Wang C-R (1991) Annu Rev Immunol 9:351-372
Flaherty L (1976) Immunogenetics 3:533-539
Flaherty L, Elliott E, Tine JA, Walsh AC, Waters JB (1990) CRC Crit Rev Immunol 10:131-175
Geliebter J, Nathenson SG (1987) Trends Genet 3:107-112
Germana S, Shinohara N (1991) Immunology 74:578-582
Gordon RD, Simpson E, Samelson LE (1975) J Exp Med 142:1108-1120
Güssow D, Ploegh H (1987) Immunol Today 8:220-222
Harshan K, Dallas A, Cranfill R, Townend W, Potter T, Koller B, O'Brien R, Born W (1992) J Immunol, in press
Hedrich SM (1992) Cell 70:177-180
Henderson RA, Michel H, Sakaguchi K, Shabanowitz J, Appella E, Hunt DF, Engelhard VH (1992) Science 255:1264-1266
Hermel E, Grigorenko E, Fischer Lindahl K (1991) Int Immunol 3:407-412
Hermel E, Grigorenko E, Aldrich CJ, Forman J, Fischer Lindahl K (1992) In preparation
Hershberg R, Eghtesady P, Sydora B, Brorson K, Cheroutre H, Modlin R, Kronenberg M (1990) Proc Natl Acad Sci USA 87:9727-9731
Hill AVS, Allsopp CEM, Kwiatkowski D, Anstey NM, Twumasi P, Rowe PA, Bennett S, Brewster D, McMichael AJ, Greenwood BM (1991) Nature 352:595-600
Howard JC (1987) MHC organization of the rat: evolutionary considerations. In: Kelsoe G, Schulze DH (eds) Evolution and Vertebrate Immunity. The Antigen-receptor and MHC Gene Families. University of Texas Press, Austin, TX, p 397
Imani F, Soloski MJ (1991) Proc Natl Acad Sci USA 88:10475-10479
Ishitani A, Geraghty DE (1992) Proc Natl Acad Sci USA 89:3947-3951
Ito K, Van Kaer L, Bonneville M, Hsu S, Murphy DB, Tonegawa S (1990) Cell 62:549-561
Jameson SC, Tope WD, Tredgett EM, Windle JM, Diamond AG, Howard JG (1992) J Exp Med 175:1749-1757
Kastner DL, Rich RR, Chu L (1979) J Immunol 123:1239-1244
Kaufmann SHE, Rodewald H-R, Hug E, de Libero G (1988) J Immunol 140:3173-3179
Klein J (1986) Natural History of the Major Histocompatibility Complex, John Wiley and Sons, New York
Klein J (1991) Hum Immunol 30:247-258
Klein J, Figueroa F (1986) CRC Crit Rev Immunol 6:295-386
Koller BH, Geraghty DE, DeMars R, Duvick L, Rich SS, Orr HT (1989) J Exp Med 169:469-480
Kress M, Cosman D, Khoury G, Jay G (1983) Cell 34:189-196
Kurlander RJ, Shawar SM, Brown ML, Rich RR (1992) Science 257:678-679
Lalanne J-L, Transy C, Guerin S, Darche S, Meulien P, Kourilsky P (1985) Cell 41:469-478
Livingstone AM (1983) Specificity of responses against rat major transplantation antigens, Ph.D. Thesis, Cambridge University
Ljunggren H-G, Stam NJ, Öhlén C, Neefjes JJ, Höglund P, Heemels M-T, Bastin J, Schumacher TNM, Townsend A, Kärre K, Ploegh HL (1990) Nature 346:476-480
Loveland BE, Wang C-R, Yonekawa H, Hermel E, Fischer Lindahl K (1990) Cell 60:971-980
Madden DR, Gorga JC, Strominger JL, Wiley DC (1991) Nature 353:321-325
Marshak AD, Doherty PC, Wilson DB (1977) J Exp Med 146:1773-1790
Matsumura M, Fremont DH, Peterson PA, Wilson IA (1992) Science 257:927-934

Monaco JJ (1992) Immunol Today 13:173-181

Nakayama K-I, Tokito S, Jaulin C, Delarbre C, Kourilsky P, Nakauchi H, Gachelin G (1990) J Immunol 144:2400-2408

O'Brien RL, Fu Y-X, Cranfill R, Dallas A, Ellis C, Reardon C, Lang J, Carding SR, Kubo R, Born W (1992) Proc Natl Acad Sci USA 89:4348-4352

Old LJ, Boyse EA, Stockert E (1963) J Natl Cancer Inst 31:977-986

Palmer M, Wettstein PJ, Frelinger JA (1983) Proc Natl Acad Sci USA 7616:7620

Pamer EG, Wang C-R, Flaherty L, Fischer Lindahl K, Bevan MJ (1992) Cell 70:215-223

Parham P, Benjamin RJ, Chen BP, Clayberger C, Ennis PD, Krensky AM, Lawlor DA, Littman DR, Norment AM, Orr HT, Salter RD, Zemmour J (1989) Cold Spring Harbor Symp Quant Biol 54:529-543

Pereira P, Zijlstra M, McMaster J, Loring JM, Jaenisch R, Tonegawa S (1992) EMBO J 11:25-31

Powis SJ, Townsend ARM, Deverson EV, Bastin J, Butcher GW, Howard JC (1991) Nature 354:528-531

Raulet DH (1989) Nature 339:342-343

Robinson PJ (1987) Proc Natl Acad Sci USA 84:527-531

Rötzschke O, Falk K, Stefanovic S, Grahovac B, Soloski MJ, Jung G, Rammensee H-G (1992) Submitted

Schwemmle S, Bevec D, Brem G, Urban MB, Baeuerle PA, Weiss EH (1991) Immunogenetics 34:28-38

Shawar SM, Cook RG, Rodgers JR, Rich RR (1990) J Exp Med 171:897-912

Shawar SM, Rodgers JR, Cook RG, Rich RR (1991a) Immunol Res 10:365-375

Shawar SM, Vyas JM, Rodgers JR, Cook RG, Rich RR (1991b) J Exp Med 174:941-944

Singer DS, Ehrlich R, Satz L, Frels W, Bluestone J, Hodes R, Rudikoff S (1987) Vet Immunol Immunopathol 17:211-221

Singer DS, Hare J, Golding H, Flaherty L, Rudikoff S (1988) Immunogenetics 28:13-21

Stanton TH, Boyse EA (1976) Immunogenetics 3:525-531

Steinmetz M, Winoto A, Minard K, Hood L (1982) Cell 28:489-498

Stephan D, Sun H, Fischer Lindahl K, Meyer E, Hämmerling GJ, Hood L, Steinmetz M (1986) J Exp Med 163:1222-1244

Stock W, Günther E (1982) J Immunol 128:1923-1928

Strominger JL (1989) Cell 57:895-898

Stroynowski I (1990) Annu Rev Immunol 8:501-530

Stroynowski I, Soloski M, Low MG, Hood L (1987) Cell 50:759-768

Teitell M, Mescher MF, Olson CA, Littman DR, Kronenberg M (1991) J Exp Med 174:1131-1138

Townsend A, Öhlén C, Bastin J, Ljunggren H-G, Foster L, Kärre K (1989) Nature 340:443-448

Townsend A, Elliott T, Cerundolo V, Foster L, Barber B, Tse A (1990) Cell 62:285-295

Ulker N, Lewis KD, Hood LE, Stroynowski I (1990) EMBO J 9:3839-3847

Vidović D, Roglić M, McKune K, Guerder S, MacKay C, Dembić Z (1989) Nature 340:646-650

Vitetta ES, Uhr JW, Boyse EA (1975) J Immunol 114:252-254

Wang C-R, Livingstone A, Butcher GW, Hermel E, Howard JC, Fischer Lindahl K (1991a) Antigen presentation by neoclassical MHC class I gene products in murine rodents. In: Klein J, Klein D (eds) NATO ASI Series, Vol. H 59: Molecular Evolution of the Major Histocompatibility Complex. Springer Verlag, Berlin Heidelberg, p 441

Wang C-R, Loveland BE, Fischer Lindahl K (1991b) Cell 66:335-345

Wang C-R, Fischer Lindahl K (1992) Submitted

Wei ML, Cresswell P (1992) Nature 356:443-446

Wu M, Van Kaer L, Itohara S, Tonegawa S (1991) J Exp Med 174:213-218

Zinkernagel RM, Doherty PC (1975) J Exp Med 141:1427-1436

Trans-species Polymorphism of the Major Histocompatibility Complex (*Mhc*) Loci

Naoyuki Takahata, Yoko Satta[*], and Jan Klein[*+]

National Institute of Genetics, Mishima 411, Japan

INTRODUCTION

The functional major histocompatibility complex (*Mhc*) genes are highly polymorphic. The *Mhc* polymorphism is unusual in many respects, among which the *trans*-species mode of evolution is the most remarkable. It has convincingly been shown that a large number of allelic lineages at functional *Mhc* loci have persisted for tens of millions of years in various lines of species that include humans. Because of the long persistence time, some human *Mhc* (*HLA*) allelic lineages are shared not only by African apes, but also by Old World monkeys, and the divergence of alleles often predates species divergences to a great extent. This finding has raised many challenging problems for evolutionary biologists. In this paper we present our analysis of the nucleotide substitution rates as well as the mode and intensity of natural selection occurring at the primate functional *Mhc* genes. The implication of the *HLA* polymorphism for human evolution is discussed.

TRANS-SPECIES POLYMORPHISM

Sampling of orthologous *Mhc* genes from a variety of primate species has revealed that often genes from different species are more similar to one another than alleles from the same species (Mayer et al. 1988, 1992; Lawlor et al. 1988; Fan et al. 1989; Gyllensten and Erlich 1989; Kenter et al. 1992; Kupfermann et al. 1992). This finding is consistent with the proposal that genes at orthologous *Mhc* loci often diverge before the divergence of the species -- that they evolve in a *trans*-species manner (Klein 1980). In such cases, the gene genealogy does not reflect faithfully species phylogeny.

Trans-species polymorphism can only be observed when genes of a given lineage are transmitted from an ancestral to descendant species and if they appear in the tested sample. [For a discussion of the conditions under which this takes place, see Klein and Takahata (1990).] But even when an allelic lineage is lost in one species and retained in another or if it is not represented in the sample, we can still infer the divergence time of two alleles obtained from the same species, provided that we can calibrate accurately the rate of molecular evolution of the *Mhc* genes. If the time turns out to be longer than the age of this species, we can conclude that the observed polymorphism has been transmitted *trans*-specifically.

[*]Max-Planck-Institut für Biologie, Abteilung Immungenetik, 7400 Tübingen, Germany

[†]Department of Microbiology and Immunology, University of Miami School of Medicine, Miami, FL 33101, USA

Usually, to study the rate of molecular evolution, we sample orthologous genes from different species and assume that the gene divergence time is the same as the species divergence time. Obviously, this assumption does not always hold for *Mhc* genes undergoing the *trans*-species mode of evolution (Mayer et al. 1988, 1992; Lawlor et al. 1988; Fan et al. 1989; Gyllensten and Erlich 1989; Kupfermann et al. 1992; Kenter et al. 1992).

The rate of nucleotide substitutions provides valuable information on the evolutionary mechanism and functional importance of molecules or part of a molecule. Although any kind of change in DNA sequences is potentially contributory to phylogenetic study, only nucleotide substitutions can be used to accurately determine the rate (Nei 1987). Under neutrality or in the absence of natural selection (Kimura 1968), the substitution rate becomes equal to the mutation rate. Therefore, if we know which genes or which parts of a gene are likely to be neutral, we can estimate the mutation rate of a gene. In the presence of natural selection, the substitution rate becomes either higher or lower than the mutation rate (Kimura 1983). Higher substitution rates are an indication that natural selection has favored certain mutations, whereas lower substitution rates imply that natural selection has tended to maintain the *status quo* of a molecule so as not to disturb its established function.

There indeed exist three distinct DNA regions in a functional *Mhc* gene in terms of action of natural selection: neutral, positively selected, and negatively selected. In the coding part of an *Mhc* gene, the candidate for the neutral region is the synonymous site at which nucleotide changes do not result in amino acid replacements. The positively and negatively selected regions are the nonsynonymous (amino acid replacement) sites in the peptide-binding region (PBR) and those in the non-PBR, respectively. Mutations at a nonsynonymous site in the non-PBR can be divided into two categories: neutral and deleterious. Deleterious mutations do not contribute to the substitution rate at all (Kimura 1991) so that only the remaining neutral mutations matter in this region. The amino acid residues in the PBR determine the repertoire of peptides bound by an Mhc molecule (Bjorkman et al. 1987) and they are therefore considered to be the main target of positive selection (Hughes and Nei 1988). The number of peptide-binding amino acid residues of a class I molecule is larger than that of a class II molecule (Brown et al. 1988).

The relative importance of natural selection acting on the nonsynonymous changes in the PBR and non-PBR region can be evaluated without knowing allelic divergence times. Under tight linkage, all the nucleotide sites of a pair of alleles must have had the same evolutionary history. Therefore, for a given pair of alleles, we can use, as a measure of positive and negative selection, the ratio of the number of nonsynonymous substitutions per site to the number of synonymous substitutions per site. We denote this ratio as γ in the PBR and as f in the non-PBR. The finding of $\gamma > 1$ and $f < 1$ by Hughes and Nei (1988) demonstrated clearly that natural selection has played an important part in *Mhc* evolution (see also Satta 1992).

The calibration of absolute substitution rates is, however, more desirable than estimating ratios because, as we shall see later, we can infer not only mutation rates and divergence times of genes from substitution rates, but also selection intensity (s) and population size (N_e). There are two statistical problems associated with the estimation of the substitution rates. One problem occurs when even two neutral genes are sampled from different species. The divergence time between these genes necessarily exceeds the species divergence time (T) by $2N_e$ generations on average (Takahata and Nei 1985). Hence, unless $T \gg 2N_e$, we cannot use T as the gene divergence time. This problem is more serious when one deals with *Mhc* genes (whose divergence time may be much longer than T) than when dealing with other genes (whose divergence time may approximate T). To avoid this problem we can compare two species with large T. However, genes of such highly diverged species may have evolved at quite different substitution rates (Li et al. 1985) and there may be a statistical difficulty in dealing with extensive multiple-hit substitutions in the aligned DNA sequences.

Table 1. Estimates of the number of nonsynonymous substitutions in the PBR ($\hat{K}_B$), the degree of selective constraint ($1-\hat{f}$) in the non-PBR nonsynonymous sites, and the enhanced nonsynonymous substitution rate ($\hat{\gamma}$) in the PBR relative to the synonymous substitution rate.

| Parameter | HLA loci | | | |
	A	B	C	DRB1
$\hat{K}_B$	25.3 (15.4)	28.9 (18.3)	10.5 (8.4)	19.4 (15.1)
$\hat{f}$	0.41	0.38	0.37	0.33
$\hat{\gamma}$	· 4.50	5.49	1.98	6.06

Estimators used are $\hat{K}_B = \dfrac{(1 + K_S + K_N) K_B(m)}{1 + m}$, $\hat{f} = \dfrac{K_N L_S}{L_N K_S}$, and $\hat{\gamma} = \dfrac{\hat{K}_B L_S}{L_B K_S}$

where $K_B(m)$ is the number of nonsynonymous substitutions in the PBR between alleles when the number of other types of substitutions is m. The value of m is 10 to 20, depending on the locus. K and L stand for the mean number of substitutions in all pairwise comparisons and the number of nucleotide sites in the region specified by the subscript. Subscripts s and N stand for the synonymous and non-PBR nonsynonymous sites. The K_B values in all pairwise comparisons (in parentheses) are smaller than $\hat{K}_B$, suggesting that multiple-hit substitutions in the PBR are underestimated by the standard method.

In a collection of *trans*-specifically evolving, orthologous *Mhc* genes from related species of the same zoological order, the distribution of divergence times among the various gene pairs can be expected to be random. In this random distribution, genes in some pairs can be expected to have diverged before the divergence of the species and others about the time of species divergence. If we could identify the latter gene pairs, we could estimate from the number of substitutions by which they differ the substitution rate of the *Mhc* genes because the divergence time of genes in each pair would, in this case, be approximately given by the divergence time of the species. The method of identifying these gene pairs was developed by Satta et al. (1991). The application of this method to the extensively studied primate *DRB1* genes indicates that the *DRB1* pair whose divergence must have been closest to species divergence are genes from human and macaque. Using these genes and $T = 23$ million years as the divergence time between human and macaque, Satta (1992) estimated the synonymous substitution rate (μ) of the *DRB1* gene as $1.2 \pm 0.4 \times 10^{-9}$ per site per year and regarded μ as the mutation rate in the *Mhc* region. Were positive selection absent at the PBR nonsynonymous sites, the neutral mutation rate per gene could be computed from $\mu(L_S + fL_N + L_B)$, where L_S, L_N, and L_B are the numbers of synonymous, non-PBR nonsynonymous, and PBR nonsynonymous sites, respectively. The mutation rates for class I and class II genes then become approximately 7.8×10^{-7} and 3.5×10^{-7} per year, respectively. In discussing the extent of polymorphism, however, an appropriate time unit is not one year, but one *generation* of breeding individuals. If we take the generation time (g) in the human lineage as 15 years, the per-generation mutation rates (v) per class I and class II locus become 1.2×10^{-5} and 0.5×10^{-5}, respectively.

These mutation rates are clearly incompatible with the hypothesis of high mutability at *Mhc* loci for the following reason: Calculations based on most non-*HLA* loci suggest that the effective population size, N_e, of humans has been about 10^4 for the last one million years (Takahata 1993). Under neutrality and in a population of $N_e = 10^4$, the probability that two randomly chosen genes at a locus of $v = 1.2 \times 10^{-5}$ are different from each other is 32% (Kimura and Crow 1964). This expected probability is much smaller than the actual values found at most functional *HLA* loci (Klein 1986), implying that the evolution of *HLA* genes has been driven by positive selection.

POSITIVE SELECTION

Positive selection appears to act on amino acid residues in the PBR. An unresolved question concerns the cause, mode, and intensity of such selection. Klein and coworkers (1992) have argued that the universal cause of selection must be pathogens (Klein 1991). There are two opposing views on the mechanism of selection. Assuming that the primary function of Mhc molecules (peptide presentation to T cells and triggering of the host immune response) is the main target of selection, the selection would confer heterozygote advantage (overdominant selection). This is because heterozygotes expressing different Mhc molecules would have a wider repertoire for binding of foreign peptides than respective homozygotes. On the other hand, if Mhc molecules are mimicked by pathogens or if blind spots of the T-cell repertoire (Klein 1986) are important in reducing viability of individuals, it would be heterozygotes that have a selective disadvantage (= underdominant selection). The blind spots could arise through the inactivation of certain T-cell clones in the process of self-tolerance induction so that the T-cell repertoire of heterozygotes would lack more functional T cells than that of homozygotes. However, the latter mechanism alone cannot explain why amino acid changes have been enhanced in the PBR ($\gamma > 1$). Moreover, underdominant selection does not generate stable polymorphism so that other maintenance mechanisms such as frequency-dependent selection would have to be invoked.

Most of the frequency-dependent selection models, including the molecular mimicry model studied by Takahata and Nei (1990), assume one-to-one interaction between an *Mhc* allele and a pathogen. However, the actual interaction seems more complex. An *Mhc* allele is capable of binding not only a variety of epitopes from different parasites but also peptides derived from various gene products of a single parasite. In such a situation, a change in the frequency of a particular *Mhc* allele seems unlikely to cause a systematic change in the frequency of a particular antigen, or *vice versa*. We assume therefore that the complex interactions lead to overdominant selection. The correctness of this view needs to be tested experimentally, but we could show that a model of overdominant selection is in surprisingly good agreement with most features of *Mhc* polymorphism (Takahata et al. 1992). The overdominance model also allows us to estimate s and N_e for the human population (Takahata et al. 1992).

Selection intensity, s, under overdominant selection is given by the extent of viability reduction in homozygotes relative to heterozygotes. From the overdominance model and the *HLA* gene sequence data available, we can estimate the ratio of s to $u = \mu g L_B$ at each of the *HLA* loci, u being the per-generation mutation rate at the PBR nonsynonymous sites. Since $u = 2.4 \times 10^{-6}$ for class I and 0.7×10^{-6} for class II approximately, s is only a few percent at most (Table 2). This finding in turn implies that selection is less efficient in a small population of $N_e < 10^3$ because $N_e s \gg 1$ must be satisfied in order for selection to be more important than random drift (Kimura and Crow 1964; Kimura 1968). In other words, when selection is this mild, *Mhc* polymorphism cannot be maintained at high levels in small populations. Extensive *Mhc* polymorphism can occur only in relatively large populations.

The value of N_e estimated from *HLA* polymorphisms is about 10^5 (Table 2), whereas that estimated from non-*HLA* polymorphism is 10^4. This discrepancy is probably due to the different time scales of these polymorphisms (Takahata 1990). We suggest that N_e in the human lineage was 10^5 for most of its evolutionary history but decreased to 10^4 approximately one million years ago and remained at that size until the agricultural revolution 10,000 years ago. A population of $N_e = 10^5$ and $s = 0.01$ is expected to be able to transmit *trans*-specifically about 25 alleles at a class I locus and 21 alleles at a class II locus (Kimura and Crow 1964; Takahata 1990). In an equilibrium population of $N_e = 10^4$, this number reduces to about 7 or 6 unless the duration time of the reduced population size is short (less than 1 my years). The more drastic the reduction in N_e (bottleneck effect), the more *trans*-specific alleles can be expected to be lost. The degree of loss of polymorphism depends on the duration (t) in terms of generations of the bottleneck. A single-generation bottleneck ($t = 1$) in which the population has been reduced to $N_e = 2$ has the same effect on persistence of alleles as a bottleneck with $t = 10$ and $N_e = 20$: both result in a substantial loss of alleles.

Table 2. Estimates of selective intensity (s) at *HLA* loci and human population size (N_e).

| Parameter | *HLA* loci | | | |
	A	B	C	DRB1
$\hat{s}$	1.1%	2.3%	0.04%	2.5%
$\hat{N}_e$	1.3×10^5	0.7×10^5	1.8×10^5	1.1×10^5

It is assumed that $u = 2.4 \times 10^{-6}$ per class I PBR per generation and $u = 7.2 \times 10^{-7}$ per class II PBR per generation. $M = N_e u$ and $S = 2N_e s$;

$$\hat{S} = \hat{\gamma}\hat{K}_B^2/\sqrt{2} \text{ and } \hat{M} = \sqrt{\hat{S}/16\pi} \exp(-\gamma/\sqrt{2}).$$

There is a controversy about the origin of modern humans and part of it concerns the size of the founding population. However, the assumption of severe bottlenecks in the evolution of the human lineage or a view that assumes small founding populations when new species emerge (Mayr 1970) is clearly contradicted by the *HLA* polymorphism. Our results indicate that for a sample of 20 alleles at the *DRB1* locus, there must have existed a dozen or so distinct allelic lineages some 5 million years ago and that the allelic genealogy goes well back to the early Miocene. The *trans*-species mode of *HLA* polymorphism suggests that the number of breeding individuals in the human lineage has been considerable at any stage of its evolutionary history and that any hypothesis assuming a severe bottleneck is unacceptable.

ACKNOWLEDGMENTS. We thank Ms. Lynne Yakes for editorial assistance

REFERENCES

Bjorkman PJ, Saper MA, Samraoui B, Bennett WS, Strominger JL, Wiley DC (1987) Nature 329:506-512

Brown JH, Jardetzky T, Saper MA, Samraoui B, Bjorkman PJ, Wiley DC (1988) Nature 332:845-850

Fan W, Kasahara M, Gutknecht J, Klein D, Mayer WE, Jonker M, Klein J (1989) Hum Immunol 26:107-121

Gyllensten UB, Erlich HA (1989) Proc Natl Acad Sci USA 86:9986-9990

Hughes AL, Nei M (1988) Nature 335:167-170

Kenter M, Otting N, Anholts J, Leunissen J, Jonker M, Bontrop RE (1992) Immunogenetics 36:71-78

Kupfermann H, Mayer WE, O'hUigin C, Klein D, Klein J (in press) Hum Immunol

Kimura M (1968) Nature 217:624-626

Kimura M (1983) The neutral theory of molecular evolution. Cambridge University Press, Cambridge

Kimura M (1991) Jpn J Genet 66:367-387

Kimura M, Crow JF (1964) Genetics 49:725-738

Klein J (1980) Generation of diversity at MHC loci: implications for T cell receptor repertoires. In: Fougereau M, Dausset J (eds) Immunology 80. Academic Press, London, p 239

Klein J (1986) Natural history of the major histocompatibility complex. Wiley, New York

Klein J (1991) Hum Immunol 30:247-258

Klein J, Takahata N (1990) Immunol Rev 113:5-25

Klein J, Satta Y, O'hUigin C, Takahata N (in press) Immunol Rev

Lawlor DA, Zemmour J, Ennis PP, Parham P (1988) Nature 335:268-271

Li WH, Luo CC, Wu CI (1985) Evolution of DNA sequences. In: MacIntyre RJ (ed) Molecular evolutionary genetics. Plenum Publishing Corporation, New York, p 1

Mayer WE, Jonker M, Klein D, Ivanyi P, van Seventer G, Klein J (1988) EMBO J: 7:2765-2774

Mayer WE, O'hUigin C, Zaleska-Rutzcynzska Z, Klein J (in press) Immunogenetics

Mayr E (1970) Population, species, and evolution. Harvard University Press, Cambirdge, Massachusetts

Nei M (1987) Molecular evolutionary genetics. Columbia University Press, New York

Satta Y (in press) Balancing selection at *HLA* loci. In: Takahata N, Clark AG (eds) Molecular paleo-population biology. Japan Science Society Press/Springer, Tokyo

Satta Y, Takahata N, Schönbach C, Gutknecht J, Klein J (1991) Calibrating evolutionary rates at major histocompatibility complex loci. In: Klein J, Klein D (eds) Molecular evolution of the major histocompatibility complex. Springer, Heidelberg, p 51

Takahata N (1990) Proc Natl Acad Sci USA 87:2419-2423

Takahata N (in press) Mol Biol Evol

Takahata N, Nei M (1985) Genetics 110:325-344

Takahata N, Nei M (1990) Genetics 124:967-978

Takahata N, Satta Y, Klein J (1992) Genetics 130:925-938

A Cellular Travelogue Major   Highways, Secret Hideouts, Spa Cuisine

Eli. E. Sercarz

University of California, Los Angeles, Department of Microbiology and Molecular Genetics, 5304 Life Sciences Building, 405 Hilgard Avenue, Los Angeles, CA 90024-1489, USA

Any account of progress in the area of antigen processing and presentation is necessarily a story of the past few years. In 1987, when Bjorkman, Wiley and their colleagues solved the mystery of class I MHC structure (Bjorkman et al, 1987) they simultaneously provided a rationale for the study of a large set of problems. How are native antigens processed to the peptides that are found in the antigen-binding groove of the MHC molecule? What are the differences between processing for class II-restricted vs class I-restricted responses? Where are the sites of molecular unfolding and proteolysis? Are the sites different for these steps in processing? What is the length, the numbers and the diversity of peptides found on display? The answers to some of these questions will be found throughout this review in the succession of Tables, which represent recent advances in specified areas.

THE MAJOR HIGHWAYS OF PROCESSING: The MHC can be viewed as a peptide-presentation device for the immune system, allowing the presentation of both endogenous and exogenous antigenic determinants to T cells, restricted by MHC-I and MHC-II respectively. Since T cells only recognize MHC-bound antigen, by separating the sources of peptide that are bound to MHC-I and MHC-II molecules, the system achieves a functional dichotomy. The peptides that are bound to MHC-I are derived from the cytosol, or the ER or mitochondria, while those bound to MHC-II are derived from the endocytic-lysosomal system which collects molecules from the cell surface and the medium.

Since CD8$^+$ T cells are generally cytotoxic and presumably respond to MHC-I viral peptide complexes which have been synthesized internally, it is sensible that these T cells be prevented from reacting with the pool of peptides garnered from external sources. Therefore, in order to maintain functional separateness, two distinct strategies have been employed.

The external milieu is sampled continually by the cells in the organism that bear, or can be induced to bear class II molecules. Here the first strategy involves uptake of an external molecule, its processing into a form able to bind to MHC-II, and then interaction with class II molecule in a specialized set of compartments within the acidic endosomal-lysosomal vesicular system. The peptide-loaded MHC-II is then sent to the cell surface for display. On the other hand, class I molecules are loaded with small (8 or 9 amino acid peptides) in the neutral pH endoplasmic reticulum (ER), and then transported through the Golgi stack to the cell membrane. In the following brief account of the major highways shown in Fig. 1, we will emphasize only the

---

Table 1. PROTEASOME - LMP COMPLEX

- The proteasome is a 650 Kd proteolytic complex with distinct peptidase subunits in an hexagonal array.
- It is widespread in phylogeny, and highly abundant in cell extracts.
- The 19S proteasome is the core of a multicatalytic 26S complex which also contains at least 2 low molecular weight proteins (LMP), coded for within the class II gene region of the MHC complex.
- Multiple types of proteasomes exist of differing subunit constitution.
- The LMP genes are polymorphic and may designate unique protein cleavages.
- The LMP may serve to optimize the delivery of peptide from the cytosol to the peptide transporter (TAP).

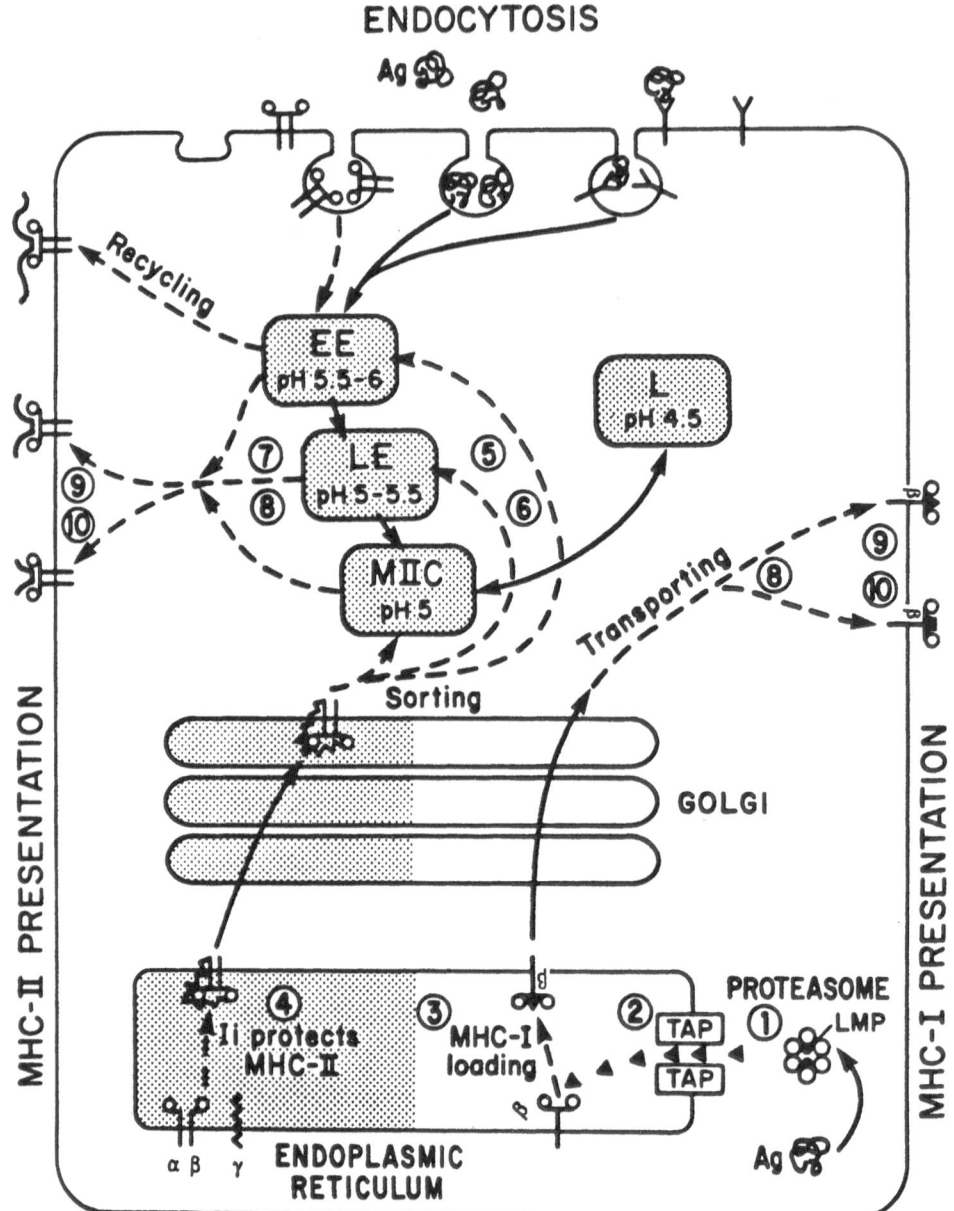

Fig. 1 Processing and Presentation - MHC-I and MHC-II. In this depiction of major pathways, on a B lymphoma cell, shaded areas refer to MHC-II related events. The circled numbers refer to Tables 1-10. Dashed arrows refer to a degree of uncertainty in the pathway. Pi symbols represent MHC-II, while T symbols with an attached β_2-microglobulin structure represent MHC-I. The 3 vesicles at the top of the Figure contain empty class II, antigen and antigen-receptor complexes, from left to right.

Table 2. TRANSPORT FOR ANTIGEN PROCESSING (TAP)

- Antigenic peptides are required for the stable assembly of MHC Class-I heavy chain and β2-microglobulin.
- Mutant cells (RMA-S, .134 and .174) lacking a complete TAP transport system express greatly reduced amounts of MHC Class-I. Addition of the peptide normalizes surface expression of MHC Class-I.
- Few ER peptides are produced from proteins resident in the ER, except for signal sequences: thus, any peptides in the ER have arrived via the peptide transporter.
- Polymorphisms exist in the peptide transporter which can lead to an altered spectrum of bound peptides.

broad landscape while some more specific comments will be listed in the Tables, referred to by circled numbers in Fig. 1. Citations in most cases will be limited to recent reviews from which the original literature can be retrieved.

CLASS I ASSEMBLY: (Monaco, 1992; Neefjes et al, 1992; Neefjes et al, 1991): Although it had been clearly visualized at the time of the 7th International Congress of Immunology that the class I and II pathways were distinct, the mechanistic details were sketchy or unknown. During this intervening period, several key observations have been made describing how the octamers and nonamers that bind to MHC-I are produced, loaded within the endoplasmic reticulum (ER) and then transported to the cell surface.

(a) Proteasome-low molecular weight protein-(LMP) complexes: (Brown et al, 1992; DeMars and Spies, 1992; Driscoll and Finley, 1992; Monaco, 1992; Yang et al, 1992): Proteins in the cytosol are processed via a phylogenetically ancient and complex proteolytic apparatus which is particularly dedicated to processing cytosolic proteins (Table 1). Importantly, its products are oligopeptides rather than amino acids. The proteasome may be connected to the ubiquitin-dependent proteolytic system: after N-terminal ubiquitination, antigenic proteins are presented more efficiently. The LMP subunits that associate with the proteasome may help in docking the proteasome at the transporter, or change the assortment of peptides produced (through LMP polymorphism) or somehow increase the level of peptide production. Interestingly, interferon-γ is a potent regulator of the association of proteasomal units and thereby plays a role in modulating the nature of the specific peptides produced by the complex.

(b) Transport for antigen processing (TAP): (see Elliott, 1991; Kelly et al, 1992; Peterson et al, this volume) Clear evidence for a peptide transporter comprised of a dimer of TAP1 and TAP2 proteins, mapping within the class II region of MHC near the 2 mapped LMP genes, has accumulated recently from several labs. An illuminating and elegant set of experiments with mutants of mouse cell line RMA-S clarified two important issues in class I peptide

Table 3. PEPTIDES ARE INTEGRAL COMPONENTS OF MHC PRESENTATION STRUCTURES

- Newly synthesized MHC-I and MHC-II are structurally unstable. Peptide binding induces the final stable conformation, acting as an intricate component of the assembly process.
- Unfilled MHC-II are formed in the ER and then protected by Ii, but gain stability after binding to antigenic peptide in the low pH endocytic system.
- A fast on and off rate exists for the floppy MHC-II state at pH 4.5-5, and as a more neutral pH is attained, a long-lived compact complex of MHC-II with trapped peptide forms, which is resistant to dissociation in sodium dodecyl sulfate.
- The stable complexes exist as long as the MHC molecule itself.
- MHC-I association occurs at neutral pH in the ER with preformed short peptides, probably aided by ER chaperones.

Table 4. INVARIANT CHAIN:(Ii)

- Acts as a chaperone to class II to help the molecule fold properly.
- Blocks antigen binding site of class II.
- Directs class II to localize in an endosomal compartment.
- In the endocytic compartments, Ii is cleaved stepwise and then dissociates from the class II molecule.
- Different molecular versions of Ii have distinct functions, e.g. in directing intracellular transport.
- Antigen-specific effects of Ii are frequently found.

presentation. This mutant produces unstable class I molecules at 37° which are not displayed on the cell surface unless the temperature is lowered, at which time stabilized molecules appear amply on the cell surface. Provision of the correct peptide reverses the 37° "lesion". The existence of empty molecules permits detailed study of the characteristics of the binding event. The mutants (and a variety of mutants have now been isolated, e.g. as seen in Table 2) are defective in a component of the peptide transport from the cytosol to the ER. Accordingly, it became evident that peptide was crucial in the proper assembly of MHC-I as well as for its stability (Table 3). In fact, the coming together of peptide, heavy chain and β_2-microglobulin in the ER (or cis-Golgi) should be considered a 3 body event, and there is much current study about the kinetics of pair interactions between these molecules .

CLASS II ASSEMBLY-(c) Invariant chain: (Anderson and Miller, 1992; Teyton et al, 1990; Teyton et al, 1992) The class II molecule is prevented from binding its ligand in the ER by the interesting device of providing a dedicated chaperone, the invariant chain (Ii), (Table 4). This maneuver delays the class II binding event until arrival at an acidic endocytic compartment where Ii dissociates and MHC-II becomes available for binding to ligands from the extracellular milieu. Thus, when Ii has completed its distinctive job of targeting class II to one of the endosomal compartments, Ii detaches and is destroyed, probably piecemeal, so that it very rarely appears on the cell surface. Transport through the Golgi occurs in the form of a 9-chain polymer of 3 $\alpha\beta$Ii trimers that Ii directs to the processing compartments. Actually, the p31 Ii is hardly invariant, and 3 variants can be detected: with a 15 a.a. N-terminal addition (p33), or a large insert (p41) or a truncation (p25), and each has other post-translational modifications. These different forms of Ii alter the distribution of $\alpha\beta$Ii complexes. The N-terminal 15-mer on p31 directs complexes to the antigen processing compartment, while p33 bears an ER retrieval signal which will shuttle the complex between the Golgi and the ER.

Other details of Ii structure and function remain to be learned, but the last few years have revealed the intricate complexity resident in this molecule underlying its 2 major functions of

Table 5. MHC-GUIDED PROCESSING OF ANTIGEN

Definition:
- Binding to MHC-II can occur when a previously compact native molecule such as HEL is unfolded and reduced, rendering high affinity determinants available for binding.
- The first determinant to stably bind will be the dominant one; it is protected from proteolysis while buried in the MHC groove.
- Trimming then occurs, leaving ragged ends and end products of ~12-25 amino acids.
Evidence:
- Unfolded proteins can bind to MHC-II without fragmentation.
- Determinant usage can depend on regions of the antigen distant or nearby, but not necessarily at the site of T cell recognition.
- In chimeric peptides, made from 2 determinants with different MHC-II restriction, competitive effects predominate.

Table 6. ANTIGEN PROCESSING AND PRESENTATION BY CLASS II MOLECULES

Exogenous antigen
- Internalization
- Denaturation or minimal degradation
- Binding to class II molecules
- Proteolytic trimming
- Surface expression of complexes
- Recognition by T cells

---

protection and guidance of MHC-II. With regard to the chaperone function in Ii's initial binding to class II, it is pertinent that class II folds differently in the absence of Ii, losing reactivity with several specific mAbs, and becomes aberrantly glycosylated.

(d) MHC-guided processing: (Sercarz et al, 1986; Sercarz, 1989; Sette et al, 1989) The state of the antigenic component of the class II molecule upon binding to MHC is still very unclear. In analogy with MHC-I, it has often been assumed that the antigen is first processed into small peptides which are the only entities able to bind. In fact, studies that are mentioned in Table 5 have shown that full length, but at least partially unfolded molecules presumably with at least one readily available agretope (MHC-binding site) can bind readily to the MHC.

Subsequently, the dangling ends of the molecule can be trimmed to the final state of 13-22 amino acid peptides, which are recoverable from the MHC-II groove. We have found that one MHC-II molecule, such as E^d in the mouse, can compete with a different MHC-II molecule such as A^d to prevent the utilization of any A^d-restricted antigenic determinant. Likewise, in immunizing with chimeric peptides prepared from parent peptides restricted to different MHC-II molecules, competition was always observed, with the dominant determinant exclusively activating the appropriate T cells. This behavior would not be expected under the assumption that small peptides are generated first, as for class I binding, but it would fit the notion that simple local unfolding will usually permit the most available determinant to bind to one of the class II MHC types: that determinant will eventually be dominant. Subsequently, we predict that trimming of the protruding ends will occur, eventually leading to the 13-25 amino acid peptides that are the minimal peptides which are extractable from the MHC molecules. By this postulated scenario, the MHC-II molecule guides processing and dominance is a natural consequence. The postulated order of events in MHC-II processing and presentation is shown in Table 6, although entries 3 and 4 in the Table can surely switch positions in the most acidic endocytic sites.

(e) The "secret hideouts" where class II and exogenous ligand meet: (Brodsky, 1992; Harding et al, 1991; Neefjes et al, 1991; Sadegh-Nasseri and Germain, 1992) The four possible secret hideouts where antigenic determinants finally meet up with class Ii molecules are

---

Table 7. PEPTIDE BINDING TO MHC-I AND II

MHC-I	CHARACTERISTICS	MHC-II
• 8-, 9- mers	size of peptide eluted	• 13→22 mers
• 8, 9- mers	size of peptide bound	• full length: (unfolded) medium: 25-50 aa small: 8, 9-25 aa
• ~ 1-3	required residues for binding	• ~ 2-3
• Endoplasmic reticulum, pH7	trysting place	• Endosomal-lysosomal system, pH 6-4,5
• Cytosolic proteasome	major proteolytic system	• Endosomal proteases
• MHC-I trimming enzymes?	extra proteolytic systems	• MHC-II trimming exo- and endopeptides

Table 8. WHAT IS THE CUISINE DELIVERED TO THE TABLE?

- $\sim 10^2$ of the same peptides is needed to activate T cell
- $\sim 10^5$ MHC-II or MHC-I molecules are displayed
- So, roughly 1000 determinants are "on the menu"
- Certain determinants are very well represented--(for example, to the level of 10% of all bound peptides) while a larger group is at a lower, near minimal frequency for T cell activation.
- For each determinant core in the MHC-II groove, there are variations at both the N and C termini.
- MHC-I displays 8,9-mer peptides, usually firmly anchored at both ends to pockets in the floor of the MHC groove.

represented in Figure 1 as discrete entities, but the early endosome (EE), the late endosome (LE) and the MIIC, an MHC-rich compartment with unique markers, can also be regarded as a continuous set of interacting vesicular compartments. EE, LE and MIIC are of increasing acidic character (see Fig. 1) and even the lysosome (L in Fig. 1) has been advocated as a major site for interaction with antigen. It is possible that Ii delivers the class II molecule to more than one vesicular endocytic site.

In any event, under acidic conditions at about pH 5, the class II molecule assumes a more open, floppy conformation enhancing the binding of potential ligands. Upon return towards neutrality, as the complex is transported towards the membrane, MHC-II adopts its final, compact conformation which is displayed to the T cell (Table 3).

(f) What is the fate of the MHC-II complex?: (Davidson et al, 1991; Harding et al, 1991) Recent evidence shows that the tightly bound peptide can remain associated with the class II molecule for the entire life of the latter. However, some empty class II molecules reach the cell surface without bound peptide and these, as well as molecules with more loosely bound peptide, can recycle. Although papers in several systems describe recycling of MHC in B cells, there is discordance in the current literature about this issue. In B cells, it is the recently synthesized class II molecules that bind to endocytosed antigen. It may be that the recycling endosomes (EE?) represent a separate class, distinct from those more acidic endosomes in which antigen processing proceeds. Meanwhile, exogenous peptide may have access to empty class II molecules on the cell surface and upon endocytosis of these complexes, exchange may occur within this recycling population.

Partially unfolded and degraded derivatives of the native antigen in the various endocytic compartments may require more acidic conditions or a more proteolytic environment in order to bind effectively to MHC-II, so that the MIIC or even the lysosomal compartment may be necessary for the conversion of the antigen to a bindable form (but see section (d) above).

Table 9. FACTORS INFLUENCING AMOUNT OF AN MHC-PEPTIDE COMPLEX ON CELL-SURFACE

- Affinity of peptide to MHC groove
  - Fit of anchor residues
  - Possible hindering residues
  - Prolines can allow kinks
- Proteolytic events
  - Sensitivity to proteases
  - Sensitivity of disulfide bonds to reduction
- Transporter mechanism efficiency
- Downstream competition by determinants with higher affinity
- Enhancement or hindrance of binding by flanking residues

Table 10. DOMINANT AND CRYPTIC DETERMINANTS

- Each self protein, as each foreign protein, presents a small minority of dominant determinants (1/5-1/3).
- These well-displayed self determinants are involved in negative selection in the thymus, and the organism is tolerant of them.
- The poorly displayed majority of subdominant and cryptic determinants do not induce tolerance (at least not complete tolerance), so that a large cadre of self-reactive T cells exists.
- Any sustained disruption leading to the heightened presentation of cryptic determinants could lead to autoimmune disease.

SPA CUISINE: - (g) The nature of the peptides served up to CD4[+] and CD8[+] T cells: (Hunt et al, 1992; Janeway et al, 1993; Rammensee et al, 1993) In Central Europe, there is little need to explain the special health requirements imposed at local spas, re: "taking the waters" or carefully prescribed diets. Since the last Congress, a wealth of information has been learned about the spa cuisine offered to T cells, and Tables 7 to 9 outline some of this information.

Acidelution from MHC-I and II affinity columns followed by microsequencing has revealed the nature of the end products of natural processing. The acid distinction between MHC-I and MHC-II is clear, and summarized in Table 7. Nonamers bind to MHC-I orders of magnitude better than longer peptides, and nonamers are eluted from these MHC molecules. Thus, the bound peptide becomes protected and the MHC-I molecule becomes stabilized simultaneously after peptide loading in the ER.

MHC-II binding has many fewer constraints: peptides of all sizes can bind and intramolecular competition for binding is a common feature with longer peptides containing multiple MHC determinants. Some quantitative estimates are shown in Table 8. An important feature of MHC-II binding is that some peptides are "more equal than others" and appear at high frequency. However, myriads of peptides can be harvested from MHC molecule that bind at lower levels.

(h) Dominant and cryptic determinants; the crypticity of self: (Gammon et al, 1992; Sercarz, 1989): The abundance of a particular peptide as well as other factors will be closely related to its MHC-binding affinity. These features will influence the immunodominance of particular determinants, which in turn, has important physiological repercussions (Table 10). We have established that readily processed, dominant determinants on self molecules induce tolerance, while poorly processed, cryptic determinants do not. Accordingly, non-tolerized self-reactive T cells, provide a source of potential autoaggression against each self molecule.

One of the issues we have not mentioned in this brief review is that of endogenous processing for MHC-II which in the cases that it has been compared, is considerably more efficient than exogenous porcessing for MHC-II. In fact, a majority of peptides extracted from MHC-II are endogenously derived. This feature will make an impact on the relative ease of tolerance induction to exogenous and endogenous self-antigens in the thymus, and therefore has serious implications for autoimmunity.

References

Anderson MS, Miller J (1992) Invariant chain can function as a chaperone protein for class II major histocompatibility complex molecules. Proc Natl Acad Sci USA **89**:2282-2286.

Bjorkman PJ, Saper MA, Samraoui B, Bennett W, Strominger JL, Wiley DC (1987) The foreign antigen binding site and T cell recognition regions of class I histocompatibility antigens, Nature **329**:512-518.

Brodsky FM (1992) Antigen processing and presentation: close encounters in the endocytic pathway. Trends in Cell Biol **2**:109-114.

Brown MG, Driscoll J, Monaco JJ (1992) Structural and serological similarity of MHC-linked LMP and proteasome (multicatalytic proteinase) complexes. Nature **353**:355-357.

Davidson HW, Reid PA, Lanzavecchia A, Watts C (1991) Processed antigen binds to newly synthesized MHC class II molecules in antigen-specific B lymphocytes. Cell **67**:105-116.

DeMars R, Spies T (1992) New genes in the MHC that encode proteins for antigen processing. Trends in Cell Biol **2**:81-86.

Driscoll J, Finley D (1992) A controlled breakdown: antigen processing and the turnover of viral proteins. Cell **68**:823-825.

Elliott T (1991) How do peptides associate with MHC class I molecules? Imm Today **12**:386-388.

Gammon G, Sercarz EE, Benichou, G (1991) The dominant self and the cryptic self: shaping the autoreactive T cell repertoire. Immunol Today **12**: 193-195.

Germain RN, Hendrix LR (1991) MHC class II structure, occupancy and surface expression determined by post-endoplasmic reticulum antigen binding. Nature **353**:134-139.

Harding CV, Collins DS, Slot JW, Geuze HJ, Unanue ER (1991) Liposome-encapsulated antigens are processed in lysosomes, recycled, and presented to T cells. Cell **64**:393-401.

Hunt DF, Michel H, Dickinson TA, Shabanowitz J, Cox AL, Sakaguchi K, Appella E, Grey HM, Sette A (1992) Peptides presented to the immune system by the murine class II major histocompatibility complex molecule I-A^d. Science **256**:1817-1820.

Janeway CA, Mamula MA, Rudensky AY (1993) Rules for peptide presentation by MHC class II molecules. Intl Revs Immun, in press.

Monaco JJ (1992) A molecular model of MHC class-I-restricted antigen processing. Imm Today **13**:173-179.

Neefjes JJ, Ploegh HL (1992) Intracellular transport of MHC-II molecules. Imm Today **13**:179-184.

Neefjes JJ, Schumacher TNM, Ploegh HL (1991) Assembly and intracellular transport of major histocompatibility complex molecules. Curr Opin in Cell Biol **3**:601-609.

Rammensee H-G, Rötzschke O, Falk K (1993) Self tolerance of natural MHC class I ligands. Intl Revs Immun, in press.

Sadegh-Nasseri S, Germain RN (1992) How MHC class II molecules work: peptide dependent completion of protein folding. Immunol Today **13**:43-46.

Sercarz EE, Wilbur S, Sadegh-Nasseri S, Miller A, Manca F, Gammon G, Shastri N (1986) The molecular context of a determinant influences its dominant expression in a T cell response hierarchy through "fine processing". Prog in Immunol **6**:227-237.

Sercarz EE (1989) The architectonics of immune dominance: The aleatory effects of molecular position on the choice of antigenic determinants. Chem Immunol **46**:169-185.

Sette A, Lamont A, Buus S, Colon SM, Miles C, Grey HM (1989) Effect of conformational propensity of peptide antigens in their interaction with MHC class II molecules. J Immunol **143**:1268-1273.

Teyton L, O'Sullivan D, Dickson PW, Lotteau V, Sette A, Fink P, Peterson PA (1990) Invariant chain distinguishes between the exogenous and endogenous antigen presentation pathways. Nature **348**:39-44.

Teyton L, Peterson PA (1992) Invariant chain - a regulator of antigen presentation. Trends in Cell Biol **2**:52-56.

Kelly A, Powis SH, Kerr L-A, Mockridge I, Elliott T, Bastin J, Uchanska-Ziegler B, Ziegler A, Trowsdale J, Townsend A (1992) Assembly and function of the two ABC transporter proteins encoded in the human major histocompatibility complex. Nature **355**:641-644.

van Bleek GM, Nathenson SG (1992) Presentation of antigenic peptides by MHC class I molecules. Trends in Cell Biol **2**:202-207.

Yang Y, Waters JB, Früh K, Peterson PA (1992) Proteasomes are regulated by interferon γ: Implications for antigen processing. Proc Natl Acad Sci USA **89**:4928-4932.

Acknowledgments

    I thank H. Jun for the preparation of this manuscript, Drs. V. Kumar and A. Miller for critical comments, M. Kowalczyk for the Figure, and grants from the NIH and ACS.

Peptide Binding and Intracellular Transport of MHC molecules

Jacques J. Neefjes

The Netherlands Cancer Institute, Plesmanlaan 121, 1066CX Amsterdam, The Netherlands

The function of MHC molecules is to present fragments of antigen in the form of peptides to T cell receptors. In general, MHC Class I molecules will present fragments derived from mitochondrial, cytoplasmic or ER molecules. MHC Class II will present fragments from antigens that are degraded in endosomes and lysosomes. This includes cell surface proteins and soluble proteins present in the medium. These introductory statements illustrate the central cell biological questions related to MHC Class I and II molecules: how and where are the antigenic peptides generated and how and where do MHC molecules associate with them? These questions are of some relevance because the type of peptides that bind to MHC molecules will determine the outcome of any T cell response. Although the binding of peptides by MHC Class I molecules has a number of aspects in common to peptide binding by MHC Class II molecules, they differ in other aspects of peptide binding and, in particular, intracellular transport.

I will first discuss peptide binding and intracellular transport of MHC Class I molecules.

Assembly and peptide binding by MHC Class I molecules.

MHC Class I molecules are assembled in the ER from a glycosylated H-chain of 45 kd and a non-glycosylated protein, β_2-microglobulin. The elucidation of the 3-D structure of a MHC Class I molecule, HLA-A2,has revealed that the polymorphic residues located in the H-chain are the major constituents of a sort of groove in which the antigenic fragment (peptide) bind (Bjorkman et al., 1987a,b). Townsend et al. (1985) had already shown that a number of nuclear and cytosolic antigens were presented in the form of peptides by MHC Class I molecules. However, how are these peptides translocated over the ER membrane in order to appear in the lumen of the ER where they can associate with MHC Class I molecules? The description of mutant cell lines that synthesized normal amounts of Class I H-chains and β_2-microglobulin but failed to assemble them, solved this question. Townsend et al. (1990) and we (Schumacher et al., 1991) showed that peptides were essential for the stable assembly of the Class I H-chain and β_2-microglobulin. A lack of peptides in the ER could thus explain the observed phenotype in the mutant cells. A number of studies has shown now that these mutant cell lines are mutated in one or two multimembrane spanning molecules that presumably pump peptides over the ER-membrane (reviewed by Monaco 1992). Another important concept derived from the study of these mutant cells is that MHC

Class I molecules should not be considered a heterodimer but a heterotrimer (H-chain, β_2-microglobulin and peptide). Peptide is an essential subunit of Class I molecules and determines its stability.

Analyzing the assembly and intracellular transport of MHC Class I molecules were analyzed in an EBV-transformed B cell line, we observed locus- and allele-specific differences (Neefjes and Ploegh, 1988). Within one cell line certain human Class I molecules assembled very efficient (mainly HLA-B locus products), whereas others assembled very inefficient (HLA-C locus products). As a result, HLA-C locus products are hardly expressed at the cell surface and are usually not considered to be important in transplantations and for the presentation of antigen. There appeared to be a correlation between the rate of assembly of MHC Class I molecules and the rate of intracellular transport. Since, in that study, the different H-chains were isolated from the same cells and were thus exposed to the same concentrations of β_2-microglobulin and, presumably, also peptide, the above described observations may be explained either by translocation of peptides over the ER membrane that associate more preferentially with HLA-B and not with -C locus products, and/or, when the amount of peptides is limiting in the ER, by a higher affinity for peptides (and β_2-microglobulin) of HLA-B locus products. The functional consequences of these differences in assembly and intracellular transport are as yet unclear although they suggest that, in particular, B-locus products may be very efficient in presentation of antigen. Recent population studies have suggested this as well (Howard 1992).

Can every peptide be presented by MHC Class I molecules? Initially, long peptides (longer then 12 aminoacids) were used to sensitize target cells. However, analysis of the peptides that really bound to Class I molecules revealed that the peptide was only 8-9 aminoacids long. Target cells were sensitized by either a contamination (a smaller peptide) in the original peptide synthesis (Schumacher et al., 1991), or the correct sized peptide was generated by proteases in the medium and/or proteases derived from cells (Neefjes et al., 1992b). The elucidation of the structure of HLA-B27 at high resolution revealed that only small peptides can bind to MHC Class I molecules because the peptide binding groove of Class I molecules is actually closed on two sites (Madden et al., 1991) prohibiting the extension of the peptide outside the Class I peptide binding groove. Still some heterogeneity with respect to the length of the peptides that bind to a particular MHC molecule does exist. Deres et al. (1992) showed that especially the proline content of a peptide may determine whether an 8-mer, 9-mer or even a 10-mer peptide will bind to a particular MHC molecule.

The sensitization of target cells with peptides rely on the fact that Class I molecules that can be loaded with peptide do exist at the cell surface. How do these "empty" Class I molecules arise? Some "empty" Class I molecules may exit the ER, and may be stable enough to survive transport and surface exposure. This is observed by culturing cells at low temperature (Ljunggren et al., 1990; Schumacher et al., 1990). Alternatively, some Class I molecules release their peptide at the cell surface, after which these, now empty, Class I molecules can bind another peptide. We have followed the binding of a Sendai virus NP derived peptide (SV9) to H-2K^b molecules. This peptide contains a tyrosine (anchor)residue at position 5, that interacts with pockets in the MHC Class I peptide binding groove. SV9 was radiolabeled on tyrosine either by iodine or by tritium. Whereas the binding of SV9 and iodinated SV9 to H-2K^b molecules was equally efficient, tritiated SV9 remained stably associated with H-2K^b molecules, but iodinated SV9 dissociated in a temperature-dependent fashion (Neefjes et al, ms in prep.). Thus, certain peptides can dissociate from MHC Class I molecules after initial binding, resulting in empty MHC Class I molecules.

Empty MHC Class I molecules at the cell surface render the risk of sensitization of naive bystander cells when peptides are released from MHC molecules on infected cells. We do not consider this a problem since the half life of empty MHC Class I molecules is short at physiological temperature. Empty Class I molecules dissociate and the resulting Class I H-chains are rapidly degraded (Neefjes et al. 1992b). Furthermore, the released peptides will be rapidly degraded by cell- and serum-derived proteases.

Intracellular transport and peptide binding by MHC Class II molecules.

Intracellular transport of MHC Class II molecules is more complex then that of MHC Class I molecules, since it involves both the normal constitutive route and, since MHC Class II molecules have to interact with internalized antigen, the endocytic pathway. Furthermore, MHC Class II molecules can bind fragments of antigen in endosomal structures. Since MHC Class II molecules are assembled in the ER, they can also bind peptides from the same peptide pool of which MHC Class I molecules extract their peptides. One similarity between MHC Class I and II molecules is that they are in principle heterodimers or, when peptide is included, heterotrimers. However, the essential difference between MHC Class I and II molecules is a molecule that transiently associates with the Class II $\alpha\beta$ dimer during biosynthesis: the γ or invariant chain.

Early during biosynthesis MHC Class II molecules are assembled as an $\alpha\beta\gamma$ heterotrimer, that trimerizes in a nine-subunit complex (Roche et al., 1991). This complex is transported from the ER through the Golgi to the trans-Golgi reticulum or network. Here, it is decided whether molecules are transported straight to the cell surface (like MHC Class I molecules), to late endosomes (like the mannose-6-phosphate receptor), or whether they follow the secretory route. We showed that MHC Class II molecules enters the endocytic route during intracellular transport to the cell surface (Neefjes et al., 1990).

Where do MHC Class II molecules enter the endocytic route? This question is relevant, because it defines the proteases that may be involved in the generation of proteolytic fragments. Harding et al. (1991) already demonstrated that antigenic fragments are generated in lysosomes. If MHC Class II molecules are transported through lysosomes, the model is simple: antigens are degraded in lysosomes where the appropriate peptides are bound by MHC Class II molecules. We indeed showed that MHC Class II molecules are transported through lysosomal structures (Peters et al.,1991). Others (Pieters et al., 1991, Guagliardi et al., 1990) have suggested transport of Class II to early endosomal structures. We have been unable to show MHC Class II molecules in early endosomes in EBV-transformed B-cells. Furthermore, delivery of MHC Class II $\alpha\beta$[γ-transferrin receptor hybrid] to early endosomes does not result in functional Class II $\alpha\beta$ dimers, which suggest that the proteolytic activity in early endosomes is not appropriate to efficiently remove the MHC Class II associated γ-transferrin receptor hybrid (M. Nijenhuis et al., ms in prep.). This result points again to a major involvement of late endocytic structures in efficient transport of MHC Class II molecules and the generation of presentable antigen.

MHC Class II molecules are, after removal of the γ chain, transported to the cell surface. How Class II molecules are transported to the cell surface is unknown. Peptide loading is, in contrast to the situation for MHC Class I molecules, not nessecary for transport of MHC Class II molecules to the cell surface (Mellins et al., 1991). Removal of the MHC Class II associated γ chain is essential, inhibition of endosomal breakdown of the γ chain will inhibit Class II transport to the cell surface (Neefjes and Ploegh, 1992a).

Once MHC Class II molecules have arrived at the cell surface, they can exchange their associated peptide by recycling. We have never observed recycling, however, recycling of MHC Class II molecules has been claimed to occur (Reid and Watts, 1990; Salemero et al., 1990). Lanzavecchia et al (1992) compared the half-life of MHC Class II molecules and associated peptide and found it to be identical. This suggests that, if recycling of MHC Class II molecules occurs, it is of minor importance.

The main difference at the protein chemical level between MHC Class I and -II molecules is the γ chain which associates transiently with MHC Class II molecules. We, and others, have determined at least five different functions of the MHC Class II-associated invariant or γ chain.

1. When the MHC Class II $\alpha\beta$ dimer is assembled in the ER, it can bind the same

The study of the CD8/ζ truncation mutants showed that the association of ZAP-70 and ζ correlates well with the signaling capability of the chimera. Since PTKs are involved in the initial events of T-cell activation, the identification of kinases which interact with the ζ chain has been one of the important goals for the study of this signaling pathway. The isolation and expression of the ZAP-70 cDNA clearly showed this molecule is a novel tyrosine kinase that can interact with the ζ chain. Interestingly, in COS cells, ZAP-70 associates with the ζ chain only when co-expressed with lck or fyn. This raises the question of what regulates the ZAP-70 interaction with lck/fyn and the ζ chain. Preliminary results demonstrate that tyrosine phosphorylation of ZAP-70 is induced by lck and fyn in the absence of the ζ chain in COS cells. However, this phosphorylated ZAP-70 does not form a complex with exogenous ζ chains when lysates from different cells are mixed. This suggests that the interaction between lck/fyn and the ζ chain is essential for the ζ:ZAP-70 association. The results obtained with J.CaM1 also showed that lck is involved in the induction of ζ:ZAP-70 association. Additionally, following TCR stimulation, ZAP-70 was detected in immunoprecipitates with only the tyrosine phosphorylated form of the ζ chain.

The study of J.CaM 1 demonstrates that lck is indispensable for the PTK activity associated with the ζ chain. However, attempts to show the association between lck and ζ chain have not been successful. This implies lck may not require a stable association with the ζ chain for its function. Instead, this kinase may function by forming transient or weak associations with the ζ chain. The induction of ζ:ZAP-70 association by lck may also reflect an additional signaling role by CD4 and CD8 TCR co-receptors. The complex formation by CD4/8-MHC-TCR would bring the cytoplasmic portion of ζ and lck into close proximity. This could raise the specific kinase activity by increasing the local concentration of the substrate and the kinase.

The association of fyn and the ζ chain has been previously detected in T cells (Samelson et al. 1990) However, when fyn was transfected into COS18 cells the kinase activity associated with the ζ chain was very low. From such transfectants, no increase of the ζ-associated PTK activity or *in vivo* tyrosine phosphorylation of the ζ chain was observed after crosslinking CD8/ζ. The functional importance of the association of fyn with the ζ chain remains to be determined.

Co-expression of fyn and ZAP-70 induced an equivalent degree of *in vivo* phosphorylation and association of ZAP-70 with ζ when compared to the lck-ZAP-70 co-transfectants. Therefore, as far as the induction of the ζ:ZAP-70 association is concerned, we were unable to distinguish the functional difference between fyn and lck. This may be due to the artificial nature of our experimental conditions. Indeed, unlike in T-cells, ζ:ZAP-70 association does not require the crosslinking of CD8/ζ chimera in COS18 cells. This suggests that the ζ:ZAP-70 association may be negatively regulated in T cells. By imposing different negative regulatory constraints, T cells may be able to control the activity and specificity of these kinases. This could also explain why fyn is not capable of compensating for the lck defect in J.CaM1, which expresses normal levels of fyn.

The isolation and characterization of these molecules and mutant cells has given us clues to elucidate the complex mechanism of TCR signal transduction. It appears that this system involves very complex machinery which is finely regulated to perform the intricate function of TCR. It is now essential to identify other components of this complex system and their molecular interactions in order to understand TCR signal transduction.

on MHC Class II restricted antigen presentation. Thus, the rate of transport of MHC Class II molecules is related to the rate of endosomal breakdown by the degradation of the γ chain. The γ chain acts as an chaparonin that retains MHC Class II molecules in endosomal structures and that has to be removed in order to allow further transport of MHC Class II molecules.

5. The γ chain may have imprinted its presence on the structure of MHC Class II molecules. In normal cells, the MHC Class II α chain contains two N-linked glycans, one of the high-mannose - and one of the complex type. In the absence of the γ chain, the Class II α chain contains two complex type carbohydrates (M. Nijenhuis, ms in prep.). Since carbohydrates are located around the peptide binding groove, different types of carbohydrates may affect the extent of an immune response (Neefjes et al., 1990b, Boog et al., 1990).

If one concept has become clear from the study of MHC Class II molecules, is that the MHC Class II associated γ chain has many different functions in intracellular transport of MHC Class II molecules and, therefore, peptide binding by MHC Class II molecules. I have not discussed the different forms of the γ or invariant chain, which shows that the invariant chain is all but invariant.

References.

Bakke O, and Dobberstein B. (1991) Cell

Bjorkman PJ, Saper MA, Samraoui B, Bennett WS, Strominger JL and Wiley DC (1987) Nature 329: 506-512

Bjorkman PJ, Saper MA, Samraoui B, Bennett WS, Strominger JL and Wiley DC (1987) Nature 329: 512-518

Boog C, Neefjes JJ, Boes J, Ploegh HL, and Melief CJM (1990) Eur. J. Immunol. 19: 537-542

Deres K, Schumacher TNM, Wiesmuller KH, Stevanovic S, Greiner G, Jung G and Ploegh HL (1992) Eur. J. Immunol. 22: 1603-1608

Guagliardi LE, Koppelman B, Blum JS, Marks MS, Cresswell P, and Brodsky FM (1990) Nature 343: 133-138

Harding CV, Collins DS, Slot JW, Geuze HJ and Unanue ER (1991) Cell 64: 393-402

Howard J (1992) Nature 357: 284-285

Lanzavecchia A, Reid PA, and Watts C (1992) Nature 351: 249-252

Ljungren HG, Stam NJ, Ohlen C, Neefjes JJ, Hoglund P, Heemels MT, Bastin J, Schumacher TNM, Townsend A, Karre K, and Ploegh HL (1990) Nature 346: 476-480

Lotteau V, Teyton L, Peleraux A, Nilsson T, Karlsson L, Schmid SL, Quaranta V, and Peterson
 PA (1990) Nature 348: 600-605

Madden DR, Gorga JC, Strominger JL, and Wiley DC (1991) Nature 353: 321-325

Mellins E, Smith L, Arp B, Cotner T, Celis E and Pious D (1991) Nature 343: 71-74

Monaco JJ (1992) Immunol. Today 13: 173-179

Neefjes JJ and Ploegh HL (1987) Eur. J. Immunol. 18: 203-210

Neefjes JJ, DeBruijn MLH, Boog CJP, Nieland JD, Boes J., Melief CJM, and Ploegh HL (1990)
 J. Exp. Med. 171: 583-589

Neefjes JJ, Stollorz V, Peters P, Geuze H, and Ploegh HL (1990) Cell 61: 171-183

Neefjes JJ and Ploegh HL (1992a) EMBO J. 11: 411-416

Neefjes JJ, Smit L, Gehrmann M, and Ploegh HL. (1992b) Eur. J. Immunol. 22: 1609-1614

Reid PA and Watts C. (1990) Nature 345: 655-657

Peters PJ, Neefjes JJ, Oorschot V, Ploegh HL, and Geuze H (1991) Nature 349: 669-676

Pieters J, Horstmann H, Bakke O, Griffith G, and Lipp J (1991) J. Cell Biol. 155: 1213-1224

Roche PA and Cresswell P (1990) Nature 345: 615-619

Salamero J, Humbert M, Cosson P and Davoust J (1990) EMBO J. 9: 3489-3496

Schumacher TNM, Heemels MT, Neefjes JJ, Kast WM, Melief CJ and Ploegh HL (1990) Cell
 62: 563-567

Schumacher TNM, DeBruijn MLH, Vernie LN, Kast WM, Melief CLM, Neefjes JJ and Ploegh
 HL Nature (1991) 350: 703-706

Townsend ARM, Gotch FM and Davey J (1985) Cell 42: 457-467

Townsend A, Ohlen C, Bastin J, Ljunggren HG and Karre K (1989) Nature 340: 443-448

The Kinetics of Association of Peptides with MHC Class 1 Molecules

V. Cerundolo, T. Elliott, A. Townsend

Institute of Molecular Medicine, John Radcliffe Hospital, Oxford, UK

INTRODUCTION

MHC class I molecules present peptides, derived from degradation of intracellular proteins, to cytotoxic T lymphocytes (CTL) (Townsend 1985; Townsend 1986). Direct sequencing of peptides eluted from MHC class I molecules demonstrated that the majority of naturally produced class I bound peptides share certain properties. They are homogeneous in their length of 8-9 amino-acids (Rotzschke 1990; Van Bleek, 1990; Jardetzky, 1992), and have conserved residues which determine the class I allele binding specificity (Falk 1991; Jardetzky 1992).

The human mutant T2 cell has a phenotype similar to that of RMA-S (Townsend 1989), consistent with loss of a mechanism required for loading class I molecules with high affinity peptides *in vivo* (Cerundolo 1990). H-2D^b (D^b) heavy chains synthesized in transfected T2, form a relatively stable complex with human ß-2m (Cerundolo 1991). These properties led us to use D^b molecules to measure the kinetics of peptide binding *in vitro*. A series of different lengthed peptides, based on the influenza nucleoprotein (NP) sequence 365-79, were synthesized replacing the N terminal residue with a tyrosine, which could be radioiodinated. We demonstrated that peptides of various lengths can bind to D^b molecules, but only the 9 amino acid peptide NP Y367-74, equivalent to the peptide isolated from influenza infected cells (Rotzschke 1990), forms a stable complex with D^b (Cerundolo 1991). Addition or deletion of a single residue at the -C and -N termini of Y367-74 greatly reduced its capacity to form a stable complex with D^b. The length dependance for the formation of stable peptide-class I complexes correlated with the ability of the optimal lengthed peptide to induce a conformational change of D^b molecules in the absence of ß-2m (Elliott 1991).

We previously measured the K_a (binding affinity constant) and k_{off} (dissociation rate constant) for a set of peptides related to NP 365-379 (Cerundolo 1991). The law of mass action defines the K_a as the ratio of the forward reaction rate constant (k_{on}) to the backward reaction rate constant (k_{off}) ($K_a = k_{on}/k_{off}$). If the reaction between peptides and the D^b class I molecule is simple, of the kind $A + B <=> AB$, the k_{on} should be predicted from the k_{off} and the K_a. We have therefore measured the k_{on} for peptides of various lengths, and compared the measured values with that predicted from the measured K_a and k_{off}. We find that for peptides extended beyond the C-terminal anchor residue, which have rapid off rates, the observed values of k_{on} are close to those expected. This suggests that the reaction between D^b and long peptides conforms with the simple model. In contrast the forward reaction of the 9 mer peptide occurs at a rate 100-200 fold faster than expected from the independently measured K_a and k_{off}. The rapidity with which equilibrium is reached suggests that the initial reaction of the 9 mer with D^b involves rapid forward and backward reaction rates (k_{on} ~5500 $M^{-1}s^{-1}$, k_{off} ~ 3×10^{-4} s^{-1}, K_a ~ 1.8×10^7 M^{-1}), that slow down drammatically during the interaction of D^b with peptide (k_{on} ~ 20 M^{-1} s^{-1}, k_{off} ~ 8.9×10^{-7} s^{-1}, K_a ~ 2.2×10^7 M^{-1}). This result suggests that peptides of optimum length may induce a conformational change in the D^b binding site that results in trapping of the bound peptide.

PEPTIDES LONGER THAN THE OPTIMUM LENGTH BIND TO MHC CLASS I MOLECULES WITH A SIMPLE KINETICS

We have previously shown that the peptide NP Y367-79, which is longer than the optimum length, forms a short lived complex with D^b molecules (half life at 4° C of ~ 2.5 hrs) (Cerundolo 1991). We have now measured the rate of the forward reaction of Y367-79 binding to D^b and derived values for k_{on} and k_{off} that described the binding curve with time.

In each experiment we did an equilibrium binding study to establish the binding affinity constant (K_a) and the concentration of D^b binding sites in the cell extract, as described (Cerundolo 1991).

The K_a was derived from Scatchard plots of the data. In a series of four experiments the plots were consistently linear (r^2 values .93 - .99) with a mean K_a value of 5.46×10^6 M^{-1}.

We have previously (Cerundolo 1991) obtained a value for the dissociation rate constant for this peptide of 7.59×10^{-5} s^{-1} ($t1/2 = 2.54$ hrs). The predicted k_{on} derived from the independently measured K_a and k_{off} was therefore 414 $M^{-1}s^{-1}$ (S.E.M. 78) .

For the measurements of the forward reaction, concentrations of peptide and D^b were chosen to give approximately 50% saturation at equilibrium. The binding curves obtained were analysed by comparison to model curves derived by integration of the mass action equation. Values for k_{on} and k_{off} that best fitted the binding data were then derived by minimizing variance.

The mean values for k_{on} and k_{off} obtained by this method were k_{on} = 572 $M^{-1}s^{-1}$ (S.E.M. 175) , k_{off} = 1.44×10^{-4} s^{-1}(S.E.M. $.38 \times 10^{-4}$) . These values were within 2 Standard Errors of the Mean of those predicted from the measured K_a and k_{off}. The results were therefore consistent with a simple reversible reaction between long peptides and D^b class I molecules.

OPTIMUM LENGTH PEPTIDES BIND TO MHC CLASS I MOLECULES WITH KINETICS NOT CONSISTENT WITH A SIMPLE MODEL

We have previously established that the peptide NP Y367-74 forms a complex with D^b molecules that is almost completely stable a 4 C ($t1/2 \sim$ 214 hrs), k_{off} 8.9×10^{-7} s^{-1}.

In a series of 4 experiments we observed a mean K_a of 2.33×10^7 M^{-1} The Scatchard plots were again linear (r^2 values .93 - .99) as described (Cerundolo 1991) . These figures predicted a k_{on} of 20.7 M^{-1} s^{-1} (S.E.M. 4.6) which would be expected to give rise to a very slow approach of the reaction to equilibrium. When followed with time the reactions between the 9 mer Y367-74 and D^b reached an equilibrium point expected from the measured K_a. However, the approach to equilibrium was much quicker than predicted by the independently measured K_a and k_{off} (half times of the reaction from 180 to 490 fold shorter than expected). The kinetic constants from the 4 experiments that best fitted the binding curves with time were: k_{on} 5585 M^{-1} s^{-1} (S.E.M. 834), k_{off} 2.89×10^{-4} s^{-1} (S.E.M. $.62 \times 10^{-4}$), K_a 2.17×10^7 M^{-1}.

These results were therefore not consistent with a simple model of binding. They suggested that the D^b molecule, as obtained from the mutant T2 cells, can interact with the optimum lengthed peptides with 2 sets of kinetics. At the beginning of the reaction, the kinetics of both the forward and backward reactions appear comparably rapid, but after contact of D^b with peptide the kinetics change and give rise to stable complexes with very slow off rates.

DISCUSSION

We have previously shown that both long and short peptides can bind to class I molecules. Long peptides form unstable complexes which dissociate rapidly at 4 C, whereas optimum lengthed peptides form a stable complex (Cerundolo 1991). In this paper we have measured the rates of binding of peptides of various lengths to D^b class I molecules . We have found that for peptides longer than the optimum length (Y367-79), and which have rapid off rates, the measured association rate constants and half times for the forward reactions, are close to those predicted from the previously measured affinity constants and off rates. These results are consistent with a simple reversible reaction between long peptides and the class I binding site of the kind A + B <=> AB.

In contrast, the optimum lengthed peptide (Y367-74), that forms extremely stable complexes at 4C, approached equilibrium from 180 to 490 fold faster than predicted from the measured affinity constant and off rate. These results suggested that either the class I binding site or the peptide ligand may change conformation during the binding reaction. The fact that this effect is specific for the optimum lengthed peptide, and correlates with its ability to induce a conformational change in the free heavy chain (Elliott 1991), suggests that it is a change in the class I binding site that is responsible. The optimum lengthed peptide may trigger a conformational change of the antigen binding site which results in a shift from rapid to slow peptide binding kinetics . The complex of heavy chain and beta-2m formed in the absence of peptide may have a more open conformation that allows rapid access of the anchor residues of the peptide to the specificity pockets in the antigen binding site (Saper 1991). The conformational change induced by these interactions may then trap the peptide and allow the formation of multiple stabilizing hydrogen bonds with the peptide's N- and C-termini (Madden 1991).

The result of this effect is that peptides with relatively low binding affinities (compared to many antibodies (Mason 1980) can form extremely stable complexes rapidly, via the formation of an unstable intermediate. A related phenomenon has been described for peptides binding to class II molecules (Sadegh-Nasseri 1989; Sadegh-Nasseri 1991). This mechanism may have evolved to cope with the requirement for each class I and class II molecule to form stable complexes with many different peptide ligands.

We have recently extended these experiments to the interaction of an NP peptide with HLA AW68.1 expressed in T2 cells (Cerundolo 1991b), and find a similar phenomenon (manuscript in preparation).

REFERENCES

Cerundolo V, Alexander J, Lamb C, Cresswell P, McMichael A, Gotch F, Townsend A (1990) Presentation of viral antigen controlled by a gene located in the MHC. Nature 345:449-451.

Cerundolo V, Elliott T, Elvin J, Rammensee HG, Townsned A (1991a) The binding affinity and dissociation rates of peptides for class I major histocompatibility complex molecules. Eur. J. Immunol. 21:2069-2075

Cerundolo V., Tse A, Salter R, Parham P, Townsend A (1991b) CD8 independence and specificity of cytotoxic T lymphocytes restricted by HLA-Aw68.1 Proc. R. Soc. Lond. B 244:169-177.

Elliott T, Cerundolo V, Elvin J, Townsend A, (1991) Peptide-induced conformational change of the class I heavy chain. Nature 351:402-406.

Jardetzky TS, Lane WS, Robinson RA, Madden DR Wiley DC, (1992) Identification of self peptides bound to purified HLA-B27. Nature 353:326-329.

Maden D, Gorga J, Strominger J, Wiley D (1991) The structure of HLA-B27 revelas nonamer self-peptides bound in an extended conformation Nature 353:321-325.

Mason D, Williams A (1980) The kinetics of anibody binding to membrane antigens in solution and at the cell surface Biochem, J. 87:1-20.

Rotzschke O, Falk K, Deres K, Schild H, Norda M, Metzger J, Jung G,

Rammensee HG, (1990) Isolation and analysis of naturally processed viral peptides as recognized by cytotoxic T cells. Nature 348:252-254.

Sadegh-Nasseri S, McConnell HM (1989) A kinetic intermediate in the reaction of an antigenic peptide and I-E^k. Nature 337:274-276.

Sadegh-Nasseri S, Germain R (1991) A role for peptide in determining MHC class II structure Nature 353:167-170.

Saper M, Bjorkman P, Wiley D (1991) Refined structure of the human histocompatibility antigen HLA-A2 at 2.6 A resolution. J. Mol. Biol. 219:277-319.

Townsend A R, Gotch FM, Davey J (1985) Cytotoxic T cells recognize fragments of influenza nucleoprotein. Cell 42:457-467.

Townsend AR,Rothbard J, Gotch FM, Bahadur D, Wraith D, McMichael AJ, (1986) The epitopes of influenza nucleoprotein recognized by cytotoxic T lymphocytes can be defined with short synthetic peptides Cell 44:959-968.

Townsend A, Ohlen VC, Bastin J, Ljunggren H-G, Foster L, Karre K (1989) Association of class I major histocompatibility heavy and light chains induced by viral peptides Nature 340:443-448.

Van Bleek GM, Nathenson SG, (1990) Isolation of an endogenously processed immunodominant viral peptide from the class I H-2 K^b molecule. Nature 348:213-216.

Rules for the Presentation of Peptides by Class II Molecules of the Major Histocompatibility Complex

C. A. Janeway, Jr., P. Preston-Hurlburt, B. Al-Ramadi, J. Rothbard, D. B. Murphy, E.-P. Reich, and A. Y. Rudensky

Section of Immunobiology, Yale University School of Medicine, and the Howard Hughes Medical Institute, New Haven, CT 06510 USA

INTRODUCTION

As virtually all adaptive immune responses involve the activation of CD4 T cells by specific antigen, a central question in immunology is the nature of the ligands recognized by these cells. It is now well established that CD4 T cells respond to peptide fragments of foreign proteins bound to a single peptide binding site of a class II molecule encoded in the major histocompatibility complex (MHC). Thus, determining the nature of peptides bound by MHC class II molecules is essential for an understanding of adaptive immunity. Moreover, as tolerance of CD4 T cells is crucial for maintaining self tolerance of B cells (Lin et al, 1991), CD8 cytolytic T cells (Guerder and Matzinger, 1992), and the CD4 T cells themselves, understanding the nature and distribution of self peptides bound to MHC class II molecules is also critical to understanding self tolerance. Finally, self peptides have been implicated in positive selection of T cells in the thymus for self MHC recognition capability. For all these reasons, we have sought to characterize self peptides bound to MHC class II molecules, their generation and distribution in various tissues, and T cell responses to these self peptides. In this paper, we summarize our findings and speculate about their implications for self tolerance and self MHC recognition.

RULES FOR PEPTIDE PRESENTATION BY MHC CLASS II MOLECULES

We have sought to analyze peptide presentation by MHC class II molecules by a direct examination of the peptides bound normally to these molecules. To this end, we have purified several different MHC class II molecules, eluted their bound peptides with acid, purified them by reversed phase high performance liquid chromatography, and determined their amino acid sequences. The peptide sequence was then checked against a data base of sequences to define the protein of origin. From this analysis, we have determined several rules that govern presentation of peptides by MHC class II molecules (Rudensky et al, 1991a, 1992).

The Length of Peptides Bound to MHC Class II Molecules

Peptides bound to MHC class II molecules are highly variable in length, ranging from 12 to 23 amino acids long in sequences determined to date. Moreover, the same core peptide has been recovered in different peaks eluted from a single MHC class II molecule, with differences of one, two, or three residues at the amino and especially the carboxyl terminus.

These results contrast with findings on peptide binding to MHC class I molecules MHC class I molecules bind peptides of 8-9 amino acids.

These peptides bind in the peptide binding cleft of the MHC class I molecule, and their amino terminal amine and carboxy terminal carboxyl groups are bound to conserved sets of residues at the ends of the peptide binding groove (Madden et al, 1991). These residues are not present in MHC class II molecules, which by modelling appear to have peptide binding grooves that are open at both ends (Margolies, 1992). The open grooves should allow length variation at the ends of peptides, and indeed, in the one case we have examined in detail, these length differences are not discriminated by CD4 T cells specific for the dominant 17 amino acid length variant of this peptide. Variants of this peptide that are longer or shorter at the amino and/or the carboxy termini bind readily to the specific MHC class II molecule and stimulate specific T cells, suggesting that peptides differing in length do not tolerize distinct sets of T cells. Moreover, these findings suggest that T cell receptors are interacting primarily with amino acids in the central region of the peptide, equivalent to the peptide bound by MHC class I molecules. Evidence supporting this comes from studies in which the same peptide was systematically mutated at each position along its sequence. Only the central 8-10 amino acids contributed to T cell recognition by this analysis. Thus, despite the differences in the length of peptides bound by MHC class I and MHC class II molecules, probably reflecting differences in the structure of these two molecules, the ligand recognized by T cell receptors specific for MHC class I or MHC class II is probably quite similar, as would be expected from the use of the same receptor genes for both types of recognition.

Proteins Donating Peptides to MHC Class II Molecules Are Abundant in Endosomes

When the peptides eluted from MHC class II molecules were sequenced, it was shown by comparing peptide sequences with protein sequence data bases that they derived from known proteins in most cases. Most peptides matched known proteins across their entire length. Similar results have been obtained by Hunt et al (1992a) and Newcomb and Cresswell (1992). These proteins derive from extracellular medium or from cell surface molecules, including the MHC class II associated invariant chain (Ii) and the MHC class I and especially the MHC class II molecules themselves. One peptide derived from the I-Eα chain was of particular interest, as it was also recognized by a monoclonal antibody when bound to the MHC class II molecule, I-A^b, from which it was originally eluted. This monoclonal antibody, called Y-Ae, because it recognizes a peptide of I-E bound to I-A^b, can be used to examine the distribution of processed self peptide presented on the cell surface, and it will be discussed in detail later in this paper (Murphy et al, 1989, 1992; Rudensky et al, 1991b). Thus, one feature of peptides bound to MHC class II molecules is that they derive from proteins that are abundant in the acidified vesicles of a cell. MHC class II molecules deliver peptides of such proteins to the cell surface, distinguishing them from MHC class I molecules, which bind and display at the cell surface peptides derived from proteins synthesized and degraded in cytosol (Jardetzky et al, 1991; Hunt et al, 1992b).

Sequence Motifs of Peptides Bound to MHC Class II Molecules

Peptides eluted from MHC class I molecules have been shown to display specific amino acid residues at particular positions in the sequence (Falk et al, 1991; Jardetzky et al, 1991). These have further been shown to be bound in deep pockets that line the peptide binding groove of MHC class I molecules (Madden et al, 1991; I. Wilson, personal communication). The structure of these binding pockets is affected by polymorphisms in MHC molecules, such that each MHC class I molecule binds a distinct set of peptides showing a unique motif of amino acid residues. The length variation in MHC class II associated peptides due to the apparently open ends of the peptide binding cleft makes it difficult to discern such motifs in peptides binding MHC class II molecules. Nevertheless, by adjusting the position of the amino terminus, it is possible to find particular amino acids that occur at particular positions in the sequence of most or all peptides that bind to a particular MHC class II molecule. As is true for peptides binding MHC class I molecules, related amino acid residues are often interchangable in such positions. The peptide motifs defined to date are shown in Fig. 1.

Fig. 1 Peptide motifs associated with MHC class II alleles in mouse

MHC Class II	Motif in single letter code*	Reference
I-A^s	X X X **I T** X X X X **H** X X X X	1
I-A^b	X X X **N** X X X X **T P** X X X X	1
I-A^d	X X X **Q** X X X **B A** X X X X X	2
I-A^{g7}	X X **K** X X X X X X X **R** X X X X†	3
I-E^b	X X **Y L** X X X X X **R** X X X X	1
DR	X X **Y** X X X X X X **K** X X X X	4

Refs: 1: Rudensky et al, 1992; 2. Hunt et al., 1992a; 3. Reich et al, 1992: 4: Jardetzky et al, 1989.
† This residue is acidic in several peptides.
* Bold letters give most frequent amino acid at positions where identical or chemically similar residues are found in all peptides.
B Bulky side chain.

As can be seen from the above motifs, no two MHC class II molecules would be expected to bind the same set of peptides, because they interact with distinct amino acid side chains in a peptide. Moreover, one should be able to identify candidate immunogenic peptides within these proteins using these sequence homologies, as has been carried out using similar motifs for MHC class I associated peptides (Pamer et al, 1991; Rötzschke et al, 1991). A protein lacking the appropriate motif for binding to a given MHC class II molecule will be unable to elicit a response in strains expressing only that molecule, giving rise to Ir gene non-responder strains of mice. Finally, it is interesting to note that peptides binding to I-A^{g7}, but not to other MHC class II molecules, frequently have an acidic residue in the C terminal end of the peptide. We believe this position lies near the ß chain residue 57, which is aspartic acid in all other genotypes, but is serine in I-A^{g7}. I-A^{g7} is associated with susceptibility to diabetes, as is the absence of an aspartic acid at this position in HLA-DQ; it is possible that this alteration allows binding of peptides that would not bind to other haplotypes, or that aspartic acid at position 57 prevents binding of peptides that have an acidic residue at this position, thus preventing diabetes. This is being tested using site-directed mutants of MHC class II molecules and peptides.

The Degradability of Peptides in the Processing Compartment Affects Presentation by MHC Class II Molecules

For a peptide to be presented by MHC class II molecules, it must be generated in the processing compartment of acidified intracellular vesicles in sufficient abundance to form numerous complexes with MHC class II molecules. Peptides that are not readily cleaved from their donor protein, or peptides that are readily degraded once liberated from the protein, may not be able to be presented to T cells. Such epitopes have been termed cryptic epitopes by Ametani et al (1989). When a self protein contains a cryptic epitope, then one might expect the individual not to be tolerant to that self peptide. In studying the T cell response to mouse cytochrome c, we have shown that to be the case.

Mice are tolerant to their own cytochrome c, as shown by their failure to mount a response to mouse cytochrome c emulsified in complete Freund's adjuvant. However, we (Lin et al, 1991) were able to immunize mice to mouse cytochrome c by mixing it with human cytochrome c. This procedure appears to improve presentation of mouse cytochrome c by activated, human and mouse cytochrome c cross-reactive B cells, since B cells from mice primed with human cytochrome c will transfer the ability to stimulate mouse cytochrome c specific CD4 T cells to naive mice. The response is directed at a peptide in the middle of the mouse cytochrome c molecule, and not at the 81-104 region that dominates mouse responses to heterologous cytochromes c. This shows that T cell tolerance to mouse cytochrome c can be broken, and that murine antigen presenting cells can process and present mouse cytochrome c. No responses to the 81-104 fragment have ever been observed in mice immunized with intact mouse cytochrome c. Thus, it was surprising when immunizing mice with a synthetic 81-104 peptide of mouse cytochrome c triggered potent T cell responses. These responses could be elicited with synthetic or with cyanogen bromide-cleaved peptide fragments of mouse cytochrome c, suggesting that the response was not directed at a contaminating synthetic peptide. However, these T cells could not respond to native mouse cytochrome c (Mamula et al, 1992). From this, it appears that mice are not tolerant to, and cannot respond against, the 81-104 fragment of mouse cytochrome c due to an inability to generate this peptide in a form that can be bound and transported to the cell surface by MHC class II molecules (Mamula, 1992). Thus, peptides must be able to be generated in sufficient quantity and be sufficiently stable in the processing compartment to get to the cell surface and be recognized by T cells.

Summary

These studies have allowed us to define four rules for peptide presentation by MHC class II molecules. First, the peptide must come from a protein that is abundant in the processing compartment of acidified cellular vesicles. Second, it must contain a sequence motif that allows strong binding to the MHC class II allelic variants present in the cell. Third, it can vary in length around the core MHC class II binding motif without qualitatively affecting recognition by T cells. And finally, the peptide must be able to be cleaved in abundance from the donor protein and be stable until it binds to the MHC class II molecule. These rules, while largely self-evident, are now supported by solid experimental data derived from direct structural analysis of peptide:MHC class II complexes.

THE IMPACT OF PEPTIDE:MHC CLASS II COMPLEXES ON THE T CELL RECEPTOR

Peptides bound to MHC class II molecules have been implicated in a variety of immune responses. First, they are the form in which foreign protein antigens are recognized by the T cell receptor. Second, self peptides bound to non-self MHC molecules have been implicated in responses to non-self MHC molecules (Panina-Bordignon et al, 1991). Third, they have been implicated indirectly in the intrathymic positive selection of T cells for self MHC recognition (Nikolic-Zugic and Bevan, 1990). Finally, self peptides bound to self MHC molecules lead to deletion of T cells whose receptors recognize these complexes during negative intrathymic selection(vonBoehmer et al, 1989). Thus, to understand the biology of T cell responses, the distribution and effect of self peptides bound to MHC molecules must be analyzed. We have used the novel monoclonal antibody Y-Ae to explore these issues.

The Distribution of a Self Peptide Affects the T Cell Repertoire

The Y-Ae antibody that recognizes a peptide fragment of the I-Eα chain bound to I-A^b allows us to determine the distribution of a self peptide:self MHC complex in situ. Studies by ourselves (Murphy et al, 1989, 1992) and by Surh et al (1992) show that this epitope is abundantly expressed on cells of hemopoietic origin, but is poorly expressed on thymic epithelial cells. Nevertheless, it is clear that the complex is expressed by thymic cortical and especially medullary epithelial cells. We have explored the impact of this self peptide on the selection of T cells by making radiation bone marrow chimeric mice that express the peptide only on bone marrow derived cells, only on thymic epithelial cells, both, or neither. These mice were then immunized with the specific peptide and their T cell responses to the peptide quantitated. We have also used isolated thymic epithelial cells to examine cell surface expression in terms of ability to stimulate T cells specific for the complex.

Our findings are clear-cut. Expression of the peptide:MHC complex on bone marrow derived cells leads to complete unresponsiveness to the peptide, as expected. However, expression only on thymic epithelial cells does not induce tolerance; indeed, on average, such mice respond better to the peptide than do mice that lack the peptide donor gene and therefore do not express the peptide on thymic epithelial cells at all. The expression of the peptide:MHC class II complex on thymic epithelial cells isolated as nurse cells and on thymic dendritic cells isolated as T cell rosettes has been confirmed. However, it should be noted that even in mice that do not express the Eα gene that encodes the peptide, thymic epithelial cells have some ability to stimulate T cell hybrids specific for the peptide:MHC class II complex, suggesting that a range of peptides may have this ability. This system should allow a careful exploration of the role of self peptide:self MHC complexes in the events of thymic selection of the T cell receptor repertoire.

Analogues of a Peptide Can Inhibit T Cell Responses to the Peptide

Many amino acid substitutions in a peptide prevent its ability to stimulate T cells. Some of these changes prevent binding to the specific MHC class II molecule, thus making it impossible for the peptide to bind. However, others bind to the MHC molecule but fail to engage the T cell receptor in such a way as to signal the T cell

(Rothbard and Gefter, 1991). In general, it has been assumed that such peptides make complexes with the MHC class II molecule that bind less well to the T cell receptor and thus fail to signal the T cell. However, recent studies by de Magistris et al (1992) strongly challenge this notion, in that some peptides closely related to the stimulating peptide can inhibit the response of specific T cells to the stimulating peptide. This phenomenon of peptide antagonism has important implications for therapy of immunological diseases, for antigen recognition by T cells, and for thymic selection of T cells. We have therefore explored the ability of peptide analogues of the Eα 54-66 peptide to inhibit T cell responses.

Our studies exploit the Y-Ae antibody that allows us to detect specific peptide:MHC complexes independent of T cell responses. We find that certain modifications of this peptide not only fail to stimulate T cells specific for the native peptide, but actually inhibit responses to the native peptide. In these cases, we can show that the amount of stimulatory peptide bound to the MHC class II molecule is unaltered by the competing analogue. Moreover, the peptide does not inhibit responses of other T cells to the same MHC molecule. Thus, the analogue must interact with the T cell receptor and prevent activation of the T cell by the peptide:MHC complexes present on the same antigen presenting cell. How this occurs is not known; a possible hypothesis is discussed in the next section.

Cross-Linking and Conformational Change in T Cell Activation

Some years ago, we used antibodies to a single T cell receptor to demonstrate that ligation and cross-linking alone did not lead to activating signals; conformational change in the receptor appeared to be required for effective signalling to occur (Rojo and Janeway, 1988; Rojo, Janeway and Saizawa, 1989; Janeway et al, 1989). If these data are correctly interpreted, they may also explain the effects of analogue peptides on inhibition of T cell responses.

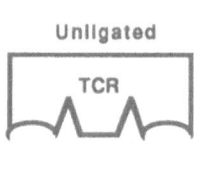
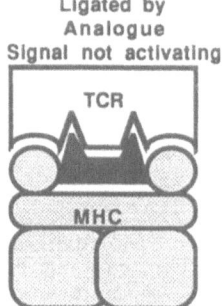
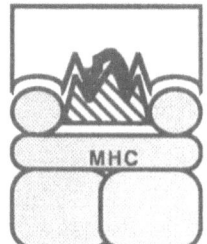

Fig. 2. The T cell receptor can bind either its stimulating peptide or a series of analogue peptides bound to self MHC. Only stimulating peptides signal the cell efficiently, because only these peptides induce a conformational change in the T cell receptor. See text for details.

The model we propose is shown in Fig. 2. We propose that stimulating peptide:MHC complexes signal by binding to multiple T cell receptors and also by inducing conformational change in them. The conformational change is required for effective signalling via the T cell receptor. Its presence may account for the low affinity measured for T cell receptor binding to stimulatory ligands (Matsui et al, 1991;Weber et al, 1992), as such induced conformational changes are usually energetically unfavorable. Analogue peptide:MHC complexes would bind the same T cell receptor without conformational change, thus signalling ineffectively or not at all. If one further postulates that two or more receptors, each undergoing conformational change, must be aggregated to deliver an effective signal, it is clear that excess analogue peptide may occupy enough receptors to prevent any signalling complexes from being generated, giving the observed inhibition. Some of these interactions of the T cell receptor with analogue peptide:MHC complexes may be of high affinity; our earlier studies with antibodies showed that signalling capacity was related to the epitope on the T cell receptor to which the antibody bound and not to its affinity for the receptor (Rojo and Janeway, 1988). Such high affinity competitors could be potent inhibitors of T cell activation, and should allow binding of T cell receptors to their ligands to be measured directly.

This hypothesis may explain thymic selection and some forms of alloreactivity as well. If one assumes that alloreactivity can involve direct binding of the T cell receptor to MHC molecules, it may generate high multiplicity cross-linking without conformational change, sufficient to signal. Likewise, positive selection must signal T cells through the receptor without delivering a stimulatory signal, which is known to induce apoptosis in immature thymocytes. This could be accomplished by extensive cross-linking of T cell receptors without conformational change by direct binding to self MHC molecules and their associated peptides. Provided these interactions only cross link and do not induce conformational changes, they will be inhibitory of activation, although some signal must be transduced to effect positive selection. Activation in the periphery will occur only when a peptide can both bind and induce conformational change in the receptor, and overcome any inhibitory effect of self peptide:self MHC complexes on the receptor. Proving this model will require direct measurement of T cell receptor binding to its ligands, including conformational changes in the T cell receptor, a search for signals generated by cross-linking alone, and demonstration of these signals in developing thymocytes undergoing positive selection.

SUMMARY

In this paper, we have described rules for the presentation of self peptides by MHC class II molecules, and have examined the impact of these self peptide:MHC class II complexes on the T cell receptor repertoire. We suggest that self peptide:self MHC class II complexes normally engage the T cell receptor, not only in the thymus but also in the periphery, and that these interactions are normally inhibitory of T cell activation. In the thymus, this signal could lead to positive selection of T cells able to recognize peptides bound to self MHC molecules. In the periphery, we propose that T cells are activated when a peptide bound to a self MHC molecule not only binds the T cell receptor but also induces a conformational change in that receptor. This signalling behavior may explain many mysterious aspects of T cell receptor behavior, such as low affinity binding to stimulatory ligands, the ability to bind self MHC molecules, and the effects of analogue peptides on T cell responses.

REFERENCES

Ametani A, Apple R, Bhardwaj V, Gammon G, Miller A, Sercarz E (1989) Cold Spring Harb Symp Quant Biol 54:505-511

De Magistris MT, Alexander J, Coggeshall M, Altman A, Gaeta FCA, Grey HM, Sette A (1992) Cell 68:625-634

Falk K, Rotzschke O, Stevanovic S, Jung G, Rammensee H-G (1991) Nature 351:290-296

Guerder S, Matzinger P (1992) J Exp Med 176:553-564

Hunt DF, Henderson RA, Shabanowitz J, Sakaguchi K, Michel H, Sevilir N, Cox AL, Appella E, Engelhard VH (1992b) Science 255:1261-1263

Hunt DF, Michel H, Dickinson TA, Shabanowitz J, Cox AL, Sakaguchi K, Appella E, Grey H M, Sette A (1992a) Science 256:1817-1820

Janeway Jr CA, Dianzani U, Portoles P, Rath S, Reich E-P, Rojo J, Yagi J, Murphy DB (1989) Cold Spring Harb Symp Quant Biol 54:657-666

Jardetzky TS, Gorga JC, Busch R, Rothbard J, Strominger JL, Wiley DC (1990) EMBO J 9:1797-1803

Jardetzky TS, Lane WS, Robinson RA, Madden DR, Wiley DC (1991) Nature 353:326-329

Lin R-H, Mamula MJ, Hardin JA, Janeway Jr CA (1991) J Exp Med 173:1433-1439

Madden DR, Gorga JC, Strominger JL, Wiley DC (1991) Nature 353:321-325

Mamula MJ (1992) Submitted

Mamula MJ, Lin R-H, Janeway Jr CA, Hardin JA (1992) J Immunol 149:789-795

Margulies DH (1992) Curr Biol 2:211-213

Matsui K, Boniface JJ, Reay PA, Schild H, Fazekas De St Groth B, Davis MM (1991) Science 254:1788-1791

Murphy DB, Rath S, Pizzo E, Rudensky AY, George A, Larson JK, Janeway Jr CA (1992) J Immunol 148:3483-3491

Murphy DB, Lo D, Rath S, Brinster RL, Flavell RA, Slanetz A, Janeway Jr CA (1989) Nature 338:765-768

Newcomb JR, Cresswell P (1992) Submitted

Nikolic-Zugic J, Bevan MJ (1990) Nature 344:65-67

Pamer EG, Harty JT, Bevan MJ (1991) Nature 353:852-855

Panina-Bordignon P, Corradin G, Roosnek E, Sette A, Lanzavecchia A (1991) Science 252:1548-1550

Reich E-P, von Grafenstein H, Janeway Jr CA (1992) In preparation.

Rojo J, Janeway Jr CA (1988) J Immunol 140:1081

Rojo J, Saizawa K, Janeway Jr CA (1989) Proc Natl Acad Sci 86:3311

Rothbard JB, Gefter ML (1991) Ann Rev Immunol 9:527-565

Rötzschke O, Falk K, Stevanovic S, Jung G, Walden P, Rammensee H-G (1991) Eur J Immunol 21:2891-2894

Rudensky AY, Preston-Hurlburt P, Al-Ramadi B, Rothbard J, Janeway Jr CA (1992) Nature Submitted

Rudensky A Y, Preston-Hurlburt P, Hong S-C, Barlow A, Janeway Jr CA (1991a) Nature 353:622-627

Rudensky AY, Rath S, Preston-Hurlburt P, Murphy DB, Janeway Jr CA (1991b) Nature 353:660-662

Surh CD, Gao E-K, Kosaka H, Lo D, Ahn C, Murphy DB, Karlsson L, Peterson P, Sprent J (1992) J Exp Med 176:495-505

von Boehmer H, Teh HS, Kisielow P (1989) Immunol Today 10:57-60

Weber S, Traunecker A, Oliveri F, Gerhard W, Karjalainen K (1992) Nature 356:793-796

Peptide Binding to MHC Class II Molecules in Living Cells: Role of Class II Synthesis and Stability of Peptide/Class II Complexes

Antonio Lanzavecchia[1] and Colin Watts[2]

[1]Basel Institute for Immunology, Grenzacherstrasse 487, Postfach CH-4005, Basel, Switzerland
[2]Department of Biochemistry, Medical Sciences Institute, University of Dundee, Dundee DD1 4HN, UK

INTRODUCTION

There is increasing evidence that the biosynthetic pathway of class II molecules intersects the endocytic pathway (Cresswell 1985, Neefjes et al. 1990) and that at this level the newly synthesized class II molecules have access to peptides derived from processing of incoming antigens. As shown schematically in fig. 1, this encounter is facilitated by the sorting and protecting action of the Invariant (I) chain (Bakke and Dobberstein 1990, Lotteau et al. 1990, Roche and Creswell 1990) and may require the action of other genes which map in the MHC and are missing in mutants that fail to correctly assemble antigenic peptides on class II molecules (Mellins 1991 and Sette, personal comunication). Stable peptide/class II complexes are then transported to the cell surface. Once on the cell surface, class II molecules recycle in peripheral endosomes (Reid and Watts 1990).

One can thus envisage two ways by which peptides can bind to class II molecules in living cells: first they can bind to newly synthesized molecules on their secretory pathway, second they can bind to mature molecules during recycling, possibly by a peptide exchange mechanism (Harding et al. 1989, Adorini et al. 1989). Indeed some surface class II molecules are empty and therefore there is some binding capacity even in the absence of peptide exchange (Shimonkevitz et al. 1983).

While there has been no general consensus on which of these two pathways may be predominant for peptide loading, it is clear that the two mechanisms of loading would lead to very different kinetics of antigen presentation. On the one hand exchange would lead to rapid increase in complex formation by providing a higher capacity, but in the absence of a continuous supply of antigen would lead to a rapid loss of the antigenic complexes by exchange with self peptides. On the other hand, loading on newly synthesized molecules would lead to a more gradual increase in complexes that may be preserved by the cell in the absence of exchange.

We have studied the loading of MHC class II molecules with antigenic peptides as well as the stability of peptide/class II complexes in living cells and conclude that in physiological conditions class II molecules bind peptides only once during biosynthesis and these peptides remain bound irreversibly for their lifetime. We review the experimental evidence and discuss the relevance of these findings for the physiology of antigen presentation.

---

* The Basel Institute for Immunology was founded and is supported by F. Hoffmann-La Roche Ltd., Basel, Switzerland. This work was partially supported by the Wellcome Trust and the Medical Research Council.

When antigen-specific B cells are pulsed with antigen, stimulatory complexes appear only after a lag of approximately 60 minutes, indicating that antigen processing and class II association requires a much longer time than that taken by class II molecules to recycle in peripheral endosomes (Roosnek et al. 1988). These results suggested that peptides are acquired in an intracellular compartment different from that involved in recycling of membrane receptors.

It has been shown that class II restricted presentation of native antigen may require new protein synthesis, as demonstrated by the inhibitory effect of brefeldin A and cycloheximide (Adorini et al. 1990, St Pierre and Watts 1990), but it was not clear whether this inhibition reflects the requirement for new synthesis of class II molecules or of other molecules necessary for antigen processing. We therefore looked for a biochemical assay that would detect the loading of processed antigen fragments on class II molecules.

When antigen-specific B cells are pulsed in the cold with ^{125}I-antigen and transfered to 37°C, it is possible to follow antigen degradation and the appearance of complexes iodinated peptides bound to class II molecules. These peptide/class II complexes are first observed at approximately 60 minutes, increase with time and are still present after 40 hours (Davidson et al., unpublished), indicating that they are very stable. Using this direct biochemical readout to measure loading of class II molecules with processed antigen and lectin chromatography to distinguish newly synthesized from mature class II molecules, we found that iodinated antigen fragments bind only to the newly synthesised class II molecules (Davidson et al. 1991). In the same experiments exogenously added peptides bound preferentially to surface class II molecules. Thus, while surface class II molecules have some binding capacity for peptides, they cannot bind processed antigen probably because the rapid recycling kinetics precludes access to the antigen processing compartment.

These results show that in human antigen-specific B cells, peptide binding to class II molecules is a biosynthetic event and that antigen presentation requires new synthesis of class II molecules.

CLASS II MOLECULES BIND PEPTIDES IRREVERSIBLY IN LIVING CELLS

Our results indicated that a peptide exchange mechanism is not involved in binding processed antigenic peptides, but still did not rule out the possibility that peptide exchange may take place during recycling of mature class II molecules. We reasoned that if there is peptide exchange the halflife of complexes should be much lower than the halflife of class II molecules. If there is no exchange, the halflifes should be the same.

We therefore measured the halflife of peptide class II complexes using both functional and biochemical assays and related it to the halflife of the class II molecules themselves (Lanzavecchia et al. 1992). To measure the functional halflife of peptide/MHC complexes we pulsed various types of APC with different concentrations of synthetic peptides corresponding to known T cell epitopes and measured at different time points the level of stimulatory complexes in a T cell proliferation assay. Using this method, a long halflife of approximately 25 hours was estimated on EBV-B cells, activated T cells and peripheral blood mononuclear cells using four different peptide epitopes.

An independent and direct estimate of complex stability was obtained by measuring the lifetime of radioiodinated peptide/MHC complexes. APC were pulsed with radioiodinated peptides and after different times at $37°C$ we immunoprecipitated class II DR molecules and measured the labelled peptide bound. The lifetime of these complexes was in good agreement with that indicated by functional assays, i.e. 30-40 hrs.

We then asked whether preformed peptide class II complexes may be displaced in living cells by adding high concentrations of a competing peptide. Although we could easily demonstrate competition for binding we were not able to displace iodinated peptide that were already bound to class II molecules even when a large excess of cold peptide was added.

Thus both functional and biochemical assays using different epitopes and different APC indicate that peptide/MHC class II complexes are very stable in living cells with a suprisingly reproducible halflife of ~30 hours. This suggested that the dissociation of the complexes might be negligible and their persistence might be determined essentially by the lifetime of the class II molecules themselves. Indeed we found that the halflife of class II molecules in our EBV-B cells is ~30 hrs. We conclude that in living cells antigenic peptides are bound irreversibly to class II molecules, are not displaced under physiological conditions and their turnover reflects the turnover of class II molecules themselves.

These epitopes however represent only a minor fraction of all class II molecules and were selected because of their immunogenicity. Thus their behavior may not reflect the bulk of the class II bound peptides. Using a novel technique based on diagonal SDS/SDS urea gels we were able to display low Mr peptides from disrupted class II αβ complexes (Lanzavecchia et al. 1992). The results showed that the halflife of this low molecular weight material, representing self peptides labeled on the cell surface, was strikingly similar to that of the α and β chains from which it was eluted, i.e. ~30 hours on EBV-B cells.

In summary, using both functional and biochemical readouts on three different cell types we measured the halflife of the complexes of DR with three synthetic peptides as well as with presumptive endogenous peptides and found that class II molecules can bind peptides irreversibly, in the sense that the halflife of the complex is determined by the halflife of the class II molecules themselves.

DISCUSSION

Our results allow to assess the relative importance of loading during biosynthesis versus peptide exchange on mature molecules. Two experimental findings indicate that in human APC the major pathway is the biosynthetic one. First, processed antigen binds to newly synthesized class II molecules. Second, the halflife of peptide/class II complexes is the same as that of class II molecules themselves, indicating that the molecules are not reused.

We would like to discuss some of the consequences of this finding and suggest that biosynthetic loading and complex stability indeed better suits the requirement of the immune system, since it allows accumulation of antigen in APC and presentation at a distant site, perhaps several days after antigen encounter.

An important working principle emerges from these results: an APC for class II is best defined not as a cell with high levels of surface class II molecules, but as one that is actually synthesizing these molecules. Class II biosynthesis is tightly regulated in APC such as B cells, macrophages and dendritic cells. Following appropriate stimulation of synthesis by lymphokines increased numbers of MHC molecules will be synthesized and made available for binding incoming antigens. The preexisting MHC molecules on the cell surface may indeed not only be useless but even play a negative role by distracting the CD4 molecules or even th T cell receptor from the productive interaction. In fact, the best APC may be those expressing a high "specific complex density", irrespective of the total level of class II molecules.

We suggest that the persistence of peptide/MHC complexes is a critical factor for the immune response since it allows the APC to keep the memory of antigen encountered over a short time window, allowing presentation at a distant site perhaps several days later. This is best illustrated by the behaviour of dendritic cells that pick up and process antigen in nonlymphoid organs such as the skin or solid tissues, where they display high capacity to capture antigen and have high rate of class II synthesis. After antigen capture and processing the dendritic cells cease class II synthesis, acquire co-stimulatory capacity and move to the T cell areas of secondary lympoid organs where they wait for antigen specific T cells. In these organs they have essentially lost the capacitiy to present new antigens but can retain long lived antigen/MHC complexes in a highly immunostimulatory form (reviewed by Steinman, 1991). Thus dendritic cells can be viewed as dedicated APC that pick up and process antigen in the periphery and present it at a distance even several days later, a function which is critically dependent from the stability of peptide/class II complexes.

Another example that illustrates the relationship between class II biosynthesis and antigen presenting function comes from activated T cells. It is known that human activated T cells are class II positive and can present peptide antigens and even protein antigens if these bind to cell surface receptors (Lanzavecchia et al. 1988), but have been reported to be relatively inefficient APC, since long term T cell lines are unable to present native protein antigens. On the light of these new findings we have now tested the antigen presenting capacty of T cells and found that recently-activated T cells that rapidly divide and synthesize class II molecules are indeed capable of presenting soluble antigens (AL and CW, unpublished results). We interpret these findings as an indication that class II biosynthesis is one of the limiting factors in presentation of antigen by activated T cells. Thus, in the case of T cells soon after activation, presentation of soluble antigen may well take place and play a physiological rolein the immune response.

We have previously shown that in antigen-specific B cells, a decrease in antigen concentration can be compensated by increasing the time of exposure to antigen,. indicating that there is net accumulation of antigen with time (Lanzavecchia 1987). We have referred to this phenomenon as a "vacuum cleaner effect" meaning that B cells can concentrate antigen by continuous capture of low levels of antigen and transformation into stable complexes until a threshold can be reached sufficient to trigger T cells. We can now explain these findings in term of sustained capture, biosynthetic loading and persistence of peptide/class II complexes (fig. 1).

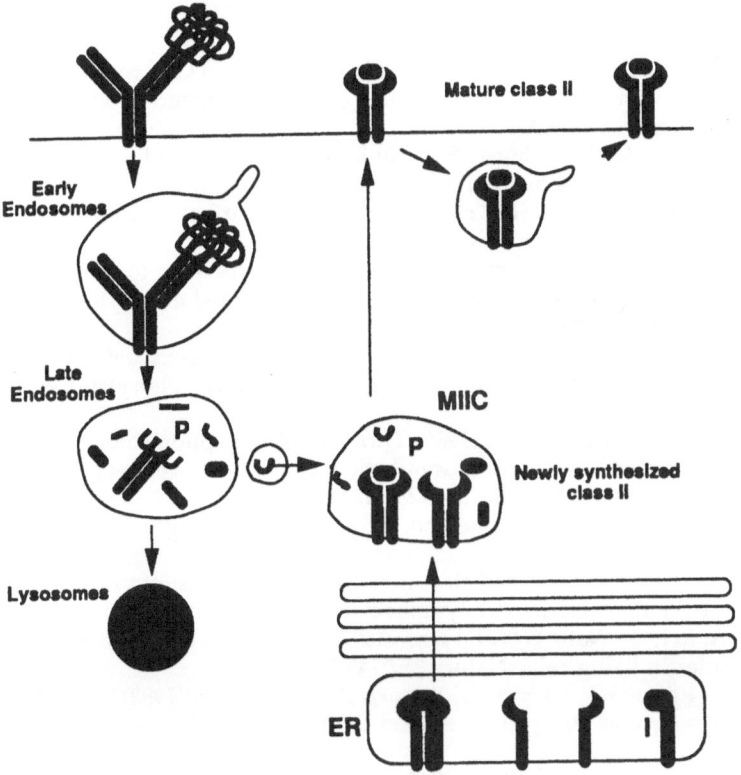

Fig. 1. Peptide binding to MHC class II molecules as a biosynthetic event.
The biosynthetic pathway of class II intersects the endocytic pathway in antigen-specific B cells. Antigen bound to mIg is rapidly internalysed (half-life on cell surface ~10 min.). Some complexes recycle, but some are delivered to the processing compartment. The antigen antibody complexes are degraded by proteases (P) and eventually completely destroyed (half-life ~2 hrs). Newly synthesized class II molecules associate in the endoplasmic reticulum with the I chain which protects the peptide binding groove and sorts these molecules in a specialised compartment (MIIC). Here the I chain is removed by proteases and newly synthesized class II molecules with free peptide binding sites

come into contact with processed antigen. The mechanism of this intersection has not been yet clarified and may involve specific signals for vescicle fusion. This "window of opportunity" decides which antigenic peptides can be loaded on class II molecules. The newly formed complexes are then transported to the cell surface from which they can recycle in the peripheral endosomal compartment. The half-life of the peptide/class II complexes is the same as the half-life of the class II molecules thamselves (~30 hrs). A small fraction (~2%) of surface class II molecules are empty and can bind exogenous peptides, but not processed antigen, since they do not reenter in the processing compartment.

REFERENCES

Adorini L, Appella E, Doria G, Cardinaux F, Nagy Z.A. (1989) Competition for antigen presentation in living cells involves exchange of peptides bound by class II mHC molecules. Nature 342:800-803

Adorini L, Ullrich SJ, Apella E, Fuchs S (1990) Inhibition by Brefeldin A of presentation of exogenous protein antigens to MHC-restricted T cells. Nature 346:63-67

Bakke O, Dobberstein B (1990) MHC class II-associated Invariant chain contains a sorting signal for endosomal compartments. Cell 63:707-713

Cresswell P (1985) Intracellular class II HLA antigens are accessible to transferrin-neuraminidase conjugates internalised by receptor-mediated endocytosis. Proc Natl Acad Sci USA 82:8188-8192

Davidson HW, Reid PA, Lanzavecchia A, Watts C (1991) Processed antigen binds to newly synthesized class II molecules in antigen-specific B lymphocytes. Cell 67:105-116

Harding CW, Roof RW, Unanue ER (1989) Turnover of Ia-peptide complexes facilitated in viable antigen presenting cells: biosynthetic turnover of Ia versus peptide exchange. Proc Natl. Acad Sci USA 86: 4230-4234

Lanzavecchia A (1987) Antigen uptake and accumulation in antigen-specific B cells. Immunol Rev 99:39-51

Lanzavecchia A (1990) Receptor mediated antigen uptake and its effect on antigen presentation to class II restricted T lymphocytes Annu Rev Immunol 8:773-793

Lanzavecchia A, Reid P, Watts C (1992) Irreversible association of peptides with class II MHC molecules in living cells. Nature 357:249-252

Lanzavecchia A, Roosnek E, Gregory T, Berman P, Abrignani S. (1988) T cells can present antigen such as HIV gp120 targeted to their own surface molecules. Nature 334: 530-532

Lotteau L Teyton L peleraux A Nilsson T Karlsson L Schmid SL Quaranta V Peterson PA (1990) Intracellular transport of class II MHC molecules directed by Invariant chain. Nature 348:600-605

Mellins E, Kempin S, Smith L, Monji T, Pious D (1991) A gene required for class II restricted antigen presentation maps to the Major Histocompatibility Complex. J Exp Med 174:1607-1615

Neefjes JJ, Stollorz V, Peters PJ, Geuze HJ, Ploegh HL (1990) The biosynthetic pathway of MHC class II but not class I molecules intersects the endocytic route. Cell 61:171-183

Reid PA, Watts C (1990) Cell surface MHC glycoproteins cycle through primaquine sensitive intracellular compartments. Nature 346:655-657

Roche PA, Cresswell P (1990) Invariant chain association with HLA-DR molecules inhibits immunogenic peptide binding. Nature 345:615-618

Roosnek E, Demotz S, Corradin G, Lanzavecchia A (1988) Kinetics of MHC-antigen complex formation on antigen presenting cells. J Immunol 140:4079-4082

Shimonkevitz RJ Kappler J Marrack P Grey HM (1983) Antigen recognition by H2-restricted T cells. I. Cell free antigen processing. J Exp Med 158:303-316

Steinman RM (1991) The dendritic cell system and its role in immunogenicity. Annu Rev Immunol 9:271-296

St. Pierre Y, Watts TH (1990) MHC class II-restricted presentation of native protein antigens by B cells is inhibitable by cycloheximide and brefeldin A. J Immunol 145:812-818

4. Signal Transduction Pathways

B Cell Antigen Receptor Ligation Activates a Receptor Associated Protein Tyrosine Kinase

J. C. Cambier and K. S. Cambier

Division of Basic Sciences, Departement of Pediatrics, National Jewish Center of Immunology and Respiratory Medicine, 1400 Jackson Street, Denver, Colorado 80206, USA

INTRODUCTION

B lymphocytes express cell surface antigen receptors which function in antigen uptake for subsequent presentation to T cells, and in signal transduction leading to cell activation. The receptor is hetero-oligomeric, being composed of a membrane immunoglobulin molecule noncovalently associated with disulfide linked heterodimers of Ig-α and Ig-β or Ig-γ (for review see Cambier and Campbell in press) These subunits are encoded by mb-1 (Ig-α) and B29 (Ig-γ and Ig-γ) genes, and function in signal transduction and receptor transport to the cell surface. Ig-α/β and -α/γ complexes have significant cytoplasmic structure which appears to act as a docking site for cytoplasmic enzymes involved in signal propagation. The enzymes include members of the src family of tyrosine kinases, phosphatidylinositol-3 kinase and other as yet unidentified molecules of 40, 42 and 38 kDa Mr (Cambier and Campbell in press; Clark et al in press). Studies utilizing fusion proteins containing Ig-α and Ig-β cytoplasmic tail sequences indicate that these proteins bind to a ~17 amino acid motif designated ARH1 (Clark et al in press). This motif, when derivated from TCRζ and CD3ε chains, has been shown to contain all structural information necessary for signal transduction. The B cell antigen receptor complex is also associated with the syk tyrosine kinase, apparently via direct interaction with the mIgM heavy chain (μ). Thus available evidence indicates that multiple tyrosine kinases associate with the antigen receptor. The dynamics of this receptor-kinase association and kinase activation following antigen receptor crosslinking is not well defined.

Studies utilizing tyrosine inhibitors indicate that tyrosine phosphorylation is of central importance in signal transduction (Cambier and Campbell in press). Further, increased tyrosine kinase activity is found associated with receptor following ligand stimulation (Burkhardt et al 1991; Yamanashi et al 1992). It is unclear, however, to what extent this reflects kinase recruitment into the receptor complex or kinase activation. To begin to approach this question we have studied ligand stimulation of tyrosine kinase activity in whole B cells, isolated membranes and in immune complexes using peptide substrate based kinase assays.

Antigen Receptor Crosslinking Stimulates Phosphorylation of receptor associated Ig-α and Ig-β

In order to confirm the ability of receptor binding ligands to stimulate protein phosphorylation in our experimental system, splenic B cells were permeabilized, loaded with ^{32}P-ATP and then stimulated for 2, 5 or 10 minutes with monoclonal anti-IgM (Bet 2) antibody. Cells were then lysed with 1% digitonin, immunoprecipitated by addition of excess anti-μ and protein G Sepharose, and analyzed by SDS-PAGE under reducing conditions. As can be see in Fig.1, stimulation for 2 minutes led to maximal stimulation of phosphorylation of receptor associated Ig-α and β. Integration of radioactivity revealed a six-fold increase in labeling of Ig-α and Ig-β proteins following stimulation.

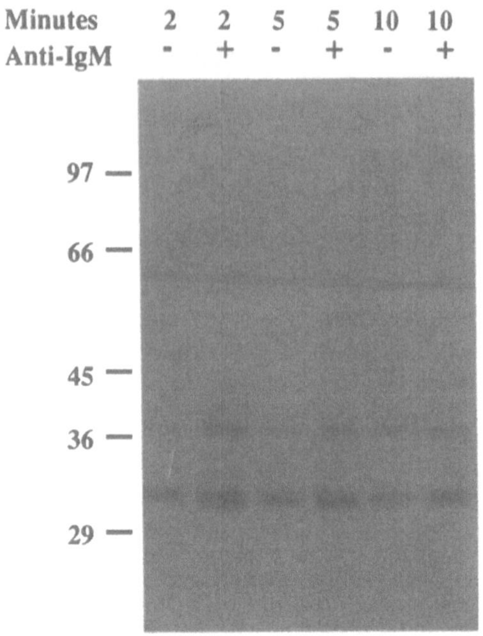

Minutes	2	2	5	5	10	10
Anti-IgM	-	+	-	+	-	+

97 —

66 —

Figure 1. Ligation of mIgM with antibody induces phosphorylation of the associated α, β and γ components. Splenic B cells were isolated, permeabilized and loaded with α-^{32}P-ATP, and then stimulated with anti-IgM (Bet-2) antibody (+) or PBS (-) and lysed at 2, 5 or 10 min. Anti-IgM antibody (Bet-2) was added to each lysate and immune complexes were isolated with protein G-Sepharose. Immunoprecipitates were fractionated using SDS-PAGE and autoradiographed.

45 —

36 —

29 —

It is possible that the kinase responsible for receptor phosphorylation is directly associated with the receptor, as discussed earlier, or free in the cytoplasm. Since splenic B cells express mIgM and mIgD which both transduce signals leading to Ig-α and Ig-β phosphorylation, it is possible to approach this question indirectly by assessing the effect of stimulation through one receptor on phosphorylation of the other receptor. Splenic B cells were loaded with ^{32}P-ATP as before and stimulated of 1 or 2 minutes with anti-IgM (Bet-2) or anti-IgD (JA12.5) antibodies. Cells were then lysed with 1% digitonin and total mIg was immunoprecipitated with sheep anti-mouse Ig and protein G Sepharose. SDS-PAGE and autoradiographic analysis revealed that anti-μ induced the phosphorylation of mIgM associated Ig-α (Mr 32 kDa) but not mIgD associated Ig-α (Mr 33 kDa) (Fig. 2). Conversely, anti-IgD stimulated phosphorylation of mIgD associated α chain but not mIgM associated α chains. Since mIgM and mIgD associated α chains are identical in cytoplasmic sequence, these findings indicate that phosphorylation is mediated by a kinase which is associated with the receptor and/or functions over a very short distance.

Figure 2. Ligation of mIgM and mIgD leads to phosphorylation of Ig-α and Ig-β associated with the homologous but not the heterologous receptors. The experiment was conducted essentially as described in figure 1 using anti-IgM (Bet-2) and anti-IgD (JA12.5) as stimuli with the following modifications. Cells were stimulated for 1 or 2 minutes before lysis in 1% digitonin and precipitation with excess sheep anti-mouse immunoglobulin.

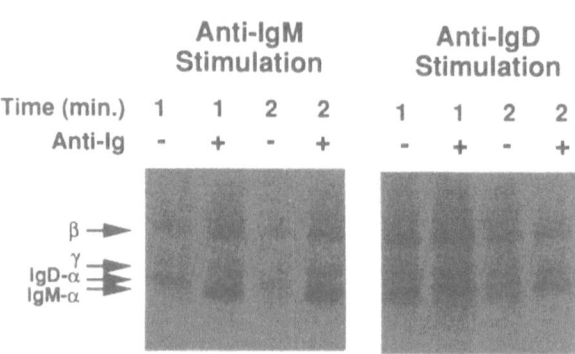

Antigen Receptors are Associated with a Kinase Which Phosphorylates the YEGLN Sequence Contained Within Ig-α and -β Chain Cytoplasmic Tails

We then undertook formal analysis of tyrosine kinase activity before and after antigen receptor ligation, utilizing the synthetic peptides shown in Fig. 3 as kinase substrates (Heasley and Johnson 1989; Cambier et al 1990). The substrates were constructed based on the amino acid sequences of the cytoplasmic tails of Ig-α (MB-1) (Sakaguchi et al 1988) and Ig-β (B29) (Hermanson et al 1988) and contain the YEGLN sequence which is uniquely found in these proteins. In view of this distribution, the tyrosine within YEGLN is a likely target for ligand induced α and β chain phosphorylation in vivo. Furthermore, mutational analysis of the ARH1 motif in TCRζ, CD3ε chains have shown that conservative replacement of this tyrosine with phenylalanine ablates receptor signaling function (Wegener et al 1992; Letourneur et al 1992). The nature of this effect is most consistent with a need to phosphorylate this residue to sustain signal transduction.

B29 peptide (residues 186-200)   RKAGMEEDHTYEGLNI

MB-1 peptide (residues 167-187)   RKFGVDMPDDYEDENLYEGLNL

Figure 3. Sequences of MB-1 and B29 used as peptide substrates in the tyrosine kinase. A basic arginine (bold "R") was added to the amino terminus of each sequence to enhance binding to P81 phosphocellulose paper in acidic conditions. Sequences are numbered according to the original description of the gene products (Hermanson et al 1988; Sakaguchi et al 1988).

We determined whether B cell membranes and receptor complexes contain a kinase which can phosphorylate this sequence and whether receptor ligation activates this kinase. B cell membranes were prepared by nitrogen cavitation and differential centrifugation and incubated with peptide in the presence of ^{32}P-ATP (see Fig. 4 legend). After incubation for 3-18 minutes, the reaction was stopped and labeled peptide was captured on P81 phosphocellulose paper, which was then washed and counted. As shown in Fig. 4 phosphorylation of the peptides was easily detected and was linear over a 20 minute period. Further, phosphorylation of both peptides was membrane dose dependent. These results indicate that the assay is appropriately sensitive, easily detecting two-fold differences in enzyme activity, to detect ligand mediated kinase activation.

Figure 4. Phosphorylation of the MB-1 and B29 peptides by B cell membranes. B cell membranes were prepared by nitrogen cavitation of purified murine splenic B cells (density > 1.066 from BDF1 mice). Nuclear/cytoskeletal debris was removed by slow speed centrifugation and membranes were subsequently pelleted by high speed centrifugation. The method of Heasley and Johnson (1989) was used to assay kinase activity (Cambier et al 1990). Membranes (1-6 μg membrane protein) were mixed with 2 mM peptide in 40 μl of assay buffer (50 mM HEPES, pH 7.2, 20 mM paranitrophenylphosphate, 8 mM MgCl₂, 2 MM MnCl₂, mM EGTA, 2 mM sodium orthovanadate, 0.1% Brij 96 detergent, 25 μM ATP-Mg salt, and 4 μCi [^{32}P]-ATP at 30 º for 3-18 minutes. Reactions were stopped by addition of 10 μl of 25% trichloroacetic acid (TCA) and entire volumes were spotted on P81 phosphocellulose filter paper. Filters were washed x 4 with 75mM of phosphoric acid, washed x 1 with acetone, and subjected to scintillation counting. Average cpm results from duplicate samples lacking peptide were subtracted from average results from duplicate samples containing peptide. Counts from duplicate samples were within 15% of each other. A. Kinetics of the phosphorylation of the B29 peptide. 4 μg of membrane protein was added to each sample. B. Comparison of membrane dose responses in phosphorylating the MB-1 and B29 peptides. Assays were performed for 6 minutes. When a peptide based on the YED containing sequence of MB-1 was used as substrates, background phosphorylation (0-5000 CPM) was seen at all timepoints.

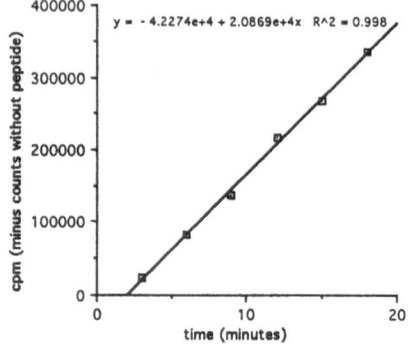

B29 Peptide Phosphorylation Time Course

$y = -4.2274e+4 + 2.0869e+4x \quad R\char`^2 = 0.998$

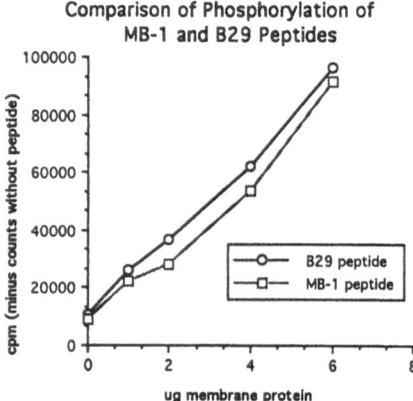

Comparison of Phosphorylation of MB-1 and B29 Peptides

The association of Ig-α/Ig-β kinase activity with the antigen receptor complex was then assessed by analysis activity in anti-Ig (SAMIG) immunoprecipitates from 1% digitonin lysates of membranes. As shown in Fig. 5. enzyme activity was readily detectable in immunoprecipitates.

Figure 5. Kinase activity is associated with mIg and is stimulated by crosslinking of the receptor. Kinase activity was assayed in immunoprecipitates and membranes as described in figure 4. Results are all the averages of cpm from duplicate samples from which the average of samples lacking peptide were subtracted from those as in Fig. 4. A. Phosphorylation of the B29 peptide by kinase(s) associated with mIg immunoprecipitates. The antigen receptor was immunoprecipitated with Sepharose conjugated sheep anti-mouse immunoglobulin (SAMIg) from 1% digitonin lysates of 0-50 million murine splenic B cells (density > 1.066 from BDF1 mice) (prepared as described in Ratcilffe and Julius 1982). Immunoprecipitates were mixed with assay buffer for 10 minutes at 30° C, stopped with TCA, spotted on phosphocellulose paper, washed, and counted as previously described. B. Crosslinking of mIg on B cell membranes stimulates phosphorylation of B29 peptide. Murine splenic B cell membranes (4 μg of membrane protein per sample) were mixed with phosphate buffered saline (□) or monoclonal anti-κ (187.1, 10 μg), (●), for 1-9 minutes prior to the addition of kinase assay buffer. The kinase reaction was allowed to proceed for 10 minutes at 30° C prior to stopping with TCA and samples were then processed as previously described.

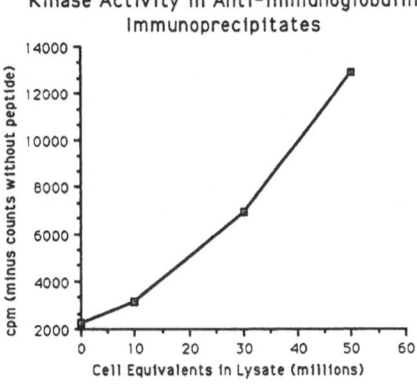

Kinase Activity in Anti-Immunoglobulin Immunoprecipitates

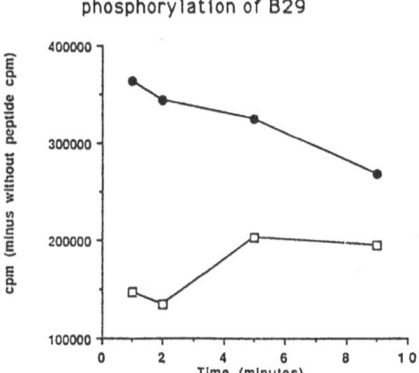

Anti-kappa stimulates tyrosine phosphorylation of B29

Antigen Receptor Ligation Stimulates Ig-α/Ig-β Kinase Activity

Finally, the effect of anti-Ig stimulation on the activity of the Ig-α/Ig-β kinase in membranes was assessed. As shown in Fig. 5b, stimulation of membranes with a monoclonal anti-κ antibody (187.1) for as little as one minute led to a four increase in detectable activity. This increased activity declined gradually over the subsequent 10 minutes.

Conclusions

The studies described above demonstrate that the antigen receptor is associated with a tyrosine kinase activity which phosphorylates the cytoplasmic tails of Ig-α and β. This activity is increased by >4 fold within 60 seconds of receptor ligation. The role of this activity in signal transduction and receptor desensitization as well as the identity of the responsible kinase is currently under study.

REFERENCES

Burkhardt AL, Brunswick M, Bolen JB, Mond JJ (1991) Proc Natl Acad Sci USA 88:7410-7414
Cambier JC, Fisher Cl, Pickles H, Morrison D (1990) J Immunol 145:13-19
Cambier JC, Campbell KS (in press) FASEB Journal
Clark MR, Campbell KS, Kazlauskas A, Hertz M, Potter, T Pleiman C, Cambier JC (in press) Science
Heasley LE, Johnson GL (1989) J Biol Chem 264: 8646-8652
Hermanson GG, Eisenberg, D, Kincade, PW, Wall R (1988) Proc Natl Acad Sci USA 85: 6890-6894
Hutchcroft JE, Harrison ML, Geahlen RL (1992) J Biol Chem 267: 8613-8619
Letourneur F, Klausner RD (1992) Science 255:79-82
Ratcliff MJH, Julius MH (1982) Eur J Immunol 12:634-641
Sakaguchi N, Kashiwamura S, Kimoto M, Thalmann P, Melchers F (1988) EMBO J. 7:3457-3464
Wegener A-M, Letourneur F, Hoeveler A, Brocker T, Luton R, Malissen B (1992) Cell 68:83-95
Yamanashi Y, Fukui Y, Wongassant W, Kinoshita Y, Ichimori Y, Toyoshima K, Yamamoto T (1992) Proc Natl Acad Sci USA 89:1118-1122

Molecular Dissection of T-Cell Antigen Receptor Signal Transduction

M. Iwashima[1], A. C. Chan[2], B. Irving[3], D. Straus[1], and A. Weiss[1, 2, 3]

Howard Hughes Medical Institute[1], Department of Medicine[2], and Department of Physiology[3], University of California, San Francisco, U426, 3rd and Parnassus Ave., San Francisco, Ca, USA

INTRODUCTION

Antigen recognition by mature and immature T-cells induces a cascade of events which ultimately leads to a variety of cellular responses (Weiss 1991). One of the major features of this signal transduction pathway is the tyrosine phosphorylation of several cytoplasmic and membrane proteins. Both kinetic studies and analyses using tyrosine kinase inhibitors showed that one of the earliest events in T-cell activation involves tyrosine kinases (June et al. 1990 a, b). PLC-γ1 is one of the substrates of this tyrosine kinase activity induced by T cell antigen receptor (TCR) stimulation (Weiss et al 1991b) This phosphorylation activates the enzyme and induces the catalysis of phosphatidylinositol 4,5-bisphosphate (PIP2) into diacylglycerol and inositol 1,4,5-trisphosphate (IP3), which activate PKC and mobilize intracellular calcium, respectively (Rhee et al 1989; Nishibe S et al 1990).

This activation process of T-cells can be mediated by a single subunit of the multi-component TCR, namely the ζ chain. Chimeric proteins were constructed by fusing the membrane and extracellular portion of CD8 or CD4 and the cytoplasmic tail of ζ. When crosslinked, these chimeras induce various events of T-cell activation in a manner indistinguishable from activation through the whole TCR complex (Irving et al, 1991; Romeo and Seed, 1991). It was also shown that the CD8/ζ chimera is associated with a protein tyrosine kinase (PTK) and this association increases after stimulation (Chan A, et al. 1991).

Here, we present three novel aspects of TCR signaling: First, we have defined the regions of ζ chain cytoplasmic tail necessary for TCR activation. Second, we have identified a new tyrosine kinase, ZAP-70, which forms a complex with the ζ chain after TCR stimulation. Third, we present genetic evidence that the lck protein tyrosine kinase plays a critical role in TCR signal transduction.

RESULTS

The motif structure of the ζ chain cytoplasmic tail

The cytoplasmic tail of the ζ chain can transduce signals indistinguishable from the entire TCR complex. (Irving 1991, Romeo 1991). This region contains three repeats of a motif that is observed among other signaling receptors expressed in hematopoietic cells. The amino acid sequence of this motif, originally noted by Reth (Reth, 1989), is shown in Fig.1. The motif consists of a charged amino acid and a repeat of the conserved sequence of *YXXL* with a spacer of 7-8 amino acids.

h ζ 1	N	Q L	Y	N E	L	N L G R R E E	–	Y	D V	L	D
h ζ 2	E	G L	Y	N E	L	Q K D K M A E A	Y	S E	I	G
h ζ 3	D	G L	Y	Q G	L	S T A T K D T	–	Y	D A	L	H
hCD3 γ	D	Q L	Y	Q P	L	K D R E D D Q	–	Y	S H	L	Q
hCD3 δ	D	Q V	Y	Q P	L	R D R D D A Q	–	Y	S H	L	G
hCD3 ε	N	P D	Y	E P	I	R K G Q R D L	–	Y	S G	L	N
rFc ε RI–β	D	R L	Y	E E	L	– H V Y S P I	–	Y	S A	L	E
rFc ε RI–γ	D	A V	Y	T G	L	N T R N Q E T	–	Y	E T	L	G
h B–1	E	N L	Y	E G	L	N L D D C S M	–	Y	E D	I	S
mB29	D	H T	Y	E G	L	N I D Q T A T	–	Y	E D	I	V
concensus	*D.*		*Y.*		*L.*			*Y.*		*L*
	E				*I*					*I*
	N									

Fig. 1. Comparison of amino acid sequences of proteins containing the YXXL based motif. The conserved amino acids are shown by the boxes. Modified from Reth, 1989. Abbreviations: h, human; r, rat; m, mouse.

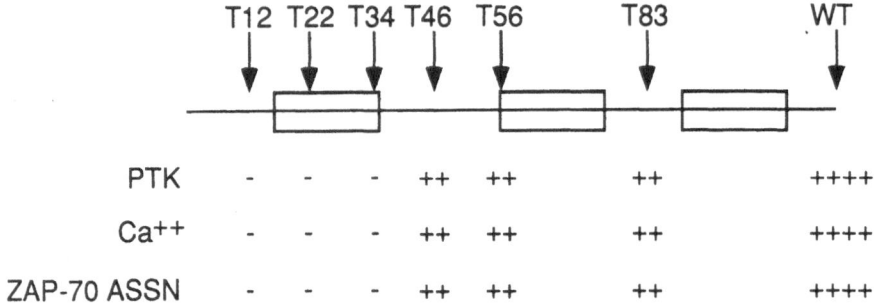

Fig. 2 A schematic representation of truncations of the CD8/ζ chimera and their signaling functions. The arrows indicate the sites of truncations. Each box represents a single copy of the ζ motif. The functional effects of each truncated chimera are summarized. These are: induction of PTK activity measured by anti-phosphotyrosine Western blot and ζ immunoprecipitates *in vitro* kinase assay (PTK), free cytosolic Ca^{++} increase (Ca^{++}), and the association of ZAP-70 and ζ determined by anti-phosphotyrosine blot of ζ immunoprecipitates (ZAP-70 ASSN) . The results are presented by comparing the response observed with the chimera with the full length cytoplasmic tail (WT).

We addressed two questions regarding this motif. First, does the motif structure correspond to a functional signaling unit? Second, if so, does the triplication of the motif signal better than a single or double copy of the motif? To answer these questions, a series of truncation mutants of the ζ cytoplasmic portion was made from the CD8/ζ chimera. As summarized in Fig.2, the deletion of two motifs does not abolish the signaling ability of the chimera (T46, T56). However, the deletion of the leucine residue at the C-terminus of the first motif (T34) results in a non-functional chimera and defines the C-terminal requirement of the motif. This supports the idea that the one motif can function as a signaling unit. Recently, the CD3 ε cytoplasmic tail was also demonstrated to be capable of transducing signals (Letouneur 1992). Since there is little similarity in the primary structures of CD3 ε and ζ chain except the conserved motif, it is likely that the motif itself is the key element for this signaling function.

It should be noted that the construct that contains only one or two motifs showed reduced signaling capability compared to the wild type. Conversely, we have made constructs containing three motifs or three copies of the first motif (data not shown) and have found that these have signaling capability more comparable to the wild type. This demonstrates that the repeat of this motif enhances the signal transduced. Such enhancement may be due to recruitment of more secondary signaling molecules (quantitative effect). Alternatively, the interaction between secondary molecules and the motif might strengthen cooperatively by dimer or a trimer formation or recruitment of different signaling components (qualitative effect).

ZAP-70, a PTK that associates with the ζ chain

We have previously identified a 70 kD tyrosine phosphoprotein ZAP-70 (ζ chain associated protein of 70 kD) which associates with ζ following stimulation of the TCR (Chan 1991). This association was also observed with crosslinking of the CD8/ζ chimera. Thus the association can be mediated by the ζ cytoplasmic domain. If the association of this protein is an important event in the cascade of TCR signaling, this association should correlate with the signaling capability of the truncated CD8/ζ chimeras. As shown in Fig. 2, this is indeed the case. Deletion of any portion of the motif resulted in the loss of the ZAP-70:ζ association. Conversely, the inclusion of only one motif preserved this association.

To elucidate the primary structure of ZAP-70, we have partially purified the ZAP-70 protein from immunoprecipitates of the CD8/ζ chimera from activated cells. The material was further purified by preparative SDS polyacrylamide gel electrophoresis and electroeluted. The purified protein was digested with CNBr and trypsin. Peptides were purified by HPLC and their amino acid sequences were determined by microsequencing. Using these amino acid sequences, we synthesized several degenerative oligonucleotides which were used as PCR primers. One of the combinations of these oligonucleotides yielded an amplified product of approximately 220 bp which was used to screen a cDNA library derived from the T-cell leukemia line PEER. Out of 4.5×10^5 clones, we were able to isolate 22 positive clones and determined the entire sequence of two clones. One of them (10.1) contained the full length cDNA. The deduced amino acid sequence from this cDNA clone predicts a 69.9 kD protein. The C-terminus contains a highly conserved tyrosine kinase domain. In addition, there are two putative SH2 domains located at the N-terminus. ZAP-70 showed the highest overall amino acid homology to the porcine spleen derived tyrosine kinase syk (Taniguchi, T et al, 1991). As shown in Fig. 3, the homologous regions are most concentrated around the SH2 and the kinase domains.

Since it was reported that syk is expressed in the thymus, it appeared that there was a possibility that ZAP-70 is human syk. To test this, we have determined the partial nucleotide sequence of human syk from PCR products amplified by oligonucleotide primers containing porcine syk sequences. These sequence comparisons showed that ZAP-70 is not the human homologue of syk. Additionally, Northern blot analysis of RNA from various sources revealed that ZAP-70 mRNA is exclusively expressed in T-cells and NK cells while syk is predominantly found in B-cells.

To confirm that ZAP-70 is a tyrosine kinase, the full length cDNA was expressed transiently in COS cells. Analysis using *in vitro* kinase assays with ZAP-70 immunoprecipitates showed a phosphorylated 70 kD protein specific to ZAP-70 transfected cells. Phospho-amino acid analysis of this protein revealed that the *in vitro* phosphorylation occurs exclusively on tyrosine residues.

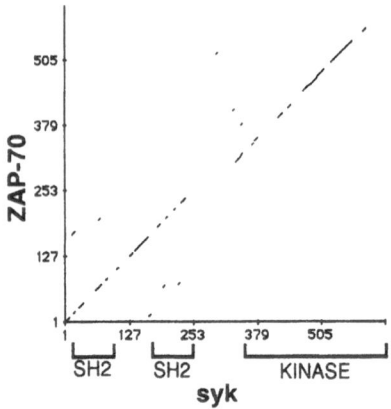

Fig. 3 Dot plot analysis of the homology between human ZAP-70 and porcine syk. Each dot represents a homology of 3 amino acids out of 5 between the two polypeptides. Sequences are from (Taniguchi, et al 1991) and this study.

A deficiency of lck prevents the activation of Jurkat T-cells

If the association between ZAP-70 and ζ chain plays a pivotal role in the TCR signaling system, any mutation that blocks such association would lead to the loss of signaling. We have previously reported the isolation of three somatic cell mutants derived from the Jurkat T cell leukemic line which fail to increase cytoplasmic free Ca^{++} (reviewed in Weiss, 1991). These three belong to different complementation groups. All of the defects are due to one of the early components of the signaling pathway since these cell lines show complete or partial impairment in the induction of tyrosine phosphorylation. The defect in one of these mutants, J.CaM1, has been recently identified. Stimulation of the

TCR in J.CaM1 fails to induce tyrosine phosphorylation of PLC. This defect does not reside in the TCR complex, since the CD8/ζ chimera does not rescue the phenotype. This indicates that the mutation of this cell line is somewhere distal to the ζ chain and blocks the pathway leading to the activation of PLC γ1. Since no tyrosine phosphorylation is induced in J.CaM1, two src family kinases that have been implicated in TCR signal transduction, fyn and lck, were examined. Analysis using *in vitro* kinase assays revealed that J.CaM 1 lacks lck kinase activity. Further analysis showed that the molecular basis for the functional deficiency of lck is caused by an aberrant splicing of lck transcripts. The lck mRNA from J.CaM1 is missing exon 7 which encodes part of the kinase domain including the putative nucleotide binding site (GxGxxG). The reason for such aberrant splicing is currently under study. To confirm that the defect of J.CaM1 is solely due to the lack of lck expression, we have introduced an exogenous murine lck cDNA. The transfectants were selected for their drug resistance and expression of the lck protein was determined by Western blotting. Cells with cDNA derived lck protein reconstitute TCR-mediated signaling events indistinguishable from the wild-type Jurkat cells.

In J.CaM1, the kinase activity associated with the ζ chain is diminished. This activity was restored in lck transfectants, and yet we were not able to identify an association between lck and the ζ chain. One possible explanation is that there is an undetectable level of lck associated with ζ chain and that ζ chain is a good substrate for lck. Anther possibility is that lck controls the activity and/or the association of another kinase with the ζ chain. One of the best candidates for such kinase is ZAP-70. Indeed, we found that ZAP-70:ζ association is impaired in J.CaM1 and this association is restored by expression of functional lck.

Induction of the association between ZAP-70 and the ζ chain.

If the cDNA clone we obtained encodes the ZAP-70 polypeptide, one would expect that the association of 70kD protein with the CD8/ζ chimera would occur when the chimera is crosslinked in the cells expressing this cDNA clone. To test this, we have transiently expressed the full length ZAP-70 cDNA (10.1) in a COS cell line (COS18) that was stably transfected with the CD8/ζ chimera. Surprisingly, we were not able to detect any ZAP-70 association with the chimera. This led us to hypothesize additional factors may be required for the association of ZAP-70 and ζ. Since both lck and fyn have been implicated in TCR signaling, COS18 cells were co-transfected with 10.1 and lck or fyn. 72 hours after the transfection, each transfectant was lysed and the CD8/ζ complex was immunoprecipitated The immune complexes were then analyzed in an *in vitro* kinase assay (Fig.4). Prominent kinase activity was precipitated when 10.1 and lck or fyn were co-transfected. However, none of these kinases alone were able to induce kinase activity above basal level. The phosphorylation of the bands corresponding to ZAP-70 and CD8/ζ chimera were predominantly on tyrosine residues since the treatment of the gel with KOH did not remove these phosphorylated bands. It should also be noted that although ζ and ZAP-70 were heavily phosphorylated in all the co-transfectants, there was no detectable change in the phosphorylation of proteins in the 50-60kD. This implies that either lck or fyn are not substrates for this ζ associated kinase activity, or very little of lck or fyn protein is contained in this complex.

In addition, Western blot analysis showed that, when ZAP-70 and lck or fyn were co-transfected, ZAP-70 and ζ form a complex and are strongly phosphorylated on tyrosine residues *in vivo* . The single transfection with lck or fyn increased the tyrosine phosphorylation of ζ slightly. On the other hand, ZAP-70 alone did not induce any tyrosine phosphorylation of ζ. Interestingly, the co-transfection of ZAP-70 and lck or fyn induced tyrosine phosphorylation of a large number of other cellular proteins in addition to ζ and ZAP-70 (data not shown).

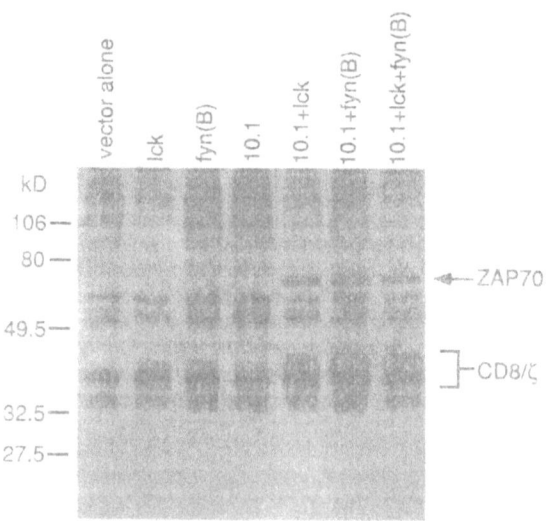

Fig. 4 *In vitro* kinase assay of the CD8/ζ chimera associated kinases from COS 18 transient transfectants. The cDNA used for each transfection is shown above the lane. fyn(B) is the brain form of fyn. We have confirmed that, in this assay system, the thymic form of fyn functions identically to the brain form (data not shown).

DISCUSSION

In order to understand how the TCR transduces the signal of antigenic recognition, we characterized several molecules which are involved in early events of this signaling process. Among the invariant chains involved in the TCR complex, the ζ chain has been of a particular interest. The cytoplasmic portion of this molecule is capable of transducing signals which are indistinguishable from those of the entire TCR complex. We have further defined a 17 amino acids motif in ζ that is sufficient to carry out the same function as the intact receptor. Our results are in a good agreement with those of others (Romeo et al, 1992). How, then, does such a motif structure transduce the signal? The molecules that interact with this structure are the essential key to understanding the process.

The study of the CD8/ζ truncation mutants showed that the association of ZAP-70 and ζ correlates well with the signaling capability of the chimera. Since PTKs are involved in the initial events of T-cell activation, the identification of kinases which interact with the ζ chain has been one of the important goals for the study of this signaling pathway. The isolation and expression of the ZAP-70 cDNA clearly showed this molecule is a novel tyrosine kinase that can interact with the ζ chain. Interestingly, in COS cells, ZAP-70 associates with the ζ chain only when co-expressed with lck or fyn. This raises the question of what regulates the ZAP-70 interaction with lck/fyn and the ζ chain. Preliminary results demonstrate that tyrosine phosphorylation of ZAP-70 is induced by lck and fyn in the absence of the ζ chain in COS cells. However, this phosphorylated ZAP-70 does not form a complex with exogenous ζ chains when lysates from different cells are mixed. This suggests that the interaction between lck/fyn and the ζ chain is essential for the ζ:ZAP-70 association. The results obtained with J.CaM1 also showed that lck is involved in the induction of ζ:ZAP-70 association. Additionally, following TCR stimulation, ZAP-70 was detected in immunoprecipitates with only the tyrosine phosphorylated form of the ζ chain.

The study of J.CaM 1 demonstrates that lck is indispensable for the PTK activity associated with the ζ chain. However, attempts to show the association between lck and ζ chain have not been successful. This implies lck may not require a stable association with the ζ chain for its function. Instead, this kinase may function by forming transient or weak associations with the ζ chain. The induction of ζ:ZAP-70 association by lck may also reflect an additional signaling role by CD4 and CD8 TCR co-receptors. The complex formation by CD4/8-MHC-TCR would bring the cytoplasmic portion of ζ and lck into close proximity. This could raise the specific kinase activity by increasing the local concentration of the substrate and the kinase.

The association of fyn and the ζ chain has been previously detected in T cells (Samelson et al. 1990) However, when fyn was transfected into COS18 cells the kinase activity associated with the ζ chain was very low. From such transfectants, no increase of the ζ-associated PTK activity or *in vivo* tyrosine phosphorylation of the ζ chain was observed after crosslinking CD8/ζ. The functional importance of the association of fyn with the ζ chain remains to be determined.

Co-expression of fyn and ZAP-70 induced an equivalent degree of *in vivo* phosphorylation and association of ZAP-70 with ζ when compared to the lck-ZAP-70 co-transfectants. Therefore, as far as the induction of the ζ:ZAP-70 association is concerned, we were unable to distinguish the functional difference between fyn and lck. This may be due to the artificial nature of our experimental conditions. Indeed, unlike in T-cells, ζ:ZAP-70 association does not require the crosslinking of CD8/ζ chimera in COS18 cells. This suggests that the ζ:ZAP-70 association may be negatively regulated in T cells. By imposing different negative regulatory constraints, T cells may be able to control the activity and specificity of these kinases. This could also explain why fyn is not capable of compensating for the lck defect in J.CaM1, which expresses normal levels of fyn.

The isolation and characterization of these molecules and mutant cells has given us clues to elucidate the complex mechanism of TCR signal transduction. It appears that this system involves very complex machinery which is finely regulated to perform the intricate function of TCR. It is now essential to identify other components of this complex system and their molecular interactions in order to understand TCR signal transduction.

A Diversity of Protein Tyrosine Phosphatases Expressed by T Lymphocytes

M. L. Thomas, D. B. Bowne, E. Cahir McFarland, E. Flores,
R. J. Matthews, J. T. Pingel, G. Roy, A. Shaw and H. Shenoi

Howard Hughes Medical Institute and the Departement of Pathology, Washington University Medical School, 660 South Euclid, St. Louis, Missouri 63110, USA

INTRODUCTION

Protein tyrosine phosphorylation is used by cells as a regulatory mechanism, controlling events such as activation and proliferation. Both protein tyrosine kinases (PTKases) and protein tyrosine phosphatases (PTPases) are required to balance the state of tyrosine phosphorylation and, therefore, the regulation of PTKase and PTPase enzymatic activity is necessary to control signal transduction. The PTKases are an extensive set of enzymes and are well characterized due to their oncogenic potential (Hanks et al.1988). The molecular identification of PTPases has emerged more recently.

The leukocyte-common antigen, CD45, is a transmembrane glycoprotein uniquely expressed by all nucleated cells of hematopoietic origin (Thomas, 1989). The cloning and sequencing of CD45 supported the biochemical data of a transmembrane protein with a heavily glycosylated exterior domain, a single membrane spanning region and a large cytoplasmic domain of approximately 80 kDa (Thomas et al.1985). The cytoplasmic domain could be divided into two subdomains of 300 amino acids that share 33% sequence similarity. The isolation and sequence analysis of a 35 kD intracellular protein tyrosine phosphatase from human placenta indicated, surprisingly, approximately 35% sequence similarity to each of the CD45 cytoplasmic subdomains, suggesting that CD45 was a transmembrane protein tyrosine phosphatase (Charbonneau et al.1988). This observation was confirmed by the demonstration that CD45 will dephosphorylate tyrosine substrates (Tonks et al.1988). The PTPase family, similar to the PTKase family, exists as both transmembrane and intracellular forms. The existence of transmembrane PTPases implies that they will interact with external stimuli which will effect the PTPase activity and, ultimately, cell behavior. To further understand how PTPases function in controlling T cell differentiation and activation, we pursued the identification and characterization of PTPases expressed by T cells.

MOUSE PTPASE SEQUENCES

Identification of sequences with high similarity to the PTPase subdomains of CD45 was performed by using degenerate oligonucleotides to sequences that were conserved between the two CD45 subdomains. The oligonucleotides were used to amplify cDNAs with similar sequences by polymerase chain reaction (PCR) and the resulting fragments were subcloned and sequenced. The region of analysis covered approximately half of a PTPase domain, approximately 130 amino acids (Fig. 1). Although this analysis does not cover the entire PTPase domain, it provides a window in which the diversity and conservation can be analyzed. Analysis of protochordate PTPases by this method identified twenty-seven fragments with significant sequence similarity to CD45 (Matthews et al.1991). Certain "hallmark" residues were found to be conserved in all PTPase domains, while other regions are significantly divergent. Importantly, certain sequences were highly similar to each other indicating that the PTPase family could be further divided into subfamilies. The breadth of sequences with similarity to CD45 indicates that the PTPase family is likely to be extensive.

A search for mouse PTPases using brain, spleen or T cell RNA revealed sequences identical to CD45, as well as 7 additional sequences (Fig. 1). From brain, three additional sequences were found, DEL, MBR and SHP; from spleen, one additional sequence, PEP; and from the T cell clone L3, three additional sequences, EPI, BET and MTC. Several of this sequences were highly similar to previously identified human and mouse sequences (Krueger et al.1990; Shen et al.1991; Gebbink et al.1991). DEL is 99% identical to human HPTPδ, SHP is highly similar to PTP1C and there are 5 differences between EPI and human HPTPε.Therefore, DEL, SHP and EPI are is likely to be the species homologues. There are 27 differences between MBR and mouse RPTPμ. The per-cent similarity is higher than with other PTPases and, therefore, MBR and RPTPμ are likely to be members of a subfamily. BET is more distantly related to human HPTPβ, 52% identity. It is not possible, therefore, to ascertain whether BET is a species homologue or a subfamily member. PEP and MTC are novel. Based on sequence similarities, DEL, EPI, BET and MBR are likely to be transmembrane proteins.

PTPASES EXPRESSED BY MOUSE T LYMPHOCYTES

The identification of multiple PTPases from a given tissue raises the question of how many PTPases will be expressed by a single cell type. CD45 is expressed by all T cells, as is the transmembrane PTPase, LRP. EPI, BET and MTC were all isolated from RNA derived from the T cell clone L3. However, surprisingly, none could be convincingly detected by northern blot analysis. This indicates that while PCR is an efficient means of isolating cDNAs encoding PTPases, caution must be used interpreting whether the cDNA is primarily expressed by the tissue or cell. It is possible that the mRNAs are regulated with respect to activation or cell cycle and thus more detailed experiments are needed to determine the expression.

Examination of the other PTPases fragments derived by PCR is more revealing (Table 1). Both SHP and PEP are expressed abundantly in thymus, spleen, lymph node and bone marrow. They are also expressed by T cell clones. This pattern is not surprising for PEP since it was derived from spleen RNA. However, SHP was derived from brain RNA, a tissue that has undetectable levels of SHP mRNA by northern blot analysis. In contrast, DEL is expressed abundantly in kidney and MBR is expressed abundantly in brain and weakly in kidney, liver, and lung. Neither are expressed abundantly in hematopoietic derived tissues. Therefore, CD45, LRP, SHP and PEP are expressed by T cells.

A.

DFWRMIWE   HCSAGVGR

B.
```
CD45 QKATVIVMVTRCEEGNRNKCAEYWPSMEEGTRAFKDIVVYINDHKRCPDYIIQKL  NVAHKKEKAT
DEL  QRSATVVMMSKLEERSRVKCDQYWP  SRGTETHGLVQVTLLDTVELATYCVRTF   AFYKNGSSE
MBR  EQSACIVMVTNLVEVGRVKCYKYWP  DDTEVYGDFKVTCVEMEPLAEYVVRTF    TLERRGYNE
EPI  QRSATIVMLTNLKERKEEKCYQYWP  DQGCWTYGNIRVCVEDCVVLVDYTIRKFCIHPQLPDSCKA
BET  KNVYAIVMLTKCVEQGRTKCEEYWP  SKQAQDYGDITVAMTSEVVLPEWTIRDF   VVKNMQNSE
SHP  ENTRVIVMTTREVEVEKGRNKCVPYWP EVGTQRVYGLYSVTNSREHDTAEYKLRTL QISPLDNGDL
PEP  YRILVIVMACMEFEMGKKKCERYWAEPGETQLQFGPFSISCEAEKKKSDYKIRTL   KAKFNNE
MTC  QNSTVIAMMTQEVEGEKIKCQRYWP  SILGHNHHGQREAALALLRMQQLKGFIDAGDGPGGYSDRE


CD45 GREVTHIQFTSWPDHGGVPEDPHLLLKL RRRVNAFSNFFSGPIVV
DEL  KREVRQFQFTAWPDHGVPEHPTPFLAFL RRVKTCNPPDAGPMVV
MBR  IREVKQFHFTGWPDHGVPYHATGLLSFI RRRVKLSNPPSAGPIVV
EPI  PRLVSQLHFTSWPDFGVPFTPIGMLKFL KKVKTLNPSHAGPIVV
BET  SHPLRQFHFTSWPDHGVPDTTDLLINFRYLVRDYMKQIPPESPIVV
SHP  VREIWHYQYLSWPDHGVPSEPGGVLSFLDQINQRQESLHHAGPIIV
PEP  TRIIYQFHYKNWPDHDVPSSIDPILQLI WDMRCYQEDDCVPICI
MTC  VRHISHLNFTAWPDHDTPSQPDDLLTFI  SYMRHIRRSGPVIT
```

Fig. 1. (A) Schematic diagram indicating the region of a PTPase
domain that was amplified by PCR. The sites of the oligonucleotides
used are indicated by arrowheads. (B) Corresponding amino acid
sequences of the derived cDNAs.

This is not an exhaustive list and EPI, BET, MBR and MTC may be
expressed at low levels or under appropriate circumstances. The
expression of multiple PTPases by T lymphocytes, indicates that a
variety of functions will be regulated by these enzymes. Furthermore,
it is certain that more phosphatases will be identified by either low
stringency hybridization or further PCR analysis.

A DIVERSITY OF T CELL PTPASE STRUCTURES

With the exception of CD45, the function of the PTPases expressed by T
lymphocytes is unclear. However, the structures have provided some
surprising insights. CD45 contains regions rich in O-linked
glycosylation and LRP is predicted to be rich in O-linked
glycosylation as well. The carbohydrate regions of both proteins are
implicated in their function. The length of the O-linked region of
CD45 is controlled by the splicing of exons encoding this region. The
smallest isoform contains a 13 amino acid amino-terminal O-linked
domain while the largest isoform contains an approximately 200 amino
acid domain. This divergence in size as well as the observation that
the size is precisely controlled in differentiation and activation
suggests that the O-linked region of CD45 is critical to function.
CD45 has been shown to be required for signal transduction elicited by

Table 1 _____ Expression of Mouse PTPases _____

PTPase	Tissue Expression	mRNA Size	T cells Expression
CD45	Leukocytes	4.8-5.2 kb	yes
LRP	Widely	3.2 kb	yes
EPI	Bone Marrow	2.5 kb	PCR[a]
DEL	Kidney	8.0 kb	no
BET	Liver, Kidney, Leukocytes	7.0 kb	PCR
PEP	Leukocytes	2.7 kb	yes
SHP	Leukocytes	2.1 kb	yes
MTC	Widely	4.6 kb	PCR
MBR	Brain, Liver, Kidney	4.7 kb	no

[a]PCR indicates that the fragment encoding the PTPase domain was derived from a T cell clone but the mRNA is undetectable by northern blot analysis.

antigen but how the carbohydrate residues of CD45 serve to modulate CD45 function is unclear.

The LRP exterior domain is also predicted to be heavily glycosylated by N- and O-linked structures and development of a monoclonal antibody to LRP, WULRP-1, has confirmed this prediction. Indeed, the extent of glycosylation may preclude protein interaction and the exterior domain may exclusively interact via the carbohydrate structures. The carbohydrate interactions of CD45 and LRP may present a novel mechanism by which enzymatic activity is regulated.

The two intracellular PTPases, SHP and PEP, both have unique structures (Matthews et al.1992). SHP contains two amino-terminal SH2 domains. Since SH2 domains have been shown to interact with phosphorylated protein residues, it is possible that SHP may interact with phosphorylated proteins to modulate signal transduction.

PEP contains a large carboxy-terminal extension of approximately 500 amino acids. The carboxy-terminal 250 amino acids is rich in proline, glutamic acid, serine and threonine residues (PEST). This type of sequence is contained within proteins that are rapidly degraded (Rogers et al.1986). Furthermore, within the carboxy-terminal domain of PEP are putative nuclear localizing sequences. PEP, therefore, is a candidate nuclear PTPase that is rapidly degraded.

CONCLUSION

There are currently five different subfamilies for both the transmembrane and intracellular PTPases. T cells express a subset of the known PTPases however the complete repertoire expressed has not been determined. The few complete structures that are known indicates that the complexity of interactions is greater than was previously appreciated from studies focusing on the PTKases. To fully understand T cell activation and differentiation it will be important to determine which enzymes are activated or regulated by a stimulus and determine which substrates are effected.

PTKase and PTPases both exist as transmembrane and intracellular forms. Many of the PTKases are transmembrane growth factor receptors, binding soluble ligands and ultimately leading to the initiation of cell cycle. However, in T cells, the PTKases implicated in antigen-specific T cell activation are both intracellular. Lck and Fyn, two members of the Src family, are associated with two important T cell transmembrane glycoproteins involved in the transduction of antigen induced signals, CD4 and the T-cell antigen receptor respectively (Rudd et al.1988; Samelson et al.1990). CD45 has been implicated in the activation of Src-family members and thus may explain the inability of CD45 deficient T cells to respond to antigen (Mustelin et al.1989; Ostergaard et al.1989; Ostergaard and Trowbridge, 1990). The interaction of PTPases and PTKases provides an important mechanism for regulating cell growth and differentiation. Understanding how the interactions of the PTPases are regulated will provide new insights in cell biology.

ACKNOWLEDGMENT

We thank our colleagues for many helpful discussions and criticisms. Anti-peptide antisera to Src-family members was kindly provided by Dr. Joe. Bolen. Our work is supported by the Howard Hughes Medical Institute, U.S. Public Health Service Grant AI 26363, grants from the Council for Tobacco Research and Grant-in-Aid from the American Heart Association. RJM is the recipient of a Cancer Research Postdoctoral Fellowship, EF was supported by a Patricia Roberts Harris Fellowship and the Division of Biological and Biomedical Sciences, Washington University, ECM, JM, HS, and EW was supported by a NIH training grant 5T32 AI07163and MLT was the recipient of an Established Investigator Award from the American Heart Association.

REFERENCES

Charbonneau H, Tonks NK, Walsh KA Fischer EH (1988) Proc Natl Acad Sci USA 85:7182-7186
Gebbink MBG, van Etten I, Hateboer G, et al (1991) FEBS Letters 290:121-130
Hanks SK, Quinn AM Hunter T (1988) Science 241:42-52
Krueger NX, Streuli M Saito H (1990) EMBO J 9:3241-3252
Matthews RJ, Flores E Thomas ML (1991) Immunogenetics 33:33-41
Matthews RJ, Bowne DB, Flores E Thomas ML (1992) Mol Cell Biol 12:2396-2405

The diversity of structures of T cell PTPases suggests a variety of regulation not previously found with the PTKases. Furthermore, a comparison of the PTPase domains indicates that they are only approximately 35% similar. This would suggest that the substrates with which they interact are also distinct. The biochemical pathways which they serve to regulate have not been determined but the structures indicate that they participate in different aspects of cell function.

THE REGULATION OF SRC-FAMILY MEMBER KINASE ACTIVITY BY CD45.

To examine CD45 activity, we have developed T cell clones that are deficient in the expression of CD45 (Pingel and Thomas, 1989; Weaver et al.1991). These cells lose the ability to respond to antigen even though the expression of the T cell antigen receptor/CD3 complex, as well as other cell surface molecules are not affected. This implies that the CD45 transmembrane PTPase is required to transduce the signal elicited by antigen and indicates that rather than function to negatively regulate cell growth, CD45 is required for cell growth.

Immunoblot analysis of the $CD8^+$ CD45-deficient T cell clone L3M-93 with anti-phosphotyrosine increased immunoreactivity with proteins of approximately 55-60 kDa. Previously, Src-family member, $p56^{lck}$, has been shown to be substrates for CD45 (Ostergaard et al.1989; Ostergaard and Trowbridge, 1990; Mustelin et al.1990). Immunoprecipitation and determination of kinase activity from L3M-93 and the parent T cell clone L3 for $p56^{lck}$, $p59^{fyn}$, $p62^{yes}$ and the ζ chain of the T-cell receptor indicated decreased kinase activity in the CD45 deficient cell line (Fig. 2). This suggests that CD45 regulates Src-family member kinase activity and provides an explanation for the inability of the CD45 deficient cells to respond to T-cell receptor stimuli. The inability of CD45 deficient cells to activate Src-family members indicates that Src-family member kinase activity is required for T-cell receptor signaling.

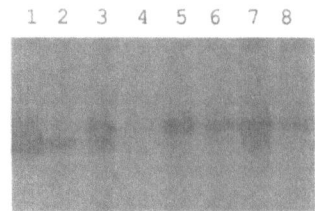

Fig. 2. Kinase activity of Src-family members from L3 and L3M-93. Cells were immunoprecipitated with antisera to either $p56^{lck}$ (lane 1 & 2), $p59^{fyn}$ (lanes 3 & 4), $p62^{yes}$ (lanes 5 & 6) or T-cell receptor ζ chain (lanes 7 & 8) and kinase activity determined. Immunoprecipitates were analyzed by SDS-PAGE and autoradiography. L3, lanes 1, 3, 5 & 7; CD45-deficient L3M-93, lanes 2, 4, 6 & 8.

Mustelin T, Coggeshall KM Altman A (1989) Proc Natl Acad Sci USA 86:6302-6306

Mustelin T, Coggeshall KM, Isakov N Altman A (1990) Science 257:1584-1587

Ostergaard HL, Schackelford DA, Hurley TR, et al (1989) Proc Natl Acad Sci USA 86:8959-8963

Ostergaard HL Trowbridge IS (1990) J Exp Med 172:347-350

Pingel JT Thomas ML (1989) Cell 58:1055-1065

Rogers S, Wells R Rechsteiner M. (1986) Science 234:364-368

Rudd CE, Trevillyan JM, Dasgupta JD, Wong LL Schlossman SF (1988) Proc Natl Acad Sci USA 85:5190-5194

Samelson LE, Phillips AF, Luong ET Klausner RD (1990) Proc Natl Acad Sci USA 87:4358-4362

Shen S-H, Bastien L, Posner BI Chrétien P (1991) Nature 352:736-739

Thomas ML, Barclay AN, Gagnon J Williams AF (1985) Cell 41:83-93

Thomas ML (1989) Ann Rev Immunol 7:339-369

Tonks NK, Charbonneau H, Diltz CD, Fischer EH Walsh KA (1988) Biochemistry 27:8695-8701

Weaver CT, Pingel JT, Nelson JO Thomas ML (1991) Mol Cell Biol 11:4415-4422

Antigen Receptors Clustering; Mobility, Size and Configurational Requirements for Effective Cellular Triggering

Idan Tamir and Israel Pecht

Department of Chemical Immunology, The Weizmann Institute of Science, Rehovot 76100, Israel

Introduction

The hallmark of immunological cell stimulation is the clustering of its respective antigen receptors [1,2]. These range from the B-cell surface anchored immunoglobulins and the different types of Fc receptors on these and other cells to the elaborate antigen recognition complex of T cells [3,4,5]. The structural homology among the different receptors for antigen became apparent over recent years and it led to naming them "Multichain Immune Recognition Receptors - MIRR's" [5]. These membranal complexes of polypeptides combine antigen recognition elements with those responsible for transmembrane signalling. Fc receptors bind the latter domains of distinct immunoglobulin isotypes, binding itself has not been found to cause the cells any stimulus. Hence Fc receptors serve to provide the cells with antigen-specific adaptor molecules. This is markedly different from the behavior of receptors for growth-factors (e.g. for EGF) which apparently undergo an allosteric transition upon binding their monovalent ligand. This causes receptor clustering which initiates the signal transduction by activation of the EGF-receptors' tyrosine-kinase domain [6]. In contrast, for signalling via the diverse forms of antigen receptors the clustering by multivalent interactions between the latter and the antigen is required [7,8,9].

In order to study the structural and temporal requirements of MIRR clustering for efficient cellular stimulation, the type 1 receptor for Fcε domains (FcεRI) on mast cells was chosen as model system. These cells promptly respond to their FcεR clustering with the secretion of mediators, which can be easily followed and quantitated [7]. Mast cells carry several hunderd thousands copies of the FcεRI on their plasma membrane. This receptor binds IgE with considerable affinity (~nM) and consists of 3 membranal polypeptides subunits: α,β and a covalent dimer of γ chains. The extracellular domains of the α subunit binds the IgE and is associated non-covalently with the other two subunits [3,10]. Clustering of FcεRI can be attained in several different ways; Physiologically, the IgE bound to the receptor is crosslinked by antigen. Alternatively, this can be attained by IgE specific antibodies, by lectins or by covalent chemical crosslinkers. The FcεRI itself may also be clustered directly by specific antibodies. All these distinct modes of FcεRI clustering lead to the cell's secretory response. Thus, the information for cellular triggering is contained within the FcεRI [11].

A mast cell line widely employed in studies of the FcεRI and its activation mechanism is RBL-2H3 [12]. It is a rat mucosal-type mast cell, which upon FcεRI stimulation exhibits distinct morphological changes, such as ruffling of the plasma membrane and marked flattening [13]. The main early biochemical events known to follow FcεRI clustering include activation of protein tyrosine kinases [14] of the src family [15] and the rapid tyrosine phosporylation of several proteins, prominently the β and γ subunits of the FcεRI [16] and phospolipase C γ1,

[17,18] the latter probably activating the enzyme, while the role of β and γ phosphorylation is still unknown.

Our aim is to gain insight into the nature of the clustering process that constitutes the very first step of the activation signal caused upon reaction between the antigen and its respective receptors. We wish to define this process in terms of the requirements for a minimal number of receptor per cluster, the distance among receptors, and the lifetime of clusters. In addition, we study the restriction of mobility and the topology of receptors within the clusters, all in relation to the eventual cellular response they produce.

In order to answer the question what the minimal size of initial receptor cluster required for signal transduction is, we have employed monoclonal antibodies specific for the α subunit of the FcεRI [19,20]. These mAbs turned out to be rather useful reagents, superior to other experimental protocols developed for that purpose (e.g. divalent haptens). FcεRI dimers emerged as the cluster size necessary and sufficient for effective cellular triggering. As expected, we have observed a cellular dose-response to the extent of FcεRI dimerization. However, marked differences were resolved in the capacity of dimers produced by the different monoclonal antibodies to induce the cellular response. In other words, the secretory response did not correlate simply with the mAbs' ability to dimerize the receptor. These data suggested that either the lifetimes of the produced dimer or the FcεRI orientation within the dimer (i.e. the topology of the subunits within this cluster) plays a crucial role in its ability to induce the secretory response [20]. We have examined in detail the kinetics of interaction between the FcεRI and several specific ligands, including the above mAbs and IgE [21]. The results showed marked differences in the expected life-times of dimer produced by the different mAbs. However, it supported the notion that the main cause for the disparate secretory response to FcεRI dimers produced by these different mAbs is their distinct topology. Namely, upon clustering a highly asymetric MIRR like the FcεRI, the relative position of each subunit of the receptor within a cluster has functional implications [19,22].

In order to further investigate the relation between the receptor mobility, size of its clusters and the cellular response it causes, we designed an experimental system where epitope density and interaction with the antigen receptor can be varied and controlled. We used a soft glass surface (e.g. microscope cover slips -CS) as a solid flat (transparent) matrix onto which specific antigenic epitopes were chemically linked. Several murine monoclonal IgEs which have different affinities towards the 2,4-dinitrophenyl (DNP) haptens are available. Hence, we chose to derivatize the above surfaces with nitroaromatic groups. The procedure for modifying glass surfaces with haptens involves first silylation with 3-aminopropyltriethoxysilane, producing pendant aminopropyl groups. This is followed either by reaction with the appropriate hapten or by introducing the appropriate spacers prior to reaction with the latter. These and analogous procedures enabled us to produce glass surfaces where the hapten-surface distance as well as its average density were controlled and correlated with the cellular response which it causes. Moreover, we have also devised surfaces to which IgE molecule could be covalently attached (by photolysis) thus limiting the lateral mobility of the IgE and therefore also that of the FcεRI. This enabled us to examine the role of receptor motion and reorganization for inducing cellular signals.

2. Results and Discussion

Conventional microscope coverslips made of regular soft glass were derivatized with 3-amino n-propyl 1-triethoxysilane as illustrated in Fig. 1. The free amine was either reacted directly with a nitroaromatic hapten, (1-fluoro,2,4-dinitrobenzene FDNB) in the presence of a non-nucleo-

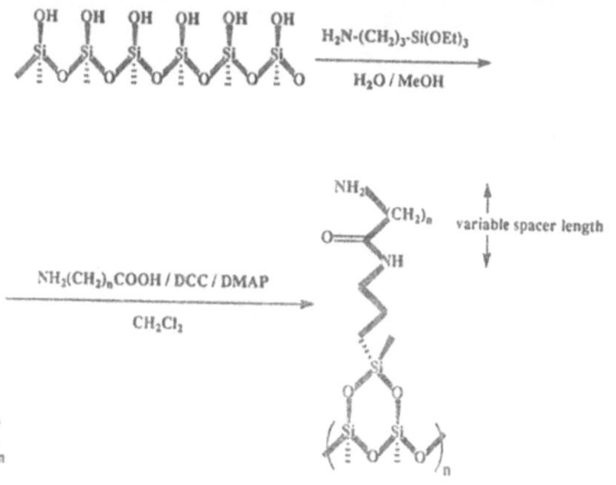

Fig. 1. Soft glass cover slips (commercially available light microscopy grade, Ø 12 mm) are employed. Following thorough cleaning, the CS are reacted with a 1% solution of 3-amino-n-propyl-1-triethoxysilane in dichloromethane yielding the amino-propyl silylated surface (APS-CS). These CS can then either be reacted directly with the appropriate antigenic epitope or with a spacer carrying it.

philic base or with an appropriately reactive spacer carrying the hapten (e.g. 6-(4-azido-2-nitrophenyl amino) hexanoic acid N-hydroxysuccinimide ester = SANPAH yielding the ANPAH epitope) (Figs. 1 and 2). Employing different concentrations of the hapten yielded a continuous range of its densities on the glass surface, while introducing the spacers of different length varied the hapten's distance from the glass surface and therefore modulated its interaction with the different monoclonal IgE molecules [23,24]. Rat mucosal mast cells of the RBL-2H3 line were first saturated with a given monoclonal IgE. Then, the cells were reacted with the appropriate CS carrying different DNP densities. This was done under conditions that were non-permissive for secretion (e.g. on ice) for 30 min. Then, the temperature was raised to 37 °C and cells were incubated for further 30 min. The resultant cellular response was determined by monitoring serotonin secretion [8]. As can be seen from Fig. 3, the cells' secretory response is dependent on the epitope density, and goes through a maximum at a certain range of DNP density. In order to identify the parameters which determine the observed cellular response, we analyzed the binding of two IgEs of distinct affinities (A2, lower [23] and SPE7 higher [24]) to the above surfaces and the secretory response of the cells reacting with the glass-carried IgE molecules. It was found that the saturation binding densities of both IgEs to these DNP carrying surfaces display a similar value, which correlates with the surface-DNP concentration causing maximal secretory response (Fig. 4). This maximum in response can be rationalized by a decrease in IgE binding at high epitope densities, caused by steric hindrance limiting the access of the IgE to adjacent DNP groups on the glass surface when their density exceeds a given threshold.
Examination of the cells' secretory response to the hapten derivatized glass surfaces carrying lower and higher affinity IgE resolved an inverse relationship between interactions with the

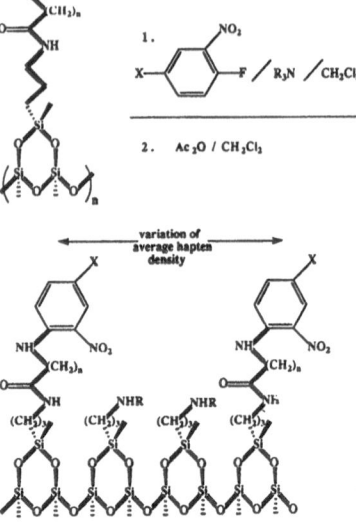

Fig. 2. Attachment of haptens to the derivatized CS. The free amino group is reacted with the hapten directly or with the spacer carrying it dissolved at the desired concentration in dichloromethane. Excess free amino groups are reacted with acetic anhydride to avoid the excess of positive charges on the CS. $R = C(=O)(CH_2)_n NHC(=o)CH_3$; $X = N_3$ or NO_2.

matrix and response. Namely, derivatized glass surfaces carrying the IgE with higher hapten binding affinity yielded a lower secretory response to stimulation by a given density of IgE than those carrying the lower affinity IgE. A similar pattern was observed when the role of spacer-length that separates the hapten from the glass matrix was examined; a longer spacer (6 carbons) enhanced the IgE binding yet lowered the cells' secretory response.

Taken together, these results led us to hypothesize that they reflect the extent of the IgE molecules mobility and hence also that of the FcεRI on the epitope carrying matrix. These findings raised the question as to what degree is receptor immobilization related to the cell stimulation it causes? In order to address this question, hapten analogues of DNP, which upon photolysis yield reactive nitrene groups forming covalent linkage of the hapten to the antigen combining site of the bound IgE, were synthesized. 4-Azido 2-nitrophenyl (ANP)haptens were thus attached via two spacers of different lengths to the glass. Photolysis with light of 360 nm converted the azide group into the nitrene which reacted with the IgE molecules arresting them to the glass (Fig. 5). ANP and ANPAH derivatized CS were incubated with each of the two DNP specific IgE and were either kept in the dark or photolysed. Then, RBL-2H3 cells were reacted with the resultant CS under non-permissive conditions for 30 min following which secretion permissive conditions were provided for another 30 minutes and secretion was assayed. One set of these experiments is illustrated in Fig. 6 and the results of the analysis of several data sets are summarized in Table 1. ANPAH, having the six carbons longer spacer is the better bound

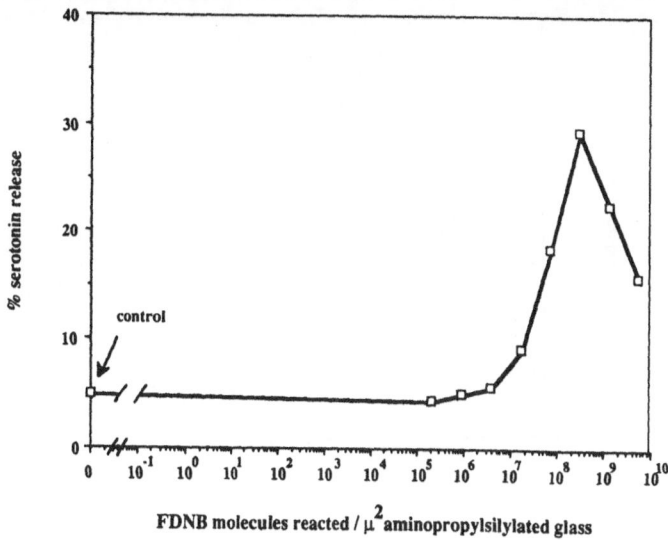

FDNB molecules reacted / μ^2 aminopropylsilylated glass

Fig. 3 - Response of A2-IgE treated RBL-2H3 cells to hapten-derivatized CS: RBL-2H3 cells were loaded overnight with ^{3}H-serotonin, harvested, washed X2 with Tyrode's buffer and incubated with 10 nM of A2-IgE in Tyrode's buffer for 1 hr. The cells are then centrifuged twice in Tyrode's and allowed to bind at 4˚C on DNP derivatized glass cover slips at a density of 1x10^6 cells/cm for 30 min. The coverslips are then washed twice with Tyrode's buffer and warmed to 37˚C in 250 microliter Tyrode's/coverslip for 30 min. Then supernatant samples are withdrawn and counted on a scintillation counter. Total serotonin was assayed by lysing the cells in 0.1% Triton-X100.

hapten. However, the secretory response it causes is the lower one with both monoclonal IgE. ANP the weaker binding hapten is causing the better response in the non-photolysed state, i.e. with non-covalently bound IgE molecules. In addition, the lower affinity IgE is also the one causing higher secretion. Photolysis, leading to covalent anchoring of the IgE to the CS glass, caused a marked decrease in the secretory response of the cells in response to all combinations of IgE and haptens. Finally, in control experiments cells bound to DNP carrying CS were also photolyzed and behaved identically to the respective non-photolyzed cells. These results clearly illustrate the requirement for mobility of the IgE-FcεRI complex in the cell's plasma membrane for attaining the secretory response. The lateral and probably rotational diffusion of this membrane complex determine its capacity to concentrate at the cell glass-surface contact zones upon binding the antigenic epitopes. This, in turn, enables the receptor microclustering required for initiation of cell stimulation. Suppression of receptor mobility by any of the above means leads to a decrease in the required receptor clustering and to the consequent reduction in cell response.

The detailed examination of the relation between epitope density and cell response is currently being pursued. Apparently, even high epitope density (i.e. comparable to that of the FcεRI on

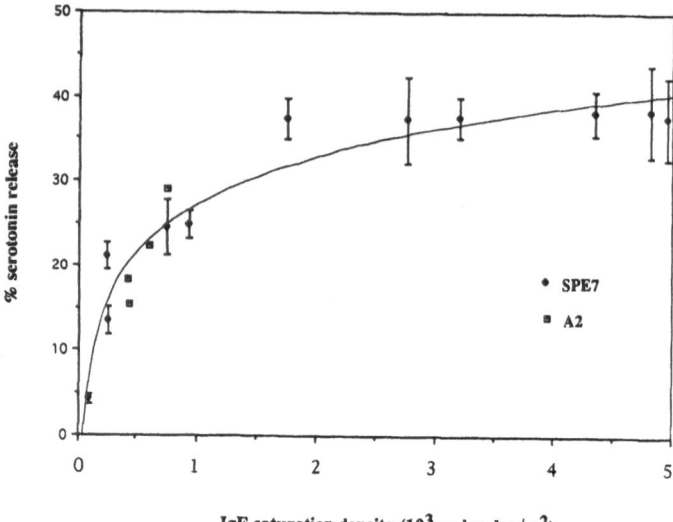

Fig. 4 - Correlation between IgE saturation density and the resultant cellular stimulation: IgE binding to CS derivatized to different hapten densities was performed by 3 hr. incubation of ^{125}I-labeled IgE with the derivatized CS in Tyrode's at 4°C. The saturation density was calculated from the fitted scatchard plot obtained. Details of the protocol for assaying the cellular response are as described in the legend to Fig. 3. ◊ DNP-specific IgE mAb SPE7; A2.

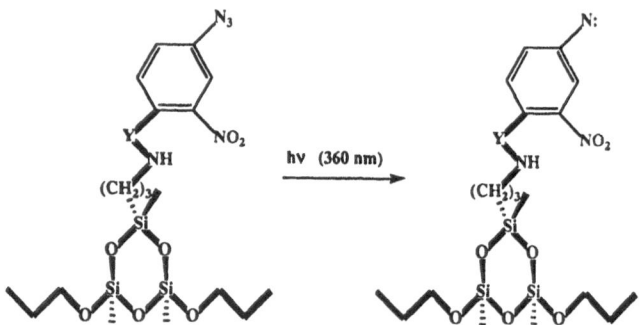

Fig. 5 - Photoactivation of haptens carried on glass surfaces: CS carrying covalently-bound 4-azido-2-nitrophenyl groups are photolysed by exposure to light of 360 nm filtered from a halogen lamp for 30 sec/CS (10 mW/cm^{-2}). The glass surfaces are incubated prior to photolysis with different concentrations of DNP-specific IgE for 3 hr at 4°C, washed twice with phosphate buffered saline and photolyzed while immersed in the latter buffer.

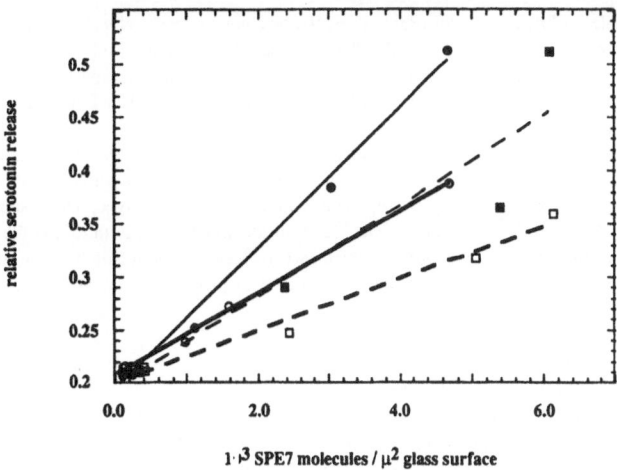

1·^{3} SPE7 molecules / µ^2 glass surface

Fig. 6 : Dose response of the RBL-2H3 cells to monoclonal DNP specific IgE (SPE7) bound to the derivatized glass surfaces: CS carrying different densities of haptens and saturated with IgE molecules were either kept in the dark or photolysed as described in the legend to Fig. 3. The cells were then incubated with these CS and the secretory response was determined as described in the legend to fig. 4 (ommiting the preincubation with IgE).
-●- ANP; -o- photolyzed ANP; -■- SANPAH; - □ - photolyzed SANPAH.

Table 1. Mast cell triggering efficiency: Modulation by IgE-matrix interactions

IgE	Surface hapten	Photolysis	Slope	Corr.
SPE7	ANP	-	0.132	0.996
SPE7	ANP	+	0.078	0.998
SPE7	SANPAH	-	0.086	0.956
SPE7	SANPAH	+	0.049	0.992
A2	ANP	-	0.193	0.980
A2	ANP	+	0.121	0.985
A2	SANPAH	-	0.187	0.971
A2	SANPAH	+	0.116	0.912

the resting mast cell) may not cause a considerable response unless the FcεRI-IgE complexes can diffuse and cluster. However, efficient stimulation would occur upon reacting cells with surfaces carrying an epitope density which is about one order of magnitude higher than the FcεRI cell-surface density, irrespective of the affinity of the IgE to the hapten carrying surface.

Thus, while prolonged immobilization of FcεRI receptors on mast cells at densities which are below or in the range of their normal cell-surface density would not result in effective cellular stimulation. At sufficiently high epitope densities, statistically some receptors are immobilized at such close proximity to be considered effectively clustered and hence initiate the cascade coupling it to the secretory response.

The topological requirements set on a stimulatory MIRR cluster are practically unknown. Even the proximity to which the cytosolic components have to be brought is still debated. Evidence is however accumulating for specificity elements present in those polypeptides responsible for signalling [26]. Thus, the common sequence motif identified by M. Reth [27] in the cytosolic domains of the latter subunits probably yield, when clustered, the required avidity to interact with further components downstream the cascade to provide the coupling signal for the respective effector function.

3. References

1.   DeLisi C (1980) Q Rev Biophys 13:201-220
2.   DeLisi C (1981) Nature 289:322-323
3.   Ravetch JV, Kinet JP (1991) Ann Rev Immunol 9:457-492
4.   Ashwell JD, Klausner RD (1990) Ann Rev Immunol 8:139-167
5.   Keegan AD, Paul WE (1992) Immunol Today 13:63-68
6.   Yarden Y, Ulrich A (1988) An Rev Biochem 57:443-478
7.   Metzger H (1978) Immunol Rev 41:186-201
8.   Balakrishnan K, Hsu FJ, Cooper AD, McConnel HM (1982) J Biol Chem 257:6427-6433
9.   Kane P, Holowka D, Baird B (1990) Imm. Invest. 19:1-25
10. Metzger H (1992) Immunol Rev 125:37-48
11. Ishizaka T, Chang TH, Taggart M, Ishizaka K (1977) J Immunol 119:1589-1596
12. Barsumian IL, Iserskiy C, Petrino MG, Siraganian R (1981) Eur J Immunol 11:317-323
13. Philips DM, Bashkin P, Pecht I (1985) J Ultrastructur Res 90:105-113
14. Benhamou M, Gutkind JS, Robbins KC, Siraganian RP (1990) Proc Natl Acad Sci USA 87:5327-5330
15. Eiseman E, Bolen JB (1992) Nature 355:78-80
16. Pauline R, Jouvin MH, Kinet JP (1991) Nature 353:855-858
17. Park DJ, Min HK, Rhee SG (1991) J Biol Chem 266:24237-24240
18. Schneider H, Dayag A, Pecht I (1992) Int Immunol 4:447-451
19. Ortega E, Schweitzer-Stenner R, Pecht I (1988) EMBO J 7:4101-4109
20. Ortega E, Hazan B, Zor U, Pecht I (1989) Eur J Immunol 19:2251-2256
21. Ortega E, Schweitzer-Stenner R, Pecht I (1991) Biochemistry 30:3473-3483
22. Pecht I, Ortega E, Jovin TM (1991) Biochemistry 30:3450-3458
23. Rudolph AK, Burrows PB, Wab MR (1981) Eur J Immunol 11:527-531
24. Eshhar Z, Ofarim M, Waks T (1980) J Immunol 124:775-779
25. Weis RM, Balakrishnan K, Smith BA, McConnel HM (1982) J Biol Chem 257:6440-6445
26. Pecht I, Schweitzer-Stenner R, Ortega E (1989) Prog Immunol 4:676-682
27. Reth M (1989) Nature 338:383-384

Acknowledgment

We gratefully acknowledge support of this research generously provided by grants from The Fritz Thyssen Foundation, F.R.G., and The Tobacco Research Council, USA.

5. Functional T Cell Subsets

Interrelationships of Peripheral T Cell Subsets

P. C. L. Beverley

Imperial Cancer Research Fund, Human Tumour Immunology Group, UCMSM, 91 Riding House Street, London, W1P 8BT England

INTRODUCTION

Although a considerable understanding of the function of peripheral T cells has been gained over the last decade, several important questions remain to be resolved. In this introduction to the conference session on "Functional T cell subsets" we will introduce these major issues and present some of our own data relating to human peripheral T cells.

Several advances in methodology have contributed to present understanding. In particular the development of monoclonal antibodies (mAbs) has allowed identification of numerous lymphocyte surface molecules, gene cloning has elucidated their structure and the development of transgenic and homologous recombinant mice has contributed greatly to understanding molecular function. Analysis of the molecular genetics of T cell receptors and of thymic mechanisms of repertoire selection have also provided an initial understanding of the potential repertoire of naive T cells in the periphery. Finally the identification and characterisation of many cytokines and of the molecular mechanisms involved in cytolysis have begun to show how T cells carry out their effector functions. In spite of these advances several important questions remain. Among those, which we consider most pressing, are the following.

It remains unclear whether all T cells are thymus processed and also whether the thymus is the only site where repertoire selection can occur. There is certainly evidence for the existence of thymus-independent $\gamma\delta$ T cells (Guy-Grand and Vassalli 1991) and in transgenic mouse models evidence that at least functional repertoire selection of $\alpha\beta$ T cells may occur peripherally. In these models, T cells capable of responding to antigens expressed only in the periphery are present but unresponsive (clonal anergy rather than clonal deletion) (Burkly et al 1989, Murphy et al 1989). However the importance of these extrathymic mechanisms under physiological circumstances remains less clear.

The relationship of T cell receptor V gene expression to disease has been much investigated recently, since expression of a dominant V gene may throw light on the nature of the disease process or disease related antigens, as well as providing a potential avenue for new therapeutic strategies. Less intensively investigated is the question of tissue specific expression of different $\alpha\beta$ T cell repertoires although for $\gamma\delta$ T cells it is clear that this is the rule (Janeway

1990). Even if αβ T cells on average are distributed homogenously throughout the body, this may not necessarily be the case among different T cell subsets.

Analysis of αβ T cells with multiple markers has revealed the existence, in addition to the well known CD4 and CD8 subsets, of extensive further heterogeneity. Recent functional studies have also shown that at least two major classes of CD4 effector cells exist in mouse and man (Mosmann and Coffman 1989, Parronchi et al 1991). These observations, together with evidence that naive and memory T cells may be identifiable, pose questions as to the relationship of phenotype and function among T cells, the ontogenetic relationships of different subsets and the stability of commitment to different functional programmes of peripheral T cells. In addition since T cell memory can be very long lived (Couch and Kasel 1983), there remain questions as to how it is maintained and what determines the pool sizes of distinct subsets of peripheral cells.

A different set of questions relate to the function of the lymphocyte molecules mainly identified with monoclonal antibodies. While many act as co-receptors in cell - cell interactions, or adhesion molecules, and the receptors for many cytokines have been defined, the function of others remains less clear. This is the case for CD45, since although the tyrosine phosphates activity of the cytoplasmic tail has been shown to be involved in signal transduction (Clark and Ledbetter 1989), the function of the large extracellular domain has not been determined nor has the functional significance of the complex alternative splicing exhibited by this part of the molecule.

This conference session will cover several of the areas outlined above, including the function of γδ T cells, cytolytic mechanisms, T effector cell function in man and the exploitation of transgenic and homologous recombinant mice in the analysis of T cell molecular function. This report will therefore concentrate on discussing the relationship of phenotype and function among T cells, the ontogeny of peripheral T cells and their lifespan.

THE ONTOGENY OF PERIPHERAL T CELLS

Cells which seed the periphery from the thymus are in the main single positive CD4 or CD8 cells. Minor subpopulations of double negative or double positive cells can be detected peripherally. Most of these are γδ T cells and will not be considered here. CD4 and 8 peripheral αβ T cells generally remain single positive, though induction of CD8 on mature CD4 cells by interleukin 4 (IL-4) has been reported (Paliard et al 1988). Thus for the present discussion the CD4 and CD8 subsets can be considered as stable sublineages of αβ T cells. However, from the time of early studies noting distinct functional properties of T_1 and T_2 cells differing in expression of Thy1 (Raff and Cantor 1971), numerous studies have explored the relationship of changes in phenotype to functional maturation of peripheral T cells. In recent years mAbs to high and low molecular weight isoforms of CD45 have proved particularly useful in several species (Arthur and Mason 1986, Lee et al 1990, Mackay et al 1990) but many other mAbs have been used, and have provided broadly similar information (Morimoto et al 1985, Budd et al 1987, Sanders et al 1988). In man mAbs reacting with isoforms of 220 and 205kDa (CD45RA), or 180 kDa (CD45RO), identify two largely non-overlapping subpopulations of T cells (Morimoto et al 1985, Smith et al 1986). Three strands of evidence strongly suggest that expression of CD45RA precedes CD45RO. Studies of cord blood T cells have shown that these mainly express CD45RA. The proportion of CD45RO cells then increases to reach adult levels around the age of 20 (Hayward et al 1989). In line with these observations, foetal blood samples have been shown to lack CD45RO T cells (Terry et al 1987). The second line of evidence is that when CD45RA T cells are stimulated in vitro, CD45RO is rapidly expressed and CD45RA lost (Akbar et al 1988). However, although this phenomenon is readily observed it is becoming clear that under certain circumstances CD45RO T cells may re-express high molecular weight isoforms in vitro (Rothstein et al 1991, Warren and Skipsey 1991). Since human cortical thymocytes also express CD45RO it seems likely that conversion to the CD45RA phenotype of the earliest peripheral T cells must occur, unless the hypothesis that only cells which do not express CD45RO survive thymic selection, is correct (Pilarski et al 1989). In any case animal data supports the idea that cells expressing low molecular weight isoforms in the periphery may subsequently express higher molecular weight forms (Bell and Sparshott 1990).

The remaining evidence for progression from the CD45RA to CD45RO phenotype is functional. Several studies have demonstrated that the ability of CD45RA cells to respond to recall antigens is poor or absent while CD45RO cells respond vigorously, both in bulk or limiting dilution cultures (Morimoto et al 1985, Smith et al 1986, Merkenschlager et al 1988, Merkenschlager and Beverley 1990). A primary response of CD45RA T cells to keyhole limpet haemocyanin has also been reported (Plebanski et al 1992). The magnitude of the response and the requirement for high density cultures, may imply a low frequency of responding CD45RA cells. In contrast preparations of malarial antigens or peptides are able to induce vigorous responses in conventional microtitre cultures from both adult and cord blood lymphocytes of non-immune donors (Jones et al 1990, Fern and Good 1992). While some of the responding cells appear to be conventional class II restricted CD4 T cells, a recent report indicates that much of the response may be due to $\gamma\delta$ cells and has features suggestive of the involvement of a superantigen (Goodier et al 1992). Thus at present the bulk of the data is in accord with the view that most CD45RA $\alpha\beta$ T cells are naive, while CD45RO cells are primed.

MEMORY AND T CELL LIFESPAN

Memory, the ability to respond more rapidly and effectively to a second exposure to antigen, is one of the most important features of the immune system but the nature of T cell memory has remained a matter of debate. Unlike B cells, T cells show no evidence of somatic mutation so that a more rapid and effective T cell response implies either increased numbers of responding cells, or a qualitative change in the ability of individual T cells to respond, or both. In addition since memory has been shown to persist in man for many years, a satisfactory account of memory must encompass mechanisms allowing for this persistence.

Our analysis of precursor frequencies for CD4 T cells able to proliferate in response to soluble recall antigens or cytotoxic precursor cells (CTLp) for EBV strongly implies that an increase in frequency is an important aspect of T cell memory (Merkenschlager et al 1988, Merkenschlager and Beverley 1990). Since the recall antigen responding cells are within the CD45RO subset, phenotypic analysis of this subset should provide information on qualitative changes in memory T cells. Several authors have documented many phenotypic differences between CD45RA and RO cells, particularly with respect to expression of adhesion and activation antigens (Sanders et al 1988, Wallace and Beverley 1990). Functional studies have documented differences in activation requirements of CD45RA and RO cells and there are major differences in the ability of the two cell types to produce cytokines (Lewis et al 1988, Salmon et al 1989). Thus there is ample evidence for qualitative differences between CD45RA and RO T cells. The detection of low levels of activation antigens on CD45RO T cells suggested that these cells might be in cycle and led us to investigate their lifespan. The results of these studies strongly suggest that the intermitotic time of CD45RO cells is short, while some CD45RA cells may persist for at least 10 years (Michie et al submitted). These data are in accord with evidence derived from in vivo labelling studies in sheep, showing that cells with memory phenotype label rapidly while those of naive phenotype do not (Mackay et al 1990). Studies of the persistence of naive cells in a transgenic model support the idea that naive cells may survive for long periods without division (Sprent et al 1991). Thus memory appears to be embodied in a population with a rapid turnover. This has led us to propose that memory consists of long lived clones rather than cells (Beverley 1990). The mechanism responsible for maintenance of memory clones in cell division remain to be adequately proven.

An additional complexity is added by recent data, which shows that cells expressing low molecular weight isoforms of rat CD45 may revert to expression of higher molecular weight forms (Bell and Sparshott 1990). In vitro human data support the idea that higher molecular weight isoforms may be re-expressed after initial down regulation (Rothstein et al 1991, Warren and Skipsey 1991). This raises the question of the role of cells which revert in vivo from memory to naive phenotype. These could represent long lived resting memory cells though we have no clear cut evidence for their existence. Alternatively during an immune response cells may be recruited non-specifically, perhaps by high local concentrations of cytokines, and at the termination of an active specific response could revert to the "naive" state while those which had responded specifically would persist as CD45R0 memory clones. Analysis of this will require serial studies of clonally marked cells during and after an immune response. This is now technically feasible using the polymerase chain reaction to identify individual T cell receptors.

COMMITMENT AND DEVELOPMENT OF EFFECTOR CELLS

Much more effort has so far been devoted to analysis of CD4 than CD8 effector heterogeneity. In mice and man two major effector types have been defined (Mosmann and Coffman 1989, Parronchi et al 1991). Th1 and Th2 CD4 cells have distinct patterns of cytokine production, IL-2 and IFN-γ being characteristic of Th1 and IL-4 and 5 of Th2. It is also clear that many cells produce both Th1 and Th2 cytokines, particularly early in immune responses (Firestein et al 1989, Street et al 1990) .

A number of factors have been shown to be important in regulating the type of effector cells which develop. Perhaps the most clear cut data relates to the effects of cytokines. Thus IL-4 present during the initial phase of culture enhances the development of Th2 effectors, while IL-2 or IFN-γ stimulate Th1 development (Gajewski et al 1989, Swain et al 1990). In rodents the nature of in vivo priming has also been shown to influence effector development. Immunisation with Brucella abortus leads to isolation of predominantly Th1 and Nippostrongylus brasiliensis to Th2 clones (Street et al 1990). A similar effect appears to operate in man, in that clones from Lyme arthritis or tuberculoid leprosy patients are Th1-like while those from allergic individuals resemble Th2 (Yssel et al 1991, Wierenga et al 1990). It is not clear how different antigens affect the development of distinct effectors but ligation of different surface molecules is known to regulate cytokine production. Micro-organisms are known to contain ligands for many leucocyte surface molecules (Sattentau et al 1986, Springer et al 1990).

A further question relates to the stage of differentiation at which commitment to develop into a specific effector type occurs. The existence of Th0 cells producing multiple cytokines has led to the idea that commitment occurs after antigen priming. However recent experiments using CD31 have suggested that the majority of CD4 CD45RA cells, which express this antigen, retain "suppressor" function even after activation and continue to produce only IL-2 (Torimoto et al 1992). In contrast others have shown a switch to helper function after activation of CD45RA cells (Clement et al 1988). A further difficulty in envisaging early commitment to a CD31+ lineage of suppressor cells is that CD31+ cells like CD45RA+ cells do not respond to specific recall antigens. Thus for the present we prefer to retain the view that CD45RA cells are precursors of CD45R0 primed or memory T cells. Table 1 summarises the phenotypic and functional properties of these two major populations of peripheral T cells. Much further heterogeneity in antigen expression remains to be investigated particularly relating to the migration pathways of T cells (Mackay et al 1990).

Table 1. Properties of CD45R-defined T cell subsets

Property		CD45RA cells	CD45RO cells
Lifespan		long	short
Alloresponse	-proliferation	+++	+++
	-cytotoxicity	+++	+++
Autoresponse		++	+
Recall response	-proliferation	-	+++
	-cytotoxicty	-	+++
Cytokine production		mainly IL-2	many
CD2 response		+/-	+++
CD3 response		++	+++
Adhesion molecules		+	+++
Activation antigens		-	+

CONCLUSIONS

Work over the last few years has indicated that after emigration from the thymus T cell differentiation is by no means complete. Furthermore it is clear that the intermitotic time of T cells in the periphery may be very long and that the peripheral T cell pool can be largely self sustaining (Bell and Sparshott 1990, Sprent et al 1991, Michie et al submitted). It is therefore of considerable importance, particularly in a long lived animal such as man, to understand the mechanisms which regulate peripheral T cell behaviour. In this review we have attempted to sketch the complexity of the peripheral T cell compartment and indicate some of the variables which affect T cell behaviour. The following papers consider several of these variables in detail.

REFERENCES

Akbar AN, Terry L, Timms A, Beverley PC & Janossy G (1988) J Immunol 140:2171-2178
Arthur RP, Mason D (1986) J Exp Med 163:774-786
Bell EB, Sparshott SM (1990) Nature 348:163-166
Beverley PCL (1990) Immunol Today 11:203-205
Budd RC, Cerottini JC, MacDonald HR (1987) J Immunol 138:1009-1013
Burkly LC, Lo D, Kanagawas O, Brinster RL, Flavell RA (1989) Nature 342:564-566
Clark EA, Ledbetter JA (1989) Immunol Today 10:225-8
Clement LT, Yamashita N, Martin AM (1988) J Immunol 141:1464-1470
Couch RB, Kasel JA (1983) Annu Rev Microbiol 37:529-549

Fern J, Good MF (1992) J Immunol 148:907-913

Goodier M, Fey P, Eichmann K & Langhorne J (1992) Int Immunol 4:33-41

Firestein GS, et al. (1989) J. Immunol. 143:518-525

Gajewski TF, Joyce J, Fitch FW (1989) J. Immunol. 143:15-22

Guy-Grand D, Vassalli P (1991) Immunol Res 10:296-301

Hayward AR, Lee J, Beverley PC (1989) Eur J Immunol 19:771-773

Janeway CJ (1990) Res Immunol 141:688-695

Jones KR, Hickling JK, Targett GA, Playfair JH (1990) Eur J Immunol 20:307-15

Lee WT, Yin XM, Vitetta ES (1990) J Immunol 144:3288-3295

Lewis DB, Prickett KS, Larsen A, Grabstein K, Weaver M, Wilson CB (1988) Proc Natl Acad Sci U S A 85:9743-9747

Mackay CR, Marston WL, Dudler L (1990) J Exp Med 171:801-817

Merkenschlager M, Beverley PCL (1990) Int Immunol 1:50-59

Merkenschlager M, Terry L, Edwards R, Beverley PC (1988) Eur J Immunol 18:1653-1661

Michie CA, McLean A, Alcock C, Beverley PCL (1992) Submitted for publication

Morimoto C, Letvin NL, Distaso JA, Aldrich WR, Schlossman SF (1985) J Immunol 134:1508-1515

Morimoto C, Letvin NL, Boyd AW, Hagan M, Brown HM, Kornacki MM, Schlossman SF (1985) J Immunol 134:3762-3769

Mosmann TR, Coffman RL (1989) Adv Immunol 46:111-147

Murphy KM, Weaver CT, Elish M, Allen PM, Loh DY (1989) Proc Natl Acad Sci USA 86:10034-10038

Paliard X, de waal Malefijt R, de Vries J, Spits H (1988) Nature 335:642-643

Parronchi P, Macchia D, Piccinni MP, Biswas P, Simonelli C, Maggi E, Ricci M, Ansari AA, Romagnani S (1991) Proc Natl Acad Sci U S A 88:4538-42

Pilarski LM, Gillitzer R, Zola H, Shortman K, Scollay R (1989) Eur J Immunol 19:589-97

Plebanski M, Saunders M, Burtles SS, Crowe S, Hooper DC (1992) Immunol 75:86-91

Raff MC, Cantor H (1971) in Progress in Immunology (B Amos, ed) pp83-91 Academic Press, New York.

Rothstein DM, Yamada A, Schlossman SF, Morimoto C (1991) J Immunol. 146:1175-1183

Salmon M, Kitas GD, Bacon PA (1989) J Immunol 143:907-912

Sanders ME, Makgoba, MW, Shaw S (1988) J Immunol 140:1401-1407

Sattentau QJ, Dalgleish AG, Weiss RA, Beverley PCL (1986) Science 234:1120-3

Smith SH, Brown MH, Rowe D, Callard RE, Beverley PCL (1986) Immunol 58:63-70

Sprent J, Schaefer M, Hurd M, Suth CD, Yacov R (1991) J Exp Med174:717-728

Springer TA (1990) Nature 346:425-434

Street NE, Schumacher JH, Fong TA, Bass H, Fiorentino DF, Leverah JA, Mosmann TR (1990) J Immunol 144:1629-39

Swain SL, Weinberg AD, English M, Huston G (1990) J. Immunol. 145:3796-3806

Terry L, Pickford A, Beverley PCL (1987) in Leucocyte Typing (eds. A.J.McMichael et al) pp225-227 Oxford University Press, Oxford

Torimoto Y, Rothstein DM, Dang NH, Schlossman SF, Morimoto C (1992) J Immunol 148:388-396

Wallace DL, Beverley PCL (1990) Immunol 69:460-467

Warren HS, Skipsey LJ (1991) Immunol 74:78-85

Wierenga EA, Snoek M, Bos JD, Jansen HM, Kapsenberg ML (1990) Eur J Immunol 20:1519-1526

Yssel H, Shanafelt M-C, Soderberg C, Schneider PV, Anzola J, Peltz G (1991) J Exp Med 174:593-601

Human Th1 and Th2 Cells:Regulation of Development and Role in Protection and Disease

S.Romagnani, G. F. Del Prete, E. Maggi, P. Parronchi, M. De Carli, R. Manetti, M-P. Piccinni, F.Almerigogna, M.G. Giudizi, R. Biagiotti and S. Sampognaro.

Division of Clinical Immunology and Allergy, University of Florence. Instituto di Clinica Medica 3, 50134 - Firenze, Italy

INTRODUCTION

In recent years it has become clear that the type of an antigen-specific immune response is determined by the selective or preferential activation of CD4+ T-cell subsets secreting different patterns of cytokines that lead to strikingly different T-cell functions. Two very distinct cytokine secretion patterns were originally defined among a panel of mouse CD4+ T-cell clones (Mosmann et al., 1986). Th1, but not Th2, cells produce interleukin (IL)-2, gamma-interferon (IFN-γ) and tumor necrosis factor (TNF)-β , whereas Th2, but not Th1, cells express IL-4, IL-5, IL-6 and IL-10 (Mosmann and Moore, 1991). Other cytokines are produced by both T-cell subpopulations. More recent data on the cytokine secretion patterns of T-cell clones and normal T cells have revealed the existence of other phenotypes. Among both mouse and human T-cell clones, a pattern of IL-2, IL-4, IL-5 and IFN-γ (Th0) has been described by several laboratories and in some cases these clones have been shown to produce all cytokines tested, including IL-3, IL-10 and granulocyte macrophage-colony stimulatory factor (GM-CSF) (Maggi et al., 1988; Paliard et al., 1988; Mosmann and Moore, 1991).

HUMAN Th1 AND Th2 CELLS

Evidence for the existence of human Th1 and Th2 cell subsets, similar to those already described in the mouse, has recently been provided by assessing the cytokine production profile of CD4+ T-cell clones specific for different antigens. The majority of allergen-specific T-cell clones derived from atopic individuals produced high amounts of IL-4 and IL-5, but no or limited IFN-γ, (production of IL-2 was variable), whereas T-cell clones specific for bacterial constituents, derived from the same donors, secreted high amounts of IL-2 and IFN-γ and only a small proportion of them were able to produce IL-4 and IL-5 (Wierenga et al., 1990; Parronchi et al., 1991). In addition, T-cell clones specific for purified protein derivative (PPD) of *Mycobacterium tuberculosis* or for *Toxocara canis* excretory-secretory (TES) antigens established from healthy individuals exhibited an opposite profile of cytokine production (Del Prete et al., 1991a). Most PPD-specific T-cell clones expressed mRNA for, and secreted, IL-2 and IFN-γ, but not IL-4 and IL-5 following stimulation with either specific antigen or phorbol myristate acetate (PMA) plus anti-CD3 monoclonal antibody. In contrast, under the same experimental conditions, most TES-specific T-cell clones expressed mRNA for, and secreted, IL-4 and IL-5, but not IL-2 and IFN-γ (Del Prete et al., 1991a). By examining the cytokine profile of T-cell clones specific for *Mycobacterium leprae* (Salgame et al., 1991; Haanen et al., 1991), *Borrelia burgdorferi* (Yssel et al., 1991), *Yersinia enterocolitica* (Lahesmaa et al., 1992) or nickel (Kapsenberg et al., 1992) a clear-cut Th1 phenotype has also been reported. These data provide firm evidence that T cells with stable Th1 or Th2 patterns exist not only in mice but also in humans. They also suggest that atopic allergens and helminth components usually expand Th2-like cells, whereas contact allergens and bacterial antigens preferentially expand Th1-like cells (Romagnani, 1991a).

The demonstration that human CD4+ T cells can be subdivided into different subsets based on their profile of cytokine production raises the question of whether such a heterogeneity exists among CD8+ T cells, as well. Two subsets of CD8+ clones specific for *Mycobacterium leprae* antigens have indeed been described. CD8+ cytotoxic clones produced IFN-γ and IL-10, but not IL-4, whereas CD8+ suppressor clones produced substantial amounts of IL-4 (Salgame et al., 1991). More recently, high proportions of CD8+ T- cell clones producing both IFN-γ and IL-5, but not IL-4, were grown from bronchial biopsy-specimens of patients with toluene-diisocyanate (TDI)-induced asthma following inhalation of TDI (Del Prete et al., submitted for publication). Finally, CD8+ clones with a clear-cut Th2 phenotype (producing IL-4 and IL-5, but not IFN-γ) were derived from the Kaposi's sarcoma skin lesions of patients with AIDS (Maggi et al., unpublished data). All these findings encourage the view that distinct CD8+ T-cell subsets can been discriminated by analysis of their lymphokine profiles.

Th1 AND Th2 CD4+ HUMAN T-CELL CLONES EXHIBIT DISTINCT FUNCTIONAL PROPERTIES

In addition to IL-2 and IFN-γ, human Th1, but not Th2, clones produce TNF-β, whereas both Th1 and Th2 clones can secrete variable amounts of IL-3, IL-6, TNF-α, GM-CSF and IL-10 (Romagnani, 1991b; Del Prete et al., submitted). Human Th1 and Th2 clones also show different responsiveness to some cytokines. IL-4 potentiated the antigen-induced proliferation and cytokine production of Th2, but not Th1, clones. In contrast, IFN-γ selectively inhibited the proliferative response and cytokine production by Th2 clones. Interestingly, human recombinant and viral IL-10 significantly inhibited the proliferation and the cytokine production of both Th1 and Th2 clones in response to either the specific antigen or phytohemagglutinin (PHA) (Del Prete et al., submitted for publication). Finally, human Th1 and Th2 clones also differ in their cytolytic potential and mode of help for B-cell immunoglobulin (Ig) synthesis. In fact, the majority of Th1, but only a minority of Th2, clones exhibited cytolytic activity in a 4h PHA-dependent assay (Del Prete et al., 1991b). All Th2 (noncytolytic) clones induced IgM, IgG, IgA, and IgE synthesis by autologous B cells in the presence of the specific antigen and the degree of response was proportional to the number of Th2 cells added to B cells. Under the same experimental conditions, Th1 (cytolytic) clones provided B-cell help for IgM, IgG, and IgA, but not IgE, synthesis with a peak response at a T-cell : B-cell ratio of 1:1. At higher T-cell : B-cell ratios, a decline in B-cell help was observed (Del Prete et al., 1991b). Interestingly, all these Th1 clones lysed EBV-transformed autologous B cells pulsed with the specific antigen and the decrease of Ig production correlated with the lytic activity of Th1 clones against autologous antigen-presenting B-cell targets (Del Prete et al., 1991b). This may represent an important mechanism for the down-regulation of antibody responses in vivo (Romagnani, 1991a).

THE PROFILE OF THE 'NATURAL' IMMUNE RESPONSE CAN DETERMINE THE Th1 OR Th2 PROFILE OF THE SUBSEQUENT SPECIFIC RESPONSE

The nature of factors that favour the development of human Th1 and Th2 clones is still unclear The presence of IL-4 in bulk cultures before cloning shifted the differentiation of PPD-specific T cells from the Th1 to the Th0, or even to the Th2, phenotype. In contrast, the addition of both IFN-γ and anti-IL-4 antibody induced allergen-, as well as TES-, specific T cells to differentiate into Th0, or even Th1, instead of Th2, clones (Maggi et al., 1992). IFN-α, transforming growth factor (TGF)-β, NK stimulatory factor (NKSF) (also named IL-12), polyinosinic acid-polycytidylic acid (poly-I-C), influenza virus and BCG also promoted the differentiation of allergen- or TES-specific T cells into Th0- or Th1-, instead of Th2-like, cells (Parronchi et al., 1992; Romagnani, 1992). In contrast, anti-IL-12 antibody favoured the differentiation of PPD-specific T cells into Th0- or Th2-, instead of Th1-, like cells (Romagnani, 1992). IFN-α and IL-12 are produced predominantly by macrophages, cells that have an important role in the presentation of antigen to Th cells. Therefore, it is reasonable to suggest that, given the capacity of viruses and intracellular bacteria to stimulate macrophage production of IFN-α and IL-12 (that induce IFN-γ production by both T cells and NK cells), Th cells may be simultaneously presented with processed antigen plus cytokines that induce them to differentiate towards the Th1 phenotype (Romagnani, 1992). Interestingly, allergen-specific T cell lines, grown in the presence of IFN-α, poly-I-C, influenza virus or BCG, contained significantly higher proportions of CD16+ cells than T-cell lines grown in the presence of allergen alone. More importantly, the removal of CD16+ cells reduced the capacity of poly-I-C to shift the differentiation of allergen-specific T cells from the Th2 to the Th0 or Th1 profile . Taken together, these data suggest that intracellular bacteria and viruses induce specific immune responses of Th1 type at least partly because they either directly stimulate NK cells to produce IFN-γ, or because they induce macrophage production of IFN-α and IL-12 which in turn, stimulate NK cell growth and IFN-γ production (Romagnani, 1992).

The production of high concentrations of IFN-γ by NK cells, although important, does not appear to be sufficient for the induction of Th1 responses. Indeed, the addition of anti-IFN-γ antibody to bulk cultures did not prevent or reverse the inhibitory effect of poly-I-C on the differentiation of allergen-specific T cells into Th2 clones. Likewise, blocking of IFN-α or IL-12 alone with specific antibodies was ineffective. In contrast, the poly-I-C-induced Th0 or Th1 differentiation of allergen-specific T cells could be driven to the Th2 profile by the simultaneous addition of IL-4 plus antibodies reactive with IFN-γ, IFN-α and IL-12 (Manetti et al., unpublished data). Thus, it is possible that poly-I-Cacts not only by inducing high IFN-γ production but also by interfering with IL-4 production by T cells. It can be concluded, therefore, that viruses and intracellular bacteria induce Th1 responses because the profile of the "natural" immmune response they evoke provides optimum conditions (high concentrations of IFN-γ and absence of IL-4) for the development of Th1 cells (Romagnani, 1992).

The reason why allergens and helminths preferentially promote the differentiation of Th cells into the Th2 phenotype is still more obscure. An absence, or low concentrations of, IFN-γ and the presence of IL-4 seem to be critical to the in vitro development of Th2 responses (Maggi et al., 1992; Parronchi et al., 1992). It is possible that allergens and helminths are poor stimulators of IFN-α and IL-2 production by macrophages and NK cells and/or that at least some atopic individuals exhibit defective macrophage or NK responses. The other component necessary for the development of Th2 cells - the presence of IL-4 - raises the important question of the cellular source of IL-4. Since T cells seem to be unable to differentiate into IL-4-producing cells in the absence of IL-4 (Paul, 1992), IL-4 production by a non-T cell should be involved. IL-4-producing non-T cells have indeed been demonstrated in both murine spleen and human bone marrow (Ben Sasson et al., 1990: Piccinni et al., 1991), but their role in the initiation and/or amplification of Th2 responses is still speculative.

The question now arises of how the cytokine profile of the "natural" immune response can influence the profile of the subsequent specific immune response. Recently, we have shown that the addition in bulk culture of IFN-α can modulate not only the cytokine profile, but also the epitope specificity, of allergen-specific T-cell clones (Parronchi et al., unpublished data). Although intriguing, this finding may reconcile different views on the factors that regulate Th1 and Th2 maturation. Local cytokines might act indeed by influencing antigen presentation and/or processing by the antigen presenting cell (APC). One possibility is that IFN-α favours the presentation of particular peptides by modulating the expression of MHC class II determinants on the APC. Another possibility is that IFN-α acts at level of antigen processing by selectively allowing the formation of a given peptide.

ROLE OF Th1 AND Th2 CELLS IN HUMAN DISEASES

Substantial evidence is available to suggest that the preferential Th1 or Th2 response not only influences the type of protection against offending agents, but is also responsible for different alterations in some immune-mediated disorders. Most CD4[+] T cells infiltrating the thyroid gland in patients with autoimmune thyroid diseases when PHA-stimulated in vitro, develop into T-cell clones producing IFN-γ, but not IL-4 (Del Prete et al., 1989). Likewise, a skew towards the Th1 phenotype was detected in the cerebrospinal fluid of patients suffering from progressive multiple sclerosis (Brod et al., 1991), suggesting that Th1 cells are involved in the pathogenesis of organ-specific autoimune disorders. Finally, nickel-specific T-cell clones derived from peripheral blood of nickel-allergic patients secreted substantial amounts of IFN-γ, IL-2, TNF-α and GM-CSF, but little or no IL-4 and IL-5 (Kapsenberg et al., 1992), indicating an important role for Th1 cells in the genesis of contact dermatitis.

A number of reports also suggest a role for Th2 responses in the genesis of allergic (IgE-mediated) disorders. Higher proportions of IL-4-producing, and lower proportions of IFN-γ-producing, T-cell clones were demonstrated in the blood of patients with severe atopic disorders than in controls (Romagnani, 1990). Moreover, most T cells infiltrating the conjunctiva of patients with vernal conjunctivitis (Maggi et al., 1991) or the skin of patients with atopic dermatitis (van der Heijden et al., 1991) were found to develop into T-cell clones producing large amounts of IL-4 but no, or limited amounts of, IFN-γ Finally, by using a different experimental approach, i.e. in situ hybridization, cells exhibiting specific signals for IL-5 and/or IL-4 mRNA, but not for IFN-γ, were found at the site of the 24-h late phase reactions in skin biopsies, of the epithelial basement membrane in endobronchial mucosal biopsies, as well as in the bronchoalveolar fluid from patients with extrinsic asthma (Kay et al., 1991; Hamid et al., 1991; Robinson et al., 1992).

More recently, to investigate whether Th2 cells present in the airway mucosa of patients with allergic respiratory disorders actually reflect a specific immune response to inhaled allergens, biopsy specimens were obtained from the bronchial or nasal mucosa of patients with grass pollen-induced bronchial asthma or rhinitis 48 hrs after positive bronchial or nasal provocation test with grass pollen extract. T-cell clones derived from these and control specimens were then assessed for surface phenotype, allergen-specificity, profile of cytokine secretion and ability to provide B cell help for IgE synthesis. Proportions ranging from 14 to 22% of CD4+ T-cell clones derived from stimulated mucosa were specific for grass allergens and most of them exhibited a clear-cut Th2 profile and induced IgE synthesis in autologous B cells. In contrast, none of T-cell clones derived from control tissues was specific for grass allergens and only a minority of them showed the Th2 phenotype (Del Prete et al., submitted). These findings provide convincing evidence that Th2 cells specific for the relevant allergen appear in the mucosa soon after allergen inhalation. These cells, because of their cytokine profile, might well play a triggering role in the initiation of the allergic cascade.

CONCLUDING REMARKS

In the last few years, a large body of evidence has been accumulated to suggest the existence of human Th1 and Th2 subpopulations, reminiscent of those described for murine T cells. Human Th1 produce IL-2, IFN-γ and TNF-β, but not IL-4 and IL-5, whereas human Th2 produce IL-4 and IL-5, but not IFN-γ, IL-2 and TNF-β. Both human Th1 and Th2 cells are able to produce TNF-α, IL-3, GM-CSF, IL-6 and IL-10. Human Th1 cells preferentially develop in response to viruses and intracellular bacteria, do not provide help for IgE synthesis and usually are cytolytic. Human Th2 cells mainly develop in response to allergens and helminth constituents, provide optimal help for IgM, IgG, IgA, and IgE syntehsis and lack cytolytic potential. The cytokine profile of the 'natural' immunity evoked by intracellular bacteria and viruses that activate both macrophages and NK cells (high IFN-γ and no IL-4 production) probably determines the phenotype of the susequent specific Th1 response. Absence or low concentration of IFN-γ and early production of IL-4 by T cells or other cell types, that occur in response to allergens and helminth components, probably favour the development of Th2 cells.

Human Th1 and Th2 cells play a different role not only in protection against offending agents, but also in immunopathology. Th1 cells are involved in organ-specific autoimmunity and contact dermatitis, whereas Th2 cells are responsible for initiation of the allergic cascade.

REFERENCES

Ben Sasson SZ, LeGros G, Conrad DH, Finkelman FD, Paul WE (1990) Proc. Natl. Acad. Sci. USA 87: 1421-1425

Brod SA, Benjamin D, Hafler DA (1991) J. Immunol. 147: 10-815

Del Prete GF, Tiri A, De Carli M, Mariotti D, Pinchera A, Chretien I, Romagnani S, Ricci M (1989) Autoimmunity 4: 267-272

Del Prete GF, De Carli M, Mastromauro C, Macchia D, Biagiotti R, Ricci M, Romagnani, S (1991a) J. Clin. Invest. 88: 346-350

Del Prete GF, De Carli M, Ricci M, Romagnani S (1991b) J. Exp. Med. 174: 809-813

Haanen JBAG, de Waal Malefyt R, Res PCM, Kraakman EM, Ottenhoff THM, de Vries RRP, Spits H (1991) J. Exp. Med. 174: 583-592

Hamid Q, Azzawi M, Ying S, Moqbel R, Wardlaw AJ, Corrigan CJ, Bradley B, Durham SR, Collins

JV, Jeffery PK, Quint DJ, Kay AB (1991) J. Clin. Invest. 87: 1541-1546

Kapsenberg ML, Wierenga EA, Stiekma FEM, Tiggelman AMBC, Bos JD (1992) J. Invest. Dermatol. 98: 59-63

Kay AB, Ying S, Varney V, Gaga M, Durham SR, Moqbel R, Wardlaw AJ, Hamid, Q (1991) J. Exp. Med. 173: 775-778

Lahesmaa R, Yssel H, Batsford S, Luukkainen R, Mottonen T, Steinman L, Peltz G (1992) J. Immunol. 148: 3079-3085

Maggi E, Del Prete GF, Macchia D, Parronchi P, Tiri A, Chretien I, Ricci M, Romagnani S (1988) Eur. J. Immunol. 144: 1045-1054

Maggi E, Biswas P, Del Prete GF, Parronchi P, Macchia D, Simonelli C, Emmi L, De Carli M, Tiri A, Ricci M, Romagnani, S (1991) J. Immunol. 146: 1169-1174

Maggi E, Parronchi P, Manetti R, Simonelli C, Piccinni MP, Santoni Rugiu F, De Carli M, Ricci M, Romagnani S (1992) J. Immunol. 148: 2142-2147

Mosmann TR, Cherwinski H, Bond MW, Giedlin MA, Coffman RL (1986) J. Immunol. 136: 2348-2357

Mosmann TR, Moore KW (1991) Immunoparasitol. Today 12: 49-53

Paliard X, de Waal Malefyt R, Yssel H, Blanchard D, Chretien I, Abrams J, de Vries J, Spits H (1988) J. Immunol. 141: 849-855

Parronchi P, Macchia D, Piccinni M-P, Biswas P, Simonelli C, Maggi E, Ricci M, Ansari AA, Romagnani, S. (1991) Proc. Natl. Acad. Sci. USA 88: 4538-14542

Parronchi P, De Carli M, Manetti R, Simonelli C, Sampognaro S, Piccinni M-P, Macchia D, Maggi E, Del Prete GF, Romagnani S (1992) J. Immunol. (in press)

Paul, W.E. (1992) IL-4 determines lymphokine producing phenotype of antigen-specific CD4 + cells. In: "Romagnani S, Mosmann TR, Abbas AK (eds) New advances on cytokines" Raven Press, New York (in press)

Piccinni, M.P., Macchia, D., Parronchi, P., Giudizi, M.G. Bani, D., Alterini, R., Grossi, A., Ricci ,M., Maggi, E. and Romagnani, S. (1991) Proc. Natl. Acad. Sci. USA 88: 8656-8660

Robinson DS, Hamid Q, Ying S, Tsicopoulos A, Barkan J, Bentley AM, Corrigan C, Durham SR, Kay AB (1992) N. Engl. J. Med. 326: 298-304

Romagnani S (1990) Immunol. Today 11: 316-321

Romagnani S (1991a) Immunol. Today 12: 256-257

Romagnani S (1991b) Int. J. Clin. Lab. Research 21: 152-158

Romagnani, S. (1992) Immunol. Today (in press)

Salgame P, Abrams JS, Clayberger C, Goldstein H, Convitt J, Modlin RL, Bloom BR (1991) Science 254: 279-281

Van der Heijden FL, Wierenga EA, Bos JD, Kapsenberg ML (1991) J. Invest. Dermatol. 97: 389-394

Wierenga EA, Snoek M, de Groot C, Chretien l, Bos JD, Jansen HM, Kapsenberg M (1990) J. Immunol. 144: 4651-4656

Yssel H, Shanafelt MC, Soderberg C, Schneider PV, Anzola J, Peltz G (1991) J. Exp. Med. 174: 593-601

ACKNOWLEDGEMENTS

The experiments reported in this paper have been supported by the National Research Council (C.N.R.) and the Italian Association for Cancer Research (A.I.R.C.). We wish to thank S. Mohapatra (University of Manitoba, Canada) for his excellent collaboration.

Differential Regulation of Murine T Lymphocyte Subsets

F. W. Fitch, M. D. McKisic, D. W.Lancki, and T. F. Gajewski

The Committee on Immunology, The Departement of Pathology, and The Ben May Institute, University of Chicago MC6027, 5841 South Maryland Avenue, Chicago, IL 60637, USA

INTRODUCTION

T lymphocytes play a central role in immune responses. Many of the effector and regulatory functions of T lymphocytes are carried out by secreted lymphokines, and the characteristics of specific immune responses are determined to a large extent by the particular arrays of lymphokines that are produced. $CD4^+$ T cells have been categorized into two distinct subsets based on the pattern of lymphokine secretion; T_H1 cells produce IL-2 and IFN-γ (and other lymphokines) but not IL-4, IL-5, IL-6, or IL-10 while T_H2 cells produce IL-4, IL-5, IL-6, and IL-10 (and other lymphokines) but not IL-2 or IFN-γ (Mosmann and Coffman 1989). These subsets also have functional differences: T_H1 cells mediate delayed-type hypersensitivity while T_H2 cells provide efficient "help" for B lymphocytes (Mosmann and Coffman 1989). The characteristics of some immune responses *in vivo* appear to relate to the dominance of particular $CD4^+$ subsets. For example, in experimental leishmania infections in mice, a T_H1-type response predominates in mouse strains that are resistant to infection, whereas a T_H2-type response predominates in mouse strains that are susceptible to progressive infection (Heinzel et al. 1989)

The developmental pathways for T_H1 and T_H2 cells, which represent polar extremes among T cell clones, are not known. $CD4^+$ T cell clones that secrete different mixtures of lymphokines are not infrequent (Gajewski et al. 1989a; Street et al. 1990; Firestein et al. 1989), and these various $CD4^+$ subsets may arise from a common precursor (Torbett et al. 1990; Rocken et al. 1992). Regulatory processes can influence selectively the development of $CD4^+$ T lymphocyte subsets. This discussion will consider those mechanisms which affect selectively T_H1 and T_H2 subsets and which may be involved in maintaining particular patterns of immune responses. However, some of these regulatory mechanisms may also influence the differentiation of uncommitted precursor cells.

Regulatory mechanisms are difficult to study using cells obtained directly from animals since lymphoid tissues consist of a heterogeneous mixture of cells. These mechanisms can be characterized more readily using clonal populations of T cells. Both murine and human T cell antigen-specific clones retain stable patterns of lymphokine production and function and maintain characteristics of non-transformed cells. These clones provide useful model systems for evaluating regulatory mechanisms. We have examined a number of murine T lymphocyte clones representing various $CD4^+$ and $CD8^+$ subsets and have found several distinct mechanisms that affect differentially the activities of these subsets. These will be described briefly below. A modified version of this presentation has been published elsewhere (Fitch et al. 1992).

1. Interferon-γ (IFN-γ) Inhibits Proliferation of but not Lymphokine Production by T_H2 cells.

IFN-γ, at relatively low concentrations (50-100 units/ml), inhibits proliferation of murine T_H2 clones stimulated with antigen, mitogens, or anti-T cell receptor (TCR) monoclonal antibody (mAb); however, secretion of lymphokines including IL-4 in response to these stimuli is not affected (Gajewski and Fitch 1988). IFN-γ also inhibits proliferation of murine T_H2 clones exposed to IL-2 or IL-4 (Gajewski and Fitch 1988; Fernandez-Botran et al. 1988). IFN-γ does not affect proliferation or lymphokine production by T_H1 or other CD4$^+$ or CD8$^+$ clones which secrete IFN-γ. Although not absolute, the inhibitory effect of IFN-γ on proliferation of murine T_H2 cells is significant and seems to be sufficient to limit the clonal expansion of such cells. While IFN-γ does not inhibit lymphokine secretion by stimulated T_H2 cells, IFN-γ does inhibit many of the agonist effects of those secreted lymphokines. For example, IFN-γ inhibits proliferation of murine bone marrow cells stimulated with IL-3, IL-4, or granulocyte-macrophage colony-stimulating factor (GM-CSF) (Gajewski et al. 1988). In addition, IFN-γ can inhibit IL-4-dependent B cell differentiation (Coffman et al. 1988; Boom et al. 1988).

The mechanisms by which IFN-γ inhibits proliferation are not clear. The inhibitory effects of IFN-γ on proliferation are observed only on those T lymphocytes which do not secrete IFN-γ. Although T_H2 clones derived by us are not affected by IL-1, many T_H2 clones derived in other laboratories are dependent on IL-1, in addition to IL-2 or IL-4, for proliferation (Chang et al. 1990). However, the inhibitory effect of IFN-γ on proliferation of T_H2 clones does not correlate with dependence on IL-1. IFN-γ exerts its effects relatively late, still causing significant inhibition of proliferation if added twenty-four hours after addition of IL-4 (T. Gajewski, unpublished observations). Thus, IFN-γ serves as an immunoregulatory molecule through which T_H1 or CD8$^+$ lymphocytes can interfere with both the clonal expansion and the effector functions of T_H2 cells.

2. IL-10 Inhibits APC-Induced Lymphokine Production by T_H1 Cells but not by T_H2 Cells.

IL-10, a lymphokine secreted by T_H2 clones, inhibits lymphokine production by T_H1 clones but not by T_H2 clones (Fiorentino et al. 1989). This inhibition is somewhat selective in that secretion of IFN-γ and IL-3, lymphokines secreted mainly at later times after stimulation, is substantially inhibited while secretion of GM-CSF and LT/TNF, lymphokines secreted mainly at early times is inhibited slightly if at all, and the effect on IL-2 production often is less than the effect on IFN-γ production (Fiorentino et al. 1989). IL-10 apparently exerts its effect indirectly by acting on APC to reduce their ability to stimulate lymphokine production by T_H1 clones (Fiorentino et al. 1991a). IL-10 acts preferentially on macrophages and does not affect the ability of B cells to stimulate lymphokine secretion by T_H1 clones; however, IL-10 does not affect the ability of either macrophages or B cells to stimulate T_H2 clones (Fiorentino et al. 1991a). The mechanism of action of IL-10 on macrophages is not certain, although IL-10 inhibits production of IL-1, IL-6, and TNF-α by activated macrophages (Fiorentino et al. 1991b). IL-10 exerts its inhibitory effect on lymphokine production by T_H1 only if metabolically active macrophages are used for stimulation (Fiorentino et al. 1991a).

3. Murine T_H1 and T_H2 Clones Proliferate Optimally in Response to Distinct Antigen-Presenting Cell (APC) Populations.

A panel of T_H1 and T_H2 murine T cell clones, derived from the same preparation of lymph node cells after priming *in vivo* with ovalbumin (OVA), secretes lymphokines and proliferates well in the presence of OVA and whole spleen cells. However, purified splenic B cells stimulate optimal proliferation of T_H2 clones while adherent spleen cells stimulate optimal proliferation of T_H1 cells. Proliferation of T_H2 stimulated with OVA and spleen cells which had been irradiated with 3300 rad is dramatically less than that observed in response to OVA and spleen cells irradiated with 1000 rad; T_H1 clones respond similarly to OVA and spleen cells exposed to either radiation dose. Hepatic non-parenchymal cells, probably Kupffer cells, stimulate proliferation of T_H1 but not T_H2 clones (Magilavy et al. 1989). Differential activation of T_H1 and T_H2 clones does not correlate with the restricting element of the major histocompatibility complex (MHC) complex or susceptibility to inhibition by anti-CD3 or anti-LFA-1 mAb (Gajewski and Fitch 1991). Differential activation of T_H1 and T_H2 clones also can not be explained by differences in antigen processing by macrophages and B cells, since similar results are obtained when either OVA or an immunogenic peptide from OVA is used (Gajewski and I itch 1991). Production of IL-3 by each subset occurs under conditions of suboptimal proliferation, indicating that both types of APC are able to induce some at least partial activation signals in each lymphocyte subset. It is not possible to restore optimal proliferation of either subset by adding either IL-1 or IL-6 to cultures (Gajewski and Fitch 1991).

This selective effect of different APC populations has not been observed in all situations. Both macrophages and resting B cells have been reported to stimulate proliferation of both T_H1 and T_H2 clones, although exogenous IL-1 was required for B cells to stimulate T_H2 cells (Fan et al. 1988; Chang et al. 1990). However, in these studies, peritoneal macrophages rather than adherent spleen cells were used as APC, and it is likely that costimulatory molecules may be differentially expressed, depending on source and prior treatment of APC populations (Liu and Janeway 1991).

Although an MHC/antigen complex usually is sufficient to stimulate T cells to secrete at least some lymphokines, additional co-stimulatory interactions are required for optimal proliferation. In the absence of additional interactions through "co-stimulatory" molecules, a state of unresponsiveness termed "anergy" is induced in T_H1 cells (Mueller et al. 1989). This occurs only in T_H1 clones; T_H2 clones do not become anergic (Williams et al. 1990). Since T_H1 and T_H2 clones respond differently in response to different kinds of APC, it seems likely that these subsets utilize different co-stimulatory receptors and ligands. The fact that these subsets also differ in their susceptibility to the induction of anergy may relate to requirements for different co-stimulatory molecules or may reflect basically different signalling events in T_H1 and T_H2 cells.

4. T_H1 and T_H2 Clones Utilize Different TCR-Associated Signaling Pathways.

Stimulation of the TCR results in hydrolysis of PIP_2 to yield diacylglycerol and IP_3 (Imboden et al. 1985), an enzymatic reaction believed to be mediated by phospholipase C. Diacylglycerol presumable activates protein kinase C, while the IP_3 generated is thought to cause an elevation of $[Ca^{2+}]_i$ (Imboden et al. 1985). Stimulation of T_H1 clones with concanavalin A (Con A) or anti-TCR mAb leads to elevated $[Ca^{2+}]_i$ and to the generation of inositol phosphates. However, these second messengers are not detected following stimulation of T_H2 clones (Gajewski et al. 1990), even though T_H2 clones can be stimulated to secrete lymphokines by treatment with active phorbol esters and calcium ionophores, agents which mimic the effects of these second messengers (Gajewski et al. 1990). $CD8^+$ CTL apparently utilized signaling pathways similar to those of T_H1 clones (Schell and Fitch

1989). The inhibitory effect of high concentrations of anti-CD3 mAb on IL-2-dependent proliferation of T_H1 but not T_H2 clones described below appears to relate to these signaling differences. Stimulation of T_H1 clones with high concentrations of anti-CD3 mAb in calcium-free medium only modestly inhibits proliferation of these clones, and both T_H1 and T_H2 clones stimulated with IL-2 in the presence of calcium ionophore exhibit markedly decreased proliferation (Gajewski et al. 1989b).

The apparent utilization of different TCR-associated signal transduction pathways by T_H1 and T_H2 clones suggests that these subsets might respond differentially to various pharmacologic agents. Lymphokine production by T_H1 clones is substantially more sensitive to the inhibitory effects of cholera toxin, cyclosporin A, and 8-Br-cAMP than is lymphokine production by T_H2 clones (Gajewski et al. 1990). Prostaglandin E_2 has been reported to inhibit production of lymphokines by T_H1 clones but not by T_H2 clones (Betz and Fox 1991).

5. High concentrations of antigen (or anti-TCR mAb) inhibit proliferation (but not lymphokine production) by T_H1 and CTL clones only.

T_H1 clones exhibit a biphasic proliferative response when exposed to increasing concentrations of antigen. Proliferation of these cloned cells is markedly inhibited at high antigen concentrations. The failure to proliferate is due to an inability to respond to IL-2 since increasing levels of lymphokines, including IL-2, are secreted at high antigen concentration (Nau, unpublished observations); similar results have been observed by other investigators (Suzuki et al. 1988). Since the affinity of the TCR appears to vary among clones, we resorted to stimulation with immobilized anti-CD3 mAb to compare the effect of varying intensity of TCR stimulation among murine T cell subsets. Increasing concentrations of anti-CD3 mAb yields increasing amounts of lymphokine production by T_H1 and T_H2 clones, and the dose-response characteristics are quite similar (Gajewski et al. 1989b). Similar results are obtained when T_H1 clones are compared with $CD8^+$ CTL clones (Nau et al. 1987). Although there are no apparent qualitative differences among the murine T cell subsets in their ability to produce lymphokines in response to stimulation with anti-CD3 mAb, there are striking differences in the proliferative response of the various subsets.

As is the case when stimulated with antigen, the dose-response curve for T_H1 clones is biphasic, and proliferation in response high concentrations of anti-CD3 mAb, either alone or with added IL-2, is profoundly inhibited (Nau et al. 1987, Gajewski et al. 1989b). Comparable results have been reported by other investigators (Williams et al. 1990; Williams and Unanue 1990). This inhibition is not due to soluble mediators, and anti-IFN-γ mAb does not reverse the effect. The proliferative response of T_H1 clones stimulated with immobilized anti-CD3 mAb in the absence of accessory cells is suboptimal; proliferation is augmented significantly when T-depleted syngeneic spleen cells are added to cultures (T. Gajewski, unpublished observations), an effect observed by other investigators (Williams et al. 1990; Williams and Unanue 1990). However, the inhibitory effect of high concentrations of anti-CD3 mAb on IL-2-induced proliferation is not reversed by the presence of accessory cells.

With T_H2 clones, optimal proliferation is observed at about the same concentration of anti-CD3 mAb that yields optimal proliferation of T_H1 clones. However, at higher concentrations anti-CD3 mAb, the proliferation of T_H2 clones remains at plateau levels. In striking contrast with the results obtained with T_H1 clones, high concentrations of anti-CD3 mAb either has no effect on IL-2-induced proliferation or actually causes augmented proliferation (Gajewski et al. 1989b). The effect of immobilized anti-TCR mAb on IL-2-induced proliferation by CD8$^+$ CTL clones is generally similar to that of T$_H$1 clones (Nau et al. 1987). However, lymphokine-independent proliferation of CD8$^+$ cells induced by immobilized anti-TCR mAb (Moldwin et al. 1986) is not affected by high concentrations of such mAb; only the IL-2-induced component of the response is inhibited (Nau et al. 1987). The striking differences in the effect of high concentrations of antigen or anti-CD3 mAb on IL-2-induced proliferation of T$_H$1 and T$_H$2 murine T cell clones is consistent with the observations reported above that at least some signalling pathways are different in these subsets.

6. Exposure of T$_H$1 Clones (but not T$_H$2 Clones or CD8$^+$ CTL Clones) to IL-2 Induces Unresponsiveness to Antigen.

Pretreatment of T$_H$1 clones for 24 to 48 hours with concentrations of IL-2, similar to those produced by TCR stimulation, renders these cells unresponsive to subsequent stimulation with antigen (Wilde et al. 1984). Cells rendered unresponsive by IL-2 pretreatment express normal levels of TCR, CD4, and LFA-1, but they do not secrete lymphokines or proliferate (Otten et al. 1986). However, they remain able to proliferate to added IL-2 and to the combination of phorbol ester and calcium ionophore (Otten et al. 1986). This unresponsive state is characterized by failure to generate lymphokine mRNA and by decreased production of diacylglycerol and inositol phosphates (Schell and Fitch 1989). The defect induced by IL-2-pretreatment appears to be proximate to the step in the postreceptor cascade which involves activation of phospholipase C.(Schell and Fitch 1989). This type of unresponsiveness cannot be induced in T$_H$2 (Gajewski et al. 1989b) or in conventional CD8$^+$ CTL (Schell and Fitch 1989).

Unresponsiveness induced by exposure to IL-2 differs from anergy in several respects even though anergy also can be induced only in T cells that secrete IL-2. Pretreatment with IL-2 inhibits production of all lymphokines (Wilde et al. 1984), while in anergy, production of IL-2 is profoundly decreased but secretion of IFN-γ and IL-3 is less affected (Mueller et al. 1989). Il-2-induced unresponsiveness apparently involves discordant impairment of TCR signalling since treatment with a calcium ionophore plus antigen restore responsiveness while phorbol ester plus antigen does not (Otten et al. 1986). In contrast, increased [Ca^{2+}]$_i$ levels apparently are required for induction of anergy; increases in [Ca^{2+}]$_i$ are observed during induction of anergy, and a calcium ionophore alone is a potent inducer of anergy (Mueller et al. 1989). Cells recover spontaneously from IL-2-induced unresponsiveness if left for several days in culture in the absence of IL-2 (Wilde et al. 1984). Anergic cells do not regain responsiveness to antigen unless maintained for several days in the presence of IL-2 (Mueller et al. 1989). Anergy and IL-2-induced unresponsiveness represent two potent regulatory processes that appear to operate through different mechanisms to render T cells that secrete IL-2 unresponsive to further antigenic stimulation.

7. T$_H$1 and T$_H$2 Clones as well as CD8$^+$ Clones can be Cytolytic, but not all T Cells use the Same Cytolytic Mechanisms.

Although CD8$^+$ cells have often been designated "cytolytic cells" and CD4$^+$ cells have been designated "helper cells", we have found that many CD4$^+$ cells have cytolytic capabilities. The majority of T$_H$1 and T$_H$2 clones lyse OVA-pulsed B lymphoma cells or anti-CD3 coated mastocytoma cells in a re-targeted assay with an efficiency similar to that of conventional CD8$^+$ clones (Lancki et al. 1991). Several cytolytic mechanisms have been suggested, including membrane injury induced by perforin molecules and a pathway that involves DNA degradation in target cells. Since sheep red blood cells (SRBC) do not have nuclei, lysis of SRBC apparently occurs by a mechanism that does not involve DNA degradation but presumable does involve membrane damage. The majority of T_H2 clones efficiently lyse SRBC coated with anti-CD3 mAb while T_H1 clones do not (Lancki et al. 1991). This is the case even though anti-CD3-coated SRBC stimulate T_H1 clones to secrete IFN-γ (Lancki et al. 1989). The ability to lyse SRBC efficiently correlates well with the ability to express perforin and CTLA-1 mRNA; mRNA for these molecules is easily induced in CD8$^+$ clones and in T_H2 clones that lyse SRBC but can not be induced in T_H1 clones (Lancki et al. 1991). However, some T_H2 clones are not cytolytic using any of the assay systems. Thus, many T_H1 and T_H2 clones have cytolytic capability, although they may not utilize the same cytolytic mechanisms. However, T_H2 clones from some mouse strains, notably those having the BALB/c background, are not cytolytic (M. McKisic, unpublished observations). This characteristic needs to be considered in evaluating the functional effects of a predominant T_H2 response in experimental infections of this mouse strain.

IMPLICATIONS OF DIFFERENTIAL REGULATION OF T LYMPHOCYTE SUBSETS

The differences in responses of T_H1, T_H2, and CD8$^+$ murine T cells to various regulatory influences indicate that the functions of these subsets can be regulated differentially. Some of these mechanisms serve to inhibit proliferation and thus limit clonal expansion of antigen-stimulated cells. Others serve to inhibit lymphokine production and thus regulate effector functions. The effects of these immunoregulatory processes on T_H1 and T_H2 clones is summarized diagrammatically in Figure 1. Some of these regulatory mechanisms may explain various immunological phenomena. For example, the inhibitory effect of high concentrations of antigen on proliferation of T_H1 clones but not T_H2 clones may account for the observation that delayed hypersensitivity reactions were favored by low daily doses of antigen while high antibody responses were favored by high daily doses of antigen (Parish 1972). Phenomena attributed to CD4$^+$ suppressor cells probably involve some of the immunoregulatory mechanisms described above.

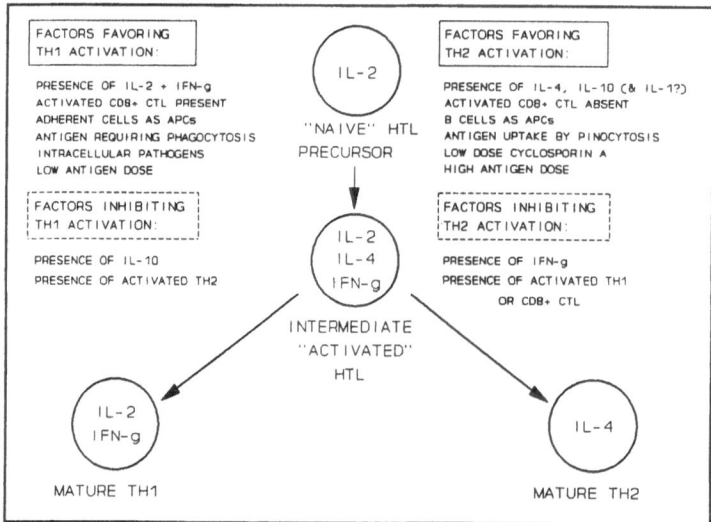

Figure 2. Factors predicted to favor or inhibit activation of T_H1 and T_H2 clones. [modified from a figure published previously (Gajewski and Fitch 1991)]

Immunoregulatory stimuli in addition to those acting directly through the TCR can affect profoundly the functions of T lymphocytes. Some of these stimuli are provided by cytokines. IFN-γ apparently acts directly on T_H2 cells to inhibit proliferation. IL-10 apparently acts indirectly by affecting macrophages to inhibit lymphokine production, and consequent IL-2-mediated proliferation of T_H1 cells. Through these mechanisms, T_H1 and T_H2 cells cross-regulate each other. IL-2, in addition to being an autocrine growth factor for T_H1 cells and some $CD8^+$ cells, acts to impair temporarily the responsiveness of these cells to antigen, an effect not observed in T_H2 cells and in $CD8^+$ cells that do not secrete IL-2. Thus, although the interaction of IL-2 with receptors for IL-2 found on all subsets of T cell leads to proliferation, this interaction has other consequences in cells that secrete IL-2. The basis for this difference is not known.

Other immunoregulatory stimuli appear to be provided by cell surface structures on APC. These include "costimulatory" molecules. Interactions through these molecules appear to be required in addition to stimulation of the TCR in order for T_H1 cells to be activated effectively. TCR stimulation in the absence of costimulation leads to the unresponsive anergic state in T_H1 cells and in $CD8^+$ cells that secrete IL-2. Anergy apparently is not induced in T_H2 cells or in $CD8^+$ cells that do not secrete IL-2 by stimulation of the TCR in the absence of costimulation. Again, the basis for this difference is not known. However, surface structures on APC appear to be able to modulate T lymphocyte functions in other ways as well. It is clear that not all APC are equally effective in stimulating all manifestations of T cell activation even though experimental results appear to vary in different laboratories. Activated B cells behave differently as APC than do resting B cells (Krieger et al. 1985), and peritoneal macrophages treated with IFN-γ function differently than do resident peritoneal macrophages (Beller 1984). It is likely that the list of cell surface molecules involved in functional interactions between T cells and APC will continue to grow.

Signaling pathways in T lymphocytes have been inadequately characterized. It is evident that differences exist among the T cell subsets. Thus, although inositol phosphates are and $[Ca^{2+}]_i$ increases after stimulation of the TCR of T_H1 and $CD8^+$ CTL, these events do not occur in T_H2 clones. Current interest is centered on the role of tyrosine phosphorylation in T cell signaling. However, information generated thus far is inadequate to explain all manifestations of T cell activation. The differences in behavior of the various T cell subsets described above indicate that there probably are multiple differences in the signalling pathways used by T lymphocyte subsets in carrying out their various functions. Understanding the mechanisms that differentially regulate T cell subsets, in addition to providing insights into the molecular events associated with activation of those subsets, should facilitate modulation of these subsets and make it possible to influence favorably the outcome of disease processes. It may be possible to exploit pharmacological means for the selective modulation of T_H1 or T_H2 functions in immunopathologic situations.

REFERENCES

Beller DI (1984) Eur J Immunol 14:138-143
Betz M, Fox BS (1991) J Immunol 146:108-113
Boom WH, Liano D, Abbas AK (1988) J Exp Med 167:1350-1363
Chang T-L, Shea CM, Urioste S, Thompson RC, Boom WH, Abbas AK (1990) J
    Immunol 145:2803-2808
Coffman RL, Seymour BW, Lebman DA, Hiraki DD, Christiansen JA, Shrader
    B, Cherwinski HM, Savelkoul HFJ, Finkelman FD, Bond MW, Mosmann TR

(1988) Immunol Rev 102:5-28

Fan X-d, Goldberg M, Bloom BR (1988) Proc Natl Acad Sci USA 85:5122-5125

Fernandez-Botran R, Sanders VM, Mosmann TR, Vitetta ES (1988) J Exp Med 168:543-558

Fiorentino DF, Bond MW, Mosmann TR (1989) J Exp Med 170:2081-2095

Fiorentino DF, Zlotnik A, Vieira P, Mosmann TR, Howard M, Moore KM, O'Garra A (1991a) J Immunol 146:3444-3451

Fiorentino DF, Zlotnik A, Mosmann TR, Howard M, O'Garra A (1991b) J Immunol 147:3815-3822

Firestein GS, Roeder WD, Laxer JA, Townsend KS, Weaver CT, Hom JT, Linton J, Torbett BE, Glasebrook AL (1989) J Immunol 143:518-525

Fitch FW, McKisic MD, Lancki DW, Schell SR, Gajewski, TF (1992) Differential regulation of T lymphocytes. In: Romagnani S (ed) Proceedings, 2nd International Congress on Cytokines: Basic Principles and Clinical Application. Ares Sereno Symposia Publications, Raven Press, New York, (In press).

Gajewski TF, Fitch FW (1988) J Immunol 140:4245-4252

Gajewski TF, Fitch FW (1991) Res Immunol 142:19-23

Gajewski TF, Goldwasser E, Fitch FW (1988) J Immunol 141:2635-2642

Gajewski TF, Joyce J, Fitch FW (1989a) J Immunol 143:15-22

Gajewski TF, Schell SR, Nau G, Fitch FW (1989b) Immunol Rev 111:79-110

Gajewski TF, Schell SR, Fitch FW (1990) J Immunol 144:4110-4120

Gajewski TF, Pinnas M, Wong T, Fitch FW (1991) J Immunol 146:1750-1758

Heinzel FP, Sadick MD, Holaday BJ, Coffman RL, Locksley RM (1989) J Exp Med 169:59-72

Imboden JB, Weiss A, Stobo JD (1985) Immunol Today 6:328-331

Krieger JI, Grammer SF, Grey HM, Chesnut RW (1985) J Immunol 135:2937-2945

Lancki DW, Kaper BP, Fitch FW (1989) J Immunol 142:416-424

Lancki DW, Hsieh C-S, Fitch FW (1991) J Immunol 146:3242-3249

Liu Y, Janeway CA,Jr. (1991) Int Immunol 3:323-332

Magilavy DB, Fitch FW, Gajewski TF (1989) J Exp Med 170:985-990

Moldwin RL, Lancki DW, Herold KC, Fitch FW (1986) J Exp Med 163:1566-1582

Mosmann TR, Coffman RL (1989) Annu Rev Immunol 7:145-173

Mueller DL, Jenkins MK, Schwartz RH (1989) Annu Rev Immunol 7:445-480

Nau GJ, Moldwin RL, Lancki DW, Kim D-K, Fitch FW (1987) J Immunol 139:114-122

Otten G, Wilde DB, Prystowsky MB, Olshan JS, Rabin H, Henderson LE, Fitch FW (1986) Eur J Immunol 16:217-225

Parish CR (1972) Transplant Rev 13:35-66

Rocken M, Saurat J-H, Hauser C (1992) J Immunol 148:1031-1036

Schell SR, Fitch FW (1989) J Immunol 143:1499-1505

Street NE, Schumacher JH, Fong TAT, Bass H, Fiorentino DF, Leverah JA, Mosmann TR (1990) J Immunol 144:1629-1639

Suzuki G, Kawase Y, Koyasu S, Yahara I, Kobayashi Y, Schwartz RH (1988) J Immunol 140:1359-1365

Torbett BE, Laxer JA, Glasebrook AL (1990) Immunol Lett 23:227-234

Wilde DB, Prystowsky MB, Ely JM, Vogel SN, Dialynas DP, Fitch FW (1984) J Immunol 133:636-641

Williams IR, Unanue ER (1990) J Immunol 145:85-93

Williams ME, Lichtman AH, Abbas AK (1990) J Immunol 144:1208-1214

T-cell Mediated Cytotoxicity

E. R. Podack, G. Deng, M. A. Bowen, Z. Wu, K. J. Olsen, M. Zakarija, M. G. Lichtenheld, D. Kägi[1], K. Bürki[2] and H. Hengartner[1]

Department of Microbiology and Immunology, University of Miami, School of Medicine, P. O. Box 016960 (R-138), Miami, Florida 33101, [1]Department of Experimental Pathology, University of Zürich, Sternwartstrasse 2, Zürich, Switzerland, [2]Sandoz AG, Basel, Switzerland

SUMMARY

Mechanistically T-cell mediated cytotoxicity uses two pathways for the destruction of a target cell. One is directed at the target membrane and the other at the target nucleus and DNA. These pathways often operate together and at least partially overlap by their common usage of perforin. Perforin mediates direct target membrane damage by pore formation and subverting the normal barrier function of the membrane. This action alone often is sufficient for target cell death. In addition perforin pores allow the entry of other molecules into the target cell either by diffusion across the pore or by uptake through repair endocytosis. Entry of Ca ions, of TNF and/or of granzymes by these mechanism may trigger apoptosis of the target cell. The membrane permeabilizing action of perforin thus also appears to be a requisite for apoptosis which is triggered by the entry of Ca and/or other factors.

Apoptosis is a target specific process. At least three constellations of apoptosis initiation are possible depending on conditions within the target cell: *i*. presence of a killer factor ready to be activated (zymogen), *ii*. presence of a protecting factor preventing killer factor formation or activity or *iii*. absence of either factor requiring gene transcription and/or translation for apoptosis. It is possible that T-cells initiate apoptosis in the target cell by interfering with the normal control of the suicide program and that the diversity of granzymes reflects different control mechanisms in different target cells.

The central role of perforin for cytotoxicity is supported by its careful transcriptional regulation and its restriction to professional killer cells. Perforin is inducible in CD8[+] and CD4[+] CTL. Gene induction requires the cell specific inactivation of perforin repressor elements and their apparent conversion to enhancers. This process in CD8[+] cells is protein synthesis independent and can be stimulated by an Il2 dependent or an Il2 independent pathway.

Perforin expression in tissue infiltrating lymphocytes in vivo correlates with the manifestation of T-cell mediated cytotoxicity. Perforin expression in vivo can occur in all T-cell subsets including CD4[+] cells. In autoimmune processes one of the largest subgroups of perforin positive cells is CD3[+], CD4[-], CD8[-] double negative.

Control of T-cell mediated cytotoxicity is achieved by apoptosis of growth factor dependent CTL through growth factor withdrawal. In addition, anti CD30 antibodies suppress cytotoxicity of the LGL line YT, possibly in a gene transcription dependent process. Since human CD4[+] CTL and Con A activated PBL express CD30 a possible role for CD30 in cytotoxicity regulation is under consideration.

MECHANISM OF T-CELL MEDIATED CYTOTOXICITY

A. MEMBRANE DAMAGE

Perforin (the name is derived from its ability to perforate) is a T-cell and NK-cell specific protein which is stored in the cytoplasmic granules of these cells (Dennert and Podack 1983; Henkart et. al. 1984; Podack and Konigsberg 1984). Specific target recognition by the killer cell results in the transmission of signals which direct the secretion of the granules towards the contact area of the killer and target cell (Yannelli et. al. 1986). The secretory process including the signalling pathway are Ca dependent and are inhibited by cyclosporin A (Havele and Paetkau 1988). The cytolytic granules are released into the intercellular space between killer and target cell. Since granules have been shown to contain the CD3 complex it is possible that the secreted material is specifically targeted to the MHC antigen complex of the target cells (Peters et. al. 1990).

Upon contact with extracellular Ca perforin binds to phospholipid headgroups and at 37° polymerizes and inserts itself into the target cell membrane (Tschopp et. al. 1989). This reaction leads to the formation of a transmembrane channel of 16nm internal diameter the wall of which is constituted of approximately 20 perforin monomers (Podack and Dennert 1983). Smaller channels are possible and commonly occur for instance when only 2 or three perforin monomers insert into the membrane. The biochemical reaction of polymerization coupled to hydrophobic membrane insertion is quite analogous to that of the complement component C9, which has been studied in greater detail (Podack 1987).

B. NUCLEAR DAMAGE

In many target cells attacked by T-cells DNA fragments are released within the first 30 min as the result of the nuclear break down (Russell 1983; Duke and Cohen 1988). The extent of DNA release is variable and is determined by the target cell rather than by the CTL. It is clear therefore that the ultimate event of DNA degradation is mediated by a target cell factor which becomes activated through the T-cell attack. The nature of the target cell factor (killer factor) becoming activated is not known. Progress has been made however in defining potential CTL factors causing the activation of DNA degradation in the target cell. In addition the role of perforin cooperating with other CTL factors triggering DNA release is becoming increasingly appreciated.

Perforin pores of any size result in the influx of Ca ions into target cells. The role of increased Ca levels preceding apoptosis is amply documented (McConkey et. al. 1989). Thus it has been argued that the sustained influx of Ca alone may be responsible for triggering DNA degradation (Hameed et. al. 1989). The influx of Ca is also the stimulus for repair endocytosis by the target cell. Endocytosis and the accompanying pinocytosis represent one pathway by which perforin triggers the uptake of external molecules independent of receptor binding. Another, probably less important pathway of uptake of external molecules results from passive diffusion across perforin pores. TNF in the presence of sublytic amounts of perforin or other pore formers dramatically increases DNA release in target cells. Even a tenfold higher amount of TNF in the absence of perforin was not cytotoxic. Since many CTL secrete TNF and lymphotoxin, a strong synergy between the two cytotoxic pathways is likely similar to what has been uncovered between perforin and granzymes. It is becoming more and more apparent that an important function of granzymes is to mediate DNA degradation in conjunction with perforin (Pasternak and Eisen 1985; Hayes et. al. 1989). Perforin allows their target entry by diffusion or by repair endocytosis. The proteolytic activity of granzymes then triggers apoptosis in an undefined mechanism. Fragmentin is (Greenberg et. al. 1992) a highly active in mediating DNA degradation in the presence of sublytic amounts of perforin. It is a member of the granzyme family produced by rat NK cells. It shows the highest degree of homology to human granzyme 3 (Hameed et. al. 1988).

In the murine system up to seven different granzymes may be present in granules together with perforin and other factors (Masson and Tschopp 1987). Each of the granzymes may gain entry into the target cell via perforin pores or repair endocytosis. The high degree of diversity among the granzymes may be required to trigger DNA degradation in different target cells that may come under T-cell attack.

APOPTOSIS

It is evident that apoptosis is controlled by the balance of at least two opposing forces: the killer factor (possibly a Ca, Mg, dependent endonuclease) and a protecting factor preventing apoptosis (e.g. bcl2). The regulation of these activities is not well understood and is subject to change with the maturation of a cell lineage. Immature thymocytes for instance require transcription and translation for the execution of apoptosis since DNA degradation is blocked in the presence of inhibitors of either (Duke and Cohen 1988). In Il2 dependent CTL, in contrast, transcriptional blockers cause apoptosis because Il2 dependent bcl2 transcription is necessary to prevent cell death (Deng and Podack, unpublished). Finally the T-cell hybridoma 2B4 is stimulated to commit suicide by DNA degradation through addition of translational, and less pronounced, transcriptional blockers (Deng and Podack, unpublished). Generally, the signals for apoptosis are coordinated with the cell cycle. This scenario suggests that most or all cells can induce their own DNA degradation under appropriate conditions. It is likely that CTL mediated apoptosis occurs by activation of the killer pathway or by inactivation of the protection pathway. The diversity of granzymes may reflect the diversity of regulatory pathways in apoptosis.

PERFORIN

Perforin is a 68 to 70kD protein with limited homology to C9 (Podack et. al. 1985; Masson and Tschopp 1985). It contains complex carbohydrates and is located in cytoplasmic granules. The signals directing it to the intracellular granules, rather than towards immediate export are not known. Unlike granzymes perforin does not use the mannose-6-phosphate pathway characteristic for lysosomal proteins.

Perforin is highly cell specific. It is synthesized constitutively by NK-cells (Lichtenheld et. al. 1988) is inducible in CD4$^+$ and CD8$^+$ (Smyth et. al. 1990) T-cells. It is not known whether perforin is induced or constitutively transcribed in double negative T-cells bearing either the $\alpha\beta$- or $\gamma\delta$-T-cell receptor. In contrast to granzymes, perforin expression has not been found in B-cells or cells of the monocyte or myelocytic series.

Perforin in human peripheral blood mononuclear cells (PBMC) is induced by Il2 dependent and by Il2 independent pathways. Il2 alone induces perforin expression in CD8$^+$ T-cells, but not in CD4$^+$ T-cells. Perforin induction by Il2 in CD8$^+$ T-cells is independent of adherent cells, independent of de novo protein synthesis and resistant to cyclosporin A (Lu et. al. [in press] 1992). Il2 induction is greatly increased in the presence of Il6 which by itself has no effect on perforin transcription (Smyth et. al. 1990). Ca-ionophores are strong inducers of perforin. Ca-signals induce perforin more rapidly than Il2 (maximum effect in 6h versus 12h for Il2) and to a higher level (15-fold versus 4-fold over basal levels). Ca signals require the presence of adherent cells for perforin induction in CD8$^+$ cells; CD4$^+$ cells remain perforin negative. Ca-signals are independent of de novo protein synthesis, however they are blocked by cyclosporin A. Activation of protein kinase C does not synergize with Ca-ionophores in the transcription of perforin. This is in contrast to other T-cell molecules such as the transcription of Il2 or granzyme B.

Extrapolating from these in vitro studies to the in vivo situation it would seem that the contact of CD8$^+$ cells with its appropriate antigen presenting cell would provide both signals for perforin induction, the Ca signal via the TCR and the accessory signal via receptors on the antigen presenting cell. Perforin transcription under these conditions is detectable within 2 hours and maximal after six hours. Importantly, this phase is helper cell independent. However it is not clear at this time whether all CD8$^+$ cells can respond in this manner. It is possible, or may be even likely, that only CD8$^+$ cells previously in contact with antigen (memory cells?) respond in this way. It is known for instance that in human PBMC among the CD8$^+$ cells only the CD11b$^+$, CD16$^+$, CD56$^+$ T cells express perforin protein and these cells may preferentially respond to Ca-signals.

PERFORIN EXPRESSION IN VIVO

In the murine model perforin expression in situ correlates with the observed local immune response (Podack et. al. 1991). Thus, perforin expression in infiltrating CD8 cells follows the tropism of viral infections in the ensuing immune response. Rejection of allografts is accompanied by perforin expression and perforin is found at sites of autoimmune reactivity. Perforin expression positively correlates with the rejection of immunogenic tumors; in contrast tumors that are not rejected are infiltrated with lymphocytes that fail to express perforin and in which perforin expression, even under in vitro stimulation, is delayed (Loeffler et. al. 1992). The case for an important role of perforin in T-cell mediated cytotoxicity is strong, notwithstanding the fact that perforin independent cytotoxic mechanisms exist. As we have succeeded in deleting the functional perforin gene by homologous recombination in transgenic mice, the biological importance of this protein will soon be established.

Perforin expression in human tissues, as far as it has been examined supports the findings of the murine findings. In heart transplants the severity of the immune rejection crisis correlates with the level of perforin expression (Hameed et. al. 1992). Viral and autoimmune disorders are accompanied by perforin expression. The majority of human CD4$^+$ CTL specific for HIV express perforin. A wide variety of T-cell subsets express perforin in Hashimoto's thyroiditis. This includes CD4, CD8 double negative T-cells with $\alpha\beta$ or $\gamma\delta$ T-cell receptors, CD8$^+$ T-cells, CD4$^+$T-cells, and CD56$^+$ and CD56$^-$ NK cells. Except for CD4$^+$ T-cells, a similar distribution of perforin expressing cells is found in cells infiltrating the thyroid in Graves' disease. Of particular interest is the high level of double negative T-cells expressing perforin in these diseases. The suggestion has been made that these cells represent T-cells that have not been selected in the thymus and that are held responsible for bone marrow rejection in irradiated, murine recipients (Kikly and Dennert 1992). It will be of interest to study the phenotype of perforin expressing cells in other autoimmune diseases to determine whether perforin containing double negative T-cells are generally prevalent.

TRANSCRIPTIONAL REGULATION OF PERFORIN

Control of perforin transcription and cell specificity is conferred to the perforin gene by 5kb of upstream sequences. It appears that transcriptional activation in CTL is primarily due to inactivation of silencer elements and their apparent conversion to enhancers (Lichtenheld and Podack J. Immunol, in press). In this sense perforin transcription relies on derepression, which may be possible only in certain lymphocytes. The details of this process are unknown but the following general picture is emerging. The first 250 nucleotides of the perforin promotor are active in lymphocytic and non lymphocytic cells independent of perforin content. If the first 800 base pairs of the promotor are tested the transcriptional activity is depressed in non lymphocytic cells but enhanced in T-cells regardless of their ability to express perforin. In constructs containing 2500 and 5000bp promotor sequence transcription is greatly enhanced in perforin expressing CTL and repressed in both non cytotoxic CD4$^+$ T-cells and non lymphocytic cells.

The perforin promotor appears unique also in that it contains very few elements known to control transcriptional responses in other genes. This organization may serve to prevent inappropriate expression of perforin in cells that are not professional killer cells and may be interpreted as an effective gun control, underlining the importance of perforin for cytotoxicity.

Experimentally the perforin promotor is well suited to direct gene products specifically to killer lymphocytes in the form of transgenes. This property of the perforin promotor is currently under study with the aim of improving the efficiency of CTL for tumor therapy.

DOWN REGULATION OF CYTOTOXICITY

CTL are generally Il2 and antigen dependent and die by apoptosis if either of the stimuli is missing (Duke and Cohen 1986). Presumably this drastic type of control is prevalent at the end of an immune response, for instance, when viral infection has been cleared by CTL and no more viral antigens are being produced. TGF-β has been shown to interfere with the induction of perforin. A true down regulation has been achieved so far only with an antibody to CD30 (C10) and YT (a human LGL tumor line). 24h preincubation of YT with anti CD30 completely blocks YT cytotoxicity for Raji cells (their best target), whereas adding the antibody to the cytotoxicity assay, without preincubation of YT, has no effect(Bowen and Podack 1992 [abstract]). Anti CD30 had no effect on perforin mRNA levels, however whether it interfered with perforin protein or secretion has not been determined. It appears that the inhibition of YT by anti CD30 requires transcriptional activity. CD30 is a T-cell activation antigen, peaking on day 3 or 4 after concanavalin A stimulation of PBMC. We have found CD30 expressed also on HIV specific mature CD4$^+$ cells. Both cytotoxic and non cytotoxic cells expressed CD30. It will be of interest to determine whether CD30 may modulate the activity of mature functional T-cells, including cytotoxicity.

PERSPECTIVES

The important role of perforin in lymphocyte mediated cytotoxicity is supported by a large number of observations from many laboratories. Perforin can be induced in all T-cell subsets and is expressed in NK-cells. B-cells apparently cannot express perforin. Perforin causes membrane damage and allows through this action the entry into the target of other molecules that may cause apoptosis. Perforin is the best marker for the presence of cytotoxic cells in situ. In autoimmunity perforin was found to be expressed in a large population of double negative T-cells.

Perforin deficient mice in conjunction with perforin transgenes and the use of the perforin promotor to direct other gene products to professional killer cells will further illuminate this aspect of T-cell mediated cytotoxicity.

REFERENCES

Bowen MA, Podack ER (1992) Faseb J 6:A1980

Dennert G, Podack ER (1983) J Exp Med 157:1483-1495

Duke RC, Podack ER (1983) Lymphokines Res 5:289-299

Duke RC, Cohen JJ (1988) The role of nuclear damage in lysis of target cells by cytotoxic T lymphocytes. In: Cytolytic Lymphocytes and Complement: Effectors of the Immune System, ER Podack (ed), CRC Press, Florida, p 235

Greenberg AH, Shi L, Aebersold R (1992) Faseb J 6:A1692

Hameed A, Lowrey D, Lichtenheld MG, Podack ER (1988) J Immunol 141:3142-3147

Hameed A, Olsen KJ, Lee MK, Lichtenheld MG, Podack ER (1989) J Exp Med 169:765-777
Hameed A, Olsen KJ, Cheng L, Fox WM, Hruban RH, Podack ER (1992) Am J Pathol 140:1025-1030
Havele C, Paetkau V (1988) J Immunol 140:3303-3308
Hayes MP, Berrebi GA, Henkart PA (1989) J Exp Med 170:933-946
Henkart PA, Millard PJ, Reynolds CW, Henkart MP (1984) J Exp Med 160:75-93
Kikly K, Dennert G (1992) 149:403-412
Lichtenheld MG, Olsen KJ, Lu P, Lowrey DM, Hameed A, Hengartner H, Podack ER (1988) Nature 335:448-451
Lichtenheld MG, Podack ER (in press) J Immunol
Loeffler CM, Smyth MJ, Longo DL, Kopp WC, Harvey LK, Tribble HR, Tase JE, Urba WJ, Leonard AS, Young HA, Ochoa AC (1992) J Immunol 149:949-956
Lu P, Garcia-Sanz JA, Lichtenheld MG, Podack ER (in press) J Immunol
McConkey DJ, Hartzell P, Nicotera P, Orrenius S (1989) FASEB J 3:1843
Masson D, Tschopp J (1985) J Biol Chem 260:9069-9072
Masson D, Tschopp J (1987) Cell 49:679-685
Pasternak MS, Eisen HN (1985) Nature 314:743-745
Peters, PJ, Van der Donk HA, Borst A (1990) Immunol Today 11:32-38
Podack ER, Dennert G (1983) Nature 314:743-745
Podack ER, Konigsberg PJ (1984) J Exp Med 160:695-710
Podack ER, Young JDE, Cohn ZA (1985) Proc Natl Acad Sci 82:8629-8633
Podack ER (1987) Perforins: A family of pore forming proteins in immune cytolysis. In: Membrane Mediated Cytotoxicity, RJ Collier B Bonavida (eds) Alan R. Liss, New York, p 339
Podack ER, Hengartner H, Lichtenheld MG (1991) Annu Rev Immunol 9:129-157
Russell JH (1983) Immunol Rev 72:97-117
Smyth MJ, Ortaldo JR, Bene W, Yagita H, Okumuta K, Young MA (1990) J Immunol 145:1156-1166
Smyth MJ, Ortaldo JR, Shinkai YI, Yagita H, Nakata M (1990) J Exp Med 171:1269-1281
Tschopp J, Schafer S, Masson D, Peitsch MC, Heusser C (1989) Nature 337:272-274
Yannelli JR, Sullivan JA, Mandell GL, Engelhard VH (1986) J Immunol 136:377-382

Development of TCRγδ Cells

J. A. Bluestone[1], A. Sperling[1], Y. Tatsumi[1], T. Barrett[1], S. Hedrick[2] and L. Matis[3]

[1]Ben May Institute and Committee On Immunology, MC 1089, University of Chicago, Chicago, IL 60637
[2]University of California, San Diego, CA
[3]Medicine Branch, National Cancer Institute, Bethesda, MD.

INTRODUCTION

Over the past several years, a novel T cell subset has been identified that, unlike conventional CD8[+] cytolytic cells and CD4[+] helper TCRαß cells, expresses a T cell receptor complex (TCR) composed of CD3-associated, disulfide-linked TCR γ and δ gene products (Bluestone et al. 1991a). The role of TCRγδ cells in immune responses is currently not well understood. TCRγδ T lymphocytes appear in enhanced numbers in skin lesions of *Mycobacterium leprae*, celiac sprue, polymyositis, listeria infections, and in the synovial fluid of joints affected by rheumatoid arthritis (Reviewed in **Immunol. Rev. Vol. 120, 1991**). These observations raise the possibility that this subset of T lymphocytes may play a role in the control of infectious processes and in autoimmune disease. The TCRγδ cells develop in waves in the thymus prior to TCRαß development with each successive wave expressing a distinct family of TCR Vγ and Vδ genes (Raulet, 1989). In fact, unlike TCRαß cells, TCRγδ cells selectively home to specific tissues including the skin, lung, reproductive tract and intestine where they reside as intraepithelial lymphocytes (IEL) (Bluestone et al. 1991b). In addition, TCRγδ cells occupy anatomically distinct regions of the lymphoid tissue, such that splenic TCRγδ cells predominate in the sinusoids while the TCRαß cells localize to the follicular areas. In spite of the great potential for diversity in this population, the TCRγδ cells have a more limited TCR repertoire than αß cells especially in the regional epithelial tissues where in some cases a single TCRγ and δ chain is expressed on resident cells. The reason for this limited TCRVγVδ pairing, restricted VDJ junctional diversity and selective tissue expression remains unclear but may reflect selection events either intrathymic or postthymic development. Yet, TCRγδ cells have been observed to recognize a variety of antigens including: MHC-encoded molecules (classical class I and class II, non-classical class I molecules [CD1, TL]); bacterial products (mycobacterial antigens, bacterial heat shock proteins and bacterial endotoxin); endogenous mammalian stress proteins such as heat shock proteins (HSP65); and a limited number of nominal antigens (G-T copolymer and tetanus toxoid) (Born et al. 1990; Del Porto et al. 1991; Holoshitz et al. 1989; Band et al. 1991). Many of the studies of murine TCRγδ cells have been limited to an analysis of TCRγδ populations localized to epithelial tissue, newborn thymus, or other non-circulating TCRγδ cells. Our approach to examining the potential repertoire and ligand specificity of TCRγδ cells, has been directed towards examining the small subset of TCRγδ cells localized in the lymphoid tissue. Although this is a minor population of TCRγδ cells, the distinct receptor usage and potential of these cells to circulate throughout the body suggested that they may have a diverse selected repertoire (Cron et al. 1990). This review summarizes our studies of the development and repertoire selection of TCRγδ cells using normal and transgenic mice.

TCR*αβ* cells recognize peptidic antigens presented in the context of self-MHC. This self/non-self discrimination is dictated by a process of thymic education known as positive and negative selection. During TCR*αβ* development in the thymus, immature T cells interact with self-MHC. If this interaction is productive, the T cells are signalled to mature and are thus positively selected. If the selected T cells interact too avidly with the self-MHC molecule, the T cells undergo apoptosis and die (clonal deletion/negative selection). Thymic education is thus a complex process in which T cells learn to distinguish self from non-self.

Previous studies from our laboratories have shown that TCR*γδ* cells can recognize a variety of class I and class II MHC antigens, including a non-classical MHC *β*2M-associated, TL-specific TCR*γδ* clone, G8 (Bluestone et al. 1988). However, the role of TCR*γδ* ligands in shaping the *γδ* repertoire had not been well understood. Therefore, we have utilized the G8 clone as a model system to study TCR*γδ* development. G8 TCR*γ* and *δ* transgenes were introduced into ligand positive (H-2^b and H-2^k) and ligand negative (H-2^d) mouse strains as a model to examine TCR*γδ* development (Dent et al. 1990).

When Tg TCR+ mice were bred to H-2^b mice, Tg TCR*γδ* cells were largely deleted from the thymus and spleen. This process begins during the first week of life in the thymus and is complete in the spleen by 6 weeks of age. Thus, it would appear that in this model, potentially self-reactive TCR*γδ* cells undergo clonal deletion. However, studies that examined the TCR*γδ* cells in the intestine in the Tg model suggest both clonal deletion and clonal anergy are involved in the tolerance evident in the Tg mice. We have shown that Tg+ TCR*γδ* cells, with normal levels of TCR expression, are found in the intraepithelial compartments of ligand-bearing H-2$^{b/d}$ mice. The TCR*γδ* cells were largely unresponsive as measured by lymphokine production, induction of IL-2 receptor (IL-2R), and proliferation even in the presence of exogenous IL-2 (Barrett et al. 1992). Thus, the maintenance of self-tolerance of TCR*γδ* cells localizing to epithelial tissues can be due to clonal anergy following antigen exposure. Therefore, a very potent antigenic stimulus in the thymus leads to clonal deletion, while a prolonged antigenic challenge in the intestine leads to clonal inactivation.

Another consequence of antigen exposure to TCR*γδ* cells has been the observation that Tg+ cells that bind antigen down regulate Thy-1 expression (Barrett, Tatsumi and Bluestone, manuscript in preparation). This down regulation is a direct result of antigen exposure and results in T cell unresponsiveness since Thy-1$^+$ cells injected into antigen-bearing but not antigen non-bearing mice downregulate Thy-1. These results have general implications for the maintenance of self-tolerance of normal TCR*αβ* and TCR*γδ* cells. In fact, normal TCR*γδ* cells down regulate Thy-1 after exposure to anti-TCR mAbs in fetal thymus organ cultures. In addition, a large percentage of TCR*αβ* and TCR*γδ* IELs express low levels of Thy-1. Previous studies have suggested that this is a result of extrathymic development (Lefrancois et al. 1990). However, our studies suggest that Thy-1 antigen expression is down-regulated as a direct result of antigen exposure and anergy induction (Fig. 1; Barrett et al., manuscript in preparation). Thus, down regulation of Thy-1 appears to be a marker of development and tolerance induction.

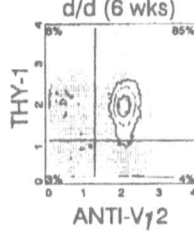 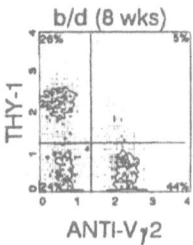

Figure 1. Tg$^+$ TCR$\gamma\delta$ IELs isolated from antigen-bearing H-2^b mice express low levels of Thy-1.

TCR$\gamma\delta$ Development-Positive Selection

Analysis of positive selection of TCR$\gamma\delta$ T cells is complicated by the current limited understanding of the $\gamma\delta$ antigen-specific repertoire. It appears that many TCR$\gamma\delta$ cells may not display the same fundamental MHC-directed specificity as TCR$\alpha\beta$ cells. Nonetheless, we have developed strong evidence for positive selection of TCR$\gamma\delta$ cells. We have found that transgenic TCR$\gamma\delta$ cells, develop in the thymus of β_2M-deficient mice (produced by disruption of the ß2M gene by homologous recombination), but cannot signal through their T cell receptor and do not exit from the thymus to populate peripheral lymphoid organs. In addition, treatment of mice with cyclosporine A blocks TCR$\gamma\delta$ development in Tg as well as normal mice (Fig. 2). In addition, several markers of TCR$\gamma\delta$ development have been identified. Using PCR analysis, we have determined that the cell surface HSA$^+$ TCR$\gamma\delta$ cells continue to transcribe the recombinase enzymes, RAG-1 and RAG-2, while mature HSA$^-$ cells do not. These results closely parallel TCR$\alpha\beta$ positive selection, and indicate that, at least some, TCR$\gamma\delta$ cells require a TCR-ligand interaction to complete differentiation.

A second approach towards examining the development of TCR$\gamma\delta$ cells has been to compare the repertoire of TCR$\gamma\delta$ splenocytes in a variety of inbred and MHC-congenic strains of mice. A panel of recently developed anti-murine TCR$\gamma\delta$ mAb has allowed for more precise studies of the peripheral lymphoid organ TCR$\gamma\delta$ repertoire. The percentages of both the Vγ2+ and Vδ4+ subsets of splenic TCR$\gamma\delta$ cells differ widely between different inbred strains of mice. Further analysis using recombinant inbred strains has demonstrated that the expression of individual γ and δ genes is positively selected and linked to the TCR δ and γ loci, respectively. In addition, we have shown that this strain-specific TCR$\gamma\delta$ development is a direct result of positive selection, not in the thymus, but post birth. This clonal expansion is not the result of environment exposure.

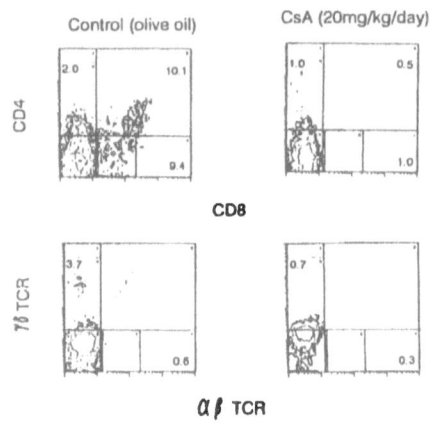

Figure 2. Flow cytofluorometric analysis of thymocytes from offspring of pregnant mice treated with cyclosporine A from day 5 to day 16 gestation. Fetuses were analyzed at day 16.

Furthermore, recent data suggests that high Vγ2 expression is linked to the δ chain usage of Vα10. Thus, this γδ pair may uniquely recognize an antigen that results in selective clonal expansion.

In summary, these studies provide a basis to understanding and determining the factors which shape the peripheral TCRγδ repertoire. It would appear that both positive and negative selection forces participate in the selection of TCRγδ cells. Like TCRαß cells, these events may be critical towards guaranteeing the selection a T cell repertoire that can distinguish self from non-self in the response to foreign antigens. If so, a fundamental characteristic of all T cell development may be the requirement for TCR/ligand interaction.

REFERENCES

Band H, Porcelli SA, Panchamoorthy G, McLean J, Morita CT, Ishikawa S, Modlin RL, Brenner MB (1991) Curr Top Microbiol Immunol 173:229-234
Barrett TA, Delvy ML, Kennedy DM, Lefrancois L, Matis LA, Dent AL, Hedrick SM, Bluestone JA (1992) J Exp Med 175:65-70
Bluestone JA, Cron RQ, Cotterman M, Houlden BA, Matis LA (1988) J Exp Med 168:1899-1916

Bluestone JA, Cron RQ, Barrett TA, Houlden B, Sperling AI, Dent A, Hedrick S, Rellahan B, Matis LA (1991a) Immunol Rev 120:5-33

Bluestone JA, Cron RQ, Rellahan B, Matis LA (1991b) Curr Top Microbiol Immunol 173:133-139

Born W, Happ MP, Dallas A, Reardon C, Kubo R, Shinnick T, Brennan P, O'Brien R (1990) Immunol Today 11:40-43

Cron RQ, Coligan JE, Bluestone JA (1990) Immunogenetics 31:220-228

Del Porto P, Mami-Chouaib F, Bruneau J-M, Jitsukawa S, Dumas J, Harnois M, Hercend T (1991) J Exp Med 173:1339-1344

Dent AL, Matis LA, Hooshmand F, Widacki SM, Bluestone JA, Hedrick SM (1990) Nature 343:714-719

Holoshitz J, Koning F, Coligan JE, De Bruyn J, Strober S (1989) Nature 339:226-229

Lefrancois L, LeCorre R, Mayo J, Bluestone JA, Goodman T (1990) Cell 63:333-340

Raulet DH (1989) Annu Rev Immunol 7:175-207

6. Cell Adhesion and Inter-Action Molecules

Cell Interaction Molecules Regulating T Cell Receptor Mediated Signal Transduction: A Comparison of Peripheral T Cell with Developing Thymocytes

K.Eichmann, T. Bartlott, A. Ehrfeld, C. Levelt, A. Potocnik, A. Reimann, A. Würch, A. Zgaga-Griesz, S. Suzuki, and K. M. Saizawa

Max-Planck-Institut für Immunbiologie, Stübeweg 51, D-7800 Freiburg, Germany

INTRODUCTION

Signal transduction through the αβ CD3 T cell receptor (TCR) can induce vastly different cellular responses at different stages of T cell differentiation. The earliest form of TCR mediated signalling is seen in immature CD4⁻8⁻ thymocytes in which ligation of CD3ε induces maturation to the CD4⁺8⁺ double positive (DP) stage (Levelt et al., 1992). At the DP stage, signal transduction through the TCR leads to negative or positive selection, corresponding either to deletion of the thymocytes or to their further maturation to CD4⁺ or CD8⁺ single positive (SP) mature thymocytes (v. Boehmer, 1988). Signal transduction through the TCR of resting peripheral T cells leads, under most circumstances, to induction of transcriptional activities and of cell division, i.e. to activation. Activated T cells are again able to respond to TCR signalling, mostly by secretion or exocytosis of preformed or de novo synthetized effector molecules (reviewed in Möller, 1989).

What are the molecular events that control this multitude of cellular responses elicited through the same receptor? Several levels of control may be considered: Firstly, the TCR or its individual components may be connected to different sets of second messengers whose composition changes with T cell maturation. Secondly, the αβ CD3 TCR complex itself may not always be of the same composition; indeed, there is evidence that the TCR complex initially expressed an immature thymocytes is composed differently from that of mature T cells (Groettrup et al., 1992). Thirdly, it is known that accessory molecules, coreceptors, and other cell interaction molecules may regulate TCR signalling (Bierer et al., 1989). It is thus possible that a differential involvement of these coregulators may in part determine the changing nature of the cellular responses during T cell differentiation (Eichmann et al., 1989). Examples for this latter mechanism will be discussed.

CD4 and CD8

CD4 and CD8 are transmembrane glycoproteins on MHC class II and class I restricted T cells, respectively (Parnes, 1989; Littman, 1987). They cooperate with the TCR in the recognition of MHC molecules presenting antigenic peptides (Emmrich et al., 1986). Cooperation has two components: Enhancement of the affinity of MHC/petide-TCR interaction, and signal transduction though the T cell membrane (Eichmann et al., 1989). CD4 is a single transmembrane polypeptide consisting of 4 extracellular Ig-like domains, a transmembrane region and a cytoplasmic tail, CD8 is a disulfide linked αβ heterodimer, each chain carrying a single N-terminal Ig domain, an elongated hinge region, a transmembrane and a cytoplasmic part (Parnes, 1989; Littman, 1987).

The cytoplasmic parts of CD4 and of the CD8α chain are associated with the protein tyrosine kinase p56^{lck} (lck) which is suspected to phosphorylate the CD3ζ chain during T cell activation (Veillette et al., 1988; Rudd et al., 1988). It is therefore reasonable to assume that the spatial relationships of the coreceptors to the TCR have a critical role in the phosphorylation of CD3ζ by lck. The associations of the cytoplasmic tails of CD4 and CD8 with lck are essential for their role in transmembrane signalling. Affinity enhancement by CD4 and CD8, on the other hand, is associated with their IgV-like domains which interact with MHC class II and class I molecules, respectively, on stimulator or target cells (Doyle and Strominger, 1987; Salter et al., 1989).

Affinity enhancement and transmembrane signalling are certainly interrelated functional activities of CD4 and CD8 in vivo. However, a host of experiments have demonstrated that both functions are in principle independent of one another.

Table 1. Functional regions on CD8 defined by monoclonal antibodies[1]

Antibody		Inhibition of	
CD8α	CD8β	activated CTL	resting T cells
53.6.72	H35-17	+	-
169.4	53.5	+	+
19/178	KT112	-	+

1) A detailed account of these experiments is published in Eichmann et al. (1991)

For CD8, these results suggest that specific target cell lysis by activated cytotoxic T cells (CTL) primarily depends on affinity enhancement, i.e. MHC binding. In contrast, activation of resting CD8$^+$ T cells depends primarily on transmembrane signalling, i.e. correct association of CD8 with the TCR (Eichmann et al., 1989). We have exploited these findings in attempts to localize regions on CD8 involved in one or the other function. Among a panel of mAb we have identified some mAb which only inhibit target cell lysis and others which inhibit only the activation of resting CD8$^+$ T cells, thus defining MHC binding regions and regions involved in interactions with the TCR, respectively (Eichmann et al., 1991). These results are summarized in Table 1. They suggest that both the α and β chains of CD8 participate in either interaction. Whereas there is no information on the localization of the CD8β epitopes, we know that all three CD8α epitopes are located within the V domain (Marion H. Brown, personal commununication).

CD4, CD8, and lck are involved in both negative and positive thymic selection (Fungleung et al., 1991). Using the collection of anti CD8 mAb described above we have investigated which of the interactions of CD8 are involved in positive selection. This was done by exposing fetal thymic organ cultures (FTOC) to these mAb and analysing their effect on the generation of CD8$^+$ SP cells. The results are summarized in Table 2 and show that all CD8 antibodies have a similar inhibitory effect, independent of their epitope specificity, and of their functional effects on peripheral T cells. These results suggest that both MHC binding and TCR association by CD8 are essential in positive thymic selection.

Table 2. Inhibition by anti-CD8 antibodies of the generation of CD8 SP cells in FTOC[1]

Antibody	α/β	% of Control[2]		maximal CD8[3] occupancy
		CD8SP	CD4SP	
53.6.72	α	19	79	75
169.4	α	11	99	100
19/178	α	5	108	100
H35-17	β	21	92	95
53.5	β	8	81	100
KT112	β	23	131	90

1) Thymic lobes were obtained from BALB/c mice at day 14 of gestation. They were incubated and treated with mAb as described by Yachelini et al. (1990). Analysis was performed by 3 color Flow cytometry using anti-CD8, anti-CD4, and anti-TCRβ antibodies.
2) SP cells expressing high TCRβ density are given as % of control FTOC cultured without mAb.
3) The maximal percent CD8 occupancy reached with each anti-CD8 mAb are given. Concentrations giving maximal occupancy varied between 7.5 µg/ml (169.4) and 200 µg/ml (53.5). Some antibodies do not
   reach 100% occupancy at even higher concentrations.

On resting peripheral T cells, we have found by immunoprecipitation experiments that CD4 and CD8 are in constant association with the CD3 complex, most notably with the CD3δ chain (Suzuki et al., 1992). We have concluded from these experiments that de novo assembly of a complex of the TCR with the coreceptors does not have to take place. Instead, we have suggested a Flip-Flop model in which upon contact of the TCR with MHC/peptide CD4 and CD8 change their position within the complex. By moving from CD3δ to CD3ζ, the coreceptor may shift lck into the correct position to achieve phosphorylation of CD3ζ. The association of the coreceptor with CD3δ may keep lck in an inactive position to maintain the resting state. Activation of resting peripheral T cells may thus be related to a proper positioning of the coreceptors within the TCR complex.

In order to investigate the spatial relationships of the TCR and coreceptors in developing thymocytes we performed immunoprecipitation experiments with purified CD4$^+$8$^+$ double positive thymocytes. Table 3 summarizes the results from a series of experiments in which the amounts of CD3 components coimmunoprecipitated with antiCD4 and antiCD8 in spleen T cells and CD4$^+$8$^+$ thymocytes are compared (K.M. Saizawa et al., manuscript in preparation).

Table 3. Comparison of CD3 components coimmunoprecipitated with antiCD3ε, antiCD4 and antiCD8 in splenic T cells and immature thymocytes

Antibody[1]		Coimmunoprecipitation of CD3 δ	ε	γ
aCD3ε	spleen T[2]	+++	+++	+++
	thymic CD4$^+$8$^+$	++	++	++
aCD4	spleen T	+++	+	+
	thymic CD4$^+$8$^+$	(+)	−	−
aCD8	spleen T	+++	+	+
	thymic CD4$^+$8$^+$	(+)	−	(+)

1) aCD3:145-2C11; aCD4:GK1.5, aCD8:169.4
2) splenic T cells were purified by Nylon wool passage; CD4$^+$8$^+$ double positive thymocytes were enriched as low density fraction after density gradient centrifugation of whole thymocytes. Cells were surface-labelled with ^{125}I using water soluble Bolten-Hunter reagent, immunoprecipitated with the indicated antibodies and subjected to 2-dimensional non-reduced, reduced SPS-PAGE gel electrophoresis (for methods see Suzuki et al., 1992). The strength of δ, ε and γ spots was judged from 6-8 experiments.

The results suggest that in CD4$^+$8$^+$ double positive thymocytes there is little if any association between the coreceptors and the TCR. We conclude that this association is established during further maturation, at some point between the two differentiation stages studied. Interpretation of these results must invoke changes in the composition of the TCR or in the structure of its members during differentiation.

The comparison of the functions of coreceptors for peripheral T cells and developing thymocytes and the analyses of their changing association with the TCR suggests an intriguing sequence of events in the maturation of the TCR/coreceptor complex. The earliest component expressed on CD4$^+$8$^+$ thymocytes appears to be CD3ε (Levelt et al., 1992). Signal transduction through this immature complex, perhaps including TCRβ homodimers (Groetrup et al., 1992) induces acquisition of the CD4$^+$8$^+$ double position state (Levelt et al., 1992) The αβCD3 complex gets expressed at this state but remains non-associated with CD4 and CD8 until mature T cells appear after positive selection. It is intriguing to speculate that the association of the TCR with the proper coreceptor gets established by interaction of both with the proper MHC molecule and is, indeed, the hallmark of positive selection. Such a mechanism is supported by our antibody inhibition studies showing that positive selection is inhibited by antiCD8 antibodies that block MHC binding as well as by those that block proper positioning of CD8. Mature T cells, in contrast, no longer require de novo assembly of a TCR/coreceptor complex but nevertheless need the proper positioning of the coreceptor. Our observation that antiCD8 antibodies blocking MHC binding do not interfere with activation of resting CD8$^+$ T cells suggest that MHC binding by the coreceptor is not essential for its proper positioning within a preformed TCR/coreceptor complex.

Other adhesion/signalling receptors

Activation of peripheral T lymphocytes by MHC/antigen depends, in addition to specific recognition, on several receptor-ligand interactions which either mediate cell adhesion alone (i.e. LFA-1

Table 4. Effects of mAb to several cell interaction molecules on peripheral T cells and on the generation of SP cells in FTOC[1]

| Antigen | Ab | Peripheral T cells[2] | | FTOC[3] | |
		inhib.	act.	CD8SP	CD4SP
CD44	I42.5	−	−	71	110
CD45	M1/9.3	+	(+)	21	18
CD2	12−15	+	(+)	16	43
Thy1	30−H12	−	(+)	24	22
CD5	53.7	−	(+)	29	75
CD11a	FD18.5	+	−	20	62
CD4	GK1.5	+	−	125	10

1) See footnote 1, Table 2.
2) Consensus results summarized. (+): Result not regularly observed or only with special mAb.
3) See footnote 2, Table 2. Inhibition of less than 30% is within the variability of the method.

with ICAM1,2) or, in addition, are suspected to have a role in signal transduction (i.e. CD2 with LFA-3) (Springer, 1990). CD45, a transmembrane tyrosine phosphatase presumably regulating autophosphorylation of lck, also may belong to the latter group although its ligand is yet unknown (Thomas, 1989; Ledbetter, 1991). Moreover, a number of prominent T cell surface molecules are suspected to have a role in cell interactions without conclusive evidence existing (i.e. Thy1, CD5). A simple way of testing the functional role of cell surface molecules is to use mAb either for direct induction of activation or for inhibition of activation of T cells by the appropriate MHC/antigen combination. A summary of data from many laboratories for several adhesion/signalling receptors on peripheral T cells is given in Table 4, (for review see Möller, 1989) and compared with results from our own experiments testing the same mAb for influences on thymocyte development in FTOC. The table contains only those antibodies that leave thymocyte development unperturbed upto the end of the DP stage, and whose effects are limited to the generation of SP cells, i.e. to positive selection. The results are surprising in several ways. Firstly, positive thymic selection appears to depend on a multitude of cell interaction systems: Inhibition is observed not only for mAb to antigens that strongly inhibit activation of peripheral T cells (LFA-1, CD2) but also for mAb to antigens that usually have little effect in the periphery (Thy1, CD5). Among the tested mAb, only anti PgP-1 (CD44) did not inhibit the generation of SP cells. Secondly, the generation of CD8 SP cells appears to depend on a greater number of cell interaction systems than that of CD4 SP cells. In particular, LFA-1 (CD11a) and CD5 appear to be required for CD8 SP but not for CD4 SP cells. The possibility that CD8 SP cells are non-specifically sensitive to rat Ig is excluded by our observation that rat anti-CD4 mAb inhibits the generation of CD4 but not of CD8 SP cells, and by the lack of inhibition by anti-PgP-1 (Table 4) and normal rat Ig (not shown). The third surprising aspect of these findings is that none of the tested mAb affected thymocyte development up to the DP stage. This suggests that earlier developmental steps proceed on the level of the individual thymocyte or depend on cell interactions by thus far unknown receptor systems.

Taken together, the results suggest that cellular interactions leading to T cell maturation and activation involve structurally and functionally highly dynamic receptor-ligand systems. The elucidation of the molecular changes in transmembrane signalling during T cell differentiation may help in the understanding of the multitute of cellular responses elicited in T lymphocytes.

REFERENCES

Bierer BE, Sleckman BP, Ratnofsky SE, Burakoff SJ (1989) Ann Rev Immunol 7:579

v Boehmer HA (1988) Ann Rev Immunol 6:309

Doyle C, Strominger JL (1987) Nature 330:356

Eichmann K, Boyce NW, Schmidt-Ulrich R, Jönsson JI (1989) Immunol Rev 109:39

Eichmann K, Ehrfeld A, Falk I, Goebel H, Kupsch J, Reimann A, Zgaga-Griesz A, Saizawa KM, Yachelini P, Tomonari K (1991) J Immunol 147:2075

Emmrich F, Strittmatter U, Eichmann K (1986) Proc Natl Acad Sci USA 83:8298

Fungleung WP, Schilham MW, Rahemtulla A, Kundig TM, Vollenweider M, Potter J, v Ewijk W, Mak TW (1991) Cell 65:443

Groettrup M, Baron A, Griffiths G, Palacios R, v Boehmer H (1992) EMBO J 11:2735

Ledbetter JA, Schieven GL, Uckum FM, Imboden JN (1991) J Immunol 146:1577

Levelt C, Ehrfeld A, Eichmann K (1992) J Exp Med, in press

Littman DR (1987) Ann Rev Immunol 5:561

Möller G Ed (1989) Immunol Rev 111

Parnes JR (1989) Adv Immunol 44:265

Rudd CE, Trevillian JM, Dasgupta JV, Wong LL, Schlossman SF (1988) Proc. Natl Acad Sci USA 85:5190

Salter RD, Norment AM, Chen BP, Clayberger C, Krensky AM, Littman DR, Parham P (1989) Nature 338:1

Springer T (1990) Nature 346:425

Suzuki S, Kupsch J, Eichmann K, Saizawa, M (1992) Eur J Immunol 22:in press

Thomas ML (1989) Ann Rev Immunol 3:143

Veilette A, Bookman MA, Horak EM, Bolen JB (1988) Cell 55:301

Yachelini P, Falk I, Eichmann K (1990) J Immunol 145:1382

The Role of Co-receptors in T Cell Activation

S. J. Burakoff[1, 3] T. Collins[1], W. C. Hahn1, J. K. Park[1],
B. P. Sleckmann[1, 5], V. Igras[1], Y. Rosenstein[1, 4], and B. E. Bierer[1, 2, 5]

[1]*Division of Pediatric oncology, Dana Farber Cancer Institute, Boston, MA, USA*
[2]*Hematology-Oncology Division, Departement of Medicine, Brigham and Women's Hospital, Boston, MA USA*
Departements of [3]*Pediatric,* [4]*Pathology and* [5]*Medicine, Harvard Medical School, Boston, MA, USA*

INTRODUCTION

Although the expression of the T cell receptor-CD3 complex (TcR-CD3) confers T lymphocyte antigen specificity, several other integral membrane proteins play significant roles during T cell-mediated cellular immune responses (Bierer *et al.*, 1989). By initiating signal transduction events and/or increasing the strength of adhesion between T cells and antigen presenting cells, these cell surface molecules cooperate with the TcR-CD3 complex to initiate biochemical signals that lead to efficient T cell activation and subsequent effector function. By binding their cognate ligands on antigen presenting cells, co-receptor molecules such as CD2, CD4 and CD43 play important roles in T cell adhesion and activation. Each of these molecules, however, contribute uniquely to the T cell response. The expression of these molecules as well as of mutated forms of these receptors in appropriate recipient cell lines has helped document the role of each of these molecules in T cell activation.

To study the role of the human co-receptors CD4, CD2 and CD43 in isolation from other human T cell surface proteins, we have expressed these molecules in an antigen-specific murine T cell hybridoma, By155.16, generated from C57Bl/6 mice primed *in vivo* with the human HLA-DR+ B cell line JY (Sleckman *et al.*, 1987). By155.16 was selected for its ability to retain antigen reactivity to stimulation with HLA-DR antigens, as assessed by IL-2 production. Human wild-type and mutated co-receptor molecules were stably introduced by DNA transfection followed by selection in the antibiotic G418. Cell lines expressing these co-receptors and equivalent levels of the TcR-CD3 complex were chosen for further analysis.

CD4 is a 55 kDa glycoprotein expressed mainly on T lymphocytes bearing TcRs which are specific for antigen in the context of MHC class II molecules. CD4 binds MHC class II antigen in the $\beta 2$ domain, a region which is similar to the CD8 binding site on class I antigens (Konig *et al.*, 1992, Cammarota *et al.*, 1992). Binding of CD4 to class II enhances T cell responsiveness to antigen (Sleckman *et al.*, 1987; Doyle *et al.*, 1987); similarly, cross-linking CD4 with the TcR/CD3 complex by monoclonal antibodies results in dramatically increased T cell stimulation compared to cross-linking of the TcR/CD3 complex alone (see Bierer *et al.*, 1989). Stimulation of the T cell by anti-TcR or anti-CD3 antibodies causes association of CD4 with the TcR/CD3 complex (Rojo *et al.*, 1989, Mittler *et al.*, 1989), suggesting that co-localization of CD4 with the TcR/CD3 complex during stimulation is necessary for optimal CD4-dependent antigen responsiveness.

In order to examine the role of CD4-mediated adhesion in TcR-dependent stimulation, we generated a glycolipid-anchored mutant of CD4 (CD4PI) which contains the entire extracellular domain of CD4 fused to the carboxyterminal 28 amino acids of the glycolipid-linked form of LFA-3 (Sleckman *et al.*, 1989). CD4PI was expressed in the murine T cell hybridoma By155.16, and the ability of this molecule to enhance the responsiveness of the hybridomas to HLA-DR was tested (Table 1). The response of the By155.16 hybridomas to HLA-DR is markedly enhanced by expression of wild-type CD4 (Sleckman *et al.*, 1988). However, of 40 independent lines expressing CD4PI, none demonstrated an increased level of IL-2 production in response to HLA-DR antigens. All cell lines were equally responsive to plate-bound anti-CD3 antibody (145-2C11), indicating that the inability to enhance an antigen response was not due to a defect in signal transduction through the TcR-CD3 complex.

To determine whether CD4PI could bind MHC class II antigen, CD4PI[+] hybridomas were tested for their ability to form conjugates with Daudi cells (Table 1). Despite their lack of CD4-dependent antigen responsiveness, the ability of the CD4PI[+] cells to form conjugates with Daudi cells via MHC class II molecules was comparable to that of cells expressing wild-type CD4. These data indicate that the enhanced level of antigen responsiveness resulting from expression of CD4 involves signal transduction mechanisms in addition to cell-cell adhesion.

Table 1. Role of CD4 cytoplasmic domain in T cell activation

Stimulation of IL-2 Production

	α-CD3	Daudi	Conjugate formation	TcR-CD3/CD4 Association (FRET)
By155.16	+	+	-	-
CD4	+	+ +	+	+
CD4PI	+	-	+	-

CD4 is noncovalently associated with the protein tyrosine kinase, $p56^{lck}$, via cysteines in the cytoplasmic domain of CD4 and in the aminoterminal end of $p56^{lck}$. $p56^{lck}$ has been proposed to be involved in CD3ζ phosphorylation (Veillette *et al.*, 1989; Nakayama *et al.*, 1989; Barber *et al.*, 1989) as well as in MAP kinase phosphorylation and activation (Ettehadieh *et al.*, 1992). It has also recently been demonstrated that $p56^{lck}$ can associate with phosphorylated GTPase activating protein (GAP) (Amrein *et al.*, 1992) and that the CD4/$p56^{lck}$ complex can associate with a GTP-binding protein (Telfer *et al.*, 1991). Thus $p56^{lck}$ appears to be an integral component of the T cell response, and it has been speculated that $p56^{lck}$ may play a role in mediating CD4 signal transduction.

To test this hypothesis we have studied the effects of mutations in CD4 which abolish its association with $p56^{lck}$ on the ability of CD4 to function as a signaling molecule (Collins *et al.*, 1992). CD4 molecules containing point mutations in the cytoplasmic domain with cysteine to serine changes at positions 420, 422, and 430 (CS420, CS422, CS430) were generated and expressed in the HLA-DR specific murine hybridoma By155.16 (Table 2). CS430 mutants continue to associate with $p56^{lck}$ whereas the mutants CS420 and CS422 are unable to associate with $p56^{lck}$.

The ability of hybridomas expressing CS420, CS422 and CS430 to co-localize with the TcR/CD3 complex using fluorescence resonance energy transfer (FRET) was assessed. Using FRET, both CS430 and wild-type CD4 co-localized with the TcR/CD3 complex while CS420 and CS422 failed to co-localize. These results suggest that association of $p56^{lck}$ with CD4 is necessary for the co-localization of CD4 with the TcR/CD3 complex. We have also assessed CS420, CS422 and CS430 expressing hybridomas for their ability to produce IL-2 in response to HLA-DR antigens expressed on the human B cell line, Daudi. Cells expressing CS420 or CS422, which lack association with p56^{lck}, failed to enhance the response of the hybridoma to HLA-DR antigens. These results are consistent with those of Glaichenhaus, *et al.*, (1991). Expression of CS430, however, caused an enhancement in antigen responsiveness which was equivalent to wild-type CD4$^+$ hybridomas. These hybridomas were also evaluated in antibody cross-linking experiments. Cells expressing CS420, CS422 or CS430 were incubated with submitogenic concentrations of either anti-TcR antibody (F23.1) or anti-CD3 antibody (145-2C11), plus saturating amounts of anti-CD4 antibody (Leu3a) then cross-linked with second antibody. CS430 expressing cells responded vigorously when both anti-CD3 and anti-CD4 mAb were cross-linked. However, CS420 and CS422 expressing cells failed to respond to such stimulation. These data indicate that p56^{lck} is involved in at least two roles in CD4-dependent stimulation: mediating the colocalization between CD4 and the TcR/CD3 complex, and enhancing T cell responses to CD4-dependent stimuli.

Table 2. Role of CD4-p56^{lck} Association in T cell Activation

| | Stimulation of IL-2 Production | | | Association p56^{lck} | TcR-CD3/CD4 Association (FRET) |
	α-CD3	Cross-linked α-CD3 + α-CD4	Daudi		
By155.16	+	-	-	-	-
CD4	+	++	++	+	+
CS420	+	-	-	-	-
CS422	+	-	-	-	-
CS430	+	++	++	+	+

The T cell glycoprotein C02, a member of the immunoglobulin superfamily, provides the T cell with a major pathway of activation and adhesion. By introducing α human wild-type and mutated CD2 molecules into the murine T cell hybridoma, By155.16, structure-function analyses of the CD2 molecule have been performed. Initial studies have demonstrated that although the cytoplasmic domain is necessary for CD2-mediated signaling events, the carboxyterminal 40 amino acids can be removed without affecting CD2-dependent Ca^{++} mobilization, cAMP production or IL-2 production (Bierer *et al.*, 1988b, 1990; Chang *et al.*, 1989; Hahn *et al.*, 1991).

Recently, we have studied the interaction of CD2 with its ligand, lymphocyte function associated antigen-3 (LFA-3, CD58). CD2$^+$ hybridomas, but not parental CD2$^-$ cells, bound immunoaffinity purified LFA-3 absorbed to polystyrene plates. MAbs to either CD2 or LFA-3 completely inhibited this binding, demonstrating the specificity of the binding assay (Hahn *et al.*, 1991, 1992). Using this model system, we have studied the effect of T cell activation on the CD2/LFA-3 interaction. Activation of purified human T cells or CD2$^+$ T cell hybridomas by cross-linking of the TcR-CD3 complex with mAbs rapidly upregulates the avidity of CD2 for purified LFA-3 two- to four-fold (Hahn *et al.*, 1992). Physiological activation of T cell hybridomas with purified MHC class II (HLA-DR) antigens resulted in increased avidity of CD2 for LFA-3. Interestingly, treatment of T lymphocytes with the phorbol ester PMA upregulates

CD2 avidity, suggesting that protein kinase C plays a direct role in the regulation of CD2 avidity.

The role of the cytoplasmic domain in TcR-mediated regulation of CD2 avidity has been assessed. CD2 cytoplasmic deletion mutants lacking 100, 40, 35, 30, 11 and 5 carboxyterminal amino acids were introduced into the murine T cell hybridoma. When these CD2 mutant expressing cell lines were tested for their ability to bind purified LFA-3, each of these CD2 deletion mutants bound LFA-3 equivalently, but none of the deletion mutants were able to upregulate CD2 avidity for LFA-3 after TcR-CD3 or PMA stimulation. Hybridomas expressing mutated CD2 molecules lacking five carboxyterminal amino acids were still capable of efficient IL-2 production in response to either stimulatory anti-CD2 mAbs or anti-CD3 mAb, indicating that TcR-mediated regulation of CD2 avidity and CD2-mediated IL-2 production are dependent upon distinct regions of the CD2 cytoplasmic domain.

Further mutagenesis of the five carboxyterminal amino acids of CD2 has identified the carboxyterminal Asn (Asn[327]) as essential for CD2 avidity regulation. Hybridomas expressing CD2 with substitutions of alanine at Pro[324] (CD2-P324A), Ser[326] (CD2-S326A), and Asn[327] (CD2-N327A) or of aspartic acid at Asn[327] (CD2-N327D) were tested for their ability to upregulate CD2 avidity. TcR-CD3 stimulation of cell lines expressing CD2 mutations at positions 324 and 326 upregulated CD2 avidity equivalently to cell lines expressing wild-type CD2. In contrast, hybridomas expressing CD2 with substitutions at Asn[327] were defective in their ability to upregulate CD2 avidity (Table 3).

Table 3. Function of hybridomas expressing single amino acid substitutions in CD2 cytoplasmic domain

| Cell line | Sequence of C-terminal AA (323-327) | Stimulation of IL-2 Production | | | TcR-regulation of CD2 avidity |
		α-CD3	α-CD2	Daudi	
By155.16		+++	-	-	-
CD2	SPSSN	+++	++	+++	+
CD2-P324A	SASSN	+++	++	+++	+
CD2-S326A	SPSAN	+++	++	+++	+
CD2-N327A	SPSSA	+++	++	+	-
CD2-N327D	SPSSD	+++	++	+	-

The response of these hybridomas to antigen stimulation was tested (Table 3). Hybridomas expressing wild-type CD2 or CD2 mutants with single amino acid substitutions in the 5 carboxyterminal amino acids were equivalent in their ability to produce IL-2 in response to stimulation of the TcR-CD3 complex via anti-CD3 mAb or of the CD2 molecule itself via stimulatory pairs of anti-CD2 mAbs. Expression of wild-type CD2 greatly enhances the ability of this hybridoma to make IL-2 in response to Daudi stimulation, demonstrating that CD2 can act as a co-receptor in TcR-mediated responses (Bierer et al., 1988a; 1988b; Hahn et al., 1992). Although wild-type CD2, CD2-P324A and CD2-S326A were equivalent in their ability to respond to Daudi stimulation, both CD2-N327A and CD2-N327D were deficient in their response to antigen stimulation (Table 3). These observations suggests that the ability to upregulate CD2 avidity for its ligand correlates directly with antigen responsiveness and thus with its ability to function as a co-receptor in TcR-mediated T cell activation.

These studies emphasize the dual role that CD2 plays during T cell interactions with antigen presenting cells. CD2-mediated signal transduction appears to play an important role in enhancing TcR signaling, perhaps augmenting or amplifying second messenger signals which lead to T cell activation. In addition, the interaction of CD2 with LFA-3 is clearly important in facilitating adhesion between T cells and their targets (Shaw *et al.*, 1986; Spits *et al.*, 1986). In murine T cell hybridomas, the introduction of CD2 greatly enhances antigen specific responses. TcR-CD3 signaling modulates the avidity with which CD2 binds LFA-3, and studies using cell lines expressing mutated CD2 molecules demonstrate that CD2 avidity regulation has a profound effect on co-receptor function of CD2.

CD43 is a heavily sialylated transmembrane glycoprotein that is defective on the T cells of patients with the Wiskott-Aldrich syndrome, a genetic deficiency. The role of the CD43 molecule in T cell activation has also been assessed. Initial studies have investigated CD43 dependent T cell stimulation with different anti-CD43 mAbs. Two anti-CD43 mAbs, B1B6 and E11B, did induce IL-2 production from human peripheral blood T cells; the addition of either mAb to cells that were co-stimulated with PMA, OKT3, ConA or PHA dramatically amplified T cell proliferation (Axelsson *et al.*, 1988). Another mAb, L10, has been shown to induce the proliferation of peripheral blood T cells without the need for additional activation signals (Mentzer *et al.*, 1987).

To assess the ability of CD43 to function as a co-receptor molecule, the dependence of the By155.16 hybridoma on the presence of accessory molecules for its activation was again exploited. By155.16 hybridoma cells expressing equivalent amounts of TcR-CD3 complex and CD43 wild-type (CD43$^+$) or of a CD43 mutant that lacked the intracytoplasmic domain of the molecule (CD43Δ^+) were screened for their ability to generate IL-2 following antigen-specific stimulation. Hybridomas that expressed the wild-type CD43 molecule were able to produce IL-2 when stimulated with Daudi cells. The CD43Δ^+ hybridomas failed to respond to Daudi stimulation although they produced IL-2 in response to anti-CD3 mAb stimulation equivalent to the wild-type CD43$^+$ cells. These data indicate that CD43 can be considered a co-receptor molecule in the activation of T cells, and that the intracellular domain of CD43 is required for co-stimulation. Furthermore, in the absence of TcR-CD3 expression, CD43 mediated signals were not sufficient to induce activation of the T cells (Park *et al.*, 1991). However, preliminary data suggest that cross-linking of wild-type CD43 and the TcR/CD3 complex by mAb results in significant enhancement of T cell activation.

The ability of affinity purified plate-bound CD43 to bind Daudi cells enabled us to search for a ligand for CD43 on Daudi cells. A panel of 56 monoclonal antibodies raised against B cell surface molecules were screened for their ability to block Daudi binding to CD43. The interaction of Daudi cells with plate-bound CD43 was consistently inhibited by anti-ICAM-1 mAbs, suggesting that ICAM-1 might be a ligand for CD43. Experiments were performed in which the binding of CD43$^+$ hybridoma cells to immobilized purified ICAM-1 was assessed. CD43$^+$, but not CD43$^-$, hybridomas bound specifically to ICAM-1 absorbed to polystyrene plates, providing additional evidence that CD43 was capable of mediating a specific interaction with ICAM-1. The ability of the two isolated molecules to interact with each other was confirmed by a conjugate assay that demonstrated the ability of beads expressing CD43 to form conjugates with beads expressing ICAM-1 (Rosenstein *et al.*, 1991). Presently, experiments are underway to determine whether ICAM-1 is the ligand used by CD43 in its role as a co-receptor in T cell activation.

In conclusion, T cell co-receptor molecules may enhance T cell activation by regulating cellular interactions and/or by transducing signals that synergize with signals mediated by the TcR/CD3 complex. Molecules such as LFA-1 and VLA-4 appear to have their major effect by increasing intercellular adhesion. CD4 and CD43 seem to exert a minimal effect on intercellular adhesion, but have a dominant effect on intracellular T cell signaling. It is noteworthy that the same ligand, ICAM-1, is apparently used by CD43 predominantly for signaling and by LFA-1 for adhesion strengthening. CD2 appears to be a multi-functional molecule, playing a role both in adhesion and in signaling. A structure-function analysis of CD2 reveals that these two functions can be localized to separable domains within the cytoplasmic tail. The complex interplay of these distinct receptor-ligand interactions regulate T cell activation and confers flexibility to the T cell immune response.

REFERENCES

Amrein KE, Flint N, Panholzer B, and Burn P (1992) Proc Natl Acad Sci USA 89:3343

Axelsson B, Youseffi-Etemad R, Hammerstrom S, and Perlmann P (1988) J Immunol 141:2912

Barber EK, Dasgupta JD, Schlossman SF, Trevillyan JM, and Rudd CE (1989) Proc Natl Acad Sci USA 86:3277

Bierer BE, Bogart RE, and Burakoff SJ (1990) J Immunol 144:785

Bierer BE, Peterson A, Barbosa J, Seed B, and Burakoff SJ (1988a) Proc Natl Acad Sci USA 85:1194

Bierer BE, Peterson A, Gorga JC, Herrmann SH, and Burakoff SJ (1988b) J Exp Med 168:1145

Bierer BE, Sleckman BP, Ratnofsky SE, and Burakoff SJ (1989) In: Paul WE, (ed) Annual Review of Immunology, Palo Alto: Annual Reviews, Inc 7:579

Cammarota G, Scheirle A, Takacs B, Doran DM, Knorr R, Bannwarth W, Guardiola J, and Sinigaglia F (1992) Nature 356:799

Chang H-C, Moingeon P, Lopez P, Krasnow H, Stebbins C, and Reinherz EL (1989) J Exp Med 169:2073

Collins TL, Uniyal S, Shin J, Strominger JL, Mittler RS, and Burakoff SJ (1992) J Immunol 148:2159

Doyle C and Strominger JL (1987) Nature 330:256

Ettehadieh E, Sanghera JS, Pelech SL, Hess-Bienz D, Watts J, Shastie N, and Aebersold R (1992) Science 255:853

Glaichenhas N, Shasti N, Littman DR, and Turner JM (1991) Cell 64:511

Hahn WC, Rosenstein Y, Burakoff SJ, and Bierer BE (1991) J Immunol 147:14

Hahn WC, Rosenstein Y, Calvo V, Burakoff SJ, and Bierer BE (1992) Proc Natl Acad Sci USA 89:7179

Konig R, Huang LV, and Germain RN (1992) Nature 356:796

Mentzer SJ, Remold-O'Donnell E, Crimmins MA, Bierer BE, Rosen FS, and Burakoff SJ (1987) J Exp Med 165:1383-1392

Mittler RS, Goldman SJ, Spitalny GL, and Burakoff SJ (1989) Proc Natl Acad Sci USA 86:8531

Nakayama T, Singer A, Hsi ED, and Samelson LE (1989) Nature 341:651

Park JK, Rosenstein Y, Remold-O'Donnell E, Bierer BE, Rosen FS, and Burakoff SJ (1991) Nature 350:706

Rojo JM, Saizawa K, and Janeway CA (1989) Proc Natl Acad Sci USA 86:3311

Rosenstein Y, Park JK, Hahn WC, Rosen FS, Bierer BE, and Burakoff SJ (1991) Nature 354:233

Shaw S, Luce GEG, Quinones R, Gress RE, Springer TA, and Sanders ME (1986) Nature 323:264

Sleckman BP, Peterson A, Jones WK, Foran JA, Greenstein JL, Seed B, and Burakoff SJ (1987) Nature 328:351

Sleckman BP, Peterson A, Foran JA, Gorga JC, Kara CJ, Strominger JL, Burakoff SJ, and Greenstein JL (1988) J Immunol 141:49

Sleckman BP, Rosenstein Y, Igras V, Greenstein JL and Burakoff SJ (1991) J Immunol 147:428

Spits H, van Schooten W, Keizer H, van Seventer G, van de Rijn M, Terhorst C, and de Vries JE (1986) Science 232:403

Telfer JC and Rudd CE (1991) Science 254:441

Veillette A, Bookman MA, Horak EM, Samelson LE, and Bolen JB (1989) Nature 338:257

Integrins and their Activation

N Hogg, C. Cabañas, J. Harvey, A. McDowall, P. Stanley, M. Stewart and R.C. Landis

Macrophage Laboratory, Imperial Cancer Research Fund, 44 Lincoln's Inn Fields, London WC2A 3PXC, UK

At present the integrin superfamily consists of 15 α and 8 β subunits found in 20 heterodimeric combinations of receptor (Hynes 1992; Pardi et al 1992; Springer 1990). Leukocytes express at least 13 of these receptors (Hogg 1991). However, it is only the β2 integrins LFA-1, CR3/Mac-1 and p150,95 and the "homing receptor" β7 integrins, α4β7 and αHML-1 β7 which are exclusively found on leukocytes. The VLA integrins of the β1 family which have specificity for extracellular matrix proteins such as fibronectin and collagen are expressed by leukocytes and also by other types of cells. This wide range of integrins renders immune cells well-equipped to enter into cell-cell adhesions and to interact with the subendothelial matrix. Many of these receptors are expressed at enhanced levels in memory T cells compared to naive T cells (Buckle and Hogg 1990; Sanders et al 1988). For example, the level of LFA-1 increases from 50-60,000 to 300,000 sites per memory T cell. Adhesion receptors are thought to have a major role in the response of memory T cells to antigen by enabling them to act rapidly through the promotion of cell-cell contact and also by facilitating movement into tissues.

The N-terminus of the integrin α subunit is characterised by 7 homologous domains (a repeating motif of ~60 amino acids) of which the last three or four are similar to EF hand type divalent cation binding sites. Integrins are known to require bound metal ions in order to function and in particular, LFA-1 makes use of Mg^{2+} (Marlin and Springer 1987; Martz 1980). All three β2 integrins together with VLA-1 and VLA-2 contain an "inserted" or "I" domain of approximately 200 amino acids situated between homologous domains II and III (Larson et al 1989). The role of this domain in integrin function remains to be elucidated.

THE LFA-1 LIGANDS

LFA-1 binds to intercellular adhesion molecule (ICAM)-1 as its principle counter receptor or ligand (Marlin and Springer 1987; Rothlein et al 1986; for review see (Hogg 1991). T cells use this receptor/ligand pair to increase the strength of adhesion to many types of targets. ICAM-1 also serves as a binding partner for a number of other molecules, namely a second β2 integrin CR3, CD43 and for the pathogens rhinovirus and *Plasmodium falciparum* in its infected erythrocyte stage. ICAM-1 uses the first two of its five domains to bind to its "ligands" with the exception of CR3 which is also bound by the third domain of ICAM-1 (Diamond et al 1991) (no information about CD43). LFA-1 binds to the front face or the CFG β sheet of domain 1 whereas malaria infected erythrocytes bind on the opposite face (Berendt et al 1992; Ockenhouse et al, 1992). Studies such as these are beginning to give us clearer ideas about the nature of molecular interactions between receptors and their ligands.

LFA-1 has two other known ligands, ICAM-2 and ICAM-3 about which less is known (de Fougerolles and Springer 1992; Staunton et al 1989). ICAM-1 and 2 are members of the Ig superfamily (IgSF) and the molecular structure of ICAM-3 remains to be established. The 3 different LFA-1 ligands may be specialised for different aspects of LFA-1 dependent leukocyte interactions. Thus, ICAM-1 which is dramatically upregulated in response to inflammatory cytokines on endothelial cells, is proposed to be important in cell migration into tissues during inflammation in the secondary immune response. In contrast ICAM-2 shows constitutive expression on endothelial cells and may be essential for normal leukocyte recirculation. ICAM-3, which is absent from endothelium, is expressed at a much higher level than ICAM-1 or 2 on resting T cells and monocytes and may be important in initiating an immune response.

The existence of a number of ligands for a single integrin is a common feature of the integrin superfamily. For example, leukocyte integrins such as VLA-4 recognise both the IgSF protein, VCAM-1 and the matrix protein fibronectin (Pulido et al 1991). So far, the β2 integrin LFA-1 appears to be unique amongst integrins in recognising only IgSF-type ligands and participating exclusively in cell-cell interactions. However the "I" domain of the β2 integrins has substantial homology to collagen binding proteins such as cartilage matrix protein and von Willebrand's factor proteins (Larson et al 1989) and the other "I" domain-containing β1 receptors, VLA-1 and VLA-2 do bind to collagen (Hemler 1990). These observations have caused speculation that collagen may serve as an additional non-IgSF ligand for the β2 integrins. We have recently investigated this possibility by both assessing the ability of intact leukocytes to bind collagen and examining T cell lysates for binding activity using collagen affinity columns (Goldman et al 1992). Three pieces of evidence indicate that β2 integrins do not mediate collagen binding. Collagen affinity columns failed to elute β2 integrins from T cell and neutrophil lysates. The ability of intact T cells (and other leukocytes) to bind to collagen is restricted to β1 integrin VLA-2. Finally neutrophils which lack β1 integrins (VLA-1, 2 and 3) do not bind to collagen. Thus the "I" domain appears not to confer collagen binding activity on β2 integrins.

Control of ligand binding may be conferred at the level of differential ligand expression as is thought to be the case for the ICAMs described above. Alternatively, receptor specificity may be determined by cell type ie VLA-2 on platelets binds collagen only, whereas on endothelial cells VLA-2 binds to collagen and laminin (Kirchhofer et al 1990). An additional point of interest from our collagen binding studies on T cells, was that VLA-1 and VLA-3, which work as collagen receptors on other cells, were expressed at levels comparable to VLA-2 but failed to bind collagen (Goldman et al 1992). Such variation in receptor specificity is not fully understood but may arise due to alternative splice variations, different membrane lipid composition or variable divalent cation usage.

LFA-1/ICAM-1 ADHESION IS TRANSIENT

A distinctive feature of the β2 integrins on leukocytes which is distinct from the extracellular matrix protein binding integrins, is the fact that they normally bind to ligand with low affinity and are not constitutively active. Activation of LFA-1/ICAM-1 interaction could be potentially controlled in several ways. One possible option would be to increase surface expression of the molecule thereby increasing the opportunity for receptor interaction. ICAM-1 levels appear to be controlled in this manner since expression is increased by the cytokines that are produced during T cell activation (for review see Hogg et al 1991). Another means of increasing the effective concentration of a membrane protein like ICAM-1 is by clustering. This appears to be the situation in some T and B cell lines, as ICAM-1 is concentrated at the ends of the cellular processes with which these cells adhere to opposing surfaces (Dougherty et al 1988; Dustin et al 1992). In contrast, LFA-1 membrane expression levels are fairly constant throughout the activation process. The initiating signal for LFA-1/ICAM-1 is determined by the LFA-1 side of the reaction in which LFA-1 is induced to bind more avidly to ICAM-1 (Dustin and Springer 1989; van Kooyk et al 1989) (see The "Inside-Out" Signal). Furthermore this increase in binding strength of receptor for ligand is transient, reaching optimal levels within 15-20 minutes of receiving the cellular activating signal and returning to basal levels by 2 hrs. The means by which transience in integrin activation is achieved is unknown but one recent suggestion has been that it might be controlled by a lipid-like factor which is readily metabolised. This lipid, named the "integrin modulating factor-1" or IMF-1, increases in neutrophils in a manner parallel to activation of the β2 integrin CR3 (Hermanowski-Vosatka et al 1992). A role for IMF-1 in activation is suggested by its ability to confer iC3b binding ability on isolated CR3 molecules.

The use of monoclonal antibodies (mAb), which can act as inducers for or indicators of the activated state, has shed much light on the activation process. We have described mAb 24 that binds to an epitope present on the α subunits of all three β2 integrins (Dransfield and Hogg 1989). Recognition of this epitope is associated with Mg^{2+} binding and, on the surface of the intact cell, expression parallels receptor activity. Therefore mAb 24 has been useful as a "reporter" of receptor activation. Recent studies have shown that mAb 24 interferes with leukocyte integrin activation in a manner which has offered some ideas about how integrins may function on leukocytes (Dransfield et al 1992). Thus mAb 24 was able to inhibit antigen presentation by monocytes to T cells and lymphokine-activated killer cell cytotoxicity which are both LFA-1-dependent processes and f-Met-Leu-Phe-stimulated chemotaxis of neutrophils when it is made to be CR3/Mac-1-dependent. These results suggested that

the epitope recognised by mAb 24 was in a functionally important region of the $\beta 2$ integrins and that exploration of the mechanism of mAb 24-mediated inhibition would lead to a better understanding of leukocyte integrin function. We found that the effects of mAb 24 could not be explained by a blockade of receptor/ligand binding but rather the reverse situation. When the mAb was added to activated cells undergoing cell-cell aggregation via LFA-1/ICAM-1, then mAb 24 had the effect of considerably enhancing the rate of aggregation. A clearer understanding of these observations was provided by experiments which used a better defined model of LFA-1/ICAM-1-dependent adhesion consisting of binding of radiolabelled T cells to purified ICAM-1 immobilised on plastic. When transient adhesion of LFA-1 on T cells was induced by cross-linking the CD3 molecule, mAb 24 had no effect on the initial phase of adhesion but prevented the "deadhesion" of LFA-1 from ICAM-1. It can be speculated that LFA-1 is unable to dissociate from ICAM-1 in the presence of mAb 24 because an essential

alteration of the α and β subunits is prevented. In other words, LFA-1 is unable to switch from an "activated" high affinity form of the receptor to an "inactive" low affinity form. *In vivo* a lack of dynamic interchange between "on" and "off" forms of the receptor could affect leukocyte function by preventing cell recycling. For example, if turnover of T cells on the presenting cell membrane is prevented, then stimulation of more T cells would not occur. Visual inspection of antigen presentation experiments, suggests that mAb 24 may be exerting just such an effect. Thus, in the presence of mAb 24, small aggregates form during the initial few hours and then remain constant in appearance for the remaining 6 days of the experiment. Similarly LAK cell killing also requires recycling of the effector cells to fresh targets. Finally it can be speculated that neutrophils responding to a chemotactic gradient would move on a substrate via an attachment-release mechanism involving CR3/Mac-1. Therefore inhibiting effects of mAb 24 on these leukocyte integrin functions can be explained by the requirement for dynamic and reversible interaction between receptor integrin and ligand and suggest a physiological relevance for the *in vitro* transient adhesion model.

Ca^{2+} AS A NEGATIVE REGULATOR OF LFA-1 ACTIVATION

LFA-1 on resting T cells expresses little 24 epitope but both expression of the epitope and binding to ICAM-1 can be induced by treating cells with Mg^{2+}, thus "short-circuiting" the activation process (Dransfield and Hogg 1989). Although LFA-1 activation is being artificially manipulated in these experiments, they suggest that control of the affinity of Mg^{2+} binding represents a potential mechanism for positive regulation of receptor function. Such speculation has recently been modified as a result of experiments which have revealed a role for a second divalent cation, Ca^{2+}, in the regulation of LFA-1 activation (Dransfield et al 1992). The role of Ca^{2+} was revealed in a series of experiments in which Mg^{2+} unexpectedly failed to induce expression of the 24 epitope or binding to ICAM-1 in the presence of Ca^{2+}. We found that in resting T cells, Ca^{2+} was bound to LFA-1 with sufficient affinity that it required removal by a short EGTA wash before it was possible to activate LFA-1 with the addition of Mg^{2+}. Moreover, this "activating" effect of Mg^{2+} could be reversed by adding Ca^{2+} back to the cells. It was therefore concluded that bound Ca^{2+} imposes a conformation on LFA-1 that is neither recognised by mAb 24 nor conducive to LFA-1-mediated binding to ICAM-1. These observations suggest that Ca^{2+} may therefore act as a negative regulator of the functional activity of LFA-1. Maintenance of the receptor in an inactive state would have the practical importance of preventing leukocytes from randomly adhering to one another in the circulation until an appropriate encounter caused the stimulation necessary to overcome the "Ca^{2+} effect" and to promote the acquisition of the "active" conformation. The speculation would be that activation causes either the release of Ca^{2+} from LFA-1 or some conformational alteration which neutralises the inhibiting role of Ca^{2+}. The nature of these initial events and how they might be manifested in LFA-1 activation is a topic of much current effort.

THE "INSIDE-OUT" SIGNAL

The initial activation requirements of LFA-1 are provided by an "inside-out" signal, which is generated from within the cell and is able to confer altered ligand binding properties to the extracellular domain of LFA-1 (Dustin and Springer 1989). This signal is triggered "remotely", by the cross-linking of cell surface molecules by specific monoclonal antibodies (mAbs) or by immobilised ligand. To date, 12 such surface molecules are apparently able to provide the first signal for LFA-1 activation (for review, see (Pardi et al 1992)). LFA-1 can also be activated directly with several special mAbs directed against LFA-1 itself (see later). There is still much uncertainty about the nature and the physiological effect of this initial "inside out" signal. Based on the ability of phorbol esters to activate LFA-1, members of the protein kinase C family are thought to be part of this activation pathway (Dustin and Springer 1989). Phosphorylation of the short cytoplasmic sequences of α and/or β subunits might cause the required changes in LFA-1 associated with activation (Valmu et al 1991). Although phorbol ester rapidly phosphorylates serine 756 on the $\beta2$ cytoplasmic tail, LFA-1 activation is unaffected by a mutation of this residue (Hibbs et al 1991). The LFA-1 α subunit "tail" has been more difficult to analyse as it is constitutively phosphorylated (Hara and Fu 1986). A possible end result of this type of signalling might be to cause integrin clustering. Increased concentration of cell membrane integrin would bring about increases in the avidity of cell-cell-interaction without the necessity for any alteration in the conformation of LFA-1. The advantage of such mechanism for "adhesion-strengthening" would be that the concentration of many low affinity receptor interactions would lead to multi-point attachment which could considerably increase the avidity of cell-cell adhesion (Kupfer and Singer 1989).

Clustering of receptor is assumed to require interaction of integrin with the cytoskeleton and there is evidence to suggest that this can occur. For example, in an antigen activated T cell line, LFA-1 was found to cocap with talin (Kupfer and Singer 1989) and the $\beta1$ subunit cytoplasmic sequence can bind to α-actinin (Otey et al 1990). In seeming conflict with this type of data are studies in which activation of primary T cells was achieved by cross-linking CD3 molecules, a procedure considered to mimic physiological activation. When receptor localisation was viewed by light microscopy, no major membrane redistribution of LFA-1 was observed (Moingeon et al 1991) [Cabañas & Hogg - submitted]. However microclusters of LFA-1 would not be detected by this methodology and to finally decide whether or not clustering of receptors is a response to the first cross-linking signal more sensitive methods of detection will be required.

ANTI-LFA-1 MONOCLONAL ANTIBODIES WHICH ACTIVATE LFA-1

Another possibility is that the first step in the activation process may be a conformational change in the receptor itself. Support for this suggestion comes from investigation of three mAbs which activate LFA-1 directly and cause enhanced binding to ICAM-1. MAb NKI-L16 recognises an epitope on the Ca^{2+} bound form of LFA-1 which is expressed at a low level on resting T cells and at increased levels on cultured cells and cell lines (Keizer et al 1988; van Kooyk et al 1991). We have now described a second mAb, MEM-83 which will also activate LFA-1 binding to ICAM-1 (Bazil et al 1990) [Landis et al - submitted]. In contrast to NKI-L16, mAb MEM-83 recognises an epitope which is constitutively expressed on LFA-1 and is not dependent on the status of bound divalent cation. A third mAb is KIM 127 which recognises an epitope on the $\beta2$ subunit and has been shown to promote both LFA-1 and CR3-dependent adhesion events (Robinson et al 1992). It can be speculated that these mAbs may be inducing the change in LFA-1 which simulates the "inside-out" signal. The fact that they bind to the extracellular domains of LFA-1 suggests that they act by inducing some allosteric alteration in the conformation which may mimic the "inside out" signal. These mAbs are effective as monovalent Fab' or Fab fragments, ruling out the possibility that they induce activation by artificially clustering receptors via bivalent antibody crosslinking. The implication of the findings with these special mAbs suggests that the initial steps in LFA-1 activation involve some alteration in the receptor itself.

ANOTHER REQUIREMENT FOR INTEGRIN ACTIVATION

Although the initial signals required for leukocyte activation derive from inside the cell, there have been recent suggestions that interaction with ligand itself might have a role in the formation of stable integrin/ligand pairing. The principal ligand of the platelet integrin GPIIbIIIa is fibrinogen and part of the recognition motif is the arginine-glycine-aspartic acid (RGD) sequence. RGD peptides will competitively inhibit GPIIbIIIa binding to fibrinogen (Frelinger et al 1988). Furthermore the binding of RGD peptides induces conformational changes in GPIIbIIIa, which give rise to neo-epitopes detected with mAbs such as PAC-1 and named ligand-induced binding sites (LIBS) (Frelinger et al 1990; Shattil et al 1985). More recently, the Ginsberg laboratory has elegantly shown that the conformational changes induced by RGD peptide in GPIIbIIIa enable this integrin to bind with high affinity to fibrinogen (Du et al 1991). This work suggests that the second signal required for stable, high affinity binding between integrin and ligand is generated by interaction with ligand itself.

Our recent studies of LFA-1 activation have amplified this suggestion. T cells will aggregate in a time dependent fashion when treated with phorbol ester or by cross-linking with CD3 associated with antigen specific T cell receptor. These aggregates are totally dependent upon the LFA-1/ICAM-1 interaction and can be dispersed with specific CD11a/CD54 mAbs. We had thought that the mAbs were acting to sterically prevent physical interaction between active LFA-1 and ligand ICAM-1. Probing this "primed" population of T cells for activation epitope 24, we have found that it was not expressed [Cabañas and Hogg - submitted]. These observations suggested that an initial interaction with ICAM-1 caused a conformational change in LFA-1 permitting expression of the 24 epitope. Thus the 24 epitope is a LIBS which characterises the receptor/ligand complex. Further experiments involving receptor fixation have shown that the change in LFA-1 induced by the inital binding also causes LFA-1 to acquire a conformation which allows stable binding to ICAM-1. Such an "induced fit" process of activation was first described by Koshland forty years ago to account for the changes in shape of binding sites of enzymes after initial encounters with substrate (Koshland and Neet 1968). It remains to be seen how common this process of ligand-induced conformational alteration is to integrin activation in general. As stated previously this stage of leukocyte integrin activation is transient and within two hours LFA-1 will have again assumed a low affinity form. How this is achieved is the next topic of interest.

REFERENCES

Bazil V, Stefanouva I, Hilgert I, Kristofova H, Vanek S, Horejsi V (1990) Folia Biologica (Praha) 36: 41-50
Berendt AR, McDowall A, Craig AG, Bates PA, Sternberg MJE, Marsh K, Newbold CI, Hogg N (1992) Cell 68: 71-81
Buckle A-M, Hogg N (1990) Eur. J. Immunol. 20: 337-341
de Fougerolles AR, Springer TA (1992) J. Exp. Med. 175: 185-190
Diamond MS, Staunton DE, Marlin SD, Springer TA (1991) Cell 65: 961-971
Dougherty GJ, Murdoch S, Hogg N (1988) Eur. J. Immunol. 18: 35-39
Dransfield I, Cabañas C, Barrett J, Hogg N (1992) J. Cell Biol. 116: 1527-1535
Dransfield I, Cabañas C, Craig A, Hogg N (1992) J. Cell Biol. 116: 219-226
Dransfield I, Hogg N (1989) EMBO J. 8: 3759-3765
Du X, Plow EF, Frelinger AL, O'Toole TE, Loftus JC, Ginsberg MH (1991) Cell 65: 409-416
Dustin ML, Carpen O, Springer TA (1992) J. Immunol. 148: 2654-2663
Dustin ML, Springer TA (1989) Nature 341: 619-624
Frelinger AL, Cohen I, Plow EF, Smith MA, Roberts J, Lam SC-T, Ginsberg MH (1990) J. Biol. Chem. 265: 6346-6352
Frelinger AL, Lam SC-T, Plow EF, Smith MA, Loftus JC, Ginsberg MH (1988) J. Biol. Chem. 263: 12397-12402
Goldman R, Harvey J, Hogg N (1992) Eur. J. Immunol. 22: 1109-1114
Hara T, Fu SM (1986) In: E. L. Reinherz, B. F. Haynes, L. M. Nadler and I. D. Bernstein (eds), Leukocyte Typing II . Springer-Verlag, New York, Berlin, Heidelberg, Tokyo, p.
Hemler ME (1990) Annu. Rev. Immunol. 8: 365-400
Hermanowski-Vosatka A, van Strijp JAG, Swiggard WJ, Wright SD (1992) Cell 68: 341-352
Hibbs ML, Xu H, Stacker SA, Springer TA (1991) Science 251: 1611-1613
Hogg N (1991) Chem. Immunol. 50: 1-12

Hogg N, Bates PA, Harvey J (1991) Chem. Immunol. 50: 98-115
Hynes RO (1992) Cell 69: 11-25
Keizer GD, Visser W, Vliem M, Figdor CG (1988) J. Immunol. 140: 1393-1400
Kirchhofer D, Languino LR, Ruoslahti E, Pierschbacher MD (1990) J. Biol. Chem. 265: 615-618
Koshland DE, Neet KE (1968) Annu. Rev. 37: 359-410
Kupfer A, Singer SJ (1989) J. Exp. Med. 170: 1697-1713
Larson RS, Corbi AL, Berman L, Springer T (1989) J. Cell Biol. 108: 703-712
Marlin SD, Springer TA (1987) Cell 51: 813-819
Martz E (1980) J. Cell Biol. 84: 584-598
Moingeon PE, Lucich JL, Stebbins CC, Recny MA, Wallner BP, Koyasu S, Reinherz EL (1991) Eur. J. Immunol. 21: 605-610
Otey CA, Pavalko FM, Burridge K (1990) J. Cell Biol. 111: 721-729
Pardi R, Inverardi L, Bender JR (1992) Immunol. Today 13: 224-230
Pulido R, Elices MJ, Campanero MR, Osborn L, Schiffer S, Garcia-Pardo A, Lobb R, Hemler ME, Sánchez-Madrid F (1991) J. Biol. Chem. 266: 10241-10245
Robinson MK, Andrew D, Rosen H, Brown D, Ortlepp S, Stephens P, Butcher EC (1992) J. Immunol. 148: 1080-1085
Rothlein R, Dustin ML, Marlin SD, Springer TA (1986) J. Immunol. 137: 1270-1274
Sanders ME, Makgoba MW, Sharrow SO, Stephany D, Springer TA, Young HA, Shaw S (1988) J. Immunol. 140: 1401-1407
Shattil SJ, Hoxie JA, Cunningham M, Brass LF (1985) J. Biol. Chem. 260: 11107-11114
Springer TA (1990) Nature 346: 425-434
Staunton DE, Dustin ML, Springer TA (1989) Nature 339: 61-64
Valmu L, Matti A, Siljander P, Patarroyo M, Gahmberg CG (1991) Eur. J. Immunol. 21: 2857-2862
van Kooyk Y, van de Wiel-van Kemenade P, Weder P, Kuijpers TW, Figdor CG (1989) Nature 342: 811-813
van Kooyk Y, Weder P, Hogervorst F, Verhoeven AJ, van Seventer G, te Velde AA, Borst J, Keizer GD, Figdor CG (1991) J. Cell Biol. 112: 345-354

CD44 in Normal Differentation and Tumor Progression

C. Tölg[1], W. Rudy[1], M. Hofmann[1], R. Arch[2], V. Zawadzki[1],
K. H. Heider[1], M. Zöller[2], S. T. Pals[3], S. Zimmer[4], H. Ponta[1], and
P. Herrlich[1]

1. *Kernforschungszentrum Karlsruhe, Institut für Genetik und Toxikologie, PO Box
3640, D-7500 Karlsruhe 1, Germany*

CD44 Represents a Polymorphic Group of Proteins

CD44, also known as HERMES antigen (Jalkanen et al. 1986) or Pgp-1 (Picker et al. 1989),
occurs in several isoforms distinguished by primary amino structure and post-
translational modification. The different sizes range from an apparent molecular weight
of 85 kD to some 250 kD. The smallest form, originally recognized by the HERMES
antibody, is the classical CD44 molecule found on many hemapoetic cells such as
granulocytes, lymphocytes and macrophages (Flanagan et al. 1989; Jalkanen et al. 1987).
Recent cloning and sequencing has shown that larger isoforms carry additional
sequences outside the transmembrane region and that the choice of these extra
sequences is quite enormous (Stamenkovic et al. 1991; Günthert et al. 1991; Hofmann et
al. 1991). Some ten additional exons can be "inserted". All CD44 molecules have the
remaining portions in common, and possess a large extracellular N-terminus, a
transmembrane region and a cytoplasmic tail.

The v6 exon as a Tumor Marker

This report concentrates on CD44 variants that express the exon v6. A number of
antibodies exist which recognize the v6 sequence of either rat or human origin. The
first of these antibodies was raised against a metastatic tumor cell line and led this
laboratory to investigate the proteins carrying the epitope, which turned out to be
members of the CD44 family (Günthert et al. 1991). Not only do several rat tumors with
metastatic properties carry the v6 exon epitope, also many progressing human tumors
are positive, for instance invasive metastatic cancers of the uterus, of the ovaries and of
the mammary glands. In an attempt to define the kinetics of expression during
carcinogenesis, a series of colorectal cancers and their precursor stages were examined
(Heider et al. 1992). Normal colon epithelium is largely negative with the exception of
some faint staining at the base of crypts. Polyp stages are heavily positive. Particularly
dysplastic areas always stain. The frequency of v6 epitope positive cells increases with
the carcinoma and the metastasis stages, suggesting that in the process of colorectal
carcinogenesis the expression once started does not discontinue. The expression appears
to be clonal and resembles the appearance of activated c-Ki-ras. We argue that the clonal
expression indicates a yet unknown growth advantage. The higher frequency of
expression with the more advanced cancers makes CD44 variant exon 6 a tumor
progression marker.

*

2 Deutsches Krebsforschungszentrum Heidelberg, Institut für Radiologie und
Pathophysiologie, Im Neuenheimer Feld 280, D-6900 Heidelberg 1, Germany
3 Academic Medical Center, Department of Pathology, Meibergdreef 9, NL-1105 AZ
Amsterdam, The Netherlands
4 Lucille P. Markey Cancer Center, Dept. of Microbiology and Immunology, 800 Rose
Street, Lexington, Kentucky 40536-0093, U.S.A.

We postulated that a protein expressed during tumor progression would also have a physiological function. We therefore examined various embryonic and adult tissues for v6 exon expression, using antibodies as probes. In the adult, the most pronounced expression level has been found in the keratinocyte layers of the skin both in humans, mouse and rat (Heider et al. 1992; Wirth et al. 1992; Tölg et al. 1992). During embryogenesis, positive areas have been found in the epithelium above dental anlagen, in fetal liver (hemapoetic cells) as well as in the ductal epithelium of the pancreas (Wirth et al. 1992; Tölg et al. unpublished results). In general, the expression of molecules carrying v6 exon sequences is by far more restricted in the adult organism than the small isoform of CD44, as is found on lymphocytes and granulocytes. Expression of the v6 epitope was almost absent in all tissues of the hemapoetic system in adult rats. We found, however, that several compartments of the hemapoetic system were stained with a v6-specific antibody, if assayed just after birth. This suggested that perhaps activated cells could be the carriers of the v6 epitope. We therefore tested whether antigenic stimulation would increase the expression in the adult hemapoetic system. Indeed, this was found: both the number of cells and the intensity of staining was dramatically increased upon antigen injection. In our initial experiments antigenic stimulation was performed by the injection of allogeneic lymphocytes. We examined which cell types were positive for v6 by using a second marker. We found the epitope expressed on T- and B-cells as well as on dentritic cells or macrophages as determined by appropriate surface markers (Zöller et al. unpublished results). That indeed antigen could trigger the expression of CD44 with the v6 sequence was confirmed by stimulation with a cross linking anti-CD3 antibody. Immediately after this treatment, peripheral human T-cells expressed the v6 epitope on their surface, prior to any DNA synthesis. When DNA synthesis started, CD44 variant 6 surface expression had already declined, and after four to six days was gone all together (Heider and Pals unpublished results). The appearance of the v6 epitope on the cell surface parallels the functional appearance of other surface molecules, such as integrins, as has been published previously (for overview see Hogg 1989). By PCR we found that two major products of CD44 are expressed, one containing the v6 sequence only, and a second splice product that carries exons v6 and v7 (Arch et al. 1992; Heider and Pals unpublished results). These experiments indicate that the CD44 variants are very early markers of lymphocyte activation. Since lymphocytes permanently express the small isoform of CD44 which does not contain additional sequences, the appearance of new isoforms indicates regulated alternative splicing.

The CD44 Variants Carrying the v6 exon are Required for Immune Responses

Using an antibody to the v6 sequence we explored whether antibody binding to CD44 molecules carrying the v6 epitope would stimulate or inhibit an immune response. After treatment of mice with an antigen such as trinitophenol (TNP) coupled to LPS or to sheep blood cells, we injected intravenously the monoclonal antibody. This resulted in a fairly severe decrease in the number of plaque-forming cells specific for TNP, as well as in the serum titer. In response to injection of allogeneic cells the appearance of cytotoxic T-cells was severely reduced. Control antibodies could not imitate this inhibition. With an antibody that recognizes the N-terminal portion of the small lymphocyte form of CD44 we found partial inhibition. Also inhibition could be achieved by Fab-fragments of the v6 specific monoclonal antibody, although the Fab-fragments were not as efficient as the complete antibody (Arch et al 1992).

These experiments tell us that activated T- and B-lymphocytes, as well as macrophages, transiently express specific variants of CD44 carrying the v6 exon. These variants are necessary in order to mount an efficient immune response in vivo.

CD44 Promoter Regulation

The lymphocytes define a cell system which expresses both the small isoform as well as one or two of the larger CD44 variants. Obviously the promoter is expressed in lymphocytes, perhaps in a tissue-specific manner. In response to stimulation of the T cell receptor complex we detect alternative splicing. In order to examine how the CD44 gene is regulated in tumor cells, in particular during tumor progression we isolated the CD44 promoter both of human and mouse origin. Interestingly, both promoters not only share elements but also these elements are classical household promoter elements found on many promoters that are expressed in proliferating cells (Zawadzki et al. unpublished results). Particularly interesting is the presence of several basal promoter elements that bind the transcription factor Sp1. In addition, the promoters from both species carry elements that bind inducible transcription factors, namely PEA3 and AP-1. Chimeric constructs carrying these promoter segments in front of a reporter gene were transfected into various types of cells. They respond strongly to ras overexpression, to phorbol esters or to jun (Zawadzki et al. unpublished results). Expression was very similar to that of collagenase promoter, where we had already shown that the decisive element for ras, for phorbol ester or for jun is the AP-1 binding site (Schönthal et al 1988). We therefore mutated the AP-1 site in the CD44 promoter at two bases known to abolish transcription factor binding. This mutant promoter construct could no longer be stimulated by either one of these three agents. Furthermore, from previous experiments we know that the adenovirus oncogene E1A interferes with AP-1 function, inhibiting collagenase promoter activity (Offringa et al 1990). Similarly, the CD44 promoter is also severely inhibited in the presence of E1A (Zawadzki 1992).

Ras Induces CD44 Variant Expression and Metastatic Behavior in CREF Cells

The finding that ras induces CD44 promoter activity made us examine cells transformed by ras for their CD44 expression and behavior as transformed cells. While CREF cells are non-tumorigenic and of course do not metastasize at all, transformation with the activated form of ras induces the ability to grow in soft agar, to form tumors in nude mice and to produce lung metastases upon injection into isogenic animals, either into the blood stream or subcutaneously (Boylan et al. 1990). The metastasis formation upon subcutaneous injection is of only moderate efficiency. While a combination of E1A and E1B (from adenovirus) confers on cells the ability to grow in soft agar, these cells are neither tumorigenic nor metastatic. Cotransformation with E1A/E1B and activated ras produces tumors in the animal and also metastases after tail-vein injection, but no metastases after implantation or injection into subcutaneous tissue (Boghaert et al. 1991). This experiment seems to support an old observation that in certain systems ras causes tumor progression. Of course this could be due to ras-induced changes in any parameter important for metastasis formation. The expression of CD44 variants is one possibility. We therefore explored whether the CD44 promoter was active and whether splice variants containing exon 6 were expressed in ras-transformed CREF cells. Indeed, the promoter was turned on and the variant exon 6 epitope was detectable on the surface, although only small amounts (Hofmann et al. 1992).

Thus it is plausible to assume that v6 exon expression correlates with metastatic behavior.

Using an inducible promoter-ras construct we can produce the ras effect on CD44 at will. By inducing ras expression, CD44 promoter activity and variant expression is elevated. We cannot decide yet whether the splice variation is simply due to leakiness under conditions of high promoter activity or whether ras influences splicing.

CD44 v6 exon Species Confer Metastatic Behavior

Two types of experiments not only correlate CD44 variant expression with metastatic behavior but also provide evidence for a causal role: overexpression in non-metastatic tumor cells, and antibody-interference of lymphogenic spread. Stable transfection of a CD44 cDNA clone encompassing either exons v4 to v7 or v6 and v7, into non-metastatic BSp73AS cells (Matzku et al 1983), in SP6 cells (Zöller et al. 1978) or rat 2/EJ transformed, induces these cells to produce metastatic tumors in vivo (Günthert et al. 1991). This amazing effect suggests that the presence of the CD44 variant on the cell surface (or internally?) triggers more than one functional change. I.V. injection of antibody directed against the v6 exon, with or briefly after the BSp73ASML cells, or after the AS cell CD44v transfectant, caused significant delay or even prevention of lymph node metastasis formation (Reber et al. 1990; Seiter et al. 1992). This localizes the role of the v6 exon sequence in CD44 to the migratory phase between site of injection and the outgrowth in the draining lymph node.

Hypotheses

That tumor cells should attempt to mimick naturally migrating cells in the body is not that surprising, and has been proposed earlier (Hart et al. 1989; Pauli et al. 1990). The striking new finding is that metastatic tumor cells and antigen stimulated lymphocytes make use of the same surface molecule, and they seem to do so in the same step, namely the expansion of cells in the draining lymph node. The expansion seems to be absolutely required for successful spreading into other sites by tumor cells. Interestingly, after contact with antigen, lymphocytes need to be activated and expanded in seemingly the same compartment, otherwise the immune response is reduced. To unravel the mechanism of CD44 action in these processes will be a challenging task. One can imagine that it will involve signal transduction, e.g. from a ligand through the CD44 protein to the cell and, as suggested by Nancy Hogg in this volume, in the opposite direction, from the carrier of CD44 to other cells.

REFERENCES:

Arch R, Wirth K, Hofmann M, Ponta H, Matzku S, Herrlich P, Zöller M (1992) Participation in normal immune responses of a splice variant of CD44 that encodes a metastasis-inducing domain. Science 257:682-685

Boghaert ER, Austin V, Zimmer SG (1991) The influence of the presence of adenovirus 5 E1a and E1b sequences on the pathology of rat embryonic fibroblasts transfected with activated c-Ha-ras and v-ras. Clinical Exper Metas 9:231-243

Boylan JF, Jackson J, Steiner MR, Shih TY, Duigou GJ, Roszman T, Fisher PB, Zimmer SG (1990) Role of the Ha-ras (Ras^H) oncogene in mediating progression of the tumor cell phenotype. Anticancer Res 10:717-724

Flanagan BF, Dalchau R, Allen AK, Daar AS, Fabre JW (1989) Chemical composition and tissue distribution of the human CDw44 glycoprotein. Immunology 67:167-175

Günthert U, Hofmann M, Rudy W, Reber S, Zöller M, Haußmann I, Matzku S, Wenzel A, Ponta H, Herrlich P (1991) A new variant of glycoprotein CD44 confers metastatic potential to rat carcinoma cells. Cell 65:13-24

Hart IR, Goode NT, Wilson RE (1989) Molecular aspects of the metastatic cascade. Biochim Biophys Acta 989:65-84

Heider K-H, Hofmann M, Horst E, van den Berg F, Ponta H, Herrlich P, Pals ST (1992) A human homologue of the rat metastasis-associated variant of CD44 is expressed in colorectal carcinomas and adenomatous polyps. J Cell Biol submitted

Hofmann M, Rudy W, Zöller M, Tölg C, Ponta H, Herrlich P, Günthert U (1991) CD44 splice variants confer metastatic behavior in rats: homologous sequences are expressed in human tumor cell lines. Cancer Res 51:5292-5297

Hofmann M, Rudy W, Günthert U, Zimmer SG, Zawadzki V, Zöller M, Lichtner R, Herrlich P, Ponta H (1992) A link between RAS and metastatic behavior of tumor cells: RAS induces CD44 promoter activity and leads to low-level expression of metastasis specific variants of CD44 in CREF cells. Cancer Res submitted

Hogg N (1989) The leukocyte integrins. Immunol Today 10:111-114

Hogg N (1992) Roll, roll, roll your leukocytes gently down the vein . . . . Immunol Today 13:113-115

Jalkanen S, Bargatze R, Herron L, Butcher EC (1986) A lympoid cell surface glycoprotein involved in endothelial cell recognition and lymphocyte homing in man. Eur J Immunol 16:1195-1202

Jalkanen S, Bargatze RF, de los Toyos J, Butcher EC (1987) Lymphocyte recognition of high endothelium: antibodies to distinct epitopes of an 85-95 kD glycoprotein antigen differentially inhibit lymphocyte binding to lymph node, mucosal, or synovial endothelial cells. J Cell Biol 105:983-990

Matzku S, Komitowski D, Mildenberger M, Zöller M (1983) Characterization of Bsp 73, a spontaneous rat tumor and its in vivo selected variants showing different metastasizing capacities. Invasion Metastasis 3:109-123

Offringa R, Gebel S, van Dam H, Timmers M, Smits A, Zwart R, Stein B, Bos JL, van der Eb A, Herrlich P (1990) A novel function of the transforming domain of E1a: repression of AP-1 activity. Cell 62:527-538

Pauli BU, Augustin-Voss HG, El-Sabban ME, Johnson RC, Hammer DA (1990) Organ-preference of metastasis. The role of endothelial cell adhesion molecules. Cancer Metast Rev 9:175-190

Picker LJ, Nakache M, Butcher EC (1989) Monoclonal antibodies to human lymphocyte homing receptors define a novel class of adhesion molecules in diverse cell types. J Cell Biol 109:927-937

Reber S, Matzku S, Günthert U, Ponta H, Herrlich P, Zöller M (1990) Retardation of metastatic tumor growth after immunization with metastasis-specific monoclonal antibodies. Int J Cancer 46:919-927

Schönthal A, Herrlich P, Rahmsdorf HJ, Ponta H (1988) Requirement for fos gene expression in the transcriptional activation of collagenase by other oncogenes and phorbol esters. Cell 54:325-334

Seiter S, Wirth K, Hofmann M, Ponta H, Herrlich P, Matzku S, Zöller M (1992) Prevention of tumor metastasis formation by anti-variant CD 44. J Exp Med submitted

Stamenkovic I, Aruffo A, Amiot M, Seed B (1991) The hematopoietic and epithelial forms of CD44 are distinct polypeptides with different adhesion potentials for hyaluronate-bearing cells. EMBO J 10:343-348

Wirth K, Arch R, Somasundaram C, Hofmann M, Weber B, Herrlich P, Matzku S, Zöller M (1992) Expression of a metastasis associated CD44 isoform in newborn and adult rats. Eur J Cancer submitted

Zöller M, Matzku S, Goerttler K (1978) High incidence of spontaneous transplantable tumors in BDX rats. Br J Cancer 37:61-66

7. Cytokines and Cytokine Receptors in Health and Disease

The Cytokine Network: Contributions of Proinflammatory Cytokines and Chemokines

J. J. Oppenheim, J-M. Wang, A. W. Lloyd, D. D. Taub, D. J. Kelvin and R. Neta[*]

Laboratory of Molecular Immunoregulation, Biological Response Modifiers Program, Division of Cancer Treatment, National Cancer Institute - Frederick Cancer and Research Development Center, Frederick, MD 21702 - 1201 Laboratory of Molecular Immunoregulation, Biological Response Modifiers Program, Division of Cancer Treatment, National Cancer Institute - Frederick Cancer and Research Development Center, Frederick, MD 21702 - 1201; [*]Department of Experimental Hematology, AFRRI, Bethesda, MD 20814

INTRODUCTION

Cytokines were initially detected beginning in 1964 as biologically active factors that appeared in supernatants of stimulated leucocyte cultures. During the past decade advances in molecular biology and biochemistry have resulted in the purification, cloning and expression of numerous cytokines. This development has provided sufficient purified recombinant cytokines for studies which have revealed the central role of cytokines in the regulation of immunological, hematopoietic, inflammatory, and reparative processes essential for host defense responses and restoration of homeostasis in vertebrate species. Over 60 cytokines are already available in recombinant form and more are yet to come.

What are Cytokines?

Cytokines are paracrine, autocrine and at times endocrine intercellular peptide signals also called "intercrine" signals that are produced by virtually all stimulated or irritated (injured) nucleated cell types with pleiotropic effects on immune, inflammatory and other receptor-bearing cell types. Cytokines, therefore, provide the means by which damaged somatic tissues can communicate with leukocytes, lymphocytes and bone marrow cells and conversely provide a means by which leukocytes and lymphocytes can marshall somatic cells such as those of the liver, vascular tissues, the skin, neuroendocrine and connective tissues to participate in defense of the host and restoration of homeostasis. Cytokines are relatively small peptide or glycoprotein mediators ranging from 6,000-60,000 kDa in molecular weight. They are exceedingly potent being active at concentrations as low as 10^{-10}-10^{-13} mol/L (1pg/ml). Consequently only very few cytokine molecules are required to modulate the growth, differentiation and functions of targeted cells (as reviewed in Thomson, 1991).

How do Cytokines Activate Cells?

Cytokines bind to complementary cell surface receptors with high affinity (kd= 10^{-9}-10^{-12}M) which results in immediate cell functions and also initiates intracellular signals that result in gene transcription and translation. Occupancy of a small minority (<5%) of cytokine receptors can be sufficient for cell activation. Cells expressing as few as 50 and up to 10^5 can be activated by cytokines. Since most nucleated cell types express multiple cytokine receptors, they can be activated by a variety of cytokines. Furthermore, multiple cell types express receptors for a given cytokine, which accounts for their pleiotropic activities (Miyazawa et al., 1992).

How are Cytokines Different from Hormones?

Endocrine hormones and neuropeptides are generally produced by specialized glands, are continually present in the circulation, act on neighboring as well as distant cells, serve to maintain homeostasis and mobilize the "flight or fight" response to stress. In contrast, cytokines generally act as paracrine or autocrine (intercrine) signals within local tissues and only following excessively potent stimulation appear in the circulation to cause systemic symptoms such as fever and other acute phase responses. Unlike hormones, cytokines are generally not constitutively produced, or stored within cells as are some neuropeptides. Cytokines are produced in response to exogenous irritants or noxious stimuli, or in response to other endogenous cytokines generated in defense against insults to the host. This cytokine cascade or network serves to rapidly mobilize and amplify inflammatory, immunological and reparative reactions in an effort to minimize damage to the host and to restore homeostasis. However, the distinctions between cytokines and those hormones and neuropeptides with major local activities such as secretin, cholecystokinin, vasoactive intestinal peptide (VIP) and testosterone are less clearcut. Similarly, some cytokines are hormone-like since they are produced constitutively and are always present in the circulation including M-CSF, erythropoietin (EPO), stem cell factor (SCF) and transforming growth factor ß (TGFß) (Thomson, 1991). Furthermore, cytokines and hormones have considerable mutual modulating effects. For example, cytokines such as IL 1, TNF, and IL 6 stimulate the hypothalamic-pituitary-adrenal (HPA) axis which leads to production of CRF, ACTH, and corticosteroids (Besedovsky et al., 1986). These cytokines induce expression of the POMC gene and consequently enhance the production of ACTH, αMSH and endorphins. Conversely, the immunosuppressive effects of corticosteroids are mediated through the suppression of production of cytokines such as IL 1, IL 2, TNF and IL 6. Corticosteroids, surprisingly, have the capacity to enhance the expression of receptors for IL 6 on hepatocytes which promotes acute phase responses (Kushner, 1993) and IL 1 on B cells (Akahoshi et al., 1988), which may account for the immune deviating effects of corticosteroids by promoting humoral immunity at the expense of cellular immunity.

What is the Cytokine Network?

Since many cytokines have multiple target cells, their activities can overlap to a great extent and they exhibit considerable redundancy (Paul, 1989). For example, even though IL 1 and TNF show no structural homology and use distinct pairs of receptors, their activities are widespread and very similar. This overlap in activities is compounded by the fact that cytokines induce the production of a cascade of other cytokines. Thus, IL 1 and TNF not only induce production of one another (e.g. IL 1 induces TNF and the reverse), but also more of themselves (e.g. IL 1 induces IL 1) as well as other cytokines such as IL 2, IL 6, IL 8, and PDGF to name a few (Neta et al., 1991). This cytokine cascade provides one means by which these signals are rapidly amplified. Further amplification is achieved through the synergistic cooperation of cytokines. Thus, low doses of IL 1, TNF and IL 6 together result in considerably greater than additive effects. This may, in part, be based on the capacity of cytokines to upregulate cytokine receptor expression. For example, IL 3 and G-CSF upregulate IL 1 receptor expression on bone marrow cells (BMC), while IL 1 upregulates the expression of receptors for G- and GM-CSF as well as those for IL 1 itself (Dubois et al., 1990). In addition, the utilization of distinct signal transduction pathways by different cytokines presumably may serve to amplify gene activation. In fact, cytokines may actually be interdependent rather than redundant. This is illustrated by studies showing that inhibition of cytokines such as IL 1, TNF or IL 6 by antibodies to any one of these cytokines can interfere with the response to one another. Thus, anti-IL 6 can markedly reduce symptoms of endotoxin induced septic shock, of IL 1 or TNF induced radioprotection or of IL 1 induced ACTH production (Neta et al., 1992). This is all the more surprising since recombinant IL 6 (which does not induce other cytokines) if administered by itself does not induce shock, nor ACTH and is not radioprotective. Thus, rather than being redundant, cytokines appear to be interdependent and their interactions are to some extent obligatory.
Finally, some cytokine interactions are characterized by antergy in that they

have down-regulating effects on one-another. For example, TGFß depresses the production of many of the proinflammatory interleukins including IL's 1, 2, 6, 8, the CSF's and TNF (as reviewed in Durum and Oppenheim, 1989). TGFß also interferes with the effects of proinflammatory cytokines in the case of IL 1 by suppressing the expression of IL 1 receptors (Dubois et al., 1990). Another very important example of antergy is exemplified by the inhibitory effects of the IFNγ produced by the subset of TH1 helper lymphocytes on the production of IL 4 and 10, while IL 10, a product of the subset of TH2 cells inhibits the production of IFNγ as well as monocyte derived cytokines such as IL 1 and TNF (Street and Mossmann, 1991). Thus, cytokines such as IL 4 and 10 also participate in deviating immune responses from the cellular to humoral type of reactions.

What are Chemokines?

I will now introduce data on the novel chemotactic cytokine family of "chemokines", that has come to light largely since 1985, to illustrate additional concepts concerning the role of cytokines in inflammatory and immunological reactions. Members of the family of chemotactic cytokines, which have been proposed to be named "chemokines" for short, are being identified as vital initiators and promulgators of inflammatory and immunological reactions (as reviewed by Oppenheim et al., 1991). The chemokines range from 8-11 kDa in MW and are produced by a wide variety of cell types. They are induced by exogenous stimulants and irritants and by endogenous mediators such as IL 1, TNF, PDGF, IL 2 and IFNγ. The chemokines can be considered "second order" cytokines because they have limited capacity to induce other cytokines, exhibit more specialized functions in inflammation and repair and at present appear to be less pleiotropic than the "first order" proinflammatory cytokines. Some of the chemokines can be assigned to an α subset based on their location of chromosome 4 (q12-32) and based on the fact that the first two of their four cysteine groups are separated by one amino acid (C-X-C). This chemokine α group consists of IL 8, melanoma growth stimulating activity (MGSA/GRO), platelet factor 4 (PF4), ß thromboglobulin (ßTG), IP-10 and ENA-78. The chemokine ß subgroup is located on chromosome 17 (q11-32), has no intervening amino acid between the first two cysteines (C-C), and includes macrophage chemotactic and activating factor (MCAF/MCP-1), RANTES, LD78 also known as human MIP-1α, ACT-2 or huMIP-1ß, and I-309 as reviewed by Schall (1991). It is surprising that chemokines responsible for mobilizing inflammatory and immune cells have been identified and cloned only over the past seven years. Initially, members of the chemokine family such as PF4, ßTG, MIP-1α and ß, IL 8, and MCAF were biochemically purified as monitored by bioassays and then cloned. More recently chemokines such as GRO, IP-10, RANTES and I-309 were discovered by subtractive hybridization and their biological activities are now being defined. This represents a major reversal in the usual sequence of events and has presented investigators with expressed cytokine products in search of biological functions.

What are the Cell Sources and Inducers of Chemokines?

I will briefly summarize the cell sources and stimulants of chemokine α and ß subsets. Suffice it to say that human IL 8, GRO, IP 10 and MCAF are produced by a wide variety of leukocytic and nonleukocytic cell types. CTAP III which is successively enzymatically cleaved to produce ß thromboglobulin (ßTG) and neutrophil attracting peptide (NAP-2) as well as platelet factor 4 (PF 4) are produced in response to stimuli that aggregate platelets and cause them to degranulate and release these two chemokines during thrombogenesis. RANTES and I-309 are lymphocyte-derived lymphokines, while macrophage inflammatory peptides (MIP) 1α and 1ß are produced by macrophages as well as lymphocytes. T cell activators induce the lymphocyte-derived chemokines, whereas noxious stimuli activate the nonlymphocytic cell types. Endogenous cytokines such as IL 1, TNF and PDGF can stimulate the production of most of the leukocyte and somatic cell-derived chemokines.

What are the Characteristics of Chemokine Receptors?

The receptors for C5a and fMLP as well as two distinct but homologous (70% at the amino acid level) receptors for IL 8 have been cloned. These
IL 8 receptors are members of the rhodopsin receptor family and have the characteristic seven transmembrane spanning regions (Holmes et al., 1991; Murphy and Tiffany, 1991). This large receptor family transduces not only immunomodulating, but also neural, hormonal, visual, olfactory and other signals (Murphy et al., 1992). Since the chemokine receptors have been difficult to purify, molecular cloning has yielded a number of rhodopsin-like receptor structures with varying degrees of homology, in search of the appropriate chemokine, neuropeptide or other ligands (Murphy et al., 1992).

The IL 8 receptors are coupled to G proteins, initiate phosphoinositide hydrolysis and are capable of rapidly elevating diacylglycerol and cytosolic Ca^{++} levels, which may lead to activation of protein kinase C (Thelen et al., 1988). IL-8 receptors are expressed by neutrophils, myelocytic lines, and basophils. Unactivated T lymphocytes express very low levels, whereas neutrophils express both types of IL 8R and their expression is upregulated by G-CSF. Biological studies including inhibition studies using pertusis toxin to inhibit G protein activation suggest that the receptors for the other chemokines also belong to the rhodopsin family. In fact, cross-utilization of the same receptors by some members of chemokine α and β subsets has been shown. Thus, GRO/MGSA and the murine homologue MIP 2 have been shown to bind the type II, but not the type I IL 8 receptor with equal affinity (Ceretti et al., unpub. observ.). In addition, NAP 2, the enzymatic cleavage product of βTG, binds the type II IL 8 receptor with 1/100 lower affinity (Leonard et al., 1991). For the chemokine β group, RANTES binding sites on monocytes can be equally competed for by MCAF and MIP 1α and β. MIP 1α and β bind equally well with one another, but only compete for 30% of the MCAF binding sites on monocytes. MCAF can compete for about 30% of MIP 1α and β binding sites on monocytes, but not for those on T cells which appear to be unique (Wang et al., 1992). A minimal model in which both MIP 1α and β and MCAF share the receptor for RANTES on monocytes, but MIP 1 and MCAF each have unique receptors as well, could account for these observations. Thus, some of the chemokines can bind to heterogeneous receptors and some of the receptors bind multiple chemokines.

What are the In Vitro and In Vivo Activities of the Chemokine α Subfamily?

The overlapping functions of the chemokines are based on the cross-utilization of common receptors. IL 8 is a major chemoattractant for neutrophils and in addition at ~10 fold higher doses stimulates neutrophil degranulation and enzyme release (Table 1). IL 8 markedly upregulates the expression of β2 integrins and promotes the adherence, penetration and tissue infiltration by neutrophils. Consequently local injections of IL 8 induce rapid local accumulation of neutrophils without erythema, edema or pain within 2-3 hours. Systemic injection of IL 8 results in neutrophilia, but neither fever nor elevation of acute phase proteins. IL 8 can easily be detected in the circulation of patients with systemic inflammatory reactions or severe trauma. IL 8 is present at inflammatory sites such as in the synovial fluid in rheumatoid arthritis, (Brennan et al., 1990) and in extracts of psoriatic skin (Sticherling et al., 1991). Thus, IL 8 is implicated as a major participant in acute as well as more prolonged inflammatory reactions.

Table 1:   Chemokine α: Cell Targets and Effects

Cytokine	Chemotactic or Haptotactic Responses	Major Other Activities
IL 8	Neutrophils Basophils Unstimulated T Cells Melanoma Cells	Activates PMN ↑Neutrophil Adhesion ↓Basophil Histamine ↑Keratinocyte Growth Acute Inflammation
GROαβγ/ muKC/ muMIP-2αβ	Neutrophils	Degranulates PMN ↑Melanoma Cell Growth ↑Fibroblast Growth Acute Inflammation
CTAP III/ βTG	Fibroblasts	↑Fibroblast Growth
βTG/NAP-2	Neutrophils	Activates PMN
PF-4	Fibroblasts	↑Fibroblast Growth Reverses Immune Suppression ↑ICAM-1 on HUVEC
IP-10/ muCRG-2	Monocytes T Lymphocytes	↑Chronic Inflammation
ENA-78	Neutrophils	Activates PMN

As predicted by the utilization of the Type II IL 8 receptor, GRO although initially detected as a melanoma growth stimulating activity (MGSA), is also a potent chemoattractant and activator of neutrophils. GRO has also been extracted from psoriatic tissues. There are three variants of GRO (α,β and γ) which exhibit about 95% homology in their amino acid sequences. They are homologues of murine macrophage derived KC, macrophage inflammatory peptide (MIP) 2α and MIP 2β respectively. Murine MIP 2α and β both compete with equal affinity for Type II receptors for IL 8 and chemoattract human as well as murine neutrophils. MIP 2 is also reported to degranulate murine neutrophils resulting in the release of lysozomal enzymes. Local in vivo injections of MIP 2 results in neutrophil accumulation and MIP 2 has been isolated from sites of wound healing. It is most likely that GRO/MIP 2 inflammatory activities overlap considerably with those of IL 8, and GRO may therefore probably also be a major acute inflammatory mediator.

PF4 and CTAP III, the precursor of βTG, are both reported to chemoattract and to stimulate fibroblasts, presumably for repair purposes (Table 1). In addition a 70 amino acid breakdown product of CTAP III/βTG known as neutrophil attracting peptide 2 (NAP 2) is a chemoattractant and activator of neutrophils, albeit at 100 fold higher concentrations than IL 8. This is consistent with the data that NAP II also binds to Type II IL 8 receptor with about one hundredth of the affinity of IL 8 (Leonard et al., 1991). However, since platelet aggregation can yield high levels of NAP 2, this chemokine presumably participates in attracting acute inflammatory cells to such sites.

IP-10 is produced by macrophages in response to IFN γ, while the murine homologue, CRG 2, can be induced by LPS, IL 1 or TNF. rhIP-10 (provided by Dr. K. Matsushima) does not compete for the IL 8, RANTES, MCAF, or MIP 1 binding sites. Antibodies to IP-10 react with many cell types present at site of delayed hypersensitivity reactions and IP-10 has been extracted from psoriatic plaques (Gotlieb et al., 1988). Thus, IP-10 can presumably be produced by many cell types and probably participates in inflammation. Our recent studies have revealed that IP-10 is a moderately potent in vitro chemoattractant of monocytes and lymphocytes, but not of neutrophils. IP-10 is also a potent promoter of adhesion of activated T lymphocytes to endothelial cells. These in vitro observations suggest that IP 10 may be a vital participant in initiating delayed type immune and cell-mediated inflammatory responses.

What is the Role of Chemokine β Subfamily Members?

Members of the chemokine β subfamily chemoattract predominantly monocytic and/or lymphocytic mononuclear cells (Table 2).

Table 2:   Chemokine β: Cell Targets and Effects

	Chemotactic or Haptotactic Responding Cells	Major Activities
MCAF (MCP-1)/ muJE	Monocytes	Macrophage Activation Basophil Histamine Release Chronic Inflammation
RANTES	Monocytes T Lymphocytes (Memory Subset)	↑T cell/HUVEC Adhesion Basophil Histamine Release Chronic Inflammation
LD78/ muMIP-1α	Monocytes Activates T Lymphocytes (>CD8 Subset)	↑BM Progenitor Stem Cells Costimulates Myelopoiesis Activates CD8 Lymphocytes ↑CD8 Adhesion to HUVEC
ACT-2/ muMIP-1β	Monocytes Activates T Lymphocytes (>CD4 Subset)	Costimulates Myelopoiesis Activates CD4 Lymphocytes ↑CD4/HUVEC Adhesion
I309/muTCA3	Monocytes	

MCAF chemoattracts and activates monocytes to release enzymes and become cytostatic for tumor cells. Binding sites for MCAF have been detected only on monocytes (Wang et al., 1992). MCAF is present in inflamed atheromatous lesions in blood vessel walls, and in the alveolar fluid of patients with pulmonary pathology and induces macrophages to accumulate by 6-18 hrs at sites of infection. In addition, MCAF is chemotactic for basophils and causes rapid degranulation of basophils resulting in histamine release (Kuna et al., 1992a). This suggests that basophils express MCAF receptors and may play an important role as a late histamine releasing factor (HRF) in the pathogenesis of allergic disorders such as atopic food allergies, asthma and chronic urticaria.
RANTES is a less potent chemoattractant for monocytes than MCAF and a potent chemoattractant for memory T cells, but not for naive T cells (Schall et al., 1990). Activated T cells respond to a greater extent to RANTES than unstimulated T cells suggesting autocrine activity. RANTES has also been detected at sites of atheromatous inflammation (Table 2). RANTES promotes the in vitro adherence

of T cells to human umbilical vein endothelial cells (HUVEC). This response is most evident when anti-CD3 activated T cells as well as IL 1 prestimulated human endothelial cells are used (Lloyd et al., unpub. observ.). In addition, RANTES, like MCAF, causes rapid basophil degranulation and histamine release (Kuna et al., 1992b).

Human MIP-1α (also known as LD 78) and huMIP-1β (also known as ACT-2) in our laboratory are equally potent chemoattractants of monocytes and also of activated T lymphocytes and promote the adhesion of T cells to HUVEC. HuMIP-1α preferentially acts on activated CD8 lymphocytes, while huMIP-1β preferentially chemoattracts activated circulating CD4 cells and T cell clones. Although natural MIP-1α and β are also reported to induce in vivo neutrophil accumulation, we could not detect any neutrophil chemotaxis in response to any of the recombinant chemokine β cytokines.

Murine MIP-1α is reported to inhibit hematopoietic stem cell replication (Maze et al., 1992), while muMIP-1β not only fails to do so, but competitively inhibits this activity of MIP-1α (Lord et al., 1992). Both muMIP-1α and β are reported to costimulate the replication of later hematopoietic progenitor cells. Therefore, although the chemokines act largely as differentiation agents on mature cell types, these observations suggests chemokines may also play a role as growth factors during the course of development.

What will the Future Hold for Cytokines?

The pathophysiological and developmental roles of many of the cytokines remain unclear. Studies of transgenic overproducers and non-producers or "knock-out" mice generated by gene targeting approaches will probably reveal cytokines to play important additional roles in embryogenesis, in normal physiology and in disease states. These approaches can be supplemented with studies utilizing cytokine inhibitors such as neutralizing antibodies, IL 1 receptor antagonists, inhibitory extra-cellular receptor domains (e.g. IL 1R, TNF-R, IL 4R, p130,, IL 6R and IL 7R), drugs such as cyclosporin A or FK 506, or by using suppressive cytokines such as TGFβ or IL 10.

The contribution of cytokine receptors to the cytokine cascade must be better defined. Inflammatory and immune responses can be markedly influenced by up or down-regulation of cytokine receptor expression as exemplified by corticosteroid enhancing IL 1R expression on B cells favoring humoral over cellular immunity and TGFβ suppression of IL 1 receptors on BM cells contributing to its suppressive effects on hematopoiesis. Furthermore, the role of shedding or alternative splicing of extra-cellular binding domains of receptors that yield soluble inhibitors or enhancers of cytokines (e.g. p80 of IL 6R), needs to be further evaluated and may provide therapeutically beneficial molecules. Sharing of receptors as shown for the chemokines, the p130 receptor chain in the case of IL 6, LIF, oncostatin M and IL 11 and of the p140 chain by GM-CSF, IL 3 and IL 5 may provide a basis for understanding the functional overlap of structurally distinct cytokines. Identification of additional cytokines that cross-utilize receptor components would be desirable. Definition of post-receptor signal transduction and regulatory sites on genes used in common by cytokines may reveal the basis for redundancy in the activities of other cytokines such as IL 1 and TNF.

Although it is dangerous to be overly optimistic, we foresee a promising future for the discipline of "intercrinology". Cytokines are crucial probes that will enable investigators to further expand our understanding of the fundamental molecular mechanisms regulating many cellular activities from the receptor to the genes. Therapeutic applications of cytokines, as with hormones, will prove to be beneficial in a variety of conditions. However, we may be in for numerous surprises and disappointments. Cytokines may prove more useful in the treatment of parasitic and viral than in neoplastic diseases. Cytokine antagonists may prove useful in therapy of autoimmune diseases and suppression of transplant rejection. Finally, gene therapy may correct deficits, modulate development, or serve to deliver cytokines in a more efficacious manner. There is much work ahead for all the investigators from the many disciplines who are interested in cytokine biology. Consequently, we foresee the evolution of studies of cytokines as culminating in the development of the discipline of "intercrinology" as a unique biomedical specialty.

REFERENCES

Akahoshi T, Oppenheim JJ, Matsushima K (1988) J Exp Med 167:924
Besedovsky H, Delkey A, Sorkin E, Dinarello CA (1986) Science 233:652.
Brennan FM, Zachariae COC, Chantry D, Larsen CG, Turner M, Maini RN, Matsushima K, Feldmann M (1990) Eur J Immunol 20:2141
Dubois CM, Ruscetti FW, Palaszynski EW, Falk LA, Oppenheim JJ, Keller JR (1990) J Exp Med 172:737
Durum SK, Oppenheim JJ (1989) Macrophage derived mediators: IL 1, TNF, IL 6, IFN and related cytokines. In: Paul WE (ed). Fundamental Immunology, Raven Press Ltd, New York, p 639
Gotlieb AB, Luster AD, Posnett DN, Carter DM (1988) J Exp Med 168:941
Holmes WE, Lee J, Kuang WJ, Rice GC, Wood WI (1991) Science 253:1271
Kuna P, Reddigari SR, Rucinski D, Oppenheim JJ, Kaplan AP (1992a) J Exp Med 175:489
Kuna P, Reddigari SR, Schall TJ, Rucinski D, Viksman MY, Kaplan AP (1992b) J Immunol 149:636
Kushner I (1993) Regulation of the acute phase response by cytokines. In: Oppenheim JJ, Rosio J, Gearing A (eds) Clinical Applications of Cytokines. Oxford Univ Press, New York, (in press)
Leonard EJ, Yoshimura T, Rot A, Noer K, Walz A, Baggiolini M, Walz DA, Goetze EJ, Castor CW (1991) J Leuk Biol 49:258
Lord BI, Dexter TM, Clements JM, Hunter MA, Gearing AJH (1992) Blood 79:2605
Maze R, Sherry B, Kwon BS, Cerami A, Broxmeyer HE (1992) J Immunol 149:1004
Miyajima A, Kitamura T, Harada N, Yokota T, Arai K-I (1992) Ann Rev Immunol 10:295
Murphy PM, Tiffany HL (1991) Science 253:1280
Murphy PM, Ozcelik T, Kenney RT, Tiffany HL, McDermott D, Franche U (1992) J Biol Chem 267:7637
Neta R, Perlstein R, Vogel SN, Ledney GD, Abrams J (1992) J Exp Med 175:689
Neta R, Sayers TJ, Oppenheim JJ (1992) Relationship of TNF to interleukins. In: Aggarwal BB, Vilcek J (eds). Tumor Necrosis Factors: Structure, Function, and Mechanisms of Action. Marcel Dekker, Inc, New York, 56:499
Oppenheim JJ, Zachariae COC, Mukaida N, Matsushima K (1991) Ann Rev Immunol 9:617
Paul W (1989) Cell 57:107
Schall TJ (1991) Cytokine 3:165
Schall TJ, Bacon K, Toy KJ, Goeddel DV (1990) Nature 347:669
Sticherling M, Bornschuer E, Schroder JM, Christophers E (1991) J Invest Dermatol 96:26
Street NE, Mossmann TR (1991) FASEB J 5:171
Thelen M, Peveri P, Kernen P, Von Tsarnev V, Walz A, Baggiolini M (1988) FASEB J 2:2702
Thomson A (1991) The Cytokine Handbook. Academic Press, London, New York, p 425.
Wang J-M, Sherry B, Kelvin D, Oppenheim JJ (1992) J Immunol (in press)

Lymphocyte Development and Immunoreactivity in IL-2 Deficient Mice

A. Schimpl, T. Hünig, I. Berberich, K. Erb, A. Elbe*, G. Stingl*, B. Sadlack, H. Schorle and I. Horak

*Institue of Virology and Immunobiology, University of Würzburg, Versbacherstrasse 7, D- 8700 Würzburg, Germany and *I. Universitätshautklinik, University of Vienna, A 1090*

Introduction

Ever since its first description, Interleukin-2 (IL-2) has been attributed a major role in the regulation of immune responses (for review see Smith 1988, 1992; Paul 1989). It is transiently produced by T cells after activation, predominantly but not exclusively by the CD4+ pre TH, TH0 and TH1 subtypes (Mosmann et al. 1989). IL-2 acts as a major autocrine and paracrine growth factor. However, it is not the only known T cell growth factor nor does it act exclusively on T cells. IL-2 promotes the growth and activity of NK cells (Trinchieri et al. 1984) and the induction of growth and differentiation of B cells (Zubler et al. 1984, Tigges et al. 1989) is a well recognized phenomenon. A role for IL-2 in intrathymic development has been suggested on the basis of the presence of IL-2 receptor bearing cells in the thymus and of inhibitory effects of anti IL-2 receptor antibodies on T cell development *in vivo* and *in vitro* (reviewed in Carding et al.1991). Other studies have, however, failed to observe such effects on the development of the major T cell subsets (Plum and de Smedt 1988). The very important role of IL-2 in vivo is also documented by the fact that three recognized cases of a congenital IL-2 defect in patients may lead to severe, and even lethal immunodeficiency (reviewed in Smith 1992).

In order to establish an experimental model in which the *in vivo* and *in vitro* role of IL-2 can be studied in detail, we have established IL-2 deficient mice by disrupting the IL-2 gene through targeted recombination (Schorle et al. 1991). Functional inactivation was achieved by insertion of the neomycin resistance gene in opposite orientation into exon 3 of the IL-2 gene. This insertion introduces several stop codons in all reading frames and leads to the loss of a cysteine essential for the biological function of IL-2 (Zurawsky and Zurawsky 1988). The mutated IL-2 gene was introduced by electroporation into embryonic stem cells derived from the 129 strain (Evans and Kaufman 1981). After identifying 2 clones carrying the properly targeted sequences, the ES cells were injected into blastocysts of C57/Bl6 mice and transferred to CD-1 fostermothers. Both ES clones gave rise to chimaeric mice which could transmit the mutated IL-2 gene in the germline and both lines are functionally equivalent. Also, litters from heterozygous or homozygous parents exhibited similar phenotypes, ruling out an effect of maternally derived IL-2.

Inactivation of the IL-2 gene was successful since we could neither detect mRNA of the correct length nor production of biologically active IL-2 by cells from IL-2 deficient animals. As determined by PCR cloning and sequencing, the IL-2 RNA made is actually shorter than expected, since the entire exon 3 is spliced out.

The first *ex vivo* analyses of homozygous IL-2 deficient animals at the age of 4 days and 4 weeks showed a surprisingly normal phenotype (see below). At later time points, however, leukocyte composition is severly disturbed and the animals get fatally ill. Data presented on the immunological status presented here were therefore derived from young animals, to reduce the possibility of secondary effects. Pathological developments later in life will be discussed separately.

Thymus Development

Intrathymic differentiation of T-cells proceeds from a T-cell receptor (TCR) negative, CD4-CD8- via a TCR^{low}, CD4+CD8+ stage towards mature TCR^{high},CD4+CD8- and TCR^{high},CD4-CD8+ phenotypes which populate the peripheral organs (reviewed in Shortman et al. 1990). Analysis of IL-2 deficient animals revealed no abnormality in the generation of these major phenotypes as defined by $\alpha\beta$–TCR, CD4 and CD8. Size and cellularity of the thymus were normal (Schorle et. al. 1991). The numbers of CD25+TCR- cells were generally also comparable to those found in controls as was the number of $\gamma\delta$ positive T-cells. We thus could not detect any significant differences between thymocytes maturing in the absence of IL-2 and those generated in control animals with respect to any of the major thymocyte subpopulations as defined by these markers. Whether there are more subtle changes in minor subpopulations and/or selection processes remains to be seen.

Presence of Dendritic Epidermal T cells

Very early in ontogeny, a wave of Vγ3+ $\gamma\delta$ cells is generated in the thymus which seems to almost exclusively home to the skin (Asarnow et al. 1988). Because of their IL-2 dependence in *in vitro* growth studies one might expect that these DETCs could be maintained by locally produced IL-2. Surprisingly, however, no difference in the density of the Thy1+Vγ3+ DETCs was seen between skin sheets of IL-2 deficient mice and of control littermates. Whether these cells are also functional has so far not been amenable to testing.

Presence of NK 1.1 Cells

A further population whose growth and activity can be influenced by culturing in IL-2 in vitro are NK cells (Trinchieri et al. 1984). Cytofluorometric analysis of spleen cells from homozygous IL-2 -/- mice and wildtype littermates with anti NK1.1 antibodies revealed approximately

3% positive cells in both groups. Functional analysis of freshly isolated NK cells was made difficult by the fact that even wildtype animals at the early age of 3-4 weeks had a very low activity on NK targets. IL-2 -/- mice and littermates which were primed with poly IC in vivo for 24 hours did, however, show measurable NK activity. In IL-2 deficient animals which already exhibited pathological changes in leukocytes composition, NK activity was very low, even after poly IC induction.

In vitro Proliferation of Peripheral T cells and Lymphokine Production

In contrast to the normal phenotype with respect to the representation of various T cell and leukocyte subsets, the consequences of IL-2 deficiency became immediately obvious when unseparated cells from thymus, spleen and lymph nodes were stimulated in vitro with the T cell mitogen Con A, with anti-CD3ε antibodies, staphylococcal superantigen SEB or allogeneic T-cell depleted stimulator cells. In all cases, DNA synthesis was markedly reduced but was restored to control values by the addition of IL-2 (Schorle et al. 1991, Schimpl et al. 1992). When T cells from IL-2 -/- mice were stimulated for 2-4 days without the addition of IL-2, the IL-2R 55KDa chain was expressed at lower levels, particularly on CD8+ cells, consistent with the known upregulation of the IL-2 receptor by interaction with its ligand (reviewed in Smith 1988, 1992)

The reduction in ^{3}H-thymidine incorporation after mitogenic stimulation of cells deficient in IL-2 production was, however, not complete. Particularly when unseparated lymph node cells or populations enriched for CD4+ cells from spleen and lymph nodes were stimulated with high doses of plastic adherent anti-CD3 antibodies, vigorous ^{3}H-thymidine incorporation was observed. This proliferation was not inhibited by anti-IL-4 antibodies, suggesting either that the antibody failed to inhibit an autocrine IL-4 pathway or that other lymphokines are involved or even that direct ligand/receptor signalling through cell interaction molecules might obviate the need for IL-2.

We also studied the lymphokines produced in the absence of signalling through IL-2 by PCR analysis of RNA isolated directly *ex vivo* or after stimulation of unseparated or CD4+ enriched spleen and lymph node cells. Essentially, mRNA for all cytokines investigated (IL-3, IL-4, IL-6, IL-7, IL-10, γ-IFN, TGFß) was easily detected and, under certain conditions of stimulation, was even overproduced in IL-2 deficient as compared to normal mice, with the exception of IL-5 which was hard to detect both *ex vivo* and after short term stimulation in both groups

Failure to Generate Cytotoxic T cells *in vitro*

In spite of residual proliferative activity in unseparated spleen and lymph node cells treated with T cell stimulating agents, we totally failed to generate cytotoxic T cells in vitro in the absence of IL-2. Neither cells polyclonally activated with ConA and tested on the 1452c11 anti-TCRε hybridoma nor cells stimulated with allogeneic T cell depleted stimulator cells assayed on allogeneic tumour cells were able to destroy their target cells. Again, the response was fully reconstituted by IL-2, added at the onset of the cultures. This indicated that precytotoxic T cells are generated in normal numbers in IL-2 deficient mice, but that their *in vitro* proliferation and/or differentiation totally depends on IL-2. Whether *in vivo* other lymphokines or ligand/receptor systems may overcome the need for IL-2 remains to be seen.

Efficient Signalling Through the CD28/B7 Ligand-Receptor System in T Cells from IL-2 Deficient Mice

It has been recognized for some time (reviewed in June et al. 1990) that costimulation with anti CD28 or the ligand of CD28, B7/BB1 leads to increased proliferation of human T cells stimulated under suboptimal conditions. At the molecular level, the costimulatory effect of anti CD28 was ascribed, at least in part, to an increase in IL-2 production, regulated at both the transcriptional and the posttranscriptional level, the latter through stabilisation of IL-2 mRNA (Lindsten et al. 1989). Recently, Allison and coworkers (Harding et al. 1992) described an anti-mouse CD28 antibody which costimulated proliferation of purified mouse CD4+ cells activated by low doses of plastic adherent anti-CD3 antibodies. Using the anti-mouse CD28 antibody provided by Dr. J. Allison we addressed the question of whether costimulation was also possible in the absence of IL-2 production. When unseparated lymph node cells from IL-2 deficient mice were treated with phorbol ester, which in the mouse does not stimulate proliferation *per se*, anti-CD28 induced an increase in ^{3}H thymidine incorporation which was comparable to that observed when cells from wildtype animals were treated in the same way. Proliferation of IL-2 -/- CD4+ T-cells was also greatly enhanced when these cells were stimulated on plastic dishes coated with low doses of anti-CD3, as was reported for normal mice by Harding et al.. An analysis of lymphokine mRNA production by PCR after costimulation with anti CD3 also showed pronounced enhancing effects on IL-3, γIFN and the residual Il-2 RNA coding for an exon3-less and, by bioassay, inactive protein. These observations clearly demonstrate that anti-CD28 costimulation provides signals which can increase T cell proliferation independently of IL-2.

Another system in which CD28/B7 interaction seems to play an important role has recently been described by Lanier and coworkers (Azuma et al. 1992). Their work showed that resting human CD4+ and CD8+ positive cells could kill B7 transfected mouse P815 cells in the presence of anti-CD3 antibodies, which crosslink target and effector cells through the TCR and the Fc receptors on P815 and delieaver an activating signal through the TCR. In contrast, only activated T cells were able to kill untransfected P815 cells in redirected lysis.

Using the B7 transfected P815 cells, provided by Dr. L. Lanier, and freshly isolated spleen and lymph node cells, we confirmed their observation for mouse effector cells and found that already at 3-4 weeks of age, IL-2 deficient mice have an increased capacity to kill these cells in the presence of anti-CD3, although, like the resting human T cells described by Azuma et al., they fail to kill untransfected P815 cells.

CD44 and Mel-14 expression on T cells from IL-2 Deficient Mice

CD44^{high}Mel-14^{low} cells have been described in the mouse as representing a phenotype which reflects previous antigen encounter of the cells (reviewed in Swain et al. 1991), similar to the CD45RO marker used in the studies by Azuma et al. For that reason and because we wondered about the activation stage that can be reached by T cells in the absence of IL-2 we investigated the levels of expression of CD44 and Mel-14 in spleen and lymph nodes of IL-2 deficient mice and their littermate controls. Already in very young animals we find an increase of CD44^{high} and of Mel-14^{low} cells in both the CD4 and CD8 subsetss. The increase is particularly drastic in the spleen, but is also observed in lymph nodes. With increasing age, almost all cells eventually carry this marker combination. This observation could explain both the increased efficiency of peripheral T cells from IL-2 deficint animals to kill the B7 transfected targets and the general increase in lymphokine production observed after anti-CD3 stimulation of purified CD4+ cells, since Swain et al. have reported that CD44^{high}Mel-14^{low} cells are mostly responsible for the production of lymphokines other than IL-2 (Swain et al. 1991).

Late phenotypic changes in IL-2 deficient mice.

When we first analysed IL-2 deficient mice, the only *in vivo* changes we observed at an early age was a certain disbalance in immunoglobulin isotypes in the serum. IgG1 and IgE were greatly enhanced as compared to those found in equally young littermates, already indicating a certain disbalance in the immune system (Schorle et al. 1991). The immune response to the T-dependent antigen TNP-KLH was normal or even enhanced (Schimpl et al. 1992) and priming for KLH-induced *in vitro* proliferative responses was also quite normal. However, when the animals get older they develop a number of disease symptoms. The first indications are splenomegaly and lymphadenopathy and signs of B lymphopoiesis outside the bone marrow. At the time of the increases in spleen size we also observe enormous increases in IgG1 and IgA secreting cells in the spleen, the two isotypes so far tested (Sadlack et al. submitted). Sera and hybridomas established from the animals contain autoantibodies of various specificities. This overstimulation of the humoral compartment of the immune system is then followed by a gradual loss of B220+ cells from spleen, lymph nodes and bone marrow.

The disease phenotypes which develop in IL-2 deficient mice vary to some degree. This may either be caused by the heterogeneous background contributed by the 129 derived ES cells and C57Bl/6 derived blastocysts or may be secondary to the individual antigenic experiences of the animals. The reason for the apparent overstimulation of the immune system is not clear, but several possibilities can be tested. Besides a direct effect of a disbalanced synthesis of lymphokines on leukocyte composition, inability to properly terminate immune responses to exogenous or to autoantigens could be the result of disrupted cytokine regulatory loops or even due the absence or overrepresentation of an as yet unidentified regulatory subset. Identification of the cells and/or products that in normal mice prevent the pathological changes observed as a result of IL-2 deficiency may provide a clue to mechanisms underlying leukocyte homeostasis.

Acknowledgments. We thank K. Borschert, H. Haber, R. Mitnacht and A. Zant for their expert technical assistance and Dr. J. Allison and Dr. L. Lanier for generously providing us with reagents and cells. This work was supported by the Sonderforschungsbereich 165 and the Fonds der Chemischen Industrie.

References

Asarnow DM, Kuziel WA, Bonyhadi M, Tigelaar RE, Tucker PW, Allison JP (1988) Cell 55: 837

Azuma M, Cayabyab D, Buck D, Philipps JH , Lanier LL (1992) CD28 interaction with B7 costimulates primary allogeneic proliferative responses and cytotoxicity mediated by small, resting T lymphocytes. J Exp Med. 175: 353

Carding SR, Hayday AC, Bottomly K (1991) Cytokines in T-cell development. Immunol Today: 12, 239

Harding FA, McArthur JG, Gross JA, Raulet DH, Allison JP (1992) CD28 mediated signalling co-stimulates murine T cells and prevents induction of anergy in T-cell clones. Nature 356: 607

Mosmann TR, Street NE, Fiorentino DF, Fong TAT, Schumacher J Jeverah JA, Trounstine M, Vieira P, Moore KW (1989) Heterogeneity of mouse helper T cells and cross-regulation of TH1 and TH2 clones. Progr Immunol. 7: 611.

Paul WE (1989) Pleiotropy and Redundancy : T cell-derived lymphokines in the immune response. Cell 57: 521

Plum J, De Smedt M (1988) Differentiation of thymocytes in fetal organ culture: Lack of evidence for the functional role of the Interleukin-2 receptor expressed by prothymocytes. Eur J Immunol.18: 795

Schimpl A, Schorle H, Hünig T, Berberich I, Horak I (1992)Lymphocyte subsets and their reactivity in IL-2 deficient mice. In "Cytokines in Health and Disease", E. Romagnani Ed.,in press

Schorle H., Holtschke T, Hünig T., Schimpl A., Horak I. (1991) Development and function of T cells in mice rendered Interleukin-2 deficient by gene targeting. Nature 352: 621

Smith KA (1988) Interleukin-2: inception, impact and implications. Science 240: 1169

Smith KA (1992) Interleukin-2. Current Opinion in Immunology 4: 271

Swain SL, Bradley LM, Croft M, Tonkonogy S, Atkins G, Weinberg AD, Duncan DD, Hedrick SM, Dutton RW, Huston G. (1991) Helper T-Cell subsets: Phenotype, function and the role of lymphokines in regulating their development. Immunol Rev 123: 115

Tigges MA, Casey LS, Koshland ME (1989) Mechanisms of Interleukin-2 signalling: mediation of different outcomes by a single receptor and transduction pathway. Science 243: 781

Trinchieri G, Matsumoto-Kobayashi M, Clark SC, Seehra J, London L., Perussia B. (1984) Response of resting human peripheral blood natural killer cells to Interleukin-2. J Exp Med 160: 1147

Zubler RH, Lowenthal JW, Erard F, Hashimoto N., Devos R., MacDonald HR (1984) Activated B-cells express receptors for, and proliferate in response to, pure Interleukin-2. J Exp Med 160: 1170

Zurawski SM, Zurawski G (1988) Identification of three critical regions within mouse Interleukin-2 by fine structural deletion analysis. EMBO J: 7, 1061

Biology of Interleukin 9 and its Receptor

J-C. Renauld, F. Houssiau, A. Vink, C. Uyttenhove, G. Warnier and J. Van Snick

Ludwig Institute for Cancer Research, Brussels Branch and Experimental Medicine Unit, Catholic University of Louvain, 74 Avenue Hippocrate, B-1200 Brussels, Belgium

INTRODUCTION

The first description of IL9 was based on the observation that supernatants of activated helper T-cell clones were capable of supporting the long term growth of certain T-cell clones in the absence of antigen and antigen presenting cells (Uyttenhove 1988). The protein responsible for this activity, designated P40 on the basis of its apparent size in gel filtration, was characterized by an elevated pI (~10) and a high level of glycosylation. Partial amino acid sequences allowed the screening of a cDNA library with oligonucleotides and the cloning of a full length clone encoding the murine P40 protein (Van Snick 1989). Subsequently, the cDNA of the human homologue was isolated by cross-species hybridization (Renauld 1990a,b), while Yang (1989) isolated the human IL9 cDNA by expression cloning of a factor stimulating the growth of the Mo7E megakaryoblastic leukemia line.

Like for most of the interleukins, it appeared soon that the biological targets of IL9 extended far beyond those initially reported. IL9-mediated activities have indeed been described recently on erythroid progenitors (Donahue 1990; Holbrook 1990), B cells (Dugas manuscript in preparation), mast cells (Hultner 1990) and fetal thymocytes (Suda 1990). In this short review, we summarize the most recent available information on the biology of IL9, with emphasis on the activity of this cytokine on human and murine T cells and the molecular characterization of its cell surface receptor.

Functional and biochemical characterization of the murine IL9 receptor was achieved by binding experiments performed with radiolabelled IL9. Scatchard analysis on a T cell line indicated the existence of a single class of binding sites exhibiting a Kd of ~100 pM. Cross-linking studies with T cell lines showed that the IL-9 binds essentially to a single 64 kDa glycoprotein, the molecular mass of which is reduced to 54 kDa on treatment with N-glycosidase F (Druez 1990).

Various cell types were found to express IL9 receptors, as assessed by binding studies. In accordance with the initial description of IL9, IL9 receptors were detected in some T cell clones and were upregulated upon stimulation. The highest number of binding sites was found on IL9-dependent T cell clones bearing 2,000 - 3,000 receptors/cells. Certain T-cell tumors such as EL4 and LBRM-33 also expressed IL9 receptors. In contrast, fresh T cells or thymocytes failed to show any significant binding. Interestingly, amongst other cell types, 2 macrophage cell lines exhibited IL9-binding sites, although the activity of IL9 on these cells is still elusive.

cDNA clones encoding the murine IL9 receptor have been identified by expression cloning in COS cells (Renauld 1992). The screening of a cDNA library from a T cell clone led to the identification of a mouse cDNA clone that was further used as a probe to isolate human IL9 receptor cDNAs. The IL9-receptor is a typical transmembrane protein with two hydrophobic regions corresponding to the signal peptide and transmembrane domain. The human and mouse protein contain respectively 468 and 522 amino acids and are 53% homologous, with 67% identity for the extracellular region.

Searches for homology revealed a significant similarity between the extracellular domain of the IL9 receptor and several other recently cloned growth factor receptors. Particularly, the presence of a WSEWS motif and of 4 cysteine residues with a fixed distance indicates that the IL9 receptor is a new member of the hematopoietin receptor superfamily. As for the cytoplasmic domain, we failed to detect any sequence suggestive of the mechanism of signal transduction. However, some sequence homology with other cytokine receptors was noticed proximally to the transmembrane domain. This segment partially fits a consensus sequence recently described by Murakami (1992) and contains a Pro-X-Pro sequence preceded by a cluster of hydrophobic residues. For the first 33 amino acids of the cytoplasmic domain, a 40% identity was observed between the human IL9-receptor and the ß chain of the IL2-receptor and a 27% identity with the Erythropoietin receptor (Fig. 1). Interestingly, these 2 receptors interact with the factors that have been shown to synergyze with IL9 for the proliferation of fetal thymocytes and erythroid progenitors, respectively.

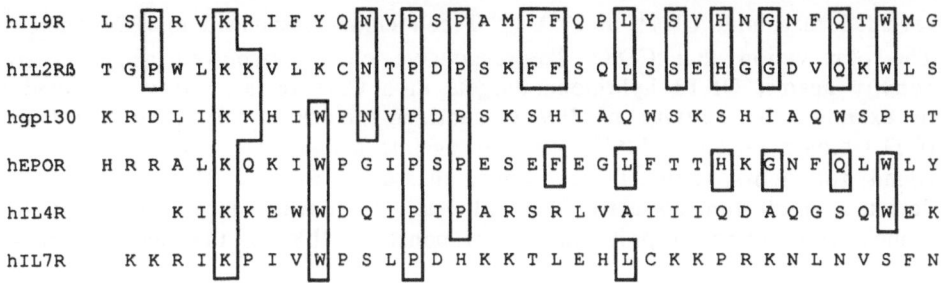

Figure 1. Sequence alignment of the cytoplasmic proximal segment of the human IL9-receptor sequence with other cytokine receptors.

RNA analysis showed that the murine IL9 receptor message consists of at least 3 transcripts of different sizes while human Northern blots showed six different bands resulting from the use of alternative polyadenylation signals or alternative splicing. In this respect, in addition to membrane-bound receptor forms, we identified a murine cDNA encoding a putative soluble form of the protein caused by the deletion of the sequences encoding the transmembrane and cytoplasmic domains (Renauld 1992).

T CELL GROWTH FACTOR ACTIVITY OF HUMAN IL9

In the mouse, the T cell growth factor activity of IL9 is apparently restricted to certain T Helper clones while fresh T cells seem unresponsive to this factor (Uyttenhove 1988; Schmitt 1989). In the human, the response of T cells to IL9 had not been studied extensively so far. No activity could be demonstrated on freshly isolated T cells but preliminary experiments indicated that the survival of some human T cell lines was enhanced by IL9 (Renauld 1990a). To address this issue further, we derived T cell lines from peripheral blood mononuclear cells (PBMC) by weekly stimulation with PHA, IL2 and irradiated allogeneic PBMC as feeders. Assuming that expression of the IL9R was a prerequisite for the response to IL9, we screened T cell lines for the presence of the IL9R at the RNA level. Interestingly, all T cell lines tested were found positive in Northern blot, as early as one week after the onset of the culture.

When stable cell lines were obtained, (after 5 to 8 weekly passages) their proliferative response to IL9 was tested. Most of the lines raised (11/12) were found responsive to IL9 (five- to fifty-fold increase in thymidine incorporation) with half-maximal proliferation being observed with as low doses as 1.5 U/ml of IL9, as measured on Mo7E cells. Cloning procedures revealed that both CD4[+] and CD8[+] T cell clones responded to IL9 (Table 1). The observation that CD8[+] IL9-responsive clones displayed a potent lytic activity against OKT3 hybridoma targets in a Cr[51] release assay prompted us to examine the activity of IL9 on antigen-specific cytolytic T lymphocyte (CTL) clones raised in MLTC cultures against melanomas or lung tumors. As found for non-specific T cell lines, messages for the IL9 receptor were detected in all tumor-specific CTL clones. Moreover, six out of eight such clones were found to proliferate in response to IL9, in the range illustrated in table 1.

Table 1. Response of human T cell clones to IL9[1]

Clone	Phenotype	Thymidine incorporation in the presence of	
			IL9
E5	CD4	5,315	29,566
B4	CD4	3,644	9,403
T5	CD8	0,974	23,915
T10	CD8	0,217	6,755
159/3	CD8 > melanoma	2,479	16,656
219B/11	CD8 > lung cancer	1,380	38,385
215/2	CD8 > melanoma	1,765	4,735

[1]: Cells were seeded in microtiter plates (1×10^5) in the presence or in the absence of IL9 (35 U/ml). Thymidine incorporations (cpm) were measured on day 2. SEM were less than 10% of the triplicate values.

It must be stressed however, that the proliferative response induced by IL9, unlike that induced by IL2, varied according to the timing after restimulation of the T cell culture. It was optimal on the 7th day after reculturing, when the cells had reached a fully blastic stage and a density of circa 1×10^6 cells/ml. When tested later in the culture, namely when the cells had already undergone a size reduction, the proliferation in response to IL9 was significantly reduced (Houssiau manuscript in preparation).

In contrast with previous data obtained in the mouse system, we failed to derive any stable IL9-dependent cell line able to grow in the absence of feeder cells or other factors. The activity of IL9 on human CTL clones also contrast with the results obtained in the mouse where IL9-responsiveness was shown to be restricted to certain T helper clones. The significance of the activity described for IL9 in these two experimental systems could thus be very different.

IL9 AND MOUSE T CELLS: A ROLE IN TUMORIGENESIS ?

In the mouse, the response of T cell clones to IL9 is gradually acquired by long-term in vitro culturing. In a first stage the clones do not proliferate with IL9. In a second stage, IL9 receptors can be detected by binding assays but the cells respond to IL9 only in the presence of another factor such as IL4 or IL3. Finally, the cells become fully responsive to IL9 and can be grown with this factor in the absence of antigen and antigen-presenting cells.

This multistep process leading to IL9-responsiveness is reminiscent of other processes leading to tumorigenic transformation, a view supported by the observation that, after transfection with the IL9 cDNA, an IL9-dependent T cell clone became capable of forming tumors in vivo (Uyttenhove 1991). This finding suggests that dysregulated production of IL9 by T cells could be part of T cell transformation. To test this hypothesis we generated transgenic mice expressing high levels of IL9 constitutively (Renauld, manuscript in preparation). Although no major morphological changes were noticed in the immune system of most mice, they show a higher susceptibility to thymic lymphoma as (i) about 5% of the mice spontaneously develop lymphomas and (ii) all transgenic mice develop the similar tumors after injection of subliminal doses of a mutagen (N-methyl nitroso-Urea).

The tumors usually involved primarily the thymus and invaded other lymphoid organs such as the spleen and the lymph nodes. Most of the tumors had rearranged the TCRß locus and expressed both the CD4 and CD8 antigens while CD3 expression varied between tumors. Noteworthy, the instability of the transgene in one of the 5 independent founders studied led to the development of chimeric mice, where part of the cells had lost the transgene. This phenomenon allowed the occurence in these mice of tumors that had lost the transgene. One of these tumors, designated 9T4, initially failed to grow in vivo after transplantation into syngeneic normal mice. However, injection of 9T4 into transgenic mice rapidly resulted in tumor formation, thereby demonstrating that IL9-requirement was not restricted to the initial steps of the oncogenesis. Injection of high numbers of 9T4 tumor cells into normal mice also resulted, in some experiments, in tumor formation after a prolonged period of time. The observation that the latter tumors are subsequently able to grow indistinctly in normal or transgenic mice suggested that they may undergo additional genetic alterations leading to an IL9-independent growth.

The occurence of thymic lymphomas in IL9 transgenic mice prompted us to investigate whether IL9 played a similar role in other models of thymic lymphomas. In this respect, we have recently found that IL9 significantly stimulates the in vitro proliferation of primary lymphomas induced either by chemical mutagenesis in DBA/2 mice or by X-ray radiation in B6 mice (Vink manuscript in preparation). Moreover, these studies have demonstrated a strong synergy between IL9 and IL2 for several lymphomas (Fig. 2), a synergy reminiscent of the activity of IL9 on murine fetal thymocytes (Suda 1990).

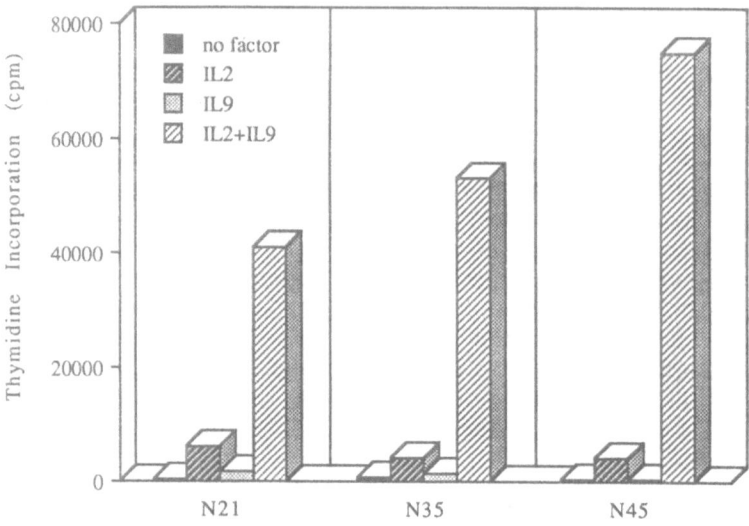

Figure 2. Synergy between IL9 and IL2 for the proliferation of NMU-induced thymic lymphomas.

CONCLUSIONS

Although IL9 was originally characterized and cloned on the basis of its T cell growth stimulatory activity, its relevance for the T cell biology is still unclear. Results recently obtained in the human demonstrate that IL9 has a broader spectrum of activity than initially observed and includes antigen-specific CTL clones. However, these studies also stress the importance of the state of activation of the T cell cultures for a response to IL9 to be observed. These results, together with the absence of IL9 responses on freshly isolated T cells, suggest that the growth factor activity of IL9 might be restricted to T cells in a particular stage of activation rather than to a specific IL9-responsive subclass of T cells.

On the other hand, the studies performed in the mouse emphasize the ability of IL9 to promote the growth of T cells undergoing transforming processes. The incidence of thymic lymphomas in IL9-transgenic mice supports the view that IL9 gene dysregulation could be involved in T cell oncogenesis. Recent observations showing constitutive IL9 expression in lymph nodes from patients with Hodgkin disease (Merz 1991), together with the demonstration of an autocrine loop for the in vitro growth of one Hodgkin cell line (Gruss 1992) suggest a potential significance for IL9 in human pathology.

REFERENCES

Donahue RE, Yang YC, Clark SC (1990) Blood 75:2271-2275

Druez C, Coulie P, Uyttenhove C, Van Snick J (1990) J Immunol 145:2494-2499

Gruss HJ, Brach AM, Drexler HG, Bross KJ, Herrmann F (1992) Cancer Res 52:1026-1031

Holbrook ST, Ohls RK, Schribler KR, Yang YC, Christensen RD (1991) Blood 77:2129-2134

Hültner L, Druez C, Moeller J, Uyttenhove C, Schmitt E, Rüde E, Dörmer P, Van Snick J (1990) Eur J Immunol 20:1412-1416

Merz H, Houssiau F, Orscheschek K, Renauld J-C, Fliedner A, Herin M, Noel H, Kadin M, Mueller-Hermelink HK, Van Snick J, Feller AC (1991) Blood 78:1311-1317

Murakami M, Narazaki M, Hibi M, Yawata H, Yasukawa K, Hamaguchi M, Taga T, Kishimoto T (1991) Proc Natl Acad Sci USA 88:11349-11353

Renauld J-C, Goethals A, Houssiau F, Van Roost E, Van Snick J (1990a) Cytokine 2:9-12

Renauld J-C, Goethals A, Houssiau F, Merz H, Van Roost E, Van Snick J (1990b) J Immunol 144:4235-4241

Renauld J-C, Druez C, Kermouni A, Houssiau F, Uyttenhove C, Van Roost E, Van Snick J (1992) Proc Natl Acad Sci USA 89:5690-5694

Schmitt E, van Brandwijk R, Van Snick J, Siebold B, Rüde E. (1989) Eur J Immunol 19:2167-2170

Suda T, Murray R, Fischer M, Yokota T, Zlotnik A (1990). J Immunol 144:1783-1787

Uyttenhove C, Simpson RJ, Van Snick J (1988) Proc Natl Acad Sci USA 85:6934-6938

Uyttenhove C, Druez C, Renauld J-C, Herin M, Noel H, Van Snick J (1991) J Exp Med 173:519-522

Van Snick J, Goethals A, Renauld J-C, Van Roost E, Uyttenhove C, Rubira MR, Moritz RL, Simpson RJ (1989) J Exp Med 169:363-368

Yang YC, Ricciardi S., A. Ciarletta A, J. Calvetti J, K. Kelleher K, and S. C. Clark SC (1989) Blood 74:1880-1884

The IL-2 Receptor and Its Target Genes in Hematopoietic Cell Cycle

Tadatsugu Taniguchi, Hiroshi Shibuya, Yasuhiro Minami, Takeshi Kono, Masanori Hatakeyama, Naoki Kobayashi, and Mitustoshi Yoneyama

Institute for Molecular and Cellular Biology, Osaka University, Yamadaoka 1-3, Suitashi, Osaka 565, Japan

Intoroduction

Cytokines are the critical requlators of proliferation and differentiation for hematopoietic cells. Interleukin-2 (IL-2) was originally identified and has been extensively studied in the context of the clonal expansion of antigen-activated T lymphocytes (T cells). In fact, antigen-specific, clonal proliferation of T cells is initiated via a process of signal transduction, wherein the specific interaction of the antigen/MHC molecule and T cell antigen receptor complex (TCR) triggers the expression of IL-2 and its homologous receptor (IL-2R). The interaction of IL-2 with IL-2R leads to the stimulation of a set of complex, yet mostly unknown, signal transduction pathways resulting in cell proliferation. The specific cell surface receptor (IL-2R) which binds IL-2 is composed of at least three distinct polypeptides, the IL-2Rα, IL-2Rβ and IL-2Rγ chains. The genes encoding IL-2 and these three receptor subunits have been cloned, and their complete primary structures have been deduced (reviewed in Greene and Leonard, 1986; Smith, 1988; Waldman,1989; Minami et al., 1993).

The molecular characterization of cytokine receptor genes have revealed a new family of receptors which permit the cytokines to stimulate responses in target cells. Importantly, members of this new receptor family lack the intrinsic protein tyrosine kinase domain that is the hallmark for the receptors that bind other growth factors such as EGF, PDGH and CSF-1 (reviewed in Bazan, 1990; Cosman et al., 1990; Miyajima et al., 1992; Minami et al., 1993).

More recently, evidence has been provided that the IL-2Rβ chain is responsible for transmitting the proliferative signal(s) via a distinct cytoplasmic region. Furthermore, the IL-2Rβ couples with a src-family protein tyrosine kinase p56[lck] both physically and functionally (Hatakeyama et al., 1989; 1991). Thus, this is the first demonstration that a cytokine receptor of this family is linked to a known signaling molecule, and suggests a mechanism by which cytokine stimulation induces protein tyrosine phoshorylation in the cytoplasm.

Here we shall provide an overview of our current knowledge of the IL-2R complex, IL-2 signal transduction and the target genes critical for the hematopoietic cell cycle progression.

Structure of the IL-2R complex

Three classes of IL-2Rs have been described on the basis of their affinity to the ligand; i.e. high-, intermediate- and low-affinity IL-2Rs. The availability of cDNAs for IL-2R subunits has made it feasable to gain insists onto the nature of the distinct classes of IL-2Rs: The high-affinity ($Kd=10^{-11}M$) IL-2R contains three distinct subunits, IL-2Rα, IL-2Rβ and IL-2Rγ. The intermediate-affinity ($Kd10^{-9}M$) IL-2R contains two distinct subunits, IL-2Rβ and IL-2Rγ. In contrast, IL-2Rα alone binds IL-2 with low affinity ($Kd=10^{-8}$). (reviewed in Smith, 1988; Waldman, 1989; Minami et al., 1993). A schematic representation of the high-affinity IL-2R complex is provided in Figure 1.

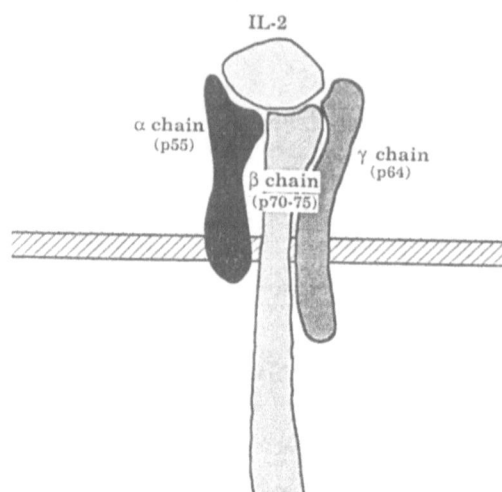

Figure 1. Schematic representation of the high-affinity IL-2R complex.

IL-2 receptor α chain (IL-2Rα)

The human IL-2Rα, originally described as the Tac antigen, was identified as a 55kDa membrane glycoprotein (p55) capable of binding IL-2 (Uchiyama et al., 1981). The deduced amino acid sequence of the human IL-2Rα from the cloned cDNA indicates a mature protein of 251 amino acids with a signal peptide of 21 amino acids in length. IL-2Rα lacks structural features characteristic of members of the immunoglobulin superfamily and does not belong to the cytokine receptor superfamily. Within this chain, regions comprised of the amino-terminal 219 a.a. residues, the internal 19 a.a. residues and the carboxy-terminal 13 a.a. residues constitute the extracellular, membrance-spanning and cytoplasmic regions, respectively.

cDNA transfection studies have revealed that IL-2Rα itself constitutes the low-affinity IL-2R(Kd,10^{-8}) in non-lymphoid cells and it participates in the formation of the high-affinity IL-2R in IL-2Rβ and IL-2Rγ positive cells. Mutation analysis of the IL-2Rα cDNA indicated that the 13 a.a. cytoplasmic region of the IL-2Rα is dispensable for IL-2 signal transduction.

IL-2 receptor β chain (IL-2Rβ)

The second IL-2R component, IL-2Rβ was originally by affinity cross-linking experiments. Structure of IL-2Rβ was elucidated through the expression cloning of the cDNA for the human IL-2Rβ, using monoclonal antibodies against IL-2Rβ (Hatakeyama et al., 1989). The full-length IL-2Rβ cDNA contains a large open reading frame which encodes a protein consisting of 551 amino acids (a.a.). From the deduced structure of the protein, the NH_2-terminal 26 amino acids apparently comprise the signal sequence, leaving 525 a.a. to make up the mature form of IL-2Rβ. Within this chain, regions of 214a.a., 25a.a. and 286 a.a. in length, constitute the extracellular, membrance-spanning, and cytoplasmic regions, respectively. Interestingly, the predicted number of amino acid residues within the extracellular region of IL-2Rβ (214a.a.) is comparable to that found in IL-2Rα (219a.a.). The cytoplasmic region of IL-2Rβ is far larger than that of IL-2Rα, which is only 13 amino acid residues in length. However, the cytoplasmic region of IL-2Rβ does not contain any apparent catalytic motifs such as a kinase consensus sequence. The cytoplasmic region of IL-2Rβ can be tentatively divided into three subregions based upon their amino acid compositions. These subregions have been designated as the "serine-rich" region, "acidic" region and the "proline-rich" region, respectively as shown in Figure 2.

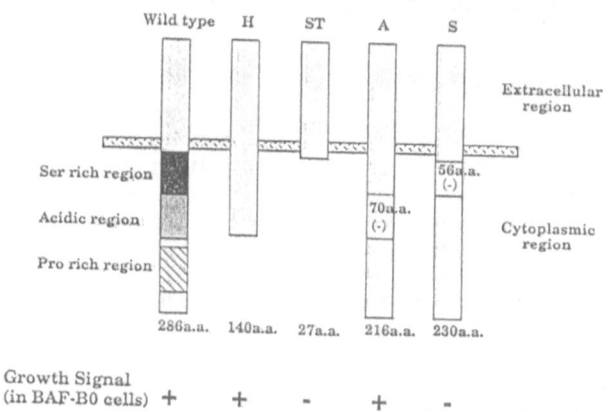

Figure 2. The IL-2Rβ mutants and their properties in growth signal transduction (*see text or Hatakeyama et al., 1989 for details*).

IL-2 receptor γchain (IL-2Rγ)

IL-2Rβ binds IL-2 with extremely low affinity (Kd=10^{-7}M) when expressed in fibroblasts such as NIH-3T3, L929 and COS-7 cells. In addition, it has been shown that IL-2 mutated at Glu-141 binds IL-2Rα and IL-2Rα/β heterodimers expressed on fibroblasts with normal affinity but are defective in binding to the same receptors on T cells. This mutant IL-2 also does not trigger the mitotic signal, suggesting the interaction of Glu-141 of IL-2 with an unidentified receptor component(s). These results suggested that a cell-type specific component [now refered to as IL-2Rγ(p64)] might be involved in the formation of the functional intermediate- and high-affinity IL-2Rs. Il-2Rγ(p64) can be co-precipitated with IL-2Rβ in the presence of IL-2 in lymphoid cells bearing the high-affinity IL-2R.

The deduced primary structure of the IL-2Rα from cloned cDNA revealed that the NH₂-terminal 22 amino acids (a.a.) apparently comprise the signal sequence, leaving 357 a.a. to make up the mature form of IL-2Rγ(22). Within this chain, regions of 232 a.a., 29a.a. and 86 a.a. in length, constitute the extracellular, membrane-spanning, and cytoplasmic regions, respectively. The cytoplasmic region of IL-2Rγ is considerably shorter than that of IL-2Rβ. Interestingly, sequences from positions 288 to 321 appears to be homologous to the Src homology region 2(SH2). IL-2Rγ belongs to the cytokine receptor superfamily. When the cDNAs for IL-

2Rα, IL-2Rβ and IL-2Rγ were introduced and expressed in fibroblasts, the high-affinity receptor for IL-2 was indeed reconstituted (22). In addition, it appears that IL-2Rγ is required for the receptor-mediated internalization of IL-2.

Other components

Other components associated with IL-2R has also been proposed based on co-immunoprecipitation of molecules that can be chemically cross-linked with IL-2R, although their biochemical natures remain to be clarified.

IL-2 signal transduction by IL-2Rβ

Among the three IL-2R component, IL-2Rβ contains the longest cytoplasmic region, suggesting the potential role of this region in the transmission of IL-2-induced proliferative signal(s). In order to elucidate further the role of the IL-2Rβ in the IL-2 signal transduction, a cDNA expression system was established in which the human IL-2Rβ transduces growth signal upon IL-2 stimulation in mouse cell lines. In fact, the cDNA linked to an expression vector was introduced in mouse mast cell progenitor cell line (IC-2) and pro-B cell line (BAF-B03). Both cell lines require IL-3 for cell growth. Although these cell lines expressed a large number of endogenous IL-2Rα, they could not respond to IL-2, indicating that the IL-2 signal is not transduced by the IL-2Rα alone. When the human IL-2Rβ was expressed in these cells by introducing the cDNA, high-affinity IL-2R was generated in conjunction with the endogenous IL-2Rα and, consequently, the cells became responsive to IL-2 as well as to IL-3. In order to identify the critical region(s) of IL-2Rβ for the growth signal transduction, mutant IL-2Rβ cDNA containing deletions within the region encoding the cytoplasmic region were generated. They were each expressed in BAF-B03 cells and the response of the cells expressing various IL-2Rβ mutants to IL-2 was examined. This study revealed that a restricted cytoplasmic region of the IL-2Rβ, i.e. the region encompassing the "serine-rich region", is most critical for IL-2 -signal transduction. Presumably, this region couples with an as yet unknown protein(s) which further drives the downstream signaling pathway(s).

IL-2R coupling with src-family kinases

It has been reported that IL-2 induces rapid phosphorylations at tyrosine residues of several proteins in the IL-2 responsive cells. Growth factor-induced tyrosine phosphorylations have been known to be essential in signal transduction by receptors containing intrinsic protein tyrosine kinase (PTK). In this regard, it has been shown that the human IL-2Rβ forms a stable complex with the lymphocyte-specific protein tyrosine kinase p56^{lck}. Specific association sites were identified in the tyrosine kinase catalytic domain of p56^{lck} and in the cytoplasmic region of IL-2Rβ which includes the "acidic" region (Figure 6). Furthermore, treatment of T cells with IL-2 promoted p56^{lck} PTK activity. Recently, we have shown that the interaction of p56^{lck} with IL-2Rβ is critical for the PTK activation induced by IL-2 (Y.M. et al, submitted for publication). These observations indicate the participation of p56^{lck} as a critical signaling molecule downstream of IL-2Rβ. Thus, this is the first demonstration that a cytokine receptor couples with a known signaling molecule. To further investigate a role of the src-family PTKs in IL-2 signalling, we analyzed a mouse pro-B cell line, BAF-BO3, in which *lck* is not expressed detectably. We observed that in this cell line, IL-2 induces activation of at least two members of the src-family, p59^{fyn} (*fyn*) and p53/56^{lyn} (*lyn*). Stimulation of this cell line with interleukin-3 (IL-3) also induces activation of these src-family PTKs. The activation of fyn or lyn seems to be selective for stimulation with IL-2 or IL-3 since stimulation with interleukin-6 (IL-6) fails to activate them. Furthermore, we have shown the physical association of *fyn* with IL-2Rβ (N.K.et al., submitted for publication). Taken together with previous results, this study suggests that different members of the src-family, each of which is expressed in a cell-type specific manner, can participate in the IL-2 signal transduction pathway (see Figure 3).

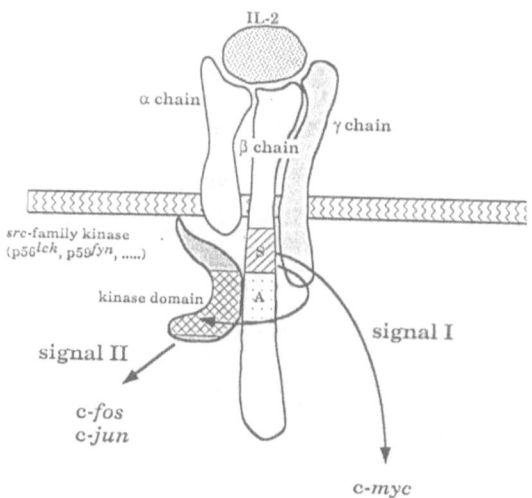

Figure 3. The IL-2Rβ coupling with src-kinases and induction of nuclear proto- oncogenes.

The target genes

As an approach to gain further insights into signal transduction pathways and the critical target genes for cytokine receptors, the properties of IL-2 and EGF receptors expressed in BAF-B03 cells were analysed with respect to their ability to induce expression of critical nuclear proto-oncogenes, the cyclins and cdc2-family kinases. The BAF-B03-delived cell lines, F7, A15 and BER2 each express wild-type IL-2Rβ, mutant Il-2Rβ lacking the "acidic region" (i.e. the primarly interaction site with src-family kinases), and wild-type EGF receptor (EGF-R). Flow cytometric analysis revealed that the growth factor-starved F7 and A15 cells were induced to transit the cell cycle by IL-2. On the other hand,The BER2 cells entered S phase following stimulation by EGF or TGF-α but they failed to progress further to G2/M phase, suggesting that a certain EGF-R signaling pathway(s) is deficient in this hematopoietic cells. The expression of the proto-oncogenes varies among the these cells after stimulation with the respective ligands. Expression of *jun*D, c-*myb* and Max, the partner of c-*myc*, at the RNA level is constitutive in these cells. Interestingly, c-*jun*, *jun*B, c-*fos* and *fos* B are ligand-inducible in BER2 and F7 cells, but not in A15 cells expressing the IL-2Rβ chain mutant that is deficient in PTK activation (Shibuya et al., 1992). Thus, inductions of these genes may be linked to a PTK pathway. Interestingly, c-*myc* gene is efficiently inducible in F7 and A15 cells but not in BER2 cells Collectively, each of the receptors analysed hare seems to deliver signals that are partly common but which are also distinct in terms of these gene induction (Figure 3). The only gene, the inducibility of which correlates with the ability of the cells to enter mitosis in c-*myc*. Furthermore, the deficiency of the BER2 cells for the EGF-induced mitotic response can be overcome or "rescued" by the ectopic expression of the human c-*myc* gene. Here, these observation suggest the role of c-*myc* in S to G2/M transition of the cell cycle.

In essence, IL-2R is linked to at least two intracellular signalling pathways. One pathway may involve a protein tyrosine kinase of the *src*-family which leads to the induction of the c-*jun*, c-*fos* genes among others (Figure 3). A second pathway, by an as yet unknown mechanism, leads to c-*myc* gene induction, which is critical for the S to G2/M cell cycle progression.

References

Bazan, J. F. 1990. Haemopoietic receptors and helical cytokines. *Immunology Today* 11: 350-354

Cosman, D., Lyman, S. D., Idzerda, R. L., Beckmann, M. P., Park, L. S., Goodwin, R. G., March, C. J. 1990. A new cytokine receptor superfamily. *TIBS* 15: 265-270

Greene, W. C., Leonard, W. J. 1986. The human interleukin-2 receptor. *Annu. Rev. Immunol.* 4: 69-96

Hatakeyama, M., Tsudo, M., Minamoto, S., Kono, T., Doi, T., Miyata, T., Miyasaka, M., Taniguchi, T. 1989. Interleukin-2 receptor β chain gene: generation of three receptor forms by cloned human α and β chain cDNA's. *Science* 244: 551-556

Hatakeyama, M., Mori, H., Doi, T., Taniguchi, T. 1989. A restricted cytoplasmic region of IL-2 receptor β chain is essential for growth signal transduction but not for ligand binding and internalization. *Cell* 59: 837-845

Hatakeyama, M., Kono, T., Kobayashi, N., Kawahara, A., Levin, S. D., Perlmutter, R. M., Taniguchi, T. 1991. Interaction of the IL-2 receptor with the *src*-family kinase p56*lck*: identification of novel intermolecular association. *Science* 252: 1523-1528

Minami, Y., Kono, T., Miyazaki, T., and Taniguchi, T. 1993. The IL-2 receptor complex; Its structure, function and target genes. Annu. Rev. Immunol., in press

Miyajima, A., Kitamura, T., Harada, N., Yokota, T., Arai, K. 1992. Cytokine receptors and signal transduction. *Annu. Rev. Immunol.* 10: 295-331

Shibuya, H., Yoneyama, M., Ninomiya-Tsuji, J., Matsumoto, K., Taniguchi, T. 1992. IL-2 and EGF receptors stimulate the hematopoietic cell cycle via different signaling pathways: Demonstration of a novel role for c-*myc*. *Cell* 70: 57-67

Smith, K. A. 1988. Interleukin-2: Inception, impact, and implications. *Science* 240: 1169-1176

Takeshita, T., Asao, H., Suzuki, J., Sugamura, K. 1990. An associated molecule, p64, with high affinity interleukin 2 receptor. *Int. Immunol.* 2: 477-480

Takeshita, T., Asao, H., Ohtani, K., Ishii, N., Kumaki, S., Tanaka, N., Munakata, H., Nakamura, M., Sugamura, K. 1992. Cloning of the γ chain of the human IL-2 receptor. *Science* 257: 379-382

Uchiyama, T., Broder, S., Waldmann, T. 1981. A monoclonal antibody (anti-Tac) reactive with activated and functionally mature human T cells. *J. Immunol.* 126: 1293-1297

Waldmann, T. A. 1989. The multi-subunit interleukin-2 receptor. *Annu. Rev. Biochem.* 58: 875-911

Zurawski, S. M., Imler, J.-L., Zurawski, G. 1990. Partial agonist/antagonist mouse interleukin-2 proteins indicate that a third component of the receptor complex functions in signal transduction. *EMBO J.* 9: 3899-3905

Biological Roles of IL-10 and the CD40 Receptor

M. C. Howard, A. W. Heath, H. Ishida, and K. W. Moore

DNAX Research Institute, 901 California Avenue, Palo Alto, CA 94304, U.S.A.

INTRODUCTION

The immune response is regulated by soluble and membrane-associated glycoproteins produced by a variety of hemopoietic and non-hemopoietic cells. We focus here on examples of each of these types of immune regulators. IL-10 is a soluble protein produced by helper T cells, macrophage/monocytes, and B cells, which exhibits a wide array of both immunosuppresive and immunostimulatory properties. CD40 is a membrane-associated receptor expressed predominantly by normal and malignant B cells, and which is an important regulator of B cell proliferation and differentiation. We consider here the biological roles of these proteins, and speculate on their clinical importance.

INTERLEUKIN 10

Discovery

The discovery of IL-10 developed from a detailed understanding of helper T cell heterogeneity and an appreciation that different T helper sub-populations (designated Th1 and Th2) regulate either antibody or cell-mediated immune responses [Mosmann and Coffman 1989]. The fact that these two effector arms of immunity are frequently mutually exclusive events lead to the concept of cross-regulation between Th1 and Th2 cells (reviewed in Mosmann and Moore 1991]). Based on this concept, Mosmann and colleagues developed an assay to identify activities expressed by Th2 cells which suppressed cytokine production by Th1 cells [Fiorentino *et al.* 1989]. This assay allowed isolation of one such activity, a novel cytokine subsequently designated IL-10 [Moore *et al.* 1990; Vieira *et al.* 1991].

IL-10, reviewed extensively elsewhere [Moore *et al.* 1992; Howard *et al.* 1992], is an acid-sensitive homodimer of 35-40 kD. It is produced by several T helper sub-populations, as well as activated B cells, macrophages, keratinocytes and mast cell lines. The single copy gene for IL-10 is located on chromosome 1 in both mouse and human species. While IL-10 is a novel mammalian protein, it is highly homologous to a previously uncharacterized ORF [termed BCRF1] in the EBV genome. EBV is indeed capable of transcribing BCRF1, and the resultant protein mediates many of the biological properties exhibited by mammalian IL-10. We have suggested that EBV has captured this immunoregulatory mammalian gene, presumably to acquire some survival advantage.

Immunosuppressive and Immunostimulatory Properties of IL-10

The immunosuppressive properties of IL-10 that have been identified to date derive primarily from its ability to inhibit cytokine production by Th1 cells, NK cells, and macrophages [Fiorentino, et al. 1989; Fiorentino *et al.* 1991, de Waal Malefyt *et al.* 1991; Hsu et *al.* 1992; Vieira, et al. 1991]. Suppression of cytokine production by Th1 and NK cells is an indirect effect reflecting IL-10 induced impairment of the accessory cell functions of macrophages and dendritic cells [Fiorentino *et al.* 1991b; de Waal Malefyt *et al.* 1991b; Hsu, et al. 1992; S. Macatonia *et al.*, submitted]. Interestingly, the accessory cell function of B cells appears to be unaffected by IL-10 [de Waal Malefyt, et al. 1991b; Fiorentino, et al. 1991b]. The mechanism of impaired accessory cell function is not fully understood, and is the subject of continuing intensive investigation. At least one component of this mechanism has been identified with the demonstration by de Waal Malefyt *et al.* that IL-10 down-regulates class II MHC antigen expression on cells of the monocyte/macrophage lineage [de Waal Malefyt, et al. 1991b]. Clearly, down-regulation of class II expression would produce serious adverse consequences on the antigen-presenting-cell capacity of these cells. However other aspects of impaired accessory cell function resulting from IL-10 exposure cannot be explained by down-regulated class II MHC antigen expression. For example, IL-10 suppresses accessory cell dependent, non-MHC restricted activation of IL-2 stimulated T cells [Fiorentino, et al. 1991b]. In addition, IL-10 inhibits monocyte/macrophage-dependent activation of cytokine synthesis by NK cells stimulated by IL-2 [Hsu, et al. 1992]. It is anticipated that IL-10 will cause modulation of co-stimulatory molecules critically involved in these latter pathways, but the identities of these co-stimulatory molecules are yet to be elucidated. In contrast, IL-10 mediated suppression of monokine production appears to be a direct effect on activated macrophages and monocytes, which leads to a striking down-regulation of numerous inflammatory monokines, such as TNFα, IL-1α, IL-6, IL-8, GM-CSF and G-CSF [de Waal Malefyt, et al. 1991a; Fiorentino, et al. 1991a]. This response does not represent a total shut-down of the macrophage protein synthesis, since production of TGFβ is unaltered and IL-1Ra expression is in fact elevated in IL-10-treated activated macrophages. In addition to its effects on production of inflammatory monokines, IL-10 also suppresses nitric oxide production by activated macrophages, a consequence of which is impaired killing of micro-organisms such as *Schistosoma mansoni* or *Toxoplasma gondii* [Gazzinelli *et al.* 1992].

Numerous immunostimulatory properties of IL-10 have also been identified. IL-10 causes up-regulation of class II expression on murine B cells, and enhances their survival *in vitro*. [Howard, et al. 1992] In addition, IL-10 is a potent co-stimulator of B lymphocyte function, augmenting growth and antibody production in cultures of human cells stimulated with cross-linked anti-CD40 antibodies [Defrance *et al.* 1992; Rousset *et al.* 1992]. In terms of stimulatory effects on other cell types, IL-10 is a growth co-stimulator for murine stem cells, megakaryocytes, mast cells, and thymocytes. Many of these stimulatory properties are discussed by D. Rennick *et al.* elsewhere in this volume.

In vivo Consequences of IL-10 Mediated Monokine Regulation

As described above, IL-10 effectively suppresses the production of inflammatory monokines such as TNFα. Conversely, continuous administration of anti-IL-10 antibodies to mice causes elevation of serum TNFα levels [Ishida *et al.* 1992b]. Two animal models were selected to explore the *in vivo* consequences of monokine regulation obtained using either IL-10 or neutralizing anti-IL-10 antibodies. Endotoxin-induced shock is a lethal inflammatory reaction mediated by monokines such as TNFα. Our data demonstrate that IL-10 effectively protects mice from endotoxin-induced shock, even when administered 30 min after the endotoxin [M. Howard *et al.*, manuscript in preparation]. We anticipate that the mechanism of this protection relates to *in vivo* down-regulation of inflammatory monokines such as TNFα together with up-regulation of the anti-inflammatory monokine IL-1Rα. The lupus-prone NZB/W F₁ mouse was selected as a second animal model, since previous studies have shown that development of autoimmunity in these mice correlates with TNFα deficiency and can be corrected by TNFα administration. Our studies demonstrated that continual IL-10 antagonism of NZB/W F₁ mice using neutralizing anti-IL-10 antibodies substantially delayed development of autoimmunity, as monitored by animal survival, development of proteinuria and pathological autoantibodies, and histological evaluation of glomerulonephritis [H. Ishida *et al.*, manuscript in preparation]. We could demonstrate that the mechanism of this anti-IL-10 antibody mediated protection was indeed related to elevation of endogenous TNFα levels, since protection could be reversed by introduction of anti-TNFα antibodies to the anti-IL-10 treated NZB/W F₁ mice. These studies together indicate that regulation of monokine production using either IL-10 or IL-10 antagonists can have important consequences in animal models of disease.

Additional insight regarding the physiological role of IL-10 *in vivo* has been derived from a series of experiments where normal mice were treated continuously from birth to adulthood with neutralizing antibodies to IL-10 [Howard, et al. 1992; Ishida, et al. 1992a; 1992b]. At 8 weeks of age, these animals appeared essentially healthy, with no change in total body weight; hematocrits; gross histology of liver, lungs, intestines, spleen or thymus; or total numbers of splenocytes, thymocytes or lymph node cells. Nevertheless, a highly-reproducible series of phenotypic changes was observed in anti-IL-10 treated mice. These consisted of increased levels of circulating IFNγ, TNFα and IL-6; reduced serum IgM and IgA; a marked depletion of peritoneal B cells; and an inability to develop *in vivo* antibody responses to two bacterial antigens. The effect of anti-IL-10 treatment on circulating IgM, IgA, and specific anti-bacterial antibody responses could be explained in large by the depletion of peritoneal B cells, a numerically small subset of murine B lymphocytes known to mediate these particular immune responses [Hayakawa and Hardy 1988]. The depletion of peritoneal B cells following anti-IL-10 treatment was an unexpected finding which was subsequently shown to be a consequence of IFN-g elevation [Ishida *et al.* 1992a]. The effects of IL-10 neutralization on non-specific and specific antibody production observed in these experiments suggest a potential role for IL-10 in augmenting antibody-mediated immunity to bacterial pathogens.

Potential Clinical Uses of IL-10 and IL-10 Antagonists

The ability of IL-10 to suppress production of inflammatory monokines e.g. TNFα and IL-1, and to elevate anti-inflammatory monokines e.g. IL-1Ra, predicts a strong anti-inflammatory role in diseases such as sepsis, rheumatoid arthritis and psoriasis. The ability of IL-10 to suppress production of cytokines by Th1 cells predicts a strong immunosuppressive role in T cell mediated autoimmune disease such as type I diabetes and multiple sclerosis, and in allograft survival. The ability of IL-10 to inhibit IL-2-induced cytokine synthesis but not LAK activity suggests its possible use with IL-2 in LAK therapy, as a means of inhibiting side effects of this treatment which may be cytokine-mediated. The ability of IL-10 to augment human B cell responses predicts a role as a vaccine adjuvant for antibody production.

Conversely, important roles for IL-10 antagonists can also be envisioned. IL-10 antagonists may selectively enhance Th1 immunity which would be of possible benefit in infectious diseases of viral origin or involving bacteria or parasites which are intracellular pathogens. In addition, IL-10 antagonists may be useful in clinical situations which benefit from TNFα administration, or in the treatment of B cell mediated autoimmune disease.

CD40

Introduction

Human CD40 is a 45-50 kd glycoprotein expressed on normal and malignant B cells, interdigitating cells, follicular dendritic cells, thymic epithelium, and some carcinomas [Clark 1990]. Agonistic antibodies to human CD40 are important co-stimulators of B cell growth and antibody production [Clark 1990]. These same antibodies can rescue germinal center B cells and induced Burkitt lymphoma cells from death via apoptosis. Armitage *et al.* [1992] have recently identified a physiological ligand for CD40 expressed transiently on activated T cells. This ligand activates B cells in a manner analogous to anti-CD40 antibodies, and appears to be the major mediator of cognate T-B collaboration. Unfortunately, little is yet known about the role of murine CD40 in murine B cell development. Murine CD40 was recently cloned by cross-hybridization [Torres and Clark 1992], and we have subsequently derived numerous reagents from this murine CD40 cDNA. These include a truncated cDNA expressing the extracellular domain of recombinant murine CD40 [designated soluble CD40], and a high-titered rat antiserum raised against soluble CD40. We discuss here the function of these reagents in murine B cell responses.

Role of CD40 on Normal and Malignant Murine B Cells

A high-titered rat anti-mouse CD40 antiserum was prepared by hyperimmunization of rats with a highly purified preparation of recombinant soluble CD40. The resultant antiserum specifically bound purified soluble CD40 coated onto microtiter plates, and L cells stably transfected with the full-length murine CD40 cDNA. The antiserum induced vigorous proliferation of both small and large murine B cells, and this proliferation was effectively suppressed by soluble CD40 [Heath *et al.*, manuscript in preparation]. Importantly, the same preparations of soluble CD40 did not adversely effect the proliferative response of normal B lymphocytes to LPS or anti-IgM antibody stimulation. The rat anti-mouse CD40 antiserum specifically bound normal and malignant B cells, but not T cells, macrophages, or a variety of other cell types [Santos-Argumedo *et al*, manuscript in preparation]. The antiserum was also capable of rescuing murine B lymphomas from apoptosis induced by anti-IgM antibody [Santos-Argumedo *et al.*, manuscript in preparation]. Thus, by most of these functional criteria, the rat anti-mouse CD40 antiserum exhibited an array of properties which closely resembled those of the previously described mouse anti-human CD40 monoclonal antibodies [Clark 1990].

In contrast to these effects on normal B cells, anti-CD40 antiserum and soluble CD40 both caused potent inhibition of the *in vitro* growth of A.20 murine B lymphoma cells [A. Heath *et al.*, manuscript in preparation]. The specificity of the former effect could be demonstrated by its reversal following pre-incubation of the anti-CD40 antiserum with soluble CD40. Similar results were obtained using two additional murine B lymphoma cell lines. One possible explanation for these data is that A.20 cell growth is dependent on a homotypic interaction between CD40 and a CD40 counterstructure also expressed on A.20 cells. Evidence in support of the existence of such a CD40 counterstructure was provided by FACS analysis demonstrating that A.20 cells specifically bound soluble CD40. Surprisingly, A.20 cells did not express the CD40 ligand expressed by activated T cells according to PCR analyses using specific probes, and immunofluorescence analyses using a monoclonal antibody specific for the T cell CD40 ligand [Noelle et al., 1992].

These data therefore suggest the existence of a second CD40 ligand which may be important in the autocrine growth of murine B lymphomas. Efforts to isolate this second CD40 ligand are currently in progress.

Potential Clinical Implications

Whatever the molecular explanation for anti-CD40 mediated inhibition of B lymphoma growth, these data implicate CD40 and its counterstructure on B cells as possible targets for therapeutic intervention in the treatment of B lymphomas. Our findings indicate that anti-CD40 antibodies and soluble CD40 produce opposing effects on normal and malignant B cells, stimulating the former and suppressing the latter. This situation is reminiscent of anti-IgM stimulation of B lineage cells, since anti-IgM antibodies induce proliferation of normal B lymphocytes, but frequently induce death by apoptosis of B lymphomas. The opposing effects of these reagents on normal and malignant B cells make them excellent candidates for therapeutic intervention in the treatment of clinical B lymphomas. Indeed, idiotype-specific anti-IgM antibodies have shown some promise in clinical trials of this nature [Levy et al. 1987]. Unfortunately, the efficacy of this approach with anti-idiotypic antibodies may be compromised by the high mutation rate of immunoglobulin genes [Levy, et al. 1987]. It is hoped that the less complex CD40 locus [Grimaldi et al. 1992] and the absolute functional requirement for a non-mutated CD40 epitope in cognate T-B collaboration [Armitage, et al. 1992], may render CD40 less susceptible to hypermutation, making this an even more attractive candidate for anti-B lymphoma therapy. Moreover, while anti-idiotypic antibodies need to be tailored for each individual patient, CD40 appears to be expressed ubiquitously on B lymphomas, and may therefore represent a universal therapeutic reagent.

To some extent, our data on murine B lymphomas appear in conflict with data reported for human B lymphomas, where anti-CD40 antibodies rescue these malignant cells from induced apoptosis [Gregory et al. 1991]. It should be appreciated, however, that such studies have been restricted to a selective subgroup of malignancy, namely Burkitt Lymphoma [BL]. BL is believed to represent the neoplastic counterpart of germinal center B cells and is characterized by a high tendency to undergo spontaneous apoptosis both in vitro and in vivo. The effective rescue of both germinal center and BL B cells from apoptosis by monoclonal antibody to human CD40 may be a feature specific for this narrow window of B cell differentiation. Perhaps a more analogous study on human CD40 is that reported by Kishimoto and colleagues showing that murine M12 B lymphoma cells became susceptible to growth inhibition by monoclonal antibody to human CD40 following their transfection and expression of the full-length human CD40 gene [Inui et al. 1990]. This study and the current report suggest that the influence of anti-CD40 antibodies on human non-Burkitt lymphomas representing different stages of differentiation is now warranted.

ACKNOWLEDGMENTS

Much of the above summary on IL-10 has been derived from the original studies of many scientists at DNAX. We wish to specifically acknowledge the role of these colleagues for their contribution to this team effort: A. O'Garra, R. de Waal Malefyt, A. Zlotnik, P. Vieira, D. Fiorentino, T. Mosmann, D. Hsu, H. Spits, J. Abrams, J. de Vries, M.G.

Roncarolo, S. Macatonia, S. Menon, R. Kastelein N. Go, D. Rennick. We also thank the following individuals for their contribution to the above studies on murine CD40: N. Harada, R. Chang, L. Santos-Argumedo, and A. Shanafelt from DNAX; J. Gordon [Birmingham, U.K.] E. Clark and R. Torres [Seattle, U.S.A.]; and R. Noelle [New Hampshire, U.S.A.]. DNAX Research Institute is fully supported by Schering-Plough Corporation.

REFERENCES

Armitage RJ, Fanslow WC, Strockbine L, Sato TA, Clifford KN, Macduff BM, Anderson DM, Gimpel SD, Davis-Smith T, Maliszewdki CR, Clark EA, Smith CA, Grabstein KH, Cosman D, Spriggs MK (1992) . Nature 357:80-82

Clark E (1990) . Tissue antigens 35:33

de Waal Malefyt R, Abrams J, Bennett B, Figdor C, de Vries J (1991a) . J Exp Med 174:1209-1220

de Waal Malefyt R, Haanen J, Spits H, Roncarolo M-G, te Velde A, Figdor C, Johnson K, Kastelein R, Yssel H, de Vries JE (1991b) . J Exp Med 174:915-924

Defrance T, Vanbervliet B, Briere F, Durand I, Rousset F, Banchereau J (1992) . J Exp Med 175:671-682

Fiorentino DF, Bond MW, Mosmann TR (1989) . J Exp Med 170:2081-2095

Fiorentino DF, Zlotnik A, Mosmann TR, Howard MH, O'Garra A (1991a) . J Immunol 147:3815-3822

Fiorentino DF, Zlotnik A, Vieira P, Mosmann TR, Howard M, Moore KW, O'Garra A (1991b) . J Immunol 146:3444-3451

Gazzinelli RT, Oswald IP, James SL, Sher A (1992) . J Immunol 148:1792-1796

Gregory CD, Dive C, Henderson S, Smith C, Williams G, Gordon J, Rickinson A (1991) . Nature 349:612

Grimaldi JC, Torres R, Kozak CA, Chang R, Clark EA, Howard M, Cockayne DA (1992) . J Immunol submitted:

Hayakawa K, Hardy RR (1988) . Ann Rev Immunol 6:197-218

Howard M, O'Garra A, Ishida H, de Waal Malefyt R, de Vries J (1992) . J Clin Immunol in press

Hsu D-H, Moore KW, Spits H (1992) . Internat Immunol 4:563-569

Inui S, Kaisho T, Kikutani H, Stamenkovic I, Seed B, Clark E, Kishimoto T (1990) . Eur J Immunol 20:1747

Ishida H, Hastings R, Kearney J, Howard M (1992a . J Exp Med 175:1213-1220

Ishida H, Hastings R, Snipes L, Howard M (1992b) . J Immunol, submitted

Levy R, Levy S, Cleary ML, Carroll W, Kon S, Bird J, Sklar J (1987) . Immunol Rev 96:43-58

Moore K, O'Garra A, de Waal Malefyt R, Vieira P, Mosmann T (1992) . Ann Rev Immunol 10:in press

Moore KW, Vieira P, Fiorentino DF, Trounstine ML, Khan TA, Mosmann TR (1990) . Science 248:1230-1234

Mosmann TR, Coffman RL (1989) . Ann Rev Immunol 7:145-173

Mosmann TR, Moore KW (1991) . Immunol Today 12:A49-A53

Noelle, R, Roy M, Shepherd D, Stamenkovic I, Ledbetter J, and Aruffo A (1992). Proc Natl Acad Sci USA 89: 6550-6554

Rousset F, Garcia E, Defrance T, Peronne C, Hsu D-H, Kastelein R, Moore KW, Banchereau J (1992) . Proc Natl Acad Sci USA 89:1890-1893

Torres RM, Clark EA (1992) . J Immunol 148:620-6

Vieira P, de Waal-Malefyt R, Dang M-N, Johnson KE, Kastelein R, Fiorentino DF, deVries JE, Roncarolo M-G, Mosmann TR, Moore KW (1991) . Proc Natl Acad Sci USA 88:1172-1176

Cytokine Receptor Directed Therapy with Genetically Engineered Monoclonal Antibodies Armed with Radionuclides

T. A. Waldmann, C. Goldman, L. Top, J. White, and D. Nelson

Metabolism Branch, National Cancer Institute, National Institutes of Health, Bethesda, MD 20892

INTRODUCTION

The hybridoma technique of Köhler and Milstein (1975) rekindled interest in the use of antibodies targeted to cell surface antigens to treat cancer patients. However, such monoclonal antibodies have been relatively ineffective (Catane and Longo 1989). There have been a number of explanations for this observed low therapeutic efficacy. One of the factors is that the murine monoclonal antibodies are immunogenic. An even more critical factor is that most of the monoclonal antibodies employed are not effective cytocidal agents against human neoplastic cells. Furthermore, in most cases, the antibodies were not directed against a vital structure present on the surface of malignant cells, such as a growth factor receptor required for tumor cell proliferation.

We readdressed this issue by: (a) using genetic engineering to create less immunogenic and more effective monoclonal antibodies, (b) arming antibodies with toxins or radionuclides to enhance their effector functions, and (c) employing the interleukin-2 receptor (IL-2R) on abnormal cells as a target for effective monoclonal antibody action (Waldmann, 1989, 1991). The scientific basis for the use of this target antigen is that resting normal cells do not express the IL-2R. Rather, this receptor is expressed by a proportion of the abnormal T cells in certain forms of lymphoid neoplasia, in select autoimmune diseases, in graft-versus-host disease, and by the T cells involved in allograft rejection (Waldmann 1989, 1991). Therapeutic trials have been initiated to exploit this difference in IL-2R expression, using an unmodified murine monoclonal antibody, termed anti-Tac, that is directed to the alpha subunit of the human interleukin-2 receptor. A disadvantage in the use of murine monoclonals is that they are highly immunogenic. In an effort to reduce its immunogenicity, anti-Tac was humanized by genetic engineering to produce an antibody that is human except for the antigen-combining regions, which are retained from the mouse (Queen *et al.* 1989, Junghans *et al.* 1990). In addition, to increase its cytotoxic effectiveness, the anti-Tac monoclonal antibody was armed with toxins or radionuclides (Lorberboum-Galski *et al.* 1988, Chaudhary *et al.* 1989, Kozak *et al.* 1986, 1989). Thus, IL-2 receptor-directed therapy provides a new approach for treating certain neoplastic diseases and autoimmune disorders and for preventing allograft rejection.

STRUCTURE AND FUNCTION OF THE MULTISUBUNIT INTERLEUKIN-2 RECEPTOR

To function as effector cells, T cells must change from a resting to an activated state. The sequence of events involved in the activation of T cells begins when a foreign pathogen encounters the antigen-specific receptor on the surface of resting T cells. This antigen-stimulated activation of these resting T cells induces the synthesis of the 15-kDa lymphokine IL-2. To exert its biologic effect, IL-2 must interact with specific high-affinity membrane receptors (Waldmann 1986, 1991, Smith 1980, Kuziel and Green 1990). Resting cells do not express high-affinity IL-2R, but they are rapidly expressed on T cells after activation with antigen or mitogen.

There are three forms of cellular receptors for IL-2: one with a very high affinity (10^{11}/M), one with an intermediate affinity (10^9/M), and another with a much lower affinity (10^8/M). We have used monoclonal antibodies and radiolabeled IL-2 in cross-linking studies to characterize chemically the multiple subunits of this receptor. Initially, a monoclonal antibody (anti-Tac) that reacts with the interleukin-2 binding site of a 55-kDa IL-2R protein (now termed IL-2R α) was identified (Uchiyama et al. 1981a, b). The receptor protein identified by anti-Tac was characterized as a glycoprotein with an apparent molecular mass of 55 kDa. Using cross-linking methods, we defined a second 70- to 75-kDa (p75 or IL-2R β) IL-2-binding protein (Tsudo et al. 1986, 1987). We proposed a multisubunit model for the high-affinity IL-2R in which both IL-2R α and IL-2R β proteins are associated in a receptor complex (Tsudo et al. 1986). In an independent study, Sharon and colleagues (1986) proposed a similar model. Recent evidence suggests a more complex subunit structure that involves peptides, in addition to the p55 and p75 IL-2-binding peptides. For example, a 65-kDa molecule, termed the gamma chain, appears to play a major role in facilitating IL-2 binding by the IL-2R β subunit (Takeshita et al. 1992).

INTERLEUKIN-2 RECEPTOR EXPRESSION IN MALIGNANCY OR AUTOIMMUNE DISORDERS

Resting T cells, B cells, or monocytes in the circulation do not display the IL-2 α receptor chain. However, most T and B lymphocytes can be induced to express this receptor subunit. Further, Rubin and coworkers (1985) showed that activated normal peripheral blood mononuclear cells and certain lines of T- or B-cell origin release a soluble form of the IL-2R α into the culture medium and that normal individuals have measurable amounts of soluble IL-2R α in their plasma. The determination of plasma levels of such IL-2R α provides a valuable noninvasive approach for analyzing both normal and disease-associated lymphocyte activation in vivo.

In contrast to the lack of IL-2R α chain expression in normal resting mononuclear cells, this receptor peptide is expressed by a proportion of the abnormal cells in certain forms of lymphoid neoplasia, in select autoimmune diseases, and in association with allograft rejection. That is, a proportion of the abnormal cells in these diseases expresses surface IL-2R α. Furthermore, the serum concentration of the soluble form of IL-2R α is elevated in the plasma of such individuals (Rubin et al. 1985). In terms of neoplasia, certain T-cell, B-cell, monocytic, and even granulocytic leukemias express the IL-2R α subunit. Specifically, virtually all of the patients with human T-cell lymphotrophic virus-I (HTLV-I)-associated adult T-cell leukemia constitutively express very large numbers of IL-2R α (Waldmann et al. 1984, Uchiyama et al. 1985). Similarly, a proportion of patients with cutaneous T-cell lymphomas expresses the Tac peptide (Waldmann et al. 1984, Schwarting et al. 1985). Further, the malignant B cells of virtually all patients with hairy cell leukemia and a proportion of patients with large- and mixed-cell diffuse lymphomas express IL-2R α (Korsmeyer et al. 1983). The IL-2R α is also expressed on the Reed-Sternberg cells of patients with Hodgkin's disease and on the malignant cells of patients with true histiocytic lymphoma (Schwarting et al. 1985). Finally, a proportion of the leukemic cells of patients with chronic and acute myelogenous leukemia express the Tac antigen (IL-2R α).

DISORDERS OF INTERLEUKIN-2 RECEPTOR EXPRESSION IN HTLV-I-ASSOCIATED ADULT T-CELL LEUKEMIA

A distinct form of mature T-cell leukemia defined by Uchiyama and coworkers (1977) was termed adult T-cell leukemia. The retrovirus HTLV-I was shown to be the primary causative agent in this leukemia (Poiesz et al. 1980). Adult T-cell leukemia is a malignant proliferation of mature CD3/CD4-expressing T cells that infiltrate the skin, lungs, and liver. Cases of adult T-cell leukemia are associated with hypercalcemia and an immunodeficiency state and usually have a very aggressive course, with a mean time to death of 20 weeks. The leukemic cells that we and others have examined from patients with HTLV-I-associated adult T-cell leukemia express high- and low-affinity IL-2R, including the IL-2R α chain (Waldmann et al. 1984, Uchiyama et al. 1985). An analysis of HTLV-I and its protein products suggests a potential mechanism for this association between HTLV-I and the constitutive IL-2R α expression. The retrovirus HTLV-I encodes a 42-kDa protein, now termed tax, that is essential for viral replication (Seiki et al. 1983, Sodroski et al. 1984). The tax protein, encoded by this retrovirus, also plays a central role in indirectly increasing the transcription of host genes, including the IL-2 and especially the IL-2R α receptor genes involved in T-cell activation and HTLV-I-mediated leukemogenesis.

INTERLEUKIN-2 RECEPTOR α AS A TARGET FOR THERAPY IN PATIENTS WITH HTLV-I-ASSOCIATED ADULT T-CELL LEUKEMIA

Unmodified Anti-Tac Monoclonal Antibody

The HTLV-I-induced adult T-cell leukemia cells constitutively express the IL-2R α chain identified by the anti-Tac monoclonal antibody, whereas normal resting cells do not. This observation provided the scientific basis for IL-2R-directed immunotherapy with this monoclonal antibody. Interleukin-2 receptor-directed immunotherapeutic agents could theoretically eliminate IL-2R α-expressing leukemic cells or abnormally activated T cells involved in other disease states while retaining the Tac-nonexpressing normal T cells and their precursors that express the antigen receptors for T-cell-mediated immune responses. In our initial studies, we administered unmodified murine anti-Tac to patients with adult T-cell leukemia (Waldmann et al. 1988). The leukemic cells of each patient with adult T-cell leukemia reacted with anti-Tac. Our goal was to inhibit the interaction of IL-2 with its growth factor receptor expressed on the malignant cells. The 20 patients treated in this study did not have untoward reactions related to the immunotherapy. Only patients undergoing a remission produced antibodies to the monoclonal antibody. Seven of the 20 treated patients had transient mixed (1), partial (4), or complete remissions (2), lasting from 1 to more than 30 months after anti-Tac therapy. This was assessed by elimination of measurable skin and lymph nodal disease, normalization of serum calcium levels, and routine hematologic and phenotypic tests of circulating cells. Further, elimination of clonal malignant cells was shown by molecular genetic analysis of HTLV-I proviral integration and the T-cell antigen receptor gene rearrangements. Thus, the use of a monoclonal antibody that prevents the interaction of IL-2 with its growth factor receptor on adult T-cell leukemia cells provides a rational approach for treating this malignancy. In most cases of the aggressive phase of adult T-cell leukemia, however, the leukemic cells no longer produce IL-2 nor do they require IL-2 for their proliferation. In this phase of the disease, the patients may not be responsive to unmodified anti-Tac therapy. Nevertheless, such cells continue to display the Tac protein. Thus, there is still a difference between the normal cells and the malignant cells that can be exploited in treatment.

Humanized Antibody to the IL-2 Receptor

There are two problems with murine monoclonal antibodies in general: their immunogenicity and their ineffectiveness at recruiting host-effector functions. We have addressed these issues by producing "humanized" antibodies. These humanized anti-Tac molecules, produced in conjunction with Cary Queen, retain the complementarity-determining region from the mouse, but virtually all the remainder of the molecule is derived from human IgG1 kappa. On the basis of computer modeling of the structure of this antibody, murine elements close to the complementarity-determining regions were identified, and those that were believed to be important to maintain the appropriate conformation of this antibody were retained (Queen et al. 1989, Junghans et al. 1990).

One primary goal in these studies is to maintain the affinity and functional capacity of the mouse monoclonal antibody. The parent anti-Tac molecule had an affinity of 9×10^9/M, whereas the hyperchimeric "humanized" version had an affinity of 3×10^9/M, still very high (Queen et al. 1989). The parent monoclonal and the humanized version showed a comparable inhibition of T-cell proliferation in response to tetanus antigen (Junghans et al. 1990). The humanized version of anti-Tac had improved pharmacokinetics when compared to the murine version, with an in vivo survival that is 2.5-fold longer (terminal $t_{1/2}$, 103 hr vs. 38 hr). In addition, humanized anti-Tac was less immunogenic than murine anti-Tac when administered to cynomolgus monkeys undergoing heterotopic cardiac allografting (Brown et al. 1991). A final goal of this project is to make an antibody that is a better effector of cell killing than is murine anti-Tac. Therefore, we were greatly encouraged by the observation that although the parent mouse anti-Tac could not function in antibody-dependent cellular cytotoxicity (ADCC) with human mononuclear cells, the hyperchimeric IgG1 anti-Tac manifests ADCC with human mononuclear cells (Junghans et al. 1990). Clinical trials have been initiated using humanized anti-Tac in the treatment of patients with leukemia/lymphoma or with graft-versus-host disease.

Monoclonal Antibody-Cytotoxic Agent Conjugates

To continue to take advantage of the difference in IL-2R expression and to improve the effectiveness of IL-2R-directed therapy, different approaches were initiated to modify the antibody for clinical purposes. For example, we have turned to β- and α-emitting isotopes as cytotoxic agents that could be conjugated to anti-Tac and are effective when bound to the surface of Tac-expressing cells. In these studies, we used the β-emitting yttrium-90 (^{90}Y) and the α-emitting bismuth-212 (^{212}Bi). Our choice of isotopes is based on the desire to have agents with a short distance of action that will act on the cell in question and on a small number of bystander cells without unwanted toxicity. In one case, we bound the β-emitting ^{90}Y to anti-Tac using chelates that did not permit elution of radiolabeled yttrium from the monoclonal antibody. Monkeys that received xenografts or allografts of cynomolgus hearts showed a marked prolongation of graft survival following administration of ^{90}Y-labeled anti-Tac (Cooper et al. 1990). Following preclinical efficacy and toxicity studies, we initiated a dose escalation trial of ^{90}Y anti-Tac for the treatment of patients with HTLV-I-associated Tac-expressing adult T-cell leukemia. At the doses utilized (5, 10, and 15 mCi per patient) 10 of the 15 patients underwent a partial (8) or complete (2) remission. Thus, anti-Tac armed with a radionuclide provides meaningful therapy for a form of leukemia that was previously universally fatal. One of the most promising directions for future development of armed monoclonal antibodies for the treatment of cancer involves the linkage of β- and α-emitting radionuclides to human or humanized monoclonal antibodies. Such conjugates may prove to be relatively non-immunogenic agents that are effective in the elimination of IL-2R-α-expressing cells.

In summary, our present understanding of the IL-2/IL-2R system opens the possibility for more specific immune intervention. The clinical applications of anti-IL-2R-directed therapy represent a new perspective for the treatment of certain neoplastic diseases, select autoimmune disorders, and graft-versus-host disease, and for the prevention of allograft rejection.

REFERENCES

Brown PS Jr, Parenteau GL, Dirbas FM, Garsia RJ, Goldman CK, Bukowski MA, Junghans RP, Queen C, Hakimi J, Benjamin W, Clark RE, Waldmann TA (1991) Proc Natl Acad Sci USA 88:2663-2667
Catane R, Longo DL (1989) Isr J Med Sci 24:471-476
Chaudhary VK, Queen C, Junghans RP, Waldmann TA, FitzGerald DJ, Pastan I (1989) Nature 339:394-397
Cooper MM, Robbins RC, Goldman CK, Mirzadeh S, Stone C, Gansow OA, Clark RE, Waldmann TA (1990) Transplantation 50:760-765
Junghans RP, Waldmann TA, Landolfi ND, Avdalovic NM, Schneider WP, Queen C (1990) Cancer Res 50:1495-1502
Köhler G, Milstein C (1975) Nature 256:495-497

Korsmeyer SJ, Greene WC, Cossman J, Hsu SM, Jensen JP, Neckers LM, Marshall SL, Bakhshi A, Depper JM, Leonard WJ, Jaffe ES, Waldmann TA (1983) Proc Natl Acad Sci USA 80:4522-4526

Kozak RW, Atcher RW, Gansow OA, Friedman AM, Hines JJ, Waldmann TA (1986) Proc Natl Acad Sci USA 83:474-478

Kozak RW, Raubitschek A, Mirzadeh S, Brechbiel MW, Junghans R, Gansow OA, Waldmann TA (1989) Cancer Res 49:2639-2644

Kuziel WA, Greene WC (1990) J Invest Dermatol 94:27S-32S

Lorberboum-Galski H, Kozak R, Waldmann T, Bailon P, FitzGerald D, Pastan I (1988) J Biol Chem 263:18650-18656

Poiesz BJ, Ruscetti FW, Gazdar AF, Bunn PA, Minna JD, Gallo RC (1980) Proc Natl Acad Sci USA 77:7415-7419

Queen C, Schneider WP, Selick HE, Payne PW, Landolfi NF, Duncan JF, Avdalovic NM, Levitt M, Junghans RP, Waldmann TA (1989) Proc Natl Acad Sci USA 86:10029-10033

Rubin LA, Kurman CC, Fritz ME, Biddison WE, Boutin B, Yarchoan R, Nelson DL (1985) Immunol 135:3172-3177

Schwarting R, Gerdes J, Stein H (1985) J Clin Pathol 38:1196-1197

Seiki M, Hattori S, Hirayama Y, Yoshida M (1983) Proc Natl Acad Sci USA 80:3618-3622

Sharon M, Klausner RD, Cullen BR, Chizzonite R, Leonard WJ (1986) Science 234:859-863

Smith KA (1980) Immunol Rev 51:337-357

Sodroski JG, Rosen CA, Haseltine WA (1984) Science 225:381-385

Takeshita T, Asao H, Ohtani K, Ishii N, Kumaki S, Tanaka N, Munakata H, Nakamura M, Sugamura K (1992) Science 257:379-382

Tsudo M, Kozak RW, Goldman CK, Waldmann TA (1987) Proc Natl Acad Sci USA 84:4215-4218

Tsudo M, Kozak RW, Goldman CK, Waldmann TA (1986) Proc Natl Acad Sci USA 83:9694-9698

Uchiyama T, Broder S, Waldmann TA (1981a) J Immunol 126:1393-1397

Uchiyama T, Hori T, Tsudo M, Wano Y, Umadome H, Tamori S, Yodoi J, Maeda M, Sawami H, Uchino H (1985) J Clin Invest 76:446-453

Uchiyama T, Nelson DL, Fleischer TA, Waldmann TA (1981b) J Immunol 126:1398-1407

Uchiyama T, Yodoi J, Sagawa K, Takatsuki K, Uchino H (1977) Blood 50:481-492

Waldmann TA (1991) J Biol Chem 266:2681-2684

Waldmann TA (1991) Science 252:1659-1662

Waldmann TA (1989) J Natl Cancer Inst 81:914-923

Waldmann TA (1986) Science 232:727-732

Waldmann TA, Goldman CK, Bongiovanni KF, Sharrow SO, Davey MP, Cease KB, Greenberg SJ, Longo D (1988) Blood 72:1805-1816

Waldmann TA, Greene WC, Sarin PS, Saxinger C, Blayney DW, Blattner WA, Goldman CK, Bongiovanni K, Sharrow S, Depper JM, Leonard W, Uchiyama T, Gallo RC (1984) J Clin Invest 73:1711-1718

Blocking Interleukin-1 in Disease

Charles A. Dinarello

Tufts University School of Medicine and New England Medical Center, 750 Washington Street, Boston, Massachusetts 02111 USA

GENERAL CONCEPTS

Numerous studies implicate a role for cytokines in the pathogenesis of disease. Recent studies using specific cytokine <u>antagonism</u> have shed considerable light on which cytokines appear to be playing a critical role. This summary will focus on interleukin-1 (IL-1) as a cytokine of primary and strategic importance to the outcome of disease.

To begin this overview, a distinction is made between the local effects of IL-1 and the consequences of systemic blood levels. The ultimate function of the host defense system is the elimination of the invading organism, foreign antigen or neoplastic cell whether by phagocytosis and antibody formation as is the case in most infections, or the induction of cytotoxic T-cells for elimination of transformed cells. Inflammation is the price the host pays for an efficient defense system. In the extreme, death is the price paid for an effective host response. In the case of IL-1, high systemic blood levels have not been a characteristic of patients with sepsis or severe inflammation compared to other cytokines, for example tumor necrosis factor (TNF). Nevertheless, IL-1 is a potent inducer of hypotension and shock alone and together with TNF, can be lethal in experimental animals. Humans are particularly sensitive to the hypotensive property of IL-1; a single intravenous injections of IL-1 of 300 ng/kg is the maximal dose tolerated because of severe fall in blood pressure. Blood levels of IL-1 in humans injected with exogenous IL-1 who develop hypotension are comparable to those measured in patients with sepsis.

The last 10 years of IL-1 research has been focused on the structure and biological properties of this cytokine and the control of its synthesis; these have recently been reviewed in detail (Dinarello, 1991) and are not the subject of this review. The focus on IL-1 research has now shifted to speccific anti-IL-1 strategies in order to antagonize the inflammatory and life-threatening aspects of this molecule. These include methods for limiting its transcription, synthesis, secretion, processing, interaction with its cell-bound receptors or post-receptor events. However, for any anti-IL-1 strategy, blockade of IL-1 receptors appears to offer specificity and effectiveness. The concept presented in this review is that blocking IL-1 can be a life-saving clinical strategy as well as an adjunct in treating acute exacerbations of chronic disease.

IL-1 AS A MEDIATOR IN DISEASE

One assumption has been that microorganisms produce lethal toxins which, upon entrance into the circulation cause hypotension, decrease perfusion of vital organs, acidosis and death. It made no difference whether these were endotoxins from Gram negative bacteria or enterotoxins from Gram positive Staphylococci (Ikejima et al., 1988). A significant breakthrough came when antibodies to TNF blocked death in mice to a lethal endotoxin challenge (Beutler et al., 1985). This experiment clearly established that blocking a cytokine would prevent a host-mediated self-destructive process. The conclusion made from that experiment was that infectious organisms (or their toxins) induce the host to make a lethal amount of TNF. Subsequent studies showed that blocking TNF with monoclonal antibodies reduced deaths in baboons given a letha injection of *E. coli* organisms (Tracey et al., 1987). However, similar data show that blocking IL-1 receptors using the IL-1 receptor antagonist (IL-1Ra) also prevents lethal shock in mice, rabbits and baboons (Alexander et al., 1991; Fischer et al., 1991b; Ohlsson et al., 1990; Wakabayashi et al., 1991). In Phase II clinical trials, IL-1Ra have been effective in reducing deaths in humans with sepsis. This antagonist is presently in Phase II trials for arthtritis, inflammatory bowel disease, asthma, leukemia and in Phase III trials for sepsis.

PREVENTING IL-1 EFFECTS

Specific Receptor Blockade Using the IL-1Ra.   Naturally occurring substances which *specifically* inhibit IL-1 activity have been detected in the serum of human volunteers injected with bacterial LPS (Dinarello et al., 1981), supernatants of human monocytes adhering to IgG coated surfaces (Arend et al., 1985) and urine of patients with monocytic leukemia (Seckinger and Dayer, 1987). The "IL-1 inhibitor" (Arend et al., 1985; Arend et al., 1989; Seckinger and Dayer, 1987; Seckinger et al., 1987) is a 23-25 kDa protein purified from the urine of patients with monocytic leukemia (Mazzei et al., 1990; Seckinger and Dayer, 1987; Seckinger et al., 1987). Natural IL-1 inhibitor blocked the ability of IL-1 to stimulate synovial cell PGE2 production, thymocyte proliferation, and decreased insulin release from isolated pancreatic islets (Balavoine et al., 1986; Dayer-Metroz et al., 1989; Seckinger and Dayer, 1987; Seckinger et al., 1987). It appears that the IL-1-specific inhibitory activities found in the serum during endotoxemia (Dinarello et al., 1981) is likely the IL-1Ra (Granowitz et al., 1991b) but the IL-1 specific inhibitor for the M20 myelomonocytic cell line (Barak et al., 1986) does not share identity with the IL-1Ra (Barak et al., 1991).

The purified naturally occurring IL-1 urinary inhibitor competes with the binding of IL-1 to its cell surface receptors (Seckinger et al., 1987). Molecular cloning of this inhibitor revealed a high degree of amino acid sequence homology to IL-1 itself and because of its mechanism of action, the IL-1 inhibitor was was re-named as IL-1Ra. Antibodies produced to the recombinant human IL-1Ra recognize the purified urinary IL-1 inhibitor of Seckinger and Dayer establishing that the IL-1 inhibitor and IL-1Ra are the same molecule (Seckinger et al., 1990).

The IL-1Ra blocks IL-1 activity *in vitro* and *in vivo*. To date, attempts to show agonist activity of IL-1Ra on a variety of cells *in vitro* have failed (Dripps et al., 1991). Humans have been injected with large amounts of IL-1Ra and symptoms or signs of agonist properties have not been observed. In a Phase I trial of the IL-1Ra, increasing doses from 1 to 10 mg/kg of IL-1Ra were injected intravenously into healthy volunteers. Blood levels were in excess of 25 µg/ml and there were no indications of altered homeostatic parameters (Granowitz et al., 1991a).

Rabbits (Ohlsson et al., 1990) or baboons (Fischer et al., 1991a) injected with IL-1 develop hypotension which is prevented by prior administration of the IL-1Ra. However, a more challenging experiment is the effect of IL-1 blockade in models of acute or chronic disease where several cytokines are produced. This, in fact, has been studied employing several animal models of disease and the results demonstrate that IL-1 receptor blockade significantly reduces the severity of disease, including those associated with infections This topic has recently been reviewed (Arend, 1991; Dinarello and Thompson, 1991). Table 1. lists the effects of IL-1Ra administration in animal models of disease. In many of these disease models, local inflammation plays a key role. The ability of IL-1Ra to block IL-1- and LPS-induced IL-8 production (Porat et al., 1992) may be a major component of the anti-inflammatory properties of IL-1Ra.

In mice and rabbits injected with lethal doses of LPS, prior administration of IL-1Ra reduces death (Alexander et al., 1991; Ohlsson et al., 1990). Decreased hypotension was observed in baboons treated with IL-1Ra and then given *E. coli* (Fischer et al., 1991a). We have studied the effects of IL-1Ra in two models of septic shock in the rabbits: Gram-negative sepsis due to *E. coli* and Gram-positive infections due to *Staphylococcus epidermidis*. In the first model, we observed no deaths in rabbits receiving *E. coli* plus the IL-1Ra whereas in control rabbits receiving *E. coli* plus a saline control, a 50% mortality was noted (Wakabayashi et al., 1991). Following the injection of *E. coli*, mean blood pressure fell in both groups and this fall takes place with the presence of TNF in the circulation. On the other hand, blood pressure returned to pre-*E. coli* levels in rabbits treated with IL-1Ra. We concluded that IL-Ra was blocking the IL-1 effects which take place associated with the IL-1 plasma peak at 180 minutes. On the other hand, we examined the effect of IL-1Ra in the model of Gram-positive sepsis and observed nearly complete block of hypotension, including the early fall in blood pressure associated with the TNF levels (Aiura et al., 1991). Since IL-1Ra blocks shock in rabbits due to either Gram-positive or Gram-negative organisms, it is not surprizing that deaths in humans with various causative microbial agents has been similarly reduced.

In humans with various forms of sepsis, significantly reduced mortality from 45% to 16% (p<0.015) was observed using 133 mg/hour constant infusion of IL-1Ra for 72 hours (personal comunication, Charles Fisher, Jr. et al, presented at Society of American Chest Surgeons, San Francisco, November, 1991). In that study, three doses of IL-1Ra were used and a dose-dependant reduction in mortality was observed. However, at the highest doses (133 mg/kg for 72 hours), the total amount of IL-1Ra appraoched 10 grams of the antagonist. A Phase III trial is presently under way. Assuming that this and other trials of IL-1Ra show improvement in disease outcome, blocking IL-1 will become an accepted therapeutic modality in several diseases. Blocking IL-1 receptors is safe, at least in the short term.

Antibodies to IL-1 Receptors. There are two IL-1 receptors (IL-1R) which are both members of the immunoglobulin superfamily. Each has a single transmembrane region. The type I receptor (IL-1RtI) is found on endothelial cells, hepatocytes, fibroblasts, keratinocytes and T lymphocytes whereas the type II receptor (IL-1RtII) is found on B lymphocytes, monocytes and neutrophils. The extracellular regions of the type I and II receptors share 28% amino acid homology but the IL-1RtII has only a short cytosolic segment.

TABLE 1   REDUCTION IN SEVERITY BY HUMAN IL-1RA IN ANIMAL
MODELS OF VARIOUS DISEASES

Death in Rabbits or Mice due to LPS or *E. coli*
(Alexander et al., 1991; Ohlsson et al., 1990; Wakabayashi et al., 1991)
Death in Newborn Rats from *Klebsiella pneumoniae*
(Mancilla et al., 1991)
Hemodynamic Shock in Rabbits and Baboons from *E. coli*
(Fischer et al., 1991b; Wakabayashi et al., 1991)
Hemodynamic Shock in Rabbits from *Staphylococcus epidermidis*
(Aiura et al., 1991)
Cerebral Malaria in Mice
(van der Meer et al., 1991)
Streptococcal Wall-induced Arthritis in Rats
(Schwab et al., 1991)
Collagen-induced Arthritis in Mice (Wooley et al., 1990)
Inflammatory Bowel Disease in Rabbits
(Cominelli et al., 1990)
Onset of Spontaneous Diabetes in BB Rats
(Dayer-Metroz et al., 1992)
Hypoglycemia and CSF Production in Mice following Endotoxin
(Henricson et al., 1991)
Proliferation and CSF Production of Acute Myeloblastic and Chronic Myelogenous Leukemia Cells
(Estrov et al., 1991; Rambaldi et al., 1991)
Neutrophil Accumulation in Inflammatory Peritonitis
(McIntyre et al., 1991)
Sciatic Nerve Regeneration in Mice
(Guenard et al., 1991)
Graft *versus* Host Disease in Mice
(McCarthy et al., 1991)
Experimental Enterocolitis in Rats
(Sartor et al., 1991)
LPS-induced Pulmonary Inflammation in Rats
(Ulich et al., 1991)

Antibodies have been produced to the IL-1RtI on murine cells (Chizzonite et al., 1989; Lewis et al., 1990). These have been used to block IL-1 effects in mouse cells *in vitro* and in murine models of disease *in vivo*. For example, in animal models of infection and inflammation, anti-IL-1RtI antibodies have reduced disease severity and use of these antibodies reveals that systemic responses of animals to IL-1 is *via* the IL-1RtI. Mice given intraperitoneal injections of inflammatory substances develop peritonitis with large numbers of neutrophils; however, prior treatment with anti-IL-1RtI prevents the influx of neutrophils, synthesis of serum amyloid A protein and circulating IL-6 levels (Chizzonite et al., 1989). Anti-IL-1RtI also blocks the neutrophil influx in response to endotoxin by 50%. Mice given an intramuscular injection of turpentine manifest several acute phase changes typical of inflammation such as decreased food intake, weight loss (lean and fat loss), IL-6 production, hepatic synthesis of amyloid P component, and elevated corticosterone levels. When anti-IL-1RtI was given prior to the inflammatory event, 80-90% of the intensity of the responses was reduced with the exception of elevated corticosterone levels (Gershenwald et al., 1990). The protective effect of IL-1 on lethal radiation appears to be due to the type I receptor since anti type I receptor antibodies blockthe protective response induced by LPS (Neta et al., 1990). These studies demonstrate that other cytokines induced by the turpentine inflammation or LPS are secondary to the production and activity of IL-1 via the type I receptor.

Soluble IL-1 Receptors. The extracellular domain of the IL-1RtI has been expressed and shown to bind both forms of IL-1. When the recombinant soluble IL-1RtI was given to mice undergoing heart transplantation, survival of the heterotopic allografts was increased. Lymph nodes directly injected with allogeneic cells have reduced hyperplasia with the use of the soluble IL-1RtI (Fanslow et al., 1990). Administration of soluble IL-1R to rats with autoimmune encephalomyelitis reduced the severity of the paralysis and delayed the onset of neurologic disease (Jacobs et al., 1991). However, it is unclear from these experiments how much of the effects of the soluble type I receptor is due decreased inflammation rather than decreased immuno-responsiveness. There are no data suggesting that the type I IL-1R is naturally shed; however, conditioned media from the IL-1RtII-bearing Raji cells contain the soluble form (35-45 kDa) of the IL-1RtII (Giri et al., 1990). Recently, an IL-1 binding protein (soluble form of an IL-1R) was found circulating in humans with inflammatory disease and appears to be related to the type II receptor (Symons et al., 1991).

CONCLUSIONS

In this review, the role for IL-1 in mediating the consequences of infection, particularly bacterial injection, has been presented. The best example of this is the model of septic shock where hypotension can be a lethal host response. However, despite the evidence that IL-1 can produce a shock-like state and that IL-1's biological properties are consistent with the host's response to infections, only by specifically blocking IL-1 has the critical role for this cytokine in infections as well as in other disease states been revealed. Indeed, the effectiveness of IL-1 receptor blockade using the IL-1Ra, particularly in septic humans, has raised some questions concerning the role of TNF in mediating the lethal consequences of infection. It is likely that the pathophysiologic events of infectious or inflammatory diseases is due to a synergism between cytokines, for example, IL-1 and TNF. Therefore, blocking either cytokine reduces the severity of the disease. Since blocking IL-1R appears to be safe and effective, the remaining question is whether total or prolonged IL-1R blockade will deprive the host of some essential role for in host defense.

Acknowlegements: These studies are supported by NIH Grant AI 15614. The author thanks Drs. K. Aiura, B. D. Clark, J. G. Cannon, J. A. Gelfand, M. Goldberg, E. V. Granowitz, T. Ikejima, G. Kaplanski, E. Lynch, J. Mancilla, L. C. Miller, D. Poutsiaka, R. Porat, L. Shapiro, E. Vannier, Sheldon M. Wolff, G. Wakabayashi, and K. Ye. In addition, I thank J-M Dayer, W. P. Arend, R. C. Thompson, D. E. Tracey, and S. Gilles.

REFERENCES

Aiura K, Gelfand JA, Wakabayashi G et al (1991) Cytokine 3:498 (abs)
Alexander HR, Doherty GM, Buresh CM et al (1991) J Exp Med 173:1029-1032
Arend WP (1991) J Clin Invest 88:1445-1451
Arend WP, Joslin FG, Massoni RJ (1985) J Immunol 134:3868-75
Arend WP, Joslin FG, Thompson RC et al (1989) J Immunol 143:1851-1858
Balavoine JF, de Rochemonteix B, Williamson K et al (1986) J Clin Invest 78:1120-4
Barak V, Peritt D, Flechner I et al (1991) Lymphokine Cytokine Res 10:in press
Barak V, Treves AJ, Yanai P et al (1986) Eur J Immunol 16:1449-1452
Beutler B, Milsark IW, Cerami A (1985) Science 229:869-871
Chizzonite R, Truitt T, Kilian PL et al (1989) Proc Natl Acad Sci USA 86:8029-8033
Cominelli F, Nast CC, Clark BD et al (1990) J Clin Invest 86:972-980
Dayer-Metroz MD, Duhamel D, Rufer N et al (1992) Eur J Clin Invest in press (abs):
Dayer-Metroz MD, Wollheim CB, Seckinger P et al (1989) J Autoimmun 2:163-71

Dinarello CA (1991) Blood 77:1627-1652

Dinarello CA, Rosenwasser LJ, Wolff SM (1981) J Immunol 127:2517-2519

Dinarello CA, Thompson RC (1991) Immunol Today 12:404-410

Dripps DJ, Brandhuber BJ, Thompson RC et al (1991) J Biol Chem 266:10331-10336

Estrov Z, Kurzrock R, Wetzler M et al (1991) Blood 78:1476-1484

Fanslow WC, Sims JE, Sassenfeld H et al (1990) Science 248:739-42

Fischer E, Marano MA, Barber AE et al (1991a) Am J Physiol 261:R442-R449

Fischer E, Marano MA, van Zee KJ et al (1991b) J Clin Invest in press:

Gershenwald JE, Fong YM, Fahey TJ3 et al (1990) Proc Natl Acad Sci USA 87:4966-70

Giri J, Newton RC, Horuk R (1990) J Biol Chem 265:17416-17419

Granowitz EV, Porat R, Gelfand JA et al (1991a) Cytokine 3:501 (abs)

Granowitz EV, Santos A, Poutsiaka DD et al (1991b) Lancet 338:1423-1424

Guenard V, Dinarello CA, Weston PJ et al (1991) J Neurosci Res 29:396-400

Henricson BE, Neta R, Vogel SN (1991) Infect Immun 59:1188-91

Ikejima T, Okusawa S, van der Meer JW et al (1988) J Infect Dis 158:1017-1025

Jacobs CA, Baker PE, Roux ER et al (1991) J Immunol 146:2983-2989

Lewis C, Mazzei G, Shaw A (1990) Eur J Immunol 20:207-13

Mancilla J, Garcia P, Dinarello CA (1991) Cytokine 3:502 (abs)

Mazzei GJ, Seckinger PL, Dayer JM et al (1990) Eur J Immunol 20:683-9

McCarthy PL, Abhyankar S, Neben S et al (1991) Blood 78:1915-1918

McIntyre KW, Stepan GJ, Kolinsky DK et al (1991) J Exp Med 173:931-939

Neta R, S.N. V, Plocinski JM et al (1990) Blood 76:57-62

Ohlsson K, Bjork P, Bergenfeldt M et al (1990) Nature 348:550-552

Porat R, Poutsiaka DD, Miller LC et al (1992) FASEB J in press:

Rambaldi A, Torcia M, Bettoni S et al (1991) Blood 78:3248-3253

Sartor RB, Holt LC, Bender DE et al (1991) Gastroenterology 100:A613 (abs)

Schwab JH, Anderle SK, Brown RR et al (1991) Inf Immun 59:4436-4442

Seckinger P, Dayer JM (1987) Ann Inst Pasteur/Immunol 138:461-516

Seckinger P, Klein-Nulend J, Alander C et al (1990) J Immunol 145:4181-4184

Seckinger P, Lowenthal JW, Williamson K et al (1987) J Immunol 139:1546-1549

Symons JA, Eastgate JA, Duff GW (1991) J Exp Med 174:1251-1254

Tracey K, Fong Y, Hesse DG et al (1987) Nature 330:662-664

Ulich TR, Yin SM, Guo KZ et al (1991) Am J Pathol 138:521-4

van der Meer JWM, Curfs JHAJ, Thompson RC et al (1991) Cytokine 3:497(abs)

Wakabayashi G, Gelfand JA, Burke JF et al (1991) FASEB J 5:338-343

Wooley PH, Whalen JD, Chapman DL et al (1990) Arthritis and Rheumat 33:S20 (abs)

IL-4 is a Major Determinant of T Cell Lymphokine-Producing Phenotype

William E. Paul, Robert A. Seder, Jane Hu-Li, Toshio Tanaka and S. Z. Ben-Sasson,

Laboratory of Immunology, National Institute of Allergy and Infectious Diseases, National Institutes of Health, Bethesda, MD 20892 U.S.A. and Lautenberg Center for General and Tumor Immunology, Hebrew University - Hadassah Medical Center, Jerusalem, 91010 Israel

ABSTRACT

Interleukin-4 present during priming causes naive T cells from T cell receptor transgenic mice to develop into cells capable of producing IL-4 upon secondary challenge. It also suppresses the capacity of such cells to produce IL-2 and interferon gamma (IFNγ). This effect is not mediated through the action of IL-10. Priming for IL-4 production is opposed by IFNγ, but only at sub-optimal concentrations of IL-4. Different types of antigen-presenting cells display different potencies for priming but all require IL-4 to cause the development of IL-4-producing cells. Administration of anti-IL-4 at the time of *in vivo* priming with keyhole limpet hemocyanin (KLH) diminishes development of T cells that produce IL-4 in response to *in vitro* challenge with KLH for up to 75 days after immunization and diminishes the capacity of T cells to produce IL-4 after secondary *in vivo* challenge. By contrast, anti-IL-4 administered at the time of secondary challenge has no effect on subsequent production of IL-4. Used acutely, *in vitro*, IL-4 blocks accessory cell-dependent, receptor mediated IL-2 and IFNγ production. Analysis of mechanism indicates that the activated T cell is the target of IL-4 activity and that inhibition is not due to blockade of the CD28 - B7 axis or its signalling pathway. It is concluded that IL-4 is a major physiologic regulator of the differentiation of CD4+ T cells into lymphokine-producing cells. The determination of the source and the clarification of the regulation of such acute IL-4 production are key to understanding the factors that determine whether immune responses will be dominated by IL-4 or by IFNγ-producing T cells.

INTRODUCTION

Much of the regulatory and effector activity of T cells is mediated by a set of small secreted proteins (lymphokines) whose synthesis is induced as a result of receptor engagement in the context of a cognate cellular interaction. A growing body of work has established that long term CD4+ T cell clones tend to segregate into those that produce IL-2 and IFNγ (T$_{H1}$ cells) and those that produce IL-4 and IL-5 (T$_{H2}$ cells) as their "signature" lymphokines (Mosmann & Coffman, 1989). Moreover, in immune responses to infectious agents and to allergens, responses often polarize into those dominated by production of IFNγ and those dominated by IL-4 production. In the case of infections, these distinctive phenotypes are often associated with different outcomes (Heinzel et al. 1989; Romani et al. 1992).

The recognition of the importance of the lymphokine-producing phenotype has naturally led to great interest in the factors that regulate the decision of CD4+ to develop into IL-4-producers or into IFNγ-producers. It has been shown that among naive T cells the frequency of cells that produce IL-4 in response to stimulation with anti-CD3 or with concanavalin A (Con A) is very low, often <1/1000 (Seder et. al., 1991), and that relatively little IFNγ is produced by naive CD4+ T cell populations (Swain et al. 1990; Seder et al. 1992). Culture of cells for 2 to 4 days with immobilized anti-CD3 or with Con A leads to cells that produce IL-4 only if IL-4 is added to the initial culture, strongly suggesting that this lymphokine plays a key role in directing the differentiation of naive CD4+ T cells (Le Gros et. al., 1990; Swain et. al., 1990).

IL-4 IS ESSENTIAL FOR *IN VITRO* PRIMING OF CD4+ T CELLS TO BECOME IL-4 PRODUCERS AND INHIBITS PRIMING FOR IFNγ PRODUCTION

In order to determine whether the activity of IL-4 in regulating priming of T cells for lymphokine production was a general one and, in particular, applied to priming of T cells to respond to conventional antigens, we have utilized a system that allows the *in vitro* priming of cells from naive donors. In this instance, T cells were obtained from mice transgenic for T cell receptor α and β genes that specify receptors for the cytochrome C peptide 88-104 and I-E^k. These mice were prepared by Barbara Fazekas de St. Groth and Mark Davis, Stanford University School of Medicine, using T cell receptor genes derived from the T cell clone 5C.C7 and the experiments using them were carried out in close collaboration between our groups (Seder et al. 1992)

Dense T cells were prepared from lymph nodes of transgenic donors that were H-2$^{a/a}$. They produced substantial amounts of IL-2 but little IL-4 or IFNγ upon immediate *in vitro* challenge with peptide 88-104 and various I-E^k+ antigen-presenting cells (APC). These cells were cultured for 3 - 4 days with peptide and a source of APC and then washed and challenged with peptide and fresh APC. The transgenic T cells responded to such primary *in vitro* immunization with a striking expansion in cell number. Upon restimulation, they produced substantial amounts of IL-2 and IFNγ but little IL-4. However, if IL-4 (1000 U/ml; 0.4 x 10^{-10} M) was added to the priming culture, the cells produced IL-4 upon secondary challenge but little or no IL-2 or IFNγ.

The effect of IL-4 appeared to be direct rather than mediated through an induction of IL-10 production. Thus, monoclonal anti-IL-10 antibody did not block the effect of IL-4 in priming for IL-4 production or in inhibiting priming for IFNγ production nor did IL-10, by itself, cause priming for IL-4 production or inhibit priming for IFNγ production.

IFNγ also played an important role in the priming process. In the presence of intermediate concentrations of IL-4 (100 U/ml; 0.4 x 10^{-11} M), IL-4 enhancement of priming for IL-4 production was partially inhibited by the presence of IFNγ. However, IFNγ had no detectable effect when a higher concentration of IL-4 was used for priming. IFNγ did not enhance priming for IFNγ production nor did it block the capacity of IL-4 to inhibit priming for IFNγ production.

To test the potential role of different types of APC in the priming process, we used dendritic cells, activated B cells, T-depleted spleen cells, and fibroblasts that expressed I-E^k as a result of stable transfection of I-E^k α and β chain genes. We observed that the different types of APC differed in their potency, with dendritic cells being more effective than activated B cells and activated B cells being superior to I-E^k+ fibroblasts. However, priming for IL-4 production required the addition of IL-4 to the priming culture and the presence of IL-4 during priming suppressed priming for IFNγ production with each of these APC as well as with T-depleted spleen cells.

These results thus indicate that IL-4 plays a central role, *in vitro*, in the process through which antigen-specific T cells develop into IL-4 or IFNγ producers.

ANTI-IL-4 TREATMENT INHIBITS DEVELOPMENT OF IL-4 PRODUCING CELLS IN IMMUNIZED MICE

These *in vitro* results lead one to ask whether IL-4 plays an equivalent role in the determination of lymphokine producing phenotype in the course of *in vivo* priming. To study this issue, mice were immunized with KLH, emulsified in complete Freund's adjuvant (CFA), and treated with monoclonal anti-IL-4 antibody at the time of priming. Among T cells from animals tested at 6-7 days after immunization, IL-4 production in response to *in vitro* challenge with KLH was diminished by more than two-fold in 4/4 experiments as a result of treatment with anti-IL-4 at priming. In addition, anti-IL-4 treatment resulted in a 2-fold or greater enhancement of IFNγ production in the majority of experiments. In this group, low density T cells gave the most reliable responses. High density T cells produced IL-4 in only two of four experiments but in both instances anti-IL-4 treatment at priming diminished production of IL-4 upon *in vitro* challenge.

In the group studied at 30 to 75 days after immunization, both low and high density T cells responded in all cases with IL-4 and with IFNγ production to *in vitro* challenge with KLH. In 4/4 experiments, high density T cells from donors that had been primed in the presence of anti-IL-4 displayed a two-fold or greater inhibition of IL-4 production while low density cells showed such inhibition in 3/4 experiments. By contrast, enhancement of IFNγ production was only observed in a minority of the experiments.

Some mice received a secondary challenge with KLH at 60 to 120 days after priming. In mice that had received anti-IL-4 at the time of priming, but not at the time of secondary challenge, IL-4 production in response to *in vitro* challenge with KLH was diminished in 3/3 experiments. By contrast, in mice that were primed in the absence of anti-IL-4 but were challenged in its presence, there was no diminution in IL-4 production.

These results thus imply a major *in vivo* role for IL-4 in priming for IL-4 production and indicate that the results observed *in vitro* have physiologic significance.

IL-4 CAUSES PROMPT INHIBITION OF ACCESSORY CELL-DEPENDENT RECEPTOR MEDIATED STIMULATION OF T CELL IL-2 AND IFNγ PRODUCTION

The striking effects exhibited by IL-4 in inducing the differentiation of T cells to develop into IL-4 producers and in blocking their production of IFNγ and of IL-2 led us to investigate the effects of IL-4 on IL-2 and IFNγ production in short term cultures of T cells. Others had shown that IL-4 suppressed production of IL-2 and IFNγ in human T cell cultures (Peleman et al. 1989; Gaya et al. 1991. By contrast, we have reported that IL-4 enhanced IL-2 production by mouse T cells stimulated with immobilized anti-CD3 antibodies (Tanaka et al. 1991). We have now verified that IL-4 inhibits IL-2 and IFNγ production by mouse T cells stimulated with soluble anti-CD3 and T-depleted spleen cells as APC but has no effect on such production in response to stimulation with phorbol ester and calcium ionophore. This result indicates that IL-4 inhibition of IL-2 and IFNγ production is limited to accessory cell-dependent, receptor-mediated T cell stimulation.

Our results indicate that IL-4 treatment causes a diminution in mRNA for IL-2 and IFNγ beginning at ~12 hours after stimulation, but has little effect prior to that time. Since IL-2 and IFNγ production is inhibited in both CD44-, LAM-1+ and CD44+, LAM-1- T cells, we interpret these results to indicate that IL-4 blocks a late amplification in IL-2 and IFNγ mRNA rather than blocking lymphokine production in only a subset of cells. Measurements of IFNγ mRNA half-life at 24 hours after stimulation indicate that T cells from mice treated with soluble anti-CD3, APC and IL-4 have less IFNγ mRNA than mice treated without IL-4 but that the half-life is similar in both cases and is relatively long (>3 hours). These results suggest that IL-4 does not achieve its effects by blocking the stabilization of IFNγ mRNA and thus imply that its effect is at the level of transcription of IL-2 and IFNγ genes.

Treatment of T cells with IL-4 prior to activation does not impair their subsequent production of IL-2 and IFNγ upon challenge but treatment with IL-4 for 16 hours in association with activation with anti-CD3 and APC strikingly diminishes their subsequent production of these lymphokines upon challenge with anti-CD3 and APC in the absence of IL-4. This effect of IL-4 appears to be on the activated T cell. This conclusion is based on an experiment in which T cells were stimulated with anti-CD3 and APC for 24 hours, in the absence of IL-4. The T cells were then purified by cell sorting and cultured for 24 hours with IL-4 but without APC or anti-CD3. Upon restimulation with anti-CD3 and APC, their production of IL-2 and IFNγ was strikingly inhibited.

The observation that IL-4 blocked accessory cell-dependent responses to soluble anti-CD3 but not accessory cell-independent responses to immobilized anti-CD3 raised the possibility that IL-4 might act by inhibiting the delivery of accessory signal(s) or by blocking the intracellular pathways that transmitted these signals. To test this, we examined the effect of IL-4 on augmentation of lymphokine production by anti-CD28 antibody and on inhibition by CTLA4Ig, which blocks the CD28 - B7 interaction (Linsley et al. 1991). Using either immobilized anti-CD3 or soluble anti-CD3 and dense B cells as APC, we observed that anti-CD28 caused a striking increase in IL-2 production. Treatment with IL-4 had no effect on that increase when immobilized anti-CD3 was used as a stimulant and caused only a modest inhibition when the stimulant was soluble anti-CD3 and dense B cells. In the latter case, the inhibition was comparable, in percentage terms, to that seen in the absence of anti-CD28. When the relative inhibitory effects of IL-4 and CTLA4Ig on IL-2 production in response to soluble anti-CD3 and T-depleted spleen were studied, it was observed that the two effects were independent of one another. These results thus suggest that the effect of IL-4 in blocking IL-2 and IFNγ production in accessory cell-dependent T cell activation is mediated by its effect on the T cell receptor-generated signal rather than on the accessory cell effect and indicate that immunosuppression achieved by inhibition of the CD28/B7 axis and the immunologic diversion caused by IL-4 are independent phenomena. The combined use of IL-4 and inhibitors of the CD28/B7 axis may lead to long lasting inhibition of IFNγ and IL-2 production and may have important implications in transplantation.

The question may be raised as to whether this acute effect of IL-4 on production of IL-2 and IFNγ has any relevance to the action of IL-4 in determining lymphokine producing phenotype. A strong argument in favor of their relatedness is the observation that the inhibitory effect of IL-4 is observed on normal T cells stimulated with soluble anti-CD3 and APC but not on cloned T cell lines that produce IL-2 and IFNγ in response to antigenic challenge. IL-4 has little or no effect on the lymphokine-producing phenotype of established T cell clones. This suggests that the IL-4 effect occurs at the time the decision is made as to the lymphokine producing phenotype of the stimulated cell.

IS IL-4 THE DOMINANT PHYSIOLOGIC REGULATOR OF THE LYMPHOKINE-PRODUCING PHENOTYPE

The results described here show that IL-4 inhibits IL-2 and IFNγ production through a novel mechanism, that it blocks priming for IL-2 and IFNγ production and that it strikingly enhances priming for IL-4 production. Furthermore, IL-4's effects are observed *in vivo* as well as *in vitro*. Thus, IL-4 appears to be a key factor in the appearance of T cells capable of producing IL-4 and in suppression of those capable of producing IFNγ. Nonetheless, it is important to determine if IL-4 is acutely induced by the priming procedure or is present at some basal level and, if it is induced, if the degree of induction determines whether a response will be dominated by IL-4 or IFNγ-producing T cells. It will also be necessary to determine the factors that determine the level of this "acute" IL-4 production and what its cellular source is.

We have suggested three potential sources of the IL-4 that may be involved in priming. They are previously primed T cells responsive to antigens cross-reactive with the antigen used for immunization, recent thymic emigrants (Bendelac and Schwartz 1991; Bendelac et al. 1992), and FcεRI+ cells (Ben-Sasson et al. 1990). Efforts to determine which of these are involved are now underway. The immunization variables that are important in determining which lymphokine-producing phenotype occurs are less clear. Antigen dose has been reported to be a possible factor, with low doses of antigen said to favor priming for IFNγ production and high doses favoring priming for IL-4 production (Pfeiffer et al. 1991; Bretscher et al. 1992). We have preliminary results from the study of T cell transgenic mice that imply that priming for IFNγ production is inhibited at high antigen concentration but these results have not been observed in every such experiment suggesting the action of other variables that are not yet understood. Similarly, although no striking effects of distinct APC have been seen in *in vitro* experiments, tissue distribution of antigen and chronicity of exposure to antigen may also play important roles in controlling the priming and continued activity of T cells producing IL-4 or IFNγ. Further, it seems most likely that other lymphokines will play roles in regulating the priming process and that the determination of outcome will depend upon a complex interplay between several of these factors.

ACKNOWLEDGMENT

We thank Dr. Barbara Fazekas de St. Groth (University of Sydney) and Dr. Mark Davis (Stanford University) for their provision of TCR transgenic mice and their collaboration in experiments involving cells from these mice; Dr. Peter Linsley (Bristol Meyers Squibb) for providing us with CTLA4Ig; and Dr. James Allison (University of California, Berkeley) for providing monoclonal anti-mouse CD28 antibodies. The expert editorial assistance of Ms. Shirley Starnes is gratefully acknowledged.

REFERENCES

Bendelac A, Schwartz RH (1991) CD4+ and CD8+ T cells acquire specific lymphokine secretion potentials during thymic maturation. Nature (London) 353:68

Bendelac A, Matzinger P, Seder RA, Paul WE, Schwartz RH (1992) Activation events during thymic selection. J Exp Med 175:731

Ben-Sasson SZ, Le Gros G, Conrad DH, Finkelman FD, Paul WE (1990) Cross-linking Fc receptors stimulate splenic non-B, non-T cells to secrete interleukin 4 and other lymphokines. Proc Natl Acad Sci USA 87:1421

Bretscher BA, Wu G, Menon JN, Bielefeldt-Ohmann H (1992) Establishment of stable, cell-mediated immunity that makes "susceptible" mice resistant to Leishmania major. Science 257:539

Gaya A, de la Calle O, Yague J, Alsinet E, Fernandez MD, Romero M, Fabregat V, Martorell J, Vives J (1991) IL-4 inhibits IL-2 synthesis and IL-2-induced up-regulation of IL-2R alpha but not IL-2R beta chain in CD4+ human T cells. J Immunol 146:4209

Heinzel FP, Sadick MD, Holaday BJ, Coffman RL, Locksley RM (1989) Reciprocal expression of interferon gamma or interleukin 4 during the resolution or progression of murine leishmaniasis. Evidence for expansion of distinct helper T cell subsets. J Exp Med 169:59

Le Gros G, Ben-Sasson SZ, Seder R, Finkelman FD, Paul WE (1990) Generation of interleukin 4 (IL-4)-producing cells in vivo and in vitro: IL-2 and IL-4 are required for in vitro generation of IL-4 producing cells. J Exp Med 172:921

Linsley PS, Brady W, Urnes M, Grosmaire LS, Damle NK, Ledbetter JA (1991) CTLA-4 is a second receptor for the B cell activation antigen B7. J Exp Med 174:561

Mosmann TR, Coffman RL (1989) TH1 and TH2 cells: Different patterns of lymphokine secretion lead to different functional properties. Ann Rev Immunol 7:145

Peleman R, Wu J, Fargeas C, Delespesse G (1989) Recombinant interleukin 4 suppresses the production of interferon gamma by human mononuclear cells. J Exp Med 170:1751

Pfeiffer C, Murray J, Madri J, Bottomly K (1991) Selective activation of Th1 and Th2-like cells in vivo-specific to human collagen IV. Immunol Rev 123:65

Romani L, Mencacci A, Grohmann U, Mocci S, Mosci P, Puccetti P, Bistoni F (1992) Neutralizing antibody toointerleukin 4 induces systemic protection and T helper type 1-associated immunity in murine candidiasis. J Exp Med 176:19

Seder RA, Le Gros G, Ben-Sasson SZ, Urban J Jr, Finkelman FD, Paul WE (1991) Increased frequency of interleukin 4-producing T cs as a result of polyclonal priming. Use of a single-cell assay to detect interleukin 4-producing cells. Eur J Immunol 21:1241

Seder RA, Paul WE, Davis MM, Fazekas de St. Groth B (1992) The presence of IL-4 during in vivo priming determines the lymphokine producing potential of CD4+ T cells from T cells receptor transgenic mice. J Exp Med in press

Swain SL, Weinberg AD, English M, Huston G (1990) IL-4 directs the development of Th2-like helper effectors. J Immunol 145:3796

Tanaka T, Ben-Sasson SZ, Paul WE (1991) IL-4 increases IL-2 production by T cells in response to accessory cell-independent stimuli. J Immunol 146:3831

Interleukin 10 and the Inflammatory Response

D. Rennick, S. Hudak, B. Hunt, G. Holland, D. Berg, K. Rajewski[1], R. Kuhn[1], and W. Muller[1]

Department of Immunology, DNAX Research Institute, 901 California Avenue, Palo Alto, CA 94304, USA
[1]*Universitat Koln, Weyertal 121, 5000 Koln, 41, Germany*

INTRODUCTION

A major advance in the study of immunity has been the identification of cytokines which mediate various aspects of acute and chronic inflammation. In vitro and in vivo studies have begun to elucidate the actions of cytokines and their importance in the generation, recruitment and effector function of different inflammatory and immune-reactive cells. In this regard, two types of helper T cells (designated Th1 and Th2) were identified based on their cytokine production. These discrete cytokine patterns have been correlated to the ability of Th1 cells to provide cell- mediated immunity and Th2 cells to provide superior help for humoral immunity. Th2 cell responses have also been linked to immediate hypersensitivity reactions (reviewed in Mosmann 1991). Because cellular and humoral immunity are mutually exclusive in certain disease states, evidence for cross regulation by Th1 and Th2 cells was sought and found. IL-10 was identified as a Th2 product which inhibits Th1 cytokine production (Fiorentino 1989; Moore 1990).

Since the initial description, IL-10 has also been shown to inhibit the production of cytokines by NK cells (Hsu 1992). The ability of IL-10 to suppress cytokine production by Th1 and NK cells is indirectly mediated via its' negative regulation of macrophage accessory cell function (de Waal Malefyt 1991; Fiorentino 1991; Hsu 1992). IL-10 also strongly inhibits the production of numerous pro-inflammatory monokines (e.g. IL-1, IL-6, IL-8, TNFa, and various CSFs) by activated macrophages (de Waal Malefyt 1991a; Fiorentino 1991a). However, it increases their production of IL-1 receptor antagonist (de Waal Malefyt 1991a). These studies suggest that IL-10 is a potent inhibitor of inflammation as well as specific cell-mediated responses.

IL-10 SELECTIVELY COSTIMULATES THE GENERATION OF INFLAMMATORY AND IMMUNE- REACTIVE CELLS

Although IL-10 has impressive immunosuppressive properties, it also has many stimulatory properties. Amongst these are its' effects on lymphocytes. IL-10 costimulates B cell proliferation, differentiation and antibody production (IgG, IgM & IgA) (Howard 1992; Defrance 1992; Rousset 1992). The effects of IL-10 on early stages of B cell development such as pre-B cells is not yet known. IL-10 appears to act on early and late stages of T cell development. It enhances the proliferation and differentiation of murine thymocytes and T

---

† Universitat Koln, Weyertal 121, 5000 Koln, 41, Germany

cells in the presence of a variety of other factors (i.e. IL-2, IL-4 and IL-7) (Suda 1990; MacNeil 1990). Furthermore, IL-10 not only augments the IL-2-dependent expansion of cytotoxic T cell precursors, it also increases their lytic activity (Chen 1990).

Effects of IL-10 on Myeloid and Erythroid Progenitors

The ability of IL-10 to stimulate the growth and differentiation of myeloid and erythroid precursors was assessed in bone marrow colony forming assays (Rennick 1992). Although IL-10 alone fails to stimulate any progenitor growth, it selectively enhances the factor-dependent growth of certain lineages. We found that IL-10 has no effect on precursors of neutrophilic granulocytes, eosinophils or monocytes. The absence of stimulatory effects for these inflammatory cells is entirely consist with the proposed role of IL-10 as a potent anti-inflammatory agent. On the other hand, IL-10 weakly potentiates the growth of mature erythroid progenitors called CFU-e and showed strong megakaryocyte potentiating activity. The functions carried out by erythrocytes and platelets are essential to almost every phase of the immune response including wound healing. Perhaps, that is why increasing numbers of cytokines have been found to directly or indirectly enhance their generation.

IL-10 also has marked effects on mast cell development. This has been demonstrated with cultured bone marrow cells from normal mice, however, the most convincing data were obtained with mesenteric lymph node cells of mice infected with *Nippostrongylus brasiliensis* (Thompson-Snipes 1991). IL-10, similar to IL-4, enhances the growth of IL-3-stimulated growth of mast cell progenitors. Optimal mast cell growth was observed when both IL-4 and IL-10 were combined with IL-3. The resulting cells were found to display the characteristics of mucosal mast cells (MMC). Mast cells play a critical role in immediate hypersensitivity reactions. The potent stimulus provided by the interaction of IL-10 with IL-3 and IL-4 accounts for the association of Th2 cells with this type of response.

IL-10 Costimulates Primitive Stem Cell Proliferation

The ability of IL-10 to affect the earliest stage of hemopoiesis was assessed using purified pluripotential stem cells selected for their expression of low levels of Thy1 antigen and the expression of stem cell antigen 1 (Sca1). Previous studies have shown that $Thy^{lo}Sca1^{+}$ stem cells are quiescent (G_0) and require multiple factor signaling in order to initiate proliferation (Heimfeld 1991). Numerous factor combinations have been identified which support their growth. These include IL-3 plus stem cell factor (SCF), IL-1 or IL-6. It was found that IL-10 is also capable of costimulating the IL-3-dependent growth of $Thy1^{lo}Sca1^{+}$ stem cells (Rennick 1992). The costimulatory effects of IL-10 were not diminished in the presence of the other cofactors such as SCF or IL-1. To the contrary, IL-10 was capable of complementing their actions and even greater stem cell proliferation was observed. During an intense inflammatory response, this type of cooperative signaling could determine if and to what extent these primitive cells will be involved in meeting the demand on bone marrow for increased cell production.

In vitro characterizations suggest that IL-10 has many bioactivities that could influence the host's ability to mount an appropriate immune response. However, none of these bioactivities are unique to IL-10. Therefore, it is not clear how or when IL-10 plays a critical role *in vivo*. An ideal system for assessing physiological relevance is the IL-10 gene-depleted (IL-10D) mouse generated by Muller and colleagues. During the early characterization of IL-10D mice, they found many were smaller than normal controls. Nevertheless, they appeared to be healthy and had normal numbers of thymocytes, mature T cells, conventional B cells and Ly1 B cells. They also had normal levels of serum immunoglobulin (unpublished). With time, however, increasing numbers of the IL-10D mice became sick and died. When it was discovered that these mice had become anemic, there was a question about their ability to generate red blood cells.

Analysis of Hemopoietic Function

Possible nemopoietic defects were investigated by analyzing the ability of IL-10D bone marrow and spleen to produce stem cells and progenitor cells. Our results showed that IL-10D mice have normal hemopoietic capabilities including the ability to generate red blood cell progenitors. In fact, the spleens of these mice showed signs of increased extramedullary hemopoiesis. Myeloid and erythroid progenitors were increased 5- to 20- fold above those of control mice. Histologically, the splenic follicles appeared normal. In contrast, the red pulp contained huge numbers of early erythroblasts and less than normal numbers of more mature erythrocytes. Further evaluation suggested that the final stages of erythrocyte maturity were defective because of depleted iron stores. One cause of iron deficiency anemia is the continuous loss of blood. The possibility of bleeding was supported by the finding that IL-10D spleens showed increased megakaryocyte production and blood platelets were elevated. Finally, the IL-10D spleens also contained abnormally high numbers of immature and mature neutrophilic granulocytes. The possibility of an inflammatory response coupled with blood loss suggested trouble in the gastrointestinal track.

Physical and Histological Examination of the Gastrointestinal Track

Upon close examination, several abnormalities were observed. Tests for occult blood were sporadically positive. After physical restraint, several IL-10D mice developed a temporary rectal prolapse. The distended tissue was red and swollen. Histological examination showed an inflammatory process in the rectum which extended proximally into the colon. The mucosa was heavily infiltrated with granulocytes as well as significant numbers of eosinophils, plasma cells and mononuclear cells. There was a decrease or absence of goblet cells in heavily infiltrated areas. Multiple ulcerations of the surface epithelium were observed and sites of previous bleeding were identified by the presence of blood cells entrapped in a fibrin mesh. Inflammation was also evident in the submucosa, however, it rarely involved the muscle wall. In addition, the inguinal and mesenteric nodes were dramatically enlarged. Unlike the spleen, there was no evidence of extramedullary hemopoiesis. The numerous cortical nodules contained well defined germinal centers and implied that these animals were actively mounting an immune response. Because histological examination of heart, liver, kidney and lung showed these tissues were normal, it appeared that the gut was the major site of inflammation.

Table 1: Comparison of inflammatory bowel disease in man and IL-10D mice

	Crohn's	Ulcerative colitis	IL-10D
Histology			
site of inflammation	anywhere	rectum 95% colon 67%	rectum 100% colon 100%
structures involved			
mucosa	+	+	+
submucosa	+	+	+
muscle wall	+	rare	rare
fistulae	+	−	−
infiltrates/cell types			
neutrophils	±	++	++
eosinophils	±	+	+
plasma cells	++	+	+
lymphocytes	+	+	+
monocytes	+	+	+
granulomas	+	rare	not seen
Lymph node involvement	+	rare	+
Clinical features			
diarrhea	+	±	not seen
bleeding	+	+	+
malabsorption	±	−	??
weight loss	+	±	+
anemia	+	+	+

358

Inflammatory Bowel Disease of Mice and Men

The finding that the IL-10D mice suffer from some form of chronic inflammatory bowel disease (IBD) prompted a search for parasites and pathogenic bacteria, but none were found. Idiopathic IBD occurs in man and has been intensively studied. Two forms, ulcerative colitis and Crohn's disease, have been distinguished based on their different clinical and histological features. The major characteristics of these two diseases are listed in Table 1 along with those manifested by the IL-10D mice. IBD in man is often associated with inflammation of other organs and this has lead to the belief that an altered immune mechanism may be involved. There is strong evidence that the tendency to develop IBD is inherited although environmental influences may contribute to its onset. When onset occurs in childhood, it can retard growth. This is of particular interest in view of the runting syndrome observed in the IL-10D mice. Furthermore, the similarities between ulcerative colitis and the inflammatory process in IL-10D are striking. This raises the possibility that IBD in man may be related to defective IL-10 production.

CONCLUDING REMARKS

The development of chronic inflammatory bowel disease in mice devoid of IL-10 is entirely consistent with the view that this cytokine plays a protective role against unnecessary and persisting inflammatory responses. However, there are reasons to think that the ability of IL-10 to help stimulate effective humoral immunity and depress inappropriate cell-mediated responses may also play a key role in maintaining the integrity of this tissue. The GI track is constantly stimulated with non-infectious agents as well as bacteria. These antigens are effectively dealt with by neutralizing antibodies. T cell- or NK-mediated cytotoxicity does not appear to play a major role in the defense of intestinal mucosa (MacDermott 1980; Kanof 1988). In fact, over-induction of cell-mediated immunity with strong lytic processes could cause senseless damage to the gut mucosa. Establishing specific effector cell defects in this mouse model should provide valuable insight into mechanisms underlying IBD in man. They may also suggest potential uses of IL-10 in the treatment of human disease.

ACKNOWLEDGMENTS

DNAX Research Institute is fully supported by Schering Plough Corporation.

REFERENCES

Chen WF, Zlotnik A (1991) J Immunol 147:528-534
Defrance T, Vanbervliet B, Briere F, Durand I, Rousset F, Banchereau J (1992) J Exp Med 175:671-682
de Waal Malefyt R, Haanen J, Spits H, Roncarolo M-G, te Velde A, Figdor C, Johnson K, Kastelein R, Yssel H, de Vries JE (1991) J Exp Med 174:915-924
de Waal Malefyt R, Abrams J, Figdor C, de Vries J (1991a) J Exp Med 174:1209-1220
Fiorentino DF, Bond MW, Mosmann TR (1989) J Exp Med 170:2081-2095
Fiorentino DF, Zlotnik A, Vieira P, Mosmann TR, Howard M, Moore KW, O'Garra A (1991) J Immunol 146:3444-3451

Fiorentino DF, Zlotnik A, Mosmann TR, Howard MH, O'Garra A (1991a) J Immunol 147:3815-3822

Howard M, O'Garra A, Ishida H, de Waal Malefyt, de Vries J (1992) J Clin Immunol in press

Hsu D-H, Moore KW, Spits H (1992) Internat Immunol 4:563-569

Kanof ME, Strober W, Fiocchi C, Zeitz M, James SP (1988) J Immunol 141:3029-3036

MacDermott RP, Franklin GO, Jenkins KM, Kodner IJ, Nash GS, Weinrieb IJ (1980) Gastroenterology 78: 47-56

MacNeil I, Suda T, Moore KW, Mosmann TR, Zlotnik A (1990) J Immunol 145:4167-4173

Moore KW, Vieira P, Fiorentino DF, Trounstine ML, Khan TA, Mosmann TR (1990) Science 248:1230-1234

Mosmann TR, Moore KW (1991) Immunol Today 12:A49-A53

Rennick D, Hunte B, Dang W, Thompson-Snipes L, Hudak S (1992) Exp Hematol submitted

Rousset F, Garcia E, Defrance T, Peronne C, Hsu D-H, Kastelein R, Moore KW, Banchereau J (1992) Proc Natl Acad Sci USA 89:1890-1893

Suda T, O'Garra A, MacNeil I, Fischer M, Bond M, Zlotnik A (1990) Cell Immunol 129:228-240

Thompson-Snipes L, Dhar V, Bond M, Mosmann TR, Moore KW, Rennick D (1991) J Exp Med 173:507-51

Cytokine Gene Transfer into Tumor Cells and its Application to Human Cancer

Felicia M. Rosenthal, Kathryn Cronin, Rita Guarini and
Bernd Gansbacher

*Department of Hematologic Oncology, Memorial Sloan Kettering Cancer Center,
New York. N. Y. 10021, USA*

INTRODUCTION

The success of conventional tumor therapy depends on the elimination of all tumor cells. Toxicity, inaccessibility or resistance to chemotherapy, however causes residual malignant cells to survive and ultimately to kill the host. Several new strategies attempt to induce host anti-tumor responses. Theoretically two approaches seem promising:
1. Activation of the afferent arm of the immune system by enhancing immunogenicity of tumor cells, e.g. by enhancing expression of MHC molecules, adhesion molecules or by transfection of tumor-specific antigens.
2. Activation of effector cells of the immune system known to be capable of killing tumor cells in vitro.

In recent years a great deal of expectation has been placed on cytokine-based immunotherapies. Cytokines are pleiotropic soluble factors secreted by diverse cell types that are responsible for communication between cells of the immune system. Those which function by recruiting, activating and/or expanding effector cell populations with anti-neoplastic activity, have been subjected to numerous clinical trials. Initial trials focused on systemic administration of Interleukin-2, because of its ability to stimulate cytotoxic T cells, NK (natural killer) and LAK (Lymphokine activated killer) cells, all of which can exert anti-tumor effects (Rosenberg 1985, 1988, West 1987, Fischer 1988, Dutcher 1989, Veelken 1992). Because of the short serum half-life of recombinant IL-2, continuous infusions and high doses were used leading to major side effects (Lotze 1986). Overall, response rates achieved by IL-2 treatment with or without the addition of ex vivo activated LAK cells or TIL (tumor infiltrating lymphocytes) were modest and only transient (Rosenberg 1985, 1988, West 1987, Fischer 1988, Dutcher 1989, Veelken 1992).

To circumvent the problems associated with systemic injection of cytokines and to mimic the physiological state in which the cytokine is released at the site of action, a new approach has emerged in recent years. By introducing the gene encoding for a cytokine into tumor cells, one can induce constitutive local secretion of the cytokine at the site where effector cells encounter their target. In this way cytotoxic effector cells at the tumor site will get activated and enriched in number. Moreover, chemotactic potentials of certain cytokines can thus be exploited. Of all gene transfer techniques, retroviral mediated gene therapy is the most suitable approach for transducing genes into cells for clinical use. The high transduction efficiency of this system allows the introduction of genes into a large proportion of replicating target cells. Further advantages are the stable integration into cellular DNA and a broad host range that not only makes the infection of adherent cells possible, but also of many suspension cells including lymphoid, myeloid and hematopoietic stem cells.

CYTOKINE GENE TRANSFER IN THE MURINE SYSTEM

Several laboratories have shown that the introduction of cytokine genes, such as IL-2 (Gansbacher 1990, Fearon 1990, Ley 1991), IL-4 (Tepper 1989, 1992, Golumbek 1991), IL-6 (Porgador 1992), IL-7 (Hock 1991, Aoki 1992), IFN-γ (Watanabe 1989, Gansbacher 1990), TNF-α (Asher 1991, Blankenstein 1991) and G-CSF (Colombo 1991, 1992) into tumor cells leads to their rejection by an immunologically competent host in syngeneic animal models. Rejection of gene modified tumor cells is mediated by different mechanisms depending on tumor and cytokine used. Secretion of G-CSF or IL-4 by tumor cells has led to localized inflammatory responses at the tumor site without apparent induction of immunological memory. IL-2, interferon-γ, and TNF-α on the other hand have induced long lasting cellular immune responses that led to rejection of potentially lethal tumor challenges at later time points. We have shown that introduction and expression of IL-2 in a murine fibrosarcoma led to the generation of tumor specific CTL, induced immunologic memory and prevented a normally seen immunosuppressive state from establishing itself (Gansbacher 1990). A similar approach showed that introduction and expression of IFN-γ led to upregulation of MHC class I molecules on the tumor cells and subsequent rejection of the lethal tumor challenges (Gansbacher 1990). Fearon et al introduced the IL-2 gene into a poorly immunogenic murine colon carcinoma (Fearon 1990). The IL-2-secreting tumors were rejected, and subsequently induced an effective cytolytic T cell response. Depletion of CD8+ cells in vivo abrogated the cellular antitumor response, implicating a pivotal role for Class I MHC restricted CD8-positive cells. The expression of IL-7 in the plasmacytoma cell line J558L induced CD4+ T cell-dependent tumor rejection, demonstrating again that different cytokines lead to tumor rejection involving different cellular mechanisms (Hock 1991).

The transduction of a number of cytokine genes into tumor cells has been successful in enhancing host anti-tumor responses. Yet, up to now the rejection of already established tumor has been reported in only one tumor model using IL-4 in a mouse renal cell carcinoma line (RENCA) (Golumbek 1991). Recently Tepper et al have reported the transfection of IL-4 in murine plasmacytoma and melanoma cell lines (Tepper 1992). Although they were able to show a potent local anti-tumor activity, mainly mediated by eosinophils, they were unable to demonstrate that IL-4 release induced growth inhibition of unmodified tumor at a distant site nor were they able to detect T cell memory. This contradicting data demonstrates that the effect of cytokines is not only determined by their intrinsic potential but also depends upon the tumor type used.

Since rejection of already established tumor is the ultimate goal in cancer therapy, we focused on optimizing the anti-tumor activity achieved with single cytokine transduced cells. To investigate the possible synergistic effects of cytokines, we have constructed retroviral vectors that carry both the IL-2 and the IFN-γ cDNA. IFN-γ is a cytokine known to upregulate molecules that participate in processing and presentation of antigens: proteasomes, peptide transporters and MHC I molecules (Basham 1983, Skoskiewicz 1985, Hammerling 1987, Brown 1991, Kleijmeer 1992). By transfection of the IFN-γ gene into tumor cells - which often are poor antigen presenters - Ag presentation to T cells can be enhanced. Clonal expansion of T cells is greatly enhanced if the antigen presenting cell also expresses co-stimulatory activity, as recently shown by Liu and Janeway (Liu 1992). Thus, by simultaneously delivering IL-2, a potent T cell growth factor as co-stimulatory signal, activated cytotoxic T cells should be clonally expanded.

Our present results demonstrate that transduction of IL-2/IFN-γ into a murine fibrosarcoma results in increased expression of MHC I molecules on the cell surface and synergistic effect on the host immune system. By selecting tumor cells secreting IL-2 in amounts to induce growth retardation but not rejection additional release of IFN-γ by these tumor cells induced tumor rejection (Gansbacher 1992)

CYTOKINE GENE TRANSFER IN THE HUMAN SYSTEM

Although the data obtained in the murine system can not be directly translated into the human system, evidence exists that cytokine gene therapy might have beneficial effects in cancer patients. Recent work demonstrates that some human melanoma cells express tumor specific antigens that can be recognized by cytotoxic T cells (CTL)(Livingston 1979, Darrow 1989, Van der Bruggen 1991). Moreover, circulating tumor specific CTL precursors can be found in the blood of some melanoma patients (Coulie 1991). Those patients whose tumor cells express CTL-recognized tumor specific antigens and who have circulating tumor specific CTL precursors are most likely to benefit from cytokine gene therapy. By injecting tumor cells that express the same antigen as the parental tumor and secrete IL-2, clonal expansion of the tumor specific CTL should be induced. A 2-4-fold enrichment of tumor specific CTL was seen in a mouse plasmacytoma model modified to secrete IL-2 (Ley 1991). CTL, which have access to systemic circulation, should be able to reach residual tumor cells at a site distant to the modified, cytokine secreting tumor cells.

Using retroviral mediated gene transfer, we and others have transfected several human cell lines and primary tumor cells with the IL-2 cDNA. IL-2 was constitutively expressed and was biologically active in vitro and in nude mice in vivo. Furthermore, we have shown that lymphokine secretion was not abolished until 3-4 weeks after lethal irradiation of the cells. This finding is important for clinical application as injection of living tumor cells into patients can be avoided (Gansbacher 1992).

The advances made in murine tumor models and in preclinical studies with retrovirally transfected human cells have led to the development of several clinical gene therapy trials in man. Some of these protocols use in vitro genetically marked cells to obtain more information about the in vivo behavior of the cells after reintroducing them into the patient: Studies with neomycin-transfected TIL have demonstrated the feasibility and safety of using retroviral vectors for the transduction of human cells. In patients that undergo autologous bone marrow transplantation for neuroblastoma, chronic myelogenous leukemia or childhood acute myelogenous leukemia, neomycin resistance-marked bone marrow cells are used to investigate the biology of marrow reconstitution and the mechanisms of relapse.

Several other federally approved protocols use cytokine transduced tumor cells to induce cellular antitumor responses. At Memorial Sloan-Kettering Cancer Center we soon will start a trail using allogeneic HLA-A2-positive melanoma cells and renal cell carcinoma cells expressing the human IL-2 gene. HLA-A2 positive cells have been chosen because HLA-A2 expressed on human melanoma cells has been shown to present shared tumor associated antigens that can be recognized by T cells. After lethal irradiation, these IL-2 secreting tumor cells will be injected into HLA-2 positive patients with metastatic melanoma or renal carcinoma. Preliminary in vitro studies showed that without the addition of any exogenous IL-2 the cytokine secreting tumor cells induced tumor specific CTL. Safety studies documented the absence of replication competent helper virus. Pending approval of RAC and FDA the protocol will be opened to patients in the fall of 1992.

REFERENCES

Aoki T, Tashiro K, Miyatake S, Kinashi T, Nakano T, Oda Y, Kikuchi H, Honjo T: Expression of murine interleukin-7 in a murine glioma cell line results in reduced tumorigenicity in vivo. Proc Natl Acad Sci USA 89(9):3850-3854, 1992
Asher AL, Mul'e JJ, Kasid A et al. Murine tumor cells transduced with the gene for tumor necrosis factor-alpha: evidence for paracrine immune effect of tumor necrosis factor against tumors. J Immunol 146:3227-34;1991.
Basham TY, Merrigan TC: Recombinant gamma interferon increases HLA-DR synthesis and expression. J Immunol 130:1492-1504, 1983
Blankenstein T, Qin Z, Ueberla K, Mueller W, Rosen H, Volk H-D, Diamantstein T: Tumor suppresion after tumor cell-targeted tumor necrosis factor α gene transfer. J Exp Med 173:1047-, 1991
Brown MC, Driscoll J, Monaco JJ: Structural and serological similarity of MHC-linked LMP and proteasome (multicatalytic proteinase) complexes. Nature 353: 738, 1991

Colombo MP, Ferrari G, Stoppacciaro A et al. Granulocyte colony-stimulating factor gene transfer suppresses tumorigenicity of a murine adenocarcinoma in vivo. J Exp Med 173:889-97;1991.

Colombo MP, Lombardi L, Stoppacciaro A, Melani C, Parenza M, Bottazzi B, Parmiani G: Granulocyte colony stimulating factor (G-CSF) gene transduction in murine adenocarcinoma drives neutrophil-mediated tumor inhibition in vivo. Neutrophils discriminate between G-CSF-producing and G-CSF-nonproducing tumor cells. J Immunol 149:113-119, 1992

Coulie, P.G., Somville, M., Lehmann, F., Hainaut, P., Rasseur, F., Devos, R., and Boon, T. Precursor frequency analysis of human cytolytic T lymphocytes directed against autologous melanoma cells. Int J Cancer, 50: 289-297, 1992.

Culver K, Cornetta K, Morgan R, et al: Lymphocytes as cellular vehicles for gene therapy in mouse and man. Proc Natl Acad Sci USA 88:3155-3159, 1991

Darrow TL, Sligluff CL: The role of HLA class I antigens in recognition of melanoma cells by tumor-specific cytotoxic T lymphocytes: Evidence for shared antigens. J Immunol 142:3329-3335, 1989

Dutcher JP, Creekmore S, Weiss GR, Margolin K, Markowitz AB, Roper MA, Parkinson D: A phase II study of interleukin-2 and lymphokine activated killer cells in patients with metastatic malignant melanoma. J. Clin. Oncol. 1989 (7):477-485

Fearon ER, Pardoll DM, Itaya T et al. Interleukin-2 production by tumor cells bypass T helper function in generation of an antitumor response. Cell 60:397-403;1990.

Fischer RI, Coltman CA, Doroshow JH, Rayner AA, Hawkins MJ, Mier JW, Wiernik P, McMannis JD, Weiss GR, Margolin KA, Gemlo BT, Hoth DF, Parkinson DR, Paietta E: Metastatic renal cell cancer treated with interleukin-2 and lymphokine-activated killer cells. A phase II clinical trial. Ann Int. Med. 1988 (108):518-523

Gansbacher B, Zier K, Daniels B et al. Interleukin-2 gene transfer into tumor cells abrogates tumorigenicity and induces protective immunity. J Exp Med 172:1217-24;1990.

Gansbacher B, Bannerji R, Daniels B et al. Retroviral vector-mediated gamma interferon gene transfer into tumor cells generates potent and long lasting antitumor immunity. Cancer Research 50:7820-5;1990.

Gansbacher B, F. Rosenthal, A.Guarini and D. Golde. Retroviral vectors carrying both the IL-2 and the IFN-gamma gene induce potent anti-tumor responses in murine tumors Abstract and oral presentation AACR meeting, San Diego, 1992, Proceedings of the American Association for Cancer Research, vol 33, 351, 2095.

Gansbacher B, K. Zier, C.Cronin, P.A. Hantzopoulos, B.Bouchard, A. Houghton, E. Gilboa and D. Golde. Retroviral gene transfer induced constitutive expression of IL-2 or IFN-gamma by irradiated human melanoma cells. 1992. 1992. Blood. In press.

Golumbek PT, Lazenby AJ, Levitzky HI, Jaffee LM, Karasuyama M, Baker, Pardoll DM: Treatment of established renal cancer engineered to secrete Interleukin-4. Science, 254: 713, 1991

Hammerling GJ, Klar D, Pulm W, et al: The influence of major histocompatibility complex class I antigens on tumor growth and metastasis. Biochem Biophys Acta 907:245-259, 1987

Hock H, Dorsch M, Diamantstein T, Blankenstein T: Interleukin-7 induces CD4+ T cell-dependent tumor rejection. J Exp Med 174:1291-1298, 1991

Kasid A, Morecki S, Aebersold P, et al: Human gene transfer: characterisation of human tumor-infiltrating lymphocytes as vehicles for retroviral-mediated gene transfer in man. Proc Natl Acad Sci USA 87:473-477, 1990

Kleijmeer MJ, Kelly A, Geuze HJ, Slot JW, Townsend A, Trowsdals J: Location of MHC-encoded transporters in the endoplasmatic reticulum and cis-Golgi. Nature 357: 342, 1992

Ley V., Langlade-Demoyen P, Kourilsky P., Larson-Sciard E.L. Interleukin-2 dependent activation of tumor specific cytotoxic T lymphocytes in vivo. Eur. J. Immunol. 21, 851, 1991.10.

Liu Y, Janeway CA: Cells that present both specific ligand and costimulatory activity are the most efficient inducers of clonal expansion of normal CD4 T cells. Proc Natl Acad Sci USA 89: 3849, 1992

Livingston PO, Shiku H: Cell-mediated cytotoxicity for cultured autologous melanoma cells. Int J Cancer 24:34-44, 1979

Lotze MT, Matory YL, Rayner AA, Ettinghausen SE, Vetto JT, Seipp CA, Rosenberg SA: Clinical effects and toxicity of interleukin-2 in patients with cancer. Cancer 1986 (58):2764-2772

Miller AD, Blaese RM, Anderson WF: Gene transfer into humans-immunotherapy of patients with advanced melanoma, using tumor-infiltrating lymphocytes modified by retroviral gene transduction. New Engl J Med 323:570-578, 1990

Porgador A, Tzehoval A, Katz A, Vadai E, Revel M, Feldman M, Eisenbach L: Interleukin 6 gene transfection into Lewis lung carcinoma tumor cells suppresses the malignant phenotype and confers immunotherapeutic competence against metastatic cells. Cancer Research:3679-3686, 1992

Rosenberg SA, Aebersold D, Cornetta K, et al: Gene transfer into humans - Immunotherapy of patients with advanced melanoma, using tumor-infiltrating lymphocytes modified by retroviral gene transduction. N Engl J Med 323:570-578, 1990

Rosenberg SA, Mule JJ, Spiess PJ, Reichert CM, Schwarz SL: Regression of established pulmonary metastases and subcutaneous tumor mediated by the systemic administration of high-dose recombinant interleukin-2. J. Exp. Med. 1985 (161):1169-1188

Rosenberg SA, Lotze MT, Muul LM, Leitman S, Chang AE, Ettinghausen SE, Matory YL, Skibber JM, Shilari E, Vetto JT, Seipp CA, Simpson C, Reichert CM: Observations on the systemic administration of autologous lymphokine-activated killer cells and recombinant interleukin-2 to patients with metastatic cancer. New Engl. J. Med. 1985 (313):1485-1492

Skoskiewicz MJ, Colvin RB, Schneeberger EE, Russel PS: Widespread and selective induction of major histocompatibility complex-determined antigens in vivo by γ Interferon. J Exp Med 162: 1645, 1985

Tepper RI, Coffman RL, Leder P: An eosinophil-dependent mechanism for the antitumor effect of interleukin-4. Science 257:548-551, 1992

Tepper RI, Pattengal PK, Leder P et al: Murine interleukin-4 displays potent anti-tumor activity in vivo. Cell 57:503;1989.

Van Der Bruggen P., Traversari C., Chomez C., Lurquin C., De Plean E., Van Den Einde B., Knuth A., Boon T. A gene encoding an antigen recognized by cytolytic T Lymphocytes on a human melanoma. Science, 254, 1643, 1991.

Veelken H, Rosenthal FM, Schneller F, v Schilling C, Guettler IC, Herrmann F, Mertelsmann R, Lindemann A: Combination of Interleukin-2 and Interferon-α in renal cell carcinoma and malignant melanoma: A phase II clinical trial. Biotechnology Therapeutics 3:1-14, 1992

Watanabe Y, Kuribayashi K, Miyatake S et al. Exogenous expression of mouse interferon gamma cDNA in mouse neuroblastoma C1300 cells: results in reduced tumorigenicity by augmented anti-tumor immunity. Proc Natl Acad Sci USA 86: 9456;1989.

Weber JS, Jay G, Tanaka K, et al: Immunotherapy of a murine tumor with interleukin-2. Increased sensitivity after MHC class I gene transfection. J Exp Med 166:1716-1733, 1987

West WH, Tauer KW, Yanelli JR, Marshall GD, Orr DW, Thurman GB, Oldham RK: Constant-infusion recombinant interleukin-2 in adoptive immunetherapy of cancer. New. Engl. J. Med. 1987 316:898-905

Molecular Mechanisms for the Regulation of Inflammation

K. Yamamoto, K. Nakayama, H. Shimizu, K. Mitomo and K. Fijimoto

Departement of Molecular Pathology, Cancer Research Institute, Kanazawa University, 13-1 Takaramachi, Kanazawa, Japan 920

INTRODUCTION

The term "inflammation " is used to collectively describe primitive but complex and coordinated host responses to various forms of tissue injuries caused by infection, trauma, necrosis, burn, malignant neoplasm and immunological abnormality. This is an important host defense mechanism we experience frequently, and is typically characterized by fever, pain, leukocytosis, increased vascular permeability, alterations in plasma metal and hormone concentrations, and drastic increases in the levels of some plasma proteins known as acute phase proteins. Although the details of mechanisms underlying inflammation are not yet understood completely, cytokines released by activated macrophages, namely IL-1, IL-6 ,and TNF-α, appear to be primary mediators of inflammatory reactions. These cytokines are produced in response to a variety of noxious inflammatory stimuli such as virus, bacterial lipopolysaccharides , double-stranded RNA, phorbol ester, UV and oxidative stress. They have diverse biological activities related to inflammatory reactions and secondarily activate many other inflammatory genes(Akira et al 1990). Thus, to understand the control mechanisms for inflammation at molecular levels, it is important to study molecular mechanisms controlling the expression of these cytokine genes as well as of genes whose expression is activated by these cytokines.

NF-κB AS A GENERAL INTRACELLULAR MEDIATOR OF INFLAMMATORY REACTIONS

The results of recent studies on inflammatory gene expression, including ours on the IL-6 and SAA genes(Shimizu et al 1990), indicate that a NF-κB transcription factor plays a major role in the regulation of inflammatory reactions. NF-κB was originally identified as a B-cell-specific nuclear factor binding to an enhancer element called κB of the κ immunoglobulin light-chain gene(Sen and Baltimore 1986) It is now clear that NF-κB is present in many different types of cells and is involved in the inducible expression of various cellular genes mostly related to host defense through various κB sequences(Table 1)(Baeuerle 1991). Although NF-κB is constitutively present in the nucleus of B cells, NF-κB is an inducible nuclear factor in other types of cells(Sen and Baltimore 1986). Subsequent studies have shown that NF-κB preexists in the cytoplasm in an inactive form complexed with inhibitory proteins termed IκB(Baeuerle and Baltimore 1988). A variety of inflammatory stimuli described above appears to activate NF-κB through dissociation of the IκB/NF-κB complex and subsequent translocation of NF-κB to the nucleus where NF-κB activates target genes. For these reasons, NF-κB may be defined as a general mediator of intracellular signal transduction for inflammatory reactions.Therefore, the regulation of NF-κB is of central importance in the inflammatory process, and we discuss how NF-κB is activated and regulated in response to inflammatory stimuli.

Table 1 Target genes for NF-κB and their κB motifs

target gene		kB motif
cytokine:	IL-6	G G G A T T T T C C
	IL-2	G G G A T T T C A C
	β-interferon	G G G A A A T T C C
	γ-interferon	T G G A A A A T T C
	TNF-α	G G G G C T T T C C
	IL-8	T G G A A T T T C C
	G-CSF	G G G G A A T C T C
	GM-CSF	G G G A A C T A C C
acute phase protein:	SAA	G G G A C T T T C C
	SAA3	T G G A A A T G C C
	angiotensinogen	G G G A T T T C C C
	factor B	G G G A T T C C C C
cell surface molecule:	ELAM-1	G G G G A T T T C C
	MHC class I	G G G A T T C C C C
	β2-microglobulin	G G G A C T T T C C
	IL-2 receptor α	G G G A G A T T C C
	T cell receptor β	G G G A G A T T C C
immunoglobulin κ		G G G A C T T T C C
transcription factor:	p50	G G G G C T T C C C
	IRF-1	G G G G A A T C C C
virus:	SV40	G G G A C T T T C C
	HIV	G G G A C T T T C C
	cytomegalovirus	G G G A C T T T C C
	adenovirus	G G G A C T T T C C
consensus		G G G R N N Y Y C C

A NF-κB/REL FAMILY

NF-κB contains two proteins of 50 and 65 kDa termed p50 and p65, respectively. Cloning and sequencing of cDNA clones encoding p65 and a 105-kDa precursor (p105) of p50 have revealed an extensive sequence homology between these proteins. Furthermore, the predicted sequences of these proteins are also highly homologous to those of the c-rel proto-oncogene and of the Dorosophila maternal morphogen dorsal gene. In particular, p65 is related to c-Rel more closely than to p50(Bours et al 1990,Ghosh et al 1990,Kieran et al 1990,Mayer et al 1991, Nolan et al 1991,Rubin et al 1991). c-rel is the cellular cognate of v-rel , the transforming gene of highly oncogenic reticuloendotherialiosis virus strain T that induces an acute fatal lymphoma in young birds(Gilmore 1990). More recently, cDNA clones for the second member of p50/p105 (p49/p100)(Neri et al 1991,Schmid et al 1991) and the third member of p65/Rel (RelB or I-Rel)(Rubin et al 1992,Ryseck et al 1992) have been isolated. Thus, the NF-κB/Rel transcription factor family currently consists of eight members including v-rel and an alternatively spliced deriverterive of p65(Fig. 1)(see below).

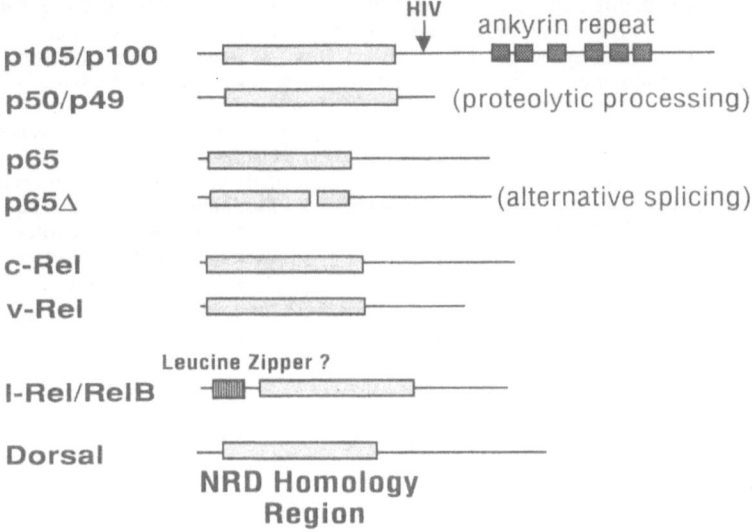

Fig 1. A schematic diagram of the structures of the NF-κB/Rel/Dorsal family members.

THE CONTROL OF DNA BINDING OF NF-κB/REL BY HOMO- OR HETERO-DIMERIZATION

The members of NF-κB/Rel family show an extensive homology in their N-terminal regions spanning about 300 amino acid residues, though their C-terminal portions are completely divergent. This region termed NF-κB/Rel/Dorsal homology region (NRD) mediates homo- or hetero-dimerization and DNA binding(Fig. 1). Despite of their homology in NRD, however, they show distinct DNA binding and sequence specificities for various κB motifs, presumably due to sequence variations within NRD. For example, the p50 homodimer shows a high affinity for symmetrical κB motifs such as the κB sequence of the MHC class I gene promoter. By contrast, asymmetric prototype κB motifs found in the enhancers of the κ immunoglobulin gene and in the SV40 and HIV viral enhancers are weak binding sites for the p50 homodimer but have a high affinity for the p50/p65 heterodimer (NF-κB). The c-Rel protein also shows an unique DNA binding and sequence specificity: Rel homodimers preferentially recognize AT-rich κB motifs such as those found in the IL-6 and β-interferon promoters(Table 1 and 2)(Nakayama et al 1992).

Table 2   Summary of DNA-binding and in vivo activities of three typical κB motifs

κB motifs	sequences	DNA binding				in vivo response	
		p50	NF-κB	c-Rel	c-Rel complex	L-TK	Jurkat
IL-6κB (AT-rich)	GGGATTTTCC	++	++	+++	++	++	-
Ig/HIV/SV40κB (asymmetrical)	GGGACTTTCC	+	++	±	-	++	++
MHCκB (symmetrical)	GGGATTCCCC	+++	++	+	±	++	++

In addition, a diverted κB motif found in the γ-interferon gene is recognized by c-Rel but not by p50. Thus, the association of p50, p65 and c-Rel in various combinations theoretically should provide the complexes with diverse but distinct DNA-binding specificities and affinities for various κB motifs and possibly with different transcriptional activities. It is not surprising,therefore,that the in vivo activities of various κB elements are differentially regulated in different types of cells, through variations in relative nuclear levels of these factors. It is not, however, clear at present whether all of possible homo- or heterodimeric complexes of these factors are physiologically present in vivo. The results of experiments with antisera for these proteins indicated that the NF-κB activities from Hela, Jurkat, Hep 3B and U937 cells are largely the complex of p50 and p65, and do not contain a significant amount of c-Rel(Nakayama et al 1992). However, Inoue et al(1991) showed that c-Rel is a part of NF-κB in pre-B cells. Furthermore, there presents a lymphoid cell-specific nuclear factor containing c-Rel but not p50(see below). In addition, there is evidence that p50 homodimers and p65/c-Rel heterodimers are physiologically present and play some roles in vivo(Kieran et al 1990, Hansen et al 1992). Thus, while a major inducible form of NF-κB/Rel complexes is the heterodimer of p50 and p65(namely, a classical NF-κB), there might be a considerable heterogeneity in the composition of the NF-κB/Rel complexes binding to various κB motifs in different types of cells.This heterogeneity may explain a diverse role for the NF-κB/Rel family in host defense.

THE CONTROL OF TRANSCRIPTIONAL ACTIVITY OF NF-κB/REL BY OTHER CELLULAR FACTORS

As discussed above, the association of c-Rel with p50 is not readily detectable in lymphoid cells, though c-Rel proteins are generally more abundunt in lymphoid cells and can readily associate with p50 in vitro. This suggests that c-Rel and p50 form discrete complexes in vivo. By using IL-6κB probes which have a high affinity for c-Rel homodimers, we indeed found the presence of a lymphoid cell-specific nuclear factor(s) (termed c-Rel complex) that contains the c-Rel homodimer but not p50 and specifically recognizes AT-rich κB motifs such as IL-6 and β-interferon κBs(Table 2). An apparent size of the c-Rel complex estimated by gel filtration(>660 kDa) and electrophoretic mobility shift assay is considerably larger than that of the c-Rel homodimer, indicating that c-Rel homodimers are in a complexed form with other cellular factors in vivo(Nakayama et al 1992). These observations are also in agreement with the results of previous studies showing that c-Rel proteins associate with other cellular factors in avian lymphoid cells(Davis et al 1990,Kochel et al 1991,Morrison et al 1989). It is clearly important to characterize and identify cellular factors interacting with c-Rel.

Another important question is what are biological functions for the c-Rel complex. The following observations indicate that this factor might function as a repressor specific for AT-rich κB motifs such as IL-6κB. The IL-6κB motif is a most important promoter element in the induction of IL-6 gene expression by IL-1 and TNF-α. In fact, IL-6κB functions as a potent IL-1/TNF-α responsive cis-element in nonlymphoid cells when inserted at the 5' end of the IL-6 TATA element(Shimizu et al 1990). However, IL-6κB is virtually inactive in TNF response in Jurkat T cells, though other κB motifs such as those found in the SV40/HIV/Ig enhancers and MHC promoter respond to cytokines in both nonlymphoid and lymphoid cells. This unresponsivenessof IL-6κB in Jurkat cells cannot be ascribed to a lack of reactivity of IL-6κB with Jurkat NF-κB, since IL-6κB and SV40/HIV/IgκB sites have comparable affinities for Jurkat NF-κB. The results of further studies indicate that the ability of various κB sequences to bind the c-Rel complex is correlated to their unresponsiveness to cytokines in T cells. Thus, the c-Rel complex appears to function as a constitutive IL-6κB specific repressor. We postulate that , because of the higher affinity of the c-Rel complex for IL-6κB sites, NF-κB cannot displace the c-Rel complex from IL-6κB sites and cannot activate transcription(Fig.2)(Nakayama et al 1992).

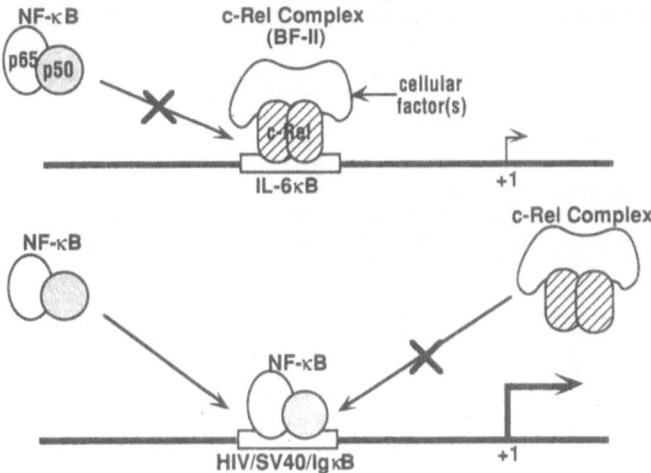

Fig.2 Model for mechanisms of differential regulation by c-Rel complexes(IL-6κB BFII) in IL-6κB- and HIV/SV40/IgκB -dependent transcriptional activation in lymphoid cells. The c-Rel complex is depicted as a repressor complex in which c-Rel homodimers function as a DNA-binding subunit of this complex.

Since the c-Rel complex recognizes other AT-rich κB motifs such as β-interferon κB, the expression of other inflammatory genes having IL-6κB-related κB motifs may also be under similar negative control mechanisms. Although the c-Rel complex is depicted as a repressor in our model, others have shown that c-Rel is a transcriptional activator(Inoue et al 1991). We speculate that the transcriptional activity of c-Rel is differentially regulated in different types of cells through interactions with other cellular factors.

The transcriptional activity of NF-κB/Rel is also modulated by other transcription factors binding to adjacent or overlapping cis-elements as shown in the following example. Serum amyloid A (SAA) proteins are most prominent acute phase reactants in mice and are encoded by four genes (Yamamoto et al 1986). The expression of one of SAA genes(SAA3) is highly IL-1-inducible. The results of characterization of the SAA3 promoter indicate:(1) both of atypical κB and downstream CEBP binding sites are essential for IL-1 induction but neither alone is sufficient for induction when inserted at the 5' end of TATA elements (by contrast, a typical κB such as IL-6κB alone is sufficient for induction as mentioned above);(2) the mutation of a sequence overlapping the κB element, which is homologous to those of IL-6 responsive elements (IL-6 RE) found in the rat α2-macroglobulin gene(Ito et al 1989, Hattori et al 1990) greatly enhances the IL-1 induction of the SAA3 gene, suggesting that IL-6 RE binding factors inhibit the NF-κB activity in SAA3 gene activation. In other examples, while both of CEBP and NF-κB motifs appear to be required for IL-8 gene expression(Mukaida et al 1990), CEBP inhibits NF-κB binding in the angiotensinogen gene (Brasier et al 1990). Thus these observations indicate that other transcription factors may function to synergize or repress the transcriptional activity of the NF-κB/Rel family by binding to adjacent or overlapping cis-elements, or by interacting directly with NF-κB/Rel.

IκB

IκBs were originally defined as cytoplasmic factors which complex with NF-κB and inhibit DNA binding(Baeuerle and Baltimor 1988). They may also have a role in cytoplasmic anchoring of NF-κB/Rel. Two different forms of IκB (α and β) have been purified from cellular extracts. They differ in their molecular weights and in the specific target proteins they can interact with: while IκBα (37 kDa) interacts only with p65, IκBβ (43 kDa) inhibits the DNA binding of both p65 and c-Rel. In addition, their inhibitory activities are abolished upon phosphorylation by protein kinase A and C(Ghosh and Baltimore 1990, Kerr et al 1991). Recently cDNAs coding for IκBβ have been cloned from human (MAD-3)(Haskil et al 1991) and avian (pp40)(Davis et al 1991) origins: pp40 was previously characterized as a cellular factor associated with c- or v-Rel in avian lymphoid cells. Interestingly, the predicted amino acid sequences of these factors show the presence of about 33 amino acids long sequences repeated in five times. These repeated sequences are so-called as ankylin repeat motifs, as they were originally found in the human erythrocyte ankyrin molecule. Surprisingly, other molecules containing ankyrin repeats (the p50 precursor and bcl-3) were subsequently found to show an IκB activity, indicating that the ankyrin repeats are an important structural element for interactions of IκB with NF-κB/Rel(Table 3).

p50 is synthesized as a105-kDa precursor protein, which is proteolytically processed in the cytoplasm to generate p50: p50 having the N-terminal NRD region moves to the nucleus and functions as a transcription factor. Although the C-terminal portion of p105 contains ankyrin repeats(Fig. 1) and has an IκB activity when translated in vitro, it is not clear whether the physiological precessing of p105 stably generates an IκB-like molecule in vivo. More recently, it has been shown that the alternative splicing or different promoter usage of the p105 gene results in the stable expression of the C-terminal portion of p105 in lymphoid cells. This protein (70 kDa) is a physiologically relevant IκB molecule derived from p105 and is designated IκBγ(Inoue et al 1992)(Table 3).

The product of bcl-3, a candidate proto-oncogene, is another molecule containing ankyrin repeats and showing an IκB activity (tentatively referred as IκBδ)(Hatada et al 1992). It was identified in a chronic lymphocyte leukemic cell line, which overproduces bcl-3 proteins and has a chromosomal translocation involving the bcl-3 gene(Ohno et al 1990). This suggests a direct role for bcl-3 in human leukemogenesis, and an important physiological function for the IκB family in general(Table 3).

Table 3   A family of IκB proteins containig ankyrin repeats

| | molecular weights (numbers of amino acids) | numbers of ankyrin repeats | inhibition of DNA binding | | |
			p65	c-Rel	p50
IκBα	37 kDa (?)	?	+	-	-
IκBβ (pp40)	40 kDa (318 a.a.)	6	+	+	-
IκBγ (p105 derived)	70 kDa (607 a.a.)	8	+	+	+
IκBδ (bcl-3)	50 kDa (446 a.a.)	7	+	?	+

THE MODULATION OF DNA BINDING OF NF-κB BY CELLULAR REDUCTION/OXIDATION(REDOX) SYSTEMS

We have recently found that the DNA binding activity of recombinant p50 proteins is greatly enhanced by a cellular factor(s). The results of subsequent characterization indicate that this cellular stimulating factor is a protein but distinct from protein kinases. Since the DNA binding activity of p50 as well as the enhancement of p50 DNA binding by this factor is blocked by sulfhydryl modifying agents such as N-ethylene maleimide and iodoacetamide, a free sulfhydryl group(s) of p50 is essential for DNA binding and a cellular factor appears to enhance DNA binding of p50 by a reduction mechanism. However, thioredoxin, a cellular enzyme that catalyzes the reduction of cystein residues, alone is unable to stimulate DNA binding of p50, suggesting that other cellular factors are required for this reaction. It has been shown recently that a similar mechanism modulates the DNA binding activity of Jun/Fos transcription factors(Xanthoudakis and Curran

1992) which are also involved in the regulation of expression of some inflammatory genes. Thus, cellular redox mechanisms play an important role in the regulation of inflammation.

OTHER CONTROL MECHANISMS

Although the activation of NF-κB involves the dissociation of NF-κB from IκB and subsequent translocation of NF-κB to the nucleus and therefore does not require *de novo* protein synthesis, the expression of all of the NF-κB/Rel family genes is induced in response to inflammatory stimuli(Bours et al 1990,1992,Bull et al 1989, Mayer et al 1991). Since NF-κB/Rel itself appears to be involved in the regulation of these genes(McDonnell et al 1992,Ten et al 1992), the activation of NF-κB/Rel may be positively autoregulated. However, it is not clear how this autoregulation is controlled. In addition, the expression of the c-rel gene is much higher in lymphoid cells than in nonlymphoid cells(Brownell et al 1988). Thus, relative nuclear levels of members of the NF-κB/Rel family are modulated by cell-type-specific mechanisms as well as by extracellular stimuli.

An additional control mechanism involves the alternative splicing of the p65 gene which results in the production of a derivertive of p65 termed p65Δ. p65Δ lacking amino acid residues 222 to 231 is unable to bind NF-κB motifs presumably due to its failure to interact with itself or p50(p65Δ weakly associates with p65), and hence lacks the ability to stimulate transcription through NF-κB sites. Although it is not clear at present what physiological role of p65Δ is, or how alternative splicing is regulated, its relative abundance in lymphoid cells at various differentiation stages suggests some roles for this protein in cell differentiation(Narayanan et al 1992).

CONCLUSION

We now begin to understand the complexity and redundancy in (and possibly cross-talks between)regulatory mechanisms for NF-κB/Rel. A next obvious question is how these mechanisms are regulated or dysregulated in various disease states. There are some examples of dysregulation in the NF-κB/Rel system. First, it has been reported that chromosomal translocations involving the p49/p100 and bcl-3 genes are associated with some lymphoid leukemia(Neri et al 1991, Ohno et al 1990). Second, v-rel, a modified form of c-rel, is a viral oncogene related to lymphomas in birds(Gilmore 1990).Thid, NF-κB is the target for the Tax oncoprotein of HTLV-1 virus(Leung et al 1988) which causes T-cell leukemia in human. Thus, the functions of the NF-κB/Rel and IκB families are not limited to the regulation of inflammation and they may play a more general role in the regulation of host defense.

REFERENCES

Akira S.,Hirano T.,Taga T.,Kishimoto T. (1990) FASEB J. 4:2860-2867
Baeuerle P. A. (1991) Biochim. Biophys. Acta 1072:63-80
Baeuerle P. A. ,Baltimore D. (1988) Science 242:540-546
Bours V.,Burd P. R.,Brown K.,Villalobos J.,Park S.,Ryeck R-P.,Bravo R.,Kelly K.,Siebenlist U. (1992) Mol. Cell. Biol. 12:685-695
Bours V.,Villalobos J.,Burd P. R.,Kelly K.,Siebenlist U. (1990) Nature (London) 348:76-80
Brasier A. R.,Ron D.,Tate J. E.,Habener J. F. (1990) EMBO J. 9:3933-3944
Brownell E.,Mittereder N.,Rice N. R. (1989) Oncogene 4:935-942
Bull P.,Hunter T.,Verma I. M. (1989) Mol. Cell. Biol. 9:5239-5243
Davis J. H.,W. Bargmann,Bose H. R. Jr. (1990) Oncogene 5:1109-1115
Davis N.,Ghosh S.,Simmons D. L.,Tempst P.,Liou H-C.,Baltimore D.,Bose H. R. Jr. (1991) Science 253:1268-1271
Ghosh S.,Baltimore D. (1990) Nature (London) 344:678-682
Ghosh S.,Gifford A. M.,Riviere L. R.,Tempst P.,Nolan G. P.,Baltimore D. (1990) Cell 62:1019-1029
Gilmore T. D. (1990) Cell 62:841-843

Hansen S. K.,Nerlov C.,Zabel U.,Verde P.,Johnsen M.,Baeuerle P. A.,Blasi F. (1992) EMBO J. 11:205-213

Haskil S.,Beg A. A.,Tompkins S. M.,Morris J. S.,Yurochko A. D.,Sampson-Johannes A.,Mondal K.,Ralph P.,Baldwin A. S. Jr. (1991) Cell 65:1281-1289

Hatada E. N.,Nieters A.,Gregory Wulczyn F.,Naumann M.,Meyer R.,Nucifora G.,McKeithan T. W.,Scheidereit C. (1992) Proc. Natl. Acad. Sci. USA 89:2489-2493

Hattori M.,Abraham L. J.,Northemann W.,Fey G. H. (1990) Proc. Natl. Acad. Sci. USA 87:2364-2368

Inoue J.,Kerr L. D.,Kakizuka A.,Verma I. M. (1992) Cell 68:1109-1120

Inoue J.,Kerr L. D.,Ransone L. J.,Bengal E.,Hunter T.,Verma I. M. (1991) Proc. Natl. Acad. Sci. USA 88:3715-3719

Ito T.,Tanahashi H.,Misumi Y.,Sakaki Y. (1989) Nucleic Acids Res. 17:9425-9435.

Kerr L. D.,Inoue J.,Davis N.,Link E.,Baeuerle P. A.,Bose H. R. Jr.,Verma I. M. (1991) Genes Dev. 5:1464-1467

Kieran M.,Blank V.,Logeat F.,Vandekerckhove J.,Lottspeich F.,Bail O. L.,Urban M. B.,Kourilsky P.,Baeuerle P. A.,Israel A. (1990) Cell 62:1007-1018

Kochel T.,Mushinski J. F.,Rice N. R. (1991) Cell 6:615-626

Leung K.,Nabel G. J. (1988) Nature (London) 333:776-778

Mayer R.,Hatada E. H.,Hohmann H. P.,Haiker M.,Bartsch C.,Rothlisberger U.,Lahm H. W.,Schlaeger E. J.,Van Loon A. P. G. M.,Scheidereit C. (1991) Proc. Natl. Acad. Sci. USA 88:966-970

McDonnell P. C.,Kumar S.,Rabson A. B.,Gelinas C. (1992) Oncogene 7:163-170

Mukaida N.,Mahe Y.,Matsushima K. (1990) J. Biol. Chem. 265:21128-21133

Nakayama K.,Shimizu H.,Mitomo K.,Watanabe T.,Okamoto S.,Yamamoto K. (1992) Mol. Cell. Biol. 12:1736-1746

Narayanan R.,Klement J. F.,Ruben S. M.,Higgins K. A.,Rosen C. A. (1992) Science 256:367-370

Neri A.,Chang C-C.,Lombardi L.,Salina M.,Corradini P.,Maiolo A. T.,Chaganti R. S. K. ,Dalla-Favera R. (1991) Cell 67:1075-1087

Nolan G. P.,Ghosh S.,Liou H. C.,Tempst P.,Baltimore D. (1991) Cell 64:961-969

Ohno H. ,Takimoto G.,McKeithan T. W. (1990) Cell 60:991-997

Ruben S. M.,Dillon P. J.,Scherck R.,Henkel T.,Chen C. H.,Maher M.,Baeuerle P. A.,Rosen C. A. (1991) Science 251:1490-1493

Ruben S. M.,Klement J. F.,Coleman T. A.,Maher M.,Chen C-H.,Rosen C. A. (1992) Genes Dev. 6:745-760

Ryseck R-P.,Bull P.,Takayama M.,Bours V.,Siebenlist U.,Dobrzanski P.,Bravo R. (1992) Mol. Cell. Biol. 12:674-684

Schmid R. M.,Perkins N. D.,Duckett C. S.,Andrews P. C.,Nabel G. J. (1991) Nature (London) 325:733-736

Sen R.,Baltimore D. (1986) Cell 47:921-928

Sen R.,Baltimore D. (1986) Cell 46:705-716

Shimizu H.,Mitomo K.,Watanabe T.,Okamoto S.,Yamamoto K. (1990) Mol. Cell. Biol. 10:561-568

Ten R. T.,Paya C. V.,Israel N.,Bail O. L.,Mattei M-G.,Virelizier J-L.,Kourilsky P.,Israel A. (1992) EMBO J. 11:195-203

Xanthoudakis S.,Curran T. (1992) EMBO J. 11:653-665

Yamamoto K.,Shiroo M.,Migita S. (1986) Science 232:227-229

Liver as a Target of Inflammatory Mediators

H. Baumann, S. Pajovic, S. P. Campos, V. E. Jones, and
K. K. Morella

*Roswell Park Cancer Institute, Department of Molecular and Cellular Biology,
Buffalo, New York 14263 USA*

INTRODUCTION

Any type of systemic tissue injury, infection or inflammation causes the liver to undergo a complex pattern of metabolic changes, referred to as the hepatic acute phase response (Kushner 1982). A prominent manifestation of this liver response is the coordinate stimulation of a set of plasma proteins called the acute phase plasma proteins (APP) (Koj 1974). The biological relevance of the liver response lies in part in the functions of the individual APPs (Koj 1985). In all vertebrate species analyzed thus far, these proteins play an essential role in controlling four processes: 1) inhibition of extracellular proteases; 2) blood clotting and fibrinolysis; 3) modulation of immune cells; and 4) neutralization and clearance of harmful components from the circulation.

Since tissue injury distant to the liver affects the expression of APP genes, the participation of humoral inflammatory signals has been assumed (Koj 1974). Studies over the past several years in many laboratories, including ours, have focussed on these issues: i) what is the biochemical nature of the mediators; and ii) what are their cellular and molecular modes of action (reviewed by Fey and Gauldie 1990; Baumann and Gauldie 1990). The qualitative and quantitative changes in the expression of specific plasma protein genes served as guides for the identification of the potential mediators. A major advance in these studies was made with the establishment of tissue culture cell systems which reproduced with high fidelity the hepatic acute phase response (reviewed by Won et al. 1992).

A variety of primary cultures of hepatocytes and established hepatoma cell lines has successfully been applied for measuring specific regulatory effects of hormones on selected APP genes. However, only a few hepatoma cell lines proved to be suitable for defining the pleiotropic action of potential inflammatory mediators. We have identified clonal lines of the rat H-35 hepatoma cells (Baumann et al. 1989) which are able not only to express all positive APPs, but also to regulate these in response to an exceptionally wide range of cytokines and endocrine hormones. The same cells proved useful for defining cis- and trans-acting elements that control the expression of APP genes (Baumann et al. 1990b).

The specific and immediate response of hepatic cells to several multi-functional cytokines has made these cells one of the favored experimental systems for the study of cytokine receptor function and for the identification of the signal transduction pathways that control functions of differentiated cells.

THE REGULATION OF APP GENES IN RAT HEPATOMA CELLS IS AN INDICATOR OF THE ACTION OF INFLAMMATORY MEDIATORS

Treatment of H-35 and Fao rat hepatoma cell lines with combinations of IL-1, IL-6 and dexamethasone, leads to the specific stimulation of the major positive APPs (representing example in Fig 1). Albeit substantial quantitative differences between the two cell lines, identical sets of APPs are regulated (Table 1).

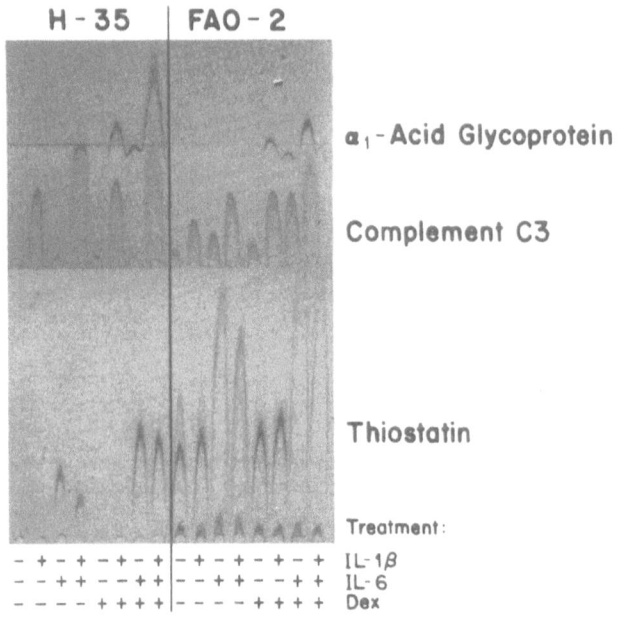

Fig. 1. Stimulation of APP synthesis in rat hepatoma cells. Confluent monolayers of H-35 (clone T-7-18) and Fao (clone 2) were treated for 24 h with serum-free medium containing human recombinant IL-1ß and IL-6 (100 units/ml) and/or dexamethasone (1 µM). Equal aliquots of the culture medium were analyzed by rocket immunoelectrophoresis for the indicated plasma proteins.

The salient features of the rat hepatoma cell response can be summarized as follows. IL-1, like TNFα, stimulates the expression of type I APPs but not type II proteins. Conversely, IL-6 primarily stimulates type II proteins but increases type I proteins as well. The combination of IL-1 and IL-6 leads to an additive to synergistic action on type I protein, in particular on α_1-acid glycoprotein (AGP) (Prowse and Baumann 1988) whereas IL-1 reduces the IL-6 effect type II proteins. Dexamethasone enhances the stimulatory effects of the cytokines on many APPs and is most prominently seen for AGP and α_2-macroglobulin (Prowse and Baumann 1988; Gehring et al. 1987). Equivalent analyses of primary cultures of hepatocytes have yielded qualitatively identical and quantitatively similar results and attest to the representative nature of hepatoma cell regulation.

The cytokine-specific regulation of the various acute phase genes provided a highly sensitive diagnostic system to assess the potential liver action of other cytokines. Although IL-1 and TNFα control the same sets of proteins, they seem to function additively (Fig. 2). This and the fact that their relative effects on the target genes vary, suggest separate signal transduction mechanisms.

Table 1 Cytokine-specific Regulation of Plasma Protein Genes

Plasma protein	Acute phase protein *in vivo*:	Stimulated in tissue culture by:
Type I Acute Phase Plasma Proteins		Interleukin 1α
Hemopexin	++	Interleukin 1β
		Tumor necrosis factor α
Haptoglobin (Rat)	++	Interleukin 6
		Interleukin 11
Complement C 3	(+)	Oncostatin M
		Leukemia inhibitory factor
		Ciliary neurotophic factor
Serum amyloid A (Rat)	-	Glucocorticoids
		(* strong synergism with
α-1-Acid glycoprotein*	+++	cytokines)
Type II Acute Phase Plasma Proteins		
		Interleukin 6
α-1-Antitrypsin	+	Interleukin 11
α-1-Antichymotrypsin (Contrapsin)	++	Oncostatin M
		Leukemia inhibitory factor
		Ciliary neurotophic factor
Fibrinogen (α,β, γ-subunits)	++	
		Glucocorticoids
Thiostatin	++	(* strong synergism with
		cytokines)
α-2-Macroglobulin*	+++	

By utilizing H-35 cells as an assay system, additional cytokines with IL-6-like activities have been discovered. The three structurally-related factors, leukemia inhibitory factor (LIF) (Baumann and Wong 1989), oncostatin M (Richards et al. 1992), and ciliary neurotrophic factor (CNTF) (Baumann and Gearing, in preparation) and the structurally distinct IL-11 (Baumann and Schendel 1991) regulate the same set of proteins as IL-6. The identical action of LIF, oncostatin M and CNTF is ascribed to the use of the same LIF receptor (Gearing et al. 1992; Gearing unpublished). Although the current model of the IL-6-related cytokine receptor action proposes the involvement of a common signal transduction subunit (Murakami et al. 1991; Gearing et al. 1992; Ip et al. 1992), the H-35 cell response still reveals characteristic differences. The stimulation of thiostatin by IL-6 receptor agonist is strongly enhanced by dexamethasone (Fig. 1), whereas stimulation by the IL-11 receptor agonist is not appreciably changed (Baumann and Schendel 1991), and that by the LIF receptor agonists is reduced (Baumann et al. 1992a). These distinct responses have only been observed in H-35 cells and still need to be verified in other hepatoma cells and primary hepatocytes.

CELL LINE SPECIFIC DIFFERENCES IN REGULATION OF APPS

The regulation of APP genes in various hepatic cell systems from humans, rabbits,

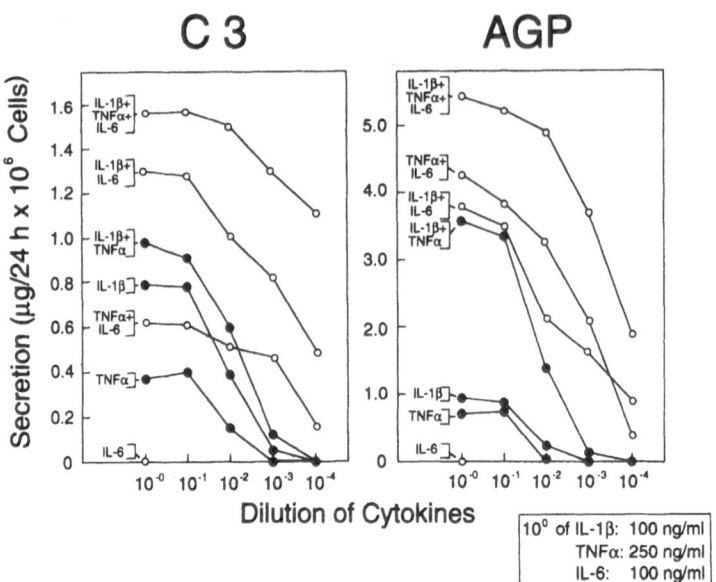

Fig. 2. Additive and synergistic action of IL-1, TNFα, and IL-6 on type 1 APPs. H-35 cells were treated for 24 h with 10-fold serially-diluted human recombinant IL-1ß and TNFα. One set of each dilution series was also included, 100 ng/ml (=100 U/ml) IL-6. The production of complement C3 and AGP was quantitated by immunoelectrophoresis.

rats, and mice shows a strikingly high degree of similarity. This conservation is especially apparent for those genes encoding APPs that are common to all mammals and include the three chains of fibrinogen, haptoglobin, AGP, and α_1-antichymotrypsin (Kushner and Mackiewicz 1987).

Differences in the pattern of APPs are in general explained by the presence of species-specific APPs, e.g., C-reactive protein in human and thiostatin and α_2-macroglobulin in the rat (Koj 1985; Fey and Gauldie 1990). There are notable variations in the responsiveness to a given cytokine or hormone among various cell lines (Won et al. 1992). In some instances, the lack of receptor function for the partial factor is recognized as the cause: HepG2 cells lack IL-11 receptors, and Hep3B cells lack receptors for IL-11, LIF, and oncostatin M. These receptor-deficient hepatoma cell lines now serve as targets for functional reconstitution of cloned cytokine receptors (such as the receptor for rat IL-6 (Baumann M et al. 1990), and LIF (Baumann and Gearing, in preparation).

APP GENES ARE REGULATED AT THE TRANSCRIPTIONAL LEVEL

In H-35 (Baumann et al. 1990b) and HepG2 cells (Baumann et al. 1990a), cytokines and dexamethasone increase transcription rates of the APP genes proportional to the changes seen at the level of mRNA and protein secretion. Although APP gene regulation is primarily a transcriptional event, in the systems studied the possible contribution of posttranscriptional regulatory processes to the overall level of APP production could not be ruled out (Birch and Schreiber 1986). The cytokine activation of most APP genes in H-35 cells (=early APP genes) is maximal within 30 min of treatment and does not require new protein synthesis. Only transcriptional stimulation of AGP and α_2-macroglobulin genes (=late APP genes) is delayed by 2 to 4 h and requires ongoing protein synthesis (Baumann et al. 1991).

We determined the major regulatory elements (RE) that are responsive to IL-1, IL-6, and dexamethasone in the genes encoding rat and mouse AGP and haptoglobin and rat ß-fibrinogen (Baumann et al. 1990a,b; Won and Baumann 1990). These elements were located within the 5' flanking regions at different positions relative to each other and relative to the transcriptional start site. The functionally-defined REs of the three genes contain binding sites for common transcription factors that include C/EBP isoforms, IL-6REBP (Hattori et al. 1990), and glucocorticoid receptors. The presence of common REs and shared transcription factors has been predicted to be responsible in part for the coordinate regulation of genetically unlinked APP genes (Fowlkes et al. 1984). C/EBPß (NF-IL-6, IL-6DBP) has been proposed as one target for cytokine regulation, since it binds to several cytokine REs and prominently mediates transactivating function (Akira et al. 1990; Poli et al. 1990). Since cytokine treatment or acute phase *in vivo* stimulate the expression of C/EBPß and C/EBPδ in hepatic cells (Alam et al. 1992; Baumann et al. 1992b), these C/EBP isoforms seem to represent indirect mediators for the cytokine signal to the APP genes. The mechanism of immediate activation of APP genes and the molecular participants are still unknown (Baumann et al. 1992b).

REGULATION OF APP IS SUBJECT TO MODULATION BY GROWTH FACTORS

Not only are cytokines and adrenocorticoids elevated during a systemic acute phase reaction, other inflammatory factors and endocrine hormones are also increased, including catecholamines, eicosanoids, thyroxine, insulin, glucagon, and various growth factors (Kushner 1982). Liver cells have receptors for many of these factors. Treatment of hepatic cells with these factors individually, however, did not result in any appreciable changes in APP synthesis. Only very recently, the role of seemingly inactive factor has been reconsidered in a physiologically relevant context of cytokines (IL-1, TNFα and IL-6) and glucocorticoids (Koj et al. 1992). Although studies are still in progress, the following findings have been made: T3 and all factors acting via cyclic nucleotides and Ca^{2+} do not significantly modulate the regulation of APP genes. Minor stimulatory or inhibitory effects have been achieved with phorbol ester activated protein kinase C (Baumann et al. 1988), and growth factors, including insulin (Campos and Baumann 1992) hepatocyte growth factor, acidic fibroblast growth factor, transforming growth factors α and ß (TGFß), and mast cell growth factor (Campos et al., in preparation). Insulin and TGFß produce the most prominent effects. While insulin generally reduces both IL-1 and IL-6 stimulation of the APP gene by a factor of two, TGFß enhances IL-6 response two-fold. The molecular mechanism of insulin inhibition is still unknown (Campos and Baumann 1992). The effect of TGFß seems in part to be mediated by enhanced expression of the IL-6 receptor (Campos et al., in preparation). It remains to be shown whether the inhibitory action of growth factors

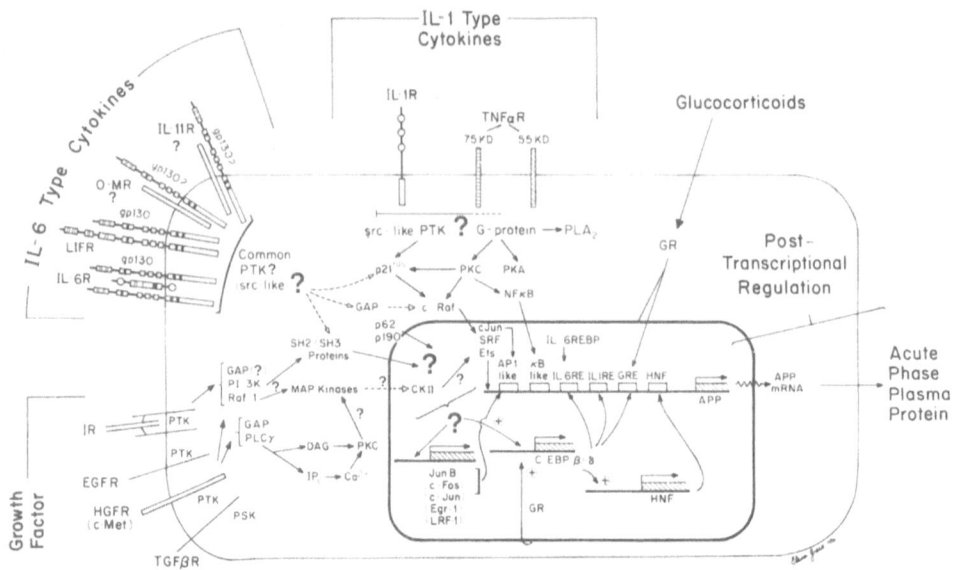

Fig. 3. Working model of potential intracellular regulatory pathways controlling expression of APP genes.

and protein kinase C is involved in the down regulation of APP genes at the late stage of the hepatic acute phase reaction (Baumann et al. 1988).

A working model for the signalling events in hepatic cells has been assembled, integrating findings of various cell systems (Fig. 3). The model obviously cannot be complete or correct for all pathways, since it takes into consideration a) contradictory reports (e.g., linkage of IL-1 and TNF receptor to protein kinase A, pertussis toxin sensitive G-proteins and src-like protein tyrosine kinases; or relative position of p21 ras to c-Raf and GAP), and b) assumptions that some pathways (or reactions) are not cell-type specific (e.g., modification of SH-2- and SH-3-domain containing proteins, nuclear action of regulatable casein kinase II and cytoplasmic to nuclear translocation of transcription factors). The emphasis of this model is that APP genes are controlled by at least four separate types of receptors (IL-1- and IL-6-type cytokines, growth factors, and glucocorticoids). Each plasma membrane receptor type must affect a discrete set of intracellular signalling molecules. Even though the involvement of common factors is noted or predicted, if not demonstrated, each of these factors cannot independently mediate the complete hormone-specific response. The unraveling of the actual regulatory mechanisms will be one of the major goals in the studies of the hepatic acute phase response.

WHAT IS THE RELEVANCE OF THE TISSUE CULTURE DATA FOR THE UNDERSTANDING OF APP GENE REGULATION *IN VIVO* ?

The expression and regulation of APP gene *ex vivo* is strongly influenced by the culture environment and by the unavoidable phenotypic changes of the liver cells in culture (Won et al. 1992). Despite these caveats, tissue culture experiments were critical in gaining conclusive evidence for the direct hepatic action of cytokines (listed in Table 1). Tissue culture analyses have not only provided information about factors which can potentially function as inflammatory mediators, but have also allowed predictions about the types of factors that are necessary for achieving an acute phase response. It has been generally accepted that *in vivo* the concerted action of cytokines of the IL-1- and IL-6-type, and glucocorticoids is responsible for the observed pattern of acute phase plasma protein changes.

The tissue culture analyses have yielded a catalog of gene-specific effects that are exerted by cytokines and endocrine hormones (e.g., Figs. 1 and 2 and Table 1). Based on this information, *in vivo* changes following administration of these factors can be predicted and can be determined accordingly. Current examples include the treatment of mice with recombinant human IL-11 (Y-C Chang, personal communication) and human LIF (D Gearing, personal communication). In both cases, a prominent stimulation of type II APP and a little change in type I APP is expected. The specific action of the administered factors can be verified by taking advantage of the known combined effects of cytokines and endocrine hormones. The test can vary from co-administration of glucocorticoids, IL-1 and TNFα, to the use of neutralizing antibodies against cytokines or cytokine receptor inhibitors. The confirmation of predicted changes in APP gene expression is especially of interest in pathologic states known to affect the systemic concentrations of cytokines and/or endocrine hormones, such as systemic rheumatoid arthritis, malignancy, or autoimmune disorders.

Lastly, the qualitative and quantitative composition of plasma proteins in disease states can serve as indicators of the type and action of circulating cytokines or hormones that may be involved. Examples include the restricted type II APP stimulation during chronic schistosomiasis (Isseroff and Baumann, unpublished) and during subcutaneous growth of transplantable sarcomas (Ebener and Baumann, unpublished).

ACKNOWLEDGEMENTS

We are greatly indebted to past and current collaborators who have made factual, material and/or intellectual contributions to our understanding of APP regulation in tissue culture and *in vivo*. We thank Genetics Institute, Immunex Corp., and Genentech, Inc. for their generous supply of cytokines, and Marcia Held for secrétarial assistance. Work in the authors laboratory is supported by NIH grants CA26122 and DK33886.

REFERENCES

Akira S, Isshiki H, Sugita T, Tanabe O, Kinoshita S, Nishio Y, Nakajima T, Hirano T, Kishimoto T (1990) EMBO J 9:1897-1906

Alam T, An, MR, Papaconstantinou J (1992) J Biol Chem 267:5021-5024

Baumann H, Gauldie J (1990) Mol Biol Med 7:147-159

Baumann H, Isseroff H, Latimer JJ, Jahreis GP (1988) J Biol Chem 263:17390-17396

Baumann H, Jahreis GP, Morella KK (1990a) J Biol Chem 265:22275-22281

Baumann H, Jahreis GP, Morella KK, Won K-A, Pruitt SC, Jones VE, Prowse KR (1991) J Biol Chem 266:20390-20399

Baumann H, Marinkovic-Pajovic S, Won K-A, Jones VE, Campos SP, Jahreis GP, Morella KK (1992a) The Ciba Foundation Symposium 167:100-124

Baumann H, Morella K, Campos SP, Cao Z, Jahreis GP (1992b) J Biol Chem 267: in press

Baumann H, Morella KK, Jahreis GP, Marinkovic S (1990b) Mol Cell Biol 10:5967-5976

Baumann H, Prowse KR, Marinkovic S, Won K-A, Jahreis GP (1989) Ann NY Acad Sci 557:280-297

Baumann H, Schendel P (1991) J Biol Chem 266:20424-20427

Baumann H, Wong GG (1989) J Immunol 143:1163-1167

Baumann M, Baumann H, Fey GH (1990) J Biol Chem 265:19853-19862

Birch HE, Schreiber G (1986) J Biol Chem 261:8077-8080

Campos SP, Baumann H (1992) Mol Cell Biol 12:1789-1797

Fey G, Gauldie J (1990) The acute phase response of the liver in inflammation. In: Popper H, Schaffner F (eds) Progress in liver disease, vol 9. WB Saunders Co, Philadelphia, p 89

Fowlkes D, Mullis NT, Comeau CM, Crabtree GR (1984) Proc Natl Acad Sci USA 81:2313-2316

Gearing DP, Comeau MR, Friend DJ, Gimbel SD, Thut CJ, McGourty J, Brasher KK, King JA, Gillis S, Mosley B, Ziegler SF, Cosman D (1992) Science 255:1434-1437

Gehring MR, Shield BR, Northemann W, de Bruijn MHL, Kan C-C, Chain AC, Noonan DJ, Fey GH (1987) J Biol Chem 262:446-454

Hattori M, Abraham LJ, Northemann W, Fey GH (1990) Proc Natl Acad Sci USA 89: 2364-2368

Ip NY, Nye SH, Boulton TG, Davis S, Taga T, Li Y, Birren SJ, Yasukawa K, Kishimoto T, Anderson DJ, Stahl N, Yancopoulos KD (1992) Cell 69:1121-1132

Kinoshita S, Akira S, Kishimoto T (1992) Proc Natl Acad Sci USA 89:1473-1476

Koj A (1974) Acute phase reactants. In: Allison AC (ed) Structure and function of plasma proteins, vol 1. Plenum Press, London, New York, p 73

Koj A (1985) Definition and classification of acute phase proteins. In: Gordon AH, Koj A (eds) The acute phase response to injury and infection, vol 5. Elsevier, Amsterdam, p 139

Koj A, Gauldie J, Baumann H (1992) Biological perspectives of cytokine and hormone networks. In: Mackiewicz A, Kushner I, Baumann H (eds) Acute phase proteins, molecular biology, biochemistry, clinical applications, CRC Press, Inc., in press

Kushner I (1982) Ann NY Acad Sci 389:39-48

Kushner I, Mackiewicz A (1987) Dis Markers 5:1-11

Murakami M, Narazaki M, Hibi M, Yawata H, Yasakawa K, Hamaguchi M, Taga T, Kishimoto T (1991) Proc Natl Acad Sci USA 88:11349-11353

Poli V, Mancini FP, Cortese R (1990) Cell 63:643-653

Prowse KR, Baumann H (1988) Mol Cell Biol 8:42-51

Richards CD, Brown TJ, Shoyab M, Baumann H, Gauldie J (1992) J Immunol 148:1731-1736

Won K-A, Baumann H (1990) Mol Cell Biol 10:3965-3978

Won K-A, Campos SP, Baumann H (1992) Experimental systems for studying hepatic acute phase response. In: Mackiewicz A, Kushner I, Baumann H (eds) Acute phase proteins, molecular biology, biochemistry, clinical applications, CRC Press, Inc., in press

Interleukin - 8 and Related Cytokines

M. Baggiolini and B. Moser

Theodor - Kocher Institute, University of Berne, Switzerland

INTRODUCTION

Our understanding of the pathophysiology of neutrophil
recruitment into diseased tissues is based on Metschnikoff's
concept of inflammation as a protective process, in which
neutrophils act as scavengers of microorganisms and other
unwanted material. Neutrophil enzymes (Metschnikoff's cytases)
bring about the intracellular digestion, but also lead to tissue
damage when they escape from the phagocytic vacuoles. The study
of phagocyte activation and migration became possible with the
identification of chemotactic agonists. The existence of tissue-
derived chemoattractants was postulated for quite some time, and
cytokines and other cell-derived factors with apparent
chemotactic properties were studied extensively (reviewed by
Baggiolini et al 1992). The discovery of a novel neutrophil-
activating protein, which is now called interleukin-8 (IL-8),
and of several related cytokines (Baggiolini et al 1989,
Baggiolini and Sorg 1992) has greatly widened our understanding
of the mechanisms of neutrophil immigration into infected or
otherwise injured tissues.

INTERLEUKIN-8

The gene of IL-8 encodes a 99-amino acid protein that is
secreted after cleavage of a leader sequence of 20 residues.
Amino-terminal processing after secretion yields variants of
increasing biological activity consisting of 77, 72, 70 and 69
amino acids (Lindley et al 1988, Clark-Lewis et al 1991). IL-8
contains four cysteines linked to disulfide bridges. Nuclear
magnetic resonance spectroscopy shows that IL-8 has a short,
flexible N-terminal domain that is anchored by the two disulfide
bonds to the core of the molecule made up of three antiparallel
ß-strands followed by a prominent C-terminal alpha-helix. Native
IL-8 is remarkably resistant to inactivation by peptidases and
proteinases, and a variety of denaturing treatments. Reduction
of the disulfide bonds, however, rapidly leads to a complete
loss of biological activity (Peveri et al 1988).

In neutrophils IL-8 induces shape change and chemotaxis,
exocytosis and the respiratory burst (Baggiolini et al 1989,
Baggiolini et al 1992). *The shape change* reflects the activation
of the contractile cytoskeleton, which enables the neutrophils
to adhere to endothelial cells and to migrate. *Exocytosis* is a

complex response involving the release of enzymes and other soluble proteins from several subcellular storage compartments, as well as the remodeling of the plasma membrane by fusion with subcellular membranes. The latter process enhances the surface expression of integrins and complement receptor type 1 (Detmers et al 1990, Paccaud et al 1990), and the ability of the neutrophils to bind to endothelial cells and the extracellular matrix (Carveth et al 1989). *The respiratory burst* is the most characteristic response of stimulated phagocytes. It is due to the transient activation of the NADPH-oxidase which forms superoxide and H_2O_2. These responses are typical for the classical chemotactic agonists like fMet-Leu-Phe or C5a, as is the mechanism of signal transduction in neutrophils stimulated with IL-8, which depends on *Bordetella pertussis* toxin-sensitive GTP-binding proteins and the activation of a phosphatidylinositol-specific phospholipase C. The phospholipase delivers two second messengers, IP_3 and diacylglycerol; IP_3 induces a rise in cytosolic free calcium, and diacylglycerol activates protein kinase C. Treatment with *pertussis* toxin, depletion of mobilizable calcium, and exposure to inhibitors like staurosporine and wortmannin can be used to modulate signal transduction and to show that the responses elicited by IL-8, fMet-Leu-Phe and C5a are controlled by a similar mechanism (Baggiolini et al 1992).

The effects of IL-8 in vivo were studied most thoroughly in rabbits, although human IL-8 induces comparable responses in mice, rats, guinea-pigs, dogs and other animals. After intradermal injection the local infiltration is rapid, restricted to neutrophils and long-lasting (Colditz et al 1989). The unusually long duration of action suggests that IL-8 is inactivated and cleared from the tissues only slowly. IL-8 injection into human skin induces a similar kind of inflammatory reaction with massive perivenular infiltration of neutrophils, no changes of lymphocytes and no participation of basophils, eosinophils or monocytes. IL-8 causes no wheal and flare, itching or pain, suggesting that it does not induce histamine release from skin mast cells (Leonard et al 1991).

IL-8 RELATED PROTEINS

Several proteins that are similar to IL-8 in terms of structure and biological activity were discovered during the last few years. GROα was originally described as "melanoma growth stimulatory activity" (MGSA) (Richmond 1988), and later shown to be a powerful neutrophil chemoattractant (Moser et al 1990). Two other GRO proteins, GROβ and GROγ, were identified subsequently (Haskill et al 1990). They are highly homologous to GROα and to each other (80-90% sequence identity) and have similar neutrophil-activating properties (Dewald and Clark-Lewis, unpublished results). *ENA-78*, was discovered in the culture supernatants of a type-II alveolar cell line (Walz et al 1991) and appears to be produced preferentially by epithelial cells. *Neutrophil-activating peptide-2* (NAP-2), the first IL-8 homologue to be discovered, differs from the other IL-8 related proteins in that it derives from the N-terminal processing of known proteins, platelet basic protein or connective tissue-activating peptide-III, which lack neutrophil-activating properties (Walz and Baggiolini 1990).

IL-8 and its five related chemotactic cytokines arise from different genes clustered on chromosome 4 (Modi et al 1990). In pathological conditions, it is likely that several of them are generated concomitantly as suggested by experiments with lung epithelial cells which release, in addition to ENA-78, IL-8, GROα and GROτ (Walz et al 1991), and by numerous observations that IL-8 and GROα are produced by the same cells in response to the same stimuli (Baggiolini et al 1992).

IL-8 RECEPTORS

In view of the occurrence of several cytokines with biological effects similar to those of IL-8, it was of primary importance to establish whether they act via the same receptors using a similar signal transduction mechanism. A simple way to approach this question is to look for desensitization of the cells upon repeated stimulation. It was shown early on that neutrophils remain responsive to IL-8 after stimulation with fMet-Leu-Phe, C5a, PAF or LTB_4, and that, in turn, IL-8 does not affect the responsiveness of the cells to the other agonists (Peveri et al 1988), suggesting the existence of a selective receptor. More recent studies showed mutual desensitization between IL-8, NAP-2 and GROα (Moser et al 1991), IL-8 and ENA-78 (Walz et al 1991) as well as among the three forms of GRO (Geiser et al. in preparation), indicating that all six related cytokine share the receptors. IL-8 receptors were also evidenced by binding studies. In general agreement with other reports, we found on average 64,500 ± 14,000 receptors per neutrophil with an apparent K_D of 0.18 ± 0.07 nM. Competition studies with IL-8, NAP-2 and GROα revealed the existence of two types of receptors on neutrophils: One with high affinity for all three ligands (K_D: 0.1-0.3 nM), and the other with high affinity for IL-8, but low affinity for NAP-2 and GROα (K_D: 100-130 nM) (Moser et al 1991). Two membrane proteins that selectively bind IL-8, NAP-2 and GROα were also demonstrated by crosslinking experiments (Moser et al 1991).

The work of Boulay et al (1990) showing that fMet-Leu-Phe binds to a seven-transmembrane-domain receptor opened the way to molecular cloning, and two IL-8 receptor cDNAs were recently reported and shown to encode similar transmembrane proteins (Holmes et al 1991, Murphy et al 1991). The cloning of two distinct cDNAs is in agreement with the biochemical evidence for two IL-8 receptors.

Using chemically synthesized N- and C-terminal truncation analogs (Clark-Lewis et al 1991) and muteins (Hébert et al 1991) it was recently established that the N-terminal sequence Glu-Leu-Arg (ELR) that precedes the first cysteine is the binding site of IL-8 to its receptor. All three residues, Arg in particular, are highly sensitive to modification. The same is likely to apply to all IL-8 related chemotactic cytokines, which in contrast to proteins that have little or no chemotactic activity like PF4 and IP10, share the ELR motif. The tripeptide ELR itself and oligopeptides containing the ELR motif are inactive, suggesting that either a particular conformation of the tripeptide, or interactions of other domains of the IL-8 molecule with the receptor are required.

DETERMINANTS OF IL-8 PATHOPHYSIOLOGY

Several features of IL-8 and its related proteins must be considered in the context of potential functions of these cytokines in physiology and pathology.

(i) Ubiquitous generation. IL-8 was originally identified in the medium of stimulated human blood monocytes (Baggiolini et al 1989). It was then realized, however, that many different cells can produce IL-8 when appropriately stimulated. The expression of IL-8 mRNA and the release of the biologically active cytokine was observed in endothelial cells, fibroblasts from different tissues, keratinocytes, synovial cells, chondrocytes, several types of epithelial cells as well as some tumor cells (Baggiolini et al 1989, Leonard and Yoshimura 1990). Even neutrophils themselves synthesize IL-8, and may thus intensify their own recruitment (Bazzoni et al 1991). Interleukin-1 (IL-1) and tumor necrosis factor (TNF) are universal stimuli, since they were found to induce IL-8 expression and secretion in all cells studied so far (Baggiolini et al 1992). Endotoxin is very effective on phagocytes and endothelial cells, but inactive on mesenchymal cells. Monocytes and macrophages generate IL-8 upon stimulation with IL-1α, IL-1β, TNFα, IL-3, GM-CSF, endotoxin, lectins, phorbol esters, immune complexes and phagocytosis (Seitz et al 1991) and some other agents (Baggiolini et al 1992). Opsonized particles are also a prime stimulus for IL-8 production by neutrophils (Bazzoni et al 1991).

(ii) Resistance to inactivation and slow clearance. IL-8 is a very sturdy protein. Unless its disulfide bonds are reduced, it remains biologically active even under the most drastic denaturing treatments (Peveri et al 1988), a property that has been of great advantage for purification. The long-lasting chemoattractant effect upon intradermal injection indicate that IL-8 is not degraded by extracellular peptidases and has a tendency to remain at the site of injection, and, by inference, within the immediate environment of the cells from which it may be released. Charge interaction with acidic tissue matrix macromolecules could restrict its diffusion. Recent studies in our laboratory show that native IL-8 is resistant to the neutrophil carboxyl-, thiol- and metallo-proteinases, and is degraded only slowly by the serine-proteinases, elastase, cathepsin G and proteinase 3 (Padrines et al., in preparation). When IL-8 is denatured by reduction and alkylation, however, it is quickly digested by several of these proteinases. Other chemoattractants like fMet-Leu-Phe, C5a, LTB$_4$ and PAF, by contrast, act more transiently as they are inactivated rapidly by oxidation or hydrolysis.

(iii) Selectivity for neutrophils. Experiments in vivo, in animals and human volunteers, have shown that IL-8 elicits the selective recruitment of neutrophils (see above). We could not confirm lymphocyte recruitment in the rat (Zwahlen, unpublished observation), as described by others (Larsen et al 1989). In vitro, however, IL-8 was shown to elicit some responses in basophils, eosinophils, monocytes and lymphocytes (Baggiolini et al 1992). In spite of these effects, IL-8 appears to be more selective than other chemotactic agonists, such as fMet-Leu-Phe, C5a, LTB$_4$ and PAF, and the pathophysiological significance of its action on other leukocytes is still questionable.

(iv) Redundancy. Five distinct chemotactic proteins that share structural and biological similarity with IL-8 and bind to the IL-8 receptors have been identified so far. This unusual redundancy may be taken to suggest that these cytokines fulfill an essential role in biology that may require multiple backing up for assuring proper function even in defective situations. The expression of IL-8 and its related cytokines often appears to be regulated in a similar way as shown for instance by numerous studies on the expression of IL-8 and GROα (Baggiolini et al 1992). On the other hand, differential expression in dependence of the tissue and the stimulus has been reported for the three *gro* genes (Haskill et al 1990). Several of these chemotactic cytokines are likely to concur in recruiting neutrophils, but may also fulfill distinguishing biological functions, possibly related to cellular growth (Sager et al 1992) in addition to neutrophil activation.

CLINICAL ASPECTS

The ability to attract and activate neutrophils qualified IL-8 from the beginning as an inflammatory mediator (Baggiolini et al 1989). Meanwhile this hypothesis has been amply verified, and a few examples shall be briefly presented for inflammatory skin and joint diseases. Considerable IL-8 immunoreactivity is found in the skin of patients with psoriasis and palmoplantar pustulosis (Sticherling et al 1991, Nickoloff et al 1991, Anttila et al 1992) , where expression of IL-8 mRNA is prominent in keratinocytes (Gillitzer et al 1991). IL-8 accumulates in the synovial fluid of arthritic joints (Brennan et al 1990, Seitz et al 1991) and is released by stimulated synovial cells (Golds et al 1989) and chondrocytes (Van Damme et al 1990). Mononuclear cells from the blood or synovial fluid of patients with rheumatoid arthritis release much higher amounts of IL-8 than cells from healthy controls after stimulation with LPS, IL-1, TNF and immune complexes (Seitz et al 1991).

The study of the of expression of IL-8, the three GRO proteins and ENA-78 is extending to cover more and more clinical conditions, and it is hoped that information on gene regulation,

on chemotactic cytokine antagonists and neutralizing antibodies
will soon have impact in the therapy of inflammatory diseases.

REFERENCES

Anttila HSI, Reitamo S, Erkko P, Ceska M, Moser B, Baggiolini M
(1992) Interleukin-8 immunoreactivity in the skin of healthy
subjects and patients with palmoplantar pustulosis and
psoriasis. J Invest Dermatol 98:96-101

Baggiolini M, Dewald B, Walz A (1992) NAP-1/IL-8 and related
chemotactic cytokines. In: Inflammation, Basic Principles and
Clinical Correlates, 2nd Edition. Gallin JI, Goldstein, IM,
Snyderman, R, eds, Raven Press New York, in press

Baggiolini M, Sorg C (1992) Interleukin-8 (NAP-1) and related
chemotactic cytokines. Cytokine vol 4 Karger, Basel, p 164

Baggiolini M, Walz A, Kunkel SL (1989) Neutrophil-activating
peptide-1/interleukin 8, a novel cytokine that activates
neutrophils. J Clin Invest 84:1045-1049.

Bazzoni F, Cassatella MA, Rossi F, Ceska M, Dewald B, Baggiolini
M (1991) Phagocytosing neutrophils produce and release high
amounts of the neutrophil-activating peptide 1/interkeukin 8.
J Exp Med 173:771-774.

Boulay F, Tardif M, Brouchon L, Vignais P (1990) Synthesis and
use of a novel N-formyl peptide derivative to isolate a human
N-formyl peptide receptor cDNA. Biochem Biophys Res Commun
168:1103-1109

Brennan FM, Zachariae COC, Chantry D, Larsen CG, Turner M, Maini
RN, Matsushima K, Feldmann M (1990) Detection of interleukin 8
biological activity in synovial fluids from patients with
rheumatoid arthritis and production of interleukin 8 mRNA by
isolated synovial cells. Eur J Immunol 20:2141-2144

Carveth HJ, Bohnsack JF, McIntyre TM, Baggiolini M, Prescott SM,
Zimmerman GA (1989) Neutrophil activating factor (NAF) induces
polymorphonuclear leukocyte adherence to endothelial cells and
to subendothelial matrix proteins. Biochem Biophys Res Commun
162:387-393.

Clark-Lewis I, Schumacher C, Baggiolini M, Moser B (1991)
Structure-activity relations of interleukin 8 determined using
chemically synthesized analogs: Critical role of N-terminal
residues and evidence for uncoupling of chemotaxis, exocytosis
an receptor binding activities. J Biol Chem 266:23128-23134

Colditz I, Zwahlen R, Dewald B, Baggiolini M (1989) In vivo
inflammatory activity of neutrophil-activating factor, a novel
chemotactic peptide derived from human monocytes. Am J Pathol
134:755-760.

Detmers PA, Lo SK, Olsen-Egbert E, Walz A, Baggiolini M, Cohn ZA
(1990) Neutrophil-activating protein 1/interleukin 8
stimulates the binding activity of the leukocyte adhesion
receptor CD11b/CD18 on human neutrophils. J Exp Med 171:1155-
1162.

Gillitzer R, Berger,R Mielke, V Müller, C Wolff K, Stingl G
(1991) Upper keratinocytes of psoriatic skin lesions express
high levels of NAP-1/IL-8 mRNA in situ. J Invest Dermatol
97:73-79

Golds EE, Mason P, Nyirkos P (1989) Inflammatory cytokines
induce synthesis and secretion of gro protein and a neutrophil
chemotactic factor but not ß2-microglobulin in human synovial
cells and fibroblasts. Biochem J 259:585-588

Haskill S, Peace A, Morris J, Sporn, S.A., Anisowicz, A., Lee,
S.W., Smith, T., Martin, G., Ralph, R, Sager, R (1990)
Identification of three related human GRO genes encoding
cytokine functions. Proc Natl Acad Sci USA 87:7732-7736.

Hébert CA, Vitangcol RV, Baker JB (1991) Scanning mutagenesis of
interleukin-8 identifies a cluster of residues required for
receptor binding. J Biol Chem 266:18989-18994

Holmes WE, Lee J, Kuang WJ, Rice GC, Wood WI (1991) Structure
and functional expression of a human interleukin-8 receptor.
Science 253:1278-1280

Larsen CG, Anderson AO, Appella E, Oppenheim JJ, Matsushima K
(1989) The neutrophil-activating protein (NAP-1) is also
chemotactic for T lymphocytes. Science 243:1464-1466.

Leonard EJ, Yoshimura T, Tanaka S, Raffeld M (1991) Neutrophil
recruitment by intradermally injected neutrophil
attractant/activation protein-1. J Invest Dermatol 96:690-694.

Leonard, EJ, Yoshimura, T (1990) Neutrophil
attractant/activation protein-1 (NAP-1 (interleukin-8)) Am J
Respir Cell Mol Biol 2:479-486

Lindley I, Aschauer H, Seifert JM, Lam C, Brunowsky W, Kownatzki
E, Thelen M, Peveri P, Dewald B, von Tscharner V, Walz A,
Baggiolini M. (1988) Synthesis and expression in Escherichia
coli of the gene encoding monocyte-derived neutrophil-
activating factor: Biological equivalence between natural and
recombinant neutrophil-activating factor. Proc Natl Acad Sci U
S A 85:9199-9203.

Modi WS, Dean M, Seuanez HN, Mukaida N, Matsushima K, O'Brien SJ
(1990) Monocyte-derived neutrophil chemotactic factor
(MDNCF/IL-8) resides in a gene cluster along with several
other members of the platelet factor 4 gene superfamily. Hum
Genet 84:185-187.

Moser B, Clark-Lewis I, Zwahlen R, Baggiolini M (1990)
Neutrophil-activating properties of the melanoma growth-
stimulatory activity. J Exp Med 171:1797-1802.

Moser B, Schumacher C, von Tscharner V, Clark-Lewis I, Baggiolini M (1991) Neutrophil-activating peptide 2 and gro/melanoma growth-stimulatory activity interact with neutrophil-activating peptide 1/interleukin 8 receptors on human neutrophils. J Biol Chem 266:10666-10671.

Murphy PM, Tiffany HL (1991) Cloning of complementary DNA encoding a functional human interleukin-8 receptor. Science 253:1280-1283

Nickoloff BJ, Karabin GD, Barker JNWN, Griffith CEM, Sarma V, Mitra RS, Elder JT, Kunkel SL, Dixit VM (1991) Cellular localization of interleukin-8 and its inducer, tumor necrosis factor-α in psoriasis. Am J Pathol 138:129-140

Paccaud J-P, Schifferli JA, Baggiolini M (1990) NAP-1/IL-8 induces upregulation of CR1 receptors in human neutrophil leukocytes. Biochem Biophys Res Commun 166:187-192.

Peveri P, Walz A, Dewald B, Baggiolini M (1988) A novel neutrophil-activating factor produced by human mononuclear phagocytes. J Exp Med 167:1547-1559.

Richmond A, Balentien E, Thomas HG, Flaggs, G., Barton, D.E., Spiess, J., Bordoni, R., Francke, U (1988) and Derynck, R. Molecular characterization and chromosomal mapping of melanoma growth stimulatory activity, a growth factor structurally related to ß-thromboglobulin. EMBO J 7:2025-2033.

Sager R, Anisowicz A, Pike MC, Beckmann P, Smith T (1992) Structural, regulatory, and functional studies of the GRO gene and protein. In Cytokines Vol.4. Interleukin-8 (NAP-1) and related chemotactic cytokines. Baggiolini M, Sorg C, eds, Karger, Basel 96-116.

Seitz M, Dewald B, Gerber N, Baggiolini M. (1991) Enhanced production of neutrophil-activating peptide-1/interleukin-8 in rheumatoid arthritis. J Clin Invest 87:463-469.

Sticherling M, Bornscheuer E, Schröder JM, Christophers E (1991) Localization of neutrophil-activating peptide-1/interleukin-8 -immunoreactivity in normal and psoriatic skin. J Invest Dermatol 96:26-30

Van Damme J, Bunning RAD, Conings R, Graham R, Russell G, Opdenakker G (1990) Characterization of granulocyte chemotactic activity from human cytokine-stimulated chondrocytes as interleukin 8. Cytokine 2:106-111

Walz A, Baggiolini M (1990) Generation of the neutrophil-activating peptide NAP-2 from platelet basic protein or connective tissue-activating peptide III through monocyte proteases. J Exp Med 171:449-454.

Walz A, Burgener R, Car B, Baggiolini M, Kunkel SL, Strieter RM (1991) Structure and neutrophil-activating properties of a novel inflammatory peptide (ENA-78) with homology to IL-8. J Exp Med 174:1355-1362

8. Molecular and Cellular Characteristics of Allergic Reactions

New Trends in Allergy

B. M. Stadler

Institute of Clinical Immunology, University of Bern, Sahlihaus, Inselspital, 3010 Bern, Switzerland

INTRODUCTION

During the last years, allergy research has evolved into an exciting and fast moving field indistinguishable from basic research in basic immunology. While in the late eighties many of the proposed regulatory moieties were mere phenomenological factors that regulated IgE synthesis or influenced effector cells of the allergic response, distinct factors have now finally been found and molecularly characterized.

The field of allergy represents nowadays an challenging area for studying basic mechanisms in immunology such as isotype switching as well as for applying basic knowledge to the clinical field. Allergy is the most frequent immunological disease and has created a very profound awareness also amongst layman all over the world. Especially the genetic predisposition for atopy seems to be known to everybody. However, there is until today no genetic marker for this disease. Thus, the claim that there exists a linkage between atopy and chromosome 11q13 (Young 1992) has obtained much attention from the public. However, this matter does not seem to be settled yet as an other group did not find such a linkage between this chromosome and asthma or atopy (Lympany 1992).

ALLERGENS

The hypothesis that the clue for developing as well as for treating allergies may lie in the molecular structure of allergens is as old as the definition of allergy. Due to the complexity of allergen extracts their biochemical characterization has taken a long time. In 1989 there were only four major allergens cloned but since then minimally 15 additional allergens from mites or plants have been molecularly cloned. This list is continuously growing (Greene 1992). The cloning of allergens has also lead to the detection of profilins as being a novel family of functional plant pan-allergens (Valenta 1992). This finding explains many of the observed cross reactions between different allergens in allergy diagnosis and may also be a step towards a better understanding of the ontogeny of allergic immune responses. It also seems that this molecular approach has generated now materials for a better standardization of allergens as well as the tools for more precisely studying allergen specific immune responses in the allergic individual.

The dominant cytokine IL-4

The cloning of IL-4 ended an era when allergologists had dealt with many different phenomenological factors that were believed to be involved in IgE synthesis or allergic responses. All of a sudden, there was one defined molecule that even in human systems reproducibly induced IgE synthesis, regulated T-cell proliferation as well as cytokine production and modulated surface receptors such as CD23 (de Vries 1992). This dominant role of IL-4 in the regulation of IgE-synthesis was shown most convincingly in in vivo models. Finkelmann (Katona 1991) had shown that neutralizing antibodies to IL-4 injected into mice completely inhibited the in vivo production of IgE. The other convincing report was by Kühn (1991), who inserted a stop codon in front of the interleukin 4 gene and created thereby IL-4 deficient mice. These mice showed a decreased IgG_1 production but no measurable IgE levels.

The T_h1 and T_h2 concept

Besides IL-4 also other cytokines were shown to be involved in the regulation of IgE synthesis. From the beginning, interferon gamma has attracted much attention for being an antagonist of IL-4 (Pène 1988). Later, it has become clear that other cytokines especially also IFN-γ can be strong inhibitors of IgE synthesis. IL-5, IL-6 and TNF-α all seem to enhance IgE synthesis (de Vries 1992).

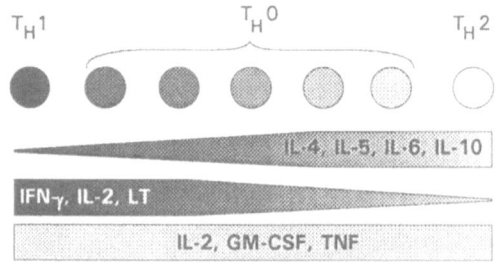

Fig. 1. Human T helper cell subsets producing different patterns and quantities of cytokines.

Thus, based on such findings the T_h1/T_h2 concept of Mosman (1986) was quickly accepted as a possible basis for explaining a differential immunoglobulin isotype regulation by T cells. He showed in mice that there existed two populations of T helper cells that secreted different patterns of cytokines. T_h2 cells were of special interest to allergologist as this cell type secreted more interleukin 4 and very little IFN-γ, while T_h1 clones produced more IFN-γ and less IL-4. For a while it seemed that this concept may only be true for the mouse. Recently, several authors (Romagnani 1991, Kay 1991) have shown that similar populations of T helper cells also exist in the human. As shown in Figure 1, the general assumption is nowadays that the T_h1 and T_h2 cells represent normally relatively rare populations, while the major subset of T helper cells can be classified as T_h0, the population that is capable of producing most cytokines and only varies in its production pattern according to the type of stimulation. T_h1 and T_h2 may only represent rather extreme poles of different T helper cells with clearly different cytokine production patterns.

MOLECULAR MECHANISMS OF IgE CLASS SWITCHING

As shown in figure 2, another interesting characteristic of IL-4 was its capacity to induce germline ε transcription (Qiu 1990, Gauchat 1990). Even though this germline ε transcript is preceding the productive ε transcripts its role cannot yet be clearly associated with the molecular events of isotype switching (Qiu 1990). There are numerous B cell lines and even two monocyte cell lines that are capable of expressing the germline ε mRNA (Stadler 1992), all of which we have so far not been able to switch to IgE synthesis.

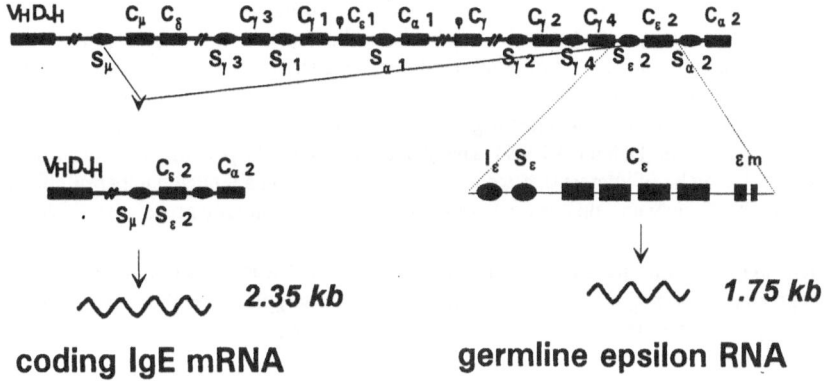

Fig. 2. Germline ε transcripts and productive ε transcripts are found in stimulated human B cells.

These new insights into the molecular mechanisms of isotype switching have still not yet addressed one of the major questions in the regulation of human IgE production. Namely, it is still puzzling that IL-4 is capable to induce IgE synthesis in vitro in the human system without adding an additional stimulus. There is no need to add a first signal, e.g., the allergen! On the basis of the present data, one would have to assume that a mere quantitative balance of different cytokines, e.g., IL-4, IFN-γ and other cytokines would produce such a specific isotype regulation in vivo resulting in 10'000 times greater IgG than IgE serum levels. It would practically mean that in vivo T_h2 helper cells are exclusively responsive to allergens and not to antigens. Even though there are data supporting this view that allergens may preferentially induce T_h2 cell clones (Romagnani 1991, Yssel 1992), it would still be difficult to imagine that such a regulation by non-specific factors leads to the observed differences in quantities of IgE and other immunoglobulin isotypes in vivo.

Human B-cell immunopoiesis and the CD40 ligand

While IL-4 finally allowed to reproducibly induce IgE synthesis in the human, the levels were still marginal and by not to be compared to the levels of other isotypes that can be found in stimulated PBMNC culture supernatants. This changed dramatically with the observation that antibodies to the CD40 antigen in the presence of IL-4 induced extremely high levels of IgE (Banchereau 1991). Thus, it was clear that the search for a ligand of the CD40 molecule got a high priority for regulating B cell immunopoiesis. The murine ligand has now been cloned (Armitage 1992) and the future will show whether it represents a molecule that also helps to understand the events leading to human IgE class-switch. It may be doubtful whether the CD40 surface structure and its ligand are involved in a positive manner. The major question still remains how the IgE isotype switch is so strict negatively controlled or suppressed in vivo.

CYTOKINES AND EFFECTOR CELLS IN ALLERGY

While cytokines were for a long time regarded as being cellular mediators that induced cell growth or differentiation or were very potent stimulators of cellular functions, it was especially interesting for the field of allergy that cytokines can be response modifiers. Growth factors such as GM-CSF, IL-3, and IL-5 were shown to be capable of priming basophils or eosinophils and render them more susceptible to secondary stimulations like IgE mediated triggering or triggering by chemotactic peptides or complement components (Dahinden 1989). The pleiotropic effect of cytokines has always been recognized but it may be of interest that some cytokines act only on a limited number of effector cells in allergy, e.g., NGF, only acts on basophils and the C-kit ligand primarily acts on mast cells.

Chemotactic agonists from the family of the intercrines, such as MCP-1, Rantes, MIP-1-α and to a weaker degree interleukin 8, also deserve attention from the allergologist as they seem to be capable of *directly* acting on some of the effector cells of the allergic response, such as inducing mediator release from basophils and eosinophils.

The findings mentioned above may also be in part why the histamine releasing factor was difficult to characterize biochemically. The different cytokines that prime or act directly on the effector cells may have been observed as a single phenomenological factor. In general the role of cytokines in allergic inflammation has become widely accepted but perhaps most importantly chronic mucosal inflammation is considered now to represent an important key in the pathogenesis of asthma (Kay 1991).

Basophils and mast cells: More efficient T_h2 alike?

The finding that mast cells in the mouse and eventually basophils in the human can produce cytokine patterns relatively similar to the patterns that are produced by T_h2 cells may be just as important for allergic inflammation as the T_h1/T_h2 concept. Namely, it has been claimed that on a per cell basis a mast cell may even produce more cytokines than a T_h2 T cell in the mouse (Plaut 1991). While the T cell can only be triggered by the antigen by its T cell receptor, the mast cell armed by many different specific IgE molecules could be triggered by a multiplicity of different antigens (allergens). As discussed above, mast cells and basophils may also be more easily triggered non-specifically by cytokines and it is well known that they can be triggered by complement components or certain drugs.

Thus, effector cells of allergy may even represent one of the major sources for perpetuating the inflammatory process in allergy.

IGE RECEPTORS

Fc epsilon receptor I (FcεRI)

During the last years, the knowledge on the high affinity IgE receptor on basophils and mast cells has constantly grown (Kinet 1990). This receptor consists of four chains. The α-chain that is the binding site for IgE, the β-chain, and two γ-chains as a homodimer that is probably involved in signal transmission. Most interestingly, the γ-chain seems to belong to a family of molecules that also includes the ζ and η chains of the T-cell receptor. The knowledge on FcεRI is presently at a level where different groups are thinking of eventually using parts of the α-chain as a therapeutic agent in atopy.

Fc epsilon receptor II (FcεRI) / CD23

The low affinity IgE receptor (CD23) once called an IgE binding factor is expressed on T an B lymphocytes, monocytes and eventually also on eosinophils and platelets. CD23 has a partial homology to animal lectins and seems to bind IgE via this lectin domain (Bettler 1989). We could show that for both, FcεRI and FcεRII the CεH3 domain of the IgE molecule represents the binding site (Fig. 3).

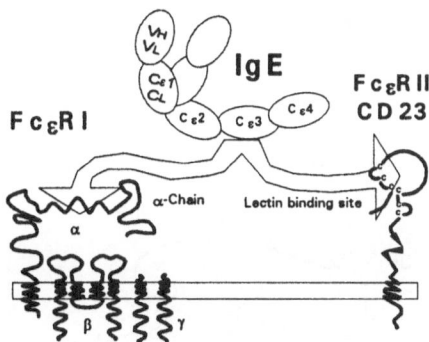

Fig. 3. The binding site of IgE to the high (FcεRI) and low affinity IgE receptor (FcεRII).

CD23 can be cleaved at two distinct sites from the surface of positive cells and it is claimed that cleaved molecules have B cell growth promoting activity as well as an effect on IgE synthesis, albeit much less than IL-4. Thus, the exact function of this molecule still remains largely unknown and it is questionable whether it is a receptor that can transmit signals.

Two independent groups have recently shown that Langerhans cells can also express FcεRI (Bieber 1992, Wang 1992). The existence of high affinity IgE receptors on other cells than mast cells in the skin also multiplies the possibilities for triggering IgE mediated immune reactions in the skin. Thus, this finding may be very important for understanding the local immunity in the skin and especially for allergic skin reactions.

ANTI-IgE AUTOANTIBODIES AS IMMUNOREGULATORY FACTORS

Anti-IgE autoantibodies have been detected in sera already soon after the detection of IgE, but for many years now their role seemed not to be understood. Many different authors have shown that the levels of such autoantibodies correlated with certain allergic diseases.

Anti-IgE autoantibodies and effector cells of the allergic response

During the last years we could show that naturally occurring anti-IgE autoantibodies can be used similarly as hetorologous anti-IgE antibodies (Stadler 1992). They seem to comprise a mixture of different isotope specificities that can either induce mediator release from basophils and mast cells or inhibit the fixation of IgE either to both either FcεRI or II. These different functions of anti-IgE autoantibodies seem to be linked to their epitope specificity. By constructing different ε immunoglobulin constant chain domains, we could show that mostly antibodies against the CεH3 domain inhibited CD23 IgE-binding. Antibodies against the second domain enhanced the binding to CD23. In the case of FcεRI there seem to exist neutralizing anti-IgE antibodies that prevent the binding of IgE to the receptor.

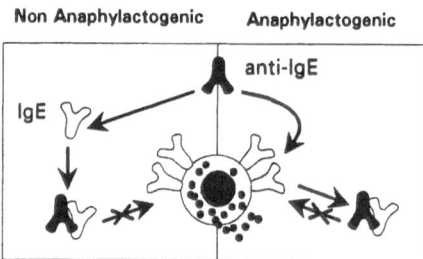

Fig. 4. A different role for anti-IgE autoantibodies depending on their epitope specificity, but even anaphylactogenic antibodies can become blocking antibodies after mediator release.

Interestingly also anaphylactogenic antibodies removed surface IgE after a successful trigger and formed immune complexes with IgE to prevent IgE to resensitize basophils (Fig. 4). Thus, it seems that autoantibodies to IgE can play a dual role depending on their epitope specificities by being either anaphylactogenic or by neutralizing the biological activity of IgE. Thus, anti-IgE autoantibodies may represent the a real blocking antibody in allergy.

Anti-IgE autoantibodies and the regulation of IgE synthesis

There is an obvious lack in isotype specificity by the available models on human IgE regulation. The activity of cytokines is too pleiotropic and the role of the allergen has not yet been shown clearly. A suppressor molecule at the level of the CD40 ligand would already represent a better and more strict regulatory moiety as one must assume from the observed serum IgE levels in vivo. In this respect, anti-IgE autoantibodies may represent a specific feedback arm for IgE synthesis. They could eventually be used for clonal deletion of cells that express surface IgE, as such antibodies may induce apoptosis or eventually mediate antibody mediated killing. There are indeed in vitro and animal models that support this hypothesis. What has not been observed yet is whether anti-IgE antibodies would not only be capable of inducing apoptosis but would eventually be capable of inducing IgE synthesis similar to anti-μ antibodies.

CONCLUSION AND A CLINICAL OUTLOOK

The significant progress in allergy research has lead to different hypothesis on the basic regulatory mechanisms of the allergic response. At present it is questionable whether the models that are exclusively based on a differential action of cytokines will deliver also the necessary drugs for the treatment of atopic diseases. Thus, other hypothesis that put more emphasis on the specific branch of the allergic response may be more promising. At the time of writing this condensed review there are new therapeutic approaches being tested in different industrial laboratories to utilize more specific tools. Approaches aiming on the one hand at the blocking of IgE receptors and on the other hand at the neutralization of the IgE molecule by chimeric or humanized anti-IgE antibodies are undertaken for a possible treatment of allergies.

REFERENCES

Armitage RJ, Fanslow WC, Strockbine L, Sato TA, Clifford KN, Macduff BM, Anderson DM, Gimpel SD, Davis-Smith T, Maliszewski CR, et-al. (1992) Molecular and biological characterization of a murine ligand for CD40. Nature 357: 80-2

Banchereau J, de Paoli P, Vallé A, Garcia E, Rousset F (1991) Long-term human B cell lines dependent on interleukin 4 and anti-CD40. Science 251: 70-2

Bettler B, Maier R, Ruegg D, Hofstetter H (1989) Binding site for IgE of the human lymphocyte low-affinity Fc epsilon receptor (FC epsilon RII/(CS23) is confined to the domain homologous with animal lectins. Proc Natl Acad Sci USA 86: 7118-22

Bieber T, de la Salle H, Wollenberg A, Hakimi J, Chizzonite R, Ring J, Hanau D, de la Salle C. (1992) Human epidermal Langerhans cells express the high affinity receptor for immunoglobulin E (Fc epsilon RI). J Exp Med 175: 1285-90

Dahinden CA, Kurimoto Y, de Weck AL, Lindley I, Dewald B, Baggiolini M (1989) The neutrophil-activating peptide NAF/NAP-1 induces histamine and leukotriene release by interleukin-3 primed basophils. J Exp Med 170:1787-92

De Vries JE, Aversa GG, Punnonen J, Bennett B, Gauchat JF (1992) Regulation of IgE synthesis by cytokines. In: Godard Ph, Bousquet J, and Michel FB (eds) Advances in Allergology and Clinical Immunology. The Parthenon Publishing Group, New Jersey, p 59

Gauchat JF, Lebman DA, Coffman RL, Gascan H, de Vries JE (1990) Structure and expression of germline epsilon transcripts in human B cells induced by interleukin 4 to switch to IgE production. J Exp Med 172: 463-73

Greene WK, Thomas WR (1992) IgE binding structures of the major house dust mite allergen Der P I. Mol Immunol 29: 257-62

Katona IM, Urban JF Jr, Kang SS, Paul WE, Finkelman FD (1991) IL-4 requirements for the generation of secondary in vivo IgE responses. J Immunol 146: 4215-21

Kay AB, Durham SR (1991) T lymphocytes, allergy and asthma. Clin Exp Allergy 21 suppl. 1: 17-21

Kinet JP (1990) The high affinity receptor for immunoglobulin E. Curr Opin Immunol 2: 499-505

Kühn R, Rajewsky K, Müller W (1991) Generation and analysis of interleukin-4 deficient mice. Science 254: 707-10

Lympany P, Welsh, K, Mac Cochrane G, Kemeny DM, Lee TH (1992) Genetic analysis using DNA polymorphism of the linkage between chromosome 11q13 and atopy and bronchial hyperresponsiveness to methacholine. J Allergy Clin Immunol 89:619-28

Mosmann TR, Cherwinski H, Bond MW, Giedlin MA, Coffman RL (1986) Two types of murine helper T-cell clones. I. Definition according to profiles of lymphokine activities and secreted proteins. J Immunol 136: 2348-57

Pène J, Rousset F, Brière F, Palliard X, Chrétien I, Banchereau J, de Vries JE (1988) IgE production by normal human B cells induced by alloreactive T cell clones is mediated by IL-4 and suppressed by IFN-γ. J Immunol 141: 1218-24

Plaut M, Pierce JH, Watson CJ, Hanley-Hyde J, Nordan RP, Paul WE (1989) Mast cell lines produce lymphokines in response to cross-linkage of FcεRI or to calcium ionophores. Nature 339: 64-7

Qiu G, Gauchat JF, Vogel M, Mandallaz M, de Weck AL, Stadler BM (1990) Human IgE mRNA expression by peripheral blood lymphocytes stimulated with interleukin 4 and pokeweed mitogen. Eur J Immunol 20: 2191-9

Romagnani S (1991) Human TH1 and TH2 subsets: doubt no more. Immunol Today 12: 256-7

Stadler BM, Vogel M, Miescher S, Aebischer I, Stämpfli MR, Furukawa K, Holzner ME, Yu Y, Qiu G (1992) Immunoregulation by anti-IgE autoantibodies. In: Godard Ph, Bousquet J, and Michel FB (eds) Advances in Allergology and Clinical Immunology. The Parthenon Publishing Group, New Jersey, p 3

Valenta R, Duchene M, Ebner C, Valent P, Sillaber C, Deviller P, Ferreira F, Teikl M, Edemann H, Kraft D (1992) Profilins constitute a novel family of functional plant pan-allergens. J Exp Med 175: 377-85

Wang B, Rieger-A, Kilgus O, Ochiai K, Maurer D, Fodinger D, Kinet JP, Stingl-G. (1992) Epidermal Langerhans cells from normal human skin bind monomeric IgE via Fc epsilon RI. J Exp Med 175: 1353-65

Young RP, Sharp PA, Lynch JR, Faux JA, Lathrop GM, Cookson WO, Hopkin JM (1992) Confirmation of genetic linkage between atopic IgE responses and chromosome 11q13. J Med Genet 29: 236-8

Yssel H, Johnson, KE, Schneider PV, Wideman J, Terr A, Kastelein R, De-Vries JE (1992) T cell activation-inducing epitopes of the house dust mite allergen Der p I. Proliferation and lymphokine production patterns by Der p I-specific CD4+ T cell clones. J Immunol 148: 738-45

T Lymphocyte/Eosinophil Interactions in Atopic Asthma: Critical Role for Interleukin-5

A. B. Kay, Q. Hamid, D. D. Robinson, A. M. Bentley, A. Tsicopoulos, Sun Ying, R. Moqbel, C. J. Corrigan, S. R. Durham

Departement of Allergy and Clinical Immunology, National Heart & Lung Institute, Dovehouse Street, London, SW3 6LY, UK

INTRODUCTION

Asthma is characterized by infiltration of the bronchial mucosa with large numbers of activated eosinophils and the presence of elevated concentrations of eosinophil-derived proteins such as major basic protein (Wardlaw et al. 1988) and the eosinophil cationic protein (De Monchy et al. 1985). The degree of the eosinophilia has been shown to correlate with the severity of airways hyperresponsiveness (Wardlaw et al. 1988, Durham et al. 1985). In a primate model of asthma, inhibition of the eosinophilic response to allergen substantially blocked the development of airways hyperresponsiveness (Wegner et al. 1990). Eosinophil-derived mediators have potential for producing many of the pathological features of asthma. For example, epithelial shedding might possibly occur through the cytotoxic effects of secreted eosinophil granule proteins (Gleich et al. 1979) and mucus hypersecretion and bronchoconstriction possibly through the release of platelet-activating factor (PAF) and leukotriene C_4 (Holgate et al. 1989). Recent studies have also suggested a role for T lymphocytes in asthma. T lymphocyte infiltration is a feature of the late-phase response to allergen in atopic individuals in both the skin (Frew and Kay 1988) and lung (Metzger et al. 1987). Increased numbers of activated T cells and concentrations of their products have been observed in the peripheral blood of acute severe asthmatics (Corrigan et al. 1988). By specific immunostaining of bronchial mucosal biopsies obtained via the fibreoptic bronchoscope, we demonstrated a significant increase in numbers of IL-2 receptor bearing cells in the airways of mild, steady-state asthmatics (Azzawi et al. 1990). This was associated with an elevation in the numbers of EG2[+] cells (EG2 is a monoclonal antibody which recognizes the secreted form of the eosinophilic cationic protein) (Tai et al. 1984).

It has been known for several years that T lymphocytes play a central role in eosinophil production and function through the release of soluble mediators. A number of cytokines with selective actions on eosinophils have now been sequenced and cloned. One of the most important is interleukin-5 (IL-5) which promotes terminal differentiation of the committed eosinophil precursor (Clutterbuck et al. 1988) as well as enhancing the effector capacity of mature eosinophils (Lopez et al. 1988). IL-5 also prolongs the survival of eosinophils *in vitro* (Yamaguchi et al. 1988). Furthermore, antibodies against IL-5 ablate the eosinophilic response to helminthic infection in mouse (Coffman et al. 1989) and an IL-5-like substance has been found in the serum of patients with hypereosinophilia (Owen et al. 1989). Interleukin-5 increases eosinophil, but not neutrophil, adhesion to vascular endothelium (Walsh et al. 1990) and also primes eosinophils for enhanced locomotory responsiveness to PAF and other chemoattractants (Sehmi et al. 1992). Release of IL-5 at the site of allergic inflammation could explain, in part, the specific eosinophil accumulation seen in these conditions. Similarly release of IL-5, in and around the bronchial mucosa, by activated T lymphocytes could lead to the specific recruitment of eosinophils, enhanced eosinophil cytotoxicity and prolonged eosinophil survival.

For these reasons we have attempted to determine whether IL-5 is associated with, and possibly regulates, the asthma process. Several approaches have been used. These include measurements of serum concentrations of IL-5 in chronic severe asthma and the identification of messenger RNA for IL-5 in bronchial biopsies and bronchoalveolar lavage (BAL) cells in patients with milder disease. We have been able to investigate steady state asthma, provoked asthma in the clinical laboratory and asthmatics before and after treatment with corticosteroids.

EXPRESSION OF MESSENGER RNA FOR INTERLEUKIN-5 IN MUCOSAL BRONCHIAL BIOPSIES IN ASTHMA

Identification of mRNA for IL-5 in the bronchi of asthmatics might provide important evidence of local IL-5 generation as well as emphasizing the possible link between eosinophils and T lymphocytes in this disease. We therefore used the technique of in situ hybridization using an IL-5 complementary RNA probe to investigate the expression of IL-5 mRNA and the pattern of distribution of IL-5-producing cells in bronchial tissue obtained from biopsies of asthmatics and normal individuals (Hamid et al. 1991). We also attempted to relate the expression of IL-5 mRNA to the severity of the disease and the degree of infiltration of the airway mucosa with eosinophils and activated T lymphocytes.

Bronchial biopsies were obtained from 10 asthmatics and 9 non-atopic normal controls. A radio-labelled cRNA probe was prepared from an IL-5 cDNA and hybridized to permeabilised sections. These were washed extensively before processing for autoradiography. An IL-5-producing T cell clone derived from a patient with the hyper-IgE syndrome was used as a positive control. As a negative control, sections were also treated with a "sense" IL-5 probe. Specific hybridization signals for IL-5 mRNA were demonstrated within the bronchial mucosa in 6 out of the 10 asthmatic subjects. Cells exhibiting hybridization signals were located beneath the epithelial basement membrane. In contrast, there was no hybridization in the control group. No hybridization was observed with the sense probe.

The six IL-5 mRNA positive asthmatics tended to have more severe disease than the negative asthmatics, as assessed by symptoms and lung function and showed a significant increase in the degree of infiltration of the bronchial mucosa by secreting (EG2+) eosinophils and activated (CD25+) T lymphocytes. Within the subjects who showed positive IL-5 mRNA, there was a correlation between the numbers of IL-5 mRNA+ cells and the number of CD25+ and EG2+ cells and total eosinophil count.

Therefore, this study provided evidence for the cellular localization of IL-5 mRNA in the bronchial mucosa of asthmatics and supported the concept that this cytokine regulates eosinophil function in bronchial asthma.

To examine whether mRNA for interleukin-5 was expressed by T cells, cytocentrifuge preparations of BAL cells from three subjects with asthma were fixed in 4% paraformaldehyde and washed in 15% sucrose in phosphate-buffered saline (Robinson et al. 1992a). Cells were incubated simultaneously with IL-5 cRNA probes labelled with uridine triphosphate-biotin and with monoclonal antibody to CD3 directly conjugated to fluorescein isothiocyanate. The conditions for *in situ* hybridization were as previously described (Giaid 1989), and positive hybridization of probe to cytokine mRNA was detected using streptavidin-Texas red staining. Controls were IgG1-fluorescein isothiocyanate with sense probes and RNase pre-treatment. Cells were quantified by fluorescence microscopy, and the percentages of cells expressing both CD3 and cytokine mRNA as well as the percentages of singly labelled cells were evaluated by counting at least 200 positive cells.

Dual fluorescence for CD3 and mRNA for IL-5 in BAL cytospin preparations showed that a mean of 91% of the cells positive for IL-5 mRNA were positive for CD3 (89%, 92%, and 91% in the three subjects), and 58% of the CD3$^+$ cells expressed IL-5 mRNA.

To examine further whether T cells were the source of IL-5 mRNA, BAL cells from five subjects with asthma were incubated with immunomagnetic beads covalently bound to monoclonal antibody to CD2 in a ratio of three beads to one lymphocyte for 20 minutes at 4°C. Cells were separated with a magnetic cell separator. CD2$^+$ cells were washed four times in phosphate-buffered saline with 0.1% bovine serum albumin, and cytospin preparations were made from both positively and negatively separated cells as well as from unseparated cells for differential counts and *in situ* hybridization for IL-5 mRNA. Hybridization was detected and quantified in cells as a percentage of BAL cells in positively or negatively separated and unseparated samples. These results were compared with the percentages of lymphocytes on cytospin slides as determined on May Grünwald Giemsa staining. There was a clear association between cytokine mRNA expression and CD2$^+$ cells. Binding of anti-CD2-coated beads to mRNA$^+$ cells in cytospin preparations identified these cells as T lymphocytes.

IL-5 mRNA AND THE LATE-PHASE ASTHMATIC REACTION

Inhalation challenge of atopic asthmatic subjects with allergen provokes an immediate or early asthmatic response detectable within minutes as a decrease in forced expiratory volume in one second (FEV$_1$). This may be followed by a second fall in FEV$_1$ between 4-8 hours after allergen exposure which may persist for 3-8 hours and is associated with an increase in non-specific bronchial responsiveness which may persist for several days (Booij-Nord et al. 1971, Robertson et al. 1974, Cartier et al. 1982). Whilst the early asthmatic response appears to be dependent on IgE triggering of mediator release from mast cells resulting in bronchoconstriction and airway oedema (Liu et al. 1991, Casale et al. 1987), the late asthmatic response is associated with an influx of inflammatory cells, particularly eosinophil leucocytes (De Monchy et al. 1985, Metzger et al. 1987, Diaz et al. 1989) into the bronchial mucosa. Eosinophil infiltration of the bronchial mucosa is a feature of the airways of patients dying of asthma (Dunnill 1978), and is present in bronchial biopsies (Bradley et al. 1991) and BAL (Wardlaw et al. 1988, Bousquet et al. 1990) of patients with mild disease. Thus, although not entirely reproducing the pattern of natural exposure, allergen challenge may provide a model for investigating allergen-induced inflammatory events in atopic asthma (O'Byrne et al. 1987).

Participation of T lymphocytes in allergen-induced asthmatic responses was suggested by the finding of lymphocyte and eosinophil infiltration in BAL after local allergen challenge (Metzger et al. 1987) and changes in CD4 and CD8 T cells in both peripheral blood and BAL after allergen challenge (Gonzalez et al. 1987, Gerblich et al. 1984, 1991). We, therefore, hypothesized that allergen inhalation challenge of atopic asthmatics results in local activation of IL-5 producing lymphocytes with subsequent eosinophilia and that this in turn contributes to the associated changes in airway calibre and hyperresponsiveness.

In order to test this hypothesis we examined the expression of IL-5 mRNA in BAL cells (Robinson et al. 1992b) and bronchial biopsies (Bentley et al. 1992) in sensitized atopic asthmatics 24 hours after inhalation challenge with either allergen or diluent control. Compared with diluent there were significant increases after allergen challenge in the numbers of cells expressing mRNA for IL-5 from both BAL and biopsies. There were also increases in eosinophils in both bronchial wash and BAL after allergen but not when compared to diluent challenge. Close associations were observed between the numbers of CD25[+] bronchoalveolar lavage CD4[+] T cells after allergen challenge, cells expressing IL-5 mRNA, and eosinophils. There was also a correlation between the numbers of cells expressing mRNA for IL-3 and IL-5 and BAL eosinophils on the allergen day. Bronchoalveolar lavage and bronchial wash eosinophilia also closely correlated with maximal late fall in FEV_1 after allergen challenge. Similarly, the numbers of EG2[+] cells in biopsies and mRNA IL-5[+] cells also correlated. We concluded, therefore, that IL-5 may contribute to late asthmatic responses in the airway by mechanisms which include eosinophil accumulation.

THE EFFECT OF CORTICOSTEROIDS ON IL-5 SYNTHESIS AND SECRETION IN ASTHMA

The beneficial effects of corticosteroids in the treatment of asthma are well documented (Walsh et al. 1966), and the use of inhaled corticosteroids in chronic asthma is widely accepted (British Thoracic Society Guidelines 1990, NHBLBI Panel Report 1991). It has been suggested that corticosteroids mediate their beneficial role in the reversal of airway obstruction and bronchial hyperreactivity through an effect on T cell associated inflammatory processes (Reed 1991). However, the mechanism of action of corticosteroids in asthma remains unclear.

We hypothesised that inhibition of T cell cytokine production (particularly IL-5) in vivo may underlie, at least in part, the anti-inflammatory actions of corticosteroids in chronic asthma. To examine the effect of prednisolone on IL-5 gene expression in symptomatic chronic asthma we performed fibreoptic bronchoscopy with BAL and endobronchial biopsies before and after 2 weeks of therapy in a double-blind placebo-controlled parallel group study (Robinson et al. 1992c). To assess the clinical relevance of cellular changes, the response to treatment was followed closely by measurements of lung function and bronchial responsiveness.

Clinical improvement in the patients receiving prednisolone was shown by decreases in airflow obstruction and in bronchial responsiveness to inhaled methacholine, which were not seen in patients receiving placebo. Between group comparison showed a significant fall in numbers of BAL cells per 1000 with positive *in situ* hybridization signals for mRNA for IL-5 with prednisolone treatment. There was also a reduction in BAL eosinophils in prednisolone-treated patients when compared with those receiving placebo. Immunohistology of bronchial mucosal biopsies revealed a significant decrease in the numbers of T cells (CD3+) and EG2+ eosinophils in those patients receiving prednisolone together with a significant reduction in numbers of tryptase only (MC_T) but not tryptase/chymase positive (MC_{TC}) mast cells by double sequential immunostaining.

Finally, we also obtained evidence that in chronic severe asthma IL-5 was detectable in the serum and decreased after treatment with prednisolone (Corrigan et al. 1992). Peripheral blood mononuclear cells and serum were obtained, on two occasions, from 15 asthmatic patients who required oral glucocorticoid therapy for moderate to severe disease exacerbations. Samples were obtained immediately before commencement of oral glucocorticoids (day 1) and again after 7 days of treatment (day 7). Samples were also isolated on two occasions 7 days apart from a group of 7 untreated normal volunteers. Serum concentrations of IL-5 were measured using an ELISA technique. Interleukin-5 was detectable in the serum of 8 of the asthmatic patients on day 1, but in none of these patients on day 7. Serum IL-5 was undetectable in all the control subjects on both occasions. The numbers of activated ($CD4^+/CD25^+$) T helper cells were also elevated in the asthmatics and decreased after treatment with prednisolone. These observations are consistent with the hypothesis that exacerbations of asthma are associated with activation of CD4 T lymphocytes which secrete IL-5, and that glucocorticoid therapy results in reduction of the activation status of these cells concomitant with inhibition of IL-5 secretion.

CONCLUSIONS

Eosinophils are implicated as major pro-inflammatory cells in the asthma process. A number of eosinophil functions are regulated by interleukin-5. These include terminal differentiation of the committed eosinophil precursor, increased adhesiveness of eosinophil to vascular endothelium and survival of eosinophils *in vitro*. Here we present evidence that IL-5 mRNA transcripts are detectable in T lymphocytes in ongoing steady state asthma as well as asthma provoked by inhalational challenge (i.e., the late-phase reactions). Furthermore, the numbers of mRNA+ cells in BAL in asthma and circulating concentrations of serum IL-5 decrease after corticosteroid therapy. Taken together, these results suggest that IL-5 plays a critical role in eosinophil-induced bronchial inflammation in asthma. It also suggests that agents which selectively target IL-5 may be of value in the treatment of this disease.

REFERENCES

Azzawi M, Bradley B, Jeffery PK, Frew AJ, Wardlaw AJ, Assoufi B, Collins JV, Durham SR, Knowles GK, Kay AB (1990) Identification of activated T lymphocytes and eosinophils in bronchial biopsies in stable atopic asthma. Am Rev Respir Dis 142:1407-1413.

Bentley AM, Qiu Meng, Robinson DS, Hamid Q, Kay AB, Durham SR (1992)
    Increases in activated T lymphocytes, eosinophils and cytokine
    messenger RNA expression for IL-5 and GM-CSF in bronchial
    biopsies after allergen inhalation challenge in atopic
    asthmatics. Submitted.
Booij-Nord H, Orie NGM, deVries K (1971) Immediate and late bronchial
    obstructive reactions to inhalation of house dust and protective
    effect of disodium cromoglycate and prednisolone. J Allergy
    Clin Immunol 48:344-353.
Bousquet J, Chanez P, Lacoste JY, Barneon G, Ghavanian N, Enander I,
    Venge P, Ahlstedt S, Simony-Lafontaine J, Godard P, Michel FB
    (1990) Eosinophilic inflammation in asthma. New Engl J Med
    323:1033-1039.
Bradley BL, Azzawi M, Jacobson M, Assoufi B, Collins JV, Irani A-M,
    Schwartz LB, Durham SR, Jeffery PK, Kay AB (1991) Eosinophils,
    T-lymphocytes, mast cells, neutrophils and macrophages in
    bronchial biopsy specimens from atopic subjects with asthma:
    Comparison with biopsy specimens from atopic subjects without
    asthma and normal control subjects and relationship to bronchial
    hyper-responsiveness. J Allergy Clin Immunol 88:661-674.
British Thoracic Society. Guidelines for management of asthma in
    adults. Statement by the research unit of the Royal College of
    Physicians of London, King's Fund Centre, National Asthma
    Campaign (1990) Br Med J 142:434-457.
Cartier A, Thomson NC, Frith PA, Roberts R, Hargreave FE (1982)
    Allergen-induced increase in bronchial responsiveness to
    histamine: relationship to the late asthmatic response and
    change in airway caliber. J Allergy Clin Immunol 70:170-177.
Casale TB, Wood D, Richerson HB, Zehr B, Zavala D, Hunninghake GW
    (1987) Direct evidence of a role for mast cells in the
    pathogenesis of antigen-induced broncho-constriction. J Clin
    Invest 80:1507-1511.
Clutterbuck EJ, Hirst EMA, Sanderson CJ (1988) Human interleukin-5
    (IL-5) regulates the production of eosinophils in human bone
    marrow cultures: comparison and interaction with IL-1, IL-3, IL-
    6 and GM-CSF. Blood 73:1504-1513.
Coffman LR, Seymour WB, Hudak S, Jackson J, Rennick D (1989) Antibody
    to interleukin 5 inhibits helminth-induced eosinophilia in mice.
    Science 245:308-310.
Corrigan CJ, Haczku A, Gemou-Engesaeth V, Doi S, Kikuchi Y, Takatsu
    K, Durham SR, Kay AB (1992) CD4 T-lymphocyte activation in
    asthma is accompanied by increased serum concentrations of
    interleukin-5: effect of glucocorticoid therapy. Submitted.
Corrigan CJ, Hartnell A, Kay AB (1988) T-lymphocyte activation in
    acute severe asthma. Lancet i:1129-1132.
De Monchy JGR, Kauffman HK, Venge P, Koeter GH, Jansen HM, Sluiter
    HJ, De Vries K (1985) Bronchoalveolar eosinophilia during
    allergen-induced late asthmatic reactions. Am Rev Respir Dis
    131:373-376.
Diaz P, Gonzalez C, Galleguillos FR, Ancic P, Cromwell O, Shepherd D,
    Durham SR, Gleich GJ, Kay AB (1989) Leukocytes and mediators in
    bronchoalveolar lavage during allergen-induced late-phase
    asthmatic reactions. Am Rev Respir Dis 139:1383-1389.
Dunnill MS (1978) The pathology of asthma. In: Middleton E, Reed CE,
    Ellis EF, eds. Allergy; principles and practice. CV Mosby, St
    Louis, Missouri, 678-686.
Durham SR, Kay AB (1985) Eosinophils, bronchial hyperreactivity and
    late-phase asthmatic reactions. Clin Allergy 15:411-418.

Frew AJ, Kay AB (1988) The relationship between infiltrating CD4+ lymphocytes, activated eosinophils and the magnitude of the allergen-induced late phase cutaneous reaction in man. J Immunol 141: 4158-4164.

Gerblich AA, Campbell AE, Schuyler MR (1984) Changes in T lymphyocyte subpopulations after antigenic bronchial provocation in asthmatics. New Engl J Med 310: 1349-1352.

Gerblich AA, Salik H, Schuyler MR (1991 Dynamic T cell changes in peripheral blood and bronchoalveolar lavage after antigen bronchoprovocation in asthmatics. Am Rev Respir Dis 143: 533-537.

Giaid A (1989) Non-isotopic RNA probes. Histochemistry 93:191-196.

Gleich G J, Frigas E, Loegering DA, Wassom DL, Steinmuller D (1979) Cytotoxic properties of the eosinophil major basic protein. J Immunol 123:2925-2927.

Gonzalez MC, Diaz P, Galleguillos FR, Ancic P, Cromwell O, Kay AB (1987) Allergen-induced recruitment of bronchoalveolar helper (OKT4) and suppressor (OKT8) T cells in asthma. Am Rev Respir Dis 136:600-604.

Hamid Q, Azzawi M, Sun Ying, Moqbel R, Wardlaw AJ, Corrigan CJ, Bradley B, Durham SR, Collins JV, Jeffery PK, Quint DJ, Kay AB (1991) Expression of mRNA for interleukin-5 in mucosal bronchial biopsies from asthma. J Clin Invest 87:1541-1546.

Holgate ST, Abraham WM, Barnes PJ, Lee TH (1989) Pharmacology and treatment. In: Holgate ST, Howell JBL, Burney PGJ, Drazen JM, Hargreave FE, Kay AB, Kerrebijn KF, Reid LM, eds. The Role of Inflammatory Processes in Airway Hyperresponsiveness. Blackwell Scientific Publications, Oxford, 179-221.

Liu MC, Hubbard WC, Proud D, Stealey BA, Galli SJ, Kagey-Sobotka A, Bleeker ER, Lichtenstein LM (1991) Immediate and late inflammatory responses to ragweed antigen challenge of the peripheral airways in allergic asthmatics. Am Rev Respir Dis 144:51-58.

Lopez AF, Sanderson CJ, Gamble JR, Campbell HR, Young IG, Vadas MA (1988) Recombinant human interleukin-5 is a selective activator of human eosinophil function. J Exp Med 167:219-224.

Metzger WJ, Zavala D, Richerson HB, Moseley P, Iwamota P, Monick M, Sjoerdsma K, Hunninghake GW (1987) Local allergen challenge and bronchoalveolar lavage of allergic asthmatic lungs. Description of the model and local airway inflammation. Am Rev Respir Dis 135: 433-440.

National Heart, Lung and Blood Institute Information Center. Expert panel report guidelines for diagnosis and management of asthma. Bethesda, Maryland, 1991.

O'Byrne PM, Dolovich J, Hargreave FE (1987) Late asthmatic responses. State of art. Am Rev Respir Dis 136:740-751.

Owen WF, Rothenberg ME, Petersen J, Weller PF, Silberstein D, Sheffer AL, Stevens RL, Soberman RJ, Austen KF (1989) Interleukin-5 and phenotypically altered eosinophils in the blood of patients with the idiopathic hypereosinophilic syndrome. J Exp Med 170: 343-348.

Reed CE (1991) Aerosol steroids as primary treatment of mild asthma. New Engl J Med 325:425-426.

Robertson DG, Kerrigan AT, Hargreave FE, Chalmers R, Dolovich J (1974) Late asthmatic responses induced by ragweed pollen allergens. J Allergy Clin Immunol 54:244-254.

Robinson DS, Hamid Q, Sun Ying, Bentley AM, Barkans J, Durham SR, Kay AB (1992b). Activated T helper cells and interleukin-5 gene expression in bronchoalveolar lavage from atopic asthma. Relationship to symptoms and bronchial responsiveness. Submitted.

Robinson DS, Hamid Q, Sun Ying, Bentley AM, Assoufi B, North J, Qui Meng, Durham SR, Kay AB (1992c) A double blind placebo controlled trial of prednisolone treatment in bronchial asthma. Clinical improvement is accompanied by reduction in bronchoalveolar lavage eosinophilia and modulation of IL-4, IL-5 and IFN-gamma cytokine gene expression. Submitted.

Robinson DS, Hamid Q, Sun Ying, Tsicopoulos A, Barkans J, Bentley AM, Corrigan CJ, Durham SR, Kay AB (1992a) Predominant T_{H2}-type bronchoalveolar lavage T-lymphocyte population in atopic asthma. New Engl J Med 326:298-304.

Sehmi R, Wardlaw AJ, Cromwell O, Kurihara K, Waltmann P, Kay AB (1992) Interleukin-5 (IL-5) selectively enhances the chemotactic response of eosinophils obtained from normal but not eosinophilic subjects. Blood (in press).

Tai P-C, Spry CJF, Peterson C, Venge P, Olsson I (1984) Monoclonal antibodies distinguish between storage and secreted forms of eosinophil cationic peptide. Nature 309:182-184.

Walsh GM, Hartnell A, Wardlaw AJ, Kurihara K, Sanderson CJ, Kay AB (1990) IL-5 enhances the *in vitro* adhesion of human eosinophils but not neutrophils, in a leucocyte integrin (CD11/18)-dependent manner. Immunology 71:258-265.

Walsh SD, Grant IWB (1966) Corticosteroids in the treatment of chronic asthma. Br Med J 2:796.

Wardlaw AJ, Dunnette S, Gleich GJ, Collins JV, Kay AB (1988) Eosinophils and mast cells in bronchoalveolar lavage in subjects with mild asthma. Relationship to bronchial hyperreactivity. Am Rev Respir Dis 137:62-69.

Wegner CD, Gundel RH, Reilly P, Haynes N, Letts GL, Rothlein R (1990) Intercellular adhesion molecule-1 (ICAM-1) in the pathogenesis of asthma. Science 247:456-459.

Yamaguchi Y, Hayashi Y, Sugama Y, Miura Y, Kasahara T, Kitamura S, Torisu M, Mita S, Tominaga A, Takatsu K, Suda T (1988) Highly purified murine interleukin-5 (IL-5) stimulates eosinophil function and prolongs *in vitro* survival. J Exp Med 167: 1737-1742

Cytokines in allergic inflammation

C. A. Dahinden, T. Brunner, M. Krieger, S. C. Bischoff and
A. L. de Weck

Institute of Clinical Immunology, Inselspital, CH-3010 bern, Switzerland

Introduction

In the recent years our knowledge about the pathogenesis of allergic diseases has increased considerably. This progress is due to a large degree to rapid developments in fields which at first sight are not directly related to allergy such as cellular immunology, molecular biology and inflammation research. There is now a general consent that asthma and allergic rhinitis are inflammatory pathologies and that the extent of the inflammatory response determines by large the severity of allergic disease. This concept has already led to clear changes in the therapeutical management of allergic diseases in that anti inflammatory therapy is used as a first-line treatment while short acting bronchodilators are rather used to relieve the remaining symptoms on a "on demand" basis. However, an inflammatory response is a rather stereotypical albeit very complex reaction of the macroorganism to tissue damage of different causes. In general, inflammatory reactions start with a vascular phase (plasma exudation / edema) followed by infiltration of the tissue by leukocytes (cellular exudate /tumor) and finally tissue repair, remodeling or fibrosis. The cellular phase of inflammation leads itself to tissue damage by activation of infiltrating effector cells and release of secondary inflammatory mediators and cytotoxic products (basic proteins, proteases, oxygen radicals etc.). Inflammatory reactions of different etiologies probably display more common features than dissimilarities, and all leukocyte types - and even platelets and tissue cells - participate to various degrees in a complex network of cellular interactions. These cellular interactions and effector functions are controlled by a very large number of mediators and cytokines. IgE-mediated allergic reactions are classically divided into an immediate response involving edema and smooth muscle contraction due to the release of mediators by mast cells and the late phase reaction characterized by cellular infiltration and tissue damage, although, in clinical practice, the continuous exposure to allergens results in a chronic inflammatory process. It is evident from clinical observations that the presence of allergen specific IgE does not predict the severity of allergic disease. This is due to the fact that the severity of allergic disease is primarily determined by the extent of cellular infiltration and activation of eosinophils and basophils in the late-phase reaction which can occur in an antigen-independent manner. Both, the immediate reaction, and in particular the activation in late phase reactions, are strongly influenced by certain cytokines. Cytokines have been primarily characterized on the basis of their growth factor and/or differentiating properties for leukocytes of the myeloid and lymphoid lineage and are thus important regulators of the immune response. However, more recent studies have clearly established that many of these cytokines also regulate the function of mature inflammatory effector cells, such as neutrophils, monocytes, eosinophils and basophils, and even resident cells such as mast cells.

In this paper we discuss the hypothesis that, through the release of different profiles of cytokines, the adaptive immune system (lymphocytes) profoundly modulates inflammatory processes (myeloid effector cells) of the innate immune system - and vice versa.

Cytokines regulating allergic inflammation

Proinflammatory cytokines

Before discussing cytokines of particular importance in allergic inflammation, the general proinflammatory effect of IL-1, IL-6 and tumor necrosis factor (TNF) will be emphasized. These cytokines are not only produced by Ag-activated T-cells, but also by monocytes and many tissue cells upon unspecific activation. A particular important function of these proinflammatory cytokines is the up-regulation of adhesive proteins (integrins, intercellular adhesion molecules = ICAM, etc.) in leukocytes as well as in tissue cells. Furthermore, IL-1 or TNF stimulates endothelial and other tissue cells to produce chemotactic cytokines (chemokines), such as IL-8. Thus, these cytokines may play a key role in the process of attachment of different leukocytes to the endothelium and their subsequent sequestration and activation in the tissue. Consistent with this view is the fact that anti-ICAM antibodies inhibit allergic late phase reactions *in vivo*. However, in allergic inflammation these cytokines probably act only locally (Brown, 1991) and a more abundant production of IL-1, IL-6, and TNF may lead to a pyogenic type of inflammation.

Effector cell priming by cytokines

Several cytokines, most of which have been initially identified as hematopoietic growth factors, not only regulate the growth and differentiation of immature leukocyte precursors, but also modulate and amplify inflammatory reactions by their action on mature effector cells. These "priming" cytokines, which we also termed "cell response modifiers", do not directly trigger effector cell functions but rather enhance the responsiveness of different leukocytes towards stimulation by all triggering molecules examined so far. All the effector functions which have been tested are strongly potentiated (i.e. chemotaxis, release of preformed and de novo synthesized mediators or cytotoxic products, cell adhesiveness, cytotoxicity). This priming phenomenon has been observed in all myeloid cell types (neutrophils, eosinophils, basophils, monocytes) and even in tissue mast cells. However, the target cell profile is distinct for the different modulatory cytokines. In addition to these rapid effects on cell function, prolonged exposure of mature effector cells to appropriate "priming" cytokines can lead to increased cell survival and further functional and phenotypic alterations by *de novo* gene expression and protein synthesis.

Priming of Effector Cells by Cytokines

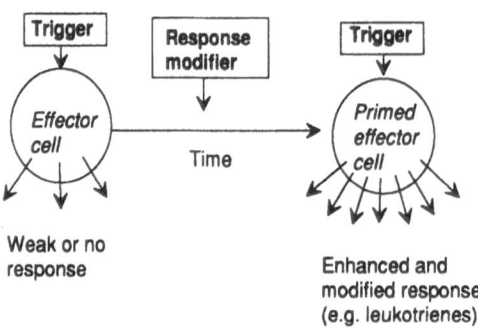

Weak or no response

Enhanced and modified response (e.g. leukotrienes)

Priming cytokines as regulators of leukotriene formation

In allergic reactions such as asthma, Leukotriene C4 (LTC4) is not only found immediately after allergen challenge formed by IgE-dependent mast cell activation, but particularly high levels are found in fluids of allergic late phase reactions. Obviously, infiltrating leukocytes, such as basophils and eosinophils, must be capable of producing LTC4 in an antigen-independent manner in response to activation by soluble cell-derived and/or humoral agonists. However, no soluble endogenous bioactive molecule has yet been identified capable of directly promoting the formation of leukotrienes in any myeloid cell type. Several studies from our laboratory, first performed with neutrophils, and more recently with mature human basophils and eosinophils, showed that appropriate hematopoietic growth factors are critical regulators of lipid mediator formation (LTB4 and platelet activation factor (PAF) in neutrophils; LTC4 in basophils; LTC4 and PAF in eosinophils). In fact, cytokine primed inflammatory effector cells produce considerable amounts of lipid mediators in response to most, otherwise inactive, soluble cell agonists when acting as a second signal. We thus propose that the presence of priming cytokines determines whether mature effector cells produce lipid mediators, while the target cell profile of the cytokines present determines which cell types respond.

Priming of basophil and mast cell function by cytokines

Basophils and mast cells are particularly potent sources of LTC4. In fact, basophils produce 10-1000 times more LTC4 than eosinophils when sequentially stimulated with a priming cytokine and different cell agonists, at least in vitro (Takafuji, 1992). Thus basophils should be considered as important effector cells of allergic late phase reactions, even if their number compared to eosinophils is lower in an allergic exudate. In basophils, three related cytokines, IL-3, IL-5 and GM-CSF, have been found to enhance the releasability of the cells (Kurimoto, 1989; Bischoff, 1990). More recently, a neurotrophic cytokine, nerve growth factor (NGF), has been added to the list of potent basophil priming cytokines, suggesting a modulation of allergic inflammation by the nervous system (Bischoff 1992c). With the exception of a weak mediator release enhancing effect of IL-1, a large list of other cytokines was inactive in modulating basophil function. Exposure of basophils to IL-3, IL-5, GM-CSF or NGF for 5-10 minutes augments the release of preformed histamine and the generation of LTC4 in response to IgE-receptor stimulation and primes cells to produce large amounts of leukotrienes in response to C5a or monocyte chemotactic peptide 1 (MCP-1), IgE-independent agonists which by themselves induce degranulation only (Kurimoto, 1989; Bischoff, 1992d). Primed basophils even become responsive to a large number of cell derived (PAF, IL-8/NAP-1, and other intercrines/chemokines, see below) or humoral (C3a) triggers which do not induce any mediator release in unprimed cells (Dahinden, 1989; Bischoff, 1990; Brunner, 1991). Thus these cytokines do not only enhance the cellular responsiveness but lead to a qualitative alteration in the behavior of the cells.

Clinical observations in asthma and rhinitis indicate that also the allergic immediate response to allergen is not solely dependent on the degree of sensitization and the level of antigen-specific IgE, suggesting that mast cell function may be regulated by factors other than IgE-receptor crosslinking. However, human mucosal lung mast cells do not seem to release mediators in response to the many IgE-independent agonists including those triggering human basophils. Recently, C-KIT-ligand (KL), a hematopoietic stem cell growth factor, which can be produced by tissue cells in either a transmembrane or soluble form, was found to be a unique potentiator of mediator release by lung mast cells in response to IgE-receptor stimulation. KL increases the sensitivity of mast cells to IgE-R-crosslinking and also increases the amount of mediators formed in response to maximal IgE-receptor stimulation. It is still unclear weather KL acts mainly as a priming agent or also as a direct agonist since, depending on the experimental conditions, KL can by itself induce the release of considerable amounts of mast cell mediators. In rodents, and probably also in the human system, KL is also an important mast cell growth and differentiating factor. By contrast, a large number of cytokines, including basophil priming cytokines and the mast cell growth factors / agonists in rodents (IL-3, IL-4, IL-9, NGF), do neither modulate nor induce mediator release by lung mast cells, indicating that observations made in the mouse cannot be easily transferred to humans, and that mouse mucosal mast cells resemble functionally more human basophils than human mucosal lung mast cells. Thus, in humans, the extent of expression of KL by tissue cells may regulate both, the number and as well as the function of lung mast cells. However, no information about KL upregulation in asthmatic patients is yet available.

Priming of eosinophil function by cytokines
Eosinophils and basophils are developmentally and functionally closely related cell types, both with regards to the cytokine profile enhancing their responsiveness and the cell agonists to which they respond, with few exceptions: NGF, does not prime eosinophil function, while tumor necrosis-factor (TNF) which does not affect basophils, increases LTC4 formation by eosinophils (Takafuji, 1992). It is interesting that a particular set of cytokines, IL-3, IL-5 and GM -CSF, primes both cell types in a similar manner, indicating that these cytokines are of particular importance in allergic late phase reactions (Takafuji, 1991). The similarity of action of IL-3, IL-5 and GM-CSF can now be explained in molecular terms, since the different ligand-α chain receptor complexes of these three cytokines interact with a common β-chain which is necessary for high affinity binding and signal transduction. The design of "β-chain antagonists" inhibiting the effect of whole groups of cytokines can now be envisioned, opening new approaches of immunomodulation and of more specific anti-allergic therapy (Bischoff, 1992).

Priming of neutrophils by cytokines
Neutrophils are the principal effector cells of acute inflammation and defense against bacterial infection, and are thought to be of lesser importance in allergic late phase reactions. The only cytokine identified so for to prime for chemotactic agonist induced lipid mediator formation (LTB4 and PAF) is GM-CSF (Dahinden, 1988; Wirthmueller, 1989). IL-1 and TNF enhance other cell functions such as oxygen radical release and degranulation.

Cytokines acting as chemotactic cell agonists (chemokines)

Most of the earlier discovered humoral and cell-derived agonists, such as C5a, FMLP, PAF, are rather pleiotrophic inducing the migration and activation of different myeloid effector cell-types. In the recent years, a large superfamily of homologous cytokines (platelet factor-4 / intercrine / chemokine superfamily) has been gradually identified, and other members of this family may still be discovered. Accordingly, the nomenclature of these cytokines is not yet settled and many synonyms for these cytokines are in use. With regards to their bio-activity, they resemble more other cell-derived or humoral inflammatory mediators, such as C5a, however with a more restricted target cell profile. Indeed, the information about the receptors for chemokines (IL-8 receptors), or the signal transducing pathway they activate (IL-8, MCP-1), indicates that many cytokines of the chemokine superfamily may interact with pertussis toxin-sensitive, G-protein-coupled, seven-transmembrane receptors, like other chemotactic cell agonists.

α-chemokines

Members of the C-X-C branch (according to the position of the first two cysteines in the conserved motif), also termed IL-8 or α-chemokine family, attract and activate mainly neutrophils, and are thus important mediators of acute inflammatory processes. Nevertheless, IL-8 receptors are also expressed on basophils and eosinophils (Dahinden, 1989; Krieger, 1992b). In contrast to neutrophils, however, these cell-types need to be primed by IL-3 or IL-5 in order to appreciably respond to IL-8 stimulation. Other α chemokines are probably inactive on basophils and eosinophils at physiological consentrarions.

β-chemokines

The biological activities of the members of C-C branch or β-chemokine family of cytokines are less well defined. They do not seem to activate neutrophils, but rather attract certain mononuclear cells. RANTES, macrophage inflammatory protein MIP-1α and MIP-1β are chemotactic for distinct lymphocyte subpopulation, while MCP-1 is a particularly potent monocyte chemotactic factor. Even less is known about the bioactivity of the other family members.

Very recent, largely unpublished, work from our laboratory and other groups revealed that distinct members of the β-chemokines are potent agonists for basophils and eosinophils, indicating that this group of cytokines may be particularly important mediators of allergic inflammation. MCP-1 is the most potent cell-derived basophil agonist identified so far, inducing the release of large amounts of histamine by itself and promoting leukotriene generation by cytokine primed basophils (Bischoff, 1992b). MCP-1 does not activate or attract eosinophils thus representing the first recognized basophil trigger incapable of activating eosinophils. RANTES and MIP-1α, however, activate both cell types in an identical manner, in contrast to MIP-1β which is inactive. All these agonists activate basophils and eosinophils by a similar mechanism inducing pertussis toxin-sensitive changes in intracellular calcium concentrations of identical kinetics, similar to the chemotaxin C5a. Cross-desensitization experiments indicate that these cytokines (including the α-intercrine IL-8) activate the cells though separate specific receptors although the MIP-1α receptor may also recognize RANTES with lower affinity. Despite these similarities in signal transduction, the fine specificities for activating the different effector functions is distinct even in the same cell type: For example, RANTES attracts basophils with higher efficacy than MCP-1, despite a much lower mediator releasing capacity which is similar to that of IL-8 / NAP-1.

The distinct and relatively restricted target cell profiles of the members of the β-chemokine family of cytokines, and the fact that they promote preferentially distinct cellular functions even in the same cell type, indicates that these cytokines regulate the fine tuning of chronic inflammatory processes of diverse etiologies. In allergic inflammation, MCP-1 may be particularly important for activating basophil mediator release. RANTES is a potent chemotactic factor for T-helper memory cells, eosinophils and basophils (but not neutrophils) and may thus mediate the selective attraction of these effector cells in allergic late phase reactions.

Thus, the specificity of allergic inflammation may not only be determined by the profile of cytokines enhancing the cellular responsiveness of particular effector cells to pleiotophic cell agonists, as suggested earlier, but also by the profile of the members of the chemokine family which act as relatively oligotrophic chemotactic agonists.

Cytokines and Effector Cells in Allergy

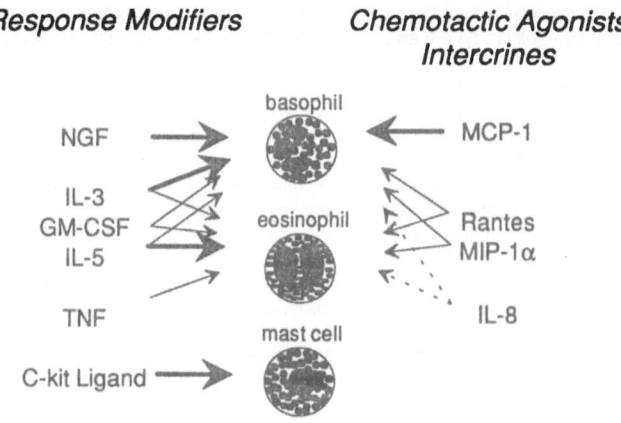

Mechanism of priming and cell activation

Importance of time and sequence of action of modulatory and activating factors
Only the sequential action of a response modifier and a cell agonist leads to an enhancement of cellular functions such as degranulation and can induce leukotriene generation. No other combination has any effect with the exception of some additive actions of sequential stimulation with certain cell agonists (Dahinden, 1989). These observations not only indicate that the time of appearance of a cytokine or agonist is of equal importance for the outcome of inflammatory reactions as the profile of the factors produced, but also suggest that the mechanism of action of cell response modifiers must be completely different from that of cell agonists.

Signal transduction for mediator release
The fundamental difference in the mode of action between response modifiers and triggers is paralleled by the different classes of receptors with which they interact (cytokine receptor- or tyrosine-kinase receptor superfamilies, and G-protein-coupled receptor superfamily, respectively), and by different intracellular signaling pathways utilized by these ligands: All basophil triggers promote a transient rise in cytoplasmatic free calcium [Ca]i, while response modifiers do not. Interestingly, the [Ca]i response is independent on cytokine priming even for incomplete agonists, and is similar for MCP-1 and IL-8 and other chemokines, although only MCP-1, in contrast to IL-8, efficiently induces degranulation in unprimed basophils. IgE-independent histamine release, but not basophil priming by growth factors, is dependent on extracellular Calcium and pertussis toxin sensitive-G-proteins, while both cellular responses are independent of PKC-activation. Thus, basophil histamine release requires Calcium and a second yet unknown signal. The IL-3 signal and IL-3-dependent LTC4 synthesis is: 1) independent of PKC activation; 2) independent of pertussis toxin-sensitive G-proteins; 3) dependent on tyrosine phosphorylation; 4) antagonized by PKC-activation. These studies indicate that lipid mediator generation and cytokine priming of basophils is regulated by tyrosine and serine/threonine phosphorylation events in an antagonistic manner (Krieger, 1992a).

Production of cytokines by lymphocytes

T-helper-1 versus T-helper-2 cells
Almost all cytokines can be produced by activated T-cells. However, T-cells can differentiate into subtypes with distinct profiles of cytokine expression, as first discovered in the mouse (Mosmann, 1991) and now clearly established in humans (Rotteveel, 1988; Maggi, 1992). It has been proposed (Mosmann, 1991) that TH-1 cells producing IL-2 and γ-INF, but no IL-4, IL-5, lead to delayed type hypersensitivity (type IV) reactions (by activation of cytotoxic T-cells and macrophages), while TH-2 cells producing IL-4, IL-5, but no γ-INF, IL-2, lead to immediate hypersensitivity (type I) diseases (by inducing the production of IgE). A mixed cytokine pattern would lead to humoral immunity or type II and III hypersensitivity (by the induction of IgG synthesis)

Hypersensitivity reactions based on cytokine profiles
Although the complete pattern of cytokines by human TH-1 and TH-2 cells has not yet been determined, it is interesting that the same profiles of cytokines also enhance the responsiveness of the respective inflammatory effector cells of the different types of inflammatory reactions in an antigen-independent manner. We thus propose an alternative classification of hypersensitivity reactions which is based on the profile of modulatory cytokines rather than on the antigen-recognizing molecules involved (T-cell receptor, IgG, IgE) as in the classification according to Coombs and Gell. "TH-2 cytokines", such as IL-3, IL-4, IL-5, should lead to allergic inflammation (type I), while "TH-1 cytokines" by the production of macrophage and NK-cell activating cytokines should lead to a granulomatous or mononuclear inflammatory response (type IV), even in the absence of antigen. A mixed profile, together with high levels of proinflammatory cytokines, such as IL-1 and TNF, should lead to neutrophilic acute inflammation (type II-III). Such a classification on cytokine profiles would also allow to include different inflammatory reactions of various etiologies other than just hypersensitivity reactions.

Cytokine profiles produced in vivo

Most importantly, the cytokine profile found in vivo in allergic late phase reactions or in chronic asthmatic patients, either by *in situ* hybidization or directly in pathological fluids (Kay, 1991; Walker, 1991), is exactly that known to prime the effector cells of allergic inflammation. It also represents the pattern produced by TH-2 cell clones, despite the fact that the immune response in vivo is, of course, polyclonal. Surprisingly, the cytokine pattern in broncho-alveolar lavage fluids of "intrinsic" asthmatic patients is different: Together with TH-2 cytokines, such as IL-5, γ-INF is also present, while IL-4 is generally lacking (Walker, 1991). Thus in different diseases, a large variety of cytokine profiles may be produced, and the TH-1 / TH-2 concept is probably an oversimplified model. However, these observations further support the utility of classifying inflammatory reactions by cytokine profiles, since very similar pathologies are found in the presence of the effector cell priming cytokines, IL-3, IL-5 and GM-CSF, in extrinsic and intrinsic asthma, despite differences of expression of IL-4 or γ-INF which mainly act on the specific immune system.

Cytokines produced by inflammatory effector cells

Inflammatory cytokines:
Proinflammatory cytokines such as TNF and IL-1, certain chemokines such as IL-8 and the pleiotrophic priming cytokine GM-CSF, are formed by many inflammatory effector cells and tissue cells and may thus amplify inflammatory processes in an unspecific manner.

Hematopoietic growth factors
The hematopoietic "priming" cytokines with a more restricted target cell profile are, in addition to lymphocytes, also generally produced by the target cells themselves, e.g. eosinophils produce IL-3, IL-5 and GM-CSF, at least upon stimulation with xenobiotics. This may lead to an amplification of a particular type of inflammatory reaction by an autocrine and paracrine mechanism.

Immunomodulatory cytokines
The expression of cytokines regulating predominantly T- and B-cell functions such as IL-4, IL-2 and γ-INF, is rather restricted to lymphocytes. In the mouse, TH-2 cytokines are expressed by mast cells and mast cell lines in response to IgE-R stimulation (Galli, 1991). We recently found that mature human basophils can produce considerable amounts of IL-4, but not IL-2 and γ-INF, by IgE-dependent and IgE-independent mechanisms (Brunner, unpublished results). By contrast, in our hands we have been yet unable to find IL-4 production by human lung mast cells, although preliminary studies by others indicate that human mast cells may also produce IL-4.

Regulation of the specific immune system by inflammatory effector cells

Inflammatory effector cells may also strongly regulate the response of the specific immune system by the production of different cytokine profiles. The production of IL-4 by basophils which regulate may be of particular importance in the pathogenesis of allergic diseases. It is interesting that IL-4 added early in cell culture in vitro results in the generation of exclusively TH-2 cells, independent of the blood donor or the nature of the antigen (Maggi, 1992), while γ-INF has an opposite effect (generation of TH-1 cells). This studies suggest that the time of appearance of IL-4 or γ-INF may be at least as important as the proportions of these two cytokines, representing a possible other example of the dimension "time and sequence" in the cytokine network. Since basophils appear to produce IL-4 more rapidly than lymphocytes, this cell type may not only represent an important effector cell of allergic late phase reactions, but also regulate the type of the immune response of the specific immune system (TH-2 respons) when a new antigen is encountered within the context of an allergic inflammatory exudate.

Conclusions

The cells of the specific immune system responsible for the recognition of antigens are connected to the inflammatory (mainly myeloid) effector cells of the innate immune system in an antigen-independent, and probably bi-directional, manner, though the action of a large numbers of cytokines and mediators. The complexity of this cytokine network makes it difficult to assess the role of an individual cytokine. Here, we stressed the possible importance of profiles of cytokines and the sequence of time of action in the cytokine network as a hypothesis to explain the different types of immune and inflammatory reaction in general and the allergic reaction in particular.

Bidirectional Interaction between Lymphocytes and Inflammatory Effector Cells in Allergy

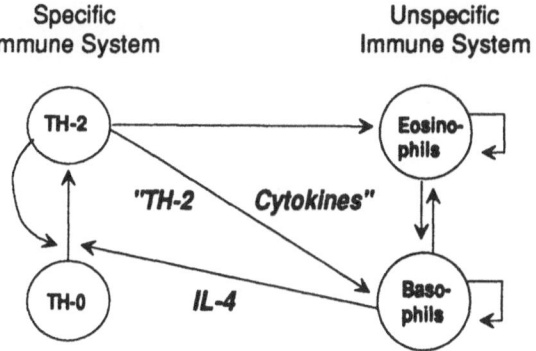

REFERENCES

Bischoff, SC, Brunner, T, de Weck, AL, Dahinden, CA (1990) J Exp Med 172:1577-1582

Bischoff, SC, Dahinden, CA (1992a) J Exp Med 175:237-244

Bischoff, SC, deWeck, AL, Dahinden, CA (1992b) Lymphokine and Cytokine Res 11:33-37

Bischoff, SC, Dahinden, CA (1992c) Blood 79:2662-2669

Bischoff, SC, Krieger, M, Brunner, T, Dahinden, CA (1992d) J Exp Med 175:1271-1275

Brown, PH, Crompton, GX, Greening, AP (1991) Lancet 338:590-592

Brunner, T, de Weck, AL, Dahinden, CA (1991) J Immunol 147:237-242

Dahinden, CA, Kurimoto, Y, de Weck, AL, Lindley, I, Dewald, B, Baggiolini, M (1989) J Exp Med 170: 1787-1792

Dahinden, CA, Bischoff, SC, Brunner, T, Krieger, M, Takafuji, S, de Weck, AL (1991) Int Arch Allergy Appl Immunol 94:161-164

Galli, SJ, Gordon, JR, Wershil, BK (1991) Current Opinion in Immunology 3:865-873

Kay, AB, Sun, Y, Varney, V, Gaga, M, Durham, SR, Moqbel, R, Wardlaw, AJ, Hamid, Q (1991) J Exp Med 173:775-778

Krieger, M, von Tscharner, V, Dahinden, CA (1992a) Eur J Immunol, in press

Krieger, M, Brunner, T, Bischoff, SC, von Tscharner, V, Walz, A, Moser, B, Baggiolini, M, Dahinden, CA (1992b) J Immunol, in press

Maggi, E, Parronchì, P, Manetti, R, Simonelli, C, Piccinni, MP, Santoni Rugiu, F, de Carli, M, Ricci, M, Romagnani, S (1992) J Immunol 148:2142-2147

Mosmann, TR, Moore, KW (1991) Immunology Today 12:A49-A53

Rotteveel, FTM, Kokkelink, I, van Lier, RAW, Kuenen, B, Meager, A, Miedema, F, Lucas, CJ (1988) J Exp Med 168:1659-1673

Takafuji, S, Bischoff, SC, de Weck, AL, Dahinden, CA (1991) J Immunol 147:3855-3861

Takafuji, S, Bischoff, SC, de Weck, AL, Dahinden, CA (1992) Eur J Immunol 22:969-974

Walker, C, Virchow, JC, Bruijnzeel, PLB, Blaser, K (1991) J Immunol 146:1829-1835

Wirtmueller, U, de Weck, AL, Dahinden, CA (1989) J Immunol 142:3213-3218

Control of IgE and Cytokine Responses

Fred D. Finkelman

Departement of Medicine, Uniformed Services University of the Health Sciences, 4301 Jones Bridge Road, Bethesda, Maryland, USA

INTRODUCTION

Our present understanding of Ig isotype selection began to develop over ten years ago with the demonstration that culture supernatants of some activated T cells contain a factor that induces activated mouse B cells to differentiate *in vitro* into IgG1 secreting cells (Isakson 1982). Subsequent *in vitro* studies identified this factor as IL-4 (Vitetta 1985), demonstrated that IL-4 also induces activated B cells to differentiate into IgE secreting cells, and showed that IFN-γ inhibits IL-4-stimulation of IgE secretion (Coffman 1986, 1986a). Studies in several laboratories have demonstrated that IL-4 and IFN-γ also have stimulatory and suppressive effects, respectively, on the induction of IgE secretion by human B lymphocytes (Pene 1988; Sarfati 1988; Vercelli 1989; Del Prete 1988; Thyphronitis 1989).

IN VIVO REGULATION OF MURINE IGE RESPONSES

Subsequently, IL-4 and IFN-γ were shown to have critical roles in the regulation of IgE responses *in vivo*. Anti-IL-4 and anti-IL-4 receptor monoclonal antibodies were found to strongly suppress the primary polyclonal IgE responses induced by nematode infections or by a foreign antibody to mouse IgD (Finkelman 1988, 1991). IgE was specifically suppressed in these systems, as serum IgG1 production was either not affected or only partially suppressed, while serum IgG2a production was either not affected or enhanced. Studies with mice that lack a functional IL-4 gene confirmed these results, although IgG1 production was suppressed more than in mice treated acutely with IL-4 antagonists (Kühn 1991). Recent experiments, in which anti-IgD antibody-injected mice mice have been treated with complexes of IL-4 and soluble CD23 (low affinity IgE receptor), demonstrate that the complexes, which slowly release IL-4 (Fernandez-Botran 1991), substantially upregulate IgE production (Sato, submitted). Thus, IL-4 is required for the generation of primary IgE responses *in vivo*, and delivery of exogenous IL-4 can enhance *in vivo* IgE responses.

In Vivo Expression of Germline ε Chain mRNA

With few exceptions, lymphocytes that secrete an Ig isotype other than IgM have first undergone a DNA recombinational event in which a productively arranged VDJ gene segment is repositioned to a site immediately 5' of the expressed C_H gene, with looping out and deletion of intervening C_H genes (Honjo 1983; Schwedler 1990). Stavnezer and others established that such deletional switching is preceded by transcription of the germline C_H gene for the isotype that will later be expressed, that the germline transcript includes the promoter-containing I region as well as C_H, but not the intervening switch region, and that IL-4 stimulates germline Cε expression (Stavnezer-Nordgren 1986; Stavnezer 1988; Rothman 1988). To investigate whether similar events occur *in vivo*, Thyphronitis has studied germline and mature ε gene expression in anti-IgE antibody-injected mice, using a quantitative PCR approach. Results indicate that increased germline ε gene expression follows increased IL-4 gene expression by approximately one day and precedes expression of mature ε chain mRNA and IgE secretion by approximately two days. IL-4 antagonists inhibit the expression of both the germline and mature forms of the ε gene transcript. Thus, the the relationship between cytokine production, germline gene transcription, and mature gene transcription that has been described *in vitro* holds true *in vivo*.

IL-4 Regulation of Secondary IgE Responses

Secondary IgE responses might theoretically be derived from B cells that had already translocated VDJ to Cε, in which case IL-4 might not be required to stimulate IgE secretion, or might be derived predominantly from B cells that undergo VDJ-Cε recombination during the course of the secondary response, in which case IL-4 would presumably be required to induce IgE secretion. *In vivo* studies with mice primed and boosted with TNP-KLH on alum support the second possibility, in that anti-IL-4 antibody selectively suppressed a secondary, TNP-specific IgE response (Finkelman 1988). *In vitro* studies of IgE secretion by peripheral blood lymphoid cells from filariasis patients (King 1990) demonstrated that while anti-IL-4 had no effect on "spontaneous" IgE secretion, it inhibited antigen-induced IgE secretion. Thus, it seems likely that both human and mouse secondary IgE responses are derived, in large measure, from cells that have not previously switched to IgE expression. These observations suggest that IL-4 antagonists might cause already established IgE responses to diminish, as IgE-secreting cells die off. This suggestion is supported by experiments in which the treatment of nematode-infected mice with IL-4 antagonists, initiated only after large IgE responses had already been generated, caused serum IgE levels to decline considerably (Katona 1991).

To test the belief that secondary IgE responses derived from B cells that had already switched to IgE expression might not require IL-4, mice were first primed by immunization with monoclonal mouse allo-anti-IgD antibodies, then rested to allow serum IgE levels to return to baseline levels, and boosted by injection of a goat anti-mouse IgE antibody. This antibody would presumably selectively stimulate IgE-expressing cells. Indeed, this treatment triggered a large, rapid, T cell-dependent IgE response. IL-4 antagonists, at doses that completely blocked anti-IgD antibody- or helminth-induced IgE responses, had little effect on the anti-IgE antibody-induced response. Taken together with our other observations, this result suggests that IL-4 is not required to stimulate B cells that already express IgE to secrete this isotype, but that most IgE secreted during a normal secondary response is derived from B cells that do not express IgE at the start of that response (Katona 1991).

Effects of IL-4 on Mucosal Mast Cells

Inasmuch as atopy is dependent upon IgE-mediated mast cell degranulation, and IL-4 stimulates mast cell growth *in vitro* (Mosmann 1986), we examined the role of this cytokine in the generation of *in vivo* mucosal mast cell responses. Both anti-IL-4 and anti-IL-3 antibodies partially inhibited the development of mucosal mastocytosis in mice infected with the nematode parasite *Nippostrongylus brasiliensis* (Nb), and both antibodies together inhibited mastocytosis by approximately 85% (Madden 1991). In contrast, injection of IL-4—anti-IL-4 antibody complexes, or IL-3—anti-IL-3 complexes that have agonist effects stimulated the development of a strong mucosal mast cell response in uninfected mice (Finkelman, unpublished data). Thus, both cytokines contribute to *in vivo* mast cell responses, and IL-4 antagonists can be expected to inhibit allergic responses by inhibiting mastocytosis as well as by inhibiting IgE production.

In Vivo Inhibitors of IgE Production

IFN-γ, IFN-α, and prostaglandin E (PGE) have all been reported to inhibit IgE production *in vitro*, although PGE has also had stimulatory effects on some IgE responses (Coffman 1986; Pene 1988; Roper 1990). Each of these agents has been tested for ability to inhibit *in vivo* IgE responses. Injection of anti-IgD antibody-stimulated mice with recombinant IFN-γ inhibits IgG1 and IgE responses, while it stimulates IgG2a responses. Anti-IFN-γ antibody has the opposite effect, indicating that sufficient IFN-γ is produced in these mice to affect Ig isotype selection (Finkelman 1988a). Exogenous recombinant IFN-α also inhibits anti-IgD antibody-induced IgE production and enhances IgG2a production, although it has little independent effect on IgG1 production. Agents, such as polyinosinic acid.polycytidylic acid, that induce endogenous IFN-α production, have the same effect; inhibition of IgE secretion by this agent is reversed by a neutralizing anti-IFN-α/β antibody (Finkelman 1991a). Selective suppression of IgE secretion was also observed when anti-IgD antibody-stimulated mice were treated with an analog of PGE$_2$ (Manning, submitted). These observations indicate that while one agent (IL-4) is critical for the induction of an IgE response, many different agents can suppress this response.

Not all cytokines that might have been expected to contribute to, or inhibit IgE secretion, based on *in vitro* studies, have had similar *in vivo* effects. *In vivo* studies with anti-IL-1 receptor antibody, anti-IL-2 and anti-IL-2 receptor antibodies, anti-IL-5 antibody, anti-IL-6 and anti-IL-6 receptor antibodies, and anti-IL-9 antibody have had no consistent effect on IgE responses in anti-IgD antibody-immunized or helminth-inoculated mice (Finkelman, unpublished data). While these results might be due to incomplete cytokine neutralization or receptor blockade, studies in IL-10 gene-knockout mice clearly indicate that this cytokine is not required to generate IgE (Kühn, unpublished data). CD23, another molecule that might have been expected to play a major role in the generation of IgE responses, based on *in vitro* studies with human lymphoid cells, also has no clear role in the generation of mouse IgE responses. An anti-CD23 mAb that modulates most CD23 from the B cell surface and blocks the ability of residual CD23 to bind IgE, has no effect on anti-IgD antibody-, anti-IgE antibody-, conventional antigen-, or nematode-induced primary or secondary, polyclonal or antigen-specific IgE responses (Katona, unpublished data).

PATTERNS OF *IN VIVO* CYTOKINE GENE EXPRESSION

Mosmann and Coffman, and Bottomly and colleagues, revolutionized the cytokine field with their observations that cytokine production by mouse CD4+ T cells tended to fall into two patterns: TH1, characterized by the production of IL-2, IFN-γ, and lymphotoxin, but not IL-4, IL-5, IL-6, IL-9, or IL-10,and TH2, characterized by the opposite cytokine profile (Mosmann 1986a, 1989; Bottomly 1988). TH1 cytokines were particularly associated with cell-mediated immune responses, while TH2 cytokines were particularly associated with allergy and antibody responses. To a major extent, similar observations have now been made with human CD4+ T cells (Parronchi 1991; Del Prete 1991; Romagnani 1991), although T cells that have a mixed pattern of cytokine production have been identified in both species (Palaird 1988; Firestein 1989; Street 1990). In addition, it has been suggested that the appearance of classic TH1 or TH2 cells is preceded first, by cells that secrete only IL-2, and next, by cells that secrete a mixed TH1/TH2 cytokine pattern (Swain 1988; Powers 1988; Weinberg 1990).

To investigate *in vivo* patterns of cytokine gene expression, William Gause and collaborators have used a quantitative PCR technique to examine three murine systems: 1) a model in which infection of mice with the nematode *Heligmosomoides polygyrus* (Hp) induces polyclonal IgE and IgG1 responses, but no IgG2a production (Finkelman 1990); 2) a model in which inoculation of mice with killed *Brucella abortus* (BA) stimulates a polyclonal IgG2a response, but little or no IgG1 and IgE production (Finkelman 1988a); and 3) a model in which injection of goat anti-mouse IgD antibody (GaMδ) stimulates a predominantly IgG1 response, with a large percentage increase in IgE and a smaller increase in IgG2a production (Finkelman 1988; Svetic´ 1991). By virtue of the previously-mentioned association of Ig isotypes with specific cytokines, it was expected that these three systems would demonstrate a TH2 cytokine pattern, a TH1 cytokine pattern, and a mixed cytokine pattern, respectively. This result was indeed observed with regard to IL-4 and IFN-γ gene expression. However, there were some interesting and important discrepancies between current dogma and the results of these studies: 1) IL-4 and IL-5 gene expression are dissociated, in that IL-4 gene expression by splenic CD4+ T cells increases over 100-fold in GaMδ-injected mice, while no increase in IL-5 gene expression by these cells has been observed; 2) No increase in IL-10 gene expression was seen in the Hp-infected mice, although there were large increases in IL-4, IL-5, and IL-9 gene expression in these mice; 3) Although a transient increase in IL-2 gene expression was seen in the GaMδ-injected mice, and preceded expression of the IL-4, IL-6, IL-10, and IFN-γ genes, no consistent early or late increase in IL-2 gene expression was seen in either the BA-injected or the Hp-infected mice; 4) The only two cytokine genes, of the group tested, that were expressed in increased quantity by T cells from BA-injected mice, were IFN-γ (a TH1 cytokine) and IL-10 (a TH2 cytokine); and 5) A mixed (Th0) pattern of cytokine gene expression was not observed in any of the three models prior to more specific cytokine gene expression patterns. This absence of evolution of cytokine gene expression patterns from TH0 to TH1 or TH2 makes it impossible to draw conclusions about whether T cells that secrete different patterns of cytokines are derived from common or from distinct precursors.

IFN-γ, IL-4, AND RESISTANCE TO DISEASE

Studies from the laboratories of Locksley and Sher pointed out the importance of different cytokines in the control of *Leishmania major* infections in mice (Locksley 1991; Sher 1992). In general, it was observed that production of IFN-γ keeps the infection from disseminating, and allows its eventual cure, while production of IL-4 leads to disseminated infection and death. It is likely that the differential cytokine effects are mediated through their different effects on macrophage activation, inasmuch as IFN-γ (with TNF-α) activates macrophages to produce the nitrites that are required to kill ingested *L. major* amastigotes, while IL-4 inhibits the activation of nitrite production by macrophages (Liew 1990; Locksley 1991).

In contrast to the response to *L. major* and several other intracellular parasites for which IFN-γ production appears to be critical (Gazzinelli 1991; Rose 1989; Ungar 1991), protective immunity to at least some gastrointestinal nematodes appears to require the production of IL-4 and is inhibited by IFNs. IL-4 antagonists have little effect on adult worm number and survival in mice inoculated with Nb or Hp larvae, but substantially increase egg production by adult worms. In contrast, treatment of Hp-inoculated mice with IL-4—anti-IL-4 antibody complexes that have IL-4 agonist activity considerably inhibits egg production by adult worms (Urban, unpublished data). In mice that have been primed by an initial Hp infection, then cured pharmacologically and reinfected, IL-4 antagonists increase both adult worm survival and egg production (Urban 1991). The mechanism by which IL-4 contributes to protection against reinfection with Hp is uncertain, and does not seem to depend upon IL-4 induction of an IgE response or mastocytosis (Urban, unpublished data). IFN-α and IFN-γ have effects opposite to those of IL-4 in mice inoculated with Nb, increasing egg production and adult worm survival (Urban, unpublished data).

Experiments performed by Else and Grencis in mice infected with the whipworm *Trichuris muris*, have yielded results consistent with those observed in the Nb and Hp systems. BALB/k mice infected with *T. muris* make an IL-4 response and spontaneously cure their infections, while *T. muris*-infected B10.BR mice make an IFN-γ response and develop a chronic infection (Else 1991). Treatment of B10.BR mice with anti-IFN-γ antibody prevents the development of a chronic *T. muris* infection, while treatment of BALB/k mice with anti-IL-4 receptor antibody leads to a chronic infection (Else, unpublished data). Thus, specific cytokines neither protect against nor exacerbate infection in general. Inasmuch as different effector mechanisms protect the host against different parasites, and different cytokines stimulate or inhibit the induction of different effector mechanisms, the host must have the ability to select the proper cytokine response to any given parasite to optimally protect itself against infection.

REGULATION OF CYTOKINE EXPRESSION

Some cytokines have been shown to influence patterns of T cell cytokine secretion *in vitro*. IFN-γ inhibits proliferation of TH2, but not TH1 cells (Gajewski 1988). The same cytokine enhances IFN-γ secretion and inhibits IL-4 secretion when added to cultures of stimulated T cells (Gajewski 1989), while IL-4 has the opposite effect (Swain 1990; Le Gros 1990). IL-10 inhibits activation of NK cells and macrophages, including the ability of macrophages to present antigen to and stimulate TH1, but not TH2 cells (Mosmann 1991). *In vivo* studies of *L. major* infected mice also point to roles for IFN-γ and IL-4 in directing T cell cytokine expression in opposite directions, but provide little evidence for a role for IL-10 in cytokine regulation (Locksley 1991; Sher 1992). Inasmuch as cytokines themselves direct T cell cytokine responses, it seems reasonable to hypothesize that the immune system recognizes certain qualities common to different pathogen types that induce the production of cytokines, such as IFN-γ and IL-4, that regulate subsequent T cell cytokine production. Such hypothetical triggers for an appropriate cytokine response would optimally be ones critical to parasite function, so that the different parasite classes could not easily modify characteristics that stimulate their control by the immune system. These triggers might also be expected to rapidly cause production of the appropriate cytokines by non-T cells, so that the controlling cytokines would be present early in the course of T cell differentiation.

In vivo studies indicate that IFN-α and IFN-β, in addition to IFN-γ, can induce T cells to differentiate in a "TH1 direction," enhancing IFN-γ secretion and inhibiting IL-4 secretion (Finkelman 1991a). Importantly, viral polynucleotides and bacterial LPS induce IFN-γ secretion by NK cells and IFN-α/β secretion by macrophages (Havell 1983). Inasmuch as the great majority of viruses and gram negative bacteria induce a T cell response that is characterized by considerable IFN-γ secretion and little IL-4 secretion, it seems likely that LPS and viral polynucleotides trigger a TH1 response, and that these triggers work by stimulating macrophage and NK cell IFN secretion.

Less is known about possible nematode triggers for a TH2 response. One possibility is that proteolytic enzymes, which are produced in abundance by nematode parasites and are required for parasite penetration and migration through host tissues, as well as for parasite maturation (McKerrow 1988, 1989; Hotez 1992), are such triggers. In accord with this possibility are the observations that: 1) the major allergen produced by the dust mice is a cysteine protease (Chua 1988), and 2) the association of papain, another cysteine protease, with frequent allergic reactions (Novey 1979). To test this possibility, my colleagues and I examined cytokine gene expression in the footpads and popliteal lymph nodes of mice after footpad injection of active or specifically inactivated papain. One hour after active papain injection, large increases were observed in mRNA levels for IL-4, IL-5, and IL-9, but not IL-2 or IFN-γ. Active papain induced considerably greater expression of TH2-associated cytokine genes than did inactive papain. While we have not yet demonstrated IL-4 secretion by cells from papain-injected mice, IL-4 secretion is suggested by the observation that serum IgE levels increase to a considerably greater extent in mice injected with active papain than in mice injected with inactive papain (Finkelman, unpublished data). The cell type(s) responsible for the increased IL-4 gene expression in papain-inoculated mice have not been identified; however, preliminary experiments suggest that T cells, B cells, and mast cells may not be required, as the increase has been seen in SCID mice and in mast cell-deficient W/W^v mice.

The concept that certain parasite characteristics may trigger a specific cytokine response neither restricts possible triggers to those that have been mentioned above, nor diminishes the possibility that other, totally distinct mechanisms also regulate cytokine expression. In addition, this concept does not easily explain why some mouse strains respond to infection with a particular parasite by generating a TH1 response, while other mouse strains infected with the same parasite generate a TH2 response. One possible explanation of this observation is that such parasites espress triggers for both types of response and that different mouse strains are more or less sensitive to these triggers, with such differential sensitivity determining the final outcome of the response. This possibility is in accord with Scharton's observation (1992) that NK cell activity, and thus the ability to make a rapid IFN-γ response to an LPS-like molecule, is greater in footpads of a mouse strain that makes a TH1 response to infection with *L. major* than in footpads of a strain that makes a TH2 response.

CONCLUSIONS AND SPECULATIONS

If parasites commonly express both TH1 and TH2 trigger molecules, differences in host sensitivity to the different trigger molecules may have evolved in response to the parasite class that constituted the greater threat. Individuals threatened predominantly by intracellular parasites may have been selected for greater sensitivity to TH1 trigger molecules, while individuals threatened predominantly by gastrointestinal nematodes may have been selected for greater sensitivity to TH2 trigger molecules. The infectious disease consequences of such differences in predisposition towards the development of a particular cytokine response are, in fact, observed in different mouse strains. C57BL mice, which are relatively resistant to intracellular parasites, tend to be more sensitive to infection with gastrointestinal nematodes, while mouse strains that are relatively resistant to nematode infection, such as BALB, tend to be more sensitive to infection with intracellular parasites. Thus, the genetic predisposition towards the development of atopy may represent a bias in the regulation of effector mechanisms that protect the host against different pathogen classes. An immune system, such as that of the BALB/c mouse or the atopic human, that has, perhaps, been selected for greater resistance to helminths and related pathogens, is unfortunately likely to be inappropriately set off by non-living environmental stimuli that either share characteristics with viable parasites, or enter the body at sites that have high concentrations of cytokines that direct T cell differentiation to a TH2 pattern. The possibility of pharmacologic regulation of production or effect of IL-4 and the IFNs, the cytokines that appear to be most involved in regulating T cell differentiation in a TH1 or TH2 direction, holds promise as a means for guiding the immune system to the most effective response against particular pathogens and for diminishing untoward allergic responses.

ACKNOWLEDGEMENTS

US Naval Medical Research Research and Development Command Contract Number N0007592WR0024 and NIH grants AI-26150 and AI-21328 provided support for much of the research described here. Many of the observations reported here were performed predominantly by colleagues from other laboratories: the use of cytokine-

cytokine receptor complexes with agonist activity was predominantly the work of T. Sato, C. Maliszewski, and colleagues at Immunex Research Corp.; studies with a PGE₂ analog were conducted by J. Manning and A. Levine at Monsanto Corp.; most of the parasitologic studies described were predominantly the work of J. Urban, Jr., at the USDA; G.Thyphronitis (USUHS) defined germline and productive ε mRNA expression *in vivo*; W. Gause, A. Svetic´, S. Morris, and Y. Jian (all from USUHS) designed and performed studies of *in vivo* cytokine gene expression; I. Katona and K. Madden (USUHS) were responsible for many of the studies of IgE production and mast cell responses; and K. Else and R. Grencis (University of Manchester) gave me permission to mention their unpublished observations with *T. muris*. I am deeply grateful to all of these individuals, as well as to I. Gresser (CNRS), R. Coffman and T. Mosmann (DNAX), M. P. Beckmann, K. Grabstein, and K. Schooley (Immunex), A. Cheever, W. Paul, and A. Sher (NIH), P. Trotta (Schering-Plough), and C. Snapper and J. Holmes (USUHS) for participating in, guiding, and supporting my efforts.

REFERENCES

Bottomly K (1988) Immunol Today 9:274-277
Chua KY, Stewart GA, Thomas WR, Simpson RJ, Dilworth RJ, Plozza TM, Turner KJ (1988) J Exp Med 167:175-182
Coffman RL, Cary J (1986) J Immunol 136:949-954
Coffman RL, Ohara J, Bond MW, Carty J, Zlotnik A, Paul WE (1986a) J Immunol 136:4538-4541
Del Prete GF, Maggi E, Parronchi P, Chretien I, Tiri A, Macchia D, Ricci M, Banchereau J, de Vries J, Romagnani S (1988) J Immunol 140:4193-98
Del Prete GF, De Carli M, Ricci M, Romagnani S (1991) J Exp Med 174:809-813
Else KJ, Grencis RK (1991) Immunology 72:508-513
Fernandez-Botran R, Vitetta ES (1991) J Exp Med 174:673-681
Finkelman FD, Katona IM, Urban JF,Jr, Holmes J, Ohara J, Tung AS, Sample JvG, Paul WE (1988) J Immunol 141:2335-41
Finkelman FD, Katona IM, Mosmann TR, Coffman RL (1988a) J Immunol 140:1022-1027
Finkelman FD, Holmes, J, Katona IM, Urban JF, Jr, Beckmann MP, Park LS, Schooley KA, Coffman RL, Mosmann TR, Paul WE (1990) Annu Rev Immunol 8:303-333
Finkelman FD, Urban JF, Jr, Beckmann MP, Schooley KA, Holmes JM, Katona IM (1991) Int Immunol 3:599-607
Finkelman FD, Svetic´A, Gresser I, Snapper C, Holmes J, Trotta PP, Katona IM, Gause WC (1991) J Exp Med 174:1179-1088
Firestein GS, Roeder WD, Laxer JA, Townsend KS, Weaver CT, Hom JT, Linton J, Torbett BE, Glasebrook AL (1989) J Immunol 143:518-525
Gajewski TF, Fitch FW (1988) J Immunol 140:4245-4252
Gajewski TF, Joyce J, Fitch FW (1989) J Immunol 143:15-22
Gazzinelli, RT, Hakin FT, Hieny S, Shearer GM, Sher A(1991) J Immunol 146:286-292
Havell EA, Spitalny GL (1983) J Reticuloendothel Soc 33:369-380
Honjo T (1983) Ann Rev Immunol 1:499-528
Hotez P, Haggerty J, Hawdon J, Milstone J, Gamble HR, Schad G, Richards F (1990) Infect Immun 58:3883-3892
Isakson PC, Pure E, Vitetta ES, Krammer PH (1982) J Exp Med 155:734-748
Katona IM, Urban JF, Jr, Kang SS, Paul WE, Finkelman FF (1991) J Immunol 146:4215-4221
King CL, Ottesen EA, Nutman TB (1990) J Clin Invest 85:1810-15
Kühn R, Rajewsky K, Müller W (1991) Science 254:707-710
Le Gros G, Ben-Sasson SZ, Seder R, Finkelman FD, Paul WE (1990) J Exp Med 172:921-929
Liew FY, Li Y, Milott S (1990) J Immunol 145:4306-4310

Locksley RM, Scott P (1991) Immunoparasitology Today. Elsevier Trends Journals, Cambridge, pp. A58-A61

Madden KB, Urban JF, Jr, Ziltener, HJ, Schrader JW, Finkelman FD, Katona IM (1991) J Immunol 147:1387-91

Manning JA, Finkelman FD, Collins PW, Levine AD (Submitted for publication)

McKerrow JH, Doenhoff MJ (1988) Parasitology Today 4:334-339

McKerrow JH (1989) Exp Parasitol 68:111-115

Mosmann TR, Bond MW, Coffman RL, Ohara J, Paul WE (1986) Proc Natl Acad Sci USA 83:5654-5658

Mosmann TR, Cherwinski H, Bond MW, Giedlin MA, Coffman RL (1986a) J Immunol 136:2348-2357

Mosmann TR, Coffman RL (1989) Annu Rev Immunol 7:145-73

Mosmann TR, Moore KW (1991) Immunoparasitology Today. Elsevier Trends Journals, Cambridge, pp. A49-A53

Novey HS, Marchioli LE, Sokol WN, Wells ID (1979) J Allergy Clin Immunol 63:98-103

Palaird X, de Wall Malefijt R, Yssel H, Blanchard D, Chretien I, Abrams J, de Vries J, Spits H (1988) J Immunol 141:849-55

Parronchi P; Macchia D; Piccinni M-P; Biswas P; Simonelli C; Maggi E; Ricci M; Ansari AA; Romagnani; S (1991) Proc Natl Acad Sci USA 88:4538-4542

Pene J, Rousset F, Briere F, Chretien I, Bonnefoy JY, Spits H, Yokota T, Arai K, Banchereau J, de Vries J (1988) Proc Natl Acad Sci USA 85:6880-6884

Powers GD, Abbas AK, Miller RA (1988) J Immunol 140:3352-3357

Romagnani S (1991) Immunol Today 12:256-257

Roper RL, Conrad DH, Brown, DM, Warner GL, Phipps RP (1990) J Immunol 145:2644-2651

Rose ME, Wakelin D, Hesketh P (1989) Infect Immun 57:1599-1603

Rothman P, Lutzker S, Cook W, Coffman RL, Alt FW (1988) J Exp Med 168:2385-2389

Sarfati M, Delespesse G (1988) J Immunol 141:2195-2199

Sato TA, Widmer MB, Finkelman FD, Madani H, Grabstein KH, Maliszewski CR (Submitted for publication)

Scharton T, Scott P (1992) FASEB J 6:A1687

Schwedler U, Jack H-M, Wabl M (1990) Nature 345:452-456

Sher A, Coffman RL (1992) Annu Rev Immunol 10:385-410

Stavnezer J, Radcliffe G, Lin Y, Nietupski J, Berggren L, Sitia R, Severinson E (1988) Proc Natl Acad Sci USA 85:7704-7708

Stavnezer-Nordgren J, Sirlin S (1986) EMBO J 5:95-102

Street NE, Schumacher JH, Annie T, Fong T, Bass H, Fiorentino DF, Leverah JA, Mosmann TR (1990) J Immunol 144:1629-1639

Svetic´A, Finkelman FD, Jian YC, Dieffenbach, CW, Scott DE, McCarthy, KF, Steinberg AD, Gause WC (1991) J Immunol 147:2391-2397

Swain SL, McKenzie DT, Weinberg AD, Hancock W (1988) J Immunol 141:3445-3455

Swain SL, Weinberg AD, English M, Huston G (1990) J Immunol 145:3796-3806

Thyphronitis G, Tsokos GC, June CH, Levine AD, Finkelman FD (1989) Proc Natl Acad Sci USA 86:5580-5584

Ungar BLP, Kao T-Z, Burris JA, Finkelman FD (1991) J Immunol 147:1014-1022

Urban JF, Jr, Katona IM, Paul WE, Finkelman FD (1991) Proc Natl Acad Sci USA 88:5513-5517

Vercelli D, Jabara HH, Arai K-I, Geha RS (1989) J Exp Med 169:1295-1308

Vitetta ES, Ohara J, Myers C, Layton J, Krammer PH, Paul WE (1985) J Exp Med 161:734-748

Weinberg AD, English M, Swain SL (1990) J Immunol 144:1800-1807

Molecular Analysis of House Dust Mite Allergens

W. R. Thomas, and K. Y. Chua

The Western Australian Research Institute for Child Health, Princess Margaret Hospital, GPO Box D184, Perth, Western Australia, Australia 6001

Allergic sensitisation to common inhalant allergens poses many questions about the steady-state immune processes as well as its role in the production of inflammatory diseases, especially asthma. There are a number of mechanisms which can produce tissue damage including the immediate release of histamine, prostaglandins and leucotrienes from IgE armed mast cells, the infiltration of eosinophils and the release of proinflammatory cytokines from epithelial and fibroblastic cells, all initiated or regulated by T cells (Frew and O'Hehir 1992; O'Hehir *et al* 1992). From the immunological perspective the questions are why doesn't the whole population develop allergy to the large number of ubiquitous allergens (Sporik *et al* 1991) and what are the mechanisms which induce some people to produce IgE and T cells releasing, for example, IL-4 and IL-5 which lead to sensitisation and inflammation. While the weight of information has shown that allergens are proteins or modified proteins without any special properties, some are major allergens in that they dominate the allergic responses of many people. That is, although allergen sources such as mites present many allergens, the bulk of the response of most patients is directed towards a limited number of components. Also, not uncommonly some patients produce IgE say to 5-6 components of rye grass pollen but not to any of the other allergens in their environment. Another problem discussed in this article is the interaction which occurs during exposure to closely related molecules produced as polymorphisms within a species or by related species. This is very common for plants, especially grasses but is also important for house dust mites and other arthropod allergens. In the case of plants, several gene families producing related molecules within the species have been described.

Allergy to the house dust mite affects 10-15% of most populations and is the allergen most associated with asthma. It is particularly significant in all except the driest climates (Platts-Mills and Chapman 1987) and has little seasonality. Modern home constructions and living styles also provide ideal indoor conditions for these scavengers, which abound in soft furnishings and warm humid environments. The most abundant domestic mites are the pyroglyphid species *Dermatophagoides pteronyssinus* and *D. farinae* (Platts-Mills and Chapman 1987). A third pyroglyphid mite *Euroglyphus maynei* although less widespread can reach at-risk levels in 25% of homes in some environments, as shown in a recent survey in Western Australia (Colloff *et al* 1991). Australia, the United Kingdom and New Zealand primarily have *D. pteronyssinus* whereas the United States, several European countries and Japan have both *D. pteronyssinus* and *D. farinae* which, as will be discussed below, could have important implications for the interpretation of investigative results, diagnosis and immunotherapy.

Allergic responses to house dust mite as measured by IgE reactions to components in house dust mite extracts are to a very large extent directed to two components termed the group I and group II allergens. Several quantitative studies (van der Zee *et al* 1988; Heymann *et al* 1989) have shown that over 50% of the responses in 50% of patients are directed to these allergens and the responses of most other patients to these allergens are also substantial. Depending on the population studied and the method of detection, from 80-100% of allergic people have been shown to react with these purified allergens. Another allergen *Der p* III also elicits responses in nearly all allergic patients (Stewart *et al* 1992), but although detailed quantitative studies were not reported, the degree of reactivity in a solid phase binding assay appeared considerably less than the group I allergens. As judged by IgE reactive bands on Western blotting a further 3-4 components react with IgE in 40-50% of patients and a further 20 show sporadic binding. The consensus of studies with Western blotting shows IgE antibody from patients generally recognises 2-6 allergens (Bengtsson *et al* 1986; Ford *et al* 1989) but a substantial proportion recognise a high number, say 10-14. Some studies of Tovey *et al* (1989) appear biased to this end of the scale presumably due to the stated use of high titre sera. In all these studies, however, there is a great need for a panel of purified allergens to perform quantitative T and B cell experiments. There is a need for detailed analysis because, in plants and some mite studies (Thomas *et al* 1992), it is clear many bands detected in Westerns are from related products. The problems with Western blotting are only too well illustrated by comparing the reactivity of the group I and group II house dust mite allergens. Antibodies titres to the group I allergens when measured with purified allergens are typically 5-10 fold larger than group II allergens and moreover this allergen is usually 3-5 fold more abundant in mite extracts. The typical Western blotting result shows, however, that serum from many patients have a very large Mr 14 K band for the group II and weak group I reactions around Mr 25-27 K. This has led to the confusion that the group II allergens are more important but is probably more related to the finding of Tovey *et al* (1989) that group II allergens can increase binding reactivity in the presence of neutral detergents and that the group I allergens are denatured during the electrophoresis. Having said this, however, studies by O'Brien *et al* (1992) have shown that *Der p* I and *Der p* II induce the same degree of polyclonal T cell proliferation and in this respect may be equally important in allergic disease.

Our studies have been to analyse and produce the major allergens by recombinant technology to provide material for investigation, more informative diagnostics and to manufacture proteins or peptides for safer and more effective immunotherapy. Although the early priorities are on investigation, even if effective therapy can be developed only for the major allergens cloned at present, this still could be effective for half of the 10-15% of the world population who seek treatment for this allergy. The work, however, is rapidly being extended to other mite allergens and perhaps an effective cocktail will be developed to replace or complement the extracts being used which vary enormously in different batches, especially in the relative amounts of allergens. These probably have many allergens at suboptimal concentration given the 20 µg/ml requirement for group I allergens as a benchmark to reliably produce optimal skin reactions (Platts-Mills and Chapman 1987).

The international nomenclature system used for allergens is to designate each allergen with the first three letters of genus followed by the first letter of the species and a number. The number is given in order of either its importance or its order of characterisation. Importantly, however, equivalent allergens from related species have the same number so it is possible to have groups. The allergens to be primarily discussed here are therefore *Der p* I and *Der p* II from *D. pteronyssinus* and *Der f* I and *Der f* II from *D. farinae*.

The group I allergens were the first analysed because of their relative abundance in mite extracts and spent mite media and by their high frequency and degree of reactivity with serum from patients allergic to house dust mites. They attracted a great deal of interest when it was found that 95% of the group I allergens is in the faecal pellets or dung balls of the mites. These particles contain 0.1 ng in a 20 μ sphere which has been calculated to give a very high concentration of 10 mg/ml of group I allergen per ml of dung (Tovey *et al* 1981). The first achievement of the cDNA investigations was to establish by sequence homology that *Der p* I was a cysteine protease (Chua *et al* 1988) belonging to the same family as papain and actinidin as well as the mammalian cathepsins B and H. About 50 enzymes belonging to this group from a variety of sources have now been sequenced. Thus, the group I allergens are probably enzymes released into the gut for digestion, a theme shown by Stewart *et al* (1992) to recur for other allergens especially trypsin and amylase, now designated group III and group IV. The 222 residue *Der p* I protein had a derived molecular weight of 25,000, expected from immunochemical analysis and a single N-glycosylation site. Sequencing of *D. farinae* homologue Der *f* I by Dilworth *et al* (1991) revealed a 223 residue protein with only 80% homology to *Der p* I and the same glycosylation site was conserved. The conservation of the N-glycosation site despite several codon changes is evidence that the allergen is indeed a glycoprotein. The amino acid differences between the proteins from the different species were notable in that they were located at the N and C termini residues 1-20 and 200-222 and in a central region from 90-130 which is a connecting loop. Based on the x-ray crystallographic structure of the homologues papain and actinidin few of the hydrophobic core residues required for the condensing of the two domains of this molecule are in the N-terminal region and there is only one conserved in residue 90-130. The C-terminal is quite different between *Der p* I and papain and actinidin and is truncated. Thus the conserved regions 20-90 and 130-200 contain 24/29 of the hydrophobic core contain residues described for actinidin. Only 2 of the remaining 5 are in fact conserved in *Der p* I. Although most of the 34/41 of amino acid differences between *Der p* I and *Der f* I are conserved with respect to their estimated ability to conserve tertiary structures many would be expected to severely affect T cell epitopes which are presented as peptides. Experiments with mice and with human T cell clones (O'Hehir *et al* 1987) have shown these allergens indeed often do not cross react at this level. Studies of polyclonal human T cell responses in areas where only one species exists have not been reported and the degree of cross reactivity would be of some interest. The level of IgE cross reactivity between and *Der p* I and *Der f* I is also not completely resolved with the studies of Heymann *et al* (1986) showing about 80% cross absorption using allergen attached to a solid phase. In the studies of Lind *et al* (1988) using fluid phase inhibition it was found that usually 10-1,000 fold more *Der f* I than *Der p* I was required to inhibit anti *Der p* I binding. These studies are not necessarily at odds and probably reflect affinity versus total absorption. Also natural exposure must be taken into account since exposure to both species may encourage the proliferation of cross reactive B cells. From our cDNA studies we have been able to use the reactivity of rabbit antiserum with a series of recombinant peptides to show the presence of cross reactive epitopes in residues 34-72, 60-72 and 166-194 and species specific epitopes from residues 82-99 and 112-140 (Greene *et al* 1990). These species specific epitopes correlate with sequence divergence between *Der f* I and *Der p* I in a flexible structure discussed above, which can be identified as an outside loop connecting the two domains. As will be outlined below this structure also appears to be very antigenic in humans so perhaps peptides could be engineered which could be used for serological diagnosis or skin tests distinguishing sensitisation to *D. farinae* and *D. pteronyssinus*. If immunotherapy works by modulating T cell reactivity it would be very important to first determine the species causing the allergic sensitisation.

Our attempts to express the recombinant *Der p* I in an unfused form in *E. coli* have not been successful, apparently due to toxicity to the bacteria. Some success has been obtained with pGEX vectors expressing the allergens or peptides as fusion on the C-terminus of a glutathione transferase molecule. Although most of the protein is insoluble, sufficient, 200-300 µg/l, is soluble and can be isolated from glutathione coupled agarose. Approximately 50% of allergic sera tested reacted, to these fusion constructs and absorption showed that the recombinant allergen could remove about 50% of the IgE reacting to the natural allergen (Greene *et al* 1991; Greene and Thomas 1992). Reactivity to a series of overlapping fusion peptides showed that providing the constructs contained more than 30 residues, reactivity across the whole molecule could be obtained. The most consistently high binding peptides were 53-99 and 98-140 which correspond to residues in the divergent loop already described. Evidence was also obtained to show flanking sequences were required for some peptides to express antigenicity and that in one case antibodies could be cross absorbed by sequentially different but structurally adjacent peptides (Greene and Thomas 1992). This indicates considerable homogeneity in the IgE antibodies which recognise the peptides, a conclusion, however, which cannot be extended to antibodies against the natural molecule at this stage. Jeannin and colleagues (1992) have now reported histamine release tests using selected synthetic peptides and find reactivity in about 40% of patients with peptides 52-71 and 117-133, the latter being in the connecting loop.

To obtain higher IgE binding the recombinant allergen has been expressed in yeast *Saccharomyces cerevisiae* as a small fusion with a few additional residues from the vector. In the results reported so far (Chua *et al* 1992), using a construct which contained a 23 residues from the truncated proenzyme region of *Der p* I the yeast product was insoluble but could be solubilised in urea and renatured in physiological buffers. A proportion of this reacted with monoclonal antibodies and IgE almost as well as the natural allergen. As well as providing clonal proof that the group I allergens did react with most allergic sera these experiments also suggested the proenzyme region may not be required for correct folding. More recent studies with a complete proenzyme of 80 residues and constructs beginning at the exact N-terminal of *Der f* I have shown improvements in yield and antigenicity. The studies pose an interesting problems not only in the expression of allergen but in determining the mechanisms for folding and proenzyme cleavage in this class of thiol proteases.

The elucidation of sequences of the allergens have also made it possible to ascertain the polymorphism of the allergens. As a start to this we have sequenced 5 clones of *Der p* I all from the same commercial source (CSL Laboratories, Melbourne). All sequences have been different usually by 1-2 residues and changes in 5 residues have been detected. The substitutions were all structurally conserved but would affect T cell epitopes. Indeed, 4/5 of the polymorphisms were in the sequences of T cell epitopes reported to date. The veracity of these changes is supported by the finding that all changes were to residues found in *Der f* I and their high frequency in a closed mite colony suggest that the level of environmental variation may be high and important. It also indicates that because of this that some commercial mite extracts may not be appropriate for particular regions for T cell studies.

The analysis of cDNA encoding the 14 kD group II allergens by Chua *et al* 1990; Trudinger *et al* 1991, and Yuuki *et al* 1991 has differed from that of the group I in two respects. Firstly, the group II molecules are highly conserved between the mite species with only 3/15 of the differences found between the 129 residue *Der p* II and *Der f* II being nonconserved (pro/asp 19; pro/thr 70; ile/ala 88). Even the nonconservative changes are, however, still sufficient for T cell clones to distinguish between the molecules as found by O'Hehir *et al* 1987. We have also found 2 high responding patients which distinguish between *Der p* II and *Der f* II in the level of the antigen induced polyclonal T cell proliferative responses (O'Brien *et al* 1992). The molecules *Der p* II and *Der f* II are said to be serologically very similar although cross absorption or inhibition experiments with IgE to formally examine this have not been published. As well, in the interspecies conservation our experience with 3 cDNA clones of *Der p* II and 4 PCR amplification products of *Der f* II and the data from Yuuki *et al* (1991) with *Der f* II show few polymorphic changes except for substitution of aliphatic moieties. To date now no homologues of the group II allergens exist in the data bases and the function of this 129 residue 6 cysteine protein is a matter of speculation. Since the relative concentration is higher in mite bodies than spent mite media it may not be a digestive enzyme.

The second noteworthy difference from the group I allergens in our studies has been the ability of *Der p* II and *Der f* II cDNA to express a product with almost all of the IgE binding capacity of the natural allergen (Chua *et al* 1990). For example, our experience with pGEX fusions show yields of 200-300 µg/l of culture of soluble fusion which reacts with all *Der p* II reactive sera and can absorb out almost all IgE reactivity to natural *Der p* II. Although this soluble material has been useful to date, experiments with yeast and mammalian expression are continuing in order to develop systems to provide larger quantities of molecules for further investigations. The group II allergens are less abundant in mite extracts and spent mite medium than the group I, so the isolation of natural allergen is more of a problem and hence the recombinant more useful. Studies with peptides produced from cDNA fragment have shown that the high reactivity of recombinant *Der p* II is not due to high reactions from small stretches of peptide. The peptides 1-69 and 70-129, which are half molecules, are usually unable to bind IgE (Chua *et al* 1991) as well as 42-117 which spans this boundary. Some patients can bind smaller peptides in keeping with the data of van't Hof *et al* 1992, although this is a small percentage of the reactivity in the sera of a small percentage of the patients.

cDNA cloning has rapidly led to the characterisation and production of recombinant allergens which are now being used in several avenues of research, particularly in studying T cell regulation. An application for immunotherapy can be predicted at least for the high percentage of patients with allergic responses directed to the class I and class II molecule. Since modification of T cell function will be targeted it is anticipated that systematic reports of human epitopes will be made to complement those described from some lines (Yssel *et al* 1992; O'Hehir *et al* 1991). As well as studying the allergic responses to these molecules in more depth there is a need to perform similar studies with further significant mite allergens. With the recent characterisation of the group III allergens as trypsin (Stewart *et al* 1992) and the importance of amylase group IV (Lake *et al* 1991) recognised, progress should be swift. Other significant allergens are the 14 kD component recognised by Tovey from IgE immunoassays of a cDNA library (Tovey *et al* 1989) and chymotrypsin (Yaseuda *et al* 1991) which are each recognised by about 40% of allergic sera. Finally, it should be said that similar developments are occurring for grass, tree, weed and mammalian allergens amongst others and that our understanding of responses and ability to investigate allergic sensitisation is being greatly increased.

ACKNOWLEDGEMENT

This work was supported by the National Health and Medical Research Council of Australia, the Asthma Foundation of Western Australia and ImmuLogic Pharmaceuticals Corp., Massachusetts.

REFERENCES

Bengtsson A, Karlsson A, Rolfsen W, Einarsson R (1986) Int Arch Allergy Appl Immunol 80:383-390

Chua KY, Stewart GA, Thomas WR, Simpson RJ, Dilworth TM, Plozza TM, Turner KJ (1988) J Exp Med 167: 175-182

Chua KY, Doyle CR, Stewart GA, Turner KJ, Simpson RJ, Thomas WR (1990) Int Arch Allergy Appl Immunol 91:118-123

Chua, KY, Dilworth RJ, Thomas WR (1990) Int Arch Allergy Appl Immunol 91:124-129

Chua KY, Greene WK, Kehal P, Thomas WR (1991) Clin Exp Allergy 21:161-166

Chua KY, Kehal P, Thomas WR, Vaughan PR, Macreadie IG (1992) J Allergy Clin Immunol 89:95-102

Colloff MJ, Stewart GA, Thompson PJ (1991) Clin Exp Allergy 21:225-230

Dilworth RJ, Chua KY, Thomas WR (1991) Clin Exp Allergy 21:25-32

Ford SA, Tovey ER, Baldo BA (1989) Clin Exp Allergy 20:27-31

Frew AJ, O'Hehir RE (1992) J Allergy Clin Immunol 89:783-788

Greene WK, Chua KY, Stewart GA, Thomas WR (1990) Int Arch Allergy Appl Immunol 92:30-38

Greene WK, Cyster JG, O'Brien RM, Chua KY and Thomas WR (1991) J Immunol 147:3768-3773

Greene WK, Thomas WR (1992) Mol Immunol 29:257-262

Heymann PW, Chapman MD, Platts-Mills TAE (1986) J Immunol 137:2841-2847

Heymann PW, Chapman MD, Aalberse RC, Fox JW, Platts-Mills TAE (1989) J Allergy Clin Immunol 83:1055-1067

Jeannin P, Didierlaurent A, Gras-Masse H, Elars AA, Delneste Y, Cardot E, Joseph M, Tartar A, Vergoten G, Pestal J (1992) Mol Immunol 29:739-749

Lake FR, Thompson PJ, Ward LD, Simpson RJ, Stewart GA (1991) J Allergy Clin Immunol 87:1035-1042

Lind P, Hansen OC, Horn N (1988) J Immunol 140:4256-4262

O'Brien RM, Thomas WR, Whotton A (1992) J Allergy Clin Immunol 89:1021-1031

O'Hehir RE, Young DB, Kay AB, Lamb JR (1987) Immunology 62:635-640

O'Hehir RE, Garman RD, Greenstein J, Lamb JR (1991) Ann Rev Immunol 9:67-95

Platts-Mills TAE, Chapman MD (1987) J Allergy Clin Immunol 80:755-775

Sporik R, Holgate ST, Platts-Mills TAE, Cogswell JJ (1991) New Eng J Med 323:502-507

Stewart GA, Ward LD, Simpson RJ, Thompson PJ (1992) Immunology 75:29-35

Thomas WR, Chua KY, Smith WA (1992) Exp Appl Acarol (in press)

Tovey ER, Chapman MD, Platts-Mills TAE (1981) Nature 289:592-593

Tovey ER, Ford SA, Baldo BA (1989) Electrophoresis 10:243-249

Tovey ER, Johnson MC, Roche AL, Cobon GS, Baldo BA (1989) J Exp Med 170:1457-1462

Trudinger M, Chua KY, Thomas WR (1991) Clin Exp Allergy 21:33-40

van der Zee JS, Van Swieten P, Jansen HM, Aalberse RC (1988) J Allergy Clin Immunol 81:884-889

van't Hof W, Driedijk C, Van den Berg M, Beck-Sickinger A, Jung G, Aalberse RC (1991) Mol Immunol 28:1225-1232

Yaseuda H, Mita H, Shida T, Ando T, Sugiyama S, Yamakawa H (1991) Allergy Clin Immunol News Suppl 1, Abstract 643 XIV ICACI Congress

Yssel H, Johnson KE, Schneider PV, Wideman J, Terr A, Kastelein R, De Vries JE (1992) J Immunol 148:738-745

Yuuki T, Okumura Y, Ando T, Yamakawa H, Suko M, Haida M, Okudairo H (1990) Jap J Allergy 39:557-561

Recent Advances in the Field of High Affinity IgE Receptors: the Connection to Signal Transduction

Martin Adamczewski, Rossella Paolini, Marie-Helene Jouvin, and Jean-Pierre Kinet

From the Molecular Allergy and Immunology Section, National Institute for Allergy and Infectious Diseases, National Institutes of Health, 12441 Parklawn Drive, Rockville, MD 20852, USA

INTRODUCTION

Mast cells and some other cell types bear on their surface receptors which bind antibodies of the immunoglobulin E class (IgE) with high affinity and specificity. The binding of IgE itself to the receptor does not activate mast cells, however, crosslinking of receptor-bound IgE by multimeric antigen leads to biochemical signals including activation of phospholipase C-γ1 (PLC-γ1), an increase in intracellular calcium levels and ultimately the release of mediators of allergic reactions such as histamine. The high affinity IgE-receptor (FcϵRI) is a tetramer of transmembrane proteins and consists of one IgE-binding α chain, one β chain and two identical disulfide-linked γ chains. In the previous volume of this series (Kinet et al. 1989) we have reviewed the structure, cloning, and expression of this receptor. In this presentation we will review recent advances of our knowledge of the connection of this receptor to signal transduction.

A FAMILY OF RECEPTORS

The review of FcϵRI in the previous volume of this series had noted the similarity of the FcϵRI γ subunit to the ζ subunit of the antigen receptor on T cells (TCR). Since then, it has become evident that the high affinity IgE receptor, together with Fc receptors for immunoglobulin G and the antigen receptors on T and B cells, forms a family of receptors defined by structural and functional similarities. The γ subunit of FcϵRI actually takes the place of TCR ζ in the antigen receptor in a subset of T cells and it is also found as part of one of the low affinity receptors for IgG (FcγRIII, CD16) on macrophages (Blank et al. 1989, Ra et al. 1989, Hibbs et al. 1989, Kurosaki and Ravetch 1989, Anderson et al. 1990, Lanier et al.1989, Orloff et al. 1990). In addition to the overall similarity between FcϵRI γ and TCR ζ there is a common motif (DxxYxxLxxxxxxxYxxL) found in the cytoplasmic domains of FcϵRI β and γ, CD3 γ, δ, ϵ, TCR ζ and η and in the IgM-associated proteins B29 and mb1 (Reth 1989). Truncation of the cytoplasmic domain of γ abolishes FcϵRI-mediated activation in a reconstituted system (Alber et al. 1991). In addition, chimeric molecules containing the cytoplasmic part of γ induce signaling upon aggregation in transfected T cell or basophil cell lines (Letourneur and Klausner 1991). These motifs therefore represent functionally relevant segments for coupling the receptors to other elements of the signal transduction chain.

ACTIVATION OF TYROSINE AND SERINE/THREONINE KINASES

Engagement of FcεRI with IgE and multivalent antigen leads to an immediate increase of phosphorylation of receptor β (on tyrosine and serine) and γ (on tyrosine and threonine) subunits. Therefore, at least two different types of kinases, a tyrosine and a serine/threonine kinase, are activated (Paolini et al. 1991). The src-related tyrosine kinases src, lyn (in the mast cell-like RBL cell line) and yes (in the mast cell-like PT-18 cell line) are activated upon receptor crosslinking and lyn and yes co-immunoprecipitate with the receptor (Eisemann and Bolen 1992). Kinase activation also leads to phosphorylation of tyrosine on substrates which are not detected by co-immunoprecipitation with the receptor. These include a prominent protein of molecular mass 72 kD with unknown function (Benhamou et al. 1990) and PLC-γ1 (Park et al. 1991), which is activated by this phosphorylation (Kim et al. 1991, Weiss et al. 1991, Mustelin et al. 1991, Nishibe et al 1990).

ACTIVATION OF TYROSINE AND SERINE/THREONINE PHOSPHATASES

The phosphorylation status of the receptor subunits reflects the balance of the competing actions of kinases and phosphatases upon them. In fact, FcεRI offers a unique opportunity to visualize phosphatase activity in vivo: when monomeric hapten is used to interrupt stimulation by polymeric antigen, the receptor and cellular proteins are immediately dephosphorylated by as yet uncharacterized phosphatases. All biochemical signaling reactions and the release of mediators are inhibited (Paolini et al. 1991)

The role of phosphatases in signal transduction is, however, not limited to returning the system to its non-activated state. To the contrary, the tyrosine-specific transmembrane phosphatase CD45 (leukocyte common antigen) is required for activating the system. Evidence for this comes from studying FcεRI-mediated signal transduction in Jurkat T cells transfected with all three subunits for the receptor and their CD45-deficient mutant counterparts. In Jurkat transfectants, FcεRI-mediated activation reactions such as phosphatidylinositol metabolism and calcium influx are faithfully reconstituted, but in the mutants they do not occur (M. Adamczewski, G.A. Koretzky, and J.-P. Kinet, unpublished observations).

We have further exploited the approach of only partially reconstituting the signal transduction system as in the case of the CD45 mutants to dissect the activation pathways coupled to the receptor. Transfection of FcεRI into non-hematopoietic cell lines adresses the question whether there are cell type-specific elements involved in signaling. CHO cells (a fibroblast-like cell line) express src, but none of the other src-related tyrosine kinases, and do not express CD45. When these cells are transfected with all three subunits of the receptor, and stimulated by IgE and antigen, there is no increase in either phosphatidylinositol metabolism or cytoplasmic free calcium. However, the receptor β and γ subunits are phosphorylated on serine/threonine and tyrosine much as in mast cells (M.-H. Jouvin and J.-P. Kinet, unpublished observations). This indicates that receptor phosphorylation does not demand a particular mast cell-specific or hematopoietic-specific element, although one or more subsequent activation steps apparently do.

These genetic approaches are complemented by the use of inhibitors of signal transduction. Phenylarsine oxide is a thiol-reactive chemical which in micromolar concentrations is a potent inhibitor of tyrosine-specific phosphatases including CD45, although at millimolar concentrations it inhibits numerous other enzymes. Phenylarsine oxide abolishes the receptor-mediated activating phosphorylation of PLC-γ1 and subsequent activation reactions including mediator release. However, it does not inhibit receptor phosphorylation and phosphorylation of cellular proteins detected in anti-phosphotyrosine blots (M. Adamczewski, R. Paolini, and J.-P. Kinet, J. Biol. Chem., in press). We interpret this as evidence for two distinct phosphorylation pathways: one that leads to phosphorylation of receptor subunits and most cellular substrates, and the other, either completely independent or branching off from the first, leading to phosphorylation of PLC-γ1.

RECEPTOR PHOSPHORYLATION/DEPHOSPHORYLATION: A COUPLING/UNCOUPLING MECHANISM

What is the role of receptor phosphorylation? The phosphorylation signal is restricted to activated receptors and is immediately reversible upon receptor disengagement. The rapidity and reversibility of the receptor phosphorylation signal suggested to us that it might represent a coupling/uncoupling mechanism. A detailed study of proteins associated with the receptor reveals at least five different polypeptides which become phosphorylated in sequence with the β and γ chains. Like β, pp180, pp48, pp42, and pp28 are phosphorylated on serine and tyrosine, while pp125 is only phosphorylated on serine. Furthermore, the physical association between pp125 and the receptor is quantitatively affected by receptor phosphorylation/dephosphorylation, thereby demonstrating

Figure 1: KINASES / PHOSPHATASES PATHWAYS

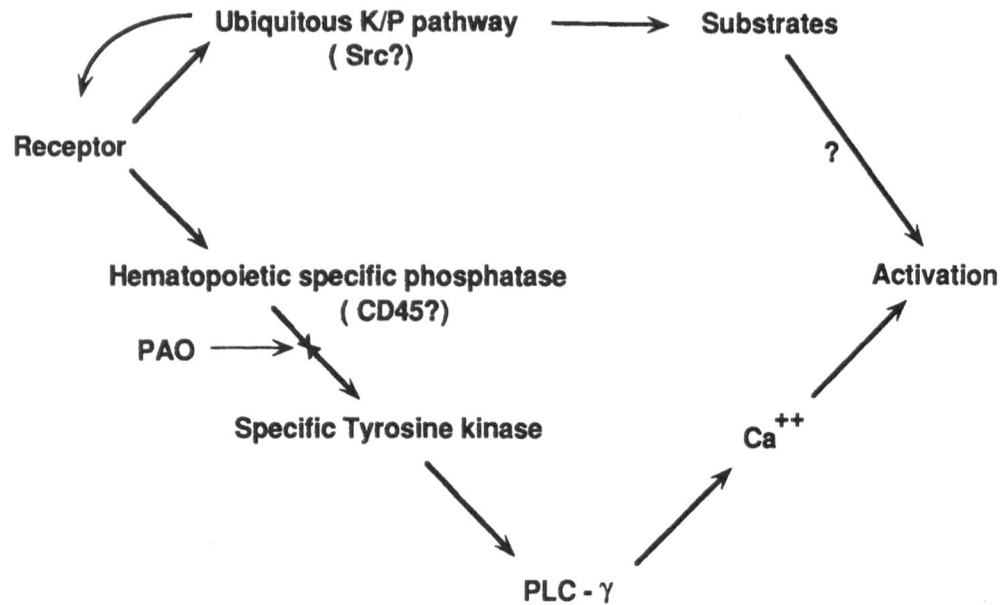

a coupling/uncoupling mechanism. (R. Paolini, R. Numerof, and J.-P. Kinet, Proc. Natl. Acad. Sci. (USA), in press).

A MODEL FOR FCεRI-MEDIATED MAST CELL ACTIVATION

Although we believe that we are still far from understanding signal transduction by FcεRI or related receptors completely, we wish to put forward a tentative model which accounts for the observations described above (Figure 1). Its salient features include the activation of kinases and phosphatases in two pathways, one hematopoietic-specific and possibly controlled by CD45, the other ubiquitous. The interaction of members of the pathways is not rigidly specific (as evidenced for example by the activation of different members of the src-family of kinases); rather, cell type specificity and appropriateness of the response are achieved by the interplay of multiple, partly specific links in the chain.

REFERENCES

Alber, G., Miller, L., Jelsema, C.L., Varin-Blank, N., and Metzger, H. (1991) J. Biol. Chem. 266, 22613-22620

Anderson, P., Caligiuri, M., O'Brien, C., Manley, T., Ritz, J., and Schlossman, S.F. (1990) Proc. Natl. Acad. Sci. (USA) 87, 2274-2278

Benhamou, M., Gutkind, J.S., Robbins, K.C., and Siraganian, R.P. (1990) Proc. Natl. Acad. Sci. (USA) 87, 5327-5330

Blank, U., Ra, C., Miller, L., Metzger, H., and Kinet, J.-P. (1989) Nature 337,187-189

Eiseman, E. and Bolen, J.B. (1992) Nature 355, 78-80

Hibbs, M.L., Selvaraj, P., Carpen, O, Springer, T.A., Küster, H., Jouvin, M.-H.E., and Kinet, J.-P. (1989) Science 246, 1608-1611

Kim, H.K., Kim, JW., Zilberstein, A., Margolis, B., Kim, J.G., Schlessinger, J., and Rhee,S.G. (1991) Cell 65, 435-441

Kinet, J.-P., Metzger, H., Alber, G., Blank, U., Huppi, K., Jouvin, M.-H., Küster, H., Miller, L., Ra, C., Rivera, J., and Varin-Blank, N. (1989), in: Progress in Immunology VII (Melchers, F., ed.), Springer, Berlin, pp.716-723

Kurosaki, T., and Ravetch, J.V. (1989) Nature 342, 805-807

Lanier, L.L., Yu, G., and Phillips, J.H. (1989) Nature 342, 803-804

Letourneur, F., and Klausner, R.D. (1991) Proc. Natl. Acad. Sci. (USA) 88, 8905-8909

Mustelin, T., Coggeshall, K.M., Isakov, N., and Altman, A. (1990) Science 247, 1584-1587

Nishibe, S., Wahl, M.I., Hernandez-Sotomayor, S.M.T., Tonks, N.K., Rhee, S.G., and Carpenter, G. (1990) Science 250, 1253-1256

Orloff, D.G., Ra, C., Frank, S.J., Klausner, R.D., and Kinet, J.-P. (1990) Nature 347, 189-191

Paolini, R., Jouvin, M.-H.E., and Kinet, J.-P. (1991) Nature 353, 855-858

Park, D.J., Min, H.K., and Rhee, S.G. (1991) J. Biol. Chem 266, 24237-24240

Ra, C., Jouvin, M.-H.E., Blank, U., and Kinet, J.-P. (1989) Nature 341, 752-754

Reth, M. (1989) Nature 338, 383-384

Weiss, A., Koretzky, G., Schatzman, R.C., and Kadlecek, T. (1991) Proc. Natl. Acad. Sci. (USA) 88, 5484-5488

9. Fc Receptors and Ig
Binding Factors

The Regulation and Functional Significance of Fc Receptors on Murine Lymphoid Cells: The Special Case of T Cells

R. G. Lynch, M. Sandor, R. Nunez, M. Hagen, P. Teeraratkul, N. Noben and R. Sacco

Department of Pathology, University of Iowa College of Medicine, Iowa City, Iowa 52242

INTRODUCTION

Fc receptors (FcR) specific for each of the five major classes of murine Ig heavy chains occur on the surface membranes of all the major lineages of murine lymphoid cells. The number of different classes of FcR expressed, their concentration and level of functional activity, and the developmental window in ontogeny during which the FcR are expressed vary considerably amongst the subsets of lymphoid cells. The regulation of FcR expression on lymphoid cells is a multi-factorial process that involves pre-programmed genetic events and local environmental factors (Lynch and Sandor, 1990).

On normal B-lineage lymphoid cells FcR are constitutively displayed and each FcR class has a characteristic developmental window during which the receptor is present (Lynch and Sandor, 1990). Once expressed, each class of FcR can be up- and down-regulated by interleukins, cytokines, immunoglobulin ligands, and cellular activation. During B-cell development the only stage at which all five classes of FcR are simultaneously expressed is the mature, resting, virgin, sIgM+/sIgD+ B-lymphocyte. In contrast, on normal murine T-lineage cells the five major classes of FcR are not constitutively expressed, but they are inducible. The major inducer of FcR expression on T-lymphocytes is cellular activation via the TCR complex (Sandor et al, 1990, 1992). It is interesting that on B-lineage cells all classes of FcR are prominently displayed on the resting immunocompetent cell prior to activation, while on T-lineage cells the FcR are expressed only after the resting immunocompetent cell has been activated by antigen.

RELATIONSHIPS BETWEEN FcR AND TCR

There is a sizable literature on the occurrence of FcR on T cells. Most reports of FcR on T cells have involved studies of activated T cells or of T cells, such as transformed or lymphoma cells, that exhibit some of the characteristics of activated T cells. Recently, Sandor et al (1990; 1992) showed that activation of cloned, antigen-specific T cells via their TCR complex induced the expression of FcR. In those studies it was shown that all five classes of FcR could be induced on some CD4+ T cell clones, but typically only two or three classes of FcR were induced on a given clone. In repeated cycles of activation/rest/re-activation, FcR were induced upon activation, disappeared during rest, and re-appeared upon re-activation. Those studies identified the close linkage of FcR expression and cellular activation via the TCR complex, and showed that FcR induction occurred early in the

G1 phase of the cell cycle in response to anti-CD3 antibody, anti-TCR clonotypic antibody, or specific antigen presented by MHC-matched accessory cells. In vivo confirmation of the linkage of FcR expression to TCR-mediated cellular activation was obtained in mice with Schistosomiasis where FcR were present on the activated T cells within hepatic granulomas but not on the resting T cells in other tissues. The linkage of FcR expression and cellular activation via the TCR complex has been shown for both alpha/beta and gamma/delta T cells (Sandor et al 1990; 1992).

In addition to the conspicuous functional linkage of FcR and TCR, recent studies have identified several interesting structural and physiological relationships between FcR and TCR. These include the findings that: a) TCR and, with one exception, FcR are members of the Ig superfamily; b) TCR and some FcR use an identical molecular subunit (Ravetch and Kinet, 1991); c) the zeta chain signal-transducing subunit of the TCR complex is structurally related to the gamma chain signal-transducing subunit of some FcR and these subunits can interchange (Lanier et al., 1991); d) TCR and FcR associate in situ with src-family and other types of protein kinases (Sugie et al., 1992); e) TCR and FcR co-migrate in the surface membrane of activated T cells (unpublished data); f) TCR and FcR use similar signal transduction pathways; and g) some classes of FcR are expressed on thymocytes during fetal development (unpublished observations).

The nature and multiplicity of the associations between TCR and FcR provides a possible explanation for the very restricted period of time during which FcR are expressed on T cells. Compared to other lymphoid cells such as B-cells, NK-cells and macrophages, it is as if T-cells avoid expressing FcR. The sharing of molecular structural components and signalling pathways by TCR and FcR might require the high stringency that characterizes FcR expression on T-cells in order to avoid inappropriate activation of the T-cells. The avoidan of FcR expression by resting T-cells could be viewed as an adaptation that prevents T-cell activation by a mechanism that would bypass the TCR and MHC restriction. If FcR were expressed on resting T-cells, antigen-antibody complexes conceivably might generate a T-cell activation signal because of the shared structural components and signalling pathways of FcR and TCR. If resting T cells did express FcR, then antigen:antibody complexes might possibly trigger a non-specific, non-MHC- restricted, polyclonal activation of T cells. However, since FcR appear on T-cells after they have been activated by antigen, antigen:antibody complexes, if present, would bind only to the activated T-cells and any regulatory effect that resulted from the binding of the complexes would appear to be antigen-specific.

In retrospect, much of the difficulty and controversy in the literature over the years about the occurrence of FcR on T-cells can now be accounted for by three special characteristics of FcR on T-cells: 1) FcR are expressed on T cells during a very restricted period of time following activation via the TCR complex and with the single, interesting exception discussed below, FcR have not been found on normal, resting murine T-cells; 2) FcR on T-cells are present at much lower concentrations than FcR on other cells such as B-cells; and 3) there is considerable variation in expression of the different classes of FcR on different subsets of T-cells.

FUNCTIONAL SIGNIFICANCE OF FcR ON T CELLS

Previous studies from several laboratories suggested that the FcR on T-cells, or released from T-cells, participate in a number of normal and pathologic immune functions. Most of these will not be considered here because recent reviews have covered the functions proposed for FcR on T-cells (Lynch and Sandor, 1990) and for the soluble forms of FcR (Fridman and Sautes, 1990).

A number of investigators have proposed that, as was previously found for B-cells, a major functional significance of FcR on T-cells is immunoregulatory. It is well-established that the IgG-FcR (CD32) on B-cells can transduce inhibitory signals that block B-cell activation, and recent studies (Waldschmidt and Tygrett, 1991) have shown that the IgE-FcR (CD23) on B-cells can transduce activation signals. The latter finding is of particular interest since CD23 has recently been shown by Sugie and colleagues (1992) to physically associate with p59fyn, a src-family tyrosine kinase. Although there is no direct proof that the FcR on T cells transduce regulatory signals, several observations make this an interesting possibility. We have recently found that the IgM-FcR on CD4+/Th2 cells can bind to the sIgM on B-cells and contribute to the formation of antigen-specific, MHC-restricted T:B cytoconjugates (Teeraratkul and Lynch, 1991). Interestingly, the IgM-FcR on the T-cell and the sIgM on the B-cell migrate in concert to form an intense zone of complimentary molecules located at the point of contact between the cells (Sandor et al, in press). These findings establish that the IgM-FcR on CD4+ T-cells and the sIgM on B-cells comprise a functional receptor:ligand pair with the properties of accessory adhesion molecules. In view of these results it becomes particularly interesting that in the studies of FcR expression on CD4+ T-cells by Sandor et al (1990), it was the Th2 subset, but not the Th1 subset, that expressed IgM-FcR upon activation via the TCR complex. Since CD4+/Th2 cells are the classic helper T-cells for antibody responses, the activation-linked expression of IgM FcR on these cells provides a molecular mechanism for the helper T cell to recognize that the conjugate B-cell expresses sIgM. Conceivably, this recognition event could result in a signal that modulates cytokine production by the T-cell. It is of additional interest that in the studies of Sandor et al (1990), IgD-FcR were almost always co-induced with IgM-FcR on the Th2 cells, in principle providing a mechanism by which Th2 cells could discriminate between sIgM+ and sIgM+/sIgD+ B-cells.

The expression of FcR on regulatory T cells is not restricted to the CD4+ subset. In a sequence of studies (reviewed by Lynch, 1987) we detected large numbers of FcR+/CD8+ T cells in mice bearing plasmacytomas and hybridomas. In each instance, the specificity of the FcR on the CD8+ T cell matched the Ig-heavy chain class of the monoclonal Ig produced by the tumor. The presence of a plasmacytoma or hybridoma appears to activate host CD8+ T cells resulting in the activation-linked induction of multiple classes of FcR. Several lines of evidence indicate that the high circulating concentration of monoclonal Ig maintains the expression of the corresponding FcR on the CD8+ T cells by ligand-induced up-regulation, while the other classes of FcR initially displayed on the activated T cells are only transiently expressed. The nature of the stimulus that activates the T-cells, whether it results in a polyclonal activation or a more restricted T cell response, and the basis for the dominance of the CD8+ subset of T cells are interesting issues that remain to be addressed.

An FcR-specific immunoregulatory function for the CD8+ T-cells in mice with Ig-secreting tumors is suggested by the investigations of Mathur et al (1990) and Nelms et al (1991). In those studies mice bearing IgE-secreting hybridomas were found to develop large numbers of CD8+ T cells that express IgE-FcR and which participate in a multicellular mechanism that down-regulates IgE synthesis in the hybridoma cells in vivo and in vitro. Analysis of the mechanism of inhibition of IgE production showed that it was effected at the level of epsilon-heavy chain gene transcription by a trans-acting element that blocked the activity of the Ig-heavy chain intronic enhancer (Nelms et al, 1991). While additional studies are necessary to identify and characterize all of the cellular and molecular events that lead to the specific turn-off of IgE production in this model system, we have established that the inhibition of IgE production in the hybridoma cells requires that the CD8+ T cells that express IgE-FcR be allowed to physically contact the target hybridoma cells.

EVIDENCE FOR ADDITIONAL LEVELS OF CONTROL IN THE REGULATION OF IgE-FcR EXPRESSION IN MURINE LYMPHOCYTES

As discussed above, the different classes of FcR are constitutively expressed on B-lineage cells and each class of FcR has a characteristic developmental window during which the receptor is present. This statement is based on studies that assessed FcR expression on the surface of the B-cell. In the case of the IgE-FcR (CD23) studies of normal B-cells in mice (Waldschmidt et al., 1988) and humans (Kikutani et al., 1986) have established that CD23 is expressed only on mature, immunocompetent B-lymphocytes and is not expressed on B-cells that have undergone heavy chain class switch. Recent findings raise the possibility that CD23 may be expressed at other times during B-cell ontogeny. We examined a large panel of murine B-lineage lymphoma clones for the presence of CD23 transcripts and found that many clones that did not express cell surface CD23, nonetheless, contained CD23 transcripts. The transcripts were present in many post-switched B-cell clones and at almost comparable levels in clones that expressed and did not express the CD23 protein. Nucleotide sequence analysis of PCR-amplified cDNA showed that the structure of the CD23 transcripts was identical in all of the clones. These findings raise the possibility that there may be circumstances in which CD23 is expressed in B-lineage cells at times other than the stage of the resting, mature, immunocompetent B-lymphocyte.

In the introduction section it was stated that FcR are not constitutively expressed on T-cells; there is at least one interesting exception to that generalization. We have found that the epidermal dendritic gamma/delta T-cells of murine skin constitutively express IgE-FcR (CD23) and IgG-FcR (CD16). When these cells are activated via the TCR complex, CD23 and to a lesser extent CD16 are down-regulated, and IgA-FcR and IgM-FcR are induced. This exception to the rule that FcR are not expressed on resting T-cells could be related to the unconventional usage of a monomorphic, unrestricted TCR by this subset of T-cells. One implication of the constitutive expression of CD23 by cutaneous T-cells is the possibility that T-cell activation via CD23 might occur in the presence of antigen and specific IgE antibodies. If this occurred, it would be an example of a mechanism in which antibodies triggered T-cell effector functions while bypassing the TCR and MHC restriction.

In other studies we have found alterations in the expression of CD23 in a number of pathologic conditions in which there is a state of polyclonal B-cell dysfunction. In mice with plasmacytomas we have found that the growth of the tumor is accompanied by a marked decrease in the expression of CD23 on host B-lymphocytes and that this appears to be mediated by TGF-beta (Berg and Lynch, 1991). Plasmacytomas in mice and multiple myeloma in humans are accompanied by a severe, polyclonal B-cell immunodeficiency syndrome (Jacobson and Zolla-Pazner, 1986). We have also found that the B-lymphocytes in mice infected with Trypanosoma brucei show a marked reduction in the expression of CD23 (R. Sacco, J. Donelson, R.G. Lynch, unpublished data) and that CD23 expression is blocked on normal murine B-lymphocytes that are cultured in vitro with the promastigote form of Leishmania donovani (Noben et al. 1991). Both of these parasitic infections are accompanied by polyclonal B-cell dysfunction. While the physiological function of CD23 on normal B-lymphocytes is still unproven, it has been proposed that CD23 is involved in some aspect of the regulation of cell cycle progression (Cairns and Gordon, 1990). It is interesting that in each of the three pathologic states characterized by polyclonal B-cell dysfunction that we have examined, CD23 expression by B-lymphocytes has been markedly reduced or absent. Further investigations of these conditions may provide insight into the physiological function of CD23 on normal B-lymphocytes.

EVIDENCE FOR MOLECULAR ISOFORMS OF THE IgE-FcR IN MURINE LYMPHOCYTES

In recent studies (R. Nunez and R.G. Lynch, submitted for publication) we searched a panel of murine T- and B-lymphoid cells for evidence of CD23 transcripts using RT-PCR and nested PCR. DNA sequence analyses of the cloned PCR products showed that CD23 transcripts were present in several cloned T- and B-lymphoma cell lines. We found that some murine T-cells, as well as B-cells, contained the full-length CD23 transcript originally isolated from murine B-cells and sequenced by Bettler and colleagues (1989). In addition we isolated from several T- and B-cell lymphomas a novel truncated transcript which was shown by DNA sequence analysis to have an in-frame deletion of the third exon of CD23. The third exon of the murine CD23 gene encodes the entire transmembrane segment and all of the cytoplasmic domain except for the first six amino acids which are encoded by the second exon. The truncated CD23 transcripts lacking the third exon sequences were reproducibly isolated from several T- and B-cell lymphomas, and from some antigen-specific T-cell clones and from thymocytes. The new splice junction between the second and fourth exons changes the histidine at residue 48 to glutamic acid, but except for that change this transcript encodes a form of CD23 whose predicted amino acid sequence is identical to the published sequence of murine CD23 but lacking the entire transmembrane segment and residues 7 through 22 of the cytoplasmic domain. Whether this truncated CD23 transcript is translated in the T- and B-cells remains to be determined.

In recent studies done in collaboration with J. Yodoi and M. Matsui (unpublished observations) we have identified truncated, CD23 transcripts in human T- and B-cell clones. As we initially found in the mouse, the truncated human CD23 transcripts contain in-frame deletions of the third exon, and the new splice junction between the second and fourth exons changes histidine to glutamic acid at the first amino acid of the extracellular domain. Interestingly, DNA sequence analyses identified truncated CD23 transcripts for both the alpha and beta isoforms of human CD23 originally described by Yokota and colleagues (1988). By DNA sequence analysis the truncated transcripts we isolated appear to be intact transcripts; whether they are translated in human T- and B-cells remains to be determined.

SUMMARY

During the past decade there has been a growing interest in the biology and pathology of lymphocyte Fc receptors. The development of new knowledge has been catalyzed by the application of molecular biological approaches to the investigation of Fc receptor structure and function. There is selectivity in the pattern of expression of different classes of FcR on different lymphoid cell populations. The expression of FcR on lymphocytes is highly regulated and the receptors appear to participate in a multiplicity of immune functions. The occurrence of alternative isoforms of FcR provides a molecular basis for multiple functions. The expression of FcR on activated T-cells may reflect a fundamental regulatory event that transiently provides a mechanism to link the antigen-recognizing molecules of classic humoral immunity with the antigen-recognizing cells of classic cellular immunity. Viewed from this perspective the linkage of T-cell FcR expression to cognate-driven activation of T-cells becomes a tightly regulated mechanism where, for brief and precisely determined periods of time, the Fc receptor functions as a molecular adapter that interfaces T-cell immunity with B-cell immunity.

REFERENCES

Berg DJ, Lynch RG (1991) J Immunol 146: 285-2872

Bettler B, Hofstetter H, Rao M, Yokoyama WM, Kilchherr R, Conrad DH (1989) Proc Natl Acad Sci USA 86:7566-7570

Cairns JA, Gordon J (1990) Eur J Immunol 20:539-543

Fridman W, Sautes C (1990) Immunoglobulin Binding Factors. IN: Metzger H (ed) Fc Recpetors and the Action of Antibodies, American Society for Microbiology, Washington, DC, p. 335

Jacobson DR, Zolla-Pazner S (1986) Semin Oncol 13:282

Kikutani H, Suemura M, Owaki H, Nakamura H, Sato R, Yamasaki K, Barsumian EL, Hardy RR, Kishimoto T (1986) J Exp Med 164:1455-1469

Lanier LL, Yu G, Phillips JH (1991) J. Immunol 146:1571-1576

Lynch RG (1987) Adv Immunol 40:135-151

Lynch RG, Sandor M (1990) Fc Receptors on T and B Lymphocytes. IN: Metzger H (ed) Fc Receptors and the Action of Antibodies, American Society for Microbiology, Washington, DC, p. 305.

Mathur A, Van Ness BG, Lynch RG (1990) J Immunol 145:3610-3617

Nelms K, Van Ness BG, Lynch RG, Mathur A (1991) Molec Immunol 28:599-606

Noben N, Wilson M, Lynch RG (1991) FASEB J 5:A1002

Ravetch JV, Kinet J-P (1991) Ann Rev Immunol 9:457-92

Sandor M, Gajewski T, Thorson J, Kemp JD, Fitch FW, Lynch RG (1990) J Exp Med 171:2171-2176

Sandor M, Houlden B, Bluestone J, Hedrick SM, Weinstock J, Lynch RG (1992) J Immunol 148:2363-2369

Sugie K, Kawakami T, Maeda Y, Kawabe T, Uchida A, Yodoi J (1992) FASEB J 6:A1011

Teeraratkul P, Lynch RG (1991) FASEB J 5:A1464

Waldschmidt T, Tygrett L (1991) FASEB J 5:A610

Waldschmidt TH, Conrad DH, Lynch RG (1988) J Immunol 140:2148-2154

Yokota A, Kikutani H, Tanaka T, Sato R, Barsumian EL, Suemura M, Kishimoto T (1988) Cell 55:611-618

Identification of Functionally Active Regions of FcγRII and Fcε RI

P. M. Hogarth, M. D.Hulett, E. Witort, A. J. Quilliam,
I. F. C.McKenzie, F. Ierino, M. S. Powell and R. Brinkworth

Schutt Laboratory for Immunology, Austin Research Institute, Heidelberg 3084,
Australia and Centre for Drug Design, University of Queensland

INTRODUCTION

The binding of antigen to antibody sensitized IgE Fc receptors (FcεR) or of immune complexes to low affinity IgG Fc receptors (FcγR) results in biologically important sequelae including the degranulation of mast cells and release of mediators; the activation of respiratory burst in neutrophils, the introduction of macromolecules into antigen presentation pathways in macrophages and the attenuation of B cell responses (reviewed in Metzger et al 1986, Unkeless et al 1988, Van de Winkel and Anderson 1991). Yet despite much analysis of FcR little is known about the nature of the regions with FcRs that are responsible for the interaction with Ig.

Since high affinity IgE receptors, FcεRI, and low affinity IgG receptors, FcγRII, play such important biological roles we have attempted to define the Ig interactive regions of these receptors and to use recombinant receptors to modify these biological responses.

APPROACHES FOR THE DEFINITION OF Ig INTERACTIVE REGIONS OF FcR

Several different strategies can be used to define Ig binding sites of Fc receptors.

Monoclonal antibodies that inhibit function or identification of genetic polymorphisms that influence Ig binding are useful guides in the search for Ig binding regions. A knowledge of the epitopes recognized by the Mab or the biochemical basis of the polymorphisms will implicate these regions in the active areas of the FcR (e.g. the Ly-17 polymorphism of FcγRII (Lah et al 1990). However, such studies imply but do not prove a role in function but do not prove it.

Definitive evidence comes from studies that directly identify Ig interactive regions. The use of dimeric receptors wherein regions of structurally related but functionally distinct receptors are exchanged (e.g. FcγRII and FcεRI) is a very powerful tool for analyzing the effect on function and indeed we have successfully used this approach (Hulett et al 1991, Hogarth et al 1992).

INDIRECT STRATEGIES: MONOCLONAL ANTIBODIES AND GENETIC POLYMORPHISMS.

In the mouse the Ly-17 alloantigen system defines a polymorphism of FcγRII and antibodies detecting the Ly-17.1 or Ly-17.2 specificities block Fc binding (Hibbs et al 1985). Sequence analysis of Ly-17.1 and Ly-17.2 proteins indicated that amino acids 116 and 161 were involved in the Ly-17 epitope and it is a possibility that they are within or near the Ig binding site (Lah et al 1990). It should be noted that they are located in the FcγRII second domain.

A polymorphism of human FcγRIIa (monocyte specific form) has been identified by differences in the capacity of FcγRIIa allomorphs to bind to mouse IgG or human IgG2 (Abo et al 1984, Tax et al 1984). Allomorphs bearing arginine at position 131 bind mIgG1 but fail to bind human IgG2 whereas those containing histidine at 131 have markedly reduced but not ablated IgG1 binding but avid mIgG2 binding (Warmerdam et al 1990, Tate et al 1992). Position 131 is located in FcγRII second domain. The use of chimeric FcR has also allowed us to map the epitope detected by several blocking anti-FcγRII antibodies to the second domain. The antibodies IV-3 (Rosenfeld et al 1985) and 8.26 (Ierino unpublished) bound to chimeric FcR containing D2 of FcγRII and D1 of FcεRI. These antibodies did not bind to FcεRI or chimeric FcR containing FcγRII domain I with FcεRI domain II (Table 1).

Studies of IgE binding to FcεRI using these approaches are less well developed. However, Riske et al (1992) have produced an anti-peptide antibody that inhibits IgE binding. This antibody recognizes a sequence that is present in the second domain which is consistent with the results of the analysis of FcγRII. Clearly the evidence described above strongly implicates FcγRII second domain as a major IgG binding region.

Table 1   Binding of Immune Complexes and MAb to Chimaeric FcR

Chimaeric FcR* (D1, D2)	Immune Complex Binding**		MAb (8.26) Binding+
	IgG	IgE	
(diagram)	+	−	+
(diagram)	−	+	−
(diagram)	+	+	NT
(diagram)	−	+	NT
(diagram)	+	+	NT
(diagram)	+	+	NT

Positions marked: 1, 86, 128, 145, 169

* ▨ FcγRII derived sequence; ☐ FcεRI derived sequence
** Binding determined using mouse IgG1 or IgE sensitized erythrocytes or human IgG1 dimers.
\+ Binding of MAb 8.26 determined by immunofluorescence

DIRECT STRATEGIES: CHIMERIC FcR AND POINT MUTATIONS.

FcγRII and FcεRI are homologous receptors but have very distinct binding characteristics, i.e. FcγRII has a low Ka for IgG and binds only IgG complexes whereas FcεRI binds IgE with Ka=10^{-10} but not IgG (Metzger et al 1986, Unkeless 1988, Hibbs et al 1988, Kinet and Metzger 1990). Thus, the construction of chimeric receptors and their transfection into cos cells enabled the direct localization of Ig interactive regions (Hogarth et al 1992). Using this strategy our experiments showed IgG bound to chimeras that contain FcγRII D2 and binding could be localized to sequences within the segment 146 - 169. This was clearly demonstrated with the placement of these residues in FcεRI which then bound IgG and when the corresponding sequence of FcεRI was used to replace only those residues in FcγRII, this resulting in the ablation of IgG binding (Table 1) (Hogarth et al 1992). However, IgE binding whilst also being localized to FcεRI D2 could be further localized to 3 regions in FcεRI D2 (Table 1).

Clearly the evidence indicates that D2 of FcγRII and FcεRI have major Ig binding sites. Point mutagenesis of residues within the Ig binding regions of FcγRII D2 have allowed a more precise localization of the Ig binding regions. These studies indicate that residues 155, 156, 157 and 161 play substantial roles in the interaction with IgG. The use of chimeric receptors directly identified Ig binding regions and localized them to D2 which is consistent with data from MAb and genetic studies (Table 2).

THREE DIMENSIONAL VISUALIZATION OF D2

The above studies indicate that residues from seemingly distant parts of FcγRII are involved in binding. Presumably the folding of D2 is such that these regions are brought together in the mature receptor. Indeed our published model of FcγRII and FcεRI (based on the solved CD-4 D2 structure) predicts this to be the case (Hogarth et al 1992). Thus, the polymorphic Ly-17 amino acids would be juxtaposed and the human equivalent of the mouse Ly-17 polymorphism, residue 159,would be present in the IgG binding amino acids 147-169. Residue 131 would be found in the C'/E loop and be present on the same face as residue 147-169

Table 2   Summary of Ig Binding Analysis[1]

Techniques		Region identified/implicated
(i) MAb blocking/binding		
hu FcγRII (chimeric FcR)	D2	
hu FcεRI (anti-peptide Abs)	D2	
(ii) Polymorphisms		
mFcγRII (Ly-17)	D2	residues 116, 161
huFcγRII (HR/LR)	D2	residue 131
(iii) Chimeric FcR (FcγRII/FcεRI)		
FcγRII - IgG binding	D2	residues 147-161
FcεRI - IgE binding	D2	residues 87 -128
		129-145
		146-169

[1] See text for references

RECOMBINANT SOLUBLE FcR INHIBITS THE ARTHUS REACTION.

A knowledge of the Ig binding sites of FcR may be helpful in the manipulation and use of recombinant FcR proteins in human disease. Indeed we have now demonstrated that soluble recombinant FcγR has potential use in the treatment of immune complex mediated hypersensitivities. Using a reverse passive model of the Arthus reaction in rats the co-administration of recombinant soluble FcR profoundly inhibited the induction of the arthus reaction with greater than 84% reduction in the inflamed area compared to buffer or irrelevent protein treated controls.

CONCLUSION

As the analysis of FcγR continues, the elucidation of the Ig binding sites is an important step in our understanding of the very earliest steps in the Ig/FcR induced cellular activation. Such information is also essential in the development of effective FcR therapy of disorders where FcR or Fc regions have a major pathophysiologic role, e.g. immune complex disorders or type I hypersensitivity.

REFERENCES

Abo T (1984) J Exp Med 160:303-309
Hibbs ML, Hogarth PM, McKenzie IFC (1985) Immunogenetics 22:335-348
Hibbs ML, Bonadonna L, Scott BM, McKenzie IFC, Hogarth PM (1988) Proc Natl Acad Sci 85:2240-2244
Hulett MD, Osman N, McKenzie IFC, Hogarth PM (1991) J Immunol 147:1863-1868
Lah M, Welch K, Deacon N, McKenzie IFC, Hogarth PM (1990) Immunogenetics 31:202-206
Metzger H, Alcaraz G, Holman R, Kinet J-P, Pribluda V, Quarto R (1986) Ann Rev Immunol 4:419-470
Riske F, Hakimi J, Mallamaci M, Griffin M, Pilson B, Tobkes N, Lin P, Danho W, Kochan J, Chizzonite R (1991) J Biol Chem 266:11245-11251
Rosenfeld SI, Looney RJ, Leddy JP, Phipps PC, Abraham GN, Anderson CL (1985) J Clin Invest 76:2317-2322
Tate BJ, Witort E, McKenzie IFC, Hogarth PM (1992) Immunol and Cell Biol in press
Tax WJM, Hermes FFM, Willems HW, Capel PJA, Koene EAP (1984) J Immunol 133:1185-1189
Unkeless JC, Scigliano E, Freedman VH (1988) Ann Rev Immunol 6:251-282
Van de Winkel JGJ, Anderson CL (1991) J Leuk Biol 49:511-524
Warmerdan PA, Van de Winkel JGJ, Grosselin EJ, Capel PJA (1990) J Exp Med 172:19-25

Structures Involved in the Activities of Murine Low Affinity Fc γ Receptors (FcγR)

C. Sautès, C. Bonnerot, S. Amigorena, J.L. Teillaud, M. Daëron and W. H. Fridman

Laboratoire d'immunologie cellulaire et clinique, INSERM U255, Institut Curie, 26 rue d'Ulm, 75231 Paris Cedex 05

INTRODUCTION

Low affinity receptors for the Fc portion of IgG (Fcγ Receptors FcγR) are an heterogeneous group of membrane glycoproteins which bind immune complexes and are expressed on most hematopoietic cells (Ravetch and Kinet, 1991). Two classes of low affinity FcγR, FcγRII and FcγRIII have common structural characteristics. In mouse, both possess nearly identical extracellular (EC) regions composed of two Ig-like domains. Soluble forms have been described for both receptors (Fridman, 1992), with antigenic determinants and IgG-binding specificities similar to those of the EC regions of the receptors. However, FcγRII and FcγRIII differ in important properties: 1) FcγRII is made of a single chain, whereas FcγRIII is a multimeric receptor composed of three chains, an α ligand-binding subunit, and a homodimer of γ chains, that is also part of the high affinity receptor for the Fc portion of IgE (FcεRI) (Ra, 1989), 2) the transmembrane (TM) and intracytoplasmic (IC) regions of FcγRII and of the α chain of FcγRIII display no structural homology, 3) different isoforms of FcγRII (IIb1 and IIb2) generated by alternative splicing exist, but not of FcγRIII. They differ by an insert of 47 aminoacids in the IC region of FcγRIIb1, and 4) FcγRII and FcγRIII have a distinct cellular distribution: FcγRII (b1 isoform) is found preferentially on B lymphocytes, NK cells express exclusively FcγRIII, and other hematopoietic cell types bear both FcγRII (b1 and b2) and FcγRIII.

The low affinity FcγR mediate important effector and regulatory functions, that depend on the cell type on which the receptors are present. The crosslinking of surface immunoglobulin (sIg) with FcγR blocks B cell activation. On macrophages, FcγR enable phagocytosis of IgG-coated particles, endocytosis of immune complexes, killing of IgG-sensitized target cells, a property shared by NK cells, and augment antigen presentation. In mast cells, the triggering of FcγR by immune complexes induces degranulation, the release of preformed

---

1.This work was performed with the collaboration of: A. Astier, C. Bouchard, A. Galinha, D. Lankar, S. Latour, A. Lynch, O. Malbec, M.A. Marloie, V. Mateo, E. Tartour, and N. Varin (INSERM U.255), R. Spagnoli and V. Mazieres (Roussel-Uclaf, Romainville, France), D. Choquet (Institut Pasteur, Paris, France), H. and C. de la Salle, D. Hanau (CRTS, Strasbourg, France), J. Drake, W. Hunziker and I. Mellman (Yale University, New Haven, USA), P.M. Hogarth (University of Melbourne, Australia) and with financial supports of INSERM, Institut Curie and Institut Scientifique Roussel.

mediators and cytokine production. In NK cells, cytokine production is induced after FcγR-crosslinking.

In the work described herein, we have addressed the question of the structural bases of these diverse FcγR functions. Since most cell types express both FcγRII and FcγRIII, we have used cells expressing recombinant FcγR. Cell lines of B lymphocyte, basophilic or fibroblastic origin were transfected with cDNA encoding wild type or mutated FcγRII and FcγRIII. The ability of recombinant receptors to mediate FcγR effector and regulatory functions has been investigated and the structures involved in these different activities have been mapped.

FcγRII FUNCTIONS

The IIA1.6 cell line is a somatic variant of the A20/2J B lymphoma that has a deletion in the 5' part of the FcγRII gene and does not express FcγR. This cell line bears membrane IgG2a and responds to anti-Ig mediated signaling by protein tyrosine kinase (PTK) activation, rise of intracellular Ca^{++} and IL-2 production. Different cDNA encoding wild type FcγRIIb1 or FcγRIIb2 or deletion mutants of these receptors were transfected by electroporation in IIA1.6 cells. Individual cell lines expressing permanently comparable levels of wild type or mutant FcγR ($1-4 \times 10^5$ receptors per cell) were then selected and tested for FcγR functions. FcγRIIb2 has an IC region of 47 aminoacids encoded by the IC2 and IC3 exons and FcγRIIb1 an additional insert of 47 aminoacids encoded by the IC1 exon. Mutants were prepared by deleting part of the IC region of FcγRIIb2 ("CT18" and "CT31", with IC regions of 18 and 31 aminoacids respectively) or of FcγRIIb1 ("CT53", with a 53 aminoacid-long IC region lacking b2-specific sequences) or with a complete deletion of IC sequences ("CT1") (Fig. 1).

Crosslinking of sIg and FcγR blocks B cell activation via sIg. Fab'2 fragments of anti-sIg, but not entire anti-sIg antibodies, stimulate B cells. Transfected cell lines were incubated with intact or F(ab')2 fragments of rabbit anti-mouse IgG and activation was assessed by the rise of intracellular Ca^{++} in single cells or by the production of IL-2. The results of the two assays were similar (Fig. 1). Whereas F(ab')2 fragments of anti-mouse Ig elicited B cell activation in FcγRIIb1 and FcγRIIb2 transfectants, the corresponding IgG failed to do so in both cell lines. Inhibition of B cell activation was next assessed on IIA1.6 cells transfected with the various FcγRII deletion mutants. Upon incubation with anti-mouse IgG2a antibodies, activation was inhibited on cells expressing the FcγRIIb2 CT31 deletant. In contrast, cells expressing FcγRIIb2 CT18 were stimulated by intact anti Ig antibodies as were those transfected with CT1 and FcγRIIb1 CT53. Thus, the region responsible for inhibition of B cell activation was localized in a 13 aminoacid-long sequence, between aminoacids 19 and 31 of the IC region of FcγRIIb2 and the b1 specific region is not necessary for this function (Amigorena, 1992a).

The ability of FcγR to form caps at one pole of the cell was next analyzed. After incubation at 37°C with the monoclonal anti-FcγR mAb 2.4G2 followed by anti-rat Ig, FcγRIIb1 transfectants exhibited typical cap structures whereas the FcγRIIb2 ones failed to do so, but rather internalized the fluorescent complexes. Efficient capping was correlated with the presence of the FcγRIIb1 specific insertion since CT53, but not CT1, mutants formed caps upon crosslinking of FcγR (Fig.1). FcγRIIb2 internalized their ligand and may thus be involved in presentation of antigens complexed with IgG antibodies as well as in endocytosis. B cell transfectants were examined for their ability to present a peptide of the λ repressor to the 24.4 T cell hybridoma line. Transfected cells were cultured with 24.4 T cells in the presence of various doses of λ repressor in the presence or in the absence of IgG anti-λ repressor antibodies. Whereas IIA1.6 cells presented soluble and IgG-complexed antigen at comparable high concentrations of repressor, cells expressing FcγRIIb2 or FcγRIIb2 CT31 presented complexed antigen at lower doses – two to three orders of magnitude – than soluble antigen. The region in FcγRIIb2 responsible for the enhancement of antigen presentation was localized between aminoacids 19 and 31 since the FcγRIIb2 CT18 deletant behaved like IIA1.6 cells. FcγRIIb1 was unable to facilitate soluble repressor presentation in the presence of antibodies, as were FcγRIIb1 CT53 and CT1 deletion mutants. The ability of transfected cell lines to endocytose immune complexes made of peroxydase-antiperoxydase IgG was also determined and the results were similar (Amigorena, 1992a) (Fig. 1).

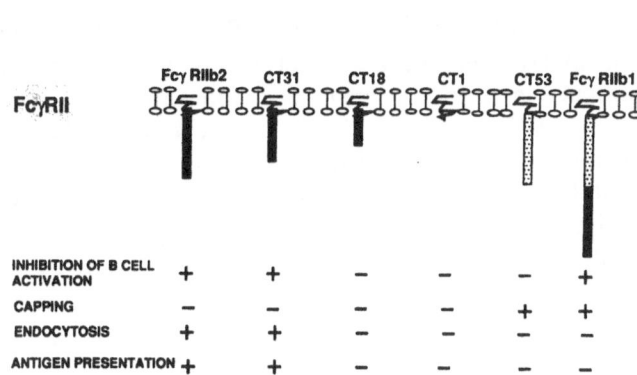

Fig. 1: Biological activities of II A1.6cells transfected with cDNA encoding wild type and deleted FcγRII. The EC region of FcγRII is not illustrated.

	FcγRIIb2	CT31	CT18	CT1	CT53	FcγRIIb1
INHIBITION OF B CELL ACTIVATION	+	+	−	−	−	+
CAPPING	−	−	−	−	+	+
ENDOCYTOSIS	+	+	−	−	−	−
ANTIGEN PRESENTATION	+	+	−	−	−	−

These results show that regions responsible for inhibition of B cell activation, antigen presentation and endocytosis are localized in a 13- aminoacid long region located in the IC region of FcγRIIb2. This sequence allows also localization of FcγRIIb2 in coated pits (Miettinen, 1989). It is present in

FcγRIIb1 but endocytosis is prevented by the cytoskeleton-binding sequence of the b1-specific insert.

FcγRIII FUNCTIONS

FcγRIII is coexpressed with FcγRIIb2 in macrophages which play a major role in antigen presentation to T lymphocytes. Since we have shown that FcγRIIb2 facilitates this process when antigen is complexed with IgG (see above), we investigated the role of FcγRIII and of its α and γ subunits in this function as well as in endocytosis, by using the same strategy as for FcγRII. Several stable cell lines with similar expression levels of recombinant α chains were prepared by transfection of IIA1.6 cells with wild type cDNA encoding α and γ chains (αγ2 cell line) or with cDNA mutated by introducing a stop codon at position 5 of the cytoplasmic domain of the α chain (αct4 mutant, which allows surface expression of the α chain without γ) or with hybrid cDNA with sequences of the EC/TM regions of FcγRII and of the IC region of the γ chain (β/ctγ chimera) (Fig. 2). Antigen presentation of IgG-complexed λ repressor to the T cell hybridoma line 24.4 was enhanced in αγ2 transfectants, the complexed antigen being presented at doses 30-to-100 fold lower than the antigen alone. This phenomenon required the γ chain and was independent on the cytoplasmic domain of the α subunit. Chimeric receptors (β/ctγ) increased antigen presentation, IC region of γ chain being required. Internalization of FcγR by crosslinking with anti-FcγR antibodies or with antigen-IgG antibody complexes-followed the same rules as antigen presentation. Internalization signals were mapped in the IC domain of the γ chain (Amigorena, 1992b) (Fig. 2).

In human NK cells, crosslinking of FcγRIII triggers Ca^{++} increase and lymphokine secretion. We thus investigated the ability of mouse FcγRIII and of the mutated receptors to transduce activation signals and compared it to that of FcγRIIb1. In contrast to FcγRIIb1, crosslinking of FcγRIII (on α γ2 transfectants) by 2.4G2 and mouse anti-rat Ig triggered the same activation signals as sIg when cross linked by F(ab')2 fragments of anti-IgG2a, i.e. intracellular Ca^{++} increase, tyrosine phosphorylation and IL2 secretion. The α ct4 transfectants were unable to respond to crosslinking by anti-FcγR antibodies, in contrast to the α ct4/γ2 ones, demonstrating that cell activation via FcγRIII requires the coexpression of the γ chain and is independent of the IC region of the α chain. The IC region of the associated γ chain is sufficient to trigger cell activation since β/ctγ responded positively (Fig. 2). Therefore, FcγRIII transduces activation signals via associated γ chains. The γ chain of FcγRIII, sIg associated α/β heterodimer as well as the TcR ζ chain share a common aminoacid motif (D-(X)7-D-(X)2-Y-(X)2-L-(X)7-Y-(X)2-L) which is involved in the activation of protein-tyrosine kinases. Transfectant cells were obtained with β/ctγ molecules in which the tyrosine in position 32 of the cytoplasmic domain of the γ chain was replaced by a leucine (β/ctγ-Y-L-32).

Crosslinking of this construct did not trigger activation signals while sIgG were still efficient, showing that tyrosine 32 is of importance in these processes (Fig. 2). The role of this motif in receptor internalization and antigen presentation was also investigated by using β/ctγ-Y-L-32 mutant as well as chimeric receptors in which tyrosine 21 was replaced by valine. Both mutants were unable to internalize antigen-IgG antibody complexes in contrast to the original chimeric receptors. These results suggest that the tyrosine-containing motif that transduces cell activation signals also determines internalization and antigen presentation via FcγRIII (Bonnerot, 1992).

FcγRIII are expressed in mast cells, together with FcγRII. Mast cells respond to FcγR mediated activation signals by degranulation, serotonin release and cytokine (TNFα) production. The role of FcγRIII and FcγRII in these processes was assessed, by transfecting rat basophilic leukemia cells (RBL-2H3) with the various wild type and mutated cDNA. Murine FcγRIII, but not FcγRIIb1 or b2, triggered release of granule (serotonin) and lipid (LTC4) mediators and the production of cytokines. The substitution of the IC domain of FcγRII for that of γ chain (in β/ctγ receptors) but not that of α chain (in β/ctα receptors) conferred the ability to induce all three activities (Daëron, 1992, Latour, 1992) (Fig. 2). The deletion of the IC region of FcγRIII α chain did not alter the ability to activate RBL cells. Thus, the ICα region is neither necessary nor sufficient to transduce these signals.

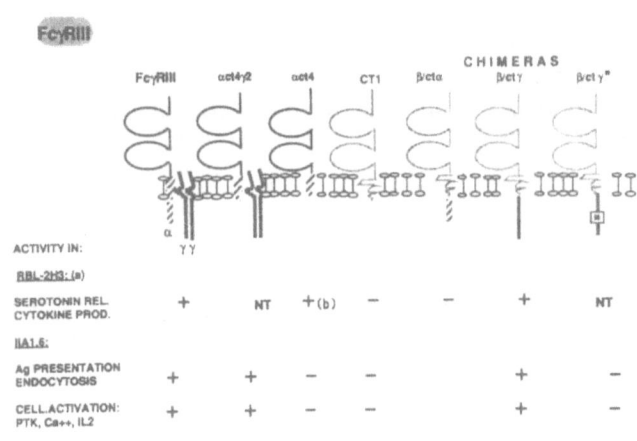

Fig. 2: Biological activities of RBL-2H3 cells and IIA1.6 cells transfected with cDNA encoding wild type, deleted, or chimeric FcγR and with cDNA encoding wild type or mutated (Y-V-21 or Y-L-32) γ chains. The EC and TM regions of α chains are represented in black and those of FcγRII in grey.
(a) Expresses endogeneous FcεRI
(b) Positive because the endogeneous γ associates with the deleted α chain

In conclusion, FcγRIII, but not FcγRII, activates B cells and mast cells. Signals are transduced via the cytoplasmic region of the γ chain. Within this region, tyrosine residues, which are part of a conserved motif involved in the activation of protein tyrosine kinase via antigen T and B cell receptor, play a key role. Internalization of immune complexes, as well as antigen presentation via FcγRIII, also requires the cytoplasmic region of the γ chain and the same tyrosine residues.

SOLUBLE Fcγ RECEPTORS (sFcγR)

Soluble FcγR have been found in supernatants of cells expressing FcγR and in biological fluids such as serum and saliva. In mouse, they bind IgG1, IgG2a, IgG2b and not IgG3, and react with antibodies directed against the EC regions of low affinity FcγR (Fridman, 1992). When deglycosylated, sFcγR produced by lymphocytes resolves in a 18 kDa polypeptide. These sFcγR are produced either by cleavage of the EC part of FcγRII, proteolysis occuring between aminoacids 165 and 180 (Sautès, 1991), or of FcγRIII (Fridman, 1992). sFcγR isolated from serum or from supernatants of macrophagic cell lines (P388D1 and J774) are different. In addition to a 18kDa polypeptide present in minor amounts, a polypeptide of larger size (25 kDa) was also found (Lynch, 1992). The 18 kDa polypeptide corresponds to the ·EC domains of cleaved sFcγRII and/or III. The analysis of the reactivities of the larger size sFcγR with antibodies directed against the EC (2.4G2 mAb and rabbit anti-NH2 terminal peptide antibodies) or the IC regions of FcγRII or FcγRIII α (anti-COOH terminal peptides), as well as one dimensional peptide mapping, showed that the 25kDa polypeptide contains the EC1, EC2, IC2 and IC3 regions of FcγRIIb2 but lacks the TM and IC1 region. The existence of mRNA encoding such a new type of FcγRII was shown by PCR amplification followed by Southern blot analysis. Hence these data demonstrate the existence of a new type of soluble FcγRII made by splicing of exon encoding the TM region. We propose to name it FcγRIIb3 (Table 1). In man a sFcγRII encoded by a transmembrane-deleted FcγRIIA mRNA present in Langherans cells was also described (Astier et al.,submitted)

Table 1. Murine soluble FcγR

sFcγR	Structure	Size a (kda) (Protein)	Present in serum	Producing cells (in vitro)	Mechanism of production	Biological role
IIb3		2 5	Yes	Macrophages Langherans cells	Alternative Splicing	NT
sFcγRII		1 9	Yes b	Lymphocytes Macrophages Langherans cells	Proteolysis of FcγRII b1 and FcγRII b2	Inhibition of Ab production
sFcγRIII		1 9	Yes b	NK cells	" of FcγRIII	"

a determined by SDS PAGE;.   b sFcγRII and/or sFcγRIII;   c not tested

☐ EC   ▨ IC   ▨ TM

Whether made by alternative splicing of FcγRII or by proteolysis of FcγRII or III, murine sFcγR share almost identical sequences in the EC regions. The question of the functional role(s) of this common structure was addressed by using the same strategy as for membrane receptors. Mouse FcγRII cDNA was mutated by introduction of a stop codon at position 175 and used to transfect fibroblastic L cells and Baby Hamster Kidney (BHK) cells. Concentrations of 32 mg sFcγRII/liter were reached upon culture of the cell lines in bioreactors. After purification to homogeneity recombinant, sFcγRII inhibited secondary IgG anti-SRBC responses *in vitro* (Varin, 1989) as well as polyclonal IgM and IgG responses of small resting B cells stimulated by LPS or by F(ab')2 fragments of anti-IgM in the presence of cytokines (IL2, IL4 and IL5). A similar suppressive role of sFcγRII on antibody production was found *in vivo,* in mice injected with endotoxin-free recombinant product (Fridman, 1992). These results show that 1) sFcγRII inhibits B cell responses, 2) the two EC domains of sFcγRII are sufficient for this activity and 3) sFcγRII can act on surface IgG negative B cells suggesting that it may bind to other ligand(s) than sIgG.

CONCLUSION

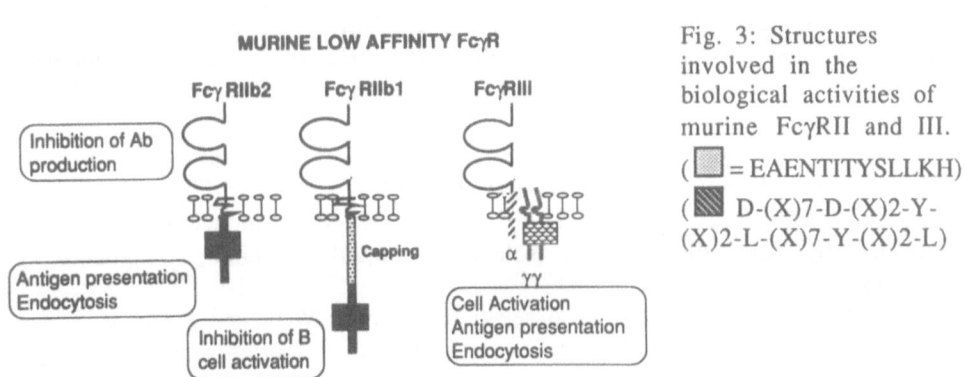

MURINE LOW AFFINITY FcγR

Fig. 3: Structures involved in the biological activities of murine FcγRII and III.

(▢ = EAENTITYSLLKH)

(▧ D-(X)7-D-(X)2-Y-(X)2-L-(X)7-Y-(X)2-L)

As illustrated in Fig. 3, our work shows that the EC and the IC regions of low affinity FcγR mediate specific functions. Beside their binding of IgG, the EC domains inhibit antibody responses *in vitro* and *in vivo*. Circulating sFcγR corresponding to these EC domains only, or with additionnal IC region of FcγRIIb2, have been described. The latter sFcγR, that we propose to call IIb3 in mouse, are made by alternative splicing of FcγRII. For membrane FcγR, the IC regions of FcγRII, or FcγRIII γ chains, are necessary to trigger the cell machinery. In the IC region of FcγRIIb2, a sequence located between aminoacids 19 and 31 is involved in endocytosis and antigen presentation and in inhibition of B cell activation. This sequence is also present in FcγRIIb1 but endocytosis is prevented by the cytoskeleton-binding sequence of the b1-specific insert. This latter sequence allows capping of FcγRIIb1 on B cells. The multimeric FcγRIII, in contrast to the monomeric FcγRII, is involved in cell activation. Cross linking of FcγRIII induces protein phosphorylation, a rise in intracellular Ca^{++}, and cytokine secretion. The sequences involved in cell activation, endocytosis and antigen presentation were mapped in the IC region of the associated γ chains. Tyrosine residues, which are part of a conserved motif involved in the activation of protein tyrosine kinase via antigen, play a key role in these processes.

REFERENCES:

Amigorena S, Bonnerot C., Drake J.R., Choquet D., Hunziker W., Guillet J.G., Webster P., Sautès C., Mellman I., Fridman W.H., (1992) Cytoplasmic Domain Heterogeneity and Functions of IgG Fc Receptors in B Lymphocytes, Science 256: 1808-1812

Amigorena S., Salamero J., Davoust J., Fridman W.H. and Bonnerot C., (1992) Tyrosine-containing motif that transduces cell activation signals also determines internalization and antigen presentation via type III receptors for IgG, Nature 358: 337-341

Bonnerot C,.Amigorena S., Choquet D., Pavlovich R., Choukroun V. and Fridman W.H., (1992) Role of associated γ-chain in tyrosine kinase activation via murine FcγRIII, Embo J. 11: 2747-2757

Daëron M., Bonnerot C., Latour S. and Fridman W.H., 1992, Murine recombinant FcγRIII, but not FcγRII, trigger serotonin release in rat basophilic leukemia cells, J Immunol. In Press

Fridman W.H., Bonnerot C., Daëron M., Amigorena S., Teillaud J.L. and Sautès C., (1992), Structural Bases of Fcγ Receptor Functions, Immunol. Rev. 125: 49-76

Latour S., Bonnerot C., Fridman W.H. and Daëron M., 1992, Induction of TNFα production by mast cells via FcγR. Role of the FcγRIII γ subunit, J. Immunol. In Press

Lynch A., Tartour E., Teillaud J.L., Asselain B., Fridman W.H. and Sautès C., (1991), Increased levels of soluble low-affinity Fcγ receptors (IgG-binding factors) in the sera of tumour-bearing mice, Clin. exp. Immunol. 87: 208-214

Miettinen H.M., Matter K., Hunziker W., Rose J.K., and Mellman.I. (1991) J. Cell Biol. 116: 875-885

Ra C, Jouvin M.HE, Blank U., Kinet J.P. (1989) A macrophage Fcγ receptor and the mast cell receptor for IgE share an identical subunit, Nature (Lond) 341: 752-755

Sautès C., Varin N., Teillaud C., Daëron M., Even J., Hogarth P.M. and Fridman W.H. (1991) Soluble Fcγ receptors II (sFcγRII) are generated by cleavage of membrane FcγRII, Eur. J. Immunol. 21: 231-234

Varin N.,Sautès C., Galinha A., Even J., Hogarth P.M. and Fridman W.H., (1989) Recombinant soluble receptors for the Fcγ portion inhibit antibody production *in vitro*. Eur. J. Immunol. 19:2263-2268

The CD23 Multifunctional Molecule and its Soluble Fragments (IgE-Binding Factors or Soluble CD23)

M. Sarfati, S. Fournier, H. Ishihara, M. Armant and G. Delespesse

Notre-Dame Hospital Research Center, 1560 Sherbrooke Street East, Montreal, Quebec, H2L 4M1 Canada (University of Montreal)

The low affinity receptor for IgE (FcεRII), recently identified as CD23 antigen, is a structurally and functionally unique immunoglobulin receptor primarily expressed on B lymphocytes. The CD23 is a 45 KD type II integral membrane glycoprotein (carboxy-terminus outside), with a short intracytoplasmic tail (23 residues), a single transmembrane domain (20 residues) and a large extracellular domain (277 residues) that shares a significant homology with calcium dependent (C-type) animal lectins (Ludin 1987; Ikuta 1987; Kikutani 1986a). Nearby the carboxy-end of the molecule, is an "RGD" sequence in a reverse configuration. RGD is a common recognition site of the integrin receptor. Three consensus sequences of 21 residues each, are located between the N-glycosylation site (position 63) and the lectin homology domain. Within this repetitive region, there are five heptadic repeats of leucine or (isoleucine) forming a "leucine zipper" motif, that may be involved in the formation of CD23 dimers (Beavil 1992). Unlike other Fc receptors, CD23 does not belong to the Ig superfamily but it is a member of a novel supergene family of type II membrane proteins displaying a lectin motif. This family includes the liver asialoglycoprotein receptors 1 and 2 (the archetype of the family), Ly49, A, B, C... (NK cell antigens), CD72 (the ligand of CD5) and A1 (a T cell antigen) (Delespesse 1992). The CD23 lectin homology domain spans from cysteine 163 to cysteine 282 and contains four highly conserved (positions 191, 259, 273 and 282) and two partially conserved cysteines (positions 163 and 174). CD23 binds IgE via its lectin-like domain, the IgE-binding site is comprised of two discontinuous segments (between residues 165-180 and 224-256) (Bettler 1989a, 1992). The binding of IgE to CD23 is calcium-dependent; it involves protein-protein interactions and probably also carbohydrate-protein interactions inasmuch as it is specifically inhibited by fucose-1-phosphate (Delespesse 1992; Vercelli 1989; Richards 1991). The CD23 lectin motif is also expressed on several membrane proteins where it was shown to be directly involved in Ca^{2+} dependent binding to carbohydrates; this is the case, for the selectins, a family of adhesion molecules (Stoolman 1989). Two isoforms of CD23 have been described (named A and B); they are generated by alternative transcription of the same CD23 gene (Yokota 1988). They only differ in their 5' untranslated sequences and in the six N-terminal amino-acids in the intracytoplasmic domain. Mouse CD23 is very similar to its human counterpart from which it differs by (a) the expression of 2 instead of 1 N-glycosylation sites (b) the presence of 4 instead of 3 consensus repeats (c) the deletion of the inverted RGD sequence at the carboxy-terminus and (d) the absence of CD23 type B isoform (Conrad 1990; Bettler 1989b). According to recent studies, multiple isoforms of CD23 type A as well as truncated transcripts encoding a secreted form of CD23 exist in murine T-lineage lymphocytes (Nunez 1992).

An intriguing feature of CD23 is its autocatalytic processing at the cell surface, leading to the release soluble fragments (soluble CD23/sCD23) of various sizes (37 KD, 33KD, 25 KD and 16 KD) retaining the ability to bind IgE (IgE-binding factors) (Letellier 1990). The rate of cleavage of CD23 is reduced by IgE and increased by N-glycoslation inhibition (Delespesse 1992).

After a brief summary of the current knowledge regarding the cellular distribution and the regulation of CD23 expression reviewed in detail very recently by Delespesse (1992), we will discuss recent data on the biological activities ascribed, to cell-bound and to soluble CD23.

The CD23 antigen may be found on a large variety of cells including B cells, T cells, monocytes, eosinophils, platelets, follicular dendritic cells and epithelial cells (Delespesse 1992). Of note, in the mouse, CD23 is exclusively found in B cells and perhaps on a subset of CD8[+] T cells (Conrad 1990). CD23 Ag is primarily defined as a B cell differentiation marker, because its expression is restricted to B cells coexpressing surface IgM and IgD, and it is lost after isotype switching (Kikutani 1986b). Given that the expression of CD23 is strikingly increased upon B cell activation in the context of IL-4 or by signals delivered by T-B cell interactions (DeFrance 1987; Crow 1989), it can also be viewed as a B cell activation marker. However, staphylococcus aureus Cowan I (SAC) or activation in the context of IL-2 does not significantly alter CD23 expression on B cells (Walker 1986; Hivroz 1989). Freshly isolated B cells exclusively express CD23 type A while type B CD23 isoform is found on all the other cell-types capable of expressing CD23 (Kikutani 1986b). The expression of CD23 on B cells is not constitutive inasmuch as CD23 protein and mRNA disappear upon incubation at 37°C. This results in the accumulation of soluble CD23 in the culture supernatant. IL-4 strikingly induces CD23 expression and, on B cells this enhancing effect is inhibited by IFNα, IFNγ, TGFβ and glucocorticoids (Delespesse 1992). Engagement of surface Ig, CD40 or CD72 increases IL-4-induced CD23 expression. Autacoids like LTB4 or PAF, moderately enhance CD23 expression in the presence of IL-4 (Delespesse 1992). Taken together, these observations indicate that the regulation of CD23 expression on B cells is tight and complex. Our recent data demonstrate that the two CD23 isoforms can be differentially regulated on normal B cells: (i) phorbol esters (PMA) selectively induce CD23 type B while IL-4 strikingly induces CD23 type B and upregulates type A isoform, and (ii) IL-4-induced CD23 mRNA type B but not type A is sensitive to cycloheximide treatment, indicating that the transcription of the two CD23 isoforms have differential requirements for de novo protein synthesis (S. Fournier and M. Sarfati, manuscript in preparation). These data suggest that CD23 type A and type B could have different functions. It was recently reported that CD23 type A mediates endocytosis on B cells, whereas CD23 type B mediates phagocytosis on monocytic cell lines (Yokota 1992). These observations are in keeping with the notion that normal murine as well as EBV-transformed human B cells are very efficient in IgE-dependent Ag presentation to T cells (Kehry 1989; Pirron 1990). This activity is totally blocked by anti-CD23 mAb. On inflammatory cells, CD23 (type B) is involved in protective immunity against parasites and in IgE-mediated inflammatory responses (e.g., IgE-dependent cytotoxicity and phagocytosis of IgE-coated particles) (Delespesse 1992). Signaling through CD23 on monocytes induces the release of IL-1, TNFα and of autacoids such as PGE2 or TXA2. Although the precise and respective role of the two CD23 isoforms remains to be delineated on B cells, several studies proposed a link between CD23 and B cell activation. Previous reports indicate that CD23 molecule is specifically expressed on Epstein-Barr virus transformed B cells (Kintner 1981) or after activation of the B cell by mitogens such as phorbol esters (PMA) (Thorley-Lawson 1985). Subsequently, one anti-CD23 mAb (MHM6) was shown to act in early G1 phase of the cell cycle by promoting progression of PMA-activated B cells through S phase, by enhancing CD25 expression and DNA synthesis in response to IL-2 (Gordon 1986). Most of the other anti-CD23 mAbs have been reported to be non-mitogenic. However, cross-linking of CD23 by IgE-immune complexes (as well as by some anti-CD23 mAbs) suppresses DNA synthesis by anti-IgM and IL-4 stimulated B cells (Luo 1991). It is proposed that mAbs directed against distinct epitopes of the same molecule may transduce positive or negative signal for B cell growth. Because the isoform preferentially induced in the above models (phorbol esters and IL-4 activation) is CD23 type B, it is possible that the regulatory signal of B cell growth is secondary to the engagement of this CD23 isoform. This would explain why the cross-linking of CD23 on LPS- and IL-4 stimulated mouse B cells has no effect on their proliferation (Conrad 1990; Bettler 1989). As already mentioned, mouse B cells do not express the equivalent of the human CD23 type B isoform. Finally, the ability of CD23 molecule to signal SAC and IL-4 activated B cells via the phosphoinositide pathway (Kolb 1990) as well as the association of Fyn tyrosine kinase to CD23 (Sugie 1991) further support a role for CD23 in lymphocyte activation. To directly assess the respective role of the two CD23 isoforms in B cell activation and proliferation, we have used anti-sense oligonucleotides specifically directed to CD23 mRNA type A or type B. The premise was that antisense oligonucleotides might block specifically the expression of CD23 protein. Unexpectedly, our results indicate that membrane CD23 protein is increased rather than decreased upon exposure to CD23 antisense but not to sense oligonucleotides. The data further show that enhanced CD23 expression following exposure to CD23 B antisense is associated with an increased B cell DNA synthesis in response to anti-IgM and IL-4 stimulation. This indicates that modulation of CD23 mRNA type B promotes B cell entry into the cell cycle. Indeed, a parallel increase in the proportion of $S+G_2/M$ cells is observed in the same cultures. Also, CD23 anti-sense oligonucleotides increase the expression of CD25 Ag as well as the frequency of G_1 cells in phorbol esters-activated B cells (S. Fournier and M. Sarfati, manuscript in preparation). As already mentioned, membrane CD23 is cleaved by autoproteolysis into soluble fragments of various sizes (sCD23). A role for sCD23 in B cell proliferation has been proposed following the observations that purified native soluble CD23 increases DNA synthesis of normal preactivated B cells (Swendeman 1987). However, these results could not be reproduced with recombinant sCD23 of various sizes by two independent groups of investigators (Luo 1991; Uchibayashi 1989). In spite of these in vitro contradicting data, in vivo and in vitro studies in chronic lymphocytic leukemia (CLL) lead us to further postulate that membrane and/or soluble CD23 have a role B cell proliferation.

CLL is characterized by the accumulation of slow dividing and long-lived CD5[+] B cells arrested at the G_0/G_1 phase of the cell cycle. These monoclonal CD5[+] B cells coexpress surface IgM, IgD and CD23 Ag. Hence, we and others proposed that CD23 Ag is a hallmark of CLL disease (Sarfati 1992). We first reported that the sera of CLL patients contain 3 to 500 fold more sCD23 than sera from patients with other B lymphoproliferative disorders (Sarfati 1988). The serum sCD23 level is weakly correlated with the clinical stage of the disease and strongly with the size of the tumor burden. Preliminary statistical analysis from ongoing prospective studies (on 110 CLL patients monitored over a 5 years period) indicate that serum sCD23 level is a prognostic marker and that the increase of serum sCD23 level may precede the progression of the clinical stage (M. Sarfati, personal data). We have next reported that abnormally high serum CD23 levels result not only from the increased pool of CD23$^+$ B cells but also from an overexpression of CD23 on the leukemic B cells (Sarfati 1990). Moreover, it was recently shown that the expression of membrane CD23 on CLL B cells has a significant pronostic value which is independent of the clinical stage (Geisler 1991). We have next examined the mechanisms underlying CD23 overexpression in CLL disease and reported that (i) in contrast to normal B cells, freshly isolated B-CLLs express the two CD23 isoforms (Fournier 1991) (ii) CD23 protein and gene are abnormally regulated inasmuch as IFNγ and IFNα significantly enhance IL-4-induced CD23 expression on B-CLLs while they display the reverse effect on normal B cells (Fournier 1992) and (iii) TNFα, which is detected in the sera of CLL patients, is enhancing CD23 on B-CLLs, whereas it has no effect on normal B cells (Fournier 1992). Signaling through CD23 Ag via the cross-linking of the molecule negatively influences the proliferation of B-CLL cells and inhibits CD23 type B (Fournier 1992). The following observations tend to indicate that the two CD23 isoforms may contribute to the physiopathology of CLL disease. We show that the ability of a given cytokine to enhance DNA synthesis of a B-CLL clone (e.g. IL-2, IFNα) correlates with its selective upregulation of CD23 type B; conversely cytokines with no proliferative activity (e.g. IL-4, IFNγ) specifically increase CD23 type A. We therefore propose that the modulation of CD23 A/B ratio toward the expression of type B isoform may determine the aptitude of a B-CLL cell to proliferate. This view is supported by recent data showing that CD23 antisense oligonucleotides enhance DNA synthesis as well as the number of B-CLL cells entering the cell cycle in response to suboptimal concentrations of IL-2. The antisense oligonucleotides have no effect on B-CLL proliferation in the absence of IL-2 (S. Fournier and M. Sarfati, manuscript in preparation).

At the present time, it is not known how CD23 contributes to B cell activation, but we conclude that CD23 plays a role in normal and leukemic B cell growth.

In addition to its role in IgE regulation, soluble CD23 appears to exert multiple functions on various cell types including (i) the maturation of early thymocytes (ii) the differentiation of myeloid cell precursors and (iii) the prevention of entry into apoptosis of germinal center B cells (Delespesse 1992). The significance of these observations is underlined by the in situ detection of CD23 producing cells in the thymus (on a subset of epithelial cells) in the bone marrow (on stromal cells) and in the light zone of the germinal centers (on follicular dendritic cells) (Delespesse 1992). Because these biological activities ascribed to sCD23 are clearly IgE independent, it was reasonable to postulate that sCD23 interacts with unidentified ligand(s) on these various cell types. Our recent data indicate that recombinant sCD23 binds to several proteins on a T cell line, as detected by Western blot analysis; this binding activity can be adsorbed on anti-CD23 mAb affinity column from which it can be recovered by acid elution (M. Sarfati, personal observations). Finally, we have attempted to relate the structure of sCD23 to its cytokine-like activities namely (i) regulation of IgE synthesis and (ii) myeloid precursors proliferation (Mossalayi 1992). We conclude from these experiments that the lectin-like domain spanning from cysteine 162 to 287 is sufficient for IgE-binding, while cysteine 288 is absolutely required for biological activities. These data further

FIGURE 1
Activation via soluble CD23

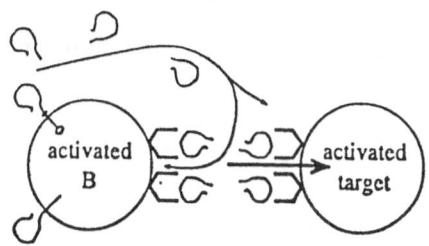

Activation via membrane CD23

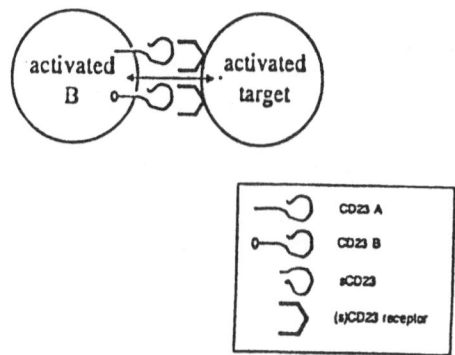

suggest that the binding of sCD23 to its counterstructure on the target cell may utilize either protein-protein or carbohydrate-protein interactions. The proposed model depicted in Fig. 1 shows our current hypothesis on the respective role of CD23 and sCD23 in B cell activation. Direct cell contact between membrane CD23 on B cells and its ligand on activated cells initiates bidirectionnal biochemical signals in both cell-types. It is proposed that the CD23 type B isoform present on activated B cells is preferentially involved in B cell activation. On the other hand, CD23 is the membrane precursor of a soluble molecule which might be capable of delivering autocrine signal(s) to the B cell as well as paracrine signal(s) to other target cells. Although not emphasized in this model, sCD23 might interact with various ligands which may be different according to the cell type on which they are expressed. The identification of the natural CD23 ligand(s) is required to both ascertain and analyze the mode of action of CD23 and sCD23 on various cell populations.

Acknowledgement
M. Sarfati is an MRC Scholar, S. Fournier is an MRC student and G. Delespesse is an MRC Associate. This work was supported by an MRC Grant. We wish to thank N. Del Bosco for her secretarial assistance.

REFERENCES

Beavil, A.J., Edmeades, R.L., Gould, H.J., and Sutton, B.J. (1992) Proc.Natl.Acad.Sci. USA 89:753.

Bettler, B., Maier, R., Ruegg, D. and Hofstetter, H. (1989a) Proc.Natl.Acad.Sci. USA 86:7118.

Bettler, B., Hofstetter, H., Rao, M., Yokoyama, W. M., Kilchherr, F. and Conrad, D. H. (1989b) Proc. Natl. Acad. Sci. USA. 86:7566.

Bettler, B., Texido, G., Raggini, S., Ruegg, D. and Hofstetter, H. (1992) J.Biol.Chem. 267:185.

Conrad, D.H. (1990) Ann.Rev.Immunol. 8:623.

Crow, M. K., Kushner, B., Jover, J.A., Friedman, S.M., Mechanic, S.E. and Stohl, W. (1989) Cell. Immunol. 121:99.

DeFrance, T., Aubry, J.P., Rousset, F., Vandervliet, B., Bonnefoy, J.-Y., Arai, N., Takebe, Y., Yokota, T., Lee, F., Arai, K., DeVries, J. and Banchereau, J. (1987) J.Exp.Med. 165:1459.

Delespesse, G., Sarfati, M., Wu, C.Y., Fournier, S. and Letellier, M. (1992) Immunological Reviews. 125:77.

Fournier, S., Trans, I. D., Suter, U., Biron, G., Delespesse, G. and Sarfati, M. (1991) Leukemia Research 15:609.

Fournier, S., Delespesse, G., Rubio, M., Biron, G. and Sarfati, M. (1992) J.Clin.Invest. 89:1312.

Geisler, C. H., Larsen, J.K., Hansen, N.E., Hansen, M.M., Christensen, B.E., Lund, B., Nielsen, H., Plesner, T., Thorling, K., Andersen, E. and Andersen, P.K. (1991) Blood 78:1795.

Gordon, J.,Rowe, M., Walker, L. and Guy, G. (1986) Eur.J.Immunol. 16:1075.

Hivroz, C., Valle, A., Brouet, J.C., Banchereau, J. and Grillot-Courvalin, C. (1989) Eur. J. Immunol. 19:1025.

Ikuta K., Takami, M., Kim, C. W., Honjo, T., Miyoshi, T., Tagaya, Y., Kawabe, T. and Yodoi, J. (1987) Proc. Natl. Acad. Sci. USA. 84:819.

Kehry, M. R., and Yamashita, L. C. (1989) Proc. Natl. Acad. Sci. USA. 86:7556.

Kikutani, H., Inui, S., Sato, R., Barsumian, E. L., Owaki, H., Yamasaki, K., Kaisho, T., Uchibayashi, N., Hardy, R. R., Hirano, T., Tsumasawa, S., Sakiyama, F., Suemura, M. and Kishimoto, T. (1986a) Cell. 47:657.

Kikutani, H., Suemura, M., Owaki, H., Nakamura, H., Sato, R., Yamasaki, K., Barsumian, E. L., Hardy, R. R. and Kishimoto, T. (1986b) J. Exp. Med. 164:1455.

Kintner, C. and Sugden, B. (1981) Nature. 294:458.

Kolb, J. P., Renard, D., Dugas, B., Genot, E., Petitkoskas, E., Sarfati, M., Delespesse, G. and Poggioli, J. (1990) J.Immunol. 145:429.

Letellier, M., Nakajima, T., Pulido-Cejudo, G., Hofstetter, H. and Delespesse, G. (1990) J.Exp.Med. 172:693.

Ludin, C., Hofstetter, H., Sarfati, M., Levy, C. A., Suter, U., Alaimo, D., Kilchherr, E., Frost, H. and Delespesse, G. (1987) EMBO J. 6:109.

Luo, H., Hofstetter, H., Banchereau, J. and Delespesse, G. (1991) J.Immunol. 146:2122.

Mossalayi, M.D., Arock, M., Delespesse, G., Hofstetter, H., Bettler, B., Dalloul, A.H., Kilchherr, E., Ouaaz, F., Debré, P., and Sarfati, M. (1992) Cytokine effects of CD23 mediated by an epitope distinct from the IgE-binding site. EMBO J. In Press.

Nunez, R., Hagen, M., and Lynch, R. (1992) FASEB J. No. 5, 6:A2695.

Pirron, U., Schlunch, T., Prinz, J. C. and Rieber, E. P. (1990) Eur.J.Immunol. 20:1547.

Richards, M.L., and Katz, D.H. (1991) Critical Reviews in Immunology 11(2):65.

Sarfati, M., Bron, D., Lagneaux, L., Fonteyn, C., Frost, H. and Delespesse, G. (1988) Blood. 71:94.

Sarfati, M., Fournier, S., Christoffersen, M. and Biron, G. (1990) Leuk. Res. 14:47.

Sarfati, M. (1992) Interleukin-4 and the two isoforms of CD23 antigen: their possible contribution to the physiopathology of chronic lymphocytic leukemia disease. The Lancet. Submitted for publication.

Stoolman, L. M. (1989) Cell 56:907.

Sugie, K., Kawakami, T., Maeda, Y., Kawabe, T., Uchida, A. and Yodoi, J. (1991) Proc.Natl.Acad.Sci. USA 88:9132.

Swendeman, S. and Thorley-Lawson, D.A. (1987) EMBO J. 6:1637.

Thorley-Lawson, D. A., Nadler, L. M., Bhan, A. K. and Schooley, R. T. (1985) J. Immunol. 134:3007.

Uchibayashi, N., Kikutani, H., Barsumian, E.L., Hauptmann, R., Schneider, F.J., Schwendenwein, R., Sommergruber, W., Spevak, W., Maurer-Fogy, I., Suemura, M. and et al. (1989) J. Immunol. 142:3901.

Vercelli, D., Helm, B., Marsh, P., Padlan, E., Geha, R. S. and Gould, H. (1989) Nature 338:649.

Walker, L., Guy, G., Brown, G., Rowe, M., Milner, A.E. and Gordon, J. (1986) Immunology. 58:583.

Yokota, A., Yukawa, K., Yamamoto, A., Sugiyama, K., Suemura, M., Tashiro, Y., Kishimoto, T. and Kikutani, H. (1992) Proc.Natl.Acad.Sci. USA 89:5030.

Yokota, A., Kikutani, H., Tanaka, T., Sato, R., Barsumian, E. L., Suemura. M. and Kishimoto, T. (1988) Cell. 55:611.

Signal Transduction by the Receptor with High Affinity for IgE

H. Metzger, V. S. Pribluda, U. M. Kent, S.-Y. Mao und G. Alber

Section on Chemical Immunology, Arthritis and Rheumatism Branch, National Institute of Arthritis and Musculoskeletal and Skin Diseases, National Institutes of Health, Bethesda, MD 20892, USA

INTRODUCTION

Since the last International Congress of Immunology in 1989, there has been a gratifyingly productive burst of activity in the field of Fc receptors. Advances in two aspects that have particularly interested our own laboratory are the relationship of the structure of the receptors to their function, and on the molecular phenomena that account for the early cellular events that are stimulated by these receptors. Here we describe our own research on these two subjects and review briefly the work of others. We focus on the receptor with high affinity for IgE (FcεRI), but where appropriate consider other Fc receptors also.

STRUCTURAL/ FUNCTIONAL RELATIONSHIPS

Cloning of Genes and cDNAs.

Sequencing of the cDNAs of various FcR has provided a sound stuctural foundation for pursuing functional questions. Interest in such explorations have been intensified by the recognition that the γ chain --originally discovered as a subunit of FcεRI (Perez-Montfort et al. 1983) --is also a subunit of FcγRIII (Ra et al.1989), and is closely related to the ζ and η subunits of the clonotypic antigen receptors of T lymphocytes forming a new family (Miller et al. 1989; Orloff et al. 1990).

Experimental Approaches and Results

The strategy now commonly used by investigators of structural /functional relationships is to manipulate genetically the structure and to transfect the modified cDNAs into appropriate cells. One tactic that has been particularly productive has been to engineer chimeric proteins in which a discrete portion of the receptor has been grafted onto an otherwise irrelevant protein. Particularly informative studies were reported by the laboratories of A. Weiss (Irving and Weiss, 1991), B.Seed (Romeo and Seed, 1991; Romeo et al. 1992), and R. Klausner (Letourneur and Klausner, 1991). Their studies demonstrated that aggregation of the cytoplasmic domains (CD) of the γ (or ζ) chains --and even of an 18 residue portion of these subunits (Romeo et al.1992)-- were sufficient to trigger both early and late cellular responses that resembled closely the phenomena initiated by aggregation of the wild-type receptors. Amigorena et al. (1992b) reported related studies just as this paper was being prepared .

We have compared the responses initiated by aggregation of wild-type receptors with the responses mediated by incomplete receptors. Both COS cell transfectants as well as transfectants of the murine P815 mastocytoma have been examined, and a variety of receptor-related events have been monitored. Some of these events are phenomena common to most if not all membrane proteins. For example we examined the basal mobility of the unaggregated receptors, the reduction in mobility secondary to aggregation, the interaction of aggregated receptors with the cytoskeleton, and endocytosis of the aggregates (Mao et al. 1991; Mao et al. 1992a; Mao et al. 1992b). In addition we monitored several early cellular phenomena that are more specifically (but not uniquely) associated with aggregation of FcεRI: phosphorylation of tyrosine residues on a variety of cellular proteins, hydrolysis of phosphoinositides, and an increase in intracellular Ca^{2+} (Miller et al. 1990; Alber et al. 1991a).

Briefly, we found the following (Table 1.):
1. Receptors missing certain cytoplasmic domains behaved indistinguishably from the wild-type receptors. Thus, receptors assembled from truncated α chains, or from β chains with a truncated NH_2-terminal cytoplasmic domain, behaved normally by all the criteria we employed.
2. Receptors missing other cytoplasmic domains (or entire subunits) behaved abnormally compared to the wild-type FcεRI. Thus, in all the assays, receptors containing β subunits whose COOH-terminal cytoplasmic domain had been truncated were functionally deficient. They did not appear to be totally inactivated, but they were clearly defective compared to the wild-type receptor. This result was curious in view of our findings with the receptors composed of *human* α and γ subunits alone. Such receptors can be studied either in COS cells cotransfected with both α and γ (Mao et al. 1992b), or in P815 mastocytoma cells which, by utilizing their endogenous γ chains, can express human α chains when transfected with cDNA for the latter chains alone. Although some of the responses were not as robust as those mediated by the wild-type receptors (Miller et al. 1990; Alber et al. 1991a) they were nonetheless considerably more vigorous than those initiated by aggregation of the β_{CT} mutant. It was interesting that transfecting the cells with the COOH-terminally truncated β also impared the responses of the FcγRIII on those same cells (Alber et al. 1991b; Alber et al. 1992). This suggested that the β chains could interact with FcγRIIIα as well as with FcεRIα. Furthermore, as discussed in detail elsewhere, it indirectly suggested that FcγRIII but not FcγRII initiated the transmembrane signals on mast cells. Both suggestions have proven to be true: Kurosaki et al. (1992) have actually found evidence for the association of FcγRIII with β when the former is transfected into mast cells, and we and others have both direct and indirect evidence that FcγRIII but not FcγRII stimulate a variety of early and late signals in a variety of

Table 1. Responses of cells transfected with variant receptors[a]

	wt	α_{CT}	$\beta_{CT(N)}$	$\beta_{CT(C)}$	γ_{CT}	$\alpha_H\gamma_R$	α-GPI
mobility of unagg.receptor	+	+	+	+	+	+	NT
immobil. of agg. receptor	+	+	+	±	+	+	NT
cytoskel. association	+	+	+	±	+	+	+
internalization (coated pit)	+	+	+	±	+	+	-
transmembrane signaling	+	+	+	±	-	+	NT

[a] Key: wt = cells transfected with wild-type subunits; X_{CT}= cells transfected with wild-type subunits except for the indicated subunit whose cytoplasmic domain was truncated; (N)= NH$_2$-terminal; (C)= COOH-terminal; $\alpha_H\gamma_R$= cells transfected with human α subunit and receptors assembled with either an endogenous (or cotransfected) rodent γ; α-GPI= cells transfected with a chimeric cDNA tin which the ectodomain of FcϵRIα subunit has been substituted for the corresponding portions on human FcγRIIIαB. When processed this leads to the expression of the ectodomains anchored to the membrane by a glycosylphosphoinositol moiety; NT= not tested.

cells (Fridman et al. 1992; Amigorena et al. 1992a; Alber et al. 1992; Katz et al. 1990; Katz et al. 1991). All the studies suggest that it is the γ subunit rather than the α subunit of FcγRIII that mediates the transmembrane signaling by this receptor. The ability of the cytoplasmic domain of the γ chains to initiate cellular responses makes it rather reasonable that P815 cells transfected with a truncated γ chain should be inactive (Alber et al. 1991a). However, this finding proved not to be as straight forwardly interpretable as we at first thought. When we examined the receptors on such transfectants it appeared that virtually all of them had been assembled from the wild-type γ chains which these cells make endogenously. How the truncated γ chains interfere with the function of the transfected FcϵRI$\alpha\beta$ (and with the endogenous FcγRα (Alber et al. 1992)) without apparently being assembled with them, is an interesting riddle that remains to be solved.

EARLY MOLECULAR EVENTS

Phosphorylation of Protein Tyrosine Residues in Intact Cells

The increasing evidence for the role of phosphorylation of protein tyrosines in cellular responses initiated by receptors in general, and the report of Benhamou et al. (1990) about such reactions stimulated by FcεRI in particular, has inspired numerous investigators to join in the further exploration of these phenomena. Some of these studies have been reviewed relatively recently (Benhamou and Siraganian, 1992), but additional important findings have since been reported (Eiseman and Bolen, 1992; Li et al. 1992).

The principal aspects that are being explored by various groups are the identification of the components that become phosphorylated, the determination of the functional significance of these phosphorylations, and the nature of the pertinent kinases. Although the identity of most of the components whose phosphotyrosine content increases is still unknown, at least three of them have been identified: the receptor itself (Paolini et al. 1991; Eiseman and Bolen, 1992; Li et al. 1992; Pribluda and Metzger, 1992), certain tyrosine kinases (Eiseman and Bolen, 1992) and phospholipase C $\gamma 1$ (Park et al. 1991; Li et al. 1992). All three are implicated in the earliest phenomena stimulated by aggregation of the receptor.

Studies on Broken Cells

Prior studies: For several years our laboratory has attempted to develop relatively simple preparations from the RBL line of rat mast cells. We have been able to prepare cytoplasts (Dreskin et al. 1989) and later lysed cytoplasts ("ghosts") (Dreskin and Metzger, 1991) which partially retain the ability to exhibit some of the early receptor-initiated cellular events. However, attempts to resolve further the essential components, e.g. by disrupting the preparations as a first step towards isolating functional membranes, led to a complete loss of activity (Dreskin et al. 1989; Dreskin and Metzger, 1991).

Recent studies: Because more receptor-proximal phenomena are more likely to be observable on broken-cell preparations, we re-examined such preparations to see whether they would manifest one or more of the tyrosine phosphorylation reactions that are stimulated by the receptors on intact cells. Extracts of *unstimulated* intact cells exhibit a large number of proteins containing phosphotyrosine, the number depending on how sensitive one makes the assay and whether one incubates the cells with phosphatase inhibitors such as phenylarsine oxide. If the cells have been sensitized with IgE and antigen is added to aggregate the FcεRI, the level of phosphorylated tyrosines increases substantially. Most if not all of the components that are heavily phosphorylated are the same ones that exhibit low levels of phosphotyrosine prior to aggregation of the receptor. Again, the degree to which this is observed depends significantly on how sensitive one makes the assay.

The sonicated broken cell preparations a different pattern was seen. First, there appears to be considerably more basal phosphorylation. However, this was also highly dependent on the sensitivity of the assay, the presence or absence of phosphatase inhibitors (we use vanadate), and the concentration of Mn^{2+} and or Mg^{2+}. Upon addition of antigen to the sonicate, superficial inspection suggested no change in such preparations regardless of how the intensity of the basal phosphorylation was manipulated. However, upon closer inspection it was apparent that two components, at $\approx$ 33 kDa and 23 kDa respectively, reproducibly exhibited an enhanced level of phosphotyrosine after addition of antigen.

We next isolated the solubilized FcεRI by immunoprecipitation from detergent extracts of intact cells or sonicates (Fig. 1).

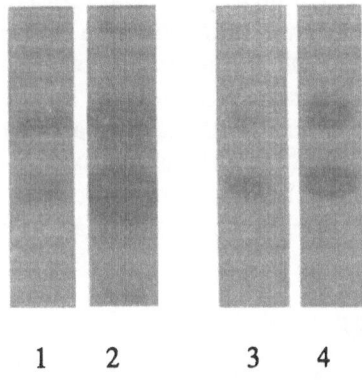

1   2   3   4

Fig. 1 Phosphorylation of tyrosine residues on the β and γ subunits of FcεRI in intact cells and partially purified membranes. RBL cells were sensitized with anti-DNP IgE and a portion of them was sonicated. The sonicated mixture was centrifuged under conditions that remove all intact cells and visible particulates. After an additional high-speed centrifugation, the membranes were re-suspended in a buffer containing 0.5mM vanadate, 0.05mM Mn^{2+}, and 2mM ATP. The mixture was brought to 37 °C and antigen was added. After 2 min. the membranes were dissolved with detergent under conditions that maintain the subunit structure of FcεRI and the latterwere then immunoprecipitated with anti-IgE and Protein A-Sepharose. The latter was extracted with sodium dodecyl sulfate and the eluted material was electrophoresed on polyacrylamide gels. The proteins on the gel were electrotransferred to nitrocellulose, reacted with horse radish peroxidase-conjugated anti-phosphotyrosine and the peroxidase activity was detected by a chemiluminescence technique which generates the autophotograph illustrated.

The specimens derived from the stimulated intact cells (lanes 1 and 2) show that among the components whose phosphotyrosine content is increased after aggregation of the FcεRI are two that by a variety of criteria, can be shown to be the β subunit and the dimer of γ subunits of the receptor. Similar results have been reported by others (Paolini et al. 1991; Eiseman and Bolen, 1992; Li et al. 1992). When this analysis was performed on the stimulated sonicate, a similar result was obtained (Fig. 1, lanes 3 and 4). Details of these studies are being reported elsewhere (Pribluda and Metzger, 1992), so here we only summarize the principal results:

1) As on the intact cells (Paolini et al. 1991), phosphorylation of the subunits of FcεRI on the stimulated broken cell preparations is strictly dependent upon aggregation of receptors: The response is proportional to the amount of aggregation and is prevented or ablated by the addition of a disaggregating stimulus (hapten in our case) before or during the course of the response. Significantly, the phosphorylation stimulated by aggregation is observed *only on those receptors that are themselves aggregated.*

2) The response is also elicited by stimulating partially purified membranes prepared either by repeated centrifugation at high speeds and resuspension by brief sonication (Pribluda and Metzger, 1992) or by centrifugation on sucrose gradients (U.M. Kent, unpublished observations). Such membranes retain both kinase(s) and phosphatase(s) capable of modifying the tyrosine residues of the receptor, and the basal phosphorylation is strongly influenced by the relative activity of these enzymes.

3) Initial experiments suggest that the responsiveness of washed membranes can be enhanced by the high-speed supernatant from sonicated cells. We are currently exploring the possibility that such supernatants may be useful for isolating the relevant kinases.

CONCLUDING STATEMENT

There has been considerable progress in the elucidation of the structure of Fc receptors, in developing methods to dissect the early receptor-mediated events, and provocative findings about which molecules participate in these early events. Therefore, one can see much more clearly than previously possible, the tasks for the next period. The molecular interactions that have been implicated, such as receptor:kinase, receptor:kinase:kinase substrate need to be more solidly confirmed, and the molecular basis for these interactions and the subsequent activations, clarified. Likewise, the identity and mechanism of action of the tyrosine phosphatases that are likely to be critical control elements need to be defined. Using the standard approaches of biochemical pharmacology as well as the newer methods of transfection, gene targeting and transgenic animals, the functional significance of these molecular events need to be elucidated. With respect to the latter, the *quantitative* aspects of both the early and late cellular phenomena must be carefully assessed. With regard to the FcεRI in particular, it is now apparent that in addition to the specificity for IgE encoded in the ectodomains of the α subunit, what sets this receptor apart from all the others, is the β subunit. It is reasonable to assume that it serves some special function; determining that role is an interesting challenge which we are pursuing.

REFERENCES

Alber G, Miller L, Jelsema C, Varin-Blank N, Metzger H (1991a) Structure/function relationships in the mast cell high-affinity receptor for IgE (FcεRI): Role of cytoplasmic domains. J Biol Chem 266:22613-22620

Alber G, Miller L, Varin-Blank N, Metzger H (1991b) Structure/function relationships in the mast cell high-affinity receptor for IgE (FcεRI): Role of cytoplasmic domains. FASEB J 5:A1675

Alber G, Kent UM, Metzger H (in press) Functional comparison of FcεRI, FcγRII and FcγRIII in mast cells. J Immunol

Amigorena S, Bonnerot C, Drake JR, Choquet D, Hunziker W, Guillet J-G, Webster P, Sautes C, Mellman I, Fridman WH (1992a) Cytoplasmic domain heterogeneity and functions of IgG Fc receptors in B lymphocytes. Science 256:1808-1812

Amigorena S, Salamero J, Davoust J, Fridman WH, Bonnerot C (1992b) Tyrosine-containing motif that transduces cell activation signals also determines internalization and antigen presentation via type III receptors for IgG. Nature 358:337-341

Benhamou M, Gutkind JS, Robbins KC, Siraganian RP (1990) Tyrosine phosphorylation coupled to IgE receptor-mediated signal transduction and histamine release. Proc Natl Acad Sci USA 87:5327-5330

Benhamou M, Siraganian RP (1992) Protein tyrosine phosphorylation: an essential component of FcεRI signaling. Immunol Today 13:195-197

Dreskin SC, Pribluda VS, Metzger H (1989) IgE receptor-mediated hydrolysis of phosphoinositides by cytoplasts from rat basophilic leukemia cells. J Immunol 142:4407-4415

Dreskin SC, Metzger H (1991) FcεRI-mediated hydrolysis of phosphoinositides in ghosts derived from rat basophilic leukemia cells. J Immunol 146:3102-3109

Eiseman E, Bolen JB (1992) Engagement of the high-affinity IgE receptor activates src protein-related tyrosine kinases. Nature 355:78-80

Fridman WH, Bonnerot C, Daëron M, Amigorena S, Teillaud J-L, Sautès C (1992) Structural basis of Fcγ Receptor Functions. Immunol Rev 125:49-76

Irving BA, Weiss A (1991) The cytoplasmic domain of the T cell receptor ζ chain is sufficient to couple to receptor-associated signal transduction pathways. Cell 64:891-901

Katz HR, Arm JP, Benson AC, Austen KF (1990) Maturation-related changes in the expression of FcγRII and FcγRIII on mouse mast cells derived in vitro and in vivo. J Immunol 145:3412-3417

Katz HR, Kaye RE, Austen KF (1991) Mast cell biochemical and functional heterogeneity. Transplant Proc 23:2900-2904

Kurosaki T, Gander I, Wirthmeller U, Ravetch JV (1992) The β subunit of FcεRI is associated with the FcγRIII on mast cells. J Exp Med 175:447-451

Letourneur F, Klausner RD (1991) T-cell and basophil activation through the cytoplasmic tail of T-cell-receptor zeta family proteins. Proc Natl Acad Sci USA 88:8905-8909

Li W, Deanin GG, Margolis B, Schlessinger J, Oliver JM (1992) FcεRI-mediated tyrosine phosphorylation of multiple proteins, including phospholipase Cγ1 and the receptor βγ2 complex, in RBL-2H3 rat basophilic leukemia cells. Mol Cell Biol 12:3176-3182

Mao S-Y, Varin-Blank N, Edidin M, Metzger H (1991) Immobilization and internalization of mutated IgE receptors in transfected cells. J Immunol 146:958-966

Mao S-Y, Alber G, Rivera J, Kochan J, Metzger H (1992a) Interaction of aggregated native and mutant IgE receptors with the cellular skeleton. Proc Natl Acad Sci USA 89:222-226

Mao S-Y, Pfeiffer JR, Oliver JM, Metzger H (submitted) Internalization of mutated IgE receptors and their localization in coated pits.

Miller L, Blank U, Metzger H, Kinet J-P (1989) Expression of high-affinity binding of human immunoglobulin E by transfected cells. Science 244:334-337

Miller L, Alber G, Varin-Blank N, Ludowyke R, Metzger H (1990) Transmembrane signalling in P815 mastocytoma cells by transfected IgE receptors. J Biol Chem 265:12444-12552

Orloff DG, Ra C, Frank SJ, Klausner RD, Kinet J-P (1990) Family of disulfide-linked dimers containing the zeta and eta chains of the T-cell receptor and the gamma chain of Fc receptors. Nature 347:189-191

Paolini R, Jouvin M-H, Kinet J-P (1991) Phosphorylation and dephosphorylation of the high-affinity receptor for immunoglobulin E immediately after receptor engagement and disengagement. Nature 353:855-858

Park DJ, Min HK, Rhee SG (1991) IgE-induced tyrosine phosphorylation of phospholipase C–γ1 in rat basophilic leukemia cells. J Biol Chem 266:24237-24240

Perez-Montfort R, Kinet J-P, Metzger H (1983) A previously unrecognized subunit of the receptor for immunoglobulin E. Biochemistry 22:5722-5728

Pribluda VS, Metzger H (in press) Transmembrane signaling by the high affinity IgE receptor on membrane preparations. Proc Natl Acad Sci U S A

Ra C, Jouvin M-HE, Blank U, Kinet J-P (1989) A macrophage Fcγ receptor and the mast cell receptor for IgE share an identical subunit. Nature 341:752-754

Romeo C, Seed B (1991) Cellular immunity to HIV activated CD4 fused to T cell or Fc receptor polypeptides. Cell 64:1037-1046

Romeo C, Amiot M, Seed B (1992) Sequence requirements for induction of cytolysis by the T cell antigen/Fc receptor ζ chain. Cell 68:889-897

10. Complement System, Structure and Function

Tissue Specific Complement Gene Expression: Evidence for Novel Functions of the Complement Proteins

H.R. Colten

Professor and Chairman, Departement of Pediatrics, Washington University School of Medicine, St. Louis, Missouri 63110 USA

INTRODUCTION

Not since the first decade following the discovery of complement, approximately 100 years ago, has the accumulation of knowledge about the complement system been as rapid, broad and deep. Table 1 lists the complement proteins and receptors and references in recent reviews (Colten and Gitlin, in press; Reid and Day 1989) summarize many of these advances. This remarkable conceptual progress has been primarily due to the application of molecular cloning methods to studies of complement genes and gene products. This symposium highlights a few areas in which these advances have already yielded practical as well as fundamental information.

The primary structure of complement proteins and genes (derived from cDNA and genomic sequencing) has been ascertained. This was a virtual impossibility 15 years ago because the soluble complement proteins present in trace amounts and complement receptors could not be analyzed using classical protein chemical methods. More recently, higher order structure of the complement proteins and receptors has been deduced and verified because bulk production of several complement proteins in vitro has been achieved. This detailed knowledge of structure and the capacity for site directed mutagenesis has led to extraordinary insights regarding structure-function relationships and in turn a better understanding of the interaction between complement proteins and pathogenic microorganisms. Interesting comparisons among the complement genes and with other protein families have facilitated an understanding of function especially for newly recognized complement receptors and regulatory proteins. Genetic variants, the phylogeny of complement and many of the genetic deficiencies of complement proteins are now understood at a molecular level. Perhaps most surprising among the recent advances is the accumulating evidence for previously unsuspected functions of complement and the suggestion that complement proteins may serve a role in reproductive biology, fat metabolism and B lymphocyte maturation, etc. These possibilities have been stimulated by discovery of the highly regulated, extrahepatic expression of many complement genes, which contributes little if any to the complement protein system in plasma.

Table 1
Proteins, Regulators, and Receptors of the Complement System

Complement Protein	Molecular Mass (daltons)	Subunit Structure	Serum Concentration (ug/ml)	mRNA[a,b] Size (kb)
C1q	410,000	6A; @24,000 6B; @23,000 6C; @22,000	70	A; 1.4 B; 1.4 C; 1.5
C1r	95,000	single chain	35	2.0
C1s	87,000	single chain	35	2.0
C2	110,000	single chain	25	2.9
B	93,000	single chain	200	2.6
D	24,000	single chain	1	1.0
C3	185,000	α; 110,000 β; 75,000	1,500	5.2
C4	200,000	α; 93,000 β; 78,000 γ; 33,000	400-600	5.3
C5	190,000	α; 115,000 β; 75,000	75	5.5
C6	115,000	single chain	75	N.D.
C7	115,000	single chain	65	3.9
C8	163,000	α; 64,000 β; 64,000 γ; 22,000	55	α; 2.5 β; 2.6 γ; 1.0
C9	71,000	single chain	60	2.4

Soluble Control Proteins

C1-INH	104,000 (34% CHO)	single chain	150	1.8
P	~224,000	6; 56,000	25	1.6
C4-bp	500,000	7; 70,000 (6α; 70,000) (1β; 45,000)	~150	2.5
I	88,000	46,000 39,000	35	2.4
H	155,000	single chain	500	4.4
S-Protein	~80,000	single chain	500	1.6

Membrane Receptors and Regulators

DAF	70,000	single chain	–	3.1 2.7 2.0
MCP	58,000-63,000	single chain	–	4.2
HRF	65	single chain	–	
CR1	A; 190,000[c] B; 220,000 C: 160,000 D; 250,000	single chain	–	A; 8.6 B; 11.6 C; 7.3 D; 12.8
CR2	140,000	single chain	–	5.0
CR3	260,000	α; 165,000 β; 95,000	–	α; 6.0 β; 3.0
C5aR	~45,000	N.D.	–	2.2
C1qR	~70,000	N.D.	–	N.D.

[a] Approximate values
[b] Only major mRNAs are shown
[c] Allotypes

Expression of complement in these extrahepatic sites is highly tissue- and in some cases species-specific, is developmentally regulated, and is responsive to acute phase stimuli, multiple cytokines, steroid hormones, growth factors, etc. Moreover, expression of soluble complement proteins which upon activation yield biologically active ligands is often spatially and temporally related to expression of the corresponding complement receptors in these tissues. This chapter and the other contributions to this symposium will briefly summarize these provocative findings and suggest several potential directions for future research. For instance, Drs. Takahashi and Campbell outline the molecular genetics of this system with emphasis on evolution of structure and regulated expression. Drs. Frank and Kazatchkine discuss the host-parasite relationship in molecular and cellular detail, emphasizing strategies employed by microorganisms (bacterial, fungal and viral agents including HIV) to subvert the elaborate host defenses mediated by the complement system.

Extensive studies by Melchers and collaborators (reviewed in Lernhardt and Melchers, 1988) have provided evidence for production of a truncated C3 mRNA species by a T cell hybridoma. This mRNA may program synthesis of a C3 polypeptide that serves a role in B cell proliferation. The product (bound C3 fragments) of these and other cells producing α-B cell growth factor(s) promotes entry of activated B cells into the S phase of the cell cycle by interaction with the complement receptor Cr2. The molecular details of this phenomenon and the interaction of Cr2 with other membrane glycoproteins important in transduction of these signals are discussed in this symposium by Dr. Fearon. Data similar to that generated with C3, but less fully developed, support a role for factor B derived cleavage products in B cell proliferation/maturation (Peters et al. 1988). Both lines of evidence are also consistent with the observation of deficits in antibody response (isotype switch) to low dose antigen administration in complement deficient guinea pigs (Berger et al. 1986). This area has received much attention recently and is likely to yield interesting insights into the nature of the early immune response to microbial products.

The complement effector and control proteins are synthesized in liver by hepatocytes but, as suggested by Thorbecke et al. (1965) many years ago, cells of ecto-, meso-, and endodermal origin in many tissues also produce complement proteins. The complement receptors and membrane bound complement regulatory proteins are also expressed in many tissues and cell types. For some complement proteins, extrahepatic synthesis is quantitatively most important. For instance, factor D is principally expressed in adipocytes (White et al. 1992) and to a lesser extent in mononuclear phagocytes, but not in hepatocytes. The peculiar tissue distribution, the marked effect of differentiation from pre-fat to fat cell, and the impact of obesity and starvation on expression of factor D (Wilkison et al. 1990; Rosen et al. 1989) has suggested a role for this protein in energy metabolism. The adipocyte is also a source of factor B and C3, the other constituents of the alternative pathway of complement activation. Finally, recent preliminary data from the Spiegelman laboratory (unpublished) suggests an effect of C3 cleavage products on regulation of gene expression in brown fat. A rate limiting function for complement in fat metabolism is at the present only speculative, but further work in this area is likely to be forthcoming and fruitful.

Another provocative example is provided by the potential role of complement in reproductive biology. For instance, uterine epithelium is striking among the many extrahepatic sites of complement synthesis because in that tissue, its expression is highly regulated by the sex hormones estrogen and progesterone (Sundstrom et al. 1990). The change in C3 expression (~25 fold) in uterine epithelium exceeds by more than an order of magnitude changes in plasma C3 during pregnancy or following in vivo hormone administration (Isaacson et al. 1989). In addition, at least three prominent C3 "regulatory" proteins, membrane cofactor protein (CD46), DAF (CD55) and homologous restriction factor (CD59) (Cervoni et al. 1991) are expressed in distinct membrane layers of sperm. It has been speculated that these protect the sperm from lytic secretory host defenses within the female genital tract but an equally attractive possibility is that the complement C3 in uterine epithelium and C3 binding proteins in sperm are more active in facilitating reproductive function. This set of observations warrants further investigation.

COMPLEMENT BIOSYNTHESIS

At the turn of the century Ehrlich and Morgenroth (1900) provided clear evidence that the liver was the prime source of serum complement. Many other investigators confirmed this (Alper et al. 1980; Morris et al. 1982) and using several different approaches, including quantitative assessments of synthetic rates, established that the hepatocyte was the major site of serum complement synthesis. Subsequent studies (Falus et al. 1987; Passwell et al. 1988) demonstrated the capacity for complement production at extrahepatic sites and rather marked regulation of this extrahepatic expression during the course of tissue injury or inflammation. This problem has been of great interest in our laboratory and in others because the evidence suggests that a distinct host advantage is afforded by peripheral sites of complement production. In addition, the potential for local complement production in the pathogenesis of several rheumatological disorders is also facilitated by local complement production at sites of inflammation (Ruddy and Colten 1974; Ahrenstedt et al. 1990).

The systemic acute phase response consists of physiological changes that include fever, leukocytosis, alteration in protein, energy and heavy metal metabolism. In addition, the concentrations of many plasma proteins are modulated, each with characteristic magnitude, direction, kinetics and duration. Among the complement proteins, some such as C3 and factor B increase 2-3 fold within several days after the acute phase stimulus and remain elevated for more than a week thereafter. The plasma concentration for others, such as C2 and C5, is unaltered by an acute phase stimulus. However, acute phase stimuli can induce local modulation of proteins (e.g.: C2) not classified as classical acute phase reactants based on changes in serum concentration.

Several central questions posed by these observations have been investigated during the past ten years. To wit: (a) What are the extracellular signalling mechanisms that regulate complement gene expression following an acute phase stimulus; (b) What are the factors that govern tissue specificity of the response; (c) How are the signals transduced; (d) What are the gene specific transcriptional controls (cis and trans elements) that mediate these acute phase changes?

SIGNALS REGULATING COMPLEMENT GENE EXPRESSION

Each of the cytokines associated with upregulation of other acute phase genes has been recognized to modulate expression of several of the complement genes also. For instance, in the case of factor B, the complement gene most broadly responsive to cytokine control, IL-1α, IL-1β, TNF-α, IL-6 and IFN-γ are able separately and together to increase transcription of the Bf gene (reviewed in Colten 1992). The C3 gene is also regulated by these cytokines, though its response to IFN-γ is species specific and even within a species such as mice, strain specific. The effect of these cytokines is manifest in many different tissues and cell types, including hepatocytes, fibroblasts, macrophages, enterocytes, neuroglia, endothelial cells and other cell types that have been less well studied. A detailed analysis of this area reveals quantitative differences in the C3 and/or Bf response among different cell types, but the basis for these differences has not yet been ascertained. In some tissues, the effect of various combinations of these cytokines is additive, in others [e.g., TNF-α and IFN-γ in renal epithelial cells (Ault,

Kelly and Colten, unpublished)] a synergistic effect is observed. Whether this is a receptor or post receptor mediated phenomenon is not known at the present time.

Some complement genes, including C4, a prominent plasma acute phase reactant, C2, C1r and C1s, and complement regulators (e.g., C1 inhibitor), are responsive only to IFN-γ and not to IL-1, IL-6 or TNF (Perlmutter and Colten 1988; Ripoche et al. 1988). This specificity of cytokine responsiveness and counter-regulation (see below) raises interesting questions about non host defense functions of these proteins. Though difficult to rigorously prove because of its known capacity to induce cytokine production, lipopolysaccharide (LPS) also directly upregulates Bf and C3 expression. However, endotoxin (LPS) stimulation of fibroblasts upregulates Bf translation as well as transcription (Katz, Cole and Strunk 1988), the former an effect thus far not observed with any of the cytokines. Secondly, the maturation of the LPS response is developmentally delayed by comparison to the cytokine responsiveness of the Bf gene (St. John Sutton et al. 1986). These observations and the tissue (cell) specificity of response to LPS which is not identical with the pattern of cytokine responsiveness strongly argue for a direct effect of LPS on these genes. The mechanisms for recognition of and transducing the LPS signal, as well as the nuclear/cytoplasmic effector events, are not yet understood. These offer interesting research opportunities.

Counter-regulation of the effects of these cytokines by the growth factors IL-4, FGF, PDGF, and EGF (Circolo et al. 1990) has prompted questions regarding the mechanisms by which the acute phase response is terminated. In other words, is the return to constitutive expression of these genes simply a result of less stimulation or is it an active extinction (downregulation) of gene expression? In any case, the effects of the growth factors are cytokine specific, gene specific and are independent of (precede) their effect on the proliferative response. As a result, work following these initial observations may be of broad fundamental interest. Not only growth factors have the capacity to counter-regulate cytokine mediated induction of complement. For instance, LPS completely counters the upregulation of C4 by IFN-γ, though it augments the effect of IFN-γ on Bf expression (Kulics, Colten and Perlmutter 1990). Again, these differences offer an advantage in working out the mechanisms responsible for the interesting set of phenomena.

TISSUE SPECIFICITY OF EXTRAHEPATIC COMPLEMENT GENE EXPRESSION

In addition to the aforementoned striking preponderance of factor D and of properdin expression (Nolan and Reid 1990) at extrahepatic sites, qualitative differences in tissue specificity of complement expression have been observed. For example, at least two murine C2 transcripts have been identified (Ishikawa et al. 1990), one of which exhibits a deletion of a seven-amino acid peptide within the serine proteinase-binding pocket. This transcript is expressed at low levels in all tissues thus far examined except in heart, where about equal amounts of the two C2 transcripts have been detected.

Two factor B transcripts resulting from alternative transcriptional initiation sites have been identified in murine kidney and intestine (Falus et al. 1987; Nonaka et al. 1989). The larger of the two is expressed in amounts equal to the shorter message only in those two tissues; in liver, the long transcript represents <5% of Bf mRNA (Ishikawa et al. 1990). No obvious promoter sequence has been recognized at the upstream initiation site, but mutations near this site, which parallel differences in expression of the Bf long transcript in different mouse strains have defined the specific sequences essential for upstream transcriptional initiation (Garnier 1992). Preliminary data by Garnier et al. in our laboratory suggest that expression of these two transcripts has significant impact on factor B synthesis; i.e., the phenomenon is biologically relevant. With few exceptions, studies of these differences have been largely descriptive thus far, but several laboratories are beginning to explore the interesting and biologically important mechanisms governing tissue specific extrahepatic complement gene expression.

SIGNAL TRANSDUCTION

Virtually nothing is known in detail about the mechanisms by which cell surface signals (cytokines, growth factor, hormones or LPS) are transduced to effect a change in complement gene expression. Thus far, it appears that no unique pathways are employed in regulation of complement so that little attention has been given to this line of research.

TRANSCRIPTIONAL CONTROL

An examination of the cis and trans elements regulating factor B and C3 expression has been undertaken in only a few laboratories thus far. Factor B and C3 have been chosen for initial studies because they are key effector proteins of the complement cascade, they are expressed in many tissues and are responsive to most of the signals regulating complement expression. Moreover, genetic variants in Bf and C3 expression provide useful natural mutations focusing attention on critical sequences for further study. For both genes, specific cis elements less than 500 bp 5' to their respective promoters are necessary and sufficient for IL-1 and IL-6 responses

(Wilson et al. 1990; Kawamura et al. 1992; Wu et al. 1987; Nonaka, Gitlin, and Colten 1989). Binding of members of the C/EBP and NFκB families of proteins to these sequences is required for transcriptional upregulation by IL-1 and IL-6. The elements upstream of the Bf and C3 genes share sufficient sequence homology so that each can specifically compete for the nucleoprotein binding to the corresponding element in the other gene (Kawamura et al., unpublished). In addition, the CATAA box (promoter region of the Bf gene) competes effectively with the C/EBP binding domain of the upstream cytokine response element (Garnier et al. 1992).

Preliminary evidence for different transacting factors governing tissue specificity of the Bf gene response have been obtained and similar studies for C3 will be of considerable interest. An interferon response element has been identified 5' to the Bf gene, but little additional information is available about the trans elements interacting with this region. Many interesting questions have been raised by these initial studies and no doubt more data will be forthcoming.

CONCLUSION

We have entered a new era in which fundamental biomedical questions can be answered in the context of complement research. Recent studies suggest these may even be of practical importance as well. This symposium has been organized to suggest some of these new directions.

REFERENCES

Ahrenstedt O, Knutson L, Nilsson B, Nilsson-Ekdahl K, Odlind B, Hallgren R (1990) Enhanced local production of complement components in the small intestines of patients with Crohn's disease. N Engl J Med 322:1345-1349

Alper CA, Raum D, Awdeh Z, Petersen BH, Taylor PD, Starzl TE. (1980) Studies of hepatic synthesis in vivo of plasma proteins including orosomucoid, transferrin, alpha-1-antitrypsin, C8 and factor B. Clin Immunol Immunopathol 16:84-89

Burger R, Gordon J, Stevenson G, Ramadori G, Zanker B, Hadding U, Bitter-Suermann D (1986) An inherited deficiency of the third component of complement, C3, in guinea pigs. Eur J Immunol 16:7-11

Cervoni F, Oglesby TJ, Nickells M, Atkinson JP, Hsi BL (1991) The expression of complement (C) regulatory proteins on human spermatozoa. Compl Inflamm 8:134

Circolo A, Pierce GF, Katz Y, Strunk RC (1990) Anti-inflammatory effects of polypeptide growth factors. Platelet-derived growth factor, epidermal growth factor, and fibroblast growth factor inhibit the cytokine-induced expression of the alternative complement pathway activator factor B in human fibroblasts. J Biol Chem 265:5066-5071

Colten HR (1992) Tissue-specific regulation of inflammation. J Appl Physiol 72:1-7

Colten HR, Gitlin JD (1992) Immunoproteins. In: Handin RI, Lux SE, Stossel TP, eds., Blood: Principles and Practice of Hematology, J.B. Lippincott, Philadelphia, in press

Ehrlich P, Morgenroth J (1900) Ueber Haemolysine. Berl Klin Wochenschr 37:453-458

Falus A, Beuscher HU, Auerbach HS, Colten HR (1987) Constitutive and IL-1 regulated murine complement gene expression is strain and tissue specific. J Immunol 138:856-860

Garnier G, Ault B, Kramer M, Colten HR (1992) Cis and trans elements differ among mouse strains with high and low extrahepatic complement factor B gene expression. J Exp Med 175:471-479

Isaacson KB, Coutifaris C, Garcia CR, Lyttle CR (1989) J Clin Endocrin Metab 69:1003-1009

Ishikawa N, Nonaka M, Wetsel RA, Colten HR (1990) Murine complement C2 and factor B genomic and cDNA cloning reveals different mechanisms for multiple transcripts of C2 and B. J Biol Chem 265:19040-19046

Katz Y, Cole FS, Strunk RC (1988) Synergism between interferon-gamma and lipopolysacharide for synthesis of factor B, but not C2, in human fibroblasts. J Exp Med 167:1-14

Kawamura N, Singer L, Wetsel RA, Colten HR (1992) Cis and transacting elements required for constitutive and cytokine regulated expression of the mouse complement C3 gene. Biochem J 283:705-712

Kulics J, Colten HR, Perlmutter DH (1990) Counter-regulatory effects of interferon-γ and endotoxin on expression of the human C4 genes. J Clin Invest 85:943-949

Lernhardt W, Melchers F (1988) The role of C3 and its fragments in the control of S phase entry of activated mouse B lymphocytes via the complement receptor type 2. Expl clin Immunogenet 5:115-122

Morris KM, Aden DP, Knowles BB, Colten HR. (1982) Complement biosynthesis by the human hepatoma derived cell line HepG2. J Clin Invest 70:906-913

Nolan KF, Reid KB (1990) Complete primary structure of human properdin: a positive regulator of the alternative pathway of the serum complement system. Biochem Soc Trans 18:1161-1162

Nonaka M, Gitlin JD, Colten HR (1989) Regulation of human and murine complement: Comparison of 5' structural and functional elements regulating human and murine complement factor B gene expression. Mol Cell Biochem 89:1-14

Nonaka M, Ishikawa N, Passwell J, Natsuume-Sakai S, Colten HR (1989) Tissue specific initiation of murine complement factor B mRNA transcription. J Immunol 142:1377-1382

Passwell J, Schreiner GF, Nonaka M, Beuscher HU, Colten HR. (1988) Local extrahepatic expression of complement genes C3, factor B, C2 and C4 is increased in murine lupus nephritis. J Clin Invest 82:1676-1684

Perlmutter DH, Colten HR (1988) Complement: molecular genetics. In: Gallin JI, Goldstein IM, Snyderman R, eds., Inflammation: Basic Principles and Clinical Correlates, Raven, New York, pp 75-88.

Peters MG, Ambrus JL, Fauci AS, Brown EJ (1988) The Bb fragment of complement factor B acts as a B cell growth factor. J Exp Med 168:1225

Reid KB, Day AJ (1989) Structure-function relationships of the complement components. Immunol Today 10: 177-180

Ripoche J, Mitchell A, Erdei A, Madin C, Moffatt B, Mokoena T, Gordon S, Sim RB (1988) Interferon gamma induces synthesis of complement alternative pathway proteins by human endothelial cells in cultures. J Exp Med 168:1917-1922

Rosen BS, Cook KS, Yaglom J, Groves DL, Volanakis JE, Damm D, White T, Spiegelman BM (1989) Adipsin and complement factor D activity: An immune-related defect in obesity. Science 244:1483-1487.

Ruddy S, Colten HR (1974) Rheumatoid arthritis: biosynthesis of complement proteins by synovial tissues. N Engl J Med 290:1284-1288

St. John Sutton MB, Strunk RC, Cole FS (1986) Regulation of the synthesis of the third component of complement and factor B in cord blood monocytes by lipopolysaccharide. J Immunol 136:1366-1372

Sundstrom SA, Komm BS, Xu Q, Boundy V, Lyttle CR (1990) The stimulation of uterine complement component C3 gene expression by antiestrogens. Endocrinology 126:1449-1456

Thorbecke GJ, Hochwald GM, von Furth LR, Muller-Eberhard HJ, Jacobson EB (1965) Problems in determining the sites of synthesis of complement components. In: Wolstenholme GEW, Knight J, eds., Complement, Churchilll, London, pp 99-119 (Ciba Found Symp)

White RT, Damm D, Hancock N, Rosen BS, Lowell BB, Usher P, Flier JS, Spiegelman BM (1992) Human adipsin is identical to complement factor D and is expressed at high levels in adipose tissue. J Biol Chem 267:9210-9213

Wilkison WO, Min HY, Claffey KP, Satterberg BL, Spiegelman BM (1990) Identification of distinct nuclear factors binding to single- and double-stranded DNA. J Biol Chem 265:477-482

Wilson DR, Juan TSSC, Wilde MD, Fey GH, Darlington GJ (1990) A 58-base pair region of the human C3 gene confers synergistic inducibility by interleukin-1 and interleukin-6. Mol Cell Biol 10:6181-6191

Wu LC, Morley BJ, Campbell RD (1987) Cell-specific expression of the human complement protein factor B genes: Evidence for the role of two distinct 5' flanking elements. Cell 48:331-342

Mediating B Lymphocyte Activation by the CR2/CD19 Complex

R.H. Carter and D. T.Fearon

Johns Hopkins University School of Medicine, Division of Molecular and Clinical Rheumatology, Baltimore, Maryland 21205, USA

The growth and differentiation of lymphocytes are controlled in part by antigen receptors. These receptors, especially in naive B cells, have broad reactivity but relatively low affinity. This leads to several problems. B cells must have mechanisms for responding to low concentrations of antigen, because waiting until pathogenic "ligands" approach the Kd of the appropriate antigen receptors would be lethal for the host. In addition, mechanisms must exist that alter the outcome of ligation of this receptor, depending on the nature (self versus non-self) and/or context (bone marrow, tissue, lymphoid follicle) of the antigen to elicit different cellular responses such as escape from the bone marrow, antigen presentation, proliferation, maintenance of viability in a germinal center and differentiation into antibody secreting cells.

CD19 is a B cell-specific member of the immunoglobulin superfamily that is expressed from the pre-B cell stage until the final differentiation into a plasma cell (Nadler 1983). The 242 amino acid cytoplasmic tail makes it a good candidate for a primary signal transduction molecule (Stamenkovic 1988; Tedder 1989). Although neither its ligand nor its role in vivo are known, the existing data suggest that CD19 helps the B cell overcome the problem of low affinity antigen receptors in antigen-induced responses.

As CD19 is known to form a complex with Complement Receptor Type 2 (see below), the clues to the potential role of CD19 start with the demonstration that complement proteins are required for normal antibody responses to low doses of antigen. Mice depleted of C3 by cobra venom factor have suppressed antibody responses after challenge with limited amounts of antigen (Pepys 1974). Guinea pigs with genetic deficiency of C2 or C4 have reduced IgM production and fail to switch to IgG after primary and secondary challenges, respectively, with low concentrations of a T-dependent antigen (Bottger 1986). More recent data suggest that the effect of complement deficiencies on antibody production is mediated by the complement receptors. Treatment of mice with antibody which recognizes both murine CR1 and CR2 blocked the response to suboptimal doses of T dependent antigen, whereas antibody to murine CR1 did not (Heyman 1990). To block interaction with ligand without binding the receptor itself, recombinant DNA technology was utilized to attach covalently the C3dg binding site of CR2 to the heavy chain of an irrelevant IgG, which is secreted as a soluble, dimeric molecule capable of competing with the native receptor for C3dg. This CR2/IgG chimera inhibited the production of antibody after intraperitoneal injection of limited doses of the T-dependent antigens, sheep erythrocytes or keyhole limpet hemocyanin (Hebell 1991). Thus, the enhancement of specific antibody responses to limited antigen by complement requires interaction with CR2.

To determine how CR2 enhances B cell activation, the effect of CR2 on intracellular free calcium concentration ($[Ca^{2+}]_i$) was tested. Changes in $[Ca^{2+}]_i$ were assayed by flow cytometric analysis of fluorescence of indo-1 loaded tonsil B lymphocytes after crosslinking of bound monoclonal anti-receptor antibodies. CR2 alone had no effect in this system. However, crosslinking CR2 together with mIg synergistically enhanced the rise in $[Ca^{2+}]_i$ (Carter 1988). An amount of mIg which would be only minimally active when crosslinked alone gave a near maximal response when co-aggregated with CR2. No synergy was observed between CR1 and mIgM, demonstrating that the effect of CR2 was not simply due to improved efficiency of binding of limited mIg through crosslinking with any other surface structure. As synergy was observed after chelation of extracellular calcium with EGTA, the mechanism involved the enhanced release of calcium from intracellular stores.

The relevance of the complement receptor data to CD19 because apparent when immunoprecipitates prepared with anti-CD19 or anti-CR2 from digitonin lysates of B lymphoblastoid cells co-precipitated a complex which included not only CD19 and CR2 but also TAPA-1, a 26kd protein with four transmembrane domains, and other membrane proteins (Matsumoto 1991; Takahashi 1990). The association between CR2 and CD19 was confirmed by co-precipitation after transfection of both receptors into K562 cells (Matsumoto 1991). The artificially created complex from the K562 cells transfected with CD19 alone or with CR2 and CD19 also contained the additional proteins initially identified in precipitates from B cells. Thus CD19, a B-cell specific member of the immunoglobulin superfamily, associated with other proteins of more general expression (Takahashi 1990) and, presumably, functions. The complement system regulatory protein, CR2, has evolved to use this CD19 complex as a mechanism for the regulation of B cell function by activation fragments of C3.

To test whether CR2 might signal by association with CD19, the function of the latter protein when crosslinked with mIg was assessed. CD19 was found to share with CR2 the ability to enhance synergistically the rise in $[Ca^{2+}]_i$ after ligation with mIgM (Matsumoto 1991). Furthermore, this function of CD19 was intact in a CR2-deficient subclone of Ramos B lymphoblastoid cells (Carter 1991), suggesting that CR2 serves primarily as a ligand binding subunit (the cytoplasmic region of CR2 has only 34 amino acids) and functions by transducing signals via CD19 or other members of the complex. This reasoning suggests that the complex has other ligands, particularly as it is expressed throughout B cell ontogeny, while CR2 is expressed only after Ig expression. Furthermore, the existence of another means of ligating the CD19 complex and mIg in addition to C3-containing complexes would explain why animals incapable of generating C3dg properly have normal total antibody (except IgG4) levels despite the deficiencies outlined above (Heyman 1990).

To determine how binding of the CD19/CR2 complex could induce synergy, the biochemical pathways of signal transduction by mIgM and CD19 were investigated. Published reports suggested they might use a common mechanism, as both stimulate release of intracellular calcium, and ligation of either inhibits the calcium response after subsequent ligation of the other (Pezzutto 1987; Rijkers 1990). CD19 and mIg colocalize after binding of mIg (Pesando 1989). Our initial studies defined the mechanism of activation of phospholipase C (PLC) by mIgM (Carter 1991). Inositol phosphate generation and the rise in $[Ca^{2+}]_i$ stimulated by mIg could be suppressed by each of three protein tyrosine kinase (PTK) inhibitors; genistein, herbimycin, and tyrphostin. Ligation of mIgM induced a rapid increase in anti-phosphotyrosine-precipitable PLC activity, suggesting that PLC is a substrate for the PTK activated by mIg. Western blot analysis revealed induction of tyrosine phosphorylation of the gamma1 and gamma2 isoforms of PLC within 60 sec of addition of anti-IgM to B cells (Carter and Fearon, unpub.).

The mechanism of PLC activation by CD19 was then compared to that by mIg. Like mIg, the rise in $[Ca^{2+}]_i$ induced by CD19 was suppressed by the PTK inhibitors, indicating that tyrosine phosphorylation is necessary for signalling by CD19. Ligation of CD19 stimulated PTK activity, as demonstrated by new protein tyrosine phosphorylation on Western blot analysis. However, the pattern of substrates was different from that induced by IgM, suggesting linkage with a PTK different from that associated with mIgM. In addition, the regulation of PLC activation by CD19 also differed. The kinetics of inositol phosphate generation were slower, but continued to rise over 30 minutes without the plateau seen after ligation of mIgM. Activators of protein kinase C, which inhibited the increase in $[Ca^{2+}]_i$ induced by mIgM, caused a delay in the initial rise but did not suppress the magnitude of the increase in $[Ca^{2+}]_i$ by CD19. Finally, CD19 ligation did not stimulate tyrosine phosphorylation of PLC, as assessed by either PLC activity in anti-phosphotyrosine immunoprecipitates or Western blot (Carter 1991). Thus, the mechanism of activation of PLC by CD19 is unique, requiring PTK activity but without tyrosine phosphorylation of PLC. Like CD19 alone, the synergistic increase in PLC activity induced by co-ligation of CD19 and mIgM persisted in the presence of PMA, and CD19/mIgM synergy did not enhance tyrosine phosphorylation of PLC beyond that induced by mIgM alone, suggesting that the CD19 component is crucial to the biochemical interaction which produces this enhanced response (Carter 1991).

Recently, the relevance of the synergy in short term biochemical assays induced by crosslinking mIg and CD19 together for cellular activation was tested on later B cell responses requiring new gene transcription. CD19 had previously been reported to suppress later mIg-induced responses, but these experiments had bound mIg and CD19 independently (Pezzutto 1987). Preliminary studies using anti-CD19 and anti-IgM immobilized together on Sepharose or on plastic culture wells demonstrated a two-fold enhancing effect of CD19 on IgM-induced proliferation. A more profound effect was discovered when mitomycin-treated fibroblasts transfected with the low affinity Fc receptor for IgG ($Fc_{gamma}IIR-L$ cells) were used to crosslink anti-receptor antibodies bound to purified peripheral blood B cells. The presence of anti-CD19 resulted in a two log reduction in the amount of anti-IgM need to produce a given level of DNA synthesis, as measured by incorporation of tritiated thymidine over the last 16 hr of a 2.5 day culture. Anti-CD19 reduced to threshold concentration of anti-IgM for induction of 100 the number of mIgM percell that had to be ligated for proliferation to occur; this represents 0.03% of total percell mIgM. To test the effect of independent binding of CD19, $F(ab')_2$ anti-CD19 was added together with intact anti-IgM. In contrast to the enhancing effect of intact anti-CD19, $F(ab')_2$ anti-CD19 suppressed proliferation induced by anti-IgM (Carter 1992).

The effect of coaggregation of other B cell membrane proteins with limited mIg was compared to that of CD19. Only CD19 enhanced DNA synthesis to a level greater than that stimulated by optimal mIg alone, but anti-MHC Class II antibodies also had an effect. In a direct comparison of anti-CD19 and anti-MHC Class II, only anti-CD19 had an enhancing effect at concentrations of anti-IgM near the threshold for induction of DNA synthesis.

Thus, the complement system, through CR2, is linked to a membrane protein of the B cell, CD19, that augments the cellular response to low levels of mIgM ligation, accounting for the capacity of the complement system to enhance the immune response to low doses of antigen. The analysis of the biochemical mechanism by which CD19 enhances signaling may uncover novel aspects of cellular activation by receptors coupled to non-receptor PTKs.

REFERENCES

Anderson D, Koch CA, Grey L, Ellis C, Moran MF, Pawson T (1990) Science 250:979
App H, Hazan R, Zilberstein A, Ulrich A, Schlessinger J, Rapp U (1991) Mol Cell Biol 11:913
Bottger EC, Hoffmann T, Hadding U, Bitter-Suermann D (1986) J Clin Invest 78:689
Campbell MA, Sefton BM (1990) EMBO J 9:2125
Cantley LC, Auger KR, Carpenter C, Duckworth B, Graziani A, Kapeller R, Soltoff S (1991) Cell 64:281
Carter RH, Fearon DT (1992) Science 256:105
Carter RH, Park DJ, Rhee SG, Fearon DT (1991) Proc Natl Acad Sci USA 88:2745
Carter RH, Spycher MO, Ng YC, Hoffman R, Fearon DT (1988) J Immunol 141:457
Carter RH, Tuveson DA, Park DJ, Rhee SG, Fearon DT (1991) J Immunol 147:3663
Costa TE, Franke RR, Sanchez M, Misulovin Z, Nussenzweig MC (1992) J Exp Med 175:1669
Coughlin SR, Escobedo JA, Williams LT (1989) Science 243:1191
Dasgupta JD, Granja C, Drucker B, Lin LL, Yunis EJ, Relais V (1992) J Exp Med 175:285
Desai D, Newtom NE, Kadlecek T, Weiss A (1990) Nature 348:66
Dymecki SM, Niederhuber JE, Desiderio SV (1990) Science 247:332
Fantl WJ, Escobedo JA, Martin GA, Turck CW, del Rosario M, McCormick F, Williams LT (1992) Cell 69:413
Gold MR, Chan VW-F, Turck CW, DeFranco AL (1992) J Immunol 148:2012

Gold MR, Law DA, DeFranco AL (1990) Nature 345:810
Graber M, Bockenstedt LK, Weiss A (1991) J Immunol 146:2935
Graziadei L, riabowol K, Bar-Sagi D (1990) Nature 347:396
Hebell T, Ahearn JM, Fearon DT (1991) Science 254:102
Hempel W, Schatzman R, DeFranco AL (1992) J Immunol 148:3021
Heyman B, Wiersma EJ, Kinoshita T (1990) J Exp Med 172:665
Kamps MP, Buss JE, Sefton BM (1985) Proc Natl Acad Sci USA
    83:3624
Kazlauskas A, Cooper JA (1989) Cell 58:1121
Kazlauskas A, Kashishian A, Cooper JA, Valius M (1992) Mol Cell
    Biol 12:2534
Klausner RD, Samelson LE (1991) Cell 64:875
Klickstein LB, Bartow TJ, Miletic V, Rabson LD, Smith JA, Fearon
    DT (1988) J Exp Med 168:1699
Kolch W, Heidecker G, Lloyd P, Rapp UR (1991) Nature 349:426
Koretzky GA, Picus J, Schultz T, Weiss A (1991) Proc Natl Acad
    Sci USA 88: 2037
Matsumoto AK, Kopicky-Burd J, Carter RH, Tuveson DA, Tedder TF,
    Fearon DT (1991) J Exp Med 173:55
Morrison DK, Kaplan DR, Escobedo JA, Rapp UR, Roberts TM,
    Williams LT (1989) Cell 58:649
Morrison DK, Kaplan DR, Rapp U, Roberts TM (1988) Proc Natl Acad
    Sci USA 85:8855
Nadler LM, Anderson KC, Marti G, Bates M, Park E, Daley JF,
    Schlossman SF (1983) J Immunol 131:244
Park DJ, Rho HW, Rhee SG Proc Natl Acad Sci USA 88:5453
Pepys MB (1974) J Exp Med 140:126
Pesando JM, Bouchard LS, McMaster BE (1989) J Exp Med 170:2159
Pezzutto A, Dorken B, Rabinovitch PS, Ledbetter J, Moldenhauer G,
    Clark EA (1987) J Immunol 138:2793
Rijkers GT, Griffioen AW, Zegers BJM, Cambier JC (1990) Proc Natl
    Acad Sci USA 87:8766
Risso A, Smilovich D, Capra MC, Baldissarro I, Yan G, Bargellesi
    A, Cosulich ME (1991) J Immunol 146:4105
Siegel JN, Klausner RD, Rapp UR, Samelson LE (1990) J Biol Chem
    265:18472
Smith MR, Lun YL, Kim H, Rhee SG, Kung HF (1990) Science 247:1074
Stamenkovic I, Seed B (1988) J Exp Med 168:1205
Takahashi S, Doss C, Levy S, Levy R (1990) J Immunol 145:2207
Tedder TF, Isaacs CM (1989) J Immunol 143:712
Turner B, Rapp U, App H, Greene M, Dobashi K, Reed J (1991) Proc
    Natl Acad Sci USA 88:1227
Ullrich A, Schlessinger J (1990) Cell 61:203
Varticovski L, Daley GQ, Jackson P, Baltimore D, Cantley LC
    (1991) Mol Cell Biol 11:1107
Weiss A, Koretzky G, Schatzman RC, Kadlecek T (1991) Proc Natl
    Acad Sci USA 88:5484
Whitman M, Kaplan DR, Schaffhausen B, Cantley L, Roberts TM
    (1985) Nature 315:239
Williams NG, Roberts TM, Li P (1992) Proc Natl Acad Sci USA
    89:2922
Yamanashi Y, Fukui Y, Wongsasant B, Kinoshita Y, Ichimori Y,
    Toyoshima K, Yamamoto T (1992) Proc Natl Acad Sci USA 89:1118
Zhou LJ, Ord DC, Hughes AL, Tedder TF (1991) J Immunol 147:1424

Molecular Mimicry in the Pathogenicity of Microorganisms

Michael M. Frank

Duke University Medical Center, Box 3352, Durham, North Carolina 27710 USA

In 1964 Damian coined the term "molecular mimicry." (Damian, 1989) He used this term to refer to sharing of antigenically cross reactive groupings on microorganisms with those on host cells and other tissue elements. He pointed out that such mimicry might lead to a series of consequences. 1) The development of autoimmunity. An immune response directed at a microbial antigen that cross reacts with host antigens might lead to host tissue damage. 2) Failure to recognize and eliminate a microbial pathogen. In this case, tolerance to ones own antigens would preclude an adequate immune response directed at similar microbial antigens. 3) Pressure within the host species to generate polymorphic antigens and to select those that do not cross react with microbial antigens. The first experiments were performed with polyclonal antimicrobial antisera. With the advent of monoclonal antibodies it became possible to perform epitope mapping, carefully detailing areas of antigenic similarity between host and microbe. The revolution in molecular biology made it possible to compare the sequence of these antigenically, functionally cross reacting antigens and to determine whether the similarity between host and microbe extends to the genetic level. It also became possible to compare literally tens of thousand of sequences, base pair by base pair, to identify new examples of mimicry not dependent on the laborious process of developing appropriate monoclonal antibodies. It has become clear that mimicry exists at every level, from genetic mimicry to structural or functional mimicry with no clear genetic similarity. It has also become clear that microbes do indeed take advantage of that mimicry to protect themselves from host defense mechanisms. In a few defined cases, mimicry has not only been used for protection against host defense mechanisms, but has also come to facilitate the invasive or infectious process. In this paper, I will discuss a number of the mechanisms employed by various microbes that exhibit functional or structural mimicry of complement components or complement control proteins. The various examples illustrated will emphasize the work of my own laboratory, but will also include the work of others.

In 1982, Cines et al reported that Herpes simplex type 1 infected cells express receptors for the Fc fragment of IgG and for C3b (Cines et al, 1982). By 1984 it was recognized that these infected cells express on their surface two proteins gC and gE that act as the complement and IgG Fc receptors respectively (Baucke et al, 1979; Friedman et al, 1984). The proteins that participate in the receptor activities are not required for infectivity in cell cultures. Nevertheless, they are found in almost all clinical isolates suggesting that they are important for viral pathogenesis in vivo. Of these two proteins, the IgG Fc receptor has been studied in greatest detail. The first herpes protein to be identified as part of the herpes IgG receptor complex was gE. It has no sequence homology to any of the known classes of IgG Fc receptors, but functions as a similar IgG Fc binding protein. Recent evidence suggests that gE by itself has low but clearly definable affinity for the IgG Fc fragment. As such it acts like a low affinity IgG receptor binding aggregated IgG most efficiently (Bell et al, 1990; Dubin et al, 1990; Hanke et al, 1990). In the presence of a second membrane protein, coded for by the virus and expressed in the membrane of infected cells, gI, a complex is formed and the receptor functions as a high affinity Fc receptor. A number of studies have shown that the presence of the receptor on Herpes simplex infected cells clearly protects the cells from immunologic attack (Adler et al, 1978; Dowler et al, 1984; Frank et al, 1989). Antibody dependent cellular cytotoxicity is clearly reduced on cells that express the IgG receptor (Dubin et al, 1991). It has been suggested that the Fc receptor complex binds to the Fc fragment of specific anti-herpes antibody that is bound via the Fab combining site to herpes antigens. The Fc receptor binding renders such antibody less likely to activate complement via C1q binding to the IgG Fc fragment and in addition inhibits ADCC (Frank et al, 1989). In addition, the binding of specific immunoglobulin and complement by the virus is partially blocked by non-specific immunoglobulin bound to the Fc receptor, presumably by steric hindrance.

The complement receptor gC also has been studied in some detail. It has clear specificity for the C3 activation fragments C3b and C3bi (Fries et al, 1986; Kubota et al, 1987; Tal-Singer et al, 1991) and some monoclonal antibodies to the C3b complement receptor CR1 block its activity (Kubota et al, 1987). Purified gC but not another virally coded membrane glycoprotein, gD, demonstrate dose dependent decay acceleration of the alternative pathway C3 convertase, C3bBb, and induce a dose dependent time independent depression of the overall efficiency of C3bBb sites (Fries et al, 1986). Thus, it inhibits lysis of cells via the alternative complement pathway. Glycoprotein C is able to bind to C3b, depressing the ability of preformed C5b6 to initiate reactive lysis. It has no effect on the ability of C8 or C9 to interact with the C5b67 lytic site and has no effect on the reactive lysis of erythrocytes sensitized with antibody alone. Glycoprotein C did not accelerate the decay of SAC14b2a and did not depress classical pathway efficiency in the presence of excess C5. This viral protein had no factor I cofactor activity.

Herpes simplex virus type 2 has been found to code for an antigenically similar protein, although this protein is not functionally expressed in the membranes of infected cells (McNearney et al, 1987). Sequence analysis has shown that much of the protein is up to 70% identical to herpes simplex virus type 1 gC (Swain et al, 1985). Purified gC-2 binds C3 in direct binding assays using purified protein, suggesting functional as well as antigenic similarity to gC type 1. However, there are clear differences in functional activity between these two proteins. In systems using purified gC2, it was found that this protein does not increase the rate of decay of the alternative pathway convertase and has no effect on the activity of C5b-9. In fact, the protein stabilizes the C3 convertase, inhibiting the normal decay (Eisenberg et al, 1987). GC-2 linked oligosaccharides are clearly different from gC-1. Endo F and endo H both abrogate binding of C3b to gC2, but not gC-1. When the gene coding for the protein is transfected into mammalian cells, these cells demonstrate newly acquired C3b binding activity. It proved difficult to map the specific regions on gC-1 that were responsible for C3b binding, but this could be accomplished with the related glycoprotein gC-2 (Seidel-Dugan, 1990).

In one series of experiments, in frame linker insertional mutants of gC-2, with each linker comprised of up to 4 additional insertional amino acids, added at sites across the protein, were examined. Each protein was expressed on the surface of transfected cells. The proteins were processed to contain N linked oligosaccharides and bound one, two, or usually three monoclonal antibodies recognizing discontinuous epitopes of gC-2. This suggested that at least regions of the inserted protein were folded into a native configuration. Three distinct regions of the protein were found to be required for C3b binding. The three regions were in residues 102-107, 222-279 and 307-379 of 480 amino acid residues. There were multiple mutants examined at each critical site and no clear relation between the nature of the amino acid inserted and the loss of C3b binding activity. It could not be determined whether these three regions come together to form a binding site or whether they affect the folding at distant regions. In spite of their different functional activities both gC-1 and gC-2 when inserted into HSV-1 and expressed, protect against viral neutralization. In addition, gC-1 has been shown to play a protective role at the infected cell surface (Harris et al, 1990; Hidaka et al, 1991; McNearney, 1987).

Although gC1 and gC2 have limited sequence homology with mammalian members of the C3b binding family, there is a membrane motif consisting of four cysteine residues, conserved tryptophan and glycine residues and a site for N linked glycosylation between cysteine residues 4 and 7 of the amino acid sequence raising a question of structural mimicry if this is found to be the true site of C3b binding (Seidel-Dugan, 1990).

The herpes viruses demonstrate even more striking examples of genetic mimicry. Nemerow et al have found a region of sequence homology in the 350 kD major envelope protein of Epstein-Barr virus, consisting of a 9 amino acid primary sequence epitope at the amino terminus of this protein, and C3d (Nemerow et al, 1989). It appears that the virus gains entry into B lymphocytes by binding to the receptor for C3d, (CR2 or CD21), and is internalized during the course of receptor triggering. This is a true example of mimicry leading not only to escape from host defense mechanisms, but subversion of a normal physiologic process to the ends of the viral pathogen. An EBV protein also acts as a regulatory molecule of complement function but, interestingly, the set of functions mediated by this regulator differs somewhat from gC (Mold et al, 1988). This molecule has factor I cofactor activity and like CR1 mediates the cleavage of C3b to iC3b and the further cleavage to C3dg. It also acts as a cofactor for the factor I mediated cleavage of C4b to iC4b with further cleavage to C4c and C4d. This protein like gC accelerates the rate of decay of the alternative pathway convertase C3bBb, but not the classical pathway convertase.

Recent studies have demonstrated striking mimicry shown by pox viruses (Kotwal et al, 1988; Kotwal et al, 1990). In this case, mimicry is structural, functional and in nucleotide sequence. The protein studied is not expressed in the membrane of virally infected cells as is gC-1 and 2 and the EBV viral envelope protein, but is a secreted product of the virally infected cells. Cells infected by vaccinia virus synthesize a number of proteins. One of the most abundant is a secreted product predicted to have a molecular weight of 28.6 kD on the basis on amino acid sequence, but with an observed weight of 35 kD on SDS gels. The protein is not required for viral replication.When the relevant gene was sequenced and compared to published sequences in genebank, it proved to have 38% sequence identity to a portion of the C4 binding protein, one of the complement regulatory proteins. C4 binding protein has 7 polypeptide chains, each with 8 short consensus repeat regions of 60 to 70 amino acids.

The vaccinia protein contained 4 consensus repeats, 38% identical to the amino half of C4bp and 28% identical to the carboxyl 4 consensus repeats. This finding was perhaps surprising since vaccinia is the most studied example of an organism whose spread is controlled by active cellular immunity. The vaccinia C4 binding protein homologue has been isolated and studied in some detail. Its binding to C4b has been confirmed and it has been found to bind as well to C3b (McKenzie et al, 1992). Thus, it binds to both of the proteins that function in the formation of the C3 convertase. This protein termed VCP (vaccinia control protein) has been shown to react rapidly with the convertase as does gC and the EBV envelope protein, the effect being seen at "zero time" in kinetic studies. It induces a dose dependent acceleration of decay of both the classical and alternative pathway convertases. Like the EBV envelope protein, it acts as a cofactor for the cleavage of C3b to iC3b and C3dg and as a cofactor for the cleavage of C4b to C4c and C4d. In initial experiments, VCP inhibited the lysis of sheep erythrocytes sensitized with rabbit antibody via the classical complement pathway and could not be shown to inhibit the lysis of these cells in an alternative pathway dependent system (Kotwal et al, 1990). Quantitative factors proved to be critical for this effect and there is clear evidence that the alternative pathway is affected as well (Isaacs et al, 1992). In fact, in direct tests of viral neutralization in the fluid phase, in the presence of excess antibody and complement, the alternative pathway showed maximal activity in viral neutralization and little activity of the classical pathway was demonstrated. In this test system, C4 deficient guinea pig serum was as effective as C4 sufficient guinea pig serum in neutralization of virus and factor B deficient human serum had no neutralizing activity when compared to normal human serum. Repletion of factor B deficient serum with factor B restored full neutralizing activity. As part of these studies, viral mutants were prepared that lacked the gene for the VCP. The supernatant fluids of virally infected cells lacked the regulatory activities listed above.

Studies were performed to determine whether the presence of the C4/C3 binding protein played a role in protection of the virus from the host defense process. Rabbits and C4 sufficient and deficient guinea pigs were injected intradermally with the wild type and mutant vaccinia virus. There was no difference in induration or swelling during the first two days following injection, suggesting that the virus replicated normally and produced the same degree of initial tissue destruction. Thereafter, and particularly by day 5 of infection, there was a clear difference between the wild type and mutant form. The mutant virus produced smaller and less indurated lesions that cleared more rapidly. This suggested that control of C4/C3 played a role in viral pathogenesis. Interestingly, other pox viruses like ectromelia have a similar protein. It will be of interest to see whether that similar protein effects the pathogenicity of this important mouse virus.

The third example of molecular mimicry of complement factors studied in my laboratory is found not in viruses, but in the fungus Candida albicans. It is known that the mycelial phase rather than the yeast phase is the form most associated with tissue pathology. It is also clear that one important aspect of fungal pathogenicity is the ability of the fungus to adhere to host tissues. There is a close correlation between adhesion and pathogenicity and mutants of the fungus that demonstrate low level adhesiveness to cells like umbilical cord endothelial cells in culture have low pathogenicity. During the course of studies of candidal binding of C3b, C3bi and C3d coated red cells in my laboratory, we noted that organisms bound EAC3bi and EAC3d avidly, but not EAC3b (Edwards et al, 1986). Further study demonstrated the presence of a membrane protein present on the surface of the mycelial form of Candida albicans, but not present on the surface of the non-pathogenic Candida stellatoidia that bound C3bi. We further noted that several monoclonal antibodies to CR3 and CR2 bound to the surface of cultured Candida albicans. Only the binding of Mo-1 was observed by fluorescence microscopy. Use of a monoclonal antibody could partially block C3d, but not C3bi binding to the candidal surface. The range of monoclonals that bound to the Candida mycelia was limited in these initial studies. Search of the literature revealed that Dierich and his colleagues had published similar studies of C3bi binding, although not of monoclonal antibody binding, in abstract and the papers describing these effects were published thereafter (Heidenreich et al, 1985; Eigentler et al, 1989). A number of individuals have now explored this phenomenon in some detail. Hostetter using FACS analysis noted that both the yeast and mycelial forms of the organism bound monoclonal anti CR3 antibodies, although the mycelial phase bound more (Gilmore et al, 1988). She noted that temperature and especially the presence of glucose markedly upregulated the expression of this receptor (Hostetter et al, 1990). She confirmed the fact that non-pathogenic forms of the organism failed to show surface expression of the receptor. The receptor protein has been purified and reported to be a glycoprotein of molecular weight on gels varying from 165-190, (Eigentler et al, 1989; Hostetter et al, 1990) although this point remains controversial (Calderone et al, 1991). A related protein has been cloned and sequenced and has 26% homology with the α chain of CR3. Non-pathogenic forms of the organism have the gene for the protein, but show no surface expression (Bendel et al, 1992; Herman et al, 1992). The reasons for this are unknown. Presence of the protein has been reported to decrease the effectiveness of C3bi in opsonization of the yeast form of the organism, implying a protective effect in the host defense process. Thought to be even more important is the ability of the protein to facilitate adhesion to cultured endothelial cells, as shown by the ability of monoclonals and complement fragments to block the adhesion (Gustafson et al, 1991). It is suggested that yeast have a number of such proteins, but that one principal protein for Candida albicans is this ß2 integrin homologue. Hostetter and colleagues find that TNF treatment of endothelial cells does not increase binding of the yeast phase organism. Since such treatment upregulates expression of ICAM-1, they suggest that ICAM-1 is not the endothelial cell protein recognized by the candidal surface protein.

A number of parasites have in addition been shown to express proteins that mimic complement control sequences. The trypomastigote form of Trypanosoma cruzi expresses a developmentally regulated molecule with sequence homology to DAF, gp160, that binds C3b and blocks alternative pathway C3 convertase assembly (Norris et al, 1991). This and other less well characterized complement regulatory molecules go far to explain the serum resistance of the parasite forms exposed to blood. Also, recently described is a 75kD molecule with antigenic similarity with C9 that can induce pore formation at low pH. It is suggested that this molecule facilitates exit of the organism from the acidified endocytic vacuole into the cytoplasms of infected cells (Andrews et al, 1990). Plasmodium falciparum contains proteins with thrombospondin repeats like those in C6-9 that may play a role in pathogenesis (Goundis et al, 1988). The infectious metacyclic form of Leishmania expresses a major surface glycoprotein, gp63, that binds to CR3 (Russell et al, 1989).

It should be noted that we have emphasized proteins in each of these discussions. Polysaccharides mediate adhesion as well and the repertoire of polysaccharides is much more limited. Therefore, overlap between parasite and host structures can be expected. Lectin like binding of polysaccharide containing structures to molecules of functional importance is well demonstrated. Thus, it is not surprising that molecular mimicry at this level is also demonstrated. Thus, Bullock and Wright have demonstrated that unopsonized Histoplasma capsulatum uses surface carbohydrate to bind to ß 2 integrins to facilitate phagocytic cell uptake (Bullock et al, 1987). This organism can resist destruction in such phagocytes and continue to propagate. Whether this is true molecular mimicry remains to be determined.

Each of the examples studied in my laboratory has interesting parallels. In each case, the protein in question helps avoid complement dependent host defense processes. In each case, the protein in question is not required for division of the microorganism; forms of the organism without the regulatory protein are capable of normal growth in culture. In each case, natural forms of the organism lacking the regulatory factor are of much lower pathogenicity than forms containing the factor and mutants with specific gene deletions of the factor have lower pathogenicity in animal models. In some cases like vaccinia, the protein is one of the major secreted products of viral infection. In other cases, like in herpes and candida, it is a membrane bound structure. Taken together these facts suggest that this set of factors plays an important role in pathogenicity. For those reasons, it is perhaps surprising that complement is usually not thought to play a major role in host defense against fungal or viral infection. Moreover, it has not been appreciated that patients lacking complement proteins have a high incidence of infection with these groups of organisms. It is not truly satisfying to add the disclaimer that there are many mechanisms of host defense, including those mediated by antibody, lymphocytes, phagocytes, etc., and that the failure of one mechanism will in most cases not be critically important because of the presence of compensatory mechanisms. Thus, there is considerable work required in understanding how these proteins function in the normal host defense process.

We have reviewed examples of molecular mimicry associated with complement and complement regulatory molecules. How frequent are examples of molecular mimicry and how do they come about? Notkins, Oldstone and colleagues studied the binding of a battery of monoclonal antibodies to a battery of 14 viruses representing most of the viral classes (Srinivasappa et al, 1986; Oldstone, 1989). In approximately 700 binding studies they demonstrated 4% cross reaction between microbial and host antigens. Their feeling was that structural mimicry of this sort is surprisingly common. When one considers the fact that the most homologous sequences may be the least likely to generate monoclonal antibodies, that figure is even more striking. With the availability of computer searches of nucleotide sequence banks, examples of molecular mimicry are easier to find and remain surprisingly common. In some cases, these represent single or a limited number of gene mutations leading to a short stretch of homology in a rather different protein. That such could lead to structural or functional advantageous homology, which is then expanded by selective pressures, is quite reasonable. In the case of vaccinia there appears to be duplication of a homologous cassette. Occasionally, as is the case with the candidal protein, the homology is extensive and occupies much of the homologous protein. Such homology raises the possibility that genetic material has been exchanged in an ancestral form and retained during evolution. It has been suggested that exon cassettes are much more limited than currently recognized and that frequent homologies are to be expected (Dorit et al, 1990). Whatever the reason, it is clear that molecular mimicry is a frequent occurrence and plays an important role, only beginning to be understood, in the pathogenicity of microbial infection.

REFERENCES

Adler R, Glorioso J, Cossman J, Levine M (1978) Infect Immun 21:442-447

Andrews NW, Abrams CK, Slatin SL, Griffiths G (1990) Cell 61:1277-87

Baucke RB, Spear PG (1979) J Virol 32:779-789

Bell S, Crange M, Borysiewicz L, Minson T (1990) J Virol 64:2181-86

Bendel CM, Hostetter MK (1992) Ped Res 30

Bullock WE, Wright SD (1987) J Exp Med 165:195-210

Calderone RA, Braun PC (1991) Microbiol Reviews 55:1-19

Cines DB, Lyss AP, Bina M, Corkey R, Kefalides NA, Friedman HM (1982) J Clin Invest 69: 123-128

Costa J, Rabson AS, Yee C, Tralka TS (1977) Nature 269: 251-252

Damian RT (1989) Curr Topics in Microbial and Immunol 145:101-115

Dorit RL, Schoenbach L, Gilbert W (1990) Science 250:1377-1382

Dowler K, Veltri R (1984) J Med Virol 13:251-259

Dubin G, Frank I, Friedman HM (1990) J Virol 64:2725-2731

Edwards JE, Gaither TA, O'Shea JJ et al (1986) J Immuno 137:3577-83

Eigentler A, Schulz TF, Larcher C, Breitweiser, E-M, Myones BL, Petzer AL, Dierich MP (1989) Infect Immunity 57:616-622

Eisenberg RJ, Ponce de Leon M, Friedman HM, Fries LF, Frank MM, Hastings JC, Cohen GH (1987) Microb Pathog 3:423-435

Frank I, Friedman HM (1989) J Virol 63:4479-4488

Friedman HM, Cohen GH, Eisenberg RJ, Seidel CA, Cines DB (1984) Nature 309:633-635

Fries LF, Friedman HM, Cohen GH, Eisenberg RJ, Hammer CH, Frank MM (1986) J Immunol 137:1636-1641

Goundis D, Reid KBM (1988) Nature 335:82-85

Gustafson KS, Vercellotti GM, Bendel CM, Hostetter MK (1991) J Clin Invest 87:1896-1902

Hanke T, Graham FL, Lulitanand V, Johnson DC (1990) Virol 177:437-444

Harris SL, Frank I, Yee A, Cohen GH, Eisenberg RJ, Friedman HM (1990) J Infect Dis 162:331-337

Heidenreich F, Dierich MP (1985) Infect Immun 50:598-600

Herman DJ, Kendrick KE, Bendel CM, Tao N, Hostetter MK (1991) Clin Res 39:244A

Hidaka Y, Sakai Y, Yasushi T, Mori R (1991) J Gen Virol 72:915-921

Hostetter MK, Lorenz J, Preus L, Kendrick KE (1990) J Infect Dis 161:761-68

Issaacs SN, Kotwal GJ, Moss B (1992) Proc Nat Acad Sci USA 89:628-632

Johnson DC, Frame MC, Ligas MW, Cross AM, Stow ND (1988) J Virol 62:1347-1354

Kotwal GJ, Isaacs SN, McKenzie R, Frank MM, Moss B (1990) Science 250:827-830

Kotwal GJ, Moss B (1988) Nature 33:176-178

Kubota Y, Gaither TA, Cason J, O'Shea JJ, Lawley TJ (1987) J Immunol 138:1137-1142

McKenzie R, Kotwal GJ, Moss B, Hammer CH, Frank MM (1992) J Infect Dis, In Press

McNearney TA, Odell C, Holers M. Spear PG, Atkinson JP (1987) J Exp Med 166:1525-1535

Mold C, Bradt BM, Nemerow GR, Cooper NR (1988) J Exp Med 168:949-969

Nemerow GR, Houghten RA, Moore MD, Cooper NR (1989) Cell 56:369-377

Oldstone MBA (1989) Current topics in Microbiol Immunol 145:127-135

Russell DG, Talamas-Rohana P (1989) Immunol Today 10:328-333

Seidel-Dugan C, Ponce De Leon M, Friedman HM, Eisenberg RJ, Cohen GH (1990) J Virol 64:1897-1906

Sjöblom I, Lundstrom M, Sjögren-Jansson E, Glorioso JC, Jeansson S, Olofsson S (1987) J Gen Virol 68:545-554

Srinivasappa J, Saegusa J, Prabhaker BS, Gentry MK, Bucheneier MJ, Witkor TJ, Koprowski H, Oldstone MBA, Notkins AL (1986) J Virol 57:397-401

Swain MA, Peet RW, Galloway DA (1985) J Virol 53:561-69

Tal-Singer R, Seidel-Duagn C, Fries LF, Huemer HP, Eisenberg RJ, Cohen GH, Friedman HM (1991) J Infect Dis 164:750-753

The Role of Complement in Enhancing Infection of Target Cells with Human Immunodeficiency Virus

C. Delibrias, N. Thieblemont, V. Boyer*, E. Fischer, N. Haeffner-Cavaillon, C. Desgranges* and M. D. Kazatchkine

*INSERUM U 28, Hospital Broussais, 75014 Paris and *INSERUM U 271, 69 424 Lyon Cedex 03, France*

INTRODUCTION

The CD4 glycoprotein mediates the high affinity binding of the gp120 envelope glycoprotein of human immunodeficiency virus (HIV) to target cells expressing CD4. CD4 serves as the primary receptor for HIV on T lymphocytes, monocyte/macrophages and dendritic cells. Infection of CD4[+] cells in vitro has been shown to be facilitated by anti-HIV antibodies through Fc receptor-mediated enhancement (Takeda 1988, Homsy 1989) and by serum from HIV-infected subjects through mechanisms involving both Fc and complement receptors (Robinson 1988, Tremblay 1990, June 1991, Bakker 1992). Recent studies have shown that a number of CD4[-] cells are susceptible to HIV infection in vivo and in vitro, indicating that alternative pathways to CD4 exist for viral penetration and that the host cell range of HIV is broader than originally described. The present review focuses on the enhancing role of complement, whether alone or in association with anti-HIV antibodies, on infection of CD4[+] and CD4[-] target cells with HIV.

COMPLEMENT ACTIVATION BY HIV

The human retroviruses HIV (Banapour 1986) and HTLV-1 resist complement-mediated lysis although the viruses efficiently activate complement. The molecular mechanisms that are responsable to the resistance of HIV to lysis by complement of human but not of animal origin have not yet been identified.
Intact HIV (Ebenbichler 1991), recombinant gp160 of HIV (Thieblemont 1992) and HIV-infected cells (Sölder 1989, Spear 1990) have been shown to activate complement in human serum resulting in the cleavage of C3 and subsequent deposition of C3 fragments on the viral surface or on the surface of infected cells. Complement activation by HIV occurs through the classical pathway following the binding of C1q to conserved peptidic sequences of gp41 and direct activation of the C1 complex (Ebenbichler 1991). The major C1-binding site in gp41 is contained within a region of the molecule that becomes exposed following the binding of gp120 to CD4 (Sattentau 1991). Although activation of C1 by gp41 and by gp160 may occur in the absence of antibodies, it is significantly enhanced by anti-HIV envelope IgG (Fig 1). Under optimal

experimental conditions, classical pathway activation by mammalian-derived recombinant gp160 results in the deposition of one molecule of C3b/iC3b per molecule of gp160. Deposition of C3 fragments on the envelope glycoprotein complex provides the basis for antibody-dependent and -independent complement-mediated enhancement of infection of C3 receptor-bearing cells. The deposition on gp160 and gp120 of C3b and iC3b allows complement-opsonized HIV to interact with CR1(CD35), CR2(CD21) and CR3(CD11b/CD18).

Complement is chronically activated in HIV-infected individuals (Perricone 1987; Senaldi 1990; Tausk 1986). HIV-infection is also associated with acquired changes in the membrane expression of complement regulatory molecules, including a decreased expression of CR1 on erythrocytes and leukocytes (Tausk 1986; Jouvin 1987; Cohen 1989), a decreased expression of CR2 on peripheral blood CD4[+] T lymphocytes (June 1992), a defective expression of DAF (CD55) on leukocytes (Lederman 1989) and of the CD59 membrane inhibitor of complement-mediated cytolysis on T lymphocytes of infected patients (Weiss 1992). The latter alterations could explain the increased susceptibility of lymphocytes of AIDS patients to homologous complement-mediated lysis (Lederman 1989).

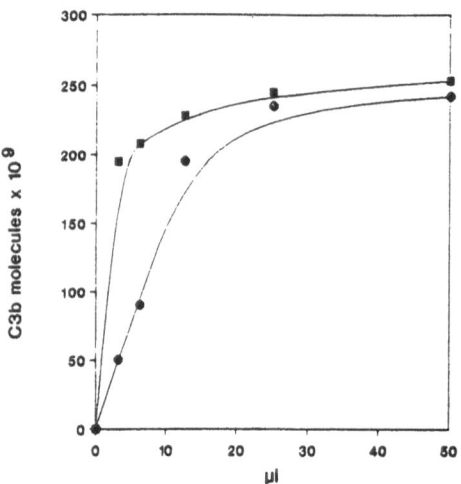

Figure 1: Complement activation by glycosylated recombinant gp160 of HIV-1. ELISA plates coated with rgp160 were incubated with increasing amounts of normal HIV seronegative serum (closed circles) or first incubated with anti-HIV IgG and then incubated with the seronegative serum (closed squares). Deposition of C3b on rgp160 was assessed using a monoclonal antibody recognizing a neo epitope expressed by the C3b and the iC3b fragment of C3. Results were expressed as numbers of C3b molecule bound per well.

COMPLEMENT-MEDIATED ENHANCEMENT OF INFECTION OF T LYMPHOCYTES

Recent studies have identified the presence of CR1 and CR2 on subsets of human peripheral blood T lymphocytes (Wilson 1983, Rodgaard 1991, Fischer 1991) and of CR2 on thymocytes (Tsoukas 1988). CR1 is a polymorphic glycoprotein of 160-250 kDa which serves as the receptor for C3b and C4b. CR1 is expressed on approximately 15% of peripheral blood CD4+ and CD8+ T cells (Wilson 1983). CR1+ T lymphocytes also express Fc gamma receptors as assessed by their ability to bind fluoresceinated aggregated IgG (Wilson1983, Cohen 1989). Most CR1+ T cells also express CR2 (Delibrias 1992). CR2 is a 145 KDa glycoprotein that functions as a receptor for C3dg/C3d and iC3b and as the receptor for EBV on B lymphocytes. CR2 is expressed on 50% of both CD4+ and CD8+ peripheral blood T lymphocytes at an approximately 10 fold lower density than on B cells. On the human T cell line HPB-ALL, CR2 and CR1 are co-internalized when cross-linked with anti-receptor antibodies; CR2 is capable of signal transduction (Delibrias 1992).

The first evidence for the role of complement in enhancing HIV infection of T cells came from the observation by Robinson (Robinson 1988, Robinson 1989) that sera from more than 80% of HIV antibody-positive individuals were capable of increasing the release of virus from the MT2 T cells infected with HIV. Enhancement of CD4+ CR2+ MT2 cells infection by seropositive serum required the presence of complement, anti-HIV antibodies and the CD4 molecule. We have subsequently found that complement alone (i.e. in the absence of anti-HIV antibodies) was capable of enhancing infection of MT2 cells with suboptimal amounts of HIV-1 (Boyer 1991); productive viral infection of MT2 cells with low inputs of HIV that had been pre-opsonized with normal human HIV-seronegative serum occurred through the interaction of opsonized virus with CR2, since infection was blocked by cross-linked mAb against CR2 but not by a mAb against the gp120 binding site of CD4 (Boyer 1991) (Fig 2). Complement alone was also shown to enhance infection of normal peripheral blood T lymphocytes as assessed by its ability to enhance infection of PHA-stimulated seronegative peripheral blood leukocytes cocultured with leukocytes from HIV-infected individuals (Boyer 1991).

Little is known at present on the role of CR1 in facilitating the infection of T cells with complement-opsonized HIV. Preliminary evidence indicates that blocking of CR1 partially decreases CD4-dependent infection of the CD4+ CD8+ CR1+ CR2+ HPB-ALL cell line with complement-opsonized HIV. CR1 on T cells could either function as a receptor mediating viral entry into the cells or as a cofactor for the cleavage of C3b into iC3b and C3dg to allow the interaction of opsonized virus with CR2.

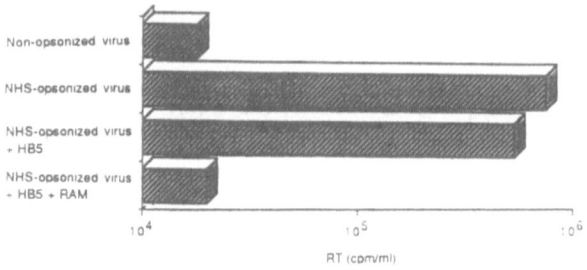

Figure 2: CR2 mediates infection of MT2 cells with complement-opsonized HIV-1 (Boyer 1991). MT2 cells were infected with low amounts of HTLV-RF that had been preopsonized with normal seronegative human serum. Before infection, MT2 cells were preincubated with saturating amounts of anti-CR2 antibody HB5 followed or not by rabbit F(ab')2 anti-mouse Ig antibodies (RAM). RT activity was assessed at day 9 of culture. Nonopsonized virus: cells infected with virus in the absence of opsonization with serum; NHS-opsonized virus: cells infected with virus that had been preincubated with seronegative human serum; NHS-opsonized virus + HB5: cells preincubated with anti-CR2 mAb HB5 and infected with complement-opsonized virus; NHS-opsonized virus + HB5 + RAM: cells preincubated with anti-CR2 mAb HB5 crosslinked with RAM and infected with complement-opsonized virus.

COMPLEMENT-MEDIATED ENHANCEMENT OF INFECTION OF MONOCYTE/MACROPHAGES.

Monocyte/macrophages represent a major reservoir of virus in HIV-infected individuals. The cells express CD4 at low density on the membrane. Depending on their state of maturation, cells of the monocytic lineage also express various amounts of Fc gamma RI (CD64), Fc gamma RII (CD32) and Fc gamma RIII (CD16) receptors, LFA-1 (CD11a/CD18), CR1 (CD35) and CR3 (CD11b/CD18). Complement enhances in vitro infection of monocyte/macrophages with HIV in the presence (Robinson 1988) and in the absence (Reisinger 1990) of antibodies. The binding of fluorescein-labeled purified HIV-1 to human monocytes was shown to be increased when virus was pre-opsonized with antibodies and complement or with complement alone (Bakker 1992). Infection of the human monoblastoid cell line U937 with low amounts of HIV was facilitated by complement without need for anti-HIV antibodies; the enhancing effect of complement infection of U937 cells was reduced in the presence of antibodies to CD4 or antibodies to CR3 and totally abrogated with a combination of these antibodies (Reisinger 1990). We have recently demonstrated that complement enhances the productive infection of cultured normal peripheral blood monocytes, of the promonocytic cell line THP-1, and of the monocytic cell line Mono Mac 6 with HIV-1 and HIV-2. The cells and cell lines express CR1 and CR3. Cultured monocytes and THP-1 cells express low amounts of CD4 whereas Mono Mac 6 cells do not express CD4 antigen nor CD4 transcript. Thus, the enhancing effect of complement on infection of cells of the monocytic lineage may occur independently of CD4, indicating that the interaction of opsonized virus with complement receptors may be sufficient to mediate penetration of HIV into monocytes. Preincubation of target monocytic cells with F(ab')2 fragments of anti-CR1 antibodies or of monoclonal antibodies directed against the alpha chain of CR3 (but not against the alpha chain of LFA-1) suppressed infection of the cells with complement-opsonized HIV.

COMPLEMENT-MEDIATED ENHANCEMENT OF INFECTION OF B LYMPHOCYTES

EBV-transformed human B cell lines, EBV⁻ cell lines and EBV-carrying normal human B lymphocytes (Laurence 1991) are susceptible to HIV-infection in vitro. B lymphocytes are activated in HIV infected-individuals. There is no evidence, however, that normal B lymphocytes may be infected with HIV in vitro or in vivo.
Lymphoblastoid B cell lines express CR2 and some of them also express CR1. The expression of CR2 is upregulated by EBV in EBV-transformed B cell lines (Cohen 1987). All normal mature B lymphocytes express CR1 and CR2 (Tedder 1983, Tedder 1984). Triggering of CR1 on pre-activated B cells enhances their differentiation into antibody-secreting cells, whereas triggering of CR2 induces B cell proliferation.
Early studies have shown that EBV-transformed human B cell lines are susceptible to HIV infection in vitro (Montagnier 1984, Salahuddin 1987). Depending on the cell line and on the strain of HIV, infection of EBV+ B cells has resulted in cell lysis or in persistent non-productive or productive infection (Dahl 1987, Dahl 1990, Tremblay 1988, De Rossi 1990). HIV has also been shown to infect EBV⁻ Burkitt lymphoma B cell lines, indicating that susceptibility to HIV infection is not strictly dependent on the presence of the EBV genome (Monroe 1988). An enhancing role of complement on infection of EBV+ B cells was first shown in an experimental system wherein a CD4+ B lymphoblastoid cell line was used as the target for infection in the presence of complement and anti-HIV antibodies (Tremblay 1990). A recent report indicated that antibody and complement-dependent enhancement of infection of the EBV+ IC-1 B cell line was mediated by CR2 and CD4, as infection was blocked either by anti-CR2 or anti-CD4 antibodies (Gras 1991). Opsonization with complement would enhance infection by increasing the number of viral particles attaching to CR2-bearing cells and by subsequently facilitating the entry of the virus through CD4 on cells expressing low amounts of the

molecule. Alternatively, one may speculate that CR2 could directly mediate the penetration of HIV into the cells through endocytosis following initial binding of the virus to CD4.

There is also evidence in the litterature that infection of B cell lines with HIV may proceed independently of CD4 (Salahuddin 1987, De Rossi 1990). We have recently demonstrated that CR2 may mediate on its own the productive infection of cells of the Raji B lymphoblastoid cell line with complement-opsonized HIV (Boyer 1992). Infection of the cells occurred independently of antibodies and of CD4 molecule since HIV-seronegative serum was used for infection and since the cells lack the expression of CD4 surface antigen and of CD4 transcript.

There has been increasing evidence that complement may facilitate infection with HIV of target cells expressing receptors for fragments of cleaved C3. Since HIV is not or only poorly lysed by human complement, activation of complement by the virus serves to facilitate viral entry into cells rather than to neutralize and eliminate the virus. Complement-mediated enhancement of infection may depend on the sole interaction of C3-bearing viral particles with C3 receptors or on a synergistic cooperation of C3 receptors, Fc receptors and/or CD4 on target cells. Although C3 receptors may mediate viral penetration and although HIV may activate complement in the absence of antibody, the facilitating role of complement is probably greatly enhanced by antibodies in seropositive individuals in vivo. Little is known however at this time on the complement-activating potential of "neutralizing" and "facilitating" anti-HIV antibodies and on how the binding of C3 to HIV may interfere with the subsequent binding of anti-HIV antibodies to the viral envelope. Enhancement of HIV-infection by complement may contribute to extend the range of target cells for the virus and to increase the rate of infection in patients with a low viral load. Complement-dependent enhancement also allows infection of cells independently of the monocytotropic or lymphocytotropic characteristics of the infective strain. The role of complement in facilitating HIV-infection should be taken into consideration in the design of vaccines and of therapeutic trials.

REFERENCES

Bakker LJ, Nottet HS, De Vos NM, De Graaf L, Van Strijp JAG, Visser MR, Verhoef J (1992) AIDS 6:35-41
Banapour B, Sernatinger J, Levy JA (1986) Virology 152:268-270
Boyer V, Desgranges C, Trabaud MA, Fischer E, Kazatchkine MD (1991) J Exp Med 173:1151-1158

Boyer V, Delibrias C, Noraz N, Fischer E, Kazatchkine MD, Desgranges C (1992) Scand J Immunol in press

Cohen JH, Fischer E, Kazatchkine MD, Lenoir GM, Lefevre-Delvincourt C, Revillard JP (1987) Scand J Immunol 25:587-598

Cohen JH, Aubry JP, Revillard JP, Banchereau J, Kazatchkine MD (1989) Cell immunol 121:383-390

Cohen JHM, Geffriaud C, Caudwell V, Kaztchkine MD (1989) AIDS 3:397-399

Dahl K, Martin K. Miller G. (1987) J Virol 61:1620-1628

Dahl KE, Burrage T, Jones F, Miller G (1990) J Virol 64:1771-1783

Delibrias C, Fischer E, Bismuth G, Kazatchkine MD (1992) J Immunol in press

De Rossi A, Roncella S, Calabro ML, D'Andrea E, Pasti M, Panozzo M, Mammano F, Ferrarini M, Chieco-Bianchi L (1990) Eur J Immunol 20:2041-2049

Ebenbichler CF, Thielens NL, Vornhagen R, Marshang P, Arlaud GJ, Dierich MP (1991) J Exp Med 174:1417-1424

Fischer E, Delibrias C, Kazatchkine MD (1991) J. Immunol. 146:865-869

Gras GS, Dormont D (1991) J Virol 65:541-545

Homsy J, Meyer M., Tateno M., Clarkson S, Levy J.A (1989) Science 244:1357-1360

Jouvin MH, Rozembaum W, Russo R, Kazatchkine MD: AIDS 1987, 1:89-94.

June RA, Schade SZ, Bankowski MJ, Kuhns M, Mc Namara A, Lint TF, Landay A L, Spear GT (1991) AIDS 5:269-274

June RA, Landay AL, Stefanik K, Lint TF, Spear GT (1992) Immunology 75:59-65

Laurence J. Astrin SM (1991) Proc Natl Acad Sci 88:7635-7639

Lederman M.M, Purvis SF, Walter EI, Carey JT Medof ME. (1989). Proc natl Acad Sci 86:4205-4209

Monroe J E, Calender A, Mulder C (1988) J Virol 62:3497-3500

Montagnier L, Gruest J, Chamaret S, Dauguet C, Axler C, Guetard D, Nugeyre MT,Barre-Sinoussi F, Chermann JC, Brunet JB, Klatzmann D, Gluckman JC (1984) Science 225:63-66

Perricone R, Fontane L ,De Carolis C, Carini C, Sirianni MC, Aiuti F (1987) Clin. Exp. Immunol 70: 500-507

Reisinger EC, Vogetseder W, Bersow D, Köfler D, Bitterlich G, Lehr HA, Wachter H, Dierich MP (1990) AIDS 4:961-965

Robinson WE, Montefiori DC, Mitchell WM (1988) Lancet i:790-794

Robinson WE, Montefiori DC, Gillespie DH, Mitchell WM (1989) J of Acq. Imm Defic Synd 2:33-42

Rodgaard A, Christensen LD, Thomsen BS, Wiik A, Bendixen G (1991) Complement Inflamm 8:303-309

Salahuddin SZ, Ablashi DV, Hunter E.A, Gonda MA, Sturzenegger S, Markham PD, Gallo RC (1987). Int J Cancer 39:198-202

Sattentau QJ, Moore JP (1991) J Exp Med 174:407-415

Senaldi G, Peakman M, McManus T, Davies ET, Tee DEH, Vergani D (1990) J Infect Diseases 162:1227-1232

Sölder BM, Schulz TF, Hengster P, LÖwer J, Larcher C, Bitterlich G, Kurth R, Wachter H, Dierich MP (1989) Immunol Lett 22 (2):135-146

Spear GT, Landay AL, Sullivan BL, Dittel B, Lint TF (1990) J Immunol 144:1490-1496

Takeda A, Tuazon CU, Ennis FA (1988) Science 242:580

Tausk FA, McCutchan JA, Spechko P, Schreiber RD, Gigli I (1986) J Clin Invest 78:977-982

Tedder TF, Fearon DT, Gartland GL, MD Cooper (1983) J Immunol 130:1668-1673

Tedder TF, Clement LT, MD Cooper (1984) J Immunol 133:678-683

Thieblemont N, Haeffner-Cavaillon N, Weiss L, Maillet F, Kazatchkine MD (1992) J. AIDS in press

Tremblay M, Fitz-Gibbon L, Wainberg MA (1988) Leukemia 2:233-240

Tremblay M., Meloche S, Sekaly R.P, Wainberg M.A (1990) J Exp Med 71:1791-1796

Tsoukas CD, JD Lambris (1988) Eur J Immunol 18:1299-1302

Wilson JG, Tedder TF, Fearon DT (1983) J Immunol 113:684-689

Weiss L, Okada N, Haeffner-Cavaillon N, Hattori T, Faucher C, Kazatchkine MD, Okada H (1992) AIDS 6:379-385

Evolution and Genetic Regulation of Complement C3 and C4 Genes

M. Takahashi, J.-H. Zengh, S. Natsuume-Sakai, N. Yamaguchi, and M. Nonaka

Departement of Immunobiology, Cancer Research Institute, Kanazawa University, Kanazawa, and Departement of Immunology, Kanazawa Medical University, Uchinada, Ishikawa, Japan

PHYLOGENETIC ORIGIN OF COMPLEMENT

Evidence has been accumulated that the complement system predated the immune system as the body defence mechanism in the vertebrates. Although the occurrence of immunoglobulines inducible by antigenic stimulation in the most primitive of the extant vertebrates has been widely accepted as a solid fact since the classical work of Marchalonis and Edelman (1968), more recent works conducted utilizing the techniques of molecular biology pointed to the other conclusion(Ishiguro et al. 1992).

In contrast to the ambiguity surrounding the occurrence of immunoglobulins in the cyclostomes, the presence of a complement system albeit primitive has been firmly established in two species of the two most primitive vertebrates, hagfish and lamprey, at the protein level and this conclusion was confirmed more recently at the molecular biological level. It is interesting to note that the cyclostome complement system is at a more primitive stage and consists of a fewer number of complement components. Cyclostome complement appears to be activated only through the alternative pathway and functions as the mediator of phagocytosis but lacks the cytolytic activity(Nonaka et al. 1984, Fujii 1992).

cDNA cloning for lamprey C3

cDNA clones coding for the C3-like molecules have been isolated from hagfish and lamprey (Ishiguro et al. 1992, Nonaka et al. 1992). The C3-like molecule of both hagfish and lamprey showed a stronger homology to mammalian C3 than to any complement components but it shared a three chain structure with mammalian C4. We have isolated lamprey C3 cDNAs from lamprey liver library using the sequence for the internal thioester bond of mammalian α_2M family as probe. They together covered the entire region encoding the putative lamprey pro-C3. The deduced amino acid sequence of this protein contained 1660 amino acids and showed 31%, 22%, 23% and 16% amino acid sequence identity with mouse C3, C4, C5 and human α_2M, respectively. The distributions of cysteine residues were completely identical between mouse C3 and lamprey C3 except that the lamprey sequence had two additional cysteine residues in the α chain. The probable $\beta-\alpha$ and $\alpha-\gamma$ processing sites in the lamprey and hagfish sequences were found exactly at the same positions as in the mammalian C4 sequences. The results suggest that the putative lamprey and hagfish C3 retains a close similarity to the common ancester of the mammalian C3 and C4 which appear to have a three subunit chain structure.

As expected, the sequence for the internal thioester bond was completely conserved in the cyclostome sequences. The neighboring sequences determing the binding specificity of the thioester bonds of lamprey and hagfish C3 were more similar to human C3 and human C4B rather than to human C4A, suggesting that the thioester bond of these cyclostome complement proteins had a stronger propensity for transesterification rather than transamidation. Comparison of the lamprey C3 with the mammalian C3, C4 and C5 revealed the presence of other conserved regions of interest. One example is the region corresponding to the C3a, C4a and C5a fragments of the mouse proteins. The arginine residues at the C-terminal end of mouse C3a, C4a and C5a were conserved in the lamprey protein, suggesting that lamprey C3 is also activated by proteolytic cleavage by serine proteases analogous to mammalian C1s, C2 and B. This observation agreed well with the results of our previous analysis of lamprey complement at the protein level, which showed that C3b-like and C3bi-like fragments were generated from lamprey C3 during its activation. The putative lamprey C3a fragment contained 76 amino acid of which 19, 22 and 27 residues were identical with mouse C3a, C4a and C5a, respectively.

cDNA cloning for lamprey Bf

Subsequently, we isolated cDNA clones coding for the lamprey counterpart of factor B/C2. By comparing the sequences of mammalian factor B and C2, we identified 2 stretches of conserved sequences about 70 amino acid residues apart in the

serine protease region of these complement proteins. Two
oligonucleotides were synthesized according to the conserved
sequences and utilized as primers for amplification of the
cDNA prepared from lamprey liver mRNA. The amplified DNAs
were cloned into plasmid and used as probe for screening
lamprey liver cDNA library.

An isolated clone of 2.6kb insert codes for a factor B/C2 like
mosaic protein consisting of short concensus repeats, von
Willebrand domains and serine protease domains. The putative
lamprey protein showed a homology to mouse factor B and C2 of
31% and 29%, respectively, and probably represents a
primordial form of factor B.

As our previous studies on rainbow trout clearly showed that
complement system in teleost is in a highly developed state
comparable to human and other mammalian counterpart(Nonaka et
al. 1981). Rainbow trout complement can be activated via the
classical as well as alternative pathway. It contained late-
acting components similar to human C5, C6, C7, C8 and C9
which together function as the cytolytic proteins apparenty by
assemblying into the membrane attack complexes.

TRANSCRIPTIONAL- AND POST-TRANSCRIPTIONAL REGULATION OF MOUSE
C4 GENE

In human and mouse, C4 (along with C2 and B) is MHC class
III molecule and is present in a duplicated form. Steroid 21-
hydroxylase gene or its pseudogene is positioned 2 kilobases
centromeric to each C4 gene, and the extant organization of
these genes is presumed to reflect tandem duplication of an
ancestral compound DNA unit 30-50kb long and containing a
single 21-hydroxylase gene and a single C4 gene.

In mouse, C4 and Slp represent the duplicated C4 genes. Mouse
sex-limited protein(Slp) is the isotype of mouse C4, shares
95% sequence identity with C4 but has no complement activity.
Slp is nonfunctional because it is not cleaved by the
activated form of complement protease C1s which
proteolytically activates C4 in the classical complement
pathway.
Slp is also distinct from C4 in that its expression in many

mouse strains is under testosterone control. Although C4 and
Slp genes showed a high degree of overall homology(95%) in the
coding regions as well as in the 5' upstream region up to
1.9kb from the transcription initiation site, their
transcriptional activity was signigicantly different(Nonaka
et al. 1986). Within the promoter region of C4 gene but not
Slp gene, we demonstrated two specific binding sites for
nuclear factors of HepG2 hepatoma and mouse liver by gel
retardation and DNase I footprinting. Nucleotide sequences of
these binding site were very similar to the recognition motifs
for NF.kB and NF-1. Of the two sites, the latter seemed to
play a major role in transcriptional activity of the C4 gene.

Insertion of B2 sequence into C4^{k}gene is the molecular basis for the low C4 production in mouse strains of H-2^{k}origin

The availability of many mutants of C4 gene expession has
served as an ideal system for studying gene regulation at
molecular level.

An extraordinary type of gene regulation was identified for C4
gene at mouse strains of H-2^k origin. The serum level of C4 in
mice carrying C4 gene derived from H-2^k mice is very low, only
5-10% of that of non-H-2^k mice. The transcriptional activities
of C4 genes of high and low C4-producer strains were shown to
be equivalent.

Hypothesizing abnormal RNA processing as the molecular basis
for the low steady-state level of C4 mRNA in low C4-producers,
we have analyzed C4 mRNA from these mouse strains and found
abnormal C4 mRNA which contained a 200bp insertion between
sequences derived from exon 13 and exon 14 (Pattanakitsakul
et al. 1992). The 5' 148bp and 3' 50bp of this insert were
derived from the B2 sequence, short interspersed repeats
abundant in mouse genome, and the central part of intron 13,
respectively. Sequence analysis of intron 13 of the C4^k gene
showed the presence of a complete copy of B2 consensus
sequence (Fig. 1).

The B2 sequence seemed to be inserted into C4^k gene by
retrotransposition, because duplicated direct repeats of the

target sequence was detected and a long adenosine-rich stretch was present at the 3' extremity of the inserted B2 sequence. The structure of the aberrant C4 mRNA indicated that the possible 3' splice site in the B2 sequence and the cryptic 5' splice site in intron 13 were used. Both insertion of the B2 sequence into intron 13 and the presence of aberrant mRNA in the liver were specific to $C4^k$-carrying mice, suggesting that the aberrant splicing due to the B2 insertion is the basis for low C4 expression in the liver of these mice.

To demonstrate directly that the B2 insertion truly caused abnormal RNA processing and low C4 production in $H-2^k$ mice, we constructed $C4^k$ gene without B2 sequence and $C4^{w7}$ gene with B2 insertion by exchanging a part of intron 13 segment between these two genes. Transfection of the intact $C4^{w7}$ gene or the chimeric $C4^k$ gene without the B2 insert into HepG2 cells resulted in the production of only normal C4 mRNA at normal levels. In contrast, the transfection of intact $C4^k$ gene or chimeric $C4^{w7}$ gene with inserted B2 sequence directed abnormal processing and decreased level of normal C4 mRNA. These results clearly showed the causal relationship between B2 insertion into C4 gene and low C4 production in $C4^k$-carrying mice.

---------------------------------→

Fig. 1.

Insertion of B2 sequence into intron 13 of $C4^k$ gene. Nucleotide sequence of intron 13 and part of adjacent exons (exons 13 and 14) of the gene of C3H/HeJ is shown. A shaded sequence indicates the B2 sequence. Two underlined sequences before and after the B2 insert indicate the terminal direct repeats. The sequence of the corresponding region of the C4 gene of B10.WR is aligned below the C3H/HeJ sequence in which identical nucleotides are shown by dots and deletion is indicated by horizontal bar. B2 insertion into intron 13 of C4 gene is observed for all of $C4^k$ carrying strains, but none of non-$C4^k$ carrying strains (Pattanakitsakul et al. 1992).

Fig. 1

Pvu II
CAGCTGGCTCCCTCGTTCTACTTTGTGGCTTACTTCTATCACCAAGGACACCCGGTGGCCAA
....
                                    exon13
CTCTCTGCTCATCAACATCCAATCCAGGGACTGTGAGGGCAAGGTGACTGGGTCTGTGGGCT
....

GTGGTGTGTGTGTGTGTGTGCTTGTGTGCATGTGTGTGCACATGCCTGTTAGCGTGCATATTCA
....

TGTGCATGTGTGTACACATGTGCAGATGTCTGTATGTACATGTGTGCTTGAATGCATGTGTA
....

CTTCCCTATGCAGGTGTGCATGTATGCTTGTGTGTGTGCATGTTCAAGTGTGCATGTGCACG
....

CAGGGGGTT *CAGAGGAGGACAG* GGGGCTGGTGAGATGAGCTCACAGATAAGAGCACCCGACTG
....

ATGTTCTGAAGGTCCTGAGTTCAATTCCCAGCAACCACATGGTGGCTCACAACCATCCCTAA

CAAGATCTGGCGCCCTCTTCTGGAGTGTCTGAAGACAGCTACAGTGTACTTACATATAATAA

ATAAATAAATCTTTTAAAAAAAAAAAAAAAAAAAAAAA *CAGAGGAGGACAG* GATGACTGGTTTAG
                                      ....
CCACTGAGTGTGCATGGGGCTGGAGGTGAGAGTTCACTCTTTTTTCTTTTTTCCTCTTCATGC
....

ATGTGAGGTGGGCATGTGTGTGTGCACATTTTGCATGCTGCATGGGTGCATGTGCTTGTAGA
....

GAGCTGGGGGTCATCTGTCTTACTCTTCAGCCTTAGTCATTGGGGCAGTTTCTTTCACTCAA
....

GCCTATAGCTCACTGATAAGCCTCATCTTCTAACCACCTTGTTCTGGGGATCCCACCCACCC
....

ACCCTGTCTCTGCCTTTCCAGGCTGGAATGACAGGCATGCCCTCACCCCCACCTGGATTCTG
....

GGGTTCAGTCCTCATGCATGGATGGCACACCTCTTAATTACTGACCCATGTCCCCAGCCCAC
....
                                          exon14
TCTCCATGTGCTGATTGTCTTTTTCCTCCCATTTCTGCCAGCTGCAATTGAAGGTGGATGGT
....
                                    Eco RI
GCCAAGGAGTATCGTAATGCGGACATGATGAAGCTCCGAATTC
....

It is interesting to note that C4 production in $H-2^k$ mice is controlled differently in hepatocytes and peritomeal macrophages, two main sources of serum C4. $C4^k$-carrying mice produce reduced level of C4 mRNA in the liver but normal level of C4 mRNA in the peritoneal macrophages. We analyzed C4 mRNA in the various organs of $H-2^k$ mouse strains by PCR and RNase protection assay and have shown that $C4^k$ transcripts are normally processed in some extrahepatic tissues including the peritoneal macrophages.

In conclusion our observation clearly represents the first demonstration of the modification of gene expression due to the insertion of B2 sequence into gene.

REFERENCES

Ishiguro HI, Kobayashi KU, Suzuki MA, Tiani KO, Tomonaga SU, Kurosawa YO (1992) Isolation of a hagtish gene that encodes a complement component. EMBO J 11:829-837

Fujii TA, Nakamura TO, Sekizawa AY, Tomonaga SU (1992) Isolation and characterization of a protein from hagfish serum that is homologous to the third component of the mammalian complement. J Immunol 148:117-123

Marchalonis JO, Edelman GA (1968) Phylogenetic origins of antibody structure III. Antibodies in the primary immune response of the sea lamprey, Petromyzon marinus. J Exp Med 127:891-914

Nonaka MA, Natsuume-Sakai SH, Takahashi MO (1981) The complement system of rainbowtrout (salmo gairdneri) II. Purificaton and chariacterization of the fifth component(C5). J Immunol 126:1495-1498

Nonaka MA, Fuji TA, Kaidoh TO, Natsuume-Sakai SH, Nonaka MA, Yamaguchi NO, Takahashi MO (1984) Purification of a lamprey complement protein homologous to the third component of the mammalian complement system. J Immunol 133:3242-3249

Nonaka MA, Kimura HI, Yu D-Y, Yokoyama SH, Nakayama KO, Takahashi MO (1986) Identification of the 5'-flanking regulatory region responsible for the difference in transcriptional control between mouse complement C4 and Slp gene. Proc Natl Acad Sci USA 83:7883-7887

Nonaka MA, Takahashi MO (1992) The complete cDNA sequence of the third component of complement(C3) of lamprey : implication for the evolution of thioester-containing proteins. J Immunol (in press)

Pattanakitsakul SA, Zheng Z-H., Natsuume-Sakai SH, Takahashi MO, Nonaka MA (1992) Aberrant splicing caused by the insertion of the B2 sequence into an intron of the complement C4 gene is the basis fow low C4 production in $H-2^k$ mice. J Biol chem 11:7814-7820

Complement System Genes and the Structures of the Proteins They Encode

R. D. Campbell[1,2] and A. J. Day[2]

[1]MRC Immunochemistry Unit, [2]Departement of Biochemistry, South Parks Road, Oxford, OX1 3QU, UK

1 Introduction

The complement system plasma proteins and most of the known cell surface receptors and control proteins have now been cloned at both the cDNA and genomic levels, and chromosomal assignments are available for the majority of the genes (Table 1) (for review see Reid and Campbell 1992). Genes encoding some of the complement control proteins and receptors such as Factor H, C4BP, CR1, CR2, DAF and MCP are present as a cluster on human chromosome 1q32 and these constitute the Regulators of Complement Activation (RCA) cluster (Pardo-Manuel et al. 1990, Hourcade et al. 1992). Another cluster of genes is found in the Major Histocompatibility Complex on human chromosome 6p21.3 where the genes encoding C4, C2 and Factor B are located within a 120 kb segment of DNA (Carroll et al. 1984, Milner and Campbell 1992). However, apart from pairs of genes encoding C1r and C1s (on 12p13), C6 and C7 (on 5q), and the C8α and C8β chains (on 1p34), and three closely-linked genes encoding the A, B and C chains of C1q (on 1p34-36), the genes encoding the other complement proteins and receptors are distributed on a large number of different chromosomes (Table 1). These cloning studies have also resulted in the determination of the primary amino acid sequences of the corresponding proteins. This has revealed that the majority of these are mosaic proteins, being built up from a number of different types of modular structures generally 40-110 amino acids in length (Reid and Day 1989, Day and Baron 1991). The mosaic nature of complement proteins is illustrated in Fig. 1. At the gene level many of the modules are encoded by discrete exons with symmetrical phase boundaries suggesting that the complement proteins have arisen through gene duplication and exon shuffling events.

The mosaic proteins of the complement system are of two major types. Proteins such as Factor H and CR1 are composed of a single type of module, in this case the Complement Control Protein (CCP) module (also referred to as Short Concensus Repeat or SCR) (Reid and Day 1989). Others such as C6, C7, C8 and C9 contain more than one type of module. For example C6 is composed of at least 5 different types of module (Fig. 1). Some of the complement proteins such as C3, C4, C5, C1q, C1-INH, the C8γ chain, and CD59, however, are not mosaic in structure, though they do share homologies with other non-complement proteins and as such are members of protein superfamilies.

In this review we will focus on the modular complement proteins and give a brief description of the structure of these proteins, relating this to what is known about the corresponding gene structures. In particular we will focus on the two most common structures found in the complement system, the CCP module and the Thrombospondin Repeat (TSR) (Reid and Day 1989, Day and Baron 1991).

2 The RCA gene cluster proteins

Physical mapping using Pulse Field Gel Electrophoresis (Carroll et al. 1988, Pardo-Manuel et al. 1990) and genomic cloning in Yeast Artificial Chromosomes (YACs) (Hourcade et al. 1992) has established the organisation of the RCA gene cluster in man. Six of the genes are clustered in a ~700kb segment of DNA in the order 5'C4BPα-C4BPβ-DAF-CR2-CR1-MCP3'. In addition a C4BP-like gene has been mapped adjacent to the 3' end of the C4BPβ gene, a CR1-like gene has been mapped adjacent to the 5' end of the MCP gene, while an MCP-like sequence has been mapped between the CR1-like and CR1 genes. All of the genes are orientated in the same 5' to 3' direction, which is likely to be a result of evolution of the RCA gene cluster by extensive gene duplication. Although YAC clones have also been isolated that include the Factor H gene, at least two Factor H-related homologues and the Factor XIIIb gene there is no information yet on the precise location of the Factor H gene relative to the MCP, CR1,

Table 1 Chromosomal localisation of proteins and receptors associated with the activation and control of the human serum complement system

Plasma Components	Chromosomal localisation	Regulatory Proteins	Chromosomal localisation
Classical pathway		Plasma	
C1q (A, B and C chains)	1p34-1p36.3	Properdin	Xp11.23-Xp11.3
C1r	12p13	C1 INH	11p11.2-11q13
C1s	12p13	Factor I	4q24-4q26
C4	6p21.3	C4BP (α and	1q32
C2	6p21.3	β chains)	
		Factor H	1q
Alternative pathway		S protein	ND[a]
		Sp40,40	ND
Factor B	6p21.3	Anaphylatoxin	
Factor D	ND	inactivator	ND
C3	19		
		Membrane bound	
Terminal pathway			
		DAF (CD55)	1q32
C5	9q32-9q34	MCP (CD46)	1q32
C6	5q	HRF (65kDa)	ND
C7	5q	CD59 (protectin, 11p	
C8 (α and β linked)	1p34	HRF20, MACIF)	
C8γ	9q		
C9	5p13		
Complement Receptors			
CR1 (CD35)	1q32	CR5	ND
CR2 (CD21)	1q32	C5a receptor	ND
CR3 (CD11b/CD18)	16p11-16p13.1	C3a receptor	ND
	(CD11a,b,c)		
CR4 (CD11c/CD18)	21q22.1 (CD18)	C1q receptor	ND

[a]ND, not determined.

CR2, DAF or C4BP genes.

A feature of the proteins of the RCA cluster is that they are almost entirely composed of variable numbers of CCP modules (Fig. 1). This module is approximately 60 amino acids in length and contains 4, generally invariant, cysteines disulphide bonded in the pattern Cys1-Cys3 and Cys2-Cys4, and several other highly conserved residues (Reid et al. 1986, Janatova et al. 1989, see Fig. 2). This structure is also present in other non-RCA complement proteins such as C1r, C1s, C2, Factor B and the terminal components (Fig. 1), as well as a wide variety of non-complement proteins such as Factor XIII, β2-glycoprotein I, ELAM-1, and the interleukin-2 receptor. The module was first described in Factor B (Morley and Campbell 1984) and therefore is sometimes called a B-type module (Patthy 1987).

CCP modules are generally encoded by single discrete exons at the genomic level (Fujisaki et al. 1989, Wong et al. 1989, Post et al. 1990, 1991, Hillarp et al. 1991, Rodriguez de Cordoba et al. 1991). However, a number of exceptions do exist. In the C4BPβ-chain gene CCP 3 is split over two exons (exons 5 and 6) (Hillarp et al. 1991), and this is also true of CCP 2 of the C4BPα-chain (Rodriguez de Cordoba et al. 1991), CCPs 2, 6, 9, 13, 16, 20, 23, and 27 of CR1(Wong et al. 1989), CCPs 4, 8, and 12 of CR2 (Fujisaki et al. 1989), CCP 3 of DAF (Post et al. 1990), and CCP 2 of MCP (Post et al.

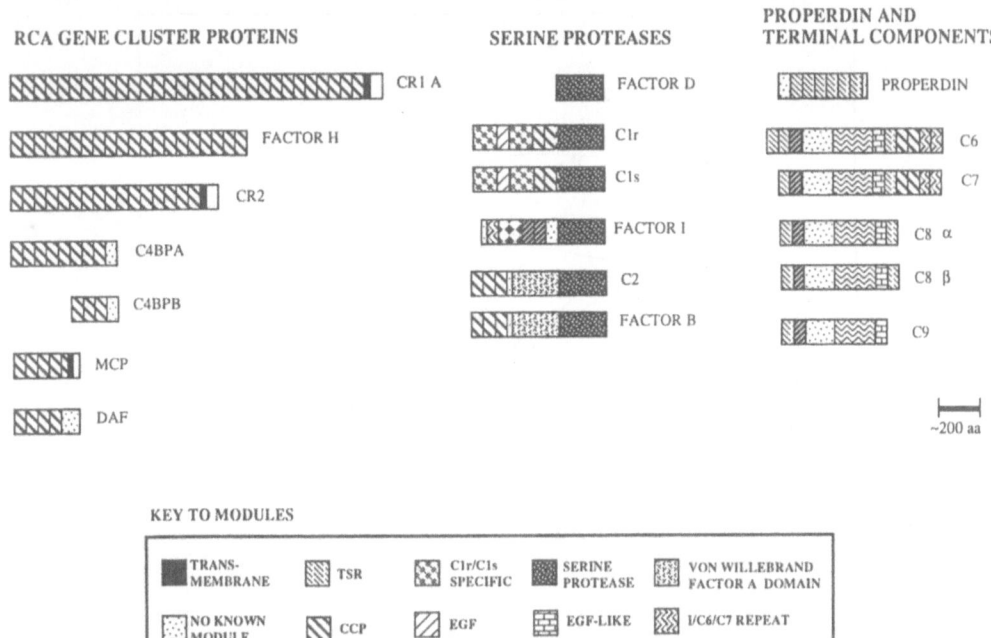

Fig. 1 The modular nature of the complement system proteins. A schematic representation of the primary structures of human complement proteins illustrating those that contain small modular structures. The various modules are defined in the key.

1991). In these cases the splice junction which interrupts the CCP module is located after the second nucleotide of the codon for a concensus Gly residue shown in Fig. 2. In addition in CR1, CR2, MCP and DAF a single exon can encode 2 CCPs.

Recently, the three-dimensional structure of the sixteenth CCP module of human factor H has been determined in solution by proton nuclear magnetic resonance spectroscopy (Barlow et al. 1991, Norman et al. 1991). The CCP module is based on a β-sandwich arrangement with one face being made up of three β-strands (see Fig. 2 - strands a, b/c, and d/e) hydrogen-bonded to form a triple-stranded region at its centre and the other face formed from two separate β-strands. Both faces of the sandwich contribute hydrophobic side chains that form a compact core. An analysis of the sequence alignment of approximately 150 CCP modules (Day AJ, unpublished) reveals a high degree of conservation of residues of structural importance (i.e., hydrophobic core residues) whereas, insertions and deletions in various repeats are accommodated in loops between elements of secondary structure. Therefore, it is thought that many members of the CCP module superfamily can be modelled on the basis of this structure (Norman et al. 1991). The solution structure of a second CCP module, CCP 5 of Factor H, has now been determined by NMR. The tertiary structure of CCP 5 reveals a compact hydrophobic core wrapped in β strand and sheet, and bears much overall resemblance to CCP 16. The dimensions of an individual CCP module are approximately 3.8 x 2.0 x 1.0 nm. This is consistent with physical studies on proteins containing many CCP modules (i.e. Factor H and CR1) which are very long and their structure likened to strings of beads (Sim and Perkins 1990). Therefore in the cell-surface molecules CR1 and CR2 CCP modules may act as spacer units, projecting the binding domain away from the cell surface.

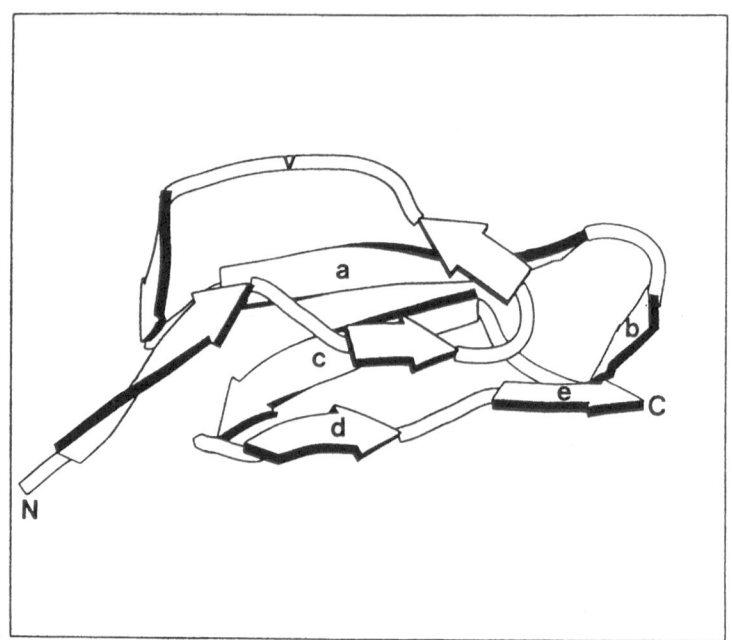

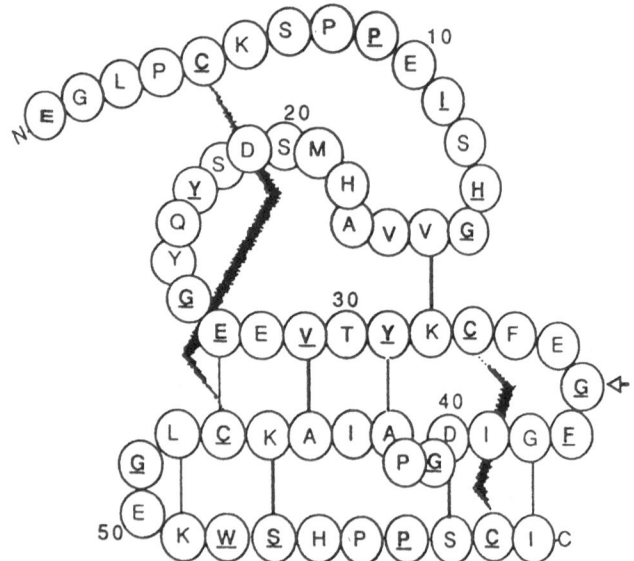

Fig.2 Tertiary (top) and secondary/primary (bottom) structure of CCP module 16 of Factor H (from Barlow et al. 1991 and Norman et al. 1991). In the upper half of the figure a MOLSCRIPT (Kraulis et al. 1991) representation of the tertiary structure is shown with arrows denoting β-strands and tubes denoting loops, turns and regions of less well defined structure. The N- and C-terminal ends are indicated by N and C, respectively, while v indicates the variable loop. In the lower half of the figure the secondary structure of this module, along with the primary amino acid sequence and disulphide bond organisation, is shown. The amino acids underlined correspond to the concensus sequence of a CCP module. The Gly residue, whose codon is sometimes split over two exons, is indicated by an arrow.

A clear function has not been ascribed to the CCP module in most of the complement and non-complement members of the CCP superfamily and in many cases may perform a purely structural role. However, this module may be important in a wide range of binding specificities, being a structural building block onto which function-related sequence elements are superimposed. In this regard the complement control proteins have been the most studied to date. In Factor H, which has 20 CCP modules (see Fig. 1) preliminary evidence suggests that the C3b-binding site is located within CCP modules 3-5 (Steinkasserer et al. 1991). In CR1 the C4b binding domain has been localised to CCPs 1 and 2, and a C3b binding domain has been localised to CCPs 8 and 9 (Klickstein et al. 1988). Protein engineering studies (Krych et al. 1991) have identified three residues within CCP 2, and a stretch of five residues in CCP 9, which are important for C4b and iC3 binding, respectively. These studies also indicated the necessity for two contiguous CCPs for the interaction of CR1 to take place with its ligands.

3 The serine proteases of the complement system

Apart from Factor D which consists only of a serine protease domain, the other serine proteases of the complement system, C1r, C1s, C2, Factor B, and Factor I, are all typical examples of mosaic proteins with various modules N-terminal to their serine protease domain (Fig. 1). These proteins have probably evolved from a common ancestor. It is believed that this ancestral serine protease had a phase-1 intron at the boundary of its protease domain with its signal-peptide. This intron was the recipient of incoming class 1-1 modules (i.e., modules encoded by exons, or cassettes of exons, with a phase-1 intron at either end - see Patthy 1991). The phase-1 intron/exon boundary 5' to the serine protease domain has been highly conserved in evolution, whereas, the number of exons and the intronic phase differs greatly within different serine proteases. This has allowed many different module types, with different binding specificities, to be combined with a serine protease domain. This is not only true for the complement system but also for the blood coagulation and fibrinolytic systems (Patthy 1985, Furie and Furie 1988).

Both the C1r and C1s genes are approximately 10-12 kb long and initial studies on the intron/exon boundaries of the C1r gene indicates that each of the modules 1 to 5 in the non-catalytic A chain may be encoded by separate exons while the entire catalytic B chain, the serine protease domain, is encoded by a single exon (Tosi et al. 1989b). The human C1s gene contains 12 exons as gauged by electron-microscopy analysis of heteroduplex molecules produced by hybridisation of a full length genomic clone and a full length cDNA clone (Tosi et al. 1989a). An unusual feature of the gene structure is the finding that the two CCPs in C1s are both encoded by two exons rather than a single exon. As for C1r, the catalytic B chain of C1s is encoded by one exon which again is unusual since in other vertebrate serine protease genes, which have been examined, the serine, aspartic acid and histidine residues which form the catalytic site are usually found in separate exons. This is particularly true of C2 and Factor B where the serine protease domain is encoded by 8 exons (Campbell and Porter 1983, Ishii et al. 1991). Amino acids 229-444 of C2 and Factor B share significant homology with the A domains of von Willebrand factor (vWF), and homologous domains in the α chain of some integrins, (called the "I" domain) and the cartilage matrix protein (CMP). Recently, the genes encoding vWF (Mancuso et al. 1989), CMP (Kiss et al. 1989) and the α chain of the integrin p150,95 (Corbi et al. 1990) have been reported. In the vWF gene the A3 domain is encoded by 4 exons, while the A1 and A2 domains are encoded by a single exon. The I domain of the α chain of p150,95 is also encoded by 4 exons, and in CMP the homologous domains are encoded by 2 exons, while in C2 and Factor B this domain is encoded by 5 exons. Comparison of the genomic organisation of this region of the Factor B gene with the corresponding regions of the vWF gene, the CMP gene and the α chain gene of the p150,95, has shown that there is little consensus on the position of the splice sites within the domain. However, the splice sites at the beginning and end of the domain are conserved as phase 1 introns in each of the genes. This supports the idea that this domain has been inserted as a unit into a diverse range of proteins during evolution, and then has subsequently undergone various modifications which presumably included duplications and intron insertions/deletions, as well as splice junction sliding.

The three CCP modules at the N-terminal ends of C2 and Factor B are each precisely encoded in separate exons (Campbell et al. 1984, Ishii et al. 1991).

4 Properdin and the terminal components

Properdin, which is a positive regulator of the alternative complement pathway, is composed almost entirely of 6 contiguous modules each of ~60 amino acids (Nolan et al. 1991) that were originally described in thrombospondin, a 420 kDa adhesion protein secreted from activated platelets (Lawler and Hynes 1986). The TSR 60-amino-acid-long consensus sequence is based primarily upon the presence of six cysteine and three tryptophan residues and it seems likely that these cysteines will form intramodular disulphide bridges thus ensuring that each TSR sequence yields an independently folding unit. There are no tertiary structural data on the TSR module. However, secondary structure prediction and Fourier transform infrared spectroscopy indicate that TSR modules are composed mainly of β-sheet and β-turn structure (Smith et al. 1991). Electron microscopy studies on properdin indicate that it has an elongated structure (2.5 x 26 nm) (Smith et al. 1984) suggesting therefore that each TSR module may have similar dimensions to the CCP module.

C6 (DiScipio et al. 1989, Haeflinger et al. 1989), C7 (DiScipio et al. 1988), C8α (Rao et al. 1987), C8β (Haeflinger et al. 1987, Howard et al. 1987) and C9 (DiScipio et al. 1984, Stanley et al. 1985) all contain TSR modules as shown in Fig. 1. The terminal components are homologous to each other and share approximately 21%-26% identity between any pair of sequences. A striking feature which emerges is the mosaic nature of the structures of these proteins. In addition to the TSR module all five chains contain a large internal domain, almost free of Cys residues, which is similar to that found in perforin (Shinkai et al. 1988), along with an EGF-like motif and a domain homologous to those found in the low density lipoprotein receptor (the LDLr module). C6 and C7 also contain CCP modules and two motifs as yet found only in C6, C7 and Factor I (Fig. 1). C6 is an excellent example of this mosaic structure (Fig. 1) being composed of two TSR modules at the N-terminal end followed by an LDLr module, the perforin/C9 domain, an EGF-like domain, a third TSR module, two CCP modules, and finally at the C-terminus two C6/C7/I modules.

The TSR modules in thrombospondin and properdin, like the CCP module superfamily, are usually encoded by discrete exons in their structural genes (Wolf et al. 1990, Nolan et al. 1992). However, this is not the case for the TSR in C9 which is encoded by two exons with the second exon also encoding half of the following LDLr domain (Marazziti et al. 1988). Fig. 1 illustrates that the terminal components probably evolved from a common ancestral C9 protein by the addition of additional modules.

5 Conclusion

Apart from Factor D (Narayana et al. 1991) none of the other complement proteins have had their complete structures determined by X-ray crystallography. This is due to the difficulties that have been encountered in crystallising these proteins, in part because of their relatively large sizes, and in part because of their mosaic nature which probably yields very mobile structures. However, given that a large number of the complement system proteins contain different types of modules, determination of the structure of a single module, or pairs of modules, by NMR in solution will make it possible to deduce concensus structures for these proteins by model building. This is now true for the RCA cluster proteins which are composed of a single type of module, the CCP module, whose solution structure has been determined. As the structures of other modules found in the complement system proteins become available this will increase our understanding of the inter-relationships between the structures and functions of these proteins, and will provide new strategies for the control of complement activation.

Acknowledgements

We thank Dr D. Norman (Oxford) for providing the MOLSCRIPT diagram of the H16 module.

References

Barlow PN, Baron M, Norman DG, Day AJ, Willis AC, Sim RB, Campbell ID (1991) Secondary structure of a complement control protein module by two-dimensional ^{1}H NMR. Biochemistry 30:997-1004

Barlow PN, Norman DG, Steinkasserer A, Horne TJ, Pearce J, Driscoll PC, Sim RB, Campbell I (1992) Solution structure of the fifth repeat of Factor H: a second example of the complement control protein module. Biochemistry 31:3626-3634

Campbell RD, Bentley DR, Morley BJ (1984) The factor B and C2 genes. Phil Trans R Soc Lond B 306:367-378

Campbell RD, Porter RR (1983) Molecular cloning and characterisation of the gene coding for human complement protein factor B. Proc Natl Acad Sci USA 80:4464-4468

Carroll MC, Alicot EM, Katzman PJ, Klickstein LB, Smith JA, Fearon DT (1988) Organisation of the genes encoding complement receptors type 1 and 2, decay-accelerating factor, and C4-binding protein in the RCA locus on human chromosome 1. J Exp Med 167:1271-1280

Carroll MC, Campbell RD, Bentley DR, Porter RR (1984) A molecular map of the human major histocompatibility complex class III region linking complement genes C4, C2 and factor B. Nature 307:237-241

Corbi AL, Garcia-Aguilar J, Springer TA (1990) Genomic structure of an integrin α subunit, the leukocyte p150,95 molecule. J Biol Chem 265:2782-2788

Day AJ, Baron M (1991) Structure/function inter-relationships of mosaic proteins. J Biomed Sci 1:153-163

DiScipio RD, Hugli TE (1989) The molecular architecture of human complement component C6. J Biol Chem 264:16197-16206

DiScipio RG, Chakravarti DN, Müller-Eberhard HJ, Fey GH (1988) The structure of human complement component C7 and C5b-7 complex. J Biol Chem 263:549-560

DiScipio RG, Gehring MR, Podack ER, Kan CC, Hugli TE, Fey GF (1984) Nucleotide sequence of cDNA and derived amino acid sequence of human complement component C9. Proc Natl Acad Sci USA 81:7298-7302

Fujisaku A, Harley JB, Frank MB, Gruner BA, Frazier B, Holers VM (1989) Genomic organisation and polymorphisms of the human C3d/Epstein-Barr virus receptor. J Biol Chem 264:2118-2125

Furie B, Furie BC (1988) The molecular basis of blood coagulation. Cell 53:505-518

Haefliger JA, Tschopp J, Naedelli D, Wahli W, Kocher HP, Tosi M, Stanley KK (1987) Complementary DNA cloning of complement C8b and its sequence homology to C9. Biochemistry 26:3551-3556

Haefliger JA, Tschopp J, Vial N, Jenne DE (1989) Complete primary structure and functional characterization of the sixth component of the human complement system. Identification of the C5b-binding domain in complement C6. J Biol Chem 264:18041-18051

Hillarp A, Rodriguez de Cordoba S, Dahlback B (1991) Organisation of the human C4b-binding β-chain gene. Complement & Inflammation 8:160

Hourcade D, Garcia AD, Post TW, Taillon-Miller P, Holers VM, Wagner LM, Bora NS, Atkinson JP (1992) Analysis of the human regulators of complement activation (RCA) gene cluster with yeast artificial chromosomes (YACs). Genomics 12:289-300

Howard OMZ, Rao AG, Sodetz JM (1987) Complementary DNA and derived amino acid sequence of the β-subunit of C8: identification of a close structural and ancestral relationship to the α-subunit and C9. Biochemistry 26:3565-3570

Ishii Y, Zhu ZB, Macon KJ, Volanakis JE (1991) Structure of the gene for human complement component C2. Complement & Inflammation 8:167

Janatova J, Reid KBM, Willis AC (1989) Disulfide bonds are localized within the short consensus repeat units of complement regulatory proteins: C4b-binding protein. Biochemistry 28:4754-4761

Kraulis PJ (1991) MOLSCRIPT: a program to produce both detailed and schematic plots of protein structures. J Appl Crystallogr 24:946-950

Kiss I, Deak F, Holloway RGJ, Delius H, Mebust KA, Frimberger E, Argraves WS, Tsonis PA, Winterbottom N, Goetinck PF (1989) Structure of the gene for cartilage matrix protein, a modular protein of the extracellular matrix. Exon/intron organization, unusual splice sites, and relation to α chains of β_2 integrins, von Willebrand factor, complement factors B and C2, and epidermal growth factor. J Biol Chem 264:8126-813

Klickstein LB, Bartow TJ, Miletic V, Rabson LD, Smith JA, Fearon DT (1988) Identification of distinct C3b and C4b recognition sites in the human C3b/C4b receptor (CR1, CD35) by deletion mutagenesis. J Exp Med 168:1699-1717

Krych M, Hourcade D, Atkinson, JP (1991) Sites within the complement C3b/C4b receptor important for the specificity of ligand binding. Proc Natl Acad Sci USA 88:4353-4357

Lawler J, Hynes RO (1986) The structure of human thrombospondin, an adhesive glycoprotein with multiple calcium-binding sites and homologies with several different proteins. J Cell Biol 103:1635

Mancuso DJ, Tuley EA, Westfield LA, Worrall NK, Shelton-Inloes BB, Sorace JM, Alevy YG, Sadler JE (1989) Structure of the gene for human von Willebrand Factor. J Biol Chem 264:19514-19527

Marizzitti D, Eggertsen G, Fey GH, Stanley K (1988) Relationships between the gene and protein structures in human complement component C9. Biochemistry 27:6529-6534

Milner CM, Campbell RD (1992) Genes, genes and more genes in the human major histocompatibility complex. Bioessays 14:(In Press)

Morley BJ, Campbell RD (1984) Internal homologies of the Ba fragment from human complement component factor B, a class III MHC antigen. EMBO J 3:153-157

Narayana SVL, Carson M, De Lucas L, Moore D, Bugg CF, Kilpatrick JM, Volanakis JE (1991) Three dimensional structure of complement Factor D. Complement & Inflammation 8:198

Nolan KF, Kaluz S, Higgins JMG, Goundis D, Reid KBM (1992) Characterisation of the human properdin gene. Biochem J 286:(In Press)

Nolan KF, Schwaeble W, Kaluz S, Dierich MP, Reid KBM (1991) Molecular cloning of the cDNA coding for properdin, a positive regulator of the alternative pathway of human complement. Eur J Immunol 21:771-776

Norman DG, Barlow PN, Baron M, Day AJ, Sim RB, Campbell ID (1991) Three-dimensional structure of a complement control protein module in solution. J Mol Biol 219:717-725

Pardo-Manuel F, Rey-Campos J, Hillarp A, Dahlbäck B, Rodriguez de Cordoba S (1990) Human genes for the α and β chains of complement C4b-binding protein are closely linked in a head-to-tail arrangement. Proc Natl Acad Sci USA 87:4529-4532

Patthy L (1985) Evolution of the proteases of blood coagulation and fibrinolysis by assembly from modules. Cell 41:657-663

Patthy L (1987) Intron-dependent evolution: preferred types of exons and introns. FEBS Letts 214:1-7

Patthy L (1991) Modular exchange principles in proteins. Current Opinion in Structural Biology 1:351-361

Post TW, Arce MA, Liszewski MK, Thompson ES, Atkinson JP, Lublin DM (1990) Structure of the gene for human complement protein decay accelerating factor. J Immunol 144:740-744

Post TW, Liszewski MK, Adams EM, Tedja I, Miller EA, Atkinson JP (1991) Membrane cofactor protein of the complement system: alternative splicing of serine/threonine/proline-rich exons and cytoplasmic tails produces multiple isoforms that correlate with protein phenotype. J Exp Med 174:93-102

Rao AG, Howard OMZ, Ng SC, Whitehead AS, Colten HR, Sodetz JM (1987) Complementary DNA and derived amino acid sequence for the α subunit of human complement protein C8: evidence for the existence of a separate α subunit mRNA. Biochemistry 26:3556-3564

Reid KBM, Campbell RD (1992) Structure and organisation of complement genes. In: Whaley K (ed) Complement in health and disease. Kluwer Academic Publishers, Lancaster, (In press)

Reid KBM, Day AJ (1989) Structure-function relationships of the complement components. Immunology Today 10:177-180

Reid KBM, Bentley DR, Chung LP, Sim RB, Kristensen T, Tack BF (1986) Complement system proteins which interact with C3b or C4b. Immunology Today 7:230-234

Rodriguez de Cordoba S, Sanchez-Corral P, Rey-Campos J (1991) Structure of the gene coding for the α polypeptide chain of the human complement component C4b-binding protein. J Exp Med 173:1073-1082

Shinkai Y, Takio K, Okumura K (1988) Homology of perforin to the ninth component of complement (C9). Nature 334:525-527

Sim RB, Perkins S (1990) Molecular modelling of C3 and its ligands. Curr Top Microbiol Immunol 153:209-222

Smith CA, Pangburn MK, Vogel CW, Müller-Eberhard HJ (1984) Molecular architecture of human properdin, a positive regulator of the alternative pathway of complement. J Biol Chem 259:4582-4588

Smith KF, Nolan KF, Reid KBM, Perkins SJ (1991) Neutron and X-ray scattering studies on the human complement protein properdin provide an analysis of the thrombospondin repeat. Biochemistry 30:8000-8008

Steinkasserer A, Barlow P, Norman DG, Kertesz Z, Campbell ID, Day AJ, Sim RB (1991) Synthesis of functional recombinant protein modules. Complement & Inflammation 8:225-226

Tosi M, Duponchel C, Meo T, Couture-Tosi E (1989a) Complement genes C1r and C1s feature an intronless serine protease domain closely related to haptoglobin. J Mol Biol 208:709-714

Tosi M, Journet A, Duponchel C, Couture-Tosi E, Meo T (1989b) Human complement C1r and C1s proteins and genes: studies with molecular probes. Behring Inst Mitt 84:65-71

Wolf FW, Eddy RL, Shows TB, Dixit VM (1990) Structure and chromosomal localization of the human thrombospondin gene. Genomics 6:685-691

Wong WW, Cahill JM, Rosen MD, Kennedy CA, Bonaccio ET, Morris MJ, Wilson JG, Klickstein LB, Fearon DT (1989) Structure of the human CR1 gene. Molecular basis of the structural and quantitative polymorphisms and identification of a new CR1-like allele. J Exp Med 169:847-863

11. Primary Immuno-Deficiencies

Antibody Deficiencies Reflect Abnormal B Cell Diffrentiation

M.D. Cooper, N. Nishimoto, K. Lassoued, C. Nunez, T. Nakamura, H. Kubagawa and J.E. Volanakis

The Howard Hughes Medical Institue, and the Divisions of Developmental and Clinical Immunology and Rheumatology, University of Alabama at Birmingham, Alabama 35294 USA

INTRODUCTION

Analysis of the genetically-determined immunodeficiencies has contributed greatly to our present view of lymphocyte differentiation pathways and their functional significance. A good example is the selective susceptibility to bacterial infections that is seen in antibody deficiencies due to inherent abnormalities of B cell differentiation versus the broader susceptibility to infection with both intracellular and extracellular pathogens when T cell development is impaired (Rosen et al. 1984, 1992). Elucidation of the genes responsible for these immunodeficiency diseases will provide further insight into immune system function.

Most primary antibody deficiencies reflect genetically-determined abnormalities in the B cell pathway of lymphocyte differentiation. One form of severe combined immunodeficiency (SCID), in which B and T cell development is aborted early in their respective differentiation pathways, is the consequence of abnormal V(D)J recombinase activity (Schuler et al. 1986; Schwarz et al. 1991). In X-linked agammaglobulinemia, which features an isolated deficiency of B cells, the arrest occurs later in pre-B cell differentiation (Pearl et al. 1978). In common variable immunodeficiency (CVID) and IgA deficiency (IgA-D), which appear to be related disorders (Schaffer et al. 1989), B cells are produced in normal numbers but fail to undergo plasma cell differentiation in response to antigens and T cell help. In all of these primary immunodeficiencies, some of which are relatively common, we still lack sufficient understanding of the molecular events in B cell differentiation to be able to pinpoint the precise defect.

This brief essay concerns current ideas about the production of B lineage cells in hemopoietic tissues and attempts to summarize our present view of the defects underlying some of the classical antibody deficiency disorders.

DEVELOPMENT OF B-LINEAGE CELLS

Normally we produce around 10^{10} B cells each day in our bone marrow. This process begins when hemopoietic stem cell progeny are influenced by neighboring stromal cells and their soluble products to become lymphoid progenitors. Environmental factors that contribute to the initial commitment to lymphoid differentiation may include hemopoietic growth factors, such as IL-3 and stem cell factor (SCF), and interactions with their respective receptors on lymphoid progenitors. By analogy with other differentiation systems we suppose that the signals transduced as a consequence of these ligand-receptor interactions activate second signal transmission pathways that ultimately result in the binding of transcription-activating proteins to the regulatory DNA sequences for particular genes. Transcription of these early genes in progenitor B cells may initiate a cascade of gene activation leading to cell growth and progression of the differentiation process. Cells thus influenced to differentiate along the B cell axis are marked by their expression of the transmembrane CD19 molecule (Dörken et al. 1989).

The elements involved in immunoglobulin (Ig) gene rearrangement and expression provide reliable indicators of cellular progression along this differentiation pathway (Cooper 1987; Alt et al. 1992). In the progenitor B cells, transcription of the recombinase activating genes, RAG1 and RAG2, is necessary for initiation of the Ig V(D)J rearrangements (Oettinger et al. 1990). Bone marrow pro-B cells that are undergoing VDJ rearrangements in the heavy chain (HC) locus express terminal nucleotidyl transferase (TdT). This enzyme is responsible for insertion of non-germline encoded nucleotides (N) into the VDJ joints, thereby adding to the combinatorial diversification of the B cell repertoire (Desiderio et al. 1984). The presence of nuclear TdT marks a large subpopulation of pro-B cells (Campana et al. 1985). Definitive pre-B cells are marked by the cytoplasmic expression of μHC, most of which is retained in the endoplasmic reticulum by a protein called Bip (Haas and Wabl 1983). When productive VJ rearrangement occurs in the κ or λ loci, Bip binds transiently to the light chains (LC) as well (Knittler and Haas 1992), but is displaced when the HC and LC bind to form IgM molecules that traverse the Golgi system en route to the cell surface (Hendershot et al. 1987).

By themselves the IgM molecules cannot reach the cell surface and, even if they could, their short intracytoplasmic tails would not allow them to transmit signals on their own (Reth et al. 1991). The Ig receptors require associated transmembrane molecules both to reach the cell surface and to transduce signals after cross-linkage by antigens. The Ig-associated α and β chains, encoded by the mb-1 (Sakaguchi et al. 1988) and B29 genes (Hermanson et al. 1988), are covalently linked to each other. These $\alpha\beta$ heterodimers are non-covalently linked to the Ig HC to form the Ig receptor complex.

The discovery of surrogate (ψ) LC genes that do not require rearrangement to become transcriptionally active in pro-B cells has altered this view of B cell development (Sakaguchi and Melchers 1986; Chang et al. 1986; Kudo and Melchers 1987). Current hypothetical models suggest that ψLC proteins encoded by the λ5/14.1 and Vpre-B genes may pair with either ψHC or μHC to form receptors which signal the progressive differentiation of pro-B, pre-B and immature B cells (Rolink and Melchers 1991). However, when we used monoclonal antibodies to examine ψLC expression during differentiation, we did not find the expression pattern expected from these models. While ψLC are produced during several developmental stages, their cell surface expression is restricted to a relatively late stage in normal pre-B cell differentiation (Lassoued et al. 1992).

An important functional role for the ψLC/μHC receptors on pre-B cells is implied by a sharp reduction in the rate of B cell production in λ5 gene-deleted mice (Kitamura et al. 1992). We know that cross-linkage of this receptor leads to its down-modulation and induces transient increase in intracellular calcium levels (Takemori et al. 1990; Ohno et al. 1990; Lassoued et al, unpublished observations). One hypothesis holds that ligand interaction with this receptor initiates a signal for VJ rearrangement in the LC loci (Tsubata et al. 1992). Alternatively, this signaling could favor pre-B cell survival and differentiation in another way. Our inability to inhibit pre-B cell growth or B cell differentiation by treatment with the anti-ψLC antibodies suggests that ψLC/μHC receptors are not involved in negative selection of the B cell repertoire. Still we do not know the ligand(s) for this receptor or the details of the presumably positive pre-B cell response.

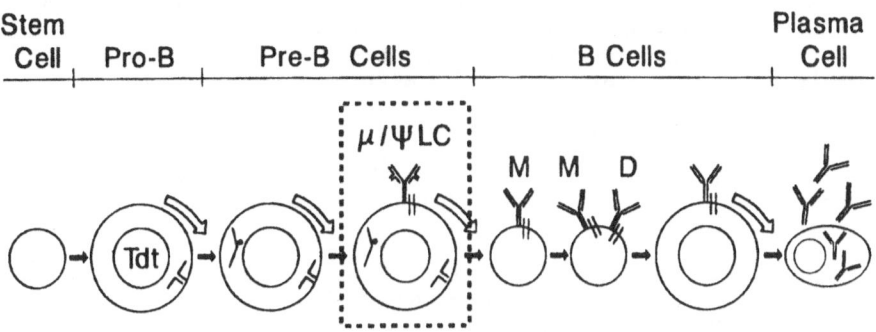

Fig. 1. Model of the B cell pathway indicating differentiation defects that underly severe combined immunodeficiency (SCID) due to abnormal V(D)J recombinase activity, X-linked agammaglobulinemia (XLA), IgA deficiency (IgA-D) and common variable immunodeficiency (CVID). The open arrows outside the circles indicate cell proliferation. The receptor isotypes indicated in the diagram are all associated with α and β chains which are represented by the two short parallel lines. The little closed circle attached to the μ heavy chain in the first pre-B cell represents the immunoglobulin-binding protein Bip. The two L shapes that are present in the cytoplasm of pro-B and pre-B cells and attached to the μ heavy chain on the surface of late pre-B cells, represent the surrogate light chain (ψLC) proteins encoded by the λ5/14.1 and Vpre-B genes.

PATHOGENESIS OF X-LINKED AGAMMAGLOBULINEMIA (XLA)

XLA, the first immunodeficiency to be recognized (Bruton 1952), is the prototypic antibody deficiency disorder. Recurrent bacterial infections are the consequence of an almost total inability to mount an antibody response. The cellular basis for the antibody deficiency is a severe deficit in B cells and their plasma cell progeny. Affected individuals have normal numbers of T cells and intact cellular immunity which allows them to clear most viral and fungal infections. In female carriers of the XLA defect, all of the circulating B cells use the normal X chromosome, while T cells and other cell lineages exhibit normal random utilization of both X chromosomes (Conley et al. 1986). This finding implies that the abnormal gene, presently mapped to the Xq 21.3-22 region (Hendriks and Schuurman 1991), is responsible for an inherent B cell defect.

On finding relatively normal numbers of μ^+ pre-B cells in the bone marrow of 11 consecutive boys with XLA, it was proposed that the differentiation arrest immediately precedes B cell development (Pearl et al. 1978). The defect is a leaky one in that newly-formed B lymphocytes are always found in small numbers in these patients. In some patients, this limited supply of immature B cells (Conley 1985; Tedder et al. 1985) can give rise to mature antibody producing cells in numbers sufficient to yield substantial levels of immunoglobulins of diverse isotypes (Goldblum et al. 1974). This demonstrated capacity for proliferation, isotype switching and plasma cell maturation suggests the defect lies in the limited level of B cell production and argues against a qualitative defect in the B cells that are produced.

Ideas abound as to exactly where the bottleneck occurs in XLA B cell production. Analysis of bone marrow samples from eight XLA adults revealed normal numbers of CD19$^+$ pro-B cells with nuclear TdT, but μ^+ pre-B cells were found in only four of these patients (Campana et al. 1990). This led to the suggestion of a maturational block in the pro-B to pre-B cell transition. Truncated DJμ transcripts and their V-less μ chain products have been reported in XLA pre-B cells, suggesting the idea of a V-DJ recombinase defect (Schwaber et al. 1983), and truncated μ chain transcripts which include a leader sequence directly spliced to Cμ RNA and thus lack VDJ sequences have been reported recently (Schwaber 1992).

Other recent investigations provide evidence indicating that XLA reflects faulty progression of pre-B cell differentiation rather that a qualitative defect in Ig gene rearrangement and expression (Nishimoto et al. 1991). While the CD19$^+$ Ig$^-$ lymphocyte compartment is normal in size, the pro-B/pre-B ratio is inverted in XLA patients. Instead of the normal ratio of approximately 40/60, we find that approximately 70% of the CD19$^+$ cells in XLA bone marrow are Tdt$^+$ pro-B cells and only 20% are μ^+ pre-B cells. A subpopulation of the latter express ψLC/μHC receptors. The μ chains produced by the XLA pre-B cells have the same relative molecular mass (76 kDa) as the full-length μ chains made by normal B cells, as noted before for EBV-transformed pre-B and B cells from XLA patients (Fu et al. 1980). Complementary results have been obtained in an analysis of the Ig gene transcripts in XLA bone marrow cells, in which normal VDJ-Cμ, full-length kappa, and Vpre-B and λ5/14.1 transcripts were found (Fougereau M, personal communication). Sequence analysis revealed that while most of the V$_H$ gene families were expressed, limited diversity of the V$_H$ genes and CDR3 region complexity resembled the fetal repertoire. These results confirm and extend results obtained in previous studies of V$_H$ diversity in EBV-transformed clones of XLA B cells (Anker et al. 1989). We can thus conclude that progression of pre-B cell differentiation and maturation of the repertoire is retarded, but we do not understand why and must still identify the responsible gene.

A significant clue may lie in the observation of a dramatic reduction in the numbers of μ^+ pre-B cells entering the S phase in XLA patients, whereas the pro-B cells appear to proliferate normally (Pearl et al. 1978; Campana et al. 1990). It is estimated that clonal expansion normally occurs largely in this differentiation compartment of B-lineage cells in the bone marrow (Osmond 1991). Understanding the molecular basis for this growth spurt could help to define the XLA defect.

RELATED PATHOGENESIS OF IgA-DEFICIENCY (IgA-D) AND COMMON VARIABLE IMMUNODEFICIENCY (CVID)

IgA-D and CVID have long been considered to be unrelated and very heterogeneous disorders. These two entities, which are the most frequently diagnosed primary immunodeficiencies, are identified primarily on the basis of serum immunoglobulin levels (Rosen et al. 1992). When IgA is below 5-10 mg/dl, and IgM and IgG levels are normal in individuals without obvious T cell abnormalities, the diagnosis is

IgA-D. When IgM and IgG levels are also low, the diagnosis is CVID. However, the distinction between the two is often blurred. IgA-D individuals can be deficient in the IgG2 and IgG4 subclasses (Oxelius et al. 1981), and patients with the diagnosis of CVID may have reasonable levels of IgM, suggesting the two immunodeficiencies may represent polar ends of a continuous spectrum. Indeed IgA-D individuals may develop CVID (Schroeder HS, unpublished). As a rule the severity of the clinical manifestations reflects the scope of the immunoglobulin isotype deficit.

A differentiation arrest at an immature B cell level is seen in both IgA-D and CVID, the chief distinction being in the Ig isotypes involved in the failure of plasma cell maturation. T cells typically are present in normal numbers in individuals with CVID or IgA-D, and consistent abnormalities of T cell function have not been found.

One of the earliest clues suggesting these immunodeficiencies might be related entities came from their recognition in members of the same Scandinavian families (Wollheim and Williams 1965). Largely ignored for several years, this point has been made forcefully by the subsequent recognition of multiplex families in which several members with either CVID or IgA-D are seen in successive generations (Volanakis et al. 1992). An underlying genetic predisposition is especially obvious for the milder, more common IgA-D (Reviewed in Schaffer et al. 1991). In Caucasian populations in Europe, North America and Australia the incidence of IgA-D is approximately 1 in 700 individuals, whereas in Japan the incidence is 1 in 18,500 individuals, more than 25 times less (Kanoh et al. 1986). The incidence also appears to be relatively low in Malaysians (Yadav and Iyngkaran 1979) and Afroamericans.

The genetic predisposition does not lie in the immunoglobulin genes. These appear structurally normal, and all of the Ig isotypes are expressed in affected individuals at the B cell surface (Hammarström et al. 1985; Cooper and Lawton 1972; Conley and Cooper 1981). Moreover, the 'silent' IgA allotypes of IgA-D individuals can be expressed normally in their offspring (Hammarström et al. 1987).

Location of the underlying susceptibility gene(s) to the MHC region is suggested by the frequent occurrence of particular MHC haplotypes in both CVID and IgA-D patients. Initially, MHC associations between IgA-D and the class I genes were noticed (Ambrus et al. 1977), then class I and II gene combinations (Oen et al. 1982; Hammarström and Smith 1983) and, finally, class III genes of certain MHC haplotypes (Wilton et al. 1985). Our studies then revealed that the same MHC haplotypes are frequently associated both with CVID and IgA-D (Schaffer et al. 1989). Importantly, these ancestral haplotypes are rare among Asians (French et al. 1991) in whom IgA-D (and probably CVID) are relatively rare. Linkage-disequilibrium, the phenomenon in which the highly polymorphic MHC genes tend to be inherited as a block, adds to the difficulty of mapping the susceptibility gene(s). One susceptibility gene candidate lies in the DQ locus, specifically alleles of the β-chain gene encoding a neutral amino acid at position 57 (Olerup et al. 1990). The other currently-favored candidate, a class III MHC gene, is suggested from an analysis of the genes lying in the most frequently conserved portion of the MHC haplotypes in IgA-D and CVID patients and their families (Volanakis et al. 1992). The C4A gene, which is often deleted in affected patients, is thus the focus of study in our laboratory.

While the class II genes are involved in T and B cell interactions, the C4A gene product indirectly enhances B cell-antigen triggering. C4A binds preferentially to antigen-antibody complexes (Law et al. 1984) and facilitates their binding to CR1 receptors on B cells. Enzymatic generation of C3bi and C3dg fragments then allows the antigen-antibody complex to become attached to CR2 (Ahearn and Fearon 1989). The CR2 molecules then interact with CD19 to trigger a signal transduction pathway that can enhance by 100-fold the antigen triggering of B cells via their Ig receptors (Carter et al. 1988; Carter and Fearon 1992). Probably for this reason, individuals with inherited complement deficiencies including deficiency of C4A have diminished antibody responses and are defective in isotype switching (Jackson et al. 1979). In C4A-deficient guinea pigs, the defect in antibody responsiveness can be repaired by administration of human C4A but not C4B (Finco et al. 1992).

For these reasons we conclude that deletion or mutations of the C4A gene could predispose to IgA-D and CVID, and our current efforts are designed to test this hypothesis.

CONCLUSION

Only a few years ago elucidation of the genetic basis of the antibody repertoire was considered the ultimate goal in immunology. Although we now know the genetic basis for antibody diversity, and that of T cell receptor diversity as well, possession of this valuable information still leaves us well short of explaining immune system development and function and of understanding what can go wrong with it. An excellent yardstick of our progress in understanding immune system development and function, the immunodeficiency diseases can offer new clues to this complex biological system and their solution challenges both our investigative and clinical skills.

REFERENCES

Ahearn JM, Fearon DT (1989) Structure and function of the complement receptors CR1 (CD35) and CR2 (CD21). Adv Immunol 46:183-219

Alt FW, Oltz EM, Young F, Groman J, Taccioli G, Chen J (1992) VDJ recombination. Immunol Today 13:306-314

Ambrus M, Hernadi E, Bajtai G (1977) Prevalence of HLA-A1 and HLA-B8 antigens in selective IgA deficiency. Clin Immunol Immunopathol 7:311-314

Anker R, Conley ME, Pollok B (1989) Clonal diversity in the B cell repertoire of patients with X-linked agammaglobulinemia. J Exp Med 169:2109-2119

Bruton OC (1952) Agammaglobulinemia. Pediatrics 9:722-727

Campana D, Farrant J, Inamdar N, Webster ADB, Janossy G (1990) Phenotypic features and proliferative activity of B cell progenitors in X-linked agammaglobulinemia. J Immunol 145:1675-1680

Campana D, Janossy G, Bofill M, Trejdosiewicz LK, Ma D, Hoffbrand AV, Mason DY, LeBacq AM, Forster HK (1985) Human B cell development. I. Phenotypic differences of B lymphocytes in the bone marrow and peripheral lymphoid tissues. J Immunol 134:1524-1530

Carter RH, Fearon DT (1992) CD19: Lowering the threshold for antigen receptor stimulation of B lymphocytes. Science 256:105-107

Carter RH, Spycher MO, Ng YC, Hoffman R, Fearon DT (1988) Synergistic interaction between complement receptor type 2 and membrane IgM on B lymphocytes. J Immunol 141:457-463

Chang H, Dmitrovsky E, Hieter PA, Mitchell K, Leder P, Turoczi L, Kirsch IR, Hollis G (1986) Identification of three new Ig-λ like genes in man. J Exp Med 163:425-435

Conley ME (1985) B cells in patients with X-linked agammaglobulinemia. J Immunol 134:3070-3074

Conley ME, Brown P, Pickard AR, Buckley RH, Miller DS, Raskind WH, Singer JW, Fialkow PJ (1986) Expression of the gene defect in X-linked agammaglobulinemia. N Engl J Med 315:564-567

Conley ME, Cooper MD (1981) Immature IgA B cells in IgA-deficient patients. N Engl J Med 305:495-497

Cooper MD (1987) B lymphocytes: Normal development and function. N Engl J Med. 317:1452-1455

Cooper MD, Lawton AR (1972) Circulating "B" cells in patients with immunodeficiency. Am J Pathol 69:513-528

Desiderio SV, Yancopoulos GD, Paskind M, Thomas E, Boss HA, Landau H, Alt FW, Baltimore D (1984) Insertions of N regions into heavy chain genes is correlated with expression of terminal deoxytransferase in B cells. Nature 311:752-755

Dörken B, Moller P, Pezzutto R, Schwartz A, Moldenhove G (1989) B cell antigens: CD19. In: Knopp W, Dörken B, Gilks WR, Rieber EP, Schmidt RE, Stein H, Vondem Born AEG (eds) Leukocyte Typing IV, Oxford University Press, Oxford, pp 34-36

Finco O, Li S, Cuccia M, Rosen FS, Carroll MC (1992) Structural differences between the two human complement C4 isotypes affect the humoral immune response. J Exp Med 175:537-543

French M, Dawkins R, Christiansen FT, Zhang W, Degli-Esposti MA, Saueracker G (1991) Reply. Immunol Today 12:135-136

Fu SM, Hurley JN, McCune JM, Kunkel HG, Good RA (1980) Pre-B cells and other possible precursor lymphoid cell lines derived from patients with X-linked agammaglobulinemia. J Exp Med 152:1519-1526

Goldblum RM, Lord RA, Cooper MD, Gathings WE, Goldman AS (1974) X-linked B lymphocyte deficiency. I. Panhypo-γ-globulinemia and dys-γ-globulinemia in siblings. J Pediatr 83:188-191

Haas IG, Wabl M (1983) Immunoglobulin heavy chain binding protein. Nature 306:387-389

Hammarström L, Carlsson B, Smith CIE, Wallin J, Wieslander L (1985) Detection of IgA heavy chain constant region genes in IgA deficient donors: evidence against gene deletions. Clin Exp immunol 60:661-664

Hammarström L, de Lange GG, Smith CIE (1987) IgA2 allotypes determined by restriction fragment length polymorphism in IgA deficiency. Re-expression of the silent A2m(2) allotype in the children of IgA-deficient patients. J Immunogen 14:197-201

Hammarström, L, Smith CIE (1983) HLA-A, B, C and DR antigens in immunoglobulin A deficiency. Tissue Antigens 21:75-79

Hendershot L, Bole D, Kearney JF (1987) The role of immunoglobulin heavy chain binding protein in immunoglobulin transport. Immunol Today 8:111-114

Hendriks RW, Schuurman RKB (1991) Genetics of human X-linked immunodeficiency diseases. Clin Exp Immunol 85:182-192

Hermanson GG, Eisenberg D, Kincade PW, Wall R (1988) A member of the immunoglobulin gene superfamily exclusively expressed on B-lineage cells. Proc Natl Acad Sci USA 85:6890-6894

Jackson CG, Ochs HD, Wedgwood RJ (1979) Immune response of a patient with deficiency of the fourth component of complement and systemic lupus erythematosus. N Engl J Med 300:1124-1129

Kanoh T, Mizumoto T, Yasuda N, Koya M, Ohno Y, Uchino H, Yoshimura K, Ohkubo Y, Yamaguchi H (1986) Selective IgA deficiency in Japanese blood donors: Frequency and statistical analysis. Vox Sang 50:81-86

Knittler MR, Haas IG (1992) Interaction of BiP with newly synthesized immunoglobulin light chain molecules: cycles of sequential binding and release. EMBO J 11:1573-1581

Kitamura D, Kudo A, Schaal S, Müller W, Melchers F, Rajewski K (1992) A critical role of λ5 protein in B cell development. Cell 69:823-831

Kudo A, Melchers F (1987) A second gene, Vpre-B in the λ5 locus of the mouse, which appears to be selectively expressed in pre-B lymphocytes. EMBO J 6:2267-2272

Lassoued K, Nunez C, Kubagawa H, Cooper MD (1992) Analysis of human surrogate light chain with monoclonal antibodies. FASEB J 6:4448

Law SKA, Dodds AW, Porter RR (1984) A comparison of the properties of two classes, C4A and C4B, of the human complement component C4. EMBO J 3:1819-1823

Nishimoto N, Kubagawa H, Cooper MD (1991) Comparison of pre-B cell differentiation in normal and X-linked agammaglobulinemic (XLA) individuals. Fed Proc 5:1346

Oen K, Petty RE, Schroeder ML (1982) Immunoglobulin A deficiency: genetic studies. Tissue Antigens 19:174-182

Oettinger MA, Schatz DG, Gorka C, Baltimore D (1990) RAG-1 and RAG-2, adjacent genes that synergistically activate V(D)J recombination. Science 248:1517-1523

Ohno T, Cooper MD, Sekar MC, Burrows PD, Hendershot LM, Kubagawa H (1990) Biochemical and functional characterization of the μ heavy chain and the surrogate light chain complex expressed on human pre-B cell lines. Fed. Proc. 4:A1846

Olerup O, Smith CIE, Hammarström L (1990) Different amino acids at position 57 of the HLA-DQβ chain associated with susceptibility and resistance to IgA deficiency. Nature (Lond) 347:289-290

Osmond D (1991) Mechanisms of B cell neoplasia 1991. Workshop at the Basel Institute for Immunology, Roche, Basel Switzerland, pp 1-19

Oxelius VA, Laurell AB, Lindquist B, Golebiowska H, Axelsson U, Bjorkander J, Hanson LA (1981) IgG subclasses in selective IgA deficiency. Importance of IgG2-IgA deficiency. N Engl J Med 304:1476-1477

Pearl ER, Vogler LB, Okos AJ, Crist WM, Lawton AR, Cooper MD (1978) B lymphocyte precursors in human bone marrow: An analysis of normal individuals and patients with antibody-deficiency states. J Immunol 120:1169-1175

Reth M, Hombach J, Wienands J, Campbell KS, Chien N, Justement LB, Cambier JC (1991) The B cell antigen receptor complex. Immunol Today 12:196-201

Rolink A, Melchers F (1991) Molecular and cellular origins of B lymphocyte diversity. Cell 66:1081-1094

Rosen FS, Cooper MD, Wedgwood RJP (1984) The primary immunodeficiencies. N Engl J Med Part I. 311:235-242, 1984 - Part II. 311:300-310, 1984

Rosen FS, Wedgwood RJP, Eibl M, Griscelli C, Seligmann M, Aiuti F, Kishimoto T, Matsumoto S, Khakhalin LN, Hanson FA, Hitzig WH, Thompson RA, Cooper MD, Good RA, Waldmann TA (1992) Primary immunodeficiency diseases. Report of a WHO Scientific Group. Immunodeficiency Reviews, 3:195-236

Sakaguchi N, Kashiwamura S, Kimoto M, Thalmann P, Melchers F (1988) B lymphocyte lineage restricted expression of mb-1. A gene with CD3-like structural properties. EMBO J 7:3457-3464

Sakaguchi N, Melchers F (1986) λ5, a new light chain-related locus selectively expressed in pre-B lymphocytes. Nature 324:579-582

Schaffer FM, Monteiro RC, Volanakis JE, Cooper MD (1991) IgA deficiency. Immunodefic Rev 3:15-44

Schaffer FM, Palermos J, Zhu ZB, Barger BO, Cooper MD, Volanakis JE (1989) Individuals with IgA deficiency and common variable immunodeficiency share polymorphisms of major histocompatibility complex class III genes. Proc Natl Acad Sci USA 86:8015-8019

Schwaber J (1992) Evidence for failure of V(D)J recombination in bone marrow pre-B cells from X-linked agammaglobulinemia. J Clin Invest 89:2053-2059

Schawber J, Molgaard H, Orkin SH, Gould HJ, Rosen FS (1983) Early pre-B cells from normal and X-linked agmmaglobulinaemia produce Cμ without an attached V_H region. Nature 304:355-358

Schuler W, Weiler IJ, Schuler A, Phillips RA, Rosenberg N, Mak TW, Kearney JF, Perry RP, Bosma M (1986) Rearrangement of antigen receptor genes is defective in mice with severe combined immunodeficiency. Cell 46:963-972

Schwarz K, Hansen-Hagge TE, Friedrich W, Kleinhauer E, Bartram CR (1991) Severe combined immunodeficiency (B-SCID) patients show an altered recombination pattern at the JH locus. In: Chapel HM, Levinsky RJ, Webster ADB (eds) Progress in Immune Deficiency III, Royal Society of Medicine Services International Congress and Symposium Series No. 173, Royal Society of Medicine Services Limited, pp 159-165

Takemori T, Mizuguchi J, Miyazoe I, Nakanishi M, Shigemoto K, Kimoto H, Shirasawa T, Maruyama N, Taniguchi M (1990) Two types of μ chain complexes are expressed during differentiation from pre-B to mature B cells. EMBO J 9:2493-2500

Tedder TF, Crain MJ, Kubagawa H, Clement LT, Cooper MD (1985) Evaluation of lymphocyte differentiation in primary and secondary immunodeficiency diseases. J Immunol 135:1786-1791

Tsubata T, Tsubata R, Reth M (1992) Crosslinking of the cell surface immunoglobulin (μ-surrogate light chains complex) on pre-B cells induces activation of V gene rearrangement at the immunoglobulin κ locus. Inter Immunol 4:637-641

Volanakis JE, Zhu Z-B, Schaffer FM, Macon KJ, Palermos J, Barger BO, Go R, Campbell RD, Schroeder HW, Cooper MD (1992) Major histocompatibility class III genes and susceptibility to immunoglobulin. A deficiency and common variable immunodeficiency. J Clin Invest 89:1914-1922

Wilton AN, Cobain TJ, Dawkins RL (1985) Family studies in IgA deficiency. Immunogenetics 21:333-342

Wollheim FA, Williams Jr RC (1965) Immunoglobulin studies in six kindreds patients with adult hypogammaglobulinemia. J Lab Clin Med 66:433-445

Yadav M, Iyngkaran N (1979) Low incidence of selective IgA deficiency in normal Malaysians. Med J Malaysia 34:145-148

X-Linked Agammaglobulinemia: Updated Criteria for Diagnosis

Mary Ellen Conley and Ornella Parolini

Departement of Pediatrics, University of Tennessee College of Medicine, and Departement of Immunology, St. Jude Childrens Research Hospital, 332 North Lauderdale, Memphis TN 38105 USA

INTRODUCTION

X-linked agammaglobulinemia (XLA), was one of the first immunodeficiencies described (Bruton 1952) and it is often thought of as a prototype immunodeficiency because the clinical and laboratory findings are felt to be distinctive. Affected boys have the onset of recurrent bacterial infections in the first few years of life. The serum immunoglobulins are profoundly decreased and the number of B cells is markedly reduced; although T cell number and function appears to be normal. Similarly affected male are seen in the maternal lineage (Ochs 1989; Lederman 1985). However, the prototype is evolving. Over the years, there has been a gradual change in the criteria required to make the diagnosis of XLA, and in the approach to identifying the underlying genetic defect.

NEW MUTATIONS OF THE GENE FOR XLA

Probably less than half of the boys who are suspected of having XLA fulfill all of the diagnostic criteria listed above (Lederman 1985). The characteristic that is most frequently absent is a family history of hypogammaglobulinemia in male relatives. However, in 1935 Haldane pointed out that one should expect a high rate of new mutations in X-linked lethal disorders. Without the constant infusion of new mutations, the gene defect would rapidly disappear from the population. Between a third and half of patients with any X-linked disorder that is lethal without medical intervention, are likely to be the first manifestation of a new mutation that occurred either in the egg that gave rise to the patient or in a gamete from an earlier generation. Therefore, a positive family history should not be a requirement for the diagnosis of XLA.

Because the incidence of an X-linked lethal disorder is dependent on the rate of new mutations, the incidence of the disorder provides some hints about the nature of the defective gene. Compared with other X linked disorders, XLA is relatively rare, occurring with a frequency of 2-5 per million (Conley 1992). In contrast, Duchenne muscular dystrophy, which is due to a defect in the dystrophin gene which is encoded by a 2300 kb gene with a transcript of 14 kb, occurs with a frequency of 200-350 per million (Darras 1990) and hemophilia A, which is due to a defect in the 186 kb Factor VIII gene with a transcript of 9 kb (White 1989) occurs with a frequency of 100 per million. The frequency of new mutations depends on the size of the gene, the occurrence of duplications in and around the gene, the cytosine content and perhaps other undefined factors. The low incidence of XLA suggests that the gene may be small.

New mutations may occur at varying sites in the XLA gene with different mutations having different physiologic consequences. In as many as 20 to 35% of families with an X-linked history of hypogammaglobulinemia, some or all of the affected males have higher concentrations of serum immunoglobulins or a higher percentage of B cells than are usually seen in patients with typical XLA (Buckley 1968; Goldblum 1974; unpublished observations). In a small number of other families, the affected males have isolated growth hormone deficiency as well as hypogammaglobulinemia (Fleisher 1980; Conley 1991; Monafo 1991). These unusual families may have allelic variants of the gene for XLA or they may have unrelated gene defects that result in clinical and laboratory findings that are similar to those seen in typical XLA.

LINKAGE ANALYSIS IN XLA

Two recently described characteristics of XLA can be used to help clarify the nature of the gene defect in XLA and to evaluate unusual families that may carry the gene for XLA. First, the gene for typical XLA has been mapped by linkage analysis to the mid-portion of the long arm of the X at Xq22 as shown in Fig. 1. The probe p212, at DXS178, has shown no recombination with XLA in approximately 200 meiotic events (Guioli 1989; Kwan 1990; unpublished observations). Flanking markers that have shown recombination can be used to limit the segment of DNA within which the gene for XLA must lie. Published flanking markers, DXS3 and DXS94 (Guioli 1989; Kwan 1990), are approximately 10 to 15 cM apart. (One cM equals 1% chance of recombination per meiotic event.) Studies from our laboratory have pared down this segment by replacing DXS3 as a proximal flanking marker with a marker that is closer to DXS178, DXS442 (Parolini, submitted). The distance between DXS442

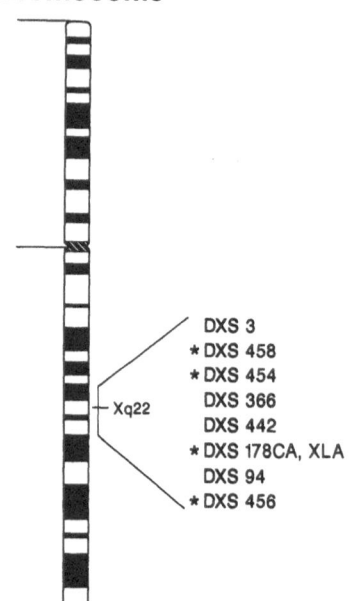

X Chromosome

Fig. 1: The gene for XLA has been mapped Xq22 as shown on this schematic diagram. No recombinations have been seen between XLA and the probe at DXS178, whereas crossovers have been seen between XLA and the probes flanking XLA, indicating that these flanking probes can be used in carrier detection or prenatal diagnosis. Loci that can be identified by PCR are marked by asterisks.

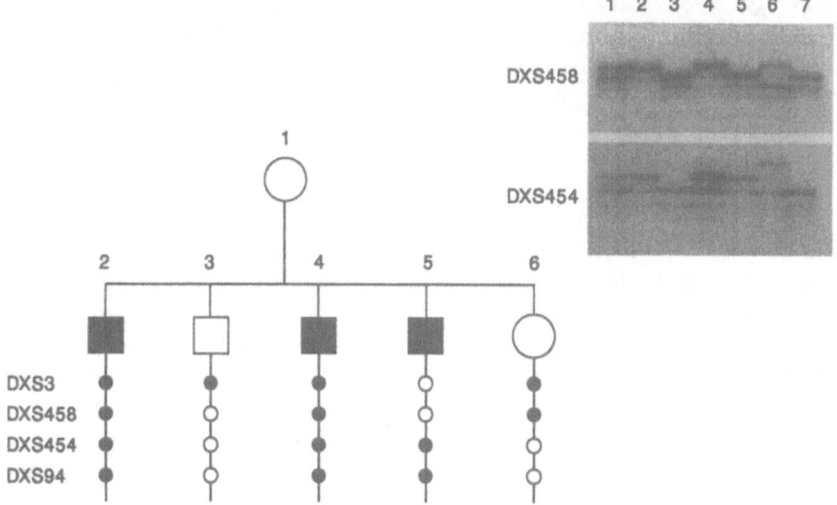

Fig. 2: *Linkage analysis and X chromosome inactivation studies were used to evaluate a family in which three males had the clinical and laboratory findings of XLA (filled squares). B cells from the mother of the affected males (subject 1) demonstrated non-random X chromosome inactivation in a series of somatic cell hybrids. B cell hybrids from the sister (subject 6) demonstrated random X chromosome inactivation, indicating that she was not a carrier of XLA. Below the symbol for each of the offspring is a diagram of that individual's maternally derived X chromosome in the region around the gene for XLA. Only loci at which the mother was heterozygous are shown. Alleles from the maternal X chromosome bearing the mutation (defined as the alleles on the X chromosome not used as the active X in B cell hybrids), are shown as filled circles. The alleles from the non-mutant X chromosome are shown as open circles. The autoradiographs demonstrating the results of PCR analysis at the loci DXS458 and DXS454 are shown on the right. Lanes 1 through 6 contain DNA from the individual with the corresponding number in the pedigree. The DNA in lane 7 is from a B cell hybrid containing the non-mutant X chromosome from subject 1. Fainter "shadow" bands are seen below each allele. In heterozygous women the smaller allele is always more intense than the larger allele. Note that the paternally derived allele in the sample from the sister (subject 6) is different in size from either of the maternal alleles.*

and DXS94 is 3.5 - 4.0 cM. Originally, two allele polymorphisms, detected by Southern blotting, were used to track the XLA gene in individual families. More recently, there has been a shift towards systems using short tandem repeat polymorphisms detected by PCR. Throughout the genome there are short tandem repeats (STRs) consisting of two to five nucleotides that are repeated a variable number of times (Weber 1989; Luty 1990). Oligonucleotide primers that flank the STR can be used in PCR reaction mixtures containing ^{32}P labelled nucleotides to amplify the intervening DNA. The polymorphic PCR products are separated on a sequencing gel and analyzed by autoradiography. Fig. 2 shows the use of these PCR polymorphisms to track the gene in a family with typical XLA and several crossovers between DXS3 and the gene defect.

There is no evidence at this time that there is more than one gene on the X chromosome that is involved in typical XLA. The gene in all families with typical disease maps to Xq22. If the families with X-linked hypogamma-globulinemia, but higher concentrations of serum immunoglobulins than are usually seen in XLA, and families with X-linked hypogammaglobulinemia and growth hormone deficiency have defects in the gene responsible for typical XLA, one would expect the defect in these families to map to Xq22. Linkage analysis of four families with X-linked hypogammaglobulinemia and two families with X-linked agammaglobulinemia and growth hormone deficiency localize the defects in these families to the same site as the gene for typical XLA (Conley 1991 and unpublished observations).

The features that are consistent in all patients with X-linked hypogamma-globulinemia, including those with higher concentrations of serum immuno-globulins than are usually seen in XLA, are early onset of disease, reduced numbers of peripheral blood B cells with small lymph nodes and tonsils, failure to make antigen specific antibody after antigen challenge and normal T cells.

X CHROMOSOME INACTIVATION ANALYSIS IN XLA

The second characteristic of XLA that has become part of the evolving definition of this disorder, is that the obligate carriers of XLA demonstrate non-random X chromosome inactivation in B cells, but in no other cell lineages (Conley 1986; Fearon 1987; Conley 1988a). As a mechanism to compensate for the fact that the female has two copies of every gene on the X chromosome, one X chromosome is inactivated in each somatic cell early in the embryogenesis of the female. In each cell there is an equal chance that the maternally or paternally derived X will be silenced by mechanisms that are not yet well understood. However, once the X chromosome is inactivated, the same X is inactivated in all the progeny of that cell. The inactive X is condensed compared to the other chromosomes, it is hypermethylated, and it replicates later in cell division than the active X.

As a result of this random X inactivation, the normal female is a mosaic. On the average, half the cells in any tissue have the maternally derived X as the active X and half have the paternally derived X as the active X. However, if one of the two X chromosome carries a mutation that is detrimental to the proliferation, maturation or survival of cells of a particular lineage, then all the cells of that lineage will be derived from cells that have the other X, the normal X as the active X.

We first showed in 1986 that B cells but not T cells, neutrophils or platelets from obligate carriers of XLA, the mother's of affected boys, demonstrate non-random X

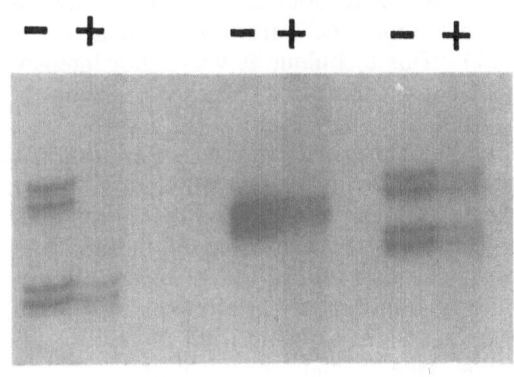

Fig. 3: DNA from early passage EBV transformed B cell lines was used to evaluate X chromosome inactivation patterns in three women who had sons with hypogamma-globulinemia and reduced numbers of B cells but who had no family history of immunodeficiency. In paired samples of DNA, one aliquot was untreated (-) and the other was digested with the methylation sensitive enzyme Hpa II (+). Primers that flank a region in the first exon of the androgen receptor gene containing a variable methylation site and a short tandem repeat were used to amplify both aliquots in a PCR mixture containing ^{32}P labelled nucleotides. The PCR products were separated on a denaturing polyacrylamide sequencing gel and analyzed by autoradiography. Because both strands of DNA are labelled, each allele is seen as a duplex. The DNA from the sample on the left clearly shows non-random X chromosome inactivation with the larger allele completely absent in the Hpa II digested DNA. The sample in the middle is more difficult to interpret because the lower band of the larger allele coincides with the upper band of the smaller allele. However, three bands are seen in the undigested DNA whereas only the two larger bands are seen in the Hpa II digested DNA, indicating non-random X chromosome inactivation. Normal, random X chromosome inactivation is seen in the DNA from the sample on the right

chromosome inactivation (Conley 1986). As in the affected boys, B cell precursors that have the mutant X as the active X do not differentiate into B cells. By default, all of the B cells in the obligate carriers are derived from precursor that have the normal X as the active X.

The non-random pattern of X chromosome inactivation in B cells from carriers of XLA indicates that the gene product is intrinsic to the B cell lineage. If the gene product were produced by stromal cells in the bone marrow microenvironment or by T cells, the normal gene product could induce the proliferation and differentiation of B cell precursors that had the mutant X as the active X. In addition to providing information about the nature of the gene defect, non-random X chromosome inactivation in B cells from the mother of a boy with hypogammaglobulinemia and reduced numbers of B cells, but no family history of immunodeficiency, can help confirm the diagnosis of XLA in the affected child.

Several techniques have been used to evaluate patterns of X chromosome inactivation. The early studies, described above, were performed using the protein polymorphism for the X encoded enzyme glucose-6-phosphate dehydrogenase (Conley 1986). However, this protein polymorphism is uncommon in the caucasian population and there are no other well described X-linked protein polymorphisms in man. We have also used a technique that involves the production of a series of somatic cell hybrids that selectively retain the active X chromosome from B cells from the obligate carrier (Conley 1988a; Conley 1988b; Conley 1991). This technique is very labor intensive but it has the advantage that it identifies the X chromosome that carries the mutation.

Other techniques to evaluate patterns of X chromosome inactivation take advantage of the fact that the active and inactive X chromosomes differ in methylation (Fearon 1987). Allen et al. (in press) have described a PCR based technique that is rapid and can be performed on a small number of cells. In the first exon of the androgen receptor gene, which is encoded on the proximal long arm of the X, there is a short tandem repeat that is less than 100 bases away from a site that is methylated on the inactive X but unmethylated on the active X. DNA from purified B cells can be digested with a methylation sensitive enzyme, Hpa II, and then amplified with PCR primers that flank both the STR and the methylation site. If there is a random pattern of X chromosome inactivation, a portion of both the maternally and paternally derived alleles will be intact after Hpa II digestion. However, if there is non-random X chromosome inactivation, the allele on the active X will be completely digested and only the other allele will be amplified.

Using this assay we examined X chromosome inactivation patterns in B lineage cells from three women who had sons with the clinical and laboratory features of XLA, but who had no previous family history of immunodeficiency. As shown in Fig. 3, B cells from two of these women exhibited a non-random pattern of X chromosome inactivation, like carriers of typical XLA; thus confirming the diagnosis of XLA in their affected sons. B cells from the third woman demonstrated a random pattern of X inactivation. There are several possible explanations for this finding. The affected child may have XLA as the result of a new mutation that occurred in the maternal gamete. An alternative explanation is that the child may have an autosomal recessive disorder that is phenotypically identical to XLA.Females with an immune deficiency that is phenotypically identical to XLA have been previously described (Hoffman 1977). In a series of 36 patients with early onset hypogammaglobulinemia, markedly reduced numbers of B cells and chronic enteroviral encephalitis, 3 were girls (McKinney 1987). We recently studied two girls with antibody deficiencies that were indistinguishable from XLA (Conley 1992). To determine whether subtle abnormalities of the X chromosome might be responsible for the disease in these patients, detailed genetic studies were performed on one of these girls. Cytogenetic studies did not reveal any deletions, inversions or translocations of the X chromosomes. Both maternal and paternal alleles could be identified at the loci flanking the gene for XLA at Xq22. Both of the patient's X chromosomes could function as the active X in somatic cell hybrids.

The lack of any abnormalities in the X chromosomes from a female with a disorder phenotypically indistinguishable from XLA suggests that her disorder is not due to a defect on the X chromosome, but instead is the result of an autosomal recessive

Table	CRITERIA FOR DIAGNOSIS OF XLA

Onset of recurrent bacterial infections in the first few years of life in a male child
Hypogammaglobulinemia with failure to make antigen specific antibody after challenge
Markedly reduced numbers of B cells; less than 2% of peripheral blood lymphocytes, with normal T cell numbers and function
Family history of similarly affected male relatives in the maternal lineage; or non-random X chromosome inactivation in B cells but not other cell lineages from the patient's mother

disorder. If as many as 5-15% of patients with the clinical and laboratory findings of XLA are females with an autosomal recessive disorder, one can expect that an equal proportion of males will have their disease as the result of the same autosomal recessive disorder.

If there is an autosomal recessive disorder that is phenotypically identical to XLA, it is likely that the gene products of the autosomal recessive and X-linked forms of the disease are functionally related. On can postulate that they might be members of a dimer or a receptor ligand pair, or part of a protein cascade in which the product of one gene precedes that of the other.

SUMMARY

The use of linkage analysis and X chromosome inactivation studies have permitted a refinement in the definition of X-linked agammaglobulinemia and have suggested updated criteria that can be used to make the diagnosis of XLA (Table). These two approaches can also be used to provide carrier detection and prenatal diagnosis in families at risk of carrying the gene for XLA, particularly when the recently described PCR based techniques are employed. In addition, these approaches have provided clues about the nature of the gene defect in XLA. The gene that is defective in XLA is probably a relatively small gene that is expressed early in the differentiation of B lineage cells but is not expressed in other cell lineages.

ACKNOWLEDGEMENTS

These studies were supported in part by grants from the National Institutes of Health AI 25129, NCI CORE grant P30 CA21765, the Fondazione Camillo Golgi, Medical Research Council of Canada MA 9337, American Lebanese Syrian Associated Charities and by funds from the Federal Express Chair of Excellence.

REFERENCES

Allen RC, Zoghbi HY, Moseley AB, Rosenblatt HM, Belmont JW (1992) Am J Hum Genet In press

Bruton OC (1952) Pediatrics 9:722-728

Buckley RH, Sidbury Jr,JB (1968) Pediatr Res 2:72-84

Conley ME (1992) Annu Rev Immunol 10:215-238

Conley ME, Brown P, Pickard AR, Buckley RH, Miller DS, Raskind WH, Singer JW, Fialkow PJ (1986) N Engl J Med 315:564-567

Conley ME, Burks AW, Herrod HG, Puck JM (1991) J Pediatr 119:392-397

Conley ME, Puck JM (1988a) J Pediatr 112:688-694

Conley ME, Puck JM (1988b) Immunol Invest 17:425-463

Conley ME, Sweinberg SK (1992) J Clin Immunol 12:139-143

Darras BT (1990) J Pediatr 117:1-15

Fearon ER, Winkelstein JA, Civin CI, Pardoll DM, Vogelstein B (1987) N Engl J Med 316:427-431

Fleisher TA, White RM, Broder S, Nissley SP, Blaese RM, Mulvihill JJ, Olive G, Waldmann TA (1980) N Engl J Med 302:1429-1434

Goldblum RM, Lord RA, Cooper MD, Gathings WE, Goldman AS (1974) J Pediatr 85:188-191

Guioli S, Arveiler B, Bardoni B, Notarangelo LD, Panina P, Duse M, Ugazio A, de Saint Basile G, Mandel JL, Camerino G (1989) Hum Genet 84:19-21

Haldane JBS (1935) J Genet 31:317-326

Hoffman T, Winchester R, Schulkind M, Frias JL, Ayoub EM, Good RA (1977) Clin Immunol Immunopathol 7:364-371

Kwan S-P, Terwilliger J, Parmley R, Raghu G, Sandkuyl LA, Ott J, Ochs H, Wedgwood R, Rosen F (1990) Genomics 6:238-242

Lederman HM, Winkelstein JA (1985) Medicine 64:145-156

Luty JA, Guo Z, Willard HF, Ledbetter DH, Ledbetter S, Litt M (1990) Am J Hum Genet 46:776-783

Lyon MF (1974) Proc R Soc Lond 187:243-268

McKinney Jr,RE, Katz SL, Wilfert CM (1987) Rev Infect Dis 9:334-356

Monafo V, Maghnie M, Terracciano L, Valtorta A, Massa M, Severi F (1991) Acta Paediatr Scand 80:563-566

Ochs HD, Wedgwood RJ (1989) Immunologic disorders in infants and children. Edited by Stiehm, E.R. Philadelphia: W.B. Saunders Company p. 226-256.

Weber JL, May PE (1989) Am J Hum Genet 44:388-396

White GC, Shoemaker CB (1989) Blood 73:1-12

Gene Therapy for Immunodeficiency and Cancer

R. Michael Blaese, Kenneth W. Culver, A Dusty Miller[*], W. French Anderson

National Cancer Institute and National Heart, Lung and Blood Institute, National Institutes of Health, Bethesda, MD
[*]*Fred Hutchinson Cancer Center, Seattle, WA, USA*

INTRODUCTION

Gene transfer technology has enormous potential for the treatment of both inherited "genetic" disorders as well as diseases not ordinarily thought of as "genetic." Since initial studies would be restricted to *ex vivo* gene transfer, our strategy for bringing this technology to the clinic has been to develop gene transfer protocols in the setting of clinical treatments which employ infusion or implantation of various types of cultured cells. Retroviral vectors are the current method of choice for clinical gene transfer because they provide relatively high efficiency gene transfer and usually give stable integration of the inserted genes. However, retroviruses only stably transfer genes into proliferating cells so that many potential target tissues (eg., brain, muscle) are not receptive to gene transfer with these vectors. This limitation has so far prevented the effective use of retroviral vectors for gene therapy of many of the disorders originally thought to be prime candidates for this treatment, disorders involving the totipotent bone marrow stem cell, since these cells are usually not cycling and thus are not receptive to this technique of gene transfer.

T LYMPHOCYTES FOR GENE THERAPY

As an alternative, T lymphocytes seemed to have several characteristics which suggested that they might be useful as cellular vehicles for the clinical application of genetic modification. Extensive preclinical studies in mice and monkeys demonstrated that cultured T cells could be efficiently transduced with retroviral vectors and that these gene-modified cells survived and functioned normally when reintroduced *in vivo*. As an initial clinical application of this technology, we initiated studies using introduced genes, not for therapeutic purposes, but rather as cell labels to assist in the determination of the distribution and survival of tumor infiltrating lymphocytes (TIL) being used in the immunotherapy of human melanoma. These studies demonstrated that the gene-marked TIL survived for weeks *in vivo* and that they accumulated preferentially at tumor sites. These studies also provided us with the critically important information that retroviral-mediated gene-transfer in man was safe enough that its application to other serious diseases could be tried.

GENE THERAPY OF ADENOSINE DEAMINASE DEFICIENCY

As a first attempt at gene therapy, we used retroviral-mediated gene-transfer to introduce a normal ADA gene into autologous T cells in 2 children with ADA deficiency SCID and then treated these children with periodic infusions of their own gene-corrected T cells. The first patient is now nearly 2 years into this treatment. She has developed an ADA level in her circulating T cells of ~25% of normal, has achieved normal T cell counts, and has developed normal antibody responses and positive delayed hypersensitivity skin tests to environmental antigens. Each child has clearly experienced significant immune reconstitution during this treatment without any observed complication of the therapy. We conclude that gene therapy can be a successful treatment for some patients with this very severe inherited immunodeficiency disease.

GENE THERAPY FOR CANCER

Another clinical application of gene transfer is the attempt to enhance the body's defense mechanisms against cancer. The introduction of the TNF gene into TIL is being evaluated, although results of this approach are not yet available. Another related approach is to produce a "tumor vaccine" by gene-modifying tumors by insertion of cytokine genes which could then recruit an immune response when these modified tumors were reinjected into patients. Extensive studies in mice have demonstrated that tumors engineered to express various cytokines including IL1, IL2, IL4, IL6, TNF, γIFN and GM-CSF are capable of enhancing anti-tumor immunity and clinical studies testing this idea have recently been started.

These approaches are principally indirect, using gene transfer to enhance the capacity of normal host defense mechanisms to attack the malignancy. Eventually the development of techniques allowing targeted *in vivo* gene transfer directly into growing tumors *in situ* should provide a mechanism for testing many different approaches for the control of malignancy. Then transfer of specific tumor suppressor genes, genes to encode production of a particular toxin, or genes encoding a process to induce apoptosis specifically in the tumor cells might become feasible. Our initial approach to this problem has been to attempt to exploit what had been viewed as a disadvantage of retroviral vectors, their inability to transfer genes into cells that are not proliferating. This characteristic could be used to help us target gene delivery specifically to growing tumors while sparing non-proliferating normal tissue in the vicinity of the tumor. Rats with a cerebral glioma were given an intratumoral stereotactic injection of murine fibroblasts engineered to produce a herpes simplex thymidine kinase (HS-tk) retroviral vector. Since the only mitotically active cells in the brain are the tumor cells, they will be the only cells to integrate the HS-tk gene. After time to allow the *in situ* produced retroviral vector to transduce the neighboring proliferating glioma with the HS-tk gene, the tumor bearing rats were treated with the anti-herpes drug ganciclovir (GCV). The gliomas in control animals grew to large size while the GCV treated rats demonstrated complete regression of the tumors both macroscopically and microscopically.

One of the most striking lessons from the early studies directed at developing clinical applications of gene transfer is the realization that his technology offers tremendous promise for the development of entirely new approaches for the treatment of many disease processes, not only those disorders that are classical genetic diseases such as ADA deficiency.

1. Blaese RM. (1991) Progress toward gene therapy. *J. Clin Immunol Immunopath* 61: S47-S55

2. Culver K, Cornetta K, Morgan R, Morecki S, Aebersold P, Kasid A, Lotze M, Rosenberg SA, Anderson WF, Blaese RM.(1991) Lymphocytes as cellular vehicles for gene therapy in mouse and man. *Proc Natl Acad Sci USA* 88:3155-59.

3. Culver KW, Morgan RA, Osborne WRA, Lee RT, Lenschow D, Able C, Cornetta K, Miller AD, Anderson WF, Blaese RM (1990) In vivo expression and survival of gene-modified rhesus T lymphocytes. *Human Gene Therapy* 1:399-410.

4. Rosenberg SA, Aebersold P, Cornetta K, Kasid A, Morgan RA, Moen R, Karson EM, Lotze MT, Yang JC, Topalian SL, Merino MJ, Culver K, Miller AD, Blaese RM, Anderson WF (1990). Gene transfer into humans: immunotherapy of patients with advanced melanoma using tumor infiltrating lymphocytes modified by retroviral gene transduction. *New Engl J Med* 323:570-78.

5. Blaese RM, Anderson WF, Culver KW (1990). The ADA human gene therapy clinical protocol. *Human Gene Therapy* 1:327-62.

6. Culver KW, Ram Z, Wallbridge S, Ishii H, Oldfield EH, Blaese RM (1992). In vivo gene transfer with retroviral vector producer cells for treatment of experimental brain tumors. *Science* 256:1550-1552.

Severe Combined Immunodeficiencies in Humans

J. P. de Villartay, G. de Saint Basile, C. Soudais, F. Le Deist,
C. Hivroz, F. Rieux-Laucat, M. Cavazzana-Calvo, J. Disanto,
S. Markievicz, B. Lisowska-Grospierre, A. Fischer

INSERM U 132, Hospital des Enfants Malades, 149. rue de Sèvres, 75730 Paris Cédex 15, France

Severe combined immunodeficiencies (SCID) form a heterogeneous group of genetic disorders characterized by intrinsic defects in T cell differentiation and function causing profound abnormalities of cellular and humoral immunity (Table 1) (Arnaiz-Villena 1992b, Conley 1992, Fischer 1991, De Saint Basile 1991b). We would like to emphasize in this manuscript recent findings on the molecular aspects of some of the SCID syndromes.

1. Autosomal recessive SCID

About 1/5 patients born without mature T lymphocytes present also with absence of mature B lymphocytes although adenosine deaminase activity is detectable. This form of SCID is inherited as an autosomal recessive trait. Lymphocytopenia is present but not total since natural killer cells are present and functional. This phenotype is reminiscent of the scid mouse model where T/B cell differentiation blockade has been assigned to faulty formation of coding joints during V(D)J elements of T and B antigen receptor rearrangements. Signal joining is in contrast normal. The gene responsible for this recombinase defect is unknown although it has been assigned to the short arm of chromosome 16. It differs from the rag 1 and 2 genes (Bosma, 1991). Two other characteristics of the scid model include the

Table 1. SCIDs

Human disease	Animal model
AR SCID	scid mouse
Adenosine deaminase deficiency	
XL SCID	XL SCID in dogs
CD3 γ / ε deficiencies	
MHC class II deficiency	MHC class II(-) mice[*]
Abnormal signal transduction in T cells	
Defective synthesis of IL2 ± other cytokines (NF-AT defect)	
Purine nucleoside phosphorylase deficiency	

[*] as achieved by homologous recombination

concomitant finding of an abnormal DNA double strand break repair (Fulop 1991, Biedermann 1991) and the existence of a leaky phenotype (Bosma, 1991). The former anomaly has been detected not only in hematopoietic cells but also in fibroblasts.

It is not strictly proven that a same gene encodes for a protein with both related functions or whether 2 linked genes are defective (deleted). Leakiness occurs in aged scid mice as oligoclonal normal T and B cell populations emerged following reverse mutation or mutation bypassing. Recently, several pieces of evidence have suggested the existence of a similar disease in humans. Indeed, Schwartz et al. (1991) have elegantly demonstrated the existence of aberrant D to J rearrangement attempts in marrow pre-B cells from AR SCID patients as shown by PCR amplification of DQ52 to JH rearrangements.

In two families, it was found that one child had typical absence of T and B lymphocytes while another presented with erythrodermia, a condition known as Omenn's syndrome. The latter is characterized by the presence of large numbers of activated T cells in blood, skin and gut while lymphoid organs are depleted. These T cells derive from a small number of T cell clones as shown by the detection of discrete bands after hybridization of T cell DNA with cß, $J\delta$ or $J\varsigma$ probes (De Saint Basile, 1991c). Clones within one given patient and in different patients differ in terms of TCR $\lambda\beta$ vs $\gamma\delta$ usage and also in terms of Vß family usage. These findings have been recently confirmed by using anchor PCR to evaluate Vß repertoire usage (Rieux-Laucat, 1992).

Finally, as in scid mouse, granulocyte-macrophage progenitors (CFU-GM) as well as fibroblasts from AR SCID patients and patients with Omenn's syndrome but not from patients with XL SCID exhibit a 2-3 fold increased radiosensitivity (Cavazzana-Calvo, 1992). Together these findings indicate that the human AR SCID and the Omenn's syndrome condition correspond to an ID caused by a defect in the recombinase system that is also likely involved in DNA repair. Definitive proof for this analogy may come from the study of artificial substrate rearrangements after transfection in patient's fibroblasts with RAG-1 and RAG-2 cDNAs. Radiosensitivity could be used as functional complementation assay for gene(s) identification.

XL SCID

Half of human SCID are characterized by the absence of mature (and immature) T lymphocytes whereas B lymphocytes are present and partially functional (Conley, 1992). This syndrome is most often inherited as an X-linked condition. It's curability by bone marrow transplantation reflects the intrinsic nature of the defect. By using RFLP determining markers in the study of X-linked pedigrees providing a number of meiotic events, it has been possible to localize the XL SCID gene to Xq1.3. No other localization has been found (De Saint Basile 1987, Puck, 1989, 1990). The locus lies between DXS132 and DXS447 as shown by deletion analysis. The estimated genetic distances between newly generated flanking markers (polymorphic tandem repeats) is approximately 5 cmorgans.

The X-chromosome inactivation pattern in obligate carriers of XL SCID is usually non random in T and B lymphocytes but not in other hematopoietic cells (Conley, 1988). This has been shown by somatic hybridization with hamster cells and the addition of a methylation sensitive restriction enzyme in Southern blot analysis using polymorphic X-chromosome-associated marker. Using the former method, Conley et al. (1988) have found that the skewed pattern of X-inactivation is more pronounced in mature B cells. These results are surprising given the XL SCID phenotype. It suggests either that the XL SCID gene product is also involved in B cell differentiation (although to a lesser extent than in T cell differentiation) or that the gene is expressed at an early step of differentiation that precedes T/B lineage commitment.

A partial T cell deficiency has been recently described in two families with X-linked inheritance (Brooks 1990, de Saint Basile 1991d). There is a quantitative T cell defect and T cell functions deteriorate with time. Suggestion that this condition reflects a different mutation of the XL locus stems from two observations 1) in obligate carriers, X-chromosome inactivation pattern is identical to the one of carriers of typical XL SCID and 2) RFLP studies assign the affected gene in the very same region (i.e. Xq1.3) where the XL SCID locus maps. One cannot however rule out the hypothesis of a closely linked distinct locus. Of note is the observation of mild phenotypes for other XL IDs such as XL agammaglobulinemia or the Wiskott-Aldrich syndrome. A canine XL SCID has recently been described (Jezyk, 1989). It is characterized by thymic hypoplasia with presence of some immature CD4(-) CD8(-) thymocytes. Further study will delineate whether canine XL SCID and human XL SCID genes are related.

T-cell ID with defective expression of the T cell receptor (TCR)/CD3 complex

Mutations affecting either the CD3 γ or the CD3 ϵ subunit have recently been described.
In two siblings from one family, heterozygous mutations of the CD3 γ encoding gene have been characterized following PCR amplification of CD3 γ cDNA and DNA (Araniz-Villena, 1992a).
Mutations occurred for one allele as a simple base substitution (A$\longrightarrow$G) in the initiation codon, and for the second as a G$\longrightarrow$C mutation in the last nucleotide of intron 2. This created an abnormal splice within exon 3 inducing deletion of 117 nucleotides. This combination of mutations resulted in the absence of normal CD3 γ protein. TCR/CD3 complex can however be detected at the T cell surface with a 50 % expression level compared to normal. This TCR/CD3 complex does not include CD3 γ chain. Such complexes can trigger T cell activation by antigens but not by lectins and anti-CD3 antibodies (Allarcon, 1988). A CD3 γ^- CD4 T cell line has been derived from one of the patient and showed normal proliferation following TCR/CD3 triggering (Perez-Aciego, 1991). In contrast CD8$^+$ CD3 γ^- T cells poorly proliferated in response to lectin and IL2. The authors claim that CD3 γ might be involved in molecular interaction with CD8, also because CD8$^+$ T cells are reduced in one of these patients. It is worth noting that despite the lack of CD3 γ, one of the patient is healthy while the other died from viral pneumonitis (Arnaiz-Villena, 1992b). This observation underlines recent evidence for distinct activation modules within the TCR/CD3 complex leaving some T cell activation with the TCR $\alpha\beta$ CD3 $\delta\epsilon\zeta$ complex.

In another patient, independent CD3 ϵ mutations have been characterized following the same approach : One mutation affects the second nucleotide of the splice site that follows exon 7 (transmembrane exon) : T$\longrightarrow$C and creates abnormal splicing deleting the transmembrane exon in most of the cases. Minor normal splice events occur allowing production of a small amount of normal CD3 ϵ subunit. The second mutation is a G$\longrightarrow$A substitution at codon 59 of exon 6. This mutation creates a stop codon and could account for the reduced CD3 ϵ mRNA level found in patient and his father (Soudais 1992). These mutations led to reduced ($\approx$ 10 % of normal) expression of the TCR/CD3 complex as recognized by anti-CD3 ϵ antibody (Thoenes, 1992). Low expression of the TCR/CD3 complex prevents activation through anti-CD3, anti-CD2 antibodies and lectins but not by certain antigens. Also the CD28 activation pathway appears preserved (Le Deist, 1991). It is remarkable to observe that a very low expression of normal TCR/CD3 complex does not preclude normal T cell differentiation including presumably positive and negative selection events and a relatively normal shaping of the T cell PCR repertoire as determined by usage of an anchor-PCR method (Rieux-Laucat, 1992). This is in correlation with the mild clinical consequences of this T-cell ID, i.e. mild bacterial lung infections. This observation underlines the redundancy of the immune system and the possibility that coactivation signals delivered for instance by B7-CD28 interaction may compensate for low number of TCR/peptide/MHC interaction.

Many other molecular abnormalities are going to be recognized as causing profound T cell dysfunctions. For intance, abnormal signal transduction in T cells that can by corrected by phorbol ester is associated in one case with a reduced pattern of tyrosine phosphorylation following TCR/CD3 triggering as shown in immunoblotting with an phosphotyrosine antibody (Le Deist, 1992). This anomaly may be caused by a defect in tyrosine kinases (p56lck or p59fyn) associated with the TCR/CD3 complex although function of both is normal in vitro.

ACKNOLEDGEMENTS

This work has been supported by INSERM and AFM.

REFERENCES

Alarcon B, Regueiro JR, Arnaiz-Villena A, Terhorst C (1988) N Engl J Med 313:1203-1207.

Arnaiz-Villena A, Timon M, Corell A, Perez-Aciego P, Martin-Villa JM, Regueiro JA (1992) N Engl J Med, in press.

Arnaiz-Villena A, Timon M, Rodriguez-Gallego C, Perez-Blas M, Corell A, Martin-Villa JM, Regueiro JR (1992) Immunol Today 13:259-265.

Bosma MJ (1991) Ann Rev Immunol 9.

Cavazzana-Calvo M, Le Deist F, De Saint Basile G, Papadopoulo D, De Villartay JP, Fischer A (1992) submitted.

Conley ME (1992) Ann Rev Immunol 10:215-238.

De Saint Basile G, Arveiler B, Oberlé J (1987) Proc Natl Acad Sci (USA) 84:7576-7579.

De Saint Basile G, Le Deist F, De Villartay JP, Griscelli C, Fischer A (1991) J Clin Invest 87:1352-1359.

De saint Basile G, Fischer A (1991) Immunol Today 12:456-461.

De Saint Basile G, Le Deist F, Caniglia M, Lebranchu Y, Griscelli C, Fischer A (1991) J Clin Invest 89:861-866.

Fischer A (1991) Immunodef Reviews 3:83-100.

Fulop GM, Philipps RA (1990) Nature 347:479-482.

Griscelli C, Durandy A, Virelizier JL, Ballet JJ, Daguillard F (1978) J Pediatr 93:404-411.

Jezyk PF, Felsburg PO, Haskins E, Patterson DF (1989) Clin Immunol Immunopathol 52:173-189.

Le Deist F, Thoenes G, Corado J, Lisowska-Grospierre B, Fischer A (1991) Eur J Immunol 21:1641-1647.

LeDeist F, Hivroz C, Buc H, Griscelli C, Fischer A (1992) submitted.

Perez-Aciego P, Alarcon B, Arnaiz-Villena A, Terhost C, Timon M, Segurado O, Regueiro JA (1991) J Exp Med 174:314-326.

Puck JM, Nussbaum R5L, Smead DL, Conley ME (1989) Am J Hum Genet 47:724-730.

Puck JM, Stewart CC, Conley ME, Nussbaum RL (1990) Am J Hum Genet 47:765.

Rieux-Laucat F, Selz F, Fischer A, De Villartay JP (1992) submitted.

Schwartz K, Hausen-Hagge TE, Knobloch C, Friedrich W, Kleihauer E (1991) J Exp Med 174:1039-1048.

Soudais C, De Villartay JP, Le Deist F, Fischer A, Lisowska-Grospierre B (1992) submitted.

Thoenes G, Soudais C, Le Deist F, Griscelli C, Fischer A, Lisowska-Grospierre B (1992) J Biol Chem 267:487-493.

Knock out Mice Models for Immunodeficiency Diseases

Werner Haas[1] and Ralf Kühn[2]

[1]Hoffmann-La Roche Inc. 340 Kingsland Street Bldg.102/Room 116 d Nutley, New Jersey 07110-1199 USA
[2]Institute für Genetik Weyertal 121 D-5000 Köln Germany

Gene knock out mice

The impressive progress in transgenic techniques in mice within the past ten years has opened up valuable new experimental approaches to immunologists. For many research purposes it is of great interest to eliminate a specific gene or at least to prevent the normal function of its product. A well known example are transgenic mice with rearranged antigen receptor genes preventing the diversification of whole lymphocyte subsets via allelic exclusion. Such antigen receptor transgenic mice have been successfully used to study T- and B-cell development (Storb 1987, Goodnow et al 1988, von Boehmer 1990). The direct modification of genes in the mouse germline could be achieved in the past only by insertional mutagenesis using transgenes or retroviruses, and chemical- or radiation-induced mutagenesis (for review see Jaenisch 1988). The limitation of these strategies is that the gene to be mutated and the exact nature of the mutation cannot be predetermined. The accuracy by which mutations can be introduced into the mouse genome has been greatly improved with the development of gene targeting techniques within the last four years allowing the modification of a particular gene in a predetermined manner (for review see Capecchi 1989, Gridley et al 1991, Bradley et al 1992). This development was made possible by the establishment of totipotent embryonic stem (ES) cell lines that can give rise to any cell type including germ cells when introduced back into a blastocyst from in vitro culture, generating a chimeric animal. The avaibility of large numbers of ES cells made it possible to select for rare cells in which a particular gene was modified by gene targeting. Targeting a modification to a particular gene is achieved by the introduction into ES cells of a mutated version of this gene designed to undergo a homologous recombination event with its chromosomal counterpart. Negative and positive selection techniques enrich for the rare cells carrying the recombinant genome (Bradley et al 1992). Whereas the first generation of targeting vectors was usually used for gene inactivation and deletion, more recently refined methods have been developed which allow the introduction of subtle changes such as single nucleotide exchanges without any further modification of the locus (Hasty et al 1991) deletions of fragments as large as 15 kb (Mombaerts et al 1991) or gene substitutions (Le Mouellic et al 1990).

Immunologists have already knocked out many genes.

To date more than 25 mouse mutants have been generated by the knock out of genes related to lymphocyte development or function, exceeding by far the few cases of naturally occuring mutants where the underlying genetic defect has been identified (see Fung-Leung and Mak 1992, Rajewsky 1992 and Tables 1-3). Conceivably many more "knock outs" will be reported at this meeting. Most of the initial targets were genes encoding proteins that are more or less directly involved in antigen recognition. Neither immunoglobulin nor T-cell antigen receptor genes could be properly assembled in RAG-1 or RAG-2 deficient mice (Mombaerts et al 1992, Sinkai et al 1992). B-cell development was drastically reduced or completely blocked in mice which cannot produce $\lambda 5$ (Kitamura et al 1992b) or the transmembrane form of IgM respectively (Kitamura et al 1991, 1992a). Mice which lack either $\alpha\beta$ T-cells, $\gamma\delta$ T-cells, CD4 $\alpha\beta$ T-cells or CD8 $\alpha\beta$ T-cells have been generated by knocking out TCR α, β or δ genes (Philpott et al 1992, Mombaerts et al 1992, Itohara and Tonegawa, personel communication), MHC class I (Zijlstra et al 1990) or class II genes (Cosgrove et al 1991 and 1992, Grusby et al 1991, Koentgen et al 1992 personal communication)), the CD4 (Rahemtulla et al 1991) or the CD8 gene (Fung-Leung 1991a,b). The development of T-cells was also drastically reduced in p56lck deficient mice (Molina et al 1992). IL-2 deficient mice show a defect in B-cell development that becomes obvious only after several weeks of life (Schorle et al 1991, Schimpl, see Paul 1992). IL4 deficient mice show reduced IgG1 and no IgE antibody responses (Kuhn et al 1991, Kopf et al 1992) and IFNγ (Dalton et al 1992) or IFNγ-receptor deficient mice (Huang et al 1992) show defects in macrophage activation and IgG2a antibody production. So far no major abnormalities of the immune system have been recognized in Thy1 (Stewart and Silver 1992, personal communication) IgD or IL10 deficient mice (Rajewsky 1992, Paul 1992). Targeting experiments which do not result in a clear mutant phenotype reveal quickly what a particular gene is not required for and the thorough analysis of the phenotypes of gene knock out mice can be more demanding than the generation of a mutant mouse strain itself.

Gene knock out mice as disease models

Animal models for human diseases have been useful to study pathogenic mechanisms and to test therapeutic approaches. Many strains of mice and other animals with inherited disorders or predispositions to particular diseases are used in basic and applied research. With the advent of methods that allow the genetic engineering of mice it has become possible to generate mouse strains that can serve as models for human diseases. The first "transgenic disease" was generated in mice by introducing a mutant α1(I) collagen gene into the germ line (Stacey et al 1988). The mutations were generated in vitro at the same sites at which human mutations were found in patients with osteogenesis imperfecta and the transgenic mice showed the same dominant lethal phenotype that is characteristic of the human disease. In contrast, the initial attempt to reconstruct a human genetic disorder using mutated ES cells failed: HPRT deficient mice are normal in contrast to HPRT deficient patients who suffer from very severe neurological disease (Hooper et al 1987, Kuehn et al 1987). The failure is thought to be due to a difference between the purine metabolism of mice and humans with mice possessing the enzyme urate oxidase, which metabolizes the toxic products that accumulate as a consequence of HPRT deficiency (Wu et al 1989). More recently ES cells in which the gene encoding glucocerebrosidase was disrupted were successfully used to generate mice with a disease closely resembling type 2 Gaucher disease that leads to death shortly after birth (Tybulewicz et al 1992). It is conceivable that many mouse strains will be generated in the near future with genetic defects analogous to those causing human heritable diseases and that in many cases the mice and the human patients with the same genetic defects will have the same or very similar symptoms e.g. cystic fibrosis (Koller et al 1991). Such animal models will be valuable tools to test for therapeutic approaches and might have a major impact on medical research. This will include immunodeficiency diseases with known underlying genetic defects such as adenosine deaminase deficiency or purine nucleotide phosphorylase deficiency. However, in most murine and human primary immunodeficiency syndromes the defective genes remain to be determined (Hong 1989, Fischer 1990, Spickett et al 1991, Arnaiz-Villena et al 1992, Greiner et al 1992, Matsumoto et al 1992). With the continuously increasing density of genetic markers in the mouse and human genome it may become easier to identify the disease causing alleles. In addition the phenotypes of already available gene knock out mice may guide clinicians in their search for the molecular cause of human immunodeficiency diseases.

Legend to tables 1-3
The tables present brief descriptions of gene knock out mice or natural genetic defects in mice (*) or humans (**) in which the development and/or function of lymphocytes is impaired. Because of the limited space only the first autors are cited in the reference column. **Abbreviations**: CTL: cytotoxic T lymphocyte, CFA: complete Freund's Adjuvants, Ig: immunoglobulin, IL: interleukin, IFN: interferon, M-CSF: macrophage colony stimulating factor, KL: Kit ligand, LCMV: lymphocytic choriomeningitis virus, MHC: major histocompatibility complex, mab: monoclonal antibody, MBP: myelin basic protein, TCR: T-cell receptor, DN, DP and SP: double negative, double positive and single positive refers to expression of CD4 and CD8, VSV: vesicular stomatitis virus, wt: wild type

Table 1 ·Defect in genes encoding plasma membrane proteins

Defective gene	Mutant phenotype	Reference
β2m	- lack of expression of MHC class I and class I like proteins; D^b is expressed at low levels - lack of mature CD8 αβ T-cells	Zijlstra 1990
	- ConA blasts from mutant mice are lysed by NK cells from normal mice; mutant mice lack NK-cells which lyse mutant target cells	Liao 1991
	- hemopoietic progenitor cells from mutant mice cannot reconstitute irradiated normal mice because they are rejected by NK-cells - NK-cells from mutant mice do not reject progenitors from mutant mice	Bix 1991
	- influenca virus is eliminated from the respiratory tract	Eichelberger 1991
	- all mice recover from vaccinia virus infection - less efficient IgG production	Spriggs 1992
	- few mice recover from LCMV infection, most are killed by anti-LCMV CD4 T-cell responses	Müller 1992
	- lack of positive selection of cells expressing transgenic γδ TCR specific for a TL region encoded protein	Wells 1991 Pereira 1991
Aβ^b	- lack of class II MHC protein expression (AαB, EαB, EαAβ) no evidence for expression of AαEβ - lack of mature CD4 T-cells	Cosgrove 1991, Grusby 1991
	- few CD4^low / CD44^high T-cells in B-cell follicles not in germinal centres - decreased serum IgG1 levels - decreased antibody responses to T-dependant antigen	
	- allografts are rejected more slowly - transgenic Eβ expression in the cortex but not in the medulla of the thymus restores CD4 T-cell development	Cosgrove 1992

in- variant chain	- highly reduced expression of MHC class II proteins - reduced numbers of CD4 T-cells - reduced antibody responses to proteins	Mathis 1992 unpublished
CD4	- lack of CD4 T-cells - normal development of CD8 T-cells - poor antibody response to sheep red cells in vivo - effective CTL response to LCMV and vaccinia in vivo	Rahemtulla 1991

CD8α
(Lyt2)
	- lack of cells epressing CD8α or CD8β (Lyt3)
	- normal development of CD4 T-cells
	- normal IgM and IgG antibody responses to VSV
	- no CTL response to LCMV in vivo; i.v. injected LCMV
	  is cleared more slowly; intracerebral injection of
	  LCMV kills all wt mice but only some CD8 deficient
	  mice
	- EAE induced by MBP in CFA: lower death rate over
	  first 30 days but higher incidence of relapse
		Fung-Leung 1991a
		
		Fung-Leung 1991b
		
		Koh 1992

IgM	- lack of membrane associated but not secreted IgM
	- B-cell development blocked at Pre-B-cell (B220^{dull})
	- lack of allelic exclusion: some B-cells of μ^a mutant /
	  μ^b heterozygous mice express both heavy chains in
	  the cytoplasma
	- Vκ-Jκ rearrangements do occur albeit at 1/20th of
	  normal rate
		Kitamura 1991
		
		Kitamura 1992a

IgD	- normal lymphocyte development and function	Roes 1991

λ5	- block in B-cell maturation; some B-cells develope
	  probably because of occasional L-chain gene
	  rearrangements before V$_H$ to DJ rearrangements
	- allelic exclusion intact
	- the number of conventional and CD5 B-cells is
	  drastically reduced in young mice; in older mice
	  CD5 B-cells accumulate more efficiently
	- serum IgM levels are slightly increased at 5 weeks
	  IgG3 and IgG2b antibody responses are decreased
		Kitamura 1992b

TCRα	- thymus: normal numbers of DN and small IL2R⁻ DP
	  but almost no SP cells
	- β rearrangements are as extensive as in wt mice
	- α rearrangments occur at somewhat reduced level
	- β and α transcription is normal ; α transcripts are
	  larger due to he neomycin resistance gene insert
	- 2 to 7% of thymocytes stain with anti αβTCR mab,
	  these cells express β chain without α chain
		Mombaerts 1992

TCRβ	- 6 to 60 fold lower number of thymocytes
	- the number of DP thymocytes is reduced to 50%
	- no β rearrangments, even though the mutant allele
	  retains intact Dβ1, Jβ1.1 and Jβ1.2 (loss of cis acting
	  rearrangement promoting element?)
	- α rearrangements occur at somewhat reduced level
	- no TCR β transcripts, low levels of α transcripts
		Mombaerts 1992

TCR δ	- lack of γδ T-cells - normal development of αβ T-cells and B-cells	Itohara 1992 unpublished

CD3ε **	- mutant CD3ε lacks transmembrane region - no CD3 γδε detected in T-cell lysates - αβ TCR and CD3ζ expressed at low levels - severe combined immunodeficiency	Fischer 1992 see Matsumoto 1992
CD3γ **	- independant point mutations on both alleles - severe combined immunodeficiency	Terhorst 1992 see Matsumoto 1992
Thy1	- normal lymphocyte development and function	Stewart and Silver, unpubl.
IFNγ-R	- embryonal fibroblasts are sensitive to anti-viral activity of IFNα/β but not IFNγ - macrophages produce nitric oxide to IFNα/β but not IFNγ - normal CTL responses against vaccinia and LCMV - normal levels of neutralizing IgG anti-VSV antibodies - marked deficiency in IgG2a and to a lesser extend IgG2b and IgG3 - increased susceptibility to infection by Listeria monocygotenes and vaccinia	Huang 1992 submitted
Kit (W)*	- 78 amino acid deletion including the transmembrane region (original W mutation) or single amino acid exchanges in the tyrosine kinase domain (other W mutants) - some W mutants encoding receptors with intact extracellular domains but nonfunctional tyrosine kinase domains have dominant negative effects in heterozygous mice - defect in hemopoietic cells, germ cells, melanocytes the severity of the defects varies with the type of mutation	Nocka 1990
Kit **	- human Kit mutations are associated with piebaldism but so far no hematological abnormalities have been observed in heterozygous individuals with one null allele	Spritz 1992
Fas (lpr)*	- in lpr mice Fas is not expressed, in lpr^cg mice a point mutation in the cytoplasmic domain renders Fas nonfunctional - lymphadenopathy and autoantibody production	Watanabe-Fukunaga 1992

Table 2   Defects in genes encoding intracellular proteins

Defective gene	Mutant phenotype	Reference
RAG-1	- block in TCR and Ig gene rearrangements - B and T-cell development arrested at early stage - no serum IgM	Mombaerts 1992

RAG-2	- block in TCR and Ig gene rearrangements - B and T-cell development arrested at early stage - transcripts of λ5, germline V_H, IL7-R, TdT	Shinkai 1992
c-abl	- abl^2 mutation abolishes kinase activity - phenotypes are very variable - high mortality in first week of life, runting (95%) - defects of T- and B-cell development in some mice	Tybulewicz 1991 Schwartzberg 1991
c-src	- delayed eye opening, reduced body size - most mutant mice die at 3 to 4 weeks of age - broader faces, domed heads, shorter long bones - osteoporosis due to decreased osteoclast function	Soriano 1991
p56 lck	- T-cell development is impaired, atrophic thymus - spleen and LN contain normal number of lymphocytes: 90-95% are B-cells, very few are αβ or γδ T-cells - residual T-cells show normal proliferative response to anti-CD3/TCR mab's - serum IgM increased, IgG1 decreased	Molina 1992
c-fyn	- normal thymocyte development - reduced response of thymocyte but not of peripheral T-cells to anti-CD3 mab	Soriano unpubl (see Fung-Leung 1992)
c-myb	- prenatal lethality due to anemia - fetal hemopoiesis is normal in yolk sac until day 15, then impaired in fetal liver, all lineages appear to be affected except megakaryocytes	Mucenski 1991
NF-AT **	- NF-AT abnormal in electrophoretic mobility shift assay - severe combined immunodeficiency - defect in production of IL2, IL3, IL4, and IL5	Castigli 1992 see Matsumoto 1992
GATA-1	- GATA-1 deficient ES cells were tested for their ability to contribute to different tissues in chimaeric mice - selective block in erythroid differentiation	Pevny 1991

Table 3 Defect in genes encoding extracellular proteins

Defecti- ve gene	Mutant phenotype	Reference
IL2	- normal T- and B-cell development - poor T-cell response to Con A or anti-CD3 mab - serum IgM levels normal; IgG1, IgG2a, IgG2b levels elevated	Schorle 1991
	- lymphadenopathy and splenomegaly in older mice - loss of B220$^+$ cells first from BM, later from spleen - death within 2 to 6 months	Schimpl 1992 (see Paul 1992)

IL4	- normal T- and B-cell development	Kühn 1991
	- serum IgM, IgG2a, IgG2b, IgA normal; IgG1 reduced IgE undetectable	
	- no predominance of IgG1 in T-cell dependant antibody responses; no IgE response to nematode infection	
	- Nippostrongylus brasiliensis infection: no upregulation of MHC class II protein and CD23, impaired or reduced production of IL3, IL5 and IL10, reduced eosinophilia	Kopf 1992 submitted
	- impaired development of TH2?	

IFNγ	- excessive proliferation of a subpopulation of antigen or mitogen activated T-cells	Dalton 1992 submitted
	- cytolytic activity increased for activated CTL, decreased for freshly isolated NK cells	
	- macrophages from BCG infected mice fail to express increased levels of MHC class II proteins, fail to produce nitric oxide and show decreased superoxide production	

IL10	- reduced body weight	Rajewsky 1992
	- no evidence for overproduction of IFNγ	

KL* (steel)	- the steel-dickie mutation encodes KL lacking transmembrane and cytoplasmic domains of the steel factor	Brannan 1991
	- defect in hemopoietic cells, germ cells, melanocytes	

M-CSF* (op)	- deficiency in osteoclasts, monocytes, macrophages	Yoshida 1990
	- osteopetrosis due to osteoclast defect	

References

Arnaiz-Villena A., Timon M., Rodriguez-Gallegon C., Perez-Blas M., Corell A., Martin-Villa J.M. and Regueiro J.R. (1992) Immunology Today 13: 259-265

Bix M., Liao Nan-Shih, Zijlstra M., Loring J., Jaenisch R. and Raulet D. (1991) Nature 349: 329-331

von Boehmer H. (1990) Annu. Rev. Immunol. 8: 531-556

Bradley A., Hasty P., Davis A. and Ramirez-Solis R. (1992) Biotechnology 10: 534-538

Brannan C.I., Lyman S.D., Williams D.E., Eisenman J., Anderson D.M., Cosman D., Bedell M.A., Jenkins N.A. and Copeland N.G. (1991) Proc. Natl. Acad. Sci. USA 88: 4671-4674

Capecchi M.R. (1989) Trends in Genetics 5: 70-76

Cosgrove D., Gray D., Dierich A., Kaufman J., Lemeur M., Benoist C. and Mathis D. (1991) Cell 66: 1051-1066

Cosgrove D., Chan S.H., Waltzinger C., Benoist C. and Mathis D. (1992) Int. Immunol. 4: 707-710

Dalton D., Pitts-Meek S., Figari I., Keshav S., Bradley A. and Stewart T. (1992) submitted

Eichelberger M., Allan W., Zijlstra M., Jaenisch R. and Doherty P.C. (1991) J. Exp. Med. 174: 875-880

Fischer A. (1990) Current Opinion in Immunol. 2: 439-444

Fung-Leung W.-P., Schilham M.W., Rahemtulla A., Kündig T.M., Vollenweider M. Potter J., van Ewijk W. and Mak T.W. (1991a) Cell 65: 443-449

Fung-Leung W.-P., Kündig T.M., Zinkernagel R.M. and Mak T.W. (1991b) J. Exp. Med. 174: 1425-1429

Fung-Leung W.-P. and Mak T.W. (1992) Current Opinion Immunol. 4: 189-194

Goodnow C.C., Crosbie J., Adelstein S., Lavoie T.B., Smith-Gill S.J., Brink R.A., Pritchard-Briscoe H., Wotherspoon J.S., Loblay R.H., Raphael K. and Basten A. (1988) Nature 334: 676-682

Greiner D.L., Rajan T.V. and Shutz L.D. (1992) Immunology Today 13: 116-117

Gridley T. (1991) The New Biologist 3: 1025-1034

Grusby M.J., Johnson R.S., Papaionnou V. E. and Glimcher L.H. (1991) Science 253: 1417-1420

Hasty P., Ramirez-Solis R., Krumlauf R. and Bradley A. (1991) Nature 350: 243-246

Hooper M., Hardy K., Handyside A., Hunter S. and Monk M. (1987) Nature 326: 292-295

Hong R. (1989) in "Human Immunogenetics, basic principles and clinical relevance", Litwin S.D. Ed., Marcel Dekker, INC New York

Huang S., Hendriks W., Hemmi S., Bluethmann H., Kamijo R., Vilcek J., Zinkernael R.M. and Aguet M. (1992) submitted

Jaenisch R. (1988) Science 240: 1468-1474

Kitamura D., Roes J., Kuhn R. and Rajewsky K. (1991) Nature 350: 423-426

Kitamura D. and Rajewsky K. (1992a) Nature 356: 154-156

Kitamura D., Kudo A., Schaal S., Muller W., Melchers F. and Rajewsky K. (1992b) Cell 69: 823-831

Koh D.-R., Fung-Leung W.-P., Ho A., Gray D., Acha-Orbea H. and Mak T.W. (1992) Science 256: 1210-1213

Koller B.H., Kim H.S., Latour A.M., Brigman K., Boucher R.C., Scambler P., Wainwright B. and Smithies O. (1991) Proc.Natl.Acad.Sci. USA 88: 10 730-10 734

Kopf M., LeGros G., Lamers M., Bluethmann H. and Köhhler G. (1992) submitted

Kuehn M.R., Bradley A., Robertson E.J. and Evans M.J. (1987) Nature 326: 295- 298

Kühn R., Rajewsky K. and Muller W. (1991) Science 254: 707-710

Le Mouellic H., Lallemand Y. and Brulet P. (1990) Proc.Natl.Acad.Sci. USA 87: 4712-4716

Liao Nan-Shih, Bix M., Zijlstra M., Jaenisch R. and Raulet D. (1991) Science 253: 199-202

Matsumoto S., Sakiyama, Ariga T., Gallagher R. and Taguchi (1992) Immunology Today 13: 4-5

Molina T.J., Kishihara D.P., Siderovski D.P., van Ewijk W., Narendran A., Timms E., Wakeham A., Paige C.J., Hartmann K.-U., Veillette A., Davidson D. and Mak T.W. (1992) Nature 357: 161-164

Mombaerts P., Clarke A.R., Hooper M.L. and Tonegawa S. (1991) Proc.Natl.Acad.Sci. USA 88: 3084-3087

Mombaerts P., Iacomini J., Johnson R.S., Herrup K., Tonegawa S. and Papaioannou V.E. (1992a) Cell 68: 869-877

Mombaerts P., Clarke A.R., Rudnicki M.A., Iacomini J., Itohara S., Lafaille J.J., Wang L., Ichikawa Y., Jaenisch R., Hooper M.L. and Tonegawa S. (1992b) submitted

Mucenski M.L., McLain K., Kier A.B., Swerdlow S.H., Schreiner C.M., Miller T.A., Pietryga D.W., Scott W.J. and Potter S.S. (1991) Cell 65: 677-689

Muller D., Koller B.H., Whitton L., LaPan K.E., Brigman K.K. and Frelinger J.A. (1992) Science 255: 1576-1578

Nocka K., Buck J., Levi E. and Besmer P. (1990) EMBO J. 9: 3287-3294

Paul W.E. (1992) Nature 357: 12-13

Pevny L., Simon M.C., Robertson E., Klein W.H., Tsai S.-F., D'Agati V., Orkin S.H. and Costantini F. (1991) Nature 349: 257-270

Pereira P., Zijlstra M., McMaster J., Loring J.M., Jaenisch R. and Tonegawa S. (1992) EMBO J. 11: 25-31

Philpott K.L., Viney J., Kay G., Rastan S., Gardiner E.M., Chae S., Hayday A.C. and Owen M.J. (1992) Science 256: 1448-1452

Rahemtulla A., Fung-Leung W.-P., Schilham M.W., Kundig T.M., Sambhara S.R., Narendran A., Arabian A., Wakeham A., Paige C.J., Zinkernagel R.M., Miller R.G. and Mak T.W. (1991) Nature 353: 180-184

Rajewsky K. (1992) Science 256: 483

Roes J. and Rajewsky K. (1991) Int. Immunol. 3: 1367-1371

Schorle H., Holtschke T., Hünig T., Schimpl A. and Horak I. (1991)
Nature 352: 621-624

Schwartzberg P.L., Stall A.M., Hardin J.D., Bowdish A.M., Humaran T.,
Boast S., Harbison M.L., Robertson E.J. and Goff S.P. (1991)
Cell 65: 1165-1175

Shinkai Y., Rathbun G., Lam K.-P., Oltz E.M., Stewart V., Mendelsohn
M., Charron J., Datta M., Young F., Stall A.M. and Alt F.W. (1992)
Cell 68: 855-867

Soriano P., Montgomery C., Geske R. and Bradley A. (1991)
Cell 64: 693-702

Spickett G.P., Misbah S.A. and Chapel H.M. (1991)
Lancet 337: 281-284

Spritz R.A., Giebel L.B. and Holmes S.A. (1992)
Am. J. Hum.Genet.50: 261-270

Spriggs M., Koller B.H., Sato T., Morrisey P.J., Fanslow W.C.,
Smithies O., Voice R.F., Widmer M.B., and Maliszewski C.R. (1992)
Proc.Natl.Acad.Sci. USA 89: 6070-6074

Stacey A., Bateman J., Choi T., Mascara T., Cole W. and Jaenisch R.
(1988) Nature 332: 131-136

Storb U. (1987) Annu. Rev. Immunol. 5: 151-174

Tybulewicz V.L.J., Crawford C.E., Jackson P.K., Bronson R.T. and
Mulligan R.C. (1991) Cell 65: 1153-1163

Tybulewicz V.L.J., Tremblay M.L., LaMarca M.E., Willemsen R.,
Stubblefield B.K., Winfield S., Zablocka B., Sidransky E., Marti
B.M., Huang S.P., Mintzer K.A., Westphal H., Mulligan R.C. and Ginns
E.I. (1992) Nature 357: 407-500

Watanabe-Fukunaga R., Brannan C.I., Copeland N.G., Jenkins N.A. and
Nagata S. (1992) Nature 356: 314-317

Wells F.B., Gahm S.-J., Hedrick S.M., Bluestone J.A., Dent A. and
Matis L.A. (1991) Science 253: 903-905

Wu X., Lee C.C., Gruzny D.M. and Caskey C.T. (1989)
Proc.Natl.Acad.Sci. USA 86: 9412-9416

Yoshida H.S., Hayashi T., Kunisada M., Ogawa S., Nishikawa S.,
Okumura H., Sudo T., Shultz L.D. and Nishikawa S. (1990)
Nature 345: 442-445

Zijlstra M., Bix M., Simister N.E., Loring J.M., Raulet D.H. and
Jaenisch R. (1990) Nature 344: 742-746

Acknowledgments: We thank Drs H. Bluethmann, D. Dalton, S. Itohara,
D. Mathis, P. Mombaerts, C.L. Stewart, S. Tonegawa for providing
unpublished manuscripts and Drs C.L. Stewart, P. Mombaerts and K.
Rajewsky for helpful suggestions.

12. Autoimmunity

T Cell Epitopes of Synthetic Antigens and of Antigens Related to Autoimmune Diseases

M. Sela, E. Zisman, E. Mozes

Department of Chemical Immunology, Weizmann Institute of Science, Rehovot, 76100 Israel

Introduction

The initial observation that synthetic polypeptides may be immunogenic in experimental animals led to the use of synthetic antigens for the elucidation of the molecular basis of various immunological phenomena (1-4) and to the distinct realization of the determinant-specific genetic control of immune response (5,6) and its correlation with the main histocompatibility locus of the species (7). In many of these studies use was made of the synthetic antigen "(T,G)-A--L", a branched polyamino acid in which peptides of L-tyrosine and L-glutamic acid are attached to side chains of poly(DL-alanine), themselves attached to a backbone of poly(L-lysine) (8). The N-terminal peptides of tyrosine and glutamic acid were prepared by polymeric techniques, and thus both the detailed amino acid composition and their sequence varied from one polymeric side-chain to another.

In recent years we have investigated this synthetic antigen to learn about processing requirements, leading to T cell activation, as well as about the nature of its T cell epitopes. We have also obtained and studied T cell hybridomas specific for a thymus-independent branched polymeric antigen, composed exclusively of D-amino acids.

Mention will be made also of Cop 1, a linear copolymer of L-alanine, L-lysine, L-glutamic acid and L-tyrosine, which serves as a candidate drug to the exacerbating-remitting type of multiple sclerosis. Cop 1 may be considered an immunomodulatory vaccine, and we shall describe here its relationship, at the T cell epitope level, to the basic protein of myelin.

Direct binding of a synthetic multichain polypeptide to class II molecules

T cell activation involves the recognition of foreign antigens as a complex with self-major histocompatibility complex (MHC) proteins on the surface of antigen-presenting cells (APC). Protein antigens usually require uptake by the APC and processsing that results in the generation of peptide fragments. The branched synthetic polypeptide (Tyr,Glu)-Ala--Lys was chosen as a mo del antigen to follow the processing requirements, leading to T cell activation. It has been demonstrated, by using fixed APC and various inhibitors of proteases, that (Tyr,Glu)-Ala--Lys has to be processed to stimulate a (Tyr,Glu)-Ala--Lys-specific T-cell line of C3H.SW (H-2^b) origin to proliferate (9) To determine whether processing of (Tyr,Glu)-Ala--Lys is required to allow its association with the MHC class II molecules, biotin was covalently attached to it. Binding of the biotinylated (Tyr,Glu)-Ala--Lys to

MHC class II gene products on the surface of intact normal APC was directly detected by phycoerythrin-streptavidin. The specificity of the binding was confirmed by its inhibition with antiI-A^b antibodies as well as with excess of nonlabeled (Tyr,Glu)-Ala--Lys. Furthermore, introducing several inhibitors of proteases to the binding assay, we could substantiate that the proteolysis of (Tyr,Glu)-Ala--Lys is required to allow association of the resulting peptide T cell epitopes with the MHC class II molecules themselves. The presence of the biotin moiety in the resulting peptides suggests that the T cell epitopes of (Tyr,Glu)-Ala--Lys contain the N-terminal portion of the side chains of the branched polypeptide. An apparent K_d of 8.05×10^{-8} M was determined, and optimal binding was detected after 10 hr of incubation with the antigen. The latter phenomenon is not due to slow uptake, since uptake of (Tyr,Glu)-Ala--Lys occurs mainly during the first 30 min of incubation, but rather reflects the events of processing that precede MHC interaction (9).

Two related tetrapeptides and their diverse immune behaviour.

In view of the intensity of studies with the above polymer, we were interested in elucidating its major B and T cell epitopes. We investigated two tetrapeptides, TyrTyrGluGlu and TyrGluTyrGlu. Both were attached to multichain branched poly(DL-alanine). Even though the two resulting synthetic immunogens are essentially identical in their molecular weight, size, shape and composition, and differ chemically only in the sequence of the tetrapeptide epitopes, the immunological differences observed were profound. Antibodies in the two systems do not cross-react. The major B cell epitope of "(T,G)-A--L" has been established as TyrTyrGluGlu (10), whereas the major T cell epitope has now been found to be TyrGluTyrGlu (11), as described below. The TyrTyrGluGlu polymer not only cross-reacts best with the random "(T,G)-A--L" but is under the same genetic control (12) and exhibits cross-tolerance (13). The TyrGluTyrGlu does not cross-react at antibody level, is under a different genetic control, and is thymus-independent (14) in contrast to the TyrTyrGluGlu polymer.

B cell activation requires specific antigen-recognition by T cells, which by themselves recognize the antigen in the context of MHC class II molecules on the surface of APC (15). It was, therefore, important to define the T cell epitopes of "(T,G)-A--L" and its ordered analogs. To investigate the cross-reactivities of TyrTyrGluGlu and TyrGluTyrGlu at the T cell level, we established T cell lines specific to "(T,G)-A--L", denoted TPB1-3 (9, 16), to TyrTyrGluGlu polymer (designated TTn), and to TyrGluTyrGlu polymer (designated TG7) from lymph-node cells of C3H.SW (H-2^b) mice that were immunized with the relevant polypeptides two weeks earlier. The fine specificities of these lines showed no cross-reactivity either of TG7 with the TyrTyrGluGlu polymer or of TTn with the TyrGluTyrGlu polymer, although both lines proliferated specifically in response to "(T,G)-A--L". The absence of cross-recognition at the T cell level resembles the previously described lack of cross-reactivity of the antibodies specific to the two ordered polypeptides (10).

Surprisingly, the "(T,G)-A--L" specific line did not proliferate in response to the TyrTyrGluGlu polymer, which bears the major B cell epitope of "(T,G)-A--L", whereas the TyrGluTyrGlu polymer that shares only minor B cell determinants with "(T,G)A--L" could specifically stimulate this T cell line to proliferate (11, 17). Thus, the "(T,G)-A--L" specific line, TPB3, and the TyrGluTyrGlu polymer specific line, TG7, shared T cell epitopes. In addition, since the "(T,G)-A--L" specific T cell line, TPB3, did not respond to the TyrTyrGluGlu polymer, the latter polypeptide might represent a cryptic T cell epitope of "(T,G)-A--L" (17).

Now, the point I want to make is that we always suspected that there must be very great differences between these two tetrapeptides, even though on paper they seem so close, and we tried several physicochemical techniques, but the one which gave the clearest result is the photochemically induced dynamic nuclear polarization (CIDNP) (18).

The CIDNP spectroscopic approach chosen for the study of local conformations of Tyr combines both specificity and high sensitivity. It has been applied previously in the studies of conformations of side chains of Tyr, Trp, and His of peptides and proteins in solution. It provides unique opportunities for the study of protein-protein and protein-ligand binding, and for conformational (accessibility) mapping.

The CIDNP spectroscopic study of the conformation of the two closely related synthetic polypeptide immunogens showed that, despite their far-reaching molecular similarity, they differ strongly in their intra-epitope aromatic interactions (17). The CIDNP results show that phenolic groups in TyrGlu-TyrGlu interact with each other, whereas they are far apart in TyrTyrGluGlu. This is reflected in the broad CIDNP peak of the aromatic Tyr protons of the TyrGluTyrGlu polymer to be compared with the fully resolved Tyr aromatic proton signals of the TyrTyrGluGlu polymer. Apparently, such interactions are directly involved in the mechanisms which control immune recognition.

MHC-Antigen - T cell interactions for thymus-dependent and thymus-independent antigens

We have previously demonstrated (19) that synthetic branched polypeptides composed of L-amino acids, although bearing repeating sequences, are thymus-dependent (L-TD), whereas the same polymer composed of D-amino acids is thymus-independent ((D-TI). Yet lymph node cells of Balb/c mice immunized with such a D-TI (multi-poly-(DPhe,DGlu)-polyDPro--polyDLys), proliferates to it in vitro. To follow T cell activation by D-TI, we established T cell hybridomas to the above D-TI, and to its stereoisomeric L-TD (multi-poly-(LPhe,LGlu)poly-L-Pro--polyLLys) for comparison (20).

The T cell hybridomas express membranal $\alpha\beta$ T cell receptors, and secrete IL-2 upon stimulation with their antigen. In addition, D-TI specific hybridomas are stimulated, to a lesser extent, by the L-TD Ag, whereas only some L-TD specific hybridomas recognize D-TI. Thus, we wanted to determine whether T cell activation involves presentation of D-TI by MHC class II gene products (Ia) on APC. Using a direct binding assay that we previously designed, we could detect binding of biotinylated analogs of D-TI and L-TD to splenic adherent cells of BALB/c mice. Binding was inhibited by excess of non-biotinylated L-TD, suggesting that both polypeptides bind to the same sites. Inhibition of the binding was also achieved with peptide p259-271 of the human acetylcholine receptor α-subunit, which binds to Ia molecules of BALB/c mice without prior processing. Radioimmunoassay of APC lysates, following incubation of the APC with biotinylated D-TI and L-TD, demonstrated that the biotin-Ag moiety is associated with Ia molecules. Apparent dissociation constants at equilibrium are quite similar, 5×10^{-8} and 3×10^{-8} M for D-TI and L-TD, respectively. On the other hand, D-TI has faster kinetics of binding than L-TD. Moreover, stimulation of the specific T cell hybridomas can occur in the presence of either the latter inhibitors or aldehyde-fixed APC. Hence, we have demonstrated an MHC class II mediated T cell response to a thymus-independent antigen.

T Cell epitopes of a synthetic antigen relevant in multiple sclerosis

A synthetic copolymer of L-alanine, L-glutamic acid, L-lysine and L-tyrosine in a residue molar ratio of 6.0 : 1.9 : 4.7 : 1.0, denoted Cop 1, has been shown by us to suppress experimental allergic encephalomyelitis in guinea pigs, rabbits, mice, rhesus monkeys and baboons (21), and to help patients with the exacerbating-remitting form of multiple sclerosis (22).

We observed already almost 20 years ago a slight cross-reaction between rabbit anti-Cop 1 antibodies and bovine myelin basic protein (MBP) (23). At the cellular level, a marked cross-reaction was observed both in vivo in the delayed hypersensitivity skin-test, and in vitro by measuring lymphocyte transformation. This cross-reactivity has been recently confirmed and investigated in detail, making use of monoclonal antibodies to MBP and to Cop 1 (24). About a third of anti-rat MBP monoclonal antibodies and most of anti-mouse monoclonal antibodies cross-reacted with Cop 1. In addition, several anti-Cop 1 monoclonal antibodies cross-reacted with MBP. Moreover, some anti-MBP and anti Cop 1 monoclonals reacted in a heteroclitic manner, and favoured the cross-reactive antigen over the immunogen.

The suppressive activity of Cop 1 correlates well with immunological manifestations. Thus an analog of Cop 1 built exclusively of D-amino acids does not cross-react at all with MBP, and indeed exhibits no suppressive activity (25). Moreover, Cop 1 does not suppress unrelated antibody formation or cellular responses, nor does it affect other autoimmune diseases.

After some preliminary studies with patients, a clinical double blind trial has been carried out (22). The participants had to have at least three attacks in the previous two years. The 23 patients on placebo had during the trial period 64 attacks, whereas the 25 patients on Cop 1 had only 16 attacks. The 13 less advanced patients had only 4 attacks (instead of the 39 expected). Thus, we have here a macromolecular drug candidate for multiple sclerosis, a drug prepared by polymeric techniques.

Of course, it is of great interest to learn more about the mechanism of action of Cop 1 in experimental allergic encephalomyelitis. One possibility is the induction of suppressor cells specific to MBP by Cop 1 (26). More recently we have shown an additional immunological mechanism. Using MBP specific mouse T cell lines and clones with various H-2 restrictions and antigen specificities, we have shown that, specifically, Cop 1 could competitively inhibit T cell responses to MBP (27). The effect of Cop 1 on both the proliferative response and IL-2 secretion induced by MBP was followed in 8 T cell lines and clones. In seven of them Cop 1 specificaly inhibited the response to MBP, and in one, Cop 1 was able to induce proliferation. Inhibition of the response was shown to be specific to Cop 1, and only T cell line responses to MBP were affected by Cop 1. These results suggest that Cop 1 or Cop 1-derived peptides can bind to the relevant MHC molecules and competitively inhibit the binding of MBP. Consequently, activation of MBP effector cells is blocked, while T cells which cross-react with Cop 1, e.g. suppressor cells, are activated. Thus, Cop 1 may be effective in suppression of EAE not only because of selective stimulation of suppressor T cells, but also by specific inhibition of MBP-specific effector T cells.

The above findings were recently extended to the human MHC (28). Cop 1 competitively inhibited the proliferative responses and interleukin 2 secretion of six BP specific T-cell lines and 13 clones of several DR restrictions and epitope specificities. Conversely, BP inhibited - albeit to a lesser extent - the response of all the Cop 1 specific T cell lines and clones, irrespective of their DR restrictions. Another random copolymer of tyrosine, glutamic acid, and alanine, had no effect on these lines. Neither Cop 1 nor MBP inhibited the response of PPD specific lines and clones. Cop 1 and BP exerted their cross-inhibitory effects only in the presence of antigen presenting cells. These results suggest that Cop 1 can compete with BP for the binding to human major histocompatibility complex molecules. In view of recent studies implicating BP reactivity in multiple sclerosis, these findings suggest a possible mechanism for the beneficial effect of Cop 1 in this disease. Cop 1 and MBP exerted their cross-inhibitory effects only in the presence of antigen presenting cells, while incubation in their absence, resulted in unresponsiveness to the homologous antigen. Thus, Cop 1 competes with MBP for the binding to the MHC, but not to the T cell receptor.

It thus seems that the polymeric candidate drug against multiple sclerosis is essentially an immunomodulatory vaccine. We shall now have more and more cases where such immunomodulatory vaccines will be used against autoimmune diseases, along the vaccines against infectious diseases.

Concluding remarks

We could learn a lot about T cell epitopes by making us of synthetic polypeptide antigens, Thus, by measuring their binding to class II antigens on the living cell, we established Kd in the range of $3-8 \times 10^{-8}$ M. We have obtained and studied T cell hybridomas against thymus-independent antigens. In this case the antigens do not seem to need processing. Finally, the study of specific T cells may help our understanding of the mode of action of Cop 1, a polymeric drug which is a candidate immunomodulating vaccine against multiple sclerosis.

References

1. Sela M (1966) Adv Immunol 5:29
2. Sela M (1969) Science 166:1365
3. Sela M (1983) Biopolymers 22:415
4. Sela M (1987) Ann Rev Immunol 5:1
5. McDevitt H.O, Sela M (1965) 122:517
6. McDevitt H.O, Sela, M (1967) 126:969
7. McDevitt H.O, Benacerraf B (1969) Adv Immunol 11:31
8. Sela M, Fuchs S, Arnon R (1962) Biochem J 85:223
9. Zisman, E, Sela M, Mozes E (1991) Proc Natl Acad Sci USA 88:9738
10. Mozes E, Schwartz M, Sela M (1974) J Exp Med 140:349
11. Mozes E, Zisman E, Kirshner S, Katz-Levy Y, Sela M (1991) Abstr Intl Congr Biochem (Jerusalem) (1991) p. 120
12. Schwartz M, Mozes E, Sela M (1975) Eur J Immunol 5:866
13. Schwartz M, Parhami B, Mozes E, Sela M (1979) Proc Natl Acad Sci USA 76:5286
14. Schwartz M, Geiger B, Hooghe RJ, Bar-Eli M, Gallily R, Mozes E, Sela M (1979) 35:849
15. Ashwell JD, Schwartz RH (1986) Nature 320:176
16. Axelrod O, Mozes E (1986) Immunobiology 172:99
17. Sela M, Mozes E, Zisman E, Muszkat KA, Schechter B (1992) Behring Inst Res Comm 91:54
18. Muszkat KA, Khait I, Hayashi K, Tamiya N (1984) Biochemistry 23:4913
19. Sela M, Mozes E, Shearer GM (1972) Proc Natl Acad Sci USA 69:2696
20. Zisman E, Sela M, Mozes E Abstr. Symp on T Cell Repertoire, September 1992 Arad, Israel, p P-33
21. Sela M, Arnon R, Teitelbaum D (1990) Bull Inst Pasteur 88:303
22. Bornstein MB, Miller A, Slagle S, Weitzman M, Crystal H, Dexler E, Keilson M, Merriam A, Wassertheil-Smoller S, Spada V, Weiss W, Arnon R, Jacobson I, Teitelbaum D, Sela M (1987) New Engl J Med 317:408
23. Webb C, Teitelbaum D, Arnon R, Sela M (1973) Europ J Immunol 3:279-
24. Teitelbaum D, Aharoni R, Sela M, Arnon R (1991) Proc Natl Acad Sci USA 88:9528
25. Webb C, Teitelbaum D, Herz A, Arnon R, Sela M (1976) Immunochemistry 13:333
26. Lando Z, Dori Y, Teitelbaum D, Arnon R (1981) J Immunol 127:1915-
27. Teitelbaum D, Aharoni R, Arnon R, Sela M (1988) Proc Natl Acad Sci USA 85:9724
28. Teitelbaum D, Milo R, Arnon R, Sela M (1992) Proc Natl Acad Sci USA 89:137

Autoimmunity to a Heat Shock Protein Probes the Immunological Homunculus

Irun R. Cohen

Departement of Cell Biology, The Weizmann Institute of Science Rehovot, Israel

THE IMMUNOLOGICAL HOMUNCULUS

The term immunological homunculus was coined as a short-hand way of designating organized natural autoimmunity (Cohen 1989; Cohen 1991a; Cohen and Young 1991; Cohen 1992a,b). It is reasonable to conclude that natural autoimmunity is organized because each human and each mouse contain populations of T cells, B cells and antibodies that recognize a discrete set of dominant self antigens. These self antigens are immunologically dominant because of the autoimmunity that is centered around them naturally; in other words, the immune system selects them as preferred antigens. The response to homunculus self antigens is thus a recall response rather than a primary response.

Dominant self antigens are not dominant because they differ in their chemical nature from the self antigens that are not dominant. The chemical building blocks - amino acids, sugars, lipids - are intrinsically the same for dominant and non-dominant antigens. Just as the immune system itself defines what is an antigen, so it is the immune system itself that determines what is a dominant antigen.

How does the immune system decide upon which molecules to focus its attention? How is immunological dominance fashioned? Although antigens are recognized by antigen receptors and antibodies, which are created by somatic recombination and mutation, it is probably the genes carried in the germ line that determine immunological dominance in general and the immunological homunculus in particular. The somatically evolved receptor repertoire is molded by the germ-line genes controlling the expression of self peptides in the thymus (Kourilsky et al. 1989) and the mechanism of positive and negative selection (van Boehmer 1990). Germ-line genes continue to control expression of the receptor repertoire in the periphery by encoding the machinery of antigen uptake, processing and presentation (Steinman 1991) including the preference of certain molecular motifs (Falk et al. 1991). Thus, the choice of a particular self antigen (even of a self epitope) for inclusion in the immunological homunculus of dominant natural autoimmunity is made by the germ line in two ways: in the way it has encoded the structure and expression of the candidate antigens and in the way it determines the deployment of the receptor repertoire. Note, if the immunological homunculus is carried even in part in the germ line, then it is likely to have adaptive value. It could not be otherwise if mice and humans have similar homunculus maps despite 80 million years of evolutionary divergence. The homunculus is shared. But because it includes only certain self antigens and ignores the others, the homunculus can be said to be a distorted representation of the individual mapped into his or her immune system.

In addition to autoimmune cells of particular specifities, the immunological homunculus includes sets of regulatory T cells (and perhaps B cells) that recognize the natural autoimmune T cells (and perhaps B cells). Some of these regulatory T cells are anti-idiotypic (Lider et al. 1988). Because the natural autoimmune lymphocytes, together with their regulatory lymphocytes, create a kind of distorted picture of the self, the term immunological homunculus, which means "little man", is apt.

The aim of this brief review is to illustrate some features of the immunological homunculus that have been revealed by investigation of immunity to the 65 KDa heat shock proteins (hsp65) of Mycobacteria (MT-hsp65) and of humans (H-hsp65).

MODEL SYSTEMS: INSULIN DEPENDENT DIABETES MELLITUS (IDDM)AND ADJUVANT ARTHRITIS (AA)

Immunity to hsp65 has been implicated in two different autoimmune models: AA in Lewis rats and IDDM in NOD strain mice (Cohen 1991c). AA is an acute to chronic polyarthritis inducible by immunization to mycobacteria (MT). It was found that a T cell clone capable of adoptively transferring AA recognized a peptide (position 180-188) in the sequence of MT-hsp65 (van Eden et al. 1988). Moreover, treatment of rats with MT-hsp65 (van Eden et al. 1988) or with its 180-188 peptide (Feige and Cohen 1991) induced resistance to AA.

The 180-188 peptide of MT-hsp65 is not a conserved part of the molecule (see Jindal et al., 1989) and this peptide is not cross-reactive with the H-hsp65 molecule. Thus, the development of AA is not related to cross-reactivity between microbacterial and mammalian hsp65. But cross-reactivity between MT-hsp65 and cartilege proteoglycan may be a factor in AA (van Eden et al. 1985; Cohen 1988).

IDDM, in contrast to AA, is a spontaneous disease which does seem to involve immunity to H-hsp65 (Elias et al. 1991). A 24 amino acid of the H-hsp65 sequence, designated p227, is the target of diabetogenic T cells and immunization to H-hsp65 can induce IDDM. Moreover, treatment with the p277 peptide leads to down-regulation of the spontaneous autoimmunity to H-hsp65 and aborts development of IDDM (Elias et al. 1991). Thus, immunity to p277 is both necessary and sufficient for IDDM in NOD mice.

H-HSP65 IS AN HOMUNCULUS SELF ANTIGEN

Despite its involvement in autoimmune diseases, immunity to hsp65 is natural autoimmunity. Healthy humans have T cells that respond to at least some epitopes of H-hsp65 (Munk et al. 1989). Indeed, humans are born with T cell immunity to hsp65 demonstrable in cord blood (Fischer et al. 1992). Hsp65 is expressed normally in the thymus of mice (Zipris & Cohen, in preparation) so that the self antigen is available during T cell differentiation and could positively select T cells (von Boehmer 1990). Thus, hsp65 is a natural self immunogen. However, if autoimmunity to H-hsp65 can cause autoimmune diseases such as AA and IDDM, what is the difference between benign homunculus autoimmunity and autoimmune disease?

REGULATORY NETWORKS

It seems that autoimmunity to hsp65 is kept benign by the activities of anti-idiotypic T cells that recognize the anti-hsp65 T cells. NOD mice manifest natural anti-idiotypic T cell reactivity early in life, before the onset of the insulitis that destroys their insulin-producing beta cells (Elias et al., in preparation). This anti-idiotypic reactivity seems to be specific for a common idiotope formed by the CDRIII region of the beta chain of the T cell receptor (Kourilski et al., 1989) specific for the p277 epitope of H-hsp65 (Tikochinski et al., in preparation). This idiotope, which we have called the C9 idiotope, is present in different mice. As long as the anti-C9 reactivity remains high, there is no insulitis. The onset and progression of insulitis is accompanied by a spontaneous fall in the anti-idiotypic T cell reactivity to C9.

The induction of AA too seems to involve a fall in natural anti-idiotypic T cell reactivity to an anti-MT-hsp65 T cell idiotope (Karin and Cohen, in preparation). If we may generalize from these two instances of natural anti-idiotypic immunity, it would seem that the biologic expression of homunculus autoimmunity is influenced by natural anti-idiotyptic networks.

How these networks originate and develop is still unknown, but it seems that they can be consolidated and strengthened by contact with microbes in the environment. Rats or mice raised early in life in isolation from microbial contamination are much more susceptible to AA or to IDDM. We have found that lack of immunological experience with infectious agents is associated with poorly developed natural anti-idiotypic networks (Karin and Cohen, in preparation). The high degree of conservation of hsp65 and its immunological dominance makes it likely that any infection can boost natural hsp65 autoimmunity and consequently can boost the regulatory cells that respond to the anti-hsp65 cells (Cohen 1991c). Thus, some forms of natural autoimmunity and its regulation are consolidated by immunologic experience with the environment (Cohen 1992a,b).

Autoimmunity to hsp90 has been shown to protect against infection with candida (Matthews et al. 1991) and it is likely that autoimmunity to hsp65 could also help fight infection. The regulatory cells that control hsp65 autoimmunity are required to prevent the autoimmune diseases that might arise from the controlled use of natural hsp65 autoimmunity to fight infection (Cohen 1992a,b). It is fitting therefore that infection itself strengthens the regulation needed to keep hsp65 autoimmunity benign.

Note that not only hsp65 but many other highly conserved, self-like antigens expressed on microbes are dominant microbial antigens (Cohen and Young, 1991). True, the immune response may choose to focus on microbe-specific, non-shared epitopes on the self-like molecule. Nevertheless, the "selfness" of a molecule, its membership to even a partial degree in the homunculus set, may endow the microbial molecule with immunological dominance (Cohen and Young, 1991). Thus, in dealing with the outside world of invaders, the immune system is self-referential, it has an interest in foreign molecules with self epitopes.

INTERPRETATION OF CONTEXT

How does the immune system know when and how to deploy hsp65 autoimmunity in the service of resistance to infection? The answer to this question is not resolved in any molecular detail but it is clear in practice; the immune system is sensitive to the context in which an antigen is recognized and it is the context that instructs the system how to deal with the antigen (Cohen 1992a,b).

The immune response producing AA for example is induced by immunizing Lewis rats with MT, but the disease will occur only if a sufficient amount of MT is administered in oil, if the MT-oil mixture is not emulsified, and if the injection is into the skin. The same amount of MT given without oil, or emulsified, or given intramuscularly will hardly induce disease. In fact, MT given under those conditions tends to induce life-long resistance to any further attempt to induce AA in the standard way. Here is a paradox; all the MT antigens are given in both the classic or variant forms of immunization, but the ancillary details - the context - make the difference between disease and its opposite, protection.

The effect of context is illustrated in IDDM too. Administering the p277 peptide alone seems to cure IDDM (Elias et al. 1991); however, administering the same self peptide conjugated to a foreign carrier molecule (such as ovalbumin or bovine serum albumin) can actually induce IDDM in non-diabetic mice (Elias et al., in preparation). The p277 peptide presented in the cleft of an MHC molecule directs the specificity of the immune response but does not determine its biologic effect. The response is protective without the carrier and destructive with the carrier. Again, we see that the context is critical.

What is the meaning of context in immunological terms? This is still an open question but I believe that the interpretation of context is assisted or determined by the antigen presenting cells and by the helper cells which express various cytokines and adhesion molecules as a consequence of the presence of "carriers", "adjuvants", anatomic site of contact and so forth. Autoimmune recognition is built into the system as the homunculus, but how to deploy this recognition in practice is a function of how the system interprets the context (Cohen 1992a,b).

AUTOIMMUNE DISEASE - MISINTERPRETATION

The above account of autoimmunity asserts that natural autoimmunity is built into the immune system. A corollary of the homunculus idea is that autoimmune disease occurs when natural autoimmunity is amplified leading to inappropriate deployment of effector mechanisms producing recurrent and/or chronically progressive damage.

A causal relationship between benign natural autoimmunity and noxious autoimmune disease could explain why autoimmune diseases are quite limited in their diversity and monotonous in their expression. The natural autoimmunity from which the autoimmune diseases arise is itself limited to the set of antigens encoded within the homunculus. Most autoimmune patients can be counted within a handful of clinical entities and the immunology of each disease is fairly fixed (see Shoenfeld and Isenberg 1989). Indeed many healthy persons harbor the same sorts of specific autoreactive lymphocytes as can be isolated from patients (Cohen 1991a, 1992a,b). for this reason, disease is less a function of the immunologic specificity of self recognition and more a function of the quality, quantity and kinetics of the biological effects resulting from self recognition. Space does not allow elaboration of specific detail, but in general one can view the transition of natural autoimmunity to disease as resulting from two factors: a misinterpretation of context and a failure of regulation. Misinterpretation of context (perhaps it would be more accurate to say inappropriateness of context) can come about when a body tissue expresses adjuvant signals such as MHC molecules, cell adhesion molecules, or cytokines that call forth an effector response. The wrong context for a self antigen can also be supplied by an invading bacterium or virus that carries a self antigen along with the signals of infection.

However, the context of inflammation and infection only triggers the transition of natural autoimmunity to an autoimmune effector mode. Development of clinical disease also requires a lapse in the regulatory networks that should operate to turn off the autoimmune effector response when it is no longer needed to fight the infection. The clinical manifestation of the disease depends on weak homunculus regulatory network, networks that fail to reinstate' harmony once the alarm is past (Cohen 1993).

VACCINATION - TEACHING THE HOMUNCULUS

The logical cure for autoimmune disease is to strengthen the regulatory department of the homunculus so that the noxious autoimmune process can be held in check.

As I mentioned above, the regulatory network of the homunculus, at least in IDDM and in AA, can be consolidated by natural immune experience with infectious agents in the environment. In addition, the homunculus can be strengthened naturally by suffering a bout of the autoimmune disease itself. This is evident in the many experimental autoimmune diseases that, like autoimmune encephalomyelitis (EAE), are often self limited and can be induced only once. Experiencing EAE is a very effective way of generating regulatory cells that can produce life-long resistance to further induction of EAE (Ben-Nun and Cohen 1982).

Therapeutic strengthening of homunculus regulation can also be achieved by vaccinating an individual with the autoimmune T cells responsible for the disease or with the peptide epitope recognized by these T cells. T cell vaccination (Ben-Nun et al. 1981) makes use of the autoimmune T cells as "immunogens" to activate the anti-idiotype T cells that should function naturally to control the natural potential for autoimmune disease present in the homunculus (Cohen 1986; Lider et al. 1988; Cohen 1991a). T cell vaccination can be used to enhance resistance to induced disease models, but it is also effective in curing spontaneous IDDM (Elias et al.1991).

Peptide vaccination is less well characterized, but administration of the p277 peptide without a carrier (Elias et al. 1991) does seem to activate the anti-idiotypic T cells that recognize the autoimmune T cells that recognize p277 (Elias et al. in preparation). The regulatory system is indeed a real network.

The proposed approach to understanding natural autoimmunity, the concept of the immunological homunculus, is only one element in a general revision of the paradigm describing the immune system (Varala and Coutinho 1991; Cohen 1992a,b). But the homunculus paradigm bears new approaches to therapy.

REFERENCES

Ben-Nun A, Cohen IR (1982) J Immunol 128:1450-1457

Cohen IR (1986) Immunol Rev 94:5-21

Cohen IR (1988) Sci Am 258:52-60

Cohen IR (1989) in Theories of Immune Networks (eds. Atlan H & Cohen IR) 6-12 (Springer-Verlag, Berlin)

Cohen IR (1991a) in Molecular Autoimmunity, N Talal Ed. (Academic Press, New York), pp 437-453.

Cohen IR (1991b) J Int Med 230:471-477

Cohen IR (1991c) Annu Rev Immunol 9: 567-589

Cohen IR (1992a) (in press) Immunol Today 13:November

Cohen IR (1992b) (in press) Immunol Today 13:December

Cohen IR (1993) in Autoimmunity, Coutinho A & Kazatchkine MD, eds., Wiley & Liss, New York, in press

Cohen IR, Young DB (1991) Immunol Today 12:105-110.

Elias D, Reshef T, Birk OS, van der Zee R, Walker MD, Cohen IR (1991) Proc Natl Acad Sci USA 88: 3088-3091

Falk K, Rotzschke O, Stevanovic' S, Jung G, Rammensee H-G (1991) Nature 351:290-296

Feige U, Cohen IR (1991) Springer Semin Immunopathol 13:99-113

Fischer HP, Sharrock CEM, Panayi GS (1992) Eur J Immunol 22:1667-1669

Jindal S, Dudani AK, Harley CB, Singh B, Gupta ES (1989) Mol Cell Biol 9:2279-2283

Kourilsky P, Claverie J-M, Prochnicka-Chalufour A., Spetz-Hagberg A-L, Larsson-Sciard E-L (1989) Cold Spring Harbor Symp Quant Biol 54: 93-103

Lider O, Reshef T, Beraud E, Ben-Nun A, Cohen IR (1988) Science 239, 181-183

Matthews RC, Burnie J P, Howat D, Rowland T, Walton F (1991) Immunology 74:20-24.

Munk ME, Schoel B, Modrow S, Karr RW, Young RA, Kaufmann SHE (1989) J Immunol 143, 2844-2849

Steinman RM (1991) Annu Rev Immunol 9:271-296.

van Eden W, Holoshitz J, Nevo Z, Frenkel A, Klajman A, Cphen IR (1985) Proc. Natl. Acad. Sci. USA 82:5117-5120

van Eden W, Thole J, van der Zee R, Noordzij A, Embden JDA, Hensen EJ, cohen IR, de Vries RRP (1988) Nature 331:171-173

Varela FJ, Coutinho A (1991) Immunol Today 12:159-166.

von Boehmer H   (1990) Annu Rev Immunol 8:531-556

ACKNOWLEDGEMENT

I thank Mrs. Vivienne Laufer for preparing the manuscript.   The experimental results which generated these ideas were supported by grants from the National Institutes of Health, Institut Merieux, Kabi, the Juvenile Diabetes Foundation International, the Minerva Foundation and Mr. Rowland Schaefer.   I am the incumbent of the Mauerberger Chair of Immunology.

Inhibition of Autoimmune T cells by "Competitor-Modulator" peptides

W. van Eden, M.H.M. Wauben, C. J. P. Boog and R. van der Zee

Departement of Infectious Diseases and Immunology, Faculty for Veterinary Medicine, University of Utrecht. Yalelaan 1, P.O. Box 80165, 3508 TD. Utrecht, The Netherlands.

Introduction:

Appreciation of the nature of autoimmunity has changed with the advent of techniques enabling the cloning of antigen specific T lymphocytes. In experimental animal models T cells have been cloned and maintained by stimulation with antigens, such as myelin basic protein (MBP), thyroglobulin and many other antigens known to cause experimental autoimmunity upon appropriate immunization. The inevitable conclusion from this was that within the healthy immunological repertoire B cell and T cell self-reactivity was a physiological component. Subsequent demonstration that such cloned T cells were capable of transferring experimental autoimmune diseases, has made the question of how these cells remain safely contained within the normal functioning repertoire into one of the most pressing issues of modern immunology (Cohen, 1986). Along with the studies aimed at furthering our understanding of this point, it has become clear that antigen specific manipulation of autoimmunity may become a reality.

Prevention or even in some instances, treatment of experimental autoimmunity has been shown to be possible, making use of the antigen or selected determinants of antigens critical to the disease. In such cases of antigen specific therapy, the prevention or treatment achieved is usually permanent and specific for the disease involved. These two qualities make such approaches into a tantalising prospect for application in spontaneous autoimmune diseases in humans. The search for relevant autoantigens in human diseases has now become one of the popular areas of immunological research. The technical possibilities to identify such antigens, are for several reasons limited. One alternative possibility, circumventing the need to identify critical autoantigens, is the inhibition of MHC presentation of critical antigens by MHC binding competitor peptides (Wraith et al, 1989; Adorini et al, 1988). This approach would be semi-specific, since the peptides are designed to bind to only one of the available MHC molecules, presumably the ones known to occur in association with raised susceptibility to the target disease. Besides this relative non-specificity, it maybe foreseen that this approach would result in only temporary disease inhibiting activity, requiring continuous and permanent infusion of competitor peptides.

Recently, we have begun to explore the possibility of designing immunogenic peptide analogues of disease associated T cell determinants in the rat model of experimental allergic encephalomyelitis (EAE) and adjuvant arthritis (AA). We have studied their qualities as MHC binding competitor peptides and have discovered that one of the AA associated peptide analogues has superior activity in inhibiting AA due to the combined qualities of MHC competition and specific immunomodulation, present within a single nonamer peptide (Wauben et al, in press). It is possible that by the use of such, so-called "Competitor-Modulator" peptides (Wauben et al, 1992) the advantages of specific active immuno-modulation can be combined with MHC competition. Attractive as this seems, application in the human situation will depend on the successful identification of antigens critical to disease. As the disease associated antigen in rat AA is the mycobacterial 65kD heat-shock protein (van Eden et al, 1988), we will discuss briefly the existing evidence that hsp60 may be one of the antigens critical to human arthritic conditions.

Heat-shock proteins and arthritis

The various forms of human reactive arthritis and Lyme arthritis, demonstrated that of all forms of clinical autoimmunity, it is especially chronic arthritis that has been seen in association with microbial immunization. Therefore, it is of no surprise that under experimental conditions, immunization with bacterial antigens has been found to lead to arthritis. One of the favourite models of experimentally induced arthritis is adjuvant arthritis, where intracutaneous injection of heat-killed mycobacteria, suspended in mineral oil, leads to an aggressive T cell mediated disease. We have used the arthritogenic T cell clone A2b (Holoshitz et al, 1984), selected by its recognition of whole M.tuberculosis to define the mycobacterial hsp65 as an antigen of mycobacteria with the capacity to activate arthritogenic T cells (van Eden et al, 1988). These findings have stimulated a renewed interest in the potential role of mycobacteria in the origin of human chronic arthritis. Although none of the findings, has been so far conclusive in itself, the overall picture that emerged is increasingly suggestive that mycobacteria or other (slow?) bacterial infections may play a role (Rook and Stanford, 1992).

As far as heat-shock proteins are concerned, it was reasoned that through sequence homologies - mycobacterial hsp65 is a member of the family of hsp60 chaperonins, exhibiting in order of 50% identity at the amino acid level with mammalian hsp60 - immune responses elicited by bacterial hsp60 would directly exhibit autoreactivity directed at its mammalian host counterpart hsp60. Although in the rat AA model no direct evidence for this is available as yet (van Eden, 1991) - A2b was found to recognise a relatively non-conserved area of the molecule and no rat hsp60 has been available so far for carrying out further critical experiments - in the human studies the principle has appeared to be essentially correct. With the use of peptides synthesized on the basis of conserved sequences and the recently cloned and expressed recombinant human hsp60, T cell cross-reactivity between bacterial and human hsp60 has been demonstrated in various instances. Although this has been shown in patients, also T cells obtained from healthy individuals have also been shown to exhibit such double specificity (Lamb et al, 1989). Thus, like the self-MBP reactive cells obtainable from healthy animals, human T cells with specificity for self hsp seem to be tolerated and persist within the immune repertoire. Despite the fact that cell-surface expression of hsp's has not been proven unequivocally to occur so far, one has to accept that hsp's, like any other intracellular self-antigen will be processed and presented in the context of MHC molecules. Alternatively, hsp's may form a target for self reactivity when, as a chaperonin having transient interactions with other proteins, it becomes co-expressed together with cell-surface molecules. Indeed, experiments have indicated that this may be the case. For instance, in a case of reactive arthritis following Yersinia infection cloned synovial T cells have been shown to display reactivity to Yersinia antigens, mycobacterial hsp65 and human hsp60 (Hermann et al, 1992). The same T cell clone was shown to lyse autologous stressed macrophages. The presence of T cells with such specificity in the synovial compartments of an arthritic joint may be significant, because of the known raised expression of hsp's in inflamed synovial lining cells.

Careful histological analyses have shown that raised expression of hsp's is a feature of synovitis both in experimental models and in human rheumatoid arthritis. In children with juvenile rheumatoid arthritis (JCA) raised synthesis of hsp60 has been shown to be one of the earlier events in the development of arthritis, as shown using an antibody (LK1) with specificity for the mammalian hsp60 and not being cross-reactive with bacterial hsp's (Boog et al, 1992). In the same children T cells were collected from their synovial fluids. These T cells were found in more than half of the cases to proliferate in the presence of purified recombinant human hsp60 (De Graeff-Meeder et al, 1991). Interestingly, the majority of the responding patients were suffering from the relatively mild oligo articular form of JCA. These patients are known to remit spontaneously. The polyarticular or systemic onset juvenile arthritis patients and the HLA-B27 positive oligoarticular patients tended to be non-responders instead. Furthermore, a control group of adult patients with advanced RA did not respond (De Graeff-Meeder et al, 1991).

Also by others it has been demonstrated that the frequency of responders among adult RA patients, may be low (Res et al, 1990) with possible exception of patients very early in their disease. In some early cases, T cell clones were obtained with specificity for conserved hsp sequences (Quayle et al, 1992). Remarkably, none of the clones obtained so far have been found to be HLA-DR4 restricted. Even in DR4 homozygous patients clones were found to be HLA-DQ or -DP restricted (Gaston et al,

1992). Altogether it seems that responses to hsp60 are most prominent in either self-limiting forms of JCA or in very early cases of RA in adults. In the latter cases responses are not found in the context of the HLA molecule known to be associated with a bad course of the disease. From this, it is tempting to speculate that in human patients responses directed against hsp's are part of protective regulatory responses rather than part of responses which are detrimental to the diseased host. Furthermore, such a speculation seems to be supported by the findings made in various experimental models of arthritis. In most of these models such as AA and streptococcal cell-wall induced arthritis, but also in oil induced (pristane) arthritis (Thompson et al, 1990) and collagen arthritis (Ito et al, 1991) immune responses to hsp have been seen to develop. In addition preimmunization with mycobacterial hsp65 has been shown to protect against arthritis induction. Thus, raised immunity to hsp's seems to reinforce resistance to arthritis rather than to precipitate disease. Since this is the case not only in bacteria induced model systems, but in others - pristane, collagen - as well, development of immunity to hsp's could well be a physiological anti-inflammatory response of a general nature (Fig. 1). As shown in the various models, the artificial manipulation of such a response, might well lead to inhibition of arthritis. This would foster optimism for the possibility of designing a specific immunotherapy based on hsp antigens.

However, suppression of active disease by administering mycobacterial hsp65 i.p. or i.c. has not been achieved, at least not in the AA model. Aggravation of disease has been seen instead (Hogervorst et al, 1992). Nevertheless, successful suppression of disease has been achieved using a recombinant vaccinia virus, expressing mycobacterial hsp65, as late as 7 days after Mt immunization (Hogervorst et al, 1991), indicating that alternative routes of immunization or immunizations with selected determinants of the molecule, might possibly lead to treatment of disease.

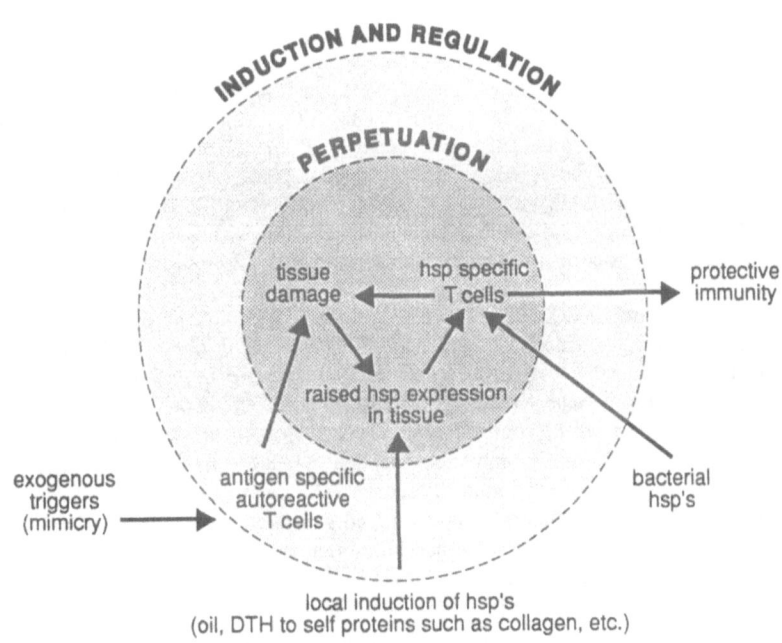

Fig. 1: In various forms of both experimentally induced and spontaneously occurring arthritis, immunity to hsp seems to evolve and probably is part of the perpetuation of the process. Evidence has been collected in favor of various possible ways to trigger or modulate such a process.

Definition and testing of competitor peptides

Lewis rats are susceptible to several experimental models, such as AA, EAE and EAU (experimental autoimmune uveoretinitis). In these models antigens have been defined and T cell lines have been selected with the capacity to passively transfer disease. So far, these disease inducing T cells, have been shown to recognise their respective antigens in the context of RT1B[1] (the I-A equivalent in Lewis rats). This would imply that disease associated peptides as determined with these T cells and some of their modified analogues would be MHC binding peptides with the potential of functioning as MHC competitors in more than one disease model. In the AA model, the arthritogenic clone A2b was found to recognize the 180-188 sequence of the mycobacterial hsp65. Other experiments have indicated that this sequence might be of crucial significance in the AA model. It was found that AA resistant Fisher rats were non-responders to 180-188, (Hogervorst et al, 1991). This was unexpected, especially since Lewis rats responded to 180-188 and Fisher rats are MHC Class II identical with Lewis rats, as also indicated by the fact that Fisher APCs may present 180-188 to Lewis T cells. Fisher rats are relatively resistant to AA, a resistance that seems determined by prior microbial exposure, as germ-free Fisher rats are susceptible. We have seen that in the relatively rare occurrences of AA in Fisher rats, responses to 180-188 do develop and first experiments have indicated that germ-free Fisher rats may respond as well (Broeren, pers. comm.). It is concluded that 180-188 can be a crucial antigenic determinant in AA. Ala-substituted peptide analogues of 180-188 were tested in a functional competition assay by adding them in increasing concentrations to A2b, antigen presenting cells and 180-188. The peptide-analogue with an Ala-substitution at position 183 [A183], was found to be an efficient inhibitor of A2b responses (Wauben et al, in press). When this inhibitor was added to Mt, and used as a co-immunogen in the protocol for induction of AA, a very marked reduction in disease development was noticed. Only a small fraction of the animals developed disease, and this disease was mild. Such an effect, reached by simply adding a single synthetic peptide analogue to a complex antigen such as whole mycobacteria, was unexpected and probably also unprecedented. Was this effect fully due to MHC competition or was an immunological relationship with native 180-188 responsible?

To check for its competitor qualities, the peptide was tested in the EAE model. First, functional competition was tested, as was done earlier with A2b, using the encephalitogenic T cell Z1a. Subsequently in vivo competition in peptide MBP 72-85 induced EAE was tested (Wauben et al, in press). In both instances the A183 peptide behaved as a powerful MHC competitor, capable of reducing development of EAE with good efficiency. Furthermore, draining lymph node responses to the MBP 72-85 peptide after disease induction clearly were reduced in animals coimmunized with A183, as compared to control animals immunized with MBP72-85 alone. Thus, the fact that priming in vivo of MBP 72-85 reactive cells was reduced by adding A183, substantiated the evidence that A183 behaved as an efficient MHC competitor peptide. For control purposes, in addition, competitor peptides were developed in the EAE model itself. Peptide MBP 72-85, being the target determinant of encephalitogenic T cell line Z1a, was used to synthesize Ala substituted analogues as was done for 180-188 earlier. From these substituted peptides, MBP72-85D$_{81}$-->A [also called by its number of synthesis:1028], was found to compete with peptide induced stimulation of both Z1a and A2b. Moreover, peptide 1028 fully inhibited peptide induced EAE. However, when tested at concentrations at which A183 was found to reduce AA dramatically, peptide 1028 was not found to cause any reduction of AA. This discrepancy urged us to determine relative MHC binding affinities of our peptide analogues, to exclude the possibility that the superior activity of A183 in suppressing AA was due to superior binding affinity.

For this purpose we used a novel direct MHC binding assay, as devised by Joosten et al (submitted). In this assay affinity purified Lewis rat RT1B molecules were incubated with biotinylated MBP 72-85 peptide and dose ranges of the non-labelled peptide competitors 1028 and A183. Subsequently MHC-peptide mixtures were separated by SDS-PAGE and blotted on nitrocellulose. The presence of labelled peptide bound to MHC products was then detected by enhanced chemiluminescence. This assay demonstrated that the AA related A183 peptide had no superior binding affinity. On the contrary, its binding turned out to be inferior in comparison with the MBP related competitor.

Furthermore, this conclusion was supported by more recent findings, showing that an unrelated OVA specific T cell clone (donated by Reske, Mainz) was inhibited more easily by the EAE related competitor than by A183. From all this we came to conclude that the remarkable efficiency of our AA related competitor to inhibit disease <u>in vivo</u>, had to be due to more than MHC blocking alone.

Immunomodulatory qualities of the AA related competitor peptide

First clues revealing the special nature of the immunological consequences of administering peptide A183, came from the analysis of lymph node responses in Mt/peptide immunized AA rats (Wauben et al, in press). It is known that responses to 180-188, present at the time of AA development around day 14, have subsided by this a later time point (Hogervorst et al, 1992). However, animals immunized with Mt in the presence of A183 responded to 180-188 and in addition to A183 itself. This indicated, besides the immunogenicity of A183, that A183 was stimulating responses to the native peptide 180-188. How could we reconcile A183 induced inhibition of AA with the peptide potentiating, and certainly not inhibiting, responses to the critical 180-188 sequence? This would only make sense, if it could be shown that cells differing from arthritogenic A2b were involved. To enable studies on A183 reactive T cells a new T cell line, ATL, with specificity for A183 was generated by <u>in vivo</u> immunization with A183 in CFA and subsequent restimulations <u>in vitro</u> using the same A183 peptide. Subcloning of this ATL T cell line revealed the presence of clone ATL11, which had a double specificity for A183, and the original peptide hsp180-188.

Thus, cells with a new specificity were activated by stimulation with A183, a specificity distinct from that of A2b, a clone that did not respond to A183. It remains to be seen whether T cells, such as ATL are responsible for the disease inhibiting activity of A183. Given their closely related antigenic specificities, it is possible that ATL and A2b, although distinct, have related T cell receptors, and that ATL is stimulating regulatory T cells in the network, similar to what has been shown previously for attenuated A2b itself (Lider et al, 1987).

Competitor peptide A183 activates T cell-line ATL with specificity for a cryptic hsp65 determinant

The regulatory activity described above could not be initiated by immunization with Mt, and therefore could not inhibit the development of AA, for the following reason. Upon testing, ATL was found not to be stimulated by Mt or mycobacterial hsp65. In other words, although ATL responded to 180-188 of hsp65, it did not respond to the protein itself and therefore ATL was an example of a T cell that recognized a cryptic determinant. Such cryptic determinants have been described previously for various other proteins by Sercarz (Gammon and Sercarz, 1989). In contrast to dominant determinants, cryptic determinants are not seen by T cells after stimulation with the complete protein. Only after stimulation with peptides have such T cell specificities been demonstrated. Crypticity or dominancy of T cell determinants were considered to be qualities of a certain determinant in the context of the complete protein. ATL has now demonstrated us that one and the same determinant, may be both cryptic and dominant, and that these respective qualities may depend on the recognizing T cell (Wauben et al, submitted). By analysis of a complete amino acid replacement network of peptide 180-186, marked differences were found between A2b and ATL. Whereas A2b permitted some substitutions at positions 180, 181 and 186 and no substitutions at 182-185, clone ATL11 permitted substitutions at 181-185 and hardly any at positions 180 and 186. So, in this respect A2b and ATL11 appeared to be almost mirror-images of one another. In contrast to A2b, ATL11 appeared to focus its attention on the sides of the critical epitope, and therefore theoretically ATL could be more sensitive in its TCR interactions to hinderance by additional flanking amino acids. This reasoning was correct, since the addition of two extra residues, as they naturally occurred in the hsp65 sequence, at the N terminus of the peptide, fully inhibited recognition by ATL11 but not A2b. This left us with the likely possibility that during processing of hsp65 such a fragment is being generated, extended at the N-terminal side of 180. This would explain the cryptic nature of the ATL specificity. To our knowledge, this is the first description of a clone with cryptic specificity generated by immunization with a peptide with rather drastic immunomodulatory - in our case disease inhibiting - qualities. So, at this point there is little to say about the generality of these findings. Nevertheless, the possibility of triggering cells with peculiar specificities by using artificially designed synthetic peptide-analogues, is expected to become an area of further intensive research. The more so as it now appears, that, possibly in contrast to earlier expectations, the majority of such artificially designed MHC binding peptides turn out to be quite immunogenic. The findings concerning peptide A183 may well demonstrate unforeseen potential of such "artificial" immunity. That the action of A183 in inhibiting AA surpassed its capacity of simply competing for MHC binding, as in the EAE model, was further illustrated by the fact that preimmunization with peptide A183 as long as 7 days prior to AA induction led to an almost complete prevention of the disease (Wauben et al, in press). Thus, besides MHC competition, the immunogenicity of peptide A183 endowed this peptide with a capacity of inducing disease protective, presumably immunomodulatory, qualities. The combined action of MHC blockade and such immunomodulation may well explain the efficiency of A183 in inhibiting AA.

The "competitor-modulator" concept

Experimental models of autoimmunity have indicated to us that resistance against the disease may be developed for relatively long periods of time both following spontaneous remission of the disease, such as in AA, and following artificial immunization, such as with hsp65 in AA. Thus, even in the susceptible host, mechanisms are present which are capable of perpetuating and somehow consolidating, protective immuno-regulatory events. This occurs not only in the contrived model systems, such as AA, but in addition in spontaneous models, such as that of diabetes in the NOD mouse. In the latter model it has been shown that after an artificially provoked, transient mild form of early disease, done with a defined hsp65 peptide, full protection against development of disease later in life can be achieved (Elias et al, 1991). Thus, it seems that exposure of the immune system to antigens crucially associated with disease either naturally during active disease or by artificial preimmunization, can trigger protective, disease-suppressing mechanisms. For the development of protective immunomodulatory agents, the critical issue is, in which form or composition and when such antigens should be administered during active disease, in order to achieve the desired disease suppressing activity and to avoid stimulating those effector mechanisms that may lead to aggravation of clinical disease instead. Whatever solution to this problem will be found, antigen-specific manipulation of autoimmunity may remain risky and may have counterproductive effects deleterious to the unfortunate host.

For this reason, one might wish to combine antigen-specific immuno-modulation with some immunosuppressive regimen. The latter could be applied prior to antigenic stimulation, in order to first silence the aggressive elements in the disease process. Alternatively, immunosuppressive drugs could be given at the same time as the immunomodulatory agent, provided that the immunosuppression would not counteract the action of the immunomodulatory agent. What about a situation in which our immunosuppressant would be the desired immunomodulator itself? To have both qualities brought together into one single immunogenic peptide? Would that not be the best of all possible worlds?

Immunosuppression by MHC blockade is attractive, relatively non-toxic and may be targeted selectively to certain disease associated MHC products. Furthermore, such suppression is transient, likely to be without long-term side effects, and it may be very effective as we and others have demonstrated using the model of EAE. Furthermore, the combination of immunomodulator and competitor in one peptide could well ensure a perfect timing of both effects. The triggering of pathogenic effector cells is countered by MHC competition just at the time the regulatory disease suppressing cells are being activated by the immunomodulation. This perfect timing would endow such "Competitor-Modulator" peptides (Table 1) with effectiveness based on real synergism between both built-in properties, in other words effectiveness that would exceed the sum of both individual qualities. The dramatic effectiveness of the A183 peptide in AA is likely to be due to such synergism. So far, none of our other peptides with proven competitive activity has been seen to have a similar capacity to inhibit AA.

Table 1: "Competitor-Modulator" peptides

---

* Compete for MHC binding

* Immunogenic

* Modulate responses to disease associated
  T-cell determinants

* SYNERGISM of passive (competition) and
  active (regulatory T cells) immunomodulation

Whether the immunomodulatory effects of A183 resulted from indirect regulatory events, such as through a cell like ATL with network regulatory events due to relatedness with arthritogenic A2b, or otherwise, has not been proven yet. Alternatively, A183 could have more direct effects on cells relevant to the disease, for instance by inducing anergy due to its relatively superior MHC binding as compared to 180-188. However, in experiments using A2b as a target cell, A183 has not been observed to induce any anergy or paralysis, whereas 180-188 did. As a second alternative possibility, one may think of A183 acting as a TCR antagonist, as recently postulated by De Magistris et al. (1992). Although in preliminary experiments no evidence for this was obtained with A2b, we can not exclude this possibility as yet. Whatever the exact mechanisms of the immunomodulatory qualities of A183 may turn out to be, the disease protective effect seen with preimmunization is suggestive of induction of an active immunomodulatory event. This augurs well for the induction, by such means, of a possible self-maintaining and perhaps permanent state of protective resistance against auto-immune disregulation. As said before, the successful implementation of the "Competitor-Modulator" peptide in the armamentarium against human autoimmune diseases, will remain dependant on the successful identification of relevant antigens. With heat-shock proteins we might have obtained our first candidate antigens as far as human chronic arthritic conditions are concerned.

Acknowledgement: We thank Mrs. Jona Gianotten for the speedy and smooth processing and Dr. Steve Anderton for critically reading this manuscript.

References

Adorini L, Muller S, Cardinaux F, Lehman PV, Falcioni F, Nagy ZA (1988) Nature 334:623-625

Boog CJP, De Graeff-Meeder ER, Lucassen MA, Van der Zee R, Voorhorst-Ogink MM, Van Kooten PJS, Geuze HJ, Van Eden W (1992) J. Exp. Med. 175:1805-1810

Cohen IR (1986) Immunol. Rev. 94:5-21

Elias D, Reshef T, Birk OS, Van der Zee R, Walker MD, Cohen IR (1991) PNAS 88:3088-3091

De Graeff-Meeder ER, Van der Zee R, Rijkers GT, Schuurman HJ, Kuis W, Bijlsma JWJ, Zegers BJM, Van Eden W (1991) Lancet 337:1368-1372

De Magistris MT, Alexander J, Coggeshall M, Altman A, Goeta CFA, Grey HM, Sette A (1992) Cell 68:625-634

Gammon J and Sercarz EE (1989) Nature 342:183-185

Gaston JSH, Life PF, Van der Zee R, Jenner PJ, Colston MJ, Tonks S, Bacon PA (1991) Intern. Immunol. 3:965-972

Hermann E, Lohse AW, Van der Zee R, Van Eden W, Mayet WJ, Probst P, Poralla T, Meyer zum Buschenfelde KH, Fleisher B (1991) Eur. J. Immunol 21:2139-2143

Hogervorst EJM, Wagenaar JPA, Boog CJP, Van der Zee R, Van Embden JDA, Van Eden W (1992) Int. Immunol. 7:719-727

Hogervorst EJM, Schouls L, Wagenaar JPA, Boog CJP, Spaan WJM, Van Embden JDA, Van Eden W (1991) Infect. Immunit. 59:2029-2035

Hogervorst EJM, Boog CJP, Wagenaar JPA, Wauben MHM, Van der Zee R, Van Eden W (1991) Eur. J. Immunol. 21:1289-1296

Holoshitz J, Matitiau A, Cohen IR (1984) J. Clin. Invest. 73:211-215

Ito J. Krço C, Yu D, Luthra HS, David CS (1991) J. Cell. Biochem. 15(A):284

Joosten I, Wauben MHM, Holewijn MC, Reske K, Hensen EJ, Buus S (submitted)

Lamb JR, Bal V, Mendez-Sampiero P, Mehlert A, So A, Rothbard J, Jindal S, Young RA, Young DB (1989) Intern. Immunol. 1:191-196

Lider O, Karin N, Shinitzky M, Cohen IR (1987) PNAS USA 84:4577-4580

Quayle AJ, Wilson KB, Li SG, Kjeldsen-Kragh J, Oftung F, Shinnick T, Sioud M, Førre Ø, Capra JD, Natuig JB (1992) Eur. J. Immunol. 22:1315-1322

Res PCM, Telgt D, Van Laar JM, Oudkerk-Pool M, Breedveld FC, De Vries RRP (1990) Lancet 336:1406-1408

Rook GAW, Stanford JL (1992) Immunol. Today 13:160-164

Thompson SJ, Rook GAW, Brealey RJ, Van der Zee R, Elson CJ (1990) Eur. J. Immunol. 20:2479-2484

Van Eden W, Thole J, Van der Zee R, Noordzij A, Van Embden JDA, Hensen EJ, Cohen IR (1988) Nature 331:171-173

Van Eden W (1991) Imm. Rev. 121:5-28

Wauben MHM, Boog CJP, Van der Zee R, Van Eden W. (1992) J. Autoimm. 5A:205-208

Wauben MHM, Boog CJP, Van der Zee R, Joosten I, Schlief A, Van Eden W (in press) J. Exp. Med.

Wauben MHM, Van der Zee R, Boog CJP, Van Dijk AMC, Holewijn MC, Meloen RH, Van Eden W (submitted)

Wraith DC, Smilek DE, Mitchell DJ, Steinmann L, McDevitt HO (1989) Cell 59:247-254

Cellular Levels of Expression of Genes Coding for Organspecific Autoimmune Diseases

J. F. Bach and H. J. Garchon

INSERM U 25, Hôopital Necker, 161 rue de Sèvres, 75015 Paris, France

Organ-specific autoimmune diseases (AID) are multifactorial polygenic diseases. The analysis of the predisposing genes is rendered difficult by the variable and often limited penetrance of each gene and by the multiplicity and variable impact of the genes in question. The identification of these multiple genes is however a problem of crucial importance both for the understanding of disease mechanisms and for clinical applications (early diagnosis, prediction and new therapeutical approaches).

HOW MANY GENES ?

The role of genetic factors in organ specific AID is examplified by the disease concordance rate observed in monozygotic twins in all the diseases where the information is available (Table 1). The high percentage of discordant twins indicates a major role for environmental factors even if one cannot exclude that the somatic diversification of B and T cell receptor genes may generate different genes in each of the monozygotic twins. Even at this level, the role of environmental factors (e.g. retroviruses) cannot be ruled out.

The difference between concordance rates in monozygotic twins and non twin siblings is considerable. Its amplitude depends on several factors primarily on the overall incidence of the disease and on the number of predisposing genes. In the best documented case of insulin-dependent diabetes mellitus (IDDM) (Spielman, 1989), the high disease frequency in siblings of diabetic patients (approximately 7%) is not compatible with a very large number of major genes (whatever the number and role of minor genes). One can evaluate the influence of environmental factors from concordance rates in monozygotic twins (which provides a minimal evaluation since twins usually share a number of environmental factors). Models for recessive or dominant genes indicate figures of the order of 3 to 5, if one assumes that the allele frequency in the general population is less than 5%. HLA genes obviously represent one (or more) of these genes as shown by the disease concordance rate in HLA identical siblings, in IDDM (12-15% versus 7% in overall siblings and 35-50% in monozygotic twins).

In the case of IDDM, if we estimate the concordance rate in monozygotic twins is approximately 50%, we can assume that the overall penetrance is about 50% (probably an overestimation if one realizes that the real concordance rate is lower and that monozygotic twins share a number of environmental factors). In these conditions, the corrected concordance rate in siblings would be close to 15%. Assuming a gene frequency less than 20%, forgetting minor genes and giving an equal weight to major genes (including HLA genes) one observes that this figure of 15% is compatible with two genes (one dominant, one recessive) or three genes (three dominant). Three genes not all dominant would not give rise to lower sibling concordance rates (e.g. 3 recessives genes would provide calculated figures of 1-5%). If one integrates HLA genes as a major codominant gene, one starts then from a corrected concordance rate of 25%. Whether or not major predisposing genes indeed include dominant HLA plus one or two unknown gene(s), it remains that most likely all other genes either represent minor genes (genes bringing a significant but limited contribution to disease predisposition) or alleles widely distributed in the general population and shared between most siblings of a given family.

HLA GENES

The association between HLA genes and organ specific AID has been known for a long time. It is interesting to note that in spite of the major role of HLA genes suggested by concordance rates in families, as just discussed, the relative risks associated with the most predisposing alleles is generally modest even if it is higher in twins carrying the alleles showing the highest relative risks (DR3/DR4 diabetic twins). The weakness of the relative risks is particularly striking in myasthenia gravis, multiple sclerosis or Graves disease. Several explanations can be proposed for this paradox. First, most often, the disease presents genetic heterogeneity as observed by considering age- or sex- defined subgroups (Table 2) (Caillat, 1992). In some of these subgroups significantly higher relative risks are observed. Second, the locus coding for the predisposition molecule is not always well defined and could not be the HLA locus under consideration. Strong indication of this has been provided for DR3-associated diseases where the predisposing locus on the A1B8 DR3 ancestral haplotype could be a class III locus (perhaps between HLA-B and TNF) according to Dawkins (1983). In other cases the predisposing genes could be peptide transporters, proteases, complement factors, or any other MHC gene to be identified.

The low HLA disease association could also relate to the fact that T cells with high affinity for the HLA-autoantigen peptide complex have been eliminated during intra thymic negative selection after binding autoantigenic peptide presented by HLA molecules, only leaving T cells binding with weak affinity to autoantigen.

These observations suggest that depending on the patients' group, different HLA loci could be involved (e.g. DR3 and DR4 in IDDM, DR3 and DQB1-6O4 in myasthenia gravis) including non class I or class II loci. At the cellular level, MHC genes could then either intervene at the level of peptide presentation (Ir genes) or at other levels associated with the amount of peptide presented in a given organ or with immunoregulation.

Another important feature of HLA gene association with organ specific AID is the observation of haplotypes providing a major protection from the disease. This has been particularly well studied in the case of IDDM and more recently in myasthenia gravis (Table 3). In fact, the statistical significance of the relative risks associated with these protection alleles is often superior to that of predisposing alleles. The mechanisms underlying this protection are still unclear. They might involve capture of the autoantigen peptides by HLA molecules that do not present them to T cells in an immunogenic fashion as suggested by Nepom (Nepom & Erlich, 1991). Alternatively they might code for some regulatory mechanisms that lead to inhibition (suppression) of the autoimmune response. This possibility is suggested by the observation of an absence of response of some mouse strains to synthetic peptides (such as GAT) that can be overcome by cyclophosphamide treatment which is known to abrogate some T cell mediated suppressor mechanisms. It is also suggested by data obtained in I-A transgenic Non Obese Diabetic mice which are made diabetic by cyclophosphamide treatment or whose lymphoid cells can transfer protection. Table 4 summarizes these findings as well as the other possible levels of intervention of disease genes.

OTHER CANDIDATE PREDISPOSITION GENES.

Candidate genes are selected among genes coding for molecules presumably involved in the pathogenesis of the disease. They include in the first place all the molecules used by the protagonists of this autoimmune response such as the autoantigen, immunoglobulins, the T cell receptor, cytokines or complement factors.

Autoantigens.

Converging arguments suggest that organ-specific AID are for most of them due to autoantigen driven immune responses. Thus autoantibody production or T-cell reactivity do not appear or are eliminated by renewal of the autoantigen as can be achieved by thyroidectomy in the obese chicken or alloxan treatment in NOD mice. One can induce in normal animals an experimental AID very similar to the human one by administering the target autoantigen incorporated in complete Freund's adjuvant. Lastly, organ specific autoantibody genes usually show somatic hypermutation in the complementarity determining regions (CDR) as one would expect from an antigen driven antibody response. Only few data are yet available on the genetic polymorphism of autoantigens. We have recently described such a polymorphism for the human acetylcholine receptor gene. We first identified two microsatellites within the AChR α-subunit gene and showed the existence of significant polymorphism with respectively 7 and 6 alleles. Preliminary data obtained in a series of 26 patients indicate that there may be an association between certain alleles and myasthenia gravis. If this observation is confirmed on a larger series of patients, it would provide the first direct evidence that genes coding for target autoantigen represent AID predisposition genes.

One may also mention at this level previously obtained data showing a genetically controlled abnormality of iodine uptake in the obese strain of chicken preceding the onset of thyroiditis which potentially contributes to the disease pathogenesis (Kroemer, 1988).

One may also quote as a putative role for autoantigen genes the data reported for the insulin gene in human IDDM. It has been shown by G.I.. Bell that one of the major haplotypes in the region of the insulin gene is associated with sporadic IDDM. The strength of the association is increased in HLA DR4 patients (Julier, 1991). In French IDDM multiplex-families, linkage of the disease with the insulin gene is conditioned by the inheritance of the HLA-DR4 haplotypes from the father. This finding that needs to be replicated is highly suggestive of an heterogeneity of the disease.

Antigen lymphoid cell receptors.

The role of lymphoid cell antigen receptors (immunoglobulin and antigen T-cell receptors) was suggested by several population studies and was recently reviewed (Garchon & Bach, 1991; Garchon, 1992). However, negative results were also reported and linkage studies are still missing.

Complement.

A number of associations of complement factor polymorphisms with organ specific AID has been reported in several diseases. All these associations involved MHC encoded genes and it remains to be proven that the association is not thus only explained by linkage desequilibrium with class I or class II or other class III MHC genes. If it was not the case, complement factors could then represent a significant predisposing element whatever the underlying mechanisms.

SYSTEMATIC SEARCH FOR PREDISPOSITION GENES THROUGH SEGREGATION STUDIES.

The identification of genes coding for monogenic diseases can now be approached in a systematic fashion by segregation studies in which the pathologic trait is followed in multiplex families in parallel to the polymorphism of a large number of genes distributed all over the genome. These genes are selected in a way that any investigative gene will be in linkage disequilibrium with one of the genetic markers. The main markers are now microsatellites and to a lesser extent variable number tandem repeats (VNTR). Microsatellites are base repeats (mostly CA) of variable length present in large number in the genome (total ~ 50.000). The polymorphism is based on the length of the repeat which is a stable trait. The microsatellite length is evaluated by gel electrophoresis after polymerase chain reaction (PCR) (Garchon & Bach, 1992). This methodology is difficult to apply directly to human AID because of the low number of multiplex families, the multiplicity of genes involved and their low penetrance. One may foresee however that this could be made easier by increasing the number of informative individuals in patients' families by considering healthy subjects presenting with some components of the disease (e.g. autoantibodies in their serum).

Alternatively, one may use experimental models of AID as has been recently accomplished in insulin-dependent diabetes with the NOD mouse. Several diabetes associated genes have been described by J. Todd and collaborators. These genes are located respectively on chromosome 1 (ch1), ch3, and ch11 (Todd, 1991 ; Cornall, 1991). These genes have not been definitively identified and their function is still unknown. It should be noted though that the ch1 gene is close to the interleukin 1 receptor gene and that the ch11 gene is only expressed in spontaneously diabetic NOD mice, not in cyclophosphamide induced diabetes, suggesting a role for this gene in the control of cyclophosphamide sensitive (presumably suppressor) T cell function. On the other hand using F2 crosses (and not backcrosses like J. Todd and coll.) and morphologic criteria (insulitis and its first phase periinsulitis) in addition to diabetes (rare in F2), we have identified another gene close to the Bcl-2 protooncogene (Garchon, 1991). Interestingly, other pathologic manifestations of NOD mice also segregate with Bcl-2 namely sialitis (mononuclear cellular infiltration of the salivary glands) and increased serum IgG levels. These data tend to suggest that Bcl-2, itself could be the gene involved in disease susceptibility. This gene could be associated with protection from apoptosis leading to long lymphocyte survival, abnormal homing to some tissues and/or hyperactivation. Indeed Bcl-2 transgenic mice show numerous signs of autoimmunity with a lupus-like syndrome (Strasser, 1991). Studies in progress will tell if there is a human equivalent of these NOD mouse IDDM predisposing genes.

CONCLUSIONS.

The complete identification of genes predisposing to organ specific AID is a central issue in immunopathology. It is yet at its beginning but the information obtained with candidate genes such as HLA, immunoglobulins, TCR or acetylcholine receptor are encouraging. Still more promising are the results obtained by segregation studies in the NOD mouse. One may hope that the knowledge of all the genes will allow to dissect in a stage-specific fashion the various anomalies that predispose to the triggering, the development and the expression of the pathogenic immune responses that characterize the various organ-specific AID.

Table 1. Genetic heterogeneity of autoimmune diseases

IDDM		Myasthenia Gravis	
	DR3 and/or DR4		DR3
children	109/112 (97 %)	young females	30/70 (43 %)
late onset IDDM (>30)	79/107 (74 %)	patients >40 years	0/12
		thymomas	0/18

Table 2. Hereditary transmission of organ specific autoimmune diseases

	IDDM		MG		Graves' disease	MS
	disease	ICA	disease	AChR ab		
Monozytotic twins	35-50 % (RR : 200)	25-44 %	40 % (RR : 4000)	NK	30-50 % (RR : 50)	20 % (RR : 200)
Siblings	7 % (RR : 30)	3-10 %	2 % (RR : 200)	18.6 % (RR : 30)	3 % (RR : 5)	1.2 % (RR : 12)
HLA identical siblings	10-15 % (RR : 70)	6-12 %	NK	30 % (RR : 60)	7 % (RR : 12)	NK

NK : not known

Table 3. HLA class II protective haplotypes in autoimmune diseases

IDDM (n = 402)			RR
DR2 DQB 602			0.15
DR2 DQB 502			> 1
DR13 DQB 603			0.23
Asp DQβ57			0.16
Myasthenia gravis (n = 115)			
DR1-	DQB1*0501-	DQA1*0101	0.3
Mechanisms			
Peptide capture			
Suppression			

Table 4. Putative levels of expression of genes predisposing to organ-specific autoimmune disease (AID)

Autoantigen	AChR/MG
HLA	All known AID
structural genes	
regulatory genes	
transporter genes	
proteasome genes	
TCR	MS, MG (?)
Immunoglobulins	IDDM, MG, ...
Complement components	IDDM, MG, ...
Cytokines and cytokine receptors	?
Regulatory cells	Ch 11 gene (NOD mice)
Lymphocyte survival (apoptosis)	Bcl-2 (NOD mice)

REFERENCES.

Caillat S et al. (in press) J Clin Invest

Cornall RJ, Prins J-B, Todd JA, Pressey A, DeLarato N, Wicker LS, Peterson (1991) Nature 353:262-265

Dawkins R. et al. (1983) Immunol Rev 70:5-22

Garchon HJ, Bach JF (1991) Human Immunol 32:1-30

Garchon HJ, Bedossa P, Eloy L, Bach JF (1991) Nature 353:260-262

Garchon HJ (in press) Current Opinion in Immunol

Garchon HJ, Bach JF (1992) Current Biology 2:268-270

Julier C, Hyer RN, Davies J et al. Nature 354:155-159

Kroemer G. et al. (1988) Eur J Immunol 18:1499-1505

Nepom GT, Erlich H (1991) Annu Rev Immunol 9:493-525

Spielman RS, Baur MP, Clerget-Darpoux F (1989) Genet Epidemiol 6:43-58

Strasser et al. (1991) Proc Natl Acad Sci USA 88:8661-8665

Todd JA, Aitman TJ, Cornall RJ et al. Nature 351:542-547

Physiopathology of Autoimmunity: The Reactivities of Natural Antibodies Define the Boundaries of the Immunological Self

A. Coutinho, A. Sundblad, M. A. R. Marcos, M. Haury and A. Nobrega

Unite d'Immunobiologie, CNRS URA 359, Institut Pasteur, 25 rue du Docteur Roux, 75724 Paris Cedex 15, France

INTRODUCTION

In modern Immunology, autoimmune diseases (AID) have a profound heuristic value that is not often considered. Every third year, in International Congresses such as this one, we are all exposed to the ever accelerating pace of progress in the analysis of components in the immune system (IS). All the more sobering, therefore, to contemplate AID and realise the extent of our current ignorance on the organization and operation of the IS. Thus, in contrast with other areas of modern medicine, we are today unable to recognise a diseased IS, before the target organ or tissue has been damaged. In other words, detection (and diagnosis ?) of AID is currently done on the basis of non-immunological parameters. Furthermore, we can of course not predict whether or not an IS will develop into an autoaggressive mode of operation, probably because, in this case, the diagnosis would have to be purely immunological. Finally, we know of no scientifically based procedure to correct immunodisfunction and bring a diseased IS back to the normal operation that characterises physiology. Thus, all established therapeutic strategies in use today aim at nonspecifically suppress all immune activities, with little consideration as to the detailed mechanism of disease, and none at all as to the specificities involved, and as to the origin of the process.

There are at least three types of explanations for this contradiction, between the apparently large volume of fundamental knowledge on immunity, and the quite complete ignorance we demonstrate in the field of AID. These are important to distinguish, for they principally determine the approaches we chose, the experiments we do, and the clinical strategies we adopt in dealing with autoimmune patients. One view defends that, in spite of much progress on the structure and function of the immune components and mechanisms, we do not know enough this far. It is necessary to know more, perhaps much more, before we can ever reach an understanding of the whole situation. This view is optimistic on several accounts. First, it satisfies the overwhelming majority of todays' immunologists, well represented here by the Introductory lecture of our Chairman in this Symposium. Second, it defends that we should just continue as we have been doing, without worrying about alternatives and basic questions of approaches, because it expects that if we accumulate enough knowledge on details, these will come together at some point, and will crystallize on a general understanding. Third, it predicts that the larger the current investment in component analysis, the quicker we will come to a solution.

A second category of views, that we heard today from our co-Chairman, contests that position, arguing that this is precisely the approach followed over the last 20 years, which has lead us nowhere. It is contended that we face here a problem of paradigms and theoretical frameworks, which determine our ways of thinking, even if many experimentalists do not realise it. Current views and approaches start from the central assumption that natural tolerance is based on immunological ignorance of self components, and that autoimmunity is necessarily associated with disease. While this has been the undisputed dogma for many years, this view defends that empirical and theoretical evidence exists today to question that paradigm, and to propose instead, that normal immune physiology is actually based on autoreactivity, natural tolerance being the result of self recognition and assertion. It is argued that with little or no knowledge of autoimmune physiology, we will be for ever unable to understand the corresponding pathologies. Moreover, the proposed new paradigm suggests that AID results from deficit, rather than from exacerbation, of autoimmune recognition, and it therefore suggests therapeutic strategies aimed at specific (or nonspecific) stimulation of autoimmunity, instead of its suppression. This view has made its way over the last 10 years or so, but it remains in clear minority among us. Thus, if today it has become acceptable to work on the physiology of autoreactivities, I also think that we are overrepresented in this Symposium, with 33% of the total time.

A third class of explanations for the present contradiction has an epistemological bases, and it addresses the levels of organization and description of biological objects, arguing that we face here a problem of true complexity. This view defends that natural tolerance or AID are global properties of a whole system or organism, that cannot be reduced to any of its components or groups of components. It argues that the current mistake is to attempt to understand higher levels of organization (system or organism) by "local" properties that are explanatory at lower levels of description (molecular or cellular), but cannot be the basis for understanding global behaviours. If tolerance is indeed a systemic property that "emerges" from the organization of the individual components, it follows that progress in this area will necessarily come from approaches that address that organization, namely the structure and function of interclonal relationships, and their impact on the dynamics of immune components at equilibrium with body structures. Although considering them insufficient, this position gives a great importance to component analyses, for it defends that knowledge on local rules is an important step for deriving reasonable propositions in the necessary global approaches.

My colleagues and I take a definite stand within the last two propositions, and have been concerned with deriving global approaches for the analyses of physiological autoreactivities in normal animals (1 - 3). I shall discuss today, some aspects of our work on natural autoantibodies.

SOME REASONS TO STUDY NATURAL ANTIBODIES

Over the last 10 years, we have paid particular attention to murine "natural antibodies" (NAbs), that is, circulating immunoglobulins (Ig) of normal, unimmunized mice. Several reasons justify this concern, but I shall only indicate two of them. The first is purely empirical and it relates to the dramatic therapeutic effects of NAbs, if injected at high doses to autoimmune patients (IVIG). This is the only nonaggressive (nonimmunosuppressive) strategy we have at hand today, and it strongly indicates that it is worthwhile to study the regulatory potential of NAbs and the respective mechanisms. The second reason for our interest in NAbs is theoretical, and it derives from our current conviction that the pool of NAbs in the serum must reflect the global operation of all components in the system, and their relationship with the rest of the body (4, 5). If we would know how to properly analyse the composition and dynamics NAbs, we could perhaps "read" all what is or is going on inside the body, as reflected in the NAbs pool.

We have been concerned with the biology of the cells secreting NAbs (natural plasma cells), their origin, selection, and V-region repertoires, with the reactivities of NAbs in relation to self and nonself structures, with their dynamic behaviours, and with their role in regulating B and T cell repertoires (6 - 12). More recently, we have also paid attention to the alterations of NAbs in infected animals and in AID strains. We have also been experimenting on the mechanisms of action of the therapeutic effects of IVIG (13 - 14). We shall summarise here, some of our results and current interpretations. This is not a final paper on all these questions, but rather a progress report, also containing putative conclusions that must await further experimental proof.

THE BULK OF NATURAL IgM ANTIBODIES, EVEN IN NORMAL INDIVIDUALS, RESULTS FROM INTERNAL, AUTOIMMUNE STIMULATION

One of the very first questions we have addressed concerns the nature of the stimuli driving NAb production. The conventional view was that all serum antibodies result from stimulation by environmental antigens, for autoreactivities were not supposed to be productive but rather, to lead to elimination or inactivation of autoreactive cells. We have taken, therefore, to compare concentrations of serum IgM and numbers of IgM-secreting cells in normal SPF mice, in germ-free animals and in germ-free mice kept for several generations in low molecular weight diets. Antigen- or germ-free mice are severely deprived of environmental antigenic stimuli, as they produce no IgG or IgA in the serum or plasma cells, and are essentially devoid of organized lymphoid tissue in the gut and of alpha-beta T cells in the intestinal epithelium and lamina propria (15, 16). Yet, the levels of IgM production in all three groups of mice are comparable (15). This observation led us to conclude that normal concentrations of IgM NAbs are produced upon stimulation by "internal antigens", be them somatic self structures or other V-regions operating in a functional network.

More recently, we have used a novel immunoblot assay that quantitatively scores antibody reactivities in normal serum against hundreds of self proteins, present in extracts of different organs or tissues and separated in SDS gels (17). Modifications introduced in current methods, allow for the automated reading and data processing of a large number of gels, in manners that are quantitative, and corrected for technical variability. The same method has also been used for scoring reactivities on nonself proteins obtained from bacterial extracts. Generally, the approach is global and sensitive enough to score more than half of all unselected monoclonal antibodies isolated from normal animals, and to distinguish the unique reactivity patterns of essentially all the positive monoclonal antibodies tested so far (Nobrega, unpublished observations). Moreover, the method readily distinguishes individual supernatants from polyclonally stimulated cultures of 10^5 or less normal "unselected" B cells. Data from a number of gels for each serum are statistically treated, as to define patterns of serum NAb reactivities with up to several hundred parameters.

The conclusions from these studies were quite striking (Nobrega et al, submitted). While NAb reactivity patterns in sera from different mouse strains, or even from single mutants and the respective wild-type mice, can be easily separated, we fail to segregate the pattern of reactivities of NAbs from germ-free and conventional mice. We conclude that the bulk of antibody reactivities in both cases is largely the same, and therefore, that even in normal animals exposed to bacterial colonization, most serum IgM NAbs are produced by autologous stimulation. Interestingly, preliminary analyses also suggest that the antigenicity of the diet might have a detectable impact in the composition of NAb reactivity patterns.

THE "IMMUNCULUS" OR THE RELEVANT OCCUPATION OF SHAPE SPACE BY SOMATIC MOLECULAR STRUCTURES

An interesting initial observation using this experimental systems was a clear support for Cohen's suggestion of the "immunological humunculus" (18). We already knew, since Medawar, that the immunological self is not a genetic listing, and the notion was established that "self" is the biochemical listing of the antigenic structures that are present in the organism and available to immune recognition during development (19). As once proposed, for the same molarity, every self structure should count as much in the establishment of the immunological identity (20). This is wrong. As Cohen has forcefully argued, molecular self structures are not treated equally, and the immunological self (the point of view of the immune system) is very distorted, as compared to the biochemical listing. Our experiments show that, from the point of view of NAbs at least, a distorted "immunculus" does exist. Thus, the reactivities of NAbs on a panel of proteins from a given organ, bear no correlation to the abundance of the proteins: some of the self proteins are very well recognized, while others of equivalent size and concentration are paid very little or no attention.

Cohen's proposition further postulates that the "immunculus" is evolutionarily established (18). Our experiments have thus far not addressed that issue, and for us it is equally plausible that the establishment of the areas of "shape space" meaningfully occupied by self structures (that is, where V-region complementarities are present) is an ontogenic and epigenetic process (19).

AUTOREACTIVITY IS A PARAMOUNT CHARACTERISTIC OF NATURAL ABS IN BOTH THE IgM AND THE IgG CLASSES

The above techniques are useful to analyse global reactivity patterns and to compare large "ensembles" on the basis of many parameters, regardless of the nature of the antigenic determinants employed. Because of possible denaturation of the antigenic proteins and their accessibility *in vivo*, the method is not appropriate to ascribe autoreactivities with certainty. Other methods (ELISA and RIA, FACS, immunohistochemistry), however, have been used by us and others to demonstrate that NAbs very frequently react with self structures. This conclusion has been derived for NAbs present in normal serum, and for antibodies secreted by hybridomas isolated from normal unimmunized mice. The notion that autoreactivity is not necessarily associated with pathology, but a constitutive property of normal immune systems is now accepted by most specialists, and this is perhaps the major shift in the dominant trends in Immunology over the last 10 years. Notably, however, other colleagues who are not directly dealing with NAbs, have not yet apprehended the new common sense, and continue to consider the absence of autoreactivities as the basis of natural tolerance.

Much remains to be done, however, in the physiology of NAbs autoreactivities, on their putative roles of neutralizing, redistributing, removing or eliminating self structures that normally turn-over. Much discussion has centered in the past on the overall significance of IgM antibodies of relatively low affinities, that also differ from selected "immune" antibodies by showing a much higher frequency of "multireactivity" with structurally unrelated antigens. [It should be noted, however, that monoclonal "natural antibodies", whatever the extent of their multireactivity, paradoxically show exquisite fine specificities, enough to identify all those studied so far. In other words, multireactive antibodies are very specific, and may distinguish, for example, DNP and TNP, while binding equally well to DNA, actin, serum albumin and other proteins and haptens.]

Many have suggested that those characteristics of natural IgM autoantibodies are incompatible with functional activity, and in this manner explained their lack of pathogenic effects. Such a conventional perspective of natural tolerance, as due to the absence of all functional self-recognition, is still frequent today. Rather than proposing *ad hoc* explanations, we have preferred to directly test for the functional significance of autoreactivities, by conducting a set of simple experiments. Their rational was the following: if the autoreactivities we can score *in vitro* are of no functional significance to the lymphocytes producing the respective antibodies, these should be in a stage of activation that is comparable to that of other B lymphocytes expressing reactivities to nonself antigens, absent from the animal. In contrast, if autoreactivity is physiological, then such lymphocytes must be activated in normal individuals. The alternatives were tested by limiting dilution analysis of normal B cells, separated *ex vivo* into small, resting cells and activated blasts (8). The results showed that several autoreactivities studied were nearly exclusively found among naturally activated B cells, and actually depleted from the resting compartment (8, 21). Testing of congenic strains and transgenic lines of mice ascertained that recruitment of autoreactive B lymphocytes into the activated cell compartment did require the corresponding self-antigen (22). We have interpreted these results as to show that natural plasma cells are indeed activated by productive interactions with the self molecular environment, newly produced B cells with the corresponding reactivities being continuously recruited into activation and NAb secretion. More recent findings have altered some aspects of this interpretation, but not the conclusion that NAb autoreactivity is functionally relevant for the lymphocytes which express it.

This contention has recently been supported by the finding that NAb autoreactivities are also abundantly found among the IgG class in normal serum. Moreover, the work of Avrameas and Kazatchkine and colleagues elegantly demonstrated that natural IgG autoreactivities are largely inhibited by V-region interactions with natural IgM and the IgG present in the same serum (23, 24). These observations oppose the notion that IgG autoantibodies are necessarily associated with autoimmune disease, and extend the physiology of autoantibodies to all isotypes (and affinities ?). The finding of IgG NAbs rises several other questions. First, as IgG production is predominantly T cell dependent, does the finding imply the functional competence of autoreactive helper T cells, as other observations also suggest? Given that germ-free animals do not produce natural IgG antibodies, however, it would seem as if the signals required for isotype switch of natural plasma cell precursors necessarily originated via recognition of non-self. In this case, how are autoreactive B cells activated, given that a stringent screening for specificity takes place in the germinal centers, where isotype switches occur in predominance? Are these conventional cross-reactivities? If so, what is the difference between such normal IgG autoantibodies and the cross-reactive IgG antibody responses to microorganisms that have been implicated in a variety of autoimmune diseases? Finally, the finding that IgM NAbs in the same individual specifically inhibit IgG binding to autoantigens (23) suggests that, at least from the "point of view" of natural IgG repertoires, a variety of self structures are "reproduced" or mimicked by IgM V-regions (1). This might have far-reaching implications for the pathogenesis of autoimmune diseases. As substantiated in other observations by those authors, disease might result from the absence of "neutralizing anti-idiotypic" antibodies, rather than from the production of IgG autoantibodies. This is actually a suggestion first made at the Pasteur Institute by Besredka in 1904, then often considered by "network regulation" models of autoimmune disease, and extensively supported by the therapeutic successes of normal NAbs administered at high doses (24).

AUTOIMMUNE DISEASE AS THE BLURRING OF SELF BOUNDARIES AND THE LOSS OF SELF IDENTITY

We have recently initiated the analysis of NAbs from animals suffering from spontaneous AID, namely lpr/lpr and NOD mice. The results from preliminary observations are suggestive of a few comments. First, systemic disease, such as the murine lupus-like syndromes, seems to be initiated by an expansion of the physiological "immunculus". In other words, no novel reactivities are consistently found, but normal patterns are exaggerated. At these stages, however, there are few, if any, clinical signs of disease. Later in time, when AID is established, a variety of "disease-associated" reactivities appear. Their most striking characteristic is precisely the fact that such reactivities are entirely "outside" the normal "immunculus", and often seem to parallel the abundance of a given autologous protein that is not a physiological component of the immunological self. Aside from the obvious implications for the development of new diagnostic tests, these observations might solve a paradox, standing ever since autoreactive NAbs were identified in normal individuals. If autoantibodies are present in normal individuals, how to explain the significance of diagnostic autoantibodies, useful for many years in clinical practise ? The task is now to precisely delineate the boundaries of self, and the set of reactivities present in physiological N(auto)Abs, such that those appearing "outside the immunculus" can be given the appropriate clinical significance. Interestingly, however, established disease seems to be associated as well, with the loss of certain autoantibody reactivities characteristic of normal individuals. This would in turn suggest that multiparametric approaches to diagnostic are likely to offer the means that the scoring of one or a few autoantibodies did not provide. Moreover, these observations suggest putative mechanisms at the origin of AID. It would seem that autoreactivities falling inside the developmentally established, physiological "immunculus" are organized such that they are included in a mode of operation that results in nonaggressive dynamics and classes of effector functions. In contrast, immune recognition of autologous proteins outside the immunologically defined "self" (the "immunculus") would seem to develop as conventional immune responses to foreign antigenic structures. If, in some respects, this hypothetical conclusion satisfies conventional views, it is rewarding to realise that it is also another way of expressing the notion that "self" is ascertained by the immune system in normal individuals. Perhaps a general solution for AID would be, in this perspective, to use appropriate manipulations (vaccination or passive transfers of NAbs) in order to include a maximum of autologous structures in the "immunculus".

REFERENCES

1 - Coutinho A, Forni L, Holmberg D., Ivars F and Vaz N (1984) From an antigen-centered, clonal perspective of immune responses to an organism-centered, network perspective of autonomous activity in a self-referential immune system. Immunol. Rev., 79 : 151-168

2- Coutinho A (1989) Beyond clonal selection and network. Immunol. Rev. 110 : 63-87.

3- Varela F and Coutinho A (1991) Second generation immune networks. Immunol.Today. 12 : 159-165.

4- Coutinho A and Avrameas S (1992) Speculations on immunosomatics : potential diagnostic and therapeutic values of concepts on immune homeostasis.Scand. J. Immunol., in press.

5- Varela F, Coutinho A and Stewart J (1992) What is the immune network for ? W. Stein and. F. Varela. Thinking about Biology : an invitation to current theoretical biology. Addisson Wesley (SFI Series on Complexity) in press

6- Thomas-Vaslin V, Coutinho A and Huetz F (1992) Origin of natural IgM-secreting cells : reconstitution potentiel of adult bone-marrow, spleen and peritoneal cells. Eur. J. Immunol. 22 : 1243-1251.

7- Freitas A, Pereira P, Huetz F, Thomas-Vaslin V, Pena-Rossi C, Andrade L., Sundblad A, Forni L and Coutinho A (1989) B cell activities in normal unmanipulated mice. P. Del Guercio and J. M. Cruse. B lymphocytes : Function and Regulation. Karger 11 : 1-26.

8- Portnoï D, Freitas,A, Bandeira A, Holmberg D and Coutinho A (1986) Immunocompetent autoreactive B lymphocytes are activated cycling cells in normal mice. J. Exp. Med. 164 : 25-35.

9- Viale AC, Coutinho A and Freitas AA (1992) Differential expression of VH-gene families in peripheral B cell repertoires of newborn or adult IgH congenic mice. J. Exp. Med 175 : 1449-1456.

10- Lundkvist I, Coutinho A, Varela F and Holmberg D (1989) Evidence for a functional network amongst natural antibodies in normal mice. Proc. Natl. Acad. Sci. USA 86 : 5074-5078.

11- Coutinho A, Marquez C, Araujo PMF, Pereira P, Toribio ML, Marcos MAR and Martinez-A. C (1987) A functional idiotypic network of T helper cells and antibodies limited to the compartment of "naturally" activated lymphocytes in normal mice. Eur. J. Immunol. 17 : 821-825.

12- Freitas AA, Viale AC, Sundblad A, Heusser C and Coutinho A (1991) Normal serum immunoglobulins participate in the selection of peripheral B cell repertoires. Proc. Natl. Acad. Soc. 88 : 5640-5644.

13- Sundblad A, Huetz F, Portnoi D and Coutinho A (1991) Stimulation of B and T cells by in vivo high dose immunoglobulin administration in normal mice. J. Autoimmunity 4 : 325-339.

14- Sundblad A, Marcos MAR, Huetz F, Freitas A, Heusser C, Portnoï D and Coutinho A (1991) Normal serum immunoglobulins regulates the numbers of bone marrow pre-B and B cells. Eur. J. Immunol. 21 : 1155-1161.

15- Pereira P, Forni L, Larsson EL, Cooper MD, Heusser C and Coutinho A (1986) Autonomous activation of B and T cells in antigen-free mice. Eur. J. Immunol. 16 : 685-688.

16- Bandeira A, Mota-Santos T, Itohara S, Degermann S, Heusser C, Tonegawa S and Coutinho A (1990) Localization of gamma/delta T cells to the intestinal epithelium is independent of normal microbial colonization. J. Exp. Med. 172 : 239-244.

17- Haury M, Grandien A, Sundblad A, Coutinho A and Nobrega A (1992) Immuno-blot as tool for the analysis of the antibody repertoire, submitted for publication.

18- Cohen IR and Yound DB (1991) Autoimmunity, microbial immunity and the immunological homunculus. Immunol. Today 12 : 105-110.
19- Coutinho A, Coutinho G, Grandien A, Marcos MAR and Bandeira A (1992) Some reasons why deletion and anergy do not satisfactorily account for natural tolerance. Res. Immunol. 143 : 345-354.
20- Vaz N, Martinez-A C and Coutinho A (1984) The uniqueness and boundaries of the idiotypic self. H. Köhler, Cazenave PA and Urbain J. Idiotypy in Biology and Medicine. Academic Press, p. 43-59.
21- Huetz F, Larsson-Sciard EL, Pereira P, Portnoï D and Coutinho A (1988) T cell dependence of "natural" auto-reactive B cell activation in the spleen of normal mice. Eur. J. Immunol. 18 : 1615-1622.
22- Pena-Rossi C, Pereira P, Portnoï D and Coutinho A (1989) MHC-linked and T cell-dependent selection of antibody repertoires. Quantitation of I-E-related specificities in normal mice. Eur. J. Immunol. 19 : 1941-1946.
23- Adib MJ, Ragimbeau S, Avrameas S and Ternynck T (1990) IgG autoantibody activity in normal mouse serum is controlled by IgM. J. Immunol. 145 : 3807-3813.
24- Rossi F, Dietrich G and Kazatchkine M (1989) Anti-idiotypes against autoantibodies in normal immunoglobulins : Evidence for network regulation of human autoimmune responses. Immunol. Rev. 110 : 135-150.

13. Therapeutic Application of Regulatory Networks

Regulation of Regulatory T Cells

M. C. Brunner, D. Caput*, M. R. Helbert, N. A. Mitchison, K. Simon,
J. Sieper and P.Wu

*Deutsches Rheumaforschungszentrum Berlin, c/o Robert Koch Institut, Nordufer 20,
D-1000 Berlin, Germany*
Laboratoire de Biologie Moléculaire, Sanofi Elf BioRecherches, Labége, France

This introductory paper reviews new ideas concerning regulatory T cells, in the fields of hyper-reactivity, cytokine balance, and desequestration of self-antigens. It also summarises an older and more troubled theme in network research, the influence on these cells of IgV regions.

Hyper-reactivity

The sub-division of CD4 T cells according to their CD45 isoform expression has attracted much attention. It is still widely believed that the transition from CD45RA to CD45R0 phenotype represents a one-way street, leading from "naive" to "memory" cells. That view is changing, under the impact of data from animal experiments (Bell and Sparshot 1990, Lightstone et al 1992). These support the alternative view that is summarised in Figure 1: activation causes T cells to become transiently hyper-reactive to further stimulation, in which condition they express a range of activation markers including CD45R0, up-regulated MHC class II (in man and rat, but not mouse), and IL-2 receptors. The acquisition and loss of this condition occurs progressively, with varying kinetics for the different markers. As might be expected, different species have substantially different kinetics: CD45R0 cells probably take months to return to quiescence in man, while in rats they do so in weeks and in mice in days. One crucial experiment is still missing. Human CD45RA T cells do not respond *in vitro* to recall antigens (as encircled for emphasis in the figure), although mouse cells of the same phenotype can do so *in vivo*. We therefore predict that recall-antigen memory, of clonal-expansion rather than hyper-reactivity type, will prove detectable *in vitro* in human CD45RA T cells, but only if they are first activated by one of the non-specific stimuli indicated in Figure 1.

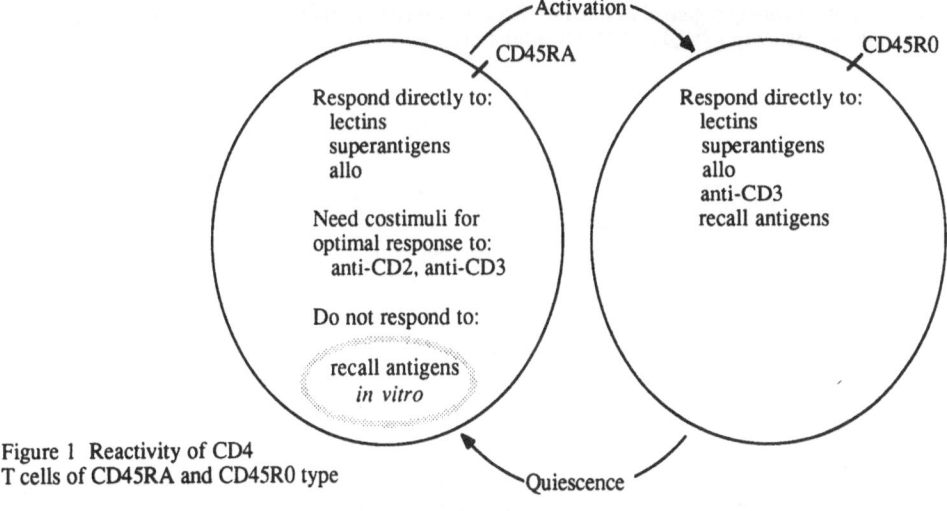

Figure 1 Reactivity of CD4
T cells of CD45RA and CD45R0 type

The cyclic behavior shown in this figure raises many important questions, such as the nature of the cell biology responsible for hyper-reactivity. These ideas may help explain HIV-induced immunodeficiency (Helbert et al 1992). Failure of CD4 T cells to receive adequate stimulation could well account for the progressive loss of these cells, first in the CD45RO pool and then in the CD45RA one. Such a failure would result from a decline in antigen-presentation, brought about by loss or functional impairment of presenting cells, or by an environmental failure such as of glutathione-depleted T cells to respond.

Cytokine balance

Remarkable success has been obtained in extending the concept of balance between TH1 TH2 cells from mouse to man. Increased TH2 activity underlies atopy associated with enhanced IgE and eosinophil levels (Parronchi et al 1992), and cutaneous hypersensitivity reactions (Kay et al 1991). PCR data (Yamamura et al 1991) and proteins secreted by cloned T cells (Salgame et al 1991) indicate that TH1 activity underlies tuberculoid leprosy and TH2, lepromatous leprosy. In arthritis, T cell cloning indicates that TH1 cells predominate in Lyme arthritis (Yssel et al 1991), and in Yersinial (Lahesmaa et al 1992) and Chlamydial (Simon et al 1992) reactive arthritis. This response is clearly appropriate for combatting infections by intracellular organisms, where the arthritis presumably represents a rare and unfortunate over-response.

These studies identify IL-4 as the critical marker of the TH2 population in man. Unfortunately it is expressed at a lower level than, for instance, IFNγ, perhaps because of its coordinate expression with the linked genes for IL-3, IL-5 and GM-CSF. Additional sources of bias are the common use of IL-2 in cellular cloning procedures, and the high concentration of antigen used to drive the clones. For this reason we regard *in situ* hybridization as an essential component of such studies.

Like others, we have encountered problems with mis-identified DNA probes. For that reason we verify their identity by the procedure outlined in Figure 2.

Recombinant DNA is currently used to detect cytokines by Northern blotting and PCR, neither of which can be regarded as a quantitative procedure, and to make measurements by the new procedure of quantitative PCR (Überla et al 1991). We are attempting to develop an RNAase-protection assay in order to provide an accurate picture of the overall cytokine pattern.

In summary, the introduction of new and more quantitative technologies is changing the face of cytokine research, so that methods in common use are rapidly becoming unacceptable. Cytokine balance is now widely regarded as the most important mode of regulation of the immune response, although the sad story related in the next section must sound a note of caution.

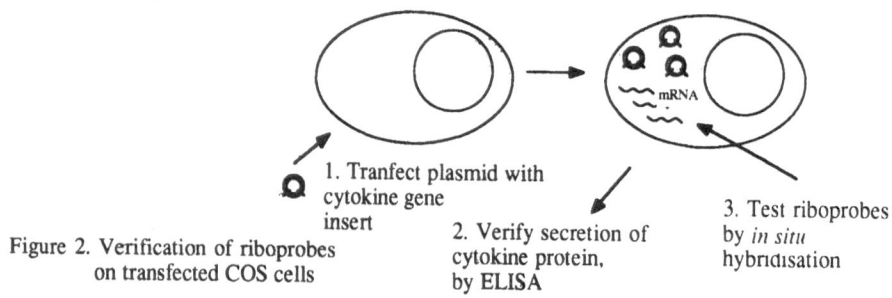

1. Tranfect plasmid with cytokine gene insert

2. Verify secretion of cytokine protein, by ELISA

3. Test riboprobes by *in situ* hybridisation

mRNA

Figure 2. Verification of riboprobes on transfected COS cells

Influence of IgV's

The main question about idiotype-anti-idiotype networks is whether they represent a physiologically important mechanism, i.e. one which is important in the normal working of the immune system while fulfilling its main function of defense, or whether they represent only an unfortunate but inescapable consequence of clonal selection, which can sometimes be demonstrated in manipulated systems but is of no physiological importance. Between these two opposing views lie various compromise positions, such as that networks are important in moulding an effective defensive repertoire even if they do not operate in short-term responses, or that they usefully constrain those self-reactive components of the T cell repertoire that are not eliminated by negative selection in the thymus. No doubt various arguments in favour of physiological significance could be cited, but surely none has attracted more attention than the contention that the T cell repertoire is profoundly influenced by immunoglobulin idioypes. That flagship argument has had its ups and downs over the past two decades. It has attracted so much effort that if it floats, then the ultimate therapeutic value of the network approach cannot be doubted. But if it sinks, then the physiological significance of the network is in jeopardy. From the following brief historical account we conclude that the flagship has passed through stormy seas but has not yet foundered.

The story starts in 1975, when Eichmann and Rajewsky (1975) demonstrated induction of Th and Ts by carefully adjusted treatment of mice with anti-idiotypic (IgV) antibody. At the time, this finding supported the belief, then widely held, that T cell receptor and immunoglobulin variable regions were encoded by the same genes.

Next, two groups independently discovered molecules made by T cells which appeared to be encoded by immunoglobulin VH genes, as judged by formal genetic analysis with IgVH congenic mouse strains (Weinberger et al 1979, Eardley et al 1979). Soon after their first discovery, by means of genetic mapping, these molecules were found to occur on the surface of suppressor T cells, where they could serve as targets for attack by anti-idiotypic antibodies, and also on suppressor factors secreted by these cells. These IgV-restricted molecules then became the subject of intensive study over the next few years. Thus five, seven, three, three, five and five substantial papers emerged in successive years from Benacerraf's laboratory. Along the way many remarkable claims were made, such as of suppressor factors assembling from two peptide chains produced by separate cells, one MHC-restricted and the other IgV-restricted. In a genetic analysis involving minimum perturbation of the immune system, Sherman (1985) found (to her surprise, as she told one of us at the time) that IgV genes exerted almost as profound a control over the TcR repertoire as do MHC genes.

Eventually, molecular cloning of the T cell receptor in 1983 excluded the possibility of encoding the T cell receptor and the immunoglobulin variable regions by the same genes. Nevertheless for a while evidence continued to accumulate in favour of immunoglobulin variable regions influencing the TcR repertoire. Thus in 1984 at least three groups independently reported that B cell depletion, by means of anti-IgM antibody, had such an influence. That year Martinez et al (1984) found an anti-Ig-id antibody that would inhibit the activity of certain Th cells.

Once the original hypothesis of common encoding had been abandoned, the interpretations of this "regulatory influence" became more cautious. Indeed one might have expected investigators to have returned to some of their earlier claims, but that has not happened, apart from the exception noted below. What, realistically, are we left with now? Our review at the last International Congress entertained the possibility that a TcR might recognise directly, without peptide processing, the combining site of another TcR or immunoglobulin (Figure 2 in Mitchison 1989). The information which has since accumulated concerning MHC-peptide-TcR trimolecular interactions surely excludes that possibility. That leaves the following possible explanations: (i) contamination, for instance of T cells with B cells, particularly in the *in vitro* experiments, of T cell products ("TsF","ThF") with APC and B cell products, and of streptococcal carbohydrate with protein; (ii) data selection, particularly when small subjective differences are involved as in DTH and PFC assays; (iii) mistaken inclusion of non-network effects, particularly from anti-IgM treatment; (iv) effects mediated by IgV-derived (and TcR-derived) peptides; (v) effects mediated by anti-IgV-idiotytpic antibodies cross-reacting with TcR's; (vi) novel regulatory pathways, novel Ts receptor molecules. The first three of these explanations amount to mistakes, and we (unlike some eminent immunologists) doubt if they alone could account for all the data. That leaves the latter, more interesting trio, each of which has grave problems. For instance, if commonly occurring cross-reactive anti-IgV-idiotypic antibodies are postulated, it becomes hard to explain how T cell proliferation assays can run in the presence of "normal" serum.

Searching for physiological significance is a challenging task, with the added deterrent in the present case of entrenched and divergent opinions. Perhaps our flagship, with its huge freight of ancient data, will finally be towed into port by some bright young immunologist interested in fundamental mechanisms. Or perhaps exploration of network-oriented therapeutic manipulations (for instance as described at this Congress) will eventually disclose the basis of these B-to-T influences.

A final word, for which N.A.M. alone takes responsibility, concerning the vexed issue of the impact of an immunoglobulin transgene on the repertoire (Weaver et al 1986). The claim was made, but later withdrawn on technical grounds, that such a transgene can profoundly perturb the network. Dr Imanishi-Kari (pers. comm.) has continued to study this system, and now believes that the effect may rather be mediated by recombination between host Ig genes and the transgene, although sequence data to support this possibility are so far lacking. Three comments are appropriate: (i) here we have yet another doubt raised about the impact of IgV's on the network, additional to those mentioned above, (ii) it is good to see doubtful data being reworked by the investigator personally involved, and (iii) we see the internal verification process of science operating as it should. Politicians, journalists, and lawyers should understand that science is an articulated structure of hypotheses, that has no place for, or need of, absolute truth.

Desequestration

Two recent studies suggest that desequestration of self-proteins occurs in animal models of auto-immunity (Lin et al 1991, Lehmann et al 1992). In collaboration with Eva Rajnavölgyi, we have recently obtained direct evidence of this process occurring in collagen-induced arthritis. Lymphocytes taken from mice immunised with chicken or bovine type II collagen are transferred into irradiated, syngeneic hosts, where they can produce antibody specific for the foreign protein.

This they do only if boosted with at least 1 μg of the foreign protein, as has been found with many other transferred responses. In contrast, the transferred cells also include a component which makes antibody specific for self-type II collagen, which goes ahead with production largely independently of boosting. This component, we conclude, can desequestrate collagen from the host's own cartilage. Furthermore this burst of self-driven production soon terminates, presumably as result of the desequestration being brought under control, possibly by regulatory T cells.

Desequestration is an intriguing mechanism, which may play an important roll in autoimmunity triggered by infection. Lehmann et al (1992) regard epitope spreading (Mitchison 1990, in their term "determinant spreading") as a hall-mark of autoimmunity, and we see the testing of these ideas in human auto-immune disease as a major challenge for future research. A minor but intriguing question is whether triggering of antisperm antibody by chlamydial infection (Cunningham et al 1991), presumably one result of desequestration, overlaps in its distribution reactive arthritis, perhaps another result of the same process.

Overlap of regulatory functions

None of the four functions discussed here could be expected to overlap any of the others completely. CD45R0, for example, if it marks hyper-reactivity in the way proposed here, could not possibly identify TH2 cells alone. It is unlikely that the cells which either mediate or prevent desequestration belong uniquely to either cyotkine-secreting set, or are uniquely locked into the network. On the other hand highly significant associations do occur: for example, the predominantly TH1 cells of arthritic synovia mentioned above also predominantly express CD45R0. Other associations (e.g. of CD45R markers with induction or prevention of autoimmunity, of network control with prevention of autoimmunity) are discussed elsewhere in these Proceedings. An important task for future research is to understand these associations, and eventually to exploit them for purposes of therapeutic manipulation.

REFERENCES

Bell EB, Sparshot SM (1990) Interconversion of CD45R subsets of CD4 T cells in vivo. Nature 348: 163-165.

Lightstone EB, Marvel J, Mitchison NA (1992) Memory in helper T cells revealed in vivo by alloimmunizations in combination with Thy1 antigen. Eur J Immunol 22:115-122.

M.R. Helbert, J. L'Age-Stehr and N.A.Mitchison (1992) Consequences of HIV mediated damage to the antigen presenting system. AIDS, submitted.

Parronchi P, De Carli M, Manetti R, Simonelli C, Piccinni MP, Macchia D, Maggi E, Del Prete G, Ricci M, Romagnani S (1992) Aberrant Interleukin (Il)-4 and Il-5 production in vitro by Cd4+ helper t cells from atopic subjects. Eur J Immunol 22: 1615-1620

Kay AB, Sun Ying, Varney V, Gaga M, Durham SR, Moqbel R, Wardlaw AJ, Hamid Q (1991) Messenger RNA expression of the cytokine gene cluster, Interleukin 3 (Il-3), Il-4, Il-5 and Granulocyte / Macrophage Colony-Stimulating Factor in allergen-induced late-phase cutaneous reactions in atopic subjects. J Exp Med 173: 775-778

Yamamura M, Uyemura K, Deans RJ, Weinberg K, Rea TH, Bloom B, Modlin RL (1991) Defining protective responses to pathogens: cytokine profiles in leprosy lesions. Science 254: 277-279

Salgame P, Abrams JS, Clayberger C, Goldstein H, Convit J, Modlin RL, Bloom B (1991) Differing lymphokine profiles of functional subsets of human CD4 and CD8 T cell clones. Science 254: 279-282

Yssel H, Shanafelt MC, Soderberg C, Schneider P, Anzola J, Peltz G (1991) Borrelia burgdorferi activates a T helper type 1-like T cell subset in Lyme Arthritis. J Exp Med 174: 593-601

Lahesmaa R, Yssel H, Batsford S, Luukainen R, Möttönen T, Steinman L, Peltz G (1992) Yersinia enterocolitica activates a T helper type 1-like T cell subset in reactive arthritis. J Immunol 148: 3079-3085

Simon K, Sieper J, Wu P, Seipelt E, Braun J (1992) Unpublished results.

Überla K, Platzer C, Diamantstein T, Blankenstein T (1991) Generation of competitor DNA fragments for quantitative PCR. PCR Meth Appl 1:136-141

Eichmann K, Rajewski K (1975) Induction of T and B cell immunity by anti-idiotypic antibody. Eur J Immunol 5:661-666

Weinberger JZ, Germain RN, Ju ST, Greene MI, Benacerraf B, Dorf ME (1979) Hapten-specific T-cell responses to 4-hydroxy-3-nitrophenyl acetyl. II. Demonstration of idiotypic determinants on suppressor T cells. J Exp Med 150:761-776

Eardley DD, Shen FW, Cantor H, Gershon RK (1979) Genetic control of immunoregulatory circuits. Genes linked to the Ig locus govern communication between regulatory T-cell sets. J Exp Med 150:44-50

Sherman L (1982) Genetic linkage of the cytolytic T lymphocyte repertoire and immunoglobulin heavy chain genes. J Exp Med 156:294-299

Martinez-A C, Pereira P, Bernabe R, Bandeira A, Larsson E-L, Cazenave P-A, Coutinho A (1984) Internal complementaries in the immune system: regulation of the expression of helper T-cell idiotypes. Proc Natl Acad Sci USA 81:4520-4523

Mitchison NA (1989) Is genes in the mouse. Progr. in Immunol. 7:845-852

Weaver D, Reis MH, Albanese C, Costantini F, Baltimore D, Imanishi-Kari T (1986) Altered repertoire of endogenous immunoglobulin gene expression in transgenic mice containing a rearranged mu heavy chain gene. Cell 45: 247-259

Lin, R-H., Mamula, M.J., Hardin, J.A., Janeway, C.A. (1991) Induction of autoreactive B cells allows priming of autoreactive T cells. J Exp Med 173:1433-1439

Lehmann PV, Forsthuber T, Miller A, Sercarz EE (1992) Spreading of T-cell autoimmunity to cryptic determinants by an autoantigen. Nature 358:155-157

Mitchison NA (1990) Unique features of the immune system: their logical ordering and likely evolution. Burger MM, Sordat B, Zinkernagel RM (eds): Cell to Cell Interaction. Basel, Karger, pp 201-214

Cunningham DS, Fulgham DL, Rayl DL, Hansen KA, Alexander NJ (1991) Antisperm antibodies to sperm surface antigens in women with genital tract infection. Am J Obstet Gynecol 164:791-796

Therapeutic Anti-clonotypic Vaccines

H. Kohler[1], S. Muller[2], M. Chatterjee[3], and K. A. Foon[4]

[1]IDEC Pharmaceuticals Corporation, La Jolla, CA 92037
[2]San Diego Regional Cancer Center, San Diego, Ca 92121
[3]Roswell Park Cancer Center, Buffalo, NY 14263
[4]Scripps Clinic and Research Foundation, La Jolla, CA 92037

INTRODUCTION

The original network hypothesis of Niels Jerne (1974) has a built-in therapeutic application as a vaccine which was first recognized by Eichmann and Rajewsky (1975). Subsequently, a formal approach to the preventive or therapeutic utility of anti-idiotypic antibodies was developed which is best known as the so-called "internal image antigen" concept (Jerne 1982 and Bona 1984). A great deal of experimental support for this concept has been obtained in animal models (as summarized in Kohler 1989). Recently, the concept has been moved to clinical settings in therapeutic trials of certain cancers (Herlyn 1987; Kageshita 1988; Mittelman 1992) and HIV-1 infection (Kang 1992).

We have recently revisited the anti-id as an antigen surrogate and proposed a revised concept (Kohler 1989). We would like to continue the process of re-thinking the internal image concept with an emphasis on clinical application. This appears to be justified by the continued interest in therapeutic interventions using anti-id antibodies for active immunization protocols.

The anti-id therapy approach is rooted in an immunobiological view of the immune response. It does not depend on the high-technology of modern biotechnology rather it aims to harness the natural potential which is built into the immune system. By doing this, it will avoid many side-effects and potential dangers often associated with biotechnology techniques such as gene and lymphokine therapies.

The Problems with the Old Internal Image Concept.

The evidence against the internal image concept comes from two types of findings. The first to be discussed is biological in nature and derived from experiments using anti-ids as antigen. One of the earliest observations made using anti-id was the failure of certain anti-ids to induce specific immunities which otherwise carried all the typical immunochemical properties of an internal image. Our findings, that in a murine tumor system, only one out of seven anti-ids was able to induce anti-tumor immunity, is an example (Raychaudhuri 1989). Similar experiences were made by other investigators and, because of negative results, often were not reported. Another example is the failure of anti-CD4 antibodies to induce anti-HIV antibodies against the CD4 binding site epitope of the HIV-1 envelope (Healey 1992). This example is particularly instructive, since the anti-CD4 used must be considered to be an internal image of the CD4-site epitope because CD4 is a potent inhibitor of HIV-1 infection. While some so-called internal image anti-ids failed as antigens, other anti-ids which did not type as internal image were able to induce specific immunities (Huang 1986 and Schick 1987). These anti-ids recognize near-binding site idiotopes because antigen is only a partial inhibitor of their binding; they were previously termed Ab2gamma, distinguishing them from the internal image Ab2beta and the non-internal image Ab2alpha (Bona 1984). Both kinds of results with Ab2alpha, Ab2beta and Ab2gamma clearly demonstrate that the distinction of different Ab2s is biologically not useful.

The failure of certain Ab2s as antigens may also be selective for different species. A given Ab2, for example, can induce the designated Ab3 response in rabbits but not in mice. Similar selective responses may be observed in outbred populations showing individual variations of the Ab3 response. This species or individual response variability is additional evidence for the non antigen character of the anti-ids.

Ab2s have been shown to induce cell-mediated immunities, such as DTH (Nelson 1987 and Raychaudhuri 1987) or T cell proliferation (Bogen 1986 and Saeki 1989) which are MHC restricted. These results would appear to support the concept of Ab2 as antigen, particularly since the T cell idiotope was recognized at the peptide level in one study (Bruck 1986). However, in other systems, the T cell response induced by Ab2 looked more like an anti-clonotypic antibody induced non-MHC restricted response (Ertl 1984).

Finally, there are observations of antigen specificity differences between the Ab1 and the Ab2 induced Ab3. For example, we have observed that the reactivity of a CEA-specific Ab1 differs from a monoclonal Ab3 induced by Ab2 in staining of tumor tissue sections (Bhattacharya-Chatterjee 1990). Similar epitope shifts from the original Ab1 to the Ab3 were seen in other systems (Viale 1989). These findings cannot be easily explained by Ab2 mimicking antigen.

Taking these observations into consideration, the biological response to immunization with Ab2s is not compatible with an internal antigen mimicry by anti-id. In this manuscript we will briefly review structural data and considerations which are equally difficult to reconcile with simple antigen mimicry.

Suppose an Ab2 folds like the original antigen; one would expect that the fit of Ab2 and Ab1 would be identical to the fit of antigen and Ab1. Thus, the fit between Ab1 and Ab2 would be a measure for the fidelity in antigen mimicry by Ab2. Accordingly, Ab2s which bind with high affinity to Ab1 would be expected to be a better internal image antigen. In our animal tumor system we have compared the affinities between Ab1 and different Ab2s and have not seen correlations of the affinities with the biological activity of the Ab2 as antigen (Raychaudhuri 1990). A weak or even absent molecular mimicry of antigen by Ab2s further supported the rather weak sequence homologies between the Ab2 and antigen in the few cases where such data was available (Bruck 1986 and Raychaudhuri 1990). An understanding of the underlying molecular contacts involved in an imperfect mimicry by Ab2 can be obtained from the recent x-ray studies of an id-anti-id complex in a lysozyme binding system (Amit 1986). This particular Ab2 made contact to Ab1 involving 13 amino acids in its variable regions; only seven of these 13 Ab1/Ab2 contacts were also involved in binding of the Ab1 to its antigen lysozyme. The imperfect overlap between the lysozyme contacting residues and the anti-id-contacts may be unique to this particular Ab2 and other Ab2s may be making contacts more resembling the contacts to antigen; however, the structure of the anti-lysozyme Ab1/Ab2 complex may also show the limitations of possible molecular mimicry of antigen by Ab2s.

In another recently published study (Garcia 1992) on the structure of an antibody complex with the small peptide hormone angiotensin II led to some interesting speculations on the molecular aspects of internal image anti-ids. The anti-peptide antibody was an Ab3 raised against a so-called Ab2beta, which immunochemically mimicked the peptide antigen. A comparison of the variable sequence regions of the Ab1 and the Ab3, which were nearly identical, indicated that they were derived by somatic mutations from common V genes. These data was interpreted as evidence that the Ab2 functioned as the true antigen image of the angiotensin. While this was a reasonable conclusion important information was missing: (i) the structure of the Ab1-peptide complex was not available and therefore the binding of the peptide to Ab1 and Ab3 could not be compared. (ii) the structure of the Ab2 was also not available, preventing a comparison of the Ab2 to the structure of the antigen; and (ii) the conformation of the peptide in solution was likely linear and not defined which showed that the binding to antibodies of the Ab3 or Ab1 type was the result of induced fit. Assuming that the same energy requirements which allow the induced fit of a random peptide conformation into the deep and tight antigen binding site fit also apply for the internal image Ab2, one would have to assume a very loose and flexible CDR region to be capable of an induced fit into the Ig receptor on B cells. Typically CDRs are structurally well defined and form a binding site in concert with other CDR, i.e. they are not free-standing and independent structures. Disregarding these objections to the internal image possibility of an Ab2 mimicking a peptide antigen, it might be possible that CDRs can mimic antigenic conformations of small peptide antigens, while similar mimicry of large protein antigens remains highly unlikely, as pointed out by Garcia, et. al (1992).

Whatever limits nature may have imposed on antibodies to mimic antigens, it appears from the studies on the Ab2-induced immune responses that structural restrictions in mimicry are of only secondary importance to the potential of anti-ids to induce desired immune responses. In the following we will develop a concept which by-passes the question of internal image and addresses the problem strictly at the level of population dynamics of immunocompetent lymphocytes.

The Concept of Anti-clonotypic Stimulation.

If structural internal image qualities of anti-ids are not required for inducing Ab3 responses with identical or similar antigen specificity as the original Ab1 and anti-ids can function as surrogate antigens or vaccines, what then are the underlying mechanisms for such biological effects? Anti-ids function as anti-clonotypic antibodies, like anti-clonotypic TCR antibodies, recognizing three-dimensional idiotope structures on Ig receptors. Two requirements must be satisfied before a significant antigen-specific Ab3 response can be induced. First, the Ab2 targeted idiotope must be expressed on Ig receptors (or antibodies) either being part of the binding site or being co-expressed with the binding site. Second, B cell clones with this particular combination of idiotope positive and antigen-specific receptors must have a significant clone size in the responding individual. If the size of this clone is small compared to other clones expressing the idiotope without association with the antigen specificity, the Ab3 response which is also antigen-specific will remain a minute fraction of the total Ab3 response.

It follows from these considerations that the clonal composition in the Ab3 response is the critical factor determining the so-called internal image response induced by an Ab2. Exceptions to this may occur in the few instances where a true internal image exists along the line of linear peptide epitope, as discussed above.

From these considerations it is imperative in the selection of effective Ab2s to aim for anti-ids which exhibit a high degree of cross-reactivity in the disease target population. For example, we have recently described (Muller 1991 and Wang 1992) a broadly cross-reactive idiotope in HIV-1 infected individuals which is found in approximately 70% of seropositive individuals. However, often it is difficult to obtain such broadly shared anti-ids. In these situations it might be possible to pre-select a therapy target subpopulation on the basis of pre-existing circulating id+ antibodies. Stimulation of such B cell clones already expanded by the disease should produce high titers of antibodies expressing the target idiotope and reacting with the given antigen.

Examples of Clinically Used Anti-clonotypic Antibodies

Tumor-specific Anti-Id: T-cell tumor gp37 target: Acute lymphoblastic leukemia (T-ALL) and cutaneous T cell lymphoma (CTCL) are human cancers affecting children and adults. Advanced CTCL is basically incurable, while T-ALL can be cured in children with chemotherapy. Among different tumor-associated antigens for these tumors a gp37 antigen has been characterized using the monoclonal antibody SN2 (Bhattacharya-Chatterjee 1987).   gp37 is highly specific for T-cell leukemia/lymphoma and was not detected on normal tissue or T lymphocytes. Two anti-Ids were generated against SN2 and used to induce Ab3 in rabbits and mice. A monoclonal Ab3 was also prepared. Monoclonal Ab1 and Ab3 were compared and no differences were detected in staining characteristics of tumor and normal tissue sections. Among six CTCL specimens tested, five were reactive to both the Ab1 and the Ab3 (Bhattacharya-Chatterjee).

A small clinical trial was initiated with the 4DC6 Ab2 which involved twelve patients. Two patients have been immunized with an alum precipitated Ab2 and both produced Ab3 responses. In one patient the cutaneous tumors began to regress after the 3rd immunization reaching a 95% reduction in tumor after the 7th and final injection.   The patient has remained stable with a 95% PR for more than nine months without any additonal therapy.

Tumor-specific Anti-Id: Carcinoembryonic Antigen (CEA) Target: Another human tumor-associated antigen, CEA, was selected as the target for an anti-idiotype therapy approach. CEA is expressed in a broad spectrum of human tumors and has been used to monitor tumor progression. There exists several anti-CEA antibodies and we selected the 8019 antibody (Bhattacharya-Chatterjee 1990) for generating Ab2s. Two Ab2s were identified which reacted with paratope-related idiotopes of the 8019 Ab1 (Bhattacharya-Chatterjee 1990). The 3H1 Ab2 was used to immunize mice and monoclonal Ab3s were obtained. The Ab3 precipitated CEA (180,000 MW) material in Western blots. The immunoreactivity of Ab1 and Ab3 were compared with tumor tissues. Interestingly, differences in the staining with Ab1 and Ab3 were observed. The 8019 Ab1 stained tumor cells and secreted mucinous material, while the Ab3 only stained tumor cells and not mucin. This epitope shift in Ab3 avoiding binding to circulating CEA would be therapeutically advantageous in the Ab3 response of Ab2 treated tumor patients.

An approach to improve the clinical response to Ab2 active immunotherapy was briefly mentioned above and entails a pre-selection of the patient population on the basis of pre-existing idiotype positive Ig in the serum. This approach is an attempt to match the anti-Id antibody with the patient disease-expanded idiotype repertoire (Idiotype Matching). We assume, that in patients who express a corresponding matching Id, Ab2 stimulation would be able to further expand B or T cells expressing the matching Id.

In this regard, we made the interesting observation that in twenty-five out of 116 randomly selected colon carcinoma patients' sera a significant level of Ig reactive with Ab2 could be detected. In only two out of 100 normal sera were similar findings were seen. Since the binding of patients' Ig was to the Fab fragment of the Ab2 it must be considered idiotype-specific. This finding of Id-matched Ab2 reactions in colon carcinoma patients suggest that they might be suitable candidates for active anti-clonotypic therapy. A clinical trial with 3H1 is currently being initiated.

Tumor-specific Anti-Id: Melanoma-associated Proteoglycan Target: Melanoma-associated proteoglycan is expressed on the majority of melanoma tumor cells and to some extent on keratinocytes and endothelial cells. Antibodies against MGP have anti-tumor effects. Anti-Ids were generated against a murine anti-MPG (Chattopadhyay 1992) and shown to induce MPG specific Ab3 responses in rabbits and non-human primates. Of interest are the studies in cynomolgus monkeys because their Ab3 response would be indicative for the Ab3 response in patients undergoing anti-Id immunotherapy. Ab3 antibodies bound to MPG expressing human cells lines, like Colo38, but not to MPG negative lines. Furthermore, purified Ab3 could precipitate MPG material similar to Ab1 precipitated MPG. Ab3 also had the interesting biological activity of inhibiting the invasion of tumor cells into a basement membrane matrix, which is representative for *in vivo* tumor invasion and tumor metastasis. The Ab2 in these monkeys was a far better antigen then a human MPG positive cell line which failed to induce MPG specific antibodies. Thus, the Ab2 was able to break a natural tolerance to MPG. Currently clinical trials are underway with the MPG specific Ab2s in melanoma patients.

HIV-1 Specific Anti-Ids: Besides cancer, infectious diseases are also experimentally tested targets for anti-clonotypic therapy. A large number of animal and human infections with viral, bacterial and parasitic infectious agents have been used to demonstrate a potential clinical applicability of the idiotype approach. We will briefly review our experiments with HIV-1 infection.

Two types of idiotopes associated with anti-HIV-1 antibodies were identified. The first type consists of a family of restricted idiotopes found on anti-gp120 antibodies identified by a panel of different Ab2s (Chamat 1992). The other type is represented by one Ab2 which recognizes a broadly shared and common idiotope associated with antibodies against different HIV-1 proteins (Muller 1991 and Wang 1992).

Restricted, gp120 CD4-site Associated Target: The aim for generating these Ab2s was to have tools for identifying gp120 epitopes which are conserved on different HIV-1 strains. Epitopes in or around the attachment site for CD4 were chosen because of their functionally imposed restriction on the degree of permissible variability. Polyclonal anti-gp120 antibodies were isolated from sera of healthy HIV-1 infected individuals (Kang 1991). These anti-gp120 were then separated into CD4 site-specific and non-CD4 site antibodies. Anti-Ids were generated against the CD4 site-specific Ab1. Using these anti-Ids as immunoabsorbents Id binding Ig was purified from Id positive sera. These purified Id+ anti-gp120 antibodies were then characterized for epitope specificity and biological activity as virus neutralizing antibodies. Most of the Id+ anti-gp120 antibodies showed broadly neutralizing activities (Kang 1991). Because of this biological activity associated with Id+ Ig from infected individuals attempts were made to use one of these Ab2s as antigen surrogate to induce neutralizing anti-gp120 antibodies. Rabbits and monkeys were immunized with 3C9 Ab2. Anti-gp120 antibodies were detected in the 3C9 immune sera which had broadly neutralizing activities (Kang 1992). A randomized clinical trial is currently underway to evaluate the safety and the potential of 3C9 to boost and broaden the neutralizing antibody titers in healthy seropositive individuals.

It is important to understand, that this therapeutic Ab2 was not selected as an internal image of the CD4 site epitope(s). Actually the 3C9 Ab2 does not bind to CD4, demonstrating that it is not an internal image of the CD4 attachment site of the viral envelope. Instead, the 3C9 was selected because it had biological activity to induce neutralizing antibodies in two species. Thus, the 3C9 functions as anti-clonotypic antibody for B cells committed to produce virus neutralizing antibodies.

Common-shared HIV-1 infection associated idiotope target: A totally different approach was used here to produce an extremely common anti-Id which could be used to identify antibodies induced by HIV-1 infection. A pool of high anti-gp120 titer sera from healthy infected donors was used to induce monoclonal anti-Ids. Among several Ab2 1F7 was isolated and characterized further (Muller 1991 and Wang 1992). 1F7 detects idiotopes on anti-HIV-1 antibodies with different specificities such as anti-gp120, anti-p24 and anti-RT. The 1F7 marker is highly specific for HIV-1 infected individuals and not detected in normal sera.

The therapeutic value of this broadly shared Ab2 needs to be explored. Several possibilities can be envisioned. Since approximately 2/3 of seropositive sera contain 1F7 positive anti-HIV antibodies one could use 1F7 as immunoabsorbent to purify anti-HIV-1 antibodies. Such anti-HIV-1 could then be used in passive immunotherapy of infected patients. Alternatively, the 1F7 Id marker may be associated with anti-HIV-1 antibodies which fail to neutralize the patient's own virus strains, because of the emergence of neutralizing escape variants. In this situation it might be advantageous to break the established B cell dominance (Kohler 1992) by anti-idiotypic suppression using the 1F7 anti-Id. Experiments to test these modalities are underway.

CONCLUSION

In the preceding sections we have summarized our experimental data using anti-idiotypic antibodies in active immunotherapy approaches. We have described the generation of therapeutic Ab2s in several cancers and in HIV-1 infection. Preliminary and limited observations in clinical trials are encouraging and merit continuation and expansion of such trials.

We have also discussed the theoretical and biological mechanisms which produce the specific immunological responses and therapeutic effects. Our interpretation favors the concept of anti-clonotypic stimulation of committed B cell clones over the conventional "internal image" of anti-Id action. Clonotypic stimulation by anti-Id can also be effective for T cells expressing clonotypic markers.

Within the clonotypic stimulation model the number of B cells and their expansion potential becomes a critical factor for inducing an effective immune response. Therefore, we propose that in the selection of the anti-clonotypic antibody for a given disease the degree of clonotypic reactivity (idiotypic match) is the most important criteria for achieving a therapeutically effective immune response.

REFERENCES

Amit AG, Mariziuzza RA, Phillips SEV and Poljak RJ. (1986) Three-dimensional structure of an antigen-antibody complex at 2.8 A resolution. Science 233:747-753

Bogen B, Malissen B, Haas W (1986) Idiotype-specific T cell clones that recognized syngeneic immunoglobulin fragments in the context of class II molecules. Eur J Immunol 16:1373-1378

Bona CA and Kohler H (1984) Anti-idiotypic antibodies and internal images In: Probes for Receptors Structure and Function 4:141-150

Bruck C, Co MS, Slaoui M, Gaulton GN, Smith T, Fields BN, Mullins JI, Greene MI (1986) Nucleic acid sequence of an internal image-binding monoclonal anti-idiotype and its comparison the sequence of the external antigen. Proc Nat Acad Sci USA 83:65788-6582

Bhattacharya-Chatterjee M., Chatterjee S.K., Vasile S., Seon B.K. and Kohler H. (1988) Idiotype vaccines against human T cell acute lymphoblastic leukemia. II.

Generation and characterization of a monoclonal idiotype cascade (Ab1, Ab2 and Ab3). J. Immunol. 141:1398-1403

Bhattacharya-Chatterjee M, Mukerjee S, Biddle W, Foon, KA and Kohler H (1990) Syngeneic monoclonal anti-idiotype antibody as a potential network antigen for human carcinoembryonic antigen. J Immunol 145:2758-2765

Bhattacharya-Chatterjee M, Pride MW, Seon BK and Kohler H (1987) Idiotype vaccines against human T cell acute lymphoblastic leukemia. I. Generation and characterization of biologically-active monoclonal anti-idiotopes. J. Immunol. 139:1354

Chamat, S, Nara P, Berquist L, Whalley, AP, Morrow WJW, Kohler, H. and Kang C-Y. (1992) Epitope Diversity in the CD4 Attachment Site of gp120 influences neutralizing antibody activity. J Immunol 149: 649-654

Chattopadhyay P, Starkey J, Morrow JW and Raychaudhuri S. (1992) Murine monoclonal anti-idiotypic antibody breaks unresponsiveness and induces a specific antibody response to human melanoma-associated proteoglycan antigen in cynomolgus monkeys. Proc Natl Acad Sci USA 89:2684-2688

Eichmann K and Rajeswsky K (1975) Induction of T and B cell immunity by anti-idiotypic antibody. Eur J Immunol 5:661-666.

Ertl HCJ and Finberg RW (1984) Sendai virus-specific T-cell clones: Induction of cytolytic T cells by an anti-idiotypic antibody directed against a helper T-cell clone. Proc Natl Acad Sci USA 81:280-2854

Garcia KC, Desiderio SV, Ronco PM, Verroust PJ and Amzel LM (1992) Recognition of Angiotensin II: Antibodies at different levels of an idiotypic network are superimposable. Science 257:528-531

Garcia KC, Ronco PM, Verroust PJ, Brunger AT and Amzel LM (1992) Three-dimensional structure of an Angiotensin II-FAb complex at 3A: Hormone recognition by an anti-idiotypic antibody. Science 257:502-507

Healey DG, Dianda L, Beverley PCL (1992) A "network antigen" for human CD4: A murine monoclonal anti-idiotype to Leu-3a induces an anti-CD4 response in naive mice J Immunol 148:821-826

Herlyn D, Wettendorf M, Schmoll E, Ilopoulos D, Schedel D, Dreikhausen U, Jaschke AH, Scriba M and Koprowski H (1987) Anti-idiotype immunization of cancer patients: Modulation of the immune response. Proc Natl Acad Sci USA 84:8055-8059

Huang J-H, Ward RE and Kohler H. (1986) Idiotope antigens (Ab2alpha and AB2beta) can induce *in vitro* B cell proliferation and antibody production. J Immunol 137:770

Jerne NK (1974) Towards a network theory of the immune system. Ann Immunol (Paris) 125C:373-389.

Jerne NK, Roland J and Cazenave P-A (1982) Recurrent idiotypes and internal images EMBO J 1:243-248

Kageshita T, Chen ZJ, Kim J-W, Kusama M, Kekish UM, Trulillo T, Temponi M, Mittelman A, and Ferrone S (1988) Murine anti-idiotypic monoclonal antibodies to syngeneic antihuman high molecular weight-melanoma associated antigen monoclonal antibodies: Development, characterization, and clinical application. Pigment Cell Res 1:185-191

Kang C-Y, Nara P, Chamat S, Caralli V, Chen A, Nguyen M-L, Yoshiyama H, Morrow, J, Ho, D and Kohler, H. (1992) Monoclonal anti-clonotypic antibody elicit broadly neutralizing antibodies in monkeys. Proc Natl Acad Sci USA 89:2546-2550

Kang C-Y, Nara P, Chamat S, Caralli V, Ryskamp T, Haigwood N, Newman R and Kohler, H (1991) Evidence for non-V3 specific neutralizing antibodies that interfere with gp120/CD4 binding in human immunodeficiency virus-1 infected humans. Proc Natl Acad Sci USA 88:6171-6175

Kang, C-Y, Nara, P, Morrow, WJW, Ho, H and Kohler, H (1992) Anti-idiotype monoclonal antibody elicit broadly neutralizing anti-gp120 antibodies in monkeys. Proc Natl Acad Sci USA 89:2546-2550

Kohler H, Goudsmit J. and Nara P (1992) Clonal Dominance in HIV-1 Infection. J AIDS. in press

Kohler H, Kaveri S, Kieber-Emmons T, Morrow WJW, Muller S and Raychaudhuri S (1989) Idiotypic networks and nature of molecular mimicry: An Overview. Methods Enzymol 178:3-35

Kohler H, Kieber-Emmons T, Srinivasan S, Kaveri, S., Morrow, W.J.W., Muller S, Kang C. and Raychaudhuri S (1989) Revised Immune Network Concepts. Clin Immunol. and Immunopathol. 52:104,

Mittelman A, Chen Z.J, Yang H, Wong, G.Y, Ferrone, S. (1992) Human high molecular weight melanoma-associated antigen (HMW-MAA) mimicry by mouse anti-idiotypic monoclonal antibody MK2-23: Induction of humoral anti-HM W-MAA immunity and prolongation of survival in patients with stage IV melanoma. Proc Natl Acad Sci USA 89:466-470

Muller S, Wang H-T, Kaveri S, Chattoppadhyay S and H Kohler (1991) Generation and specificity of monoclonal anti-idiotypic antibodies against human HIV-specific antibodies. J Immunol. 147: 933-941

Nelson KA, George E, Swenson C, Forstrom JW, Hellstrom I and Hellstrom K (1987) Immunotherapy of murine sarcomas with auto-anti-idiotypic monoclonal antibodies which bind to tumor specific T cells J Immunol 139:2110-2114

Raychaudhuri S, Kang C-Y, Kaveri S-V, Kieber-Emmons T and Kohler, H (1990) Tumor idiotypic vaccines VII. Analysis and correlation of structural, idiotypic and biological properties of protective and non-protective Ab2s. J Immunol 145:760-767

Raychaudhuri S, Kohler H, Saeki Y, and Chen J-J. (1989) Potential role of anti-idiotype antibodies in active tumor immunotherapy. Critical Reviews in Oncology/Hematology 9(2):109-124

Raychaudhuri S, Saeki Y, Chen J-J. and Kohler, H. (1987) Tumor specific idiotype vaccine III. Induction of T helper cells by anti-idiotype and tumor cells. J Immunol 139:2096

Saeki Y, Chen J-J, Shi L. and Kohler H. (1989) Characterization of "regulatory" idiotope-specific T cell clones to a monoclonal anti-idiotypic antibody mimicking a tumor-associated antigen (TAA) J Immunol 142:1046-1052

Schick M.R, Dreeseman G.R, Kennedy R.C (1987) Induction of an anti-hepatitis B surface antigen response in mice by non internal image (Ab2beta) anti-idiotypic antibodies. J Immunol 138:3419-3425

Viale G, Flamini G, Grassi F, Buffa R, Natali PG, Pelagi M, Leoni F, Menard S, Siccardi AG (1989) Idiotypic replica of an anti-human tumor-associated antigen monoclonal antibody. J. Immunol.143:4338-4344

Wang H-T, Muller S and Kohler H (1992) Human Anti-HIV antibodies with different epitope specificity share common clonotypic markers, Eur J Immunol 22:1749-1755

Suppression of Organ-Specific Autoimmune Diseases by Oral Administration of Autoantigens

Howard L. Weiner, Ariel Miller, Samia J.Khoury, Z. Jenny Zhang, Ahmad Al-Sabbagh, Stanley A. Brod, Ofer Lider, Paul Higgins, Raymond Sobel, Makoto Matsui, Mohamed Sayegh, Charles Carpenter, George Eisenbarth, Robert B. Nussenblatt, and David A. Hafler

Center for Neurologic Diseases and Laboratory of Immunogenetics and Transplantation Brigham and Women's Hospital and Harvard Medical School, 75 Francis Street, Boston, Massachusetts 02115 and Laboratory of Immunology, National Eye Institute, National Institutes of Health, Bethesda Maryland 20205 and The Joslin Diabetes Center, One Joslin Place, Boston, Massachusetts 02215

One of the primary goals in developing effective therapy for autoimmune diseases is to specifically suppress autoreactive immune processes without affecting the remainder of the immune system. Autoimmune diseases involve the presence of autoreactive clones that have not been deleted in the thymus and thus these cells must be inactivated in the periphery. We have been investigating antigen-driven peripheral immune tolerance as a means to suppress autoimmune processes using the oral route of antigen-exposure to the immune system because of its inherent clinical applicability. An effective and long-recognized method of inducing immunologic tolerance is the oral administration of antigen, which was first demonstrated by Wells for hen's egg protein.[1] The mechanism by which orally administered antigen induces tolerance most probably relates to the interaction of protein antigens with gut-associated lymphoid tissue (GALT) and the subsequent generation of regulatory or suppressor T cells.[2] The two primary points of contact of orally administered antigen are Peyer's patches and gut epithelial cells, the latter of which overlie intraepithelial lymphocytes. Investigators have reported that specific suppressor cells can be found in the Peyer's patches following oral administration of antigen and that such cells then migrate systemically.[2] Intestinal epithelial cells express class II antigens on their surface and thus have the capacity to function as antigen-presenting cells.[3] Furthermore, it has been shown that human gut epithelial cells preferentially stimulate CD8+ cells *in vitro* which can function to suppress *in vitro* immune responses.[4] Although most investigators have reported that the generation of antigen-specific suppressor T cells is the primary mechanism responsible for mediating oral tolerance, other reported mechanisms include anti-idiotypic antibodies, immune complexes and biologically filtered antigen (reviewed in REF. 5).

In order to test whether feeding an autoantigen could suppress an experimental autoimmune disease, the Lewis rat model of experimental autoimmune encephalomyelitis (EAE) was studied.[6] Animals were fed increasing amounts of myelin basic protein (MBP), either once or three times prior to immunization with MBP in complete Freund's adjuvant. With increasing dosages, the incidence and severity of disease was suppressed. In addition, proliferative responses of lymph node cells to MBP was also suppressed. Antibody responses to MBP were decreased but not as dramatically as proliferative responses. Thus, it appears that oral tolerance to MBP preferentially suppresses cellular immune responses. EAE is associated with inflammatory cells that accumulate in the central nervous system. In animals fed myelin basic protein, there was a marked decrease in the number of cells infiltrating the nervous system. In order to determine the length of protection following feeding, animals were fed three times prior to immunization and then immunized at weekly intervals. Animals were protected for approximately 2-3 months after this feeding regimen.

It is known that there are specific regions of MBP that are encephalitogenic in the Lewis rat. In order to determine whether nonencephalitogenic portions of MBP could suppress EAE, both fragments of MBP and synthetic peptides were orally administered.[6] Suppression of disease occurred by feeding fragments or synthetic peptides prior to immunization with BP/CFA. There was some suggestion that nonencephalogenic fragments were more potent in generating suppression than encephalitogenic fragments, although more investigation is needed in this area. We also noted that feeding bovine MBP was able to suppress EAE in the Lewis Rat and in the strain 13 guinea pig, showing cross-species tolerization.[7] Nonetheless in a recent series of experiments, it appears that homologous MBP is a more potent oral tolerogen for EAE than heterologous MBP.[8]

The majority of studies related to mechanisms of oral tolerance suggest that active suppression is generated following exposure of antigen via the gut.[5] In order to test this mechanism in the EAE model, mesenteric lymph nodes and spleen cells were adoptively transferred from animals fed myelin basic protein into naive animals that were then immunized with MBP/CFA. We found that protection could be adoptively transferred and that such protection was dependent on CD8+ T cells.[9] Splenic and mesenteric T cells from fed animals were also able to suppress both *in vitro* proliferative responses and antibody production by MBP-primed popliteal lymph node cells. *In vitro* suppression was also mediated by CD8+ T cells. The suppression was antigen specific in that adding T cells from animals fed MBP suppressed MBP responses, but not responses to mycobacteria.

To further study the mechanism of oral tolerance in the EAE model, we studied the ability of cells from MBP fed animals to suppress proliferative responses of an MBP or OVA line using a transwell system. We found that cells from MBP tolerized animals could suppress either an MBP or an OVA line across the transwell provided that the cells from the fed animals were triggered *in vitro* with the oral tolerogen.[10] The factor responsible for the suppression was identified as TGFβ. CD8 cells from animals fed MBP release TGFβ *in vitro* when stimulated with the fed antigen. Furthermore, *in vivo* administration of anti-TGFβ antibody abrogates oral tolerance. In addition, natural recovery in untolerized animals is prolonged in animals treated with anti-TGFβ antibody.[11] Detailed immunohistology was performed in animals orally tolerized with myelin basic protein and in animals naturally recovering from EAE.[12] Brains from OVA fed animals at the peak of disease showed perivascular infiltration with activated mononuclear cells which secreted the inflammatory cytokines IL-1, IL-2, TNF-α, IFN-γ, IL-6 and IL-8. Inhibitory cytokines TGFβ and IL-4 and prostaglandin $E_2(PGE_2)$ were absent. In MBP orally tolerated animals there was a marked reduction of the perivascular infiltrate and downregulation of all inflammatory cytokines. In addition, there was upregulation of the inhibitory cytokine TGFβ. In MBP + LPS orally tolerized animals[13, see below], in addition to upregulation of TGFβ and reduction of inflammatory cytokines there was an enhanced expression of IL-4 and PGE_2, presumably secondary to activation of an additional population of immunoregulatory cells. In control-fed animals have recovered (day 18) staining for inflammatory cytokines diminished and there was an appearance of TGFβ and IL-4. These results suggest that the suppression of EAE either induced by oral tolerization or that which occurs during natural recovery related to the secretion of inhibitory cytokines or factors that actively suppress the inflammatory process in the target organ.

EAE in the Lewis rat is usually an acute monophasic illness. In order for oral administration of autoantigens to have clinical applicability, it must be effective in patients in whom the disease process has expressed itself and in whom activated autoreactive cells already exist. Thus, experiments were performed in relapsing models of EAE.[7] A relapsing Lewis rat model of EAE occurs following injection of spinal cord homogenate plus adjuvant. Feeding MBP to animals following recovery from their first attack significantly suppressed the second attack and decreased histologic manifestations of the disease. In addition, cell-mediated immunity as measured by DTH and anti-myelin antibody responses were also suppressed. A more chronic model of EAE occurs in the strain 13 guinea pig. In a series of experiments, guinea pigs were injected with white matter homogenate plus adjuvant and upon recovering from the first attack, were fed 10 mg of bovine myelin or BSA, three times weekly over a three-month period. In animals fed the bovine myelin preparation there was a diminution in frequency of attacks, and a decrease in demyelination in the spinal cord and certain portions of the white matter. These results demonstrate that oral administration of myelin antigens can suppress chronic relapsing EAE and have direct relevance to the therapy of human demyelinating disorders such as multiple sclerosis.

The generation of an immune response in animals often requires concomitant administration of an adjuvant to enhance antibody production and to generate T cell-mediated responses. Antigens administered per os generally result in systemic hyporesponsiveness even though there may be local stimulation of IgA antibody.[5] We thus initiated a series of experiments to determine whether a tolerogenic adjuvant could be found for oral tolerance to MBP. It has been suggested that colonization of the gastrointestinal tract by LPS-producing bacteria is one of the requirements for oral tolerization as LPS converts germ-free mice to sensitivity to oral tolerance induction.[14] Furthermore, LPS-nonresponsive C3H/HeJ mice are unable to be orally tolerized to sheep red blood cells, [15,16] and LPS and dextran sulfate have been reported to enhance oral tolerance for DTH responses to picryl chloride in mice,[17] although LPS did not affect DTH responses to ovalbumin.[18] We found that the oral administration of LPS enhanced the suppressive effects of myelin basic protein on EAE.[13] LPS given without MBP had no effect on EAE and LPS given subcutaneously with orally administered MBP tended to abrogate oral tolerance. The enhanced suppression of EAE associated with oral administration of LPS was also associated with a decrease in DTH responses to MBP, but not with decreased anti-MBP antibody responses. Further experiments demonstrated that it was the lipid A moiety of LPS that was active in enhancing oral tolerance to MBP in the EAE model.[13] The mechanism of action of LPS is unknown but as demonstrated in our studies of cytokines in the brains of MBP+LPS fed animals [12] presumably relates to enhanced generation of cellular suppressive mechanisms by gut associated lymphoid tissue.

In addition to oral exposure to antigen, the body is constantly exposed to inhaled antigen which contacts the mucosal immune system at the level of the bronchial associated lymphoid tissue. In order to determine the effect of this route of antigen exposure on EAE, MBP was aerosolized to Lewis rats on days -10, -7, -5, -3 prior to immunization with MBP in Freund's adjuvant and on days 0, +2, +4 following immunization. Five ml of PBS containing 5 mg/ml of MBP was aerosoled to a group of 5 rats in an airtight plastic cage over a 10-minute period. Aerosolization of MBP completely abrogated clinical EAE: incidence in controls, 20/20; in treated group, 0/20. CNS inflammation and DTH and antibody responses to MBP were also significantly reduced in aerosol-treated animals. Aerosolization of histone, a basic protein of similar weight to MBP had no effect. Disease was also suppressed with one aerosol treatment on day -3. Aerosolization was more effective than oral administration of MBP over a wide dose range (0.005-5 mg) suggesting that protection via aerosolization was not merely secondary to gastric absorption of aerosolized antigen. Splenic T cells isolated from aerosoled animals adoptively transferred protection to naive animals immunized with MBP. Aerosolization of MBP to animals with relapsing EAE after recovery from the first attack decreased subsequent attack severity, and MBP antibody and DTH responses. Thus, aerosolization of an autoantigen is a highly potent method to downregulate an experimental T cell-mediated autoimmune disease and suggests that exposure of antigen to lung mucosal surfaces preferentially generates immunologic tolerance.[16]

In order to further assess oral tolerization as a method to treat autoimmune diseases, studies were performed in experimental autoimmune uveitis and in adjuvant arthritis. Oral administration of S-antigen (S-Ag), which is a retinal autoantigen that induces experimental autoimmune uveoretinitis (EAU), prevented or markedly diminished the clinical appearance of S-Ag-induced disease as measured by ocular inflammation.[20] Furthermore oral administration of S-Ag also markedly diminished uveitis induced by the uveitogenic M and N fragments of the S-Ag. Oral administration of S-Ag did not prevent MBP-induced EAE. *In vitro* studies demonstrated a significant decrease in proliferative responses to the S-Ag in lymph node cells draining the site of immunization from fed versus nonfed animals. Furthermore the addition of splenocytes from S-Ag-fed animals to cultures of a CD4+ S-Ag-specific cell line profoundly suppressed the cell line's response to the S-Ag, whereas these splenocytes had no effect on a PPD-specific cell line. The antigen-specific *in vitro* suppression was blocked by anti-CD8 antibody demonstrating that suppression was dependent on CD8+ T cells.

Oral tolerance was also tested in the adjuvant arthritis model. Previous investigators have demonstrated suppression of collagen-induced arthritis by feeding collagen type II.[21-22] We studied adjuvant arthritis (AA), another well-characterized and more fulminant form of experimental arthritis.[23] Adjuvant arthritis is induced by injection of Mycobacterium tuberculosis (MT) into the base of the tail. Attempts to suppress adjuvant arthritis by oral administration of MT were not successful. Nonetheless we found that oral administration of chicken collagen type II (CII) given at a dose of 3 mg per feeding on days -7, -5, and -2 before disease induction consistently suppressed the development of AA. A decrease in delay-type hypersensitivity responses to CII was also observed that correlated with suppression to AA. AA was optimally suppressed by 3 and 3 mg CII, variably by 300 mg, and not by 0.3 mg or 1 mg. Oral administration of collagen type I also suppressed AA; only minimal effects were seen with collagen type III. Suppression was antigen specific in that feeding collagen type II did not suppress EAE, and feeding MBP did not suppress AA. Suppression of AA could be adoptively transferred by T cells from CII-fed animals and could be obtained when CII was fed after disease onset. These results suggest that autoimmunity to CII may have a pathogenic role in AA. Alternatively, suppression of AA by type II collagen may be related to the phenomenon of antigen-driven bystander suppression.[10] Bystander suppression in which oral administration of an antigen different than the immunizing antigen has been demonstrated in the EAE model in which orally administered MBP suppresses PLP induced disease in the SJL mouse.[28]

NOD diabetic mice spontaneously develop an autoimmune form of diabetes associated with insulitis. This is a naturally occurring disease and the autoimmune nature of the disease is suggested by lymphocytic infiltration of the islets of Langerhans which precedes the destruction of insulin producing beta cells. A variety of immunomodulatory treatments have been studied in the NOD mouse and immunosuppressive therapy that affects T cell function has been successful. To test oral tolerance as a mode of therapy we administered porcine insulin at a dose of 1 mg orally twice a week for five weeks and then weekly until one year of age.[24] The severity of lymphocytic infiltration of pancreatic islets was reduced by oral administration of insulin and there was a delay in the onset of diabetes and a decreased incidence of diabetes in animals followed for one year. As expected, orally administered insulin had no metabolic affect on blood glucose levels. Furthermore, splenic T cells from animals orally treated with insulin adoptively transferred protection against diabetes, demonstrating that oral insulin generates active cellular mechanisms that suppress disease. Additional studies have demonstrated the ability to suppress insulitis administering insulin peptides or the A or B chain of insulin. Given the mechanism of antigen-driven bystander suppression and the role of TGFβ in oral tolerance, our results do not definitively implicate autoreactivity to insulin as a pathogenic mechanism in the NOD mouse. Oral insulin may act by inducing T cells via the gut which migrate to the pancreas and release TGFβ or other suppressive cytokines when triggered by insulin at the target organ.

In order to further test oral tolerance as a mechanism to suppress immune reactions, models of allografts were tested. In an initial series of experiments, we fed splenocytes from Wistar Furth rats to Lewis rats and studied the accelerated allograft rejection model. We found that oral tolerance to spleen cells prevents sensitization by skin grafts and transforms accelerated rejection of vascularized cardiac allografts to an acute form typical of unsensitized recipients. In addition, the mixed lymphocyte response *in vitro* and delayed-type hypersensitivity responses were suppressed following oral administration of antigen.[25] In a second series of experiments we have found induction of immunity and oral tolerance with polymorphic class II major histocompatability complex allopeptides in the rat. Inbred Lewis rats were immunized or fed class II synthetic MHC allopeptides. In vivo these animals developed delayed-type hypersensitivity responses. Furthermore, oral administration of the allopeptide mixture daily for 5 days before immunization reduced DTH responses both to the allopeptide mixture and to allogeneic splenocytes. This reduction was antigen-specific.[26] Thus, oral tolerance may be of benefit in downregulating alloreactivity associated with transplantation.

In order to further study antigen-driven tolerance we compared oral tolerance to intravenously administered MBP in both actively induced EAE and EAE adoptively induced by the transfer of an MBP reactive T cell line. Spleen cells from orally tolerized animals suppressed adoptively-transferred EAE when co-transferred with encephalitogenic cells or when injected into recipient animals at a different site at the time encephalitogenic cells were transferred. This suppression was mediated by CD8+ T cells, correlated with suppression of DTH responses to MBP, and was associated with decreased inflammation in the spinal cord. Unlike oral tolerization, spleen cells from IV tolerized animals did not suppress adoptively-transferred EAE when co-transferred with encephalitogenic cells. MBP peptides were then utilized to further characterize differences between IV and oral tolerization in the actively-induced disease model. Both orally and intravenously administered MBP suppress actively-induced EAE. However, EAE was only suppressed by prior IV tolerization with the encephalitogenic MBP peptide 71-90, but not with the non-encephalitogenic peptide 21-40, whereas prior tolerization with 21-40 did suppress actively-induced EAE when administered orally. These results suggest that the dominant mechanism of suppression associated with IV tolerization to MBP in EAE is related to the elicitation of clonal anergy whereas oral tolerization suppresses primarily by the generation of active suppression.[27] Depending on the manner in which MBP is administered orally, anergy may be generated.[29] Large doses of MBP given in a fashion that prevents digestion in the stomach, may induce anergy by direct absorption of MBP into the bloodstream via the gut.

To determine whether oral administration of a foreign protein could induce suppression of immune responses to that fed antigen in humans, two healthy subjects were given 30 mg of keyhole limpet hemocyanin (KLH) orally every other day for 2 to 3 months. 7-day proliferative responses to KLH and the precursor frequency of KLH-reactive cells were serially studied over a 6 month period. To test for suppressor factors released by cells following oral tolerization, supernatants were collected from blood mononuclear cells (MNC) cultured with media, KLH or tetanus toxoid (TT). Precursor frequencies of KLH-reactive cells were decreased in association with feeding, while 7-day proliferation assays varied with time. Supernatants from KLH but not TT or media stimulated MNC suppressed autologous and allogeneic responses to another antigen (mumps). This suppressor activity was dose-dependent and attenuated with anti-TGFβ antibody.[30]

CONCLUSIONS

1. Oral administration of autoantigens suppresses experimental autoimmune diseases (EAE, EAU, AA, CIA, NOD diabetes) in a disease and antigen-specific manner.

2. Suppression can be adoptively transferred by CD8+ T cells which act by releasing TGFβ following antigen-specific triggering. TGFβ is present in the target organ of the fed antigen.

3. "Antigen-driven tissue-directed" suppression occurs following oral administrastion of an antigen from the target organ even if it is not the disease-inducing antigen (bystander suppression).

4. Synthetic peptides can induce oral tolerance. Tolerogenic epitopes of MBP may be different from the encephalitogenic epitope.

5. Active suppression can be demonstrated in Peyer's patches 48 hours after feeding. Disease protection lasts approximately 2 months.

6. Passage of antigen through the stomach may facilitate and be necessary for oral tolerance.

7. Intravenous tolerance with MBP is mechanistically different than oral tolerance and may represent clonal anergy as opposed to active suppression.

8. Tolerance following oral administration of antigens may also involve anergy due to passage of oral antigen into the bloodstream.

9. Graft rejection and the MLR can be suppressed by oral administration of allogenic cells or MHC peptides.

10. Chronic relapsing EAE can be suppressed by oral administration of myelin antigens after disease expression.

11. LPS given orally enhances oral tolerance to MBP in EAE.

12. Aerosol administration of MBP suppresses EAE and of collagen type II suppresses AA and CIA.

13. Oral administration of KLH to humans suppresses cell-mediated responses to KLH.

14. Pilot clinical trials in multiple sclerosis, rheumatoid arthritis and uveitis have been initiated. No toxicity or exacerbation of disease was observed. Decreased cellular reactivity to MBP and S-antigen occurred in multiple sclerosis and uveitis. Double-blind trials are currently in progress to establish clinical efficacy in patients with early relapsing remitting multiple sclerosis at the Brigham and Women's Hospital in Boston, in rheumatoid arthritis utilizing orally administered collagen type II (David Trentham, Beth Israel Hospital, Boston) and in uveitis using S-antigen (Robert Nussenblatt, National Eye Institute, NIH).

REFERENCES

1. Wells H (1911) Studies on the chemistry of anaphylaxis. III. Experiments with isolated proteins, especially those of hen's egg. J Infect Dis 9:147
2. Mattingly J & B Waksman (1978) Immunologic suppression after oral administration of antigen. 1. Specific suppressor cells found in rat Peyer's patches after oral administration of sheep erythrocytes and their systemic migration. J Immunol 121:1878
3. Santos LMB, O Lider, J Audette, SJ Khoury & HL Weiner (1990) Characterization of immunomodulatory properties and accessory cell function of small intestinal epithelial cells. Cell Immunol 127:26-34
4. Mayer L & R Shlien (1987) Evidence for function of Ia molecules on gut epithelial cells in man. J Exp Med 166:1471
5. Mowat, A (1987) The regulation of immune responses to dietary protein antigens. Immunol Today 8:193
6. Higgins, PJ & HL Weiner (1988) Suppression of experimental autoimmune encephalomyelitis by oral administration of myelin basic protein and its fragments. J Immunol 140:440-445
7. Brod SA, A Al-Sabbagh, RA Sobel, DA Hafler & HL Weiner Suppression of experimental autoimmune encephalomyelitis by oral administration of myelin antigens. IV. Suppression of chronic relapsing disease in the Lewis rat and strain 13 guinea pig. Ann Neurol In press
8. Miller A, O Lider, A Al-Sabbagh & HL Weiner (1992) Suppression of experimental autoimmune encephalomyelitis by oral administration of myelin basic protein. V. Hierarchy of suppression by myelin basic protein from different species. J Neuroimmunol 39:243-250
9. Lider O, LMB Santos, CSY Lee, PJ Higgins & HL Weiner (1989) Suppression of experimental autoimmune encephalomyelitis by oral administration of myelin basic protein.II. Suppression of disease and in vitro) immune responses is mediated by antigen-specific CD8+ T lymphocytes. J Immunol 142:748-752
10. Miller A, O Lider & HL Weiner (1991) Antigen-driven bystander suppression following oral administration of antigens. J Exp Med 174:791-798
11. Miller A, O Lider, A Roberts, MB Sporn & HL Weiner (1992) Suppressor T cells generated by oral tolerization to myelin basic protein suppress both in vitro and in vivo immune responses by the release of TGF-ß following antigen specific triggering. PNAS 89:421-425,12
12. Khoury SJ, WW Hancock & HL Weiner (in press) Oral tolerance to myelin basic protein and natural recovery from experimental autoimmune encephalomyelitis are associated with down-regulation of inflammatory cytokines and differential upregulation of TGF-β, IL-4 and PGE expression in the brain. J Exp Med
13. Khoury SJ, O Lider, A Al-Sabbagh & HL Weiner (1990) Suppression of experimental autoimmune encephalomyelitis by oral administration of myelin basic protein. III. Synergistic effect of lipopolysaccharide. Cell Immunol 131:302-310
14. Wannemuehler MJ, H Kiyono, JL Babb, SM Michalek & JR Mcghee (1982) Lipopolysaccharide (LPS) regulation of the immune response: LPS converts germfree mice to sensitivity to oral tolerance induction. J Immunol 129:959-965
15. Kiyono, JR Mcghee, MJ Wannemuehler & SM Michalek (1982) Lack of oral tolerance in C3H/HeJ mice. J Exp Med 155: 605-610
16. Michalek SM, H Kiyono, MJ Wannemuehler, LM Mosteller & JR Mcghee (1982) Lipopolysaccharide (LPS) regulation of the immune response: LPS influence on oral tolerance induction. J Immunol 128:1992
17. Newby TJ, CR Stokes & FJ Bourne (1980) Effects of feeding bacterial lipopolysaccharide and dextran sulphate on the development of oral tolerance to contact sensitizing agents. J Immunol 41:617-621

18. Mowat AM, MJ Thomas, S Mackenzie & DMV Parrott (1986) Divergent effects of bacterial lipopolysaccharide on immunity to orally administered protein and particulate antigens in mice. Immunology 58:677

19. Weiner HL, A Al-Sabbagh & R Sobel (1990) Antigen driven peripheral immune tolerance: suppression of experimental autoimmune encephalomyelitis (EAE) by aerosol administration of myelin basic protein. FASEB J (Abstr.) 4(7):2102

20. Nussenblatt RB, RR Caspi, R. Mahdi, C-C Chan, R Roberge, O Lider & HL Weiner (1990) Inhibition of S-antigen induced experimental autoimmune uveoretinitis by oral induction of tolerance with S-antigen. J Immunol 144:16891695

21. Nagler-Anderson C, LA Bober, ME Roslnson, GW Siskind & GJ Thorsecke (1986) Suppression of type II collagen-induced arthritis by intragastric administration of soluble type 11 collagen. Proc Natl Acad Sci USA 83:7443

22. Thompson HSG & NA Staines (1986) Gastric administration of type 11 collagen delays the onset and severity of collagen-induced arthritis in rats. Clin Exp Immunol 64:581

23. Zhang JZ, CSY Lee, O Lider & HL Weiner (1990) Suppression of adjuvant arthritis in Lewis rats by oral administration of type II collagen. J Immunol 145:2489-2493

24. Zhang ZJ, L Davidson, G Eisenbarth & HL Weiner (1991) Suppression of diabetes in NOD mice by oral administration of porcine insulin. PNAS 88:10252-10256

25. Sayegh MH, ZJ Zhang, WW Hancock, CA Kwok, CB Carpenter & HL Weiner (1992) Down-regulation of the immune response to histocompatibility antigens and prevention of sensitization by skin allografts by orally administered alloantigen. Transplantation 53:163-166

26. Sayegh MH, SJ Khoury, WW Hancock, HL Weiner & CB Carpenter (in press) Induction of immunity and oral tolerance with polymorphic class II MHC allopeptides in the rat. PNAS

27. Miller A, AJ Zhang, M Prabhu Das, RA Sobel & HL Weiner (1992) Active suppression vs. clonal anergy following oral or I.V. administration of MBP in actively and passively induced EAE. Neurology 42(Suppl 3):301

28. Al-Sabbagh A, A Miller, RA Sobel & HL Weiner (1992) Suppression of PLP induced EAE in the SJL mouse by oral administration of MBP. Neurology 42(Suppl 3):346

29. Whitacre CC, IE Gienapp, CG Orosz & D Bitar (1991) Oral tolerance in experimental autoimmune encephalomyelitis. III. Evidence for clonal anergy. J Immunol 147:2155-2163.

30. Polanski M, M Matsui, A Miller, SJ Khoury, DA Hafler & HL Weiner (1992) Oral tolerization to keyhole limpet hemocyanin in humans. FASEB 6(5):1700

TCR Peptide Therapy in Autoimmunity[1]

Arthur A. Vandenbark[*], George Hashim[**], and Halina Offner[*].

[*]Neuroimmunology Research 151D, V.A. Medical Center, and Depts. of
Microbiology and Immunology and Neurology, Portland, OR, 97201
[**]Dept. of Microbiology and Surgery, St. Luke's Roosevelt Hospital Center and
Columbia University, New York, NY.

INTRODUCTION

T cell-mediated autoimmunity may be viewed as the selection and expansion
of "self" reactive clones that are not deleted during thymic maturation.
The thymic negative selection process presumably does not function because
organ specific autoantigens are virtually absent, and one may assume that
potentially pernicious T cells are continually replenished from stem cells
throughout the life of an individual. The presence of autoreactive T
cells in the circulation of clinically normal individuals indicates that
natural mechanisms effectively regulate these T cells. The goal of this
manuscript is to describe and evaluate one such regulatory mechanism
directed at epitopes found on germline T cell receptor V region sequences.

TREATMENT OF ACUTE EAE

The paralytic disease, Experimental Autoimmune Encephalomyelitis (EAE), is
mediated by T lymphocytes specific for central nervous system (CNS) myelin
antigens, including basic protein (MBP) and proteolipid protein (PLP). In
Lewis rats, the pathogenic cells tend to utilize a common V region gene
(Vβ8.2) and thus common germline sequences in their T cell receptor (TCR)
for antigen. A synthetic peptide corresponding to residues 39-59 of rat
Vβ8.2 was highly immunogenic, and could induce anti-TCR peptide T cells
and antibodies that could protect against EAE (Vandenbark et al. 1989).
When injected after onset of EAE, Vβ8.2-39-59 or Vβ8-44-54 prevented
disease progression and speeded recovery. This rapid treatment effect was
due to a boosting of a natural anti-TCR peptide response that was induced
as a consequence of TCR over-expression related to the EAE disease process
(Offner et al. 1991).

Lewis rat T cells specific for the major encephalitogenic epitope of MBP,
residues 72-89, over-express Vβ8.2 and utilize a common ASP-SER motif in
the third complementarity determining region (CDR3)(Gold et al. 1991). T
cells specific for a secondary encephalitogenic epitope, residues 85-99,
over-express Vβ6 and utilize a common ARG-GLY motif in CDR3 (Gold et al.
1992). The biased expression of Vβ8.2 or Vβ6 was most pronounced in the
CNS among activated, IL-2 responsive T cells, but was weakly reflected in
the cerebrospinal fluid (CSF). The regulatory idiotope found on Vβ8.2-39-
59 was also present on Vβ6-39-59, thus allowing either peptide to regulate

1 Supported by The Department of Veterans Affairs, USA, DHHS grants
NS23221, NS23444, NS21466, and XOMA Corporation.

both encephalitogenic T cell specificities. Our working hypothesis is that the regulatory T cells and antibodies recognize naturally processed TCR epitopes that are expressed in association with MHC class I or II molecules on the encephalitogenic T cell surface (Offner et al. 1992). Prevention or treatment of EAE with Vβ8-44-54 reduced the frequency of MBP-72-89 specific T cells in the periphery and the CNS as well as CNS inflammation (Vandenbark et al. In press), but not the percentage of Vβ8.2+ T cells. These findings support the hypothesis that anti-TCR immunity regulates the function but does not delete T cells bearing Vβ8.2.

TREATMENT OF CHRONIC-RELAPSING EAE

The regulatory effects of TCR peptides were also evaluated in the SJL/J mouse, in which EAE induced by a PLP peptide (residues 139-151) is relapsing, progressive and demyelinating. The T cell response to PLP-139-151 is more heterogeneous than in rats or most other mouse strains, partially due to the deletion of nearly half of the Vβ genes, including Vβ8. Encephalitogenic T cells bearing Vβ4 and Vβ17 specific for PLP-139-151 have been reported, and on this basis, SJL/J mice were treated with Vβ4 + Vβ17 peptides on the first day of onset of EAE. As in Lewis rats, the severity of the initial episode of EAE in SJL/J mice was reduced. Moreover, the treated mice had fewer relapses, minimal progression, and a reduced demyelination. These data demonstrate that T cell responses involving multiple V genes and complex histopathology can be regulated using combinations of TCR peptides.

HUMAN T CELL RESPONSES TO MBP

The human disease multiple sclerosis (MS) may also involve encephalitogenic T cells directed at myelin antigens. If so, it may be feasible to utilize TCR peptides to regulate these responses. Because of its documented immunogenicity in humans and its widespread encephalitogenic activity, MBP was chosen for our initial studies as a target antigen in progressive MS patients. Our intention was to compare T cell responses to MBP in MS patients and controls, with the hope that meaningful differences could be established in the frequency and TCR V gene usage of MBP reactive T cells. Biased expression of TCR V genes would provide the rationale to apply the principles of TCR peptide therapy described in rats and mice with EAE.

Elevated Frequency of Myelin Reactive T Cells in MS

In animals, the severity of passively transferred clinical EAE is T cell dose dependent. In vivo, the degree of sensitization can be quantitated using the limiting dilution assay which determines the number of antigen-specific T cells in a mixed cell population. Using this assay, we observed that the estimated frequency of MBP-specific T cells increased from approximately 0.1/100,000 cells in blood or lymph nodes of unimmunized rats, to approximately 1/100,000 cells in rats paralyzed with EAE (Vandenbark et al. In press). In the CNS, the MBP-specific T cell frequency rose from undetectable levels to >50/100,000 cells just prior to disease onset, and then declined rapidly during recovery.

The average MBP T cell frequency in MS blood was estimated to be 0.61/100,000, a level 4-5 times higher than in normal donors, neurologic controls, and rheumatoid arthritis patients (Chou et al. 1992 and Fig. 1

top). The frequency of PLP peptide-reactive cells in blood was approximately 0.1/100,000 cells in all the patient groups, suggesting lack of sensitization, whereas the frequency of Herpes virus specific T cells was approximately 10/100,000 cells in each group (Fig. 1). These data demonstrate that MBP-reactive T cells occur at a higher frequency in the blood of MS patients than in control subjects. However, this frequency was lower than for recall antigens, explaining perhaps, the variable results obtained in previous studies. Although it is unknown whether circulating MBP-reactive T cells are clinically important, it is striking that the estimated level in MS patients approximates the frequency described above in the blood of Lewis rats with EAE.

In MS CSF, we found at least 37% of activated IL-2/IL-4 responsive T cells isolated were specific for MBP or a PLP peptide, unlike OND subjects in whom only 5% of isolates were myelin antigen reactive (Chou et al. 1992). The relative frequency of MBP reactive T cells was estimated as 22/100,000 CSF cells, almost 20 times higher than in OND patients (Fig. 1 bottom). Both MS patients and OND possessed similar frequencies (13-15/100,000) of T cells responsive to MBP fragments but not intact MBP ("cryptic" epitopes). The significance of this CSF subpopulation is unknown, but similar MBP peptide-specific T cell clones from rats were not encephalitogenic. The relative frequency of PLP peptide-reactive T cells

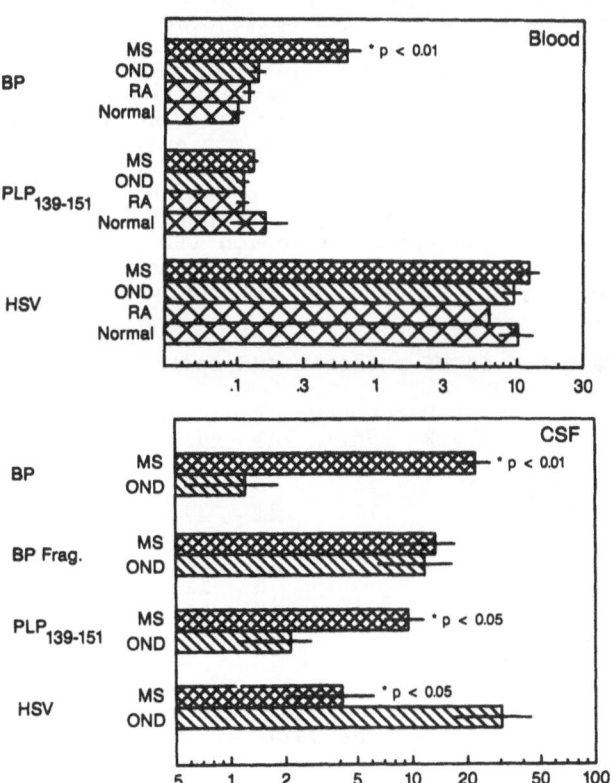

Figure 1. Estimated frequencies of MBP and PLP-peptide specific T cells in blood and CSF.

was estimated as 10/100,000 cells, 5 times higher than in OND patients (Fig. 1). Again, it is striking that the frequency of MBP reactive T cells in MS CSF, proximal to the affected tissue, was similar to the frequency of blood T cells stimulated by vigorous vaccination protocols (eg. to Tetanus toxoid) and to the frequency of MBP reactive T cells found in the CNS of rats with EAE. In two recent studies using a novel technique based on IFN-γ production (Olsson et al. 1990 and Sun et al. 191), even higher frequencies of MBP- and PLP-reactive T cells were detected in CSF.

In summary, MS patients appear to have an increased frequency of MBP-specific T cells in their blood and, more importantly, activated MBP- and PLP-specific T cells in the CSF. While these results do not establish a definite role for MBP- and PLP-specific T cells in MS, they are consistent with the hypothesis that MS is an autoimmune disease caused by myelin-reactive T cells.

TCR Vβ Gene Expression.

The preferential use of Vβ8.2 and Vβ6 by encephalitogenic rodent T cells raises the important question as to whether V region gene biases can also be detected in human MBP-specific T cells. Our analysis of MS blood T cell clones revealed an over-expression of Vβ5.2 and to a lesser degree of Vβ6.1 genes in the TCR of MBP reactive cells compared to non-MBP reactive T cells from the same patients (Kotzin et al. 1991), or to MBP-reactive T cells from normal donors that over-expressed Vα2, Vα15, Vβ7, and Vβ14 (Table 1). Vβ5.2 was expressed by T cells with different MBP epitope specificities, in accordance with the notion that the mechanism responsible for biased V gene selection is not epitope driven. The over-utilization of Vβ5.2 and Vβ6.1 is even more striking in light of recent data demonstrating a significant over-expression of TCR Vβ5.2 and Vβ6 genes in the CNS plaques of DR2/Dw2+ MS patients (Oksenberg et al. Submitted). In several instances, a common LEU-ARG-GLY CDR3 motif was noted in brain plaque-derived material and in human or rat MBP-specific T cell clones, suggesting the presence of MBP-specific clones in MS plaque tissue.

Table 1. TCR V gene use in MS patients and controls.

	N	Percentage of Clones Expressing:						
		Vβ5.2	Vβ6.1	Vβ7	Vβ14	Vβ18	Vα2	Vα15
MS Blood	(46)	59	20	0	0	0	14	7
Normal Blood	(42)	5	10	21	33	7	59	41
MS CSF	(41)	2	7	29	12	15	44	10
MS CSF (Cryptic)	(9)	0	0	0	11	22	44	0
ONDCSF (Cryptic)	(8)	0	0	25	13	0	63	13

Surprisingly, MBP-specific T cells from MS CSF over-expressed different V region genes, including Vα2, Vβ7, and Vβ18, but not Vβ5.2 and Vβ6.1 genes that were over-expressed by clones from MS blood (Table 1). It is interesting that Vα2 and Vβ7 were also over-expressed in MBP peptide-specific (cryptic) CSF T cell clones from OND patients, as well as in the blood of normal donors mentioned above. The differences in blood versus CSF V gene use in response to MBP in MS patients suggests that these are discrete T cell subpopulations, and it is currently unknown whether one is more clinically relevant than the other. This problem is currently being addressed in both the Lewis rat and SJL/J mouse EAE models.

A recent study by Wucherpfennig et al. 1990, suggested that T cells from HLA-DR2+ donors specific for the 84-102 epitope of MBP tended to utilize $V\beta17$, whereas T cells specific for the 149-170 epitope tended to utilize $V\beta12$ and to a lesser extent, $V\beta14$. A second study by Ben-Nun et al. 1991, found different TCR V gene biases ($V\beta2$, $V\beta12$ and $V\beta15$) that varied among HLA types. Additionally, Richert et al. (personal communication) observed a $V\alpha8.2$ bias in a single MS patient. Still other reports have not detected TCR V region gene biases in MBP-specific T cells (Martin et al. 1992, Wekerle, personal communication). At present, it is unclear why the results from different laboratories are so divergent, although some of the differences can be accounted for by patient selection or technical procedures.

In summary, MS patients are similar to animals with EAE in their response to MBP. The frequencies of MBP reactive T cells in blood and CSF appear to be sufficient to induce inflammation, and there is over-utilization of $V\beta5.2$ and $V\beta6.1$ in the TCR of MBP-specific T cells from the blood of MS patients. Although MBP cannot be implicated directly as the target antigen in MS, the increased levels of MBP specific T cells may have encephalitogenic potential.

IMMUNITY TO TCR PEPTIDES IN PATIENTS WITH MS

The over-utilization of $V\beta5.2$ and $V\beta6.1$ genes by MS T cells specific for MBP suggested the possibility that TCR peptide boosting might be applicable to human autoimmunity. Eleven patients with progressive multiple sclerosis were injected intradermally with two TCR $V\beta$ peptides, $V\beta5.2-39-59$ and $V\beta6.1-39-59$, to test antigenicity and side effects. T cell responses to the TCR peptides were measured by frequency analysis, and antibody responses by ELISA. Frequencies were estimated by the χ-squared minimization test that allows assessment of significant changes in frequency. In 3 patients, prior studies had established a $V\beta5.2/V\beta6.1$ bias in MBP-reactive T cell clones from blood. The remaining 8 patients were selected without prior evaluation of V gene bias.

Five patients had a significantly increased T cell frequency to both TCR peptides after respective boosting with each peptide. In addition, one patient responded only after injection with $V\beta5.2-39-59$, and one patient responded only after injection of $V\beta6.1-39-59$. In many cases, increased T cell frequencies were corroborated by positive delayed type hypersensitivity responses at the injection site. On average, the frequency of $V\beta5.2-39-59$ T cells in the responders increased from 1.2 to 3.9 cells/million, and the responses could be maintained for about 16 weeks; the frequency of $V6.1-39-59$ specific T cells increased from 1.0 to 5.9/million, and the response could be maintained for about 27 weeks (Fig. 2). Responses occurred at doses ≤ 300 μg, with higher doses tending to induce non-responsiveness. Non-responders did not increase in their T cell frequency, even at doses as high as 3 mg. Only one patient had increased antibody response to the $V\beta6.1-39-59$ peptide. Side effects were minimal at doses ≤300 μg.

Nineteen T cell clones responsive to the $V\beta5.2$ peptide were subsequently isolated from a single responder. These clones were all found to be CD4+, interestingly, the response to $V\beta5.2$ peptide was restricted either by HLA-B7 (class I) or HLA-DR2 (class II). As is noted above, rat T cells specific for $V\beta8.2-39-59$ peptide were also CD4+, but the response was mainly MHC class I restricted. We would speculate that these differences

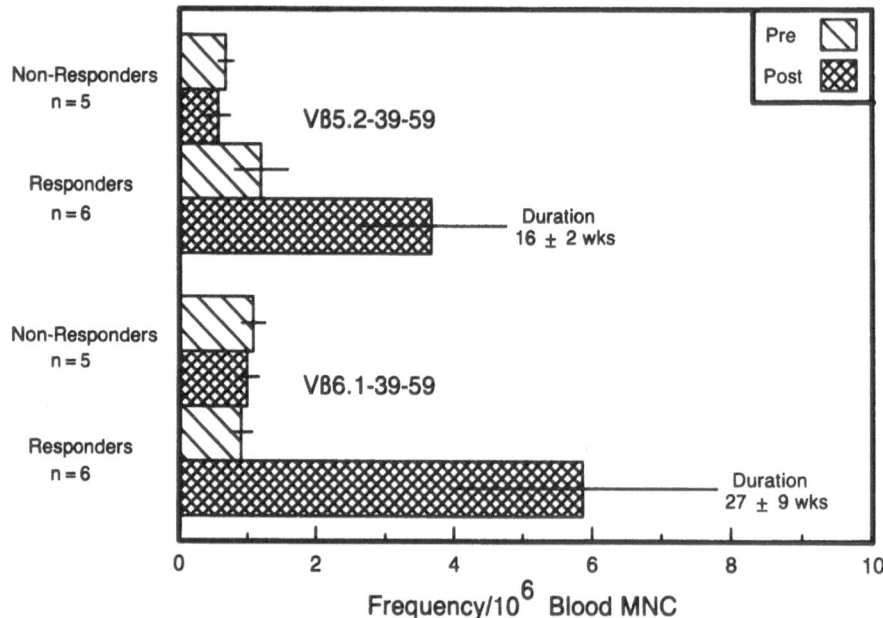

Figure 2. TCR Vβ peptides boost T cell frequency in MS patients.

in MHC restriction may reflect opposing regulatory functions, an issue currently under further investigation. Additionally, 31 T cell clones specific for the Vβ6.1 peptide were isolated from a different responder. With the exception of 1 CD8+ clone, these clones had characteristics similar to the Vβ5.2 reactive clones.

T cell frequencies to MBP were episodic. Longitudinal studies showed that MBP frequencies approximated or exceeded levels observed in animals with EAE in 9/11 patients at least once during the course of the study. Increased MBP responses did not necessarily result in clinical worsening, similar to the occurrence of MRI lesions. Increased responses to TCR peptides did not prevent increased responses to MBP, although MBP responses in all of the treated patients eventually returned to baseline.

Clinically, the 3 patients in whom the V gene bias for MBP-specific T cells had been established improved or remained stable. Two of the other 4 TCR peptide responders were stable, and 2 were clinically worse. Similarly, 2 of the 4 non-responders to the TCR peptide injections were stable, and 2 were clinically worse.

These results indicate that TCR peptides are immunogenic and relatively safe for use in humans, although it is still not clear to what extent immunity to TCR peptides can regulate MBP responses. It would appear to be important to use TCR V region peptides based on an established TCR V gene bias, however. Long term regulation of T cell responses characterized by V gene bias using TCR peptides may allow a critical test of the hypothesis that myelin reactive T cells participate in the pathogenesis of MS.

REFERENCES

Ben-Nun A, Liblau RS, Cohen L, Lehmann D, Tournier-Lasserve E, Rosenzweig A, Jingwu Z, Raus JCM, Bach MA (1991) Restricted T-cell receptor Vβ gene usage by myelin basic protein-specific T-cell clones in multiple sclerosis: Predominant genes vary in individuals. Proc Natl Acad Sci USA 88:2466

Chou YK, Bourdette DN, Offner H, Whitham R, Wang RY, Hashim GA, Vandenbark AA (1992) Frequency of T cell specific for basic protein and proteolipid protein in multiple sclerosis. J Neuroimmunol 38:105-114

Gold DP, Offner H, Sun D, Wiley S, Vandenbark AA, Wilson DB (1991) Analysis of T cell receptor β chains in Lewis rats with experimental allergic encephalomyelitis: Conserved complementarity determining region 3. J Exp Med 174:1467

Gold DP, Vainiene M, Celnik B, Wiley S, Gibbs C, Hashim GA, Vandenbark AA, Offner H (1992) Characterization of the immune response to a secondary encephalitogenic epitope of basic protein in Lewis rats. II. Biased TCR Vβ expression predominates in spinal cord infiltrating T cells. J Immunol 148:1712

Kotzin BL, Karuturi S, Chou YK, Lafferty J, Forrester JM, Better M, Nedwin GE, Offner H, Vandenbark AA (1991) Preferential T cell receptor Vβ gene usage in myelin basic protein reactive T cell clones from patients with multiple sclerosis. Proc Natl Acad Sci USA 88:9161

Martin R, Utz U, Coligan JE, Richert JR, Flerlage M, Robinson E, Stone R, Biddison WE, McFarlin DE, McFarland HF (1992) Diversity in fine specificity and T cell receptor usage of the human CD4+ cytotoxic T cell response specific for the immunodominant myelin basic protein peptide 87-106. J Immunol 148:1359

Offner H, Hashim GA, Vandenbark AA (1991) T cell receptor peptide therapy triggers autoregulation of experimental encephalomyelitis. Science 251:430

Offner H, Hashim G, Chou YK, Bourdette D, Vandenbark AA (1992) Prevention, suppression, and treatment of EAE with a synthetic T cell receptor V region peptide. In: Molecular Mechanisms of Immunological Self - Recognition. Alt FW, Vogel MG. Eds. Academic Press

Oksenberg JR, Panzara MA, Begovich AB, Mitchell D, Erlich HA, Murray RS, Shimonkevitz R, Sherritt M, Rothbard J, Bernard CCA, Steinman L (Submitted) Selection of T cell receptor Vβ-Dβ-Jβ gene rearrangements with specificity for an epitope of myelin basic protein in brain lesions of HLA-DRB1*1501, DQB1*0602 patients with multiple sclerosis.

Olsson T, Wang WZ, Hojeberg B, Kostulas V, Yu-Ping J, Anderson G, Ekre HP, Link H (1990) Autoreactive T lymphocytes in multiple sclerosis determined by antigen-induced secretion of interferon-γ. J Clin Invest 86:981

Ota K, Matsui M, Milford EL et al. (1990) T-cell recognition of an immunodominant myelin basic protein epitope in multiple sclerosis. Nature 346:183

Sun JB, Olsson T, Wang WZ, Xiao BG, Kostulas V, Fredrikson S, Ekre HP, Link H (1991) Autoreactive T and B cells responding to myelin proteolipid protein in multiple sclerosis and controls. Eur J Immunol 21:1461

Vandenbark AA, Hashim G, Offner H (1989) Immunization with a synthetic T-cell receptor V-region peptide protects against experimental autoimmune encephalomyelitis. Nature 341:541

Vandenbark AA, Vainiene M, Celnik B, Hashim G, Offner H (1992) TCR peptide therapy decreases the frequency of encephalitogenic T cells in the periphery and the central nervous system. J Neuroimmunol, In press

Wucherpfennig KW, Ota K, Endo N, Seidman JG, Rosenzweig A, Weiner HS, Hafler DA (1990) Shared human T cell receptor Vβ usage to immunodominant regions of basic protein. Science 248:1016

Suppression of Autoimmunity through Manipulation of Immune Network with Normal Immunoglobulin G

S. V. Kaveri, G. Dietrich, N. Ronda, V. Hurez, V. Ruiz de Souza,
D. Rowen, T. Vassilev., and M. D. Kazatchkine

INSERM U28, Hopital Broussais, Paris, France
*Centre for Infectious and Parasitic Diseases, Sofia, Bulgaria

Expression of autoreactivity is efficiently regulated under physiological conditions to maintain the homeostasis of the immune system, thus preventing the emergence of autoaggressive T and B cell clones. Selection of preimmune repertoires and control of autoreactivity in healthy individuals involve variable (V) region interactions between antibodies and lymphocytes within a functional network (Sundblad 1991). The emergence of pathological autoimmunity could reflect a failure in the function of immune network. Following this view, therapeutic intervention in autoimmunity should primarily be aimed at stimulating the molecular and cellular mechanisms involved in physiological regulation of autoreactivity. We propose that these concepts provide a basis for understanding the therapeutic effects of the infusion of pooled normal polyspecific IgG (intravenous immunoglobulins : IVIg) in patients with autoimmune diseases. In this review, we discuss the evidence supporting the hypothesis that the immunoregulatory effect of IVIg in autoimmune disease is dependent on the selection of recipient's immune repertoires by V region reactivities of infused IgG.

There is now ample evidence that B and T cells reactive with a wide range of self antigens are present in healthy individuals, that autoreactive B cell clones are positively selected during ontogeny and that a functional network regulating the expression of the autoreactive repertoire is established during prenatal development (Martinez 1988; Sundblad 1989). In healthy adults, natural autoantibodies of both the IgM and IgG isotypes represent at least up to 50% of immunoglobulins in serum (Avrameas, personal communication). Natural autoantibodies are encoded by germline genes and connected through V region-complementary interactions. The autoantibodies, often polyreactive, recognize intracellular and membrane components and circulating plasma proteins. Some of the antigens recognized by natural autoantibodies are also targets of autoantibodies in autoimmune diseases eg., thyroglobulin, neutrophil cytoplasmic antigens, Fcg, glomerular basement membrane, intrinsic factor and factor VIII (Dietrich 1992a). Some of the natural IgM and IgG autoantibodies present in normal serum function as anti-idiotypic antibodies against autoantibodies reflecting V region-dependent regulatory interactions between antibodies and lymphocytes that control expression of autoreactivity in healthy individuals (Dietrich 1992b; Dietrich 1992c).

We have recently analyzed age-related expression and control of the natural autoreactive IgG repertoire in normal human serum. Serum IgG autoreactvity is significantly lower than that of the purified IgG fraction on a panel of self antigens. In patients with autoimmune diseases, on the other hand, there is no difference between the binding activity of purified IgG and that of serum IgG (Fig. 1). Both IgG-depleted serum and purified IgM are capable of inhibiting

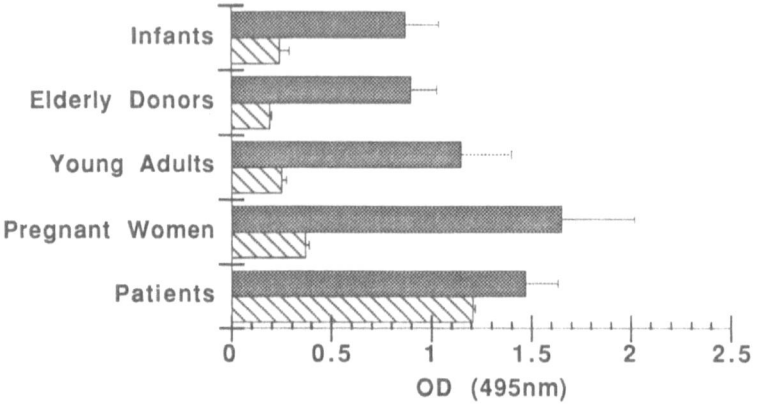

Figure 1. Comparison between the mean binding to various autoantigens. of purified IgG (black columns) and of IgG in whole serum (hatched columns) of infants, elderly donors, young adults, pregnant women and patients with autoimmune thyroiditis. Each column represents the mean OD ± SD of binding of IgG. Autoreactivity was measured by ELISA.

the binding of IgG to many self antigens. The IgM regulation of IgG autoantibody activity is mediated by idiotypic interactions as indicated by the binding of IgM to $F(ab')_2$ fragments of IgG. A further line of evidence for the role of IgM-IgG interactions in autoimmune regulation comes from the fact that, in humanized SCID mice, serum IgG is not less potent in its binding to self antigens than the purified IgG fraction. This apparent lack of control of IgG autoreactivity in the serum may be related to the low expression of human IgM in the serum of these mice. These results together with the previous studies in mice (Adib 1990) indicate that IgM plays a crucial role in regulation of physiological autoimmunity and an adequate IgG and IgM connectivity appears to be important in maintaining autoantibodies under non-pathological levels in whole serum (Hurez 1992).

The role of spontaneously generating anti-idiotypes in the recovery from certain autoimmune diseases has been well documented. Anti-idiotypic antibodies against autoantibodies have been found in remission sera of patients with myasthenia gravis, Guillain-Barré syndrome, systemic vasculitis with ANCA autoantibodies, SLE, anti-factor VIII autoimmune disease, and anti-fibrinogen autoimmune disease (Ronda 1992). In the case of patients who spontaneously recovered from anti-factor VIII autoimmune disease, Guillain-Barré Syndrome and in patients in remission from systemic vasculitis with ANCA autoantibodies, we have observed that F(ab')₂ fragments of patients' post-recovery IgG inhibited autoantibody activity in F(ab')₂ fragments of autologous IgG obtained during the acute phase of the disease (Sultan 1987; Van Doorn 1990; Rossi 1991).

Immunoglobulin populations that are naturally produced consistently manifest a definite pattern temporal fluctuations in their circulating levels (Fig. 2). The pattern of fluctuations are similar in mice and man. In autoimmune conditions, serum Ig concentrations follow dynamical patterns that tend to resemble random fluctuations or to exhibit a more marked rhythmicity than in healthy individuals. Autoimmune disease is associated with a general disturbance in the dynamics of the self-referential network rather than with a clonally localized escape of tissue-specific autoantibodies (Varela 1991a). Thus, in autoimmune

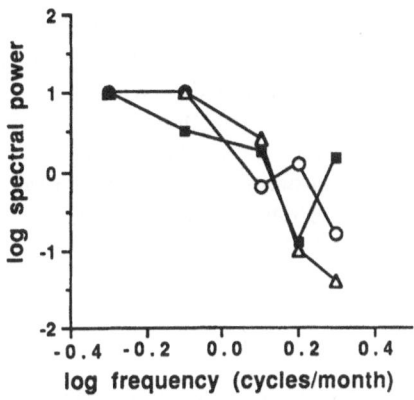

Figure 2. Frequency power spectra for the expression of anti-thyroglobulin IgG in the serum from three healthy individuals. The patterns of temporal fluctuations in autoantibody concentrations are similar to the "red-shifted" spectra that have been observed in normal mice.

disease, a basic perturbation in network regulation may affect the expression of a wide range of disease associated autoantibodies as well as natural antibodies that are not involved in pathology. As the dynamics of natural antibodies is determined by V region interactions of cell-bound and free Igs of each clone, the autoimmune behaviour could represent a generalized defect in connectivity. These concepts would support therapeutic approaches aimed at restoring the disrupted autoantibody dynamics possibly through modifications in connectivity. This might be possible by using the very components of the normal immune system, i.e., pools of natural antibodies embodying the normal connectivity levels (Varela 1991b).

Intravenous immunoglobulins (IVIg) are therapeutic preparations of normal intact IgG obtained from a pool of plasma from a large number of healthy donors. IVIg thus represent a wide spectrum of the expressed normal human IgG repertoire, including antibodies to external antigens, IgG autoantibodies and anti-antibodies. IVIg have initially been used for replacement therapy of antibody deficiency states. A beneficial effect of the intravenous administration of high-dose IVIg has now been reported in autoimmune peripheral cytopenias and in a variety of other autoimmune diseases in which there is direct or indirect evidence for the role of pathogenic autoantibodies or autoaggressive T cells (Dwyer 1992). The immunomodulatory effects of IVIg may depend on interactions between the Fc portion of infused IgG and Fc-receptors on inflammatory cells and lymphocytes and/or on the modulation or selection of the expressed antibody repertoire through the interaction of variable regions of IVIg with circulating immunoglobulins and antigen receptors on immuno-competent cells.

Several lines of evidence led us to speculate that IVIg act in autoimmune patients by restoring and/or stimulating physiological control mechanisms of autoimmunity through network interactions (Kaveri 1991). Idiotypic interactions between IVIg and disease-associated autoantibodies have been demonstrated by the following observations; i) F(ab')$_2$ fragments of IVIg neutralize the functional activity of autoantibodies and/or inhibit the binding of autoantibodies to autoantigens, ii) different disease-associated autoantibodies are retained on affinity chromatography columns of F(ab')$_2$ fragments of IVIg coupled to Sepharose, iii) IVIg do not contain detectable antibodies against the Gm1 (3), Gm1 (4), Gm1 (17), Gm1 (1) and Km (1) allotypes that are most commonly expressed in the F(ab')$_2$ region of human IgG, and iv) IVIg share anti-idiotypic reactivity towards idiotypes of autoantibodies with heterologous anti-idiotypic reagents (Dietrich 1990a; Dietrich 1990b). IVIg interact not only with disease-associated autoantibodies but also with natural autoantibodies and interfere with the regulatory function of the immune network in vivo. IVIg recognize idiotypic determinants on natural IgM and IgG autoantibodies from healthy individuals. We have demonstrated that IVIg interact with natural polyreactive IgM antibodies through idiotypic interactions by using autoreactive monoclonal IgM antibodies secreted by Epstein Barr virus-transformed normal B lymphocytes (Rossi 1990). IVIg also contain anti-idiotypes against idiotypic determinants expressed on natural IgG autoantibodies as it may be demonstrated by using affinity chromatography of F(ab')$_2$ fragments of IVIg on Sepharose-bound F(ab')$_2$ fragments of IVIg (Dietrich 1992a). By reacting with natural IgG and IgM which are components of the normal idiotypic network, IVIg could exert an influence on the expression of the available immune repertoire of an individual so as to restore a physiological control of autoimmunity.

We have recently analyzed the changes that occur in the expressed autoreactive antibody repertoire and in network organization following the infusion of normal polyspecific IgG in a patient with autoimmune thyroiditis (Dietrich

1992d). The results have allowed us to gather several lines of evidence indicating that changes in serum antibody concentrations observed after infusion of IVIg do not merely reflect passive transfer of IgG into the patient. The concentration of IgG in patient's serum after the second infusion of IVIg increased to higher levels than would be expected from the amount of transfused IgG. Accordingly, the two consecutive infusions of IVIg resulted in a cumulative increase in the production of IgM. These observations illustrate an in vivo activation of lymphocytes following infusion with normal Ig. The profile of antibody activity of IgG in serum after infusion of IVIg is not the same as would be expected from the passive transfer of the antibodies present in IVIg. For e.g., IVIg contains equivalent amounts of anti-PC and anti-gliadin antibodies, however, the titer of anti-gliadin antibody sharply increased in serum after infusion of IVIg (Fig. 3). In addition, antibody activities expressed

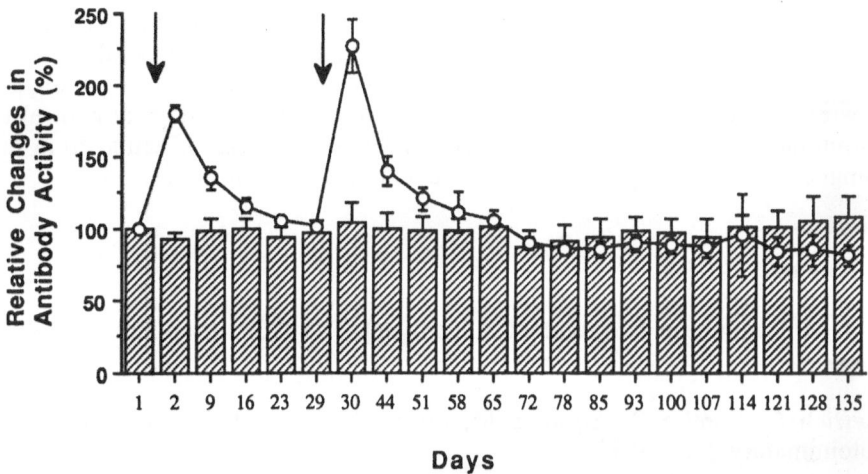

Figure 2. Anti-phosphorylcholine (hatched bars) and anti-gliadin (O) activity in serum following administration of IVIg. The Figure depicts changes in titers of IgG antibodies with time relative to pretreatment levels. Arrows denote infusions of IVIg.

by IgG in the patient's serum after infusion of IVIg differ from that observed in the patient's pre-infusion serum upon reconstitution with IVIg in vitro. Finally, we have observed that the relative changes in anti-IVIg activity of IgG in serum do not follow the changes in the concentration of total serum IgG. Together, these data suggest that infusion of Ig leads to a recruitment of subsets of B cells reactive with IVIg. Analysis of idiotypes shows regulation of certain disease-specific IVIg-reactive clones. For example, anti-TG activity within the IVIg-reactive subfraction of HT patient's IgG fluctuates in a pattern different from that of anti-TG IgG activity measured in serum, indicating an interaction between IVIg and a distinct subpopulation of anti-TG B cells. Within that population, T44 (a phenotypic marker of anti-TG autoantibodies associated with Hashimoto's thyroiditis) (Dietrich 1991) idiotype positive clones are down regulated for several weeks following infusion of IVIg. These observations may be taken as an indication that infusion of IVIg results in transient suppression of disease-specific antibody clones. In addition, the dynamical behaviour of autoantibodies in the patient prior to infusion of immunoglobulin exhibits a clearly distinct pattern with marked rhythmicity suggestive of disruptions of connectivity within the immune network. The kinetic pattern following the second infusion of IVIg, on the other hand, is similar to that seen in healthy individuals. Thus, infusion of pooled normal immunoglobulin restored in the patient a network organization of autoantibodies characteristic of the physiological conditions.

Taken together, these observations support our hypothesis that the beneficial effect of IVIg in autoimmune diseases is not merely due to the passive transfer (transfusion) of neutralizing anti-idiotypic antibodies against autoantibodies but that IVIg alters the structure, the function and the dynamics of the idiotypic network in the autoimmune patient so as to restore a physiological control of autoimmunity. IVIg thus, would clearly differ in its mode of action from the immunosuppressive approach to the treatment of autoimmune diseases.

References

Adib M, Ragimbeau J, Avrameas S, Ternynck T (1990) J Immunol 145:3807-3813

Dietrich G, Pereira P, Algiman M, Sultan Y, Kazatchkine MD (1990a) J Autoimmunity 3: 547-557

Dietrich G, Kazatchkine MD (1990b) J Clin Invest 85: 620-625

Dietrich G, Piechaczyk M, Pau B,. Kazatchkine MD (1991) Eur J Immunol 21: 811-814

Dietrich G, Kaveri SV, Kazatchkine MD (1992a) Eur J Immunol 22:1701-1706

Dietrich G, Kaveri SV, Kazatchkine MD (1992b) Clin Immunol Immunopathol 62: S73-S81

Dietrich G, Algiman M, Sultan Y, Nydegger U. Kazatchkine MD (1992c) Blood 79: 2946-2951

Dietrich G, Varela F,. Hurez V, Bouanani M, Pau B, Kazatchkine MD (1992d) (submitted for publication)

Dwyer MD (1992) N Engl J Med 326: 107-116

Hurez V, Kaveri SV, Kazatchkine MD (1992) (submitted for publication)

Kaveri SV, Dietrich G, Hurez V, Kazatchkine MD (1991) Clin Exp Immunol 86:192-198

Martinez AC, Pereira P, Toribio ML, Marcos MAR, Bandeira A, De la Hera A, Marquez C, Cazenave P-A, Coutinho A (1988) Immunol. Rev. 101: 191-215

Ronda N, Hurez V, Kazatchkine MD (1992) Vox Sang. (in press)

Rossi F, Guilbert B, Tonnelle C, Ternynck T, Fumoux F, Avrameas S, Kazatchkine MD (1990) Eur J Immunol 20: 2089-2094

Rossi F, Jayne DRW, Lockwood CM, Kazatchkine MD (1991) Clin Exp Immunol 83: 298-303

Sultan Y, Rossi F, Kazatchkine MD (1987) Proc Natl Acad Sci (USA), 84: 828-831.

Sundblad A, Marcos M, Huetz F, Freitas A, Heusser C, Portnoï D, Coutinho A (1991) Eur J Immunol 21: 1155-1161

Sundblad A, Hauser S, Holmberg D, Cazenave P-A, and Coutinho A (1989) Eur J Immunol 19: 1425-1430

Van Doorn PA, Rossi F, Brand A, Van Lint M, Vermeulen M, Kazatchkine MD (1990) J Neuroimmunology 29: 57-64

Varela F., Anderson A., Dietrich G, Sundblad A., Holmberg D., Kazatchkine MD, Coutinho A (1991a) Proc Natl Acad Sci (USA).88: 5917-5921

Varela F, Coutinho A (1991b) Immunol. Today 5: 159-166

Control of Fertility by Immunisation with Antiprogesterone Antiidiotypes

M. J. Taussig, M. W. Wang[+] , A. S. Humphreys, M. J. Sims[+],
J. Coley[*], M. Gani[*], A. Feinstein and R. B. Heap

*Structural Studies Laboratory, Departement of Immunology, AFRC Institute of
Animal Physiology and Genetics Research, Babraham, Cambridge CB2 4AT,
England*

INTRODUCTION

The induction of autoantibodies against hormones can be expected to
interfere with physiological processes under hormone control.
Reproduction is an example of such a process: active immunisation
with progesterone-protein conjugate, which induces anti-progesterone
autoantibodies, blocks pregnancy in rats (Kaushansky et al. 1977) and
rabbits (Elsaesser 1980). Similarly, we have shown that passive
administration of monoclonal antibodies (mAbs) against progesterone
prevents implantation and pregnancy in mice (Wright et al. 1982;
Ellis et al. 1988), rats (Phillips et al. 1988) and ferrets (Rider
and Heap 1986). This suggests that progesterone, the key steroid
hormone required for the establishment and maintenance of pregnancy,
is a possible target autoantigen for immunocontraception. The action
of anti-progesterone immunisation appears to cause progesterone
withdrawal by a two-fold mechanism. One effect of anti-progesterone
antibody is to bind the steroid in the circulation and prevent it
from reaching progesterone receptors in uterine tissue (Rider et al.
1985); a second may be exerted locally in the uterus itself, where
we have shown that passively administered anti-progesterone mAbs
localise specifically on the luminal and glandular epithelia around
the time of implantation (Wang et al. 1989b).

We have used polyclonal anti-idiotypes (Ab2) raised against anti-
progesterone mAbs (Ab1) to induce an anti-progesterone (Ab3) response
in mice (Wang et al. 1989a, 1991). With some anti-idiotypic
reagents, the anti-progesterone response is sufficient to cause a
period of infertility. In this article we review our studies on the
anti-idiotypic approach to fertility regulation.

[+]Current addresses: M-W.Wang, Ligand Pharmaceuticals, 9393 Towne
Centre Drive, Suite 100, San Diego, CA 92121, USA; M.J. Sims, The
Wellcome Research Laboratories, Langley Court, South Eden Road,
Beckenham, Kent BR3 3BS, England.
[*]Department of Immunology, Unilever Research, Colworth Laboratory,
Sharnbrook, Bedford MK44 1LQ, England.

A panel of anti-progesterone mAbs was raised from mice immunised with progesterone-11α-succinyl-bovine serum albumin (progesterone-BSA) (Wright et al. 1982; Ellis et al. 1988). The antibodies have affinities (Ka) in the range 10^8-10^9 M^{-1}; their specificities for a range of steroids have been described (Ellis et al. 1988). The antibodies used in this study are designated DB3, 11/32 and 11/64; both DB3 and 11/32 are IgG1, while 11/64 is an IgM. All were shown to be capable of blocking pregnancy after passive administration, though the IgG mAbs were the more effective (Ellis et al. 1988).

The variable regions of twelve anti-progesterone mAbs have been sequenced and a remarkable degree of homogeneity in VH and VL gene usage has been observed. All use a VH gene segment derived from the small VGAM3.8 family, in association with the Vk105 (VK1) light chain V segment (Stura et al. 1987; Sims et al., 1992 and in preparation). The CDR3H loops show considerable sequence variation, while maintaining a consistent length (5 amino acids in the D/N segment) and conserving residues at certain positions. Thus the first CDR3H residue (position 95) is invariably glycine, residue 97 is tyrosine in 10/12 cases (phenylalanine or tryptophan in the other two), and residue 100 is tryptophan in 8 mAbs and tyrosine in others. The intervening residues on this loop are highly variable and likely to contribute to private idiotopes. The Fab' fragment of the DB3 antibody has been crystallised both in native form and as a complex with progesterone and related steroids (Stura et al. 1987), and the structures have recently been solved at 2.7 A resolution by X-ray crystallography (Arevalo et al. 1992). This has shown that the conserved CDR3H residues are those in contact with the steroid, while the side-chains of the variable residues are directed away from the binding pocket. It has also provided an explanation for the repetitive use of the VGAM3.8 family VH segment in anti-progesterone mAbs, since the two contact residues which it contributes to the binding site (asparagine-35H and tryptophan-50H) are rarely found together in other VH gene families (Sims et al. 1992).

The mAbs used in this study differ in their CDR3H sequences partly as a result of having different JH segments (Table 1). The use of JH1 in DB3 gives rise to tryptophan at position 100H (the first JH residue), whereas both 11/32 and 11/64 express JH4 and as a result have a tyrosine at this position, with further differences at 100a, 100b and 102.

Table 1.　　Individuality of VHCDR3 regions of monoclonal anti-progesterone antibodies DB3, 11/32 and 11/64

Residue:	95	96	97	98	99	100	a	b	101	102
DB3	G	D	Y	V	N	W	Y	F	D	V
11/32	.	.	.	Y	.	Y	A	L	.	Y
11/64	.	T	.	S	D	Y	A	L	.	F

Anti-idiotypes against DB3, 11/32 and 11/64 (anti-DB3-id, anti-11/32-id, anti-11/64-id) were raised by repeated immunisation of rabbits; antisera were rendered specific by absorption with sepharose-conjugated mouse Ig and anti-ids purified by absorption and elution from columns of the specific insolubilised mAb (Taussig et al. 1986; Wang et al. 1989a, 1991). Each anti-idiotypic reagent showed a high degree of specificity for its inducing mAb, but with significant cross-reactivity (about 10%) with other members of the anti-progesterone panel. Hence, they were mainly directed against private idiotopes on DB3, 11/32 or 11/64, but with a cross-reactive component which can probably be attributed to the repetitive use of VH and VL segments among the panel. All three anti-idiotypes blocked binding of ^{125}I-labelled progesterone to the isologous mAb.

INDUCTION OF ANTI-PROGESTERONE ANTIBODIES BY ANTI-IDIOTYPIC IMMUNISATION

Female BALB/c mice were immunised with rabbit anti-idiotype by injection of 20μgm in CFA on the back, followed by boosters of equal amounts in IFA (once) or PBS (four times). They responded with production of anti-progesterone antibodies, but the level of response was dependent on the specificity of the anti-idiotypic reagent (Wang et al. 1989a, 1991). Anti-DB3-id was the most effective of the three, the anti-progesterone level increasing progressively to a peak of about 60μgm/ml after 4 inoculations (Table 2). There was a wide range in response from 5-97μgm/ml, with some animals (7/20) consistently making less than 10μgm/ml. In contrast, anti-11/32-id and anti-11/64-id were both much less effective, inducing mean responses of less than 15μgm/ml. With anti-11/32-id, only 3/10 mice made a response over 10μgm/ml, and for anti-11/64-id only 1/12 responded well. Immunisation with progesterone-BSA as a positive control induced very high levels of circulating anti-progesterone (mean of 3.8mg/ml after 5 inoculations); anti-mouse-Ig immunised mice (negative controls) did not make an anti-progesterone response.

Table 2. Anti-progesterone antibody responses and plasma progesterone levels after anti-idiotypic immunisation

No. of mice	Immunisation	Anti-progesterone response(mean±SE) μgm/ml	Plasma progesterone ng/ml
20	Anti-DB3-id	57.2±16.4	60.0±6.4
10	Anti-11/32-id	13.6±6.1	31.0±3.0
12	Anti-11/64-id	12.0±7.9	33.4±5.4
10	Progesterone-BSA	3800±900	273±61.3
11	Anti-mouse-Ig	-	29.3±3.4

The presence of anti-progesterone antibodies in the circulation leads to an increase in total plasma progesterone, due to the increased half-life of the steroid complexed with antibody. This was seen in mice immunised with anti-DB3-id or progesterone-BSA, but not after immunisation with the two other anti-idiotypes (Table 2).

ANTI-FERTILITY EFFECT OF ANTI-IDIOTYPIC IMMUNISATION

Ten days after the last inoculation of anti-idiotype or progesterone-BSA, the fertility of immunised mice was studied by mating with males of the same strain. Immunised mice were caged with males overnight and vaginal plugs checked the following morning (day 1 of pregnancy); where no plug was found, they were recaged with males on subsequent nights. Implantation sites were counted on autopsy 10 days post-coitum (p.c.). The pregnancy rate of control mice immunised with rabbit anti-mouse-Ig was over 90%, while in mice immunised with anti-DB3-id the rate fell to 30% (Table 3) (Wang et al. 1989a). All mice which became pregnant in this group had anti-progesterone levels of below 15μgm/ml; those with levels over 30μgm/ml were all infertile. Statistically, there was a clear correlation between pregnancy blocking and the level of circulating anti-progesterone induced. In contrast, immunisation with anti-11/32-id or anti-11/64-id did not block pregnancy and the number of implantation sites was normal (Wang et al. 1991). None of the mice immunised with progesterone-BSA became pregnant.

Table 3. Effect of anti-idiotypic immunisation on pregnancy

Immunisation	No. pregnant/ no. mated	Implantation sites (mean±se)	%pregnant
Anti-DB3-id	6/20	2.6±0.9	30
Anti-11/32-id	9/10	6.3±1.4	90
Anti-11/64-id	10/12	7.3±1.1	83
Progesterone-BSA	0/10	0	0
Anti-mouse-Ig	10/11	7.5±1.3	91

DURATION OF INFERTILITY

Groups of 10 mice immunised with either anti-DB3-id, anti-mouse Ig or progesterone-BSA were caged with males 10 days after the final inoculation, and the time between the first detection of a copulation plug to delivery of pups was observed.

In the control group treated with anti-mouse-Ig, pups were delivered at 21.9±1.2 days, whereas in anti-DB3-id immunised mice this was prolonged to 29.1±3.3 days and in progesterone-BSA immunised mice to 78.7±8.7 days. Taking into account the 10-day interval after immunisation and before mating, the duration of infertility was about 20 days or 4-5 oestrous cycles after anti-DB3-id immunisation, and 70 days or 16-17 cycles after progesterone-BSA immunisation.

DISCUSSION

These experiments demonstrate that polyclonal anti-idiotypes against anti-progesterone mAbs are able to mimic the antigenicity of a steroid hormone, inducing anti-progesterone antibodies without the use of steroid-protein conjugate. They also show that it is possible to regulate the activity of a hormone *in vivo* through induction of autoantibodies via the idiotypic network. With one anti-idiotypic reagent (anti-DB3-id), the anti-progesterone response was large enough to block pregnancy and to cause a period of infertility which lasted for a few oestrous cycles. The response to anti-idiotype depended on the specificity of the reagent and was in all cases much lower than that to immunisation with progesterone-BSA, suggesting that anti-idiotype selectively activates only that fraction of the total anti-progesterone B cell repertoire which carries the private idiotype recognised by the reagent; i.e. the anti-ids are acting as clonotypic or oligoclonal stimulators rather than 'internal images' of progesterone. Since the effectiveness in pregnancy blocking was determined by anti-progesterone level in the circulation, it was also of much longer duration after progesterone-BSA immunisation than after anti-idiotype.

Polyclonal Ab2 against different mAbs clearly varied in their ability to act as 'surrogate steroids' in induction of Ab3, despite having similar properties of private, site-directed specificity and the close V-gene relationship between the inducing mAbs. There was also a wide range of response in individual mice. Both observations may share a common explanation in the number of B cell clones carrying each private idiotope. The latter will depend on the frequency with which particular VDJ combinations arise in the repertoire of individual animals. As noted in Table 1, there is a significant difference between the CDR3H of DB3 and the other two mAbs, in that DB3 expresses a JH1 segment while the others use JH4. The use of JH1 is considerably more common than JH4 in the anti-progesterone mAb panel (9/12) (Sims et al., in preparation), suggesting that the frequency of rearrangements to JH4 among naive progesterone-specific B cells is lower than to JH1. Assuming that the variable nature of CDR3H contributes to the private idiotopes of the mAbs, this could explain the generally lower efficacy of anti-11/32-id and anti-11/64-id compared with anti-DB3-id, since the response they induce should be directly proportional to the frequency with which their 'target' clones are generated.

REFERENCES

Arevalo JH, Stura EA, Taussig MJ, Wilson IA (1992) Nature (London), submitted for publication
Ellis ST, Heap RB, Butchart AR, Rider V, Richardson NE, Wang M-W, Taussig MJ (1988) J Endocrinol 118:69-80.
Elsaesser F (1980) J Reprod Fert 58:213-218.
Kaushansky A, Bauminger S, Koch Y, Lindner HL (1977) Acta Endocrinol (Copenhagen) 84:795-803.
Phillips A, Hahn DW, McGuire J, Wang M-W, Heap RB, Rider V, Taussig MJ (1988) Contraception 38:109-116.

Rider V, Heap RB (1986) J Reprod Fert 76:459-470.
Rider V, McRae A, Heap RB, Feinstein A (1985) J Endocrinol 104:153-158.
Sims MJ, Krawinkel U, Taussig MJ (1992) J Immunol 149:in press.
Stura EA, Arevalo JH, Feinstein A, Heap RB, Taussig MJ, Wilson IA (1987) Immunology 62:511-521
Taussig MJ,Brown N, Ellis S, Holliman A, Peat D, Richardson NE, Heap RB, Feinstein A (1986) Immunology 58:445-452.
Wang M-W, Heap RB, Taussig MJ (1989a) Proc Nat Acad Sci USA 86:7098-7102.
Wang M-W, Sims MJ, Symington PR, Humphreys AS, Taussig MJ (1991) Immunology 73:348-355.
Wang M-W, Whyte A, King I, Taussig MJ, Heap RB (1989b) J Reprod Fert 86:211-218.
Wright LJ, Feinstein A, Heap RB, Saunders JC, Bennett RC, Wang M-Y. (1982) Nature (London) 295:415-417.

14. Immunology of Infectious Diseases and Vaccination

Viruses Escaping Immunological Surveillance

R. M. Zinkernagel, D. Moskophidis, B. Odermatt, Ch. Schalcher,
M. Battegay, D. Kyburz, M. Bründler, I. F. Ciernik, H. P. Pircher,
H. Hengartner

*Institute for Experimental Immunology, Departement of Pathology, University of
Zurich, Sternwartstrasse 2, CH-8091 Zurich, Switzerland*

INTRODUCTION

Patterns of anti-viral immunity varies from virus to virus (Mims,
1982); for recovery from some virus infections antibodies are all
important (e.g. rabies virus) whereas for others T cell immunity is
crucial. Protection against reinfection is mediated mostly by antibody dependent mechanisms. Of course, macrophages, interleukins
(particularly IFN c, TNF etc.) possibly ADCC, NK activities etc. may
play some role in both recovery from primary infection or in
protection against rechallenge, but their respective roles are far
less clear. These immunological parameters on one hand and the virological and parasitic ones on the other hand will crucialy determine
the overall outcome of an infection and of disease (summarized in
Table 1).

By definition, cytopathic infectious agents including viruses will
either kill the host or will be eliminated efficiently early enough
so that death of the host is prevented. These infectious agents will
in general define the efficiency standards of the host defences,
including immune responses; immune escape of these agents will always be detrimental to the host and therefore in the end also to the
virus. In contrast, non- or poorly cytopathic viruses that usually
do not kill the host will be able to establish varying equilibria
with the host; these agents may often be permitted to escape immune
surveillance.

MECHANISMS OF IMMUNE ESCAPE OF VIRUSES

There are many possible mechanisms by which viruses may persist in a
host (Table 2). Transplacental transmission of poorly or noncytopathic viruses from mother to offspring during fetal life or around
birth usually leeds to persistent infection of the offspring.; virus
is usually found in thymus and in many organs (e.g. LCMV (Mims,
1982; Lehmann-Grube, 1984; Johnson *et al.* 1978; Hotchin, 1962;
Zinkernagel and Doherty, 1979), hepatitis B, LDH virus, mammary tumour virus etc.). Complete T cell tolerance has been well documented
in LCMV carrier mice (Hotchin, 1962; Buchmeier *et al.* 1980; Lehmann-
Grube, 1984; Zinkernagel and Doherty, 1979). T cell tolerance in
these mice is due to clonal deletion of LCMV-specific T cells in the
thymus (Pircher *et al.* 1989).

This is not the place to summarize virological mechanisms leading to
viral persistence other than to note that several of them result in
a drastic reduction of viral antigen accessible to immune effector
mechanisms. Obviously any measure of reducing viral antigens or pro-

Table 1: Selected examples of virus-host relationships

Virus	Cytopatho-genicity	Consequences of infection with status of immunocompetance:		Efficient response required for survival of host		Incubation time for disease	Late disease? (incubation period)	Immunological escape mechanisms
		incompetent	competent	T cells	neutralizing antibodies			
POX	++	death	recovery	yes		7-10 days	none	none
Influenza A	+	death	recovery		yes	7-10 days	none	antibody escape epidemiologicaly at population level
Herpes simplex	+/-	death	control but no elimination	yes	yes	7-10 days	recurrent infection	hiding in privileged sites
Hepatitis B	-	carrier	recovery - aggr. hepatitis	no	no	weeks to months	aggr. hepatitis (weeks to years) primary liver Ca (> 30 years)	tolerance
HIV	?	? carrier	variable	? no	? no	weeks to years	immunopathology immunosuppression (weeks to years)	immunosuppression, T cell epitope escape mutants? antibody escape mutants
LCMV	-	carrier	variable	no	no	6 days to years	immunopathology immune complex disease (weeks to years)	tolerance immunosuppression T cell epitope escape mutants?

cessed peptides on cell surfaces will reduce their vulnerability to immune mechanisms. Virus integration is the most complete way for a virus to hide (e.g. HIV, retroviruses in general). Downmodulation of cellsurface antigens by capping has been postulated to play a role in measles or herpes virus infection (Oldstone *et al.* 1980). Some viruses also inhibit MHC expression so as to reduce T cell recognition.

Many viruses hide in epithelial, fewer in mesenchymal cells that all lack antigen presentation capacity. E.g. herpes, cytomegalo, measles, rubella viruses or LCMV may persist in neurons or kidney cells, papilloma viruses in keratinocytes etc. (Lehmann-Grube, 1984; Buchmeier *et al.* 1980) The important point here is that even primed memory immune effector lymphocytes may not be triggered by these infected cells, a finding that has recently been supported by studies using mice expressing LCMV-GP as a transgene in insulin producing cells (Ohashi *et al.* 1991). Whether their usually very low MHC class I expression or lack of "factors" contributes importantly to this hiding remains to be fully established.

The possibility that extensive infection by virus may lead to an inactivation of T cells, by various means has been found for MMTV transmissible by milk (Marrack *et al.* 1991; Held *et al.* 1992) and LCMV (Zinkernagel and Doherty, 1979; Ahmed *et al.* 1984). Also, immune suppression is observed during many virus infections and this may facilitate establishment of virus peristence; the example of LCMV will be discussed in greater details below. In addition to direct viral cytopathic effects on monocytes or lymphocytes, optimal noncytopathic impairment of cell function have been postulated to cause immune suppression by affecting lymphokine production as examples of blocking of "luxury" function of cells by virus infection. However, it is not established yet why and how viruses may differentiate between special and household functions (Valsamakis *et al.* 1987).

Two escape mechanisms namely mutation of either B or T cell determinants on viruses are from the immunological point of view highly sophisticated and complicated mechanisms that are evolutionarily fascinating. The capacity of influenza viruses to escape neutralization by mutating B cell epitopes stepwise (antigenic drift), and more drastically by antigenic shift, has been analysed in great detail (Webster *et al.* 1982). Of particular interest is the fact that there is a directionality to it in that sequential virus isolates express determinants that may be neutralized only by new specific antibodies but not by the preexistent ones, whereas antibodies with the "new" specificities may neutralize related epidemiologically older viruses. In fact the "new" virus often induced increases in titers against the previous viruses while the titers against the specific new determinant are slow to rise. This phenomenon has been called original antigenic sin. This rather unique phenomenon has been the target of many speculations and excellent experiments (Fazekas de St.Groth, 1981).

Recently viruses that escape CD8 T cell surveillance have been described (Pircher *et al.* 1990; Aebischer *et al.* 1991) and will be discussed in the next section (Phillips *et al.* 1991). In general antibody escape variants are relevant at the population level; in contrast T cell escape mutants may escape T cell surveillance in the infected individual but not in the population, at least not easily within short periods of time. Nevertheless accummulation of mutations may lead to the evolution of an ideal virus, that is neither cytopathic nor induces an immune response.

EXAMPLES OF IMMUNE EVASION:

A virus inducing immunosuppression by LCMV

It had been known for some time that LCMV causes immunosuppresion in mice. When reevaluating this in various mouse strains using varying LCMV-isolates, we found that an LCMV infection in mice suppressed their capacity to mount an IgM or IgG response to vesicular stomatitis virus (VSV). It also rendered them considerably more susceptible to this virus, which replicates not measurably in adult mice and is usually non-pathogenic for mice if infected subcutaneously or intravenously. The extent of immune suppression by LCMV depended upon the virus isolate and the mouse strain used. Both, the MHC haplotype and non-MHC genes played a major role (Leist et al. 1988; Roost et al. 1988; Odermatt et al. 1991; Moskophidis et al. 1992).

The following experimental results suggest that the antiviral T-cell response is responsible for immunosuppression. When LCMV-carrier mice were evaluated with respect to their immune responsiveness, they were found to mount anti-VSV IgM and IgG responses comparable to normal control mice. T-cell deprived nude mice infected with LCMV also had normal IgM responses. This indicates that LCMV alone is not immunosuppressive and that the observed immune suppression is not caused by the action of interferons on VSV. In contrast, LCMV-infected nude mice inoculated with LCMV immune cytotoxic T cells exhibited suppressed antibody responses. Also, while LCMV-infected mice failed to exhibit an antibody response, similarly infected mice treated with anti-CD8 antisera at the time of LCMV infection mounted normal IgM and IgG responses. These results are compatible with the view that antiviral cytotoxic T cells are responsible for immune suppression in this model infection. Accordingly, LCMV may infect few lymphocytes (Borrow et al. 1991) but mainly antigen presenting cells (Odermatt et al. 1991), which are involved in triggering of antibody responses; these infected cells are then in turn destroyed by anti-LCMV specific cytotoxic T cells.

The question of whether virus-induced immunosuppression includes the antibody response against the infecting virus itself was evaluated in LCMV infected mice (Moskophidis et al. 1992). All mice generated excellent T cell dependent ELISA-antibody titers against LCMV-GP or NP. Therefore, anti-LCMV antibody responses were probably induced before immune suppression was effective. If the CD8+ T cell kinetics could be accelerated then possibly an earlier CTL response may prevent also the anti-LCMV antibody response. This was in fact shown in the following model situation (Moskophidis et al. 1992): Transgenic mice expressing the T-cell receptor (TCR) specific for peptide 32-42 of lymphocytic choriomeningitis virus (LCMV) glycoprotein 1 presented by D^b reacted with a strong transgenic cytotoxic T-lymphocyte (CTL) response starting on day 3 after infection with a high dose (10^6 PFU intravenously) of the WE strain of LCMV (LCMV-WE); LCMV-specific antibody production in the spleen was suppressed in these mice. Low-dose (10^2 PFU i.v.) infection resulted in an antiviral antibody response comparable to that of the transgene-negative littermates. The induction of suppression of LCMV-specific antibody responses was specifically mediated by CD8+ TCR transgenic

CTLs, since the LCMV-8.7 variant virus (which is not recognized by transgenic TCR-expressing CTLs because of a point mutation (see below) did not induce suppression. In addition, treatment with CD8-specific monoclonal antibody in vivo abrogated suppression.

Such virus-triggered, T-cell-mediated immunopathology causing the suppression of B cells and of protective antibody responses, including those against the infecting virus itself, may therefore permit certain viruses to establish persistent infections.

Although LCMV infection in mice may differ from HIV infections in humans with respect to the kinetics of induction of immunodeficiency, the tropism, failing to integrate etc., it nevertheless exhibits several features common to findings with AIDS in humans: for example, tropism for lymphohemopoietic cells, immunosuppression, difficulties in inducing appreciable in vivo protective neutralizing antibodies, the capacity to induce wasting disease in neonatal or adult mice, and the ability to induce a virus carrier state in immunoincompetent fetuses (by vertical transmission from the mother) or in newborns as well as in immunocompetent adults.

T cell epitope escape virus mutants

An immunological game viruses may play to escape immune surveillance has been lately discovered by infection of mice with lymphocytic choriomeningitis virus (LCMV) by the demonstration that the anti-viral cytotoxic T cell response may select in vivo viruses with a mutated T cell-epitope to escape T cell immune responses in vivo and in vitro (Pircher *et al.* 1990; Aebischer *et al.* 1991). For example, lymphocytic choriomeningitis virus (LCMV) can escape the cytotoxic activity of LCMV-specific cloned CTLs by single amino acid changes within the recognized T-cell epitope defined by residues 275-289 of the LCMV glycoprotein. LCMV-infected fibroblasts (multiplicity of infection = 10^{-3}) exposed to virus-specific CTL at an effector-to-target cell ratio of 4:1 4 hrs after infection was optimal for virus mutant selection. The selections were carried out with several LCMV-GP-(275-289)-specific CTL clones but selection was also possible with LCMV-GP-(275-289)-specific cytotoxic polyclonal T cells. The most common escape mutation was an amino acid change of aspargine (AAT) to aspartic acid (GAT) at position 280; an additional mutation was glycine (GGT) to aspartic acid (GAT) at position 282. The results show that relevant point mutations within the T-cell epitope of LCMV-GP-(275-289) occur frequently and that they are selectable in vitro by CTLs.
The biological role of such T cell epitope escape mutants in vivo is not very clear as yet; but at least for LCMV it is conceivable that such mutant viruses may establish persistent infections in mice more readily.

Such T cell epitope escape mutants have also been sought in HIV patients with variable success. Whereas one paper failed to find them (Meyerhans *et al.* 1991) another study did. Philips et al. (Phillips *et al.* 1991) isolated HIV from patients over 2-4 years and tested whether HIV-specific cytotoxic T cell responses of the same patients were able to recognize the HIV variants or not. They found a few examples of HIV isolates that expressed a mutated T cell epitope, so that the new variant was not recognized by T cells specific for the corresponding epitope in previous isolates.

If CD8 T cells are involved in HIV control, true T cell epitope escape mutants may reveal two consequences dependent upon diametrically opposed points of view. If HIV is a cytopathic virus then such mutants should render the virus highly pathogenic, and accelerate disease. If as is more likely the case, HIV is not cytopathic and causes cell destruction via T cell mediated immunopathology, then such escape mutants should cause less or no disease. There are two experiments of Nature that may be very revealing. First, HIV infection of the human embryo before 7-10 weeks of gestation i.e. before immunocompetence is reached for T cells, may cause an asymptomatic carrier infection, because of immunological tolerance. Second, the isolation of an HIV that is not recognized by any CTL response of a patient or by any T cells from any HIV immune patient at all may be one ultimate proof of such successful T cell epitope HIV escape mutants and of the postulate that HIV is basically a poorly or noncytopathic virus. HLA class I antigen polymorphism which enables the species to present a great number of different peptides immunogenically is obviously a serious obstacle for such a virus to develop but the idea need not be unreasonable and the consequences not too worrying, because possibly we are observing a virus on its way to become an "ideal" virus that is finding an optimal balance with its host.

CONCLUSION

Immune escape by infectious agents that are not directly and acutely pathogenic for the host, is the result of coevolution. There is not one general mechanisms, instead there are some general principles that are used or misused by each virus or group of viruses (or infectious agents in general).

Each example has its own characteristics and the sophisticated balances and "gentlemen agreements" between virus and host reveal a very precise contour of the immune system, of the rules and efficiencies of induction of immune responses and the efficiency of effector mechanisms. Analysis of mechanisms of persistance of viruses in a host therefore provides not only excellent insights into pathogenesis of disease but also into fundamental parameters of the immune system.

Table 2. Mechanisms of viruses escaping immune surveillance

Immunological

Tolerance:	Infection during fetal life or at birth
	Mimicry
Antigenicity reduced:	Viral antigen down modulated
	mutation of B or T cell epitope
	MHC products blocked or down modulated
Immunogenicity reduced :	Viral antigen in non-APC, integration of virus
Immunosuppression:	Specific / nonspecific
	via cytopathic effects
	via impairment of "luxury" function of cell
	via immunopathology

REFERENCES

Aebischer, T., Moskophidis, D., Hoffmann Rohrer, U., Zinkernagel, R.M. and Hengartner, H. (1991) *Proc.Natl.Acad.Sci.USA*, **88**, 11047-11051.

Ahmed, R., Byrne, J.A. and Oldstone, M.B.A. (1984) *J.Virol.*, **51**, 34-41.

Borrow, P., Tishon, A. and Oldstone, M.B.A. (1991) *J.Exp.Med.*, **174**, 203-212.

Buchmeier, M.J., Welsh, R.M., Dutko, F.J. and Oldstone, M.B.A. (1980) *Adv.Immunol.*, **30**, 275-312.

Fazekas de St.Groth, S. (1981) In Steinberg, C.M.,et al. *The Immune System - The joint evolution of antigens and antibodies*. S.Karger, Basel, Munchen, Paris, Vol. 2, pp. 155-168.

Held, W., Shakhov, A.N., Waanders, G., Scarpellino, L., Luethy, R., Kraehenbuhl, J.-P., Robson MacDonald, H. and Acha-Orbea, H. (1992) *J.Exp.Med.*, **175**, 1623-1633.

Hotchin, J. (1962) *Cold Spring Harbor Symp.Quant.Biol.*, **27**, 479-499.

Johnson, E.D., Monjan, A.A. and Morse III, H.C. (1978) *Cell.Immunol.*, **36**, 143-150.

Lehmann-Grube, F. (1984) *Bacterial and viral inhibition and modulation of host defences*. Academic Press, London, pp. 211-242.

Leist, T.P., Rüedi, E. and Zinkernagel, R.M. (1988) *J.Exp.Med.*, **167**, 1749-1754.

Marrack, P., Kushnir, E. and Kappler, J. (1991) *Nature*, **349**, 524-526.

Meyerhans, A., Dadaglio, G., Vartanian, J-P., Langlade-Demoyen, P., Frank, R., Asjö, B., Plata, F. and Wain-Hobson, S. (1991) *Eur.J.Immunol.*, **21**, 2637-2640.

Mims, C.A. (1982) *Pathogenesis of infectious disease*. Academic Press, London, 2nd (ed.) pp. 10-160.

Moskophidis, D., Pircher, H.P., Ciernik, I., Odermatt, B., Hengartner, H. and Zinkernagel, R.M. (1992) *J.Virol.*, **66**, 3661-3668.

Odermatt, B., Eppler, M., Leist, T.P., Hengartner, H. and Zinkernagel, R.M. (1991) *Proc.Natl.Acad.Sci.USA*, **88**, 8252-8256.

Ohashi, P.S., Oehen, S., Bürki, K., Pircher, H.P., Ohashi, C.T., Odermatt, B., Malissen, B., Zinkernagel, R. and Hengartner, H. (1991) *Cell*, **65**, 305-317.

Oldstone, M.B.A., Fujinami, R.S. and Lampert, P.W. (1980) *Prog.Med.Virol.*, **26**, 45-93.

Phillips, R.E., Rowland-Jones, S., Nixon, D.F., Gotch, F.M., Edwards, J.P., Ogunlesi, A.O., Elvin, J.G., Rothbard, J.A., Bangham, Ch.R.M., Rizza, Ch.R. and McMichael, A.J. (1991) *Nature*, **354**, 453-459.

Pircher, H.P., Bürki, K., Lang, R., Hengartner, H. and Zinkernagel, R. (1989) *Nature*, **342**, 559-561.

Pircher, H.P., Moskophidis, D., Rohrer, U., Bürki, K., Hengartner, H. and Zinkernagel, R.M. (1990) *Nature*, **346**, 629-633.

Roost, HP., Charan, S., Gobet, R., Rüedi, E., Hengartner, H., Althage, A. and Zinkernagel, R.M. (1988) *Eur.J.Immunol.*, **18**, 511-518.

Valsamakis, A., Riviere, Y. and Oldstone, M.B.A. (1987) *Virology*, **156**, 214-220.

Webster, R.G., Laver, W.G., Air, G.M. and Schild, G.C. (1982) *Nature*, **296**, 115-121.

Zinkernagel, R.M. and Doherty, P.C. (1979) *Adv.Immunol.*, **27**, 52-142.

Immune Avoidance by Pathogenic *Neisseria*

J. E. Heckels

Molecular Microbiology Group, University of Southampton, Southampton General Hospital, Southampton SO9 4XY, UK

INTRODUCTION

The pathogenic *Neisseria* species comprise the gonococcus and the meningococcus, the causative agents of gonorrhoea and meningococcal meningitis respectively. Despite the great difference in the type of infection caused, the close genetic relationship between the two species results in similar antigenic structure, and mechanisms of pathogenesis. Gonococci and meningococci are specific human pathogens which initially colonise the mucosal surfaces of the genital and upper respiratory tracts respectively. The critical first stage of the infectious process is adhesion to non-ciliated columnar epithelial cells. Subsequent events include cell-invasion, intracellular multiplication, and exocytosis into sub-epithelial tissues. In the case of gonococcal infection, tissue penetration usually gives rise to the acute inflammatory response responsible for the typical symptoms of acute gonorrhoea, but less commonly may result in bacteremia and disseminated infection. With meningococci, invasion leads to bacteremia and subsequent spread to the CSF, resulting in the typical symptoms of meningococcal disease.

The above processes involve multiple interactions between the bacterial surface antigens and host systems, for which the *Neisseria* are particularly well adapted. In particular both species have evolved a series of complex mechanisms for immune avoidance including antigenic shift, antigen masking, molecular mimicry and antibody inactivation. This review will focus on the biological basis of immune avoidance with particular emphasis on the considerable problems this has posed for the development of effective vaccines against both gonococcal and meningococcal disease.

ANTIGENIC SHIFT

Pili

Pili are filamentous protein appendages which extend several microns from the surface of the bacteria and are predominantly composed of multiple copies of a single structural protein, pilin. They were the first *Neisseria* antigens to be associated with virulence. Freshly isolated gonococci which were infectious to human volunteers were found to be piliated, while laboratory passaged bacteria, which lost pilus expression, also lost virulence. It is generally agreed that the key role of pili is to mediate the initial adhesion between the bacteria and the mucosal epithelial cell surface, but the precise molecular basis for this interaction remains unclear (Heckels, 1989).

The association of pili with virulence initially prompted many studies to determine their potential for vaccination against gonorrhoea. Antibodies raised against purified pili in laboratory animals, have the ability to inhibit adhesion to epithelial cells, to opsonise for phagocytosis by PMN and to protect tissue culture cells against challenge by gonococci (Virji & Heckels, 1984). However the protective effects seen in these studies were usually confined to the homologous strain since pili isolated from different gonococcal strains show little antigenic cross reactivity.

Even greater antigenic diversity is generated by the fact that the pili expressed by a single strain undergo antigenic shift. During laboratory growth pilus expression is subject to phase variation with the bacteria switching between piliated (Pil$^+$) and non-piliated (Pil$^-$) phases. On reversion to the piliated phase the resultant pili are frequently composed of pilin with altered subunit molecular weight (Swanson & Barrera, 1983a). Similar variation in pilus expression can be seen during the course of both gonococcal (Zak et al. 1984) and meningococcal (Tinsley & Heckels, 1986) infections and in challenge experiments with human volunteers (Swanson et al. 1987). The variant pili so produced, exhibit very limited antigenic cross-reactivity (Virji et al. 1982), so that variation during the course of the natural infection enable the bacteria to escape the consequences of the host immune response (Zak et al. 1984).

The genetic mechanisms responsible for pilus antigenic shift have been the subject of considerable attention and a detailed description is beyond the scope of this review. In general new pilus types arise at high frequency, due to recombination between incomplete variant pilin silent sequences and the pilin structural gene (Hagblom et al. 1985; Hill et al. 1990). Further antigenic diversity can be generated through transformation with pilus gene sequences from other strains (Gibbs et al. 1989). Thus both gonococci and meningococci have evolved complex mechanisms which allow them to express an enormous, perhaps limitless, diversity of different pilus variants. Despite the apparent attraction of inducing an immune response designed to inhibit the initial colonisation of mucosal surfaces, the occurrence of widespread antigenic shift has frustrated attempts to use pili in a vaccine designed to prevent gonococcal infection (Boslego et al. 1991).

Considerable information on the structural basis of antigenic diversity has accumulated from sequencing pilin genes from variants of a number of gonococcal strains (Hagblom et al. 1985; Swanson et al. 1990; Nicolson et al. 1987). Pilins can be considered to contain three major regions, a region encompassing the first 53 amino acids which is highly conserved between strains, a semi-variable region (residues ca. 54-114) and a hypervariable region at the carboxy terminus, in which two cysteine residues form a disulphide loop. Amino acid substitutions give rise to variations within the semi-variable region, while insertions and deletions of up to four amino acids occur within the hypervariable region. Comparison of amino acid sequences of the pili from gonococci and meningococci shows that they have approximately 80% of their amino acids in common, variation in only 20% of the molecule is therefore responsible for generating the extreme antigenic diversity (Heckels, 1989).

Studies with synthetic peptides have revealed that immunisation with purified pili produces antibodies which predominantly react between residues 121 - 151, corresponding to the disulphide loop of the hypervariable region. The low levels of cross reacting antibodies generated, were directed against a weakly immunogenic determinant between residues 4860

(Rothbard *et al.* 1984). One possible strategy to overcome the problems posed by the antigenic variability of pili would be to use vaccination regimes designed to produce a response to the normally immunorecessive conserved regions of the pilus molecule, rather than the immunodominant variable regions. However the use of monoclonal antibodies has revealed that antibodies directed against the type-specific epitopes inhibit pilus adhesion to epithelial cells, but those directed against the conserved epitope within the weakly immunogenic regions 48 - 60 failed to display any protective effect (Virji & Heckels, 1984). Thus the main immune response to pili is directed against highly immunogenic hypervariable determinants and even the low levels of cross-reacting antibodies which are obtained are directed against non-protective epitopes.

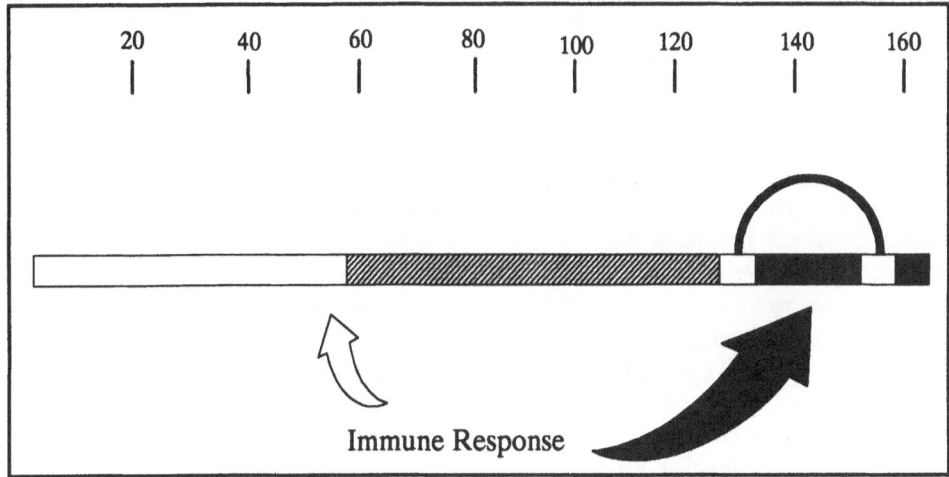

Figure 1: Immunochemistry of *Neisseria* pili. The structure of pilin molecule showing the position of conserved (open), semi-variable (hatched) and hypervariable regions (solid). The immune response is largely directed against the immunodominant variable domains. Conserved domains are immunorecessive with very limited response to the region encompassed by amino acids 48-60.

An alternative strategy to immunisation with intact pili has used synthetic peptides based on conserved pilin sequences. In one report immunisation with a peptide corresponding to the normally non-immunogenic region 69 - 84 produced antibodies which reacted with intact pili and which could inhibit adhesion of the homologous and one heterologous strain to human endometrial cells (Rothbard *et al.* 1985). However subsequent studies have revealed the occurrence of variation even within this region suggesting that such antibodies are unlikely to be widely cross-reactive.

One possibility which remains is that while the variable pilin molecule constitutes the major pilus structural unit, additional conserved accessory protein(s) may be present which act as specific cell receptors (Muir *et al.* 1988). If so, then this might represent a potential target for immunisation. However at the moment the occurrence of antigenic shift has defeated all attempts to use *Neisseria* pili for vaccination.

Opa proteins

The second stage of adhesion to the epithelial surface is mediated by outer membrane proteins designated Opa. Like pili the Opa protein is subject to phase variation, and variants which express the protein show increased adhesion to epithelial cells (Lambden *et al.* 1979). Antibodies to Opa inhibit gonococcal adhesion, opsonise for phagocytosis by PMN and protect tissue culture cells against challenge (Virji & Heckels, 1985). However Opa is also subject intra-strain antigenic shift on laboratory culture (Lambden & Heckels, 1979), during gonococcal (Zak *et al.* 1984) and meningococcal infections (Tinsley & Heckels, 1986) and in gonococcal challenge experiments with human volunteers (Swanson *et al.* 1988). At any time a single strain is capable of expressing between zero and four Opa species from a total repertoire of up to 11 (Swanson & Barrera, 1983b; Bhat *et al.* 1991) and antibodies raised against one variant protein show less than 5% cross reactivity against the other variants produced by the same strain (Heckels, 1981).

Despite the apparent similarity between the pilin and Opa systems gene expression is controlled by a quite different mechanism. Within a strain all the *opa* genes are constitutively transcribed, together with a preceeding series of repeats of the pentameric sequence CTCTT. Expression of any individual the Opa is controlled by additions or deletions to the repeat sequence which have the effect of moving the coding region in or out of frame (Stern *et al.* 1986). Sequencing of *opa* genes reveals the proteins to show considerable structural homology, with variation confined to two hypervariable and one semivariable regions (Aho & Cannon, 1988; Bhat *et al.* 1991). Experimental evidence and structural predictions indicate that the variable regions are exposed on the surface of the protein while conserved regions are located within the outer membrane (Heckels, 1989; Bhat *et al.* 1991). Thus as with pili the immune response to Opa is directed against hypervariable regions of the molecule, while conserved regions are immunorecessive.

Direct evidence for the effect of antigenic shift during the course of an infection has come from a study with serum from groups of consorts infected by gonococci. Serum from one woman recognised a 31kDa Opa but not a 29.5kDa Opa present in the isolate from her urethra, and did not recognise the Opa from the isolate of her male partner, thus confirming the specificity of the immune response. However the patient's serum did contain antibodies reactive with a 29.0kDa Opa from a second female contact of her partner, despite the absence of this protein from her own isolates (Zak *et al.* 1984). Thus a model for antigenic shift during infection is that the gonococcus produces one of several possible Opa variants, inducing an immune response which leads to the elimination of variants expressing that particular protein. The high rate of variation in Opa expression would always ensure the presence of alternate variants which would then have the selective advantage to grow and establish a new majority population. The cycle could then be repeated several times. Equivalent selection pressures would also operate during meningococcal infection and also concurrently on pilus expression.

INTERACTIONS WITH PHAGOCYTIC CELLS.

In order to cause systemic infection both meningococci and gonococci must avoid surveillence by phagocytic cells, particularly by polymorphonuclear leukocytes (PMN). Like many virulent bacteria meningococci produce an extracellular polysaccharide capsule with antiphagocytic properties (see below). In contrast gonococci do not produce a capsule and so an additional interaction of Opa with the immune system may occur. While expression Opa is associated with increased adhesion to epithelial cells, it is also associated with an increased adhesion to, and subsequent killing by PMN (Virji & Heckels, 1986; Rest *et al.* 1982). Thus the advantage of Opa expression on the mucosal surface becomes a disadvantage when the bacteria enter the circulation. However in laboratory experiments with Opa$^+$ variants, the high rate of phase variation between Opa$^+$ and Opa$^-$ ensures that a substantial minority of Opa$^-$ survive (Virji & Heckels, 1986). Thus the switch to Opa$^-$ during the natural infection would give the bacteria the ability to resist phagocytosis and subsequently recolonise other niches. This would be in accord with the observations that gonococci isolated from disseminated infection may be predominantly Opa$^-$, in contrast to genital isolates which are predominantly Opa$^+$. In addition, recent studies suggest that modification of LPS (see below) may also promote gonococcal resistance to Opa mediated phagocytosis (Rest & Frangipane, 1992).

BLOCKING ANTIBODIES

In contrast to the enormous antigenic diversity exhibited by the pili and Opa proteins, gonococci produce a highly conserved outer membrane protein PIII, and meningococci express an equivalent protein initially designated class 4 protein. Structural studies have revealed that the two proteins differ from one another by only a few amino acid residues (Klugman *et al.* 1989) and they are now designated with the common name Rmp. The prime defence against meningococcal infection is the presence of circulating antibodies which promote complement mediated bactericidal killing, similar activity is associated with immunity to the complications of gonococcal infection. An attractive strategy for immunisation would therefore be to induce an immune response to such a conserved protein which might be expected to protect against a wide range of species from both strains. However antibodies directed against the Rmp protein are not only non-bactericidal, but in addition, they inhibit the bactericidal activity of antibodies directed against other surface antigens. Thus antibodies directed against Rmp purified from convalescent sera of patients after gonococcal infection, inhibit the natural bactericidal activity of human sera (Rice *et al.* 1986). Similarly monoclonal antibodies raised against gonococcal Rmp, inhibit the bactericidal effect of monoclonal antibodies directed against range of other surface antigens on both gonococci and meningococci, and in addition inhibit killing of meningococci by normal human sera (Virji & Heckels, 1988; Munkley *et al.* 1991).

The mechanism of the blocking effect of antibodies directed against Rmp is not clear, but does not simply result from steric hindrance to antibodies directed against other surface antigens. Two monoclonal antibodies have been shown to recognise the same epitope located in a surface exposed loop on the Rmp protein, but differ in their blocking activity (Virji & Heckels, 1989). The two antibodies also differ in their ability to activate complement, only the blocking antibody being able to do so. In contrast other antibodies which activate complement but recognise different Rmp epitopes do not exert the blocking effect. Thus blocking effect is dependent on the binding of complement activating antibodies to specific Rmp epitopes suggesting that the mechanism may involve the diversion of complement components away from bactericidal sites (Joiner *et al.* 1985). Certainly the removal of Rmp protein in any potential vaccine is to be recommended, as it is likely to be antagonistic to the development of effective immunity.

Meningococcal Capsules

A major difference in antigenic structure of gonococci and meningococci is the production of an extracellular polysaccharide capsule by the latter. Like capsules of other virulent bacteria its function is to protect the meningococcus from host defences, the hydrophilic surface preventing phagocytosis by PMN. The structure of the capsule varies between strains and is responsible for serogroup specificity. Production of antibodies directed against the capsular polysaccharides is associated with the development of protective immunity. These observations have formed the basis of the development of purified polysaccharide vaccines directed against meningococci of serogroups A and C (Frasch, 1989). Unfortunately these vaccines provide only limited protection. Firstly, polysaccharides are poorly immunogenic in young children, the age group most at risk of meningococcal infection. Secondly the polysaccharide from serogroup B, the predominant cause of infection in most temperate countries is non-immunogenic even in adults. The group B polysaccharide is a polymer of α2-8 linked N-acetylneuraminic acid. Oligosaccharides containing the same structure occur in neural cell adhesion molecules found in the developing fetus and at lower levels in adult neural tissue (Finne et al. 1987). It would therefore appear that the purified group B polysaccharide capsule is not seen as a foreign antigen by the immune system and is hence non-immunogenic. One possible strategy for production of a vaccine against group B meningococcal disease would be to attempt to break tolerance by chemical manipulation of B-polysaccharide together with conjugation to a suitable carrier protein to provide Th-cell epitopes. However this approach is controversial because of the possibility of inducing antibodies with auto-immune effects.

Lipopolysaccharide

Like all Gram-negative bacteria the pathogenic *Neisseria* produce lipopolysaccharide in the outer membrane. Many of the symptoms of systemic meningococcal disease appear to result from the release into the circulation, of membrane vesicles containing LPS, which have endotoxin activity. The structure of the LPS is of the R type in which a relatively short oligosaccharide core is linked to lipid A, without the addition of polysaccharide O chains. Antibodies directed against LPS are bactericidal for both gonococci and meningococci (Virji & Heckels, 1988; Saukkonen et al. 1988). However like pili and Opa, the LPS molecule is subject to antigenic shift during infection (Apicella et al. 1987) and virulence has been associated with the expression of a core oligosaccharide which mimics human blood group precursors (Schneider et al. 1991). In addition gonococci produce an enzyme which utilises host cell CMP-NANA to modify LPS, by the addition of N-acetyl neuraminic acid to the core oligosaccharide (Mandrell et al. 1990). This sialylation is associated with the development of the resistance of gonococci to bactericidal killing by normal human sera. The sialylated epitope is also constitutively expressed by meningococci (Mandrell et al. 1991). Thus the structure of this important target for immune defences against the pathogenic *Neisseria* is subject not only to antigenic shift, but also to molecular mimicry and molecular masking.

IgA1 PROTEASE

Pathogenic *Neisseria* are mucosal pathogens and as such first encounter the immune system on the mucosal surface. Mucosal secretions from patients with gonorrhoea contain IgA antibodies directed against pili, and these have the ability to inhibit gonococcal adhesion to epithelial cells (Tramont, 1977). However both species produce a protease which is active against IgA1, cleaving the molecule at the hinge region (Plaut, 1983). The production of the enzyme is associated with virulence, since it is not secreted by the non-pathogenic *Neisseria* but is also produced by other mucosal pathogens. However the precise contribution of the enzyme in pathogenesis is not fully established since it has no activity against IgA2.

CONCLUSIONS

The pathogenic *Neisseria* are highly successful pathogens which are well adapted to colonisation of mucosal surfaces, tissue invasion and potential systemic spread. As such they encounter, and must evade both mucosal and systemic immune defences. Accordingly they produce an impressive repertoire of mechanisms which modify their antigenic properties and so permit immune avoidance. These include phase variation of antigen expression, antigenic shift, induction of blocking antibodies, molecular mimicry, molecular masking and inactivation of mucosal antibodies. These mechanisms not only play an important role in their success as pathogens but have so far hampered attempts to produce successful vaccines against both gonococcal and meningococcal infections.

Table: *Neisseria* antigens which contribute to immune avoidance

Neisseria Antigen	Biological Role	Immune Avoidance
Pili	Cell Adhesion	Phase variation: Antigenic Variation
Opa	Cell Adhesion	Phase variation: Antigenic Variation
Rmp	Unknown	Blocking Antibody
LPS	Endotoxin	Phase Variation : Molecular Masking: Molecular Mimicry
IgA1 Protease	Protease	Antibody Destruction
Meningococcal Capsule	Anti-phagocytic	Molecular Masking: Molecular Mimicry

REFERENCES

Aho EL Cannon JG (1988) Microb Path 5:391-398

Apicella MA Shero M Jarvis G *et al.* (1987) Infect Immun 55:1755-1761

Bhat KS Gibbs CP Barrera O et al (1991) Mol Microbiol 5:1889-1901

Boslego J W Tramont E C Chung R C et al (1991) Vaccine 9:154-162

Finne K Bitter-Suerman D Goridis C Finne U (1987) J Immunol 138:4402-4407

Frasch C (1989) Clin Microbiol Rev 2:S134-S138

Gibbs C Reimann BY Schultz E Kaufmann A Haas R Meyer T (1989) Nature 338:651-652

Hagblom P Segal E Billyard E So M (1985) Nature 315:156-158

Heckels J E (1981) J Bacteriol 145:736-742

Heckels J E (1989) Clin Microbiol Rev 2:S67-S71

Hill S Morrison S Swanson J (1990) Mol Microbiol 4:1341-1352

Joiner KA Scales R Warren KA Frank MM Rice PA (1985) J Clin Invest 76:1765-1772

Klugman K Gotschlich E Blake MS (1989) Infect Immun 57:2066-2071

Lambden PR Heckels JE James LT Watt PJ (1979) J Gen Microbiol 111:305-312

Lambden P Heckels J (1979) FEMS Microbiol Lett 5:263-265

Mandrell RE Lesse A Sugai JV et al (1990) J Exp Med 171:1649-1664

Mandrell RE Kim JJ John CM et al (1991) J Bacteriol 173:2823-2832

Muir LL Strugnell RA Davies JK (1988) Infect Immun 56:1743-1747

Munkley A Tinsley CR Virji M Heckels JE (1991) Microb Path 11:447-452

Nicolson IJ Perry ACF Heckels JE Saunders JR (1987) J Gen Microbiol 133:553-561

Plaut AG (1983) Ann Rev Microbiol 37:603-622

Rest RF Fisher SH Ingham ZZ Jones JF (1982) Infect Immun 36:737-744

Rest RF Frangipane JV (1992) Infect Immun 60:989-997

Rice PA Vayo HE Tam M Blake MS (1986) J Exp Med 164:1735-1748

Rothbard JB Fernandez R Schoolnik GK (1984) J Exp Med 160:208-221

Rothbard JB Fernandez R Wang L *et al.* (1985) Proc Nat Acad Sci USA 82:915-919

Saukkonen K Leinonen M Kayhty H *et al.* (1988) J Infect Dis 158:209-212

Schneider H Griffiss JM Boslego JW *et al.* (1991) J Exp Med 174:1601-1605

Stern A Brown M Nickel P Meyer TF (1986) Cell 47:61-71

Swanson J Robbins K Barrera O et al (1987) J Exp Med 165:1344-1357

Swanson J Barrera O Sola J Boslego J (1988) J Exp Med 168:2121-2129

Swanson J Morrison S Barrera O Hill S (1990) J Exp Med 171:2131-2139

Swanson J Barrera O (1983a) J Exp Med 158:1459-1472

Swanson J Barrera O (1983b) J Exp Med 157:1405-1420

Tinsley CR Heckels JE (1986) J Gen Microbiol 132:2483-2490

Tramont E C (1977) Journal Of Clinical Investigation 59:117-123

Virji M Everson JS Lambden PR (1982) J Gen Microbiol 128:1095-1100

Virji M Heckels JE (1984) J Gen Microbiol 130:1089-1095

Virji M Heckels JE (1985) Infect Immun 49:621-628

Virji M Heckels JE (1986) J Gen Microbiol 132:503-512

Virji M Heckels JE (1988) J Gen Microbiol 134:2703-2711

Virji M Heckels JE (1989) J Gen Microbiol 135:1895-1899

Zak K Diaz JL Jackson D Heckels JE (1984) J Infect Dis 149:166-174

Escape Mechanisms in Schistosomiasis

M. Capron, O. Duvaux-Miret and A. Capron

Centre d'Immunologie et de Biologie Parasitaire, Unité Mixte INSERM U167 - CNRS 624, Institut Pasteur, 1 rue du Pr A. Calmette, 59019 Lille Cédex (France).

INTRODUCTION

Survival and growth of infectious organisms in the face of the immune response mounted by their immunocompetent hosts is made possible through one of two strategies : one is to outrun rejection by fast replication and/or mutation, as observed in some bacterial or viral infections ; the other is to impair development or expression of immunity so as to evade the deleterious consequences of the host's counter-attack. Indeed, it is now recognized that in most infections, it is the infectious organism that determines whether or not to allow itself to be rejected by immune effectors. This is clearly the case of parasites which in general induce chronic infections and are thus confronted for months and often years with the immune response that they have to maintain within limits compatible with both their own survival and that of their mammalian hosts. This leads in many instances to the refined elaboration in these organisms of a complex network of processes allowing them to escape the defence mechanisms of their host. Such a complexity gives the expression of immunity in parasitic diseases a dynamic aspect reflecting the permanent balance between effector and regulatory mechanisms (Capron and Dessaint, in press).

In helminthic infections, such as schistosomiasis, the basis for acquired resistance is the development of so-called concomitant immunity, a situation in which the adult worm provides the major antigenic stimulus without it being affected by the resulting effector mechanisms. The major target of immunity in this case is the invasive (larval) stage of the parasite, i.e. the schistosomula. Among the multiplicity of adaptative mechanisms, the present review will focus on 3 main escape mechanisms well illustrated in schistosomiasis : antigenic mimicry, immunosuppression by release of parasite-derived neuropeptides and isotypic selection, essential components of the immunodeviation towards non protective immune response at the expense of putatively protective ones.

ANTIGENIC MIMICRY

A first level of escape to strong immune response is antigenic mimicry (reviewed in Capron and Dessaint, in press). Indeed, one of the simplest ideas when considering the host-parasite relation and its stability in terms of evolution, population dynamics, and individual infection, is the likely existence of structures common to parasites and their hosts which would allow a precise adaptation of their respective metabolic requirements.

In the case of schistosomes, there are cross-reactivities between the various developmental stages of the parasite and antigenic components of its invertebrate or vertebrate host. Numerous parasite genes cloned so far show high levels of nucleotide sequence homology with mammalian genes, attributable to functional conservation since the corresponding proteins are potent immunogens. Besides, antigenic disguise appears to be employed by schistosomes during their life in their vertebrate host : indeed, within hours after their penetration through the skin they acquire a masking coat of host molecules, both glycolipids and glycoproteins, and worm receptors for host molecules such as the Fc portion of IgG have been characterized (Torpier et al., 1979). An interesting observation is acquisition by schistosomes of class I and class II major histocompatibility complex products (Sher et al., 1978). Although these acquired MHC products are recognizable on the surface of worms by alloantibodies or CD8+ T cells (Sher et al., 1978, Butterworth et al., 1979), they do not appear to play a role in promoting parasite recognition by T cells or in immunoregulation. It is generally assumed that these host antigens limit access of immune effectors to target antigens since their acquisition coincides with the capacity of transforming schistosomula to resist immune damage. Indeed, from this stage onward, disguised schistosomes are resistant to immune effectors, a phenomenon contributing to concomitant immunity.

In fact, the contribution of antigenic disguise or antigenic mimicry in the reduction of surface recognition of parasites is difficult to appreciate. In schistosomes, the acquisition of so-called host antigens proceeds concomitantly with intrinsic changes in membrane susceptibility to antibody-dependent killing and with the loss of expression in the worm membrane of the major target antigens that are expressed on the schistosomulum surface. Sharing of epitopes between host and parasite may accordingly appear to be related more to phyletic convergence and adaptation to common metabolic environments than to a strategy of evasion from immune clearance. However, foreign parasitic antigens may trigger an inappropriate immune response against host self-antigens through molecular mimicry, which can be involved in the autoimmunity often associated with parasitic diseases.

IMMUNOSUPPRESSION BY RELEASE OF PARASITE-DERIVED NEUROPEPTIDES

The mechanisms underlying the induction of suppression by parasites appear in some models to be related to intrinsic properties of parasite products. Many parasite-derived factors with suppressive activity have been described, but most of these are poorly characterised at the molecular level. In the case of S.mansoni, adult worms were shown to liberate, in low osmolarity conditions, a small molecule named SDIF for Schistosome-Derived Inhibitory Factor. It exhibits several immunosuppressive properties such as inhibition of the proliferative response of lymphocytes to parasite antigens and to interleukin 2 (Mazingue et al., 1986). More indirectly, S.mansoni schistosomula were shown to bear a Fc γ receptor which allows the binding of IgG on the surface of the parasite (Torpier et al., 1979). The same larvae can liberate several proteases, and the resulting IgG hydrolysate suppresses most functions of macrophages.

This immunosuppressive effect was reproduced by a tripeptide named TKP (threonine-lysine-proline) which sequence can be deduced from the second constant domain of IgG, given the putative cleavage sites which can be identified (Auriault *et al.*, 1985). These examples illustrate clearly the ability of parasites to disturb their host defences.

More recently, evidence has been given that schistosomes could produce immunomodulatory neuropeptides (Duvaux Miret et al., 1990) which might constitute tools of immune evasion (Duvaux Miret *et al.*, 1992). Neuropeptides, originally described in the nervous system, are also expressed by cells of the immune system and exhibit a number of immunomodulatory properties. Among neuropeptides supporting the information exchange between the nervous and immune systems, opioids have been particularly studied. Although more than eight endogenous opioids have been isolated from the central nervous system, their molecular origin appears to be rather simple in mammalian species, where they derive from three different precursors. Proopiomelanocortin (POMC) differs from the two others, proenkephalin and prodynorphin, in that it contains a single copy of the opioid core Tyr-Gly-Gly-Phe-(Met or Leu), at the amino-terminal end of β-Endorphin (βE). It exhibits also the particular feature to be cleaved into several other bioactive peptides, mainly adrenocorticotropin (ACTH) and melanocyte stimulating hormones (α-, β- and γ-MSH) which present biological activities completely different from opioid effects (see review by Duvaux-Miret and Capron, 1991).

Several studies have been concerned with the effects of these neuropeptides on cells of the immune system, which will not be reviewed in detail here . The main papers focusing on ACTH show its ability to suppress antibody synthesis by B cells and IFN γ production by T cells in mouse (Johnson *et al.*, 1982; Johnson *et al.*, 1984). ACTH has also been shown to inhibit the IFN γ activation of macrophages but to enhance the proliferative response of B lymphocytes (see review in Smith *et al.*, 1991). Few MSH effects have been described; the main report demonstrated its capacity to inhibit the migration of human monocytes and granulocytes *in vivo* and *in vitro* (Van Epps and Mason, 1990). Concerning β-Endorphin, many papers have reported its various immunomodulatory properties. Some effects seem to be now established such as antibody production inhibition, diminution of T lymphocyte chemotactic factor production by monocytes, augmentation of granulocyte migration (reviewed in Duvaux-Miret *et al.*, 1992).

More recently, evidence has been given that invertebrates synthesize neurohormones and that functions of invertebrate immunocytes can be regulated through mechanisms very similar to the vertebrate neuroimmune interactions (Stefano *et al.*, 1989). Considering the immunomodulatory properties of POMC-derived peptides, and their phylogenetic conservation, we have postulated their existence in *Schistosoma mansoni* and their possible implication in the interactions with both intermediate and definitive hosts.

POMC-related peptides were identified in the main stages of the parasite life cycle (cercariae, schistosomula, adult worms and miracidia). Radioimmunoassays detected molecules related to ACTH, αMSH and βE in crude extracts of all stages (Duvaux-Miret et al., 1990 ; Duvaux-Miret and Capron, 1991). The presence of three main POMC-derived peptides led us to hypothesize the existence of a POMC gene in the parasite. Using oligonucleotide probes specific for the regions of the genes encoding POMC conserved amino-acid motifs, we could demonstrate that related sequences are present in the parasite genome (Duvaux-Miret et al., 1991). To assess our hypothesis of their implication in the interaction with the vertebrate host, the three peptides were tested for their ability to be released by S. mansoni adult worms under physiological conditions (incubation at 37°C in Minimal Essentiel Medium). We could demonstrate the release of β-E and ACTH. ACTH and β-E could also be released after incubation of S. mansoni miracidium with MEM medium, as well as in the haemolymph of S. mansoni infected snails, but not in the case of non-infected snails (Duvaux-Miret et al., 1992).

The role of the parasite POMC-derived peptides in immunosuppression in the definitive host (man) and in the intermediate host (Biomphalaria glabrata snail) has been examined. Coincubation of adult worms with human polymorphonuclear leukocytes or B. glabrata immunocytes led to the appearence of MSH in the medium. This α-MSH-resulted from conversion of parasite ACTH by a neutral endopeptidase (NEP, enkephalinase, CALLA or CD10) present on the surface of vertebrate PMN and invertebrate immunocytes (Shipp et al., 1990). α et β MSH were shown to inhibit activation of mammalian PMN and snail immunocytes in vitro (Duvaux-Miret et al., 1992). Interestingly the participation of POMC-derived factors in the inhibition of immunocytes obtained from S. mansoni infected snails was suggested by the reversal of inhibition in the presence of antibodies directed against ACTH or αMSH (Duvaux-Miret et al., 1992). These findings are highly significant because αMSH has been shown to inhibit adherence and locomotion of PMN, monocytes and invertebrate immunocytes (Stefano et al., in press). ACTH can exert the same immunosuppressive cellular effects through conversion into αMSH by means of NEP (Shipp et al., 1990).

Taken together, the results strongly favor our hypothesis regarding the role of neuropeptides, specifically ACTH and MSH in an adaptative process of the parasite to circumvent host attack. In the definitive host, liberated ACTH appears to exert a direct effect on host cells. Of particular relevance is the observation of a decrease of IFN γ and IL-2 during schistosomiasis, after maturation of the worms and oviposition (Grzych et al., 1991; Pearce et al., 1991). It is now suggested that in mouse this is due to a recruitment of Th2 and a diminution of the Th1 populations (Pearce and Sher, 1991). Decrease of IFN production by T cells is one of the described effects of ACTH. Furthermore, release of ACTH might contribute to the preferential induction of Th2 responses during schistosomiasis, since it would be expected to stimulated the adrenal gland to produce corticosterone, able to induce switching from Th1 to Th2 responses (reviewed in Pearce and Appleton, in press). The transformation of ACTH into an MSH-like molecule, which can exhibit different immunosuppressive properties, could account for the immune evasion of adult worms from many non specific effector mechanisms. The present results demonstrate the ability of schistosomes to release ACTH and β-Endorphin which can (i) act directly on immune cells of both the definitive and the intermediate hosts and (ii) be converted into immunosuppressive subtances by NEP, an enzyme present on the same cells. They constitute an example of molecular mimicry by which parasites use phylogenetically conserved molecules to interfere with the host response.

BLOCKING ANTIBODIES AND ISOTYPIC SELECTION

Observations made in our laboratory have demonstrated that anaphylatic antibody isotypes (IgE and certain subclasses of IgG) play a prominent role in conjunction with non-lymphoid cells like macrophages, eosinophils and platelets in antibody-dependent cell-mediated cytotoxicity(ADCC) against schistosomes (Capron et al., 1985). In addition to this restricted range of protective antibody isotypes, other restricted antibody isotypes are involved in the escape strategy of schistosomes. Against the carbohydrate epitopes of the major surface glycoprotein of schistosomula (gp38), two antibody isotypes can indeed be produced during the experimental infection. One, IgG2a in the rat, is an anaphylatic antibody involved in protection (Grzych et al., 1982); the other, IgG2c in the rat, is a blocking antibody as shown both in vitro by inhibiting ADCC involving eosinophils and in vivo, since injection of a monoclonal IgG2c antibody inhibits the protection conferred by passive transfer of the protective IgG2a monoclonal antibody specific for the same gp38 antigen (Grzych et al., 1984). This observation led to the identification of such blocking antibodies in human Schistosomiasis mansoni. The presence of antibodies to S. mansoni carbohydrate epitopes was shown to be highly correlated to a state of non-resistance to reinfection in human children and blocking antibodies of IgM and IgG2 isotypes were characterised in the susceptible but not in the resistant population, and progressively declined with age (Khalife et al., 1986; Butterworth et al., 1987).

The hypothesis was therefore that young children's continued susceptibility to reinfection was due to the predominant "blocking" Ab response, preventing the effects of potentially protective Ab. Although a protective role of IgE in helminth infection has been postulated for many years and clearly demonstrated in the rat model of schistosomiasis (Capron and Dessaint, 1985), the nature of the protective responses expressed in older individuals has been only recently elucidated in 3 different studies. First in a group of patients with S. haematobium infections, Hagan et al., (1991) have demonstrated a significant relationship between IgE Ab against adult worms or egg Ag and resistance to reinfection. Similar findings have been reported in an endemic area for schistosomiasis in Kenya (Dunne et al., 1992) and in Brazil (Rihet et al., 1991). These field studies demonstrated therefore for the first time, the putative protective role of IgE in man. Interestingly, it was also reported in the same studies that, in contrast to IgE, IgG4 Ab were in fact associated with susceptibility to reinfection, suggesting that resistance to schistosomiasis was depending on the balance between the protective effects of IgE and the blocking effect of IgG4. Concerning the precise effector mechanisms, IgE Ab have been clearly involved in ADCC reactions mediated by macrophages, eosinophils and platelets, in both the rat and human (Capron and Dessaint, 1985). IgG4 antibodies have been shown to inhibit IgG dependent ADCC by eosinophils (Khalife et al., 1989), whereas they could also compete with IgE in the binding to larval Ag (Rihet et al., 1991), and inhibit IgE dependent release of histamine by basophils (A. Dessein, personal communication).

More recently, the isotypic profiles of the antibody response to a protective antigen of S. mansoni, the glutathione S-tranferase (Sm28-GST) have been investigated. A relationship was established between IgE and IgG4 antibody levels to rSm28-GST and, respectively, the presence or the absence of immunity to reinfection supporting the conclusions of Hagan et al., (1991) and Rihet et al., (1991) of an association between IgE antibodies to adult worm antigens and the lack of reinfection after treatment. We have recently extended this study to an examination, in the same group of individuals, of IgA antibodies to rSm28-GST. We present evidence for an age-related association between an increase in IgA antibodies and a decrease in the level of reinfection after chemotherapeutic cure (Grzych et al., in press). In addition to the participation of IgA antibodies in ADCC mechanisms mediated by eosinophils and directed against schistosomula (Capron et al., 1988), we could also show that IgA Ab were involved in the impairment of schistosome fecundity, an evaluated by the reduction of egg laying and egg viability (Grzych et al., in press).

Taken all together, these studies have demonstrated a variety of antibody responses that correlate respectively either with susceptibility (IgM, IgG2, IgG4) or with resistance (IgE, IgA) to subsequent reinfection. This is not surprising. It might be expected that naturally developing immunity to such a complex organism would be multifactorial in nature. Moreover in a chronic transmissible disease such as schistosomiasis, whereas the antibody profile of potentially protective antibodies might not differ significantly in immune and non-immune populations, susceptibility is controlled by the development of a blocking antibody response and thus it might be crucial to characterise markers of susceptibility rather than putative indicators of protection as it was anticipated by Capron et al. (1984). Besides, the general concept that a given antigen might selectively be involved in the production of a defined effector isotype and that the same molecule may elicit both effector and blocking isotypes, is obviously of considerable importance in the framework of vaccine strategy.

Molecular mimicry, release of immunoregulatory molecules, induction of isotype selection represent only a few examples of the ingenious machinery by which schistosomes can escape effector mechanisms and can maintain their prolonged survival in their immunocompetent hosts. Recent evidence has been obtained that together with their capacity to express genes encoding host or host-like regulatory molecules, schistosomes can also express receptors for host interleukins such as TNF α for instance and that signal provided through this interaction could lead to an up-regulation of female worm fecundity (Amiri et al., 1992). It is likely that the increasing availability of refined molecular tools will allow in the future a better understanding of the permanent dialogue between parasites and their hosts.

REFERENCES

Amiri P, Locksley RM, Parslow TG, Sadick M, Rector E, Ritter D, Mc Kerrow JH, (1992) Nature 356:604-607

Auriault C, Joseph M, Tartar A, Bout D, Tonnel AB, Capron A (1985) Int J Immunopharmacol 7:73-79

Butterworth AE, Bensted-Smith R, Capron A, Duiton PR, Dunne DW, Grzych JM, Kariaki HC, Khalife J, Koech D, Muyambi M, Ouma JH, arap Siongok JK, Sturrock RF (1987) Parasitology 94:281-300

Butterworth AE, Vadas MA, Martz E, Sher A (1979) J Immunol 122:1314-1321

Capron A, Capron M, Joseph M, Dissous C, Auriault C (1984) New approaches to vaccine development, In : Bell R, Torrigiani G (eds) Schwabbe and Co Ag, Basel, p 460

Capron A, Dessaint JP (1985) Ann Rev Immunol 3:455-476

Capron A, Dessaint JP (in press) Adv Neuroimmunol

Capron M, Tomassini M, Van der Vorst E, Kusnierz JP, Papin JP, Capron A (1988) Cr Acad Sc Immunol 307:397-402

Dunne D, Butterworth AE, Fulford AJC, Kariuki HC, Langley JG, Ouma JH, Capron A, Pierce RJ, Sturrock RF (1992) Eur J Immunol 22:1483-1494

Duvaux-Miret O, Capron A (1991) Adv Neuroimmunol 1:41-57

Duvaux-Miret O, Dissous C, Gautron JP, Pattou E, Kordon C, Capron A (1990) New Biol 2:93-99

Duvaux-Miret O, Stefano GB, Smith EM, Dissous C, Capron A (1992) Proc Natl Acad Sci USA 89:778-781

Grzych JM, Capron M, Bazin H, Capron A (1982) J Immunol 129:2739-2743

Grzych JM, Capron M, Dissous C, Capron A (1984) J Immunol 133:998-1004

Grzych JM, Grezel D, Xu CB, Neyrinck JL, Capron M, Ouma JH,

Butterworth AE, Capron A (in press) J Immunol

Grzych JM, Pearce EJ, Cheever A, Caulada ZA, Caspar P, Heiny S, Lewis F, Sher A (1991) J Immunol 146:1322-1327

Hagan P, Blumenthal UJ, Dunne D, Simpson AJG, Wilkins HA (1991) Nature 349-243

Johnson HM, Smith EM, Torres BA, Blalock JE (1982) Proc Natl Acad Sci USA 79:4171-4174

Johnson HW, Torres BA, Smith EM, Dion LD, Blalock JE (1984) J Immunol 132: 246-250

Khalife J, Capron M, Capron A Grzych JM, Butterworth AE, Dunne DW, Ouma JH (1986) J Exp Med 164:1626-1640

Khalife J, Dunne DW, Richardson BA, Mazza G, Thorne KJ, Capron A, Butterworth AE (1989) J Immunol 142:4422-4427

Mazingue C, Stadler B, Quatannens B, Capron A, De Weck A (1986) Int Archs Allergy Appl Immunol 80:347-354

Pearce E, Appleton J (in press) Parasitol Today

Pearce EJ, Caspar P, Grzych JM, Lewis FA, Sher A (1991) J Exp Med 173:159-166

Pearce EJ, Sher A (1991) Exp Parasitol 73:110-116

Rihet P, Demeure CE, Bourgeois A, Prata A, Dessein AJ (1991) Eur J Immunol 21:2679-2686

Sher A, Hall BF, Vadas MA (1978) J Exp Med 148:46-52

Shipp MA, Stefano GB, D'Adamio L, Switzer SN, Howard FD, Sinisterra J, Scharrer B, Reinherz E (1990) Nature 347:394-396

Smith EM, Hughes TK, Leung MK, Stefano GB (1991) Adv. Neuroimmunol. 1:7-16

Stefano GB, Leung MK, Zhao X, Scharrer B (1989) Proc Natl Acad Sci USA 86:626-630

Stefano GB, Smith DE, Smith D, Hughes TK (in press) MSH can deactivate both TNF stimulated and spontaneously active immunocytes In: Boer H, Maat A (eds) Molluscan Neurobiology, Elsevier North Holland, Amsterdam

Torpier G, Capron A, Ouaïssi MA (1979) Nature 278:447-449

Van Epps DE, Mason MM (1990) Modulation of leukocyte migration by alphamelanocyte stimulating hormone In: Florey E, Stefano G (eds) Comparative Neuropeptide Pharmacology, Manchester University Press, UK p 335

Can New Vaccines Overcome Parasite Escape Mechanisms?

P.-H. Lambert, J.A. Louis, and G. Del Giudice

Microbiology and Immunology, World Health Organization, CH-1211 Geneva 27, Switzerland

INTRODUCTION

The development of new vaccines remains an essential goal of research dealing with the immunology of infectious diseases. Most existing vaccines are the result of efforts largely based on empiricism. However, in view of the numerous parasite escape mechanisms which characterize many non-vaccine preventable infectious diseases more logical approaches are now required to reach that goal.

ESCAPE MECHANISMS

Three types of parasite escape mechanisms may have to be overcome.

First, intrinsic parasite factors such as antigen polymorphism, parasite sequestration and surface membrane enzymatic activities are often involved.

Second, host-related escape mechanism are commonly hampering attempts to prevent or control infectious diseases. This includes specific effects on antiparasite responses such as antigenic mimicry, induction of tolerance, low immunogenicity or genetically restricted immune responses. Several agents can also generate inefficient or blocking antibody isotypes as well as unappropriate effector T cells.

Third, parasite escape may depend on the production of immunosuppressive molecules (e.g. neuropeptides), on cytopathic or functional effects on cells of the immune systems as well as on polyclonal B/T activation or superantigen effects.

SOME EXISTING VACCINES DO OVERCOME ESCAPE MECHANISMS

The possibility to overcome such escape mechanism has been evidenced by the fact that some existing vaccines do operate despite similar unfavourable conditions.

Table 1: Existing vaccines which overcome escape mechanisms

DISEASE	ESCAPE MECHANISM	VACCINATION STRATEGY
influenza	antigenic variation	monitoring of antigenic drift/shift yearly immunisation with adapted vaccine
tetanus	no immunity following infection	protection by anti-toxin Ab
Hepatitis B	perinatal infections leads to tolerance and carrier state (+cirrhosis and cancer)	immunization with HB surface Ag in alum, soon after birth, prevents tolerization and infection
Haemophilus influenzae b meningitis	no B cell response to capsular polysaccharide under 2 years of age	oligosaccharides conjugated to toxoid (D or T) or to Neisseria OMP become T dependant

DEFINITION OF AN OPTIMAL VACCINE PROFILE.:
THE EXAMPLE OF LEISHMANIASIS

A first step towards the development of new vaccines is the definition of an optimal vaccine profile through deciphering the nature of specific protective responses and of escape mechanisms specific for the target disease. Leishmaniasis provides a good example of such approaches.

At the cell level, *protection* against leishmania parasites has been shown to depend largely on the triggering of the nitric oxyde pathway (Mauel et al 1991) following the activation of infected macrophages by cytokines (Green et al 1991). This process requires CD4+ T cells of the Th1 type and/or CD8+ T cells, as demonstrated through comparing cytokine profiles in resistant and susceptible mice (Heinzel et al. 1991) and through cell transfer experiment in SCID mice(Holaday et al 1991). It also appears that CD4+ cells specific for antigens expressed only by *live* parasites are needed.(Müller and Louis 1991).

Several *escape mechanisms* have been defined: (a) hiding of parasites inside phago-lysosomes (b) generation of CD4+ Th2 cells, as shown in non-healing strains of mice ; (c) competition between antigens released from live and dead organisms and (d) reduced capacity to present antigen to MHC II in infected macrophages (see review by Louis, same volume)

On the basis of those observations, several approaches have been used to *prevent or modulate* leishmaniasis. Thus, in experimental systems, protection against *L. major* has been induced by injecting parasite antigens intravenously, which resulted in a predominant CD8+

T cell response (Farrell et al. 1989). Similarly, the disease can be modulated by treatment of mice with anti-IL4 antibodies (Sadick et al, 1990) .

APPROACHES TOWARDS VACCINES WHICH OVERCOME ESCAPE MECHANISMS

A general strategy to overcome parasite escape mechanisms can be envisaged as follows
(i) to trigger selectively appropriate antigen presentation pathways for binding to MHC I or MHC II;
(ii) to influence the balance between Th1 and Th2 subsets;
(iii) to increase the immunogenicity of potential vaccines .

Recent data indicate the feasibility of *orienting presentation pathways* of selected antigens. For example, acid-resistant liposome-encapsulated antigens undergo processing in antigen-presenting cells with a binding of resulting peptides to MHC I but not MHC II, whereas the processing of acid-sensitive liposomes results in peptide binding to both MHC II and MHC I (Harding et al. 1991). Mannan-coated liposomes have also been used to induce CD8+ T cells specific of HTLV1 *env* and *gag* sequences (Noguchi et al., 1991). Non-replicating or highly attenuated live vectors are now the most advanced candidates for the development of vaccines having the capacity to induce CTL responses . Thus, a highly attenuated virus (NYVAC) derived from vaccinia through the deletion of 18 ORF genes (Tartaglia et al., 1992) and a canary pox virus (ALVAC),which does not replicate in mammalian cells, have been used successfully as vectors for vaccines (Taylor J, 1991). Both of these vectors have been shown to induce the expression of foreign antigens and to generate high levels of antibodies and/or of CTL. The technology has been applied to rabies, and measles.(Taylor J, 1991,1992).

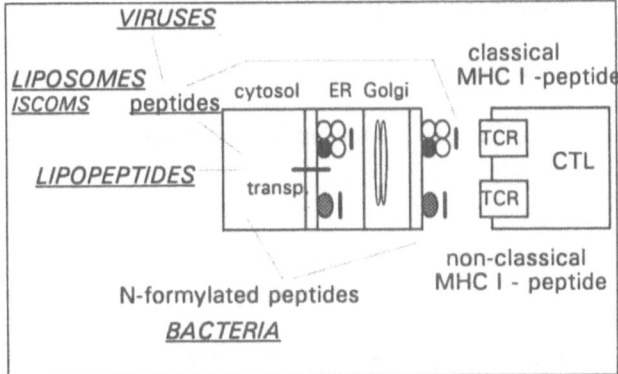

Fig. 1. The use of live vectors and of non-living vehicles for MHC I presentation of antigens

Similar CD8+ responses were obtained using as vector Salmonella strains attenuated by genetic engineering and which could not anymore survive in vivo (Aggarwal et al. 1990).The presentation of bacterial peptides might then follow different rules. Indeed, it was recently shown that N formylated peptides could bind to less polymorphic non-classical MHC I molecules and thus were recognized with a much lower degree of genetic restriction than viral peptides (Pamer et al., 1992)

There is an important potential role of new adjuvants and carriers for *inducing selectively Th1 or Th2* patterns of responses with new candidate vaccines. For example, a conjugate of BSA with a *Plasmodium cynomolgi* peptide, $NAGG_5$, was shown to induce close to 100% of IgG1 anti-$NAGG_5$ antibodies in mice, if injected in 2% squalane whereas it induces over 55% of IgG2a, IgG2b and IgG3 if injected in 2% squalane containing a non-ionic block polymer (L121) and a low-toxicity LPS (Kalish, Check and Hunter 1991). These results confirm that selected adjuvants can favour the generation of either Th1 or Th2 patterns of cytokines.

New procedures are now emerging which allow for *increasing the immunogenicity* of relatively poor immunogens using carriers or vaccine vehicles.

First, *new carriers* have been identified which may act as "promiscuous" or universal T cell epitopes recognized by most MHC II haplotypes (table 2). They appear promising for the design of peptide vaccines but are presently limited by their requirements for strong adjuvants.

Table 2. *Some promiscuous T cell carrier epitopes*

tetanus toxin	*aa 830-843*	*(Panina-Bordignon et al., 1989)*
	aa 947-957	*(Ho et al.,1990)*
influenza haemagglutinin	*aa 307-319*	*(Roche & Cresswell, 1990)*
P. falciparum CS	*aa 378-398*	*(Sinigaglia et al,, 1988)*

Second, multiple antigen peptide systems (MAP) have been shown to considerably amplify the response to synthetic peptides (Tam 1988) and in some instances to overcome genetic restriction (Pessi et al 1991).

Third, some *heat-shock proteins* from mycobacterial origin have been used as carriers for peptides and for oligosaccharides in animals primed or not with BCG. These HSP's have exhibited an exceptional carrier effect in the absence of any additional adjuvant (Lussow *et al* 1990,1991, Barrios *et al* 1992). Furthermore, one of these molecules, HSP70, was found to act as an excellent carrier in the absence of any priming with BCG, in mice, rabbits and monkeys (Barrios et al 1992).

In view of these results a hypothesis was made suggesting that the HSP carrier may exert a chaperone effect during the late stage of antigen processing (Del Giudice 1992).

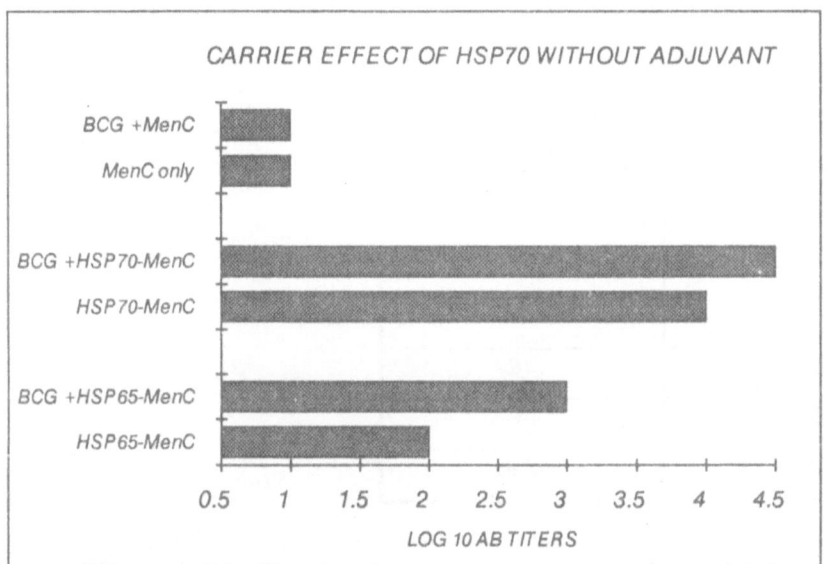

Fig. 2. Antibody response to a group C meningococcal polysaccharide conjugated to HSP70 injected in mice primed or not with BCG. Control mice received the non-conjugated oligosaccharide.

The use of *controlled release systems* for antigen delivery in vivo now appears as a very promising approach to optimize immunization procedures and it may contribute to the overcoming of parasite escape. A remarkable increase in the duration of immune responses after a single exposure to subunit vaccines (figure 3) has been obtained with such controlled-release biodegradable microspheres containing antigens (Aguado and Lambert 1992).

Therefore, the hope to do better than evolution and overcome some of the limitations within the equilibrium between infectious agent and host, which is the declared goal of vaccination strategies, is now moving towards more realistic approaches after a phase of doubts and disappointments.

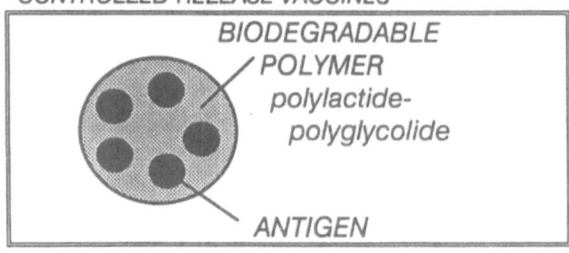

CONTROLLED-RELEASE VACCINES

lactide/glycolide ratio	time of antigen release (weeks)
50/50	4
65/35	8
85/15	16
100 D-L	30
100 L	52-85

Fig. 3. Delayed release of antigen from microspheres of different polylactide/polyglycolide ratios.

REFERENCES

Aggarwal A, Kumar S, Jaffe R, Hone D, Gross M, Sadoff J (1990) J Exp Med 172: 1083-1090
Aguado MT, Lambert PH (1992) Immunobiol 184: 113-125
Barrios C,Lussow AR, van Emden J, Van Der Zee R, Rappuoli R, Costantino P,, Louis JA,
   Lambert PH, Del Giudice G (1991) Eur J Immunol 22: 1365-1372
Del Giudice G (1992) Current Opin Immunol 4: 454-459
Farrell JP, Müller I, Louis JA (1989) J Immunol 142: 2052-2056
Green SJ, Nacy CA, Meltzer MS (1991) J Leuk Biol 50: 35-42
Harding CV, Collins DS, Kanagawa O, Unanue ER (1991) J Immunol 147: 2860-2863
Heinzel FP, Sadick MD, Mutha SS, Locksley RM (1991) Proc Natl Acad Sci USA 88: 7011-
   7015
Ho PC, Mutch DA, Winkel KD, Saul AJ, Jones JI, Doran TJ, Rzepcyk CM,(1990) Eur J
   Immunol 20: 477-483
Kalish MI, Check IJ, Hunter RL (1992) J Immunol 146: 3583-3590
Lussow AR, Del Giudice G, Renia L, Mazier D, Verhave JP, Verdini AS, Pessi A, Louis JA,

Lambert PH (1990) Proc Natl Acad Sci USA 87: 2960-2964

Lussow AR, Barrios C, van Emden J, Van Der Zee R, Verdini AS, Pessi A, Louis JA, Lambert PH, Del Giudice G (1991) Eur J Immunol 21: 2297-2302

Mauel J, Betz-Corradin S, Buchmüller-Rouiller Y (1991) Res Immunol 142: 557-580

Müller I, Louis JA (1989) Eur J Immunol 19: 865-871

Noguchi Y, Noguchi T, Sato T, Yokoo Y, Itoh S, Yoshida M, Yoshiki T, Akiyoshi K, Sunamoto J, Nakayama E, Shiku H (1991) J Immunol 146: 3599-3603

Pamer EG, Wang CR, Flaherty L, Fischer Lindahl K, Bevan MJ (1992) Cell 70: 215-223

Panina-Bordignon P, Tan A, Termitelen A, Demotz S, Corradin G, Lanzavecchia A (1989) Eur J Immunol 19: 2237-2242

Pessi A, Valmori D, Migliorni P, Tougne C, Bianchi E, Lambert PH, Corradin G, Del Giudice G (1991) Eur J Immunol 21: 2273-2276

Roche PA, Cresswell P (1990) J Immunol 144: 1849-1856

Sadick MD, Heinzel FP, Holaday BJ, Pu RT, Dawkins RS, Locksley RM (1990) J Exp Med 171: 115-127

Sinigaglia F, Guttinger M, Kilgus J, Doran DM, Matile H, Etlinger H, Trzeciak A, Gillesen D, Pink JR (1988) Nature 336: 778-780

Tam JP (1988) Proc Natl Acad Sci USA 85: 5409-5413

Tartaglia J,Perkus M, Taylor J, Norton E, Audonnet JC, Cox w, Davis SW, Van Der Hoeven J, Meignier B, Riviere M, Languet B,Paoletti E (1992) Virology 188: 217-232

Taylor J, Trimarchi C, Weinberg R, Languet B, Guillemin F, Desmettre P, Paoletti E (1991) Vaccine 9: 190-193

Taylor J, Weinberg R, Tartaglia J, Richardson C, Alkhatin G, Breidis D, Appel M, Norton E, Paoletti E (1992) Virology 187: 321-328

HIV and Human Complement: Molecular Mechanisms and Biological Consequences

Manfred P. Dierich, Peter Marschang and Clara Larcher

Institut für Hygiene, Leopold-Franzes-Universitaet and Ludwig-Boltzmann-Institut für AIDS-Forschung, Fritz-Pregl-Str. 3, A-6020 Innsbruck, Austria

In the late seventies retrovirologists and complementologists were convinced, based on a large body of data that retroviruses could be killed by complement-dependent lysis. It was for instance shown that human serum effectively destroyed avian, rodent and feline leukaemia viruses (Welsh et al. 1975, 1976). Therefore it was assumed that due to the human complement system no retroviruses were found in humans, but when in 1978/1979 HTLV-I and a few years later HIV-1 were discovered this dogma was not true anymore.

Obviously human retroviruses are not killed by human serum (Hoshino et al. 1984, Banapour et al. 1986). The same experience was made in our laboratory. In the presence of 50 % rat serum (37°C, 60') HIV-1 strain IIIB lost its infectivity completely. On the other hand, human serum did not affect the capacity of HIV-1 to induce syncytia formation of C8166 cells, when assayed on day 5 or day 10 after infection in comparison to buffer treatment (unpublished results).

These findings raised the question: *Do human retroviruses in contrast to animal retroviruses not interact with complement?*
The complement components can be activated either by the classical or by the alternative pathway. In both pathways enzymes are formed, which cleave C3, the central component of the complement system, and generate the small fragment C3a and the large fragment C3b. This fragment then in turn activates C5 and induces the generation of the C5b-C9 complex, which, when formed, typically induces lysis of the target cell or the target pathogen. The fact that human retroviruses are not lysed by human serum does not necessarily imply that these early steps of complement activation with generation of C3b and its covalent fixation do not occur on the virus or HIV-1-infected cells.

We first tested whether on HIV-infected cells treated with native serum, in comparison with cells treated with heated serum, the deposition of the third component of the complement system could be detected. This was clearly the case. HIV-infected cells carried large amounts of C3 fragments after incubation with human serum (Sölder et al. 1989a). In addition, complement-coated HIV-infected cells formed aggregates with human macrophages, which express various complement receptors, namely complement receptor type 1 and 3 (CR1, CR3). This aggregation was not observed with cells which had been pretreated with heated serum (Sölder et al. 1989a).

A detailed analysis of this phenomenon proved that the deposition of C3 on HIV-infected cells was brought about via the alternative pathway of complement activation. This is based on the following facts: Heat inactivated serum and serum containing EDTA caused no deposition of C3, both blocking the alternative as well as the classical pathway. A selective block of the classical pathway by EGTA-chelated or C4-deficient serum, however, did not inhibit C3 deposition. Uninfected cells, on the other hand, showed very little complement activation (Sölder et al. 1989a).

These findings lead to the conclusion that HIV-infected cells upon interaction with human serum activate the alternative pathway of the complement system and deposit C3 fragments on their surface. This opsonization allows interaction with cells carrying complement receptors but does not induce lysis of the infected cells.

To test whether isolated virus was also capable to activate the complement system human serum was incubated with HIV-1. It could be proven that increasing amounts of virus consumed increasing amounts of haemolytic activity. This clearly shows that *in the absence of antibody HIV-1 activates the complement system*. Since this activation could be inhibited not only by EDTA but also by EGTA, it was very likely that the classical pathway of the complement system was the mechanism of activation in the case of isolated virus (Sölder et al. 1989a).

This was tested directly using the reconstituted C1 complex which contained radiolabelled C1s. The cleavage of the proenzyme C1s is an obvious indicator of C1 activation. Immune complexes serving as a positive control were well capable of inducing transition from proenzyme C1s into its activated two-chain form. It was shown that purified HIV, obtained from the supernatants of infected H9 cells, also induces this activation process. Virus-free supernatants, prepared in parallel to the virus preparation, were unable to do so (Ebenbichler et al. 1991).

To test, whether this C1 activation was brought about by C1q binding, HIV-1 was incubated with radiolabelled C1q. Then this mixture was run on a Sephacryl S-1000-column. For comparison, HIV-1 was incubated with C1s. In this experiment HIV-1 co-eluted with C1q, indicating binding to the virions, whereas C1s eluted independently.

On the basis of these data we postulate that HIV-1 activates the classical pathway of the complement system by binding C1q directly and thus activating the C1 complex. The consecutive activation of C4 and C2 leads to the cleavage of C3, which then becomes deposited on the virus surface. The demonstration of complexes between viral proteins and human C3 provided direct evidence for this mechanism (Sölder et al. 1989a).

Presently we do not know why the virus, although it activates the complement system, is not lysed. Our hypothesis, which we are trying to prove, is the following:
It had been observed by several investigators that human cells are inefficiently lysed by human serum (Atkinson and Farries 1987, Lachmann 1991). For this effect membrane-bound restriction molecules are responsible, like the decay accelerating factor (DAF) (Nicholson-Weller et al. 1982), membrane cofactor protein (MCP) (Cole et al. 1985, Seya et al. 1986), CR1 (Medof et al. 1982), homologous restriction factor (HRF) (Schönermark et al. 1986, Zalman et al. 1986) or protectin (CD59) (Sugita et al. 1988). Our assumption is that a budding virus may take along these restriction factors from the host cell membrane, which inhibit the complement cascade at various steps (Dierich et al. 1990). Thus the virus may activate complement and deposite C3, but the efficient formation and function

of the C5b-C9 complex may be inhibited. Whether this hypothesis is true or not, has still to be proven. It is known that the virus takes along host cell proteins when budding from the cell surface, which has been nicely documented by Gelderblom. He showed that HIV-1 incorporates HLA class 2 antigens, when propagated in class 2 positive H9 cells, while it does not contain these HLA antigens after growth in MOLT-3 cells, which do not express these molecules (Gelderblom et al. 1988, 1989). Instead of relying on human host cell proteins the virus may use own, hitherto unknown, factors for protection against human complement. If so, they would behave like the human complement restriction factors.

To analyze, which viral protein was responsible for C1q binding and classical pathway activation, in a first basic experiment recombinant gp160, the precursor of the envelope proteins of HIV-1, was deposited on nitrocellulose paper and then incubated with normal human serum, serum containing EDTA or EGTA, heated normal serum or C4 deficient serum. Only in case of normal serum deposition of C3 was observed (Sölder et al. 1989a). This clearly proved that a portion of gp160 activated the classical pathway of the complement system.

In a next set of experiments gp160, gp120, gp41 and p24 were compared by ELISA for their capacity to activate the complement system. Aside from gp160 only gp41 was able to do so (Ebenbichler et al. 1990a, 1990b).
In these experiments a soluble recombinant form of gp41 (sgp41) was used, which comprises the proposed outer membrane part of the transmembrane protein, i.e. aa 539-684 of gp160. It could be proven that spg41 bound C1q dose dependently nearly as efficient as immune complexes and that this binding leads to the activation of the C1 complex.

To analyze the sites in gp41 responsible for this classical pathway activation a series of peptides, each 15 amino acids long, were tested for their capacity to activate the system and deposit C3. Two peptides covering amino acids 591-620 and another peptide covering amino acids 561-575 appeared to activate the complement system best (Ebenbichler et al. 1991).

What are the *biological consequences of complement activation by HIV-1 in the absence of antibodies against HIV?*
First of all we tested whether HIV-1 treated with human serum would show enhanced infectivity. While limiting amounts of HIV-1 within 21 days did not cause infection of U937 cells, as detected by fluorescence, virus pretreated with complement efficiently infected this promonocytic cell line. With antibodies against CR3, the iC3b receptor, this infection could be inhibited (Sölder et al. 1989b, Reisinger et al. 1990). Later Boyer and colleagues proved that infection of MT2 cells depended only on complement receptor type 2 (CR2) and pretreatment of the virus with complement; an interaction with CD4 was not necessary (Boyer et al. 1991). A similiar enhancement of infection was demonstrated by us for CR2-expressing Raji cells (unpublished results) and EBV-transformed B cells (Gras et al. 1991). The mechanisms involved are activation of complement by HIV-1, deposition of C3 on the virus and efficient uptake of the complement-coated virus by complement receptor carrying cells (Fig.1). These cells include macrophages and B-lymphocytes (Dierich et al. 1988), follicular dendritic cells (Reynes et al. 1985), and T-lymphocytes (Wilson et al. 1983, Fischer et al. 1991).

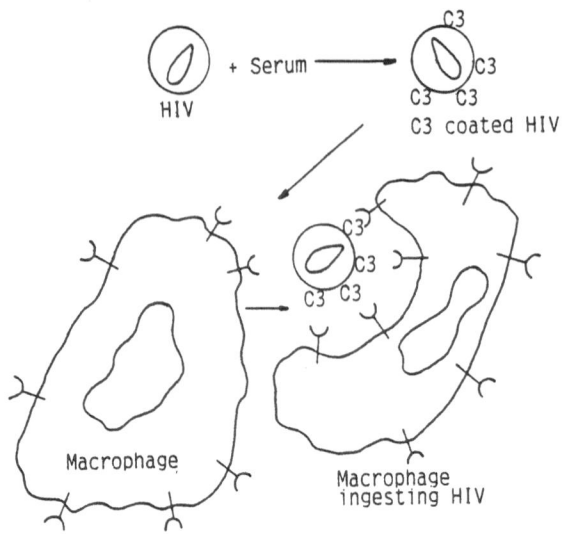

Fig. 1. Comlement-dependent enhancement of HIV-1 infectivity. HIV-1 directly activates the complement system, leading to C3 deposition on the viral surface. Opsonization of the virus increases uptake by complement receptor carrying cells, e.g. macrophages.

This concept has enormous implications. Robinson originally showed with diluted human serum that antibody-mediated enhancement of infection by HIV-1 is dependent on complement activation by antibody-coated HIV-1 (Robinson et al. 1988). Recently Jiang demonstrated that all antibodies against any peptide of gp160, except for the antibody against the V3 loop, dependent on complement enhanced the infection (Jiang et al. 1991). The antibody against V3 loop inhibited the infection. When comparing various isolates, he also could show that this held true only for some isolates, while in some cases also antibodies against the V3 loop would enhance infection in a complement-dependent way.

Thus we have to accept that HIV-1 activates the complement system in the pre-immune phase and that the virus is efficiently taken up by cells carrying complement receptors, including macrophages and T-cells. After generation of antibodies against the envelope proteins complement becomes again very important in so far as it turns antibodies into infection-enhancing mechanisms. This might have critical implications for individuals who are vaccinated with recombinant gp160 or subunits of it (Fig. 2).

Pathologists have demonstrated that HIV is found in large quantities in lymph follicles (Tenner-Racz et al. 1986, 1992, Spiegel et al. 1992). Obviously follicular dendritic cells are infected in the early phase of HIV infection.

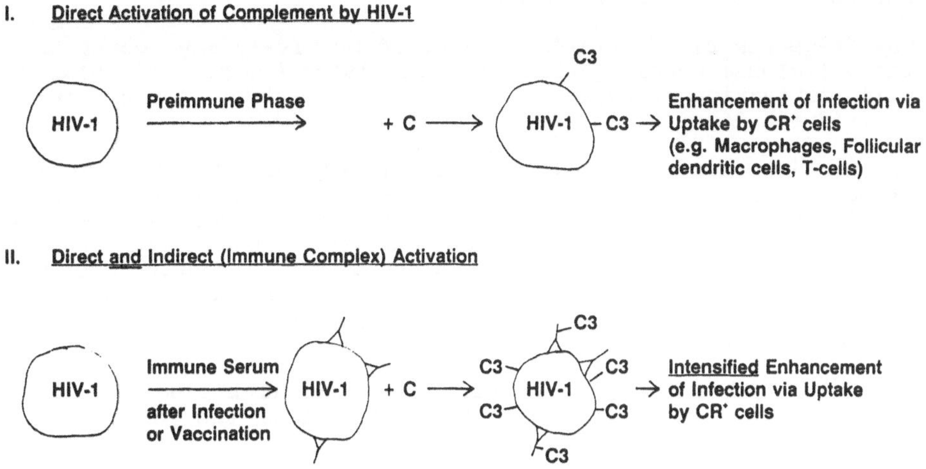

I.   **Direct Activation of Complement by HIV-1**

II. **Direct and Indirect (Immune Complex) Activation**

Fig. 2. Two mechanisms of complement-dependent enhancement of HIV-1 infection. In the early preimmune phase of infection, direct activation of complement by HIV-1 facilitates the infection of complement receptor carrying cells. After generation of anti-HIV antibodies additional activation of complement by virus-bound antibodies increases C3 deposition and leads to intensified enhancement HIV-1 infection.

This is not surprising at all, since follicular dendritic cells carry CR1, the C3b receptor, CR2, the C3d receptor and CR3, the iC3b receptor (Reynes et al. 1985). Furthermore, the complement dependence of the localization of antigens and immune complexes in lymph follicles had been shown in the early seventies (Dukor et al. 1974). Therefore one can savely assume that the position of HIV-1 in lymph follicles and in particular in follicular dendritic cells is complement-mediated.

Various groups have reported that patients with HIV infection show a constant activation of their complement system, as indicated by reduced complement levels and appearance of complement fragments (Perricone et al. 1987, Senaldi et al. 1989, Lin et al. 1988). This could be due to direct viral effects but certainly also to immune complexes, present in large quantities in these sera.

Another effect of the complement activation by HIV-1 and in particular by gp41, has been demonstrated by Füst and collagues. He demonstrated that gp41, when covered with C3, shows less

695

accessibility for antibodies against gp41 (Füst 1992). In this respect, complement might interfere with the effector phase of the antibody-based immune response.

The fragments of C3, which are induced by HIV-1, may contribute to the polyclonal B-cell (Melchers et al. 1985) and possibly also to T-cell activation (Erdei et al. 1984) observed in HIV-infected individuals (Lane et al. 1983, Fuchs et al. 1988).

Another important aspect of complement related HIV-1 biology is the effect of HIV-1 on the expression of CR2. As could be demonstrated in our laboratory HIV-1 induces down-regulation of CR2 in CR2-carrying T-cells and B-cells. In both cases, CR2 was either totally or largely lost (Larcher et al. 1990). In contrast to HIV, infection with HTLV-I is associated with an increased expression of CR2 on some T-cell lines (Schulz et al. 1986) as well as peripheral blood lymphocytes (Mc Nearney et al. 1991).

Thus, HIV-1 has numerous complement-dependent biological effects, which in all likelihood have a major impact on the pathogenesis of the disease.

Table 1. Biological consequences of the interaction between HIV-1 and human complement

o   Enhancement of infection via CR1, CR3 and CR2 independent of antibody (first phase of infection)

o   Follicular localization of HIV-1

o   Conversion of antibodies against gp120/gp41 into enhancing antibodies except for some against V3 loop

o   Complement consumption

o   Covering of epitopes

o   Immunomodulation (CR2), B cell activation

ACKNOWLEDGEMENT

This work was supported by the Ludwig Boltzmann-Gesellschaft, State of Tyrol, FWF P8287

REFERENCES

Atkinson JP, Farries T (1987) Immunol Today 8: 212–215
Banapour B, Sernatinger J, Levy JA (1986) Virol 152: 268–271
Boyer V, Desgranges C, Trabaud MA, Fischer E, Kazatchkine MD
  (1991) J Exp Med 173: 1151–1158
Cole JL, Housley GA, Dykman TR, MacDermott RP, Atkinson JP (1985)
  Proc Natl Acad Sci 82: 859–863
Dierich MP, Schulz TF, Eigentler A, Huemer H, Schwäble W (1988)
  Mol Immunol 25: 1043–1051
Dierich MP, Ebenbichler CF, Hallfeldt PH, Prodinger WM, Fuchs D,
  Wachter H (1990) Mol Immunol 27: 1349–1353
Dukor P, Dietrich FM, Gisler RH, Schuman G, Bitter-Suermann D (1974)
  In: Brent L, Holborrow J (eds), Progress in Immunology 11, vol 3,
  North-Holland, Amsterdam, p 99
Ebenbichler CF, Thielens NM, Vornhagen R, Marschang P, Arlaud GJ,
  Dierich MP (1991) J Exp Med 174: 1417–1424
Ebenbichler CF, Weyrer W, Vornhagen R, Wachter H, Dierich MP (1990)
  VIIIth Int Congr of Virol, Berlin, Abstract P58-029
Ebenbichler CF, Weyrer W, Vornhagen R, Wachter H, Dierich MP (1990)
  Complement and Inflamm 7/3: 122
Erdei A, Späth E, Alsenz J, Rüde E, Schulz TF, Gergely J, Dierich MP
  (1984) Mol Immunol 21: 1215–1221
Fischer E, Delibrias C, Kazatchkine MD (1991) J Immunol 146:
  865–869
Fuchs D, Hausen A, Reibnegger G, Werner ER, Dierich MP, Wachter H
  (1988) Immunol. Today 9: 150–155
Füst G (1992) Immunol Today 13: A23–A24
Gelderblom HR, Özel M, Hausmann EHS, Winkel T, Pauli G, Koch
  MA (1988) Micron and Microscopica 19: 41–60
Gelderblom HR, Özel M, Pauli G (1989) Arch Virol 106: 1–13
Gras GS, Dormont D (1991) J Virol 65: 541–545
Hoshino H, Tanaka H, Miwa M, Okada H (1984) Nature 310: 324–325
Jiang S, Lin K, Neurath R (1991) J Exp Med 174: 1557–1563
Lachmann PJ (1991) Immunol Today 12: 312–315
Lane HC, Masur H, Edgar LC, Whalen G, Rook AH, Fauci AS (1983)
  N Engl J Med 309: 453–458
Larcher C, Schulz TF, Hofbauer J, Hengster P, Romani N, Wachter
  H, Dierich MP (1990) J AIDS 3: 103–108
Lin RY, Wildfeuer O, Franklin MM, Candido K (1988) Int Arch
  Allergy Appl Immunol 87: 40–46
McNearney T, Ebenbichler C, Marschang P, Tötsch M, Schulz TF,
  Dierich MP (1991) Complement and Inflammation 8/3-4: 192–193
Medof ME, Iida K, Mold C, Nussenzweig V (1982) J Exp Med. 156:
  1739–1754
Melchers F, Erdei A, Schulz TF, Dierich MP (1985) Nature 317:
  264–267
Nicholson-Weller A, Burge J, Fearon DT, Weller PF, Austen KF
  (1982) J Immunol 129: 184–189
Perricone R, Fontana L, De Carolis C, Carini C, Sirianni MC,
  Aiuti F (1987) Clin Exp Immunol 70: 500–507
Reisinger EC, Vogetseder W, Berzow D, Köfler D, Bitterlich G,
  Lehr HA, Wachter H, Dierich MP (1990) AIDS 4: 961–965
Reynes M, Aubert JP, Cohen JHM, Audouin J, Tricottet V, Diebold
  J, Kazatchkine MD (1985) J Immunol 135: 2687–2694
Robinson WE, Montefiori DC, Mitchell WM (1988) Lancet i: 790–794
Schönermark S, Rauterberg EW, Shin ML, Löke S, Roelcke D,
  Hänsch GM (1986) J Immunol 136: 1772–1776
Schulz TF, Petzer A, Stauder R, Eigentler A, Dierich MP (1986)
  Immunobiol 173: 372

Senaldi G, Peakman M, McManus T, Davies ET, Tee DEH, Vergani D
  (1989) Lancet ii: 624
Seya T, Turner JR, Atkinson JP (1986) J Exp Med 163: 837-855
Sölder BM, Schulz TF, Hengster P, Löwer J, Larcher C,
  Bitterlich G, Kurth R, Wachter H and Dierich MP (1989)
  Immunol Lett 22: 135-146
Sölder BM, Reisinger EC, Köfler D, Bitterlich G, Wachter H,
  Dierich MP (1989) Lancet ii: 271-272
Spiegel H, Herbst H, Niedobitek G, Foss HD, Stein H (1992)
  J Pathol 140: 15-22
Sugita Y, Nakano Y, Tomita M (1988) J Biochem 104: 633-637
Tenner-Racz K, Racz P, Bofill M, Schulz-Meyer A, Dietrich M, Kern P,
  Weber J, Pinching AJ, Veronese-Dimarzo F, Popovic M, Klatzmann D,
  Gluckman JC, Janossy G (1986) AJP 123: 9-15
Tenner-Racz K, Racz P, Schmidt H, v.Stemm A, Fox CH (1992)
  VIII Int Conf AIDS/III STD World Congr, Amsterdam, 19-24 July,
  Abstr No PoA2130
Welsh RM, Cooper NR, Jensen FC, Oldstone MBA (1975) Nature
  257: 612-614
Welsh RM, Jensen FC, Cooper NR, Oldstone MBA (1976) Virol
  74: 432-440
Wilson JG, Tedder TF, Fearon DT (1983) J Immunol 131: 684-689
Zalman LS, Wood LM, Müller-Eberhard HJ (1986) Proc Natl Acad
  Sci 83: 6975-6979

B Cell Activation and HIV-1 Infection

A. Amadori, R. Zamarchi, M.L. Veronese, A. Veronesi,
S. Indraccolo, M. Mion, E.D. Andrea, A. Del Mistro and L. Chieco-
Bianchi

*Institute of Oncology, Interuniversity Center for Research on Cancer, University of
Padova, Via Gattamelata 64, I-35128, Padova, Italy*

INTRODUCTION

The human immunodeficiency virus type 1 (HIV-1), the causative agent
of the acquired immunodeficiency syndrome (AIDS) in man, shows
tropism for CD4+ T cells mostly, and this accounts for the intense
impairment in cellular immunity function (for review, see Rosenberg
and Fauci 1989); however, B cell function is also severely deranged
by HIV-1 infection (for review, see Amadori and Chieco-Bianchi
1990). Although the biologic properties of HIV-1, and its life cycle
in the host are well understood, a comprehensive view of the
pathogenesis of AIDS is still lacking, and it is still debated
whether the T cell deficiency depends only on the virus' cytopathic
effect, or whether other mechanisms also come into play. Among the
pathways proposed as co-factors in generating AIDS, the possibility
that B cell deregulation might be involved is intriguing. This
article reviews current knowledge on the features characterizing B
cell function during HIV-1 infection, and addresses the possible
participation of the humoral compartment in the pathogenesis of AIDS
and associated disorders.

FEATURES OF B CELL DEREGULATION DURING HIV-1 INFECTION

Abnormalities in B cell function were evidenced in AIDS patients
even before the causative agent was identified (Lane et al 1983),
and include decreased production of antibodies against recall
antigens following in vitro polyclonal activation, high numbers of
activated B cells in circulation, and spontaneous in vitro antibody
synthesis (Lane et al 1983; Birx et al 1986). While this important B
cell activation is in fact polyclonal, its target is not
indiscriminate; we (Amadori et al 1988, 1989) and others (Pahwa et
al 1989) demonstrated that most spontaneously produced antibodies in
unstimulated culture supernatants of peripheral blood mononuclear
cells (PBMC) from seropositive patients are directed against HIV-1
determinants, and that no significant re-activation of memory B

cells directed against recall antigens occurs.

This observation was further substantiated by the finding that IgG oligoclonal bands present in the serum of some HIV-1-infected patients are specific for HIV-1 determinants (Ng et al 1989; Amadori et al 1990). Thus, it is evident that HIV-1 infection induces a considerable B cell function derangement entailing the preferential expansion of HIV-1-specific cells, with the physical or functional loss of B lymphocytes that recognize previously encountered foreign antigens. Moreover, normal B cell development seems to be affected by HIV-1 infection; in fact, recent observations indicate that B lymphocytes from most seropositive patients show a maturational arrest at the level of the germinal center (Berberian et al 1991).

The mechanisms underlying spontaneous B lymphocyte activation are not fully defined. In fact, although interleukin-6 seems to play a pivotal role in sustaining its terminal steps (Amadori et al 1991a), little is known concerning the early events that promote this phenomenon. T cell responses to HIV-1 antigens or peptides are poor (Wahren et al 1987; Krowka et al 1989); while recent evidence indicated that HIV-1-infected T cells in vitro are able to provide help to normal B cells for polyclonal Ig synthesis (Macchia et al 1991), this is probably not the case in vivo, given the relatively low number of infected T cells in the peripheral blood and lymphoid organs (Schnittman et al 1990), and the relatively compartmentalized organization of the lymphoid tissues in humans. On the other hand, the spontaneous synthesis of HIV-1-specific antibodies can be down-regulated in vitro by the addition of mitogens through the activation of cytotoxic effectors (Amadori et al 1991b).

B cells in man and in rodents are not a homogeneous population, and two major subsets may be identified. One is characterized by the presence of the CD5 (Ly-1) antigen at the cell surface (for review, see Kipps 1990); while the precise function of this B cell sub-population is presently unclear, it seems to include B cells involved in the production of polyreactive autoantibodies and broadly-reacting anti-idiotypes (Casali and Notkins 1988). In view of the frequency of autoimmune phenomena in AIDS (reviewed in Amadori 1992), and the B cell derangement observed in seropositive patients, we recently addressed the phenotypic and functional behaviour of these sub-populations during HIV-1 infection. As shown in Fig. 1, an increase in the percentage of circulating CD5+ B cells was observed in seropositive patients; while this phenomenon was not strictly related to disease progression, it was mostly evident in AIDS patients. Recent evidence also demonstrated that CD5+ B cells are able to produce interleukin-10, a cytokine endowed with immunoregulatory activities (Ishida et al 1992); thus, it is also possible that the increased number of CD5+ B cells might potentiate immune impairment. From a functional point of view, most spontaneous B cell activation was found in the CD5- B cell compartment, and CD5+ B lymphocytes only produced about 10% of the total Ig spontaneously synthesized by in vivo activated B cells. In addition, only the CD5-

B cell subset showed spontaneous production of HIV-1-specific antibodies, which could not be demonstrated in unstimulated CD5+ B cell culture supernatants (manuscript in preparation). However, the specificity of the low levels of antibodies released by spontaneously activated CD5+ B cells is still obscure; they might possibly represent low-affinity, polyreactive autoantibodies, or an anti-idiotypic response to the steady production of HIV-1-specific

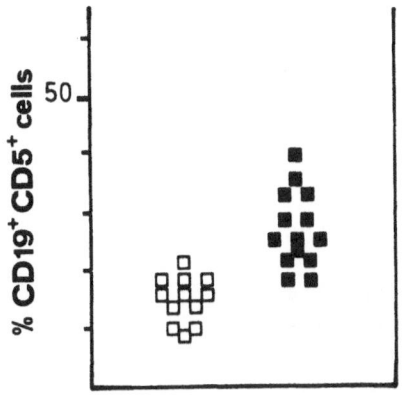

Fig. 1. Percentage of circulating CD5+ B cells in seropositive (■) and seronegative (□) subjects

Ig. This last case is particularly intriguing, as anti-idiotypic B cells raised to HIV-1-specific antibody-producing B cells might work to maintain their continuous stimulation; indeed, a putative "memory" role for CD5+ B cells has been advanced (UytdeHaag et al 1991).

We also recently addressed the effects of zidovudine (AZT) therapy in seropositive patients. AZT treatment was associated with a certain degree of humoral response normalization, since following therapy completion the patients showed significant recovery of mitogen-induced in vitro antibody production against recall antigens, including the Epstein-Barr virus (EBV). On the other hand, AZT treatment was not associated with a consistent change in the levels of spontaneous activation of HIV-1-specific B cells (manuscript in preparation).

POSSIBLE PATHOLOGIC SIGNIFICANCE OF ANTI-HIV-1 ANTIBODIES

Despite the intense and steady production of HIV-1-specific Ig, their protective role in controlling virus spread and disease progression in seropositive patients is unclear. In general, the early stages of viral infections are more efficiently controlled by antibodies, while established infections are better challenged by virus-specific cytotoxic T lymphocytes. Indeed, in the case of HIV-1 infection, anti-HIV-1 antibodies might be endowed with pathologic significance, leading to increased virus spread in the organism (Poli et al 1990; Bolognesi 1989), and generation of immunodeficiency through amplification of the virus-induced direct cytopathic effect on CD4+ T cells. Some of the pathways possibly leading to antibody-mediated immunopathologic damage (Amadori and Chieco-Bianchi 1992) are summarized in Table 1; here we will briefly discuss the possibility that gp120-containing immune complexes might down-regulate CD4+ cell function, and steady B cell activation might be involved in the pathogenesis of lymphoma development in AIDS patients.

Table 1. Possible pathways of anti-HIV-1 antibody-mediated immune damage

1. Enhancement of infection
2. Immunopathologic damage to non-infected CD4+ cells
   - molecular mimicry
   - production of anti-CD4 (anti-idiotypic) antibodies
   - gp120/anti-gp120 antibody complex formation
3. Increased risk of lymphoma development

Effect of Gp120/anti-gp120 Antibody Complexes.

We recently demonstrated that circulating CD4+ cells from HIV-1-infected patients with advanced disease are extensively covered by gp120/anti-gp120 antibody complexes adhering to the CD4 antigen, and that this phenomenon modulates their function in vitro (Amadori et al 1992). As shown in Table 2, the presence of these micro-complexes at the CD4+ cell surface might be responsible for damage to the immune system through several pathways; in particular, the possibility that gp120-containing immune complexes could mimic anti-CD4 antibody activity is worth stressing. In mice, anti-CD4 antibodies exert an intense modulating effect, in vitro as well as in vivo on T cell function

Table 2. Possible consequences of gp120/anti-gp120 antibody complex adhesion to CD4+ cells

1. Modulation of CD4 expression
2. Impaired cell-to-cell interactions
3. Complement- or cell-mediated damage
4. Functional cell impairment (death by apoptosis?)

(Goronzy and Weyand 1989; Biasi et al 1992); in addition, Newell et al (1990) demonstrated that separate ligation of CD4 in anti-T cell receptor (TCR)-stimulated CD4+ cells causes cell death through activation of the endonuclease machinery. It was recently shown that T lymphocytes from AIDS patients undergo apoptosis following in vitro culture and activation (Ameisen et al 1992). Thus, it is conceivable that the presence of gp120/anti-gp120 antibody complexes on the CD4+ cell surface might act as a "second signal" (Biasi et al 1992) promoting activation of the biochemical machinery, that ultimately leads to programmed cell death upon cell stimulation through the TCR:CD3 complex.

Lymphomagenesis in AIDS Patients

B cell lymphoma development is a major concern in seropositive patients (Pluda et al 1990). In some cases, Burkitt-like pictures are observed, with typical c-myc activation and chromosomal translocations; in others, the neoplastic cells show no genomic alterations, nor are EBV sequences consistently found. The pathogenesis of EBV positive or negative lymphomas without genomic alterations is mostly obscure; HIV-1 might play an indirect role by sustaining the steady B cell proliferation, and thus promoting the expansion of potential targets of subsequent transforming events. Indeed, serum oligoclonal IgG bands, which reflect extreme B cell clone deregulation, are mostly directed against HIV-1 determinants (Ng et al 1989; Amadori et al 1990); moreover, other human retroviruses have been shown to play an indirect promoting role in lymphomagenesis (Mann et al 1987). If this were also the case in HIV-1 infection, and in view of the prevalent orientation of the B cell activation towards HIV-1 antigens, it could be predicted that, at least in some cases, the lymphoma cells are specific for HIV-1 determinants, much as occurs in some B cell malignancies arising during HTLV-I infection (Mann et al 1987). Work is now underway in this laboratory to assess whether a part of the B cell lymphomas arising in AIDS patients derive from B cells initially directed against HIV-1 antigens.

While the view that HIV-1-specific B cell activation possibly underlies lymphoma development may be mostly attractive in the case of EBV-negative lymphomas, the role of EBV in at least some of the B cell malignancies arising in AIDS patients is clear; indeed, about 50% of these lymphomas contain EBV sequences, even though no chromosomal alterations are present (Pelicci et al 1986). To better understand the mechanisms involved in lymphoma generation in AIDS patients, we turned to a new lymphomagenesis model, the severe combined immunodeficiency (SCID) mouse. These mice are characterized by a chromosome 16 mutation that prevents correct TCR and Ig gene rearrangements, leading to virtually complete T and B cell deficiency (Bosma et al 1983); these features make the SCID mouse an appropriate recipient for lymphoid cell grafts. When these cells derive from EBV+ donors, it was shown (Mosier et al 1988) that SCID mice frequently develop lymphomas constituted by the oligoclonal proliferation of EBV-infected B cells of human origin, without c-myc deregulation or chromosomal alterations (Cannon et al 1990; Rowe et al 1991). Nonetheless, the mechanisms underlying lymphoma development in this model are not forthcoming, despite accurate phenotypic and genotypic chracterization. We recently demonstrated that the presence of functional T cells in the injected PBMC population was necessary to allow expansion of EBV-infected B cell precursors, and that factors derived from either CD4+ or CD8+ T cells undergoing activation against xenoantigens were probably involved (Veronese et al 1992). Thus, immunostimulation seems to play a central role in the lymphomagenesis process in SCID mice reconstituted with lymphoid cells from EBV+ donors. In the setting of human immunodeficiencies, including AIDS, the relative roles of the immune impairment status and of immunostimulating events are difficult to determine. Our observations in the SCID experimental model seem to suggest that the immunodeficiency status alone, while constituting a pre-requisite condition, could not be sufficient to sustain the expansion of EBV-infected B cell precursors and their progression to tumors in AIDS patients. Indeed, despite the low number of circulating CD4+ cells, a strong immune stimulation by opportunistic pathogen antigens probably takes place in these patients; CD8+ T cells, which alone are able to sustain lymphoma development in SCID mice (Veronese et al 1992) might suffice in providing the putative factor(s) involved in EBV-infected B cell precursor expansion and tumor generation. In view of the potential importance of this working hypothesis also in the setting of human allograft transplantation, further studies are needed.

CONCLUSIONS

B cell dysregulation is a prominent feature of HIV-1-infected patients; as recently suggested, one might wonder whether the acronym AIDS could not be better interpreted as "acquired immune dysregulation syndrome" (Edelman and Zolla-Pazner 1991). Nonetheless, the significance of this dysregulation within the natural history of the disease is undefined, and future research must address many open questions. In the first place, what is the source of help for such intense B cell activation? It is possible that virus-containing immune complexes bound to the surface of lymph-node dendritic cells might provide the appropriate stimuli for maintaining B cell activation. Second, are anti-HIV-1 antibodies also involved in immunodeficiency generation? Third, what are the mechanisms at play in lymphomagenesis in AIDS patients? Lymphomas are now a major cause of death among seropositive patients, especially those undergoing long-term AZT treatment (Pluda et al., 1990). While our findings that AZT treatment is not associated with a significant decrease in B cell activation rate is not in contrast with the hypothesis that this latter phenomenon is a key event favouring subsequent lymphoma development, a better understanding of the underlying mechanisms in the immunodeficient host may open new avenues of therapeutic approach to this condition.

ACKNOWLEDGMENTS

The Authors were supported in part by grants from Istituto Superiore di Sanita', AIDS Project; CNR, Target Project FATMA, Target Project on Biotechnology and Bioinstrumentation, and Target Project ACRO; Italian Association for Research on Cancer (AIRC). The precious help of Ms. Patricia Segato in preparing this manuscript is gratefully acknowledged.

REFERENCES

Amadori A (1992) AIDS and autoimmunity. In: Bona CA, Siminovitch K, Teofilopoulos AN, Zanetti M (eds) The molecular pathology of autoimmunity, Harwood, New York (in press)
Amadori A, Chieco-Bianchi L (1990) Immunol Today 11:374-379
Amadori A, Chieco-Bianchi L (1992) Int J Clin Lab Res 22:11-16
Amadori A, DeRossi A, Faulkner-Valle GP, Chieco-Bianchi L (1988) Clin Immunol Immunopathol 46:342-351
Amadori A, DeSilvestro G, Zamarchi R, Veronese ML, Mazza MR, Schiavo G, Panozzo M, DeRossi A, Ometto L, Mous J, Barelli A, Borri A, Salmaso L, Chieco-Bianchi L (1992) J Immunol 148:2709-2716
Amadori A, Gallo P, Zamarchi R, Veronese ML, DeRossi A, Wolf D, Chieco-Bianchi L (1990) AIDS Res Human Retroviruses 6:581-586
Amadori A, Zamarchi R, Ciminale V, Del Mistro A, Siervo S, Alberti A, Colombatti M, Chieco-Bianchi L (1989) J Immunol 143:2146-2152
Amadori A, Zamarchi R, Veronese ML, Panozzo M, Barelli A, Borri A, Sironi M, Colotta F, Mantovani A, Chieco-Bianchi L (1991a) J Immunol 146:57-62
Amadori A, Zamarchi R, Veronese ML, Panozzo M, Mazza MR, Barelli A, Borri A, Chieco-Bianchi L (1991b) AIDS 5:821-828
Berberian L, Valles-Ayoub Y, Sun N, Martinez-Maza O, Braun JA (1991) Blood 78:175-179
Biasi G, Facchinetti A, Panozzo M, Zanovello P, Chieco-Bianchi L, Collavo D (1992) J Immunol 147:2284-2288
Birx DL, Redfield RR, Tosato G (1986) N Engl J Med 314:874-879
Bolognesi D (1989) Nature 340:431
Bosma GC, Custer RP, Bosma MJ (1983) Nature 301:527-530

Cannon MJ, Pisa P, Fox RI, Cooper NR (1990) J Clin Invest 85:1333-1337

Casali P, Notkins AL (1988) Immunol Today 10:364-368

Edelman A, Zolla-Pazner S (1991) FASEB J 3:222-230

Goronzy JJ, Weyand CM (1989) J Immunol 142:4435-4440

Groux H, Torpier G, Monte D, Mouton Y, Capron A, Ameisen JC (1992) J Exp Med 175:331-340

Ishida H, Hastings R, Kearney J, Howard M (1992) J Exp Med 175:1213-1220

Kipps TJ (1989) Adv Immunol 47:117-185

Krowka JF, Stites DP, Jain S, Steimer KS, George-Nascimento C, Gyenes A, Barr PJ, Hollander H, Moss AR, Homsy JM, Levy JA, Abrams DI (1989) J Clin Invest 83:1198-1203

Lane HC, Masur H, Edgar LC, Whalen G, Rook AH, Fauci AS (1983) N Engl J Med 309:453-458

Macchia D, Parronchi P, Piccinni MP, Simonelli C, Mazzetti M, Ravina A, Milo D, Maggi E, Romagnani S (1991) J Immunol 146:3413-3418

Mann DL, DeSantis P, Mark G, Pfeifer A, Newman M, Gibbs N, Popovic M, Sarngadharan MG, Gallo RC, Clark J, Blattner W (1987) Science 236:1103-1106

Mosier DE, Gulizia RJ, Baird SM, Wilson DB (1988) Nature 335:256-259

Newell MK, Haughn LJ, Maroun CR, Julius MH (1990) Nature 347:286-288

Ng VL, Chen KH, Hwang KM, Kayam-Bashi H, McGrath MS (1989) Blood 74:2471-2475

Pahwa S, Chirmule N, Leombruno C, Lim W, Harper R, Bhalla R, Pahwa R, Nelson RP, Good RA (1989) Proc Natl Acad Sci USA 7532-7536

Pelicci P, Knowles DM, Arlin ZA, Wieczoreck R, Luciw P, Dina D, Basilico C, Dalla-Favera R (1986) J Exp Med 164:2049-2060

Pluda JM, Yarchoan R, Jaffe ES, Feuerstein IM, Solomon D, Steinberg SM, Wyvill KM, Raubitschek A, Katz D, Broder S (1990) Ann Intern Med 113:276-282

Poli G, Bressler P, Kinter A, Duh E, Timmer WC, Rabson A, Justement JS, Stanley S, Fauci AS (1990) J Exp Med 172:151-158

Rosemberg ZF, Fauci AS (1989) Adv Immunol 47:377-431

Rowe M, Young LS, Crocker J, Stokes H, Henderson S, Rickinson AB (1991) J Exp Med 173:147-158

Schnittman SM, Greenhouse JJ, Psallidopoulos MC, Baseler M, Salzman NP, Fauci AS, Lane HC (1990) Ann Intern Med 113:438-442

UytdeHaag F, Van der Heijden R, Osterhaus A (1991) Immunol Today 12:439-442

Veronese ML, Veronesi A, D'Andrea E, Del Mistro A, Indraccolo S, Mazza MR, Mion M, Zamarchi R, Menin C, Panozzo M, Amadori A, Chieco-Bianchi L (1992) J Exp Med (in press)

Wahren B, Morfeldt-Manson L, Biberfeld G, Moberg L, Sonnerborg A, Ljungman P, Werner A, Kurth R, Gallo R, Bolognesi D (1987) J Virol 61:2017-2023

TH1 and TH2 Type Responses in HIV Infection

Mario Clerici, Gene M. Shearer, Robert L. Coffmann[*]

Experimental Immunology Branch National Cancer Institute Bethsda, Maryland
[*]*Departement of Immunology DNAX Research Institute of Molecular and Cellular Biology, Inc. Palo Alto, California*

I. Introduction

The most dramatic and well known immunologic consequence of infection with the human immunodeficiency virus type-1 (HIV) is the severe depletion of CD4[+] T cells in the progression toward the acquired immunodeficiency syndrome (AIDS) (reviewed by Fauci, 1988). The ultimate outcome of this loss of helper cells is susceptibility to opportunistic infections and the development of neoplastic conditions such as generalized Kaposi's sarcoma and non-Hodgkin's lymphoma. Most immunologic studies of HIV infection and AIDS have focused on this CD4-deficient state. However, to understand the complex series of events that ultimately results in AIDS and the extreme immune deficient state that is its hallmark, it is necessary to investigate the earlier events of immune dysregulation that result from HIV infection.

Several laboratories have reported defects in helper cell (TH) function prior to a critical reduction in CD4+ cell numbers (CD4 count) (Lane et al, 1985; Smolen et al, 1985; Shearer et al, 1986; Garbracht et al, 1987; Giorgi et al, 1987; Miedema et al, 1988; Clerici et al, 1989a; and Petersen et al, 1989). However, it has been mainly the research of two laboratories that has persisted in a systematic study of the T helper cell dysfunction that preceeds AIDS symptoms and the decline in CD4 counts. Thus, Miedema's laboratory in Amsterdam has studied the loss of T cell proliferation by stimulation with anti-CD3 monoclonal antibody and with other T cell mitogens (Miedema et al, 1988; Schellekens et al, 1990). Our laboratory has extensively investigated TH function by interleukin 2 (IL-2) production and proliferation in asymptomatic, HIV-seropositive (HIV+) individuals by stimulation of peripheral blood leukocytes (PBL) with: a) recall antigens (REC) such as influenza A virus or tetanus toxoid; b) irradiated allogeneic PBL from HIV-seronegative (HIV-) individuals (ALLO); and c) phytohemagglutinin (PHA) (Clerici et al, 1989a; 1989b; Lucey et al, 1991; Shearer and Clerici, 1991). We observed a sequential loss of TH function such that: a) 34%of HIV+ individuals responded to REC, ALLO, and PHA, designated +/+/+, similar to TH responses seen using PBL from healthy, HIV- individuals; b) 44% failed to respond to REC but responded to ALLO and PHA, designated -/+/+; c) 10% responded to PHA only, designated -/-/+; and d) 12% failed to respond to all of the above stimuli, designated -/-/-. This frequency of the various categories has been established using more than 1000 HIV$^+$ individuals. This loss of TH function is progressive such that [+/+/+] -----> [-/+/+] -----> [-/-/+] -----> [-/-/-], and the incidence of reversals in patients not on therapy is 5.5%. This CD4-independent loss of TH function is predictive for a more rapid decline in CD4 counts (Lucey et al, 1991), as well as for more rapid progression to AIDS symptoms, and it is associated with a higher incidence of opportunistic and bacterial infections among HIV-infected children (Roilides et al, 1991). Reversals in TH function (at least for responses to ALLO and PHA) are detected in more than 50% of patients on antiretroviral drug therapy (Clerici et al, 1992a; 1992b).

The mechanism(s) responsible for the CD4-independent loss of TH function is not known. An interesting characteristic of the phenomenon is that the early lack of response is selective for CD4-mediated, HLA self-restricted responses. Thus, we observed that 44% of HIV$^+$ individuals were selectively unresponsive to recall antigens. Such responses require CD4$^+$ helper T cells and the processing and presentation of antigens by autologous antigen presenting cells (APC) (Via et al, 1990). Thus, we detected a selective loss of responses to influenza A virus, tetanus. toxoid, and synthetic peptides of HIV envelope (Clerici et al, 1989a; 1989b). It has been reported that T cells expressing memory markers are lost relatively early in the progression toward AIDS (Schnittman et al, 1990; van Noesel et al, 1990), although not all laboratories have seen a selective loss of memory cells (Giorgi et al, 1991). Since TH responses to recall antigens would involve memory cells, our data could be explained by selective loss of memory cells. However, we have seen that the CD4-mediated, self-restricted component of the ALLO response (which is not mediated by memory cells) is lost concomitantly with the loss of responses to recall antigens (Clerici et al, 1991a). Therefore, it is likely that the early loss of TH function to recall antigens is more dependent on the self-restricted aspects of TH immunity than on memory per se. It should be noted that cultures enriched for CD45RA+ lymphocytes from HIV+ individuals still do not proliferate in response to HLA self-restricted antigens (Giorgi et al, 1991). Although we failed to detect a defect in APC function in these HIV+ individuals (Clerici et al, 1991b), we did observe multiple defects in the APC of approximately 60% of patietns with AIDS (Clerici et al 1990). From co-culture experiments of PBL from HIV+ and HIV⁻ monozygotic twins, it is clear that a defect resides in the TH cells themselves. However, it is also possible that a defect not detected by this experimental approach exists in the APC population, as well.

It is interesting that the phenomenon of "immune suppression" can be demonstrated in vitro by co-culturing PBL from individuals who are HIV+ and from HIV⁻ (Laurence et al, 1984; Hofmann et al, 1986). In fact, using PBL from HIV+ and HIV⁻ monozygotic twins, one can demonstrate that a soluble factor is produced by CD8+ T cells from the HIV+ patient that selectively suppresses responses to REC generated by PBL from the HIV⁻ twin (Clerici et al, 1992c). This finding raises the possibility that soluble factors contribute to the loss of TH function prior to CD4 depletion. That factors independent of CD4 counts contribute to HIV-induced immune deficiency is also supported by the finding that anti-retroviral drug therapy can result in restoration of TH function without a concomitant increase in CD4 counts (Clerici et al, 1992a; 1992b).

II. TH1 and TH2 Helper Cells

It has been reported by several laboratories that two functionally distinct types of murine TH cell clones can be isolated: TH1, which produce IFN-γ and IL-2, and TH2, which produce IL-4, IL-5, IL-6, and IL-10 (reviewed by Mosmann and Coffman, 1987; Fiorentino et al, 1989). Exceptions to these TH1 and TH2 patterns have been reported, and other cells termed "TH0" and "THp" have been identified (Mosmann and Coffman, 1989; and Firestein et al, 1989). Futhermore, cytokine cross-regulation has been demonstrated, in which certain cytokines produced by one TH subtypet can down-regulate cytokines produced by the other TH subtype (Fiorentino et al, 1989). For example, TH1 produce IFN-γ, which can down-regulate TH2 function; and TH2 produce IL-4 and IL-10, which can down-regulate TH1 function (Mosmann and Moore, 1991) (Fig. 1.) Recently, TH1 and TH2 clones were isolated from cultures of human leukocytes, and these clones appear to exhibit cytokine production profiles similar to those reported for murine cultures (DelPrete et al, 1991; Yssel et al, 1991).

The observations that TH functional loss and restoration (summarized above) can occur independently of CD4 count, and that PBL from HIV+ individuals produce soluble factors that inhibit TH function raise the possibility that immunoregulatory cytokines contribute to the TH functional loss summarized above. Furthermore, the fact that TH function assessed by IL-2 production and proliferation can occur in HIV-infected patients who exhibit activated B cells and hypergammaglobulinema (reviewed by Kopelman and Zolla-Pazner, 1988), raises the possibility that cytokine cross-regulation may be part of the immune dysregulation seen in HIV+ individuals

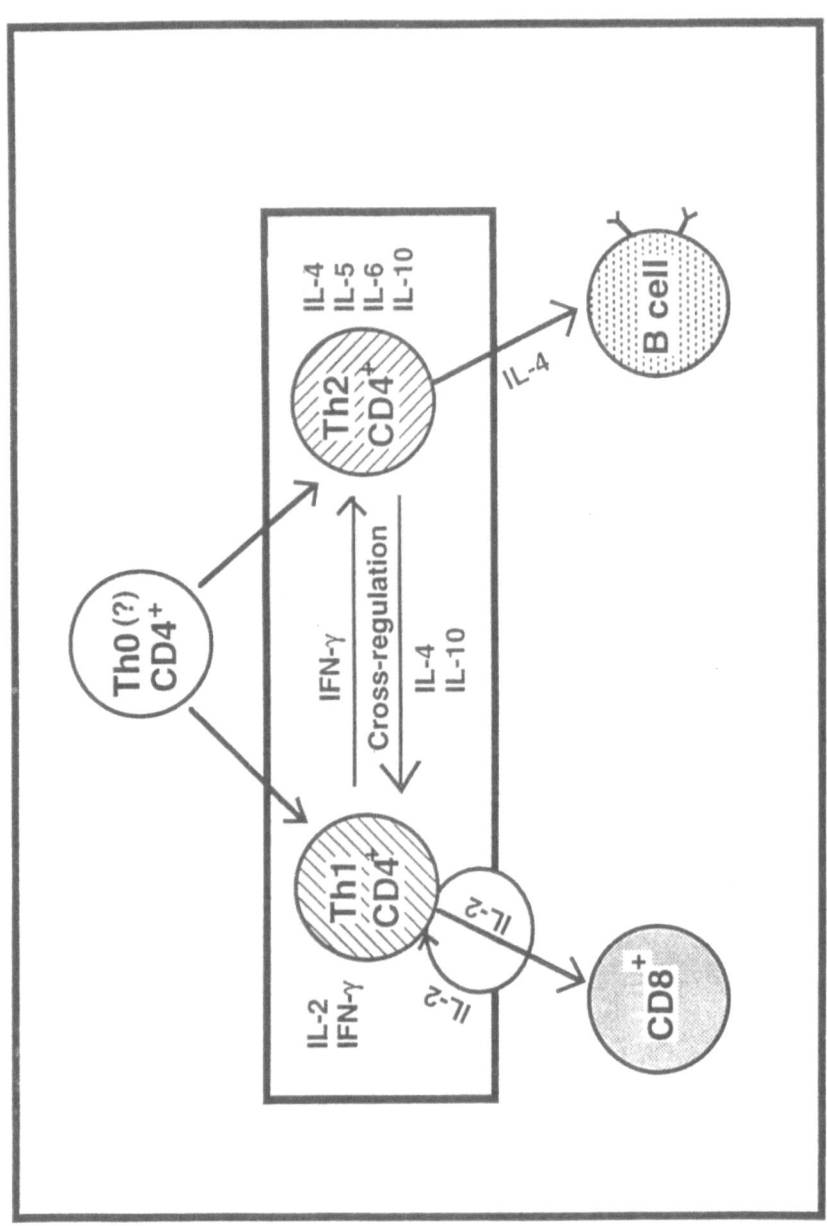

Fig. 1. Model of TH1 and TH2 development, and of TH1-TH2 cytokine crossregulation.

(Shearer and Clerici, 1992). Thus, an immunoregulatory cytokine profile that would reduce IL- 2 production and the generation of cytotoxic T lymphocyte responses, but enhance B cell activity might indicate a fundamental problem of immune dysregulation that is induced by immunoregulatory cytokines.

III. TH1 and TH2 Type Responses in HIV Infection

The recent finding that TH1 and TH2 clones can be isolated and identified from cultures of human T cells (DelPrete et al, 1991) provided an important prerequisite for our assessing of whether TH1-like and TH2-like responses could contribute to immune dysregulation after HIV infection. Thus, we have recently begun to question whether the +/+/+, -/+/+, -/-/+, and -/-/- profiles of IL-2 production could be correlated with increases in other immunoregulatory cytokines, particularly IL-4 and IL-10. Although IFN-γ may be a better indicator of TH1-type function than IL-2, we have compared IL-4 with IL-2 production because of: a) the extensive experience that we obtained assaying for IL-2 production (in more than 1000 HIV+ individuals); and b) the strong correlation that we demonstrated between the loss of TH responses by IL-2 production and progression to AIDS. As illustrated in Fig.2, IL-2 production in response to REC is lost first [(+/+/+) -----> (-/+/+)], and is accompanied by an increase in PHA-stimulated IL-4 production. The step that involves loss of the proliferation and IL-2 production in response to ALLO ([-/+/+] -----> [-/-/+]) is associated with a loss of IL-4 production in response to PHA, but an increase in PHA-stimulated IL-10 production. It appears that PHA-stimulated IL-10 production may decline concomitant with the loss in PHA-stimulated IL-2 production ([-/-/+] -----> [-/-/-]). It should be noted that all of these

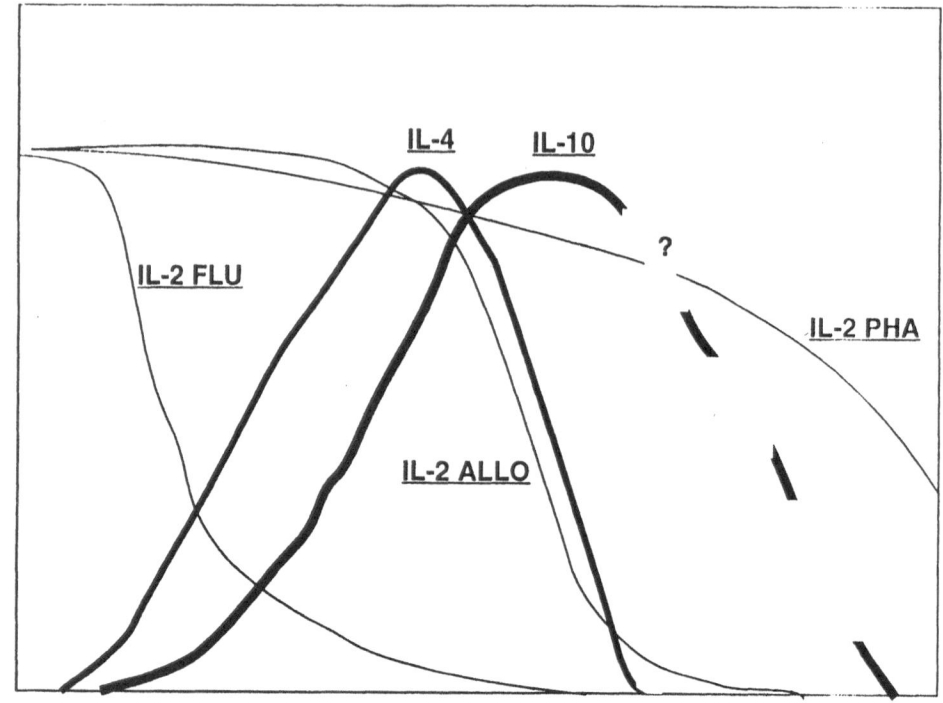

Fig. 2. Model of kinetics of cytokine production in HIV+ individuals in the progression toward AIDS. REC-stimulated IL-2 production with no PHA-stimulated IL-4 or IL-10 production is characteristic of healthy HIV- individuals and of HIV+ individuals who exhibit the "+/+/+" IL-2 profile. PHA-stimulated IL-4 production but not IL-10 production with selective loss of REC-stimulated IL-2 production is characteristic of HIV+ individuals who exhibit the " -/+/+" IL-2 profile. PHA-stimulated IL-10 production but not IL-4 production with additional loss of ALLO-stimulated IL-2 production is characteristic of HIV+ individuals who exhibit the "-/-/+" IL-2 profile. HIV+ individuals who exhibit the "-/-/-" IL-2 profile may be unresponsive to all of these stimuli for IL-2, IL-4, and IL-10 production. The decline in IL-10 is indicated as (- - - -) because we are not yet as certain of the decline in IL-10 as we are of the decline in Il-2 and IL-4.

changes in cytokine profiles: a) occur in HIV+ individuals who do not have symptoms of AIDS; and b) were detected in primary cultures of PBL, and did not require the selection and expansion of T cell clones. The changes in the cytokine profiles of these HIV+ individuals is consistent with a switch from a predominantly TH1-like pattern to one that is manily TH2-like. Table I outlines several immunologic parameters associated with a cytokine profile that would predict a predominant TH2 over TH1 pattern, and also indicates whether this pattern is observed in HIV+ individuals. Thus, there is an increase in those parameters that promote B cell activation and a decrease in the parameters that augment T effector cell function. It is noteworthy that a similar pattern of cytokine dysregulation has been recently reported in the progression of the retrovirus-induced murine model of an AIDS-like syndrome that closely resembles the immune dysregulation that we have detected after HIV infection in humans (Morse et al, 1991).

We have recently begun to study HIV-induced IL-2 production in exposed, HIV-seronegative individuals (Clerici et al, 1991c; 1992d), using a series of envelope synthetic peptides of HIV-1 that were prepared and studied in mice by Jay A. Berzofsky's laboratory (reviewed by Berzofsky, 1991). We previously studied these responses to these peptides in HIV+ individuals (Clerici, 1989b) and in human volunteers immunized with the MicroGeneSys rgp160 candidate vaccine (Clerici et al, 1991d). We have found that more than 50% of seronegative individuals in each of four at-risk groups responded to these peptides, whereas 5% of presumed unexposed individuals responded to one or more peptides. These findings indicate that HIV-specific T cell responses can be elicited in the absence of HIV antibody responses, and raise the possibility of protective T cell immunity against HIV.

Table I

Concordance between the Predicted Immune Function Profile of a TH2-like
Condition and those observed in asymptomatic HIV infection.

Immune parameter	Predicted*	Observed	Reference
B cell activation and hypergammaglobulinemia	√	√	Stahl et al., 1982
Hyper IgE and allergies	√	√	Israel-Biet et al., 1992
Reduced DTH	√	√	Reviewed by Fauci, 1988
Reduced CTL activity	√	√	Shearer et al., 1986
Reduced in vitro T cell proliferation	√	√	Reviewed by Fauci, 1988
Reduced production of IL-2	√	√	Clerici et al., 1989a
Increased production of IL-4	√	√	Clerici et al., in press
Increased production of IL-6	√	√	Nakajima et al., 1989e
Increased production of IL-10	√	√	Clerici et al., unpublished

* Predicted based on known elevated TH-2 like and reduced TH-1 like functions.

Based on the above considerations, one might expect to find HIV+ individuals whose immune responses move from a TH1-like to a TH2-like pattern as they pregress toward AIDS. In fact, we have recently made such observations in that our +/+/+ category for IL-2 production is associated with low Il-4 production, whereas the -/+/+ category for IL-2 production is acssociated with high IL-4 production (Clerici et al, 1992e).

IV. TH1 and TH2 Model of Susceptibility and Resistance to HIV Infection

Based on the findings summarized above, we have developed a model of resistance and susceptiblity to HIV infection and/or progression to AIDS (Shearer and Clerici, 1992; Sher et al, 1992). In this model, we postulate that a TH1-type immune profile (strong IFN-γ and IL-2; weak IL-4 and IL-10) is associated with protection against HIV infection and/or progression to AIDS in individuals already infected. In contrast, we suggest that a TH2-type response is indicative of susceptibility to infection and/or progression to AIDS among HIV+ individuals. Thus, individuals who exhibit a TH1-type>TH2-type response profile should be more resistant to HIV infection and/or progression to AIDS than individuals who exhibit a TH2-type>TH1-type response profile. It may be significant, therefore, that many individuals produce HIV-specific antibodies (as well as cellular immunity) and go on to develop AIDS, which contrasts with the hundreds (and probably thousands) of individuals who have been exposed multiple times to HIV (infected ?), and who generate strong HIV-specific T cell immunity but not antibody responses. A high proportion of these latter individuals

have not developed AIDS symptoms.

If the TH1-TH2 model is relevant for susceptibility and resistance to HIV infection and/or AIDS progression, it might be possible to reverse TH1-to-TH2 progression by cytokine therapy - either by antibodies against the TH2 cytokine that is produced, or by cytokine-crossregulation, in which the TH1 cytokine IFN-γ might be used to down-regulate TH2 cytokine production. We have tested the former possibility in vitro by stimulating PBL from -/+/+ individuals who were selectively unresponsive to REC with influenza A virus (FLU) in the absence or presence of an anti-IL-4 antibody. As shown in Table II, several of the -/+/+ HIV[+] individuals exhibited a restored proliferation in response to FLU when the PBL were stimulated in the presence of the anti-IL-4 antibody. The control antibody against IFN-γ had no restorative effect on the seven PBL samples tested, and the control antibody against TGF-β restored only 2/9 samples tested. These in vitro results suggest that it may be possible to reverse the TH2>TH1 cytokine profile using cytokine-based therapy.

The TH1-TH2 model of HIV resistance and susceptibility may also be important in the design of effective AIDS vaccines. The current plan for AIDS vaccines is for "high dose" immunization that will induce HIV-specific antibodies as well as HIV-specific cell mediated immunity. However, it was demonstrated many years before the discovery of TH1 and TH2 cells that cellular or humoral immunity can be selectively induced, depending on the dose of antigen used (Parish, 1972; Lagrange and MacKaness, 1974). Furthermore, it has been well established that strains of mice that are genetically susceptible to certain parasitic infections generate cell mediated immunity, whereas susceptible strains elicit antibody responses (for review see Locksley and Scott, 1991; Finkelman et al, 1991). The most recent and potentially

Table II

Patient[1]	³H Thymidine Incorporation (CPM) of FLU-Stimulated PBL			
	Medium alone	+anti-IL-4[2]	+ anti-IFN γ[3]	+ anti-TGF β[4]
1	490 ± 67	17179 ± 3489	N.T	N.T
2	955 ± 47	4018 ± 513	N.T	N.T
3	671 ± 101	5073 ± 292	N.T	375 ± 52
4	323 ± 224	7419 ± 641	N.T	2389 ± 403
5	1898 ± 88	14263 ± 1048	1021 ± 210	1297 ± 122
6	1021 ± 335	8136 ± 1689	804 ± 124	24134 ± 6467
7	2102 ± 355	10682 ± 829	2345 ± 348	9878 ± 1209
8	1903 ± 505	11862 ± 405	1019 ± 344	2098 ± 451
9	312 ± 102	9419 ± 641	1548 ± 88	1648 ± 104
10	466 ± 93	866 ± 128	N.T	912 ± 168
11	554 ± 68	1128 ± 121	898 ± 114	N.T
12	1021 ± 344	1344 ± 117	1233 ± 58	786 ± 114

[1] HIV+, -/+/+ individuals
[2] 5µg/ml anti -IL-4 antibody
[3] 5µg/ml anti -IFN γ antibody
[4] 2.5µg/ml anti- TGF β antibody
N.T. Not Tested

relevant example of this phenomenon is the work of Bretscher et al (1992), in which BALB/c mice that are susceptible to *Leishmania major*, were given a high dose or a low dose immunization against *L. major*. The low dose immunized mice generated cellular immunity to *L. major*, but not antibody and were protected against subsequent challenge. In contrast, the high dose immunized mice made antibody but not cellular immunity, and were susceptible upon challenge with the parasite (Bretscher et al, 1992).

Based on our experience in HIV immunity (outlined above), and the murine model of Bretscher el al, it is possible that the current approach for vaccine development will not be as effective as a protocol that would preferentially induce cellular immunity (Salk et al, 1992). Although neutralizing antibodies may provide some protection against HIV infection, the issue may not be whether antibodies can kill HIV, but rather whether: a) cellular immunity is more effective than humoral immunity; and b) the immune system can be made to maximally generate both cellular and humoral responses. It appears that the immune system is regulated to maximally produce one of these types of immunity at the expense of the other. If this is correct, the induction of strong antibody response could be at the expense of potent cellular immunity.

References

Berzofsky, J. A. (1991) *J. Acquired Immune Defic. Synd.* **4**, 451.

Bretscher, P. A., Wei, G., Menon, J. N., and Bielefeldt-Ohmann, H. (1992) *Science* **257**, 539.

Cierici, M., Stocks, N. I., Zajac, R. A., Boswell, R. N., Lucey, D. R., Via, C. S., and Shearer, G. M. (1989a) *J. Clin. Invest..* **84**, 1892.

Clerici, M., Stocks, N. I., Zajac, R. A., Boswell, R. N., Bernstein, D. C., Mann, D. L., Shearer, G. M., and Berzofsky, J. A. (1989b) *Nature* **339**, 383.

Clerici, M., Stocks, N. I., Zajac, R. A., Boswell, R. N., and Shearer, G. M. (1990) *Clin. Immunol. Immunopathol.* **54**, 168.

Clerici, M., Via, C. S., Lucey, D. R., Roilides, E., Piozzo, P. A., and Shearer, G. M. (1991a) *Eur. J. Immunol.* **21**, 665.

Clerici, M., Landay, A. L., Kessler, H. A., Zajac, R. A., Boswell, R. N., Muluk, S. C., and Shearer, G. M. (1991b) *J. Immunol.* **146**, 2207.

Clerici, M., Berzofsky, J. A., Shearer, G. M., and Tackett, C. O. (1991c) J. Infect. Dis. **164**, 178.

Clerici, M., Tackett, C. O., Via, C. S., Muluk, S. C., Berzofsky, J. A., and Shearer, G. M. (1991d) *Eur. J. Immunol.* **21**, 1345.

Clerici, M., Landay, A. L., Kessler, H. A., Venzon, D. J., Lucey, D. T., and Shearer, G. M. (1992a) *J. Infet. Dis.* in press.

Clerici, M., Roilides, E., Butler, K. M., DePalma, L., Venzon, D., Shearer, G. M., and Pizzo, P. A. (1992b) *Blood* in press.

Clerici, M., Roilides, E., Via, C. S., Pizzo, P. A., and Shearer, G. M. (1992c) *Proc. Natl. Acad. Sci. U. S. A.* in press.

Clerici, M., Giorgi, J. V., Gudeman, V. K., Chou, C.-C., Zack, J. A., Nishanian, P. G.,Dudley, J. P., Berzofsky, J. A., and Shearer, G. M., (1992d) *J. Infect. Dis.* **165**, 1012.

Clerici. M, Hakim, F. T., Venzon, D. J., Blatt, S., Hendrix, C. W., Wynn, T. A., and Shearer, G. M. (1992e) *J. Clin. Invest.* in press.

DelPrete, G. F., DeCarli, M., Matromauro, C., Biagiotti, R., Macchia, D., Falagiani, P., Ricci, M., and Romagnani, S. (1991) *J. Clin. Invest.* **88**, 346.

Fauci, A. S. (1988) *Science* **239**, 617.

Finkelman, F. D., Pearce, E. J., Urban, J. F. Jr., and Sher, A. *In* Immunoparasitology Today Ash, C. and Gallagher, R. B. (eds) Elsevier, Cambridge. pp. A62-A65.

Fiorentino, D. F., Bond, M. W., and Mosmann, T. R. (1989) *J. Exp. Med.* **170**, 2081.

Firestein, G. S., Roeder, W. D., Laxer, J. A., Townsend, K. S. Weaver, C. T., Horn, J. T. Linton, J., Torbett, B. E., and Glasebrook, A. L. (1989) *J. Immunol.* **143,** 518.

Garbrecht, F. C., Sisakind, G. W., and Wexler, M. E. (1987) *Clin. Exp. Immunol.* **67**, 245.

Giorgi, J. V., Fahey, J. L., Smith, D. C., Hultin, L. E., Cheng, H.-L., Mitsuyasu, R. T., and Detels, R. T. (1987) *J. Immunol.* **138**, 3725.

Giorgi, J. V. (1992) *In* Immunodeficiency in HIV Infection and AIDS. Janossy, G, Autran, B., Miedema, F. (eds) Karger, Basel. pp 1-17.

Hoffman, B. Odum, N., Jakobsen, B. K., Platz, P., Ryder, L.P., Nielsen, J.O., Gerstof, J., and Svejgaard, A. (1986) *Scand. J. Immunol.* **23**, 669.

Kopelman, R. G., and Zolla-Pazner, S.(1988) *Am. J. Med.* **84**, 82.

Lagrange, G. B., and MacKaness, T. E. (1974) *J. Exp. Med.* **139**, 528.

Lane, H. C., Depper, J. M., Greene, W. C., Whalen, G., Walsdmann, T. A., and Fauci, A. S. (1985) *N. Engl. J. Med.* **313**, 79

Laurence, J., Gottlieb, A. B., and Kunkel, H. G. (1983) *J. Clin. Invest.* **72**, 2072.

Locksley, R. M., and Scott, P. in Immoloparasitology Today Ash, C. and Gallagher, R. B. (eds) Elsevier, Cambridge. pp A58-A61.

Lucey, D. L., Melcher, G. P., Hendrix, C. W., Zajac, R. A., Goetz, D. W., Butzin, C. A., Clerici, M., Warner, R. D., Abbadessa, S., Hall, K., Jaso, R., Woolford, B., Miller, S., Stocks, N. I., Salina, C. M., Wolfe, W. H., Shearer, G, M., and Boswell, R. N. (1991) *J. Infect. Dis.* **164**, 631.

Miedema, F., Chantal-Petit, A. J., Terpstra, F. G., Eeftinek-Schasttenkerk, J. K. M., deWolf, F., Roos, M., Lange, J. M. A., Danner, S. A., Gouldsmit, J., and Schellenkens, P. T. A. (1988) *J. Clin. Invest.* **82**, 1908.

Gazzinelli, R. T.,Makino, M., Chattopadhay, S. K., Snapper, C. M., Sher, A., Higin, A. W., and Morse, H. C. III. (1992) *J. Immunol.* **148**, 182.

Mosmann, T. R., and Coffman, R. L. (1987) *Immunology Today* **8**, 223.

Mosmann, T. R., and Coffman, R. L. (1989) *Adv. Immunol.* **46**, 111.

Mosmann, T. R., and Moore, K. W. (1991) *In* Immunoparasitology Today Ash, C., Gallagher, R. B. (eds) Elsevier, Cambridge. pp A49-A61.

Parish, C. R. (9172) *Transpl. Revs.* **13**, 35.

Petersen, J., Chuarch, J., Gomperts, E., and Parkman, R. (1989) *J. Pediatr.* **115**, 944.

Roilides, E., Clerici, M., DePalma, L., Rubin, M., Pizzo, P. A., andf Shearer, G. M. (1991) *J. Pediatr.* **118**, 724.

Salk, J., and Salk, P. L. VIII International Congress on AIDS/III STD Woarl d Congress Amsterdam, The Netherlands 19-24 July, 1992 (Abstract).

Schellenkens, P. T., Roos, M. T., DeWolfe, F., Lange, J. M., and Miedema, F. (1990) *J. Clin. Immunol.* **10**, 121.

Schnittman, S. M., Lane, H. C., Greenhouse, J., Justement, J. S., Baseler, M., and Fauci, A. S. (1990) *Proc. Natl. Acad. Sci. U. S. A.* **87**, 6058.

Shearer, G. M., Bernstein, D. C., Tung, K. S., Via, C. S., Redfield, R., Salahuddin, S. Z., and Gallo, R. C. (1986) *J. Immunol.* **137**, 2514.

Shearer, G. M., and Clereici, M. (1991) *AIDS* **5**, 245.

Shearer, G. M., and Clerici, M. (1992) *Prog. Chem. Immunol.* **54**, 21.

Sher, A., Gazzinelli, R. T., Oswald, I. P., Clerici, M., Kullberg, M., Pearce, E. J., Berzofsky, J. A., Mosmann, T. R., James, S. L., Morse, H. C. III, and Shearer, G. M. (1992) *Immunol. Revs.* **127**, 123.

Smolen, J. S., Bettlehaim, P., Koller, U., McDougal, S., Grainger, W., Luger, T. A., Knapp, W., and Lechner, K. (1985) *J. Clin. Invest.*. **75**, 1828.

van Nossel, C. J. M., Gruters, R. A., Terpstra, F. G., and Miedema, F. (1990) *J. Clin. Invest.* **86**, 293.

Via, C. S., Tsokos, G., Stocks, N. I., Clerici, M., and Shearer, G. M. (1990) *J. Immunol.* **144**, 2524.

Yssel, H., Shanafelt, M. C., Soderberg, C., Schnieder, P. V., Anzola, J., and Peltz, G. (1991) *J. Exp. Med.* **164**, 593.

HIV-1 Strain-dependent CD4 T Cell Depletion in hu-PBL-SCID Mice

D. E. Mosier[1,2], R. J. Gulizia[1,2], P. D. MacIsaac[1], B. E. Torbett[1,2] and J. A. Levy[3]

[1]Immunology Division, Medical Biology Institute, La Jolla, Ca 92037, USA
[2]current address: The Scripps Research Institute, La Jolla, CA 92037, USA
[3]Cancer Research Institute, School of Medicine, University of California-San Francisco, San Francisco, CA 94143, USA

INTRODUCTION

The most devastating consequence of HIV-1 infection is the decline in CD4 T lymphocytes, which seems to play a central role in the progression of asymptomatic disease to clinical AIDS. This process typically occurs over more than a decade in infected individuals, and an animal model with accelerated CD4 cell depletion would aid in answering several critical questions about the mechanism of pathogenesis. The simplest model for CD4 cell loss would be a direct, cytopathic infection of cells by HIV, but some isolates of virus show no cytopathic effect on cultured human T cells (Cheng-Mayer, 1989), and the extent of viral infection in patients may be quite limited when CD4 cell depletion is occurring (Ho, 1989)(Clark, 1991). HIV or free gp120 envelope protein may interact with the CD4 receptor on T cells, and induce a state of activation or anergy that predisposes CD4 T cells to programmed cell death (apoptosis)(Habeshaw, 1990). While this mechanism does not require infection of every T cell that undergoes deletion, it should be closely related to the overall viral burden. These intrinsic mechanisms of CD4 T cell depletion should be experimentally distinguishable from extrinsic pathways, such as killing of HIV-infected cells by cytolytic T lymphocytes (CTL) or by antibody-dependent cellular cytotoxicity (ADCC). Finally, hypotheses for CD4 depletion must take into account the variation in pathogenicity of different HIV-1 isolates (Evans, 1989;Tersmette, 1989; Miedema, 1990), the high mutation rate of the viral genome (Mullins, 1991; Phillips, 1991), and the existence of multiple viral quasispecies in an individual at any one time (Wain-Hobson, 1989).

RESULTS AND DISCUSSION

To address these issues, we have studied the fate of human CD4 T cells in SCID mice reconstituted with adult human peripheral blood leukocytes (hu-PBL-SCID mice)(Mosier, 1988; Mosier, 1991) and infected with several well-characterized HIV-1 viruses derived from molecular clones. Following the intraperitoneal transfer of 20 x 10^6 PBL to SCID mice, human T cells survive for many months (Torbett, 1991), and the resulting hu-PBL-SCID mice are highly susceptible to infection with multiple strains of HIV-1 and HIV-2 (Mosier, 1991). Infection of hu-PBL-SCID mice with a highly T cell-tropic strain such as IIIB leads to a rapid decline in CD4 T cell numbers (Figure 1).

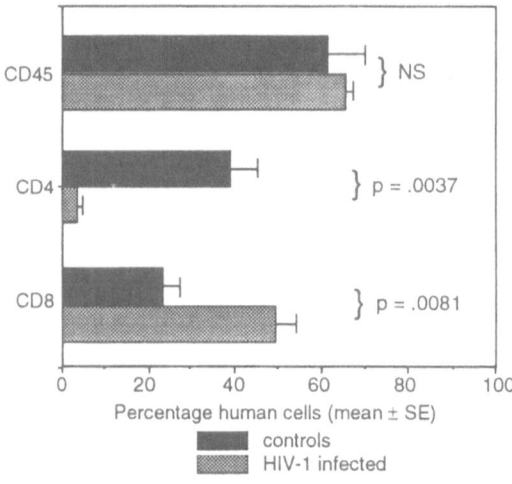

Fig. 1. Infection of hu-PBL-SCID mice with HIV-1 strain IIIB leads to a decrease in CD4 T cells and a relative increase in CD8 T cells. Data are from FACS analysis of human cells recovered from the peritoneal cavity of 5 mice/group reconstituted with 20×10^6 PBL 4 weeks previously and infected with 10^3 tissue culture infectious doses of virus 2 weeks previously. P values compare levels of significance between infected and control groups by the two-tailed t-test.

These results show that human CD4 T cells are depleted by HIV-1 infection of hu-PBL-SCID mice, and that the percentage of CD8 T cells increases following infection. This depletion occurs at a time when no CTL response or antibody formation to HIV has been detected, thus it is unlikely that these extrinsic means of CD4 cell depletion are active in this model. CD4 T cells in hu-PBL-SCID mice are not self-renewing, so that the rate of depletion is not counterbalanced by new T cell generation, as probably occurs in infected individuals. To further examine mechanisms of CD4 T cell depletion, we evaluated HIV-1 isolates that were likely to show differences in pathogenic potential.

For these experiments, we have employed molecularly cloned isolates of HIV-1, SF2, SF13, SF33, and SF162, viruses that were recovered from patients at different stages of disease and are known to have different cytopathic effects in vitro (Tateno, 1988). The isolates differ in their in vitro host range, replication rate, and cytopathic effects (Cheng-Mayer, 1988a; Cheng-Mayer, 1988b; Cheng-Mayer, 1990; Liu, 1990; York-Higgins, 1990) as outlined in Table 1. Moreover, their biologic properties have been mapped to differences in *tat* and gp120 (Cheng-Mayer, 1991).

Table 1. Biologic properties of molecular clones of HIV-1 SF2, 13, 33 and 162

Strain	Replication in		CPE	Sensitivity
	T cells	Macrophage	in CD4 cells	to neutralization
SF2	high	low	+	high
SF13	high	high	++	high
SF33	high	low	++	high
SF162	low	high	-	low

In two replicate experiments, these viruses were used at the same TCID$_{50}$ (100) to infect hu-PBL-SCID mice two weeks following PBL reconstitution. Five animals per group were examined for numbers of human CD4 and CD8 T cells at two or four weeks post-infection. Representative data are shown below in Fig. 2.

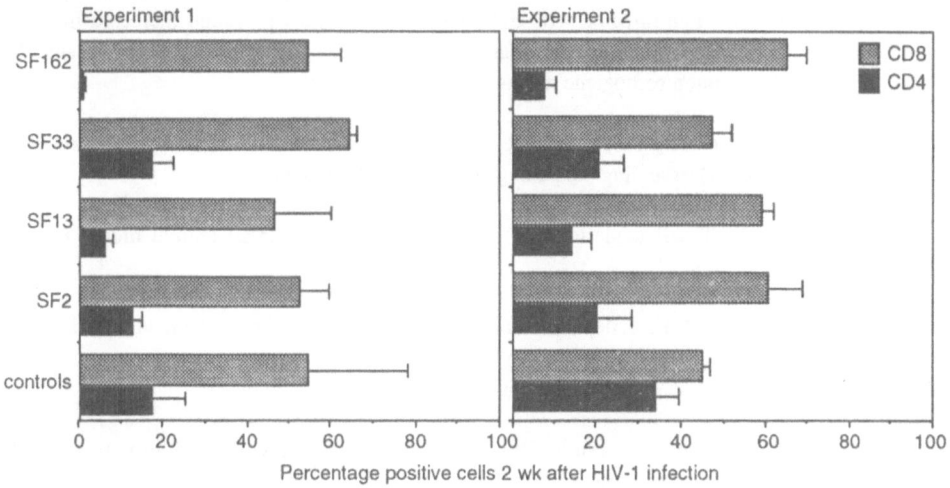

Fig. 2. Recovery of human CD4 and CD8 T cells from the peritoneal lavage fluid of hu-PBL-SCID mice infected 2 weeks previously with the indicated strain of HIV-1. Data represent the analysis of at least 10^4 cells by two-color flow cytometry. At 4 weeks post-infection, more extensive CD4 cell depletion was seen in all groups except those infected with HIV-1 SF33 (data not shown). These data are reported in more detail elsewhere (Mosier, 1992).

These results show a strain-dependent CD4 T cell depletion, with SF162 giving the most rapid and extensive depletion, and SF33 giving the slowest and least extensive depletion. Similar results were observed when hu-PBL-SCID spleen cells were examined. This result is surprising, since SF162 is a macrophage-tropic isolate with no cytopathic effect on cultured T cells. It is also noteworthy that the most macrophage-tropic strain gave the greatest depletion of CD4 cells, and that the cytopathic SF33 strain did not. Viruses recovered from hu-PBL-SCID mice infected with SF162 and SF33 were shown to have the same biologic properties as outlined in Table 1, so passage in SCID mice had not selected for a variant phenotype. In addition, quantitative PCR analysis of proviral genome copy number was performed (data not shown). These data showed that SF33 replicated to highest copy number in hu-PBL-SCID mice, that SF162 and SF13 showed similar replication, and that SF2 showed the lowest copy number. The extent of CD4 depletion thus does not seem to be directly related to the extent of viral replication. SF33 maintains its high replicative capacity in hu-PBL-SCID mice, yet this does not lead to marked CD4 T cell depletion.

Infection of hu-PBL-SCID mice with HIV-1 leads not only to CD4 T cell depletion, but also to an early polyclonal activation of human B cells that is manifested by a spike in human immunoglobulin levels (Mosier, 1991). Mice infected with SF162 showed the most profound increases in human immunoglobulin at 2 weeks post-infection (data not shown), suggesting that this change and CD4 depletion may share common pathogenic mechanisms.

These results, while examining only four molecularly cloned viruses, establish several points which should prove useful in understanding the pathogenesis of AIDS in humans:

(1) reproducible CD4 T cell depletion can be demonstrated in a small animal model for HIV infection;

(2) the rate and extent of CD4 cell depletion is highly dependent upon the infecting virus strain;

(3) CD4 depletion is not directly correlated with extent of virus replication;

(4) the extent of CD4 cell depletion in hu-PBL-SCID mice was not predicted by in vitro studies of viral replication rate or cytopathic effect;

(5) the most macrophage-tropic virus isolate caused the most CD4 cell depletion, suggesting that macrophage infection may be related to the mechanism of CD4 cell loss.

Viral variation thus would be expected to contribute to the rate and extent of CD4 T cell depletion in infected individuals as well as hu-PBL-SCID mice. A more extensive examination of both molecularly cloned viruses and patient isolates in the hu-PBL-SCID model system may give more insight into the viral determinants of pathogenesis.

ACKNOWLEDGMENTS

This work was supported by NIH grants AI29182 and AI30238 to DEM. Heidi Strobel is thanked for technical assistance. The skilled animal care of Kim VonSederholm and Rachel Mac Dowell is gratefully acknowledged. All work with HIV-1 infected hu-PBL-SCID mice was performed under biosafety level 3 containment (Milman, 1990).

REFERENCES

Cheng-Mayer C, Homsy J, Evans LA, Levy,JA (1988a) Identification of human immunodeficiency virus subtypes with distinct patterns of sensitivity to serum neutralization. Proc. Natl. Acad. Sci. USA 85:2815-2819

Cheng-Mayer C, Seto D, Tateno M, Levy JA (1988b) Biologic features of HIV that correlate with virulence in the host. Science 240:80-82

Cheng-Mayer C, Weiss, C, Seto D, Levy JA (1989) Isolates of human immunodeficiency virus type 1 from the brain may constitute a special subgroup of the AIDS virus. Proc Natl Acad Sci USA 86:8575-8679

Cheng-Mayer C, Quiroga M, Tung JW, Dina D, Levy JA (1990) Viral determinants of human immunodeficiency virus type 1 T-cell or macrophage tropism, cytopathogenicity, and CD4 antigen modulation. J Virol 64:4390-4398

Cheng-Mayer C, Shioda T, Levy JA (1991) Host range, replicative, and cytopathic properties of human immunodeficiency virus type 1 are determined by very few amino acid changes in *tat* and gp120. J. Virol. 65:6931-6941

Clark SJ, Saag MS, Decker WD, Campbell-Hill,S, Roberson JL, Veldkamp PJ, Kappes JC, Hahn BH, Shaw GM (1991) High titers of cytopathic virus in plasma of patients with symptomatic primary HIV-1 infection. N. Engl. J. Med. 324:954-960

Evans LA, Levy JA (1989) Characteristics of HIV infection and pathogenesis. Biochem. Biophys. Acta 989:237-254

Habeshaw JA, Dalgleish AG, Bountiff L, Newell AL, Wilks D, Walker LC, Manca F (1990) AIDS pathogenesis: HIV envelope and its interaction with cell proteins. Immunol. Today 11:418-425

Ho DD, Moudgil T, Alam M (1989) Quantitation of human immunodeficiency virus type 1 in the blood of infected persons. N. Engl. J. Med. 321:1621-1625

Liu Z-Q, Wood C, Levy JA, Cheng-Mayer C (1990) The viral envelope gene is involved in macrophage tropism of a human immunodeficiency virus type 1 strain isolated from brain tissue. J. Virol. 64:6148-6153

Miedema F, Tersmette J, van Lier RAW (1990) AIDS pathogenesis: A dynamic interaction between HIV and the immune system. Immunol. Today 11:293-297

Milman G (1990) HIV research in the SCID mouse: Biosafety considerations. Science 250:1152

Mosier DE, Gulizia RJ, Baird SM, Wilson DB (1988) Transfer of a functional human immune system to mice with severe combined immunodeficiency. Nature (London) 335:256-259

Mosier DE, Gulizia RJ, Baird SM, Wilson DB, Spector DH, Spector SA (1991) Human immunodeficiency virus infection of human-PBL-SCID mice. Science 251:791-794

Mosier DE, Gulizia RJ, MacIsaac PD, Torbett BE, Levy JA (1992) Human immunodeficiency virus type 1 strains differ in their ability to deplete CD4 T cells in SCID mice grafted with peripheral blood leukocytes. Manuscript submitted.

Mullins JI, Hoover EA, Quackenbush SL, Donahue PR (1991) Disease progression and viral genome variants in experimental feline leukemia virus-induced immunodeficiency syndrome. JAIDS 4:547-557

Phillips RE, Rowland-Jones S, Nixon DF, Gotch FM, Edwards JP, Ogunlesi AO, Elvin JG, Rothbard JA, Bangham CRM, Rizza CR, McMichael AJ (1991) Human immunodeficiency virus genetic variation that can escape cytotoxic T cell recognition. Nature 354:453-434

Tateno M, Levy JA (1988) MT-4 plaque formation can distinguish cytopathic subtypes of the human immunodeficiency virus. Virology 167:299-301

Tersmette M, Gruters RA, de Wolf F, de Goede REY, Lange JMA, Schellekens PTA, Goudsmit J, Huisman HG, Miedema F (1989) Evidence for a role of virulence human immunodeficiency virus (HIV) variants in the pathology of acquired immunodeficiency syndrome: studies on sequential HIV isolates. J. Virol. 63:2118-2125

Torbett BE, Picchio GR, Mosier DE (1991) hu-PBL-SCID mice: A model for human immune function, AIDS, and lymphomagenesis Immunol Revews 124:139-164

Wain-Hobson S (1989) HIV genome variability *in vivo*. AIDS 3 (Suppl. 1):S13-S18

York-Higgins D, Cheng-Mayer C, Bauer D, Levy JA, Dina D (1990) Human immunodeficiency virus type 1 cellular host range, replication and cytopathicity are linked to the envelope region of the viral genome. J. Virol. 64:4016-4020

Immune Response Regulations in Parasitic Infections

P. Perlmann, M. Troye-Blomberg, H. Perlmann, M. Kullberg and
S. Kumar*

*Department of Immunology, Stockholm University, S - 106 91 Stockholm, Sweden
*Malaria Section, NIAID, NIH, Bethesda, MD 20892, USA

INTRODUCTION

The yearly death toll in the major infectious diseases plaguing primarily tropical and subtropical areas of the world is estimated to be appr. 18 million (WHO, Geneva, reproduced in Science, 1992, 256:1135). These figures do not account for the enormous background of morbidity caused by these diseases and their disastrous consequences both for those directly affected and for the socio-economic development of the countries concerned. Seen in these perspectives, it may appear surprising that only a relatively minor fraction of worldwide research in immunology during the past decades has dealt with the immunology of infection. This situation seems, however, to be changing slowly partly because of the increasing need to fight infectious diseases by immunological manipulation, i.e. with vaccines. This has led to the recognition that construction of modern vaccines requires better insights in the basic mechanisms involved in immunity to infection. Much of the research in this area has dealt with the characterization and cloning of pathogen derived antigens expected to give rise to protective immunity in the infected host. However, more recently, attention has also been given to the various factors regulating immune protection. Some of these regulations will be exemplified and discussed in the following.

Although it may not be true in general, it is fair to state that acquisition and maintenance of protective immunity to infection in most cases is T cell dependent. This is valid both when immunity is primarily cell mediated and when it is humoral. Of the two major T cell subsets, both CD8$^+$ and CD4$^+$ T cells have regulatory and effector functions. Recently, murine CD4$^+$ T cells have been further divided into at least two different subsets, TH1 cells which upon activation produce IL-2 and IFN-γ and TH2 cells producing IL-4, 5, 6 and 10 (Mosmann and Coffman 1989; Mosmann and Moore 1991). Although the generic relationship between these cell types is not clear and, importantly, many intermediates exist (Swain et al 1991; Mosmann et al 1991) there is now good evidence that these two response patterns are crucial for the course and outcome of many murine infections. It was also initially stated that this division of murine CD4$^+$ T cells does not hold up in humans. However, recent evidence suggests that CD4$^+$ T cells differing in lymphokine secretion in a similar although perhaps not identical way also exist in humans (Romagnani 1991).

In this brief overview we will discuss regulation of both experimental and human immune responses involving CD4$^+$ T cells to some of the parasites causing major tropical diseases. Although it is now apparent that CD8$^+$ T cells may exhibit a similar functional heterogeneity as CD4$^+$ T cells (Salgame et al 1991) this discussion will be restricted to the latter for which most information is available. In a final section we will also briefly review recent data which illustrate how pathogens may interfere with immune control by utilizing some of the host's growth factors and receptors essential for the regulation of the immune response.

Leishmania Major

The biological significance of CD4$^+$ T cells differing in their cytokine secreting phenotypes is most clearly demonstrated by certain experimental parasite infections. Thus, subcutaneous infection with the protozoon *Leishmania major* causes localized cutaneous lesions in most mouse strains. The mice resolve infection within a few weeks and become resistant to reinfection. In contrast, infection of BALB/c mice with the same parasite results in fatal disease disseminating to visceral organs. The difference reflects a genetically determined difference in the immune response of the mice to the same parasite. Although CD8$^+$ T cells have been shown to play a role in resistance to *L.major* under certain circumstances (Müller *et al* 1991), the major T cell response affecting ·the course of infection in this system is a CD4$^+$ T cell response (Liew *et al* 1989). While the healer phenotype of splenic CD4$^+$ T cells in infected resistant mouse strains is characterized by elevated levels of IFN-γ mRNA and low levels of IL-4 mRNA, this situation is reversed in cells from infected BALB/c mice. When incubated *in vitro* with *L.major* antigen, CD4$^+$ T cells from the former mice produce IFN-γ while those from BALB/c produce primarily IL-4 and IL-5. Thus, it appears that a CD4$^+$ T cell response of TH1 type is controlling infection while one of TH2 type exacerbates it (Heinzel *et al* 1989; Scott *et al*, 1988). In line with this, mice which are able to control their infection respond to *L.major* with strong DTH reactions while those which are poor DTH responders form large amounts of antibodies, including IgE. Many experiments performed by different groups directly support the importance of the CD4$^+$ T cell dichotomy for the course of the infection in this system. These experiments include both adoptive transfer of *L.major* specific TH1 or TH2 type cell lines into histocompatible recipients and treatment of infected mice with cytokines or anti-cytokine antibodies. Taken together, the results indicate that the major, although not exclusive, factor responsible for the generation of a healer phenotype is IFN-γ while the one responsible for the non-healer phenotype is IL-4 (Coffman *et al* 1991).

L.major is a parasite living in macrophages and it is conceivable that lymphokines regulating macrophages are directing the course of infection by acting on macrophage effector function. However, although these cytokine activities probably play a role in this context, available evidence suggests that the main effects of IFN-γ and IL-4 on the course of *Leishmania* infection are exerted early, implying that they regulate CD4$^+$ T cell differentiation or selection rather than macrophage effector function (Scott 1991; Chatelain 1992; Sher *et al* 1992). The nature of the factors which initially direct the response to *L.major* in different mouse strains towards either TH1 or TH2 type patterns is presently not known. There is no consistent evidence that the different response patterns reflect differences in recognition of different antigens. The fact that the responses in different strains can be changed into TH1 or TH2 patterns by vaccination or administration of anti-cytokine antibodies also indicates that the strain differences seen are not due to genetic differences in TCR or antibody repertoires. Rather, they may reflect genetic differences in antigen processing or presentation. As IFN-γ and IL-4 are known to be important factors driving CD4$^+$ T cell development into different directions, it has been suggested that small amounts of these cytokines produced by other cells and to different degrees in different strains already before *L.major* infection may be responsible for the early fixation of the CD4$^+$ developmental pathway (Coffman *et al* 1991; Sher *et al* 1992; Sher and Coffman, 1992, Mosmann *et al* 1991).

Helminths

For protozoa like *Leishmania* and other parasites living intracellularly, the importance of cell mediated immune mechanisms involving IFN-γ and activated macrophages for resistance can easily be appreciated. The picture is more complicated for parasitic *helminths* which are multicellular organisms usually living extra-cellularly and not replicating in the vertebrate host. Infection of mice by the trematode *Schistosoma mansoni* may serve as example. In humans, this *helminth* causes one of the major tropical diseases characterized by chronic granulomatous inflammatory reactions to parasite eggs deposited in the liver. Characteristic for this infection are eosinophilia, mastocytosis and elevated serum levels of IgE (Sher and Coffman 1992). All these characteristics are typical reflections of lymphokines secreted by TH2 cells, eosinophilia being induced by IL-5, mastocytosis by IL-3, IL-4 and IL-10 and IgE elevation by IL-4. In agreement with this, the principal lympho-kines expressed when granuloma formation has its peak are IL-4 and IL-5. In contrast, expression of typical TH1 cytokines appears to be partially depressed by deposition of the eggs. A down-regulation of IFN-γ and IL-2 responses can also be seen when splenocytes of infected mice are stimulated *in vitro* with either mitogen or egg antigen (Pearce *et al* 1991; Grzych *et al* 1991). Importantly, this down-regulation becomes also apparent after stimulation with unrelated antigens to which infected mice were immunized, implying that infection modulates the immune response in general (Kullberg *et al* 1992). The same response pattern can also be induced by injection of *S.mansoni* eggs, suggesting the direct involvement of egg antigens. However, there is no conclusive evidence that the TH1/TH2 imbalance in this system is caused by some particular antigen released from the eggs. Rather, it may be due to the mode of antigen processing and/or presentation. In any event, the TH1/TH2 imbalance in *S.mansoni* infection probably reflects a cytokine mediated cross-regulation resulting in impaired TH1 differentiation or, more likely, inhibition of effector function such as TH1 lymphokine secretion. Key factors in these events are IL-4, IL-10 and TGF-β. An important role for IL-10 has been established through administration of anti IL-10 antibodies which upregulate IFN-γ production by CD4$^+$ T cells in *S.mansoni* infected mice (Sher *et al* 1992; Kullberg *et al* 1992). IL-10 also has important inhibitory effects on nitric oxide mediated parasite killing by blocking TNF-α synthesis required as a costimulatory signal to fully activate IFN-γ primed effector macrophages (Gazzinelli *et al* 1992).

It is generally assumed that the typical TH2 response seen in murine *S.mansoni* infection is also mediating protection (Urban *et al* 1992). However, recent evidence suggests that protective immunity to this parasite as well as other *Helminths*, induced naturally or by vaccination, may actually be associated with TH1 type responses (Sher and Coffman 1992). Although this is not uniformly true, it has given rise to speculations that TH1 type responses in infection are generally protective while TH2 responses are harmful and are induced by the pathogen to protect itself. However, such generalizations are not warranted even when considering different *helminth* infections, e.g. that of BALB/c mice with the nematode *Heligmosomoides polygyrus*, where administration of anti IL-4 antibodies has been shown to completely block protective immunity to reinfection (Urban *et al* 1992).

Experimental Malaria Infections

Investigation of other parasitic diseases does not either support the simple concept of TH1 responses being important for protection and TH2 responses for causing disease progression. Malaria caused by the protozoon *Plasmodium* may be taken to illustrate this. The form of the parasite transmitted by the mosquito to the vertebrate host, the sporozoite, develops in the liver for 1 - 2 weeks. It is then released into the blood where it grows and multiplies in red blood cells. It is the blood stages (rings, trophozoites, schizonts, merozoites) which are solely responsible for malaria morbidity and mortality. Immunity to the sporozoite and liver stages of this parasite is T cell dependent with cytotoxic T cells of both CD8$^+$ and CD4$^+$ phenotype being of major importance for preventing sporozoite development in liver (Troye-Blomberg and Perlmann 1992). In contrast, acquired immunity to the blood stages is primarily dependent on regulatory T cells of the CD4$^+$ phenotype.

P.chabaudi: The role of antibodies for blood stage immunity varies with different _Plasmodial_ species and this is also reflected by the functional activities of the CD4$^+$ T cells elicited by infection. Thus, the murine malaria parasite _P.chabaudi chabaudi_ induces in most mouse strains a non-lethal infection similar to human disease. In both susceptible and resistant strains, two phases of infection can be distinguished; an acute phase lasting appr. 10 days, and a second phase when parasitaemia becomes subpatent and recrudescenses may occur (35 - 70 days) (Langhorne _et al_ 1989 a, b). Infection gives rise to a rapid IgM anti-plasmodial antibody response while formation of IgG antibodies is slow and characterized by a relative deficiency of IgG1 antibodies in the early phases of infection. This is also true for antibodies to other antigens during the acute phase of infection. These and other findings suggest that in the acute phase of infection the major regulatory T cells are of TH1 type producing IFN-γ. However, in the later phases of infection, when clearance of parasitaemia by antibodies becomes prominent, T cells of TH2 type which help B cells to produce antibodies _in vitro_ appear to predominate. Taken together, these results imply that the major protective mechanism in early infection involves destruction of intraerythrocytic parasites by activated macrophages while T-dependent antibodies play a major role in late infection, probably by focusing parasites to proper areas of the spleen and promoting phagocytosis.

P.vinckei: It should be stressed that blood stage infection of mice of the same strains by other _plasmodial_ species may elicit completely different immune responses. Thus, _P.vinckei vinckei_, which is lethal for mice, gives rise to solid immunity in BALB/c mice which have been infected and drug cured. This long lasting immunity is CD4$^+$ T cell dependent and can develop in the absence of CD8$^+$ T cells as well as B cells (Kumar _et al_ 1989). _In vitro_ stimulation of spleen cells from immune mice with _P.vinckei_ antigen induced a strong T cell dependent release of IFN-γ for at least 5 months after the mice had been drug cured. However, while appr. 2000 per million splenic T cells from immune mice could be stimulated with _P.vinckei_ antigen to produce IFN-γ, this number was up to 3000 times higher than that of IL-4 producing cells in immune as well as non-immune mice and regardless of addition of antigen to the cultures (H. Perlmann _et al_, ms in preparation). These results indicate that protective immunity to _P.vinckei_ is due to cellular mechanisms distinct from those giving rise to immunity to _P.chabaudi_. Elucidation of the factors directing the immune response into different directions in these two related malaria systems will be of considerable interest.

REGULATION OF THE HUMAN IMMUNE RESPONSE TO _P.FALCIPARUM_ MALARIA

Functional Heterogeneity of CD4$^+$ T Cells

In humans, the most wide spread and serious malaria is caused by _P.falciparum_. Immunity to the blood stages of this parasite may be acquired but requires repeated exposure and is relatively unstable. Blood T lymphocytes from donors primed by natural exposure to _P.falciparum_ may be induced specifically to proliferate or release lymphokines _in vitro_ by exposure to parasite antigens. Although γ/δ T cells have recently been shown to be responsible for some of these activities in acutely infected donors (Langhorne _et al_ 1992), the majority of the responding cells in repeatedly exposed and clinically immune donors are CD4$^+$ T cells with α/β receptors. When stimulated with a variety of _P.falciparum_ antigens _in vitro_ these cells proliferate and secrete IFN-γ. Importantly, in individual donors, proliferation and IFN-γ production induced by malarial peptides are poorly and sometimes even negatively correlated with each other or with the expression of IL-4 which also may be induced in CD4$^+$ T cells by the same peptides. Furthermore, the expression of IL-4 but not that of IFN-γ or proliferation is usually well correlated with elevated serum antibody concentrations to the peptides used to activate the T cells (Troye-Blomberg _et al_ 1989). The direct T-helper activity of such peptide activated T cells can also be shown by induction of antibody production _in vitro_ in

T/B cooperation systems (Kabilan *et al* 1987; Chougnet *et al* 1991). In contrast, there is no association between serum antibody levels and *in vitro* induction of primed CD4$^+$ cells to produce IFN-γ. Taken together, these results indicate that functionally distinct CD4$^+$ T cell subsets similar to what has been found in mice also are involved in the regulation of the human immune response to *P.falciparum* malaria. It may be assumed that helper T cells of TH2 type are involved in the induction of *P.falciparum* specific antibodies important for parasite clearance in immune individuals. However, IL-4 production is also associated with elevated levels of serum IgE (Mosmann and Coffman 1989) and it was therefore of interest to note high serum IgE levels in some *P.falciparum* primed donors (H. Perlmann, unpublished). Moreover, preliminary results suggest a significant association between elevated serum IgE and the occurrence of cerebral malaria, the most severe and often fatal form of *P.falciparum* malaria. However, whether or not TH2 cells or IgE antibodies are directly involved in malaria pathogenesis remains to be established.

Genetic Regulation

CD4$^+$ T cell clones specific for *P.falciparum* antigens have been generated from both immune and non exposed donors (Troye-Blomberg and Perlmann 1992). Such clones are MHC class II restricted. It has therefore been assumed that high or low responsiveness to defined *P.falciparum* antigens by donors primed through natural infection may also reflect MHC class II restrictions. However, and not unexpectedly in view of the extensive MHC class II polymorphism (Olerup *et al* 1991), it has been difficult to find consistent associations between MHC and anti-*P.falciparum* responses in outbred human populations (Troye-Blomberg *et al* 1991). The involvement of other genetic factors in the regulation of some anti-*P.falciparum* immune responses has recently been shown by studying twins and their age matched siblings (Sjöberg *et al* 1992). Both antibody and T cell responses to defined epitopes (peptides) of a major *P.falciparum* blood stage antigen were found to be more concordant within monozygotic than within dizygotic twin pairs and their siblings. The immune responses measured were genetically regulated but there was no association with different MHC class II *DRB*, *DQA* or *DQB* alleles or haplotypes, suggesting that the regulation observed was due to some other genes. Such genes could act at different levels involving either B cells, T cells or antigen presenting cells (APC). The importance of the latter for determining the magnitude of the T cell response is suggested by the results of recent experiments in which T cells were stimulated by a well defined *P.falciparum* antigen in the presence of either autologous or MHC class II identical APC from unrelated donors (Troye-Blomberg, unpublished). The magnitude of the proliferative response of T cells from a given donor frequently differed depending on the origin of the APC. The results suggest that both the origin of APC and the degree of sensitization and specificity of the T cells determine the outcome of these tests. It is presently not known whether a genetic regulation at the APC level reflects differences in antigen processing and presentation or release of factors such as IL-1.

USE OF CYTOKINES AND HOST RECEPTORS BY PATHOGENS

Long standing co-evolution between parasites and their hosts has resulted in efficient adaptation of the pathogen to evade killing by the latter (and vice versa). Well known mechanisms of evasion involve the development of extended genetic repertoires making it possible for the pathogen to hide from an immune response by antigenic diversity or variation. More recently, it has been shown that a pathogen may directly affect host responses to its favour by utilizing or even producing growth factors or growth factor receptors normally involved in host response regulation. This is true for both pro- and eukaryotic pathogens. Thus, human IL-1 but not TNF or IL-4 may enhance the growth of virulent but not avirulent strains of *E.coli*, implying that this cytokine is a virulence factor for some gram negative bacteria which also may produce IL-1 or IL-1 like molecules themselves (Porat *et al* 1991). Similarly, invasion of mammalian cells *in vitro* by wild type

Salmonella typhimurium appears to involve stimulation of the receptor for epidermal growth factor (EGF), a receptor which also participates in the internalization of vaccinia virus (Galán *et al* 1992). Another recent example of cytokine utilization by the pathogen is provided by *Schistosoma mansoni* infection in mice (Amiri *et al* 1992).These parasites make use of TNF-α for granuloma formation, the most important pathogenic event in Schistosomiasis. In addition, these parasites also utilize TNF-α as a signal promoting egg-laying and -excretion from the host.

EGF also modifies the growth *in vitro* of the African trypanosome *T.brucei*, suggesting that it is one of the mammalian host factors required for parasite growth and differentiation. *T.brucei* also provides an additional example illustrating the adaptation of the parasite to the host's immune system. Growth of this parasite in rats is associated with extensive infiltration of the tissues by CD8[+] T cells. Depletion of these cells *in vivo* delays parasitaemia and significantly increases survival of the infected rats. This result does not reflect a conventional immune suppressor effect of CD8[+] T cells which also promote parasite growth *in vitro*. The trypanosomes activate CD8[+] T cells by means of a soluble factor which binds to a T cell receptor, assumed to be the CD8 molecule itself. This interaction results in the release of IFN-γ which has a direct growth promoting effect on the parasites (Olsson *et al* 1992). This host - parasite interplay is distinct from that of the American trypanosome *T.cruzi* which grows for a certain period intracellularly in macrophages and where IFN-γ has a protective role during infection, probably involving macrophage activation. In line with this, CD8[+] T cell depletion in experimental *T.cruzi* infection increases both parasitaemia and mortality (Tarleton 1990).

CONCLUDING REMARKS

The division of T cells into subsets differing in their patterns of cytokine secretion following activation was first detected by investigating T cell clones of murine origin (Mosmann and Coffman 1989). However, the examples discussed in this short review demonstrate that such a functional subdivision of CD4[+] T cells plays a major role in determining the course and outcome of various infections in mice. This is probably valid for human infections as well. Although we have only discussed some of the parasites responsible for major tropical diseases, there is little doubt that a similar case may be made for functional subdivision of T cells controlling, for example, mycobac- terial, yeast and probably many other infections (Walker *et al* 1992; Romani *et al* 1992). Moreover, although this discussion has been restricted to CD4[+] T cells it should be emphasized that a similar functional heterogeneity of CD8[+] T cells (Salgame *et al* 1991) may be of importance for some of the infections discussed herein.

For convenience, we have used the designation TH1 and TH2 to stress the fact that certain cytokines, such as IFN-γ on one hand and IL-4 or IL-10 on the other, appear to have antagonistic regulatory functions. However, the heterogeneity of CD4[+] T cells differing in cytokine secreting phenotypes is probably much greater than initially thought (Mosmann *et al* 1991). In any event, while it is obvious that it exists, the factors giving rise to such heterogeneity are incompletely understood. As illustrated herein, e.g. for *L.major* infection in different mouse strains, host genetics may play an important role in determining in which direction a CD4[+] T cell response will develop. However, the genes involved in this control are largely unknown. Moreover, as various manipula- tions such as administration of cytokines or anti-cytokine antibodies can change the type of response it can be concluded that, at least in the cases discussed in this paper, epitope specificity, TCR- selection and V-region usage appear not to be essential for the development of a particular response pattern. The relative importance of the physical form or concentration of antigen, of hormones or of antigen processing and presentation also remains to be clarified (Coffman *et al* 1991). These are major issues which have to be considered for optimizing vaccine design.

A further complication in the context of vaccine constructions lies in the recent recognition that pathogens may make use to their own benefit of some of the cytokines and cytokine-receptors involved in growth control as well as the regulation of the immune response itself. Obviously, vaccination focusing on IFN-γ induction and/or CD8$^+$ T cell activation, which may be advantageous to control infection by malaria sporozoites, would appear to be disastrous in the case of African trypanosomiasis. These aspects of the pathogen host interplay will also have to be explored and accounted for in order to be able to effectively optimize vaccine design.

REFERENCES

Amiri P, Locksley RM, Passlow TG, Sadick M, Rector E, Ritter D, McKerrow JH (1992) Nature 356:604-607

Chatelain R, Varkila K, Coffman RL (1992) J Immunol 148:1182-1187

Chougnet C, Troye-Blomberg M, Deloron P, Kabilan L, Lepers JP, Savel J, Perlmann P (1991) J Immunol 147:2295-2301

Coffman RL, Varkila K, Scott P, Chatelain R (1991) Immunol Revs 123:189-207

Galán JE, Pace J, Hayman M (1992) Nature 357:588-589

Gazzinelli RT, Oswald IP, James SL, Sher A (1992) J Immunol 148:1792-1796

Grzych JM, Pearce E, Cheever A, Caulada ZA, Caspar S, Hieny S, Lewis F, Sher A (1991) J Immunol 146:1322-1327

Heinzel FP, Sadick MD, Holaday BJ, Coffman RL, Locksley RM (1989) J exp Med 169:59-72

Hide G, Gray A, Harrison CM, Tait A (1989) Mol Biochem Parasitol 36:51-60

Kabilan L, Troye-Blomberg M, Patarroyo ME, Perlmann P (1987) Clin Exp Immunol 68:288-297

Kullberg MC, Pearce EJ, Hieny SE, Sher A, Berzofsky JA (1992) J Immunol 148:3264-3270

Kumar S, Good MF, Dontfraid F, Vinetz JM, Miller LH (1989) J Immunol 143:2017-2023

Langhorne J, Gillard S, Simon B, Slade S, Eichmann K (1989 a) Int Immunol 1:416-424

Langhorne J, Meding SJ, Eichmann K, Gillard SS (1989 b) Immunol Revs 112:71-94

Langhorne J, Goodier M, Behr C, Dubois P (1992) Immunol Today 13:298-300

Liew FY (1989) Immunol Today 10:40-45

Mosmann TR, Coffman RL (1989) Ann Rev Immunol 7:145-173

Mosmann TR, Moore KW (1991) In Ash C and Gallagher RB (eds) Immunoparasitology Today, Elsevier Trends Journals, Cambridge UK, p A49

Mosmann TR, Schumacher JH, Street NF, Budd R, O'Garra A, Fong TAT, Bond MW, Moore KWM, Sher A, Fiorentino DF (1991) Immunol Revs 123;209-229

Müller I, Pedrazzini T, Kropf P, Louis J, Milon G (1991) Int Immunol 3:587-597

Olerup O, Troye-Blomberg M, Schreuder GMT, Riley E (1991) Proc Natl Acad Sci USA 88:8480-8484

Olsson T, Bakhiet M, Kristensson K (1992) Parasitol Today 8:237-239

Pearce EJ, Caspar P, Grzych JM, Lewis FA, Sher A (1991) J exp Med 173:159-166

Porat R, Clark BD, Wolff SM, Dinarello C (1991) Science 430-432

Romagnani S (1991) Immunol Today 12:256-257

Romani L, Mencacci A, Grohman U, Mocci S, Mosci P, Puccetti P, Bistoni F (1992) J exp Med 176:19-25

Salgame P, Abrams JS, Clayberger C, Goldstein H, Convit J, Modlin RL, Bloom BR (1991) Science 254:279-282

Scott P (1991) J Immunol 147:3149-3155

Scott P, Natovitz RL, Coffman RL, Pearce E, Sher A (1988) J exp Med 168:1675-1684

Sher A, Coffman RL (1992) Ann Rev Immunol 10:385-409

Sher A, Gazzinelli RT, Oswald IP, Clerici M, Kullberg M, Pearce EJ, Berzofsky JA, Mosmann TR, James SL, Morse HC, Shearer GM (1992) Immunol Revs 127:183-204

Sjöberg K, Lepers JP, Raharimalala L, Larsson Å, Olerup O, Marbiah NT, Troye-Blomberg M, Perlmann P (1992) Proc Natl Acad Sci USA 89:2101-2104

Swain SL, Bradley LM, Croft M, Tonkonogy S, Atkins G, Weinberg AD, Duncan DD, Hedrick SM, Dutton RW, Huston G (1991) Immunol Revs 123:115-144

Tarleton RL (1990) J Immunol 144:717-724

Troye-Blomberg M, Olerup O, Larsson Å, Sjöberg K, Perlmann H, Riley E, Lepers JP, Perlmann P (1991) Int Immunol 3:1043-1051

Troye-Blomberg M, Perlmann P (1992) In Molecular Immunological Considerations in Malaria Vaccine Development, Good M, Saul A (eds) CRC Press, Boca Raton, Fl, USA (in press)

Troye-Blomberg M, Riley EM, Perlmann H, Andersson G, Larsson Å, Snow RW, Allen SJ, Houghten RA, Olerup O, Greenwood BM, Perlmann P (1989) J Immunol 143:3043-3048

Troye-Blomberg M, Riley EM, Kabilan L, Holmberg M, Perlmann H, Andersson U, Heusser CH, Perlmann P (1990) Proc Natl Acad Sci USA 87:5484-5488

Urban JF, Madden KB, Svetic A, Cheever A, Trotta PP, Gause WC, Katona IM, Finkelman FD (1992) Immunol Revs 127:205-220

Walker KB, Butler R, Colston MJ (1992) J Immunol 148:1885-1889

A Vaccine Against Schistosomiasis: Strategy and Perspectives

A. Capron

Centre d'Immunologie et de Biologie Parasitaire, Unité Mixte INSERM U167 - CNRS 624, INSTITUT PASTEUR, 1 rue du Pr A. Calmette, 59019 Lille Cédex (France)

INTRODUCTION

Schistosomiasis, the second major parasitic disease in the world after malaria affects at least 200 millions people, 500 millions being exposed to the risk of infection and is responsible for 300 to 500 000 deaths per year.

Morbidity observed in this chronic and debilitating disease is essentially related to the remarkable female worm fecundity, hundreds of eggs beeing laid every day and deposited in numerous mucous membranes and tissues. Granuloma formation around eggs, in particular in the liver, leads to the development of severe fibrotic and often irreversible lesions.

Although active drugs, such as Praziquantel are available, evidence is now accumulating that, while they can reduce the overall incidence of severe forms of the disease, they do not prevent reinfection, have little effect on already developped hepatosplenic manifestations and do not significantly affect transmission.

Unlike protozoan parasites, schistosomes, as metazoans, do not replicate in their vertebrate hosts. It is agreed, on the basis of experimental and epidemiological studies, that a significant but partial reduction, estimated around 60 percent, of the worm burden following infection would considerably reduce pathology and affect parasite transmission.

Recent epidemiological studies (Butterworth 1985, Hagan 1991) have now clearly established that protective immunity in chronically exposed human population is slowly built up and begins to be expressed after the age of puberty. Children, who get infected, as soon as they can walk, will know before they reach adolescence a long period of susceptibility to multiple reinfections and will represent both privileged targets for the development of the disease and major actors of transmission.

It therefore appears that a vaccine strategy which could lead to the anticipated induction of effector mechanisms reducing the level of reinfection and ideally parasite fecundity would deeply affect the incidence of pathological manifestations as well as the parasite transmission potentialities.

On the basis of these general principles, our strategy has aimed, in a first phase, at the identification of effector and regulatory mechanisms of the immune response to schistosomes in experimental models and in human populations.

EFFECTOR AND REGULATORY MECHANISMS

Extensive studies performed in our laboratory have attempted to a detailed analysis of the humoral components and of the cellular partners involved in *in vitro* killing of target schistosome larvae, namely schistosomula. Using the rat as a model system in parallel with cytotoxicity assays in humans and in primates, we identified novel ADCC mechanisms involving proinflammatory cell populations (macrophages, eosinophils and platelets) as cellular partners and unusual antibody isotypes such as IgE or a subclass of IgG with anaphylactic properties, like rat IgG2a in the particular case of eosinophils (reviewed in Capron *et al.* 1987). These observations of IgE dependent cell mediated killing, which could be confirmed in human schistosomiasis and in primates (Joseph *et al.* 1978, Capron et *al*. 1984) remained however limited at this stage to *in vitro* experiments thus raising the problem of their *in vivo* relevance. Indeed the essential question brought by these *in vitro* ADCC mechanisms was related to the implication, so far unsuspected, of anaphylactic antibody isotypes and specially IgE in the mechanisms of protection against metazoan parasites. The production of monoclonal antischistosome antibodies of the rat IgE and IgG2a isotype led to the demonstration of their high protective capacity by passive transfer (Grzych *et al.* 1982, Verwaerde *et al.* 1987). Together with the diminished protection passively conferred by IgE depleted immune rat serum and abrogation of immunity after anti μ and anti ε antibody treatment of neonate rats (Bazin *et al.* 1980, Kigoni et *al*. 1986), evidence was then accumulated that IgE, at least in rats, could have a more beneficial function than being mainly involved in deleterious allergic manifestations. The relevance of these experimental findings to human immunity to schistosomes have been very recently confirmed by three independent immunoepidemiological studies which have brought convergent evidence for a protective role of IgE antibody in human infection. Studying the rate of reinfection after treatment in a community exposed to *S. haematobium* infection in the Gambia, Hagan *et al.* have demonstrated a positive correlation between the specific IgE antibody response to worm antigens and the acquired resistance to reinfection (1991). Multiple regression analysis show in particular that the risk of reinfection is ten times more likely when IgE antibodies are absent or in the lowest quintile (Hagan *et al.* 1991). Similar findings have been made by Dunne *et al.* (1992) in a community exposed to *S. Mansoni* infection (Dunne et al. 1992) as well as by Rihet *et al.* in Brazil (Rihet *et al.* 1991). Without excluding the possibility of the participation of additional mechanisms, as mentionned later in this review, specific IgE antibody response appears therefore as a strong correlate of protective immunity in humans, confirming the views we have expressed for many years regarding the unsuspected functions of this class of antibody (Capron and Dessaint, 1985, Capron *et al.* 1987).

Among the various mechanisms that regulate the expression of protective immunity, one stems from isotypic regulation itself. Evidence for the selective production of defined antibody classes during the course of experimental schistosome infection in rats raised questions about the functions of other isotypes shown not to be directly implied in killing pathways. The decrease in immunity observed at certain periods of the infection in rats indeed is not related to a sharp decrease in antibody production but is concomitant with the appearance of non anaphylactic IgG subclasses. A representative IgG2c monoclonal antibody was shown to inhibit the capacity of an IgG2a monoclonal antibody both to induce eosinophil dependent killing of schistosomula and to confer passive protection *in vivo* (Grzych *et al*. 1984). The concept of blocking antibody was supported by the observation that this IgG2c monoclonal antibody can inhibit the recognition by the protective IgG2a monoclonal antibody of the carbohydrate moiety of a major surface glycoprotein of schistosomula described as gp 38 (Grzych *et al*. 1984).

The possibility that a similar phenomenon might be important in humans infected by *S. mansoni* was first indicated by the observation that susceptibility to reinfection after treatment of school children is significantly correlated with the presence of high levels of antibodies that inhibit the binding to the major gp38 schistosomulum surface antigen of the protective monoclonal IgG2a antibody. In addition, IgM and IgG2 antibodies isolated from the sera of various individuals directly block the eosinophil-dependent killing of schistosomula mediated by IgG antibodies from the same sera. IgM antibodies with specificity for schistosomulum surface antigens are present in higher levels in the young susceptible children than in the older, resistant subjects (Khalife *et al*. 1986, Butterworth *et al*. 1987). More recently, analysing the isotypes of antibodies to a recombinant protective protein (Sm28 GST) and its derived synthetic peptides, we found a significant correlation between susceptibility to reinfection to *S. mansoni* in humans and increased production of IgG4 antibodies to Sm28 and its defined B cell epitopes (Auriault *et al*. 1990). Consistently, in the framework of their studies, Hagan et al.(1991), and Dunne *et al*. (1992) have also shown a clear correlation between IgG4 antibody response to schistosome antigens and increased susceptibility to reinfection. More strikingly Dessein et al. ·(personal communication) have shown in Brazil that the association of low levels of IgE antibodies to *S. mansoni* with high IgG4 levels resulted in an increase of over 100-fold in susceptibility to reinfection. The main message from these studies is that in human schistosomiasis blocking antibodies are important components of the clinical expression of acquired resistance at its early stages, and afterwards clinical expression of immunity is positively correlated to the presence of detectable IgE antibody response to schistosomes. Such indications are obviously in the heart ot the design of defined antischistosome vaccines.

STRATEGY TOWARD VACCINE

On the basis of these concepts, we have developped during the last five years extensive investigations aiming at the identification and the molecular characterisation of potentially protective antigens against schistosomiasis. The genes encoding several *S. mansoni* proteins have now been cloned in our laboratory, among which one of them initially named P28 (Balloul *et al*. 1987) appears as a promising vaccine candidate. After its successful cloning in collaboration with Transgene, P28 was identified as a glutathion S.transferase (GST) (Taylor *et al*. 1987) and distinct in its molecular structure of a GST recently cloned from a *S. japonicum* cDNA library (Davern *et al.* 1987). Sm 28 GST has been expressed in various vectors, including *E. coli, Saccharomyces cerevisiae*, and the Vaccinia virus. Vaccination experiments performed with the highly purified native protein indicated a level of protection close to 70 % in rats, 50 % in the mouse and in hamsters (Balloul *et al.* 1987). Immunisation performed with the recombinant protein in the presence of aluminium hydroxyde confirmed the initial results and led to a mean protection of over 50 % in rats and 40 % in mice (Balloul *et al.* 1987). Several vaccination experiments were undertaken in baboons and a very significant protection up to 80 % could be obtained in some animals. However a large degree of individual variation was noticed and the mean protection observed was 42 % (Boulanger *et al.* 1991).

During these preliminary experiments, our attention was drawn to the existence, even in the very partially protected animals, of a significant decrease in the size and the volume of egg granulomas in the liver, whereas a mean reduction of 68 % of fecal egg output per female worm and per day was noticed. Similar observations were made in the monkey *Patas patas* immunized against *S. haematobium* infection. A dramatic decrease of urinary bladder lesions studied by ultrasound tomography during a period of eight months was observed in immunized animals compared to controls. More strikingly, eggs collected from vaccinated monkeys showed a marked decrease (85 %) in their hatching capacity and in the viability and infectivity (58 %) of the miracidia (Boulanger *et al.* 1991).

The use of appropriate monoclonal antibody probes to Sm 28 GST epitopes has recently allowed to relate the antifecundity effect observed after immunisation to the inhibition of expression of the GST enzymatic activity of Sm 28 GST. Indeed results obtained both *in vitro* and *in vivo* indicate that a monoclonal antibody which inhibits the enzymatic activity confers, with a significant protection against challenge, a dramatic reduction in egg laying and egg viability whereas, in contrast, another protective monoclonal antibody which does not inhibit the enzymatic activity confers protection in reducing worm burden but has no effect on egg production and viability (Xu *et al.* 1991).

The mapping of the major epitopes of the molecule has led to the identification of the major role played by the N and C terminal domains in the expression of the enzymatic activity. The construction of corresponding synthetic peptides has allowed, after immunisation, to decrease by 70 % parasite fecundity and egg viability. In this context the C. terminal epitope (190-211) appears of particular interest for the optimization of an anti parasite fecundity vaccine. In parallel, an immunodominant epitope associated with protection against challenge in experimental models and acquired resistance in human population has been identified. The immunisation with an octameric construction of the corresponding peptide (115-131) has led to significant degree of protection in rats (Wolowczuk *et al.* 1991)).

The study of the immunological mechanisms, underlying the inhibition of parasite fecundity and of egg viability has revealed the existence of an unsuspected mechanism related to the neutralising activity of IgA antibodies.

Recent studies performed in human populations have revealed a close association between the production of IgA antibodies to Sm 28 GST, their neutralising activity of the enzymatic function of the molecule and the age dependent decrease in the egg output observed in human population in parallel with acquisition of immunity (Grzych et al. in press). More recently monoclonal IgA antibodies to the N and C terminal epitopes have been thown to reproduce, by passive transfer, a dramatic inhibition of female worm fecundity (Grezel et al. submitted).

It therefore appears that in terms of vaccine strategy against schistosomiasis immunisation with Sm 28 GST, might achieve two complementary goals in human population :

a) A partial but significant reduction of the worm population resulting from infection or reinfection

b) A significant reduction of pathological consequences by a marked decrease in parasite fecundity and egg viability, this effect affecting directly transmission potentialities of the disease.

It also appears, both from the study of experimental models and of human populations that at least two distinct immunological mechanisms, for which the cellular components remain to be defined, may account for these two effects. For the first, IgE antibodies appear as a major humoral component of acquired resistance to reinfection whereas for the other, IgA antibodies appear as a major humoral factor affecting parasite fecundity and its pathological consequences. As it could be expected immunity to such complex organisms as schistosomes is obviously multifactorial in nature and there is no a priori reason to think that successful immunisation against this pathogen can be achieved though the elicitation of a single effector mechanism.

PERSPECTIVES

The relevance of our observations to vaccine strategy has been recently confirmed by vaccination experiments performed in Sudan against cattle schistosomiasis due to *Schistosoma bovis*. This model was chosen because of many common features with human infection by *S. mansoni*.

In collaboration with A. Bushara and M. Taylor (Bushara et al. submitted), we could show that immunisation of calves with *S. bovis* GST results in a dramatic reduction in egg production and tissue egg count. (over 80 %) and acquisition of resistance to a lethal infection. These results confirmed in a natural host that the major effect of immunisation with schistosome GST is to very significantly reduce parasite fecundity. They also open feasible perspectives for a veterinary vaccine against schistosomiasis in a near future and allow to consider the acceptability of Phase I trials in human populations.

At the same time, the recent crystallisation of Sm 28 GST (Trottein et al. 1992) allows the study of its 3D molecular structure and new approaches toward molecular design of an optimal vaccine. In this respect the recent cloning in our laboratory of the GST from *Schistosoma haematobium* and *Schistosoma bovis* has allowed the demonstration of the high degree of conservation of the C terminal domain among the various species of schistosomes (Trottein et al. 1992). It is not unlikely therefore that a cross specific anti fecundity vaccine might be achieved.

The demonstration of the unsuspected function of IgA antibodies in Schistosomiasis and their role on worm fecundity paves the way to new possibilities of immunisation strategy through oral route which are currently explored in our laboratory.

From experimental models to human populations, from the bench to endemic areas, studies performed during the last 15 years have revealed at the level of effector mechanisms of immunity, immunoregulation and pathogenesis, novel modalities, the interest of which extend far beyond the field of Schistosomiasis (Capron and Dessaint 1992, Capron 1992). These studies seem presently to represent a promising approach towards the possible immunological control of one of the major human parasitic disease through the identification not only of potentially protective antigens but also of the components of the immune response which vaccination should aim at inducing.

REFERENCES

Auriault C, Gras-Masse H, Pierce RJ, Butterworth AE, Wolowczuk I, Capron M, Ouma JH, Balloul JM, Khalife J, Neyrinck JL, Tartar A, Koech D, Capron A (1990) J. Clin. Microbiol. 28:1918-1924

Balloul JM, Sondermeyer P, Dreyer D, Capron M, Grzych JM, Pierce RJ, Carvallo D, Lecocq JP, Capron A (1987) Nature 326:149-153

Balloul JM, Grzych JM, Pierce RJ, Capron A (1987) J. Immunol. 138:3448-3453

Bazin H, Capron A, Capron M, Joseph M, Dessaint JP, Pauwels R (1980) J. Immunol. 124:2373-2377

Boulanger D, Reid GD, Sturrock RF, Wolowczuk I, Balloul JM, Grezel D, Pierce RJ, Otieno MF, Guerret S, Grimaud JA, Butterworth AE, Capron A (1991) Parasite Immunol. 13:473-490

Butterworth AE, (1985) Trans. R. Soc. Med. Trop. 79:393-408

Butterworth AE, Bensted-Smith R, Capron A, Dalton PR, Dunne DW, Grzych JM, Kariuki HC, Khalife J, Koech D, Mugambi M, Ouma JH, Siongok JK, Sturrock RF (1987) Parasitology 94:281-300

Capron A, Dessaint JP (1985) Ann. Rev. Immunol. 3:455-476

Capron A, Dessaint JP, Capron M, Ouma JH, Butterworth AE (1987) Science 238:1065-1072

Capron A, Dessaint JP (1992) Ann. Rev. Med. 43:209-218

Capron A (1992) Current Opinion in Immunology 4:419-424

Capron M, Spiegelberg HL, Prin L, Bennich H, Butterworth AE, Pierce RJ, Ouaissi MA, Capron A (1984) J. Immunol. 232:462-468

Davern KM, Tiu W, Morahan G, Wright MD, Garcia EG, Mitchell GF (1987) Immunol. Cell. Biol. 651:473-482

Dunne D, Butterworth AE, Fulford AJC, Kariuki HC, Langley JG, Ouma JH, Capron A, Pierce RJ, Sturrock RF (1992) Eur. J. Immunol. 22:1483-1494

Grzych JM, Capron M, Bazin H, Capron A (1982) J. Immunol. 129:2739-2743

Grzych JM, Grezel D, Xu CB, Neyrinck JL, Capron M, Ouma JH, Butterworth AE, Capron A (in press) J. Immunol.

Grzych JM, Capron M, Dissous C, Capron A (1984) J. Immunol. 133:998-1004

Hagan P, Blumenthal UJ, Dunne D, Simpson AJG, Wilkins HA (1991) Nature 349:243-245

Joseph M, Capron A, Butterworth AE, Sturrock RF, Houba V, (1978) Clin. Exp. Immunol. 33:36-45

Kigoni EP, Elsas PPX, Lenzi HL, Dessein AJ (1986) Eur. J. Immunol. 16:589-595

Khalife J, Capron M, Capron A, Grzych JM, Butterworth AE, Dunne DW, Ouma JH (1986) J. Exp. Med. 164:1626-1640

Rihet P, Demeure CE, Bourgeois A, Prata A, Dessein AJ (1991) Eur. J. Immunol. 21:2679-2686

Taylor JB, Vidal A, Torpier G, Meyer DJ, Roitsch C, Balloul JM, Southan C, Sondermeyer P, Pemble S, Lecocq JP, Capron A, Ketterer B (1988) Embo. J. 7:465-472

Trottein F, Godin C, Pierce RJ, Sellin B, Taylor M, Gorillot I, Sampaio Silva M, Lecocq JP, Capron A (1992) Mol. Biochem. Parasitol. 54:63-72

Verwaerde C, Joseph M, Capron M, Pierce RJ, Damonneville M, Velge F, Auriault C, Capron A (1987) J. Immunol. 138:4441-4446

Wolowczuk I, Auriault C, Bossus M, Boulanger D, Gras-Masse H, Mazingue C, Pierce RJ, Grezel D, Reid GD, Tartar A, Capron A (1991) J. Immunol. 146:1987-1995

Xu CB, Verwaerde C, Grzych JM, Fontaine J, Capron A (1991) Eur. J. Immunol. 21:1801-1807

Correlation of Natural Killer Cell Activation, CD4+ TH1 Cell Development and resistance in Experimental Cutaneous Leishmaniasis

P. Scott and T. Scharton

Departement of Pathobiology, School of Veterinary Medicine, University of Pennsylvania, 3880 Spruce Street, Philadelphia, Pennsylvania, 19104

INTRODUCTION

Human cutaneous leishmaniasis is a protozoal infection associated with a spectrum of clinical presentations ranging from self-limiting lesions to non-healing metastatic infections. Infections of mice with *Leishmania* also exhibit a wide range of outcomes. For example, in BALB/c mice *L. major* infections are eventually fatal, while in resistant strains of mice, such as C3H/HeN, this parasite causes a self-limiting disease. Studies from many laboratories have shown that CD4+ T cell subsets determine the outcome of infection (reviewed in (Locksley and Scott, 1991)). Thus, the susceptibility observed in BALB/c mice is associated with TH2 responses, while resistant C3H/HeN mice exhibit TH1 responses.

A major question in this field has been what factors influence which T cell subsets develop following infection with *L. major*. Results from several laboratories show that a critical component is the endogenous levels of cytokines present during the first several days of infection (Coffman, Varkila, Scott and Chatelain, 1991). In our laboratory, we have found that in vivo depletion of IFN-γ inhibits the development of TH1 cells in C3H/HeN mice, and dramatically enhances susceptibility (Scott, 1991). Furthermore, we have found that administration of IFN-γ enhances TH1 responses in BALB/c mice, and functions as a potent immunopotentiator in a vaccine model. We have recently focused on the role of natural killer (NK) cells as a source of IFN-γ during the first few days of infection, as well as their potential contribution to successful immunization against *Leishmania*.

RESULTS

Induction of NK cell activity during an active infection

As early as three days following infection, the lymph node cells from C3H/HeN mice draining the site of a *L. major* inoculation produce high levels of IFN-γ when cultured in vitro, while significantly lower levels are produced by cells taken from BALB/c mice (Scott, 1991). We have now found that correlated with this early IFN-γ response is an NK cell cytotoxic response. Thus, lymph node cells taken from C3H/HeN mice infected with 2 million purified metacyclic promastigotes exhibit high levels of cytotoxicity against Cr-^{51} labeled YAC-1 cells, while minimal cytotoxicity is observed by cells from BALB/c mice (40% specific lysis by C3H/HeN cells vs. 10% specific lysis by BALB/c cells at a 100:1 effector to target cell ratio) (Scharton and Scott, 1992b; Scharton and Scott, 1992a). The NK cell response seen in C3H/HeN mice is completely depleted by in vivo administration of an anti-asialo antisera, and moreover, such elimination abrogates approximately 75% of the IFN-γ produced by LN cells when subsequently cultured in vitro. An interesting aspect of this model is that depletion of CD4+ cells also abrogates a significant proportion of the IFN-γ response (Scott, 1991; Scharton and Scott, 1992a). These results might be interpreted to indicate that the NK cell response is dependent upon IL-2 from naive T cells, since IFN-γ production by NK cells is significantly augmented by IL-2 (Henney, Kuribayashi, Kern and Gillis, 1981; Trinchieri, 1989). However, we also observe that the immediate inflammatory response in the lymph node is dependent upon CD4+ cells, since in contrast to control infected animals, no increase in lymph node size is observed in anti-CD4+ treated C3H/HeN mice following infection (unpublished observations).

In order to determine if ablation of NK cells alters the outcome of the infection, C3H/HeN mice were depleted of NK cells by administration of anti-asialo antisera and the course of infection monitored. Abrogation of the NK cell response at the time of infection resulted in a significant increase in the number of parasites in the lesions at 2 weeks of infection. This increase was accompanied by a decrease in the level of IFN-γ produced by both spleen and lymph node cells, and an increase in the amount of IL-4 produced (Scharton and Scott, 1992a). Despite the maintenance of significantly larger lesions than controls up to 8 weeks after infection, these animals eventually healed.

IFN-γ and NK cell responses in C3H/HeN and BALB/c mice following immunization

When a soluble leishmanial antigen (SLA) preparation was injected into the footpad of either C3H/HeN and BALB/c mice, the pattern and kinetics of IFN-γ production in the popliteal lymph node was similar to that observed with inoculation of live parasites (Fig. 1). The levels of IFN-γ produced by cells from BALB/c mice were low, while high levels of IFN-γ were produced by cells from C3H/HeN mice. Since the early IFN-γ produced after an active infection is NK cell dependent, we next investigated whether SLA induces NK cytotoxic responses in the lymph nodes draining the site of antigen administration. The NK cell response was evaluated on cells harvested 2 days after antigen administration, which corresponds to the peak of IFN-γ production. As seen in Fig. 2, inoculation of SLA into the footpad induced a significant NK cell cytotoxic response in the draining lymph node of C3H/HeN mice, but not in BALB/c mice.

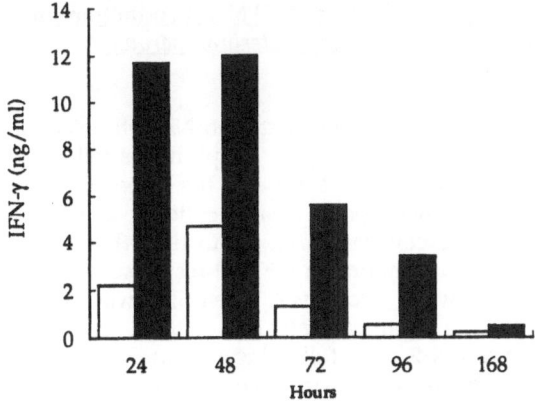

Figure 1: IFN-γ production by lymph node cells from C3H/HeN (solid bars) or BALB/c (open bars) mice injected in the footpad with SLA. Mice were injected with 50 ug of SLA in the footpad and the popliteal lymph nodes were harvested at various times after injection. Cells were cultured at 5 x 10 ^{6}/ ml with 50 ug/ml of SLA. The level of IFN-γ in supernates harvested at 72 hours was assessed by ELISA (Scott, 1991). SLA was prepared as previously described (Scott, Pearce, Natovitz and Sher, 1987).

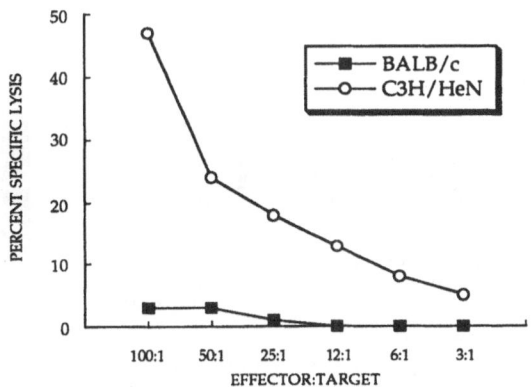

Figure 2: NK cell cytotoxic responses by lymph node cells taken 2 days following inoculation of SLA. Mice were injected with SLA as described in Fig. 1. Cells were cultured with Cr^{51}-labeled YAC-1 cells at various effector to target cell ratios, and the percent specific lysis calculated after 4 hours of incubation.

Enhancement of NK cell activity and IFN-γ production in BALB/c mice by administration of the adjuvant *Corynebacterium parvum*

In order to show that there was not a generalized NK cell defect in BALB/c mice, we assessed NK cell cytotoxic responses in the lymph nodes following footpad injection of the bacterium, *C. parvum* or the response in the spleen following intraperitoneal injection of the NK cell inducer, poly inosinic-cytidylic acid. Both of these agents induced a significant NK cell response, indicating that given the appropriate stimulus, BALB/c mice could mount an NK cell response in both the spleen and the lymph node (data not shown). Moreover, when SLA was co-administered with *C. parvum* in the footpad, a significant NK cell cytotoxic response was observed by lymph node cells harvested 2 days following immunization (Fig. 3A). Associated with the increase in NK cell cytotoxicity was an increase in the levels of IFN-γ produced by these cells, both in unstimulated cultures and even more dramatically in cultures restimulated in vitro with SLA (Fig. 3B). Since we previously demonstrated that administration of SLA and IFN-γ in the footpad also enhances subsequent IFN-γ production by lymph node cells, we assessed whether IFN-γ given with SLA could induce an NK cell cytotoxic response. As seen in Fig. 3A, in contrast to *C. parvum*, IFN-γ had little ability to stimulate an NK cell cytotoxic response.

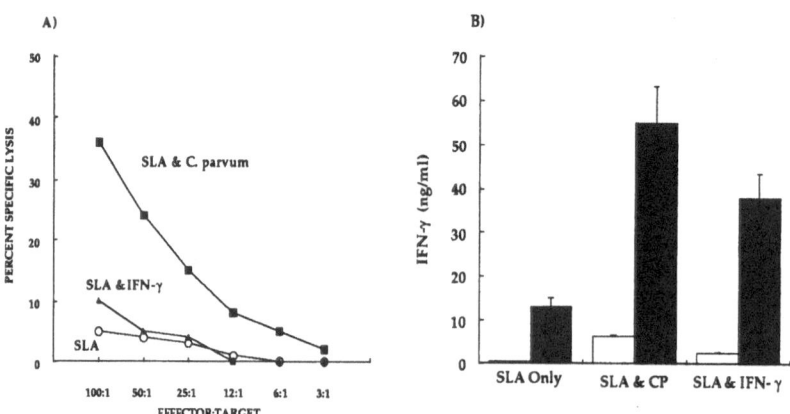

Figure 3: NK cell and IFN-γ response by BALB/c cells following injection of SLA and *C. parvum*. Popliteal lymph nodes were harvested 2 days after inoculation of SLA (50 ug) with or without *C. parvum* (100 ug) in the footpad. (A) Cytotoxic responses were assessed as described in Fig. 2. (B) IFN-γ production by cells incubated without additional in vitro stimulation (open bars) or following incubation with 50 ug/ml SLA (closed bars) is shown.

SUMMARY AND DISCUSSION

It is now firmly established that the divergent pattern of *L. major* infection observed in BALB/c and C3H/HeN mice is due to differential stimulation of CD4+ T cell subsets, and that the endogenous level of IFN-γ at the time of infection is one critical factor in T cell subset development (Scott, 1991). Thus, C3H/HeN mice depleted of IFN-γ develop TH2 responses and become susceptible to *L. major*. Studies in our laboratory currently aim to determine the source of the IFN-γ in C3H/HeN mice during the first few days of infection. Our results indicate that one source of IFN-γ in the C3H/HeN during this time period may be NK cells. Moreover, we find that depletion of NK cells using a polyclonal anti-asialo GM1 antisera significantly enhances the number of parasites in the lesions at 2 weeks. Associated with this increase is an alteration in the cytokine profile observed in both the spleen and lymph nodes, such that these animals exhibit dramatically lower levels of IFN-γ, and significant levels of IL-4, when compared with controls. These data establish the NK cell as an important source of IFN-γ capable of biasing the differentiation of naive CD4+ cells into a TH1 cell phenotype. Unexpectedly, these animals were able to eventually heal. Since administration of an anti-IFN-γ monoclonal antibody leads to a non-healing infection (although only with the appropriate high affinity monoclonal antibody), the inability to alter the eventual outcome of these infections might indicate that another cell contributes IFN-γ during the early stages of infection. However, it is also possible that the anti-asialo GM-1 antisera used in these experiments is unable to completely deplete NK cells for a sufficient period of time to permanently alter the response. In this regard, it should be noted that it is unknown how long the anti-IFN-γ monoclonal needs to be present in vivo in order to mediate its effects. Ongoing studies in the laboratory are directed at addressing these issues.

Since NK cells participate in the development of a TH1 response in C3H/HeN mice after an active infection, stimulation of these cells during immunization may also augment TH1 cell development. Initial studies to test this hypothesis examined whether NK cells are stimulated during immunization with SLA. Similar to an active infection, SLA injected in the footpad of C3H/HeN mice induces a strong NK cell response, while no response is observed in BALB/c mice. Associated with these responses was significant IFN-γ production by lymph node cells of C3H/HeN, but not BALB/c mice. The inability to induce IFN-γ in BALB/c mice is consistent with the inability to vaccinate BALB/c mice by the subcutaneous route with parasites or SLA alone (Liew, Hale and Howard, 1982)(unpublished observations). However, we have previously found that BALB/c mice can be protectively immunized via the subcutaneous route if both IFN-γ and the bacterial adjuvant, *C. parvum*, are included with SLA (Scott, 1991). Interestingly, while IFN-γ and SLA, without *C. parvum*, induced a TH1 response in the lymph node, such an immunization was unable to induce protection. We now show that one difference between these immunizations is that only the *C. parvum* injection induced an NK cell response, at least as measured by cytotoxicity. Future experiments will directly determine if the induction of NK cells is required for the development of protective immunity.

The lytic capacity of NK cells has long been the focus of study in the field of NK cell biology. However, a more critical role for these cells in infectious disease may involve their ability to predispose subsequent immune responses towards cell-mediated immunity. Furthermore, the targeting of NK cells during immunization may be a useful strategy in vaccine development. IFN-γ must be present for the development of TH1 cells following infection with *L. major* (Scott, 1991), and by extension may be required for the induction of protective TH1 cells during immunization. However, administration of IFN-γ in a vaccine may not be practical in the field. On the other hand, NK cells may provide an endogenous source of modulatory cytokines, such as IFN-γ, that might be significantly more effective than exogenously administered immunomodulators. Indeed, the success of *C. parvum* as an adjuvant in several leishmanial immunization models may, in part, be due to its capacity to stimulate NK cells.

Acknowledgements: The authors wish to thank Dr. Luis Afonso and Dr. Leda Vieira for helpful discussions and Leslie Taylor and Mary Beth Costenbader for technical assistance. This work was supported by a National Institutes of Health grant # RO1 AI30073.

REFERENCES

Coffman RL, Varkila K, Scott P and Chatelain R (1991) Immunol Rev 123:189-207

Henney CS, Kuribayashi K, Kern DE and Gillis S (1981) Nature 291:335-338

Liew FY, Hale C and Howard JG (1982) J Immunol 128:1917-1922

Locksley R and Scott P (1991) Immunol Today 12:A58-A61

Scharton T and Scott P (1992a) manuscript submitted

Scharton T M and Scott P (1992b), Immediate IFN-g production in C3H/HeN mice infected with *Leishmania major*. In: Sonnenfeld G, Czarniecki C, Nacy C, Byrne G and Degre M. (eds) Cytokines and Resistance to Nonviral Pathogenic Infections, The Biomedical Press, Augusta, GA, p in press.

Scott P (1991) J Immunol 147:3149-3155

Scott P, Pearce E, Natovitz P and Sher A (1987) J Immunol 139:221-227

Trinchieri G (1989) Adv Immunol 47:187-375

Recombinant Migration Inhibitory Factor (MIF): its Potential as an Adjuvant for Developing Vaccines

J. R. David[*+], R. G. Titus[*], C. Shoemaker[*] and W. Y. Weiser[+]

[*]*Departement of Tropical Public Health, Harvard School of Public Health and*
[+]*Departement of Medicine, Harvard Merdical School and Departement of Rheumatology, Brigham and Womens Hospital, Boston, MA 02115*

INTRODUCTION

Migration inhibitory factor (MIF) was the first lymphokine to be discovered (Bloom and Bennett 1966, David 1966). The factor was identified by its ability to prevent the migration of guinea pig macrophages out of capillary tubes in vitro. We showed that the production of MIF correlated with delayed hypersensitivity and cellular immunity in humans and animal models (Rocklin et al. 1970, David and David 1977). MIF also altered macrophage physiology (Nathan et al. 1971) and increased macrophage functions such as killing microorganisms and tumor cells (Fowles et al. 1973, Piessens et al. 1975). However, a purified or cloned lymphokine was required to demonstrate that these altered functions of macrophages were induced by MIF and not by other factors. Using functional expression cloning in COS cells, a cDNA, p7-1, encoding for a human MIF was isolated. (Weiser et al. 1989).

RECOMBINANT MIF (rMIF)

Sequence analysis

Supernatants from COS cells transfected with p7-1 cDNA contained a biologically active rMIF, a 12kD protein with an amino acid sequence differing from other known proteins (Weiser et al. 1989). It shared no sequence homology with IFNγ or IL-4 which have been reported to have MIF activity (Thurman et al. 1985, McInnes and Rennick 1988), or with other cytokines or two cDNA encoding MIF-related proteins MRP-8 and MRP-14 (Odink et al. 1987). The p7-1 cDNA codes for two potential asparagine-linked glycosylation sites and three cysteine residues (Weiser 1989). It does not have a conventional secretory leader sequence which is also lacking in IL-1 α and IL-β (March et al. 1985) and acidic and basic fibroblast growth factors (Giminez-Gallego et al. 1985, Abraham et al. 1986).

Insertional mutants

Insertional mutations of p7-1 were made (Weiser et al. 1989). p7-1-24B, contained a 14-mer, with a termination codon inserted in the unique *Pst* I site to disrupt the open reading frame. p7-1-24232 contains an inserted 99-mer at the same site to extend the open reading frame. Supernatants from COS-1 cells transfected with these had neither MIF activity nor a 12kD protein band. A new band of apparent 15.5kD was seen with the p7-1-24232 insert. These studies confirmed that the active 12kD protein was coded by p7-1. The cDNA containing the stop codon was used as a control in some experiments described below and was referred to as "stopMIF".

Detection of mRNA for rMIF in stimulated lymphocytes

The mRNA for rMIF was detected in concanavalin A stimulated blood lymphocytes but not in unstimulated cells using Northern analysis (Weiser et al. 1989). It was also detected in T cell hybridoma cells which produce MIF.

ACTIVITY OF rMIF ON HUMAN MONOCYTES/MACROPHAGES

Adhesion induced by rMIF.

In collaboration with John Caulfield, human monocytes, cultured with rMIF or supernatants from mock-transformed COS-1 cells (mock supernatants), were examined by electronmicroscopy. rMIF enhanced viability and caused the aggregation and clumping of monocytes. The membranes of many were closely juxtaposed (<20 nm) over long distances (1 μm).

Up-regulation of HLA-DR, IL-1ß, IL-6 and TNFα.

rMIF up-regulates HLA-DR and induces gene expression and protein production of IL-1β, IL-6 and tumor necrosis factor (TNFα) by human monocyte-derived macrophages. Human monocytes were cultured for 7-10 days and the resulting macrophages cultured for 6-12 hrs with rMIF, stopMIF or mock supernatants. The RNA was extracted and the mRNA was measured by Northern blot analysis. rMIF up-regulated HLA-DR, IL-1ß, and TNFα. The induced changes were comparable to that seen with interferon-γ (IFNγ). rMIF also increased the production of the protein IL-1ß (Weiser et al. 1992), TNFα (Pozzi and Weiser 1992) and IL-6 (G. Newman, unpublished) as determined by ELISA and/or biologic activity.

rMIF activates human macrophages to kill the protozoan parasite _Leishmania donovani_.

rMIF activated blood monocyte-derived macrophages in vitro suppress the growth of and kill _L. donovani_ promastigotes and amastigotes (Weiser et al. 1991). The anti-leishmanial effect, 50-77% reduction in parasites compared to controls, is maximal when macrophages have been incubated 48-72 hrs before infection, and is similar to that seen with macrophages activated by IFNγ. In five of eleven experiments, there was also a significant decrease in the number of parasites at 48 hr compared to 2 hr, suggesting cytolysis. Of interest, although IL-4 inhibits macrophage activation by IFNγ or GM-CSF, it has no effect on macrophage activation by rMIF.

rMIF activates human macrophages and monocytes to kill tumor cells.

Fresh human blood monocytes and/or derived-macrophages were incubated with rMIF for 48 or 72 hr. Then, human tumor cell lines (^{3}H-thymidine labeled A375, a melanoma line or K562, a myeloma line) were added to the wells at an effector to target ratio of 1:1. Both activated human monocytes and macrophages killed these tumor cell lines, demonstrating cytotoxicity of 34-57% compared to cells incubated in stopMIF (Pozzi and Weiser 1992).

Regulation by IL-10.

IL-10, a regulatory cytokine synthesis inhibitory factor, was shown to be an effective inhibitor of MIF at several levels. IL-10, when added to MIF-producing T cell hybridomas, inhibited the expression of MIF-mRNA as assessed by Northern blot and of MIF production as assessed by ELISA. IL-10 also inhibited the effect of MIF on macrophages, preventing inhibition of migration and macrophage activation, as measured by the ability to kill *L. donovani* (Weiser et al. unpublished).

HUMAN MIF ACTS ON MOUSE CELLS

It was of interest to determine whether human rMIF lacks species specificity, as does natural MIF, and would act on murine cells. If it did, we could carry out in vivo experiments to study the mechanism of action of rMIF.

Human rMIF inhibits the migration of mouse and guinea pig macrophages using both the agarose and capillary tube assays. Human MIF-cDNA hybridizes with mRNA from concanavalin A activated murine spleen cells (but not from unactivated spleen cells), and with EL-4 cells, a cell line that makes MIF. In studies with Richard Titus, human rMIF activated murine macrophages to kill *Leishmania major*. Further, we cloned mouse MIF by screening an EL-4 library with the human MIF-cDNA clone. The sequence of the murine MIF cDNA has 90% homology with the human MIF clone and the majority of differences are conservative (David and Shoemaker, unpublished).

Nitric oxide (NO) synthetase and NO production.

rMIF induces the production of nitric oxide (NO) synthetase and NO by murine macrophages. In collaboration with Dr. F.Y. Liew in England, we have shown that rMIF increases NO and NO synthetase in murine macrophages. The L-arginine derived toxic nitrogen intermediates have been shown to be important in the killing of Leishmania by macrophages (Green et al. 1990, Liew et al. 1990). Of special interest, TGFβ, which does not affect NO production induced by IFNγ, inhibits the NO and NO synthetase induced by rMIF. The data so far suggest that IFNγ may be regulated, in part, by IL-4, and rMIF by TGFβ. rMIF also induces H_2O_2 production by human monocytes (Pozzi and Weiser 1992).

ANTIBODY TO HUMAN MIF

Because the sequence of human and mouse MIF were over 90% homologous, we attempted to produce antibodies to human rMIF in chickens rather than in mice. Chickens were immunized at East Acres Biologicals, Southbridge, MA. with the purified rMIF-maltose binding fusion protein. We showed that IgY isolated from the eggs of these chickens bound and removed the rMIF from COS cell supernatants, and gave positive Western blots with rMIF from COS cells, with purified rMIF-glutathione-transferase fusion protein produced with the pGEX vector, with a non-fusion MIF bacterial recombinant, and with natural MIF from stimulated human blood lymphocytes and from T cell hybridomas. Using an ELISA assay, the antibody reacted with the above MIFs as well as some of the MIFs from stimulated blood lymphocytes. The antibody should be useful in delineating the role of rMIF in immune mechanisms in in vitro and in vivo experiments.

HUMAN MIF ACTS AS A POTENT ADJUVANT

The ability of human rMIF to act on murine cells allowed us to carry out the following in vivo experiments. A model antigen, bovine serum albumin (BSA), was injected into mice with rMIF-containing COS cell supernatants in incomplete Freund's adjuvant (ICFA); the draining lymph nodes were removed 7-10 days later and assayed for BSA-induced T cell proliferation. Controls included BSA in complete Freund's adjuvant (CFA), BSA with mock transfected COS cell supernatants with ICFA, and BSA with supernatants from COS cells transfected with the mutant stopMIF and ICFA. The results of 8 experiments show that T cell proliferation induced by BSA given with rMIF in incomplete Freund's adjuvant is as great as the response induced by BSA in CFA (See Figure 1).

Six experiments with BSA given with IFNγ and ICFA showed that IFNγ had little to no adjuvant effect in this system (See Figure 1). BSA given with rMIF without ICFA had no effect, probably because the rMIF rapidly diffuses away from the antigen into the surrounding tissues.

In preliminary experiments, BALB/C mice were injected with 10μg HIV gp120 in CFA, ICFA, ICFA with rMIF/or mockMIF supernatants. Lymph node lymphocytes obtained 8 days later from animals receiving rMIF showed marked proliferation to 1 and 2.5 μg of HIV gp120, almost as great as that seen with CFA and 10 fold higher than with the mock control.

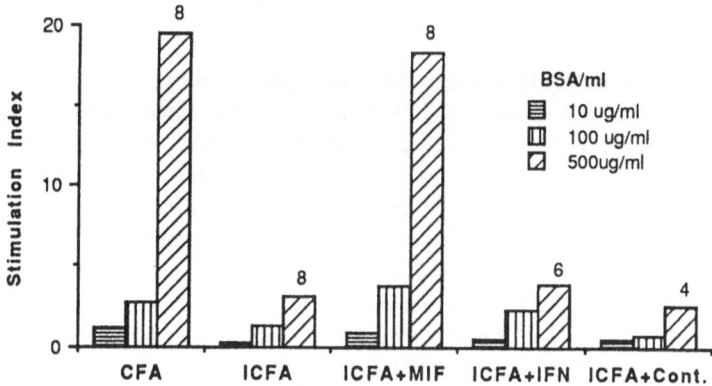

Fig 1. BSA with the adjuvants listed was injected into mice and the regional lymph nodes removed. The T cells assayed for BSA induced proliferation in vitro. The number of experiments is indicated on top of the bars. ICFA+Cont are 2 experiments with stopMIF and 2 with mock transfected COS cell supernatants. From Weiser et al. 1992.

Human rMIF encapsulated in liposomes has adjuvant-like activity.

In studies with Stephen Furlong, we encapsulated COS-cell supernatants containing rMIF or stop-MIF with BSA into phosphatidylcholine/cholesterol liposomes at a molar ratio of 1:1. Administration of these BSA-rMIF encapsulated liposomes to mice yielded lymph node T cells which exhibited a marked increase in BSA induced proliferation. The increase was almost as great as that seen by T cells from lymph nodes of mice that received BSA in CFA.

rMIF enhances *Leishmania* antigen vaccine.

Preliminary studies with Richard Titus show that rMIF greatly enhances the protection of mice given a *Leishmania* candidate vaccine antigen, cloned gp63, to subsequent challenge with *Leishmania major*. C57Bl/6 mice were each injected subcutanously in the right foot pad with an ICFA-emulsion of recombinant *L. major* gp63 (Dr. R.McMaster, Canada) and COS cell MIF or stopMIF. One week later, animals were boosted intraperitoneally with an ICFA emulsion of gp63 and COS supernatants containing rMIF or stopMIF. Two weeks after the boost, animals were challenged in the left foot pad with 10^6 *L. major* promastigotes. The average size of the lesions at 30 days for mice receiving MIF were 0.4 mm, compared to 1.3 mm for those receiving stopMIF and 1.8 mm for controls receiving nothing. At 40 days, they were 0.0 mm, 0.45 mm and 0.5 mm respectively.

SUMMARY

A cDNA encoding a human MIF was isolated through functional expression cloning in COS-1 cells. It encodes a 12 kD protein with MIF activity. Concanavalin A stimulated blood lymphocytes show increased expression of MIF-mRNA. The recombinant MIF from COS-1 cells affects the physiology and function of human monocyte-derived macrophages and murine macrophages. Human macrophages incubated with rMIF exhibit marked aggregation and clumping, show enhanced expression of HLA-DR mRNA, and enhanced expression and production of IL-1ß, IL-6, and TNFα. rMIF activates macrophages to kill *Leishmania donovani* and tumor cells. Human rMIF acts on murine cells, inducing the production of nitric oxide (NO) synthetase and NO by murine macrophages. Some of the functions of MIF are inhibited by IL-10 or TGFß but not by IL-4. Human MIF acts as a potent immuno-adjuvant, enhancing T cell proliferation to BSA and HIV gp120 similar to that of complete Freund's adjuvant when given with these antigens in oil or liposomes. In preliminary experiments, rMIF given with the leishmanial antigen gp63 to mice, enhances protection to subsequent parasite challenge. These experiments taken together suggest that rMIF may be useful as an adjuvant in developing vaccines.

REFERENCES

Bloom BR, Bennett B (1966) Science 153: 80-82

David JR (1966) Proc. Natl. Acad. Sci. USA 56: 72-77

Fowles RE, Fajardo JM, Leibowitch JL, David JR (1973) J. Exp. Med. 138: 952-964

Giminez-Gallego G, Rodkey J, Bennett C, Rios-Candelore M, Di Salvo J, Thomas K (1985) Science 230: 1385-1388

Abraham JA, Mergia A, Whang JL, Tumolo A, Friedman J, Hjerrild KA, Gospodarowicz D, Fiddes JC (1986) Science 233: 545-548

Green SJ, Crawford, RM, Hockmeyer JT, Meltzer MS, Nacy CA (1990) J. Immunol. 145: 4290-4297

Liew FY, Millot S, Parkinson C, Palmer RMJ, Moncada S, (1990) J. Immunol. 144:4794-4797.

March CJ, Moseley B. Larsen A, Cerretti DP, Braedt G, Price V, Gillis S, Henney CS, Kronheim SR, Grabstein K, Conlon P, Hopp TP, Cosman D (1985) Nature (London) 315: 641-647

McInnes A, Rennick DM (1988) J. Exp. Med. 167: 598-611

Nathan CF, Karnovsky ML, David JR (1971) J. Exp. Med. 133: 1356-1376

Odink K, Cerletti N, Bruggen J, Clerc RG, Tarcsay L, Zwadlo G, Gerhards G, Schlegel R, Sorg C (1987) Nature (London) 330: 80-82

Piessens WF, Churchill WH, David JR (1975) J. Immunol. 114: 293-299

Pozzi LM, Weiser WY (1992) Cell. Immunol. In press

Rocklin RE, Rosen F, David JR (1970) N. Engl. J. Med. 282: 1340-1343

Thurman GB, Braude IA, Gray PW, Oldham RK, Stevens HC (1985) J. Immunol. 134: 305-309

Weiser WY, Temple TA, Witek-Giannotti JS, Remold HG, Clark CC, David JR (1989) Proc. Natl. Acad. Sci. USA 86: 7522-7526

Weiser WY, Pozzi LM, David JR (1991) J. Immunol. 147: 2006-2011

Weiser WY, Pozzi LM, Titus RG, David JR (1992) Proc. Natl. Acad. Sci USA. In press

Still Learning from Leprosy

Barry R. Bloom[1], Padmini Salgame[1], Jacinto Convit[2], Masahiro Yamamura[3], and Robert L. Modlin[3]

[1]Departement of Microbiology & Immunology, Albert Einstein College of Medicine, Bronx, New York, 10461, USA
[2]Instituto de Biomedicina, Caracas, Venezuela
[3]Division of Dermatology, Departement of Microbiology & Immunology, UCLA School of Medicine, Los Angelkes, CA 90033, USA

INTRODUCTION:

Reciprocity is not a value usually associated with scientific endeavors, yet perhaps no area of biomedical science has greater possibilities for providing reciprocal insights between basic science and clinical medicine than the immunology of infectious diseases. Basic knowledge of the immune system is clearly essential to understand pathogenesis of many infectious diseases and to develop new interventions to modulate or control resistance to infectious pathogens. It is increasingly clear that study of the interactions between pathogens and the immune system is reciprocally providing extraordinarily valuable insights in the fundamental questions of function and regulation of the immune system. Leprosy provides a wonderful example of the reciprocity between basic immunology and clinical medicine.

IMMUNOLOGICAL UNRESPONSIVENESS IN LEPROSY

Great progress has been made in the treatment of leprosy, and there is now a Multi-Drug Therapy (MDT) consisting of rifampicin, dapsone and clofazimine, that brings about cure rates approaching 100% in both lepromatous and tuberculoid forms[1]. While there were estimated to be about 10.5 million cases of leprosy worldwide and, 5.4 million of registered cases in 1985 when MDP was introduced by WHO, that number has declined to 3.7 million in 1990. Almost 60% of registered cases are on MDT, and over 1 million cases have been cured. Thus there is now an effective treatment for leprosy but it is unclear, in a disease of long latency, whether treatment willl have a significant effect on transmission. Clearly, an effective vaccine is still an important research goal.

One of the fascinations of leprosy is that the disease presents not as a single clinical picture, but with a diversity of manifestations that comprise a clinical and histopathological spectrum[2]. At one end of that spectrum, tuberculoid leprosy, there is a high level of cell-mediated immunity and localized lesions with organized granulomas that ultimately self-heal, albeit often with damage to nerves. At the other end of the spectrum, lepromatous leprosy, there is an absence of cell mediated immunity, and the organisms grow to extraordinarily high numbers, up to 10^{10} acid-fast bacilli/cm^2 of skin.

One of the fundamental issues in immunology remains, mechanisms of specific immunological unresponsiveness and tolerance. Patients with lepromatous leprosy fail to exhibit positive delayed-type hypersensitivity skin reactions or *in vitro* lymphocyte transformation to antigens of *M. leprae*, although they often make high levels of antibodies. Their failure to restrict the growth of *M. leprae* indicates that protection requires appropriate cell mediated responses. Lepromatous leprosy patients who are unresponsive to antigens of *M. leprae* are almost invariably able to respond with delayed type hypersensitivity to antigens of *M. tuberculosis:* thus lepromatous leprosy provides an extraordinary model for understanding specific immunologic unresponsiveness in a human disease.

The ability of these patients to respond to antigens of *M. tuberculosis* and other recall antigens in the face of specific unresponsiveness of *M. leprae* antigens prompted us to suggest that there may be one or a small number of unique antigens associated with *M. leprae* that have the ability to produce antigen-specific suppressor T cells, which function by blocking expansion of specific or cross-reactive T cell clones[3]. Studies in which *M. leprae* was added to cultures stimulated with T cell mitogens such as Con A, that trigger the T cell receptor complex indicated that in a majority of lepromatous patients the presence of *M. leprae* significantly inhibited the proliferation of CD4 T cells in the blood of lepromatous leprosy patients but not of patients with tuberculoid leprosy or lepromin positive healthy contacts. This suppression was induced by antigens of *M. leprae*, but not from eight other mycobacteria. Thus there was evidence *in vitro* for an *M. leprae* antigen-induced suppression that was stage of disease associated. The phenotype of the suppressor cell in those studies was CD8[+], CD3[+], Fc receptor positive, HLA-DR[+], TAC[+], and CD28[-]. Immune histological characterization of the cells infiltrating lesions across the spectrum of leprosy have indicated a preponderance of CD4 T cells in tuberculoid leprosy lesions, and conversely a preponderance of CD8 T cells in lesions of lepromatous leprosy, consistent with *in vitro* cellular responsiveness data[4,5]. Of current interest, approximately 70% of the CD8 cells in tuberculoid lesions expressed CD28, characteristic of cytotoxic T cells, whereas only about 35% of the CD8 cells in lepromatous lesions express CD28[6].

CHARACTERIZATION OF T SUPPRESSOR CELLS IN LEPROSY

One of the unique attributes of leprosy is the transparency of the disease, its accessibility to study. Most lesions occur in the skin, and biopsies are necessary for diagnosis and staging. It is thus readily possible to have access to immune cells directly from lesions as well as blood[7]. When CD8 cells were separated from biopsies, and amplified for a short time in absence of antigen by IL-2, approximately half of the T cell lines obtained from lepromatous lesions were found to have suppressor activity *in vitro*, where as none of the CD8 lines from tuberculoid lesions had that activity. It was subsequently possible to clone cells from those lesions and demonstrate that all expressed T cell $\alpha\beta$ receptors, in contrast to the controversial suppressor clones reported in murine systems which lacked expression of T cell receptor α or β receptor chains.[8] After surface iodination and immunoprecipitation with specific antibodies, all CD8 T suppressor clones were found to express T cell $\alpha\beta$ receptors. Further, it was possible to show that inclusion of monoclonal antibodies specific for Vβ chains expressed on T suppressor clones markedly reduced or abolished the *in vitro* suppression, indicating that the T cell receptors were required for suppression[9]. The availability of clones of CD8 cells from lepromatous patients, and CD4 clones that respond to *M. leprae* antigens by proliferation and production of lymphokines, such as IFNγ and IL-2 enabled us directly to test whether there was an MHC restriction on suppression. When CD8 and CD4 clones were matched for MHC Class 2 *in vitro* suppression was observed. When mismatched for MHC Class 2, no suppression was seen. Using human recombinant individuals with linkage dissocations between DR and DQ, it was possible to show that the CD8 antigen specific suppressor activity in all clones studies were HLA-DQ restricted, and not HLA-DR restricted[10]. That was in marked contrast to the proliferative responses of the CD4 clones, that were in our hands, all HLA-DR restricted and none HLA-DQ restricted. This is consistent with the results of Sasazuki et al[11] who found that suppression was epistatic to responsiveness and DQ linked in family studies in man.

MECHANISMS OF SUPPRESSION IN LEPROSY

Several mechanisms could be envisioned to explain the *in vitro* suppression by the CD8 clones from lepromatous leprosy patients: i) T suppressor cells could kill antigen presenting cells or block their antigen presentation; ii) TSL's could kill CD4 T cells and an antigen or MHC restricted manner; iii) T cells could secrete a factor that blocks the CD4 response to antigen; iv) T cells could present antigen/MHC, in the absence of a costimulatory signal (veto); and v) Ts cells could negate appropriate antigen presenting signals by a novel mechanism.

To the extent possible, each of these possibilities has been experimentally tested *in vitro*[3,9]. When MHC, HLA-DR and HLA-DQ matched antigen presenting cells, either primary macrophages, SV40 transformed macrophage cell lines or EBV transformed B cell lines were ^{51}Cr-labeled, there was no evidence for chromium release or for killing of APC's. Similarly, when the CD4 clones whose proliferation was suppressed were labeled and placed in suppressor assays *in vitro*, there was similarly no release of ^{51}Cr. Further in all *in vitro* experiments with clones, a routine control was the inclusion of IL-2. All clones that failed to respond to antigens in the presence of suppressor cells plus *M. leprae* antigens were perfectly capable of proliferating in the presence of IL-2, indicating that they were not killed in the process of being suppressed. Concentration of culture supernatants failed to indicate a secreted suppressor factor. The facts that T suppressor clones in the presence of *M. leprae* antigens could block influenza specific CD4 clones, and that veto mechanisms are dependant on recognition of antigen through the same restricting element, we believe, excluded the veto mechanism. That left only a mechanism by which all appropriate signals were available to the system for producing a positive response in the CD4 clones, and that those signals were negated by the CD8 suppressor clones activated by *M. leprae* antigens.

To elucidate differences between the cells infiltrating the tuberculoid lesions and lepromatous lesions, biopsies were extracted for RNA, in RNA was reverse transcribed to produce cDNAs, and using PCR primers for 15 lymphokines and cytokines, it was possible to analyze the nature of lymphokine mRNAs present within the lymphocytes infiltrating the lesions[12]. To normalize for different numbers of T cells, the PCR products were standardized for the CD3δ chain present only on T cells. The results were quite striking. Many lymphokines were found in greater abundance in tuberculoid than lepromatous lesions, for example, IL1β, TNFα, TGFβ, GMCSF. Only three lymphokines were found in greater abundance in lepromatous than in tuberculoid lesions. They were IL-4, IL-5 and IL-10. Further, IFN-γ that was found in abundance in tuberculoid lesions was invariably absent in the biopsies from lepromatous lesions. The findings with mRNAs for lymphokines in lesions were largely confirmed by the study of lymphokine production *in vitro* by clones of CD8 T supressor cells stimulated by anti CD3 antibodies[13]. Again while CD4 clones produced IFN-γ, the CD8 suppressor clones failed to do so. However, when stimulated the CD8 clones produced IL-4 and IL-5, not IFN-γ and interestingly not IL-10. Subsequent studies on cells from lesions have indicated that IL-10 is produced by macrophages and not by T cells. Thus, this pattern is highly reminiscent of the murine TH1 and TH2 patterns, with the major difference being that the IL-4/IL-5 producing T cells in the human leprosy system are produced by CD8 rather than CD4 cells. *In vitro* data indicate that anti-IL-4 antibodies block the *in vitro* suppression. However, preliminary experiments on clones indicate that IL-4 while necessary, it appears not to be sufficient for inducing suppression. In further studies on the mechanism of suppression induced by IL-4, preliminary data indicate that IL-4 can block transcription in Jurkat cells of IL-2 mRNA[14]. Were this to be true of primary T cells as well, the possibility is suggested that one function of IL-4 would be to block the production of the lymphokine, IL-2, required for expansion of TH1 cells.

DISCUSSION

The study of T cell subsets and the development of reagents and assays that distinguish them have provided an important new level of approach to the analysis of the immunology of infectious, autoimmune and neoplastic disease. These approaches led to the understanding that the predominant cell in lesions of immunologically unresponsive patients with lepromatous leprosy is a CD8 suppressor T cell which has the capability of inhibiting the proliferation of antigen specific CD4 responding cells *in vitro*. The characterization of the lymphokines of those cells indicated that the pattern of lymphokine production was similar to that of TH2 cells in the mouse, even though their phenotype was CD8 positive. That led to a comparison of the patterns of lymphokines produced by four kinds of human T cell clones: i) Antigen-specific CD4 cells capable of producing IFN-γ analogous to murine TH 1 cells; ii) IL-4 producing human CD4 cells shown to have helper activity for immunoglobulin and antibody production *in vitro*; iii) CD8 cytotoxic T lymphocytes; and iv) CD8 T suppressor cells. These comparisons indicated that both CD4 and CD8 cells contained functional subsets that produced two distinct patterns of lymphokines. Because the CD8 cells are not thought to function as helper cells we found the nomenclature awkward. Consequently, we have suggested that CD4 and CD8 subsets be defined as Type 1 or Type 2 depending on the pattern of lymphokines produced[13]. Type I producing IFN-γ and IL-2, and Type 2 producing IL-4 and IL-5. Rather than having a fixed function, we believe that the potential functions of human T cell subsets are largely defined, in the broadest sense, by the lymphokines produced (in this context we would include perforins and other molecules associated with cytotoxicity as lymphokines)[15]. Thus, while Type 1 cells appear to have a major role to play in cell-mediated immunity, they may also serve to suppress development of antibodies. Reciprocally, Type 2 cells producing IL-4, appear to have suppressive activity for development and expression of cell mediated immunity, yet can exert helper activity for antibody production (Fig. 1).

A major concern over the next several years in cellular immunology will be to define the factors that determine the functional T cell subsets developed from the T₀ or multi-potential T cell. Some features which influence that determination, for which there is some experimental evidence, include: the nature of the antigen involved; the nature of the epitope involved; and the nature of MHC restriction elements involved and MHC clearly can play a role. Perhaps the key factor in experimental studies that influences the development of TH1 and TH2 cells is lymphokines themselves. If IL-4 does have a role in inhibiting transcription as the *in vitro* data from suppression in leprosy suggest, then it could serve both to block expansion of already existing Type 1 cells that would be required for bringing about protective immunity against this infection. Furthermore, it could also block the expansion

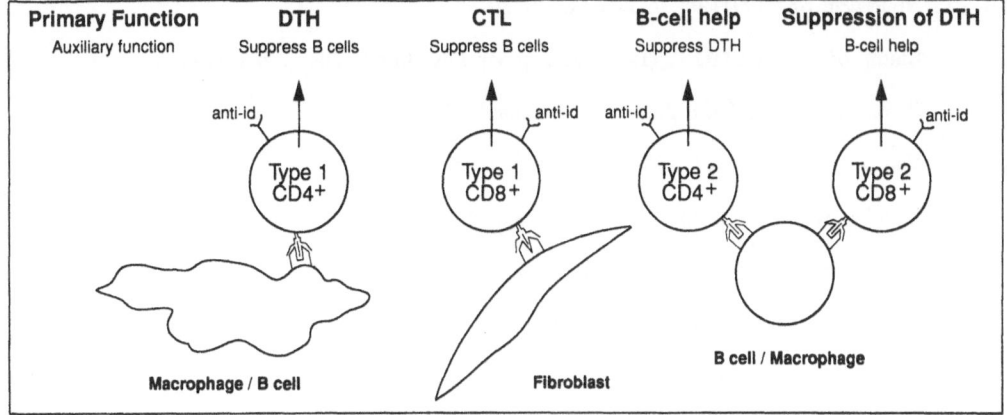

Primary Function	DTH	CTL	B-cell help	Suppression of DTH
Auxiliary function	Suppress B cells	Suppress B cells	Suppress DTH	B-cell help

Type 1 CD4+   anti-id

Type 1 CD8+   anti-id

Type 2 CD4+   anti-id

Type 2 CD8+   anti-id

Macrophage / B cell   Fibroblast   B cell / Macrophage

Fig. 1. *Model for T-cell suppression involving functional CD4+ and CD8+ T-cell subsets, differential antigen presentation and differing patterns of lymphokine production.* **From Ref 15.**

and differentiation of To cells to Type 1 cells, and thus have a pivotal role in determining the level of cell-mediated immunity in any given context. It should be noted, however, that IL-4 is clearly not the only lymphokine or cytokine involved in down-regulation of immune responses. Clearly data exist indicating that in addition to IL-4, IFN-γ, IL-10 and TGF-β have major negative regulatory roles in expansion or cytokine production by various T cell subsets.

Clonal deletion is clearly a major mechanism for explaining neonatal tolerance, although in our judgement even in that circumstance there is reason to believe that peripheral mechanisms maybe important[15]. Yet it is increasingly recognized that there are circumstances, as in the case of unresponsiveness in leprosy, which require development of peripheral tolerance. We believe that at least three mechanisms are likely to be involved. One is the case described here for leprosy in which Type 1 or Type 2 CD8 or CD4 cells secreting particular patterns of lymphokines are capable of blocking proliferation and/or lymphokine production of T cell clones. The ultimate effect of lymphokine suppression, at least from *in vitro* experiments, is indistinguishable from the second major mechanism, which is clonal anergy or veto, in which T cells expressing specific receptors are unable to respond to appropriate antigen and MHC, but are not immediately killed or deleted. Thus, both these mechanisms represent means of engendering clonal anergy. Finally, there are a number of circumstances, particularly relevant to autoimmune diseases where the numbers or antigens or epitopes involved in the pathologic response are very limited, where anti-clonotypic suppression or cytotoxicity appears to be a most appropriate and useful mechanism for inducing peripheral tolerance. It is our view that understanding these mechanisms of peripheral tolerance, including clonal anergy, veto and T cell suppression has acquired a new urgency and perhaps centrality in immunology, if we are to develop interventions for preventing or blocking inappropriate autoimmune responses and for producing more effective and specific vaccines against infectious diseases.

1. Noordeen, SK (1991) Int. J. Lepr. 62:72-80

2. Bloom, BR and Godal, T. (1983) V. Leprosy Rev. Infect. Dis. 5:765-779

3. Bloom, BR et al (1992) Ann. Rev. Immunol. 10:453-88

4. Van Voorhis, WC et al (1982) N. Engl. J. Med. 37:1593

5. Modlin, RL (1983) J. Am. Acad. Dermatol. 8:181

6. Modlin, RL et al (1988) Proc. Natl. Acad. Sci. 85:1213-1217

7. Modlin, RL et al (1986) Nature 322:459-461

8. Modlin, RL et al (1987) Nature 329:541

9. Salgame, P et al (1989) Int. Immunol. 1:121

10. Salgame, P et al (1991) Proc. Natl. Acad. Sci. 88:2598

11. Sasazuki, T et al (1989) Immunology 2:21

12. Yamamura, M (1991) Science 254:277-279

13. Salgame, P et al (1991) Science 254:279-282

14. Schwarz, EM et al (1992) *submitted*

15. Bloom, BR et al (1992) Immunol. Tod. 13:131-136

Development of Synthetic Vaccines

R. Arnon

Weizmann Institute of Science, Department of Chemical Immunology Rehovot, Israel, 76100.

INTRODUCTION

The development of vaccines has been one of the most important achievements in immunology and medicine to date. It was initiated almost two hundred years ago, by the famous trial of Jenner, who inoculated a small boy with cow pox, thereby achieving preventive immunity against the fatal small pox, a process which eventually led to the eradication of this disease. The major breakthrough, occurring just exactly one hundred years ago, was Pasteur's development of rabies vaccine, a direct result of his pioneering systematic experiments on attenuation of microorganisms, which paved the way for the development of a whole series of vaccines based on viral attenuation. The presently existing vaccines, which consist of killed or live attenuated disease-causing microbial agents or their isolated components, have definitely diminished the incidence, morbidity and mortality of a large number of infectious diseases, including major killers such as diphtheria and polio.

Notwithstanding these invaluable achievements, there are several shortcomings to the conventional vaccines, both killed and attenuated. For example: (1) The difficulty in preparation of sufficient material for vaccine production in case of some viruses which cannot be cultivated *in vitro*, such as HBV, and even more so in the case of parasites. (2) Safety considerations - the difficulty of ascertaining complete killing or adequate attenuation of the vaccine preparation, and the hazard which may be caused by exposure of both the vaccines and those involved in vaccine production. Indeed, there are several examples of calamities resulting from vaccination, as recorded in this century. This consideration is of particular consequences in case of fatal incurable diseases such as AIDS. (3) The genetic variations in viruses, or the frequent recurring variations in the antigenic components of parasites, which result in the evolution of new strains with different serological specificities, for which continuous development of new vaccines is obligatory.

For these reasons and others, new approaches are being sought for vaccine development in the future. These include: (1) Use of recombinant DNA technology for the production of the relevant microbial protein antigens in either bacterial, yeast or animal cells, for vaccine preparation. (2) Use of recombinant DNA techniques for production of live vaccines by introducing the relevant gene(s) into the genome of appropriate vectors such as vaccinia virus or *Salmonella* vaccine mutants. (3) Application of anti-idiotype antibodies, which mimic the antigenic structure of the relevant epitope(s), and thus could induce protective immunity. (4) The utilization of synthetic peptides which constitute the relevant protective epitopes of viruses, bacterial toxins or parasites, for eliciting neutralizing immune response towards the disease causing agent (5) The employment of synthetic oligonucleotides coding for such peptides as synthetic "genes" in recombinant constructs. This presentation focuses on the latter two approaches.

IDENTIFYING AND LOCALIZING EPITOPES

Before considering the use of synthetic peptides as potential immunogens for the purpose of vaccination, it is important to develop strategies for identifying and localizing those regions in the antigen, denoted epitopes, which are recognized by both antibodies and immunocompetent cells. Epitopes of protein are usually classified as either continuous (sequential) or discontinuous (conformational). The first refers to short linear peptide fragments of the antigen that bind to antibodies, whereas the latter describes a more complex epitope that is present only in the folded structure of the protein (Sela et al, 1967). There are several strategies for identifying and localizing the epitopes, including crystallographic analysis of the antigen or the antigen-antibody complexes, analysis of the reactivity of peptide fragments with polyclonal or monoclonal antibodies, identification of antibody contact residues by comparison with some protein homologs, or by the use of synthetic peptides.

A number of alternative approaches are being used for the synthesis of peptide epitopes, such as a systematic synthesis of overlapping peptides alongside the amino acid sequence of the protein antigen, usually on polyethylene pins (Geysen 1987), or the recently described "split synthesis" strategy (Lam 1991). The latter allows the synthesis of about 64x10^6 different hexapeptides produced on separate resin beads, from which the epitope could be selected by means of a monoclonal antibody and identified. Strategies also exist for increasing the level of conformational involvement in the synthesized peptide, primarily by using various approaches for predicting the location of epitopes in the protein sequence according to several structural parameters such as hydrophilicity, segmental mobility, accessibility or sequence variability. In a recent study the validity of 22 different scales for such prediction was analyzed (Pellequer 1991).

In our laboratory, early studies have demonstrated that synthetic antigens containing an immunoreactive region of a protein can give rise to a specific, and often conformation-dependent, immune response towards the intact native protein (Arnon et al, 1971). When the protein in question is a component of a virus e.g. the coat protein of MS-2 coliphage, the antibodies induced by a synthetic fragment were capable of neutralizing the viability of the phage (Langbeheim et al., 1976). These findings paved the way for the use of synthetic peptides as the basis for vaccine design. We have employed this approach for the study of three systems - the influenza virus and the bacterial toxins of cholera and shigella, as summarized in the following.

VACCINES BASED ON SYNTHETIC PEPTIDES

Cholera toxin:

The B subunit of this toxin, which expresses most of the immunodominant epitopes capable of neutralizing the biological activities was the subject of our study. Several peptides corresponding to various regions in the molecules, when conjugated to tetanus toxoid (TT), elicited antibodies that reacted to different extents with the native toxin. The most reactive peptides were CTP1 (residues 8 to 20) and CTP 3 (residues 50 to 64), which as such did not cause any toxic or other biological side effects, but elicited antibodies which inhibited the biological activities of cholera toxin (both the enterotoxicity and the induction of adenylate cyclase) to a level of 60 to 70% (Jacob et al., 1983; Jacob et al., 1984a).

In view of the high level of sequence homology between the B subunits of the cholera toxin (CT) and the heat labile toxin of *E. coli* (LT), and the immunological relationship between these two toxins (Lindholm 1983)), we studied the cross-reactivity of the above synthetic peptides with LT. This is of importance, since the LT of pathogenic strains of *E. coli* is the causative agent of diarrhea in many tropical countries and due to its wide spread, it presents probably a more serious health problem than cholera. Indeed, the antiserum elicited by CTP3 and to a lesser extent by CTP1 were cross-reactive with LT of multiple strains of human and porcine *E. coli*. Moreover, these antisera, which are inhibitory towards CT, were found equally effective in neutralizing the biological activity of the *E. coli* LT (Jacob et al., 1984b). This indicates that synthetic peptides may serve as the basis for a general vaccine against the coli-cholera family of diarrheal diseases.

Shiga toxin:

Vaccination against Shigella species, to prevent shigellosis, presents a number of problems, and as a consequence no efficient shigella vaccine is available as yet. In addition to the invasive properties of the organism, the various species of *Shigella* produce a protein toxin, which experimentally reproduces the major features of the infection. Shiga toxin (ST), isolated from *Shigella dysenteriae 1* strains, is one of the most potent of the lethal microbial toxins. It consists of A ad B subunits, and antibodies raised against the B subunit were shown to neutralize the cytotoxic effects of the toxin (Donohue-Rolfe et al., 1984). Several peptides of the B chain, corresponding to its amino and carboxy terminal regions, were synthesized and conjugated to macromolecular carriers. Rabbit antisera raised against all the conjugates or against the polymerized peptides were highly reactive with the respective homologous peptides and cross-reacted with the native Shiga toxin. More significantly, the antisera showed considerable neutralizing capacity (60%-80%) against all three biological effects of the toxin, namely cytotoxicity towards HeLa cells, enterotoxic activity in rats, as well as neurotoxic lethal effects in mice (Harari et al., 1989).

Immunization of mice and rats with the peptide conjugates led to partial protection against the detrimental effects of the Shiga toxin. Thus, for example, mice immunized with the conjugates of the peptides and subsequently exposed to lethal dose of toxin, showed up to 80%

long-term survival. Furthermore, oral immunization of rats with the synthetic peptide conjugates by intragastric feeding, following priming by parenteral immunization, led to elevation of specific IgA antibodies and the immunized rats manifested significant protection against the enterotoxic effect of Shiga toxin (Arnon., et al 1990). These findings indicate that synthetic peptide vaccines are capable of inducing both systemic and local immunity, with protective efficacy.

Influenza:

The presently available influenza vaccines consist of either attenuated or inactivated viral particles. Their effectiveness is only partial and of short duration, mainly due to the the frequent antigenic variation of the external glycoproteins of virus (Laver and Air 1979), each new strain presenting a new challenge to the host immune system. The two major antigenic components of influenza virus are the haemagglutinin and the nucleoprotein. The haemagglutinin (HA), towards which the neutralizing anti-viral antibodies are directed, occurs as trimeric spikes projecting from the viral proteolipid envelope. It is qualitatively the most important glycoprotein in the viral surface, and undergoes frequent genetic variations, denoted "shifts" and "drifts". Several synthetic peptides of the HA molecule, have been studied, all of which proved immunoreactive (Muller et al., 1982; Shapira et al 1985a). The most effective peptide consisted of 18 amino acid residues corresponding to the sequence 91-108 of the HA molecule. This region, which is common to all H3 strains, when conjugated to tetanus toxoid, elicited in both rabbits and mice antibodies that reacted with the synthetic peptide, as well as with the intact influenza virus of several type A H3 strains. These antibodies caused haemagglutination inhibition and also interfered with the in vitro growth of the virus in tissue culture. Furthermore, as shown in Fig. 1 mice immunized with the peptide conjugate were partially protected against challenge infection with several H3 strains of the virus (Muller et al., 1982).

The above results were achieved by immunization in complete Freund's adjuvant (CFA), which is a very effective adjuvant evoking high level and long lasting immunity, but is not suitable for human use. It is possible, however, to replace the CFA by a less harmful substance, the synthetic adjuvant MDP (N-acetyl-murmamyl-L-alanyl-D-isoglutamine). In a conjugate with the 91-108 peptides, this adjuvant was similar to CFA in the induction of anti-peptide antibodies and led to protection against in-vivo viral challenge. As shown in Fig. 1, the level of protection was even slightly higher than that induced in the presence of CFA (Shapira et al., 1985). It should be emphasized that this conjugate is water soluble and was administered in a physiological aqueous solution and hence constitutes a synthetic vaccine with built-in adjuvanticity, which could be suitable for use in humans.

AIDS:

At present there is no vaccine available towards the AIDS causing HIV. Furthermore, live attenuated vaccines are probably not appropriate in this case, primarily because of the high risk of reverting to virulence. Hence, the synthetic approach to vaccination seems specially attractive for this system, and consequently extensive research is currently focused on synthetic peptides of HIV in several laboratories around the world. The major component of the virus subjected to these studies is the envelope protein gp 120, in the V3 loop of which a principal neutralizing determinant (PND), as well as an immunodominant cytotoxic T cell (CTL) epitope, have been identified (Takahashi et al., 1988). Peptides derived from this region, when conjugated to an appropriate carrier, induced anti-HIV response, and when combined with a CTL epitope they were recently shown to induce highly specific CD8+ MHC Class I restricted CTLs *in vivo* (Nardelli 1992; Hart 1991). These results demonstrate the potential of synthetic peptides for production of vaccines towards HIV.

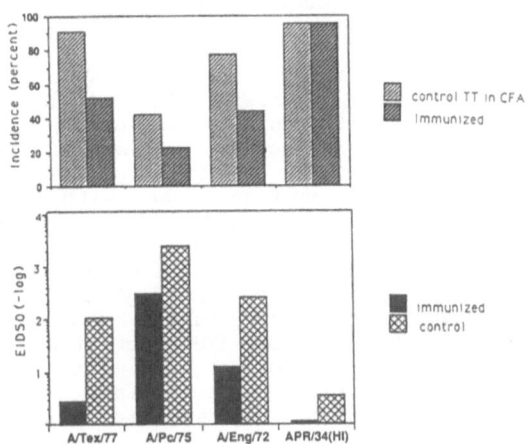

Fig. 1. Protection of mice against challenge infection with influenza virus. Mice of several strains were immunized with a conjugate of the 91-108 epitope of influenza HA and tetanus toxoid. The level of protection against infection is indicated by reduced incidence of infection in the immunized mice vs the control group (Top), and by lower titer of the virus in the lungs, as measured by the 50% egg infective dose - EID 50 - (Bottom). Mice were protected against challenge with H3 strains but not against H1 strain.

SYNTHETIC RECOMBINANT VACCINES BASED ON OLIGONUCLEOTIDES

An alternative approach to the chemical synthesis of vaccines is the use of genetic engineering. This technology is used for insertion of genes into expression vectors, for the biosynthesis of intact proteins, for the preparation of either subunit or live vaccines, as will be discussed in subsequent presentations of this Symposium. In our studies attempts were made to bridge the synthetic and recombinant DNA approaches with regard to the three systems investigated in our laboratory, namely, cholera toxin, Shiga toxin and influenza haemagglutinin, by expressing their relevant epitopes in recombinant products.

In the case of cholera toxin, plasmids containing a synthetic "gene" coding for CTP 3 were inserted in phase into the gene coding for *E. coli* β-galactosidase. Immunization with the resulting fusion protein did not lead however, to a significant titer of antibodies recognizing CT. In the case of Shiga toxin, several plasmids were prepared, by insertion of synthetic oligonucleotides coding for two N-terminal peptides (corresponding to residues 9-21 and 19-31, respectively),

and/or the carboxy terminal peptide (residues 54-67) of the B subunit.
These regions were then expressed in *E. coli* K12, or in the non-
virulent mutant of *Salmonella dublin* SL 1438.   Rabbits immunized
with the respective partially purified recombinant fusion proteins
responded towards the homologous peptides and showed very high cross-
reactivity with the intact Shiga toxin.   Furthermore, rats immunized
by the oral route of administration with the live intact recombinant
Salmonella expressing either the C-terminal peptide, or both the N-and
the C-terminal peptides of the Shiga toxin B chain were protected to
a level of 90% against the enterotoxicty of the toxin.

In more recent studies this approach was employed in the case of
influenza virus as well (McEwen 1991), by expressing the
haemagglutinin epitope 91-108 in *Salmonella* flagellin.   For that
purpose, a synthetic oligonucleotide coding for the corresponding
sequence was inserted into the plasmid pLS408 and expressed in
Salmonella dublin.   Rabbits immunized either with the live recombinant
S. dublin or with the flagellin isolated from it, showed significant
levels of IgG response against the synthetic peptide 91-108, as well
as against the intact influenza virus.   Mice immunized with the same
preparations developed influenza-specific IgG antibodies in the blood
and secreted IgA antibodies in their lungs.   Furthermore, as depicted
in Fig 2, these mice showed about 50% protection against challenge
infection with the virus (McEwen 1991).   interestingly, the most
successful results were achieved by intranasal immunization with the
isolated recombinant flagellin without the aid of adjuvant.

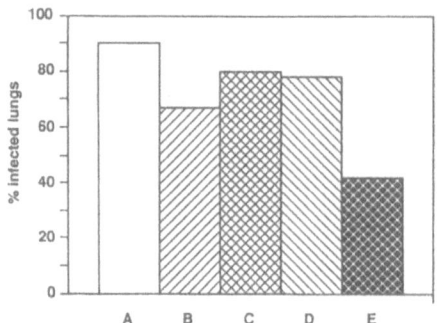

Fig. 2.   protective effect of intranasal immunization of C57B1/6 mice
against viral infection by immunization with: A, Untreated controls;
B,   *Salmonella* control; C, *Salmonella* expressing the peptide 91-108; D,
flagellin control;   E, isolated flagellin expressing the 91-108
epitope.   Protection was assessed by the percent infected lungs in the
immunized mice as compared with the respective controls.

These cumulative findings demonstrate that when relevant peptide
epitopes are defined, they can serve as the basis for protective
synthetic vaccines and, in addition, their corresponding synthetic
oligonucleotides can be expressed in suitable constructs.   In this
way, a combination of genetic engineering and peptide chemistry is
employed for the preparation of either killed or live synthetic
recombinant vaccines, offering a new concept in the development of
vaccines.

REFERENCES

Arnon R, Harari I, Keusch GT (1990) Synthetic peptide toxoid
  vaccines against Shiga dysentery. In: New Generation
  Vaccines. G.Woodrow,M.Levine,eds. Marcel Dekker, Inc,
  Publisher NY p.688-697.
Arnon R, Maron E, Sela M, Anfinsen CB (1971) Antibodies reactive
  with native lysozyme elicited by a completely synthetic
  antigen. Proc. Natl. Acad. Sci. USA 68: 1450.
Donohue-Rolfe AG, Keusch, GT, Elson, C, Thorley-Lawson D,
  Jacewicz, M. (1984) Pathogenesis of Shigella diarrhea.
  IX. Simplified high yield purification of Shigella toxin and
  characterization of subunit composition and function by the use
  of subunit-specific monoclonal and polyclonal antibodies.
  J Exp Med 160: 1767-1781.
Geysen HM, Rodda SJ, Mason TJ, Tribbick G, Schoofs PG (1987)
  Strategies for epitope analysis using peptide synthesis.
  J Immunol Meth 102: 259-274.
Harari I, Donohue-Rolfe AG, Keusch G, Arnon R (1989). Synthetic
  peptides of Shiga toxin B-subunit induce antibodies which
  neutralize the biological activities of the toxin. Infection and
  Immunity 56: 1618-1624.
Hart MK, Weinhold KJ, Scearce RM, Washburn EM, Clark CA, Palker TJ,
  Haynes BF. (1991) Priming of anti-human immunodeficiency virus
  (HIV) CD8$^+$cytotoxic T cells in vivo by carrier-free HIV
  synthetic peptides. Proc Natl Acad Sci USA 88: 9448-9452.
Jacob CO, Pines M, Arnon, R (1984b) Neutralization of heat labile
  toxin of E. coli by antibodies to synthetic peptides derived from
  B subunit of cholera toxin. EMBO J 3:2889-2893.
Jacob CO, Sela M, Pines M, Hurwitz, S, Arnon R (1984a)
  Adenylate cyclase activation by cholera toxin as well as its
  activity are inhibited with antibodies against related
  synthetic peptides. Proc Natl Acad Sci USA 8: 7893-7896.
Jacob CO, Sela M, Arnon R (1983) Antibodies against synthetic
peptides
    of the B subunit of cholera toxin: Cross-reaction and
neutralization
  of the toxin. Proc Natl Acad Sci USA 80: 7611-7615.
Lam KS, Salmon SE, Hersh EM, Hruby VJ, Kazmierski WM, Knapp RJ,
  (1991) A new type of synthetic peptide library for identifying
  ligand-binding activity. Nature 354: 82-84.
Laver WG, Air GM. Eds (1979.) Structure and variation
  in influenza virus.Elsevier North Holland, Amsterdam.
Langbeheim H, Arnon R, Sela M (1976) Antiviral effect on MS-2
  coliphage obtained with a synthetic antigen. Proc. Natl. Acad.
  Sci USA 73: 4636.
Lindholm L, Holmgren J, Wikstrom M, Karlsson V, Andersson K,
  Lycke, N (1983) Monoclonal antibodies to cholera toxin with
  special reference to cross-reactions with Escherichia coli
  heat-labile enterotoxin. Infect Immun 40:570-576.
McEwen J, Leitner M, Harari I, Arnon R (1989) Expression of
  Shiga toxin epitopes in E. coli immunological characterization.
  Immunology Letters 21: 157-164.
McEwen J, Levi R, Horwitz RJ,Arnon R. (1991) Synthetic recombinant
  vaccine expressing Influenza HA epitope leads to partial
  protection Vaccine 10: 405-412.
Muller GM, Shapira M, Arnon R (1982) Anti-influenza response
  achieved by immunization with a synthetic conjugate. Proc Natl
  Acad Sci USA 79: 569-573.
Nardelli B, Lu YA, Shiu DR, Delpierre-Defoort C, Profy AT,
  Tam JP (1992) A chemically determined synthetic vaccine model
  for HIV-1. J of Immunol 148: 914-920.

Pellequer JL, Westol, E, Van Regenmortel, MHV (1991) Overview
of methods for predicting the location of continuous epitopes
in proteins from their primary structures. Meth. Enzymol.
203: 176-201.

Sela M, Schechter B, Borek F (1967) Antibodies to sequential and
conformational determinants. Cold Spring Harbor Symp.
on Quantitative Biology 32:537-545.

Shapira M, Jolivet, M, Arnon R (1985) Synthetic vaccine against
influenza with built-in adjuvanticity. Int J Immunopharmacology
7: 719-723.

Shapira M, Misulovin, Z, Arnon R(1985a) Specificity and cross-
reactivity of synthetic peptides derived from a major antigenic
site of influenza hemagglutinin. Mol Immunol 22: 23-28.

Takahashi H, Cohen J, Hosmalin A, Cease KB, Houghten R, Cornette JL
Delisi C, Moss B, Germain RN, Berzofsky JA (1988) An immunodominant epitope of the human immunodeficiency virus envelope
glycoprotein gp 160 recognized by Class I major histocompatibility complex molecule-restricted murine cytotoxic T lymphocytes.
Proc Natl Acad Sci USA 85: 3105-3109.

Wiley DC, Wilson IA, Skehel JJ (1981) Structural identifications of
the antibody binding sites of Hong Kong influenza hemagglutinin and
their involvement in anti-genic variation. Nature 289: 373-378.

The Analysis of the Function and Specificity of T Cells Triggered During Infection with *Leishmania*, Its Importance for the Rational Design of a Vaccine

J.A. Louis[*], U. Fruth[*], J.P. Rosat[*], F. Conceicao-Silva[*], B. Perlaza[*], P. Romero, I. Müller[*], and G. Milon[o]

WHO Immunology Research and Training Centre[*], Institute of Biochemistry, U. of Lausanne, Ludwig Institute for Cancer Research[+], 1066 Epalinges, Switzerland and Institut Pasteur[o], Paris

INTRODUCTION

Infections with protozoan parasites of the genus *Leishmania* encompass a spectrum of diseases dependent upon the species of the microorganism and the host immune response. On the one hand, it is established that the elimination of these parasites from infected hosts depends on the development of T cell responses capable of activating macrophages to a parasiticidal state. On the other hand, clear evidence exists that T cell response(s) also play a role in mediating susceptibility to infection with *Leishmania*. Therefore, the rational design of a vaccine against these parasites critically depends on the recognition of the precise T cell function(s) instrumental in mediating either the resolution of lesions or the progression of disease and on devising means by which to selectively trigger protective responses after immunization.

Study of the murine model of infection with *Leishmania major (L.major)* has been particularly fruitful for the characterization of the various components of the T cell response induced by infection. In this short communication, we have summarized informations pertaining to the types of T cells triggered during infection of mice with *L.major* and their effects on the disease process.

CD4 T CELL SUBSETS

Results using the murine model of infection with *L.major* have clearly shown that polarized CD4 Th1 or Th2 cell responses were correlated with resistance and susceptibility to disease respectively. Indeed, it was demonstrated that CD4 T cells expanding during infection of genetically resistant mice (i.e. C57Bl/6) contain mainly mRNA encoding lymphokines characteristics of Th1 cells i.e. Interferon γ (IFN-γ) and Interleukin-2 (IL-2), whereas CD4 T cells from susceptible mice (i.e. BALB/c) have a higher expression of mRNA encoding for Th2-like lymphokines such as Interleukin 4 (IL-4) and Interleukin 10 (IL-10) (Heinzel *et al*, 1989, Locksley *et al*, 1991). Furthermore, using a protocol (administration of either α-CD4 or α-IL-4 mAbs) that allows the expression of a resistant phenotype in otherwise susceptible BALB/c mice, it was confirmed that IL-4 and IL-10 transcripts are present in mice with progressive disease whereas IFN-γ and IL-2 transcripts were observed in mice induced to heal by immune intervention (Heinzel *et al*, 1991).

Direct proof for the role of Th1 and Th2 type cells in mediating respectively resistance and susceptibility to *L.major* infection was obtained using cell transfer experiments in immunodeficient *scid* mice. *L.major*-specific Th1 and Th2 cell lines derived from lymph nodes draining the lesions of infected unmanipulated BALB/c or BALB/c mice capable of healing their lesions following administration of anti-CD4 mAb were adoptively transferred in *scid* mice. Compared to non-reconstituted *scid* mice which are not able to control infection, mice receiving Th1-like cells were able to restrict the parasite growth whereas those receiving Th2 like cells exhibited exacerbated disease (Holaday *et al*, 1991). IFN-γ certainly accounts, at least in part, for the beneficial role of Th1 cells during *L.major* infection, as shown by observations showing that neutralization *in vivo* of IFN-γ could significantly enhance disease progression (Belesovic *et al*, 1989; Müller *et al*, 1989). Evidence that IL-4, a product of Th2 cells, plays a role in the progression of infection also derives from observations showing that neutralization of this lymphokine *in vivo* enables normally highly susceptible mice to heal (Sadick *et al*, 1990). IL-4 has been shown to inhibit macrophage activation by IFN-γ and TNF-α *in vitro* (Liew, 1989; Scott *et al*, 1989), eventhough this effect appears to depend critically upon the sequence of addition of these lymphokines to macrophages cultures (Bogdan *et al*, 1991).

Critical to the rational design of effective anti-*Leishmania* vaccine is thus the understanding of the rules that control the maturation and activation of Th1 and Th2 cells. The bulk of available information suggest that mature Th1 and Th2 cells are derived from a common precursor and do not differ in their antigenic repertoire. The role of some lymphokines as factors controlling the differential development of CD4 T cell precursors is being suggested. IFN-γ would favor the differentiation toward a Th1 functional phenotype (Gajewski *et al*, 1989), whereas IL-4 would constitute a differentiation factor toward Th2 cells (Abehsira-Amar *et al*, 1992). In this vein, using the murine model of infection with *L.major* it has been shown that IFN-γ plays a key role in modulating the early IFN-γ response of resistant mice during infection (Scott, 1991). However, it should be mentioned that administration of IFN-γ to susceptible BALB/c mice for up to 2 weeks after infection did not modify its outcome (Sadick *et al*, 1990). Observations showing that administration of IL-4 to resistant mice, early during infection, resulted in a transient but marked shift, toward a Th2-like response, in the lymphokines released by lymph node cells upon stimulation *in vitro*, suggest that, in this model of infection, IL-4 plays a role in differentiation of CD4 cells toward a Th2 functional phenotype (Chatelain *et al*, 1991). However, IL-4 given to resistant mice for 3 weeks at the beginning of infection could not alter the healing of lesions (Sadick *et al*, 1991).

Recent results also suggest an important role of the T cell receptor (TCR) in the generation of T helper cell subsets and indicate that prolonged TCR occupancy and ligation is critical for the development of Th2-like cells (Röcken *et al*, 1992). A working hypothesis derived from these findings could be that parasitized macrophages from susceptible mice present parasite antigenic epitopes in such a way as to allow prolonged interaction with the TCR of *L.major*-specific T cells precursors, thus favoring their differentiation into Th2-like cells.

IMPORTANCE OF THE SPECIFICITY OF TH1 T CELLS ON THEIR EFFECT ON THE DEVELOPMENT OF LESIONS

Results accumulated during the last years have revealed the capacity of some *L.major*-specific Th1 lines and clones to enhance disease progression after transfer to normal syngeneic mice. The exacerbation of lesions seen after transfer of these functionally characterized parasite-specific Th1 cells did not require the participation of host's T and B cells and was related to the number of T cells adoptively transferred. The finding that other Th1-like T cells recognizing epitopes unrelated to *L.major* could also mediate exacerbation of cutaneous leishmaniasis provided that their specific epitopes were injected with the parasite inoculum strongly indicate that their activation generated the signal(s) for exacerbation (Titus *et al*, 1991). In contrast to these cells, other *L.major*-specific Th1 cloned T cells shown to protect normal syngeneic BALB/c mice against infectious challenge were able to recognize only antigen(s) associated with living parasites (Müller and Louis, 1989). This finding led to the hypothesis that, in order to be protective, Th1 cells must be capable of responding to antigens presented at the surface of parasitized macrophages i.e. the target cells of protective immune effector mechanisms. The reason(s) why infected macrophages would not present their specific parasite-derived peptide to exacerbating Th1 cells which were selected *in vitro* using a lysate of *L.major* as antigen is not known. It could be that inside macrophages, *L.major* amastigotes express antigens different from these expressed by extracellular promastigotes or that antigen presentation is modified in macrophages harbouring living parasites. Results in table I show that, compared to normal macrophages, parasitized macrophages have an impaired capacity to present unrelated proteins to specific T cells. This inhibition was related to the number of parasites used for macrophage infection *in vitro*. Similar results have been observed using macrophages infected with *L.amazonensis* (Prina *et al*, 1991). Interestingly, the ability of macrophages infected with *L.major* to present exogenously added antigen peptides to specific T cell hybrids was not reduced compared to normal macrophages.

Together with results showing that this impairment of antigen presentation by infected macrophages was not the result of a reduced uptake or catabolism of the antigen, these data might suggest that the presence of parasites in macrophages could interfere with the intracellular loading of class II MHC molecules (Fruth and Louis, manuscript submitted, 1992). This interference could simply reflect competition between various antigenic peptides for binding with MHC class II molecules within the endosomal compartment loaded with antigenic material constituted by the parasites. According to this scheme, the specific epitope of Th1 cells derived from lymphoid tissue of mice immunized with *L.major* antigens might not necessarily be presented by infected macrophages and, as a consequence, the anti-*Leishmania* activity of these T cells cannot be expressed.

Table 1.　　Inhibition of antigen presentation by macrophages
infected with *L.major*

| Macrophages | [^{3}H]TdR incorporation | |
	OVA	OVA 323-339
normal	65 995 ± 3686	35 257 ± 1128
infected	5 180 ± 265	32 895 ± 708

Bone marrow-derived macrophages were cultured overnight in the
presence of rIFN-γ (50 U/ml), distributed in microtiter wells (2.5 x
10^4/well) and incubated for 2h in fresh medium with or without *L.major*
promastigotes (20/macrophage). Then the macrophages were washed and
incubated for 3h in fresh medium, followed by a 2h incubation with
native antigen (ovalbumin [OVA]; 2 mg/ml). At the end of these
incubations the macrophages were fixed and incubated with the OVA-
specific T cell hybridoma DO11.10. After 24h supernatants were tested
for the presence of IL-2 using the IL-2-dependent cell line CTLL.
[^{3}H]TdR incorporation of CTLL is expressed as mean cpm ± SD of
triplicate cultures. The antigen-presenting capacity of normal or
parasitized macrophages was also tested after pulse with OVA 323-339
(10 µg/ml), the peptide specifically recognized by DO11.10 in the
context of I-A^d.

γδ$^+$ T CELLS

Evidence for the accumulation of γδ$^+$ T cells in lesions induced by
Leishmania has been obtained in human cases of American leishmaniasis
and results indicate that these cells were instrumental in granuloma
formation (Modlin *et al*, 1989). Furthermore, elevated numbers of γδ$^+$ T
cells in the blood of patients infected with *L.amazonensis* have also
been recently documented (Russo *et al*, 1991). Recent observations made
in our laboratory have shown an expansion of the γδ$^+$ T cell population
within lymphoid tissue of mice infected with *L.major* (Rosat, McDonald
and Louis, manuscript submitted 1992). Maximal numbers were observed
in the spleens of chronically infected BALB/c mice where γδ$^+$ T cells
represent a large proportion of CD3$^+$ T cell blasts (figure 1).

It is noteworthy that this expansion of γδ$^+$ T cells was related to the
parasite burden and that there is indirect evidence for their possible
involvement in host defense against this parasites. The mechanisms by
which γδ$^+$ T cells could participate in restricting parasite growth in
the murine model of infection with *L.major* as well as their
specificity remain important issues to be elucidated.

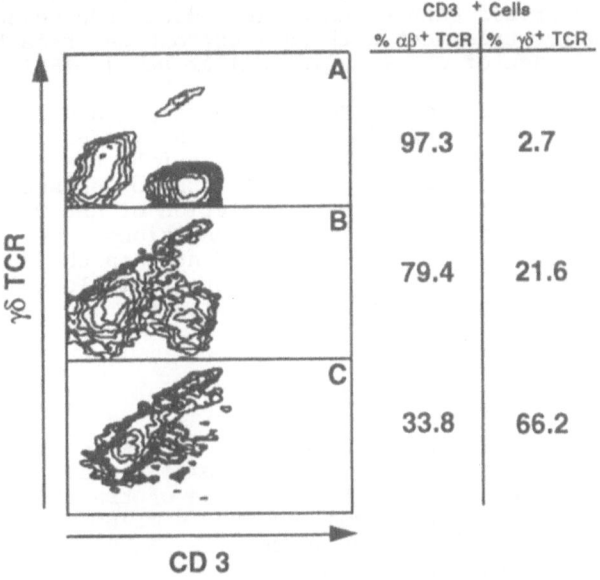

	CD3 + Cells	
	% αβ+ TCR	% γδ+ TCR
A	97.3	2.7
B	79.4	21.6
C	33.8	66.2

γδ TCR

CD 3

Figure 1 γδ+ T cells in the spleen of normal (A), *L. major* infected BALB/c mice (B) and within the blast population of infected mice (C). Cells were double stained with FITC-conjugated anti-CD3 mAb 17A2 and biotin-conjugated anti-γδ TCR mAb GL3 (revealed by AV-PE). The blast population was obtained by gating on cells with high forward and side light scatter.

CD8 T CELLS

The triggering of CD4 T cells during infection with *L.major* and their importance as anti-*Leishmania* effector cells are consistent with the localization of these parasites within the phagolysosomal compartment of their host's macrophages where antigenic peptides associate with MHC class II molecules. However, experimental evidence now exist that CD8 T cells are also involved in the development of immunity against these parasites and thus are triggered during the course of infection.

Role of CD8 T Cells in Immunity Against *L.major*

Studies of the role of CD8 T cells on the development of lesions induced by *L.major* revealed that these cells were more important for the resolution of secondary lesions developing after infectious challenge of immune mice than for the healing of primary infection. Resistance to infection which is induced by i.v. immunization of susceptible BALB/c mice with killed *L.major* was prevented by depletion of CD8 T cells *in vivo* with anti-CD8 mAb (Farrell *et al*, 1989). Small secondary lesions developping following reinfection of genetically resistant mice having spontaneously resolved a primary lesion were substantially more severe in mice given anti-CD8 mAb from the time of infectious challenge (Müller et al, 1989). The immunity to secondary infection which characterizes BALB/c susceptible mice induced to heal a primary lesion by prior treatment with anti-CD4 mAb was remarkably dependent upon the presence of CD8 T cells (Müller et al, 1991).

A contribution for CD8 T cells to the resolution of primary lesions has also been substantiated by experimental results. Indeed, the state of resistance to a primary infection which results from the administration of either anti-CD4 or anti-IL-4 mAbs to susceptible BALB/c mice could be partially overcome by depletion of CD8 T cells in vivo. Interestingly, under these experimental conditions, the protective capacity of CD8 T cells was more evident when the degree of infection was evaluated by determining the number of viable parasites in lesions rather than by measuring the size of lesions (Müller et al, 1991).

Expansion of CD8 T Cells During Infection

The first evidence for the expansion of specific CD8 cells in lymphoid cells from mice infected with L.major derives from observations showing an increased frequency of CD8 T cells able of transferring parasite-specific DTH reactions to normal syngeneic mice (Milon et al, 1985). In addition, lymphoid tissues of susceptible BALB/c mice rendered resistant by pretreatment with anti-CD4 mAb were shown to contain CD8 T cells capable of mediating L.major-specific DTH reactions and which participate in the production of cytokines endowed with anti-Leishmania effector function (Müller et al, 1991). It is noteworthy that within T cell population in the blood of these mice only parasite-specific CD8 cells could be demonstrated to transfer specific DTH responses to normal recipient mice. Finally, consistent with observations showing that resistance of immune mice to secondary infectious challenge was more dependent upon the activity of CD8 cells are our recent results showing that reinfection of immune mice elicits secondary IFN-γ responses and that CD8 T cells significantly contribute to this memory response (Müller, Kropf, Louis, manuscript submitted, 1992).

L.major-Specific CD8 T Cell Lines and Clones

Recently, it has been possible to derive in vitro L.major-specific CD8 T cell lines and clones from the spleen and lymph nodes of resistant mice that had recovered spontaneously from a primary infection. These cells displayed specific cytolytic activity against tumor and macrophage target cells sensitized with peptide digests from L.major in a MHC class I restricted fashion. In addition, these CD8 T cells release significant amounts of IFN-γ after specific stimulation in vitro. These results provide additional evidence that MHC class I restricted CD8 T cells are elicited during the course of infection with L.major in mice.

Taken together, these data showing that Leishmania-specific CD8 T cells are triggered during infection and play a role in its outcome could suggest that these cells recognize epitopes presented by macrophages harbouring parasites i.e. the target cells of protective immunity. This would imply that antigenic peptides from these exclusively intravacuolar parasites have access to the class I MHC pathway of antigen presentation. Alternatively, CD4 cells mediated activation of infected macrophages leading to parasite destruction could lead to the generation of antigenic peptides that could directly bind to the MHC class I molecules of cells in the vicinity with the consequent triggering and expansion of CD8 cells.

CONCLUDING REMARKS

The murine model of infection with *L.major* has been instrumental in deciphering the immune mechanisms accounting for resistance and susceptibility to infection. These studies combined with those aimed at a) identifying parasite molecules which are able to elicit and are the target of protective T cell responses and b) defining the rules by which to selectively trigger protective responses are prerequesite for the rational design of an effective vaccine.

In addition and importantly, basic research using this experimental model has significantly contributed to unravel the complexity of the T cell response elicited during infection.

ACKNOWLEDGMENTS

Our experimental work is supported by the Swiss National Science Fdt.,the Deutsche Forschungsgemeinschaft (U.F.), the Sandoz Research Fdt. and the World Bank / UNDP / WHO Special Programme on Tropical Diseases.

REFERENCES

Abehsira-Amar O, Gibert M, Joliy M, Thèze J, Jankovic DL (1992) IL-4 plays a dominant role in the differential development of Th0 into Th1 and Th2 T cells. Immunol 148:3820-3829

Belosevic M, Finbloom DS, VanderMeide PH, Slayter MV, Nacy CA (1989) Administration of monoclonal anti-IFN-γ antibodies *in vivo* abrogates natural resistance of C3H/HeN mice to infection with *Leishmania major*. J Immunol 143:266-272

Bogdan C, Stenger S, Röllinghoff M, Solbach W (1991) Cytokine interactions in experimental cutaneous leishmaniasis. Interleukin 4 synergizes with interferon-γ to activate murine macrophages for killing of *Leishmania major* amastigotes. Eur J Immunol 21:327-333

Chatelain R, Varkila K, Coffman RL (1991) IL-4 induces a Th2 reponse in *Leishmania major*-infected mice. J Immunol 148:1182-1187

Farrell JP, Müller I, Louis JA (1989) A role for Lyt2⁺ T cells in resistance to cutaneous leishmaniasis in immunized mice. J Immunol 142:2052-2056

Gajewski TF, Fitch FW (1988) Anti-proliferative effect of IFN-γ in immune regulation. I IFN-γ inhibits the proliferation of Th2 but not Th1 murine HTL clones. J Immunol 140:4245-4252

Heinzel FP, Sadick MD, Holaday BJ, Coffman RL, Locksley RM (1989) Reciprocal expression of interferon-γ or interleukin 4 during the resolution or progression of murine leishmaniasis. J Exp Med 169:59-72

Heinzel FP, Sadick MD, Mutha SS, Locksley RM (1991) Production of interferon-γ, interleukin 2, interleukin 4 and interleukin 10 by CD4⁺ lymphocytes *in vivo* during healing and progressive leishmaniasis. Proc Natl Acad Sci 88:7011-7015

Holaday BJ, Sadick MD, Zhi-En Wang, Reiner SL, Heinzel FP, Parslow TG, Locksley RM (1991) Reconstitution of *Leishmania* immunity in severe combined immunodeficiency mice using Th1- and Th2-like cell lines. J. Immunol 147:1653-1658

Liew FY (1989) Functional heterogeneity of CD4⁺ T cells in leishmaniasis. Immunol Today 10:40-45

Locksley RM, Heinzel FP, Holaday BJ, Mutha SS, Reiner SL, Sadick MD (1991) Induction of Th1 and Th2 CD4 subsets during murine *Leishmania major* infection. Res Immunol 142:28-32

Milon G, Titus RG, Cerottini JC, Marchal G, Louis JA (1986) Higher frequency of *Leishmania major*-specific L3T4⁺ Tcells in susceptible BALB/c mice than in resistant CBA mice. J Immunol 136:1467-1471

Modlin RL, Pirmez C, Hofman FM, Torigian V, Uyemura K, Rea TH, Bloom BR, Brenner MB (1989) Lymphocytes bearing antigen-specific γ/δ T-cell receptors accumulate in human infectious disease lesions. Nature 339:544-547

Müller I, Garcia-Sanz JA, Titus R, Behin R, Louis JA (1989) Analysis of the cellular parameters of the immune responses contributing to resistance and susceptibility of mice to infection with the intracellular parasite *Leishmania major*. Immunol Rev 112:95-113

Müller I, Pedrazzini T, Farrell JP, Louis JA (1989) T-cell responses and immunity to experimental infection with *Leishmania major*. Annu Rev Immunol 7:561-578

Müller I, Pedrazzini T, Kropf P, Louis JA, Milon G (1991) Establishment of resistance to *Leishmania major* infection in susceptible BALB/c mice requires parasite-specific CD8⁺ T cells. Int Immunol 3:587-597

Müller J, Louis JA (1989) Immunity to experimental infection with *Leishmania major*: Generation of protective L3T4⁺ T cell clones recognizing antigen(s) associated with live parasites. Eur J Immunol 19:865-871

Prina E, Antoine JC, Guillet JG, Jouanne C (1991) MHC class II molecule (Ia) expression and antigen presentation by mouse macrophages infected with *Leishmania amazonensis* amastigotes. Mem Inst Osvaldo Cruz 86(suppl.1):207 (Abstr)

Röcken M, Müller KM, Saurat JH, Müller I, Louis JA, Cerottini JC, Hauser C (1991) Central role for TCR/CD3 ligation in the differentiation of CD4⁺ T cells toward a Th1 or Th2 functional phenotype. J Immunol 148:47-54

Russo D, Barral-Netto M, Armitage R, Barraal A, Grabstein K, Reed S (1991) Characterization of circulating γ/δ T cells in patients with leishmaniasis. FASEB Jour. III.A1680 (Abstr)

Sadick MD, Heinzel FP, Holaday BJ, Pu RT, Dawkins RS, Locksley RM (1990) Cure of murine leishmaniasis with anti-interleukin-4 monoclonal antibody. Evidence for a T-cell-dependent, interferon-γ-independent mechanism. J Exp Med 171:115-127

Sadick MD, Street N. Mosmann TR, Locksley RM (1991) Cytokine regulation of murine leishmaniasis. Interleukin-4 is not sufficient to mediate progressive disease in resistant mice. Infect Immun 59:4710-4714

Scott P (1991) IFN-γ modulates the early development of Th1 and Th2 responses in a murine model of cutaneous leishmaniasis. J Immunol 147:3149-3155

Scott P, Pearce E, Cheever AW, Coffmann RL, Sher A (1989) Role of cytokines and CD4⁺ T cell subsets in the regulation of parasite immunity and disease. Immunol Rev 112:161-182

Titus RG, Müller I, Kimsey P, Cerny A, Behin R, Zinkernagel RM, Louis JA (1991) Exacerbation of experimental murine cutaneous leishmaniasis with CD4⁺ *Leishmania major*-specific T cell lines or clones which secrete interferon-γ and mediate parasite-specific delayed-type hypersensitivity. Eur J Immunol 21:559-567

15. Transplantation Immunity

MHC Restricted Allorecognition as a Modulator of the Homograft Reaction

Jon J. van Rood and Frans Claas

University Hospital Leyden the Netherlands

In preparing this symposium it was decided to concentrate on one specific topic in transplantation immunology which is receiving more and more attention: the attainment of tolerance in man. Even this restricted topic cannot be covered in full and for that reason we have (with one exception) selected authors which did not contribute to a recent excellent review (Brent 1991).

One of the core problems in attaining tolerance in man is that with current immunosuppression the transplant results of most organs have shown a steady improvement over the last years, which understandably makes clinicians hesitant to embark on largely unchartered routes to obtain even better results using highly experimental approaches. The challenge is thus to devise protocols which are less toxic then current immunosuppression and might lead to tolerance.

Obviously this will only be possible when we understand the different cellular, humeral and molecular mechanisms of the homograft reaction. As an introduction to the symposium first the modulating effect of MHC restricted T-cells on the allograft reaction will be reviewed.

After the publication on the induction of neonatal tolerance by Billingham *et al.* in 1953, it has taken a full twenty years to realize that infusion of allogeneic cells in man, although not inducing tolerance, could down-regulate the homograft reaction. Opelz and co-workers (1973) must be credited with showing that Pre Transplant Blood Transfusion (PTB) is a major down-regulating factor of the homograft reaction in man.

The main criticism of most studies that provided evidence for the importance of PTB in man was that the results were not based on prospective randomized trials. We will review studies that indicate that part of the PTB will improve graft survival and part will not, in this way negating the above-mentioned criticism. If one studies the effect of PTB, it becomes obvious that PTB appears to be effective in some individuals, but not in others.

The challenge is to identify the variables influencing the effectiveness of PTB. Lagaaij *et al.* (1989) analysing patients who had received a single blood transfusion could show that those who had received a blood transfusion sharing an HLA-DR antigen with the recipient had a 20 per cent improvement of graft survival. The patients who received a blood transfusion in which there was a complete DR antigen mismatch between donor and recipient had a survival that was similar to that of patients who had received no blood transfusion. This suggests that the sharing of at least one DR antigen by the blood transfusion donor and patient is essential for the effectiveness of the blood transfusion in downregulating homo graft sensitivity. This finding is independent of the match between patient and the kidney donor; in each match group the same difference is observed, i.e. graft survival is superior when the blood transfusion donor shares an HLA-DR antigen with the recipient.

Sharing of a DR antigen between recipient and blood transfusion donor was also found to be essential for a blood transfusion effect in cyclosprine A treated heart transplant patients. In these patients the diagnoses of rejection was made by histology of biopsies and was not indirectly based on clinical criteria (Table 1).

Table 1. Episodes of graft rejection among heart-transplant recipients followed for six months, according to status for matching for HLA-DR antigens between blood transfusion donor and recipient.

Number of episodes	Matching for one DR antigen (N = 10)	No matching for DR antigen (N = 10)	P value
	Number of patients		
At six months			
None	7	1	
One	2	1	0.0029
Two or more	1	8	

Figure 1. Estimates of Precursor Frequency in 10 Patients Receiving Blood Transfusions from Donors with Whom They Had HLA Antigens in common: a)   against these donors
b)   against third party controls (from van Twuyver et al.1991)

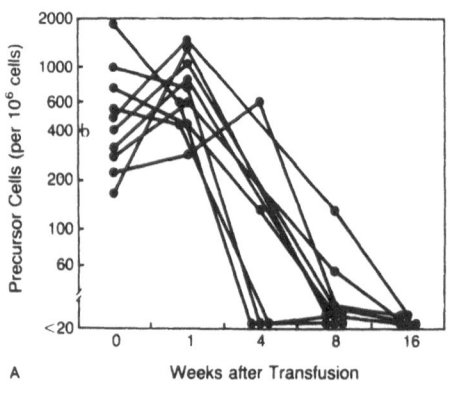
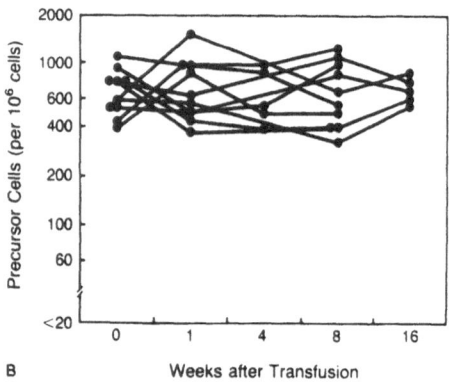

Similarly data of van Twuyver, De Waal, and co-workers (1991) show a significant decrease in cytotoxic T-cell precursor (CTLp) frequency after blood transfusion of a donor, who shared an HLA-B-DR haplotype with the recipient. Such an effect was not observed after an HLA-B-DR mismatched blood transfusion (Figure 1).

If confirmed, these findings might provide an effective approach to controlling homograft sensitivity in the clinic.

Transplantation happens to only a few of us, blood transfusion to many, but we have all been exposed to and been in contact with allogeneic cells of our mothers. What are the implications of this contact for our B- and T-cell repertoire? We were confronted with this question when studying patients who had formed broadly reactive antibodies against HLA because of previous blood transfusion, pregnancy, and/or graft rejection. Because such patients are very difficult to transplant (the crossmatch is almost always positive) a protocol was initiated to help these patients by identifying those HLA antigens to which they had not formed antibodies. A systematic study showed that these acceptable mismatches were often identical to or included in the non-inherited maternal HLA antigens (NIMA). In about half of the patients with high panel reactivity, no antibodies are formed against non-inherited maternal HLA antigens. The noninherited paternal antigens (NIPA) served as a control (Claas et al. 1988).

Zhang, Li, et al. (1991) have investigated whether a similar phenomenon could be found at the level of the T-cell allorepertoire. They found that the frequency of CTL precursors against the mother is lower than against the father. In seven of 32 children, CTL precursors against the mother were not demonstrable, while they were present against the father. These children may be tolerant for the noninherited antigens of the mother. It is possible that the same mechanism is involved in the DR Matched blood transfusion effect and the low CTLp frequency against NIMA. After all, in both instances, a part of an HLA haplotype donor is shared between donor and recipient.

In a recent analysis it was shown that those childeren who had been breastfed were especially likely to have a low CTLp frequency against the NIMA. The implication is that this (partial) tolerance against maternal mismatched antigens could be the result of oral contact of the child with maternal lymphocytes present in the colostrum (Zhang Li et al. 1991).

It was expected that a clear-cut NIMA effect would also be evident in organ and bone marrow transplantation. Although several findings do support this, there is certainly no general agreement on this point. Part of the explanation for the discrepancies noted might be the finding that only half of the childeren show a significant down-regulation of the CTLp frequency against the NIMA's. Furthermore, it might be that not only CTLp frequency but also the avidity of the CTL should be taken into account. Roelen *et al.* (1992) recently showed that, in hyperimmunized patients, CTL directed against HLA antigens, against which the patients had nevertheless not formed antibodies, could be inhibited by CD8 or cyclosporine A, while CTL against HLA antigens against which the patient had formed antibodies could not be inhibited. In other words, avidity of the CTL as measured by anti-CD8 or cyclosporine A inhibition might correlate better with the clinical situation, than the number of CTLp.

It is evident that we will only be able to handle the influence of allogeinic cells on the immune repertoire, if we understand the mechanism. Several mechanisms that can down-regulate the homograft reaction after a blood transfusion have been published, such as veto cells and the formation of a-idiotypic or enhancing antibodies, all of which have been shown to be responsible for graft prolongation in rodents.

We would like to add to the confusion by emphasizing certain findings and formulating a working hypothesis. Our findings can be summarized as follows: a blood transfusion from a donor who shares (part of a) haplotype with the recipient will down-regulate both the humoral and the cellular alloimmune response and improve graft survival. The blood transfusion effect has a number of characteristics, and our working hypothesis should be compatible with them. The characteristics are as follows. The down-regulating blood transfusion effect is:

(1) HLA-DR restricted, in other words recognition of an allopeptide in the context of self might play a role;
(2) Longlasting;
(3) Specific;
(4) Because HLA class II restriction is involved, it is likely that a CD4+ T cell plays an active role in the effect.

It is likely that a class II positive donor cell triggers the blood transfusion effect, although we do not know whether it is an antigen-presenting cell (APC), a dentritic cell, a B cell, or an activated T cell. Let us assume that it carries HLA-DR1 and HLADR2 and that some of the HLA-DR1 molecules on this donor cell carry a DR2 peptide. Most probably, this cell will activate in the recipient a CD4 + T cell that is HLA DR1 and for example, DR3 positive. The specifity of this cell is a DR1-restricted anti DR2 peptide and might, in its turn, activate CD8+ cells and the humoral immune response. Because the phenomenon is DR-restricted, we are ignoring non-restricted alloresponses for the time being, to which we will come back.

So far, it is a normal, although DR restricted, alloresponse and it does not explain how down-regulation of this alloimmune response could take place. The finding that MHC peptides can travel from donor cells to activated recipient T cells might be relevant in this context. If a DR1 molecule in the recipient's cells carries a DR2 peptide originating from the blood transfusion donor, such cells could be recognized and attacked by their own HLA-DR1, restricted HLA-DR2 peptide - specific T cells. Obviously, targets could not only be CD4+ cells but also CD8 + cells, APC, etc. as long as they carry the donor's mismatched peptide on their HLA-DR1 molecule, Unless donor cells make the recipient chimeric (which is certainly a possibility but has not been proven), the production of DR2 peptides will stop and thus the blood transfusion effect would not be longlasting; if the recipient becomes chimeric with donor cells and the production of DR2 peptides continues, where is the end? All class-II-positive cells will be attacked, paralysed, or lysed. The blood transfusion effect might then be longlasting, but is not specific. All DR+ cells carrying the allopeptide are attacked, It is clear that we need another factor that leads to the induction of the specific longlasting blood transfusion effect; for instance, the assumption that peptides can "travel" only a short distance. When, for example, a kidney is transplanted, the down-regulating effect might take place in or near the kidney, but not elsewhere in the body. Perhaps the only good thing about this working hypothesis is that it is testable. If true, DR-restricted antidonor allopeptide and perhaps antirecipient anti-allopeptide CD4+ clones ought to be demonstrable and chimerism might exist (van Rood and Claas 1990).

As a matter of fact Kaminsky et al. (1992) recently published convincing evidence for an increased cytolytic activity against autologous cells in patients which had received multiple transfusions as compared to non transfused patients and normal controls. These findings could be in agreement with the hypothesis mentioned above but so far the relation of HLA-DR sharing between blood transfusion donor and recipient has not been studied. Nevertheless they open a new interesting approach to study the influence of allogenic cells on the immune repertoire in vitro.

Concluding remarks

Although the experiments described above do not lead to a protocol to induce tolerance in man, they do make several important points. The first is that it is apparently possible to manipulate the homograft reaction in a reproducible fashion: a blood transfusion sharing a part of an HLA haplotype of the donor with the recipient down-regulates cellular and probably humeral homograft reactivity, a complete mismatch blood transfusion up-regulates it.

In the second place the findings of van Twuyver, Roelen, and Kaminsky provide us with the possibility to study the "MHC restricted blood transfusion" effect in vitro.
Probably the most important point however is that these findings make it likely that tolerance induction, with the help of pretransplant blood transfusion, can be induced far more effectively if blood transfusion donor recipient and organ donor share a part of a haplotype and would accordingly need less inductive immunosuppression.

Part of this publication will also appear in: HLA 1991. Publisher Oxford University Press, Editor Sasazuki et al..

Acknowledgements

Support was received from the Ernst Jung Stiftung, the Max Geldens Price, the Dutch Organisation for Scientific Research (NWO), and the J.A. Cohen Institute for Radiopathology and Radiation Protection (IRS).

References

Billingham R.E., Brent L., and Medawar P.B., *Nature* 172, 603 (1953).

Brent L., *Transplantation, Current Opinion in Immunology*, vol. 3, 707-758 (1991).

Claas F.H.J., Gijbels Y., van der Velden-de Munck J., van Rood J., *Science* 241, 1815 (1988).

Kaminsky E.R., Hows J.M., Goldman J.M., and Batchelor J.R., *British Journal of Haematology* vol. 81:23-26, (1992).

Lagaaij B.L., Henneman I.Ph.H., Ruigrok M., de Haan M.W., Persijn G.G., Termijtelen A., Hendriks G.F.J., Weimar W, Claas F.H.J., and van Rood J.J., *New Eng. J. Med.* 321, 701 (1989).

Opelz G., Mickey M.R., and Terasaki P.I., *Transplantation* 16, 649 (1973).

Roelen D., Datema G., van Bree S., Zhang Li, van Rood J.J., and Claas F.H.J., *Transplantation* 53: 899-903 (1992).

Rood van J.J., and Claas F.H.J., *Science* 248: 1388-1393 (1990).

Twuyver van E., Mooijaart R.J.D., Berge ten I.J.M., Horst van der A.R., Wilmink J.M., Kast W.M., Melief C.J.M., Waal de L.P., *New Eng J. Med.* 325, 1210 (1991).

Zhang L. Li. S. G., van Rood J.J., and Claas F.H.J. *Res. Immunol.*, 142, 441-445.

Zhang L. Li, S. G., van Bree J., van Rood J.J., Claas F.H.J. *Transplantation* 52: 914-916, (1991)

Quantitation and Manipulation of Alloimmune Responses

Robert Lechler[+], Tony Schwarer[*], Sarah Deacock[+], John Goldmann[*], Sid Sidhu[+], Giovanna Lombardi[+] and Richard Batchelor[+]

Departments of Immunology[+] and Haematology[*] Royal Postgraduate Medical School, Hammersmith Hospital, Du Cane Road, London W12 0NN

Introduction

The capacity of T lymphocytes to respond vigorously to allogeneic tissues has always represented the major obstacle to successful tissue transplantation. This obstacle is usually overcome in solid organ allografting by means of powerful immunosuppressive drugs, with or without attempts to minimise HLA incompatibility. Despite this, there is a significant failure rate, and pharmacological immunosuppression has a number of unwanted side effects. In the context of bone marrow transplantation, even when HLA identical sibling donors are used a significant proportion of patients develop severe graft versus host disease (GVHD). For these reasons the evolution of assays to predict the severity of alloimmune reactivity following transplantation, and the development of methods to interfere more specifically with host versus graft and graft versus host responses are highly desirable goals.

The frequency of donor anti-recipient interleukin-2-producing T cells correlates with outcome after bone marrow transplantation

In addition to achieving the best possible degree of HLA matching between donor and recipient, functional assays have been used in an attempt to predict the severity of GVHD following bone marrow transplantation. Traditionally this has relied on the mixed lymphocyte reaction (MLR), however the MLR is notoriously unreliable and insensitive. We have reported recently that the measurement of donor anti-recipient cytotoxic T lymphocyte precursor (CTLp) frequencies by limiting dilution provides a more quantitative measure of donor anti-recipient reactivity (1). In unrelated donor/recipient combinations CTLp frequency correlated with the severity of acute GVHD, however the correlation was not complete, and measurable frequencies were not detected in HLA-identical sibling pairs. For these reasons we adapted a technique for the measurement of helper T lymphocyte precursor (HTLp) frequencies first described by Bishop and Orosz (2), and went on to apply it to a series of 46 transplant pairs.

The assay involved the culturing of a fixed number of stimulator cells with serial dilutions of responder cells. After 64 hours the plates were irradiated, and the interleukin-2 (IL-2)-dependent indicator cell line, CTL-L20 was added. Proliferation of the CTL-L cells was measured by quantiation of tritiated thymidine incorporation. Initially, 82 responder/stimulator pairs were studied, with varyiing degrees of histocompatibility. The results of these comparisons showed that the mean frequency increased with increasing degrees of mismatch. It was also noted that there was a wide spread of frequencies within groups that had the same level of matching; this was most notable for the HLA-identical sibling combinations, in which frequencies ranged from 1:80,000 to 1:1000,000. This last observation was particularly encouraging, in that it suggested that it might be possible to use donor anti-recipient HTLp frequency measurement as a predictive parameter for GVHD. The phenotype of the IL-2-producing Th cells was also examined in six combinations, by measuring the HTLp frequencies in unseparated, $CD4^+$ and $CD8^+$ cells. This revealed that a significant fraction, between one third and one half of the HTLp cells were contained in the $CD8^+$ population.

In the donor/recipient transplant pairs, Th frequencies were measured in 25 HLA-identical sibling, and 21 unrelated combinations. In the sibling pairs the patients that received marrow from donors with high HTLp frequencies had a significantly higher incidence of moderate to severe (grades II-IV) acute GVHD, when compared with those that received marrow from donors with a low HTLp frequency, as illustrated in Table 1. In the unrelated donor/recipient combinations, a nonsignificant trend towards higher HTLp frequencies was seen in the donors for the patients developing moderate to severe acute GVHD, when compared to those developing trivial or no GVHD. Nine deaths occurred due to acute GVHD in the study group, and of these patients seven received marrow from donors with a high, one with an intermediate, and one with a low HTLp frequency.

Several points emerge from this study. The first is that this limiting dilution assay for HTLp measurement appears to be sufficiently sensitive to detect primary anti-minor histocompatibility antigen helper T cell (Th) responses. The responder cells for these assays had not been transfused nor had any of these donors been pregnant, so that it is unlikely that any *in vivo* priming against minor antigens had occurred. The second, and clinically most important observation was that HTLp frequency correlated with development of acute GVHD in the recipients of HLA-identical sibling bone marrow. The same trend was seen in the recipients of unrelated marrow, although statistical significance was not reached. Similarly, a correlation was seen between donor anti-recipient HTLp frequency and acute GVHD-related

Table 1.   Occurrence of acute graft versus host disease correlates with donor anti-recipient HTLp frequency in recipients of bone marrow from HLA-identical siblings.

Donor anti-recipient HTLp frequency	Percentage of patients with grades II-IV acute GVHD
High ($>1:10^5$) n = 9	70
Intermediate ($1:10^5$-$1:4 \times 10^5$) n = 8	38
Low ($<1:4 \times 10^5$) n = 8	10

Table 2. Frequency of "primed" HTLp specific for donor DR alloantigens is increased in patients that lost their grafts within one month due to early rejection, and is decreaed in patients that lost their grafts more than two years after transplantation.

Recipient	Donor	Graft loss	Stimulator	HTLp frequency[+]
DR1,w15	DR1,4	Early[*]	DR1	Not detected
			DRw15	Not detected
			DR4	1:19.0
			DRw13	1:126.8
DRw15,9	DRw15,7	Early	DRw15	1:1,304.9
			DR9	Not done
			DR7	1:24.4
			DR4	1:59.2
DRw11,w13	DRw15,6	Late	DRw11	1:1,302.4
			DRw13	1:114.7
			DRw15	1:1,242.4
			DR3	1:50.4
DRw11,-	DR1,4	Late	DRw11	1:92.2
			DR1	1:82.5
			DR4	1:924.9
			DRw15	1:23.6

[+]Frequencies are shown as $1:n \times 10^{-3}$ [*]Grafts lost "early" failed due to acute rejection within one month of transplantation; grafts lost "late" were removed two years or more after transplantation, with no evidence of acute rejection

deaths. Further studies will determine whether this assay has predictive value in clinical transplantation. The third finding was that a substantial fraction of the IL-2-producing T cells were of the $CD8^+$ phenotype. This result adds further evidence for the importance of MHC class I-restricted helper T cells.

The prolonged residence of an allograft may induce a state of hyporesponsiveness against donor alloantigens

A modified form of this assay was used to measure the frequency of sensitised interleukin-2-producing T cells in dialysis patients that had lost a previous renal allograft. The modified form of this assay involves light irradiation of the responder T cells at the start of the culture. This modification is based on the observation that IL-2 production by memory T cells is less irradiation sensitive than IL-2 production by naive responder cells. Four dialysis patients were analysed in detail, two that had lost their graft due to early, irreversible rejection within two months of transplantation, and two that had lost their graft more than two years after transplantation with no evidence of acute rejection. As illustrated in Table 2, the frequency of "memory" Th was notably higher against donor DR alloantigens in the first two patients; in marked contrast, frequencies against the donor DR types were significantly lower than those measured against third party DR types in the second two patients (3). This pilot study raises the possibility that prolonged exposure to MHC alloantigens on the surface of cells that cannot act as potent antigen-presenting cells can lead to a hyporesponsive state. Two mechanisms could be responsible for this phenomenon. One possiblity is that anti-donor alloreactive T cells are clonally deleted. The alternative is that a state of anergy is induced in anti-donor T cells, such that they may be present, but are non-responsive. These results are being followed up by the study of further patients that have lost a previous renal transplant. They also prompted a series of experiments aimed at the induction of anergy in allospecific T cells *in vitro*.

Recognition of DR alloantigens on the surface of human T cells induces a state of non-responsiveness in anti-DR alloreactive human T cell clones

A series of results have been presented over the past few years that led to the conclusion that T cell recognition can have two distinct outcomes. If recognition is accompanied by the necessary accessory signals, activation ensues. If, on the other hand, signals are transduced through the TcR:CD3 complex in the absence of co-stimulatory signals, this can lead to a state of non-responsiveness referred to as anergy. Recognition of antigen on the surface of

796

several non-professional antigen-presenting cells has been shown to indue T cell anergy. Such cells have included γ-interferon-induced, MHC class II-expressing mouse and human keratinocytes (4,5) and ß cells of pancreatic islets from transgenic mice expressing H-2E molecules (6). The ability of DR-expressing human T cells to present antigen has been a contentious point. Some years ago it was reported that presentation of high concentrations of peptide by T cells to T cells led to tolerance in a peptide-specific clone (7). More recent studies have documented the ability of T cells to present peptide antigens effectively, leading to the activation of responder T cells (8). We have investigated this for antigen-specific and anti-DR alloreactive human T cell clones. The ability of two DR1-restricted T cells, specific for the 306-324 peptide of influenza haemagglutinin, to proliferate in response to peptide in the absence of added APC was examined. T cells that had been stimulated one week earlier with APC and antigen were able to proliferate in response to peptide alone, however these preparations of T cells were also able to respond to phytohaemagglutinin without the addition of accessory cells. This suggested that the T cells were contaminated with accessory cells or accessory cell membranes that were co-purified with the T cells. In contrast, if the T cells were prepared using a protocol designed to eliminate contaminating accessory cells, no proliferation due to T cell:T cell peptide presentation was detected, as shown in Figure 1. This was followed by examining the consequences of peptide recognition on the surface of a T cell clone expressing the appropriate DR restriction element. T cells specific for the 306-324 peptide of HA were incubated overnight with the HA peptide or with an irrelevant peptide, and then were challenged the following day with fresh peptide and B cell APC. As little as 1µg/ml of specific peptide induced a state of profound non-responsiveness (data not shown). Having established that low concentrations of antigen were sufficient to induce tolerance when presented by T cells, we argued that the recognition of alloantigens on the surface of an allogeneic T cell should induce non-responsiveness in alloreactive T cells in a similar manner. The ability of three anti-DR1 T cell clones to recognise DR1 on the surface of a DR1-expressing T cell clone was tested by incubation of the alloreactive T cells with the DR1-expressing T cell, in the presence of third party peripheral blood mononuclear cells. Two of the three clones proliferated under these conditions. Pre-incubation of these three anti-DR1 clones with the DR1-expressing T cell overnight led to a marked reduction in the response to a subsequent challenge with DR1[+] B cells for the two clones that recognised the DR1 alloantigen on the T cell's surface (Figure 2) (9). No effect was seen for the third clone (data not shown).

These results suggest that human T cells cannot function as autonomous antigen-presenting

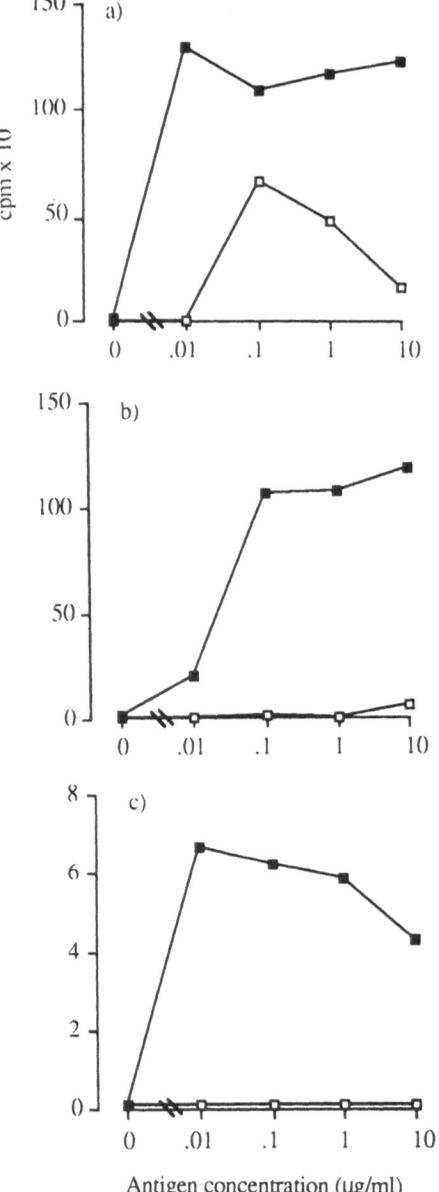

Legends to Figures

Figure 1. Accessory cell-free T cels are unable to act as autonomous antigen-presenting cells.
10^4 T cells, a) non-purified NF4 (a DR1-restricted clone specific for the 306-324 peptide of
influenza haemagglutinin), b) purified NF4, and c) purified HC1 (same specificity as NF4)
were cultured with different concentrations of peptide alone (open squares), or with 3×10^4
peptide-pulsed, irradiated B-LCL.

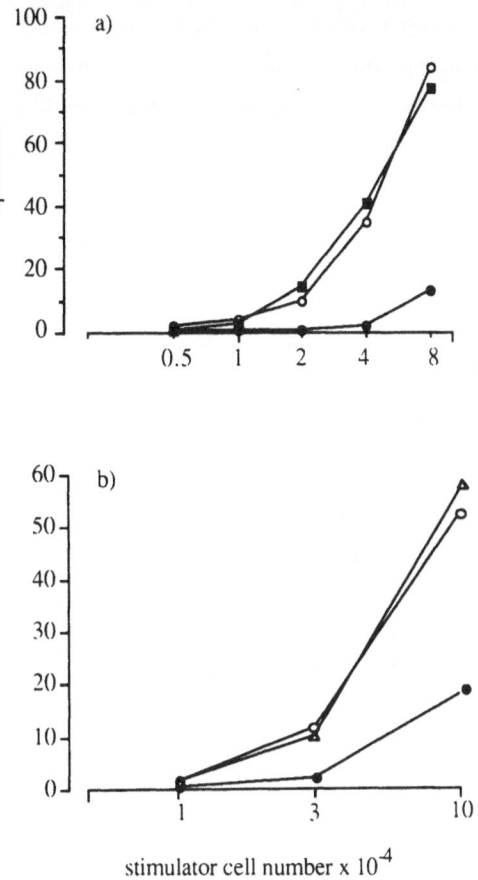

Figure 2. Recognition of DR1 alloantigens on the surface of a T cell leads to the induction of anergy in anti-DR1 T cells. Purified anti-DR1 T cells, 4×10^5, a) G3, and b) G11 were pre-cultured with 8×10^5 irradiated DR1-expressing purified T cells (NF4) (filled circles), medium alone (open circles), 8×10^5 irradiated NF4 cells in the presence of anti-DR antibody (filled squares), 8×10^5 irradiated DRw17-expressing purified T cells (open triangles). After washing, 10^4 anti-DR1 T cells were cultured with different doses of irradiated DR1-expressing B-LCL.

cells, but induce tolerance in antigen-specific and allospecific responder cells. The discrepancy between these findings and the results of others may reflect heterogeneity amongst T cell clones, or differences in the experimental systems used. More importantly, these results may have potential application for the induction of allospecific T cell tolerance *in vivo*.

References

1. Kaminski, E. Hows, J., Man S., et al. 1989. Prediction of graft versus host disease by frequency analysis of cytotoxic T cells after unrelated donor bone marrow transplantation. Transplantation 48: 608

2. Bishop, D.K. and Orosz, C.G. 1989. Limiting dilution analysis for alloreative TCGF-secreting T cells. Transplantation 47: 671

3. Deacock, S. and Lechler, R.I. 1992. Frequencies of alloreactive T helper cells in chronic renal failure patients may correlate with T cell sensitisation. Transplantation - in press

4. Gaspari, A.A., Jenkins, M.K. and Katz, S.I. 1988. Class II MHC-bearing keratinocytes induce antigen-specific unresponsiveness in hapten-specific Th1 clones. J. Immunol. 141: 2216

5. Bal, V., McIndoe, A., Denton, G.D. et al. 1990. Antigen presentation by keratinocytes induces tolerance in human T cells. Eur. J. Immunol. 20: 1893

6. Markman, J., Lo, D., Naji, R.D., et al. 1988. Antigen-presenting function of class II MHC-expressing pancreatic beta cells. Nature 336: 476

7. Lamb, J.R., Skidmore, N., Green, J.M., et al. 1983. Induction of tolerance in influenza virus-immune T lymphocyte clones with synthetic peptides of influenza haemagglutinin. J. Exp. Med. 157: 1434

8. LaSalle, J., Ota, K. and Hafler, D.A. 1991. Presentation of autoantigen by human T cells. J. Immunol. 147: 774

9. Sidhu, S., Deacock, S., Bal, V., et al. 1992. Human T cells cannot act as autonomous antigen-presenting cells but induce tolerance in antigen-specific and alloreactive responder cells. J. Exp. Med. - in press.

T Cell Subsets Resisting Induction of Mixed Chimerism Across Various Histocompatibility Barriers

Y. Sharabi[1], D. H. Sachs, and M. Sykes

Transplantation Biology Research Center, Massachusetts General Hospital, Harvard Medical School, Charlestown, MA 02129, USA

INTRODUCTION

We have previously demonstrated that a permanent state of mixed allogeneic chimerism and donor-specific transplantation tolerance can be induced in mice conditioned with a mild, non-myeloablative regimen involving host pre-treatment with anti-CD4 and anti-CD8 mAbs, sublethal (3 Gy) whole body irradiation, and local thymic irradiation (Sharabi 1989). The present studies were conducted in order to determine the minimal requirements for host T cell depletion in order to achieve mixed chimerism in the setting of defined major and minor histocompatibility barriers. We report here the results of studies involving selective depletion of host CD4$^+$ and/or CD8$^+$ T cells in the setting of isolated class I and/or class II MHC, or multiple minor histoincompatibilities.

METHODS

Conditioning and BMT

Mice were purchased from Jackson Laboratories (Bar Harbor, ME). Ascites from hybridoma lines GK1.5 (a rat anti-mouse CD4 mAb) (Dialynas 1983), and 2.43 (a rat anti-mouse CD8 [Lyt2.2] mAb) (Sarmiento 1980) were used. Mice received 0.1 ml of GK1.5 and/or 2.43 ascites intraperitoneally on days -6 and -1 before BMT. On the day of BMT, 3 Gy WBI, 7 Gy selective thymic irradiation (TI) and 15×10^6 unmanipulated donor bone marrow cells (BMC) were administered as described (Sharabi 1989).

MAbs and flow cytometry

Chimerism was assayed by FCM after staining with fluoresceinated anti-K^b mAb 5F1 (Sherman 1981) followed by biotinylated anti-D^d mAb 34-2-12 (Ozato 1981) (B10.D2→B10 combination) or anti-K^k mAb 16-1-11 (Ozato 1981) (AKR, C3H/HeJ, or CBA/J→B10 combination) plus Texas Red Streptavidin, as previously described (Sharabi 1989; Sykes 1990). Chimerism in the bm12→B6 combination was determined by staining with mAb 34-5-3 (Hansen 1981) (staining all Ia positive cells of B6 origin but not of bm12 origin) and with Y3P (Janeway 1984) (staining Ia positive cells of both strains). Non-specific staining was evaluated with an irrelevant mAb (fluoresceinated or biotinylated anti-human CD3, Leu 4, Becton Dickinson), and the percentage of cells stained with this antibody was subtracted from the values of staining with the specific Ab to determine specific staining. Fc receptors were blocked before staining with an anti-FcγR mAb, 2.4G2 (Unkeless 1979). A B10.D2 anti-DBA/2 alloantiserum prepared in our laboratory was used to distinguish B10 cells from C3H.SW cells. This antiserum binds to C3H.SW cells and not to B10 cells. Cells were incubated with antiserum, washed, then incubated with FITC-labeled goat-anti-mouse Ig. We have previously published the details of staining and FCM analysis (Sykes 1990; Sykes 1988).

---

[1]Department of Life Sciences, Bar-Ilan University, Ramat-Gan 52100, ISRAEL

Skin grafting

Skin grafting was performed by the method we have previously described (Sykes 1988).

RESULTS

Engraftment across selected histocompatibility barriers

As shown in Table 1, injection of 15×10^6 unmanipulated allogeneic BMC following conditioning with anti-CD4 plus anti-CD8 mAbs, 3 Gy WBI and 7 Gy thymic irradiation (TI) resulted in engraftment of allogeneic BMC across isolated class I MHC, isolated class II MHC, class I plus class II MHC, multiple minor histocompatibility antigen (HA), and class I plus class II MHC plus minor HA barriers. For most of these disparities, engraftment resulted in the development of mixed chimerism, with the percentage of donor PBL ranging from 10 to >90%.

Table 1. Engraftment of BMC across various genetic barriers following host conditioning with the complete non-myeloablative regimen[a]

Combination (Donor→Recipient)	Disparity	Donor Lyt 2 Allele	# of Mice Engrafted/ Total	% Donor PBL[a]
bm1→B6	MHC Class I	2	9/12	45-89
B10.MBR→B10.AKM	MHC Class I	2	5/5	83-95
bm12→C57BL/6	MHC Class II	2	8/10	10-92
C3H.SW→C57BL/10	Multiple minors	1	12/12	34-75
B10.D2→C57BL/10	MHC Class I+II	2	10/12	36-86
AKR→C57BL/10	Class I+II, minors	1	5/5	100
CBA/J→C57BL/10	Class I+II, minors	1	4/4	53-97
C3H/HeJ→C57BL/10	Class I+II, minors	1	5/6	19-100

[a] PBL chimerism was evaluated 5 weeks following BMT.

Selective T cell depletions in fully MHC-mismatched BMT

As shown in Table 2, selective depletion of only $CD4^+$ or only $CD8^+$ T cells from B10 mice treated with 3 Gy WBI and 7 Gy TI did not permit engraftment of fully MHC-disparate allogeneic marrow (B10.D2→B10). In contrast, depletion of both subsets permitted stable mixed chimerism to develop in 12 of 13 mice. Thus, both $CD4^+$ and $CD8^+$ T cell subsets are independently capable of resisting engraftment of class I plus II disparate allogeneic BMC.

T cell subsets resisting engraftment of class I- or class II- MHC-disparate marrow grafts

Studies were performed to evaluate the requirement for depletion of $CD4^+$ and/or $CD8^+$ cells for engraftment of class I or class II disparate BMC. As shown in Table 3, engraftment across an isolated class I disparity (bm1→B6) was achieved by host treatment with anti-CD4 plus anti-CD8 mAbs or by treatment with anti-CD8 alone. Six of 12 B6 mice treated only with anti-CD8 (in addition to 3 Gy WBI and 7 Gy TI) engrafted with bm1 BMC, while 9 of 12 showed engraftment after treatment with both mAbs. Engraftment after depletion of only CD8[+] T cells was similarly demonstrated in the additional

Table 2. Engraftment of class I+II MHC disparate BMC following selective depletion of host T cell subsets

Combination	mAb Treatment[a]	Chimeric Animals/Total (% Donor PBL)	
		Expt. 1	Expt. 2
B10.D2→B10	GK1.5+2.43	6/7 (47-73)	6/6 (36-86)
(Class I+II)	GK1.5	0/7	0/6
	2.43	0/7	0/6

[a] B10 recipients were pre-treated with indicated mAb on day -6 and -1, then received 3 Gy WBI, 7 Gy TI and BMT on day 0.

Table 3. T cell subsets resisting engraftment of class I vs. class II disparate allogeneic BMC

Combination	mAb Treatment	Fraction of Chimeric Animals (% Donor PBL)	
		Experiment 1	Experiment 2
bml→B6[a]	GK1.5+2.43	4/6 (45-86)	5/6 (74-78)
(Class I)	GK1.5	0/6	0/5
	2.43	3/6	3/6 (45-71)
B10.MBR→B10.AKM[b]			
(Class I)	GK1.5+2.43	5/5 (83-95)	
	GK1.5	0/4	
	2.43	4/5 (77-100)	
TCD B6→bml	2.43	4/4 (49-72)	
bm12→B6[c]	GK1.5+2.43	2/4 (10-27)	6/6 (17-92)
(Class II)	GK1.5	0/6	2/7 (29-30)
	2.43	0/6	0/5

[a] Chimerism in the bml→B6 combination was tested by staining with 5F1 (staining all B6 cells but not bml cells).
[b] Chimerism in the B10.MBR→B10.AKM combination was determined by staining with fluoresceinated 5F1 (anti-K[b]) (staining B10.MBR cells and not B10.AKM cells).
[c] Chimerism in the bm12→B6 combination was determined by staining with 34-5-3 (staining all Ia positive cells of B6 origin but not those of bm12 origin) and with Y3P (staining Ia positive cells of both

strains). Percent bm12 cells was calculated, after selecting cells in the two color plots which were positive for Y3P and negative for 34-5-3 staining, by the formula:
%bm12 cells=(% chimera Y3P$^+$ 34-5-3$^-$ cells/% B6 Y3P$^+$ cells)x100%.

B10.MBR→B10.AKM class I disparate strain combination. T cell depletion of donor marrow did not inhibit engraftment of B6 marrow in class I-disparate bm1 recipients depleted of CD8$^+$ cells (Table 3). Depletion of host CD4$^+$ cells only did not permit engraftment in either class I-disparate combination (Table 3).

Engraftment in the class II-disparate bm12→B6 combination was determined using mAb 34-5-3, which binds B6 but not bm12 Ia-positive cells (Hansen 1981), and mAb Y3P, which binds both bm12 and B6 I-A molecules (Janeway 1984). In contrast to the results obtained in the presence of isolated class I differences, host depletion with anti-CD8 alone did not permit engraftment of bm12 BMC in B6 recipients. Depletion with only anti-CD4 mAb resulted in engraftment in 2 of 13 mice. In the same two experiments, 8 of 10 recipients pre-treated with anti-CD4 plus anti-CD8 mAbs demonstrated engraftment of bm12 BMC (Table 3).

To evaluate the correlation between engraftment and tolerance induction, B6 mice transplanted with bm1 or bm12 BMC were grafted with donor-type skin. All B6 hosts with persistent chimerism were tolerant of donor-type skin grafts. These data confirm the specificity of FCM in detecting the presence of donor-type PBL.

T cell subsets resisting engraftment of minor HA-disparate, MHC-matched, allogeneic BMC

Selective depletion of CD8 cells alone resulted in engraftment of C3H.SW BMC in B10 mice (10/12), while depletion of CD4 cells alone failed to permit engraftment (0/12). Depletion of both subsets resulted in engraftment in 12 of 12 animals in the same experiment.

DISCUSSION

Previous studies have evaluated the roles of T cell subsets in skin allograft rejection and GVHD. These studies showed that GVHD and skin graft rejection are predominantly mediated by CD8$^+$ T cells in the case of class I disparity alone, by CD4$^+$ T cells in the case of class II disparity alone, and by either subset in the case of class I- and class II-mismatched transplants (Pietryga 1987; Rosenberg 1987; Sprent 1988; Sprent 1986). CD8$^+$ T cells appear to mediate GVHD alone in some strain combination. involving minor histocompatibility differences only (Korngold 1987). Results of skin grafting studies suggest that a source of both Th and CTL is needed for effective skin graft rejection to occur (Rosenberg 1987).

We have now performed studies to evaluate the T cell subsets resisting bone marrow engraftment in various strain combinations. Consistent with skin grafting and GVHD studies, our results demonstrate that CD8$^+$ T cells play a predominant role in resisting engraftment of class I-disparate BMC, whereas CD4$^+$ T cells predominate in resistance to class II-disparate BMC. In addition, CD8$^+$ T cells contribute substantial resistance to class II-disparate grafts, whereas CD4$^+$ T cells pose little or no barrier to engraftment of minor HA-disparate (in the one strain combination tested) or class I-disparate BMC. In general, our observations are consistent with those in human and animal studies indicating a predominant role for CD8$^+$ T cells in resisting allogeneic marrow engraftment across major and minor histocompatibility barriers (Bierer 1988; Bordignon 1989; Kernan 1987; Schwartz 1987).

The results of studies involving an isolated class II MHC disparity indicate that $CD4^+$ and $CD8^+$ T cells both resist marrow engraftment in this setting. The mechanism of CD8-mediated resistance to class II-disparate grafts may involve presentation of peptides derived from endogenous class II MHC molecules by class I molecules on the surface of donor BMC. Since these class I molecules are also shared by the recipients, this complex could be recognized as self class I MHC plus peptide antigen by recipient $CD8^+$ T cells, and thus target the donor BMC for CD8-mediated destruction. Presentation of class II peptides to $CD8^+$ CTL in this manner has been previously described (Shinohara 1988). In addition, some $CD8^+$ CTL are capable of recognizing intact allogeneic class II molecules (Shinohara 1988), providing another possible mechanism of CD8-mediated resistance to class II-disparate marrow grafts.

Our results have implications for the potential clinical application of this approach to inducing transplantation tolerance across MHC barriers. When performing class II MHC-mismatched transplants, regardless of the presence or absence of class I MHC disparities, it might be essential to deplete host $CD4^+$ as well as $CD8^+$ T cells in order achieve mixed chimerism and transplantation tolerance.

REFERENCES

Bierer BE, Emerson SG, Antin J, et al (1988) Transplantation 46:835
Bordignon C, Keever CA, Small TN, et al (1989) Blood 74:2237
Dialynas DP, Quan ZS, Wall KA, et al (1983 J Immunol 131:2445
Hansen TH, Walsh WD, Ozato K, Arn JS, Sachs DH (1981) J Immunol 127:2228
Janeway CA Jr, Conrad PJ, Lerner EA, Babich J, Wettstein P, Murphy DB (1984) J Immunol 132:662
Kernan NA, Flomenberg N, Dupont B, O'Reilly RJ (1987) Transplantation 43:842
Korngold R, Sprent J (1987) J Exp Med 165:1552
Ozato K, Henkart P, Jensen C, Sachs DH (1981) J Immunol 126:1780
Ozato K, Mayer NM, Sachs DH (1982) Transplantation 34:113
Pietryga DW, Blazar BR, Soderling CCB, Vallera DA (1987) Transplantation 43:442
Rosenberg AS, Mizuochi T, Sharrow SO, Singer A (1987) J Exp Med 165:1296
Sarmiento M, Glasebrook AL, Fitch FW (1980) J Immunol 125:2665
Schwartz E, Lapidot T, Gozes D, Singer TS, Reisner Y (1987) J Immunol 138:460
Sharabi Y, Sachs DH (1989) J Exp Med 169:493
Sherman LA, Randolph CP (1981) Immunogenetics 12:183
Shinohara N, Bluestone JA, Sachs DH (1986) J Exp Med 163:972
Shinohara N, Hozumi N, Watanabe M, Bluestone JA, Johnson-Leva R, Sachs DH (1988) J Immunol 140:30
Sprent J, Schaefer M, Gao E, Korngold R (1988) J Exp Med 167:556
Sprent J, Schaefer M, Lo D, Korngold R (1986) J Exp Med 163:998
Sykes M (1990) J Immunol 145:3209
Sykes M, Sheard M, Sachs DH (1988) J Immunol 141:2282
Sykes M, Sheard MA, Sachs DH (1988) J Exp Med 168:661
Unkeless JC (1979) J Exp Med 150:580

Temporary Tolerance or Suppressive Regulation Induced By Non MHC Alloantigens in Transplantation and Pregnancy*

A. Padányi, É. Gyódi, A. Horuzsko, R. Mihalik, É. Pócsik, J. Szelényi, M. Réti, I. Szigetvári[1], F. Perner[2], M. Kassai[3], B. Schmidt, Gy. G. Petrányi

National Institute of Haematology, Blood Transfusion and Immunology, Daróczi u. 24, H-1113 Budapest, Hungary

INTRODUCTION

A special type of tolerance or suppressive regulation will be reported, which can be characterized by the complementary participation of two alloantigen systems; the major histocompatibility complex, and either a minor histocompatibility or certain type of differentiation antigen (secondary, subordinated) alloantigen system of functional importance. There are representative experimental observations on this phenomena, from which one characteristic model, reported by Hutchinson and Morris (Hutchinson and Morris 1987) is explained. Prior to kidney transplantation between RTL incompatible rats recipients were transfused with blood obtained from various strains characterized by either matching or mismatching with the transfusion and organ donor or recipient strain in the minor histocompatibility system. The matching between transfusion and kidney donor and a mismatching between transfusion donor and organ recipient as regards minor histocompatibility alloantigen system has to be emphasized as a new requirement. No similar situation has been reported in human beings. However, the induction of tolerance by transfusion in certain models is a well established phenomena. The importance of class II antigen matching between transfusion donor and recipient were outlined and proved in series of clinical observations based on in vivo and in vitro parameters including cytotoxic antibody production, MLC, cytotoxic precursor cell function and kidney survival (Lagaay et al. 1989, Claas et al 1991, van Rood and Claas 1990, de Waal and van Twuyer 1991).

Recent reports attempt to clarify in human this secondary "subordinated" alloantigen system, which seems to have a mandatory function besides the relation in the major histocompatibility complex antigens between the blood transfusion donors and recipient for induction the immunsuppressive regulation. A natural and an artificial model was selected for the demonstration of the phenomena, i.e. pregnancy and transplantation. Both have analogous mechanisms involving major and secondary alloantigen systems and priming, as well as inducing and grafting events.

For better interpretation the explanation of the object will be given in order to understand the in vivo relevance of immunologic phenomena.

---

* Supported by the European Foundation for Sciences, Arts and Cultures Grant.
1. Dept.Gynacology, Postgraduate Medical School, Budapest
2. 1st Clin. of Surgery, Semmelweis Medical School, Budapest
3. Blood Center, County Hospital, Hódmezővásárhely

1) Association of immunoregulatory IgG with improved kidney survival and normal pregnancy

In the case of unrelated cadaver kidney transplantation if transfusions are routinely administered in the preoperative phase, in most of the cases Fcγ receptor blocking antibodies are detectable by various rosetting techniques termed as erythrocyte antibody inhibition (EAI) assay (Burlingham and Sollinger 1986, MacLeod et al. 1982, Sandilands et al 1990, Padányi et al 1990).

These antibodies are present either alone or together with cytotoxic anti-HLA antibodies. Figure 1 shows the presence of FcγR blocking antibodies in polytransfused patients on haemodyalisis and its absence in the non-transfused healthy individuals. It is demonstrated simultaneously that the occurrence of cytotoxic anti-HLA antibodies is much more frequent when a low level of blocking activity is demonstrated.

If the presence of FcγR blocking antibodies is compared, in a retrospective study, with kidney survival, it is remarkable that besides improved graft survival, almost no kidney was lost in the first six months after transplantation. This suggests that the short term effect of this kind of immunoregulation is conspicuous (Fig. 2).

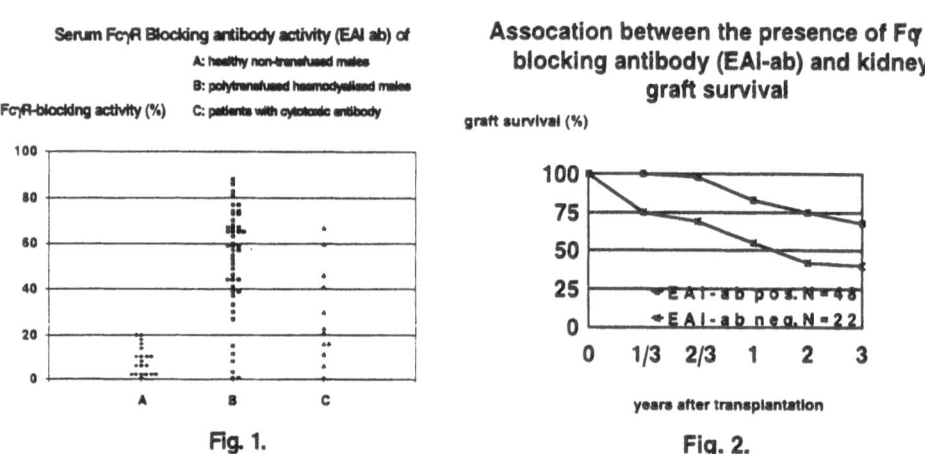

Fig. 1.

Fig. 2.

Analogous association could be found in the case of normal pregnancy in contrast to recurrent habitual abortions (Padányi et al. 1991, Takakuwa et al. 1986, Mueller-Eckhardt et al. 1989). In the latter case no FcγR blocking activity was found in the sera of the patients, however, primiparous and multiparous women possess significant blocking activity in their sera. It is widely accepted that in the case of the classical immunological type of recurrent habitual abortion the loss of the foetus is due to the immunologic rejection of the placental tissues (Editorial 1983, Johnson and Ramsden 1988).

2) Characterization of FcγR blocking IgG antibody.

It was previously reported that the FcγR blocking activity could be associated with the IgG fraction of the sera, as well as to the $(Fab)_2^1$ fragment of the IgG, and it was not joined with specificity to the FcγRII receptor (Padányi et al. 1991, Sármay et al. 1990). The FcγR blocking function could be explained as in the case of anti-beta$_2$ microglobulin, class II and anti-IgG antibodies. Namely, the antibody bound to their ligand at the cell surface indirectly affects the function of FcγR. This statement was further supported by the finding that the blocking activity can also be absorbed from the sera using FcγR receptor not expressing cell lines (Forwell et al. 1986).

The question of what the tissue distribution of antigen or receptor on the cell surface reacting with the IgG antibody is, was approached with the initiation of absorption experiments. Almost all of the various cell compartments of the peripheral blood were able to absorb the blocking activity with the exception of erythrocytes and few lymphocyte subpopulation. Trophoblast cell membrane, preparates and liophylised seminal plasma are also excellent absorbents. Determination at the molecular level resulted in finding a monomer with 66 kD Mw on trophoblasts reacting with the FcγR blocking antibody. Based on the molecular weight and the tissue distribution studies it was suggested that the FcγR blocking antibody probably reacts with a molecule analogous to the alpha chain of TLX/MCP/CD46 dimer (McIntyre and Faulk 1982, Purcell et al. 1990). However, it cannot be excluded that the target molecule of the blocking antibody is in association with FcγRII or III.

The inducibility of the FcγR blocking antibody production after one transfusion is demonstrated (Fig. 3). In almost all individuals, a very low level of FcγR blocking antibody can be found in the sera. After the first transfusion of buffy coat or platelets, however, immediate increase in the level of blocking activity can be observed that can be further strenghtened by boosting. Various lines of evidences obtained by testing the sera from one individual on a series of unrelated B cells in the EAI assay suggested the allotypic nature of the antibody.

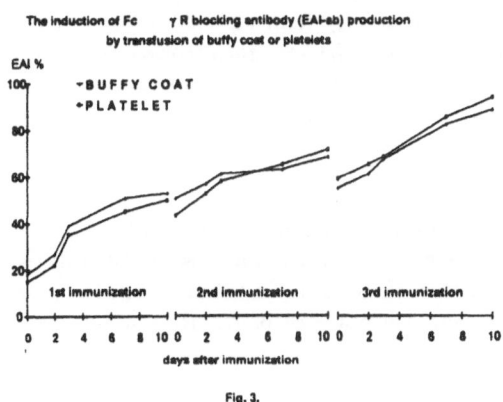

The induction of Fc γ R blocking antibody (EAI-ab) production by transfusion of buffy coat or platelets

Fig. 3.

3) The TLX/MCP/CD46 polymorphic system and its phenotypes in the population

Ten different blocking sera were available for determination of the population's allotypic pattern. B cells from each panel member were separated and tested in the EAI assay simultaneously with the ten sera. The TLX phenotype of the individuals was determined by the blocking pattern of the ten "typing" sera. It shows that at least four clusters can be established designated as TLX-B1, -2, -3, and -4 (Fig. 4).

Efforts were directed to define the possibility of TLX typing for both the donor and recipient in the transfusion and pregnancy model. It shows that in the pairs of the transfusion model, the production of the TLX-B allospecific antibodies were induced when certain alloantigen expressed in the donor were lacking in the recipient. It should be pointed out that the pairs shared a certain number of class I and class II antigens. There were few representative pairs of individuals, in whom there was matching in TLX phenotypes but no blocking antibody production occurred (Fig. 5). Thus, a prerequisite for FcγR blocking or TLX specific allo-antibody production in the transfusion model is the mismatching in the TLX polymorphic system and probably, a certain type of matching in the HLA system in addition to that.

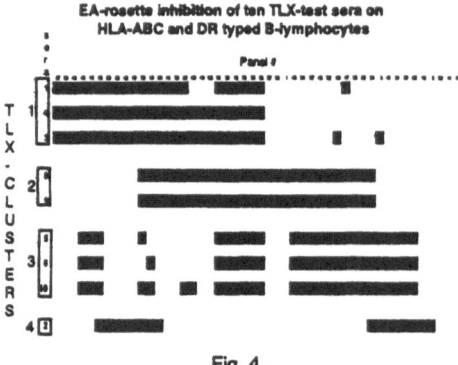

Fig. 4.

The relevance of TLX allotype in the pregnancy model were approached by means of absorption studies on trophoblasts with the simultaneous use of all typing sera. It shows that trophoblasts express at least 2 TLX allotypes inherited partially from the mother and partially from the father. Our preliminary study in collaboration with McIntyre's group showed that soluble antigen in the seminal plasma also expresses the adequate TLX allotype, though it can be concluded that either seminal plasma or the sperms the trophoblast itself can be the appropriate priming (tolerogeneic) stimuli for inducing the blocking antibody production or suppressor regulatory pathway (Torry et al. 1988, Anderson et al. 1989).

It had been shown (Purcell et al 1990, Lublin et al 1988) that TLX/MCP and probable CD46 gene is located outside MHC on the chromosome 1 short arm, close to the FcγR coding genes. Our family studies, in which first of all the allotypes were proven to be controlled by genetic rules and further the segregation of TLX type showed independence of the HLA phenotypes have strengthened this observation (Padányi et al 1990).

4) The biological importance of the TLX phenotypes in kidney transplantation cases and in habitual recurrent abortion

We put forward the question whether the previously demonstrated association between the TLX phenotypes of donor and recipient concerning blocking (TLX alloantibody) production can also be found in the two models in relation to the in vivo beneficial biological consequences on transplantation immune response. In related kidney transplantation, we have a few representative cases in which, because of extremly high MLC reactivity, a donor specific transfusion protocol with platelets was indicated prior to transplantation. In those cases a high titre of FcγR blocking antibody was produced and the mismatching between TLX phenotypes of donor and recipient were present parallel. It is not necessary to stress that in those cases haplo identity existed concerning the HLA phenotypes. The MLC reactivity decreased significantly and kidneys are surviving for more than 4 years, free from rejection crisis. In contrast to a series of successful donor-specific transfusion treatment, we also have other cases in which the donor-specific transfusion was unsuccessful parallel for inducing blocking antibodies, the MLC reactivity remained high and the TLX phenotype between transfusion donor and recipient corresponded to the theory in the majority of cases.

The model of recurrent habitual abortion in contrast to normal pregnancy shows a very clear feature. In almost all investigated cases a matching in the TLX phenotype was found between the couples, while in normal pregnancy this situation was the opposite, thus, a foreign TLX allotype observable in the husband, was lacking in the wife (Padányi et al 1991). It is worthwhile to point out that the first trimester is the most critical for the rejection or survival of the foetus. This combined with the observation in kidney transplantation that suppressive regulation was most effective in the first 3-6 months of the post-transplant period suggests the notion that this kind of temporary tolerance or suppressive regulation may be responsible for controlling early alloreactive response and may have less importance in the later or chronic phase of transplantation reactivity.

5) The probable mechanism of suppressive immune regulation induced by TLX/MCP/CD46 polymorphic system.

As far as the possible mechanism is concerned, only preliminary data, obtained by in vitro tests are available. As demonstrated previously, the sera containing the blocking antibody or IgG prepared from it, displays a powerful suppression on PHA transformation in MLC test, but less in conA stimulation and none on soluble antigen (PPD and tetanus toxoid) induced activation (Horuzsko et al 1990).

Three main characteristics could be drawn from all those experiments. 1) The suppression on lymphocyte transformation in vitro was obvious only in the case of stimulation by alloantigens and mitogens and failed by PPD and bacterial proteins. 2) The blocking activity of the TLX IgG allospecific antibody is nonspecific, third party MLCs could also be suppressed. 3) Based on an MLC panel, a certain type of random blocking pattern emerges, thus, not all MLC tests are blocked by the sera with certain type of TLX-specificity.

We were encouraged to define the background of the curious suppressive pattern found in the MLC panel study and, therefore, approached first whether the blocking IgG is effecting on the stimulator or effector part of the MLC. Figure 6 shows that isolate pretreatment of the cells from both partners resulted in blocking activity only when the effector cells were treated. In the second series of MLC investigation blocking sera or IgG with certain type of TLX specificity were used in the pretreament protocol on effector cells possessing adequate or different TLX phenotype. Eight out of ten cases were blocked by sera with specificity 1+2 in the case when the effector cells possessed the adequate TLX specificity (TLX-B1, 2), but only three of twelve were blocked if the effector cells lacked it (TLX-B3,4). We concluded that the binding of allospecific TLX blocking IgG to its ligand on the cell surface as a type of allospecific restriction may be a key event in the mechanism of suppressive regulation.

Treatment of effector and stimulator cell population
MLC with TLX blocking sera

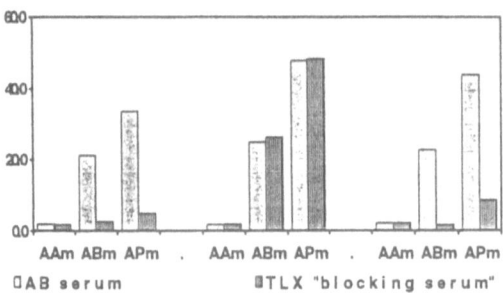

AAm ABm APm · AAm ABm APm · AAm ABm APm

◻AB serum   ▉TLX "blocking serum"

Fig. 6.

The allo-restricted suppressive regulation is, moreover, a cell mediated phenomena demonstrated by transfer experiment in vitro. Primary MLC comprising effector/responder cells possessing certain TLX specificity were treated with the appropriate allospecific blocking sera and thereafter at day five the washed cells of MLC were transferred in a secondary MLC where they displayed their suppressive regulation.

In an other line of experiments, activation markers were determined in vivo in the case of buffy coat and platelet infusion. The latter case was found as a representative model for separate induction of blocking antibodies. The activation marker expression seems to be characteristic for both different priming protocols. The most striking was that class II lower expression was found in the case of activation with platelets, in contrast to buffy coat immunization. Activation marker CD30 and CD70, CD54 were absolutely lacking parallel to this feature. This observation is only an indirect notion that probably the induction phase of sensitization to foreign MHC antigens could be altered if only class I antigen is presented parallel with the TLX foreign alloantigen. In this case, the induction of blocking IgG and probably suppressive cell-mediated regulation can be a characteristic consequence (Pócsik et al 1990).

Change in CD4, CD8 blast cell ratio in
MLC treated with blocking IgG (anti TLX)
(Analysis of double stained FACS study)

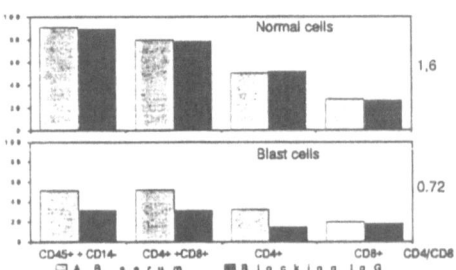

Fig. 7.

sIL2-R in MLC supernatants
2nd day

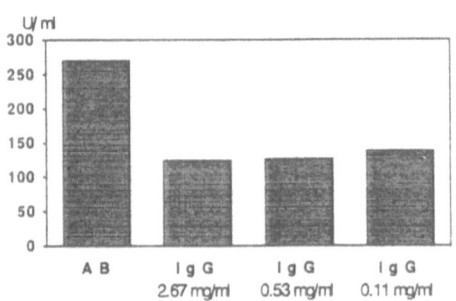

A B   I g G   I g G   I g G
           2.67 mg/ml   0.53 mg/ml   0.11 mg/ml

Fig. 8.

Considerable effort has been expanded to characterize the expression of differentiation antigen markers in MLC with or without the effect of the blocking IgG. As far as the first two preliminary experiments are concerned, there were no characteristic activation marker changes. However, focusing on the blast cells in the MLC, an alteration was found in the blocking IgG treated cell cultures documented in Fig. 7. The MLC reactivity which was significantly blocked by the immunosuppressive pooled IgG preparate showed a shift in the CD4, CD8 blast cells in the direction to CD8 cell over- representation. In contrast to the non-treated MLC in which, beside a higher number of blast cells, CD4 cells represented the majority of this population. Parallel to this shift significant decrease in the amount of soluble IL-2 receptor was found in the supernatant of the suppressive regulated cultures (Fig. 8).

Conclusion

In conclusion, temporary immunosuppressive regulation to transplantation alloreactivity could be achieved by complementary collaboration of the MHC and the secondary TLX/MCP/CD46 (FcγR) alloantigens. Conditions: priming-induction events can be characterized by MHC class I or/and class II partial matching and TLX/MCP/CD46 (FcR) mismatching between transfusion or seminal plasma donor and recipient. The immunregulatory process induced by the priming effect comprised by a) characteristic activation route; b) production of TLX allospecific/autoreactive IgG.

Allo/autoreactive IgG display powerful non-specific immunsuppressive/immunregulatory effect on T cell proliferation, which could be mediated by cells involved in alloreactivity binding the antibody to the cell surface. Immunsuppressive immunregulation is probably based on the unbalanced collaboration between monocyte, CD4- CD8- lymphocyte subpopulation and lymphokine kaskade (IL-2 production and IL-2 receptor solubilization, turn over down regulation).

In vivo relevance of the immunosuppressive regulation by the "double polymorphic" complementary mechanism could be suggested in the "beneficial effect of blood transfusion" in organ transplantation and the "lack of induction here" in recurrent habitual abortion.

REFERENCES

Anderson DJ, Michaelson JS, Johnson PM (1989) Trophoblast/Leukocyte-common antigen is expressed by human testicular germ cells and appears on the surface of acrosome-reacted sperm. Biol Reprod 41:285-293

Burlingham JW, Sollinger HW (1986) Action of donor-specific transfusions - analysis of three possible mechanisms. Transplant Proc XVIII:685-689

Claas FHJ, Lagaay EL, van Rood JJ (1991) Immunological consequences of blood transfusion. Schweiz med Wschr 121 (Suppl 43):70-

de Waal LP, van Twuyver E (1991) Blood transfusion and allograft survival: Is mixed chimerism the solution for tolerance induction in clinical transplantation? Immunol 10:417-425

Editorial (1983) Maternal blocking antibodies, the fetal allograft and recurrent abortion. Lancet ii:1175-1176

Forwell MA, Peel MG, Froebel KS, Belch JJF, MacSween RNM, Sandilands GP (1986) Transfusion-induced FcP-receptor-blocking antibodies: Spectrum of cellular reactivity. J Clin Lab Immunol 20:63-67

Horuzsko A, Gyódi É, Réti M, Mayer K, Kassai M, Petrányi Gy (1990) Selective effect of non-cytotoxic blocking alloantibodies produced after platelet transfusions on MLC mitogen- and soluble antigen induced response of lymphoytes in human. Transplantation 50:497-501

Hutchinson IV, Morris PJ (1987) The role of major and minor transplantation antigens in the blood transfusion effect. Transpl Proc 19:3087-3088

Johnson PM, Ramsden GH (1988) Recurrent miscarriage. Bailliére's Clin Immunol Allergy 2:607-624

Lagaay EL, Hennemann PH, Ruiszonetal M (1989) Effect of one-HLA-DR-antigen-matched and completely HLA-DR-mismatched blood transfusion survival of heart and kidney allografts. N England J Med 321:701

Lublin DM, Liszewski K, Post TW, Arce MA, LeBeau MM, Rebentisch MB, Lemons RS, Seya T, Atkinson JP (1988) Molecular cloning and chromosomal localization of human membrane cofactor protein (MCP). Evidence for inclusion in the multigene family of complement regulatory proteins. J Exp Med 168:181-194

MacLeod AM, Mason RJ, Stewart KN et al (1982) Association of Fc-receptor-blocking antibodies and human renal transplant survival. Transplantation 39:521-523

McIntyre JA, Faulk WP (1982) Allotypic trophoblast-lymphocyte cross-reactive (TLX) cell surface antigens. Human Immunol 4:27-35

Mueller-Eckhardt G, Garming W, Neppert J, Heine O (1989) Letters to the Editor. Maternal immune response and recurrent miscarriage. Lancet i:437

Padányi Á, Gyódi É, Horuzsko A, Réti M, Fülöp V, Petrányi GGy: (1991) Functional importance of TLX-B alloantigen system in reproductive immunity. In HLA 1991 (ed. Sasazuki et al.) Vol 12.Oxford University Press (in press)

Padányi Á, Gyódi É, Sármay G, Réti M, Mayer K, Kassai M, Petrányi GGy (1990) Functional and immunogenetic characterization of FcR-blocking antibody. Immunol Letters 26:131-138

Pócsik É, Mihalik R, Réti M, Pálóczi K, GGy Petrányi, Benczur M (1990) Activation of lymphocytes after platelet allotransfusion possessing only class I MHC product. Clin Exp Immunol 82:102-107

Purcell DFJ, Deacon NJ, Andrew SM, McKenzie JFC (1990) Human non-linaege antigen CD46 (HuLy-m5): purification and partial sequencing demonstrate structural homology with complement regulating glycoproteins. Immunogenetics 31:21

Purcell DFJ, McKenzie IFC, Lublin DM, Johnson PM, Atkinson JP, Oglesby TJ, Deacon NJ (1990) The human cell-surface glycoproteins HuLy-m5, membrane co-factor protein (MCP) of the complement system, and trophoblast leucocyte-common (TLX) antigen, are CD46. Immunology 70:155-161

Sandilands GP, Cocker JE, McMillan MA, Owsianka AM, Marsden H, Junor BJR, Briggs JD, MacSween RNM (1990) Isolation and characterization of a high molecular weight lymphocyte FcÞ-receptor blocking factor associated with renal allograft survival, Clin Exp Immunol 82:140-144

Sármay G, Iványi J, Gergely J (1980) The improvement of a preformed cytoplasmic receptor pool in the re-expression of Fc receptors following their interaction with various antibodies. Cellular Immunol 56:452-464

Takakuwa K, Kanazawa K, Takeuchi S (1986) Production of blocking antibodies by vaccination with husband's lymphocytes in unexplained recurrent aborters: The role in successful pregnancy. AJRIM 10:1-9

Torry DS, Thaler C, Faulk WP, McIntyre JA (1988) Seminal vesicles: A source of TLX in seminal plasma. Human Immunol 23:152

van Rood JJ, Claas FHJ (1990) The influence of allogeneic cells on the human T and B cell repertoire. Science 248:1388-1393

Xenotransplantation of Neovascularised Endocrine Tissue in Autoimmune Diabetes

T. E. Mandel and M. Koulmanda

Transplantation Unit, The Walter and Eliza Hall Institute of Medical Research Parkville, 3050 Victoria, Australia

INTRODUCTION

Organ transplantation is now routine therapy for the treatment for many cases of organ failure and over the past 3 decades has been responsible for saving many lives and improving the quality of many others. However, a critical shortage of donor organs and continuing problems with the long-term non-specific immunosuppression (First, 1992) has largely limited transplantation to either life-saving procedures, eg heart or liver transplantation, or to situations where organ replacement is clearly superior to alternative treatment, eg. kidney transplantation where the improved quality of life with a successful graft makes the risks of immunosuppression preferable to chronic dialysis. In addition, a shortage of organ donors also greatly limits the availability of grafts. The immune response to xenografts is still poorly understood, but if such grafts were able to be accepted by recipients and the immune response against them adequately and safely controlled, xenotransplantation could be a way of overcoming some of the problems of transplantation. In recent years there has been renewed interest in xenotransplantation and its potential as a source of tissues and organs is again being widely investigated

ENDOCRINE PANCREAS REPLACEMENT

Pancreas transplantation for the treatment of insulin-dependent diabetes mellitus (IDDM) is not life-saving and while clearly improving the quality of life of recipients (Voruganti and Sells, 1989; Corry and Zehr, 1990) is currently generally restricted to patients that also are in end-stage renal failure (ESRF) and need a renal allograft (Sutherland, 1991, Robertson, 1991). In most cases, such patients receive a concurrent kidney and pancreas graft from the same donor and rejection episodes in the kidney can be used to monitor possible rejection of the pancreas. Immunosuppression, required in any case for the kidney allograft, is also used to maintain the pancreas. Thus, the transplantation of another organ in addition to the kidney can be readily justified as it adds relatively little extra risk to the recipient.

However, patients with IDDM in ESRF also usually have other severe complications of diabetes, such as retinopathy and often severe macrovascular disease, and many are excluded by these from consideration for pancreas transplantation. Even in patients judged suitable for a pancreas graft, the advanced complications already present appear to be irreversible and, while some such as neuropathy may be improved (Kennedy et al 1990), retinopathy may, at best, be stabilised (Ramsey et al 1988). Pancreas transplants in patients without ESRF are performed in some instances but the success of these grafts is not as good as it is in combined kidney-pancreas transplantation for reasons that are still not well understood (Sutherland, 1991). Thus, pancreas transplantation in IDDM is currently palliative rather than curative. However, studies in experimental animals have indicated that islet transplantation can prevent the development of diabetic complications and this suggests that perhaps this may also be possible in humans.

There is now ample evidence that most if not all of the major organ and tissue complications of IDDM are the result of the disordered metabolic state, maily hyperglycemia, (Brownlee, 1985) and may be prevented if diabetes is adequately treated before structural changes develop in the various target organs. For example, typical diabetic changes often develop in the transplanted kidney in patients with IDDM treated with conventional parenteral insulin, (Mauer et al, 1989) and may, occasionally, result in ESRF. In contrast, kidney allografts in patients with a functioning pancreas transplant are generally protected from such changes (Bilous et al, 1989). However, it is virtually impossible to maintain constant euglycemic control with parenteral insulin and even "good" control is generally accompanied by periods of poor control during which glycosylation-induced tissue damage may occur, but a pancreas graft can maintain constant euglycemia (Robertson 1991).

The survival of a pancreas allograft requires continuing immunosuppression but currently this is too dangerous for the long-term use that would be required if endocrine pancreas replacement were to be used prophylacticaly soon after diagnosis in an attempt to cure diabetes and prevent the development of its complications. Unless an alternate form of immunosuppression is developed, ideally resulting in tolerance to the graft, early pancreas transplantation is unlikely to become an option, even if sufficient suitable tissue becomes available.

An alternative to vascularised pancreas grafts is the use of only the endocrine component of the gland, the islets of Langerhans, either as isolated islets from the adult pancreas or as fetal or immature pancreas in which the exocrine tissue is either undeveloped or can be eliminated by pretransplant organ culture. However, at present, the function of isolated adult islet grafts is far inferior to whole or even segmental pancreas grafts and even when islets allografts function, generally more than one donor pancreas is required for each recipient (Warnock et al, 1992). Thus, at present where there is already a major donor shortage, it seems unlikely that islet grafts from adult human donors will become a viable alternative to vascularised organ grafts for large numbers of patients. The use of fetal donors is also an unlikely alternative, at least in most Western countries, but fetal pancreas grafts are being widely used in China and in some other countries, with at least anecdotal reports of success (Federlin et al, 1992).

RECURRENT AUTOIMMUNE DISEASE

A separate potential and perhaps even actual problem is the recurrence of the original disease in the transplant. IDDM is an autoimmune T-cell mediated disease (Castano and Eisenbarth, 1990) and in both clinical and experimental models recurrent disease has been described. Recurrent disease developed rapidly in the graft in the absence of host immunosuppression in patients who received a pancreas from an identical twin donor, but recurrent disease was controlled with relatively low doses of immunosuppresion (Sibley and Sutherland, 1988). Recurrent disease was also described in the grafts from HLA-identical living-related donors, but not in grafts from cadaver donors (Sibley and Sutherland, 1988). It is possible that the lower amount of immunosuppresion in the former may be responsible for recurrent disease but this observation also suggests that recurrent disease may be HLA-restricted. In animal models of spontaneous autoimmune diabetes, principally in the non-obese diabetic (NOD) mouse and the BB rat, recurrent disease is also a problem when isografts are performed without immunosuppression, but there is controversy whether disease recurrence is MHC-restricted. In BB rats Prowse et al (1986) and Weringer and Like (1985) suggested that it is, whereas Hegre et al suggested that it is not. MHC-restricted recurrence has also been described in NOD mice by Terada et al (1988) but this was disputed by Wang et al (1991). It is not known whether autoimmune attack of xenogeneic islets will occur as the data so far are scarce. We have some evidence, however, that fetal pig pancreas transplanted into NOD mice may not be susceptible to disease recurrence, (Mandel et al, 1990) but there is evidence that cells from the peripheral blood of patients with recently diagnosed IDDM may respond in vitro against homogenates of fetal pig islets suggesting that these cells are sensitised and able to react specifically with antigens on the xenogeneic islets (Harrison et al, 1992), thus implying that disease recurrence may be a problem with pig islets.

XENOGRAFTS

A potential solution to donor shortage is to use xenografts. After a long period of relative neglect, xenografts are again being actively investigated (Auchincloss, 1988). Since the only major function required of a pancreas graft is the appropriate secretion of insulin, any species that produces insulin of the appropriate type and amount in response to physiological stimuli may be suitable as a graft donor. However, xenografts, particularly if used as vascularised grafts, are likely to be rejected hyperacutely due to the action of preformed natural antibody (NAb) on the endothelial cells of the transplanted organ. NAbs activate complement resulting in endothelial cell activation producing loss of anticoagulant activity, intravascular thrombosis and graft infarction (Platt et al 1990a, b, c).

Hyperacute rejection is a particular problem when "discordant" donor-recipient combinations are used but with "concordant" combinations this may not be a major concern. "Concordat" recipients, defined as those in which NAbs are not present, are generally phylogeneticaly closely related to the donor. In clinical transplantation concordant donors would probably have to be primates and possibly even higher apes; a source that is most unlikely to be available. In contrast, in "discordant" transplantation, donors are generally phylogeneticaly distantly related to the recipient and are potentially readily available. In particular, domestic animals already being raised for human consumption as food species would be readily available. Of the potentially useful species, pigs are the most likely as they produce insulin that is almost identical to human insulin and has been widely and successfully used for the past 70 years. In addition, the biochemistry of pigs is similar to humans and insulin secretion by a pig islet graft may be appropriately regulated in humans. However, rejection of xenografts will need to be adequately controlled without the need for crippling immunosuppression.

Free fetal tissue grafts that have been immunomodified prior to transplantation have been successfully used in rodents without the need for any immunosuppression, even across major histocompatibility barriers, (Collier and Mandel, 1983; Simeonovic and Lafferty, 1982). This has not, however, been achieved in large animals or in NOD mice, although there are reports of successful allografts of immunomodified neonatal islets in BB rats (Hegre et al 1990). The rationale for immunomodification is that the removal from the graft of antigen presenting cells (APC) that can present immunogenic peptides directly to the recipient renders the graft less immunogenic as this removes the source of *direct* antigen presentation by donor APCs. In the absence of graft APCs, antigen presentation is only by the host's APC's, ie it is *indirect*, and this may be less effective and perhaps easier to control with immunosuppression. Theoretically, however, xenogeneic graft APCs should not be able to present antigens directly and their depletion should have no effect on graft survival.

In addition, there are a number of other factors that can reduce the strength of an interaction between a donor's stimulator cells and recipient's responders. These include an inadequate interaction between the TcR and the MHC molecules on the xenogeneic cells, inappropriate interaction between CD4 or CD8 molecules on responder cells with MHC class I and II molecules respectively on donor target cells, and suboptimal effects of xenogeneic cytokines, as well as inadequate interaction between accessory molecules and their ligands, such as ICAM-1 and LFA-1. These may all diminish or even ablate the response between xenogeneic donor and recipient (Moses et al, 1992). All of these factors may play a role in determining the nature and avidity of a xenogeneic antigraft response over and above that produced by the presence of antibodies. Thus, if endocrine pancreas replacement is to have a future as a treatment (and potential cure) of IDDM, a number of major problems will need to be solved. These include the identification of an appropriate donor, control of rejection, prevention of recurrent disease in the graft, and avoidance of crippling immuno-suppression, particularly if islet replacement is to be a long-term procedure.

ISLET XENOGRAFTS IN NOD MICE; THE USE OF ORGAN-CULTURED FETAL PIG PANCREAS

In an attempt to address some of these issues we have been testing islet xenotransplantation in an animal model of IDDM. As recipients we have used NOD mice since, in contrast to the lymphopenic BB rats - the other major rodent model of IDDM - NOD mice are immunologically relatively normal (Bernard et al, 1992). As donors we use fetal pig pancreas since this is readily available and, as stated above, the pig may be a physiologically appropriate donor species. We are currently testing various forms of immunosuppression as well as altering the structure of the grafted tissue by pretransplant immunomodification. Our recent data suggest that xenografts of fetal pig pancreas can be successfully grafted in NOD mice and that these grafts also may be resistant to recurrent disease.

FETAL PANCREAS CULTURE

Organ culture of fetal pancreas is efficient way of enriching it for its endocrine component (Mandel et al, 1982) since in culture the endocrine tissue survives preferentially and the developing exocrine tissue dies. This has been noted in a number of species and is also true for fetal pig pancreas (Thompson and Mandel, 1990). When the pancreas fragments ($\sim$1mm^3) are placed in a gas-medium interface in 10%CO_2-90O_2% air, the peripheral rim of tissue survives while the central region of the fragment dies of ischemic necrosis. In contrast, when the fragments are placed in 90%O_2-10%CO_2, there is little or no central necrosis, but a preferential loss occurs of interstitial cells that presumably include "passenger leucocytes" and other cells that may act as APCs. Thus, specifically adapted organ culture methods may be useful in preserving the maximum amount of endocrine tissue and retaining its capacity for proliferation and differentiation, as well as potentially altering its immunogenicity. Whether this is of any consequence in xenotransplantation is debatable but our data suggest that it may be.

TRANSPLANTATION

Fetal pancreas fragments are placed beneath the renal capsule of recipient NOD mice. This site is useful because it is technically simple to transplant multiple grafts to each kidney, and because it is well vascularised it provides an ideal site for free tissue grafts that need to survive until they becomes vascularised by ingrowth of vessels from the recipient's vascular bed. Whether the subcapsular site is also relatively immuno-privileged is contentious with some reports suggesting that it is while others suggest that it is not. In contrast, islets embolised into the liver are often well preserved and the liver does seem to have at least a degree of immunoprivilage. However, intraportal grafts often fail even as auto or isografts, perhaps because of their perfusion with portal blood that is at least intermittently hyperglycemic.

IMMUNOSUPPRESSION

Rejection of xenografts is generally regarded as being dependent on CD4$^+$ T cells (Auchincloss, 1988). In contrast, MHC-mismatched allografts are rejected by both CD4$^+$ and CD8$^+$ T cells, and rejection can occur even in the absence of CD4$^+$ cells. There are indeed reports that xenografts can survive, and that even tolerance can be achieved, in mice that have been immunosuppressed with large doses of anti-CD4 MAbs (Simeonovic et al, 1990). However, this has not been achieved in NOD mice with the same treatment.

We have tested anti-CD4 treatment in NOD mice and compared organ cultured fetal pig pancreas graft survival with that of organ cultured fetal CBA mouse allografts and NOD isografts. In NOD/Wehi male mice that develop insulitis but have a low prevalence of spontaneous diabetes, there was preferential xenograft survival 28 days after transplantation following peritransplant (p/t) anti-CD4 MAb treatment when compared to the survival of MHC-mismatched allografts (Mandel et al, 1989; 1990). In these mice the isografts became infiltrated and allografts were rejected in the presence of anti-CD4 immunosuppression while the xenografts were totally free of infiltration. However, when immunosuppression was discontinued, the xenografts were also eventually rejected. Excellent prolonged (>28d) xenograft survival was also seen in CBA males and in BALB/c females after p/t anti-CD4 treatment (Mandel and Koulmanda, 1992). Interestingly, CBA mice were used for tolerance induction with anti-CD4 MAb by Simeonovic (1990), and BALB/c mice had long-term human islet xenografts after treatment of the recipient with anti-HLA $F(ab)_2$ MAb (Faustman and Coe, 1991).

In contrast, fetal pig grafts in female NOD/Lt mice that have a high incidence of spontaneous diabetes, failed to survive with p/t anti-CD4 immunosuppression. Continued anti-CD4 treatment prolonged survival but the grafts showed slow rejection with a gradual loss of islet cells, increasing graft site fibrosis and a moderate infiltration by mononuclear cells (Mandel and Koulmanda, 1991). Thus, despite the apparent dependence of xenograft rejection on the presence of CD4+ T cells, in some circumstances there seems to be an escape mechanism that allows the grafts to be destroyed even when the number of CD4+ T cells is low. Whether the rejection is mediated by the relatively few CD4+ T cells still present, or is CD4-independent, is unclear.

Since rejection was present despite anti-CD4 MAb treatment we decided to test the effect of anti-CD3 treatment. This type of immunosuppression is widely used clinically with good effect both for control of rejection episodes and for induction therapy, but can be hazardous and severe reactions are common (Chatenoud and Bach, 1990). There have been a few reports of anti-CD3 treatment with a hamster anti-CD3 MAb, 145-2C11 (Leo et al, 1987) in murine allotransplantation but some suggest that it is effective and, as in humans, side effects are common and in mice may be lethal (Hirsch et al 1988, Mackie et al, 1990).

We have used a rat IgG2a anti-CD3 MAb; KT3 (Tomonari 1988). In contrast to 145-2C11, KT3 is safe and no mice died as a result of treatment. When given ip as p/t treatment there was transient graft survival but when used p/t with additional weekly doses, 2 of 8 grafts were intact and uninfiltrated for over 12 weeks, when the experiment was terminated. However, when grafts were cultured for 3 days in $90\%O_2$, 7 of 9 were still present intact and not infiltrated at this time. When peripheral blood cells were assessed by flow cytometry for lymphoid subsets, there was a marked depletion of CD3 expression but, in addition, also a loss of CD8+ T cells and a lesser loss of CD4+ cells, so that the CD4:CD8 ratio was >10 (Mandel and Koulmanda, 1993). This finding is similar to that reported by Hirsch et al (1988). Interestingly, none of the MAb-treated mice developed diabetes although the pancreas still showed islet infiltration, suggesting that autoimmune disease could be controlled. Furthermore, the absence of infiltration of the grafts suggests that they are resistant to recurrent disease.

In a second experiment where we compared p/t plus weekly anti-CD4 treatment with similar anti-TcR (CD3) therapy, and used HiO_2 cultured grafts, we again saw excellent graft survival in the KT3-treated groups; 16 of 16 grafts were present and of these 14 were intact and not infiltrated, whereas of the 14 grafts present (of 16) in the GK1.5-treated mice, all were infiltrated and only 2 were still largely intact. In these mice too, the KT3-treated animals had marked depletion of T cells with an almost total loss of the CD8+ subset, and the remaining cells were Thy1+/CD3-.

CONCLUSIONS

Thus, the data suggest that islet xenotransplantation in the presence of an underlying autoimmune disease may be feasible if sufficient immunosuppression is used. In NOD/Lt mice, anti-CD4 treatment that seems adequate in some other strains is clearly not sufficient to maintain the grafts, but anti-CD3 treatment does appear to be effective. The continued effectiveness of this reagent also suggests that antibodies against it are not being formed, but this has not been formally tested in these studies so far. Experiments are in progress to determine whether treatment of limited duration will allow a state of long-term unresponsiveness to develop as has been demonstrated with unprimed human T cells exposed in vitro to anti-CD3 MAbs (Anasetti et al, 1990).

REFERENCES

Anasetti C, Tan P, Hansen JA, Martin PJ (1990) J Ex Med 172: 1691-1700
Auchincloss HJr (1988) Transplantation 46: 1-20
Bernard CCA, Mandel TE, Mackay IR (1992) in "Autoimmune Diseases" 2nd ed. Rose NR and Mackay IR (eds), Academic Press NY.
Bilous RW, Mauer SM, Sutherland DER et al (1989) N Engl J Med 321: 80-85
Brownlee M. "Joslins' Diabetes Mellitus", 12 ed (eds Marble A et al) Lea & Febiger. p.380, 1985
Castano L, Eisenbarth GS. (1990) Ann Rev Immunol 8: 647-679
Chatenoud L, Bach J-F (1990) Seminars Immunol 2: 437-447
Collier SA, Mandel TE (1983) Transplantation 36: 233-237
Corry RJ, Zehr P (1990) Clin Transplantation 4: 238-41.
Faustman and Coe, 1991 Science 252: 1700-1702

Federlin KF, Bretzel RG, Hering BJ (1992) in "Pancreatic islet cell transplantation" ed. Ricordi C. 462-472, Landes Co Austin USA

First MR (1992) Transplantation 53: 1-11

Harrison LC, Chu SX, DeAizpurua HJ et al (1992) J Clin Invest 89: 1161-1165

Hegre OD, Serie JR, Weinhaus AJ et al (1990) Horm Metab Res (suppl) 25, 108-116

Hirsch R, Eckhaus M, Auchincloss HJr et al (1988) J Immunol 140: 3766-3772

Kennedy WR, Navarro X, Goetz FC, et al. (1990) N. Engl. J. Med. 322: 1031-1037

Leo O, Foo M, Sachs DH et al (1987) Proc Natl Acad Sci USA 84: 1374-1378

Mackie JD, Pankewycz OG, Bastos MG et al Transplantation 49: 1150-1154

Mandel TE, Hoffman L, Collier SA et al (1982) Diabetes 31 (suppl 4) 39-47

Mandel TE, Koulmanda M, Loudovaris T et al (1989) Transplant Proc 21: 3813-3814

Mandel TE, Koulmanda M, Bacelj A (1990) Horm Metab Res suppl 25: 166-173

Mandel TE, Koulmanda M (1991) Transplant Proc 23: 583-584

Mandel TE, Koulmanda M (1992) Diabetes Metab Nutrition (in press)

Mandel TE, Koulmanda M (1993) Transplant Proc (in press)

Mauer SM, Goetz FG, McHugh LE et al (1989) Diabetes 38: 516-523

Moses RD, Winn HJ, Auchincloss HJr. (1992) Transplantation 53: 203-209

Platt JL, Lindman BJ, Chen H, et al. (1990a) Transplantation 50: 817-822

Platt JL, Vercellotti GM, Lindman B, et al.(1990b) J Exp Med 171: 1363-1368

Platt JL, Vercellotti GM, Dalmasso AP, et al. (1990c) Immunol Today 1 11: 450-456

Prowse SJ, Bellgrau D, Lafferty KJ. (1986) Diabetes 35: 110-114

Ramsay RC, Goetz FC, Sutherland DER, et al (1988) N Engl J Med. 318: 208-214

Robertson RP (1992). Diabetes 40: 1085-1089

Sibley RK, Sutherland DER (1988) Amer J Pathol 128: 151-170

Simeonovic CJ, Lafferty KJ. (1982) Aust J Exp Biol Med Sci 60: 391-395

Simeonovic CJ, Ceredig R, Wilson JD (1990) Transplantation 49: 849-856

Sutherland DER (1991) Diabetologia 34 (Suppl 1); 28-39

Terada M, Salzer M, Lennartz K, Mullen Y. (1988) Transplantation 45: 622-627

Thompson SC, Mandel TE (1990) Transplantation 49: 571-581

Tomonari K (1988) Immunogenetics 28: 455-458

Voruganti LNP, Sells RA (1989) Clin Transplantation 3: 78-82.

Wang Y, Pontesilli O, Gill RG et al (1991) Proc Natl Acad Sci USA 88: 527-531

Warnock, GL, Kneteman NM, Ryan EA et al. (1992) Diabetologia 35: 89-95

Weringer EJ, Like AA (1985) J Immunol 134: 2383-2386

16. Reproductive Immunology

Reproductive Immunology 1989-1992: Some Important Recent Advances about Feto Maternal Relationship

Gerard Chaouat, Elisabeth Menu, François David, Valentine Djian, Radslav Kinsky.

Biologie cellulaire et moléculaire de la relation materno fetale
Service de Gynécologie et d' Obstétrique. Hôpital Antoine Béclère. Clamart. 92140
France

This lecture will, as intended by the Congress organisers, review some important recent advances in the field in the last 3 years. I will deal with these in the first half of my talk, and, then, indulge myself to present some data of our laboratory. The symposium was designed as to have J. Szekeres Bartho deal with an immuno endocrine pathway, D. Clark with decidua with due emphasis on TGF beta 2, T. Wegmann with cytokine networks at the feto maternal interface, and GP Talwar report on contraceptive vaccines. I hope that this format will enable the audience to get out of this Congress with as complete a view of Reproductive Immunology as possible, and apologise for the non inclusion of a gamete Immunology report.

Without wanting to impede John Mac Intyre (the first speaker), I would mention, saluting our hosts, work from Petranyi et al. Originally dealing with kidney allograft enhancement by pretransfusions, they obtained recent data suggestive that "blocking antibodies" found in the sera of pregnant women endowed with the so said "anti Fc gamma RII blocking activity", suppress a MLR and are directed against trophoblast antigens, being seemingly absent in the serum of recurrent aborters (1).

Now, let us review other important new topics. The field has, in 3 years, seen important accomplishments. I would like to mention 6 that I believe are headlines in our discipline.

(A) Placental interferons (Martal, Imakawa, Bazer). (

(B) IL-6, TNF and production of HCG (Tanizawa, Kishimoto, Saji).

(C) A cloned T cell suppressor factor (Progesterone dependent) and fetal survival (Beaman and Hoversland).

(D) HLA-G, the only trophoblast MHC antigen (Ellis, Kovatts, De Mars).

(E) MCP and DAF on trophoblast membrane (Mac Kenzie, Johnson, Hsi, Fenichel) (F) IL-1 effects on Central Nervous system and abortion (Croy).

A)PLACENTAL INTERFERONS.

In sheep, there is no such hormone as HCG. Corpus luteum maintenance is mediated by Trophoblastin/oTP1. The substance is secreted constitutionally by the trophoblast, in large amounts, in early ovine gestation. It has been purified, cloned by French and then US groups and expressed in Coli or Yeast.

All its 5 isoforms have antiviral activity, and its AA and DNA genomic sequences show that the material is analogy with interferons alpha (2,3,4). It is more conserved accross species than within the INF-alpha , defining a new INF family . We and Imakawa et al have shown that it is cytostatic for ovine T cells, and we have shown using Dr Mackay 's MoAbs (Basel Institute for Immunology) that it acts on both CD4 and CD8 ovine T cells (5).

Thus, it could prevent in early pregnancy development of CTLs, as well as antipaternal antigens DTH reactions, which might otherwise prevent successful implantation. In synergy with PAF aceter (anti PAF aceter prevent implantation, PAF being secreted by uterine epithelial cells) it would early embryos. Considerable interest arose from the wide specy distribution of placental INFs .

The search is going on in human , Imakawa reporting recently (FASEB) a 800 Bp sequence possibly coding for human oTP like .It is also, evidently, one of the most saliently convincing immuno endocrine network so far demonstrated.

B) IL-6 and HCG.

Human Placenta secretes High level of IL-6 (6) which regulates production of HCG, another example of immuno endocrine circuitry (7).Further, IL-6 itself is regulated by TNF (7), secreted in decidua and by uterine Granular Metrial Gland cells, regulating placental growth.

C) Progesterone mediated immunosuppression and cloned T cell suppressor factor .

J. Szekeres Bartho will tell you about her own progesterone dependent pathway. I would like to mention the important work of Beaman and Hoversland. They first raised a MoAb against a T cell suppressor inducer factor (cross reacting, interestingly, with lipomodulin). Thos MoAb is abortifacient (8) without affecting embryo development in mice (9), a VERY important control NOT carried , amongst others needed, for studies done with polyclonal antisera (themselves of unchecked specificity) .

This factor has been cloned and christened J6B7. (10).The cloned material is immunosuppresive . Its mRNAs are induced in T cells only if activated , and progesterone treated.

Such elegant , well controlled studies are worth signalling, for J6B7 analogs could be of interest in Veterinary Medicine, Clinics, and because it is a 3rd, cleanly demonstrated, this time at molecular level, example of immuno-endocrine interactions . The abortive effects of anti J6B7 in pregnancy show a key role in a physiological situation of (balanced , see underneath) suppression (by T cells) .

D) HLA-G, the ONLY human trophoblast MHC antigen .

I will not detail the variations of antigenic status between various placentations, but focus on HUMAN placenta. All cells of outer layers of trophoblast lack polymorphic MHC antigens, the due to abnormal DNA hypermethylation, and possibly specific repressors(11,12). But the so called extravillous cytotrophoblast columns are labelled by W6/32 MoAb, initially postulated as expression of HLA A,B,C .

In fact, immunoprecipitations show no variations of material between individuals, and the band had lower MW than "conventional" HLA. A truncated MHC molecule has been cloned , without variation between individuals. It has been christened HLA-G (13,14). Its expression is restricted to placenta, and there are analog candidates in other species . Such a restriction in expression, and conservation in evolution, suggest a key role in fetal protection.

Indeed, elegant transfection experiments by Kovatts et al in non HLA-G expressing cells show that these are then much less sensitive to cell mediated lysis (15).

The mechanism of this effect remains unclear,.Nothing is yet known about which set of peptides would eventually HLA-G select/present, but a truncated, monomorphic HLA could "defuse or confuse " TcRs. Nevertheless, it is clear that abscence of conventional Antigen Presenting molecules on the first layer of the interface, and the presence of a monomorphic element play as important a role as decidua, placenta and immuno/endocrine pathways in assuring the success of the "fetal allograft" . Rodent placentae do express on spongiotrophoblasts low amounts of polymorphic MHCs, but they also express monomorphic like determinants.

Yet, HLA-G or analogs offer no absolute protection since both murine and human trophoblasts are sensitive to activated NKs and LAKCs mediated lysis, such cells being involved in murine models of natural abortion .

E) MCP and DAF on the placenta.

20% (1st pregnancy) to 75-80% (multiparous) women develop anti paternal IgG anti MHC alloantibodies, but, in human those are NOT restricted to the IgG1 subclass as in mice e.g. they are cytotoxic. Since an intact complement pathway is present in the uterus at early pregnancy stage, and since alloantibodies do cross the placental syncytiotrophoblasts , and gain easily access to MHC + cells the (syncytio)trophoblast neutral barrier, and since some of the sera from pregnant women do cross react with HLA-G, trophoblast could theoretically be vulnerable to Ab + C' attack.

But as a totally unexpected spin off of the search for the so called TLX antigens, presence of MCP (Membrane Cofactor Protein) and DAF (Decay Accelerating Factor) was unambiguously demonstrated on trophoblast surface (16,17,18,19). This offers another protection for the fetus,.It would also turn ANY placenta bound alloAb into a so-called "blocking" (enhancing) one.

F) IL1 and abortion.

In a sery of recent studies, Croy and colleagues observed that intracerebral injection of IL-1 causes pregnancy failure in mice, while affecting severely peripheral cellular immune responses (20). This Il-1 induced "excess of suppression" is of in my opinion of interest, because it has implications for imbalance between the immunotrophic and suppressive pathways , as well as for neuro endocrine immunocontroled pathways, and possibly parturition where IL-1 levels are enhanced dramatically.

It also cast light on stress induced hypo or unfertiltity, as well as abortions, as possibly immune mediated events.

Having dealt as far and as fairly, I hope, as I could with those topics, I did not review HILDA/LIF, and other important points close to Reproductive Immunology but dealt with somewhere else in this Congress as it should. The controversy (21) about the existence and treatment of human (immune ?) recurrent abortions of putative will be dealt with in workshop. I apologise to anyone mistreated / nonquoted. I would like to adress now some of my own work:

A) MHC,NKS,LAKCS and fetal survival.

B) Final identification of placental suppressor factor.

C) Materno fetal transmission of HIV.

A)MHC, NKS, LAKCS AND FETAL SURVIVAL.

We have designed these last years (in mice !) models of artificially induced resorbtions abortions, such as Ds RNA treated mice, high or low doses of LPS or Recombinant TNF, and ultrasound stress . A common feature of these models is that abortion/resorbtion can be prevented by alloimmunisation, a feature shared with "natural " abortion e.g. CBA x DBA/2 , B10 x B10.A systems (where we already tested lymphokine therapy (22). I will let Tom Wegmann and David Clark review the involvement of cytokines and LAKcs in such systems.

We have used Ds RNAs systems to demonstrate that abortion in Ds RNA system, as well as in spontaneous ones,can be prevented by immunisation with appropriatly H-2 K transfected L cells (23). The same is true for low doses of LPS induced abortions, as well as for stress induced ones.

We have then been able to use an H-2 K^b dominant peptide ,and used it in alloimmunising C3H mice (22). This prevented Ds RNA induced abortions. Control peptides such as HEL N terminal peptides, or influenza peptides , in IFA do not induce such effect, be it in C3H or C57BL/6, (in the latter they are well known indeed to Ts to the HEL molecule).

In close cooperation with Philippe Kourilsky, Christian Jaulin, Jean Pierre Abastado, we have examined whether mutations on the border of the pocket of the MHC AG could affect recognition by T cells ensuring optimal fetal survival. Figure 1 shows effect of immunising CBA/J with such cells, .The less there is a protective effect, the more those H-2^d mutants are seen as "H-2^k" ,e.g self, by CBA/J. Thus Figure 1 shows that indeed some of these K^d mutants are seen as "less allo" than native H-2^d by CBA/J. and that that rsults in a loss of fetal protection.

We have also raised C3H x BALB/c F1s , and immunised them with K^d transfected L cells, or K^d mutants transfected L cells, before injection of Poly I C12u and matings. As a control, we immunised against DBA/2 (Minor loci difference only, since they are H-2^d), or the allogenic C57BL/6, and pseudo immunised with BALB/c splenocytes . The results of Figure 2. show that some K^d mutations on the helixes are recognised as foreign to H-2^d ,and afford fetal protection.

Since maternal reaction against those antigens would elicit protective reaction against activated NKs cells,LAKCs and their products, these data show that one of the selective pressure for such mutations in evolution (in mammals !) could be the Reproductive advantage they offer. They explain partly the maintenance of such mutations.

B) FINAL CHARACTERISATION AND MODE OF ACTION OF HUMAN PLACENTAL SUPPRESSOR FACTOR.

_Human placenta secretes ,as do choriocarcinomas, suppressor molecules active in vivo and in vitro. The crude material blocks a variety of in vitro immunologic tests, and especially Il-2 and Il-4 driven CTLL-2 proliferation, CTLs and NKs at the effector stage (24). In cooperation with Transgene SA, and with R Raghupathy (National Institute of Immunology,India), we have demonstrated that active molecule is a very low MW material, as far as anti-proliferative assays are concerned (<1 Kd). Purified material is active in vitro in

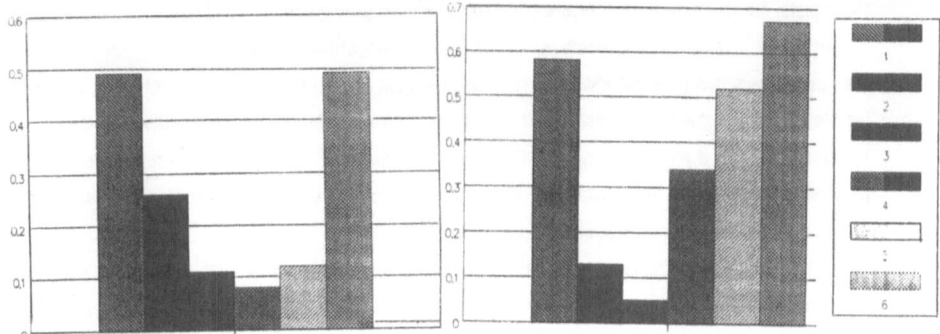

% resorbtions day 13.5 pregnancy

Left (1a) CBA x DBA/2, natural resorbtion model.

Right, (1b) C3HOH x BALB/c F1s. Ds RNA (Poly I Poly C12 U , Day 0) induced abortions.

Immunisations Day - 7 1)Nil (2) Mutant 163 (3) Mutant 97 (4) Mutant 154 (5) BALB/c splenocytes.

(6) nontransfected L cells.

Figure 1a.1b : Effects of immunising with K⁴ mutant transfected L cells (1a) CBA/J x DBA/2 matings .Mutants 129,69, 70 are compared for protection with K⁴ (BALB/c splenocytes) versus control immunisation with untransfected L cells. (1b) Effects of such mutants (129,69,70,80) in C3H x BALB/c F1s pregnant of BALB/c after Poly I Poly C12u treatment as described in (23).

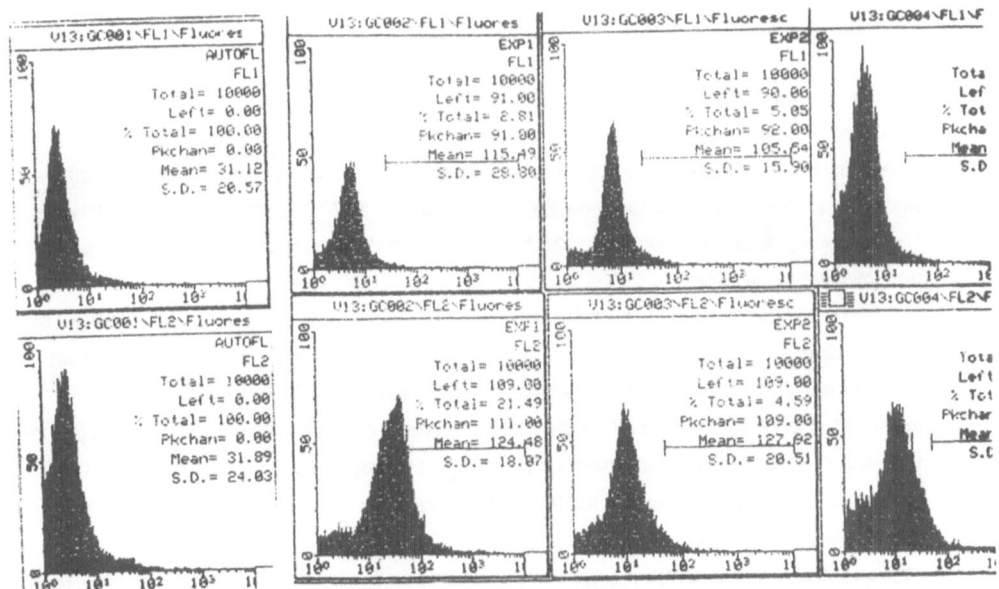

Figure 2 : FACS profile of H-2 ᵇˣᵈ (B6x D2) F1s splenic cells lethally irradiated and repopulated with H-2ᵇ cells immeditely after lethal irradiaton .B6 cells were injected with purified placental factor .Spleens were harvested 90 days after such injection. One can see that the cells that repopulate the mice bear only H-2ᵇ markers .Top line . anti H-2ᵈ.Bottom.Anti H-2ᵇ. Right FACS controls. Left 1,2,3 : profiles of experimental mice.

those tests, but the material that blocks CTL activity at effector stage, (HTC1 and HTC2 anti Thyroglobulin CTL T hybrids ,via B Texier and J Charreire, U 283 INSERM) sems of higher M.W.

We have shown that the small MW material is very resistant to a variety of chemical treatments, rendering incidentally its absorbtion per os feasible. It then corrects/prevents Natural/Induced abortion. Injected systematically (27), it prevents general GVH reaction in lethally irradiated F1s recipients repopulated with parental splenocytes, and we checked that recipient immunocompetence is due to donor cells , not host cells that would have escaped radiation effects , by FACS (Fig 2).

If cells are treated with the material plus PHA , they become unresponsive or hyporesponsive to a subsequent PHA challenge. This effect is not due to general cytotoxicity of the material, since the incubation with material alone does not induce such effects (e.g. the cells stay responsive to PHA) . The same is seen in an MLR. Thus, it acts only if the cells have been previously triggered at Tcr level. We are now investigating whether this is due to clonal deletion or anergy, both by in vivo and in vitro assays .

In the same vein, the material acts on Jurkatt cells or K562. Antiproliferative effects are reversible, providing the cells are washed from excess material before 16-24 hours. If not, the effect becomes unreversible , and cells die very quiclky. These data suggest, but do not prove, deletion by cell death (with features apoptosis like in those cells) in T cell blasts rather than anergy. The material does not affect cAMP levels, nor does it seems to act on Ca fluxes.

The previously ascribment of suppressor activities to higher MW molecules, already challenged by ourseleve using acidic Ph conditions had to be reexamined in the context of what we know now,e.g. that the purified material is a very small MW , hydrophobic molecule that will stick to carrier proteins of various MW according to the "purification" procedures. It is fair to say that, under those conditions, the material could appear to "reside" to one or the other carrier , according to the "purification" .

The christening of suppressor factors in pregnancy as "elusive" was justified in that respect, and it is fair to assume that many "purifications" were in fact dealing with such artefacts, due to recombination of the active molecule to a caarier . It is decent to say that MY own "ASF, CTL-IF, NK-IF" were likely in fact such artefacts . The low MW material acts on many cells , not only lymphocytes, probably more general inhibitor of cell growth than sole immunosuppressant.

We have also studied in cooperation with Miljenco Kapovic of Rijeka, Croatia, whose work is now hampered by the tragic situation reigning in his country, and Olaf Heine , Germany, the induction of suppressor cells by the supernatants of JAR,JEG and BeWO choricarciomas, which was also shown with Raj Raghupathy to be operational in vivo in preventing local GVH and inducing suppressor cells (26). Both CD4 and CD8 cell appear to be "induced", though "more" activity resides in CD8 cells, and the target is both CD4 proliferation and CD8 proliferation/ differentiation. Ts act via a non MHC restricted cytostatic factor (INFs ,IL-10 or TNFs are owing to their MW the most likely candidates).

The spin off of such suppressor and suppressor inducer materials for Clinics, be it organ tranplantation , bone marrow engraftment, and the amelioration of fecundity and Livestock in cattle are obvious,as is probably their place in optimising Embryo Transfer in man and cattle, and cloning of embryos in cattle.

Finally, it is worth mentioning that we have found similar low MW suppresisve materials in the S/N of several malignant lines.

The regulation of the production of such factors by cytokines and hormones is now feasible a project, as would be the design of neutralising antibodies.The reason why the data with the previously raised ones were quite inconsistant is that they were in fact adressing to pseudo epitopes on the carrier molecule(s), varying with batches of FCS etc....

C)MATERNO FETAL TRANSMISSION OF AIDS.

About 20% (in North -Western Europe) of infants born from HIV +mothers are effectively will undergo perinatal AIDS. The mechanisms leading to such a transmission during pregnancy were our main topic, since evidence of post partum transmission per milk, possibly because of colostral macrophages. is small. Three hypothesis were invoked, non mutually exclusive.

(1) Microchimerism e.g. direct passage of maternal leukocytes in the maternal circulation. Tackling this problem requires not yet available specific PCR technology (2) Uptake via placental RFcs of HIv anti HIV Abs immune complexes, (3) CD4 expression on the placenta.

We have obtained using Klinman culture technique, choriocarcinoma cell line and our own trophoblast lines, as well as by studies on cryostat tissue sections,evidence for expression of CD4 on the syncytitiotrophoblast (27), both by immunohistochemical and Molecular Biology technology , in close cooperation with F Barre Senoussi, Brigitte Autran, Catherine Vacquero, and several groups of clinicians.

That this CD4 is functional could be proved by direct infection of JAR cell line and purified trophoblast, and its blocking by appropriate anti CD4 monoclonal antibodies as well as SCD4. The FcRs were mapped on trophoblast (27).

Syncytio expresses a Poly IG receptor, whereas RFC gamma isoforms could be detected on the cytotrophoblast. Their precise molecular characteristics are now apporoached by the relevant Molecular Biology techniques. The role of such RFcs could be evidenced by the tremendous increase of JAR infection / HIV production by enhancing antibodies (28) whereas neutralising antibodies were ....neutralising.

Unfortunately, as for macrophages (indeed !), almost _every_ cytokines involved, safe TGFbeta, positively or negatively, in the materno fetal relationship , enhance HIV infection of trophoblasts. Only 20% in Europe of infants are infected suggest step by step spreading from trophoblast to fetal blood, and search is ongoing for placental reservoir cells as well as mechanisms regulating HIV production/release in the placenta. The replication rate of HIV inside trophoblasts is much lower than in lymphocytes.

It is important to note that we have delineated now in the S/N of trophoblasts a low MW material (distinct from our antireplicative material) active on HIV production by infected lymphocytes.It does not affect NF kappa B . Such materials act in vitro and possibly in vivo on various retroviral processes.

The confirmation of the presence, and final isolation of such materials, as well as modulation of trophoblast receptivity by cytokines, could help explain why only 20% of children are HIV +, whereas one could also seek there an explanation between the apparent differences between European (20% or lower) and African (about 40%) vertical transmission rates. It is hoped that from Reproductive Immunology might spread a non toxic , natural, inhibitor of some retroviral infections, explaining the placental barrier to some viruses, and putatively helping partly to solve AIDS and other retroviruses problems.

Future directions include now mastering the early steps of pregnancy, including understanding early lymphocyte redistribution and lymphokine patterns and cellular patterns (29,30,31), as well as regulation of IgG production and isotype regulation by placental factors. These goals, as well as contraceptive vaccines, could perhaps be achieved for Delhi 9th IUIS Congress. I guess the following speakers will give you now an idea of how we (the field) have advanced in 3 years as far as cytokine networks, contraceptive vaccines, local suppression and role of hormones in pregnancy has progressed in of the field, showing it should regain a popular status amongst immunologists in general.

REFERENCES.

1) Padanayi A,. Horuzko A., Gyodi E., Sarmay G., Reti M., Kassai M., Fulop V. and Petranyi G. 1992. Tiss.Antig. (in press).
2) Charpigny ,Reinaud,Huet ,Guillomot ,Charlier ,Pernollet and Martal 1988.FEBS letters.228.1.12.
3) Charlier ,Hue, Martal and Gayle.1989.Gene.77. 341-348.
4) Imakawa, Anthony, Kazemi, Marotti ,Polites, Roberts, 1987.Nature.330. 377.
5) Martal J., Charpigny C., Fillion C., Asall Meliani A. and Chaouat G.1991.Biol. cell. mol. relation materno fetale. Editions INSERM John Libbey.Paris. Pages 317-325.
6) Nishino E., Matsuzaki N., Masuhiro K., Kameda T., Taniguchi T., Takagil T., Saji F. and Tanizawa O.1990. . J. Clin End. Met.72(2) 436-440.
7) Neki R., Matsuzaki N., Masuhiro K., Taniguchi T., Shimoya K., Jo T., Li Y., Takagi T., Saji F. and Tanizawa T. 1991. 5th Jap. Soc. Reprod.Immunol meeting 1990. K Honjo and S Kasahura.JSBR and JSMR. Ed. Mure Printing. P.108-114.
8) Beaman K.D., and Hoversland R.C. 1988. J. Reprod. Fert. 1998.82. 135-139.
9) Hoversland R.C. and Beaman K.D.1991. Am. J. Reprod. Immunol. 26:84-88.
10) Lee C.K., Ghosshal K., Beaman K.D. 1990.Mol. Immunol. 1990.27.1137-1134.
11) Le Bouteiller et al. 1991 in Biol. cell. mol. relation materno fetale. Editions INSERM John Libbey.Paris. Pages 208-222.
12) Boucraut J. Phd Thesis. Uniiversity Marseille Luminy France. June 1992.
13) Kovatts S., Main E.K.,Libbrach C., Stublebline M., Fischer S.J., De Mars R.1990.Science.248 :220-223.
14) Ellis S. 19990. AM. Reprod. Immunol. 23:84-86.
15) Kovatts S., Librach C., Fisch P., Main E.K., Sondel P.M., Fisher S.J. and De Mars R. 1991. Biol. cell. mol. relation materno fetale. Editions INSERM John Libbey.Paris. pages 41-51.
16) Purcell D.F. J., Brown M.A., Russel S.M., Clark C.J., Mc Kenzie I.F.C. and Deacon N.J. 1989.J. Rep. Immunol. S. 207.
17) Roberts J.M.,Taylor C.T., Melling G.C., Kingsland C.R. and Johnson P.M. 1992.Immunology.In press.
18) Hsi B.L., Fenichel P.,and Cervoni F. 1991.Biol. cell. mol. relation materno fetale.INSERM John Libbey.Paris.Pages 3-13
19) Holmes C.H., Simpson K.L.,Wainwright S.D., Tate C.G., Houlihan J.M., Sawyer J.H., Roger I.P., Spring F.A., Amstee D.J. and Tanner M.J. 1990. J. Immunol. 144.3099-3115.
20) Croy B.A., Malashenko B.A., Poterski R., Yamashiro S. and Summerlee A. 1990. Prog. Neur.End. Immunol. 3. 242-250.
21) "Immune abortion : The controversy" .Biol. cell. mol. relation materno fetale. Editions INSERM John Libbey.Paris.
22) Chaouat G , Menu E., Dy M., Minkowski M.,Clark D.A. and WEegmann T.G.1990. J.Fert. Steril.89.447-458.
23) Chaouat G., Menu E., Davoid F., Szekeres-Bartho J., Kinsky R., Dang D.C.,Kapovic M.,Ropert S.and Wegmann T.G.1991. Periodicum Biologorum.93.1.49-54 .
24) Menu E. , Jankovic D.,Theze J, David V. and Chaouat G. 1991. Reg.Immunol.3.5.p 254-259
25) Menu E.,Djian V.,Kinsky R.,Jankovic D.,Delage G., Rosin, and Chaouat G.(1991).Biol. cell. mol. relation materno fetale. Editions INSERM John Libbey.Paris.Pages 197-205.
26) Khrihsnan L., Menu E.,Chaouat G., Talwar G.P. and Raghupathy R. 1991. Cell.Immunol. 138. 313-326..
27) David F.J.E., Autran B., Tran H.C., Menu E., Raphael M., Debré P., Barré Sinoussi F., Wegmann T.G., Hsi B., and Chaouat G. (1992).Clin. Exp.Immunol.88.10_16
28) David et al. This meeting ,and 8th Intl AIDS conference communications, and submitted forpublication/
29) Kachkache M., Acker G.M.,Chaouat G;, Noun A; and Garabedian M. 1991. Biol. Reprod.Dec 1991.
30) Mac Master M.T., Newton R.C., Sudhanski K.Dey,and AndrewsG.K. 1992.J. Immunol.148. 1699-1705.
31) Sanford R., De M. and Wood G. 1992.J Reprod. Fert.94.1. 213-220

Maternal Responses to Trophoblast Antigens: Idiotype-Antiidiotype Control

J. A. McInbtyre and R. G. Roussev

Center for Reproduction and Transplantation Immunology, Methodist Hospital for Indiana, Indianapolis, Indiana 46202 USA

INTRODUCTION

Pregnancy challenges the female to numerous alloantigens (McIntyre, 1992). We have focused our attention upon trophoblast lymphocyte crossreactive (TLX) alloantigens because trophoblast forms the allogeneic interface between mother and fetus. The maternal immune response to TLX antigens has been shown by Torry et al. (1989) to be regulated by an auto-antiidiotype network. An intact TLX network is postulated to govern immune acceptance of the conceptus by the mother. Absent, suboptimal or unchecked responses to TLX antigens may result in infertility or recurrent spontaneous abortion (RSA) (Roussev et al., 1992). A group of women suffering from secondary RSA was identified that produced high titer antipaternal lymphocytotoxic antibodies (McIntyre et al., 1984). These antibodies reacted in a non-HLA restricted pattern against third-party lymphocyte donors. The serum reactivity was removed by absorptions with some, but not other individual trophoblast membrane preparations suggesting that the antibodies were directed to TLX alloantigens (McIntyre and Faulk, 1986). TLX antigens are also expressed on platelets (Kajino et al., 1987). Lymphocytotoxic TLX antibodies can be removed from secondary aborter sera by absorption with platelet membranes (McIntyre, 1988). Platelet eluates containing purified anti-TLX IgG (idiotype or Ab1) were prepared as immunogens for producing heterologous antiidiotypic antibodies (Ab2). The Ab2 was used by Torry et al. (1991) to determine if Ab1 produced to TLX antigens by different women employed similar variable (V) region genes, detectable as cross-reactive idiotypes (CRI). Initial studies with this rabbit Ab2 (RAb2) showed that women respond to trophoblast antigens during normal pregnancy. This report summarizes and extends previous observations of allotypic TLX CRI in normal pregnancy sera and provides evidence for lack of TLX antigen recognition in a subgroup of patients designated as primary aborters.

MATERIALS AND METHODS

Patients and Sera: Three secondary aborting patients' anti-TLX sera were chosen on the basis of their differential cytotoxicity reactions on a panel of 57 lymphocyte donors. Anti-TLX IgG (Ab1) were affinity purified from each of the 3 patients by using selected third-party platelets (Torry and McIntyre 1992).

Rabbit Antiidiotype Production: Outbred New Zealand white male rabbits were immunized with 200-400 μg of each patient's Ab1 suspended in an equal volume of complete Freund's adjuvant. At 2 week intervals the rabbits received 4-6 booster injections consisting of 150-300 μg of Ab1 in incomplete Freund's adjuvant. Aliquots of these rabbit antisera were heated (56°C-30 min), absorbed at least x4 with pooled human male plasma (n=30) coupled to Sepharose and with immobilized pooled human male immunoglobulin (n=42). The absorptions continued until the RAb2 no longer reacted in the ELISA with male IgG.

ELISA: Protein A affinity purified IgG from each sera tested was coated onto microtiter wells at 200 μg/ml for 16 hr at 4°C. A predetermined working dilution (usually 1/20) of RAb2 was used for control and CRI determinations. IgG fragments were produced by immobilized papain and immobilized pepsin digestions to yield the representative Fab or F(ab')$_2$ fragments. The IgG fragments were verified by SDS-PAGE, double immunodiffusion assays and ELISA by using antihuman IgG (gamma chain Fc specific) and antihuman kappa light chain antisera. The RAb2 were incubated in PBS containing 1% BSA (PBS-BSA) for 40 min in each well. The wells were then washed x3 with PBS-Tween-20 and a peroxidase conjugated swine antirabbit immunoglobulin conjugate 1:2000 in PBS-BSA was added. Wells were washed x5 with PBS-Tween-20 and developed with 100 μl of substrate (0.42 mM tetramethylbenzidine in 0.1M Na acetate, 0.0045% H_2O_2, pH 6.0). Reactions were stopped by adding 35 μl of 2M H_2SO_4 and the optical density (OD) of each well measured at 450 nm in an automated Beckman Biomeck.

Complement Dependent Cytotoxicity (CDC) Assay: Lymphocytotoxicity assays were performed as previously described by Torry et al. (1987). Briefly, Ficoll-Hypaque prepared lymphocytes were resuspended in 0.2 ml $Na_2{}^{51}CrO_4$ for 50 min, then washed x3 in RPMI media containing 10% fetal calf serum (FCS) and adjusted to 2×10^6 cells/ml. Adjusted concentrations of secondary aborter IgG that provided half maximal $CDC^{51}Cr$ were incubated with increasing dilutions of RAb2 for 1 hr at 4°C. Incubations with normal rabbit sera (NRS) were used for controls. After these incubations 50 μl of the human-rabbit sera mixture were added in duplicate to 2.5×10^4 target cells. All tubes received 200 μl of fresh rabbit complement, incubated at 37° for 3 hr whereupon the reaction was stopped by addition of ice-cold RPMI. The tubes were centrifuged and the supernatants decanted in clean test tubes. Both pellets and supernatants were counted for 1 min in a gamma counter. Percentage ${}^{51}Cr$ release was calculated as [supernatant cpm/supernatant cpm + pellet cpm] x 100. The percent specific ${}^{51}Cr$ release was calculated as: [test ${}^{51}Cr$ release - spontaneous release/maximum ${}^{51}Cr$ release - spontaneous release] x100%.

RESULTS

Production of Antiidiotypic Antibodies to Human Anti-TLX

Rabbit Ab2 were produced by immunization with anti-TLX Ab1 prepared from 3 secondary aborter sera chosen on the basis of their differential cytotoxicity reactions on a 57 member panel of lymphocyte donors (Fig. 1). None of the cytotoxicity reactions were explainable on an HLA basis. Secondary aborter K.R. was positive with 40% of the panel donors. Secondary aborter M.H. was positive with 37% of the panel donors with a 25% overlap of cytotoxicity with K.R. Secondary aborter H.H. also was reactive with 37% of the panel but with many different donors than were K.R. (10% overlap) and M.H. (31% overlap). Together, these 3 patient antisera reacted with 80% of the cell panel.

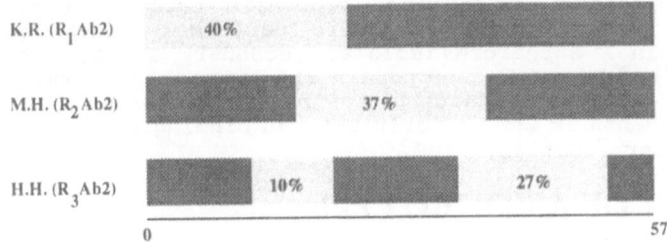

Fig.1. A graphic illustration of the cell panel reactivity and cross-reactivity of the 3 secondary aborter antisera selected for RAb2 production. Lightly shaded bars represent % of total and cross reactivity.

When tested in ELISA against purified IgG from the respective immunizing secondary aborter's serum, the 3 RAb2 (R₁Ab2; R₂Ab2; R₃Ab2) demonstrated dose dependent binding to their respective immunizing secondary aborter IgG, but not to normal male IgG or to Ab1 preparations from the other two secondary aborters (Fig. 2).

SPECIFICITIES OF ANTIIDIOTYPES

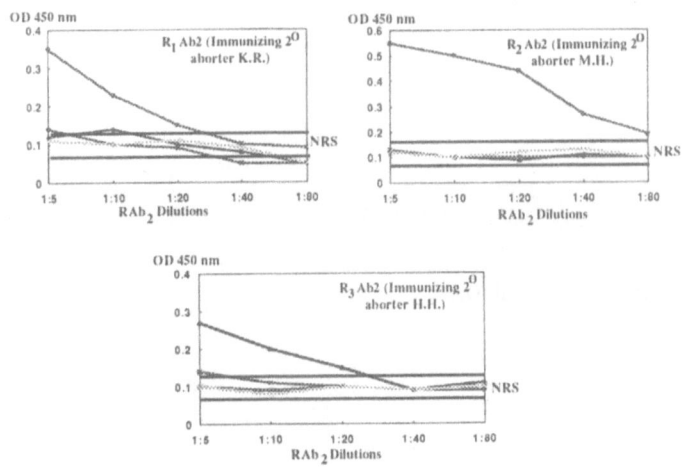

Fig. 2. Direct binding ELISA of the RAb2 versus their respective immunizing IgG isolated from secondary aborter sera. 200µg/ml IgG were coated onto wells. The top line in each graph represents the RAb2 activity against the Ab1 IgG used for immunization. The lines within the range depicted for control NRS represent the reactivity of the RAb2 with male IgG or the 2 other secondary aborters' IgG.

To establish that the RAb2 preparations recognize true idiotypic determinants, a series of experiments were performed using IgG fragments from protein A Sepharose isolated secondary aborter and normal male IgG obtained by pepsin and papain digestion. The results shown in Fig. 3 demonstrate that the RAb2 preparations were restricted to determinants in the variable antigen binding regions of the respective secondary aborter's IgG.

SPECIFICITY OF ANTIIDIOTYPE

R_1Ab2 vs $F(ab')_2$ and Fab K.R.

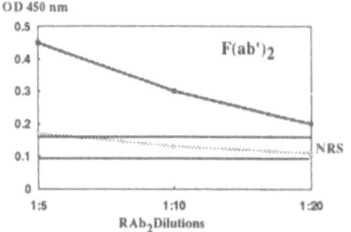
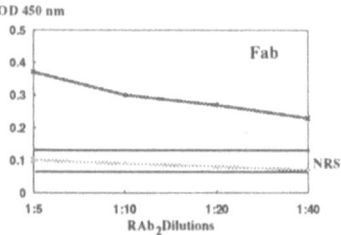

R_2Ab2 vs $F(ab')_2$ and Fab M.H.

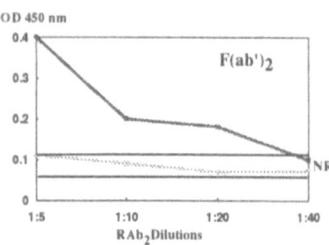
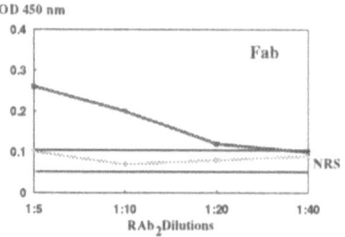

R_3Ab2 vs $F(ab')_2$ and Fab H.H.

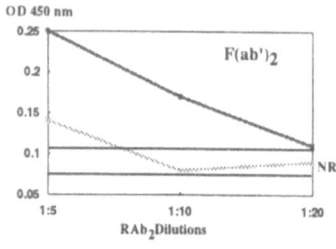
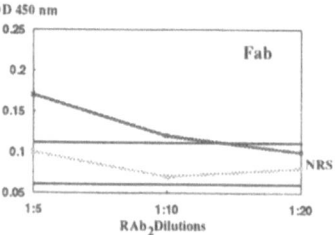

Fig. 3. Direct binding ELISA of the 3 RAb2 versus $F(ab')_2$ and Fab fragments from each respective immunizing secondary aborter IgG. 200µg/ml $(Fab')_2$ or Fab fragments were incubated in the wells. The upper darker line in each graph depicts the ELISA reactivity of the

respective RAb2 on its immunizing IgG fragment preparation. The lower line represents the RAb2 reactivity to male control F(ab')$_2$ and Fab fragments. NRS = the range of activity of the control NRS.

To test functionally if the RAb2 recognize idiotypic determinants within or close to the antigen binding site (paratope) of TLX Ab1, a competitive inhibition CDC^{51}Cr assay was employed. The RAb2 and a control NRS were preincubated for 1 hr with adjusted concentrations of IgG from the Ab1 donors to ascertain their ability to inhibit Ab1 cytotoxicity. Shown in Fig. 4, RAb2 produced significant CDC^{51}Cr inhibition of their respective immunizing Ab1. None of the RAb2 had CDC^{51}Cr reactivity against target cells (data not shown). None of the RAb2 showed inhibition when incubated with the non-immunizing secondary aborter IgG preparations (data not shown). The results confirm that the RAb2 recognize idiotypic determinants within or adjacent to the antigen binding site of their respective TLX Ab1 antibodies.

INHIBITION OF CYTOTOXICITY BY ANTIIDIOTYPE

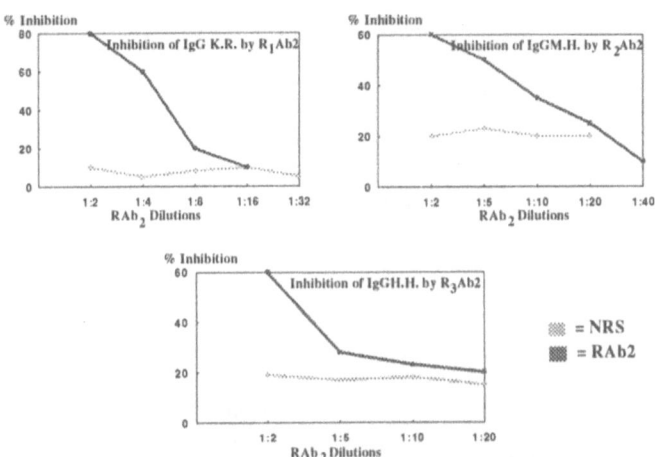

Fig. 4. Inhibition of CDC^{51}Cr activity of secondary aborter IgG by the respective RAb2. All IgG preparations were concentrated to 6mg/ml. The upper dark lines represent the CDC^{51}Cr inhibition for each RAb2. The lower stippled lines represent the inhibition observed by using NRS.

Screening Human IgG for Cross-Reactive Idiotypes (CRI) on TLX Antibodies

The ability of the RAb2 to detect CRI on anti-TLX antibodies enabled us to screen serum samples from both normal reproducing women and women suffering from unexplained repeated miscarriages. IgG was purified from 43 secondary aborters, 41 normal multiparous women, 18 primigravidae and 18 primary aborters. The reactivities of the 3 RAb2 were tested in a direct ELISA by coating each woman's IgG onto wells of microtiter plates. Positive results were determined by the

$$\text{formula: negative control} + \left[\frac{\text{positive control} - \text{negative control}}{2} \right]$$

where male IgG is the negative control value and the respective immunizing secondary aborter IgG is the positive control value. All OD readings above this calculated value were considered positive. Figure 4 shows by Venn diagrams the CRI frequencies for these 4 groups of women. The size and overlap of the circles correspond to the actual percentages of the reactions of each RAb2. Overlapping circles represent cross-reactivity of their IgG with each RAb2. Of the 120 women tested, 67, (56%) reacted with only 1 of the 3 RAb2, 14, (12%) reacted with 2 of the RAb2 and none of the women reacted with all 3 RAb2. In the primary abortion group only 1 patient serum showed reactivity with a single RAb2.

FREQUENCY OF CROSS-REACTIVE TLX IDIOTYPES

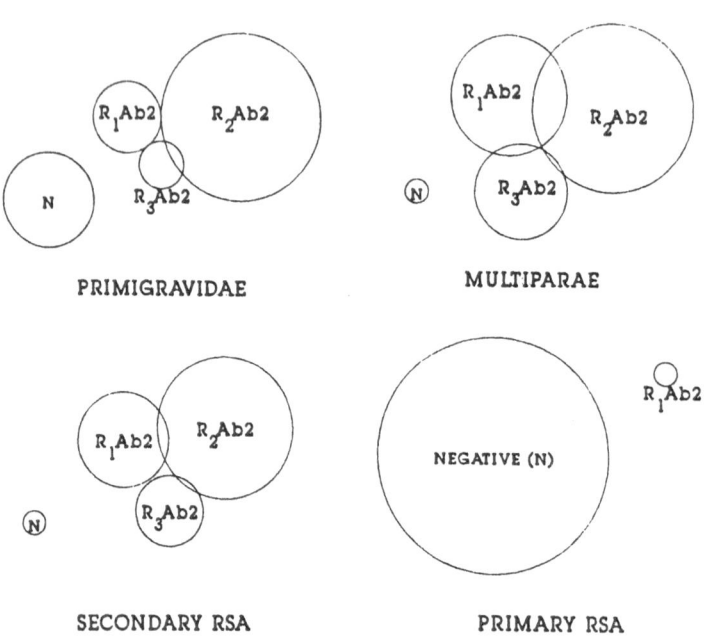

PRIMIGRAVIDAE

MULTIPARAE

SECONDARY RSA

PRIMARY RSA

Fig. 5. Venn diagrams representing the frequencies of RAb2 reactivities with IgG from normal reproducing women and women suffering RSA. The size of each circle depicts the percentage of positive, negative and cross-reactive activity.

DISCUSSION

We have prepared and tested 3 rabbit antiidiotypic antibodies (RAb2) to 3 different human anti-TLX antibodies (Ab1). Inhibition by blocking and functional assays indicated that the 3 RAb2 reagents were specific for a region in or very close to the antigen binding epitope (paratope) of the respective immunizing idiotype (Ab1). Little cross-reactivity was observed among the 3 RAb2. By using these RAb2 as probes to detect CRI, 92 of 102 (90%) multiparous, primigravid and secondary aborting women tested were found reactive with one or occasionally two of the RAb2 but never with all 3. The presence of CRI from unrelated women supports our hypothesis that normal pregnant and multiparous women respond to trophoblast alloantigens (Roussev et al., 1991). These results also indicate that a conserved variable (V) region gene, or family of genes is used to produce antibodies with TLX antigen specificities. These RAb2 data also support our hypothesis that primary aborters fail to recognize TLX alloantigens. The finding of CRI on TLX antibodies in normal pregnancy and multiparous women and lack of such antibodies in primary RSA support the rationale for the use of immunotherapy in primary abortion by maternal immunization with TLX antigens. The presence of these CRI raises the possibility of novel methods to prevent immunologically mediated primary and secondary RSA.

ACKNOWLEDGEMENTS

We thank Dr. Donald Torry, Dana Farber Cancer Institute, Boston, MA, for providing the initial R_1Ab2 antisera.

REFERENCES

Kajino T, Faulk WP, McIntyre JA (1987) Antigens of human trophoblast: Trophoblast lymphocyte cross-reactive (TLX) antigens on platelets. Am J Reprod Immunol Microbiol 14:70-87.

McIntyre JA (1992) Immune Recognition at the maternal fetal interface. Am J Reprod Immunol (in press, 1992)

McIntyre JA (1988) In search of trophoblast lymphocyte cross-reactive (TLX) antigens. Am J Reprod Immunol Microbiol 17:100-110.

McIntyre JA, Faulk WP (1986) Trophoblast antigens in normal and abnormal human pregnancy. Clin Obstet Gynecol 29:976-998.

McIntyre JA, McConnachie PR, Taylor CG, Faulk WP (1984) Clinical, immunologic and genetic definitions of primary and secondary recurrent spontaneous abortions. Fertil Steril 42:849-855.

Roussev RG, Vanderpuye OA, McIntyre JA (1992) TLX alloantigens and pregnancy. In: Immunology of Reproduction (ed. R.K. Naz) CRC Press, New York (in press).

Roussev RG, Vanderpuye OA, Wagenknecht DR, McIntyre JA (1991) A role for TLX antigens in pregnancy. Acta Europaea Fertilitatis 22:181-187.

Torry DS, Faulk WP, McIntyre JA (1991) Trophoblast immunity in human pregnancy defined by antiidiotype. Am J Reprod Immunol 251:181-184.

Torry DS, Faulk WP, McIntyre JA (1989) Regulation of immunity to extraembryonic antigens in human pregnancy. Am J Reprod Immunol 21:76-81.

Torry DS, McIntyre JA, McConnachie PR (1987) Characterization of immunoglobulin class and subclass responses in secondary aborter sera. J Reprod Immunol 10:33-42.

Torry DS, McIntyre JA (1992) Immune regulation in pregnancy: role of idiotype-antiidiotype network. In: Immunological Obstetrics (eds. C.B. Coualm, W.P. Faulk, J.A. McIntyre) W.W. Norton and Company, New York pp 333-356

Role of a Unique Species of Transforming Growth Factor Beta in Preventing Rejection of the Conceptus During Pregnancy

D.A. Clark, R.G. Lea, K.C. Flanders, D. Banwatt and G. Chaouat

McMaster University, Hamilton, Ontario, Canada (DAC, RGL, DB), Laboratory of Chemoprevention, NIH, Bethesda, MD, USA, (KCF), Batimnet de Gynecologye,/Obsterique, Univerity Peraxis-XI-Hospital Antione Beclere, Clamart, France (GC).

Suppression of rejection of the conceptus is a thesis arising from the paradigm that the fetus resulting from mating of genetically dissimilar individuals is an allograft that paradoxically is highly successful. Cells able to suppress the generation of cytotoxic T lymphocytes (CTL) against paternal Class I MHC were found in paraaortic lymph nodes draining the uterus in most strains of allogeneically mated mice and, where deficient (i.e., C57Bl10ScSn, CBA/J), were detectable in the decidua (Clark 1991). Deficient decidual suppressor cell activity was found in mice with a high rate of spontaneous resorption (Clark et al 1986, 1990a, 1991a). It is important to note that kinetic studies of suppressor cell activity in decidua during mouse pregnancy demonstrated two types of suppressor cells, a $CD8^+$ T cell population present during the pre- and peri-implantation phase of pregnancy, and a non-T non-B cell population which developed 4-5 days after implantation (Clark et al 1991b). Both were antigen non-specific, and the latter non-T non-B suppressor cell was particularly interesting as it released a soluble suppressor factor closely related to TGF-β2, and developed just before the usual time of onset of abortion in mice (Clark et al 1990b, 1991b, Lea et al 1992).

REJECTION OF THE "FETAL ALLOGRAFT" MODEL

While the semiallogeneic fetus behaves as an allograft (Clark 1991), the fetal trophoblast which envelopes the fetus and forms the maternal-conceptus interface is not susceptible to transplantation immunity (Clark 1991). Indeed, trophoblast was found to be resistant to lysis by CTL, by ADCC, and by NK and NC cells (Head 1989, King and Loke 1990). Nevertheless, for a time it was thought that local suppression at the level of the decidua in which the placenta is embedded could function to prevent sensitized maternal T cells from crossing the trophoblast into the fetus and causing allograft necrosis (Clark et al 1987). Experiments done to test the role of transplantation immunity in the CBA/J-DBA/2 mouse model of allopregnancy abortion associated with a cytotoxic cell infiltrate indicated that CTL likely appeared after the fact and antigen-specific immunity was not essential for initiation or completion of resorption (Clark 1991, Clark et al 1991a). In the DBA/2 mated-CBA/J female, abortion is a late effect of exposure to intrauterine bacterial LPS that occurs at the time of mating; asialoGM1$^+$NK-like cells have been shown to be essential (Clark 1991). In this system, production of TGF-β2-related suppressor factor at the implantation site is delayed with respect to production of TNF-α, - a "window" not seen in DBA/J-mated C3H/HeJ mice which have a low abortion rate (Clark et al 1991b). It can be shown that the spleen cells from CBA/J mice contain asialoGM1$^+$ effector cells that can kill an NK/NC/CTL TNF-α resistant trophoblast target (Be6 cell line) *in vitro*; cytokines such as TNF-α boost killing, and the decidual TGF-β blocks activation as does pretreatment of the mice with antibody to asialoGM1 (Figure 1). The TNF-α^+TGF-β^- window can be eliminated by antibiotic (tetracycline) treatment of the males and females before mating, and the rate of abortion lowered (Clark et al 1992c). Interestingly, a similar TNF-α^+TGF-β^- window has been demonstrated in C.B-17 SCID mice which have a high abortion rate, but in these mice raised under SPF-barrier conditions, cells sensitive to injection of asialoGM1 antibody do not appear to be essential for abortion (Clark et al 1989, 1992a). However, this treatment is only partially effective in depleting anti-trophoblast killer cells *in vivo* and appears more active against those effectors that are responsive to bacterial lipopolysaccharide as seen in CBA/J mice. Activation of anti-trophoblast killer cells SCID mice appears to occur by a different activation mechanism (Clark et al 1989, 1992a).

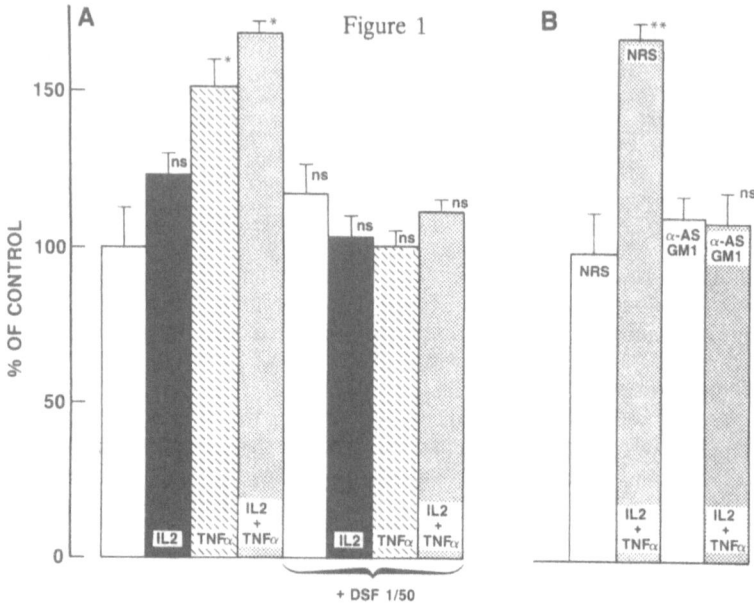

Figure 1

Footnotes: Graphs show effect of cytokines on generation of cells which kill Be6 trophoblast. One x 10^6 CBA/J spleen cells were cultured for 3 days alone with 10 u/ml murine recombinant IL-2 and or 1 u/ml TNF-α (Genzyme). (A) Effect of cytokines is expressed as % of control lysis. * denotes significant boosting of lysis per culture. HPLC-purified decidual suppressor factor (DSF) (TGF-β2-related) was added to some of the cultures when set up. Note abrogation of cytokine boosting. (B) shows cytokine boosting of spleen cells from CBA/J injected 24 hours previously with normal rabbit serum (NRS). Injection of rabbit anti-asialoGM1 antibody blocks responsiveness. Note that when CBA/J spleen cells were directly tested on Be6 targets at an E:T ratio of 50-100:1, addition of DSF reduced lysis at 18 hours by 54%; 1 ng/ml TGF-β reduced lysis by 67%

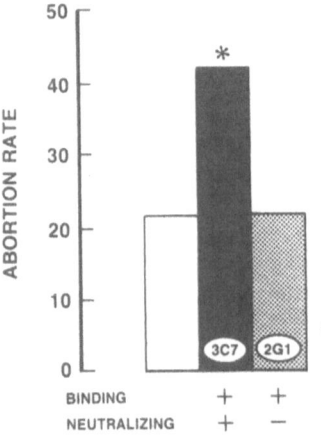

Evidence that TGF-β2-producing suppressor cells are deficient at resorption sites is not a completely convincing argument for a role in preventing abortion as these cells are activated by signals from trophoblast and activation signals could disappear with the onset of trophoblast/conceptus failure (Slapsys et al 1988). However, a deficiency of suppressor activity has been found prior to resorption (Clark et al 1986), and injection of 100 μg monoclonal anti-TGF-β2 neutralizing antibody (3C7 kindly provided by Celltrix Laboratories, Palo Alto, CA, USA) on day 8.5 of pregnancy prior to the onset of abortion boosted (Figure 2) the abortion rate; a control monoclonal which binds but does not neutralize the suppressor factor (2G1) had no effect.

Recent data indicate that the TGF-β2-related factor released from murine decidua differs from the expected 25 kD TGF-β2 molecule. Decidua supernatants separated on HPLC sieving columns at pH 2.9 in 0.1 M KC1 showed a dominant peak of activity co-purifying with the cytochrome C marker, and electroelution of PAGE gels confirmed activity in the 15-23 kDa range (Clark et al 1990b). Figure 3 shows a confirmatory result. Western blotting also confirmed the presence of a low molecular weight TGF-β2 related factor (Clark et al 1990b) . As low pH can activate TGF-β, the state of the suppressor factor in untreated supernatant was uncertain. It was not possible to test crude supernatant containing 10% fetal bovine serum by PAGE and Western blotting due to protein overloading. Recent studies using supernatants generated in serum-free medium have shown that substantial quantities of TGF-β2-like factor are released in the form of a doublet with an estimated molecular size of 28-30 kDa. The most suppressive TGF-β2-related activity occurs at molecular weights less than 25 kDa which gives a weak signal on Western blots. Anomalous high molecular weight TGF-β has also been found in human breast cancer (King et al 1989).

Figure 3

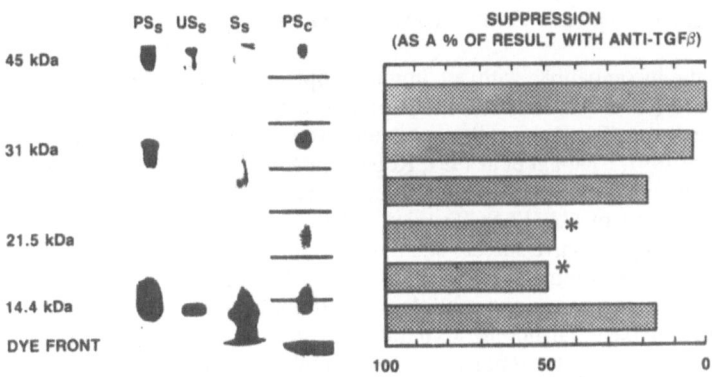

Footnotes: PS$_s$ - prestained standards - silver stained; US$_s$ - non-prestained standards visualized by silver stain; S$_s$ suppressive sample - silver stained; PS$_c$ - prestained standards, no silver stain. Horizontal lines show site of gel cuts. *significant suppression.

SIMILARITY OF MOUSE AND HUMAN PREGNANCY: SUCCESS AND FAILURE.

Biologically active TGF-β2-related suppressor activity has also been found in supernatants human pregnancy decidua and appears to be produced by the mononuclear cells (Michel et al 1989, Clark et al 1991b, 1992). We initially suspected the factor was produced by granulated null-type cells as in the mouse, and recent data confirms that CD56+ CD16- cells (the dominant lymphomyeloid cell type in first trimester decidua (Starkey 1992)) produce the TGF-β2-like suppressor factor (unpublished data - manuscript in preparation). By *in-situ* hybridization using the pcdG1G2 probe that selectively detects TGF-β2-producing suppressor cells in murine pregnancy decidua, a subpopulation of patients with recurrent pregnancy loss has been identified which lacks these suppressor cells (Clark et al 1991ab, 1992d). Further, an infiltrate of "activated" CD56+ CD16+ LGL (which exert most of the non-specific cytotoxic activity of freshly isolated or IL2-stimulated decidual CD56+ cells (Bulmer 1992, Christmas et al 1990)) has been noted in biopsies of incipiently aborting patients with the recurrent pregnancy loss problem. These findings are compatible with rejection by activated NK-lineage cells similar to that occurring in the mouse (Michel et al 1989, Kodana et al 1992, Gotoh et al 1992).

T LYMPHOCYTES IN DECIDUA AND THE IMMUNOTROPISM CONTROVERSY

Immunization of female CBA/J mice against the MHC antigens of the DBA/2 male mate prevents abortion and produces larger babies with thicker placental trophoblast (Chaouat et al 1990, Wegmann et al 1987, 1989). Further, T cell cytokines such as IL-3 and GM-CSF can prevent abortion, and *in vivo* administration of anti-CD4 + anti-CD8 augmented the abortion rate (Chaouat et al 1988, 1990). Together with the finding of T cells in decidua and Class I MHC on trophoblast, these data led to the idea that maternal decidual T cells were recognizing trophoblast Class 1 and secreting cytokines which exerted a positive immunotrophic growth effect on trophoblast (Wegmann 1987, 1989). As trophoblast is insensitive to killing by NK/NC cells and specific immune effectors (Head 1989, King & Loke, 1990), trophoblast should act as an effective barrier between the mother and fetus, and prevent abortion. The finding that the dominant T cell in murine and human decidua is $CD8^+$ is consistent with recognition of Class I MHC.

Experimental data incompatible with an immunotrophic role of maternal T cells has been reviewed elsewhere (Clark 1991). With the exception of the 2-3 days immediately after mating, IL-3, the only T-cell-specific cytokine, has not been detected, and GM-CSF is produced by non-T cells (Sanford et al 1991, Chaouat et al 1991, Robertson & Seamark 1991). SCID mice which lack functional T and B cells, as well as T cell-deficient nude mice produce GM-CSF and reproduce successfully; in the case of C.B-17 SCID mice, a heterozygous background genotype appears essential for optimal reproductive success and the abortion rate is *not* reduced by T cells as shown by the study of +/+ C.B-17 females (Clark et al 1992a). It is argued that immunotrophic effects are only needed under non-gonobiotic conditions, where there is exposure to the NK-macrophage activating stimulus of bacterial lipoplysaccharide. In the CBA/J-DBA/2 system where such a mechanism has been shown to be important under natural conditions, administration of anti-T cell antibodies boosts the abortion rate and, in this case, only anti-CD8 is effective (Clark 1991). Further, in this system, monoclonal anti-GM-CSF also boosts the abortion rate (Clark et al 1992b). However, administration of GM-CSF to mice co-treated with anti-CD8 does *not* prevent abortion as would be expected if the function of $CD8^+$ cells were to produce this cytokine. Rather, the protective effect of GM-CSF is blocked by anti-CD8 treatment. Further, the protective effect of alloimmunization is also blocked by anti-CD8 treatment (Chaouat, unpublished data). These observations are consistent with the idea that immunization and GM-CSF act in concert with or via $CD8^+$ T cells. $CD8^+$ cells in pregnancy may acquire progesterone receptors and in response to progesterone, secrete a 34 kDa factor which suppresses NK-LAK killer activity (Szekeres-Bartho et al 1990). Injection of GM-CSF also appears to downregulate endogenous killer cell activity in spleen as measured by the ability to lyse an NK/NC/CTL-resistant trophoblast cell target *in vitro* (Clark et al 1991b); this inhibitory effect of GM-CSF is blocked by concommitant administration of anti-CD8 but not anti-CD4 antibody (Clark et al 1992b). The so-called immunotrophic benefit of $CD8^+$ cells may therefore be attributable to an immunosuppressor action on a unique non NK-type of natural effector cell *in utero*. Indeed, there is a very close correlation between changes in splenic antitrophoblast killer cells and abortion in the CBA-DBA/2 system as summarized in Table 1. It is important to note that the 3C7 monoclonal (Figure 2) does *not* boost this killer activity (data not shown).

Table 1

Treatment	Abortion Rate CBA/JmDBA/2[A]	Change in Splenic Anti-Be6 Lytic Activity
NRS	No change	No change
Anti-ASIALOGM1	Down by 58%	Down by 47%
Anti-IL-2R (7D4 mAb)	Up by 67%	Up by 34%
E. coli LPS 2 μg IP	Up by 77%	Up by 43%
GM-CSF 400 u IP	Down by 69%	Down by 74%
Anti-CD8 mAb	Up by 70%	Up by 27%

Footnotes: A) Abortion rate without treatment or with medium injection ranges from 24-42% assessed on day 13.5 of pregnancy. Spleen cell lytic activity measured in an 18 hr assay with ^{51}Cr-labelled Be6 murine trophoblast cell line (LAK sensitive) at E:T ratios of 100:1 to 25:1. Lytic activity has been corrected for changes in spleen cell content. LPS = lipopolysaccharide.

A NEW PARADIGM?

What may be the relative importance of systemic regulation of anti-trophoblast killer cells in reproductive success (as suggested by Table 1), as compared to local regulation of killer cells by suppressor cell activity at the feto-maternal interface (Clark et al 1991b)? The TGF-β2-producing suppressor population in murine decidua is located deep in the decidua and not at the contact area between fetal membrane trophoblast and decidua (Clark 1991). In this respect, it is interesting that trophoblast invades human decidua and also maternal arterial walls and replaces the endothelium; similar arterial wall invasion occurs in the mouse (Redline & Lu 1989). In the human, CD56+ cells are often located near trophoblast-invaded arterioles (Bulmer 1989, 1992). In abortion in mice, trophoblast cells at the fetal membrane-decidual interface are not obviously destroyed or lysed but there are areas of necrosis in adjacent decidua. Hemorrhage is also a feature of the abortion process. Taken together these data suggest the vascular system may be the target (Clark et al 1991c). As illustrated in Figure 4, the "deep" location of suppressor cells in decidua would allow TGF-β2-like factors to be locally produced and affect activation of effector cells at the "vascular trophoblast"-blood interface. CD8+ cells may exert both a systemic and local action on effector cells, but resistance to lipopolysaccharide-induced abortion in mice appears with development of the TGF-β2-producing cells in decidua, - an observation suggesting the importance of these local suppressor cells in resistance to abortion.

Figure 4

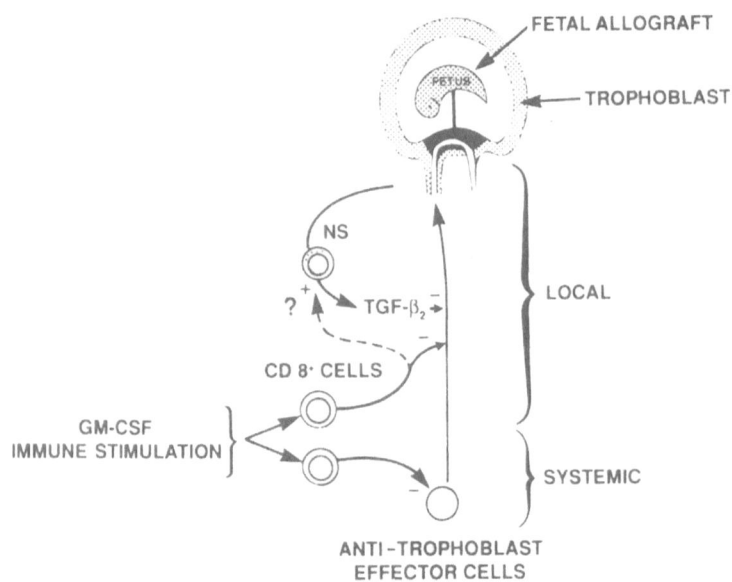

Footnotes: NS = trophoblast activated non T non B natural suppressor cell.

Immunization that prevents abortion has been found to boost the activity of TGF-β2-producing suppressor cells (Clark 1991). The true immunotrophic action of CD8+ T cells required for the protective effect of immunization may (and possibly GM-CSF) therefore be directed towards stimluating TGF-β2-producing natural suppressor cells in decidua. Testing this model should significantly enhance our understanding and exploitation of immunoregulatory mechanisms that impact on the success of mammalian pregnancy.

REFERENCES

Bulmer JN (1989) Curr Op Immunol 1:1141-1147
Bulmer JN (1992) In: Proc. 6th Annual Meeting. Japan Society for Immunology of Reproduction, Hayashi - Kobo Co.Ltd. Tokyo, p 26
Chaouat G, Menu E, Athanassakis I, Wegmann TG (1988) Reg Immunol 1:143-148
Chaouat G, Menu E, Clark DA, Minkowsky M, Dy M, Wegmann TG (1990) J Reprod Fert 89:447-458
Chaouat G, Menu E, Szekeres-Bartho J, Rebut-Bonneton C, Bustany P, Kinsky R, Clark DA, Wegmann TG (1991) In: Molecular and Cellular Immunobiology of the Materno-Fetal Interface Oxford University Press, New York, p 277
Christmas SE, Bulmer JN, Meager A, Johnson PM (1990) Immunology 71:182-189
Clark DA (1991) Crit Rev Immunol 11:215-247
Clark DA, Chaput A, Tutton D (1986) J Immunol 136:1668-1675

Clark DA, Chaput A, Slapsys R, Brierley J, Daya S, Rosenthal K, Chaouat G (1987) In: Reproductive Immunology:Materno-Fetal Relationship.INSERM Colloque, John Libbey Eurotext, Paris, Vol 154:77-87

Clark DA, Banwatt DK, Manuel J, Fulop G, Croy BA (1989) In: Current Topics in Microbiology and Immunology 152:227-234

Clark DA, Drake B, Head JR, Stedronska-Clark J, Banwatt D (1990a) J Reprod Immunol 17:253-0264

Clark DA, Flanders KC, Banwatt D, Millar-Book W, Manuel J, Stedronska-Clark J, Rowley B (1990b) J Immunol 144:3008-3014

Clark DA, Lea RG, Denburg J, Banwatt D, Manuel J, Namji N, Underwood J, Michel M, Mowbray J, Daya S, Chaouat G (1991a) In: Cellular and Molecular Biology of the Materno-Fetal Relationship INSERM Colloque, John Libbey Eurotext, Paris, 212:171-178

Clark DA, Lea RG, Podor T, Daya S, Banwatt D, Harley C (1991b) Annals NY Academy Science 626:524-536

Clark DA, Head JR, Drake B, Fulop G, Brierley J, Manuel J, Banwatt D, Chaouat G (1991c) In: Molecular and Cellular Immunobiology of the Maternal-Fetal Interface. Oxford University Press, New York, p 294

Clark DA, Banwatt D, Fulop G, Quarrington C, Croy BA (1992a) Am J Reprod Immunol 27:48-49

Clark DA, Lea RG, Chaouat G, Pearce M, Abrams J (1992b) Am J Reprod Immunol 27:48

Clark DA, Banwatt D, Manuel J (1992c) Am J Reprod Immunol 27:34

Clark DA, Lea RG, Underwood J, Michel M, Flanders KC, Harley C, Daya S, Hirte H, Beard R, Mowbray J (1992d) 5th ICRI, Serono, Rome (in press)

Gotoh M, Matsumoto K, Akiyama Y, Kusuhara K, Terashima Y (1992) In: Proc. 6th Annual Meeting. Japan Society for Immunology of Reproduction, Hayashi, Kobo Co.Ltd. Tokyo, p 102

Head JR (1989) Am J Reprod Immunol 20:100-105

King A, Loke YW (1990) Cell Immunol 129:435-448

King RJB, Wang DY, Daly RJ, Darbre PD (1989) J Ster Bioch 34:133-138

Kodama T, Okamoto E, Ohata K, Hara T, Ohama K, Kusunoki Y, Akiyama M (1992) In: Proc. 6th Annual Meeting. Japan Society for Immunology of Reproduction, Hayashi, Kobo Co.Ltd. Tokyo, p 80

Lea RG, Flanders KC, Harley CB, Manuel J, Banwatt D, Clark DA (1992) J Immunol 148:778-787

Michel M, Underwood J, Clark DA, Mowbray J, Beard RW (1989) Am J Obs Gyn 161:409-414

Redline RW, Lu CY (1989) Lab Invest 61:27-36

Robertson S, Seamark RF (1991) In: Cellular and Molecular Biology of the Materno-Fetal Relationship INSERM Colloque, John Libbey Eurotext, Paris 212:113-120

Sanford TH, De M, Andrews GA, Wood GW (1991) In: Cellular Signals Controlling Uterine Function, Plenum Publishing, New York, p 172

Slapsys RM, Younglai E, Clark DA (1988) Reg Immunol 1:182-189

Starkey PM (1992)In: The Natural Immune System: The Natural Killer Cell. IRL Press, p 206

Szekeres-Bartho J, Varga P, Kinsky R, Chaouat G (1990) Res Immunol 141:175-181

Wegmann TG (1987) Am J Reprod Immunol Microbiol 15:67-69

Wegmann TG, Athanassakis I, Guilbert L, Branch D, Dy M, Menu E, Chaouat G (1989) Transplant Proc 21:566-568

Vaccines for Control of Fertility

G. P. Talwar, Om Singh, R. Pal, S. Sad, K. Arunan, N. Chatterjee,
C. Kausic, S. N. Upadhyay, P. Sahai, M. Singh, S. Chandrasekhar,
H. Gupta, and D. Salunke

National Institue of Immunology, New-Delhi-110067, India

INTRODUCTION

Although sperm toxic antibodies were raised in a heterospecies in 1899 by two pioneers, Landsteiner (1899) and Metchinikoff (1899), it is only today that one can visualize potential vaccines for control of animal or human fertility. An injectable that sterilizes male mammals without loss of libido and decline of testosterone has already received the New Drug Authorization from the Drugs Controller of India and is being marketed through a company. A luteinizing hormone releasing hormone (LHRH) vaccine is being advanced for reversible fertility control of pets by an Australian company and an American company is proposing zona based immunization for sterilization of female animals. As regards human fertility control, an important milestone has recently been reached; a vaccine intended for females is observed to protect women from becoming pregnant. This presentation will briefly review the progress that has taken place in the field and will discuss particularly recent work from our lab, where more than one birth control vaccines (BCVs) are under development.

MULTIPLE VACCINES THEORETICALLY POSSIBLE

The principle of BCVs is to induce antibodies and/or cell mediated immunity (CMI) against one or more hormones or gamete antigens which are important for reproduction. The mammalian reproductive system is regulated by a cascade of hormones. A hormonal signal also ensues fertilization. Immunization against many of these hormones leads to impairment of fertility in experimental animals, including sub-human primates. Clinical infertility due to antibodies or CMI directed against sperm antigens is a well recognized entity and such cases have been recorded all over the world. In most cases, couples are otherwise healthy and have no other problem. Thus immunization strategies against gamete antigens have a place and active research is leading to the identification of a number of promising candidate antigens on the sperm (Primakoff et al. 1988; Herr et al. 1990; Naz and Menge 1990; Shaha et al. 1990; Goldberg 1991) and on the egg (Sacco 1987; Henderson et al. 1988; Millar et al. 1989).

In the light of the above, it follows that more than one vaccine is theoretically possible. In the long run, it may be necessary to adopt a multivaccine approach with the idea of ensuring efficacy in almost all recipients. Immune response is determined by the genetic make-up and it may be expected that no single vaccine will be able to evoke a high enough response in all individuals. By adopting a multivaccine strategy, deficiency of response against a given antigen may be compensated by adequacy of response to others in a polyvalent mixture. Vaccines directed against reproductive hormones are at present more advanced than those directed against gamete antigens. Three vaccines have reached the stage of clinical trials after completion of experimental and toxicology studies.

FSH VACCINE

This vaccine is meant for control of male fertility. It was observed that spermatogenesis in primates and hopefully in humans requires follicle stimulating hormone (FSH) stimulus on a continuing basis, in contrast to rodents in which after initiation, FSH is no longer considered necessary (Moudgal 1981). Immunization of monkeys with ovine FSH generated antibodies cross-reactive with monkey FSH. Active and passive immunization led to a fall in sperm counts (Moudgal et al. 1988). Although azoospermia was not achieved, the sperm had diminished fertilization capability. The effect was reversible with regain of fertility on decline of antibodies. Anti-FSH immunization did not cause a decrease in testosterone levels. In the coming years, we should have better information on this vaccine from the results of the Phase I clinical trial that has just started in India in 6 volunteers.

LHRH VACCINE

LHRH is a key decapeptide that regulates the synthesis of gonadotropins in both males and females. These in turn modulate gonadal function, namely, the making of the egg and the sperm and also the sex steroid hormones. Interception of LHRH by bioeffective antibodies can cut off the entire chain of events in both males and females. In early experiments, effective immunization required the use of Freund's complete adjuvant (FCA) (Arimura et al. 1973; Fraser and Gunn 1973). Two groups of investigators have made semi-synthetic vaccines that produce a bioeffective response without the use of FCA. One of these has LHRH linked through the N-terminal to diphtheria toxoid (DT) (Ladd et al. 1988) and the other, developed by us, has utilized the aminoacid at position 6 for creating a functional group for covalent linkage; Gly^6 has been replaced by D-lysine. This was in turn linked to amino caproic acid, a spacer molecule to which the carrier DT was attached via its ϵ-NH$_2$ group (Jayashankar et al. 1989). This design has a rationale. Modeling of the structure of LHRH on the basis of conformational preferences of homologous sequence stretches in the protein data banks and the known data on structure-activity relationships suggests that the LHRH molecule is folded through a β-turn defined through a weak hydrogen bond between the backbone carboxylate of Trp^3 and the amide of Gly^6. The folding brings <Glu^1, the guanidyl group of Arg^8 and the amide on the C-terminal end in close proximity, forming the probable receptor recognition site. Carrier conjugation at position 6 would not disturb this site. NMR and computer graphic studies indicate that the LHRH molecule folds in a manner to bring the N- and C-terminal aminoacids proximal to each other. We observed in previous studies that the amide group at the C-terminal was important for the recognition of the hormone by conformation reading antibodies (Talwar et al. 1985).

The vaccine impairs the fertility of male animals. It has a particularly marked effect on the prostate, which atrophies dramatically (Jayashankar et al. 1989; Giri et al. 1990). With decline of antibodies, testicular functions are restored; the prostate, along with other accessory reproductive organs, regenerates. An issue which has been concerning us is whether the regenerated prostate is of a young or old type. There are no markers available to get information on this point. The question is, however, posed and it will be of interest to determine whether following immunological surgery by the vaccine, one can replace an aging prostate by one having the characteristics of a young prostate.

After drug regulatory and ethical approvals, probing clinical trials have been conducted with the vaccine in patients of carcinoma of the prostate in two centres in India and in one centre in Austria. The available results indicate that in patients in whom adequate antibody titres were generated by the vaccine, testosterone fell to castration levels. The mass of the prostatic tissue diminished, as seen by nephrestograms and ultrasound scans. This was corroborated by a fall in the prostatic specific antigen levels. The vaccine can thus be of utility in androgen dependent prostate cancers. What is required is a sustained delivery system so that the antibodies are maintained at high level for prolonged periods. Encapsulation of the vaccine in microspheres may provide such a response (Singh et al. 1991).

Immunization against LHRH can also be useful in control of fertility in females. A bioeffective anti-LHRH monoclonal suppressed estrus in dogs and inhibited ovulation in rats (Talwar et al. 1985). These antibodies could also abrogate pregnancy in mice (Gupta et al. 1985) and prevent menstrual cyclicity in primates (Talwar et al. 1984).

In recent clinical trials, treatment of postpartum women with an LHRH agonist prevented ovulation. No untoward side effects were observed (Fraser et al. 1989). Immunization with the present LHRH vaccine can be expected to achieve the same results. Post-delivery immunization of monkeys with the vaccine has produced no adverse effects on lactation and growth of the infants. Antibodies were absent or below the limits of detection of assay (<0.1 ng/ml) in milk of monkeys and in serum samples of infants. On weaning, control monkeys resumed ovulatory cycles in course of time, whereas immunized monkeys carrying anti-LHRH antibodies continued to have suppressed ovulation. Preliminary clinical trials have been approved in postpartum women to determine whether the vaccine can indeed prolong lactational amenorrhoea without adverse effects on the mother or the suckling infant.

hCG VACCINE

Human chorionic gonadotropin (hCG) is made and secreted fairly early after fertilization. The hormone is made by pre-implantation blastocysts (Fishel et al. 1984). It is essential for the establishment of early pregnancy. Its action may reside at two points: it may have a role in implantation; marmoset embryos exposed to anti-hCG antibodies fail to implant (Hearn et al. 1988) and it may also be important for maintenance of the endometrium to receive the embryo. This is accomplished by sustained production of progesterone by the ovaries under the stimulus of hCG. The interception of hCG by circulating antibodies may thus prevent the initiation of pregnancy. Based on these facts and confirmed by experimentation in three species of primates (Hearn 1976; Stevens 1976; Talwar et al. 1980), hCG was an early choice as a target for a BCV. Another advantage in choosing hCG was the expectation that the normal physiology of the non-pregnant female will not be interfered with. In this respect, it would differ from the widely used steroidal contraceptives, which though very effective, block ovulation as well as the endogenous production of sex steroids, which are replaced by synthetic compounds.

At present two vaccines are under development against hCG. One of them uses the 37-aminoacid carboxy terminal peptide (CTP) of βhCG (Jones et al. 1988). The other uses the entire βhCG or a heterospecies dimer (HSD) created by association of βhCG with the α-subunit of ovine LH (Talwar et al. 1990). In both, the hormonal ligand is linked to TT or DT. CTPs are poor immunogens as compared to βhCG based vaccines (Ramakrishnan et al. 1979). HSD was considered a better proposition than βhCG on the grounds that it has a conformation similar to the hCG. To attain this conformation the two subunits (α and β) of the hormone have to associate; the dissociated subunits recognize poorly, if at all, receptors on target tissues. As the homologous dimer could not be used for fear of cross-reactivity with human FSH and TSH (which have a common α-subunit), we exploited the conserved potential of the subunits to associate with each other across the species. Thus the α-subunit of ovine origin readily combines with the β-subunit of hCG. The HSD thus created recognizes receptors on target tissues.

Safety and Reversibility of HSD Vaccine

Experimental observations in rats and monkeys revealed that the HSD vaccine was more immunogenic as compared to the βhCG vaccine (Talwar and Om Singh 1988), and generated antibodies of higher bioneutralization capacity (Talwar et al. 1988). Pre-clinical toxicology studies were carried out in two species of animals according to standard international protocols as specified by the Indian Council of Medical Research. These studies indicated the lack of adverse side effects. After due approvals, multicentric Phase I clinical trials were undertaken in 112 women. These studies confirmed the better immunogenicity of the HSD vaccine in comparison to the βhCG vaccine. All women immunized with the HSD vaccine produced antibodies (Om Singh et al. 1989). Antibodies were of high affinity and were capable of neutralizing the bioactivity of hCG both in vitro and in vivo (Pal et al. 1988). Antibodies declined to near zero levels in the absence of a booster injection, indicating the reversibility of immunization. All hematological and clinical chemistry parameters remained normal (Talwar et al. 1990) and no immunopathological reactivities were observed. Ovulation and menstrual cyclicity remained undisturbed (Kharat et al. 1990).

Repeated immunization with HSD or βhCG linked to a single carrier, led to a state of non-responsiveness in some women. This observation was similar to the phenomenon previously observed in mice by Herzenberg et al. (1983). It could be overcome by presentation of the hormonal ligand on an alternate carrier (Gaur et al. 1990).

Phase II Efficacy Trials

Having found that the HSD vaccine was immunogenic, made the right type of antibodies with characteristics of high avidity for hCG and the ability to inactivate its bioactivity, that immunization with the HSD vaccine was free of side effects, and that the response was reversible, clinical trials were undertaken to get answer to the crucial issue of efficacy. Phase I clinical trials for all hCG vaccines investigated so far were conducted on women who had previously undergone elective tubal ligation. For efficacy studies, women of proven fertility (with at least two live children) cohabiting with partners of proven fertility were enrolled by informed consent. Several women were hyperfertile and had undergone one or more elective termination of pregnancy. The trials were carried out in three major centres: the All India Institute of Medical Science and the Safdarjung

Hospital, New Delhi, and the Postgraduate Institute of Medical Education and Research, Chandigarh. The main objective of the trials was to determine whether antibody titres above a threshold of 50 ng/ml neutralization capacity could prevent pregnancy. The protocol required that at least 750 protected cycles be recorded to establish firmly the efficacy. The vaccine employed was HSD linked to TT or DT, used in an alternating sequence. The dose was 300 μg per injection. The primary immunization schedule consisted of three injections at six weekly intervals (as is the case for TT and DT immunization). Booster immunizations were given as and when necessary to maintain antibody levels higher than the threshold of 50 ng/ml neutralization capacity.

Observations: As on 27th July, 1992, observations had been recorded on 928 cycles, with only one pregnancy having taken place at antibody titres above 50 ng/ml. 13 women had been protected for 18-27 cycles, 24 for at least 12 cycles and 30 for 6 to 11 cycles. Immunization did not affect ovulation. Progesterone in bleeds taken in the mid-luteal phase ranged from 14-44 nM/L, which are indicative of ovulation. All women continued to menstruate regularly, about 88% of the cycles were of normal duration (22-35 days). Shortening or lengthening of cycles was not related to antibody titres, and their frequency was the same as in control women. Libido of immunized women remained normal. Post-coital test (PCT), conducted at mid-cycle in 8 vaccinated women showed that they did not become pregnant in spite of cervical scores ranging from 12-14 and sperms of high motility and number detectable.

The effect of the vaccine was reversible. Pregnancies took place below 35 ng/ml bioneutralization capacity. These observations modify previous estimates of 20 ng/ml to prevent pregnancy (Nash et al. 1980; Jones et al. 1988). The present studies are the first of their type to demonstrate unequivocally the possibility of preventing pregnancy by a BCV. They also substantiate the crucial role of hCG in the establishment of pregnancy.

While there is a regain of fertility without any problem, an important issue raised is the possible residual ill-effects of immunization on the progeny. Three immunized women were desirous of having children. They were not given booster immunizations and they conceived when their respective antibody titres were < 5, 10 and 20 ng/ml. Pregnancy proceeded to term in all three cases, resulting in normal children.

Planned Further Developments

While an important milestone has been passed in establishing safety, reversibility, and efficacy of the HSD vaccine for birth control, the vaccine at present is not ready for adoption in family planning programmes. Primary immunization demands three injections, and it takes about 3-4 months to build up antibody levels above the protective threshold. This period will be vulnerable to pregnancy and it is important to devise an approach which is compatible with the HSD vaccine for covering this lag period. It is further logistically important to deliver at one visit the multiple immunization doses required to ensure efficacy for a defined period, say 6 months, 1 year or 2 years. In fact, it may be possible to make capsules for all these time periods. The following progress has taken place to meet these demands:

Coverage of lag period by a vaccine for inducing local CMI (VILCI): A purified extract (Praneem) of an ancient Indian tree, neem (Azadirachta indica), has been observed to activate locally the cell mediated immune reactions. This has been exploited to develop a potential companion vaccine. Administration of a small amount of Praneem in the uterus blocks fertility for several months in rodents (Upadhayay et al. 1990). The treatment does not disturb ovarian functions and ovulation is maintained. The action is exercised by activation of macrophages and by the induction of MHC class II antigens on various cells in the uterus. These cells manifest an increased reactivity to sperm, causing the production of lymphokines such as τ-interferon and TNF (Upadhyay et al. 1992). These interesting observations offer not only an additional product for control of fertility, but also have promise for covering the lag period in the development of the immune response to the hCG vaccine. These observations also underline the importance of CMI for control of fertility. Hitherto, primary attention has been focused on antibodies, which may have a primary role in inactivating a hormone, but for gamete antigens CMI may be equally, if not more, important for achieving efficacy. The possible role of CMI in infertility is emerging from clinical observations (Anderson and Hill 1988). Several years ago, we achieved aspermatogenesis in male animals by essentially activating CMI locally in the testis (Talwar et al. 1979). The intervention did not lead to the formation of anti-sperm antibodies. The effect was localized and did not get transferred to the contralateral testis if the injection was given on one side. Administration of Praneem VILCI to one uterine horn does not prevent normal pregnancy in the other horn of rats (Upadhyay et al. 1990), again demonstrating a highly localized effect. Pre-clinical toxicology studies have been completed on Praneem VILCI, and with the approval of the regulatory authorities, clinical trials with the HSD vaccine ± VILCI are envisaged in the near future.

Biodegradable microspheres for vaccine delivery: Experiments are in progress to encapsulate the vaccine in polylactic and polyglycolic acid polymer microspheres to develop a formulation which can engender an antibody response for 6 month or a year after a single administration. Results show the possibility of making a six months response capsule (Singh et al. 1991). The delivery system has, however, possibility of inducing antibody response of longer duration (Stevens 1992).

A long term antibody response may be possible by the use of a live recombinant vaccine. We have cloned the βhCG gene in the vaccinia virus, which is expressed in a bioactive and immunoreactive form (Lall et al. 1988; Chakrabarti 1989). By co-expression of the βhCG in alignment with a 48 aminoacid transmembrane fragment, high and sustained antibody responses were obtained in monkeys following a single immunization with the recombinant virus and a conventional booster (Srinivasan et al. 1992).

SUMMARY & CONCLUDING COMMENTS

The BCVs are entering a decisive phase. Efficacy of one of them, the HSD vaccine which induces bioeffective antibodies against hCG has recently been demonstrated. It can prevent pregnancy in women above antibody levels of 50 ng per ml. Earlier studies have established the safety and reversibility of the vaccine. It is posed for further developments to make it suitable logistically for family planning programme. The lag period of antibody build-up to protective levels has to be covered by a compatible approach. Purified extract of neem seeds, modulating locally CMI in the uterus, has shown desirable characteristics in rats and monkeys. Praneem acts as a vaccine inducing local CMI, emphasizing that birth control vaccines need not be confined to those making antibodies, though these are necessary to inactivate hormones. Local activation of CMI in the genital tract can offer an effective method for fertility control with no adverse effect on ovarian functions.

Biodegradable vaccine delivery systems would be required to administer multiple doses at a single contact point. Microspheres encapsulating the vaccines engendering a long term response are under development. Live recombinant vaccines offer an alternate way to induce a long term response (up to two years) following a single immunization and a conventional booster. In addition, these vaccines would be cheaper and amenable to large scale production. An LHRH vaccine has entered early Phase I clinical trials for the extension of lactational amenorrhoea in postpartum women. The vaccine has also therapeutic potential and is a cheaper alternative to LHRH agonists for the treatment of carcinoma of the prostate and other disorders. The FSH vaccine is the first to enter Phase I clinical trials for fertility control in males.

ACKNOWLEDGEMENTS

Research and clinical trials discussed in this paper were supported by S & T Project of the Dept of Biotechnology, Govt. of India, IDRC of Canada and the Rockefeller Foundation. The work benefited from cooperative interaction with the International Committee for Contraception Research of the Population Council, New York.

REFERENCES

Anderson DJ, Hill JA (1988) Am J Reprod Immunol Microbiol 17:22-30
Arimura A, Sato H, Kumasaka T, Worobec RB, Debeljuk L, Dunn J, Schally AV (1973) Endocrinology 93:1092-1103
Chakrabarti S, Srinivasan J, Lall L, Rao LV, Talwar GP (1989) Gene 77:87-93
Fishel SB, Edwards RG, Evans CJ (1984) Science 223:816-818
Fraser HM, Dewart PJ, Smith SK, Cowen GM, Sandow J, McNeilly AS (1989) J Clin Endocrinol Metab 69:996-1002
Fraser HM, Gunn A (1973) Nature 244:160-161
Gaur A, Arunan K, Om Singh, Talwar GP (1990) Int Immunol 2:151-155
Giri DK, Chaudhuri MK, Jayashankar R, Neelaram G, Jayaraman S, Talwar GP (1990) Exp Mol Pathol 52:54-62
Goldberg E (1991) Lactate dehydrogenase C_4 as an immunocontraceptive model. In: Alexander N, Griffin D, Spieler J, Waites GMH (eds) Gamete interaction, prospects for immunocontraception. Wiley Liss Inc, New York, p 63
Gupta SK, Om Singh, Talwar GP (1985) Am J Reprod Immunol Microbiol 7:104-108
Hearn JP (1976) Proc R Soc Lond B 195:149-160
Hearn JP, Gidley-Baird AA, Hodges JK, Summers PM, Wibley GE (1988) J Reprod Fertil (Suppl) 36:49-58
Henderson CJ, Hulme MJ, Aitken RJ (1988) J Reprod Fertil 83:325-343
Herr JC, Flickingen CJ, Homyk M, Klotz K, John E (1990) Biol Rep 42:181-193
Herzenberg LA, Tokuhisa T, Hayakawa K (1983) Ann Rev Immunol 1:609-632
Jayashankar R, Chaudhuri MK, Om Singh, Alam A, Talwar GP (1989) Prostate 14:3-11
Jones WR, Bradley J, Judd SJ, Denholm EH, Ing RMY, Mueller UW, Powell J, Griffin PD, Stevens VC (1988) Lancet i:1295-1298.
Kharat I, Nair NS, Dhall K, Sawhney H, Krishna U, Shahani SM, Banerjee A, Roy S, Kumar S, Hingorani V, Om Singh, Talwar GP (1990)

Contraception 41:293-299

Ladd A, Prabhu G, Tsong YY, Probst T, Chung W, Thau R (1988) Am J Reprod Immunol Microbiol 17:121-127

Lall L, Srinivasan J, Rao LV, Talwar GP, Chakrabarti S (1988) Ind J Biochem Biophys 25:510-514

Landsteiner K (1899) Zentralb Bacteriol 25:546

Metchinikoff E (1899) Ann Inst Pasteur (Paris) 14:577

Millar SE, Chamow SM, Baur AW, Oliver C, Robey F, Dean J (1989) Science 246:935-938

Moudgal NR (1981) Arch Androl 7:117-125

Moudgal NR, Murthy GS, Ravindranath N, Rao AJ, Prasad MRN (1988) Development of a contraceptive vaccine for use by the human male: Results of a feasibility study carried out in adult male bonnet monkeys (Macaca radiata). In: Talwar GP (ed) Contraception research for today and the nineties. Springer-Verlag, New York, p 253

Nash H, Talwar GP, Segal S, Luukkainen T, Johannsson EDB, Vasquez J, Coutinho E, Sundaram K (1980) Fertil Steril 34:328-335

Naz R, Menge A (1990) Hum Reprod 5:511-518

Om Singh, Rao LV, Gaur A, Sharma NC, Alam A, Talwar GP (1989) Fertil Steril 52:739-744

Pal R, Om Singh, Rao LV, Talwar GP (1990) Am J Reprod Immunol 22:124-126

Primakoff P, Lathrop W, Woolman L, Cowan A, Myles D (1988) Nature 335:543-546

Ramakrishnan S, Das C, Dubey SK, Salahuddin M, Talwar GP (1979) J Reprod Immunol 1:249-261

Sacco AG (1987) Am J Reprod Immunol Microbiol 15:122-130

Shaha C, Suri A Talwar GP (1990) Int J Androl 13:17-25

Singh M, Singh A, Talwar GP (1991) Pharm Res 8:958-961

Srinivasan J, Om Singh, Pal R, Lall L, Chakrabarti S, Talwar GP (1992) A recombinant anti-fertility vaccine. In: Local immunity in reproductive tract tissues. (in press)

Stevens VC (1976) Perspectives of development of a fertility control vaccine from hormonal antigens of the trophoblast. In: Development of vaccines for fertility regulation. Scriptor, Copenhagen, p 93

Stevens VC (1992) Scand J Immunnol (in press)

Talwar GP, Das C, Tandon A, Sharma MG, Salahuddin M, Dubey SK (1980) Immunization against hCG: efficacy and teratological studies in baboons. In: Anand Kumar TC (ed) Non-human primate models for study of human reproduction. Karger, Basel, p 190

Talwar GP, Gupta SK, Singh V, Sahal D, Iyer KSN, Om Singh (1985) Proc Natl Acad Sci USA 82:1228-1231

Talwar GP, Hingorani V, Kumar S, Roy S, Banerjee A, Shahani SM, Krishna U, Dhall K, Sawhney H, Sharma NC, Om Singh, Gaur A, Rao LV, Arunan K (1990) Contraception 41:301-316

Talwar GP, Naz RK, Das C, Das RP (1979) Proc Natl Acad Sci USA 76:5882-5885

Talwar GP, Om Singh (1988) Birth control vaccines inducing antibodies against chorionic gonadotropin. In: Talwar GP (ed) Contraception research for today and the nineties. Springer-Verlag, New York, p183

Talwar GP, Om Singh, Rao LV (1988) J Reprod Immunol 14:203-212

Talwar GP, Sharma NC, Dubey SK, Salahuddin M, Das C, Ramakrishnan S, Kumar S, Hingorani V (1976) Proc Natl Acad Sci USA 73:218-222

Talwar GP, Singh V, Om Singh, Das C, Gupta SK, Singh G (1984) Pituitary and extra-pituitary sites of action of gonadotropin-releasing hormone: potential uses of active and passive immunization against gonadotropin-releasing hormone. In: Saxena BB, Catt KJ, Birnbaumer L, Martini L (eds) Hormone receptors in growth and reproduction. Raven Press, New York, p 351

Upadhyay SN, Kaushic C, Talwar GP (1990) Proc R Soc Lond B 242:175-179

Upadhyay SN, Dhawan S, Garg S, Talwar GP (1992) Int J Immunopharm (in press)

The Role of Placental IL-10 in Maternal-Fetal Immune Interactions

T. G. Wegmann, L. J. Guilbert, T. R. Mosmann, and H. Lin

Departement of Immunology, University of Alberta, 8-65 Medical Sciences Building, Edmonton, Alberta, Canada T6G 2H7

Over the past decade a fair amount of information has accumulated indicating that the maternal immune response can either harm the fetus or improve its chances to survive. Spontaneous fetal resorption in the CBA x DBA/2 mouse model was shown to be reversible following injection with allogeneic spleen cells. T cells are clearly involved in this phenomenon because removal of them by monoclonal antibody after the immunization has taken place completely abrogates the effect (Athanassakis et al 1990). That components of the immune system were causing the damage in the first place was shown by the experiments of Baines and his colleagues in which they demonstrated that local NK cells play a role in fetal resorption (Gendron and Baines 1988). This was confirmed by Kinsky and his colleagues who showed that the double-stranded RNA poly IC can induce fetal resorption, and this effect can be adoptively transferred by spleen cells from poly IC-treated animals. The adoptive transfer effect can be eliminated by treating the spleen cells prior to transfer with anti-NK antibodies (Kinsky et al 1990).

The poly IC induced fetal demise can also be reversed by immunizing with isolated class I or II MHC antigens in the form of mutant cells, transfected L cells and immunogenic peptides (Kinsky et al submitted). Over the past five years, my colleagues and I, and others as well, have been investigating the cytokine basis for these effects. We have primarily concentrated on the positive effects of the immunization procedure because, not only does alloimmunization increase fetal survival, it also leads to increased placental and fetal weight and increased placental phagocytosis. From these observations we determined that members of the CSF family, including GM-CSF, IL-3 and CSF-1 can serve as growth and differentiation factors for the trophoblast, as recently reviewed elsewhere (Wegmann TG 1991). In addition, we have found that injecting GM-CSF or IL-3 into pregnant females whose pregnancies are at risk leads to increased fetal survival, as well as enhanced feto-placental growth. On the other hand, injecting γ-IFN, IL-2 or TNF-α leads to decreased fetal survival (Chaouat et al 1990).

Until recently the regulation of these phenomena was not clear. However, a set of recent observations now suggest that IL-10 might be involved. As a background to these recent findings, Holland et al (1984), showed a number of years ago that during early pregnancy in the mouse it is difficult to induce priming for delayed type hypersensitivity reactions. More recently Dresser has shown in the mouse that the lymph nodes draining the pregnant uterus are much more prone to form plaque forming cells against sheep red blood cells than are other lymphoid organs in the pregnant female or uterine draining lymph nodes in virgin females (Dresser 1991). In the meantime, Mosmann and his colleagues have shown that IL-10 has the property of downregulating γ-IFN and thus promoting the activity of TH2 versus TH1 type cells (Mosmann and Moore 1991). We thus formulated the hypothesis that IL-10 would be released locally from the tissues of the fetal-placental unit during pregnancy in order to direct the maternal immune response from potentially harmful effects against the fetus (NK, DTH) and towards a more beneficial response (TH2).

We therefore have examined the supernatants of tissue cultures derived from placenta, decidua, draining lymph nodes and systemic immune tissues of pregnant females at all three trimesters of pregnancy for IL-10, IL-5 and γ-IFN by double monoclonal antibody sandwich ELISA assays. The results support the above conjecture in that there is a constitutive production of large amounts of IL-10 from the placenta and/or the decidua at days 6, 12 and 18 of pregnancy. There is also a lesser amount of IL-10 present in the lymph nodes draining the uterus, and none in other lymphoid tissues such as spleen and mesenteric lymph nodes. Significant amounts of IL-5 are also present in the same distribution. γ-IFN is also secreted but in very low amounts and only in the day 6 preparations (Lin et al unpublished observations). This pattern clearly fits the description given in the above hypothesis.

Additional unpublished experiments done in collaboration with Raj Ragupathy and his colleagues at the National Institute of Immunology in New Delhi provide a very interesting correlate of these observations. Placentas were taken from an aborting strain combination, namely CBA x DBA/2, or a non-aborting combination, CBA x BALB. In both cases, virgin CBA splenic lymphocytes were added to the cultures and supernatants were collected after 3 days for analysis by the above cytokines assays. The results were as expected from the above description. Thus in the mixed lymphocyte placental reaction where the aborting placenta was the target, large amounts of γ-IFN were released along with practically undetectable levels of IL-5. On the other hand, with the non-aborting placentas as the target large amounts of IL-5 were released with low levels of γ-IFN. We also recently found that IL-3 is constitutively produced at the maternal-fetal interface (Lin et al unpublished observations). Surprisingly, IL-3 was released in mixed lymphocyte cellular reactions in which the aborting placenta was the target. This is not what we would have predicted from our previous observations that IL-3 positively affects placental growth and function (Athanassakis et al 1990), but it should be noted that both TH1 and TH 2 cells release IL-3 (Mosmann and Moore 1991).

A prediction that one can make from the above results is that eliminating IL-10 during pregnancy should be harmful to the fetus. This result has been achieved by Rajewsky and his colleagues, as reported in the proceedings of this congress (Kuhn et al 1992). They found that mice rendered deficient for IL-10 by gene knockout experiments were runted at birth with an average weight only two-thirds that of their litter mate controls. This observation indicates that the probable source of the IL-10 is not from the maternal immune system, but from embryonically derived tissue, perhaps from the trophoblast itself. However, detailed tissue localization studies remain to be done. In the meantime, the implication of all the above results is that cells within the placenta, apparently of embryonic origin, direct the pattern of the maternal immune response away from harmful delayed type hypersensitivity and in particular natural killer type reactivity and towards harmless and perhaps even beneficial humoral immunity by secreting abundant quantities of IL-10. If this balance is shifted they become runted.

How can we explain the runting? One presumes that absence of IL-10 leads to an increase in γ-IFN. It is known that γ-IFN can inhibit the GM-CSF which is released from the uterine epithelial cell lining under the influence of estrogen, as shown by Robertson and her colleagues (personal communication). GM-CSF promotes the growth of the early trophoblast, as well as implantation of the embryo into the uterine epithelium. In addition, and more to the point, GM-CSF promotes trophoblast syncytialization which can lead to the increased production of placental lactogen, as shown by our group (Guilbert et al 1991) as well as by Shiverick and her colleagues in the rat. Indeed, the latter group has shown that particular families of placental lactogen are induced by this cytokine (Shiverick personal communication), and, since placental lactogen is a fetal growth factor, this provides a possible pathway for explaining the runting in the absence of IL-10. It will therefore be interesting to measure the relevant cytokine balances in the IL-10 knockout mice, to test this conjecture.

SUMMARY

We have shown that there is a constitutive release of IL-10 and IL-5 from cells of the maternal-fetal unit during all three trimesters of pregnancy. We have also shown that supernatants from these cells can downregulate γ-IFN using the IL-2 activated spleen cell assay. This can help explain why IL-10 knockout mice are runted at birth, given previous knowledge of the deleterious effects of γ-IFN on fetal growth and survival. The implications of these studies for understanding immunological aspects of fetal survival as well as intervening in a positive or negative manner will be of interest to pursue.

REFERENCES

Athanassakis I, Chaouat G, Wegmann TG (1990) Cell Immunol 129:13-21

Chaouat G, Menu E, Clark D, Dy M, Minkowski, Wegmann TG (1990) J Reprod Fertil 89:447-458

Dresser DW (1991) J Reprod Immunol 20:253-266

Gendron R, Baines M (1988) Cell Immunol 113:261

Guilbert LJ, Athanassakis I, Branch DR, Christopherson R, Crainie M, Garcia-Lloret M, Mogil R, Morrish D, Ramsoondar J, Vassiliadis S, Wegmann TG (1991) The placenta as an immune:endocrine interface. In: Wegmann TG et al (eds) Molecular and cellular immunology. Oxford University Press, pp 261-276

Holland D, Bretscher P, Russell AS (1984) Lab Immunol 14:177-179

Kinsky R, Delage G, Rosin N, Thang MN, Hoffmann M, Chaouat G (1990) Am J Reprod Immunol 23:73

Kinsky R, Kapovic M, Menu E, Jaulin C, Kourilsky P, Wegmann TG, Thang, MN, Chaouat G (submitted)

Kuhn R, Rajewsky K, Muller W (1992) IL-4 and IL-10 deficient mice. In: Abstracts, 8th International Congress of Immunology, Budapest, Hungary, August 23-28, p. 203.

Lin H, Guilbert L, Mosmann TR, Wegmann TG (unpublished observations)

Mosmann TR, Moore KW (1991) Immunol Today 12:A49-A53

Robertson S (personal communication)

Shiverick K (personal communication)

Wegmann TG (1991) Curr Op Immunol 3:759-761

Immuno-Endocrine Networks in Pregnancy

Julia Szekeres-Bartho, R. Kinsky[*], M. Kapovic[**], G. Chaouat[*],
P. Varga, T. Csiszar

Inst. Microbiol. University Med. School H-7643 Pécs, Hungary
[*]Biologie Cellulaire et Moleculaire de la Relation Materno Fetale Clamart 94120
France [**]Dept. Physiol. and Immunol. University of Rijeka, Yugoslavia

INTRODUCTION

Immunological aspects of the fetal-maternal relationship have attracted interest of reproductive biologists, endocrinologists and immunologists for many years. However, the central question of how the fetus is protected from a potentially harmful maternal immune response has not been adequately answered.

Several lines of evidence support the concept that nonspecific immunological mechanisms are involved in spontaneous pregnancy termination. The higher incidence of viral infections (Pickard, 1968) or tumors (Janerich, 1980) suggest that cell mediated immunity is altered during pregnancy.

Sex related differences in immune responsiveness and alteration of endocrine functions during pregnancy focused attention on the role of sex steroids in regulation of the immune response.

Sex steroids influence both the immune cell number and function in the uterus (Mathur et al., 1979). Progesterone is essential for the maintenance of pregnancy, blocking of progesterone binding sites by an antiprogesterone causes abortion in humans. High local concentrations of this hormone prolong the survival of xenogeneic and allogeneic skin grafts (Hansen et al., 1986). Progesterone blocks in vitro T cell activation (Stites et al 1983), inhibits IL-1 induced proliferation (Stites and Siiteri, 1983) and interferes with IL-2 receptor binding (Van Vlasselaer and Vandeputte, 1986). Relatively high concentrations of progesterone (0.5 to 20 ug/ml) were needed to affect responses of lymphocytes from healthy non-pregnant individuals.

PROGESTERONE BINDING IN PREGNANCY LYMPHOCYTES

Earlier we observed an immunological response of pregnancy lymphocytes to in vitro treatment by low concentrations of progesterone and the lack of response, in the same conditions, of non-pregnancy lymphocytes (Szekeres-Bartho et al., 1985). Lymphocytes of pregnant women displayed a significantly higher progesterone binding capacity than those of healthy donors, and progesterone binding capacity of the lymphocytes showed an inverse correlation with their cytotoxic activity (Szekeres-Bartho et al., 1983). Considering the high progesterone concentrations in pregnancy serum, *alteration of progesterone sensitivity of the lymphocytes appears to be an important factor in pregnancy-related changes of immune responsiveness.*

Specific high affinity receptors have so far not been identified in resting human lymphocytes. We observed specific binding of ^{3}[H] progesterone and that of progesterone receptor specific MoAbs in the CD8+ population of pregnancy lymphocytes, but not in those of non-pregnant individuals (Szekeres-Bartho et al., 1990). The percentage of progesterone receptor (PR) containing lymphocytes increased throughout gestation, and spontaneous pregnancy termination was associated with a fall in the number of PR positive cells.

THE IMMUNOLOGIC EFFECTS OF PROGESTERONE ARE MEDIATED BY A PROTEIN

In the presence of progesterone, PR positive lymphocytes (mainly CD8+) produce a 34 kD protein (PIBF) which inhibits the release of arachidonic acid (Szekeres-Bartho et al., 1985 a). The release of the PIBF is inhibited in the presence of PR blockers, but not affected by glucocorticoid antagonists, suggesting that functioning PRs are needed for its production (Szekeres-Bartho et al., 1990a). Immunological actions of this protein include inhibition of NK mediated lysis and antigen induced proliferation as well as generation of suppressor activity (Szekeres-Bartho et al.,1989).

THE REGULATION OF PROGESTERONE DEPENDENT IMMUNOMODULATION

The PIBF is present in the serum of pregnant women and it disappears at the termination of pregnancy (Szekeres-Bartho et al.1989a). This raises the question: which are the signals, that initiate lymphocyte PR expression at the beginning of gestation, and those resulting in the disappearance of these binding sites at the termination of pregnancy. In endocrine tissues steroid hormones regulate the actual cellular level of their receptors. *The regulation of lymphocyte PRs is hormone-independent.*

Data obtained from in vitro and in vivo studies suggest, that *PR induction, thus the production of the PIBF is related to lymphocyte activation* (Szekeres-Bartho et al.,1990, 1989b). Recent observations suggest that activation via the CD3 molecule alone is sufficient for induction of PR positivity.

Studies on transfected cell lines revealed that polymorphic Class I as well as Class II HLA antigens induce PRs in non-pregnancy lymphocytes, whereas cells transfected with a murine monomorphic Class I antigen do not. PHA-induced PR expression was downregulated in the presence of murine monomorphic MHC products or a HLA-G positive human choriocarcinoma (BeWo) cell line. Based on these observations it cannot be ruled out that *expression of monomorphic MHC products might in certain circumstances, e.g., labour induce a transient refractoriness to the activation of PRs.*

Labour is associated with a reduced number of PR positive cells in peripheral blood. This phenomenon might result from either recirculation of receptor negative lymphocytes after local downregulation of PRs, or alternatively, the message responsible for downregulation of PR expression might be present in peripheral blood. Testing the latter possibility we found, that mid-pregnancy sera did not substantially influence PHA induced PR expression. Sera obtained at the time of delivery downregulated PHA-induced PR expression. Absorption of term sera with antibody to a monomorphic Class I antigen (W6/32) abrogated this effect.

Regulation of Progesterone-Dependent Immunomodulation by Placental Cells.

The above data allow the assumption that in pregnancy antigenic stimulation by the fetus might be involved in the induction of PRs but the tissue presenting the appropriate antigen is still to be identified. Trophoblast - a tissue of embryonic origin and in close contact with maternal blood throughout gestation might be a possible candidate. Maternal immune response to paternal (Pence et al., 1975) or neonatal (Rocklin et al.,1973) or placental alloantigens is well documented. The trophoblast does not express classical polymorphic HLA antigens, but it expresses trophoblast specific antigens, (Billington and Bell, 1983) trophoblast leucocyte cross reactive antigens (McIntyre and Faulk, 1982) and monomorphic Class I HLA antigens (Redman et al.,1984; Ellis et al., 1986).

In different mammalian species the length of gestation largely varies, which obviously corresponds to the size of the animal, thus the size of the placenta. It is conceivable, that the length of gestation is related to the life span of the placenta.

In function of placental differentiation and maturation the presentation of certain antigens on trophoblast cells might change, which in turn would exert a regulatory effect on lymphocyte PR expression. In support of this hypothesis, in our hands lectin binding of the trophoblast depended on the stage of gestation, and while trophoblast of normal 1st trimester placentae showed a strong ConA positivity, no ConA binding was seen in trophoblast from spontaneous abortion of a similar gestational age. The Con A binding of the trophoblast might indicate a gain or loss of glycoprotein antigens.

Syncytiotrophoblast villous surface membranes (STPM) or trophoblast enriched cells from 1st trimester placentae induced PRs in lymphocytes of non-pregnant individuals, while those from term placentae did not, and even downregulated PHA-induced lymphocyte PR expression. This allows the assumption, that trophoblast cells in the process of ageing might begin to lose or express surface structures this way regulating PR expression.

Gestational-age related changes in membrane fluidity influence antigen presentation by trophoblast cells: The theory of immunological inertia of the trophoblast was based on its reduced susceptibility to lysis by immune effectors (Zuckerman and Head, 1987) and the lack of polymorphic HLA antigens in this tissue. However, immunization with trophoblast has ben proved to be successful, in the presence of Freund adjuvant (Davies, 1982), suggesting that trophoblast is not inert but the antigens eliciting an immune response are not, at all stages of development properly presented.

Modulation membrane fluidity alters antigenicity of certain cell types(Shinitzky et al., 1979). Data from our laboratory suggest that the membrane fluidity of trophoblast cells is related to the gestational age or the physiological state of the placenta (Szekeres-Bartho et al.,1989c).The PR inducing capacity as well as susceptibility to NK mediated lysis of trophoblast cells changes in function of membrane fluidity.

These data imply, that *the expression of structures responsible for down regulation of lymphocyte PRs is determined by membrane fluidity.*

THE BIOLOGICAL SIGNIFICANCE OF PROGESTERONE-DEPENDENT IMMUNOMODULATION

Data from in vitro studies suggest a correlation between progesterone binding capacity of the lymphocytes and the success or failure of pregnancy. In vivo experiments provided a direct evidence for the importance of the progesterone-mediated immunomodulatory pathway. These studies also revealed that:
a)The anti-abortive effect of the PIBF in vivo is manifested via blocking the NK-TNF pathway: Earlier studies by Gendron and Baines (1988) suggest an NK involvement in murine resorptions. Transfer of high NK activity spleen cells from poly (I) poly (C 12U) treated mice is abortogenic in pregnant Balb/c mice (Kinsky et al., 1990). The mechanism by which NK cells affect fetal survival remains to be elucidated. NK cells are producers of TNF, on the other hand natural cytotoxic activity has been proposed to be either a cell-associated or a secreted form of TNF (Patek and Lin, 1989). The relationship

between TNF and fetal damage is well established. Parant (1987) observed the cessation of murine pregnancy after injection of recombinant murine or human TNF alpha. TNF antibodies or TNF antagonists correct the spontaneously high resorption rates in CBA/JXDBA/2 matings. (Chaouat et al., 1991). TNF is produced by activated lymphoid cells thus the transfer of activated spleen cells might result in extremely high TNF production from an exogenous source. This treatment resulted in elevated serum and placental TNF concentrations in pregnant Balb/c, simultaneously with increased resorption rates. Extracts from damaged placentae with resorbed fetuses contained significantly higher TNF levels than intact ones. Both TNF levels (Fig.1) and resorptions were corrected by simultaneous PIBF administration.

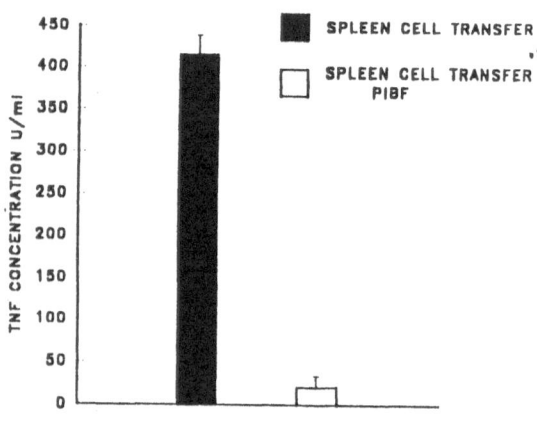

Fig.1 The effect of the PIBF on serum TNF levels in NK-mediated abortion

Whether the PIBF affects TNF release directly or inhibits its effects, remains to be established. In vitro studies suggest, that the PIBF counteracts the cytotoxic effect of TNF, but does not influence its production. Supernatants of progesterone-treated human pregnancy lymphocytes significantly inhibited the cytotoxic effect of recombinant human TNF alpha on L 929 cells. Control supernatants did not exert such effect (Fig 2).On the other hand, the PIBF did not inhibit LPS-induced TNF release by PBL. It seems clear, that *modulation of TNF activity is a step in the progesterone-dependent immunomodulatory pathway, and that the final anti-abortive effect of the PIBF is mediated by inhibition of the NK-TNF pathway.*

b) A proper stimulation of the maternal immune system is required for the operation of the progesterone-dependent immunomodulatory pathway: Data obtained in two murine models provide supportive evidence for the concept, that an appropriate antigenic stimulation is needed for lymphocyte PR induction and for the setting in of the progesterone-mediated immunomodulatory pathway.

1) Antiprogesterone treatment of pregnant Balb/c mice results in abortion, which can be prevented by simultaneous administration of the PIBF (Szekeres-Bartho et al., 1990b). Alloimmunization of mice with paternal strain type cells prior to mating results of a higher dose requirement of an antiprogesterone to induce abortion (Fig.3).

This allows the assumption that the rate of antigenic stimulation during pregnancy determines the number of progesterone receptors. The different number of cellular binding sites would result in different responsiveness to an antiprogesterone treatment.

2) When CBA/J females are mated to DBA/2 males, 25 to 40 % of the fetuses are resorbed, in contrast to the normal resorption rates (10%) observed in other combinations. Specific (Chaouat et al. 1983) or nonspecific immunostimulation (Toder et al. 1990) of the pregnant females has been shown to prevent resorption in this system. Spleen cells from pregnant mice fail, in this particular combination to produce the progesterone-dependent inhibitory substance in vitro.

Complete Freund adjuvant treatment of the pregnant females restores
this capacity, simultaneously correcting the resorption rates
(Szekeres-Bartho et al 1991).

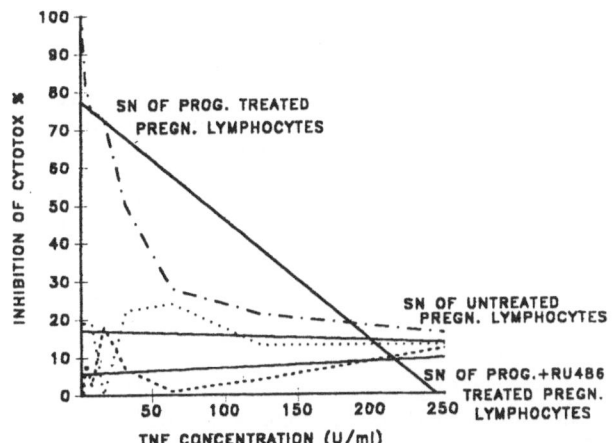

**Fig.2
The effect of the PIBF on
TNF-mediated cytotoxic-
ity.**

*Since the situation seems
to be similar in humans
these data imply, that in
pregnant women lacking
progesterone receptors in
their lymphocytes
(possibly as a result of
inadequate antigen pre-
sentation) the suppres-
sive effect of proges-
terone can be substituted
by the administration of
a progesterone-induced
blocking factor. This
might be of therapeutic
importance in certain
forms of spontaneous
abortion, offering an al-
ternative to active immu-
nization.*

**Fig.3
The effect of alloimmu-
nization on the abortive
effect of progesterone
receptor block in mice**

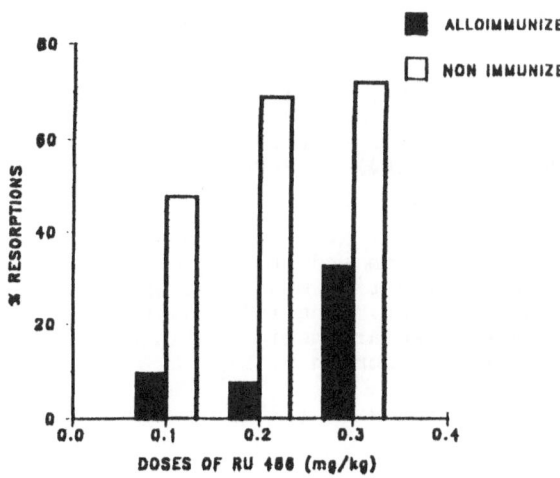

**THE PIBF IS INDISPENSABLE
FOR THE MAINTENANCE OF
GESTATION**

        The ultimate
proof, that the PIBF is
compulsory for normal
gestation would consist
in, inducing fetal loss,
by neutralizing the PIBF
activity in vivo.
        Nine to 12 days
pregnant Balb/c mice were
treated intravenously or
locally in the uterine
horns by a single injec-
tion of mouse anti-PIBF
serum. In case of local administration one of the uterine horns was
injected with a control serum. Animals treated the same way with
control sera or normal mouse serum were used as controls. Two days
later the mice were sacrificed and uteri inspected. Neutralizing the
PIBF activity with an antiserum administered either locally or
systematically to pregnant mice resulted in either an increased rate
of resorptions or complete abortion (Fig. 4).Control sera

(individual samples from normal mice, and different anti-Salmonella and anti-Shigella mouse sera) gave negative results.

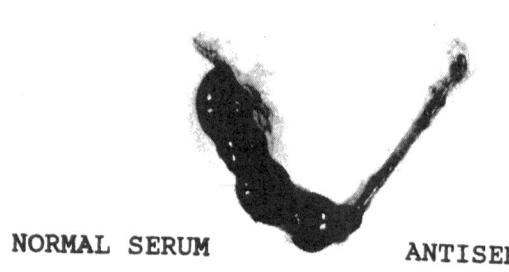

Fig.4
Fetal vastage induced by local administration of an anti-PIBF serum.

NORMAL SERUM   ANTISERUM TO PIBF

These data allow the conclusion that the progesterone-induced immunomodulatory protein is indispensable for the maintenance of normal gestation in mice.

CONCLUSIONS

Due to allogeneic stimulation during gestation lymphocytes of pregnant women develop a progesterone binding capacity. Under the effect of progesterone these lymphocytes produce a protein with immunomodulatory and anti abortive properties.

REFERENCES

Billington, W.D., Bell, S.C.(1983) Immunobiology of mouse trophoblast In: Loke Y.W., Whyte A. (eds.) Biology of trophoblast Amsterdam, Elsevier p. 571.

Butterworth, M., McClellan, B., Allansmith, M. (1967) Nature, 214: 1224.

Chaouat, G., Kiger, N., Wegmann, T.G. (1983) J. Reprod. Imunol 5: 389-392.

Chaouat, G., Menu, E., Wegmann, T.G. (1991) Role of lymphokines of the CSF family and of TNF, gamma-interferon and IL-2 on placental growth and fetal survival, studied in 2 murine models of spontaneous resorptions. In: G. Chaouat, and G. Mowbry. (eds.) Cellular and Molecular Biology of the Materno-Fetal Relationship. Colloque INSERM Vol 212, John Libbey Eurotext Ltd Paris, London,p. 91.

Davies, M., McLaughin, M.E.E., Sutcliffe, R.G. (1982) Immunology 47: 459-468.

Ellis, S.A., Sargent, I.L., Redman, C.W.G., McMichael, A.J. (1986) Immunology 59: 595-601.

Gendron, R., Baines, M. (1988) Cell. Immunol. 113: 261-268.

Hansen, P.J., Bazer, F.W., Segerson, E.C. (1986) Am. J. Reprod. Immunol. Microbiol. 12: 48-54.

Jancnich, D.T. (1989) Medical Hypotheses 6: 1149-1155.

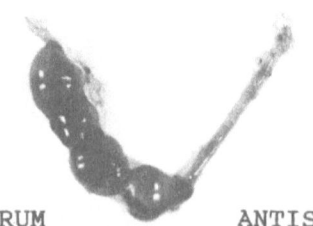

RMAL SERUM   ANTISERUM TO PIBF

Kinsky, R., Delage, G., Rosin, N., Thang, M.N., Hoffmann, M., Chaouat, G. (1990) Amer. J. Reprod. Immunol. 23: 73-77.

Mathur, S., Mathur, R.S., Goust, J.M., Williamson, H.O., Fudenberg, H.H. (1978) Clin. Immunol. Immunopathol. 13: 246.

McIntyre, J.A., Faulk, W.P.(1982) Hum. Immunol. 4: 27-35.

Parant, M. (1987) Immunobiol. 175: 26.

Patek, P.Q., Lin, Y. (1989) Immunology 67: 509-513.

Pence, H., Petty, E.M., Rocklin, R.E. (1985) J. Immunol. 114: 525-528.

Pickard, R.E. (1968) Am. J. Obstet. Gynecol. 100: 504.

Redman, C.W.G., McMichael, A.J., Stirrat, G.M. et al.(1984) Immunology 52: 457-468.

Rocklin, R.E., Zuckerman, J.R., Alpert, E. et al.(1973) Nature 241: 130-131.

Shinitzky, M., Skornick, Y., Haran-Ghera, N.(1979) Proc. Natl. Acad. Sci. USA 76: 5313-5316.

Stites, D.P., Bugbee, S., Siiteri, P.K. (1983) J. Reprod. Immunol. 5: 215-228.

Stites, D.P., Siiteri, P.K. (1983) Immunol. Rev. 75: 118-138.

Szekeres-Bartho, J., Csernus, V., Hadnagy, J., Pacsa, A.S. (1983) J. Reprod. Immunol. 5: 81-88

Szekeres-Bartho, J., Hadnagy, J., Pacsa, A.S.(1985) J. Reprod. Immunol 7: 121-128.

Szekeres-Bartho, J., Kilar, F., Falkay, G., Csernus, V., Torok, A., Pacsa, A.S. (1985 a) Am. J. Reprod. Immunol. Microbiol. 5: 15-19.

Szekeres-Bartho, J., Autran, B., Debre, P., Andreu, G., Denver, L., Chaouat, G. (1989) Cell. Immunol. 122: 281-294.

Szekeres-Bartho, J., Varga, P., Pejtsik, B. (1989 a) J. Reprod. Immunol. 16: 19-29.

Szekeres-Bartho, J., Weill, B.J., Mike, G., Houssin, D., Chaouat, G. (1989 b) Immunol. Letters 22: 259-262.

Szekeres-Bartho, J., Nemeth, A., Varga, P., Csernus, V., Koszegi, A., Paal, M. (1989 c) Amer. J. Reprod. Immunol. 19: 92-98 .

Szekeres-Bartho, J., Szekeres, Gy., Debre, P., Autran, B., Chaouat, G. (1990) Cell. Immunol. 125: 273-283.

Szekeres-Bartho, J. Philibert, D., Chaouat, G. (1990a) Amer. J. Reprod. Immunol. 23: 42-43.

Szekeres-Bartho, J., Kinsky, R., Chaouat, G. (1990 b) Amer. J. Ob. Gyn. 163: 1320-1322.

Szekeres-Bartho, J., Kinsky, R., Kapovic, M., Chaouat, G. (1991) Amer. J. Reprod. Immunol. 26: 82-83.

Toder, V., Strassburger, D., Irlin, I., Carp, H., Pecht, M., Trainin, N. (1990) Amer. J. Reprod. Immunol. 24: 63-66.

Van Vlasselaer, P., Vandeputte, M. (1986) Effect of sex steroids and trophoblast culture supernatants on the cytotoxic activity in mice. in:Han, J. (ed.) Pregnancy proteins in animals. Walter de Gruyter and Co. Berlin, New York, p.482.

Zuckerman, F.A., Head, J.R. (1987) Transplant. Proc. 1: 554-556.

17. Tumorimmunology

Genes Coding for Tumor Rejection Antigens

T. Boon, P. van der Bruggen, C. Traversari, B. van den Eynde,
B. Lethé, P. Chomez, A. Van Pel, P. Coulie, C. Lurquin, E. De Plaen,
F. Brasseur.

Ludwig Institute for Cancer Research, Brussels Branch 74 avenue Hippocrate - B-1200 Brussels, Belgium and Cellular Genetics Unit, Université Catholique de Louvain, B-1200 Brussels, Belgium

New antigens recognized on mouse tumor cells by syngeneic cytolytic T lymphocytes (CTL) can arise as a result of point mutations. Such mutations were observed in the genome of antigenic variants ("tum- variants") obtained by mutagenic treatment of mouse tumor cells. Three unrelated mutated genes coding for three different tum⁻ antigens have been identified (De Plaen et al. 1988); Sibille et al. 1990; Szikora et al. 1990). The normal equivalents of these genes are expressed in normal tissues. In all instances, the tum- mutation is located in the sequence that encodes the antigenic peptide and it results into a change of one amino-acid in the peptide. For two antigens, this confers to the antigenic peptide the capability to bind to the presenting class I MHC molecule (Lurquin et al. 1989).

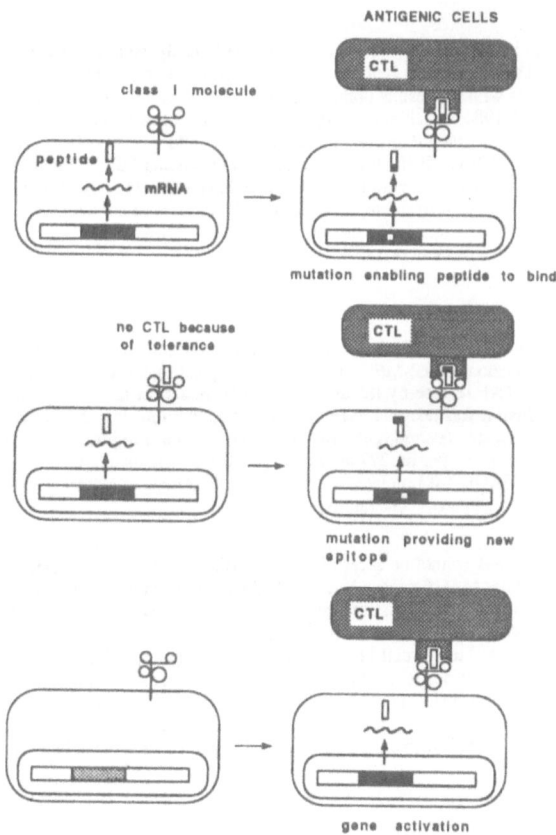

Fig. 1 Genetic processes underlying the production of new antigenic peptides. When a new epitope is provided by a mutation affecting a gene expressed in normal cells, the CTL directed against the peptide presented by these normal cells are probably eliminated or inactivated as a result of natural tolerance.

For the third, the mutation provides a new epitope to a peptide that is already capable of binding to a MHC molecule (fig. 1). In this instance, the CTL that recognize the original peptide, which should be produced in all normal cells, are presumably removed or inactivated during the establishment of natural tolerance. The mutational mechanism for the generation of new antigens would be expected to produce a vast array of different antigens, since every gene that is expressed constitutes a potential target. It is important to note that tum- antigens do not qualify as tumor rejection antigens since they are artificially induced on tumor cells by in vitro mutagenesis. But they constitute an potential model for tumor rejection antigens. In this regard, it is interesting that point mutated ras genes have recently been shown to produce specific peptides recognized by CD4+ T lymphocytes (Jung and Schluesner 1991; Peace et al. 1991).

The origin of a mouse tumor rejection antigen has been elucidated. This is antigen P815A, which is recognized on mouse tumor P815 by syngeneic CTL. The encoding gene, P1A is not expressed in normal mouse tissues, but it is expressed at a high level in tumor P815, so that an antigenic peptide is produced that binds to Ld (Van den Eynde et al. 1991) (fig. 1). Interestingly, this peptide bears two epitopes recognized by distinct CTL and, as a result of a point mutation, one of these epitopes can be lost while the other is maintained (Lethé et al. 1992). Gene P1A is activated in several mastocytoma tumors, but not in normal mast cells. The analysis of mastocytoma tumors selected from mast cell lines suggests that the activation of gene P1A is often but not always linked to the tumoral transformation (Van den Eynde and Moroni, unpublished observations).

To extend these findings to human tumors, we have systematically analyzed a number of human melanoma cell lines for their ability to stimulate in vitro lymphocytes drawn from the blood of the same patient (autologous lymphocytes). In accordance with the results obtained by several groups (Vanky and Klein 1982; Mukherji and MacAlister 1983; Knuth et al. 1984; Anichini et al. 1987), we have obtained specific anti-tumoral CTL responses. Stable CTL clones have been obtained that exert no LAK activity and that have high lytic activity against the autologous tumor (Hérin et al. 1987). Using this approach with melanoma MZ2-MEL we have obtained a panel of autologous anti-tumor CTL clones and with these clones we have selected resistant tumor cell variants. Most of these antigen-loss variants proved resistant to the selective CTL clones but not to the other
members of the panel. This analysis thus led to the conclusion that tumor MZ2-MEL presents a total of 6 distinct antigens to autologous CD8+ T cells (Van den Eynde et al. 1989).

To isolate the gene coding for antigen MZ2-E, an E- antigen-loss variant was transfected with a cosmid library prepared with the DNA of melanoma MZ2-MEL. Transfectants expressing the antigen were identified on the basis of their ability to stimulate TNF release by the anti-E CTL (Traversari et al. 1992). Such transfectants were obtained and from one of them it was possible to retrieve a cosmid that transfers the expression of the antigen at high efficiency (fig. 2). From this cosmid, gene MAGE-1 (melanoma antigen) was isolated. It comprises 3 exons, with a open reading frame coding for 277 amino-acids entirely contained in the third exon (fig. 3) (van der Bruggen et al. 1991). Gene MAGE-1 is unrelated to any previously characterized gene. It belongs to a family of more than twelve highly related genes (MAGE family).

The expression of gene MAGE-1 cannot be ascertained by Northern blotting and hybridization because probes cross-hybridize with all the other MAGE genes. But specific PCR (polymerase chain reaction) primers have been developed that distinguish different MAGE genes and that also distinguish mRNA from the genomic sequence because these primers are located in different exons. Using this PCR system we have screened a large array of normal tissues, tumor samples and tumor cell lines for MAGE-1 expression. MAGE-1 is not expressed at all in normal tissues with the exception of testis. Approximately 40% of all melanoma tumors express MAGE-1. So does a significant fraction of breast (Brasseur et al. 1992) and non-small cell lung tumors. Among melanomas, the level of expression appears to be quite variable from one tumor to another. It appears likely that some tumors express the gene at a level that is too low to produce enough antigen to be recognized by CTL (Table 1).

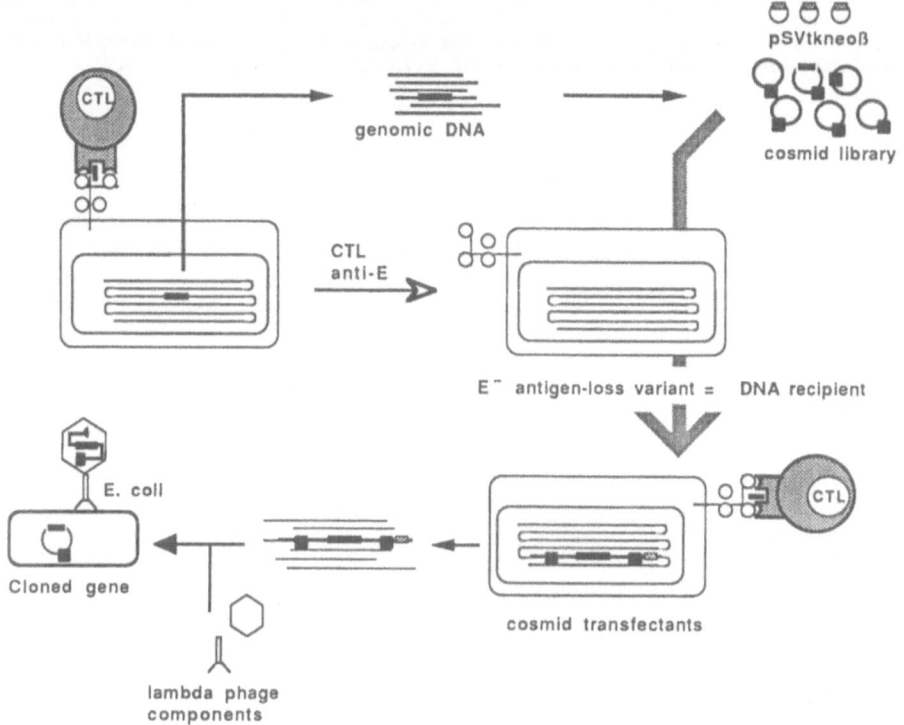

Fig. 2 Outline of the isolation of a gene encoding an antigen recognized by cytolytic T lymphocytes on a human melanoma.

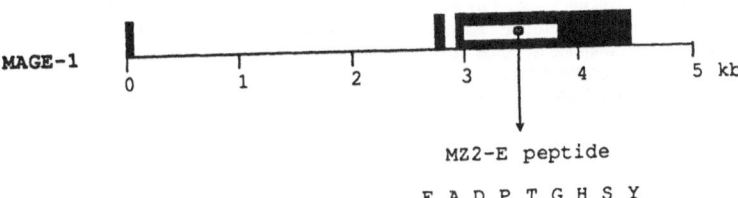

MZ2-E peptide

E A D P T G H S Y

Fig. 3 Structure of gene MAGE-1. The gene is represented with exons as black boxes and the open reading frame of exon 3 as white box. The sequence of the antigenic peptide MZ2-E encoded by gene MAGE-1 is indicated.

Antigen MZ2-E is presented by HLA-A1. By transfecting various fragments of the gene, it was possible to identify a short region coding for the antigenic peptide. Synthetic peptides sensitizing HLA-A1 cells to the anti-E CTL were obtained. A nonapeptide was defined that appears to be optimal because removal of either its N-terminal or its C-terminal amino-acid abolishes its antigenicity (fig. 3) (Traversari et al. 1992).

Table 1. Expression of gene MAGE-1 and antigen MZ2-E

	Expression of gene MAGE-1	Recognition by anti-E CTL tested by :	
		TNF release	Lysis
Cells of patient MZ2			
Melanoma cell line	++++	+	+
Tumor sample	+++		
Antigen-loss variant	-	-	-
CTL clone	-		
PHA-activated PBL	-		
Melanoma cell lines of other HLA-A1 patients			
LB34-MEL	++	+	±
LB36-MEL	+	+	-
LB45-MEL	-	-	-
MI665/2-MEL	-	-	-
MI10221-MEL	-	-	-
MI13443-MEL	+++	+	+
MI30462-MEL	-	-	-
SK33-MEL	-	-	-
SK23-MEL	-	-	-
SK24-MEL	++	+	+
LE1-MEL	++	+	±
LY1-MEL	+	+	+
LY2-MEL	+++	+	+

The results obtained with MAGE-1 lead to several positive conclusions. First, they vindicate the approach to identify human tumor rejection antigens by autologous mixed lymphocyte tumor cell cultures (MLTC) generating anti-tumor CTL. There was a risk that under the artificial conditions of these MLTC, where IL-2 is added to the culture, artefactual auto-immune response might be generated leading to CTL recognizing antigens expressed by normal cells. Our results demonstrate that genuine tumor-specific antigens can be identified by this method.

Secondly, the cytolytic T cell approach appears to lead to the identification of new genes, different from the known oncogenes, that are expressed specifically by tumor cells. The elucidation of the function of these genes may provide new insights regarding the mechanisms of tumor transformation and progression. However, we have not yet obtained any indication of the role that MAGE-1 may play in promoting the malignancy of tumor cells.

Finally, the availability of genes such as MAGE-1 opens new perspectives in cancer immunotherapy. Approximately 26% of the Caucasian population possesses the HLA-A1 gene. Melanoma patients about to receive surgery could be HLA-typed. For the HLA-A1 patients, the RNA extracted from a frozen tumor sample could then be analyzed by PCR to establish whether the tumor cells express MAGE-1. This should lead to a proportion of approximately 10% of melanoma patient that could be immunized with antigen MZ2-E with the certainty of immunizing against an antigen carried by their tumor. This approach may not lead rapidly to therapeutic advances, because several questions remain regarding our ability to immunize effectively against antigen MZ2-E. But the availability of the gene and the antigenic peptide offers many possibilities other than the injection of irradiated allogeneic cells. Cells engineered to express a high amount of antigen and perhaps interleukins may prove to be effective immunogens (Golumbek et al. 1991). Viral or bacterial engineered vaccines may also prove useful (Stover et al. 1991; Taylor et al. 1991; Taylor et al. 1992). Finally immunization with the MZ2-E peptide combined with adjuvant or peptide-presenting cells may be very effective (Deres et al. 1989; Aichele et al. 1990; Kast et al. 1991; Schild et al. 1991). It will be very important to acquire the ability to evaluate reliably the increase in anti-MZ2-E CTL precursors frequency obtained by these various modes of immunization. Improvements in limiting dilution assays may provide this (Brunner et al. 1980; Moretta et al. 1983; Coulie et al. 1992).

It is likely that the method used to identify gene MAGE-1 will soon provide the identification of new genes coding for human tumor rejection antigens. It is also possible that the direct isolation of antigenic peptides and their analysis will lead to the same result (Rötzschke et al. 1990; Wallny and Rammensee 1990; Hunt et al. 1992).

REFERENCES

Aichele, P, Hengartner, H, Zinkernagel, RM and Schulz, M (1990). Antiviral cytotoxic T cell response induced by in vivo priming with a free synthetic peptide. J. Exp. Med. 171: 1815.

Anichini, A, Fossati, G and Parmiani, G (1987). Clonal analysis of the cytolytic T-cell response to human tumors. Immunol. Today 8: 385-389.

Brasseur, F, Marchand, M, Vanwijck, R, Hérin, M, Lethé, B, Chomez, P and Boon, T (1992). Human gene MAGE-1, which codes for a tumor rejection antigen, is expressed by some breast tumors. Int. J. Cancer in press: -.

Brunner, K, McDonald, R and Cerottini, JC (1980). Antigenic specificity of the cytolytic T lymphocyte (CTL) response to murine sarcoma virus-induced tumors. II. Analysis of the clonal progeny of CTL precursors stimulated in vitro. J. Immunol. 124: 1627-.

Coulie, PG, Somville, M, Lehmann, F, Hainaut, P, Brasseur, F, Devos, R and Boon, T (1992). Precursor Frequency Analysis of Human Cytolytic T Lymphocytes Directed against Autologous Melanoma Cells. Int J Cancer 50: 289-297.

De Plaen, E, Lurquin, C, Van Pel, A, Mariamé, B, Szikora, J, Wölfel, T, Sibille, C, Chomez, P and Boon, T (1988). Tum- variants of mouse mastocytoma P815. IX. Cloning of the gene of tum- antigen P91A and identification of the tum- mutation. Proc. Natl. Acad. Sci. USA 85: 2274-2278.

Deres, K, Schild, KH, Wiesmüller, K-H, Jung, G and Rammensee, H-G (1989). In vivo priming of virus-specific cytotoxic T lymphocytes with synthetic lipopeptide vaccine. Nature 342: 561.

Golumbek, PT, Lazenby, AJ, Levitsky, HI, Jaffee, LM, Karasuyama, H, Baker, M and Pardoll, DM (1991). Treatment of established renal cancer by tumor cells engineered to secrete interleukin-4. Science 254: 713-716.

Hérin, M, Lemoine, C, Weynants, P, Vessière, F, Van Pel, A, Knuth, A, Devos, R and Boon, T (1987). Production of stable cytolytic T-cell clones directed against autologous human melanoma. Int. J. Cancer 39: 390-396.

Hunt, DF, Henderson, RA, Shabanowitz, J, Sakaguchi, K, Michel, H, Sevilir, N, Cox, AL, Appela, E and Engelhard, VH (1992). Characterization of peptides bound to the class I MHC molecule HLA-A2 by mass spectrometry. Science 255: 1261-1263.

Jung, S and Schluesner, HJ (1991). Human T lymphocytes recognize a reptide of single point-mutated, oncogenic ras proteins. J Exp Med 173: 273-276.

Kast, WM, Roux, L, Curren, J, Blom, HJJ, Voordouw, AC, Meloen, RH, Kolakofsky, D and Melief, CJM (1991). Protection agaisnt lethal Sendai virus infection by in vivo priming of virus-specific cytotoxic T lymphocytes with a free synthetic peptide. Proc. Natl. Acad. Sci. USA 88: 2283.

Knuth, A, Danowski, B, Oettgen, HF and Old, L (1984). T-cell mediated cytotoxicity against autologous malignant melanoma : analysis with interleukin-2-dependent T-cell cultures. Proc. Natl. Acad. Sci. USA 81: 3511-3515.

Lethé, B, Van den Eynde, B, Van Pel, A, Corradin, G and Boon, T (1992). Mouse tumor rejection antigens P815 A and B : two epitopes carried by a single peptide. Eur.J.Immunol. in press:

Lurquin, C, Van Pel, A, Mariamé, B, De Plaen, E, Szikora, J, Janssens, C, Reddehase, M, Lejeune, J and Boon, T (1989). Structure of the gene coding for tum- transplantation antigen P91A. A peptide encoded by the mutated exon is recognized with Ld by cytolytic T cells. Cell 58: 293-303.

Moretta, A, Pantaleo, G, Moretta, L, Mingari, MC and Cerottini, JC (1983). Quantitative assessment of the pool size and sub-set distribution of cytolytic T lymphocytes within human resting or alloactivated peripheral-blood T-cell populations. J. Exp. Med. 158: 571-585.

Mukherji, B and MacAlister, TJ (1983). Clonal analysis of cytotoxic T cell response against human melanoma. J. Exp. Med. 158: 240-245.

Peace, DJ, Chen, W, Nelson, H and Cheever, MA (1991). T Cell Recognition of Transforming Proteins Encoded by Mutated *ras* Proto-oncogenes. J Immunol 146: 2059-2065.

Rötzschke, O, Falk, K, Deres, K, Schild, H, Norda, M, Metzger, J, Jung, G and Rammensee, H-G (1990). Isolation and analysis of naturally processed viral peptides as recognized by cytotoxic T cells. Nature 348: 252-254.

Schild, H, Deres, K, Wiesmuller, KH, Jung, G and Rammensee, HG (1991). Efficiency of peptides and lipopeptides for in vivo priming of virus-specific cytotoxic T cells. Eur. J. Immunol. 21: 2649-2654.

Sibille, C, Chomez, P, Wildmann, C, Van Pel, A, De Plaen, E, Maryanski, J, de Bergeyck, V and Boon, T (1990). Structure of the gene of tum- transplantation antigen P198 : a point mutation generates a new antigenic peptide. J. Exp. Med. 172: 35-45.

Stover, CK, de la Cruz, VF, Fuerst, TR, Burlein, JE, Benson, LA, Bennett, LT, Bansal, GP, Young, JF, Lee, MH, Hatful, GF, Snapper, SB, Barletta, RG, Jacobs, WRJ and Bloom, BR (1991). New use of BCG for recombinant vaccines. Nature 351: 456-460.

Szikora, J, Van Pel, A, Brichard, V, André, M, Van Baren, N, Henry, P, De Plaen, E and Boon, T (1990). Structure of the gene of tum- transplantation antigen P35B : presence of a point mutation in the antigenic allele. EMBO J. 9: 1041-1050.

Taylor, J, Trimarchi, C, Weinberg, R, Languet, B, Guillemin, F, Desmettre, P and Paoletti, E (1991). Efficacy studies on a canarypox-rabies recombinant virus. Vaccine 9: 190-193.

Taylor, J, Weinberg, R, Tartaglia, J, Richardson, C, Alkhatib, G, Briedis, D, Appel, M, Norton, E and Paoletti, E (1992). Nonreplicating viral vectors as potential vaccines : recombinant canarypox virus expressing measles virus fusion (F) and Hemagglutinin (HA) glycoproteins. Virology 187: 321-328.

Traversari, C, van der Bruggen, P, Luescher, IF, Lurquin, C, Chomez, P, Van Pel, A, De Plaen, E, Amar-Costesec, A and Boon, T (1992). A nonpeptide encoded by human gene MAGE-1 is recognized on HLA-A1 by CTL directed against tumor antigen MZ2-E. J. Exp. Med. in press: -.

Traversari, C, van der Bruggen, P, Van den Eynde, B, Hainaut, P, Lemoine, C, Ohta, N, Old, L and Boon, T (1992). Transfection and expression of a gene coding for a human melanoma antigen recognized by autologous cytolytic T lymphocytes. Immunogenetics 35: 145-152.

Van den Eynde, B, Hainaut, P, Hérin, M, Knuth, A, Lemoine, C, Weynants, P, van der Bruggen, P, Fauchet, R and Boon, T (1989). Presence on a human melanoma of multiple antigens recognized by autologous CTL. Int J Cancer 44: 634-640.

Van den Eynde, B, Lethé, B, Van Pel, A, De Plaen, E and Boon, T (1991). The gene coding for a major tumor rejection antigen of tumor P815 is identical to the normal gene of syngeneic DBA/2 mice. J Exp Med 173: 1373-1384.

van der Bruggen, P, Traversari, C, Chomez, P, Lurquin, C, De Plaen, E, Van den Eynde, B, Knuth, A and Boon, T (1991). A gene encoding an antigen recognized by cytolytic T lymphocytes on a human melanoma. Science 254: 1643-1647.

Vanky, F and Klein, E (1982). Specificity of auto-tumor cytotoxicity exerted by fresh, activated and propagated human T lymphocytes. Int. J. Cancer 29: 547-553.

Wallny, HJ and Rammensee, HG (1990). Identification of classical minor histocompatibility antigen as cell-derived peptide. Nature 343: 275-277.

The Immunological Basis of Active Specific Immunotherapy (ASI) for Human Melanoma

Malcolm S. Mitchell, William Harel, June Kan-Mitchell, and Robert J. Deans

Kenneth Norris Jr. Cancer Center, University of Southern California School of Medicine, Los Angeles, California, USA

INTRODUCTION

Over the past 6 years we have studied a therapeutic vaccine ("theraccine") for human melanoma, while at the same time using this system as an immunological model with which to learn about the mechanisms underlying rejection of a human tumor. We have also attempted to discover which melanoma-associated antigens (MAA) are immunogenic in humans, i.e., which epitopes stimulate human T helper cells, and which epitopes are targets for cytolytic T lymphocytes.

From the outset we have assumed that there were antigens held in common among melanomas, supported by our earlier research with human monoclonal antibodies (Kan-Mitchell et al. 1986; Kan-Mitchell and Mitchell 1988). Those studies showed that a monoclonal antibody from a particular patient with melanoma invariably recognized internal (cytoplasmic or internal membrane) antigens common to a large number of melanoma cell lines. Our results with cell-mediated immune responses in melanoma have supported the view that we have immunized with antigens common to many melanomas, which are capable of eliciting rejection of autochthonous tumors.

In this paper we will summarize our clinical results, including some encouraging new findings with Interferon-alfa 2b (IFN-alfa) following theraccine, and discuss the specificity of human T cells we have derived from the blood or tumor nodules of immunized patients. Finally, we will mention the isolation of a gene derived from melanoma cells by subtractive hybridization, which appears to encode a human melanoma-associated antigen.

CLINICAL STUDIES

A mixture of mechanical lysates from two melanoma cell lines has been administered with a novel immunological adjuvant, DETOX[tm1] to nearly 200 patients with disseminated (metastatic) melanoma. We did not know which antigens were most critical in causing rejection of autologous tumors and avoided enzyme treatment or irradiation, lest we diminish the already weak immunogenicity of the antigens. We therefore decided to use mechanical lysates of melanoma cells derived from two cell lines of very different characteristics, such as size, growth rate, melanin pigmentation, and the expression of HLA and surface gangliosides GD2 and GD3.

Clinical remissions have been noted in approximately 20% of the patients, with 5% complete and 15% partial remissions (Mitchell et al. 1988, 1990; Mitchell 1991). There has been a good deal of consistency in the results with various lots of theraccine, including those produced with our cell lines in multi-institutional trials by Ribi ImmunoChem Research, Hamilton, Montana, U.S.A. More importantly than shrinkage of existing disease, the

---

[1] DETOX is a mixture of detoxified (monophosphoryl) lipid A, mycobacterial cell wall skeletons and squalane oil with Tween-80 emulsifier.

spread of melanoma to new sites has apparently been retarded, which has resulted in improved survival among the responders. From data collected by Ribi ImmunoChem, in patients who had stability of disease or a major objective response to treatment, the median survival from time of entry onto study was >21 months (Elliott et al. 1992). Most of the patients tabulated were not treated solely with theraccine, but went onto various other types of immunotherapy and chemotherapy after relapse from the theraccine. However, of our own patients who responded to theraccine for >6 months and then received repeated maintenance injections, 10 have lived at least 22 months. The median survival in the group is >31 months from time of entry onto study, with one patient alive and free of disease more than 72 months after entry (Table 1). In contrast to these figures, the usual survival in metastatic melanoma from diagnosis is only 6-12 months.

IMMUNOLOGICAL RESPONSES IN PATIENTS RECEIVING ASI

A proportion of the patients immunized to the theraccine can then react against their own tumor in vivo. Immunological studies performed on the patients has confirmed that fact. This immunological system has permitted us to study in detail the cytotoxic T lymphocytes (Tc) generated by immunization with melanoma antigens. Sixty five patients (58%) of 111 had an increased frequency of Tc in the blood during immunization, as measured by limiting dilutions weekly (Mitchell et al. 1988, 1990; Mitchell 1991). This increase was often more than 10-fold above the baseline frequency of Tc among PBL. Reactivity was measured against the melanoma cell lines used in the theraccine. When cold target competition assays were performed, inhibition by several melanoma cell lines was noted, but little or none by non-melanoma tumors such as Daudi lymphoma, squamous lung carcinoma and leukemia K562. The presence of Tc was correlated with a clinical response, such that approximately one-third of patients with an increase in Tc during ASI had an objective remission of their disease, while none of those without a rise had a clinical response (Mitchell 1991). In contrast, although antibodies were present in a substantial proportion of the patients, as measured by enzyme immunoassay, their presence was not associated with a clinical response, nor was any consistent specificity established by absorption experiments (Table 2).

Table 1: Current Status of Patients Originally Responding >6 Months to Melanoma Theraccine (as of August 5, 1992)

Patient	Original Response	Date of Last Followup	Duration of Survival	Status
L.A.	CR 7/30/86	6/92	>72 mo	Alive. No evidence of disease. Brain metastasis resected 8/21/88.
H.H.	PR 4/8/87	7/91	52 mo	Relapsed in brain 7/90. Died 7/91.
S.H.	PR 7/7/87 CR 3/8/88	6/92	>59 mo	Alive. Single, stable 8-10 mm lung nodule since 9/88.
L.R.	Stable from 9/8/87	12/90	39 mo	Relapsed in brain 1/89. Died 12/90. Metastases slowly progressive from 1/89.
P.D.	PR 12/2/87 (CR ileum)	12/89	24 mo	Died 12/89. Ileal lesions never recurred.

R.B.	CR 12/27/89	6/92	>31 mo	Alive. No evidence of disease.
K.R.	PR 11/30/89	6/92	>30 mo	Alive. 3 5 mm s.c. nodules remain. No visceral disease.
G.W.	PR 9/5/90 (CR omental mass, 10/30/90)	8/92	23 mo	Died 8/5/92. Omental mass never recurred.
J.S.	PR 6/27/90	6/92	>24 mo	Alive.
C.K.	PR 10/28/90	6/92	>22 mo	Alive. Progressed at local site 10/91.

Median survival = >31 mo.

Table 2: Immunological Responses of Patients Given Melanoma Theraccine

| Trial | Number of Patients/Total Tested | |
	pTc	Antibodies
Phase I (MAC 1+3)	12/22	5/22
Phase II (MAC 4)	14/21	11/25[a]
Phase I (MAC 5)	8/18[b]	12/17[c]
Phase II (MAC 6)	13/20	N.D.
Phase I (MAC 6 lyophilized)	8/15	N.D.
Phase II (CY + MAC 6 lyophilized)	10/15	N.D.
TOTAL	65/111 (58%)	28/64 (44%)

"MAC n" designates melanoma antigen from cultured cells, nth batch. CY = cyclophosphamide. pTc designates precursor of cytotoxic T cells.

[a] Only 2/11 specific for melanoma on absorption.
[b] Dose-response noted: titers increased in 2 of 6, 1 of 6, and 5 of 6 patients, respectively given 5, 10 or 20 x 10^6 tumor-cell-equivalents of lysate per dose.
[c] No dose-response noted; specificity not tested by absorptions.

ASSOCIATION OF CERTAIN HLA ANTIGENS WITH A CLINICAL RESPONSE

Statistical analysis of the first 77 patients treated with melanoma theraccine revealed the association of 3 HLA Class I alleles with a clinical response (Mitchell et al. 1992). Those patients with one or more of the alleles on their lymphocytes had a clinical remission, while those who lacked all three did not. Those alleles were HLA-A2 (and HLA-A28, serologically cross-reactive), HLA-B12 "split" (including HLA-B44 and B45), and HLA-C3. Those with at least 2 alleles had a response rate of 38%, while those lacking all three had a response rate of less than 10%. The results were consistent with the possibility that those who did best were patients with HLA alleles able to present antigen most effectively to Tc, since HLA-A2 (and most recently, HLA-B44) are known to be important restriction elements in the in vitro cross-reactivity of Tc to MAA. An extension of this reasoning is that certain specific epitopes presented most efficiently by those HLA Class I molecules might be crucial for facilitating rejection in vivo.

INCREASED RESPONSE RATE TO INTERFERON-ALFA AFTER THERACCINE

A group of 18 patients was given IFN-alfa 2b (Schering) 1 month after completing a course of a uniform lot of melanoma theraccine (MAC-6, lyophilized). During that course, none of the patients had achieved an objective clinical response. Fourteen of the patients were seen at our institution, while four were treated by Dr. James Jakowatz at the University of California, Irvine. IFN-alfa was given at 5 to 6 million units per m^2 3 times weekly, s.c.

Eight of 18 patients achieved an objective remission, for a response rate of 44%. The median duration of response was 11 mo. Moreover, the duration of survival is currently 16 + mo for the responders Vs. 7.75 mo for nonresponders and 10 mo for the group as a whole.

These results encourage a controlled study of ASI combined with IFN-alfa preceding, following or concomitantly. Whether the effect of IFN-alfa was due to an upregulation of HLA Class I molecules, adhesion molecules and/or tumor-associated antigens is uncertain, but is consistent with our immunological data. That is, those patients who responded to IFN-alfa after theraccine were those who were successfully immunized but somehow failed to achieve an objective remission. This supports the hypothesis that the autologous tumor cells had evaded recognition.

CLONING OF T CELLS FROM PATIENTS GIVEN ASI

We have cloned T cells from several of our patients treated with ASI to study the restriction of these unusual Tc, many of which seemed to react against a common MAA found on a variety of melanoma cell lines. CD4 and CD8 Tc clones were developed, both in the presence of additional MAA stimulation, but in the absence of APC for the CD8 Tc. By this means, we hoped to determine the true in vivo reactivity of the latter, without the possibility of inducing new specificities in vitro. Through the use of both IL-2 and IL-4 in the medium, we have been able to grow a higher proportion of CD8 Tc, and to keep most Tc clones of both CD types viable for 4 to 6 months.

Characterization of CD4 Tc Clones

CD4 Tc were cloned in the presence of autologous APC and irradiated autologous melanoma cells. We should note that in some early experiments where irradiated allogeneic melanoma cells were used to re-reducate in vitro, the specificity and HLA restriction of the clones were similar to those of cells stimulated with autologous tumor.

CD4 lymphocytes from melanoma patients undergoing ASI were found to contain both noncytotoxic (presumably T helper (Th)) and cytotoxic T cells, when isolated from FACS-sorted TIL or PBL after MLTC restimulation in vitro (Harel et al. 1990). The Th were classically MHC Class II-restricted, but the Tc often were not. In 3 independent studies from our laboratory, CD4 Tc reactive against allogeneic and autologous melanomas were identified. Seven of 16 CD4 clones cytotoxic to autologous tumor cells were exclusively Class I-restricted, while 5 others were restricted by either Class I or II, i.e., blocked by pretreatment with mAbs against Class I or II antigens. Two clones were not inhibited by either anti-Class I or -II, one in each of 2 studies. Only two exclusively Class II-restricted CD4 Tc have been identified. Thus, in the CD4 Tc response to melanoma, Class I restriction, exclusively or with some degree of Class II restriction, appears to be common.

Cytotoxicity by CD4 T cells was weaker than that of CD8, requiring 16h assays, 40:1 effector:target ratio and, most importantly, pretreatment of the melanoma with IFN-gamma, which mainly upregulated tumor antigen expression. Accessory molecules such as LFA-1 and ICAM-1 were involved in T cell-melanoma cell interaction, such that inhibiting them led to diminished lysis (Goedegebuure P et al., submitted for publication; LeMay L et al., submitted for publication). TNF production was found in nearly all CD4 Tc clones, but was similar in noncytotoxic CD4 clones, making it unlikely to be the main mediator of cytotoxicity (LeMay, L and Mitchell, MS, unpublished data).

With the collaboration of Dr. Elwyn Loh, University of Pennsylvania School of Medicine, we have characterized the T cell receptor (TCR) of 8 CD4+ Class I-restricted Tc clones. No singular usage of any single Va or Vb genes has been noted. However, several newly described Va and Vb sequences were found, including Va w24.1, w27.1 and 29.1, and Vb 23.1 and 24.1. These studies are continuing in our laboratory with methods developed by Dr. Loh.

Characterization of CD8 Tc Clones

CD8 lymphocytes were purified from tumor-infiltrating (TIL) or peripheral blood lymphocytes by fluorescence-activated cell sorting and restimulation with irradiated autologous melanoma cells in IL-2 and IL-4 but without APC. When obtained from a patient who had undergone ASI for 5 years to maintain a complete remission, 21 of 28 Tc clones lysed only the autologous melanoma, and the other 7 lysed both autologous (HLA-A2) and allogeneic (HLA-A28) melanoma targets. None lysed autologous LCL. Of the ostensibly self-restricted clones, 6 were analyzed in more detail. Two of them cross-reacted with melanomas sharing HLA-A2 with the patient's Tc, while 4 were exclusively self-MHC restricted.

TCR genes utilized by the 4 self-restricted Tc clones and the 2 HLA-A2-restricted clones were compared, in a collaboration with Drs. Lawrence Steinman and Jorge Oksenberg of Stanford University. Va genes were not amplified in 2 clones: one, self-restricted and one, A2-restricted. The combination of Va17, Vb7 was found in 2 of the other 3 self-restricted clones. (Va__, Vb17 was found in the other). Vb10 was present in one self-restricted (with Va8) and one A2-restricted clone (Va__). The final A2-restricted clone was Va3, Vb15. Thus, there was a less heterogeneous usage, particularly of Vb genes, among these clones reactive exclusively with melanoma. The utilization of Va17, Vb7 by 2 self-restricted Tc clones was particularly interesting in this regard.

ISOLATION AND SEQUENCING OF NOVEL GENES FROM MELANOMA CELLS BY MOLECULAR SUBTRACTION

With the T cell clones obtained from patients immunized with our mixture of melanoma antigens, we have been able to start to determine which immunogens are most important in stimulating anti-tumor immunity. This information will facilitate development of a wholly synthetic melanoma theraccine composed of proteins or epitopic peptides.

Genes with novel sequences have been produced by molecular subtraction (subtractive hybridization) between a squamous lung carcinoma cell line (Lu-1) and melanoma M-1, one of components of the theraccine. Lu-1 is a tumor that lacked antigens present on melanoma, as judged by its consistent insensitivity to lysis by anti-melanoma Tc (Vs. LAK cells), and its inability to block anti-melanoma lysis in cold-target competition assays. From the molecular subtraction experiments, we obtained a series of cDNA clones that were represented predominantly in melanoma cells rather than other tumors or normal tissues. Twelve of the genes were novel, i.e., not found in GenBank computer files, of which we have reported six (Hutchins et al. 1991). Gene 50 (called clone 50 in our paper) had a distribution fairly restricted to melanomas by Northern blot analysis, which made it a reasonable first candidate for investigation. Most recently, its expression in cells and tissues has been studied by a quantitative PCR analysis. Gene 50 was found to be expressed by most melanoma cells, but was not detected in a variety of normal tissues such as lung, kidney, skin, stomach as well as LCL, K562, Daudi cell lines. It is, however, expressed to varying degrees by breast carcinoma, sarcoma and colorectal carcinoma (Kan-Mitchell et al. 1992).

We have developed oligonucleotide probes for completing the molecular description of this gene. They have been used in PCR amplification strategies to orient the inserts of cDNA we have obtained with one another in a coding context. To date we have sequenced approximately one-third of the bases in gene 50.

Synthesis and Immunological Study of a Peptide Fragment of the Protein Encoded by Gene 50

Attempts to produce a fusion protein from cDNA 50 were unsuccessful. We then synthesized a 17 amino acid peptide ("peptide 50") whose sequence was deduced from the complete open reading frame of gene 50. Peptide 50 was found to be immunogenic in vitro, by its ability to stimulate proliferation of CD4 cells from an autologous melanoma-specific TIL derived from a melanoma patient, in a 7d assay together with autologous APC. The peptide caused stimulation above that of an irrelevant, cDNA-derived peptide saponin-C of approximately the same length. We then derived 32 CD4 clones from a culture of the same TIL with peptide 50. Four clones proliferated in response to the peptide 50 but not to saponin-C.

This strongly suggested that peptide 50 was immunogenic in humans, and that the complete protein encoded by gene 50 may contain other epitopes. Thus far, incubation of peptide 50 with LCL or T cells has failed to sensitize either for cytotoxicity by autologous CD8 Tc lines or Tc clones.

Responses to Peptide 50 as a Result of ASI with Melanoma Theraccine

In a pilot study, we investigated whether an increased frequency of PBL responding to peptide 50 was a consequence of ASI with melanoma theraccine. Various dilutions of PBL from 12 patients who had had an increase in their Tc frequency resulting from ASI were incubated with peptide 50 for 7d, after which 3H-thymidine incorporation was measured. The frequency of proliferating cells was then estimated by the Poisson distribution. Eight of the 12 patients had an increase in lymphocytes reactive with peptide 50. The mean number of responsive cells was $136/10^6$ lymphocytes in the entire group, and $205/10^6$ among the $8/10^6$ responders. No consistent DR or DQ phenotype restriction was obvious. Thus, a high proportion of these patients responded to the protein antigen of gene 50, as expressed on M-1 melanoma cells in the immunizing lysate.

In contrast to these results, only 3 of 10 melanoma patients who failed to be immunized with ASI (i.e., had no increase in Tc frequency), and none of 5 normal controls (not shown), responded to peptide 50.

CONCLUSIONS

These results have encouraged us to continue clinical trials with more purified versions of melanoma theraccine, including synthetic peptides derived from gene 50 and other genes isolated from melanoma cells. We have also been encouraged to continue our study of the T cells elicited by specific immunotherapy. In fact our original tenet was that immunotherapy given to human subjects must be studied at every stage of its development, not simply during Phase I trials. That has been continually reinforced by the information we have gleaned from our melanoma patients about the nature of their immune response to melanoma antigens.

ACKNOWLEDGMENTS

We would like to acknowledge with gratitude the expert collaboration of the following people in various aspects of these studies: Drs. Xiu-Qing Huang, Peter Goedegebuure and Lin LeMay in work on T cell clones, Dr. Jeff Hutchins and Ms. Sarah Weiler in work on gene 50 and peptide 50, Ms. Grace Dean, R.N. and Ms. Lucy Stevenson, R.N., and Drs. Raymond Kempf and William Boswell in the clinical trials. Drs. Lawrence Steinman and Jorge Oksenberg of Stanford University and Dr. Elwyn Loh, University of Pennsylvania have been our expert collaborators in the sequencing of the TCR on our clones.

This research was supported by USPHS Grants CA 36233 (MSM), EY 09031, EY 09427, (JKM), a grant from the Concern Foundation, and a contract with Ribi ImmunoChem Research.

REFERENCES

Elliott GT, McLeod RA, Perez J, and Von Eschen KB (1992) Proc Amer Assoc Cancer Res 33: 332 (abstract).

Harel W, Li VA, Morse AC, Kan-Mitchell J, Ewoldsen MA and Mitchell MS (1990) Proc Amer Assoc Cancer Res 31: 253 (abstract).

Hutchins JT, Deans RJ, Mitchell MS, Uchiyama C and Kan-Mitchell J (1991) Cancer Res 51: 1418-1425

Kan-Mitchell J, Deans RJ, Danenberg PV, White WL, Danenberg KD, Granada ESV, Ozbun LL, Levy S, Harel W and Mitchell MS (1992) Proc Amer Assoc Cancer Res 33: 320 (abstract)

Kan-Mitchell J, Imam A, Kempf RA, Taylor CR and Mitchell MS (1986) Cancer Res 46: 2490-2496

Kan-Mitchell J and Mitchell MS (1988) Human monoclonal antibodies for the diagnosis of tumors. In: Kupchik HZ (ed) In vitro diagnosis of human tumors using monoclonal antibodies. Marcel Dekker, New York, pp 289-304

Mitchell MS (1991) Int Rev Immunol 7: 331-347

Mitchell MS, Harel W and Groshen, S (1992) J Clin Oncol 10: 1158-1168

Mitchell MS, Harel W, Kempf RA, Hu E, Kan-Mitchell J, Boswell WD, Dean G and Stevenson L (1990) J Clin Oncol 8: 856-869

Mitchell MS, Kan-Mitchell J, Kempf RA, Harel W, Shau H and Lind S Cancer Res 48: 5883-5893

18. Late Arrivals

gp130, common signal transducer for cytokines including IL-6

Tadamitsu Kishimoto[1,2] and Tetsuya Taga[2]

1 Department of Medicine III, Osaka University Medical School, 1-1-50 Fukushima, Fukushima-ku, Osaka 553, Japan
2 Division of Immunology, Institute for Molecular and Cellular Biology, 1-3 Yamada-oka, Suita, Osaka 565, Japan

Functional pleiotropy and redundancy are characteristic features of cytokines. To understand the signaling mechanisms of such cytokines, we have proposed a two-chain interleukin-6 receptor (IL-6-R[†]) model: IL-6 triggers the association of a ligand-binding chain (IL-6-R) and a non-binding signal transducer (gp130) to form a high-affinity receptor complex, causing transmission of the signal by the cytoplasmic portion of gp130. This model would help us to explain the functional redundancy of cytokines if we were to assume that gp130 interacts with several different receptor chains. In fact, we have recently demonstrated that gp130 functions as a common signal transducer for IL-6, IL-11, oncostatin M (OM), leukemia inhibitory factor (LIF), and ciliary neurotrophic factor (CNTF), based on the results that anti-gp130 monoclonal antibodies (mAbs) completely block the biological responses induced by all of these factors, and stimulation of cells by the above cytokines rapidly induces tyrosine-phosphorylation of gp130.

INTRODUCTION

Most cytokines are characterized by their pleiotropic and redundant functions; i.e., each factor exerts multiple effects in different cells and different factors can act on the same cell to induce similar effects. Some of the biological effects induced by interleukin 6 (IL-6) (Kishimoto 1989), a typical example of such cytokines, are also mediated by IL-11, oncostatin M (OM), and leukemia inhibitory factor (LIF), e.g. acute phase protein-induction in hepatocytes (Baumann 1992, Richards 1992, Hilton 1991), suggesting a common signaling mechanism. LIF functions as a cholinergic differentiation factor in nerve cells (Yamamori 1989) as does ciliary neurotrophic factor (CNTF), another cytokine functioning in the neural system (Stockli 1989). CNTF-R, which is anchored to the membrane by a glycosyl-phosphatidylinositol linkage, is structurally very similar to IL-6-R except that it possesses no transmembrane and cytoplasmic regions (Davis 1991). This is reminiscent of the observation that the extracellular soluble form of IL-6-R (sIL-6-R) mediates the IL-6 signal through cell-surface gp130 (Taga 1989). The two-chain receptor model we have proposed (Taga 1989; Hibi 1990), in which a receptor-associated signal transducer generates the cytoplasmic signal, would be one way to explain these overlapping functions, if we assume that a signal transducer is shared by these three cytokines. In this paper, we describe that gp130 is involved in exerting the biological functions of IL-6, IL-11, OM, LIF, and CNTF.

---

† Abbreviations: IL-6, interleukin 6; LIF, leukemia inhibitory factor; OM, oncostatin M; CNTF, ciliary neurotrophic factor; IL-6-R, IL-6 receptor (-R attached to any cytokine refers to its receptor); sIL-6-R, soluble IL-6-R; mAb, monoclonal antibody.

THE SIGNAL TRANSDUCING IL-6 RECEPTOR COMPONENT, GP130

The cytoplasmic region of IL-6-R does not include sequences known to be important in signal transduction, such as tyrosine kinase domains. Furthermore, the cytoplasmic region of IL-6-R is very short (Yamasaki 1988) and can be deleted without affecting IL-6 signal transduction (Taga 1989), suggesting that an associated molecule is responsible for mediating the IL-6 signal. The existence of such an accessory signal transducing molecule was discovered: Binding of IL-6 to IL-6-R was shown to trigger the association of IL-6-R and a non-ligand-binding 130 kD signal transducing molecule, gp130 (Taga 1989). Both receptor components belong to the hematopoietic cytokine receptor family (Hibi 1990). Whereas expression of the IL-6-R cDNA in an IL-6-R negative cell resulted in the generation of only low-affinity binding sites, co-expression of the cDNAs for both IL-6-R and gp130 resulted in the formation of both high- and low-affinity binding sites (Hibi 1990). Furthermore, the addition of anti-gp130 monoclonal antibody reduced the number of the high-affinity IL-6 binding sites and inhibited transmission of the IL-6-signal, demonstrating that gp130, despite its lack of IL-6-binding capability, plays a role in both the formation of high-affinity binding sites and in signal transduction. This association takes place extracellularly, because the genetically engineered extracellular soluble form of IL-6-R (sIL-6-R) is able to associate with gp130 in the presence of gp130 and to transduce the signal (Taga 1989; Hibi 1990).

GP130 SERVES AS SIGNAL TRANSDUCER FOR IL-6, IL-11, LIF, OM, AND CNTF

The functional redundancy which is a characteristic feature of the action of many cytokines could be explained if we assumed that several different cytokine receptors were able to interact with a common signal transducing component, such as gp130 mentioned above. LIF, originally identified as a factor which inhibits the growth of a murine myeloid leukemia cell line (M1), was shown to be a multifunctional cytokine, many of whose activities overlapped with those of IL-6 (Kishimoto 1989; Hilton 1991). A cDNA encoding LIF-R was molecularly cloned, and expression of this cDNA in COS cells resulted in the generation of only low-affinity binding sites (Gearing 1991). In contrast, LIF responding cells express both high- and low-affinity LIF binding sites. This suggested the existence of an additional high affinity converting subunit of LIF-R. Subsequently, this converter was molecularly cloned and shown to be identical to the IL-6 signal transducer, gp130 (Gearing 1992b).

OM is a cytokine that was originally identified as a growth inhibitor of human melanoma cells. OM is structurally similar to LIF and IL-6, and shares multiple functions with these two factors (Rose 1991; Richards 1992), suggesting that gp130 might also be involved in the signaling processes triggered by OM. gp130 has been shown to bind OM with low intrinsic affinity, but this binding affinity increases somewhat when gp130 is co-expressed with LIF-R. The latter molecule has no OM binding capability, indicating that gp130 and LIF-R associate to form an OM receptor complex (Gearing 1992b; Liu 1992). However, some melanoma cells express OM binding sites with affinity higher than that conferred by gp130 and LIF-R, even though these cells do not express LIF-R. This observation suggests the existence of an unidentified receptor component for OM (Gearing 1992a, 1992b). Anti-gp130 mAbs completely block the OM-induced acute-phase protein production in hepatoma cells and OM-mediated growth inhibition of melanoma cells, confirming that gp130 is essential for transducing the signals of OM, in addition to those of IL-6 and LIF (Liu 1992; Taga 1992). Furthermore, stimulation of cells with LIF or OM rapidly induced tyrosine-phosphorylation of gp130 as has been observed with IL-6 (Murakami 1991; Taga 1992).

CNTF, initially identified based on its activity to support the survival of ciliary neurons, functions pleiotropically within the neural system. Its actions include the enhancement of the survival of motor neurons and an effect on the cholinergic differentiation of sympathetic neurons (Stockli 1989). Interestingly, CNTF and LIF, also known as cholinergic differentiation factor (CDF), elicit similar responses within some neuronal cells (Stockli 1989; Yamamori 1989). Molecular cloning of the CNTF-R, which is expressed exclusively in the neuronal system, revealed that it shows the highest sequence homology to IL-6-R (Davis 1991). Furthermore, CNTF-R lacks a cytoplasmic region and anchored to the membrane via a glycosyl-phosphatidylinositol (GPI) linkage. The absence of a cytoplasmic region in the CNTF-R is reminiscent of the observation that the IL-6-R can mediate the IL-6 signal even when its cytoplasmic region has been deleted (Taga 1989). Studies with CNTF-responsive cell lines have confirmed the possible involvement of gp130 in the CNTF-R system. CNTF-stimulation of a neuronal cell line (MAH) inhibited its growth and induced a tyrosine specific phosphorylation of gp130 (Ip 1992). A complex of IL-6 and sIL-6-R acts on several neuronal cell lines as well as primary cultured neurons to initiate cellular responses similar to those induced by CNTF (N. Y. Ip et al., unpublished). Conversely, soluble CNTF-R plus CNTF could induce a human erythroleukemia cell line (TF1), which does not express the CNTF-R, to initiate DNA synthesis as does IL-6 (Taga 1992). The proliferation of TF1 cells induced by soluble CNTF-R and CNTF, and CNTF-actions on neuronal cells were completely blocked by anti-gp130 antibodies, indicating that the CNTF-signaling process involves gp130 (Ip 1992; Taga 1992). The above findings clearly show that the gp130 molecule, present within each of the receptor complexes, is essential for transducing the signals of IL-6, OM, LIF, and CNTF. CNTF stimulation rapidly induces tyrosine-phosphorylation of both gp130 and a 190 kD protein. This 190 kD-phosphoprotein can be co-precipitated with gp130, indicating a physical association. This 190 kD protein is most likely LIF-R, suggesting that the functional CNTF-R complex may include LIF-R in addition to gp130 and CNTF-R (Ip 1992; Taga 1992). These multiple interactions of gp130 with the above cytokine receptors are depicted in Fig. 1.

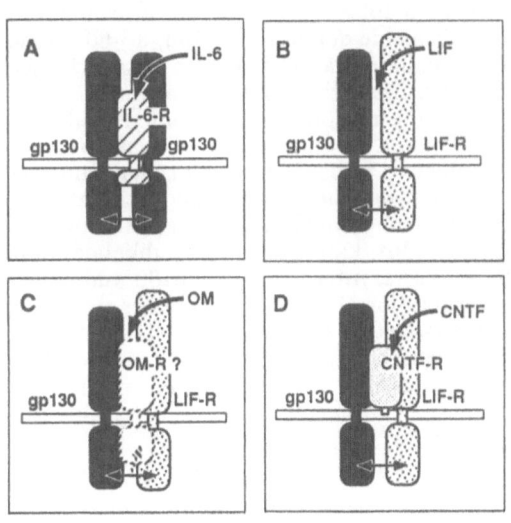

Fig. 1. Schematic models of the multisubunit receptors for IL-6 (A), LIF (B), OM (C), and CNTF (D). A signal transducing component, gp130, is shared by these receptor complexes and is essential for transmitting their respective cytokine signals. In the IL-6-R complex, IL-6 triggers the association of a low affinity ligand-binding subunit (IL-6-R) and a non-binding signal transducer (gp130) to form a high affinity complex, leading to the dimerization of gp130. In other receptor systems, ligand-induced formation of heteromeric complexes, including gp130 and LIF-R molecules, is believed to occur. Existence of an unidentified receptor component for OM is hypothesized. Oligomerization of the receptor components (indicated by the arrow at the bottom of each figure) is postulated to result in the interaction with a cytoplasmic molecule such as a tyrosine kinase.

IL-11 exerts multiple biological functions similar to those of IL-6 (Baumann 1992): IL-11 promotes proliferation of plasmacytoma cells, formation of immunoglobulin-secreting B cells, and production of acute phase proteins in hepatocytes. IL-11 also promotes the formation of hematopoietic, especially megakaryocytic, colonies in the presence of IL-3. These observations have suggested that similar signalling processes may be operating in the IL-6 and IL-11 systems. A specific receptor for IL-11 has been detected (Yin 1992), but its structure has not yet been elucidated. Anti-gp130 mAbs have been shown to inhibit the IL-11-induced TF-1 cell proliferation, suggesting that gp130 is a component of the IL-11-R complex and is essential for IL-11 signal transduction (Y.-C. Yang et al., unpublished).

It should be noted that while the various cytokines which utilize gp130 as a common signal transducers exhibit overlapping biological functions, each cytokine also exhibits some specific activities as well. Thus, while the ubiquitously-expressed gp130 is involved in mediating signals elicited by all the above mentioned cytokines, the ability of a cell to respond to each of these factors specifically appears to be regulated by the specific expression of distinct receptor chains.

INVOLVEMENT OF TYROSINE KINASE IN GP130-MEDIATED SIGNAL TRANSDUCTION

gp130 does not contain any known signal transduction sequence motifs, such as the tyrosine kinase domains, that are observed in several conventional growth factors. However, a series of studies have provided clues to the understanding of how receptors lacking the signalling motifs propagate their signals. In the case of the receptors belonging to the hematopoietic cytokine receptor family, a number of results have suggested the involvement of protein tyrosine kinases in the first step of the signalling process: Stimulation of cells, e.g, by IL-2, IL-3, IL-4, IL-5, IL-6, IL-7, GM-CSF, EPO, and CNTF has been shown to activate intracellular tyrosine kinases and induce the tyrosine-specific phosphorylation of cellular proteins (Kanakura 1990; Nakajima 1991; Murakami 1991; Ip 1992; Linnekin 1992; Otani 1992; Taga 1992). The addition of tyrosine kinase inhibitors blocks cellular responses induced by IL-6 (Nakajima 1991). Stimulation of cells with IL-6 induces the tyrosine phosphorylation of its signal transducer, gp130, and stimulated gp130 has been shown to be associated with tyrosine kinase activity (Murakami 1991; Ip 1992; Taga 1992; M. Murakami et al., unpublished). In the 277 amino acid cytoplasmic domain of gp130, a ~60 amino acid region proximal to the transmembrane domain was shown to be essential for signal transduction (Murakami 1991). Two short stretches of amino acids from this region are highly conserved among many cytokine receptors and signal transducers belonging to the cytokine receptor family. Thus, a common or structurally related signaling molecule(s), such as an intracytoplasmic tyrosine kinase(s), may interact with this region of homology. In the IL-6 signaling, the stimulation of target cells with a complex of sIL-6-R and IL-6 induces the homo-dimerization of gp130, and the tyrosine-specific phosphorylation of gp130 (M. Murakami, et al., unpublished). At present, it remains unclear which tyrosine kinase(s) interacts with gp130 and which downstream signaling molecules are the targets for this tyrosine kinase(s). Since gp130 has been shown to be essential for transduction of the IL-6, IL-11, OM, LIF, and CNTF signals, it is necessary to examine whether the gp130 molecule in each of the receptor complexes for these cytokines interacts with the same or different kinase.

Acknowledgement: This work was partly supported by the Human Frontier Science Program.

REFERENCES

Baumann H, Schendel P (1992) Interleukin-11 regulates the hepatic expression of the same plasma protein genes as interleukin-6. *J. Biol. Chem.* **266**, 20424-20427

Davis S, Aldrich TH, Valenzuela DM, Wong V, Furth ME, Squinto SP, Yancopoulos GD (1991) The receptor for ciliary neurotrophic factor. *Science* 253, 59-63

Gearing DP, Bruce AG (1992a) Oncostatin M binds the high-affinity leukemia inhibitory factor receptor. *The New Biologist* 4, 61-65

Gearing DP, Comeau MR, Friend DJ, Gimpel SD, Thut CJ, McGourty J, Brasher KK, Kin, JA, Gillis S, Mosley B, Ziegler SF, Cosman D (1992b) The IL-6 signal transducer, gp130: an oncostatin M receptor and affinity converter for the LIF receptor. *Science* 255, 1434-1437

Gearing DP, Thut CJ, Vandenbos T, Gimpel SD, Delaney PB, King J, Price V, Cosman D, Beckmann MP (1991) Leukemia inhibitory factor receptor is structurally related to the IL-6 signal transducer, gp130. *EMBO J.* 10, 2839-2848

Hibi M, Murakami M, Saito M, Hirano T, Taga T, Kishimoto T (1990) Molecular cloning and expression of an IL-6 signal transducer, gp130. *Cell* 63, 1149-1157

Hilton DJ, Gough NM (1991) Leukemia inhibitory factor: A biological perspective. *J. Cell Biochem.* 46, 21-26

Ip NY, Nye SH, Boulton TG, Davis S, Taga T, Li, Y, Birren SJ, Yasukawa K, Kishimoto T, Anderson DJ, Stahl N, Yancopoulos GD (1992) CNTF and LIF act on neuronal cells via shared signaling pathways that involve the IL-6 signal transducing receptor component gp130. *Cell* 69, 1121-1132

Kanakura Y, Druker B, Cannistra SA, Furukawa Y, Torimoto Y, Griffin JD (1990) signal transduction of the human granulocyte-macrophage colony-stimulating factor and interleukin-3 receptors involves tyrosine phosphorylation of a common set of cytoplasmic proteins. *Blood* 76, 706-715

Kishimoto T (1989) The biology of interleukin 6. *Blood* 74, 1-10

Linnekin D, Evans GA, D'Andrea A, Farrar WL (1992) Association of the erythropoietin receptor with protein tyrosine kinase activity. *Porc. Natl. Acad. Sci., USA.* 89, 6237-6241

Liu J, Modrell,B, Aruffo A, Marken,JS, Taga T, Yasukawa K, Murakami M, Kishimoto T, Shoyab M (1992) Interleukin-6 signal transducer gp130 mediates oncostatin M signaling. *J. Biol. Chem,* 267, 16763-16766

Murakami M, Narazaki M, Hibi M, Yawata H, Yasukawa K, Hamaguchi M, Taga T, Kishimoto T (1991) Critical cytoplasmic region of the IL-6 signal transducer, gp130, is conserved in the cytokine receptor family. *Proc. Natl. Acad. Sci., USA* 88, 11349-11353

Nakajima K, Wall R (1991) Interleukin-6 signals activating *jun*B and *tis*11 gene transcription in a B-cell hybridoma. *Mol. Cell. Biol.* 11, 1409-1418

Otani H, Siegel JP, Erdos M, Gnarra JR, Toledano MB, Sharon M, Mostowski H, Feinberg MB, Pierce JH, Leonard WJ (1992) Interleukin (IL)-2 and IL-3 induce distinct but overlapping responses in murine IL-3-dependent 32D cells transduced with human IL-2 receptor beta chain: involvement of tyrosine kinase(s) other than p56*lck*. *Proc. Natl. Acad. Sci. USA.* 89, 2789-2793

Richards CD, Brown TJ, Shoyab M, Baumann H, Gauldie J (1992) Recombinant oncostatin M stimulates the production of acute phase proteins in HepG2 cells and rat primary hepatocytes in vitro. *J. Immunol.* 148, 1731-1736

Rose,TM, Bruce AG (1991) Oncostatin M is a member of a cytokine family that includes leukemia-inhibitory factor, granulocyte colony-stimulating factor, and interleukin6. *Proc. Natl. Acad. Sci. USA* 88, 8641-8645

Stöckli KA, Lottspeich F, Sendtner M, Masaiakowsk P, Carroll P, Götz R, Lindholm D, Thoenen H (1989) Molecular cloning, expression and regional distribution of rat ciliary neurotrophic factor. *Nature* 342, 920-923

Taga T, Hibi M, Hirata Y, Yamasaki K, Yasukawa K, Matsuda T, Hirano T, Kishimoto T (1989) Interleukin-6 triggers the association of its receptor with a possible signal transducer, gp130. *Cell* 58, 573-581

Taga T, Narazaki M, Yasukawa K, Saito T, Miki D, Hamaguchi M, Davis S, Shoyab M, Yancopoulos GD, Kishimoto T (1992) Functional inhibition of hematopoietic- and neurotrophic-cytokines (LIF, OM, and CNTF) by blocking IL-6 signal transducer, gp130. *Proc. Natl., Acad. Sci., USA.* in press

Yamamori T, Fukada K, Aebersold R, Korsching S, Fann MJ, Patterson PH (1989) The cholinergic neuronal differentiation factor from heart cells is identical to leukemia inhibitory factor. *Science* 246, 1412-1416

Yamasaki K, Taga T, Hirata Y, Yawata H, Kawanishi Y, Seed B, Taniguchi T, Hirano T, Kishimoto T (1988) Cloning and expression of the human interleukin-6 (BSF-2/IFNβ2) receptor. *Science* 241, 825-828

Yin T, Miyazawa K, Yang YC (1992) Characterization of interleukin-11 receptor and protein tyrosine phosphorylation induced by interleukin-11 in mouse 3T3-L1 cells. *J. Biol. Chem.* 267, 8347-8351

The Tyrosine Kinases pp561ck and pp59fyn are Activated in Thymocytes Undergoing Positive Selection

Ana C. Carrera, Carrie L. Baker, Thomas M. Roberts and Drew M. Pardoll

Dana Farber Cancer Institute, 44 Binney St., Boston, MA 02115, and The Johns Hopkins University School of Medicine, Dept. of Medicine, Oncology and Molecular Biology and Genetics, Baltimore, MD 21205

ABSTRACT

Developing T cells undergo distinct selection processes that determine the T cell receptor (TCR) repertoire (1). Positive selection involves the differentiation of thymocytes bearing self-MHC TCR to mature T cells (2). In order to study the potential involvement of tyrosine phosphorylation in the mechanism of positive selection we have analyzed the activation of pp56^{lck} and pp59^{fyn} in thymocytes from transgenic mice expressing a unique TCR (anti-HY +H2D^b)(3). Thymocytes undergoing positive selection (H2D^b) displayed high specific kinase activity for pp56^{lck} and pp59^{fyn} as compared with thymocytes from nonselecting mice (H2D^d). Furthermore,the increase in kinase activity is found selectively in the CD4$^+$CD8$^+$ subpopulation of thymocytes, where the selection process is believed to occur (4,5). These data suggest that tyrosine phosphorylation is part of the intracellular signals involved in MHC I-driven positive selection from CD8$^+$CD4$^+$ to CD8$^+$CD4$^-$ cells.

Positive selection allows the maturation of only those thymocytes whose TCR recognizes foreign antigens in the context of self-MHC molecules (1,2). Significant advances in the study of thymic selection have come from the development of TCR transgenic mice in which rearranged TCR α and β genes derived from antigen-specific, MHC-restricted T cell clones are inserted into the germline (3,6,7). In these mice, a large proportion of developing thymocytes express the transgenic TCR and their developmental fate can be followed in a clonal fashion. We have analyzed the transgenic anti-HY (male antigen) + H-2D^b TCR backcrossed onto a SCID background. In these mice virtually all thymocytes express only the transgenic TCR (8).

It is generally held that the developmental consequences of thymocyte TCR engagement result from the generation of signals by intracellular second messengers. Antibody crosslinking of TCR on immature CD4$^+$CD8$^+$ thymocytes results in rises in intracellular free calcium (9-11) and phosphorylation of associated ζ homo- and heterodimers (12,13); these events have been linked to negative selection since TCR crosslinking *in vitro* causes apoptosis in immature thymocytes. Given that the earliest signaling events detectable upon TCR crosslinking appear to be activation of tyrosine kinases (14-18), and given that two src family tyrosine kinases appear to be involved in the activation of mature T cells (pp56^{lck} and pp59^{fyn}) (19-21) we decided to study the

activation state of these kinases in the TCR transgenic mouse model
described above, whose thymocytes uniformly follow a distinct
developmental pathway (positive selection or non selection in the case
of female mice) depending solely on the mouse's MHC haplotype.

The comparison of pp56^{lck} and pp59^{fyn} kinase activity in total
thymocytes from nonselecting (H-2D$^{d/d}$) versus positively selecting
(female H-2D$^{b/d}$) mice was performed by immunoisolating these kinases
(using specific antibodies) and subsequent analysis of the kinase
activity *in vitro* at 5 minutes (linear range). Figure 1A shows a
representative set of mice of 10 different couples analyzed. The
autophosphorylation signal obtained in pp56^{lck} and pp59^{fyn}
immunoprecipitates from positively selecting thymocytes was
significantly higher than the corresponding signal from nonselecting
mice. Thymocytes from normal, nontransgenic mice show an intermediate
level of activity reflecting that they are a heterogeneous mixture of
cells undergoing different developmental fates. The 56-60 kDa pp56^{lck}
bands and the 58-62 pp59^{fyn} bands (indicated in the Fig 1) reflect the
marked heterogeneity of src-family kinases that we and others have
previously described (22,23). Increased tyrosine kinase activity was
also evident when acid-denatured enolase was included in the kinase
assay as an exogenous substrate (Fig 1B).

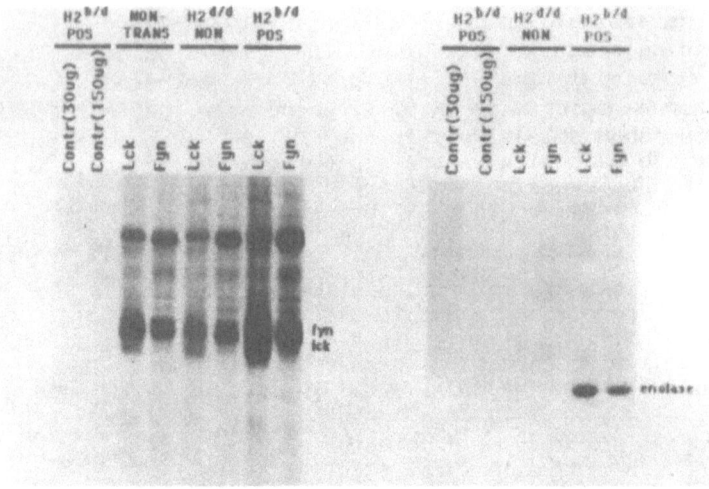

Fig. 1: Tyrosine kinase activity of pp56^{lck} and pp59^{fyn} in thymocytes undergoing positive selection.
(A) Thymocytes from H-2$^{b/d}$ scid/scid, TCR-transgenic (positively selecting) and H-2$^{d/d}$ *scid/scid*, TCR-
transgenic (nonselecting), or from control, nontransgenic C57BL6 normal mice were lysed in 1% Triton X-
100. After normalize for protein content, lysates were immunoprecipitated using 2ml of either control,
anti-pp56^{lck} or anti-pp59^{fyn} serum. Immunoprecipitates were then incubated with ^{32}P-gATP for a kinase
assay *in vitro*. Phosphorylated products were separated on a 10% SDS-PAGE gel and analyzed by
autoradiography. Autophosphorylated tyrosine kinases are identified.(B) Immunoprecipitates were
subjected to kinase assay *in vitro* in the presence of acid-denatured enolase. 30 mg of total protein
was used for pp56^{lck} immunoprecipitations and 150 mg of total protein was used for pp59fyn
immunoprecipitations. Therefore, immunoprecipitations with the control irrelevant antibody were done
with both 30 mg for pp56^{lck} (Ctr L) and 150 mg for pp59^{fyn} (Ctr.F) of protein. Non=nonselecting ,
Pos=positively selecting.

Table 1 represents the quantitative analysis of ^{32}Pi incorporated in pp56^{lck} and pp59^{fyn} bands in each of six representative experiments similar to the one described. The activity of these src-family tyrosine kinases was reproducibly increased in positively selecting thymocytes relative to non-selecting thymocytes with a mean +/- SD of 3.2 +/- 0.9 for pp56^{lck} and 6.2 +/- 4.7 for pp59^{fyn} . We describe for the first time a primary T cell in which pp56^{lck} and pp59^{fyn} seem to be specifically activated in response to a physiological induction.

Table 1: Relative activities of pp56lck and pp59fyn in positive selecting thymocytes vs non selecting thymocytes.
Thymocytes from H-2$^{b/d}$ *scid/scid*, TCR transgenic (positively selecting) and H-2$^{d/d}$ *scid/scid* orTCR-transgenic (nonselecting) mice were lysed in 1% Triton X-100 lysis buffer, immunoprecipitated with either anti-pp56^{lck} or anti-pp59^{fyn} antiserum. The immunoprecipitated proteins were subjected to a kinase assay in vitro, and the phosphorylated products were separated by SDS-PAGE (10%) as in Fig.2. The gels were placed in a betascope to quantitate the ^{32}Pi incorporated in the bands corresponding to pp56^{lck} and pp59^{fyn}.

Exp.	SELECT.	PP56^{lck}cpm**	PP59^{fyn}cpm**	PP56^{lck}POS/ PP56^{lck}NON/	PP59^{fyn}POS/ PP59^{fyn}NON/
1	POS	29.33	23.02	3.52	2.47
	NON	8.34	9.30		
2	POS	6.20	41.98	1.24	1.80
	NON	5.00	23.30		
3	POS	5.93	65.24	3.92	15.79
	NON	1.51	4.13		
4	POS	9.79	41.47	2.89	9.29
	NON	3.38	4.46		
5	POS	16.90	25.30	3.19	3.03
	NON	5.30	8.33		
6	POS	14.23	11.58	2.87	5.12
	NON	4.96	2.26		
				$\overline{X} \pm$ SD	$\overline{X} \pm$ SD
				2.9 ± 0.9	6.25 ± 5.4

**gels were placed in a betascope and cpm for pp56^{lck} and pp59^{fyn} bands were dtermined and normalized for bacground based on control lanes immunoprecipitated with irrelevant antibodies.

In order to determine whether the increased tyrosine kinase activity seen in positive selecting mice was specific or due to an increased absolute amount of tyrosine kinase, we analyzed the total content of $pp56^{lck}$ and $pp59^{fyn}$ in each lysate. This determination was performed using equivalent amounts of protein of the different lysates in several ways. First, by western blotting of total cell lysates, this could not be used for $pp59^{fyn}$ because the sera does not recognize denatured protein. Second, by immunoprecipitation (with anti-$pp56^{lck}$ or $pp59^{fyn}$ antisera) from lysates of cells labeled with ^{35}S methionine *in vivo* and third, by immunoprecipitation (with anti-$pp56^{lck}$ or $pp59^{fyn}$ antisera) from lysates labeled with ^{125}I. No significant differences in total $pp56^{lck}$ or $pp59^{fyn}$ content were observed between lysates from nonselecting and positively selecting mice of this age (Table 2).

Table 2. Relative amount of $pp56^{lck}$ and $pp59^{fyn}$ in positively selecting thymocytes vs nonselecting thymocytes.
Thymocytes from $H2^{b/d}$ scid/scid, TCR trasgenic (positively selecting) and $H2^{d/d}$ scid/scid, TCR trasgenic (non selecting) mice were lysed in 1% Triton X-100 lysis buffer and analyzed as in several ways (as indicated) for their $pp56^{lck}$ and $pp59^{fyn}$ content. Proteins were resolved on a SDS-PAGE. Bands intensities (western blot and ^{125}Iodine) were analyzed using the densitometry program included in Enhance TM 1.0.1 in a Macintosh computer or quantitated using a betascope (^{35}S-methionine). Background levels were substracted in all the analysis.

Analysis	Selection	$PP56^{lck}$(a)	$pp59^{fyn}$(a)	$PP56^{lck}POS/$(b) $PP56^{lck}NON$	$PP59^{fyn}POS/$(b) $PP59^{fyn}NON$
(c)western blot	POS	51.1±6.6	ND	0.9	ND
	NON	55.6±1.6			
(d)^{35}S methionine	POS	154.1	230.8±32.4	1.3	1.1
	NON	117.5	211.2±42.5		
(e)^{125}Iodine	POS	216.7	53.36±7.1	1.1	0.8
	NON	199.9	68.95±0.5		

(a)$\overline{X} \pm$ SD (except when only one experiment is included)

(b)Ratio of the $\overline{X}$

(c)50mg of cellular lysate from the different mice was resolved by SDS-PAGE, transfered onto nitrocellulose and probed with anti-pp56lck antiserum as described (22), pp59 fyn could not be analyzed by western blot with our antibodies.

(d)16 hours ^{35}S-methionine labeling was carried out in tissue organ culture. Lysates were normalized for their protein content and $pp56^{lck}$ and pp59 ^{fyn} analyzed by immunoprecipitation using the appropiate antibodies. The immunoprecipitated proteins were separated by SDS-PAGE.

(e)Iodination of tyrosines or lysines was carried out in the TX100 lysates of the different mice. Iodination (10mg of total cellular protein) was carried out either on tyrosines or on lysines as described (25,26).Lysates were normalized and pp561ck and pp59 fyn analyzed by immunoprecipitation using the appropiate antibodies immunoprecipitated proteins were resolved by SDS-PAGE.

The observation that thymocytes in a nonselecting environment halt their differentiation at the CD4+CD8+ stage, together with recent CD8 transgenic experiments, indicate that positive selection occur at the CD4+CD8+ stage (4,5). It was therefore important to compare directly the activation state of tyrosine kinases within purified CD4+CD8+ thymocytes from nonselecting versus positively selecting mice. Figure 2 demonstrates that purified CD4+CD8+ thymocytes from positively selecting mice display at least as great an increase in pp56^{lck} and pp59^{fyn} activity as is seen among unfractionated thymocytes. Interestingly, there is essentially equivalent tyrosine kinase activity within the CD4$^-$CD8$^-$ thymocytes from nonselecting (Fig.2 indicated as NON) and positively selecting (Fig.2 indicated as POS) mice. This result demonstrates that the differential tyrosine kinase activity is selectively found in the thymocyte subset upon which positive selection is acting. This difference in kinase activity can not be due to the presence of a small percentage of mature thymocytes in positive selecting mice because the separation was performed based in the CD4 expression, which is absent in the mature subpopulation of these mice. Therefore, the small percentage of mature thymocytes would appear in the CD4$^-$CD8$^-$ subset. These results are not due either to the treatment with anti-CD4 antibodies used in the purification process because the same treatment was carried out for non-selecting and positively selecting thymocytes. Thus, the difference in activity among the unfractionated population (Fig.1) can only be accounted for within the double positive subset (Fig.2).

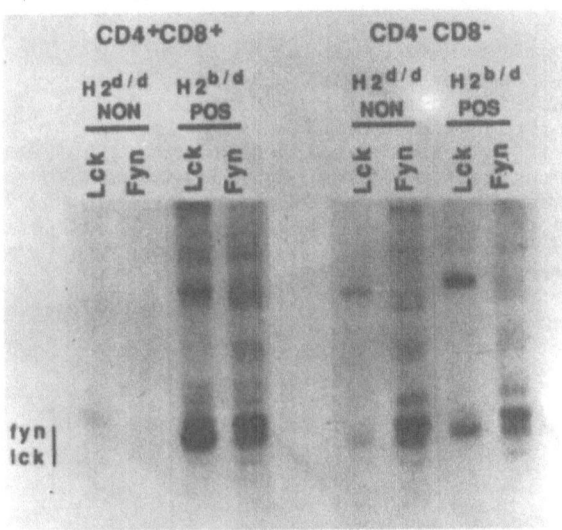

Fig. 2: Tyrosine kinase activity of pp56^{lck} and pp59^{fyn} in different thymocyte subsets. Thymocytes from H-2$^{b/d}$ *scid/scid*, TCR transgenic (positively selecting) and H-2$^{d/d}$ *scid/scid*, TCR-transgenic (nonselecting) mice were separated into CD4+CD8+ and CD4$^-$CD8$^-$ subsets. Each subset was lysed in 1% Triton X-100. 10 mg and 40 mg of each lysate were used for preparing pp56^{lck} and pp59fyn immunoprecipitates respectively. Immunoprecipitated proteins were subjected to an in vitro kinase assay and resolved by SDS-PAGE (8%). The approximately position of autophosphorylated tyrosine kinases is identified.

The fact that development from the immature $CD4^+CD8^+$ stage to the mature single-positive stage only occurs when the correct MHC allele is present, implies that TCR engagement on $CD4^+CD8^+$ thymocytes generates a differentiation signal. The preferential activation $pp56^{lck}$ and $pp59^{fyn}$ when the proper restricting MHC allele is present suggests that either these kinases are involved in triggering positive selection or that as a result of the selection process they have been induced. In both cases $pp56^{lck}$ and $pp59^{fyn}$ would have a role in the intracellular signaling generated during positive selection. An intriguing idea suggested by the above data is that $pp56^{lck}$ and $pp59^{fyn}$ might regulate each others activation state when brought into physical proximity by crosslinking driven by MHC of the membrane receptors to which they are bound ($pp56^{lck}$ with CD8 and $pp59^{fyn}$ with TCR). Preliminary experiments indeed suggest the presence of both kinases and their associated transmembrane molecules in a complex on the membrane of positively selecting thymocytes (data not shown). The involvement of $pp56^{lck}$ and $pp59^{fyn}$ in the process of thymic development is not surprising (although needed to be demonstrated) considering the previous data on ζ induced phosphorylation upon treatment with antibodies (12) and the fact that the overexpression of these kinases in transgenic mice dramatically affected the thymus phenotype ($pp56^{lck}$, 27) and responsiveness ($pp59^{fyn}$, 28); but this is the first evidence that links a triggered signal ($pp56^{lck}$ and $pp59^{fyn}$ increased specific activity) to the physiological process of positive selection.

Acknowledgments

We wish to thank Fred Ramsdell, Sue Demecki, Mark Soloski, Steve Desiderio and Doug Fearon for helpful discussions and review of the manuscript, Mitzi Baker for antibody preparation and Ping Li for technical advice and Linda Reavis. We thank Harald von Boehmer for the historical advice and the generous gift of TCR transgenic mice and anticlonotypic antibody. TMR was supported by PHS grant CA43803. ACC was supported by a fellowship from the Spanish CSIC. DMP is a recipient of the Cancer Research Institute Benjamin Jacobson Family Investigator Award and the RJR Nabisco Research Scholars Award.

REFERENCES:
1. Fowlkes BJ and Pardoll DM. Adv. Immunol.1989. 44:207-264
2. Von Boehmer H. Ann. Rev. Immunol.1988. 6:309-325
3. Kisielow P, Bluthmann H, Staerz UD, Steinmetz M, and Von Boehmer H. Nature 1987. 333:742-746
4. Robey EA, Fowlkes BJ and Pardoll DM. Seminars Immunol.1990.2:25-34
5. Robey EA, Fowlkes BJ, Gordon JW, Kioussis D, Von Boehmer H, Ramsdell F, Axel R. Cell 1991. 64:99-107
6. Berg LJ, Fazekas de St. Groth B, Pullen AM, Davis MM. Nature 1989. 340:559-562
7. Kaye J, Hsu ML, Sauron ME, Jameson JC, Gascoigne RJ, Hedrick SM. Nature 1989. 341:746-749
8. Scott B, Bluthmann H, Teh HS, Von Boehmer H. Nature 1989 338:591-593
9. Havran WL, Poenie M, Kimura J, Tsien R, Weiss A and Allison JP. Nature, 1987. 330:170-175
10. Finkel TH, McDuffie M, Kappler JW, Marrack P, Cambier JC. Nature 1987 330:(6144),179-181
11. Weiss A, Dazin PF, Shields R, Fu SM, Lanier LL. J Immunol.1987. 139: 3245-3250
12. Nakayama T, Singer A, Hsi ED, Samelson LE. Nature 1989.341:651-655
13. Vivier E, Morin P, Qingsheng T, Daley J, Blue ML, Schlossman SF, Anderson P. J. of Immunology 1991. 146:1142-1148
14. Samelson LE, Patel MD, Weissman AM, Harford JB, Klausner RD. Cell 1986. 46:1083-1090
15. Patel MD, Samelson LE, Klausner RD. J.Biol.Chem.1987. 262:(12), 5831-5838
16. Veillette A, Bookman MA, Horak EM, Samelson LE, Bolen JB. Nature 1989. 338:257-262
17. Hsi ED, Siegel JN, Minami Y, Luong ET, Klausner RD, Samelson LE. J. Biol. Chem.1989. 264:10836-10842
18. June CH, Fletcher MC, Ledbetter JA, Schieven GL, Siegel JN, Phillips AF and Samelson LE. Proc. Natl. Acad. Sci. USA 1990, 87:7722-7726
19. Marth JD, Peet R, Krebs EG, Perlmutter RM. Cell 1985. 43:393-406
20. Rudd CE, Trevillyan JM, Dasgupta JD, Wong LL, Schlossman SS. Proc. Natl. Acad. Sci. USA 1988. 85:(14),5190-5194
21. Samelson LE, Phillips AF, Luong EI, Klausner RD. Proc. Natl. Acad. Sci. USA 1990. 87:4358-4362
22. Carrera AC, Li P and Roberts TM. Inter. Immunol 1991,673-682
23. Da Silva AJ, Barber E, Zalvan CH, Dasgupta JD, Zamoyska R,Rudd CE Proc. Natl. Acad. Sci. 1990. 85:5190-5194
24. Pierres A, Naquet P, Van-Agthoven A, Bekkhoucha F, Denizot F, Mishal Z, Schmitt-Verhulst AM, Pierres M. J.Immunol.1984. 132:(6),2775-2782
25 Fraker J. and Speck JC. Biochem.Biophys.Res.Commun.1978.80:849-857
26. Bolton AE and Hunter WM. Biochem J. 1973. 133:529-538
27. Abrahamn KM, Levin SD, Marth JD, Forsbush KA, Perlmutter RM.,J.Exp Med. 1991. 173:1421-1432
28. Cooke, et al. Cell 1991. 65:281
29. Gee CE, Griffin J, Sastre L, Miller LJ, Springer TA, Piwnica-Worms H, Roberts TM. Proc. Natl. Acad. Sci. USA 1986. 83:5131-5135
30. Singer A, Munitz TI, Gress RE.Transplant-Proc.1987.19:(6,7),107-110

19. Subject Index

A

AA (adjuvant arthritis), 580,587,630
acute phase
- genes
- - regulation of, 378
- proteins, 369
- response, 297
- - and complement, 486
- - hepatic, 377,387
- - systemic, 488
ADA (adenosine deaminase) deficiency, 554
ADCC (antibody dependent cytotoxicity),
336,502
- against *Schistosome*, 679
- mediated by eosinophil, 680
- mediated by IgA, 680
- mediated by pro-inflammatory cell, 740
Adenovirus and CD44, 291
adhesion/signalling receptors, 273
adjuvant, 582
adoptive transfer, 6
affinity maturation, 5,48
- in antibody response, 5
- in germinal center, 21
AIDS (acquired immunodeficiency syndrome),
699,707
- pathogenesis, 699
alkaline phosphatase PB-76 12
ALL (acute lymphoblastic leukemia) T cell, 620
allelic exclusion, 37
- of alpha/beta TCR genes, 129,132
allelic inheritance
- of V_H allotypes in rabbits, 100

allergen, 427
- group I,II, 425
- house dust mite, 395,427
- immune response to, 395
- in asthma, 403
- plant, 395
- rye grass pollen, 427
- inflammation, 403,411
allergic
- reaction
- - immediate, 411
- - late phase, 411
- response, 427
- rhinitis, 411
allergy, 395
- activation of basophils and eosinophils in, 411
- chemokines in, 411
- cytokines in, 398,411,412
- edema in, 411
- effector cell in, 398,412
- infiltration of basophils and eosinophils in, 411
- mast cell in, 411
- mediator release in, 411,415
- signal transduction in, 415
- smooth muscle contraction in, 411
alloantigen system, 807
allograft, 841
- survival, 329
alloimmune response, 793
allorecognition, 785
alpha-actinin, 286
$alpha_2$-macroglobulin, 378,518
alternative complement pathway, 502
amylase, 429
ANCA autoantibodies, 644
anergy (functional unresponsiveness),
25,70,73,251,796
- induction, 67
- of allospecific T cells, 796
anti-clonotypic
- stimulation, 620
- vaccine, 619
antibody
- anaphylatic, 400,679
- anti-clonotypic, 3

dust mite allergen, 427

nitric oxyde pathway, 684,755
NK (natural killer) cell, 305
- antigen Ly49, 465
- cytokine production, 328,355
- in fetal resorption, 857
- in leishmaniasis, 747
- in pregnancy, 827
NK cell-TNF pathway, 863
NOD mice, 580,608,630,817
non-responsiveness in alloreactive T cell, 796
nuclear factor
- c-Rel complex, 372
- lymphoid cell specific, 372
- NF-kappa-B, 369
nucleotide substitution rate in MHC genes, 154

O

Omenn's syndrome, 558
oncogenes, 874
oncostatin M, 303,379,887
Opa protein of *Neisseria*, 670

P

p150,95, 283
PAF (platelet activating factor), 387,403,412,466
papain, 429
Papilloma virus, 661
parasite
- antigenic mimicry, 683
- enzymatic activity of, 683
- fecundity, 739
- infection TH1,2 cells in, 731
- intrinsic, 683
- of tropical disease, 731
PBR (peptide binding regoion) of MHC, 137
peptide
- binding
- - to MHC class I, 159,163

- - to MHC class II, 163,189,159
- bound to MHC proteins, 58,59
- exchange on MHC class II molecules, 190
- in ER, 161
- MHC binding competitor, 587
- presentation by MHC molecules, 161
- therapy, 636
- transporter TAP, 161
peptide/MHC complex
- lifetime, 191
perforin, 257
peripheral tolerance, 61,70
Peyer's patches sheep ileal, 121
PF4 (platelet factor 4), 299
PFGE (pulse field gel electrophoresis),
115,298,525
PGE (prostaglandin), 421
PGE$_2$ (prostaglandin E$_2$), 341,466
Pgp-1, 289
phosphorilation
- of CD3-zeta chain, 75
- of protein tyrosine, 74
- of Ig associated alpha/beta chain, 199
- pathway in IgE receptor, 437
phylogenetic trees of MHC evolution, 137
PIBF (progesteron inhibitory factor), 862
pilin filamentous protein, 667
PKC (protein kinase C), 74
- activation, 87
- family, 286
placenta
- IL-10 production, 857
- suppressor factor of, 8828
placental lactogen, 859
plasma cell, 22,51
- in immunodeficiency, 537
plasmablast, 21,25
plasmocytoma, 54
Plasmodium falciparum malaria, 283,506,734
PLC-gamma (phospholipase C-gamma), 78
PLP (proteolipid protein), 635
poly-Ig receptor of throphoblast, 831
polyarthritis, 580
POMC (pro-opiomelanocortin), 677
positive selection, 57,58

S

type 2 Gaucher disease, 562
tyrosine phosphatases, 436
- CD45, 76
tyrosine phosphorylation
- after antigen cross-linking, 30
- in signal transduction, 199

U

ubiquitin, 161
ulcerative colitis, 359

V

V gene
- hypermutation, 4,8,11
- Ig, 44
- pre-B, 52,536
V region
- complementary interaction, 643
- interaction of natural antibodies, 645
vaccination, 583
- strategy, 684
vaccine
- anti-clonotypic
- - in cancer, 619
- - in HIV-1 infection, 619
- based on oligonucleotide, 771
- controlled release, 688
- for birth control, 849
- immunomodulatory, 577
- recombinant, 772
- synthetic, 767
- vehicles, 686
vasculitis systemic, 644
vernal conjunctivitis, 242
veto cell, 788
V_H
- allotypes in rabbit, 99

- families, 93,110
V_H gene
- diversity, 110
- N-region diversity, 110
- usage in rabbit, 99
- in chicken, 124
V_H repertoire, 110
V_H segments, 93
VIP (vasoactive intestinal peptide), 298
virulence of *Neisseria*, 667
virus-host relationship, 660
V_L gene
- in chicken, 124
VLA integrins, 283
VLA-1,2, 284
VLA-4, 13,279
von Willebrand domains, 519
von Willebrand factor, 529
VSV (*Vesicular stomatitis* virus) 662
WSEWS motif, 313

X

X chromosome
- inactivation, 547,558
- methylated, 550
X-linked disorders, 545
xenograft, 817
xenotransplantation, 815
XL (X-linked) SCID, 557
XLA (X-linked aggamaglobulinemia), 537
- mutation of genes, 545

Y

YAC (yeast artificial chromosome), 91,525
Yersinia arthritis, 614
Yersinia enterocolitica, 239
Yersinia infection, 588